THESAURUS OF TEXTILE TERMS

Thesaurus of Textile Terms
covering
FIBROUS MATERIALS AND PROCESSES
SECOND EDITION

Stanley Backer and Emery I. Valko
Editors

Miriam (Liang) Chu, Hans U. Rudolf, and Signe A. Dayhoff
Information Specialists

Roger A. Roach, Robert C. Sheldon, and David A. Anderson
Computer Specialists

THE M.I.T. PRESS
MASSACHUSETTS INSTITUTE OF TECHNOLOGY
CAMBRIDGE, MASSACHUSETTS, AND LONDON, ENGLAND

First Edition Published February 1966
by Fibers and Polymers Division,
Department of Mechanical Engineering,
Massachusetts Institute of Technology

Second Edition Published February 1969
by the M.I.T. Press

This thesaurus was prepared in the Fibers and Polymers Laboratories, Department of Mechanical Engineering, M.I.T., under Contracts C.S.T.-1170 and C.S.T.-164 with the National Bureau of Standards. N.B.S. - M.I.T. liaison has been provided by Dr. F. Cecil Brenner, Chief, Textile and Apparel Technical Center, Institute for Applied Technology, National Bureau of Standards, U.S. Department of Commerce

Library of Congress catalog card number: 69-19245
Printed in the United States of America

THESAURUS OF TEXTILE TERMS

FOREWORD

This thesaurus has been prepared as a part of the textile information program conducted in the Fibers and Polymers Division, M.I.T., under sponsorship of the U. S. Department of Commerce. It is intended to serve as an aid to authors, editors, indexers, librarians, information specialists, and users of textile engineering literature. In particular it is designed to provide an effective language interface between people and manual storage systems or electronic computers capable of handling immense stores of technical information.

The problems of storage and retrieval of textile information are treated in some detail in Appendix A of this document. The ground rules of the thesaurus, governing its form and content and its usage, are cited in Appendix B. Hence we will confine this foreword to an explanation of the philosophy that guided the compilers of the thesaurus in their task. This philosophy has several facets.

First, the thesaurus is intended to provide language control and to minimize word ambiguity. The minimization of term ambiguity must be undertaken both from the point of view of the indexer (input) and of the searcher (output).

Second, this thesaurus has somewhat of a teaching function. It is designed to furnish the indexer with suggestions on relationships between terms and concepts, and it attempts to teach the searcher which concepts and systems are associated and therefore warrant consideration during the retrieval process. These associations are based on hierarchical word structures, on upstream and downstream elements in textile process flow, and on mechanistic associations based on the understanding and the technical experience of the compilers.

Third, this edition of the thesaurus should provide maximum subdivision in concepts. The degree of term refinement and subdivision of concepts should, of course, be directly related to depth of indexing. Since any decision on depth of indexing must be postponed until sufficient experience is developed in indexing and retrieval, it is desirable at first to provide maximum subdivision in this thesaurus. At a later date this refinement can be reduced automatically, if deemed advisable, with minimum intellectual input, whereas if a coarser concept structure were used first, any move towards greater refinement at a later date would necessitate repeating the intellectual input of the original indexing.

Fourth, it may be expected that different users of this thesaurus will have different requirements for concept refinement. It is therefore more logical to develop a system that can be employed in part according to user needs, rather than to design a language that must be augmented in depth by a wide range of users.

This thesaurus is designed primarily for use in storage and retrieval of information on textile materials and processes. Its coverage serves the needs of the fiber producer, the textile manufacturer and finisher, the converter, the manufacturer and distributor of clothing, the producer of textile auxiliaries and dyes, the textile machinery maker, the materials testing laboratories of government, of retailing organizations and of consumer groups, and the textile specialist in faculties of home economics.

Developing a thesaurus from start is not a task to be turned over to a series of part-time committees representing many organizations. There are simply too many day-to-day decisions that influence the continuing development of the system, and the need to check constantly with other groups geographically separated would make such a joint project entirely impractical. Each of the 8000 key words represents a positive decision, as does each of the 72,000 relationships listed in the thesaurus. Obviously, there were a proportional number of decisions not to include other words and other relationships. However, the continuing modification and expansion of an established thesaurus is well suited to cooperative activities with professional organizations, with publishers, and with processors and users of textile information. And this joint activity must cross national boundaries if it is to be really effective.

The development of a thesaurus document for technical literature is a never-ending task, for it reflects the vocabulary of that literature. Just as that vocabulary expands with the introduction of new terms and the refinement of old concepts, so must the thesaurus be expanded and continuously modified.

Concerning the state of completeness and correctness of the current edition of this document, we refer to the comment in Peter Roget's Preface to the first edition of his thesaurus in 1852, "But, in a work of this nature, where perfection is placed at so great a distance, I have thought it best to limit my ambition to that moderate share of merit which it may claim in its present form; trusting to the indulgence of those for whose benefit it is intended, and to the candor of critics who, while they find it easy to detect faults, can at the same time duly appreciate difficulties."

Stanley Backer

Cambridge, Massachusetts

September 1968

ACKNOWLEDGMENTS

This thesaurus has been compiled by staff members of the Fibers and Polymers Division, Mechanical Engineering Department, M.I.T. However, it reflects input from a large number of sources in industry, in government, and in other departments at M.I.T.

Errors and omissions of the thesaurus can be attributed only to the editors. Credit for the overall document must be shared with a much larger group of people and organizations. This note is intended to provide a detailed list of the participants in this activity with an indication of their individual roles.

The officers of the following professional societies were most cooperative in arranging for committee participation in the information activities here described. This covered any of a spectrum of activities--from provision of a sounding board for ideas, plans, and programs to active involvement in developing word relationships for particular sections of the thesaurus. Thanks are due to the American Association of Textile Technologists, the American Association of Textile Chemists and Colourists, Committee D-13 of the American Society for Testing and Materials, and The Fiber Society for their interest and cooperation. Particular mention should be made of the aid of Mr. P. J. Fynn's Committee on Information Storage and Retrieval in the A.A.T.C.C. It should be emphasized that appearance of this document marks but the beginning of liaison with many of the above named groups, for complete communication depends, to a great extent, on availability of a complete thesaurus.

Word lists used by two fiber producers were made available at the beginning of this program and we are indebted to the Celanese Fibers Company and the Chemstrand Research Center for their generous assistance. Discussions with Miss Helen Sommar of Celanese and with Mr. Daniel Mills and Dr. Bruno Roberts of Chemstrand were particularly valuable at the beginning of the M.I.T. program. We have also benefited greatly from discussions with Mrs. Darlene L. Ball of Burlington Industries, Inc. and with Dr. J. G. Van Oot of the E. I. du Pont de Nemours & Company.

To Miss Dorothy Eastman we are most grateful on two counts: first, for her direct contribution to the unending job of typing associated with the numerous drafts of this thesaurus; and second, for her responsible assumption of the additional office managerial duties related to this information project. Thanks are also due to Miss Consuelo Godfrey for typing and clerical activities in the program.

Personnel of the Fibers and Polymers Division actively engaged in compiling or editing different drafts of the thesaurus included the following doctoral candidates: Peter Popper, Harvey Plonsker, Robert Kimmel, Willard Anderson, Dieter Ender, and Aly El-Shiekh. And student programmers included David Anderson and James Bixby. We are indebted to the Director and Staff of the M.I.T. Computational Center for continuing assistance in execution of the computer phases.

During the course of this program, we sought guidance on numerous occasions from several information specialists at M.I.T. including Dr. M. M. Kessler, Associate Director, and Dr. W. H. Locke, Director of the M.I.T. Libraries, Miss R. L. Taggart, Engineering Librarian, and Mr. Charles H. Stevens of M.I.T.'s Project INTREX. And we benefited greatly from participation in the 1965 Planning Conference of Project INTREX held at Woods Hole, Massachusetts, under the direction of Dr. C. F. J. Overhage.

Financial support of this program has been made possible through a contract with the Department of Commerce. We have received full support and encouragement in this work from Mr. Robert Stern, Dr. George Gordon, and Dr. F. Cecil Brenner of the Department's Textile and Apparel Technology Center, and from Dr. Herbert F. Schiefer and Mr. Ernest Kaswell, consultants to the Center. For this support we are appreciative and we hope the progress indicated in this document satisfies their expectations.

Finally, special acknowledgment is due to Dr. Aidan B. McNamara who played a leading role in the first years of the M.I.T. Textile Information Program. The project as a whole bears the impact of his leadership and his effort. He was a principal contributor to the earlier *Thesaurus of Textile Engineering Terms* that served as the basis of the present expanded document.

NOTATIONS USED IN THESAURUS

USE (Use) -- A term followed by the instruction to "use" another term or terms indicates that the referred-to term(s) should be used instead of the term at hand.

UF (Used For) -- A term followed by this notation is shown to be the preferred term for those terms that follow the notation. That is, the referred-to terms instruct the user to "use" the term at hand. Thus the "used for" reference is the "reverse" of the "use" reference.

NT (Narrower Terms) -- A term which is followed by this notation is the name of a class of concepts that contains as members those concepts symbolized by the referred-to terms. Example: TEXTURING, as a key word with NT FALSE TWISTING listed below it. False twisting is a member of the class of processes known as Texturing. The narrower-term cross-references can be employed to make indexing or inquiries more specific.

BT (Broader Terms) -- A term which is followed by this notation is the name of a member of the class of concepts symbolized by the referred-to term. Example: FALSE TWISTING as a key word with BT TEXTURING listed below it. The class of processes known as Texturing includes the process of false twisting. The broader-term cross-references can be employed to make indexing more general.

RT (Related Terms) -- A term that is preceded by this

notation is related in an unspecifiable manner to the term at hand. It may be a near-synonym to the term or may sometimes be synonymously or hierarchically related to the term at hand; it may be mechanistically related, or it may represent upstream or downstream elements of related processes. Example: SPINNING with RT DRAFTING (STAPLE FIBER) and RT TWISTING.

TABLE OF CONTENTS

	Page Number
FOREWORD	v
ACKNOWLEDGMENTS	ix
NOTATIONS USED IN THE THESAURUS	xiii
KEY WORDS AND RELATIONSHIPS	1
LIST OF KEY WORDS	264
REFERENCES USED IN THE PREPARATION OF THE THESAURUS	332
APPENDICES	
A. PROBLEMS OF TEXTILE INFORMATION RETRIEVAL	339
B. GROUND RULES FOR USE OF THE TEXTILE THESAURUS IN INDEXING AND SEARCHING	393
C. TABLES OF HIERARCHICAL RELATIONSHIPS	411
RECENTLY RECOMMENDED KEYTERMS	447

THESAURUS OF TEXTILE TERMS

ABRASION
RT ABRASION RESISTANCE
ABRASION RESISTANCE FINISHES
ABRASION TESTING
ABRASIVES
BREAKING
CLEANING
COEFFICIENT OF FRICTION
CORUNDUM
CUTTING
DURABILITY
DYNAMIC FRICTION
EMERY
FABRIC TO FABRIC FRICTION
FATIGUE
FATIGUE RESISTANCE
FRICTION
FRICTIONAL FORCE
FROSTING
GALLING
GRINDING (MATERIAL REMOVAL)
MECHANICAL PROPERTIES
PILLS
PLOUGHING
POLISHING
RUBBING
SCORING
SCRATCHING
SCRUBBING
SCUFFING
SEAM ABRASIVE STRENGTH
SIZES (SLASHING)
SKIN IRRITATION
SLIDING
SLIPPAGE
SNAG RESISTANCE
SNAGGING
STRESS CONCENTRATION
STRETCH BREAKING
WARP TENSION
WEAR
WEAR RESISTANCE
WEAVING DYNAMICS
YARN SLIPPAGE
YARN TO METAL FRICTION
YARN TO YARN FRICTION

ABRASION MARKS
USE CHAFE MARKS

ABRASION RESISTANCE
BT END USE PROPERTIES
FINISH (PROPERTY)
MECHANICAL PROPERTIES
SURFACE PROPERTIES (MECHANICAL)
RT ABRASION
ABRASION RESISTANCE FINISHES
DURABILITY
FABRIC PROPERTIES
FABRIC PROPERTIES (MECHANICAL)
FATIGUE RESISTANCE
FRICTION
LUBRICITY
SEAM ABRASIVE STRENGTH
SIZE (SLASHING)
SNAG RESISTANCE
STOLL QM ABRASION TESTING
WEAR RESISTANCE

ABRASION RESISTANCE FINISHES
BT FINISH (SUBSTANCE ADDED)
MECHANICAL DETERIORATION PROPERTIES
RT ABRASION
ABRASION RESISTANCE
ANTISLIP AGENTS
DURABILITY
FATIGUE RESISTANCE
FRICTION
LUBRICANTS
SNAG RESISTANCE
WEAR RESISTANCE

ABRASION TESTERS
BT TESTING EQUIPMENT

ABRASION TESTING
NT SCHIEFER ABRASION TESTING
STOLL QM ABRASION TESTING
TABER ABRASION TESTING
BT TESTING
RT ABRASION
BENDING TESTING
BURST TESTING
CREEP TESTING
CUTTING TESTING
FATIGUE TESTING
IMPACT PENETRATION TESTING
IMPACT TESTING
PILLING TESTING
SNAG TESTING
STRENGTH TESTING
TEAR TESTING
TENSILE TESTING
TESTING EQUIPMENT
WEAR TESTING

ABRASIVES
NT CORUNDUM
EMERY
RT ABRASION
ASPERITIES
CERAMICS
FRICTION
GRINDING (MATERIAL REMOVAL)
GRIT
POLISHING
PUMICE
SCORING
SCRATCHING
WEAR

ABSENCE
RT APPEARANCE
CORES (CENTER)
HOLES
HOLLOWNESS
INTENSITY
INVERSION
MATRIX FREEDOM
PRESENCE
SPACE
VACUUM

ABSORBATES
RT ABSORBERS

ABSORBENCY (MATERIAL)
BT END USE PROPERTIES
PHYSICAL CHEMICAL PROPERTIES
TRANSFER PROPERTIES
RT ABSORPTION (MATERIAL)
ABSORPTIVITY (RADIATION)
CHROMATOGRAPHY
CONDITIONING
CRASH TOWELING
CRITICAL MOISTURE CONTENT
EQUILIBRIUM MOISTURE CONTENT
FABRIC PROPERTIES
FABRIC PROPERTIES (PHYSICAL EXCLUDING MECHANICAL)
FILTRATION
HYDROPHILIC PROPERTY
HYDROPHOBIC PROPERTY
HYGROSCOPICITY
INHIBITION
MERCERIZED FINISH
OSMOSIS
POROSITY
REGAIN
SORPTION OF DYES
TOWELS
WATER REPELLENCY

ABSORBERS
RT ABSORBATES
ABSORBENCY (MATERIAL)
ABSORPTION (MATERIAL)
ADSORPTION
HYDROPHILIC PROPERTY
HYGROSCOPICITY
SORPTION

ABSORPTION (MATERIAL)
NT DYNAMIC ABSORPTION (WATER)
MOISTURE ABSORPTION
STATIC ABSORPTION (WATER)
WATER ABSORPTION
BT SORPTION
RT ABSORBENCY (MATERIAL)
ABSORBERS
ABSORPTION (RADIATION)
ABSORPTION RATE
ABSORPTIVITY (RADIATION)
ACQUISITION
ADSORPTION
CHROMATOGRAPHY
DESORPTION
DIFFUSION
DYE AFFINITY
DYEING RATE
DYEING THEORY
DYE MIGRATION
EXTRACTION
HYGROSCOPICITY
IODINE ABSORPTION
LANGMUIR ISOTHERM
MOISTURE CONTENT
PARTICLE SIZE
PENETRATION
PERMEABILITY
PHYSICAL CHEMICAL PROPERTIES
POROSITY
SANITARY NAPKINS
SORPTION OF GASES
SORPTION OF LIQUIDS
SORPTION OF WATER
SWELLING
WET PICKUP

ABSORPTION (RADIATION)
RT ABSORPTION (MATERIAL)
ABSORPTION SPECTRA
BRIGHTNESS
COLOR
COLORIMETRY
HUE
LIGHTNESS
LUSTER
OPTICAL DENSITY
OPTICAL PROPERTIES
REFLECTIVITY
SORPTION
SPECTROPHOTOMETRY
THERMAL PROPERTIES
TRANSLUCENCY
TRANSMITTANCE
TRANSPARENCY

ABSORPTION EQUILIBRIUM
USE SORPTION EQUILIBRIUM

ABSORPTION ISOTHERM
USE ADSORPTION ISOTHERM

ABSORPTION RATE
BT SORPTION RATE
RT ABSORPTION (MATERIAL)
ADSORPTION ISOTHERM
ADSORPTION RATE
EQUILIBRIUM (DYEING)
SORPTION
SORPTION EQUILIBRIUM

ABSORPTION SPECTRA
BT SPECTRA
RT ABSORPTION (RADIATION)

ABSORPTION TESTERS
BT TESTING EQUIPMENT

ABSORPTIVITY (RADIATION)
BT OPTICAL PROPERTIES
THERMAL PROPERTIES
TRANSFER PROPERTIES
RT CONDUCTIVITY (THERMAL)
EMISSIVITY
RADIATION
REFLECTANCE
REFLECTIVITY
TRANSMISSIVITY
TRANSMITTANCE

ACCELERANTS (DYEING)
NT CARRIERS (DYEING)
BT DYEING AUXILIARIES
RT RETARDING AGENTS

ACCELERATION (MECHANICAL)
UF DECELERATION
RETARDATION
NT CORIOLIS ACCELERATION
RT ACCELEROMETERS
ANGULAR VELOCITY
BRAKING
CENTRIFUGAL FORCE
CONSTANT SPEED
DYNAMOMETERS
FORCE
GRAVITATION
HIGH SPEED
IMPACT
IMPULSES
INERTIA
MECHANICAL SHOCK
MOMENTUM
REVERSAL
TRANSLATION
VARIABLE SPEED
VELOCITY
VELOCITY (RELATIVE)
VELOCITY CHANGE POINT

ACCELEROMETERS
BT MEASURING INSTRUMENTS
TESTING EQUIPMENT
RT ACCELERATION (MECHANICAL)
BRAKING
CENTRIFUGAL FORCE
CORIOLIS ACCELERATION
DYNAMIC BALANCING
FORCE
GRAVITATION
IMPACT
INERTIA
MECHANICAL SHOCK
VELOCITY
VIBRATION

ACCESSIBILITY (INTERNAL)
BT FINE STRUCTURE (FIBERS)
RT AMORPHOUS REGION
CRYSTALLINITY
DIFFUSION
MOISTURE
PENETRATION
PERMEABILITY

ACCESSORIES
RT COMPONENTS
DEVICES
GADGETS
MACHINERY
MECHANISMS
STANDS

ACCIDENTS
RT DEAFNESS
DOWN TIME
OCCUPATIONAL HAZARDS
SAFETY

ACCORDION FABRICS
BT FILLING KNITTED FABRICS
KNITTED FABRICS

ACCUMULATION
RT ACQUISITION
AGGREGATION
ASSEMBLIES
COLLECTION
FILTER CAKES
FILTRATION
INTEGRATION

ACCURACY
RT ADJUSTMENTS
CALIBRATION
COMPENSATION
ARITHMETIC
CONFIDENCE LIMITS
CORRECTION
DETECTION
DRIFT
ERRORS
EXAMINATION
EXPERIMENTAL DESIGN
EXPERIMENTATION
HYSTERESIS (MECHANICAL)
LINEARITY
NUMERICAL ANALYSIS
STANDARD DEVIATION

ACETAL RESINS
BT RESINS
RT ACETALS
CATALYSTS
CROSSLINKING
FINISH (SUBSTANCE ADDED)

ACETALDEHYDE SULFOXYLATES
RT DYEING AUXILIARIES
FORMALDEHYDE SULFOXYLATES
PRINTING
REDUCING AGENTS
VAT DYEING
VAT PRINTING
ZINC FORMALDEHYDE SULFOXYLAT*

ACETALS
UF DIACETALS
NT DIMETHYL ACETAL
FORMALS
HEMIACETALS
POLY (VINYL ACETAL)
POLY (VINYL BUTYRAL)
BT REACTANTS
RT ACETAL RESINS
ALDEHYDES
ETHERS
FINISH (SUBSTANCE ADDED)
METHYLOLATION
POLYETHERS
RESIN FINISHES

ACETAMIDE
BT AMIDES

ACETATE
USE ACETATE FIBERS

ACETATE DYEING
BT DYEING (BY FIBER CLASSES)
RT ACETATE DYES
BLEND DYEING
COTTON DYEING
DOPE DYEING
RAYON DYEING
SILK DYEING
TOW DYEING
TRIACETATE DYEING
WOOL DYEING

ACETATE DYES
BT DYES (BY FIBER CLASSES)
RT ACETATE FIBERS
ACETATE DYEING
ACRYLIC DYES
CELLULOSE DYES
DISPERSE DYES
DYEING (BY FIBER CLASSES)
NYLON DYES
POLYESTER DYES
SILK DYES
WOOL DYES

ACETATE ESTERS
NT ETHYL ACETATE
VINYL ACETATE
BT ESTERS
RT ACETATE SALTS
ACETIC ACID

ACETATE FIBERS
UF ACETATE
ACETATE RAYON
NT CELACRIMP (TN)
CELAFIBRE (TN)
ESTRON (TN)
FIBROCETA (TN)
LOFTURA (TN)
BT CELLULOSE ESTER FIBERS
MAN MADE FIBERS
RT ACETATE DYES
ACETYLATED COTTON
ALKALI CELLULOSE
ARTIFICIAL SILK (ARCHAIC)
CELLULOSE ACETATE
CELLULOSE DERIVATIVES
CELLULOSE ESTERS
CELLULOSE NITRATE
CUPRAMMONIUM RAYON
CYANOETHYLCELLULOSE
CYCLOSET YARNS (TN)
TRIACETATE FIBERS
VISCOSE RAYON

ACETATE RAYON
USE ACETATE FIBERS

ACETATE SALTS
NT MERCURIC ACETATE
POTASSIUM ACETATE
SODIUM ACETATE
RT ACETATE ESTERS
ACETIC ACID

ACETIC ACID
BT CARBOXYLIC ACIDS
ORGANIC ACIDS
RT ACETATE ESTERS
ACETATE SALTS
ACETIC ANHYDRIDE
HYDROXYACETIC ACID
PERACETIC ACID

ACETIC ANHYDRIDE
BT ACID ANHYDRIDES
RT ACETIC ACID

ACETONE
BT KETONES
RT HYDROXYACETONE
SOLVENTS

ACETONITRILE
BT NITRILES

ACETYLATED COTTON
BT CHEMICALLY MODIFIED COTTON
RT ACETATE FIBERS
ACETYLATION
AMINIZED COTTON
BENZOYLATED COTTON
CARBAMOYLETHYLATED COTTON
CHEMICAL MODIFICATION (FIBERS)
CYANOETHYLATED COTTON
DS (CELLULOSE)
ESTERIFICATION

ACETYLENE
BT ALKYNES
RT VINYLATED COTTON

ACETYLATION
BT ACYLATION
RT ACETYLATED COTTON
CHEMICAL MODIFICATION (FIBERS)
ESTERIFICATION
REACTIONS (CHEMICAL)

ACETYLENE DIUREA FORMALDEHYDE
USE METHYLOL ACETYLENE DIUREA

ACETYLENIC COMPOUNDS
NT ALKYNES
BT UNSATURATED COMPOUNDS

ACETYLENIC HYDROCARBONS
USE ALKYNES

ACETYLTRIETHYL CITRATE
BT ESTERS
RT PLASTICIZERS

ACID AGING
BT AGING (STEAMING)
RT AGERS
FLASH AGING
PRINTING

ACID ANHYDRIDES
UF ANHYDRIDES
NT ACETIC ANHYDRIDE
MALEIC ANHYDRIDE
PHTHALIC ANHYDRIDE
PROPIONIC ANHYDRIDE
RT ACID CHLORIDES

ACID BINDING
BT CHEMICAL PROPERTIES
RT BASE BINDING

ACID CHLORIDES
RT ACID ANHYDRIDES
ACIDS
CHLORINE COMPOUNDS

ACID DAMAGE
USE ACID DEGRADATION

ACID DEGRADATION
UF ACID DAMAGE
BT DEGRADATION
RT ACID HYDROLYSIS
CHEMICAL PROPERTIES
COPPER NUMBER
DEGRADATION PROPERTIES
DEGREE OF POLYMERIZATION
HYDROCELLULOSE
METHYLENE BLUE NUMBER
REACTIONS (CHEMICAL)
TENDERING
VISCOSITY

ACID DYEING
BT DYEING (BY DYE CLASSES)
RT ACID DYES
ACID SHOCK METHOD
ANIONIC SITES
AZOIC DYEING
BASIC DYEING
DIRECT DYEING
DISPERSE DYEING
HYDROXYLAMINE
METALLIZED DYEING
MORDANT DYEING
NEUTRAL DYEING
OVERDYEING
PIGMENT DYEING
REACTIVE DYEING
SULFUR DYEING
VAT DYEING

ACID DYES
NT MILLING DYES
BT ANIONIC DYES
DYES (BY CHEMICAL CLASSES)
RT ACID DYEING
BASIC DYES
DYEING (BY DYE CLASSES)
DYES (BY FIBER CLASSES)
NEUTRAL DYES

ACID END GROUPS
NT CARBOXYL END GROUPS
SULFONIC END GROUPS
BT END GROUPS
RT AMINE END GROUPS
DEGREE OF POLYMERIZATION
MOLECULAR WEIGHT CONTROL
POLYMERIZATION
POLYMERS
ORGANIC ACIDS

ACID HYDROLYSIS
BT HYDROLYSIS
RT ACID DEGRADATION
ALKALI HYDROLYSIS
CHEMICAL STABILITY

ACID MILLING DYES
USE MILLING DYES

ACID SHOCK METHOD
BT DYE FIXING
RT ACID DYEING
AGING (STEAMING)
CONTINUOUS DYEING
STEAMING
THERMOFIXING (DYEING)

ACID SOLUBILITY
BT CHEMICAL PROPERTIES
DEGRADATION PROPERTIES
RT ACID SOLUTIONS
ACIDS
ALKALI SOLUBILITY
DEGRADATION
KERATIN
UREA BISULFITE SOLUBILITY
WOOL

ACID SOLUTIONS
BT SOLUTIONS (LIQUID)
RT ACID SOLUBILITY
ACIDITY
ALKALINE SOLUTIONS
DYE SOLUTIONS
PH

ACIDITY
RT ACID SOLUTIONS
ACIDS
ALKALINITY
PH

ACIDS
NT DIBASIC ACIDS
INORGANIC ACIDS
ORGANIC ACIDS
TRIBASIC ACIDS
(CONT.)

RT ACID CHLORIDES
ACID SOLUBILITY
ACIDITY
BASES
CATALYSTS
ELECTROLYTES
SALTS (GENERAL)
SODIUM BISULFATE

ACOUSTIC BAFFLING
BT INDUSTRIAL FABRICS
RT ACOUSTIC INSULATION
SOUND

ACOUSTIC INSULATION
UF INSULATION (ACOUSTICAL)
RT ACOUSTIC BAFFLING
ACOUSTIC PROPERTIES
SOUND

ACOUSTIC PROPERTIES
BT PHYSICAL PROPERTIES (EXCLUDING MECHANICAL)
RT AESTHETIC PROPERTIES
BIOCHEMICAL PROPERTIES
CHEMICAL PROPERTIES
ELECTRICAL PROPERTIES
FABRIC PROPERTIES
FABRIC PROPERTIES (PHYSICAL EXCLUDING MECHANICAL)
FIBER PROPERTIES
MECHANICAL PROPERTIES
OPTICAL PROPERTIES
PHYSICAL CHEMICAL PROPERTIES
SOUND
STRESS OPTICAL PROPERTIES
STRUCTURAL PROPERTIES
THERMAL PROPERTIES
TRANSFER PROPERTIES

ACQUISITION
RT ABSORPTION (MATERIAL)
ACCUMULATION
COLLECTION
DELIVERY
EXTRACTION
FILTRATION
SORPTION

ACRILAN (TN)
BT ACRYLIC FIBERS

ACROLEIN
UF ACRYLALDEHYDE
METHACROLEIN
BT ACRYLIC COMPOUNDS
ALDEHYDES
RT POLYACROLEIN

ACRYLALDEHYDE
USE ACROLEIN

ACRYLAMIDE
UF DIACETONE ACRYLAMIDE
BT ACRYLIC COMPOUNDS
AMIDES
RT ACRYLONITRILE
CARBAMOYLETHYLATED COTTON
CHEMICAL MODIFICATION (FIBERS)
METHACRYLAMIDE
METHYLOL ACRYLAMIDE
POLYACRYLAMIDE
REACTANTS

ACRYLAMIDO DYES
UF PRIMAZIN (TN)
PROCILAN (TN)
BT REACTIVE DYES

ACRYLIC ACID
BT ACRYLIC COMPOUNDS
CARBOXYLIC ACIDS
ORGANIC ACIDS
RT ACRYLIC ESTERS
ACRYLIC SALTS
METHACRYLIC ACID
POLYACRYLIC ACID

ACRYLIC COMPOUNDS
NT ACROLEIN
ACRYLAMIDE
ACRYLIC ACID
ACRYLIC ESTERS
ACRYLIC SALTS
ACRYLONITRILE
BUTYL ACRYLATE
ETHYL ACRYLATE
METHACRYLAMIDE
METHACRYLIC ACID
METHACRYLIC ESTERS
METHACRYLONITRILE
METHYL ACRYLATE
METHYL METHACRYLATE
METHYLOL ACRYLAMIDE
METHYLOL METHACRYLAMIDE
SODIUM ACRYLATE
BT MONOMERS
RT ALLYL COMPOUNDS
VINYL COMPOUNDS

ACRYLIC DYEING
BT SYNTHETIC DYEING
DYEING (BY FIBER CLASSES)
RT ACRYLIC FIBER BLENDS
ACRYLIC FIBERS
BLEND DYEING
NYLON DYEING
POLYESTER DYEING
WOOL DYEING

ACRYLIC DYES
BT DYES (BY FIBER CLASSES)
RT ACETATE DYES
ACRYLIC FIBERS
CELLULOSE DYES
DYEING
DYES (BY CHEMICAL CLASSES)
MODACRYLIC FIBERS
NYLON DYES
PERSULFATES
POLYACRYLATES
POLYESTER DYES
SILK DYES
WOOL DYES

ACRYLIC ESTERS
NT BUTYL ACRYLATE
ETHYL ACRYLATE
METHYL ACRYLATE
BT ACRYLIC COMPOUNDS
ESTERS
RT ACRYLIC ACID
ACRYLIC SALTS
METHACRYLIC ESTERS
POLYACRYLATES

ACRYLIC FIBER BLENDS
BT BLENDS (FIBERS)
VINYL FIBER BLENDS
RT ACRYLIC FIBERS
NYLON BLENDS
POLYESTER FIBER BLENDS
WOOL BLENDS

ACRYLIC FIBERS
UF ACRYLICS
NT ACRILAN (TN)
CRESLAN (TN)
NITRON (TN)
ORLON (TN)
ORLON SAYELLE (TN)
ZEFKROME (TN)
ZEFRAN (TN)
BT MAN MADE FIBERS
SYNTHETIC FIBERS
RT ACRYLIC DYES
ACRYLONITRILE
MODACRYLIC FIBERS
NYTRIL FIBERS
OLEFIN FIBERS
POLYACRYLATES
POLYACRYLONITRILE
POLYMETHACRYLATES
POLYVINYLS

ACRYLIC RESINS
BT RESINS
THERMOPLASTIC RESINS
RT POLYESTER RESINS

ACRYLIC SALTS
NT SODIUM ACRYLATE
BT ACRYLIC COMPOUNDS
RT ACRYLIC ACID
ACRYLIC ESTERS
POLYACRYLATES
SODIUM POLYACRYLATE

ACRYLICS
USE ACRYLIC FIBERS

ACRYLONITRILE
BT ACRYLIC COMPOUNDS
NITRILES
RT ACRYLAMIDE
ACRYLIC FIBERS
CHEMICAL MODIFICATION (FIBERS)
CYANOETHYLATED COTTON
CYANOETHYLATION
POLYACRYLONITRILE

ACTINIC DEGRADATION
USE PHOTOCHEMICAL DEGRADATION

ACTIVATED CARBON
UF DECOLORIZING CARBON
DEODORIZING CARBON
BT CARBON
RT CARBON BLACK
DEODORANTS

ACTIVATION ENERGY
BT ENERGY
PHYSICAL CHEMICAL PROPERTIES
RT DYEING
FREE VOLUME
MACRORHEOLOGY
RHEOLOGY

ACTIVATORS
USE CATALYSTS

ACTIVE SPORTSWEAR
USE ATHLETIC CLOTHING

ACTUAL DRAFT
BT DRAFT
RT BREAK DRAFT
DRAW RATIO
FALSE DRAFT

ACYLATION
NT ACETYLATION
BT REACTIONS (CHEMICAL)
RT ESTERIFICATION

ACYL ESTER SULFONATES
NT IGEPON A (TN)
BT SULFONATE SURFACTANTS

ADAPTATION
RT ADJUSTMENTS
COMPENSATION
IMPROVEMENTS
INSERTION
INSTALLATION
MODIFICATION

ADD ON
RT FINISH (SUBSTANCE ADDED)
IMPREGNATION
MANGLING
PADDING
WET PICKUP

ADDITION
BT COMPUTATIONS
RT DIVISION
MULTIPLES
MULTIPLICATION
SUBTRACTION

ADDITION POLYMERIZATION
BT POLYMERIZATION
RT ADDITION POLYMERS
CONDENSATION POLYMERIZATION
COPOLYMERIZATION
FREE RADICALS
INITIATORS
VINYL GROUPS
VINYLATED DYES

ADDITION POLYMERS
RT ADDITION POLYMERIZATION
CONDENSATION POLYMERIZATION
COPOLYMERS
POLYMERS
VINYL COMPOUNDS
VINYL FIBERS (GENERAL)

ADDITIVES (CHEMICAL)
RT BUILDERS (DETERGENTS)
CATALYSTS
DYEING AUXILIARIES
FINISH (SUBSTANCE ADDED)
TEXTILE CHEMICALS

ADHESION
UF STICKING (ADHESION)
BT MECHANICAL PROPERTIES
RT ADHESIVES
BONDING
CEMENTING
COHESION
DELAMINATING
FOAM BACKED FABRICS
FUSION (MELTING)
GLUING
HEAT SEALING
JOINING
LAMINATED FABRICS
LAMINATING
NONWOVEN FABRICS
PEELING
RESINS
SEALING
SEAMING
TACK
TACKING
TAPING
THERMOPLASTICITY
SYNTHETIC LEATHER
VINYL COATED FABRICS

ADHESIVE BACKING
BT BACKING (MATERIAL)
RT BACKING FABRICS

ADHESIVE BONDED (NONWOVENS)
UF ADHESIVE BONDED FABRICS
BT BONDED FIBER FABRICS
NONWOVEN FABRICS
RT BINDERS (NONWOVENS)
BONDING

ADHESIVE BONDED FABRICS
USE ADHESIVE BONDED (NONWOVENS)

ADHESIVES
    BT BACKING COMPOUNDS
    RT ADHESION
       BINDERS (NONWOVENS)
       BONDED YARN FABRICS
       BONDING STRENGTH
       CEMENTS
       COHESION
       GLUES
       GUMS
       LAMINATING RESINS
       NONWOVEN FABRICS
       PASTES
       SEALERS
       SEAMS
       SIZES (SLASHING)
       TAPING

ADHESIVES (FOOTWEAR)
    BT FOOTWEAR COMPONENTS

ADIPALDEHYDE
    BT DIALDEHYDES
    RT CROSSLINKING
       GLUTARALDEHYDE

ADIPIC ACID
    BT CARBOXYLIC ACIDS
       DIBASIC ACIDS
       ORGANIC ACIDS
    RT MONOMERS
       NYLON 66
       POLY (HEXAMETHYLENE ADIPAMIDE)
       POLY (M-XYLYLENE ADIPAMIDE)
       TEREPHTHALIC ACID

ADJUSTMENTS
    UF SELF ADJUSTING ACTION
    RT ACCURACY
       ADAPTATION
       ALIGNMENT
       ALTERATIONS
       BALANCE
       CALIBRATION
       CLEARANCES
       COMPARISON
       COMPENSATION
       CONTROLLING
       CORRECTION
       DRIFT
       ERRORS
       FITTING
       FLAT SETTINGS (CARD)
       FLUCTUATIONS
       FOCUSING
       INCREASE
       INSERTION
       INSTALLATION
       LEVELLING
       MAINTENANCE
       MINIMIZATION
       OPTIMIZATION
       POSITIONING
       PROCESS VARIABLES
       REVERSAL
       REVISIONS
       SETTING (ADJUSTMENTS)
       TOLERANCES

ADSORPTION
    NT PREFERENTIAL ADSORPTION
    BT SORPTION
    RT ABSORBERS
       ABSORPTION (MATERIAL)
       ADSORPTION RATE
       ADSORPTIVITY
       CHROMATOGRAPHY
       DESORPTION
       DIFFUSION
       DYE AFFINITY
       DYEING
       DYEING RATE
       DYE MIGRATION
       EXTRACTION
       LANGMUIR ISOTHERM
       PARTICLE SIZE
       POROSITY
       SITES (DYEING)
       SORPTION OF DYES
       SORPTION OF GASES
       SORPTION OF LIQUIDS
       SORPTION OF WATER
       SURFACE PROPERTIES (PHYSICAL CHEMICAL)

ADSORPTION ISOTHERM
    UF ABSORPTION ISOTHERM
    NT LANGMUIR ISOTHERM
       BET ISOTHERM
       FREUNDLICH ISOTHERM
    RT ABSORPTION RATE
       ADSORPTION RATE
       ADSORPTIVITY
       DYE AFFINITY
       DYEING THEORY
       EQUILIBRIUM (CHEMICAL)
       EQUILIBRIUM (DYEING)
       PARTITION COEFFICIENT
       SORPTION EQUILIBRIUM

ADSORPTION RATE
    BT SORPTION RATE
    RT ABSORPTION (MATERIAL)
       ABSORPTION RATE
       ADSORPTION
       ADSORPTION ISOTHERM
       EQUILIBRIUM (DYEING)
       SORPTION

ADSORPTIVITY
    RT ABSORPTION (MATERIAL)
       ADSORPTION
       ADSORPTION ISOTHERM
       SORPTION

ADVANTAGES
    RT ALTERATIONS
       IMPROVEMENTS
       INCREASE
       MINIMIZATION
       MODIFICATION
       OPTIMIZATION
       QUALITY

AERATED YARNS
    RT TEXTURED YARNS (FILAMENT)

AERESS (TN)
    BT MODACRYLIC FIBERS
    RT ACRYLIC FIBERS

AERODYNAMIC CARDING
    BT CARDING
    RT AERODYNAMICS
       CARD SLIVERS
       CARD WEBS
       CARDS
       COTTON CARDING
       DOFFERS (CARD)
       DOFFING (OF A CARD)
       DUO CARDS
       SLIVER CANS
       STRIPPERS
       STRIPPING ACTION
       WOOLEN CARDING
       WOOLEN CARDS
       WORSTED CARDING

AERODYNAMIC CARDS
    BT CARDS
    RT AERODYNAMICS

AERODYNAMIC CONFIGURATIONS
    RT AERODYNAMICS
       AIR DRAG
       AIR RESISTANCE
       BALLOON SHAPE
       BALLOON DYNAMICS
       BALLOON STABILITY
       BALLOONS
       DRAG

AERODYNAMIC HEATING
    RT AERODYNAMICS
       HEATING

AERODYNAMICS
    BT DYNAMICS
       MECHANICS
    RT AERODYNAMIC CARDING
       AERODYNAMIC CARDS
       AERODYNAMIC CONFIGURATIONS
       AERODYNAMIC HEATING
       AIR FLOW
       AIR JET TEXTURING
       AIR RESISTANCE
       BOUNDARY LAYER CONTROL
       CONTROL SURFACES
       CURRENTS
       DRAG
       FLUID FLOW
       FLUID FRICTION
       FLUID SPINNING
       HYPERSONIC FLOW
       INVISCID FLOW
       JET QUENCHING
       LAMINAR FLOW
       LIFT (FLUID)
       PARACHUTES
       PERMEABILITY
       PNEUMAFIL (TN)
       POROSITY
       PRESSURE DROP
       STABILITY
       THERMODYNAMICS
       TURBULENCE
       TURBULENT FLOW
       VISCOUS FLOW
       WIND TUNNELS

AEROSOL 18 (TN)
    BT SULFOSUCCINAMIDES

AEROSOL 22 (TN)
    BT SULFOSUCCINAMIDES

AEROSOLS
    BT SOLS
    RT ATOMIZERS
       FOAMS (FROTH)
       HYDROSOLS
       SPRAYING

AESTHETIC APPEAL
    BT AESTHETIC PROPERTIES
       FABRIC PROPERTIES
       FABRIC PROPERTIES (AESTHETIC)
    RT APPEARANCE
       BLOOM
       BRIGHTNESS
       COLOR
       DESIGN
       DRABNESS
       DRAPE
       LIVELINESS
       LUSTER
       RICHNESS
       TEXTURE

AESTHETIC PROPERTIES
    NT AESTHETIC APPEAL
       APPEARANCE
       BLOOM
       BRIGHTNESS
       COMFORT
       COMPRESSIBILITY
       DRABNESS
       DRAPE
       DULLNESS
       HAIRINESS
       HAND
       LIVELINESS
       LUSTER
       LUXURIOUSNESS
       PATTERN DEFINITION
       RICHNESS
       ROUGHNESS
       SLICKNESS
       SMOOTHNESS
       SOFTNESS
       SPARKLE
       TEXTURE
       TRANSLUCENCY
       TRANSPARENCY
    RT BIOCHEMICAL PROPERTIES
       CHARACTERISTICS
       CHEMICAL PROPERTIES
       CLEANLINESS
       COLOR
       COVER
       DESIGN
       END USE PROPERTIES
       FABRIC PROPERTIES
       FABRIC PROPERTIES (AESTHETIC)
       FABRIC PROPERTIES (STRUCTURAL)
       FIBER PROPERTIES
       FINISH (PROPERTY)
       MECHANICAL PROPERTIES
       OPTICAL PROPERTIES
       PATTERN (FABRICS)
       PHYSICAL CHEMICAL PROPERTIES
       PHYSICAL PROPERTIES (EXCLUDING MECHANICAL)
       PROPERTIES
       RESILIENCE
       SHADE
       SLUBINESS
       STITCH CLARITY
       SURFACE CONTOUR
       SURFACE FRICTION
       WEAVE

AFFINITY (DYES)
    USE DYE AFFINITY

AFTERCHROME DYEING
    USE MORDANT DYEING

AFTERCHROME DYES
    USE MORDANT DYES

AFTERTREATMENTS (GENERAL)
    NT BLEACHING
       CROSSLINKING
       DYE FIXING
       RINSING
       SCOURING
       SOAPING
    RT DRY FINISHING
       DYEING
       FINISHING PROCESS (GENERAL)
       PRETREATMENTS (GENERAL)
       PRINTING
       WET FINISHING

AFTERWASH
    USE SOAPING

AFTERWELT
    UF ANTILADDER BANDS
       ANTIRUN COURSES
       GARTER BANDS
       SHADOW WELT
    RT KNITTING
       STOCKINGS

AGERS
RT ACID AGING
AGING (STEAMING)
DYEING
FLASH AGING
PRINTING
REACTIVE PRINTING
STEAM
STEAMERS
STEAMING
VAT PRINTING

AGILON (TN)
BT CRIMPED YARNS
EDGE CRIMPED YARNS
STRETCH YARNS (FILAMENT)
TEXTURED YARNS (FILAMENT)
RT BICOMPONENT FIBER YARNS (FILAMENT)
BICOMPONENT FIBER YARNS (STAPLE)
BULKED YARNS
EDGE CRIMPING
FALSE TWIST YARNS

AGING (MATERIALS)
NT AGING (RAYON MANUFACTURE)
AGING (STEAMING)
AGING (STORAGE)
RT CHEMICAL PROPERTIES
DEGRADATION
DEGRADATION PROPERTIES
INHIBITION
STABILIZATION
TIME

AGING (RAYON MANUFACTURE)
BT AGING (MATERIALS)
RT ALKALI STEEPING
DOPE (POLYMER)

AGING (STEAMING)
BT AGING (MATERIALS)
NT ACID AGING
FLASH AGING
RT ACID SHOCK METHOD
AGERS
DYE FIXING
PRINTING
REACTIVE PRINTING
STEAM
STEAMING
VAT PRINTING

AGING (STORAGE)
UF STORAGE AGING
BT AGING (MATERIALS)
RT DEGRADATION PROPERTIES
STORAGE VESSELS
TRANSPORTING VESSELS
WEATHERING

AGITATORS
USE STIRRERS

AGRICULTURAL FABRICS
BT FABRICS (BY END USES)
RT AUTOMOTIVE FABRICS
BAGS
FABRICS (ACCORDING TO STRUCTURE)
FURNISHING FABRICS
GUNNY SACKS
HOUSEHOLD FABRICS
INDUSTRIAL FABRICS
TARPAULINS

AGGREGATION
RT ACCUMULATION
ASSEMBLIES
COAGULATING
COLLECTION
DYES
ZETA POTENTIAL

AIR
RT BLOWERS
DRYING
OXIDIZING AGENTS
OXYGEN
PNEUMATIC INSTRUMENTS

AIR BEARINGS
BT BEARINGS
RT ANTIFRICTION BEARINGS
AXLES
BALL BEARINGS
BEARING ALLOYS
BUSHINGS
CONICAL BEARINGS
FALSE TWIST SPINDLES
JOURNAL BEARINGS
LUBRICATION
MAGNETIC BEARINGS
NEEDLE BEARINGS
PLAIN BEARINGS
RADIAL BEARINGS
ROLLER BEARINGS
RT SLEEVE BEARINGS
SPINDLES
TAPERED ROLLER BEARINGS
THRUST BEARINGS

AIR BRAKES
BT BRAKES (FOR ARRESTING MOTION)
RT MAGNETIC BRAKES
MECHANICAL BRAKES

AIR BRIDGES
RT CONDENSER SCREENS
CURLATORS (TN)
NONWOVEN FABRIC MACHINES
RANDO FEEDER (TN)
RANDO WEBBER (TN)
RANDO WEBS (TN)
TUFT FORMERS
WEBS
WET PROCESS (WEB)

AIR CONDITIONING
RT COOLING
HEATING
HUMIDITY CONTROL
REFRIGERATION
TEMPERATURE CONTROL

AIR DRAG
BT FORCE
RT AERODYNAMIC CONFIGURATIONS
AIR RESISTANCE
BALLOON DYNAMICS
BALLOON TENSION
BOUNDARY LAYER
CURRENTS
DRAG
FLUID FLOW
LIFT (FLUID)
PRESSURE DROP
TERMINAL VELOCITY
WIND TUNNELS
YARN DYNAMICS

AIR FLOW
RT AERODYNAMICS
AIR PERMEABILITY
AIR RESISTANCE
AIR SPINNING
AIR STREAMS
CAUSTICAIRE SCALE
CAUSTICAIRE VALUE
CONVECTION
COUNTERFLOW PROCESSES
CURRENTS
DRYERS
DRYING
EDDIES
FLUID FLOW
FLUID FRICTION
FLUID MECHANICS
FLUID SPINNING
FLUID TWISTING
FLUIDIZED BEDS
HEATING
HOT AIR DRYING
HOT AIR HEATING
HYPERSONIC FLOW
LAMINAR FLOW
MASS TRANSFER
MICRONAIRE
PNEUMATIC CONTROL
PNEUMATIC INSTRUMENTS
TURBULENCE
TURBULENT FLOW
VISCOSITY
WIND TUNNELS

AIR FLOW TESTERS
USE AIR PERMEABILITY TESTING

AIR JET BULKED YARNS
USE AIR JET TEXTURED YARNS

AIR JET BULKERS
USE AIR JET TEXTURING MACHINES

AIR JET BULKING
USE AIR JET TEXTURING

AIR JET LOOMS
BT JET LOOMS
LOOMS
RT DRAPER SHUTTLELESS LOOMS (TN)
OUTSIDE FILLING SUPPLY
RAPIER LOOMS
SHUTTLELESS WEAVING
SULZER LOOMS (TN)
WATER JET LOOMS

AIR JET TEXTURED YARNS
UF AIR JET BULKED YARNS
NT LOOP YARNS
SKYLOFT (TN)
TASLAN (TN)
BT BULKED YARNS
FILAMENT YARNS
TEXTURED YARNS (FILAMENT)
RT BICOMPONENT FIBER YARNS (FILAMENT)
BICOMPONENT FIBER YARNS (STAPLE)
BULKING
CRIMPING
EDGE CRIMPED YARNS
FALSE TWIST YARNS
FLUID TWISTING
GEAR CRIMPED YARNS
INTERLACING (FILAMENT IN YARN)
JET QUENCHING
PRODUCERS YARNS
ROTOSET YARNS (TN)
STUFFER BOX CRIMPED YARNS

AIR JET TEXTURING
UF AIR JET BULKING
AIR TEXTURING
JET BULKING
JET CRIMPING
JET LOOPING
BT CRIMPING
TEXTURING
RT AERODYNAMICS
AIR JET TEXTURING MACHINES
BULKED YARNS
BULKING
EDGE CRIMPING
GEAR CRIMPING
JET QUENCHING
JETS
NOZZLES
STEAM JET TEXTURING
STRETCH YARNS (FILAMENT)
STUFFER BOX CRIMPING
TASLAN (TN)
TEXTURED YARNS (FILAMENT)

AIR JET TEXTURING MACHINES
UF AIR JET BULKERS
RT AIR JET TEXTURING
BULKING
CRIMPING
NOZZLES
STEAM JET TEXTURING

AIR LAY RANDOM WEB
USE RANDO WEBS (TN)

AIR PERMEABILITY
BT PERMEABILITY
RT AIR FLOW
AIR RESISTANCE
COVER FACTOR
FRAZIER PERMEOMETER
GURLEY PERMEOMETER
PACKING FACTOR
PARACHUTES
PERMEABILITY TESTING
PORES
WATER PERMEABILITY
WATER REPELLENCY
WATER VAPOR TRANSMISSION
WATERPROOFING
WIND TUNNELS

AIR PERMEABILITY TESTING
UF AIR FLOW TESTERS
BT PERMEABILITY TESTING
TESTING
RT FRAZIER PERMEOMETER
GURLEY PERMEOMETER
WATER RESISTANCE TESTING

AIR POLLUTION
BT POLLUTION
RT CONTAMINATION
WEATHERING
WEATHERING TESTING

AIR RESISTANCE
RT AERODYNAMIC CONFIGURATIONS
AERODYNAMICS
AIR DRAG
AIR FLOW
AIR PERMEABILITY
BALLOON DYNAMICS
BALLOON SHAPE
BALLOONS
CAUSTICAIRE SCALE
CAUSTICAIRE VALUE
COVER FACTOR
DRAG
FABRIC INTERSTICES
FLUID FLOW
FLUID FRICTION
MICRONAIRE
MOISTURE VAPOR TRANSMISSION
PARACHUTES
PORES
POROSITY
WATER VAPOR TRANSMISSION

AIR SPINNING
BT FLUID SPINNING
OPEN END SPINNING
SPINNING

RT AIR FLOW
AIR STREAMS
COMB SPINNING
CONTINUOUS SPINNING (STAPLE FIBER)
FLUID TWISTING
RING SPINNING

AIR STREAMS
RT AIR FLOW
AIR SPINNING
CURRENTS
FLUID FLOW
LAMINAR FLOW
TURBULENCE
TURBULENT FLOW

AIR TEXTURING
USE AIR JET TEXTURING

AIR TUCK STITCHES
BT STITCHES (SEWING)

AIR TUCKING
BT TUCKING (GARMENT MANUFACTURE)

AIR TWISTING
USE FLUID TWISTING

ALBERT TWILL
BT FLAT WOVEN FABRICS
WOVEN FABRICS
RT TWILL WEAVES

ALBUMIN
BT PROTEINS

ALCIAN BLUE (TN)
BT INGRAIN DYES
RT BASIC DYES
REACTIVE DYES

ALCOHOL SULFATES
USE ALKYL SULFATES

ALCOHOLS
NT ALKANOLAMIDES
ALKANOLAMINES
ALLYL ALCOHOL
BENZYL ALCOHOL
BUTYL ALCOHOL
ETHANOL
FATTY ALCOHOLS
HALOHYDRINS
HYDROXYACETONE
ISOPROPYL ALCOHOL
METHANOL
POLY (VINYL ALCOHOL)
POLYOLS
SORBITOL
RT CARBOHYDRATES
ESTERS
ETHERS
HYDROXIDES
PHENOLS
SOLVENTS

ALDEHYDE GROUPS
RT ALDEHYDES
ALKOXYL GROUPS
AMINO GROUPS
CARBONYL GROUPS
CARBOXYL GROUPS
END GROUPS

ALDEHYDES
NT ACROLEIN
CHLORAL
DIALDEHYDE CELLULOSE
DIALDEHYDE STARCH
DIALDEHYDES
FORMALDEHYDE
POLYACROLEIN
BT REACTANTS
RT ACETALS
ALDEHYDE GROUPS
CARBOHYDRATES
CARBONYL GROUPS
FINISH (SUBSTANCE ADDED)
KETONES
REDUCING AGENTS

ALENCON
BT HAND MADE LACES
NEEDLE POINT LACES

ALEO
BT HAND MADE LACES
PILLOW LACES (BOBBIN LACES)

ALGINATE FIBERS
BT MAN MADE FIBERS
REGENERATED FIBERS (EXCLUDING CELLULOSIC AND PROTEIN)
RT ALGINIC ACID

ALGINIC ACID
BT CARBOXYLIC ACIDS
ORGANIC ACIDS
POLYSACCHARIDES
RT ALGINATE FIBERS
PRINTING PASTES
THICKENING AGENTS

ALHAMBRA QUILT
RT JACQUARD WEAVES

ALICYCLIC COMPOUNDS
NT ALICYCLIC HYDROCARBONS
RT ALIPHATIC COMPOUNDS
ALIPHATIC HYDROCARBONS
AROMATIC COMPOUNDS
HETEROCYCLIC COMPOUNDS

ALICYCLIC HYDROCARBONS
BT ALICYCLIC COMPOUNDS
HYDROCARBONS
RT ALIPHATIC HYDROCARBONS
AROMATIC HYDROCARBONS
UNSATURATED HYDROCARBONS

ALIGNMENT
RT ADJUSTMENTS
CALIBRATION
CLEARANCES
CONTROLLING
CORRECTION
INSERTION
INSTALLATION
MAINTENANCE
ORIENTATION
POSITIONING
TOLERANCES

ALIGNMENT CHARTS
USE NOMOGRAMS

ALIPHATIC COMPOUNDS
NT ALIPHATIC HYDROCARBONS
FATTY ACIDS
FATTY ALCOHOLS
SATURATED COMPOUNDS
UNSATURATED COMPOUNDS
RT ALICYCLIC COMPOUNDS
AROMATIC COMPOUNDS
HETEROCYCLIC COMPOUNDS

ALIPHATIC HYDROCARBONS
NT ALKANES
ALKENES
ALKYNES
BRANCHED ALIPHATIC HYDROCARBONS
NORMAL ALIPHATIC HYDROCARBONS
BT ALIPHATIC COMPOUNDS
HYDROCARBONS
RT ALICYCLIC COMPOUNDS
ALICYCLIC HYDROCARBONS
AROMATIC COMPOUNDS
AROMATIC HYDROCARBONS
UNSATURATED HYDROCARBONS

ALKALI CELLULOSE
UF SODA CELLULOSE
BT CELLULOSE DERIVATIVES
RT ACETATE FIBERS
CARBOXYMETHYLCELLULOSE
CELLULOSE
CELLULOSE ESTERS
CELLULOSE ETHERS
CELLULOSE NITRATE
CELLULOSE XANTHATE
CYANOETHYLCELLULOSE
RAYON (REGENERATED CELLULOSE FIBERS)
REGENERATED CELLULOSE

ALKALI DEGRADATION
BT DEGRADATION
RT ALKALI HYDROLYSIS
CHEMICAL PROPERTIES
DEGRADATION PROPERTIES
TENDERING

ALKALI HYDROLYSIS
UF SAPONIFICATION
BT HYDROLYSIS
RT ACID HYDROLYSIS
ALKALI DEGRADATION
CHEMICAL STABILITY

ALKALI SOLUBILITY
BT DEGRADATION PROPERTIES
CHEMICAL PROPERTIES
RT ACID SOLUBILITY
ALKALIES
ALKALINE SOLUTIONS
DEGRADATION
KERATIN
SOLUBILITY
UREA BISULFITE SOLUBILITY
WOOL

ALKALI STEEPING
RT AGING (RAYON MANUFACTURE)
RAYON (REGENERATED CELLULOSE FIBERS)

ALKALI SOAPS
NT AMMONIUM SOAPS
POTASSIUM OLEATE
POTASSIUM SOAPS
POTASSIUM STEARATE
SODIUM OLEATE
SODIUM PALMITATE
SODIUM ROSINATE
SODIUM SOAPS
SODIUM STEARATE
BT SOAPS
RT AMINE SOAPS
METALLIC SOAPS
OLEATE SALTS

ALKALIES
NT CARBONATES
HYDROXIDES
BT BASES
RT ALKALI SOLUBILITY
ALKALINITY

ALKALINE SOLUTIONS
BT SOLUTIONS (LIQUID)
RT ACID SOLUTIONS
ALKALI SOLUBILITY
ALKALINITY
DYE SOLUTIONS
PH
SODIUM HYDROXIDE

ALKALINITY
RT ACIDITY
ALKALIES
ALKALINE SOLUTIONS
PH

ALKANES
UF SATURATED ALIPHATIC HYDROCARBONS
SATURATED HYDROCARBONS
BT ALIPHATIC HYDROCARBONS
SATURATED COMPOUNDS
RT ALKENES
ALKYNES

ALKANOL S (TN)
BT TETRAHYDRONAPHTHALENE SULFONATES

ALKANOLAMIDES
NT DIETHANOLAMIDES
MONOETHANOLAMIDES
BT ALCOHOLS
AMIDES
RT ALKANOLAMINES
DIETHANOLAMIDE SURFACTANTS

ALKANOLAMINE SOAPS
NT DIETHANOLAMINE SOAPS
TRIETHANOLAMINE SOAPS
BT AMINE SOAPS

ALKANOLAMINES
NT DIETHANOLAMINE
MONOETHANOLAMINE
TRIETHANOLAMINE
BT ALCOHOLS
AMINES
RT ALKANOLAMIDES
SOAPS

ALKENES
UF OLEFINIC HYDROCARBONS
NT DIENES
POLYENES
BT ALIPHATIC HYDROCARBONS
OLEFINIC COMPOUNDS
RT ALKANES
ALKYNES

ALKOXYL GROUPS
RT ALDEHYDE GROUPS
AMINO GROUPS
CARBONYL GROUPS
CARBOXYL GROUPS
END GROUPS
ETHERS

ALKYD RESINS
USE POLYESTER RESINS

ALKYL CARBAMATE FORMALDEHYDE
USE METHYLOL CARBAMATES

ALKYL DIMETHYLAMINE SURFACTANTS
BT AMINE SURFACTANTS

ALKYL ETHOXY PHOSPHATES
BT PHOSPHATE SURFACTANTS
RT ALKYL PHOSPHATE SURFACTANTS
ALKYLARYL ETHOXY PHOSPHATES

ALKYL ETHOXY SULFATES
UF ETHERALCOHOL SULFATES
BT SULFATE SURFACTANTS
RT ALKYL SULFATES
ALKYLAMIDE SULFATES
ALKYLARYL OXYETHYL SULFONATES

ALKYL GLYCEROSULFATES
UF MONOGLYCERIDE SULFATES
SULFATED GLYCERIDES
SULFOGLYCERIDES
NT LECITHIN SULFATE
BT SULFATE SURFACTANTS

ALKYL PHENOXYETHYL SULFONATES
BT ALKYLARYL OXYETHYL SULFONATES
RT ALKYLARYL ETHOXY SULFATES

ALKYL PHOSPHATE SURFACTANTS
BT PHOSPHATE SURFACTANTS
RT ALKYL ETHOXY PHOSPHATES
ALKYLARYL ETHOXY PHOSPHATES

ALKYL SULFATES
UF ALCOHOL SULFATES
FATTY ALCOHOL SULFATES
NT DODECYL SULFATE
GARDINOL (TN)
OCTYL SULFATE
SULFATED OLEFINS
BT SULFATE SURFACTANTS
RT ALKYL ETHOXY SULFATES
ALKYL SULFONATES
ALKYLAMIDE SULFATES

ALKYL SULFONATES
UF HYDROCARBON SULFONATES
PETROLEUM SULFONATES
BT SULFONATE SURFACTANTS
RT ALKYL SULFATES
ALKYLAMIDE SULFONATES
ALKYLARYL SULFONATES
ALKYLAMINO SULFONATE SURFACTANTS

ALKYLAMIDE SULFATES
BT SULFATE SURFACTANTS
RT ALKYL ETHOXY SULFATES
ALKYL SULFATES
ALKYLAMIDE SULFONATES
ALKYLAMINO SULFATE SURFACTANTS
ALKYLAMINO SULFONATE SURFACTANTS

ALKYLAMIDE SULFONATES
NT METHYLTAURIDE SURFACTANTS
BT SULFONATE SURFACTANTS
RT ALKYL SULFONATES
ALKYLAMIDE SULFATES
ALKYLBENZENE SULFONATES

ALKYLAMINE SURFACTANTS
UF ALKYLDIMETHYLAMINE SURFACTANTS
PRIMARY AMINE SURFACTANTS
SECONDARY AMINE SURFACTANTS
TERTIARY AMINE SURFACTANTS
NT ARMACS (TN)
ARMEENS (TN)
BT AMINE SURFACTANTS
RT ALKYLMORPHOLINE SURFACTANTS
AMIDOAMINE SURFACTANTS
AMINE OXIDE SURFACTANTS
IMIDAZOLINE SURFACTANTS
OXAZOLINE SURFACTANTS
POLYALKYLAMINE SURFACTANTS
POLYETHOXY ALKYLAMINE SURFACTANTS
POLYETHOXY POLYPROPOXY ETHYLENEDIAMINE

ALKYLAMINO CARBOXYLATE SURFACTANTS
BT AMPHOTERIC SURFACTANTS

ALKYLAMINO SULFATE SURFACTANTS
BT AMPHOTERIC SURFACTANTS
RT ALKYLAMIDE SULFATES
ALKYLAMINO SULFONATE SURFACTANTS

ALKYLAMINO SULFONATE SURFACTANTS
BT AMPHOTERIC SURFACTANTS
RT ALKYL SULFONATES
ALKYLAMIDE SULFATES
ALKYLAMINO SULFATE SURFACTANTS

ALKYLARYL ETHOXY PHOSPHATES
BT PHOSPHATE SURFACTANTS
RT ALKYL ETHOXY PHOSPHATES
ALKYL PHOSPHATE SURFACTANTS

ALKYLARYL ETHOXY SULFATES
BT SULFATE SURFACTANTS
RT ALKYL PHENOXYETHYL SULFONATES

ALKYLARYL OXYETHYL SULFONATES
NT ALKYL PHENOXYETHYL SULFONATES
BT SULFONATE SURFACTANTS
RT ALKYL ETHOXY SULFATES
ALKYLARYL SULFONATES

ALKYLARYL POLYETHOXY ETHERS
USE POLYETHOXY ALKYLARYL ETHERS

ALKYLARYL SULFONATES
NT ALKYLBENZENE SULFONATES
ALKYLBIPHENYL SULFONATES
ALKYLNAPHTHALENE SULFONATES
BT SULFONATE SURFACTANTS
RT ALKYL SULFONATES
ALKYLARYL OXYETHYL SULFONATES
NAPHTHALENE FORMALDEHYDE SULFONATES
TETRAHYDRONAPHTHALENE

ALKYLATION
NT ETHYLATION
METHYLATION
BT REACTIONS (CHEMICAL)
RT BENZYLATION
DIAZOMETHANE
ESTERIFICATION
ETHERIFICATION
ETHOXYLATION
THIOETHERS

ALKYLBENZENE SULFONATES
NT LINEAR ALKYLBENZENE SULFONATES
NACCONOL NR (TN)
NONYLBENZENE SULFONATES
OCTYLBENZENE SULFONATES
BT ALKYLARYL SULFONATES
RT ALKYLAMIDE SULFONATES
ALKYLBIPHENYL SULFONATES
ALKYLNAPHTHALENE SULFONATES

ALKYLBIPHENYL SULFONATES
BT ALKYLARYL SULFONATES
RT ALKYLBENZENE SULFONATES

ALKYLDI (HYDROXYETHYL) AMINE OXIDE SURFACTANTS
BT AMINE OXIDE SURFACTANTS
RT ALKYLDIMETHYLAMINE OXIDE SURFACTANTS

ALKYLDIMETHYLAMINE OXIDE SURFACTANTS
BT AMINE OXIDE SURFACTANTS
RT ALKYLDI (HYDROXYETHYL) AMINE OXIDE SURFACTANTS

ALKYLDIMETHYLAMINE SURFACTANTS
USE ALKYLAMINE SURFACTANTS

ALKYLENE CARBONATES
NT ETHYLENE CARBONATE
PROPYLENE CARBONATE
BT ESTERS
RT CARRIERS (DYEING)
SOLVENTS

ALKYLETHER SULFONATES
BT SULFONATE SURFACTANTS

ALKYLMORPHOLINE SURFACTANTS
BT AMINE SURFACTANTS
RT ALKYLAMINE SURFACTANTS
AMINE OXIDE SURFACTANTS

ALKYLNAPHTHALENE SULFONATES
NT AMYLNAPHTHALENE SULFONATES
BUTYLNAPHTHALENE SULFONATES
NEKAL A (TN)
NEKAL B (TN)
NONYLNAPHTHALENE SULFONATES
BT ALKYLARYL SULFONATES
RT ALKYLBENZENE SULFONATES
NAPHTHALENE FORMALDEHYDE SULFONATES
TETRAHYDRONAPHTHALENE SULFONATES
WETTING AGENTS

ALKYNES
UF ACETYLENIC HYDROCARBONS
NT ACETYLENE
BT ACETYLENIC COMPOUNDS
ALIPHATIC HYDROCARBONS
RT ALKANES
ALKENES

ALL OVER (LACE)
BT LACES

ALLIGATOR
BT LEATHER
RT SYNTHETIC LEATHER

ALLOTMENT
RT END BREAKAGES
FLOW (MATERIALS)
INVENTORIES
MANAGEMENT
PATROLLING
PROCESS CONTROL
PROCESS EFFICIENCY
PRODUCTION

ALLOWANCE FACTOR (WARP PREPARATION)
RT ROLLER SLIPPAGE
SLIPPAGE
WARP PREPARATION

ALLOYS
NT BEARING ALLOYS
METALLIC ALLOYS
POLYMER ALLOYS
STAINLESS STEEL
RT METALS

ALLYL ALCOHOL
BT ALCOHOLS
ALLYL COMPOUNDS

ALLYL BROMIDE
BT ALLYL COMPOUNDS
BROMINE COMPOUNDS

ALLYL CHLORIDE
BT ALLYL COMPOUNDS
CHLORINE COMPOUNDS

ALLYL COMPOUNDS
NT ALLYL ALCOHOL
ALLYL BROMIDE
ALLYL CHLORIDE
ALLYL GLYCIDYL ETHER
METHALLYL CHLORIDE
BT MONOMERS
RT ACRYLIC COMPOUNDS
VINYL COMPOUNDS

ALLYL GLYCIDYL ETHER
BT ALLYL COMPOUNDS
OXIRANE COMPOUNDS

ALLYLCELLULOSE
BT CELLULOSE ETHERS

ALPACA
BT HAIR
KERATIN FIBERS
PROTEIN FIBERS
RT CAMEL HAIR
CASHMERE

ALPACA FABRICS
BT FLAT WOVEN FABRICS
WOVEN FABRICS

ALPHA BETA TRANSFORMATION
RT CRYSTALLINITY
KERATIN
X RAY ANALYSIS

ALPHA KERATIN
USE KERATIN

ALPHA PARTICLES
RT BETA PARTICLES
RADIATION

ALPHASULFOMETHYL PALMITATE
USE SULFOFATTY ACID ESTERS

ALPHASULFOPALMITIC ACID
USE SULFOFATTY ACIDS

ALTERATIONS
UF ALTERING
NT LETTING OUT (GARMENTS)
TAKING IN (GARMENTS)
BT GARMENT MANUFACTURE
RT ADJUSTMENTS
ADVANTAGES
FITTING
GARMENTS
INCREASE
MINIMIZATION
OPTIMIZATION
REMODELLING
REPAIRS
REVISIONS
TAILORING
TOLERANCES

ALTERING
USE ALTERATIONS

ALTERNATE PICKS
RT LOOMS
PICKING (WEAVING)
WEAVING

ALTERNATING TWIST
BT TWIST
RT FALSE TWIST
FLUID TWISTING
HIGH TWIST
INTERLACING (FILAMENT IN YARN)
LOW TWIST
REAL TWIST
S TWIST
YARNS
Z TWIST
ZERO TWIST

ALUMINA
UF ALUMINUM OXIDE
NT COLLOIDAL ALUMINA
BT ALUMINUM COMPOUNDS
OXIDES
RT PIGMENTS

ALUMINUM
BT METALS
RT ALUMINUM COMPOUNDS
ALUMINUM FOIL
ALUMINUM SILICATE FIBERS

ALUMINUM ABIETATE
BT ALUMINUM COMPOUNDS
METALLIC SOAPS
RT ALUMINUM SOAPS
ALUMINUM STEARATE
ROSIN

ALUMINUM COMPOUNDS
NT ALUMINA
ALUMINUM ABIETATE
ALUMINUM SOAPS
ALUMINUM STEARATE
RT ALUMINUM
ALUMINUM SILICATE FIBERS

ALUMINUM FOIL
BT FOIL
HIGH TEMPERATURE FIBERS
INORGANIC FIBERS (MAN MADE)
METALLIC FIBERS
RT ALUMINUM

ALUMINUM OXIDE
USE ALUMINA

ALUMINUM SILICATE FIBERS
BT CERAMIC FIBERS
HIGH TEMPERATURE FIBERS
INORGANIC FIBERS (MAN MADE)
RT ALUMINUM
ALUMINUM COMPOUNDS
BORON NITRIDE FIBERS
SILICON CARBIDE FIBERS

ALUMINUM SOAPS
BT ALUMINUM COMPOUNDS
METALLIC SOAPS
RT ALUMINUM ABIETATE
ALUMINUM STEARATE

ALUMINUM STEARATE
BT ALUMINUM COMPOUNDS
METALLIC SOAPS
RT ALUMINUM ABIETATE
ALUMINUM SOAPS

AMBIENT TEMPERATURE
BT TEMPERATURE
RT DRYING
ROOM TEMPERATURE

AMBLER SUPERDRAFT
RT ACTUAL DRAFT
APRONS
BRADFORD SYSTEM PROCESSING
CASABLANCA SYSTEM
CONTROL SURFACES
DRAFT
DRAFTING (STAPLE FIBER)
DRAFTING MACHINES
HIGH DRAFT
HIGH DRAFT SPINNING
NEW BRADFORD SYSTEM PROCESSING
RING FRAMES
SPINNING
SPINNING FRAMES
TWIST CONTROL
WORSTED SYSTEM PROCESSING

AMERICAN CUT
USE CUT SYSTEM

AMERICAN EGYPTIAN COTTON
UF UPLAND COTTON
BT COTTON

AMERICAN MERINO WOOL
BT MERINO WOOL

AMERICAN RUN
USE RUN SYSTEM

AMERICAN SYSTEM PROCESSING
BT WORSTED SYSTEM PROCESSING
RT BACKWASHING
BRADFORD OPEN DRAWING
BRADFORD SYSTEM PROCESSING
CAP SPINNING
CARBONIZING
CARDING
CONTINENTAL SYSTEM PROCESSING
COTTON SYSTEM PROCESSING
GILLING
NEW BRADFORD SYSTEM PROCESSING
PICKING (OPENING)
PIN DRAFTING
SCOURING
SPINNING
TOP MAKING
TWISTING
WOOLEN SYSTEM PROCESSING
WORSTED CARDING
WORSTED COMBING
WORSTED SYSTEM SPINNING

AMERICAN UPLAND COTTON
BT COTTON
RT LONG STAPLE COTTON

AMIDES
NT ACETAMIDE
ACRYLAMIDE
ALKANOLAMIDES
BENZAMIDE
DIMETHYL ACETAMIDE
DIMETHYL FORMAMIDE
METHACRYLAMIDE
POLYACRYLAMIDE
RT AMINES
AMINO ACIDS
AMMONIUM COMPOUNDS
HYDRAZIDES
IMIDES
LACTAMS
METHYLOL AMIDES
POLYAMIDES
UREA

AMIDOAMINE SURFACTANTS
BT AMINE SURFACTANTS
RT ALKYLAMINE SURFACTANTS
IMIDAZOLINE SURFACTANTS
POLYALKYLAMINE SURFACTANTS

AMILAR (TN)
USE DACRON (TN)

AMINE END GROUPS
BT END GROUPS
RT ACID END GROUPS
AMINES
AMINO GROUPS
DEGREE OF POLYMERIZATION
MOLECULAR WEIGHT CONTROL
ORGANIC ACIDS
OVERDYEING
POLYMERIZATION

AMINE GROUPS
USE AMINO GROUPS

AMINE OXIDE SURFACTANTS
UF TERTIARY ALKYLAMINE OXIDES
TERTIARY AMINE OXIDES
NT ALKYLDI (HYDROXYETHYL) AMINE
OXIDE SURFACTANTS
ALKYLDIMETHYLAMINE OXIDE
SURFACTANTS
BT AMINE SURFACTANTS
RT ALKYLAMINE SURFACTANTS
ALKYLMORPHOLINE SURFACTANTS
AMINE OXIDES

AMINE OXIDES
RT AMINE OXIDE SURFACTANTS
AMINES

AMINE SOAPS
NT ALKANOLAMINE SOAPS
MORPHOLINE SOAPS
BT SOAPS
RT ALKALI SOAPS
AMINE SURFACTANTS
AMMONIUM SOAPS

AMINE SURFACTANTS
NT ALKYL DIMETHYLAMINE
SURFACTANTS
ALKYLAMINE SURFACTANTS
ALKYLMORPHOLINE SURFACTANTS
AMIDOAMINE SURFACTANTS
AMINE OXIDE SURFACTANTS
IMIDAZOLINE SURFACTANTS
OXAZOLINE SURFACTANTS
POLYALKYLAMINE SURFACTANTS
POLYETHOXY ALKYLAMINE
SURFACTANTS
POLYETHOXY POLYPROPOXY
ETHYLENEDIAMINE
BT CATIONIC SURFACTANTS
RT AMINE SOAPS
AMINES
ONIUM SURFACTANTS
QUATERNARY AMMONIUM
SURFACTANTS
TRIETHANOLAMINE SOAPS

AMINES
UF DIAMINES
TRIAMINES
NT ALKANOLAMINES
AMINO ACIDS
CYCLOHEXANE, 1,4-BIS
(METHYLAMINE)-
ETHYLENEDIAMINE
HEXAMETHYLENEDIAMINE
MELAMINE
PRIMARY AMINES
QUATERNARY AMMONIUM COMPOUNDS
SECONDARY AMINES
TERTIARY AMINES
BT BASES
RT AMIDES
AMINE END GROUPS
AMINE OXIDES
AMINE SURFACTANTS
AMINO ACIDS
AMINO GROUPS
AMINO RESINS
AMMONIA
AMMONIUM COMPOUNDS
AZIRIDINE COMPOUNDS
DECRYSTALLIZED COTTON
HYDROXYLAMINE
IMINES
POLYAMINES
PYRIDINE DERIVATIVES

AMINIZED COTTON
BT CHEMICALLY MODIFIED COTTON
RT ACETYLATED COTTON
ACID DYES
ANIMALIZING
BENZOYLATED COTTON
CARBOXYMETHYLATED COTTON
CYANOETHYLATED COTTON
DYEABILITY

AMINO ACIDS
NT CYSTEINE
CYSTINE
LANTHIONINE
LYSINE
THREONINE
TRYPTOPHAN
TYROSINE
BT AMINES
CARBOXYLIC ACIDS
ORGANIC ACIDS
RT AMIDES
AMMONIUM COMPOUNDS
KERATIN
LACTAMS
NYLON 6
POLYPEPTIDES
PROTEINS
SILK
WOOL

AMINO GROUPS
UF AMINE GROUPS
RT ALDEHYDE GROUPS
ALKOXYL GROUPS
AMINE END GROUPS
AMINES
ANIMAL FIBERS (GENERAL)
BASES
CARBONYL GROUPS
CARBOXYL GROUPS
CATIONIC SITES
END GROUPS
KERATIN FIBERS
POLYAMINES
PROTEIN FIBERS
PROTEINS

AMINO RESINS
UF AMINOPLASTS
NT MELAMINE FORMALDEHYDE RESINS
UREA-FORMALDEHYDE RESINS
BT RESINS
RT AMINES
DURABLE PRESS FINISHES
EASY CARE FABRICS
FINISH (SUBSTANCE ADDED)
METHYLOL COMPOUNDS
POLYAMIDES
PRECONDENSATES
PROTEINS

AMINOPLASTS
USE AMINO RESINS

AMMONIA
BT BASES
RT AMINES
AMMONIUM COMPOUNDS
AMMONIUM SULFATE
HYDROXYLAMINE

AMMONIUM COMPOUNDS
NT AMMONIUM SOAPS
AMMONIUM SULFATE
QUATERNARY AMMONIUM COMPOUNDS
RT AMIDES
AMINES
AMINO ACIDS
AMMONIA
PYRIDINE DERIVATIVES

AMMONIUM SOAPS
BT ALKALI SOAPS
AMMONIUM COMPOUNDS
RT AMINE SOAPS

AMMONIUM SULFATE
BT AMMONIUM COMPOUNDS
SULFATES
RT AMMONIA
DYEING AUXILIARIES

AMORPHOUS POLYMERS
RT AMORPHOUS REGION
CRYSTALLINE POLYMERS
DEGREE OF POLYMERIZATION
FINE STRUCTURE (FIBERS)
LONG CHAIN MOLECULES
POLYMERS
TACTIC POLYMERS

AMORPHOUS REGION
BT FINE STRUCTURE (FIBERS)
RT AMORPHOUS POLYMERS
CRYSTALLINE ORIENTATION
CRYSTALLINE REGION
DEGREE OF CRYSTALLINITY
GLASS RUBBER TRANSITION
GRAIN BOUNDARIES
LONG CHAIN MOLECULES
MATRIX (FINE STRUCTURE)
MOLECULAR STRUCTURE
WET MODULUS
WET RELAXATION
WET STRENGTH

AMPHOLYTIC SURFACTANTS
USE AMPHOTERIC SURFACTANTS

AMPHOTERIC SURFACTANTS
UF AMPHOLYTIC SURFACTANTS
NT ALKYLAMINO CARBOXYLATE
SURFACTANTS
ALKYLAMINO SULFATE SURFACTANTS
ALKYLAMINO SULFONATE
SURFACTANTS
ZWITTERIONIC SURFACTANTS
BT SURFACTANTS
RT ANIONIC SURFACTANTS
ANTIFOAM AGENTS
CATIONIC SURFACTANTS
DYEING AUXILIARIES
EMULSIFYING AGENTS
LEVELLING AGENTS
NONIONIC SURFACTANTS
WETTING
WETTING AGENTS

AMPLIFIERS
BT ELECTRONIC INSTRUMENTS
RT OSCILLATORS
RECORDING INSTRUMENTS
TRANSDUCERS

AMPLITUDE
RT AREA
CRIMP AMPLITUDE (FIBER)
CRIMP FREQUENCY
DIMENSIONS
DRAFTING WAVES
DYNAMIC RESPONSE
FREQUENCY
IRREGULARITY
LENGTH
MAGNITUDE
OSCILLATIONS
PARTICLE SIZE
PERIODICITY
RADIUS
REPEATS
THICKNESS
TRANSVERSE VIBRATIONS
VIBRATION

AMYLNAPHTHALENE SULFONATES
BT ALKYLNAPHTHALENE SULFONATES

AMYLOPECTIN
BT STARCH
RT AMYLOSE

AMYLOSE
BT STARCH
RT AMYLOPECTIN

ANALOGUES
RT COMPUTERS
EXPERIMENTATION
MODELS (MATHEMATICAL)

ANALYSIS
USE ANALYZING

ANALYSIS (MATHEMATICAL)
USE MATHEMATICAL ANALYSIS

ANALYSIS OF VARIANCE
RT DEGREES OF FREEDOM
DISTRIBUTION
EXPERIMENTAL DESIGN
STATISTICAL ANALYSIS
STATISTICAL MEASURES
STATISTICAL METHODS
STATISTICS
VARIANCE LENGTH CURVES

ANALYZING
UF ANALYSIS
RT CHEMICAL ANALYSIS
CONFIDENCE LIMITS
CORING
COUNTING
DETECTION
DETERMINATION
DIMENSIONAL ANALYSIS
EMPIRICAL ANALYSIS
ESTIMATION
EVALUATION
EXAMINATION
EXPERIMENTAL ANALYSIS
EXPERIMENTAL STRESS ANALYSIS
EXPERIMENTATION
FABRIC ANALYSIS
FIBER IDENTIFICATION
FORECASTING
INSPECTION
JOB ANALYSIS
MATHEMATICAL ANALYSIS
MEASURING
NUMERICAL ANALYSIS
PHYSICAL ANALYSIS
PREPARATION (CHEMICAL)
PRODUCTION CONTROL
REVIEWS
SAMPLING
SPECTRUM ANALYSIS
STATISTICAL ANALYSIS
STRESS ANALYSIS
STRUCTURAL ANALYSIS
SURVEYS
TESTING
THEORETICAL ANALYSIS
THEORIES
THERMAL ANALYSIS
TRIAL AND ERROR SOLUTIONS
X RAY ANALYSIS

ANATASE
USE TITANIUM DIOXIDE

ANGLE OF INCIDENCE
RT ANGLE OF WRAP
ANGULAR VELOCITY
HELIX ANGLE
IMPACT
LIGHT
REFLECTANCE
WINDING ANGLE

ANGLE OF TWIST
USE HELIX ANGLE

ANGLE OF WIND
USE WINDING ANGLE

ANGLE OF WRAP
RT ANGLE OF INCIDENCE
BELT DRIVES
BELT SLIPPAGE
BELT TENSION
COEFFICIENT OF FRICTION
FINGER TENSION
TENSION
WRAP

ANGLE STRIPPERS
BT CARD ROLLS
STRIPPERS
RT BREAKERS
CARD CLOTHING
CARD CYLINDERS
CARDING
CARDS
LICKER IN
STRIPPING ACTION
WOOLEN CARDING
WOOLEN CARDS
WORSTED CARDING
WORSTED CARDS

ANGOLA (YARN)
BT CONDENSED YARNS
WOOLEN SPUN YARNS
RT COTTON BLENDS
WOOL BLENDS

ANGOLA (FABRIC)
RT PLAIN WEAVES
TWILL WEAVES

ANGORA (RABBIT)
USE RABBIT HAIR

ANGORA (GOAT)
USE MOHAIR

ANGULAR DEFLECTION
RT COUPLES
HELIX ANGLE
TORSION (GEOMETRIC)
TORSIONAL RIGIDITY
TWIST
TWISTING
TWISTING MOMENTS

ANGULAR SEAMING
BT SEAMING
RT BACK TACKING
BASTING
CHAIN SEWING
CURVED SEAMING
MACHINE SEWING
SEAM BASTING
SEAMS
SEWING
STITCHING

ANGULAR VELOCITY
UF ROTATIONAL VELOCITY
BT VELOCITY
RT ACCELERATION (MECHANICAL)
ANGLE OF INCIDENCE
CENTRIFUGAL FORCE
CORIOLIS ACCELERATION
INERTIA
MOMENTUM
MOTION
ROTATION
VELOCITY (RELATIVE)

ANHYDRIDES
USE ACID ANHYDRIDES

ANILINE BLACK DYES
BT DYES (BY CHEMICAL CLASSES)

ANIMAL FATS
BT FATS
RT FATTY ACIDS
VEGETABLE FATS

ANIMAL FIBERS (GENERAL)
NT FUR
GUARD HAIRS
HAIR
KERATIN FIBERS
LAMBSWOOL
MOHAIR
SILK
WOOL
BT FIBERS
NATURAL FIBERS
PROTEIN FIBERS
RT AMINO GROUPS
AZLON (REGENERATED PROTEIN
FIBERS)
FOLLICLE
KERATIN
LANTHIONINE
MAN MADE FIBERS
MEDULLA
POLYPEPTIDES
SHEEP
STAPLE FIBERS
VEGETABLE FIBERS (GENERAL)

ANIMAL HUSBANDRY
NT GOAT BREEDING
SHEEP BREEDING
RT ANIMALS
COWS
GOATS
SHEEP
WOOL

ANIMALIZING
RT AMINIZED COTTON
CHEMICAL MODIFICATION (FIBERS)
COTTON DYEING
DYE AFFINITY
DYEING
WOOL DYEING

ANIMALS
NT COWS
GOATS
RABBITS
SHEEP
RT ANIMAL HUSBANDRY
SILKWORMS

ANION EXCHANGE
USE ION EXCHANGE

ANION EXCHANGE RESINS
BT ION EXCHANGE RESINS
RT CATION EXCHANGE RESINS
ION EXCHANGE

ANIONIC DYES
NT ACID DYES
DIRECT DYES
VAT DYES
BT DYES (BY CHEMICAL CLASSES)
RT ANIONIC SITES

ANIONIC SITES
BT SITES (DYEING)
RT ACID DYEING
ANIONIC DYES
CATION EXCHANGE RESINS
CATIONIC SITES
DYEABILITY
DYEING THEORY
ION EXCHANGE

ANIONIC SOFTENERS
    BT SOFTENERS
    RT ANIONIC SURFACTANTS
        CATIONIC SOFTENERS
        NONIONIC SOFTENERS
        SOAPS
        SULFATE SURFACTANTS
        SULFONATE SURFACTANTS

ANIONIC SURFACTANTS
    NT CARBOXYLATE SURFACTANTS
        PHOSPHATE SURFACTANTS
        SULFATE SURFACTANTS
        SULFONATE SURFACTANTS
    BT SURFACTANTS
    RT AMPHOTERIC SURFACTANTS
        ANIONIC SOFTENERS
        CATIONIC SURFACTANTS
        DETERGENTS
        DYEING AUXILIARIES
        FATTY ACIDS
        LEVELLING AGENTS
        NONIONIC SURFACTANTS
        OLEATE SALTS
        OLEIC ACID
        RETARDING AGENTS
        WETTING
        WETTING AGENTS

ANIONS
    BT IONS
    RT CATIONS
        FREE RADICALS
        SALTS (GENERAL)

ANISOTROPIC MATERIALS
    RT ANISOTROPY (STRESS STRAIN)
        COMPOSITES
        DEGREE OF ORIENTATION
        ISOTROPIC MATERIALS
        NONWOVEN FABRIC STRUCTURE
        ORIENTATION
        ORIENTED WEBS
        ORTHOTROPIC MATERIALS
        STRUCTURAL ANALYSIS
        STRUCTURAL MECHANICS
        STRUCTURAL PROPERTIES

ANISOTROPY (STRESS STRAIN)
    BT STRESS STRAIN PROPERTIES
        STRUCTURAL PROPERTIES
    RT ANISOTROPIC MATERIALS
        DEGREE OF ORIENTATION
        ISOTROPIC MATERIALS
        ORIENTATION

ANNEALING
    RT CRYSTALLIZATION
        DECRYSTALLIZATION
        DRAWING (FILAMENT)
        HEAT SETTING (SYNTHETICS)
        JET QUENCHING
        MELT SPINNING
        TWIST SETTERS

ANTHROPOMETRIC KINEMATICS
    UF BODY KINEMATICS
        BODY MOVEMENTS
        SKINEMATICS
    RT ANTHROPOMETRIC MEASUREMENTS
        FIT
        SKIN
        STRETCH FABRICS

ANTHROPOMETRIC MEASUREMENTS
    RT ANTHROPOMETRIC KINEMATICS
        GARMENTS
        MEASURING
        SHOE SIZES
        STRETCH FABRICS

ANTIBACTERIAL AGENTS
    USE ANTIBACTERIAL FINISHES

ANTIBACTERIAL FINISHES
    UF ANTIBACTERIAL AGENTS
        BACTERICIDES
        BACTERIOSTATIC AGENTS
    BT ANTIMICROBIAL FINISHES
        FINISH (SUBSTANCE ADDED)
        PRESERVATIVES
    RT ANTIBACTERIAL TREATMENTS
        CATIONIC SURFACTANTS
        QUATERNARY AMMONIUM COMPOUNDS
        SOIL BURIAL TEST

ANTIBACTERIAL TREATMENTS
    BT ANTIMICROBIAL TREATMENTS
    RT ANTIBACTERIAL FINISHES
        MILDEW RESISTANCE TREATMENTS

ANTICLOCKWISE
    UF COUNTERCLOCKWISE
    RT CLOCKWISE
        SENSE (DIRECTION)

ANTICREASE FINISHES
    USE WASH WEAR FINISHES

ANTIFELTING AGENTS
    NT WURLAN (TN)
    BT FINISH (SUBSTANCE ADDED)
    RT ANTIFELTING TREATMENTS
        CHLOROCYANURIC ACID
        FELTING
        PERMANGANATES
        RESIN FINISHES

ANTIFELTING TREATMENTS
    UF FELT PROOFING
    NT WURLAN (TN)
    BT WET FINISHING
    RT ANTIFELTING AGENTS
        CHLORINATED WOOL
        CHLORINATION (SHRINK PROOFING)
        DIFFERENTIAL FRICTION
        FELTING
        FELTING RESISTANCE
        FELTING SHRINKAGE
        PERMANGANATES
        SCALE MASKING
        SHRINK PROOFING

ANTIFOAM AGENTS
    UF DEFOAMERS
    RT FOAMING AGENTS
        FOAMS (FROTH)
        SILICONES
        SURFACTANT APPLICATIONS
        SURFACTANTS

ANTIFRICTION BEARINGS
    BT BEARINGS
    RT AIR BEARINGS
        NYLON BEARINGS
        TEFLON BEARINGS

ANTILADDER BANDS
    USE AFTERWELT

ANTILUBRICANTS
    USE ANTISLIP AGENTS

ANTIQUE (LACE)
    BT HAND MADE LACES
        PILLOW LACES (BOBBIN LACES)

ANTIMICROBIAL FINISHES
    UF ANTISEPTICS
        BIOCIDES
        DISINFECTANTS
        GERMICIDES
        SANITIZING AGENTS
    NT ANTIBACTERIAL FINISHES
        FUNGICIDES
        MILDEW RESISTANCE FINISHES
    BT FINISH (SUBSTANCE ADDED)
        PRESERVATIVES
    RT ANTIMICROBIAL TREATMENTS
        BACTERIAL DEGRADATION
        BACTERIAL INHIBITION
        DEGRADATION PROPERTIES
        DEODORANTS
        DEODORIZING
        FABRIC PROPERTIES
         (DEGRADATION)
        INSECT RESISTANCE FINISHES
        MERCURIC ACETATE
        MICROBIOLOGICAL DEGRADATION
        MILDEW RESISTANCE
        ORGANOBISMUTH COMPOUNDS
        ORGANOMERCURY COMPOUNDS
        ORGANOTIN COMPOUNDS
        PENTACHLOROPHENYL LAURATE
        PHENYLMERCURIC ACETATE
        PHENYLMERCURIC
         DIOCTYLSULFOSUCCINATE
        PHENYLMERCURIC SUCCINATE
        SALICYLANILIDE
        SILVER COMPOUNDS
        SOIL BURIAL TEST
        TIN COMPOUNDS
        ULTRAVIOLET STERILIZERS

ANTIMICROBIAL TREATMENTS
    UF DISINFECTION
    NT ANTIBACTERIAL TREATMENTS
        MILDEW RESISTANCE TREATMENTS
    BT WET FINISHING
    RT ANTIMICROBIAL FINISHES
        ULTRAVIOLET STERILIZERS

ANTIOXIDANTS
    BT STABILIZERS (AGENTS)
    RT HEAT STABILIZERS
        INHIBITION
        OXIDATION
        PHOTOCHEMICAL DEGRADATION
        THERMAL STABILITY

ANTIREDEPOSITION AGENTS
    RT CARBOXYMETHYLCELLULOSE
        CLEANING
        DEPOSITION
        DETERGENCY
        DETERGENTS
        LAUNDERING
        POLY (VINYL PYRROLIDONE)
        SOIL REMOVAL
        SOIL RESISTANCE FINISHES
        SUSPENSIONS
        WHITENESS RETENTION

ANTIRUN COURSES
    USE AFTERWELT

ANTISEPTICS
    USE ANTIMICROBIAL FINISHES

ANTISHRINK COMPOUNDS
    USE SHRINK RESISTANCE FINISHES

ANTISLIP AGENTS
    UF ANTILUBRICANTS
    BT FINISH (SUBSTANCE ADDED)
    RT COLLOIDAL SILICA
        LUDOX (TN)
        RESIN FINISHES
        SLIP RESISTANCE
        SLIPPAGE
        SYTON (TN)

ANTISOILING FINISHES
    USE SOIL RESISTANCE FINISHES

ANTISTATIC AGENTS
    UF ANTISTATIC FINISHES (SUBSTANCE
         ADDED)
    BT FINISH (SUBSTANCE ADDED)
    RT ANTISTATIC BEHAVIOR
        MONTMORILLONITE
        POLYAMINES
        SPIN FINISHES
        STATIC ELECTRICITY
        STATIC ELIMINATORS
        TRIBOELECTRIC SERIES
        TRIBOELECTRICITY

ANTISTATIC BEHAVIOR
    BT END USE PROPERTIES
        FINISH (PROPERTY)
    RT ANTISTATIC AGENTS
        STATIC ELECTRICITY
        STATIC ELIMINATORS
        STATIC RESISTANCE
        TRIBOELECTRIC SERIES
        TRIBOELECTRICITY

ANTISTATIC FINISHES (SUBSTANCE ADDED)
    USE ANTISTATIC AGENTS

ANTIWEDGE RINGS
    BT RINGS
    RT LUBRICATED RINGS
        PLYING
        RING FRAMES
        RING SPINNING
        SINTERED METAL RINGS
        SPINNING
        SPINNING FRAMES
        TRAVELERS
        TWISTING
        UNLUBRICATED RINGS
        WEDGING

ANTRON (TN)
    BT NYLON (POLYAMIDE FIBERS)
        NYLON 66

APO
    UF TRIS (AZIRIDINYL) PHOSPHINE
         OXIDE
    BT AZIRIDINE COMPOUNDS
        PHOSPHORUS COMPOUNDS
    RT DURABLE PRESS FINISHES
        EASY CARE FABRICS
        ETHYLENE IMINE
        FIRE PROOFING AGENTS
        FIRE RETARDANCY AGENTS
        MAPO
        PROPYLENE IMINE

APPARATUS
    NT LABORATORY APPARATUS
        MEASURING INSTRUMENTS
    RT COMPONENTS
        DEVICES
        EQUIPMENT
        INSTRUMENTATION

APPAREL
    USE GARMENTS

APPAREL DESIGN
    UF GARMENT DESIGN
    NT DRESS DESIGN
    BT DESIGN
    RT CUTTING (TAILORING)
        FABRIC DESIGN
        FASHION (APPAREL)
        GARMENTS
        MANNEQUINS
         (CONT.)

RT MODELLING
PATTERN MAKING
PATTERN (APPAREL)
SKETCHING (APPAREL DESIGN)
STYLING (APPAREL)
TAILORING

APPAREL FABRICS
UF DRESSGOODS
BT FABRICS
FABRICS (BY END USES)
RT END USE PROPERTIES
END USES
FABRICS (ACCORDING TO STRUCTURE)
GARMENTS
HOUSEHOLD FABRICS
INDUSTRIAL FABRICS

APPARENT CONTACT AREA
BT CONTACT AREA
RT COEFFICIENT OF FRICTION
CONTACT AREA
FRICTION
FRICTION THEORY
TRUE CONTACT AREA

APPARENT DENSITY
USE BULK DENSITY

APPEARANCE
BT AESTHETIC PROPERTIES
END USE PROPERTIES
FABRIC PROPERTIES
FABRIC PROPERTIES (AESTHETIC)
RT ABSENCE
AESTHETIC APPEAL
BLOOM
BRIGHTNESS
COLOR
COVER
DESIGN
DRABNESS
DRAPE
DULLNESS
FABRIC DEFECTS
FINISH (PROPERTY)
HAIRINESS
LUSTER
LUXURIOUSNESS
OPTICAL PROPERTIES
PATTERN (FABRICS)
PATTERN DEFINITION
PRESENCE
RICHNESS
ROUGHNESS
SHADE
SMOOTHNESS
SPARKLE
STITCH CLARITY
TEXTURE
TRANSLUCENCY
TRANSPARENCY
WEAVE
YARN LUSTER

APPLICATIONS
RT END USES
END USE PROPERTIES
FABRICS (BY END USES)
PRODUCTS
WEARING (OF APPAREL)

APPLIED ELASTICITY
BT ELASTICITY
RT STRENGTH OF MATERIALS

APPROXIMATIONS
RT COMPUTATIONS
ERRORS
ESTIMATION
TRIAL AND ERROR SOLUTIONS

APRON CONTROL
BT CONTROL SURFACES
RT APRON DRAFTING
ACTUAL DRAFT
CASABLANCA SYSTEM
DRAFT
DRAFTING (STAPLE FIBER)
FIBER CONTROL
SINGLE APRONS
SPINNING
VELOCITY CHANGE POINT

APRON DRAFTING
BT DRAFTING (STAPLE FIBER)
RT APRON CONTROL
CONTROL SURFACES
ROLLER DRAFTING
TWIST CONTROL

APRONS
UF CONTROL APRONS
NT DOUBLE APRONS
SINGLE APRONS
BT CONTROL SURFACES
RT AMBLER SUPERDRAFT
CASABLANCA SYSTEM
COTTON SYSTEM PROCESSING
DRAFTING (STAPLE FIBER)
DRAFTING ROLLS
GILLING
HIGH DRAFT SPINNING
ROLLS
SPINNING
WORSTED SYSTEM PROCESSING

APRONS (CLOTHING)
BT INDUSTRIAL CLOTHING
WORK CLOTHING
RT OVERALLS
SAFETY CLOTHING

AQUEOUS SCOURING
BT SCOURING
RT SOLUTIONS (LIQUID)
WATER SOLUTIONS

AQUEOUS SOLUTIONS
USE WATER SOLUTIONS

ARABEVA FABRIC (TN)
BT NONWOVEN FABRICS
STITCH BONDED FABRICS

ARABEVA PROCESS (TN)
BT ARACHNE SYSTEMS (TN)

ARACHNE FABRIC (TN)
BT NONWOVEN FABRICS
STITCH BONDED FABRICS
STITCH REINFORCED FABRICS
RT ARACHNE PROCESS (TN)
FELTS
FIBER WOVEN (TN)
MALIMO FABRIC (TN)
MALIPOL FABRIC (TN)
MALIWATT FABRIC (TN)
MALIWATT PROCESS (TN)
NEEDLE LOOMS (NEEDLING)
NEEDLE PUNCHED FELTS
RANDO WEBBER (TN)

ARACHNE PROCESS (TN)
BT ARACHNE SYSTEMS (TN)
RT ARACHNE FABRIC (TN)
KNITTING
MALI PROCESSES (TN)
MALIMO PROCESS (TN)
MALIPOL PROCESS (TN)
MALIWATT PROCESS (TN)
NEEDLED FABRICS
NEEDLING
NONWOVEN FABRIC MACHINES
STITCHING
WEAVING

ARACHNE SYSTEMS (TN)
NT ARABEVA PROCESS (TN)
ARACHNE PROCESS (TN)
ARAKNIT PROCESS (TN)
ARALOOP PROCESS (TN)
RT CROSS LAID YARN FABRICS
FABRICS (ACCORDING TO STRUCTURE)
MALI PROCESSES (TN)
NONWOVEN FABRICS
STITCH BONDED FABRICS
STITCH BONDING MACHINES
STITCH REINFORCED FABRICS
STITCHED PILE FABRICS

ARAKNIT PROCESS (TN)
BT ARACHNE SYSTEMS (TN)

ARALAC (TN)
BT AZLON (REGENERATED PROTEIN FIBERS)
PROTEIN FIBERS

ARALOOP FABRIC (TN)
BT NONWOVEN FABRICS
STITCH BONDED FABRICS
STITCHED PILE FABRICS

ARALOOP PROCESS (TN)
BT ARACHNE SYSTEMS (TN)

ARDIL (TN)
BT AZLON (REGENERATED PROTEIN FIBERS)
PROTEIN FIBERS

AREA
RT AMPLITUDE
CROSS SECTIONS
CROSS SECTIONAL AREA
DIMENSIONS
FINENESS
LENGTH
PARTICLE SIZE
RADIUS
SPACE
SURFACES
THICKNESS

AREA MOMENT OF INERTIA
USE MOMENT OF INERTIA

AREA OF CONTACT
USE CONTACT AREA

ARGYLE GIMP
BT NARROW FABRICS

ARITHMETIC
RT ACCURACY
COMPUTATIONS
DATA REDUCTION
MATHEMATICAL ANALYSIS
MULTIPLICATION

ARMACS (TN)
BT ALKYLAMINE SURFACTANTS

ARMEENS (TN)
BT ALKYLAMINE SURFACTANTS

ARMENIAN (LACE)
BT HAND MADE LACES
LACES

ARMURE

RT EMBOSSED FABRICS

ARNEL (TN)
BT CELLULOSE ESTER FIBERS
TRIACETATE FIBERS

AROMATIC COMPOUNDS
UF ARYL COMPOUNDS
NT AROMATIC HYDROCARBONS
RT ALICYCLIC COMPOUNDS
ALIPHATIC COMPOUNDS
ALIPHATIC HYDROCARBONS
BENZOIC ACID
HETEROCYCLIC COMPOUNDS
PHENOLS
TEREPHTHALIC ACID

AROMATIC HYDROCARBONS
NT BENZENE
TETRAHYDRONAPHTHALENE
TOLUENE
BT AROMATIC COMPOUNDS
HYDROCARBONS
RT ALICYCLIC HYDROCARBONS
ALIPHATIC HYDROCARBONS
UNSATURATED HYDROCARBONS

ARRAS TAPESTRIES
BT TAPESTRIES

ARRESTERS
BT MILITARY PRODUCTS
RT END USES
ENERGY ABSORPTION
IMPACT STRENGTH
PARACHUTES

ARSONIUM COMPOUNDS
BT ONIUM COMPOUNDS

ART
RT AESTHETIC PROPERTIES
DESIGN
PATTERN (FABRICS)
TAPESTRIES
WEAVE

ARTERIAL REPLACEMENTS
BT MEDICAL TEXTILES
RT HEART VALVES (FABRIC)

ARTIFICIAL DAYLIGHT
BT DAYLIGHT
LIGHT
RT FLUORESCENT LIGHT
ILLUMINATION
INCANDESCENT LIGHT
POLARIZED LIGHT
SUNLIGHT

ARTIFICIAL FURS
BT PILE FABRICS (WOVEN)
WOVEN FABRICS
RT FUR
GUARD HAIRS
SLIVER KNITTING

ARTIFICIAL SILK (ARCHAIC)
BT FIBERS
MAN MADE FIBERS
RT ACETATE FIBERS
RAYON (REGENERATED CELLULOSE FIBERS)
REGENERATED CELLULOSE
SILK
SYNTHETIC FIBERS
TRIACETATE FIBERS
VISCOSE RAYON

ARTIFICIAL SOIL
BT SOIL
RT CLAYS
DIRT

ARYL COMPOUNDS
USE AROMATIC COMPOUNDS

ASBESTOS
UF ASBESTOS FIBERS
CHRYSOTILE
BT HIGH TEMPERATURE FIBERS
MINERAL FIBERS
RT GLASS
INORGANIC FIBERS (MAN MADE)

ASBESTOS FIBERS
USE ASBESTOS

ASPERITIES
RT ABRASIVES
COEFFICIENT OF FRICTION
FRICTION
FRICTION THEORY
HOLES
ROUGHNESS
SNAG RESISTANCE
SNAGGING

ASSEMBLIES
UF SUBASSEMBLIES
RT ACCUMULATION
AGGREGATION
COLLECTION
COMPONENTS
CONFIGURATION
EQUIPMENT
FIBER ASSEMBLIES
STRUCTURES

ASYMMETRY
USE SYMMETRY

ASTRAKHAN
BT FILLING KNITTED FABRICS

ATACTIC POLYMERS
RT POLYMERS
TACTIC POLYMERS

ATHLETIC CLOTHING
UF ACTIVE SPORTSWEAR
NT BATHING SUITS
GLOVES
GUARDS (ATHLETIC CLOTHING)
HELMETS
KNICKERS
LEOTARDS
ROBES
SHINGUARDS
SHIRTS
SHORTS
SKI PANTS
SLACKS
SOCKS
SWEATSHIRTS
TRUNKS
BT GARMENTS
RT BOXER SHORTS
JACKETS
SNEAKERS
SPORTSWEAR

ATMOSPHERIC DRYERS
BT DRYERS
RT CENTRIFUGAL DRYERS
DRUM DRYERS
DRYING CANS
FLUIDIZED BED DRYERS
HOT AIR DRYERS
INFRARED DRYERS
LOOP DRYERS
VACUUM DRYERS

ATOMIZERS
RT AEROSOLS
ATOMIZING
SPRAYING
SPRAYING MACHINES

ATOMIZING
RT ATOMIZERS
COMMINUTION
GRINDING (COMMINUTION)
POWDERING
SPRAYING

AUBOSSON TAPESTRIES
BT TAPESTRIES

AUSTRALIAN MERINO WOOL
BT MERINO WOOL

AUTOCLAVES
UF PRESSURE VESSELS
BT VESSELS
RT KETTLES (HEATING)
PRESSURE DYEING MACHINES
PRESSURE KETTLES

AUTOCORRELATION
UF AUTOCORRELATION COEFFICIENT
RT CORRELATION
DRAFTING THEORY
IRREGULARITY
PERIODICITY
REGRESSION
REPEATS
STATISTICAL MEASURES
VARIATION

AUTOCORRELATION COEFFICIENT
USE AUTOCORRELATION

AUTOCOUNT (TN)
RT AUTOMATIC CONTROL
AUTOMATION
CARD WEBS
CARDING
CONTROL SYSTEMS
DOFFERS (CARD)
DOFFING (OF A CARD)
FINISHER PART
INSTRUMENTATION
IRREGULARITY
OPTICAL SCANNERS
OPTICAL SCANNING
PHOTOELECTRIC SCANNING
PROCESS CONTROL
SLUBBINGS
WEB UNIFORMITY
WEBS
WOOLEN CARDING

AUTODRAFTER
RT DRAFTING (STAPLE FIBER)
IRREGULARITY CONTROL
PROCESS CONTROL

AUTOLEVELLERS
NT RAPER AUTOLEVELLER (TN)
RT AUTOMATIC CONTROL
COMBED SLIVERS
CONTROL SYSTEMS
DRAFT CONTROL
DRAFTING (STAPLE FIBER)
DRAFTING MACHINES
EQUALIZING (DRAFTING)
EVENERS
IRREGULARITY CONTROL
PROCESS CONTROL
SLIVERS
TOPS
WEIGHT CONTROL

AUTOLEVELLING
RT ACTUAL DRAFT
AUTOMATIC CONTROL
DRAFTING MACHINES
DRAFTING THEORY
DRAWFRAMES
FEEDBACK CONTROL
HIGH DRAFT
IRREGULARITY
IRREGULARITY TESTERS
RAPER AUTOLEVELLER (TN)
SLIVERS

AUTOMATED SPINNING (STAPLE FIBER)
UF COMPLETELY AUTOMATED SPINNING
BT SPINNING
RT AUTOMATIC CONTROL
AUTOMATIC DOFFING
CONTINUOUS SPINNING (STAPLE FIBER)
COTTON SYSTEM PROCESSING
INTEGRATED SYSTEMS
OPEN END SPINNING
SLIVER TO YARN SPINNING

AUTOMATIC CONTROL
UF AUTOMATIC REGULATION
NT DIFFERENTIAL CONTROL
FEEDBACK CONTROL
INTEGRAL CONTROL
ON OFF CONTROL
OPEN LOOP CONTROL
RT AUTOCOUNT (TN)
AUTOLEVELLERS
AUTOLEVELLING
AUTOMATED SPINNING (STAPLE FIBER)
AUTOMATION
COLOR CONTROL
COMPUTERS
CONTINUOUS CONTROL
CONTROL CHARACTERISTICS
CONTROL EQUIPMENT
CONTROL POINTS
CONTROL SURFACES
CONTROL SYSTEMS
CONTROL THEORY
CYBERNETICS
DRAFT CONTROL
DYNAMIC CHARACTERISTICS
ELECTRIC CONTROLLERS
ELECTROMAGNETIC DEVICES
ELECTRONIC INSTRUMENTS
FLOW CONTROL
GUARDING DEVICES
HUMIDITY CONTROL
INDICATORS (INSTRUMENTATION)
INTEGRATED SYSTEMS
IRREGULARITY CONTROL
MEASURING INSTRUMENTS
MOISTURE CONTENT CONTROL
OPTICAL SCANNERS
OPTICAL SCANNING
PH CONTROL
PICKER EVENERS
PICKERS (OPENING)
PNEUMATIC CONTROL
PNEUMATIC INSTRUMENTS
PRESSURE CONTROL
PROCESS CONTROL
PUNCHED CARD EQUIPMENT
QUALITY CONTROL
RAPER AUTOLEVELLER (TN)
RECORDING INSTRUMENTS
REGULATORS
REMOTE CONTROL
SCHLUPP REGULATOR
SERVO DRAFT
SPEED CONTROL
STITCH LENGTH CONTROL
STOP MOTIONS
SUPERLEVELLER (DRAFTING)
TEMPERATURE CONTROL
TENSION CONTROL
WEIGHT CONTROL

AUTOMATIC DOFFING
BT DOFFING (PACKAGE)
RT AUTOMATED SPINNING (STAPLE FIBER)

AUTOMATIC KNOTTING
BT KNOTTING
RT CHEESE WINDING
CONING
DRAWING IN
END BREAKAGES
KNOTS
KNOTTERS
PACKAGES
PIECING
SLUB CATCHING
UNICONER (TN)
WINDING

AUTOMATIC LOOMS
BT LOOMS
RT BOX LOOMS
DOBBY LOOMS
HAND LOOMS
JACQUARD LOOMS
LAPPET LOOMS
LENO LOOMS
PILE FABRIC LOOMS
SHUTTLE CHANGING
SHUTTLELESS LOOMS

AUTOMATIC REGULATION
USE AUTOMATIC CONTROL

AUTOMATIC STOP MOTIONS
BT STOP MOTIONS
RT BRAKES (FOR ARRESTING MOTION)
END BREAKAGES
MAGNETIC BRAKES
MECHANICAL BRAKES

AUTOMATIC TRANSFER (KNITTING)
RT KNITTING
LOOP TRANSFER

AUTOMATIC WINDING
BT WINDING
RT DRUM WINDING
PRECISION WINDING

AUTOMATION
RT AUTOCOUNT (TN)
AUTOMATIC CONTROL
COMPUTERS
CONTROL EQUIPMENT
CONTROL SYSTEMS
CONTROL THEORY
CYBERNETICS
INFORMATION SYSTEMS
INSTRUMENTATION
PROCESS CONTROL
REGULATORS
REMOTE CONTROL
SYSTEMS ENGINEERING

AUTOMOTIVE FABRICS
BT FABRICS
FABRICS (BY END USES)
RT FABRICS (ACCORDING TO STRUCTURE)
AGRICULTURAL FABRICS
END USES
FURNISHING FABRICS
FURNISHINGS
HOUSEHOLD FABRICS
(CONT.)

RT INDUSTRIAL FABRICS
UPHOLSTERY
UPHOLSTERY FABRICS
UPHOLSTERY LEATHER
VINYL COATED FABRICS

AUXILIARY CAMS
RT KNITTING MACHINES
STITCH CAMS

AVRIL (TN)
BT CELLULOSIC FIBERS
HIGH WET MODULUS RAYON
RAYON (REGENERATED CELLULOSE FIBERS)
VISCOSE RAYON

AWL CUTTING DRILLS
BT CUTTING DRILLS

AWNING CLOTH
BT FLAT WOVEN FABRICS
WOVEN FABRICS
RT AWNINGS
DUCK
PLAIN WEAVES

AWNINGS
RT AWNING CLOTH
INDUSTRIAL FABRICS

AXIAL MOTION
BT MOTION
RT PARALLEL MOTION
PLANE MOTION
ROTATION
SURGING
TRANSLATION
VIBRATION

AXLES
RT AIR BEARINGS
BALANCING
BEARINGS
BUSHINGS
FALSE TWIST SPINDLES
LINKAGES
SHAFTS
SPINDLE ECCENTRICITY
SPINDLES
VIBRATION

AXMINSTER
USE AXMINSTER CARPETS

AXMINSTER CARPETS
UF AXMINSTER
BT CARPETS
PILE FABRICS (WOVEN)
WOVEN CARPETS
WOVEN FABRICS
RT AXMINSTER LOOMS
CHENILLE CARPETS
FLOCKED CARPETS
NONWOVEN CARPETS
TAPESTRIES
TUFTED CARPETS
VELVET CARPETS
WILTON CARPETS

AXMINSTER LOOMS
BT LOOMS
RT AXMINSTER CARPETS

AZIDES
RT CATALYSTS
CROSSLINKING
FREE RADICALS
POLYMERIZATION

AZIDOSULFONYL DYES
BT REACTIVE DYES
RT CHLOROPYRIMIDINE DYES
CHLOROTRIAZINE DYES
VINYLSULFONE DYES

AZIRIDINE COMPOUNDS
NT APO
ETHYLENE IMINE
MAPO
OCTADECYL ETHYLENE UREA
PROPYLENE IMINE
BT IMINES
REACTANTS
RT AMINES
WASH WEAR FINISHES

AZLON (REGENERATED PROTEIN FIBERS)
UF REGENERATED PROTEIN FIBERS
NT ARALAC (TN)
ARDIL (TN)
COLLAGEN FIBERS
FIBROLANE (TN)
LANITAL (TN)
VICARA (TN)
BT FIBERS
MAN MADE FIBERS
PROTEIN FIBERS
RT ANIMAL FIBERS (GENERAL)
CASEIN
GUT (COLLAGEN)
KERATIN
POLYMERS
POLYPEPTIDES
RAYON (REGENERATED CELLULOSE FIBERS)
REGENERATED FIBERS (EXCLUDING CELLULOSIC AND PROTEIN)
SYNTHETIC FIBERS
ZEIN

AZLON BLENDS
BT BLENDS (FIBERS)

AZOIC BASES
RT AZOIC DYEING
AZOIC DYES
COUPLING (CHEMICAL)
DIAZO COMPOUNDS
DIAZOBENZOYLATED COTTON
DIAZOTIZATION
NAPHTHOLS

AZOIC DYEING
BT DYEING (BY DYE CLASSES)
RT ACID DYEING
AZOIC BASES
AZOIC DYES
BASIC DYEING
COUPLING (CHEMICAL)
DEVELOPERS (DYEING)
DIAZO COMPOUNDS
DIAZOTIZATION
DIRECT DYEING
DISPERSE DYEING
METALLIZED DYEING
MORDANT DYEING
NEUTRAL DYEING
PIGMENT DYEING
REACTIVE DYEING
SOAPING
SULFUR DYEING
VAT DYEING

AZOIC DYES
UF DEVELOPED DYES
INSOLUBLE AZO DYES
NAPHTHOL DYES
BT DYES (BY CHEMICAL CLASSES)
RT AZOIC BASES
AZOIC DYEING
COUPLING (CHEMICAL)
DIAZO COMPOUNDS
DIAZOTIZATION
DIRECT DYES
DISCHARGE PRINTING
DYEING (BY DYE CLASSES)
DYES (BY FIBER CLASSES)
NAPHTHOLS
NITRITES
POTASSIUM NITRITE
SULFUR DYES
VAT DYES

B (L) CURVES
    USE VARIANCE LENGTH CURVES

BABY DELAINE (WOOL CLASS)
    BT WOOL CLASS

BABY FLANNEL
    USE FLANNEL

BACK (FABRIC)
    RT BACK FILLING
       BACKED FABRICS
       COVER
       DOUBLE KNIT FABRICS
       FABRIC STRUCTURE
       FABRICS
       FACE (FABRIC)
       KNITTED FABRICS
       WOVEN FABRICS

BACK AND FACE EFFECTS
    BT DYEING DEFECTS
       FABRIC DEFECTS

BACK COATING
    USE COATING (PROCESS)

BACK CROSSING HEDDLES
    UF BACK STANDARD
    RT DOUP
       FRONT CROSSING HEDDLES
       GROUND HEDDLES
       HARNESSES
       LENO WEAVES
       LENO WEAVING
       STANDARD HEDDLES

BACK FILLING
    BT FILLING YARNS
    RT BACK (FABRIC)
       BACK WARP
       BACKED FABRICS
       FACE YARNS
       MULTIPLE LAYER FABRICS
       WEAVING

BACK FILLING MACHINES
    RT COATING (PROCESS)
       COATING MACHINES
       IMPREGNATING MACHINES
       KNIFE COATING
       NIP COATING
       ROLLER COATING

BACK GRAY
    UF BUMP GRAY
    RT BLANKET (FINISHING)
       DYEING
       PRINTING
       PRINTING MACHINES

BACK REST (LOOM)
    USE WHIP ROLLS

BACK ROLLS
    BT DRAFTING ROLLS
       ROLLS
    RT ACTUAL DRAFT
       BREAK DRAFT
       CONTROL SURFACES
       DRAFTING (STAPLE FIBER)
       DRAFTING THEORY
       DRAFTING WAVES
       FLOATING FIBERS
       FRONT ROLLS
       GILL BOXES
       INTERSECTING GILL BOXES
       RATCH
       VELOCITY CHANGE POINT

BACK STANDARD
    USE BACK CROSSING HEDDLES

BACK TACKING
    BT GARMENT MANUFACTURE
    RT ANGULAR SEAMING
       BASTING
       CHAIN SEWING
       CURVED SEAMING
       GATHERING
       HAND SEWING
       HEMMING
       PINCHING
       PUCKERING
       RUFFLING
       SEAM BASTING
       SEAM BINDING
       SEAMING
       SEAMSTRESSES
       SEWING
       SEWING MACHINES
       SHIRRING
       STITCHING
       TACKING
       TAILORING
       TUCKING (GARMENT MANUFACTURE)

BACK TO BACK FABRICS (KNITTED)
    USE DOUBLE FACED FABRICS (KNITTED)

BACK WARP
    BT WARP ENDS
    RT BACK FILLING
       BACKED FABRICS
       FACE YARNS
       MULTIPLE LAYER FABRICS

BACKED FABRICS
    RT BACK (FABRIC)
       BACK FILLING
       BACK WARP
       BACKING COMPOUNDS
       BACKING FABRICS
       FABRIC STRUCTURE
       FLAT WOVEN FABRICS
       LAMINATED FABRICS
       MULTIPLE LAYER FABRICS
       PILE FABRICS (WOVEN)

BACKING (MATERIAL)
    NT ADHESIVE BACKING
       BACKING FABRICS
       FOAM BACKING
    RT BACKING COMPOUNDS
       FOAM BACKED FABRICS
       FOAM BACKING
       HEAT SEALING
       LAMINATED FABRICS
       LAMINATES
       UNDERLAY (CARPETS)

BACKING CLOTHS
    USE BACKING FABRICS

BACKING COMPOUNDS
    NT ADHESIVES
       FOAMS
       FOAM RUBBER
    RT BACKED FABRICS
       BACKING (MATERIAL)
       BACKING FABRICS
       CARPETS
       FOAM BACKING
       LAMINATES
       LAMINATING RESINS

BACKING FABRICS
    UF BACKING CLOTHS
    BT BACKING (MATERIAL)
    RT ADHESIVE BACKING
       BACKED FABRICS
       BACKING COMPOUNDS
       CARPET BACKING
       FOAM BACKED FABRICS
       FOAM BACKING
       HEAT SEALING

BACKSTITCHES
    BT STITCHES (SEWING)

BACKWASHING
    BT CLEANING
       SCOURING
       WORSTED SYSTEM PROCESSING
    RT BATHS
       BLEACHING
       BRADFORD SYSTEM PROCESSING
       CARBONIZING
       COMBING
       FUGITIVE DYES
       IMPURITIES
       OILING
       PREPARING (WORSTED)
       TOP MAKING

BACKWASHING MACHINES
    BT SCOURING MACHINES
    RT BACKWASHING
       SCOURING TRAINS

BACTERIA REPELLENCY
    USE BACTERIAL INHIBITION

BACTERIA RESISTANCE
    USE BACTERIAL INHIBITION

BACTERIAL DEGRADATION
    BT DEGRADATION
    RT ANTIMICROBIAL FINISHES
       BACTERIAL INHIBITION
       ENZYMATIC DEGRADATION
       ODORS
       PERSPIRATION
       PERSPIRATION RESISTANCE
       PRESERVATIVES

BACTERIAL INHIBITION
    UF BACTERIA REPELLENCY
       BACTERIA RESISTANCE
    BT DEGRADATION PROPERTIES
       END USE PROPERTIES
       FINISH (PROPERTY)
    RT ANTIMICROBIAL FINISHES
       BACTERIAL DEGRADATION
       BIOCHEMICAL PROPERTIES
       INHIBITION
       MICROBIOLOGICAL DEGRADATION
       MILDEW RESISTANCE
       ODOR CONTROL
       PERSPIRATION
       PERSPIRATION RESISTANCE

BACTERICIDES
    USE ANTIBACTERIAL FINISHES

BACTERIOSTATIC AGENTS
    USE ANTIBACTERIAL FINISHES

BAGGINESS
    UF BAGGING TENDENCY
    BT FABRIC PROPERTIES
       FABRIC PROPERTIES (AESTHETIC)
    RT COMFORT
       CREEP
       DIMENSIONAL STABILITY
       FIT
       GARMENTS
       GROWTH (FABRIC)
       LOOSENESS
       NARROWING
       PAPERMAKERS FELTS
       SKEW
       SLACKS
       STRETCH
       STRETCH FABRICS
       TROUSERS
       WIDENING

BAGGING (PACKAGES)
    USE BAGS

BAGGING MACHINES
    RT BAGS
       BALING MACHINES

       PACKAGES
       PACKAGING (OPERATION)
       PACKAGING EQUIPMENT

BAGGING TENDENCY
    USE BAGGINESS

BAGS
    UF BAGGING (PACKAGES)
       SACKS (BAGS)
    BT PACKAGES
    RT AGRICULTURAL FABRICS
       BAGGING MACHINES
       BALES
       BOXES
       CARTONS
       GUNNY SACKS
       HESSIAN
       INDUSTRIAL FABRICS

BAIZE
    RT FELTS

BAKING
    USE DRY CURING

BALANCE
    RT ADJUSTMENTS
       CALIBRATION
       COMPENSATION
       LEVELLING

BALANCED SHED
    BT SHED (WEAVING)
    RT SHEDDING
       WEAVING

BALANCED WEAVES
    RT ENDS PER INCH
       MOCK LENO WEAVES
       PICKS PER INCH
       PLAIN WEAVES
       WEAVE
       WEAVING

BALANCED YARNS
    BT YARNS
    RT DIRECTION (TWIST)
       MULTIPLIED YARNS
       NO TORQUE YARNS
       PLIED YARNS
       PLY TWIST
       PLYING
       REAL TWIST
       RESIDUAL TORQUE
       S TWIST
       SINGLES YARNS
       SNARLING TENDENCY
       TWIST
       TWIST LIVELINESS
       TWISTING
       TWO PLY YARNS
       WILDNESS
       Z TWIST

BALANCING
    NT DYNAMIC BALANCING
    RT AXLES
       SPINDLES

BALBRIGGAN
    USE PLAIN KNITTED FABRICS

BALDNESS
    RT HAIRINESS

BALE BREAKERS
    RT BALE BREAKING
        BALE PLUCKERS
        BEATERS (OPENING)
        COTTON SYSTEM PROCESSING
        STOCK (FIBER)

BALE BREAKING
    BT COTTON SYSTEM PROCESSING
    RT BALE BREAKERS

BALE PLUCKERS
    RT COTTON SYSTEM PROCESSING
        BALE BREAKERS

BALES
    NT COTTON BALES
        WOOL BALES
    BT PACKAGES
    RT BAGS
        BALING
        BALING MACHINES
        BOXES
        CARTONS
        GINNING
        STOCK (FIBER)
        STRAPPING
        STRAPPING MACHINES

BALING
    RT BALING MACHINES
        BALES
        COTTON BALES
        COTTON GINS
        COTTON SYSTEM PROCESSING
        GINNING
        PACKAGING (OPERATION)
        WOOL BALES

BALING MACHINES
    UF BALING PRESSERS
    RT BAGGING MACHINES
        BALES
        BALING
        PACKAGING EQUIPMENT

BALING PRESSERS
    USE BALING MACHINES

BALL AND SOCKET TENSION
    RT DISC TENSION
        FINGER TENSION
        GATE TENSION
        TENSION
        TENSION CONTROL
        TENSION DEVICES
        WRAP

BALL BEARINGS
    BT BEARINGS
    RT AIR BEARINGS
        CONICAL BEARINGS
        JOURNAL BEARINGS
        NEEDLE BEARINGS
        NYLON BEARINGS
        ROLLER BEARINGS
        SLEEVE BEARINGS
        TEFLON BEARINGS
        THRUST BEARINGS

BALL BURST TESTING
    UF BALL BURST TESTS
    BT BURST TESTING
    RT BIAXIAL STRESS
        BREAKING STRENGTH
        BURSTING STRENGTH
        DIAPHRAGM BURST TESTING
        MECHANICAL PROPERTIES
        MULLEN BURST TESTING

BALL BURST TESTS
    USE BALL BURST TESTING

BALL WARPS
    BT PACKAGES

BALLET TOES (KNITTING)
    RT KNITTING
        TOES

BALLING HEADS
    RT CARD SLIVERS
        COMBED SLIVERS
        COMBER LAP MACHINES
        COMBER LAPS
        COMBING
        FEED BOXES
        NOBLE COMBS
        PUNCH BALLS (COMBING)
        SLIVER CANS
        SLIVERS
        TOPS

BALLISTIC FABRICS
    RT BULLET PROOF CLOTHING
        IMPACT STRENGTH
        MILITARY CLOTHING

BALLISTIC LIMIT
    RT BULLET PROOF CLOTHING
        IMPACT TESTING
        MILITARY CLOTHING

BALLISTIC TESTING
    BT TESTING
    RT IMPACT STRENGTH
        IMPACT TESTING

BALLOON COLLAPSE
    BT BALLOON DYNAMICS
    RT BALLOON CONTROL
        BALLOON CONTROL RINGS
        BALLOON HEIGHT
        BALLOON SEPARATORS
        BALLOON STABILITY
        BALLOON TENSION
        BALLOONS
        COLLAPSED BALLOON SPINNING
        END BREAKAGES
        RING SPINNING
        SPINDLES
        SPINNING
        SPINNING BALLOONS
        TWISTING
        UNWINDING ACCELERATORS
        UNWINDING BALLOONS

BALLOON CONTROL
    UF BALLOON CONTROLLERS
    BT BALLOON DYNAMICS
    RT BALLOON COLLAPSE
        BALLOON CONTROL RINGS
        BALLOON HEIGHT
        BALLOON SEPARATORS
        BALLOON STABILITY
        BALLOON TENSION
        BALLOONS
        DOUBLE RING SYSTEM
        DOWNTWISTING
        RING SPINNING
        SPINNING
        SPINNING BALLOONS
        TWISTING
        UNWINDING ACCELERATORS
        UNWINDING BALLOONS

BALLOON CONTROL RINGS
    UF BALLOON RINGS
    RT BALLOON COLLAPSE
        BALLOON CONTROL
        BALLOON DYNAMICS
        BALLOON SEPARATORS
        BALLOON TENSION
        BALLOONS
        DOUBLE RING SYSTEM
        DOWNTWISTING
        END BREAKAGES
        RING FRAMES
        RING SPINNING
        SPINNING
        SPINNING BALLOONS
        TRAVELERS
        TWISTING
        UNWINDING ACCELERATORS

BALLOON CONTROLLERS
    USE BALLOON CONTROL

BALLOON DYNAMICS
    NT BALLOON COLLAPSE
        BALLOON CONTROL
        BALLOON SHAPE
        BALLOON TENSION
    BT SPINNING DYNAMICS
        TWISTING DYNAMICS
        WINDING DYNAMICS
        YARN DYNAMICS
    RT AERODYNAMIC CONFIGURATIONS
        AIR DRAG
        AIR RESISTANCE
        BALLOON CONTROL RINGS
        BALLOON HEIGHT
        BALLOON SEPARATORS
        BALLOONS
        BALLOON STABILITY
        DOWNTWISTING
        DRAG
        RING SPINNING
        RINGS
        SPINDLES
        SPINNING
        SPINNING BALLOONS
        SPINNING TENSION (STAPLE YARN)
        THREADLINE
        TRAVELERS
        TWISTING
        TWISTING TENSION
        UNWINDING
        UNWINDING ACCELERATORS
        UNWINDING BALLOONS
        WAVE LENGTH

    RT WINDING
        WINDING TENSION
        YARN TENSION

BALLOON HEIGHT
    RT BALLOON COLLAPSE
        BALLOON CONTROL
        BALLOON DYNAMICS
        BALLOON SHAPE
        BALLOON STABILITY
        BALLOON TENSION
        BALLOONS
        BUILDER MOTIONS
        HEIGHT
        LARGE PACKAGES
        LIFT (WINDING)
        UNWINDING BALLOONS
        YARN TENSION

BALLOON RINGS
    USE BALLOON CONTROL RINGS

BALLOON SEPARATORS
    RT BALLOON COLLAPSE
        BALLOON CONTROL
        BALLOON CONTROL RINGS
        BALLOON DYNAMICS
        BALLOON TENSION
        BALLOONS
        RING SPINNING
        SEPARATORS
        SEPARATOR SPACING
        SPINNING
        SPINNING BALLOONS
        TWISTING
        UNWINDING BALLOONS

BALLOON SHAPE
    BT BALLOON DYNAMICS
    RT AERODYNAMIC CONFIGURATIONS
        AIR RESISTANCE
        BALLOON COLLAPSE
        BALLOON HEIGHT
        BALLOON STABILITY
        BALLOON TENSION
        CENTRIFUGAL FORCE

BALLOON STABILITY
    RT AERODYNAMIC CONFIGURATIONS
        BALLOON COLLAPSE
        BALLOON CONTROL
        BALLOON DYNAMICS
        BALLOON HEIGHT
        BALLOON SHAPE
        BALLOON TENSION
        STABILITY

BALLOON TENSION
    BT BALLOON DYNAMICS
        TENSION
    RT AIR DRAG
        BALLOON COLLAPSE
        BALLOON CONTROL
        BALLOON CONTROL RINGS
        BALLOON HEIGHT
        BALLOON SEPARATORS
        BALLOON STABILITY
        BALLOONS
        END BREAKAGES
        LARGE PACKAGES
        RING SPINNING
        SPINDLE SPEED
        SPINNING
        SPINNING BALLOONS
        SPINNING TENSION (STAPLE YARN)
        THREAD TENSION
        TRAVELER BURNOUT
        TRAVELER CHATTER
        TRAVELERS
        TWISTING
        TWISTING DYNAMICS
        TWISTING TENSION
        UNWINDING
        UNWINDING ACCELERATORS
        UNWINDING BALLOONS
        WINDING
        YARN TENSION

BALLOONS
    UF YARN BALLOONS
    NT SPINNING BALLOONS
        UNWINDING BALLOONS
    RT AERODYNAMIC CONFIGURATIONS
        AIR RESISTANCE
        BALLOON COLLAPSE
        BALLOON CONTROL
        BALLOON CONTROL RINGS
        BALLOON DYNAMICS
        BALLOON HEIGHT
        BALLOON SEPARATORS
        BALLOON TENSION
        CAP SPINNING
        CAPS (SPINNING)
        DOWNTWISTING
        FLYER SPINNING
        LIFT (WINDING)
        OVEREND UNWINDING
        PRECISION WINDERS

RT RING FRAMES
RING SPINNING
RINGS
SPINDLES
SPINNING
TRAVELERS
TWISTING
UNWINDING
UNWINDING ACCELERATORS
WINDING TENSION
YARN DYNAMICS
YARN TENSION

BANCROFT PROCESS (TN)
BT FIRE RETARDANCY TREATMENTS

BAND KNIFE CUTTING MACHINES
BT CUTTING MACHINES

BANDAGES
BT MEDICAL TEXTILES
RT GAUZE

BANDING
RT TAPING

BANGING OFF
RT DROP WIRES
FEELER MECHANISMS
KNOCK OFF
LOOM STOPS
LOOMS
STOP MOTIONS
WEAVING

BANLON (TN)
BT CRIMPED YARNS
TEXTURED YARNS (FILAMENT)
RT BULKED YARNS
BULKING
CRIMPING
SPUNIZE (TN)
STUFFER BOX CRIMPING
STUFFER BOXES
TEXTRALIZED YARNS (TN)
TEXTURING

BANNOCK BURN
RT CHEVIOT TWEED
TWILL WEAVES

BAR TACKS
RT TACKING

BAR WARP MACHINES
BT LACE MACHINES
RT BARMEN MACHINES
BOBBINET MACHINES
CURTAIN MACHINES
DOUBLE LOCKER MACHINES
GO THROUGH MACHINES
LEAVERS MACHINES
MECHLIN MACHINES
ROLLING LOCKER MACHINES
SIVAL MACHINES
STRING WARP MACHINES
WARP LACE MACHINES

BARATHEA
BT FLAT WOVEN FABRICS
WOVEN FABRICS
RT HOPSACK
TWILL WEAVES

BARIUM ACTIVITY NUMBER
RT BARIUM HYDROXIDE
MERCERIZED FINISH
MERCERIZING

BARIUM COMPOUNDS
NT BARIUM HYDROXIDE

BARIUM HYDROXIDE
BT BARIUM COMPOUNDS
HYDROXIDES
RT BARIUM ACTIVITY NUMBER
BARMEN MACHINES
BT LACE MACHINES
RT BAR WARP MACHINES
CURTAIN MACHINES
GO THROUGH MACHINES
LEAVERS MACHINES
MECHLIN MACHINES
ROLLING LOCKER MACHINES
SIVAL MACHINES
STRING WARP MACHINES
WARP LACE MACHINES

BAROTOR MACHINES
BT PIECE DYEING MACHINES
PRESSURE DYEING MACHINES

BARRAS
RT SACKCLOTH

BARRE
NT DIAMOND BARRING
PIRN BARRE
BT DYEING DEFECTS
FABRIC DEFECTS
RT COCKLE (DEFECT)
FABRIC INSPECTION
FILLING STREAKS
PERIODICITY
POSITIVE FEED
QUALITY CONTROL
REPEATS
SHADE BARS
STREAKINESS
WARP STREAKS

BASE BINDING
BT CHEMICAL PROPERTIES
RT ACID BINDING

BASES
NT ALKALIES
AMINES
AMMONIA
CARBONATES
HYDROXIDES
RT ACIDS
AMINO GROUPS
CATIONS
ELECTROLYTES
SALTS (GENERAL)

BASIC DYEING
UF CATIONIC DYEING
BT DYEING (BY DYE CLASSES)
RT ACID DYEING
AZOIC DYEING
BASIC DYES
DIRECT DYEING
DISPERSE DYEING
METALLIZED DYEING
MORDANT DYEING
NEUTRAL DYEING
PIGMENT DYEING
REACTIVE DYEING
SULFUR DYEING
VAT DYEING

BASIC DYES
UF CATIONIC DYES
NT METHYLENE BLUE
BT DYES (BY CHEMICAL CLASSES)
RT ACID DYES
ALCIAN BLUE (TN)
BASIC DYEING
CATIONS
DYEING (BY DYE CLASSES)
DYES (BY FIBER CLASSES)
NEUTRAL DYES

BASKET WEAVES
BT WEAVE (TYPES)
RT FLAT WOVEN FABRICS

BAST FIBERS
NT FLAX
HEMP
HIBISCUS
JUTE
RAMIE
BT CELLULOSIC FIBERS
FIBERS
NATURAL FIBERS
RT CELLULOSE
DECORTICATING
HESSIAN
RAFFIA
RETTING
SEED HAIR FIBERS (GENERAL)

BASTARD COPS
USE QUILLS

BASTING
BT GARMENT MANUFACTURE
RT ANGULAR SEAMING
BACK TACKING
BONDING
CEMENTING
CURVED SEAMING
DRESSMAKING
FUSION (MELTING)
GATHERING
HAND SEWING
HEMMING
MACHINE SEWING
PINCHING
PUCKERING
RUFFLING
SEAM BINDING
SEAMING
SEAMS
SEAMSTRESSES
SEWABILITY
SEWING
SEWING MACHINES
SHIRRING
STITCH QUALITY
STITCHING
TACKING
TAILORING
TUCKING (GARMENT MANUFACTURE)

BATCH DRYING
BT BATCH FINISHING
DRYING
RT CONTINUOUS DRYING
FREEZE DRYING
TUMBLE DRYING
VACUUM DRYING

BATCH DYEING
BT DYEING (BY PROCESS FLOW)
RT BATCH DYEING MACHINES
BATCH FINISHING
CONTINUOUS DYEING
DYEING MACHINES
PAD DYEING
PAD JIG DYEING
PAD JIG FINISHING
PAD ROLL DYEING
PAD ROLL FINISHING
PAD STEAM DYEING
PAD STEAM FINISHING
PIECE DYEING
SEMI CONTINUOUS DYEING
SIMULTANEOUS DYEING AND FINISHING
STOCK DYEING
TOP DYEING
YARN DYEING

BATCH DYEING MACHINES
BT DYEING MACHINES (BY PROCESS FLOW)
DYEING MACHINES
RT BATCH DYEING
CONTINUOUS DYEING RANGE
DYEING
DYEING (BY PROCESS FLOW)
DYEING MACHINES (BY MATERIAL ASSEMBLY)
GARMENT DYEING MACHINES
PACKAGE DYEING MACHINES
PIECE DYEING MACHINES
SEMI CONTINUOUS DYEING RANGE
STOCK DYEING MACHINES
TOP DYEING MACHINES
YARN DYEING MACHINES

BATCH FINISHING
NT BATCH DRYING
BATCH SCOURING
BT WET FINISHING
RT BATCH DYEING
CONTINUOUS FINISHING
FINISHING MACHINERY (GENERAL)
PAD BATCH CURE PROCESS
PAD FINISHING
PAD JIG FINISHING
PAD ROLL FINISHING
PAD STEAM FINISHING

BATCH SCOURING
NT PACKAGE SCOURING
BT BATCH FINISHING
SCOURING
WET FINISHING
RT CONTINUOUS SCOURING
KIER SCOURING

BATH RATIO
UF LIQUOR RATIO
RT BATHS
DYE BATHS
FINISHING PROCESS (GENERAL)

BATHING SUITS
UF SWIM SUITS
BT ATHLETIC CLOTHING
SPORTSWEAR
RT BEACH ROBES
BIKINIS
KNITTED FABRICS
SUN SUITS

BATHS
NT DYE BATHS
SCOURING BATHS
RT BACKWASHING
BATH RATIO
BLEACHING
COAGULATING
COATING (PROCESS)
DESIZING
DYEING
FINISHING PROCESS (GENERAL)
LAUNDERING
NEUTRALIZING (FIBER EXTRUSION)
PADDING
QUENCHING (FILAMENT)
SCOURING
SLASHING
SOAKING
STRIPPING (COLOR)

BATIK DYEING
BT DYEING (FOR EFFECT)
RT BATIK DYEING MACHINES
JIG DYEING
OPEN WIDTH DYEING
RESIST PRINTING
WAXES

BATIK DYEING MACHINES
    RT BATIK DYEING
        REEL MACHINE (DYEING)

BATINES
    BT NOVELTY YARNS

BATISTE
    BT FLAT WOVEN FABRICS
        WOVEN FABRICS
    RT CORSET BATISTE
        LAWN
        POPLIN

BATT MAKING MACHINES
    RT BATTING
        NONWOVEN FABRIC MACHINES
        PAPER MACHINES
        WADDINGS
        WEB FORMING MACHINES
        WEBS

BATTENBERG
    BT HAND MADE LACES
        PILLOW LACES (BOBBIN LACES)
    RT MACHINE MADE LACES

BATTENS (LOOM)
    USE SLEYS (LOOM)

BATTING
    RT BATT MAKING MACHINES
        COMPRESSIBILITY
        FILLER
        INSULATION (THERMAL)
        MEDICAL TEXTILES
        PACKING
        PADDING (MATERIAL)
        WADDINGS
        WARMTH
        WEBS

BATTLEDRESS
    BT MILITARY CLOTHING

BAULK FINISHING
    BT WET FINISHING
    RT FULLING
        SCOURING
        TENTERING
        WOOLEN FABRICS

BAVE (SILK)
    RT SILK

BEACH ROBES
    BT SPORTSWEAR
    RT BATHING SUITS

BEAD FILLER FABRICS
    USE TIRE FABRICS

BEAD WOVEN FABRICS
    USE TIRE FABRICS

BEAD YARNS
    BT NOVELTY YARNS
    RT BOUCLES
        BUG YARNS
        FLAKE YARNS
        FLAMES (NOVELTY YARNS)
        FRILLS (NOVELTY YARNS)
        KNICKERBOCKER YARNS
        LOOPS (NOVELTY YARNS)
        NOVELTY TWISTERS
        NUB YARNS
        RATINES
        SEEDS (NOVELTY YARNS)
        SLUB YARNS
        SPIRALS (NOVELTY YARNS)
        SPLASHES
        TWISTERS

BEADING (LACE)
    BT LACES

BEAM DYEING (YARN)
    UF WARP DYEING
    BT PACKAGE DYEING
        YARN DYEING
    RT BEAM DYEING MACHINES (YARN)
        BEAMS
        CREELING
        DYE PENETRATION
        DYE UNIFORMITY
        MUFF DYEING
        WARP PREPARATION
        WARPING

BEAM DYEING MACHINES (FABRIC)
    BT PIECE DYEING MACHINES
        PRESSURE DYEING MACHINES

BEAM DYEING MACHINES (YARN)
    BT PACKAGE DYEING MACHINES
        YARN DYEING MACHINES
    RT BEAM DYEING (YARN)
        BOBBIN DYEING MACHINES
        CAKE DYEING MACHINES
        CHEESE DYEING MACHINES
        CONE DYEING MACHINES
        COP DYEING MACHINES
        HANK DYEING MACHINES
        WARP DYEING MACHINES

BEAM WARPING
    USE WARPING

BEAMERS
    USE WARPERS (MACHINE)

BEAMING
    USE WARPING

BEAMS
    UF WARP MILLS
    RT BEAM DYEING (YARN)
        BOBBINS
        CAKES
        CHEESES
        CONES
        CREELING
        FLANGES
        JACKSPOOLS
        LET OFF MOTIONS
        LOOMS
        MALIMO PROCESS (TN)
        PACKAGES
        QUILLS
        SECTIONAL WARPING
        SLASHING
        TUFTING
        WARP KNITTING
        WARP PREPARATION
        WARPERS (MACHINE)
        WARPING
        WEAVING
        WINDERS
        WINDING

BEARDED MOTES
    RT COTTON
        MOTE KNIVES
        MOTES
        SEEDS (TRASH)

BEARDED NEEDLES
    USE SPRING NEEDLES

BEARING ALLOYS
    BT ALLOYS
    RT AIR BEARINGS
        BEARINGS

BEARING STRENGTH
    BT MECHANICAL PROPERTIES
        STRENGTH
        STRESS STRAIN PROPERTIES
        (COMPRESSIVE)
    RT BONDING STRENGTH
        COMPRESSIVE STRENGTH
        HARDNESS
        IMPACT STRENGTH
        KNOT EFFICIENCY
        KNOT STRENGTH
        LOOP EFFICIENCY
        LOOP STRENGTH
        PERALTA ROLLS
        SHEAR STRENGTH
        STRAIN
        STRENGTH ELONGATION TESTING
        STRENGTH TESTING
        STRESS
        STRESS STRAIN CURVES
        TESTING EQUIPMENT
        WET STRENGTH

BEARINGS
    NT AIR BEARINGS
        ANTIFRICTION BEARINGS
        BALL BEARINGS
        CONICAL BEARINGS
        FOOTSTEP BEARINGS
        JOURNAL BEARINGS
        MAGNETIC BEARINGS
        NEEDLE BEARINGS
        NYLON BEARINGS
        PLAIN BEARINGS
        RADIAL BEARINGS
        ROLLER BEARINGS
        SLEEVE BEARINGS
        TAPERED ROLLER BEARINGS
        TEFLON BEARINGS
        THRUST BEARINGS
    RT ACCESSORIES
        AXLES
        BEARING ALLOYS
        BOLSTERS
        BUSHINGS
        COTS (ROLLERS)
        LOAD BEARING CAPACITY
        LUBRICATION
        MECHANISMS
        PACKING
        SPINDLES

BEAT UP
    USE BEATING UP

BEATERS (OPENING)
    RT BALE BREAKERS
        GRID BARS
        IMPELLERS
        OPENERS
        PICKERS (OPENING)
        SAW TOOTHED ROLLS

BEATING UP
    UF BEAT UP
    RT BOUNCE (WEAVING)
        CLOTH FELL
        CRIMP INTERCHANGE
        FILLING YARNS
        JAMMING
        LET OFF
        LOOM TAKE UP
        LOOMS
        PICKER STICKS
        PICKING (WEAVING)
        PICKS
        PICKS PER INCH
        REED MARKS
        REEDS
        SHED (WEAVING)
        SHEDDING
        SLEYS (LOOM)
        TAKE UP
        WARP TENSION
        WEAVING

BEAUVAIS TAPESTRIES
    BT TAPESTRIES

BEAVER CLOTH
    RT FACE GOODS

BECK DYEING
    UF WINCH DYEING
    BT PIECE DYEING
    RT BECKS
        OPEN WIDTH DYEING
        ROPE DYEING
        ROPE MARKS
        SEMI CONTINUOUS DYEING

BECKS
    BT PIECE DYEING MACHINES
    RT BECK DYEING
        SEMI CONTINUOUS DYEING RANGE

BEDFORD CORD
    BT FLAT WOVEN FABRICS
        WOVEN FABRICS
    RT PIQUE
        PLAIN WEAVES

BEDJACKETS
    BT NIGHTWEAR
    RT BEDSOCKS
        PAJAMAS

BEDSOCKS
    BT NIGHTWEAR
    RT BEDJACKETS
        PAJAMAS

BEDSPREADS
    NT TUFTED BEDSPREADS
    RT BLANKETS
        EIDERDOWNS
        HOUSEHOLD FABRICS
        PILLOWS

BEER
    UF PORTER
    RT REEDING
        REEDS
        SET (WOVEN FABRIC)
        WARP ENDS
        WARP PREPARATION
        WARPING
        WEAVING

BEETLING
    BT DRY FINISHING
    RT FULLING

BELFAST (TN)
    BT WASH WEAR TREATMENTS
    RT COTTON
        DIMENSIONAL STABILITY
        EASY CARE FABRICS

BELT CRIMPING
    BT CRIMPING
        TEXTURING
    RT GEAR CRIMPING
        STUFFER BOX CRIMPING

BELT DRIVES
    BT DRIVES
    RT ANGLE OF WRAP
        BELT SLIPPAGE
        BELTS

RT CHAIN DRIVES
CONVEYOR BELTS
FLUID DRIVES
GEAR DRIVES
GEARS
HYDRAULIC DRIVES
MECHANICAL DRIVES
VARIABLE SPEED DRIVES

BELT SLIPPAGE
BT SLIPPAGE
RT ANGLE OF WRAP
BELT DRIVES
COEFFICIENT OF FRICTION

BELT TENSION
BT TENSION
RT ANGLE OF WRAP
DISC TENSION
FINGER TENSION
WRAP

BELTING
USE BELTS

BELTS
UF BELTING
NT CONVEYOR BELTS
BT INDUSTRIAL FABRICS
RT ACCESSORIES
BELT DRIVES
CHAINS
DEVICES
MECHANISMS
PAPERMAKERS FELTS
SAFETY BELTS

BELTS (APPAREL)
NT GARTER BELTS
BT GARMENT COMPONENTS
GARMENTS
RT SAFETY BELTS
SUSPENDERS

BEMBERG FIBERS
USE CUPRAMMONIUM RAYON

BENCH SORTING (WOOL)
USE WOOL SORTING

BENDING
UF FLEXING
BT MODES OF DEFORMATION
RT BENDING ENERGY
BENDING LENGTH
BENDING MOMENTS
BENDING RECOVERY
BENDING RIGIDITY
BOWING
BUCKLING
CAMBER
COEFFICIENT OF FRICTION
COUPLES
CREASE RETENTION
CREASING
CURL
CURVATURE
DEFLECTION
DEFORMATION
DISTORTION
DRAPE
DUCTILITY
ELASTIC MODULUS (TENSILE)
FATIGUE
FRICTION
LOOPING
MOMENT OF INERTIA
NEUTRAL AXIS
PLEATING
RADIUS OF CURVATURE
RESILIENCE
SAG
STIFFNESS
STRAIGHTENING
STRAIN
TENSION
WEAVING DYNAMICS
WRINKLING

BENDING ENERGY
BT ENERGY
STRAIN ENERGY
RT BENDING
BENDING MOMENTS
BENDING RIGIDITY
CURVATURE
INTERNAL ENERGY
KINETIC ENERGY
POTENTIAL ENERGY
TORSIONAL ENERGY

BENDING LENGTH
RT BENDING
BENDING MOMENTS
BENDING RECOVERY
BENDING RIGIDITY
BREAKING LENGTH
DEFLECTION
DRAPE
LENGTH

BENDING MOMENT DIAGRAMS
RT BENDING
BENDING MOMENTS
LOAD DEFLECTION CURVES
MOMENTS
SHEAR FORCE

BENDING MOMENTS
BT MOMENTS
RT BENDING
BENDING ENERGY
BENDING LENGTH
BENDING RIGIDITY
BUCKLING
DEFLECTION
LOAD DEFLECTION CURVES
SHEAR FORCE
STRAIGHTENING
TWISTING MOMENTS

BENDING RECOVERY
BT MECHANICAL PROPERTIES
RECOVERY (SELF RESTORATION TO ORIGINAL CONDITION)
RT BENDING
BENDING LENGTH
CREEP RECOVERY
CRIMP RECOVERY
DELAYED RECOVERY
DRY WRINKLE RECOVERY
ELASTIC RECOVERY
FABRIC PROPERTIES
FABRIC PROPERTIES (MECHANICAL)
MOMENT CURVATURE CURVES
PERMANENT SET
RESIDUAL STRAIN
RESIDUAL STRESS
STRAIGHTENING
STRESS RELAXATION
WASH WEAR RATING
WET WRINKLE RECOVERY
WRINKLE RECOVERY
WRINKLE RESISTANCE
WRINKLES

BENDING RIGIDITY
UF BENDING STIFFNESS
FLEXIBILITY
FLEXURAL RIGIDITY
BT MECHANICAL PROPERTIES
RT BENDING
BENDING ENERGY
BENDING LENGTH
BENDING MOMENTS
BOWING
BUCKLING
COEFFICIENT OF FRICTION
COMPRESSIBILITY
COUPLES
CREASE RETENTION
CREASING
CREPING
DEFLECTION
DEFORMATION
DRAPE
ELASTIC MODULUS (TENSILE)
EXTENSIBILITY
FABRIC PROPERTIES
FABRIC PROPERTIES (MECHANICAL)
FIBER FIBER FRICTION
FLEXURAL FATIGUE
FLEXURAL STRENGTH
FRICTION
FRL DRAPE TESTER
HAND
MATRIX FREEDOM
MOMENT CURVATURE CURVES
MOMENT OF INERTIA
NEUTRAL AXIS
PAPERINESS
PILL RESISTANCE
PILLING
SIZES(SLASHING)
SOFTENERS
STARCHING
STIFFNESS
STRAIGHTENING
WRINKLE RECOVERY
YARN PROPERTIES

BENDING STIFFNESS
USE BENDING RIGIDITY

BENDING STRENGTH
USE FLEXURAL STRENGTH

BENDING TESTING
BT TESTING
RT ABRASION TESTING
BURST TESTING
FRL DRAPE TESTER
IMPACT TESTING
TEAR TESTING
TENSILE TESTING
TESTING EQUIPMENT

BENZAMIDE
BT AMIDES

BENZENE
BT AROMATIC HYDROCARBONS
RT SOLVENTS

BENZOATE ESTERS
NT BENZYL BENZOATE
BT ESTERS
RT BENZOATE SALTS
BENZOIC ACID

BENZOATE SALTS
NT SODIUM BENZOATE
RT BENZOATE ESTERS
BENZOIC ACID
PRESERVATIVES

BENZOIC ACID
BT CARBOXYLIC ACIDS
ORGANIC ACIDS
RT AROMATIC COMPOUNDS
BENZOATE ESTERS
BENZOATE SALTS
PRESERVATIVES

BENZOQUINONE
USE QUINONE

BENZOYLATED COTTON
BT CHEMICALLY MODIFIED COTTON
RT ACETYLATED COTTON
AMINIZED COTTON
CARBAMOYLETHYLATED COTTON
CARBOXYMETHYLATED COTTON
COTTON
CYANOETHYLATED COTTON

BENZYL ALCOHOL
BT ALCOHOLS
RT CARRIER DYEING
CARRIERS (DYEING)
SOLVENT ASSISTED DYEING

BENZYL BENZOATE
BT BENZOATE ESTERS

BENZYLCELLULOSE
BT CELLULOSE ETHERS
RT BENZOYLATED COTTON

BENZYL CHLORIDE
BT CHLORINE COMPOUNDS
RT BENZYLATION

BENZYLATION
BT ETHERIFICATION
RT ALKYLATION
BENZYL CHLORIDE
CHEMICAL MODIFICATION (FIBERS)
CHEMICALLY MODIFIED COTTON

BERYLLIUM CHLORIDE
BT BERYLLIUM COMPOUNDS
CHLORIDES

BERYLLIUM COMPOUNDS
NT BERYLLIUM CHLORIDE
BERYLLIUM HYDROXIDE

BERYLLIUM HYDROXIDE
BT BERYLLIUM COMPOUNDS
HYDROXIDES

BET ISOTHERM
BT ADSORPTION ISOTHERM
RT FREUNDLICH ISOTHERM
LANGMUIR ISOTHERM

BETA KERATIN
USE KERATIN

BETA PARTICLES
RT ALPHA PARTICLES
RADIATION

BETA RAYS
BT IONIZING RADIATION
RADIATION

BETAINE SURFACTANTS
BT ZWITTERIONIC SURFACTANTS
RT CARBOXYLATE SURFACTANTS
SULFOBETAINE SURFACTANTS

BHES
USE BIS (HYDROXYETHYL) SULFONE

BIAXIAL STRENGTH
BT FABRIC PROPERTIES
FABRIC PROPERTIES (MECHANICAL)
STRENGTH
RT STRENGTH TESTING
UNIAXIAL STRENGTH

BIAXIAL STRESS
BT STRESS
TENSILE STRESS
RT BALL BURST TESTING
BURST TESTING
BURSTING STRENGTH
COMPRESSION
(CONT.)

RT CRIMP INTERCHANGE
DIAPHRAGM BURST TESTING
ELASTIC STRAIN
ELASTICITY
EXTENSIBILITY
HOOP STRESS
HYDROSTATIC STRESS
MOHRS CIRCLE
POISSONS RATIO
SHEAR STRESS
STRENGTH OF MATERIALS
STRENGTH TESTING
STRESS ANALYSIS
STRESS CONCENTRATION
STRESS DISTRIBUTION
STRETCH
TENSION
TRELLIS MODEL
UNIAXIAL STRESS
WEAVING DYNAMICS

BIBLIOGRAPHY
RT HISTORY
LITERATURE SURVEYS
REVIEWS
SURVEYS

BICARBONATES
NT POTASSIUM BICARBONATE
SODIUM BICARBONATE
BT CARBONATES

BICOMPONENT FIBER YARNS (FILAMENT)
NT CANTRECE (TN)
BT BULKED YARNS
CRIMPED YARNS
FILAMENT YARNS
SELF CRIMPING YARNS
STRETCH YARNS (FILAMENT)
TEXTURED YARNS (FILAMENT)
RT AGILON (TN)
AIR JET TEXTURED YARNS
BICOMPONENT FIBER YARNS (STAPLE)
BICOMPONENT FIBERS
BULKING
CORE SPUN YARNS
CRIMP
CRIMP TENDENCY
CRIMPING
DIFFERENTIAL SHRINKAGE
EDGE CRIMPED YARNS
FALSE TWIST YARNS
FOLLOW THE LEADER CRIMP
GEAR CRIMPED YARNS
HILOW BULKED YARNS
NATURAL CRIMP
ORLON SAYELLE (TN)
PRODUCERS YARNS
STRETCH YARNS (SPUN)
TEXTURED SPUN YARNS
TEXTURING
WAVINESS
YARN CRIMP
Z CRIMP

BICOMPONENT FIBER YARNS (STAPLE)
BT CRIMPED YARNS
SELF CRIMPING YARNS
TEXTURED SPUN YARNS
RT BICOMPONENT FIBERS
HILOW BULKED YARNS
ORLON SAYELLE (TN)

BICOMPONENT FIBERS
UF BICOMPONENT FILAMENTS
FIBERS (BICOMPONENT)
NT ORLON SAYELLE (TN)
RT BICOMPONENT FIBER YARNS (FILAMENT)
BICOMPONENT FIBER YARNS (STAPLE)
BICOMPONENT STRUCTURE
BICONSTITUENT FIBERS
BLENDS (POLYMERS)
CRIMP
CRIMP TENDENCY
DIFFERENTIAL SHRINKAGE
EDGE CRIMPED YARNS
HELICAL CRIMP
FIBER CRIMP
MAN MADE FIBERS
NATURAL CRIMP
ORTHOCORTEX
PARACORTEX
TEXTURED YARNS (FILAMENT)
WOOL

BICOMPONENT FILAMENTS
USE BICOMPONENT FIBERS

BICOMPONENT SPINNING
UF CONJUGATE SPINNING
RT BICOMPONENT FIBERS
BICOMPONENT STRUCTURE

BICOMPONENT STRUCTURE
UF BILATERAL FIBER STRUCTURE
RT BICOMPONENT FIBERS
BICOMPONENT SPINNING
DIFFERENTIAL SHRINKAGE
MOMENT OF INERTIA

BICONSTITUENT FIBERS
RT BICOMPONENT FIBERS
BLENDS (FIBERS)
BLENDS (POLYMERS)
MAN MADE FIBERS

BIKINIS
BT SPORTSWEAR
RT BATHING SUITS
BEACH ROBES
KNITTED FABRICS
SUN SUITS

BILATERAL FIBER STRUCTURE
USE BICOMPONENT STRUCTURE

BILLIARD CLOTH
BT FABRICS

BINCHE
BT HAND MADE LACES
LACES

BINDER EFFICIENCY
RT BINDERS (NONWOVENS)
BONDED FIBER FABRICS
NONWOVEN FABRICS
STRENGTH EFFICIENCY

BINDER MIGRATION
BT MIGRATION (SUBSTANCE)
RT BONDED FIBER FABRICS
NONWOVEN FABRICS

BINDERS (NONWOVENS)
UF BONDING AGENTS (NONWOVENS)
NT EMULSION BINDERS (NONWOVENS)
SOLUTION TYPE BINDERS
THERMOPLASTIC POWDERS
THERMOSETTING POWDERS
RT ADHESIVE BONDED (NONWOVENS)
ADHESIVES
BONDING
LAMINATING RESINS
NONWOVEN FABRIC STRUCTURE
NONWOVEN FABRICS
SELF BONDED FABRICS (NONWOVENS)

BINDERS (PIGMENTS)
NT EMULSION BINDERS (PIGMENTS)
RT OIL IN WATER EMULSIONS
PIGMENT DYEING
PIGMENT PRINTING
THERMOSETTING RESINS
WATER IN OIL EMULSIONS

BINDING (SEAMS)
USE SEAM BINDING

BINDINGS (FOOTWEAR)
NT LACES (SHOES)
RT BOOTS
FOOTWEAR
LEATHER
LININGS (FOOTWEAR)
PUMPS (FOOTWEAR)
SHOES
SLIPPERS
SNEAKERS
SOLES
STITCHING
TONGUES
UPPERS

BINGHAM BODY
USE PLASTICOVISCOSITY

BINOMIAL DISTRIBUTION
RT CHI SQUARED DISTRIBUTION
DISTRIBUTION
F DISTRIBUTION
GAMMA DISTRIBUTION
NORMAL DISTRIBUTION
T DISTRIBUTION

BIOCHEMICAL OXYGEN DEMAND
UF BIOLOGICAL OXYGEN DEMAND
BOD
RT BIOCHEMICAL PROPERTIES
BIODEGRADABILITY
OXYGEN

BIOCHEMICAL PROPERTIES
NT BIODEGRADABILITY
RT AESTHETIC PROPERTIES
ANTIMICROBIAL FINISHES
BACTERIAL INHIBITION
BIOCHEMICAL OXYGEN DEMAND
CHEMICAL PROPERTIES
DERMATITIS
END USE PROPERTIES
FABRIC PROPERTIES
FIBER PROPERTIES
MECHANICAL PROPERTIES
PHYSICAL CHEMICAL PROPERTIES
PHYSICAL PROPERTIES (EXCLUDING MECHANICAL)
PROPERTIES
SKIN IRRITATION

BIOCIDES
USE ANTIMICROBIAL FINISHES

BIODEGRADABILITY
BT BIOCHEMICAL PROPERTIES
CHEMICAL PROPERTIES
DEGRADATION PROPERTIES
RT BIOCHEMICAL OXYGEN DEMAND
DETERGENTS
SURFACTANTS
SYNTHETIC DETERGENTS
WASTE

BIOLOGICAL OXYGEN DEMAND
USE BIOCHEMICAL OXYGEN DEMAND

BIRDS EYE WEAVES
BT WEAVE (TYPES)
RT PIQUE

BIREFRINGENCE
NT ORIENTATION BIREFRINGENCE
BT OPTICAL PROPERTIES
STRESS OPTICAL PROPERTIES
RT DEGREE OF ORIENTATION
DICHROISM
DIFFRACTION PATTERNS
EXPERIMENTAL STRESS ANALYSIS
GLASS RUBBER TRANSITION
GLASS TRANSITION TEMPERATURE
INTERFERENCE FRINGES
MELTING POINT
MOLECULAR CHAIN MOVEMENT
OPTICS
QUENCHING (FILAMENT)
REFRACTIVE INDEX
SECONDARY TRANSITION TEMPERATURES
STRESS OPTICAL COEFFICIENT
SUBGROUP MOVEMENT (MOLECULAR)
TRANSITIONS (POLYMERS)
X RAY ANALYSIS

BIS (HYDROXYETHYL) SULFONE
UF BHES
BT REACTANTS
SULFONES
RT DIVINYL SULFONE
EASY CARE FABRICS
SULFONIUM REACTANTS

BISMUTH BROMIDE
BT BISMUTH COMPOUNDS
BROMIDES

BISMUTH CHLORIDE
BT BISMUTH COMPOUNDS
CHLORIDES

BISMUTH COMPOUNDS
NT BISMUTH BROMIDE
BISMUTH CHLORIDE
ORGANOBISMUTH COMPOUNDS

BISULFATES
NT SODIUM BISULFATE
BT SULFUR COMPOUNDS
RT BISULFITES
DYEING AUXILIARIES
SULFATES
SULFURIC ACID

BISULFITES
UF MONOETHANOLAMINE BISULFITE
NT SODIUM BISULFITE
BT SULFUR COMPOUNDS
RT BISULFATES
PYROSULFITES
REDUCING AGENTS
REDUCTION
SULFITES
UREA BISULFITE SOLUBILITY

BLACKBODY
RT ABSORPTIVITY (RADIATION)
EMISSIVITY
REFLECTIVITY
THERMAL PROPERTIES
TRANSMISSIVITY

BLADES (ROLLER)
USE IMPELLERS

BLAMIRE FEED
BT INTERMEDIATE FEED
RT CARDS
FINISHER PART
HOPPER FEED
INTERMEDIATE PART
SCOTCH FEED
WOOLEN CARDING
WOOLEN CARDS

BLANC DE BLANCS (TN)
    BT NYLON (POLYAMIDE FIBERS)
        NYLON 6

BLANCOPHOR (TN)
    BT FLUORESCENT BRIGHTENERS

BLANK NEEDLES
    USE DUMMY NEEDLES

BLANKET (FINISHING)
    RT BACK GRAY
        BLANKET MARKS
        EMBOSSING
        COMPRESSIVE SHRINKAGE
        PRINTING MACHINES

BLANKET MARKS
    RT BLANKET (FINISHING)
        EMBOSSING

BLANKET SHUTTLES
    BT SHUTTLES
    RT BOAT SHUTTLES
        FLY SHUTTLES
        LOOMS
        PICKING (WEAVING)
        QUILLS
        RACE BOARDS
        REEDS
        SHEDDING
        SHUTTLE BINDERS
        SHUTTLE BOXES
        SHUTTLE MARKS
        SLEYS (LOOM)
        STICK SHUTTLES
        WEAVING

BLANKETS
    RT BEDSPREADS
        EIDERDOWNS
        HOUSEHOLD FABRICS
        PILLOWS

BLEACH HOUSE
    USE BLEACHERIES

BLEACHERIES
    UF BLEACH HOUSE
    RT BLEACHING
        BLEACHING MACHINES
        BLEACHING RANGES
        COTTON
        DYE HOUSE
        WET FINISHING

BLEACHES
    USE BLEACHING AGENTS

BLEACHING
    NT CONTINUOUS BLEACHING
        DRY IN BLEACHING
        GRASS BLEACHING
        OPEN WIDTH BLEACHING
        ROPE BLEACHING
    BT WET FINISHING
    RT BACKWASHING
        BATHS
        BLEACHERIES
        BLEACHING AGENTS
        BLEACHING MACHINES
        BLEACHING RANGES
        CHELATING
        CHLORINE BLEACHING AGENTS
        CHLORINE DAMAGE
        CHLORINE DIOXIDE
        CHLORINE FASTNESS (COLOR)
        CHLORINE FASTNESS (OF FINISH)
        CHLORITES
        CLEANING
        CLEANLINESS
        COLORFASTNESS
        CONTINUOUS SCOURING
        COTTON WAX
        DECOLORIZING
        DYE BATHS
        DYEING
        FADING
        FLUORESCENT BRIGHTENERS
        FUGITIVE DYES
        HYDROGEN PEROXIDE
        HYPOCHLORITES
        KIER SCOURING
        LAUNDERING
        LITHIUM HYPOCHLORITE
        OXIDATION
        PAPER MAKING
        PEROXIDE BLEACHING AGENTS
        PEROXIDES
        SILICATES
        PEROXYGEN COMPOUNDS
        SCOURING
        SODIUM HYPOCHLORITE
        SODIUM PEROXIDE
        STRIPPING (COLOR)
        WHITENESS

BLEACHING AGENTS
    UF BLEACHES
    NT CHLORINE BLEACHING AGENTS
        PEROXIDE BLEACHING AGENTS
    BT TEXTILE CHEMICALS
    RT BLEACHING
        BLEACHING MACHINES
        BLUEING AGENTS
        CHLORINE DAMAGE
        CLEANING
        DETERGENTS
        FINISH (SUBSTANCE ADDED)
        FLUORESCENT BRIGHTENERS
        HYPOBROMITES
        LAUNDERING
        OXYCELLULOSE
        SCOURING
        SOAPS

BLEACHING MACHINES
    RT BLEACHERIES
        BLEACHING
        BLEACHING AGENTS
        BLEACHING RANGES
        KIERS
        LAUNDERING
        SCOURING
        SCOURING MACHINES

BLEACHING RANGES
    BT WET PROCESSING RANGES
    RT BLEACHERIES
        BLEACHING
        BLEACHING MACHINES
        CASCADE WASHERS
        CONTINUOUS BLEACHING
        J BOXES

BLEEDING
    UF COLOR BLEEDING
    BT FABRIC DEFECTS
    RT BLEEDING TESTING
        DYEING
        LAUNDERING
        RESISTANCE TO BLEEDING
        WASHFASTNESS (COLOR)
        WET CROCKING

BLEEDING TESTING
    BT COLORFASTNESS TESTING
    RT BLEEDING
        LIGHTFASTNESS TESTING
        RESISTANCE TO BLEEDING
        WASHFASTNESS TESTING

BLEMISHES
    USE DEFECTS

BLEND DRAFTING
    RT BLENDING
        BLENDS (FIBERS)
        DRAFTING (STAPLE FIBER)
        INTIMATE BLENDS
        ORTHOBLENDS
        RADIAL DISTRIBUTION

BLEND DYEING
    BT DYEING (BY FIBER CLASSES)
    RT ACETATE DYEING
        BLENDS (FIBERS)
        CELLULOSE DYEING
        COTTON DYEING
        CROSS DYEING
        FUGITIVE DYES
        RAYON DYEING
        SILK DYEING
        SYNTHETIC DYEING
        TONE IN TONE DYEING
        UNION DYEING
        WOOL DYEING

BLEND LEVEL
    USE BLENDED FIBER RATIO

BLENDED FABRICS
    RT BLENDED YARNS
        BLENDED YARNS (FILAMENT)
        BLENDED YARNS (STAPLE)
        BLENDING
        BLENDS (FIBERS)
        BLENDS (POLYMERS)
        ORTHOBLENDS

BLENDED FIBER RATIO
    UF BLEND LEVEL
    RT BLENDING
        BLENDS (FIBERS)
        FIBER CONTENT
        FIBER PROPERTIES

BLENDED YARNS
    NT BLENDED YARNS (FILAMENT)
        BLENDED YARNS (STAPLE)
    BT YARNS
    RT BLENDED FABRICS
        BLENDING
        BLENDS (FIBERS)
        BLENDS (POLYMERS)
        CORE SPINNING
        CORE SPUN YARNS
    RT FIBER ASSEMBLIES
        FIBER CONTENT
        INTIMATE BLENDS
        RADIAL DISTRIBUTION

BLENDED YARNS (FILAMENT)
    BT BLENDED YARNS
    RT BLENDED FABRICS
        BLENDED YARNS (STAPLE)
        BLENDING
        COVERED YARNS
        FIBER CONTENT
        RADIAL DISTRIBUTION

BLENDED YARNS (STAPLE)
    RT BLENDED FABRICS
    BT BLENDED YARNS
    RT BLENDED YARNS (FILAMENT)
        BLENDING
        COVERED YARNS
        FIBER CONTENT
        INTIMATE BLENDS
        RADIAL DISTRIBUTION

BLENDERS
    RT BLENDING
        BLENDS (FIBERS)
        HOMOGENIZERS

BLENDING
    NT SANDWICH BLENDING
    BT MIXING
    RT BLEND DRAFTING
        BLENDED FABRICS
        BLENDED FIBER RATIO
        BLENDED YARNS
        BLENDED YARNS (FILAMENT)
        BLENDED YARNS (STAPLE)
        BLENDERS
        BLENDS (FIBERS)
        BLENDS (POLYMERS)
        BRADFORD SYSTEM PROCESSING
        CARDING
        CARDING OIL
        COMBING
        COMPOUNDING
        COTTON SYSTEM PROCESSING
        DOUBLING (DRAFTING)
        FIBER CONTENT
        INTIMATE BLENDS
        LUBRICATION
        NEW BRADFORD SYSTEM PROCESSING
        OILING
        ORTHOBLENDS
        PAPER MAKING
        PICKING (OPENING)
        STIRRING
        TINTING
        TRACER FIBERS
        WOOLEN SYSTEM PROCESSING
        WOOLEN SYSTEM SPINNING
        WORSTED SYSTEM PROCESSING
        YARNS

BLENDS (FIBERS)
    UF FIBER BLENDS
        MAN MADE FIBER BLENDS
        NATURAL FIBER BLENDS
        SYNTHETIC FIBER BLENDS
    NT ACRYLIC FIBER BLENDS
        AZLON BLENDS
        CELLULOSE BLENDS
        CELLULOSE ESTER FIBER BLENDS
        COTTON BLENDS
        INORGANIC FIBER BLENDS
        NYLON BLENDS
        OLEFIN FIBER BLENDS
        POLYESTER FIBER BLENDS
        POLYTETRAFLUOROETHYLENE FIBER BLENDS
        RAYON BLENDS
        SILK BLENDS
        SPANDEX FIBER BLENDS
        SYNTHETIC RUBBER FIBER BLENDS
        VINYL FIBER BLENDS
        WOOL BLENDS
    RT BICOMPONENT FIBERS
        BICONSTITUENT FIBERS
        BLEND DRAFTING
        BLEND DYEING
        BLENDED FABRICS
        BLENDED FIBER RATIO
        BLENDED YARNS
        BLENDERS
        BLENDING
        BLENDS (POLYMERS)
        INTIMATE BLENDS
        MAN MADE FIBERS
        ORTHOBLENDS
        RADIAL DISTRIBUTION
        SANDWICH BLENDING
        SYNTHETIC FIBERS
        TINTING
        TRACER FIBERS

BLENDS (POLYMERS)
    RT BICOMPONENT FIBERS
        BICONSTITUENT FIBERS
        (CONT.)

RT BLENDS (FIBERS)
BLENDED FABRICS
BLENDED YARNS
BLENDING
POLYMERS

BLIND STITCHING
BT GARMENT MANUFACTURE
RT BLINDSTITCHES
GATHERING
HEMMING
LOCKSTITCH BLINDSTITCHES
SEAMING
SERGING
SEWING
SINGLE THREAD BLINDSTITCHES
STITCHES (SEWING)
STITCHING
TAILORING

BLINDSTITCHES
NT SINGLE THREAD BLINDSTITCHES
BT STITCHES (SEWING)
RT BLIND STITCHING
LOCKSTITCH BLINDSTITCHES
SINGLE THREAD BLINDSTITCHES

BLOCK POLYMERS
NT POLYETHOXY POLYPROPOXY ETHYLENEDIAMINE
BT COPOLYMERS
RT CONDENSATION POLYMERS
GRAFT POLYMERS
HOMOPOLYMERS
PLURONICS (TN)

BLOCK PRINTING
BT PRINTING
RT DISCHARGE PRINTING
DUPLEX PRINTING
MELANCE PRINTING
PRINTING MACHINES
RESIST PRINTING
ROLLER PRINTING
SCREEN PRINTING
THROUGH PRINTING

BLOCK PRINTING EQUIPMENT
BT PRINTING MACHINES
RT DUPLEX PRINTING MACHINES
MELANCE PRINTING EQUIPMENT
SCREEN PRINTING MACHINES

BLONDE (LACE)
BT HAND MADE LACES
LACES

BLOOM
BT AESTHETIC PROPERTIES
FABRIC PROPERTIES
FABRIC PROPERTIES (AESTHETIC)
RT AESTHETIC APPEAL
APPEARANCE
BRIGHTNESS
COLOR
DRABNESS
DULLNESS
GLOSS
LUSTER
LUXURIOUSNESS
OPTICAL PROPERTIES
RICHNESS
SCHREINERING
SHADE
SPARKLE

BLOTCHES
RT BLOTCHINESS
DYEING

BLOTCHINESS
RT BLOTCHES
ORIENTATION UNIFORMITY
PATCHINESS
SPOTTINESS
WEB ORIENTATION
WEB UNIFORMITY

BLOUSES
RT DRESSES

BLOWERS
UF BLOWING EQUIPMENT
RT AIR
CLEANERS
TRAVELING BLOWERS
VACUUM CLEANERS

BLOWING EQUIPMENT
USE BLOWERS

BLOWN FINISH
BT DRY FINISHING
WET FINISHING

BLUE C NYLON (TN)
BT NYLON (POLYAMIDE FIBERS)
NYLON 66

BLUE C POLYESTER (TN)
BT POLYESTER FIBERS

BLUE C SPANDEX (TN)
BT SPANDEX FIBERS

BLUEING
BT WET FINISHING
RT BLUEING AGENTS
FLUORESCENT BRIGHTENERS
LAUNDERING
YELLOWING

BLUEING AGENTS
NT METHYLENE BLUE
ULTRAMARINE BLUE
BT FINISH (SUBSTANCE ADDED)
RT BLEACHING AGENTS
BLUEING
FLUORESCENT BRIGHTENERS

BLUETTE
RT TWILL WEAVES

BOARDING
UF PREBOARDING
PRESS FINISHING
BT DRY FINISHING
GARMENT MANUFACTURE
SHAPING
RT CIRCULAR HOSIERY
CIRCULAR KNITTING
DIMENSIONAL STABILITY
FABRIC RELAXATION
FASHIONING
FORMING
HALF HOSE
HEAT SETTING (SYNTHETICS)
HOSIERY
HOSIERY MACHINES
IRONING
KNITTED FABRICS
KNITTING
MACHINE KNITTING
MOLDING (TAILORING)
PRESSING
SEAMLESS HOSIERY
SEAMLESS HOSIERY MACHINES
SHRINKING
SOCKS
STOCKINGS
STRESS RELAXATION
STRETCH HOSIERY
TRIMMING (OPERATION)
WET FINISHING

BOAT SHUTTLES
BT SHUTTLES
RT BLANKET SHUTTLES
FLY SHUTTLES
LOOMS
PICKING (WEAVING)
QUILLS
RACE BOARDS
REEDS
SHEDDING
SHUTTLE BINDERS
SHUTTLE BOXES
SHUTTLE MARKS
SLEYS (LOOM)
STICK SHUTTLES
WEAVING

BOBBIN CASES
RT BOBBIN HOOKS
SEWING BOBBINS
SEWING MACHINES

BOBBIN DYEING
BT PACKAGE DYEING
YARN DYEING
RT MUFF DYEING

BOBBIN DYEING MACHINES
BT PACKAGE DYEING MACHINES
YARN DYEING MACHINES
RT BEAM DYEING MACHINES (YARN)
CAKE DYEING MACHINES
CHEESE DYEING MACHINES
CONE DYEING MACHINES
COP DYEING MACHINES
HANK DYEING MACHINES
WARP DYEING MACHINES

BOBBIN HOOKS
RT BOBBIN CASES
SEWING BOBBINS
SEWING MACHINES

BOBBIN LEAD
RT FLYER LEAD
FLYER SPINNING FRAMES
PRESSURE FINGERS

BOBBIN LOOPS
RT NEEDLE LOOPS
SEWING MACHINES

BOBBIN RINGS
UF RINGS (BOBBINS)
RT BOBBINS

BOBBIN STRIPPING MACHINES
UF QUILL STRIPPERS
RT QUILLS
STRIPPERS

BOBBIN THREADS (SEWING)
BT SEWING THREADS
THREADS
RT COTTON THREADS
HAND SEWING
LEA STRENGTH PRODUCT
LINEN THREADS
MACHINE SEWING
NEEDLE THREADS
NYLON THREADS
SEWABILITY
SEWING
SEWING BOBBINS
SEWING MACHINES
STITCHING
THREADS

BOBBINET MACHINES
UF PLAIN NET MACHINES
BT LACE MACHINES
RT BAR WARP MACHINES
BARMEN MACHINES
CURTAIN MACHINES
DOUBLE LOCKER MACHINES
GO THROUGH MACHINES
LEAVERS MACHINES
MECHLIN MACHINES
ROLLING LOCKER MACHINES
SIVAL MACHINES
STRING WARP MACHINES
WARP LACE MACHINES

BOBBINETTE LACE
BT LACES
MACHINE MADE LACES

BOBBINS
UF **REELS**
SPOOLS
NT **BOTTLE BOBBINS**
BRAIDER BOBBINS
LET OFF SPOOLS
SEWING BOBBINS
SPINNING BOBBINS
TWISTING BOBBINS
BT **PACKAGES**
RT **BEAMS**
BOBBIN RINGS
CAKES
CAPS (SPINNING)
CHEESES
CONES
CORES (CENTER)
DOFFING (PACKAGE)
EMPTY PACKAGES
FLANGES
FLYER LEG
FLYERS
FULL PACKAGES
GROOVES
HEADS (TOP)
JACKSPOOLS
LARGE PACKAGES
LIFT (WINDING)
PACKAGE HOLDERS
QUILLS
SEWING MACHINES
SHUTTLE ACCESSORIES
SPINDLES
TAPER (BOBBIN)
YARN CARRIERS

BOD
USE BIOCHEMICAL OXYGEN DEMAND

BODICES
BT GARMENT COMPONENTS

BODY KINEMATICS
USE ANTHROPOMETRIC KINEMATICS

BODY MOVEMENTS
USE ANTHROPOMETRIC KINEMATICS

BODY PLY
USE CASING PLY

BOHEMIAN (LACE)
BT HAND MADE LACES
LACES

BOIL OFF
USE BOILING OFF

BOIL OFF MACHINES
RT BOILING OFF
SCOURING MACHINES

BOILING
RT BOILING OFF
BOILING POINT
DYEING
KIER SCOURING

BOILING OFF
UF BOIL OFF
RT BOIL OFF MACHINES
BOILING
DEGUMMING
DESIZING
SCOURING
SERICIN

BOILING POINT
BT PHYSICAL CHEMICAL PROPERTIES
PHYSICAL PROPERTIES (EXCLUDING MECHANICAL)
THERMAL PROPERTIES
RT BOILING
DEW POINT
DIFFUSION
DIFFUSIVITY
ENTHALPY
ENTROPY
FLASH POINT
FREEZING POINT
HEAT CONTENT
LATENT HEAT
MELTING POINT
MOISTURE CONTENT
PRESSURE
RELATIVE HUMIDITY
SPECIFIC HEAT
SPECIFIC HUMIDITY
TEMPERATURE
THERMODYNAMICS
VAPOR PRESSURE

BOLLS
UF COTTON BOLLS
RT COTTON
COTTON GINS
GINNING
LINT (FIBER)
MOTE KNIVES
PICKING (HARVESTING)
SEED HAIR FIBERS (GENERAL)
SEEDS (TRASH)
TRASH

BOLSTERS
RT BEARINGS
SPINDLES
SPINNING FRAMES

BOLT CAMS (KNITTING)
UF PLUNGER CAMS
RT KNITTING
KNITTING MACHINES

BOLTON SHEETING
RT TWILL WEAVES

BOMBAZINE
BT FLAT WOVEN FABRICS
WOVEN FABRICS
RT PRINT CLOTH

BOND DISSOCIATION ENERGY
BT ENERGY
RT BOND ENERGY
CHEMICAL BONDS

BOND ENERGY
BT ENERGY
RT BOND DISSOCIATION ENERGY
CHEMICAL BONDS
VAN DER WAALS FORCES

BONDED FABRICS (NONWOVENS)
USE BONDED FIBER FABRICS

BONDED FABRICS (LAMINATED)
USE LAMINATED FABRICS

BONDED FIBER FABRICS
UF BONDED FABRICS (NONWOVENS)
NT ADHESIVE BONDED (NONWOVENS)
PAPER
SELF BONDED FABRICS (NONWOVENS)
SPUNBONDED (NONWOVENS)
BT FABRICS (ACCORDING TO STRUCTURE)
NONWOVEN FABRICS
RT BINDER EFFICIENCY
BINDER MIGRATION
BONDED YARN FABRICS
BONDING
BONDING STRENGTH
CROSS LAID WEBS
CURING
DRY POWDER BONDING
EMULSION BINDERS (NONWOVENS)
FELTING
FLAME BONDING
FOAM BACKED FABRICS
FOAM BONDING
FUSION (MELTING)
HEAT SEALING
LAMINATED FABRICS
MECHANICAL BONDING
NEEDLED FABRICS
NONWOVEN FABRIC STRUCTURE
NONWOVEN SCRIMS
RANDO WEBBER (TN)
RANDO WEBS (TN)
RODNEY HUNT SATURATOR
SATURATION BONDING
SATURATORS
SOLVENT BONDING
SPRAY BONDING
STITCH BONDED FABRICS
THERMOPLASTIC FIBER BONDING
THERMOPLASTIC POWDERS
THERMOPLASTICITY
THERMOSETTING POWDERS
UNIDIRECTIONALLY ORIENTED WEBS
WALDRON SATURATOR
WEBS

BONDED YARN FABRICS
BT FABRICS (ACCORDING TO STRUCTURE)
RT ADHESIVES
BONDED FIBER FABRICS
BONDED YARNS
BONDING
COMPOSITES
CROSS LAID YARN FABRICS
DIP BONDING
EMULSION BINDERS (NONWOVENS)
FELTS
FLOCKED FABRICS
LAMINATES
NEEDLED FABRICS
NONWOVEN FABRICS
NONWOVEN SCRIMS
PAPER
SELF BONDED FABRICS (NONWOVENS)
SPRAY BONDING
STITCH BONDED FABRICS
STITCH REINFORCED FABRICS
STITCHED PILE FABRICS

BONDED YARNS
BT UNCONVENTIONAL YARNS
YARNS
RT BONDED YARN FABRICS
LAMINATED YARNS
SLIT FILM YARNS
TEKJA (TN)

BONDING
NT DIP BONDING
DRY POWDER BONDING
FLAME BONDING
FOAM BONDING
HEAT BONDING
MECHANICAL BONDING
SATURATION BONDING
SOLVENT BONDING
STITCH BONDING
SPRAY BONDING
THERMOPLASTIC FIBER BONDING
BT JOINING
RT ADHESION
BASTING
BINDERS (NONWOVENS)
BONDED FIBER FABRICS
BONDED YARN FABRICS
BONDING STRENGTH
CEMENTING
CHEMICAL BONDS
COHESION
FUSION (MELTING)
GLUING
LAMINATING
MALI PROCESSES (TN)
NONWOVEN FABRICS
PEELING
SEALING
SEAMING
SELF BONDED FABRICS (NONWOVENS)
SPUNBONDED (NONWOVENS)
TAPING
THERMOPLASTICITY
WEAVING

BONDING AGENTS (NONWOVENS)
USE BINDERS (NONWOVENS)

BONDING STRENGTH
BT STRENGTH
RT ADHESIVES
BEARING STRENGTH
BONDING
CEMENTS
COHESION
GLUES
PASTES
SEALERS
SEAMS
SELF BONDED FABRICS (NONWOVENS)
SETTING TIME
STRENGTH ELONGATION TESTING
TAPING

BOOKBINDING
BT INDUSTRIAL FABRICS

BOOKS (SILK)
RT SILK

BOOSTER BOXES
RT CONTINUOUS DYEING RANGE

BOOTS
NT FELT BOOTS
INSULATED BOOTS
LEATHER BOOTS
SHOEPAC
BT FOOTWEAR
RT BINDINGS (FOOTWEAR)
GALOSHES
LACES (SHOES)
LASTS
LEATHER
LININGS (FOOTWEAR)
PUMPS (FOOTWEAR)
RAWHIDE
RUBBER FOOTWEAR
SHOE LININGS
SHOES
SLIPPERS
SNEAKERS
SOLES
SPATS
STITCHING
STOCKINGS
SYNTHETIC LEATHER
TONGUES
UPPERS
VINYL COATED FABRICS

BORON COMPOUNDS
NT BORON HYDRIDES
BORON NITRIDE
BORON TRICHLORIDE
DIBORANE
SODIUM BOROHYDRIDE
ZINC FLUOBORATE

BORON HYDRIDES
NT DIBORANE
SODIUM BOROHYDRIDE
BT BORON COMPOUNDS
RT REDUCING AGENTS

BORON NITRIDE
BT BORON COMPOUNDS

BORON NITRIDE FIBERS
BT CERAMIC FIBERS
HIGH TEMPERATURE FIBERS
INORGANIC FIBERS (MAN MADE)
RT ALUMINUM SILICATE FIBERS
SILICON CARBIDE FIBERS

BORON TRICHLORIDE
BT BORON COMPOUNDS
CHLORIDES

BOTANY WOOL
RT MERINO WOOL

BOTTLE BOBBINS
BT BOBBINS
PACKAGES

BOTTOM COVER
BT COVER
RT FABRIC COVER
FUZZ
HAIRINESS
HAND
NAP
PATTERN DEFINITION
STITCH CLARITY
TOP COVER
YARN COVER

BOTTOM FILLER (FOOTWEAR)
BT FOOTWEAR COMPONENTS

BOTTOM ROLLS
BT DRAFTING ROLLS
ROLLS
RT DRIVEN ROLLS
DRIVING ROLLS
FLUTED ROLLS
FRONT ROLLS
INTERMEDIATE ROLLS
PLAIN ROLLS
TOP ROLLS
TUMBLERS (ROLLS)

BOTTOMING
RT DYEING
SCOURING
TOPPING

BOUCLES
BT NOVELTY YARNS
RT BEAD YARNS
BUG YARNS
FLAKE YARNS
FLAMES (NOVELTY YARNS)
FRILLS (NOVELTY YARNS)
KNICKERBOCKER YARNS
LOOPS (NOVELTY YARNS)
NOVELTY TWISTERS
NUB YARNS
PLYING
RATINES
SEEDS (NOVELTY YARNS)
SLUB YARNS
SPIRALS (NOVELTY YARNS)
SPLASHES

BOUNCE (WEAVING)
UF BOUNCING
RT BEATING UP
CLOTH FELL
JAMMING
LOOMS
PICKING (WEAVING)
REED MARKS
REEDS
WEAVING

BOUNCING
USE BOUNCE (WEAVING)

BOUND WATER
RT CAPILLARY WATER
DRYING
MOISTURE
MOISTURE CONTENT
REGAIN
WATER

BOUNDARY LAYER
RT AIR DRAG
BOUNDARY LAYER CONTROL
BOUNDARY LUBRICATION
DRAG
DRYING
FLUID FLOW
FLUID FRICTION
HYDRODYNAMIC LUBRICATION
LAMINAR FLOW
SHEAR (MODE OF DEFORMATION)
TURBULENCE
TURBULENT FLOW

BOUNDARY LAYER CONTROL
RT AERODYNAMICS
BOUNDARY LAYER
FLUID FLOW

BOUNDARY LUBRICATION
BT LUBRICATION
RT BOUNDARY LAYER
DYNAMIC FRICTION
FLUID FLOW
FLUID FRICTION
FRICTION
HYDRODYNAMIC LUBRICATION
ROLLING
SLIDING
WEAR

BOURDON
BT HAND MADE LACES
LACES

BOW
UF DOUBLE BOW
BT FABRIC PROPERTIES
RT BOWED FABRICS
DISTORTION
FABRIC DEFECTS

BOW CONTROL MACHINES
USE BOW STRAIGHTENERS

BOW STRAIGHTENERS
UF BOW CONTROL MACHINES
RT BOWED FABRICS
BOWING
EXPANDERS
FILLING STRAIGHTENING MACHINES
OPTICAL SCANNERS
PROCESS CONTROL
SKEW STRAIGHTENERS

BOWED FABRICS
UF DRAWN PIECE
BT FABRIC DEFECTS
RT BOW STRAIGHTENERS
BOWED ROLLS
BOWING
BOW

BOWED ROLLS
RT BOWED FABRICS
BOWING
ROLLS

BOWING
RT BENDING
BENDING RIGIDITY
BOW STRAIGHTENERS
BOWED FABRICS
BOWED ROLLS
BUCKLING
CAMBER
CROWNS (ROLL)
DEFORMATION
DISTORTION
DRAPE
SAG
STIFFNESS

BOWLS
USE ROLLS

BOX LININGS
BT LININGS
RT CASKET LININGS
GLOVE LININGS
INTERLININGS
LINING FELT
POCKET LININGS
SHOE LININGS

BOX LOOMS
UF DROP BOX LOOMS
BT LOOMS
RT AUTOMATIC LOOMS
DOBBY LOOMS
JACQUARD LOOMS
LENO LOOMS

BOX TOES
BT FOOTWEAR COMPONENTS

BOXER SHORTS
RT ATHLETIC CLOTHING
BRIEFS
DRAWERS

BOXES
BT PACKAGES
RT BAGS
BALES
CARTONS

BRABANT
BT HAND MADE LACES
LACES

BRACE WEB
BT NARROW FABRICS
RT BRACES

BRACES
RT BRACE WEB
SUSPENDERS

BRADFORD OPEN DRAWING
UF OPEN DRAWING
BT WORSTED SYSTEM PROCESSING
RT BRADFORD SYSTEM PROCESSING
CONE DRAWING
GILL BOXES
GILLING
NOBLE COMBING
OIL COMBED TOPS

BRADFORD SYSTEM PROCESSING
BT WORSTED SYSTEM PROCESSING
RT AMBLER SUPERDRAFT
AMERICAN SYSTEM PROCESSING
BACKWASHING
BLENDING
BRADFORD OPEN DRAWING
CAP SPINNING
CARDING
COMBING
CONTINENTAL SYSTEM PROCESSING
COTTON SYSTEM PROCESSING
DRAFTING (STAPLE FIBER)
FINISHER GILLING
GILLING
LONG STAPLE SPINNING
NEW BRADFORD SYSTEM PROCESSING
NOBLE COMBING
NOBLE COMBS
OILING
PIN DRAFTING
SCOURING
SPINNING
TOP MAKING
TWISTING
WOOL
WOOLEN SYSTEM PROCESSING
WORSTED CARDING
WORSTED COMBING
WORSTED SYSTEM SPINNING

BRAGGS LAW
RT X RAY ANALYSIS

BRAID (WOOL GRADE)
BT GRADE (FIBERS)
WOOL GRADE

BRAIDED FABRICS
USE BRAIDS

BRAIDER BOBBINS
BT BOBBINS
PACKAGES
RT CAKES

BRAIDERS
UF PLAITING MACHINES (BRAIDING)
RT BRAIDS
CARRIER (BRAIDS)

BRAIDING
UF PLAITING (BRAIDING)
RT CARRIER (BRAIDS)
KNITTING
MALI PROCESSES (TN)
PLIED YARNS
PLYING
WEAVING

BRAIDS
UF BRAIDED FABRICS
FABRICS (BRAIDS)
NT CORE BRAIDS
DOUBLE BRAIDS
FLAT BRAIDS
HOLLOW BRAIDS
NO CORE BRAIDS
PARACHUTE CORDS
REGULAR BRAIDS
SOLID BRAIDS
BT FABRICS (ACCORDING TO STRUCTURE)
FIBER ASSEMBLIES
RT BRAIDERS
CARRIER (BRAIDS)
CORDAGE
CORDS
ROPES
TAPE

BRAKE DISCS
RT BRAKE DRUMS
BRAKES (FOR ARRESTING MOTION)
MAGNETIC BRAKES
MECHANICAL BRAKES

BRAKE DRUMS
RT BRAKE DISCS
BRAKES (FOR ARRESTING MOTION)
MAGNETIC BRAKES
MECHANICAL BRAKES

BRAKE LININGS
RT BRAKES (FOR ARRESTING MOTION)
LININGS
MAGNETIC BRAKES
MECHANICAL BRAKES

BRAKES (FOR ARRESTING MOTION)
UF WHEEL BRAKES
NT AIR BRAKES
ELECTRIC BRAKES
HYDRAULIC BRAKES
MAGNETIC BRAKES
MECHANICAL BRAKES
POWER BRAKES
WATER BRAKES
RT ACCESSORIES
AUTOMATIC STOP MOTIONS
BRAKE DISCS
BRAKE DRUMS
BRAKE LININGS
BRAKING
CLUTCHES
FRICTION
HYDRAULIC CYLINDERS
LET OFF MOTIONS
MECHANISMS
MOTION
RETARDERS
WEAR

BRAKING
RT ACCELERATION (MECHANICAL)
ACCELEROMETERS
BRAKES (FOR ARRESTING MOTION)
FORCE
FRICTION
INERTIA
MAGNETIC BRAKES
MECHANICAL BRAKES
MECHANICAL SHOCK
MOTION
VELOCITY

BRANCHED ALIPHATIC HYDROCARBONS
UF BRANCHED CHAIN ALIPHATIC HYDROCARBONS
BT ALIPHATIC HYDROCARBONS
RT NORMAL ALIPHATIC HYDROCARBONS

BRANCHED CHAIN ALIPHATIC HYDROCARBONS
USE BRANCHED ALIPHATIC HYDROCARBONS

BRANCHED CHAIN MOLECULES
    BT LONG CHAIN MOLECULES
       MOLECULES
    RT LINEAR CHAIN MOLECULES

BRASSIER CLOTH
    BT FLAT WOVEN FABRICS
       WOVEN FABRICS

BRASSIERES
    BT FOUNDATION GARMENTS
       GARMENTS
       UNDERWEAR
    RT BRIEFS
       GARTER BELTS
       GARTERS
       GIRDLES
       LINGERIE
       PANTIES
       SHORTS
       SLIPS
       SUSPENDERS
       VESTS

BRATTICE CLOTH
    BT INDUSTRIAL FABRICS

BREADTH
    USE WIDTH

BREAK DRAFT
    BT DRAFT
    RT ACTUAL DRAFT
       BACK ROLLS
       DRAFTING ROLLS
       DRAFTING THEORY
       FALSE DRAFT
       FINISHER GILLING
       GILL BOXES
       GILLING
       INTERSECTING GILL BOXES
       PROCESS VARIABLES
       ROLLER DRAFT

BREAK SPINNING
    USE OPEN END SPINNING

BREAKAGES
    USE BREAKS

BREAKER FABRICS
    BT INDUSTRIAL FABRICS
    RT TIRES

BREAKER PART
    USE BREAKERS

BREAKERS
    UF BREAKER PART
       SCRIBBLER PART
    RT ANGLE STRIPPERS
       BURR BEATERS
       BURR CRUSHERS
       CARD CLOTHING
       CARD CYLINDERS
       CARD WEBS
       CARDING
       CARDING ACTION
       CARDING EFFICIENCY
       CARDS
       DOFFER COMBS
       DOFFERS (CARD)
       DOFFING (OF A CARD)
       FANCY (NOUN)
       FEED ROLLS
       FINISHER PART
       HOPPER FEED
       INTERMEDIATE FEED
       INTERMEDIATE PART
       LICKER IN
       MOTE KNIVES
       PERALTA ROLLS
       SCRIBBLING
       STRIPPERS
       WOOLEN CARDING
       WOOLEN CARDS
       WORKERS
       WORSTED CARDING

BREAKING
    RT ABRASION
       BREAKING ELONGATION
       BREAKING ENERGY
       BREAKING LENGTH
       BREAKING STRENGTH
       BREAKS
       CHIPPING
       CRACK PROPAGATION
       END BREAKAGES
       FATIGUE
       FIBER BREAKAGE
       FRACTURING
       RUPTURE
       SCORING
       SCRATCHING
       SPLITTING
       STRENGTH
       STRETCHING (YARNS)
       WEAR

BREAKING ELONGATION
    UF BREAKING EXTENSION
       BREAKING STRAIN
       ELONGATION AT BREAK
       EXTENSION AT BREAK
       ULTIMATE ELONGATION
    BT STRAIN
       STRESS STRAIN PROPERTIES
        (TENSILE)
    RT BREAKING
       BREAKING LENGTH
       BREAKING STRENGTH
       BREAKS
       COUNT STRENGTH PRODUCT
       ELASTIC STRAIN
       ELASTICITY
       END BREAKAGES
       EXTENSIBILITY
       FIBER BREAKAGE
       FIBER STRENGTH
       FRACTURING
       GAUGE LENGTH
       IMPACT STRENGTH
       LEA STRENGTH PRODUCT
       NECKING (FILAMENT)
       PLASTIC STRAIN
       STRENGTH OF MATERIALS
       STRESS STRAIN CURVES
       STRETCH
       STRETCHING (YARNS)
       STRAIN
       TOUGHNESS
       UNIAXIAL STRESS

BREAKING ENERGY
    UF WORK TO BREAK
    BT ENERGY
       STRESS STRAIN PROPERTIES
       STRESS STRAIN PROPERTIES
        (TENSILE)
    RT BREAKING
       BRITTLE FRACTURE
       ELASTICITY
       IMPACT STRENGTH
       TOUGHNESS

BREAKING EXTENSION
    USE BREAKING ELONGATION

BREAKING LENGTH
    BT STRESS STRAIN/GATION (TENSILE)
    RT BENDING LENGTH
       BREAKING
       BREAKING ELONGATION

       BREAKING STRENGTH
       COUNT STRENGTH PRODUCT
       END BREAKAGES
       FIBER BREAKAGE
       FIBER STRENGTH
       GRAB STRENGTH
       IMPACT STRENGTH
       LEA STRENGTH PRODUCT
       LENGTH
       PRESSLEY TEST
       RAVEL STRIP STRENGTH
       SHOCK RESISTANCE
       STRENGTH EFFICIENCY
       STRETCH BREAKING
       TEAR STRENGTH
       TENACITY
       WEAK SPOTS
       YARN STRENGTH EFFICIENCY

BREAKING LOAD
    USE BREAKING STRENGTH

BREAKING STRAIN
    USE BREAKING ELONGATION

BREAKING STRENGTH
    UF BREAKING LOAD
       BREAKING STRESS
       BREAKING TENACITY
       BREAKING TENSION
       RUPTURE STRENGTH
       RUPTURE TENACITY
       STRENGTH AT RUPTURE
       TENSILE STRENGTH
       ULTIMATE STRENGTH
       ULTIMATE TENACITY
    NT BUNDLE STRENGTH
       COUNT STRENGTH PRODUCT
       FABRIC STRENGTH
       FIBER STRENGTH
       IMPACT STRENGTH
       LEA STRENGTH PRODUCT
       YARN STRENGTH
    BT END USE PROPERTIES
       STRENGTH
       STRESS STRAIN PROPERTIES
        (TENSILE)
    RT BREAKING
       BREAKING ELONGATION
       BREAKING LENGTH

       BREAKS
       COMPRESSIVE STRENGTH

    RT ELASTICITY
       END BREAKAGES
       FABRIC PROPERTIES
       FABRIC PROPERTIES (MECHANICAL)
       FABRIC STRENGTH
       FIBER BREAKAGE
       FIBER STRENGTH
       GRAB STRENGTH
       LOOP EFFICIENCY
       LOOP STRENGTH
       PRESSLEY TEST
       RAVEL STRIP STRENGTH
       SHOCK RESISTANCE
       SNAP BACK
       STRENGTH EFFICIENCY
       STRENGTH ELONGATION TESTING
       STRESS STRAIN CURVES
       TEAR STRENGTH
       TENACITY
       TENSILE STRESS
       TENSILE TESTING
       UNIAXIAL STRESS
       WEAK SPOTS
       YARN STRENGTH
       YARN STRENGTH EFFICIENCY

BREAKING STRESS
    USE BREAKING STRENGTH

BREAKING TENACITY
    USE BREAKING STRENGTH

BREAKING TENSION
    USE BREAKING STRENGTH

BREAKS
    UF BREAKAGES
    BT DEFECTS
    RT BREAKING
       BREAKING ELONGATION
       BREAKING STRENGTH

       CRACKS (MECHANICAL)
       END BREAKAGES
       PIECING

BREAST BEAMS
    RT LOOMS

BREAST SWIFTS
    USE CARD CYLINDERS

BRETON
    BT MACHINE MADE LACES

BRI-NYLON (TN)
    BT NYLON (POLYAMIDE FIBERS)
       NYLON 66

BRIEFS
    BT GARMENTS
       LINGERIE
       UNDERWEAR
    RT BRASSIERES
       DRAWERS
       FABRICS
       FOUNDATION GARMENTS
       GARTER BELTS
       GARTERS
       GIRDLES
       KNITTED FABRICS
       PANTIES
       SHORTS
       SLIPS
       SUSPENDERS
       TRUNKS
       VESTS

BRIGHT PICKS
    USE SHINERS

BRIGHT SPECKS
    BT FABRIC DEFECTS

BRIGHTNESS
    BT AESTHETIC PROPERTIES
       FABRIC PROPERTIES
       FABRIC PROPERTIES (AESTHETIC)
       FABRIC PROPERTIES (PHYSICAL
        EXCLUDING MECHANICAL)
       OPTICAL PROPERTIES
    RT ABSORPTION (RADIATION)
       AESTHETIC APPEAL
       APPEARANCE
       BLOOM
       COLOR
       DAYLIGHT
       DRABNESS
       DULLNESS
       FLUORESCENT BRIGHTENERS
       ILLUMINATION
       INTENSITY
       LIGHT
       LUSTER
       MUNSELL COLOR SYSTEM
       RICHNESS
       SHADE
       SMOOTHNESS
        (CONT.)

RT SPARKLE
SPECTROPHOTOMETRY
TRANSPARENCY
WHITENESS
YARN LUSTER

BRILLIANT SPOT
BT FLAT WOVEN FABRICS
WOVEN FABRICS

BRINS (SILK)
RT SILK

BRITTLE FRACTURE
BT FRACTURING
RT BREAKING ENERGY
CRACK PROPAGATION
ENERGY ABSORPTION
IMPACT
RUPTURE
SHOCK

BRITTLENESS
BT STRESS STRAIN PROPERTIES
STRESS STRAIN PROPERTIES (TENSILE)
RT BREAKING
COMPRESSIVE STRENGTH
CRACK PROPAGATION
DUCTILITY
ELASTICITY
FRACTURING
HARDNESS
IMPACT STRENGTH
PLASTICITY

BROADCLOTH
BT FLAT WOVEN FABRICS
WOVEN FABRICS
RT PLAIN WEAVES
TWILL WEAVES

BROCADES
BT FIGURED FABRICS
FLAT WOVEN FABRICS
WOVEN FABRICS
RT DAMASK

BROCATELLES
BT FIGURED FABRICS
FLAT WOVEN FABRICS
WOVEN FABRICS
RT DAMASK

BROKEN END COLLECTION
USE BROKEN END COLLECTORS

BROKEN END COLLECTORS
UF BROKEN END COLLECTION
NT PNEUMAFIL (TN)
VACUUM END COLLECTORS
RT END BREAKAGES
SPINNING
WINDING

BROKEN FIBERS
RT END BREAKAGES
FIBER BREAKAGE

BROKEN SELVEDGES
BT FABRIC DEFECTS
RT FABRIC INSPECTION

BROKEN THREADS
BT FABRIC DEFECTS
RT END BREAKAGES
FABRIC INSPECTION

BROMIDES
NT BISMUTH BROMIDE
LITHIUM BROMIDE
POTASSIUM BROMIDE
SODIUM BROMIDE
BT HALIDES
RT BROMINE
BROMINE COMPOUNDS
CHLORIDES
FLUORIDES
IODIDES

BROMINE
RT BROMINE COMPOUNDS
BROMIDES
CHLORINE
IODINE

BROMINE COMPOUNDS
NT ALLYL BROMIDE
ETHYLENE DIBROMIDE
VINYL BROMIDE
BT HALOGEN COMPOUNDS
RT BROMIDES
BROMINE
HALIDES

BRONZING
BT DYEING DEFECTS
FABRIC DEFECTS
RT DYEING
SPECKINESS
STREAKINESS

BRUISED PLACES
USE CHAFE MARKS

BRUSHING
BT DRY FINISHING
RT BRUSHING MACHINES
BULKING
BUTTON BREAKING
DOFFER BRUSHES
FACE FINISH
FULLING
NAPPING
PATTERN DEFINITION
POLISHING
SHEARING (FINISHING)
SUEDING

BRUSHING ACTION
RT CARDING ACTION
DOFFER BRUSHES
FANCY (NOUN)
STRIPPING ACTION

BRUSHING MACHINES
RT BRUSHING

BRUSSELS TAPESTRIES
BT TAPESTRIES

BUCKET SPINNING
USE POT SPINNING

BUCKLES
BT FASTENERS
RT BUTTONS
HOOK CLOSURES
JOINTS
SNAP FASTENERS
ZIPPERS

BUCKLING
NT COMPRESSIVE BUCKLING
DYNAMIC BUCKLING
TORSIONAL BUCKLING
YARN BUCKLING
BT MODES OF DEFORMATION
STRESS STRAIN PROPERTIES (COMPRESSIVE)
RT BENDING
BENDING MOMENTS
BENDING RIGIDITY
BOWING
COCKLING
COMPRESSION
COMPRESSIVE MODULUS
DEFORMATION
DISTORTION
DRAPE
DYNAMIC BUCKLING
PUCKER (DEFECT)
PUCKERS (FOR EFFECT)
RIPPLES
SEAM PUCKER
SNAP BACK
STABILITY
STRAIN
TWISTING MOMENTS
WAVINESS

BUCKRAM
BT FLAT WOVEN FABRICS
WOVEN FABRICS
RT PLAIN WEAVES

BUFFING WHEELS
RT INDUSTRIAL FABRICS

BUG YARNS
BT NOVELTY YARNS
RT BEAD YARNS
BOUCLES
FRILLS (NOVELTY YARNS)
KNICKERBOCKER YARNS
LOOPS (NOVELTY YARNS)
NUB YARNS
SEEDS (NOVELTY YARNS)
SPIRALS (NOVELTY YARNS)
SPLASHES

BUILDER FABRICS
BT INDUSTRIAL FABRICS

BUILDER MOTIONS
RT BALLOON HEIGHT
CAP SPINNING FRAMES
CAPS (SPINNING)
CLOSE WINDING
DRUM WINDERS
DRUM WINDING
EMPTY PACKAGES
FLYER SPINNING FRAMES
FULL PACKAGES
HEADS (TOP)
LAYER LOCKING
LIFT (WINDING)
NOSE BUNCHING
OPEN WINDING
PACKAGES
POT SPINNING FRAMES
PRECISION WINDING
RIBBON BREAKERS
RING FRAMES
RING RAILS
SPINDLE RAILS
SPINNING
SPINNING FRAMES
TAPER (BOBBIN)
TRAVERSE
TWISTERS
WARP WIND
WINDING

BUILDERS (DETERGENTS)
RT ADDITIVES (CHEMICAL)
CHELATING AGENTS
DETERGENCY
DETERGENTS
FOAMING AGENTS
PHOSPHATES
SURFACTANT APPLICATIONS

BULGING (FILAMENT)
RT RHEOLOGY
SPINNING (EXTRUSION)

BULK
UF LOFT
BT PHYSICAL PROPERTIES (EXCLUDING MECHANICAL)
RT SPACE
BT STRUCTURAL PROPERTIES
RT BULK DENSITY
COVER
CRIMP
DENSITY
FABRIC PROPERTIES
FALSE TWIST YARNS
HAND
MECHANICAL PROPERTIES
PACKING FACTOR
POROSITY
RESILIENCE
STRETCH
TEXTURE
TEXTURING
THICKNESS
VOLUME
WOOLEN SYSTEM PROCESSING
YARN PROPERTIES

BULK DENSITY
UF APPARENT DENSITY
BT DENSITY
STRUCTURAL PROPERTIES
PHYSICAL PROPERTIES (EXCLUDING MECHANICAL)
RT ABSORBENCY (MATERIAL)
BULK
INSULATION (THERMAL)
LINEAR DENSITY
PACKING FACTOR
POROSITY
RESILIENCE
THICKNESS

BULK MODULUS
BT MODULUS
STRESS STRAIN PROPERTIES
RT COMPRESSIBILITY
COMPRESSIVE MODULUS
DILATANCY
FREE VOLUME

BULK POLYMERIZATION
BT POLYMERIZATION
RT EMULSION POLYMERIZATION
SUSPENSION POLYMERIZATION

BULK YARNS (FILAMENT)
USE BULKED YARNS

BULKED YARNS
UF BULK YARNS (FILAMENT)
NT AIR JET TEXTURED YARNS
BICOMPONENT FIBER YARNS (FILAMENT)
CRINKLE TYPE YARNS
EDGE CRIMPED YARNS
LOOP YARNS
TEXTRALIZED YARNS (TN)
BT FILAMENT YARNS
TEXTURED YARNS (FILAMENT)
RT AIR JET TEXTURING
BANLON (TN)
BICOMPONENT FIBER YARNS (STAPLE)
BULKING
CRIMPED YARNS
CRIMPING
CRINKLE TYPE YARNS
DIFFERENTIAL SHRINKAGE
DYNALOFT (TN)

RT EDGE CRIMPED YARNS
EDGE CRIMPING
FALSE TWIST YARNS
FALSE TWISTING (TEXTURING)
FLUFLON (TN)
GEAR CRIMPING
HELANCA (TN)
HILOW BULKED YARNS
NYLOFT (TN)
SAABA (TN)
SELF CRIMPING YARNS
SPUNIZE (TN)
STRETCH FABRICS
STRETCH KNITTED FABRICS
STRETCH WOVEN FABRICS
STRETCH YARNS (FILAMENT)
STUFFER BOX CRIMPING
STUFFER BOXES
SUPERLOFT (TN)
TASLAN (TN)
TEXTURING
TYCORA (TN)
YARN CRIMP
Z CRIMP

BULKED YARNS (STAPLE)
USE TEXTURED SPUN YARNS

BULKING
BT TEXTURING
RT AIR JET TEXTURED YARNS
AIR JET TEXTURING
AIR JET TEXTURING MACHINES
BANLON (TN)
BICOMPONENT FIBER YARNS (FILAMENT)
BICOMPONENT FIBER YARNS (STAPLE)
BULKED YARNS
CRIMPED YARNS
CRIMPING
EDGE CRIMPED YARNS
EDGE CRIMPING
FALSE TWISTING (TEXTURING)
HEAT SETTING (SYNTHETICS)
LATENT CRIMP
SELF CRIMPING YARNS
STUFFER BOX CRIMPING
TEXTRALIZED YARNS (TN)
TEXTURED YARNS (FILAMENT)
THROWING
YARN CRIMP
Z CRIMP

BULKING MACHINES
BT THROWING MACHINES
RT BULKING
BULKED YARNS
TEXTURING

BULLET PROOF CLOTHING
BT MILITARY CLOTHING
RT BALLISTIC FABRICS
IMPACT STRENGTH

BUMP GRAY
USE BACK GRAY

BUMP YARNS
RT CANDLEWICK YARNS
TORCHWICK YARNS

BUNA N COATED FABRICS
BT COATED FABRICS
COMPOSITES
RT BUTYL COATED FABRICS
COATINGS (SUBSTANCES)
LACQUER COATED FABRICS
NEOPRENE COATED FABRICS
RUBBER COATED FABRICS
SBR COATED FABRICS
VINYL COATED FABRICS

BUNCHING COEFFICIENT
RT FIBER ARRANGEMENT
FIBER ORIENTATION
NONWOVEN FABRIC STRUCTURE
NONWOVEN FABRICS
ORIENTATION
ORIENTATION UNIFORMITY
PATCHINESS
SPOTTINESS
UNIT CELLS (NONWOVEN FABRICS)
WEB ORIENTATION
WEB UNIFORMITY
WEBS

BUNDESMANN TESTING
BT WATER RESISTANCE TESTING
RT WATER REPELLENCY
WATER RESISTANCE
WATERPROOFING

BUNDLE EFFICIENCY
BT EFFICIENCY (STRUCTURAL)
STRESS STRAIN PROPERTIES (TENSILE)
RT STRENGTH EFFICIENCY

BUNDLE STRENGTH
NT PRESSLEY STRENGTH
BT BREAKING STRENGTH
STRENGTH
STRESS STRAIN PROPERTIES (TENSILE)
RT COUNT STRENGTH PRODUCT
CREEP RUPTURE STRENGTH
FIBER STRENGTH
IMPACT STRENGTH
LEA STRENGTH PRODUCT
LOOP EFFICIENCY
LOOP STRENGTH
MOISTURE STRENGTH CURVES
RUPTURE
STRAIN
STRENGTH EFFICIENCY
STRENGTH TESTING
STRESS
STRESS STRAIN CURVES
TESTING EQUIPMENT
WET STRENGTH

BUNDLES
BT FIBER ASSEMBLIES
RT HANKS
SKEINS
STRANDS

BUNDLING (TAILORING)
BT GARMENT MANUFACTURE
RT CUTTING (TAILORING)
PATTERN (APPAREL)
SEWING
TAILORING

BUNTE SALTS
USE THIOSULFONATES

BUNTING
BT FLAT WOVEN FABRICS
WOVEN FABRICS
RT FLAGS
PLAIN WEAVES

BURLING
BT DRY FINISHING
RT DEFECTS
FABRIC INSPECTION
MENDING
PERCHING

BURLING IRONS
USE BURLING MACHINES

BURLING MACHINES
UF BURLING IRONS
IRONS (BURLING MACHINE)
PICKING TONGS
RT DRY FINISHING
FABRIC INSPECTION

BURMILIZED (TN)
BT TEXTURED YARNS (FILAMENT)

BURNING
RT FIRE RESISTANCE
FIRE RETARDANCY
FLASH POINT
SCORCHING MARKS
SINGEING

BURR BEATERS
RT BREAKERS
BURR CRUSHERS
BURR PICKING
BURRY WOOL
CARD ROLLS
CARD WASTE
CARDING
CLEARERS (CARD)
MOTE KNIVES
PERALTA ROLLS
WOOLEN CARDING
WORSTED CARDING

BURR CRUSHERS
UF BURR CRUSHING MACHINES
RT **BREAKERS**
BURR BEATERS
BURR PICKING
BURRS
BURRY WOOL
CARD ROLLS
CARD WASTE
CARDING
CARDS
CLEARERS (CARD)
MOTE KNIVES
NEPS
NIP PRESSURE
PERALTA ROLLS
WOOLEN CARDING
WOOLEN CARDS
WORSTED CARDING

BURR CRUSHING MACHINES
USE BURR CRUSHERS

BURR PICKING
UF BURR REMOVAL
RT BURR BEATERS
BURR CRUSHERS
BURRS
CLEANLINESS

BURR REMOVAL
USE BURR PICKING

BURRS
BT TRASH
VEGETABLE MATTER
WASTE
RT BURR CRUSHERS
BURR PICKING
BURRY WOOL
CLEARERS (CARD)
IMPURITIES

BURRY WOOL
RT BURR BEATERS
BURR CRUSHERS
BURR PICKING
BURRS
WOOL

BURST TESTERS
BT TESTING EQUIPMENT
RT BURST TESTING
BURSTING
BURSTING STRENGTH

BURST TESTING
NT BALL BURST TESTING
DIAPHRAGM BURST TESTING
MULLEN BURST TESTING
BT STRENGTH TESTING
TESTING
RT ABRASION TESTING
BENDING TESTING
BIAXIAL STRESS
BURST TESTERS
BURSTING STRENGTH
IMPACT TESTING
PERMEABILITY TESTING
TEAR TESTING
TENSILE TESTING
TESTING EQUIPMENT
WATER RESISTANCE TESTING

BURSTING
RT BURST TESTING

BURSTING STRENGTH
BT FABRIC PROPERTIES
FABRIC PROPERTIES (MECHANICAL)
STRENGTH
STRESS STRAIN PROPERTIES (TENSILE)
RT BALL BURST TESTING
BIAXIAL STRESS
BREAKING STRENGTH
BURST TESTING
BURSTING
COMPRESSIVE STRENGTH
DIAPHRAGM BURST TESTING
FABRIC STRENGTH
LOOP STRENGTH
MULLEN BURST TESTING
PRESSURE CONTAINERS
STRENGTH ELONGATION TESTING

BUSHINGS
RT AIR BEARINGS
AXLES
BEARINGS
LUBRICATION
NYLON BEARINGS
POROUS MATERIALS
SHAFTS
SLEEVE BEARINGS
TEFLON BEARINGS

BUSTS
BT GARMENT COMPONENTS

BUTADIENE
BT DIENES
RT BUTADIENE DIEPOXIDE
BUTADIENE RUBBER FIBER
COPOLYMERIZATION
MONOMERS
POLYBUTADIENE
POLYMERIZATION

BUTADIENE DIEPOXIDE
BT OXIRANE COMPOUNDS
RT BUTADIENE

BUTADIENE RUBBER FIBER
BT MAN MADE FIBERS
SYNTHETIC FIBERS
SYNTHETIC RUBBER FIBERS
RT BUTADIENE
POLYBUTADIENE
RUBBER FIBER (NATURAL)

BUTTED SEAMS
BT SEAMS
RT BUTTERFLY SEAMS
COVERED SEAMS
FELLED SEAMS
FRENCH SEAMS
FULL SEAM WIDTH
HEM SEAMS
JOINTS
LAPPED SEAMS
PINKED SEAMS
PIPED SEAMS
SANDWICH SEAMS
SEAM ABRASIVE STRENGTH
SEAM ALLOWANCE
SEAM EFFICIENCY
SEAM GRIN
SEAM HEADING
SEAM LET OUT
SEAM QUALITY
SEAM SIZE
SEAM SLIPPAGE
SEAM SLIPPAGE STRENGTH
SEAM SPECIFICATIONS
SEAM STRENGTH
SEAM TENSILE STRENGTH
SEAM THICKNESS
SEAM WIDTH
SEWING
STITCH ELASTICITY
STITCH SEQUENCE
STITCH SIZE (SEWING)
STITCH TENSION
STITCHES (SEWING)
STITCHING
SUPERPOSED SEAMS
TAILORED SEAMS
YARN SEVERANCE

BUTTERFLY SEAMS
BT SEAMS
RT BUTTED SEAMS
COVERED SEAMS
FELLED SEAMS
RT FELLING
FRENCH SEAMS
FULL SEAM WIDTH
HEM SEAMS
JOINTS
LAPPED SEAMS
PINKED SEAMS
PIPED SEAMS
SANDWICH SEAMS
SEAM ALLOWANCE
SEAM GRIN
SEAM HEADING
SEAM LET OUT
SEAM QUALITY
SEAM SIZE
SEAM SLIPPAGE
SEAM SLIPPAGE STRENGTH
SEAM SPECIFICATIONS
SEAM STRENGTH
SEAM TENSILE STRENGTH
SEAM THICKNESS
SEAM WIDTH
SEWING
STITCHES (SEWING)
STITCHING
SUPERPOSED SEAMS
TAILORED SEAMS

BUTTON BREAKERS
RT NAPPING MACHINES
SUEDING MACHINES

BUTTON BREAKING
BT DRY FINISHING
RT BRUSHING
NAPPING

BUTTONHOLES
BT GARMENT COMPONENTS
RT BUTTONHOLING
BUTTONS

BUTTONHOLING
RT BUTTONHOLES
BUTTONS
TAILORING

BUTTONS
BT GARMENT COMPONENTS
RT BUTTONHOLES
BUTTONHOLING
JOINTS
VELCRO (TN)

BUTTS (NEEDLES)
USE NEEDLE BUTTS

BUTYL ACRYLATE
BT ACRYLIC COMPOUNDS
ACRYLIC ESTERS
RT ETHYL ACRYLATE
METHYL ACRYLATE

BUTYL ALCOHOL
BT ALCOHOLS

BUTYL COATED FABRICS
BT COATED FABRICS
COMPOSITES
RT BUNA N COATED FABRICS
COATINGS (SUBSTANCES)
LACQUER COATED FABRICS
NEOPRENE COATED FABRICS
SBR COATED FABRICS
VINYL COATED FABRICS

BUTYLNAPHTHALENE SULFONATES
BT ALKYLNAPHTHALENE SULFONATES

BUTYROLACTONE
BT LACTONES

BYPRODUCTS
RT OUTPUT
PROCESS EFFICIENCY
WASTE
YIELD (RETURN)

CABLE CORDS
    NT FRENCH CREPE CORDS
       LACING CORDS

CABLE STITCH
    RT LINKS LINKS MACHINES

CABLED YARNS
    USE PLIED YARNS

CABLES
    BT FIBER ASSEMBLIES
    RT CABLING
       CORDAGE
       CORDS
       PLIED YARNS
       PLYING
       ROPES

CABLING
    RT CABLES
       CORDAGE
       CORDS
       FISHING LINES
       HAWSERS
       PLIED YARNS
       PLYING
       ROPES
       TWISTING

CADMIUM CHLORIDE
    BT CADMIUM COMPOUNDS
       CHLORIDES

CADMIUM COMPOUNDS
    NT CADMIUM CHLORIDE
       CADMIUM SELENIDE
       CADOXEN

CADMIUM ETHYLENEDIAMINE HYDROXIDE
    USE CADOXEN

CADMIUM SELENIDE
    BT CADMIUM COMPOUNDS
    RT MILDEW RESISTANCE FINISHES

CADON (TN)
    BT NYLON (POLYAMIDE FIBERS)
       NYLON 66

CADOXEN
    UF CADMIUM ETHYLENEDIAMINE
        HYDROXIDE
    BT CADMIUM COMPOUNDS
    RT CELLULOSE
       CUPRAMMONIUM HYDROXIDE
       CUPRIETHYLENEDIAMINE HYDROXIDE
       FERRIC TARTRATE
       MARSCHALL SOLUTION
       SOLVENTS
       VISCOSITY

CAKE DYEING
    BT PACKAGE DYEING
       YARN DYEING
    RT BEAM DYEING (YARN)
       BOBBIN DYEING
       CHEESE DYEING
       CONE DYEING
       COP DYEING
       HANK DYEING
       MUFF DYEING
       PRESSURE DYEING
       SKEIN DYEING

CAKE DYEING MACHINES
    BT PACKAGE DYEING MACHINES
       YARN DYEING MACHINES
    RT BEAM DYEING MACHINES (YARN)
       BOBBIN DYEING MACHINES
       CAKE DYEING
       CHEESE DYEING MACHINES
       CONE DYEING MACHINES
       COP DYEING MACHINES
       WARP DYEING MACHINES

CAKE SIZING
    BT SLASHING
    RT CAKES
       HANK SIZING
       SIZES (SLASHING)

CAKES
    BT PACKAGES
    RT BEAMS
       BOBBINS
       BRAIDER BOBBINS
       CAKE SIZING
       CHEESES
       CONES
       EXTRUSION
       LARGE PACKAGES
       POT SPINNING
       POT SPINNING FRAMES
       QUILLS
       SPINNING (EXTRUSION)
       WET SPINNING

CALCIUM CHLORIDE
    BT CALCIUM COMPOUNDS
       CHLORIDES
    RT DELIQUESCENT AGENTS

CALCIUM COMPOUNDS
    NT CALCIUM CHLORIDE
       CALCIUM HYDROXIDE
       CALCIUM SOAPS
       CALCIUM STEARATE

CALCIUM HYDROXIDE
    BT CALCIUM COMPOUNDS
       HYDROXIDES

CALCIUM SOAPS
    BT CALCIUM COMPOUNDS
       METALLIC SOAPS
    RT CALCIUM STEARATE
       LIME SOAP

CALCIUM STEARATE
    BT CALCIUM COMPOUNDS
       METALLIC SOAPS
    RT CALCIUM SOAPS
       LIME SOAP

CALCULATING
    USE COMPUTATIONS

CALCULATIONS
    USE COMPUTATIONS

CALENDER COATING
    BT COATING (PROCESS)
    RT CALENDER ROLLS
       CALENDER SPREADING
       CAST COATING
       COATED FABRICS
       DIP COATING
       EXTRUSION COATING
       FLEXIBLE FILM LAMINATING
       KNIFE COATING
       NIP COATING
       PAPER MAKING
       ROLLER COATING

CALENDER PRINTING
    USE ROLLER PRINTING

CALENDER ROLLS
    NT EMBOSSING CALENDERS
    BT CALENDERING MACHINES
       ROLLS
    RT CALENDER SPREADING
       HYDROSTATIC LOADING
       NIP PRESSURE
       PERALTA ROLLS
       PRESS SECTION
       PRESSURE ROLLS
       ROLL COVERINGS
       SUCTION PRESSES

CALENDER SPREADING
    RT CALENDER COATING
       CALENDER ROLLS
       CALENDERING

CALENDERING
    NT FRICTION CALENDERING
    BT DRY FINISHING
    RT CALENDER SPREADING
       EMBOSSING
       EMBOSSING CALENDERS
       GLAZE
       PAPER MAKING
       POLISHING
       ROLLER SLIPPAGE
       SCHREINERING

CALENDERING MACHINES
    NT CALENDER ROLLS
       EMBOSSING CALENDERS
    RT CALENDERING
       DRY FINISHING
       LUSTERING MACHINES

CALFSKIN
    BT LEATHER
    RT SYNTHETIC LEATHER

CALGON (TN)
    UF METHAPHOSPHATES
       SODIUM HEXAMETAPHOSPHATE
    BT PHOSPHATES
    RT CHELATING
       CHELATING AGENTS
       DEMINERALIZING
       DISPERSING
       EDTA
       LAUNDERING
       TRISODIUM PHOSPHATE
       TSPP
       WATER

CALIBRATION
    RT ACCURACY
       ADJUSTMENTS
       ALIGNMENT
       BALANCE
       COMPARISON
       COMPENSATION
       DRIFT
       ERRORS
       SETTING (ADJUSTMENTS)
       TOLERANCES

CALICO
    BT FLAT WOVEN FABRICS
       WOVEN FABRICS
    RT PLAIN WEAVES

CALIFORNIA COTTON
    BT COTTON

CALORIMETRY
    BT THERMAL ANALYSIS
    RT DIFFERENTIAL THERMAL ANALYSIS
       MEASURING INSTRUMENTS
       SPECIFIC HEAT

CAM ANGLES
    RT CAMS
       KNITTING NEEDLES
       NEEDLE BUTTS
       STITCH CAMS

CAM LOOMS
    BT LOOMS
    RT DOBBY LOOMS
       JACQUARD LOOMS
       WEAVING

CAMBER
    RT BENDING
       BOWING
       CURVATURE
       DEFLECTION
       DEFORMATION
       DISTORTION

CAMBRIC
    BT FLAT WOVEN FABRICS
       WOVEN FABRICS
    RT PLAIN WEAVES

CAMOUFLAGES
    BT MILITARY PRODUCTS

CAMBRIDGE EXTENSOMETER (TN)
    BT TENSILE TESTERS
    RT INSTRON TENSILE TESTER (TN)
       SCOTT TENSILE TESTER (TN)

CAMEL HAIR
    BT HAIR
       KERATIN FIBERS
       PROTEIN FIBERS
    RT ALPACA

CAMS
    NT NEEDLE CAMS
       RAISING CAMS
       SINKER CAMS
       STITCH CAMS
    RT ACCESSORIES
       CAM ANGLES
       GEARS
       LINKAGES
       MECHANISMS

CANDLE FILTERS
    RT SPINNING (EXTRUSION)

CANDLEWICK FABRICS (TUFTED FABRIC)
    BT TUFTED FABRICS

CANDLEWICK YARNS
    RT BUMP YARNS

CANOPIES
    RT INDUSTRIAL FABRICS

CANTON SILK
    BT SILK

CANS (DRYING)
    USE DRYING CANS

CANTRECE (TN)
    BT BICOMPONENT FIBER YARNS
        (FILAMENT)
       NYLON (POLYAMIDE FIBERS)
       TEXTURED YARNS (FILAMENT)
    RT NYLON 6
       NYLON 66

CANVAS
    USE DUCK

CAP SPINNING
    BT SPINNING
       WORSTED SYSTEM PROCESSING
    RT BALLOONS
       CAPS (SPINNING)
       DOWNTWISTING
        (CONT.)

RT FLYER SPINNING
MULE SPINNING
POT SPINNING
RING SPINNING
WORSTED SPUN YARNS
WORSTED SYSTEM SPINNING

CAP SPINNING FRAMES
BT SPINNING FRAMES
RT BOBBINS
BUILDER MOTIONS
CAPS (SPINNING)
CASABLANCA SYSTEM
CONTROL SURFACES
DOFFING (PACKAGE)
DRAFTING ROLLS
FLYER SPINNING FRAMES
PACKAGES
POT SPINNING FRAMES
RING FRAMES
ROVINGS
SPINDLES
SPINNING
WORSTED SYSTEM PROCESSING
WORSTED SYSTEM SPINNING

CAPACITANCE (ELECTRICAL)
UF CAPACITY (ELECTRICAL)
BT ELECTRICAL PROPERTIES
RT CAPACITANCE BRIDGES
CAPACITANCE MOTORS
CAPACITANCE TYPE IRREGULARITY TESTERS
CHARGE (ELECTRICAL)
CURRENT (ELECTRICAL)
DIELECTRICS
DIELECTRIC CONSTANT
DIELECTRIC PROPERTIES
DIELECTRIC STRENGTH
DISCHARGE (ELECTRIC)
DISSIPATION
FIELDEN WALKER IRREGULARITY TESTER (TN)
FILTERS (ELECTRICAL)
FIXED CAPACITORS
IMPEDANCE
IMPEDANCE BRIDGES
INDUCTANCE
IRREGULARITY TESTERS
LEAKAGE RESISTANCE (ELECTRICAL)
PROXIMITY TESTERS
R C CIRCUITS
REACTANCE (ELECTRICAL)
RESISTANCE (ELECTRICAL)
RESONANT FREQUENCY
STATIC ELECTRICITY
TIME CONSTANT
USTER IRREGULARITY TESTER (TN)
VARIABLE CAPACITORS
VARIANCE LENGTH CURVES
VOLTAGE

CAPACITANCE BRIDGES
RT CAPACITANCE (ELECTRICAL)
ELECTRONIC INSTRUMENTS
IMPEDANCE BRIDGES
MEASURING INSTRUMENTS

CAPACITANCE MOTORS
RT CAPACITANCE (ELECTRICAL)

CAPACITANCE TYPE IRREGULARITY TESTERS
NT FIELDEN WALKER IRREGULARITY TESTER (TN)
USTER IRREGULARITY TESTER (TN)
BT IRREGULARITY TESTERS
RT CAPACITANCE (ELECTRICAL)
IRREGULARITY
PNEUMATIC IRREGULARITY TESTERS
QUALITY CONTROL
ROVINGS
SACO LOWELL IRREGULARITY TESTER (TN)
SLIVERS
VARIANCE LENGTH CURVES
YARNS
ZELLWEGER TESTER (TN)

CAPACITY (ELECTRICAL)
USE CAPACITANCE (ELECTRICAL)

CAPACITY (GENERAL)
RT LOAD BEARING CAPACITY
OUTPUT
PLANNING
PRODUCTION CONTROL
SALES
SPACE
VOLUME

CAPILLARITY
UF CAPILLARY EFFECTS
RT CAPILLARY PRESSURE
CAPILLARY TUBES
CONTACT ANGLES
DRYING
HYGROSCOPICITY
RT PORES
POROSITY
SURFACE TENSION
WETTING
WICKING

CAPILLARY EFFECTS
USE CAPILLARITY

CAPILLARY PRESSURE
RT CAPILLARITY
CAPILLARY WATER
PRESSURE
SURFACE TENSION

CAPILLARY TUBES
RT CAPILLARITY
CAPILLARY PRESSURE
SURFACE TENSION

CAPILLARY WATER
RT BOUND WATER
DRYING
MOISTURE
SORPTION
SURFACE TENSION
WATER
WATER IMBIBITION
WATER RETENTION

CAPROLACTAM
BT LACTAMS
RT NYLON 6
POLYCAPROLACTAM

CAPROLAN (TN)
BT NYLON (POLYAMIDE FIBERS)
NYLON 6

CAPRYL DIETHANOLAMIDE
BT DIETHANOLAMIDE SURFACTANTS

CAPS (HEADGEAR)
BT HEADGEAR
WORK CLOTHING
RT HATS
RAINWEAR

CAPS (SPINNING)
RT BALLOONS
BOBBINS
BUILDER MOTIONS
CAP SPINNING
CAP SPINNING FRAMES
CONTROL SURFACES
DOFFING (PACKAGE)
DRAFTING ROLLS
FLYERS
PACKAGES
RINGS
SPINDLES
SPINNING
SPINNING FRAMES
TRAVELERS
WORSTED SYSTEM PROCESSING
WORSTED SYSTEM SPINNING

CARBAMOYLETHYLATED COTTON
BT CHEMICALLY MODIFIED COTTON
RT ACETYLATED COTTON
ACRYLAMIDE
BENZOYLATED COTTON
CHEMICAL MODIFICATION (FIBERS)
CYANOETHYLATED COTTON
DS (CELLULOSE)
ETHERIFICATION
REACTANTS

CARBODIIMIDES
BT REACTANTS
RT CROSSLINKING
IMIDES
ISOCYANATES

CARBOHYDRATES
UF LAMINARIN
NT CELLOBIOSE
CELLULOSE
HEMICELLULOSE
PECTINS
POLYSACCHARIDES
STARCH
SUCROSE
SUGARS
RT ALCOHOLS
ALDEHYDES
CHITIN
DEACETYLATED CHITIN
NATURAL POLYMERS

CARBON
NT ACTIVATED CARBON
CARBON BLACK
GRAPHITE
RT COAL TAR
GRAPHITE FIBERS

CARBON BLACK
BT CARBON
RT ACTIVATED CARBON

CARBON DISULFIDE
BT SULFIDES (INORGANIC)
RT CELLULOSE XANTHATE
SOLVENTS
VISCOSE RAYON

CARBON TETRACHLORIDE
BT CHLORINATED HYDROCARBONS
RT CARRIERS (DYEING)
CHLORINATED SOLVENTS
DRY CLEANING
DRY CLEANING SOLVENTS
PERCLENE (TN)

CARBONATES
NT BICARBONATES
MONOETHANOLAMINE CARBONATE
POTASSIUM CARBONATE
SODIUM CARBONATE
SODIUM SESQUICARBONATE
BT ALKALIES
BASES
RT HARD WATER

CARBONIZED NOILS
BT NOILS
RT CARBONIZING
WOOL

CARBONIZERS
RT CARBONIZED NOILS
CARBONIZING

CARBONIZING
BT WET FINISHING
RT AMERICAN SYSTEM PROCESSING
BACKWASHING
CLEANING
CONTINENTAL SYSTEM PROCESSING
DRYING
NEUTRALIZING
OILING
RINSING
SCORCHING MARKS
SCOURING
SOAP SODA SCOURING
SOLVENT SCOURING
SQUEEZING
SULFURIC ACID
VEGETABLE MATTER
WOOLEN SYSTEM PROCESSING
WORSTED SYSTEM PROCESSING

CARBONYL GROUPS
RT ALDEHYDE GROUPS
ALDEHYDES
ALKOXYL GROUPS
AMINO GROUPS
CARBOXYL GROUPS
END GROUPS
KETONES

CARBOXYL CONTENT
RT CELLULOSE
DEGRADATION
POLYAMIDES
POLYESTERS
PROTEINS

CARBOXYL END GROUPS
BT ACID END GROUPS
RT CARBOXYL GROUPS
CARBOXYLIC ACIDS
DEGREE OF POLYMERIZATION
END GROUPS
POLYMERS
SULFONIC END GROUPS

CARBOXYL GROUPS
RT ALDEHYDE GROUPS
ALKOXYL GROUPS
AMINO GROUPS
CARBONYL GROUPS
CARBOXYL END GROUPS
CARBOXYLIC ACIDS
END GROUPS
PROTEINS

CARBOXYLATE SURFACTANTS
NT LAMEPONS
SARCOSIDE SURFACTANTS
SOAPS
BT ANIONIC SURFACTANTS
RT BETAINE SURFACTANTS

CARBOXYLIC ACIDS
NT ACETIC ACID
ACRYLIC ACID
ADIPIC ACID
AMINO ACIDS
BENZOIC ACID
CHLOROACETIC ACID
CINNAMIC ACID
CITRIC ACID
ALGINIC ACID

NT EDTA
FATTY ACIDS
FORMIC ACID
HYDROXYACETIC ACID
ISOPHTHALIC ACID
ITACONIC ACID
LACTIC ACID
METHACRYLIC ACID
PHTHALIC ACID
POLYACRYLIC ACID
POLYMETHACRYLIC ACID
PYRUVIC ACID
SALICYLIC ACID
TANNIC ACID
TEREPHTHALIC ACID
BT ORGANIC ACIDS
RT CARBOXYL END GROUPS
CARBOXYL GROUPS

CARBOXYMETHYLATED COTTON
BT CHEMICALLY MODIFIED COTTON
RT AMINIZED COTTON
BENZOYLATED COTTON
CARBOXYMETHYLATION
CARBOXYMETHYLCELLULOSE
COTTON

CARBOXYMETHYLATION
BT REACTIONS (CHEMICAL)
RT CARBOXYMETHYLATED COTTON
CARBOXYMETHYLCELLULOSE
CELLULOSE
CHEMICAL MODIFICATION (FIBERS)
FINISHING PROCESS (GENERAL)

CARBOXYMETHYLCELLULOSE
UF SODIUM CARBOXYMETHYLCELLULOSE
BT CELLULOSE ETHERS
RT ALKALI CELLULOSE
ANTIREDEPOSITION AGENTS
CARBOXYMETHYLATED COTTON
CARBOXYMETHYLATION
CYANOETHYLCELLULOSE
LAUNDERING
POLYELECTROLYTES
PRINTING
SANITARY NAPKINS
SIZES (SLASHING)
THICKENING AGENTS
VISCOSITY

CARCASE
USE CASING PLY

CARD CLEARERS
USE CLEARERS (CARD)

CARD CLOTHING
UF FLAT CLOTHING
NT FILLET CLOTHING
METALLIC CARD CLOTHING
SHEET CLOTHING
RT ANGLE STRIPPERS
BREAKERS
CARD CLOTHING FOUNDATION
CARD COUNT
CARD CROWN
CARD CYLINDERS
CARD FLATS
CARD GAUGE
CARD GRINDING
CARD NAILING
CARD ROLLS
CARD STRIPS
CARD WIRES
CARDING
CARDING ACTION
CARDING LEATHER
CARDS
COTTON CARDING
COTTON CARDS
COUNT CROWN AND GAUGE
DOFFERS (CARD)
DOFFING (OF A CARD)
FANCY (NOUN)
FETTLING
FINISHER PART
FLAT SETTINGS (CARD)
KNEE
LICKER IN
NOGG
NOGGS PER INCH
RIB SET
STRIPPERS
STRIPPING ACTION
TWILL SET
WOOLEN CARDING
WORKERS
WORSTED CARDING

CARD CLOTHING FOUNDATION
RT CARD CLOTHING
CARD NAILING
CARD WIRES
FILLET CLOTHING
METALLIC CARD CLOTHING
RIB SET
SHEET CLOTHING
TWILL SET

CARD COUNT
RT CARD CLOTHING
CARD CROWN
CARD GAUGE

CARD CROWN
RT CARD CLOTHING
CARD COUNT
CARD WIRES
METALLIC CARD CLOTHING

CARD CYLINDERS
UF BREAST SWIFTS
CYLINDERS (CARD)
SWIFTS
BT CARD ROLLS
RT ANGLE STRIPPERS
BREAKERS
CARD CLOTHING
CARD FLATS
CARD STRIPS
CARDING
CARDING ACTION
CARDING EFFICIENCY
CARDS
CLEARERS (CARD)
COTTON CARDING
COTTON CARDS
DOFFERS (CARD)
DOFFING (OF A CARD)
FANCY (NOUN)
FETTLING
FINISHER PART
INTERMEDIATE PART
LICKER IN
PERALTA ROLLS
PERCENTAGE PLATES
STRIPPERS
STRIPPING ACTION
WOOLEN CARDING
WORKERS
WORSTED CARDING

CARD CYLINDERS (JACQUARD KNITTING)
RT JACQUARD KNITTING

CARD DOFFERS
USE DOFFERS (CARD)

CARD FLATS
UF FLATS
RT CARD CLOTHING
CARD CYLINDERS
CARD ROLLS
CARD STRIPS
CARD WASTE
CARD WIRES
CARDING
CARDING ACTION
CARDS
COTTON CARDING
COTTON CARDS
CYLINDER SCREENS
DOFFERS (CARD)
FLAT CHAINS (CARD)
FLAT SETTINGS (CARD)
FLAT STRIPPING COMBS
FLAT STRIPS
STRIPPING ACTION

CARD GAUGE
UF GAUGE (WIRE)
RT CARD CLOTHING
CARD COUNT
CARD WIRES
CARDS
FILLET CLOTHING

CARD GRINDERS
USE GRINDING ROLLS (CARDS)

CARD GRINDING
RT CARD CLOTHING
CARD CYLINDERS
CARD NAILING
CARD ROLLS
CARD WIRES
CARDS
DOFFERS (CARD)
FILLET CLOTHING
GRINDING ROLLS (CARDS)
SHEET CLOTHING
STRIPPERS
WORKERS

CARD NAILING
RT CARD CLOTHING
CARD CLOTHING FOUNDATION
CARD GRINDING
CARD ROLLS
CARD WIRES
CARDS
FILLET CLOTHING
METALLIC CARD CLOTHING
SHEET CLOTHING

CARD ROLLERS
USE CARD ROLLS

CARD ROLLS
UF CARD ROLLERS
NT ANGLE STRIPPERS
CARD CYLINDERS
DIVIDERS (CARD ROLLS)
DOFFERS (CARD)
FANCY (NOUN)
FEED ROLLS
LICKER IN
PERALTA ROLLS
STRIPPERS
WORKERS
BT ROLLS
RT CARD CLOTHING
CARD GRINDING
CARD NAILING
CARD WIRES
CARDS
FETTLING

CARD SLIVERS
UF CARDED SLIVERS
BT FIBER ASSEMBLIES
SLIVERS
RT BALLING HEADS
CARD WEBS
CARDING
CARDS
COILERS
COMBED SLIVERS
COMBER LAPS
COTTON CARDING
COTTON CARDS
CROSROL WEB PURIFIER (TN)
DOFFERS (CARD)
DOFFING (OF A CARD)
DOUBLE TRUMPETS
DOUBLING (DRAFTING)
DRAFTING (STAPLE FIBER)
FIBER ORIENTATION
FINISHER PART
IRREGULARITY
LEADING HOOKS
LINDSLEY RATIOS
NEPS
ROVINGS
SLIVER CANS
SLIVER KNITTING
SLIVER REVERSALS
SLIVER TENDERNESS
SLIVER TESTERS
SLUBBINGS
TOPS
TOW
TRAILING HOOKS
TRUMPETS
WEB UNIFORMITY
WOOLEN CARDING
WORSTED CARDING
YARNS

CARD STRIPPING
USE FETTLING

CARD STRIPS
NT FLAT STRIPS
BT CARD WASTE
RT CARD CLOTHING
CARD CYLINDERS
CARD FLATS
DOFFER STRIPS
DOFFERS (CARD)
FETTLING
FLY (WASTE)
LINT (WASTE)
STRIPPERS
TRASH
WORKERS

CARD WASTE
NT CARD STRIPS
CYLINDER STRIPS
DOFFER STRIPS
FETTLING WASTE
FLAT STRIPS
FLY (WASTE)
MOTES
RT BURR BEATERS
BURR CRUSHERS
CARD FLATS
CARD YIELD
CARDING
CARDS
CLEARERS (CARD)
COTTON CARDING
FETTLING
FLAT STRIPPING COMBS
HACKLE COMBS
KNIFE PLATES
LICKER IN SCREENS
MOTE KNIVES
NEPS
PERALTA ROLLS
WASTE CONTROL KNIVES
WOOLEN CARDING
WORSTED CARDING

CARD WEBS
BT WEBS
RT AERODYNAMIC CARDING
AUTOCOUNT (TN)
CARD SLIVERS
CARDING
CARDS
CONDENSERS
CROSS LAID WEBS
CROSS ROLLS
DOFFER COMBS
DOFFERS (CARD)
DOFFING (OF A CARD)
FELTS
FIBER HOOKS
FIBER ORIENTATION
FINISHER PART
LEADING HOOKS
NEP COUNT
NEPS
NONWOVEN FABRICS
OPTICAL SCANNING
ORIENTATION UNIFORMITY
PATCHINESS
PERALTA ROLLS
RANDO WEBS (TN)
SLIVERS
SLUBBINGS
SPOTTINESS
STRIPPING ACTION
TAPE CONDENSERS
TRAILING HOOKS
TRUMPETS
WEB ORIENTATION
WEB UNIFORMITY

CARD WIRES
RT CARD CLOTHING
CARD CLOTHING FOUNDATION
CARD CROWN
CARD FLATS
CARD GAUGE
CARD GRINDING
CARD NAILING
CARD ROLLS
COUNT CROWN AND GAUGE
FILLET CLOTHING
KNEE
METALLIC CARD CLOTHING
NOGG
NOGGS PER INCH
RIB SET
SHEET CLOTHING
STAPLES (WIRE)
TWILL SET

CARD YIELD
RT CARD WASTE
CARDING
OUTPUT
PROCESS EFFICIENCY
YIELD (MECHANICAL)

CARDED SLIVERS
USE CARD SLIVERS

CARDED YARNS
BT SPUN YARNS
YARNS
RT COMBED YARNS
COTTON SPUN YARNS
WOOLEN SPUN YARNS
WORSTED SPUN YARNS

CARDER PART
USE FINISHER PART

CARDIGANS
BT GARMENTS
KNITTED OUTERWEAR
OUTERWEAR
SWEATERS
RT CIRCULAR KNITTING
FULLY FASHIONED KNITTING
JERKINS
JUMPERS
KNITTED FABRICS
KNITTING
PULL OVERS
SLIP ONS
SPORTSWEAR

CARDING
NT AERODYNAMIC CARDING
COTTON CARDING
WOOLEN CARDING
WORSTED CARDING
RT AMERICAN SYSTEM PROCESSING
ANGLE STRIPPERS
AUTOCOUNT (TN)
BLENDING
BRADFORD SYSTEM PROCESSING
BREAKERS
BURR BEATERS
BURR CRUSHERS
CARD CLOTHING
CARD CLOTHING FOUNDATION
CARD CYLINDERS
CARD FLATS
CARD GAUGE
CARD GRINDING
CARD NAILING
CARD ROLLS
CARD SLIVERS
CARD WASTE
CARD WEBS
CARD WIRES
CARD YIELD
CARDING ACTION
CARDING DYNAMICS
CARDING EFFICIENCY
CARDS
CLEARERS (CARD)
COMBING
CONDENSERS
CONDENSING (YARN MANUFACTURING)
CONTINENTAL SYSTEM PROCESSING
COTTON CARDS
COTTON SYSTEM PROCESSING
DOFFER COMBS
DOFFERS (CARD)
DOFFING (OF A CARD)
DOUBLE TRUMPETS
ENTANGLEMENTS
FANCY (NOUN)
FEED PLATES
FEED ROLLS
FETTLING
FIBER BREAKAGE
FIBER HOOKS
FINISHER PART
GARNETTING
HOOK DIRECTION
HOOK FORMATION
HOOK REMOVAL
HOOKED FIBERS
HOPPER FEED
INTERMEDIATE PART
LEADING HOOKS
LICKER IN
LINDSLEY RATIOS
MOTE KNIVES
MOTES
NEP COUNT
NEPS
NEW BRADFORD SYSTEM PROCESSING
ORIENTATION UNIFORMITY
PARALLELIZATION
PERALTA ROLLS
PICKING (OPENING)
SCOTCH FEED
SCRIBBLING
SLIVER CANS
SLIVERS
SLUBBINGS
STRIPPERS
STRIPPING ACTION
TRAILING HOOKS
TRUMPETS
TUFTS (FIBER)
WEB ORIENTATION
WEB UNIFORMITY
WEBS
WOOLEN SYSTEM PROCESSING
WORKERS

CARDING ACTION
UF WORKING ACTION
RT BRUSHING ACTION
CARD CLOTHING
CARD CYLINDERS
CARD FLATS
CARDING EFFICIENCY
CARDS
DOFFING (OF A CARD)
ENTANGLEMENTS
FIBER BREAKAGE
FIBER WITHDRAWAL FORCE
NEP COUNT
NEPS
SCRIBBLING
STRIPPERS
STRIPPING ACTION
WORKERS

CARDING DYNAMICS
BT PROCESS DYNAMICS
RT CARDING
CARDING ACTION
DRAFTING DYNAMICS
ENTANGLEMENTS
FIBER WITHDRAWAL FORCE
KNITTING DYNAMICS
MECHANISM (FUNDAMENTAL)
SPINNING DYNAMICS
TWISTING DYNAMICS
WEAVING DYNAMICS
WINDING DYNAMICS
YARN DYNAMICS

CARDING EFFICIENCY
UF CARDING POWER
BT PROCESS EFFICIENCY
RT CARD CYLINDERS
CARD YIELD
CARDING
CARDING ACTION
CARDING DYNAMICS
COTTON CARDING
DOFFERS (CARD)
DOFFING (OF A CARD)
LICKER IN
STRIPPING ACTION
WORKERS

CARDING ENGINES
USE CARDS

CARDING LEATHER
RT CARD CLOTHING

CARDING OIL
BT LUBRICANTS
RT BLENDING
LUBRICATION
OILING
SPINNING LUBRICANTS
YARN FINISH

CARDING POWER
USE CARDING EFFICIENCY

CARDING WOOL
BT WOOL CLASS
RT PREPARING WOOL
WORSTED SPUN YARNS

CARDS
UF CARDING ENGINES
NT AERODYNAMIC CARDS
COTTON CARDS
DUO CARDS
HIGH PRODUCTION CARDS
REVOLVING FLAT CARDS
ROLLER TOP CARDS
SRRL CARD
WOOLEN CARDS
WORSTED CARDS
RT ANGLE STRIPPERS
BLAMIRE FEED
BREAKERS
BURR CRUSHERS
CARD CLOTHING
CARD CLOTHING FOUNDATION
CARD CYLINDERS
CARD FLATS
CARD ROLLS
CARD SLIVERS
CARD WASTE
CARD WEBS
CARD WIRES
CARDING
CLEARERS (CARD)
COILERS
CONDENSERS
CROSROL WEB PURIFIER (TN)
CROSS ROLLS
DOFFER COMBS
DOFFERS (CARD)
FANCY (NOUN)
FEED LATTICES
FEED ROLLS
FINISHER PART
FLAT CHAINS (CARD)
FLY (WASTE)
GARNETTING MACHINES
HACKLE COMBS
HOPPER FEED
KNIFE PLATES
LAP GUIDES
LICKER IN
LICKER IN SCREENS
MOTE KNIVES
NONWOVEN FABRIC MACHINES
PERALTA ROLLS
PICKERS (OPENING)
SCOTCH FEED
SHEET CLOTHING
SLIVER CANS
SLIVERS
SLUBBINGS
STAPLES (WIRE)
STRIPPERS
TRUMPETS
TUFT FORMERS
WASTE CONTROL KNIVES
WORKERS

CARPET BACKING
RT BACKING FABRICS

CARPET DYEING
BT PIECE DYEING
RT CARPETS
OPEN WIDTH DYEING

CARPET STUFFER YARNS
RT WOVEN CARPETS

CARPETING
USE CARPETS

CARPETS
 UF CARPETING
  RUGS
 NT AXMINSTER CARPETS
  CHENILLE CARPETS
  FLOCKED CARPETS
  INGRAIN CARPETS
  KNITTED CARPETS
  NONWOVEN CARPETS
  TUFTED CARPETS
  WILTON CARPETS
  WOVEN CARPETS
 RT BACKING COMPOUNDS
  CARPET DYEING
  DECORATIVE FABRICS
  FOAM BACKED FABRICS
  FOAM RUBBER
  FURNISHING FABRICS
  FURNISHINGS
  HOUSEHOLD FABRICS
  MULTIPLE LAYER FABRICS
  NEEDLED FABRICS
  NONWOVEN FABRICS
  PILE FABRICS (WOVEN)
  TAPESTRIES
  TUFTED FABRICS
  UNDERLAY (CARPETS)
  WOVEN FABRICS

CARRIAGES (SPINNING MULE)
 RT COPS
  COUNTERFALLERS
  DELIVERY ROLLS
  DRIVING BANDS
  END BREAKAGES
  FALLERS (MULE)
  HEADSTOCKS
  JACKSPOOLS
  MULE DRAFT
  MULE SPINNING
  MULES
  QUADRANTS (MULE)
  SLUBBINGS
  SPINDLES
  SURFACE DRUMS

CARRICKMACROSS (LACE)
 BT HAND MADE LACES
  LACES

CARRIER (BRAIDS)
 RT BRAIDERS
  BRAIDING
  BRAIDS

CARRIER DYEING
 UF EMULSION DYEING
 BT DYEING (BY ENVIRONMENTAL
  CONDITIONS)
 RT BENZYL ALCOHOL
  CARRIERS (DYEING)
  DYEING AUXILIARIES
  LOW TEMPERATURE DYEING
  PRINTING
  SOLVENT ASSISTED DYEING
  SOLVENT DYEING
  SWELLING
  VACUUM DYEING

CARRIER RODS
 RT FEEDERS
  FLOW (MATERIALS)
  FULLY FASHIONED KNITTING
  MACHINES
  KNITTING
  KNITTING MACHINES
  KNITTING NEEDLES
  YARN CARRIERS
  YARNS

CARRIERS (DYEING)
 BT ACCELERANTS (DYEING)
  DYEING AUXILIARIES
 RT ALKYLENE CARBONATES
  BENZYL ALCOHOL
  CARBON TETRACHLORIDE
  CARRIER DYEING
  CHLORINATED HYDROCARBONS
  DICHLOROBENZENE
  DYES
  EMULSIFYING AGENTS
  LEVELLING AGENTS
  PHENOLS
  SOLVENT ASSISTED DYEING
  SURFACTANTS
  SWELLING AGENTS
  TOLUENE
  TRICHLOROBENZENE
  WETTING AGENTS

CARROTING
 BT WET FINISHING
 RT FUR
  MERCURY COMPOUNDS
  NITRIC ACID
  ORGANOMERCURY COMPOUNDS

CARTONS
 BT PACKAGES
 RT BAGS
  BALES
  BOXES

CARVED PILE FABRICS
 USE SCULPTURED PILE FABRICS

CASABLANCA SYSTEM
 RT ACTUAL DRAFT
  AMBLER SUPERDRAFT
  APRON CONTROL
  APRONS
  CAP SPINNING FRAMES
  CONTROL SURFACES
  COTTON SYSTEM PROCESSING
  DOUBLE APRONS
  DRAFTING (STAPLE FIBER)
  DRAFTING MACHINES
  DRAWFRAMES
  DUO ROTH SYSTEM
  HIGH DRAFT
  HIGH DRAFT SPINNING
  RING FRAMES
  ROTH SYSTEM
  SHAW SYSTEM
  SPINNING FRAMES
  TWIST CONTROL
  WORSTED SYSTEM PROCESSING

CASCADE WASHERS
 BT SCOURING MACHINES
 RT BLEACHING RANGES
  SCOURING

CASEIN
 BT PROTEINS
 RT AZLON (REGENERATED PROTEIN
  FIBERS)

CASEMENT
 BT FLAT WOVEN FABRICS
  WOVEN FABRICS
 RT PLAIN WEAVES

CASHMERE
 BT HAIR
  KERATIN FIBERS
  PROTEIN FIBERS
 RT ALPACA
  DEHAIRING
  GUARD HAIRS

CASING PLY
 UF BODY PLY
  CARCASE
 RT TIRE CORDS
  TIRES

CASKET LININGS
 BT LININGS
 RT BOX LININGS
  FABRICS
  GLOVE LININGS
  INDUSTRIAL FABRICS
  INTERLININGS
  LINING FELT
  POCKET LININGS
  SHOE LININGS

CAST COATING
 BT COATING (PROCESS)
 RT CALENDER COATING
  COATED FABRICS
  DIP COATING
  EXTRUSION COATING
  FLEXIBLE FILM LAMINATING
  KNIFE COATING
  NIP COATING
  ROLLER COATING

CASTING OFF
 RT COMPOUND NEEDLES
  FILLING KNITTING MACHINES
  FULLY FASHIONED KNITTING
  MACHINES
  KNITTING
  LANDING LOOPS
  LAYING IN (KNITTING)
  TAKE DOWN
  TRICOT KNITTING MACHINES
  TUCKING (KNITTING)
  WARP KNITTING MACHINES

CATALYSIS
 NT REDOX CATALYSIS
 RT CATALYSTS
  INITIATION
  REACTANTS
  REACTIONS (CHEMICAL)
  RESIN FINISHES

CATALYSTS
 UF ACTIVATORS
 NT INITIATORS
 RT ACETAL RESINS
  ACIDS

 RT ADDITIVES (CHEMICAL)
  AZIDES
  CATALYSIS
  CERIUM COMPOUNDS
  CUPROUS CHLORIDE
  CURING

  DURABLE PRESS
  EASY CARE FABRICS
  ENZYMES
  FERRIC CHLORIDE
  FINISH (SUBSTANCE ADDED)
  INHIBITION
  MAGNESIUM CHLORIDE
  MAGNESIUM NITRATE
  ORGANOTIN COMPOUNDS
  PERSULFATES
  POLYMERIZATION
  REAGENTS
  REDOX CATALYSIS
  RETARDING AGENTS
  TIN COMPOUNDS
  ZINC COMPOUNDS
  ZINC FLUOBORATE
  ZINC NITRATE

CATAPHORESIS
 USE ELECTROPHORESIS

CATCH BARS
 RT LACE MACHINES

CATCHING
 RT END BREAKAGES
  PICKING (SNAGGING)
  PUNCTURE RESISTANCE
  PUNCTURING
  RUN RESISTANCE
  RUNS (KNITTING)
  SNAG RESISTANCE
  SNAG TESTERS
  SNAGGING
  SNAGS

CATION EXCHANGE
 USE ION EXCHANGE

CATION EXCHANGE RESINS
 BT ION EXCHANGE RESINS
 RT ANION EXCHANGE RESINS
  ANIONIC SITES
  ION EXCHANGE

CATIONIC DYEING
 USE BASIC DYEING

CATIONIC DYES
 USE BASIC DYES

CATIONIC SITES
 BT SITES (DYEING)
 RT AMINO GROUPS
  ANIONIC SITES
  DYEABILITY
  DYEING THEORY
  ION EXCHANGE
  POLYELECTROLYTES

CATIONIC SOFTENERS
 BT SOFTENERS
 RT ANIONIC SOFTENERS
  CATIONIC SURFACTANTS
  LUBRICANTS
  NONIONIC SOFTENERS
  QUATERNARY AMMONIUM
  SURFACTANTS

CATIONIC SURFACTANTS
 NT AMINE SURFACTANTS
  ONIUM SURFACTANTS
 BT SURFACTANTS
 RT AMPHOTERIC SURFACTANTS
  ANIONIC SURFACTANTS
  ANTIBACTERIAL FINISHES
  CATIONIC SOFTENERS
  CATIONS
  DETERGENTS
  DYEING AUXILIARIES
  EMULSIFYING AGENTS
  LEVELLING AGENTS
  NONIONIC SURFACTANTS
  RETARDING AGENTS
  WETTING
  WETTING AGENTS

CATIONS
 BT IONS
 RT ANIONS
  BASES
  BASIC DYES
  CATIONIC SURFACTANTS
  ELECTROLYTES
  FREE RADICALS
  SALTS (GENERAL)
  SURFACTANTS

CAUCHY DISTRIBUTION
 RT BINOMIAL DISTRIBUTION
  (CONT.)

RT CHI SQUARED DISTRIBUTION
DISTRIBUTION
F DISTRIBUTION
NORMAL DISTRIBUTION
T DISTRIBUTION

CAUSTIC (NOUN)
USE SODIUM HYDROXIDE

CAUSTIC POTASH
USE POTASSIUM HYDROXIDE

CAUSTIC SCOURING
BT SCOURING
RT SOAP SODA SCOURING

CAUSTIC SODA
USE SODIUM HYDROXIDE

CAUSTICAIRE INDEX
USE CAUSTICAIRE VALUE

CAUSTICAIRE SCALE
RT AIR FLOW
AIR RESISTANCE
CAUSTICAIRE VALUE
COTTON
COUNT
CURVILINEAR SCALE
FIBER DIAMETER
FIBER FINENESS
FIBER SURFACE
FINENESS TESTING
MICRONAIRE
MICRONAIRE FINENESS

CAUSTICAIRE VALUE
UF CAUSTICAIRE INDEX
RT AIR FLOW
AIR RESISTANCE
CAUSTICAIRE SCALE
COTTON
COUNT
CURVILINEAR SCALE
FIBER DIAMETER
FIBER FINENESS
FIBER SURFACE
GRADE (FIBERS)
IMMATURE FIBERS
MATURITY INDEX
MICRONAIRE
MICRONAIRE FINENESS

CAVALRY TWILL
BT FLAT WOVEN FABRICS
WOVEN FABRICS
RT TWILL WEAVES

CAVITIES
USE HOLES

CELACRIMP (TN)
BT ACETATE FIBERS
CELLULOSE ESTER FIBERS
CRIMPED YARNS
RT STUFFER BOX CRIMPING

CELLOTETRAOSE
USE POLYSACCHARIDES

CELAFIBRE (TN)
BT ACETATE FIBERS
CELLULOSE ESTER FIBERS

CELLOBIOSE
BT CARBOHYDRATES
RT CELLULOSE

CELLOPHANE
BT FILMS
REGENERATED CELLULOSE
RT CELLULOSE
LAMINATES
PACKAGING (MATERIAL)
VISCOSE RAYON

CELLULAR FABRICS
UF LENO CELLULAR FABRICS
RT FLAT WOVEN FABRICS
HONEYCOMB WEAVES
LENO WEAVES
MOCK LENO WEAVES
WEAVE (TYPES)

CELLULAR PLASTICS
USE FOAMS

CELLULAR YARNS
UF FOAM YARNS
BT YARNS
RT FOAMS
POROUS MATERIALS

CELLULASE
BT ENZYMES
RT CELLULOSE
HYDROLYSIS
CELLULOSE DERIVATIVES
METHYLENE BLUE NUMBER

CELLULOSE
NT CELLULOSE I
CELLULOSE II
CELLULOSE III
CELLULOSE IV
DEGRADED CELLULOSE
BT CARBOHYDRATES
NATURAL POLYMERS
POLYSACCHARIDES
RT ALKALI CELLULOSE
BAST FIBERS
CADOXEN
CARBOXYL CONTENT
CARBOXYMETHYLATION
CELLOBIOSE
CELLOPHANE
CELLULASE
CELLULOSE DERIVATIVES
CELLULOSE DYES
CELLULOSE ESTERS
CELLULOSE ETHERS
CELLULOSE NITRATE
CELLULOSE XANTHATE
CELLULOSIC FIBERS
COPPER NUMBER
COTTON
CUPRAMMONIUM HYDROXIDE
CUPRIETHYLENEDIAMINE HYDROXIDE
FERRIC TARTRATE
FIBRILS
HEMICELLULOSE
LEAF FIBERS
MARSCHALL SOLUTION
METHYLENE BLUE NUMBER
RAYON (REGENERATED CELLULOSE FIBERS)
WOOD
WOOD FIBERS

CELLULOSE I
BT CELLULOSE
RT CRYSTAL LATTICE

CELLULOSE II
BT CELLULOSE
RT CRYSTAL LATTICE

CELLULOSE III
BT CELLULOSE
RT CRYSTAL LATTICE

CELLULOSE IV
BT CELLULOSE
RT CRYSTAL LATTICE

CELLULOSE ACETATE
BT CELLULOSE ESTERS
RT ACETATE FIBERS
CELLULOSE ACETATE BUTYRATE

CELLULOSE ACETATE BUTYRATE
BT CELLULOSE ESTERS
RT CELLULOSE ACETATE
LACQUERS
PAINTS

CELLULOSE BLENDS
NT COTTON BLENDS
RAYON BLENDS
BT BLENDS (FIBERS)
RT CELLULOSE ESTER FIBER BLENDS

CELLULOSE DERIVATIVES
NT ALKALI CELLULOSE
CELLULOSE ESTERS
CELLULOSE ETHERS
DEGRADED CELLULOSE
DIALDEHYDE CELLULOSE
HYDROCELLULOSE
OXYCELLULOSE
RT ACETATE FIBERS
CELLULOSE
CELLULOSE ESTER FIBERS
CHEMICALLY MODIFIED CELLULOSIC FIBERS
DS (CELLULOSE)
HEMICELLULOSE
POLYMERS
RAYON (REGENERATED CELLULOSE FIBERS)
REGENERATED CELLULOSE

CELLULOSE DYEING
NT COTTON DYEING
RAYON DYEING
BT DYEING (BY FIBER CLASSES)
RT BLEND DYEING
SILK DYEING
SYNTHETIC DYEING
WOOL DYEING

CELLULOSE DYES
UF COTTON DYES
RAYON DYES
BT DYES (BY FIBER CLASSES)
RT ACETATE DYES
ACRYLIC DYES
CELLULOSE
COTTON
COTTON DYEING
DIRECT DYES
DYEING
DYES (BY CHEMICAL CLASSES)
NYLON DYES
POLYESTER DYES
RAYON DYEING
SILK DYES
VISCOSE RAYON
WOOL DYES

CELLULOSE ESTER FIBER BLENDS
BT BLENDS (FIBERS)
RT CELLULOSE BLENDS

CELLULOSE ESTER FIBERS
NT ACETATE FIBERS
ARNEL (TN)
CELACRIMP (TN)
CELAFIBRE (TN)
ESTRON (TN)
FIBROCETA (TN)
LOFTURA (TN)
TRIACETATE FIBERS
TRICEL C (TN)
BT FIBERS
MAN MADE FIBERS
RT CELLULOSE DERIVATIVES
CELLULOSE ESTERS
CUPRAMMONIUM RAYON
RAYON (REGENERATED CELLULOSE FIBERS)
VISCOSE RAYON

CELLULOSE ESTERS
NT CELLULOSE ACETATE
CELLULOSE ACETATE BUTYRATE
CELLULOSE NITRATE
CELLULOSE PHOSPHATE
CELLULOSE XANTHATE
BT CELLULOSE DERIVATIVES
ESTERS
RT ACETATE FIBERS
ALKALI CELLULOSE
CELLULOSE
CELLULOSE ESTER FIBERS
CELLULOSE ETHERS
CHEMICAL MODIFICATION (FIBERS)
CHEMICALLY MODIFIED COTTON
RAYON (REGENERATED CELLULOSE FIBERS)
TRIACETATE FIBERS

CELLULOSE ETHERS
NT ALLYLCELLULOSE
BENZYLCELLULOSE
CARBOXYMETHYLCELLULOSE
CYANOETHYLCELLULOSE
ETHYLCELLULOSE
HYDROXYALKYLCELLULOSE
HYDROXYETHYLCELLULOSE
METHYLCELLULOSE
BT CELLULOSE DERIVATIVES
ETHERS
RT ALKALI CELLULOSE
CELLULOSE
CELLULOSE ESTERS
CELLULOSE XANTHATE
CHEMICAL MODIFICATION (FIBERS)
CHEMICALLY MODIFIED COTTON
VINYLATED COTTON

CELLULOSE NITRATE
UF NITROCELLULOSE
BT CELLULOSE ESTERS
RT ACETATE FIBERS
ALKALI CELLULOSE
CELLULOSE
CELLULOSE PHOSPHATE
NITRATES
NITRO COMPOUNDS
RAYON (REGENERATED CELLULOSE FIBERS)

CELLULOSE PHOSPHATE
BT CELLULOSE ESTERS
RT CELLULOSE NITRATE
PHOSPHATES

CELLULOSE PULP
USE PULP

CELLULOSE REACTANTS
USE REACTANTS

CELLULOSE XANTHATE
BT CELLULOSE ESTERS
RT ALKALI CELLULOSE
CARBON DISULFIDE
CELLULOSE
CELLULOSE ETHERS
CYANOETHYLCELLULOSE
REGENERATED CELLULOSE

CELLULOSIC FIBERS
NT AVRIL (TN)
BAST FIBERS

NT COIR
CHEMICALLY MODIFIED CELLULOSIC FIBERS
COLORAY (TN)
CORDURA (TN)
COTTON
CUPRAMMONIUM RAYON
DURAFIL (TN)
FIBRO (TN)
FLAX
FORTISAN (TN)
FRUIT FIBERS
HEMP
HIBISCUS
HIGH TENACITY VISCOSE RAYON
HIGH WET MODULUS RAYON
JUTE
KAPOK
LEAF FIBERS
LINT (FIBER)
MANILA
POLYNOSIC FIBERS
RAFFIA
RAMIE
RAYON (REGENERATED CELLULOSE FIBERS)
SEED HAIR FIBERS (GENERAL)
SISAL
TENASCO (TN)
TENAX (TN)
VEGETABLE FIBERS (GENERAL)
VISCOSE RAYON
WOOD FIBERS
BT FIBERS
RT CELLULOSE
NATURAL FIBERS

CELON (TN)
BT NYLON (POLYAMIDE FIBERS)
NYLON 6

CEMENTING
BT JOINING
RT ADHESION
BASTING
BONDING
CEMENTS
COHESION
FUSION (MELTING)
GLUING
LAMINATING
PEELING
SEALING
SEAMING
TAPING

CEMENTS
RT ADHESIVES
BONDING STRENGTH
CEMENTING
GLUES
PASTES
SIZES (SLASHING)

CENTER GUIDES
BT CLOTH GUIDES
GUIDES
RT EDGE GUIDES

CENTER LOOPS
UF WEFT LOOPS
BT FABRIC DEFECTS

CENTER OF GRAVITY
UF CENTER OF MASS
RT MASS

CENTER OF MASS
USE CENTER OF GRAVITY

CENTER SELVEDGES
BT SELVEDGES
RT CRACKED SELVEDGES
DOG LEGGED SELVEDGES
DRESSING SELVEDGES (LACE)
LENO SELVEDGES
LOOP SELVEDGES
PULLED IN SELVEDGES
SEALED SELVEDGES
SELVEDGE WARPS
SLACK SELVEDGES
TAPE SELVEDGES
TIGHT SELVEDGES
WEAVING
WOVEN FABRICS

CENTER SHED
USE SPLIT SHED

CENTRALIZED CONTROL
RT CONTROL SYSTEMS

CENTRIFUGAL DRYERS
UF CENTRIFUGAL EXTRACTORS
BT DRYERS
RT ATMOSPHERIC DRYERS
CENTRIFUGAL FORCE
CENTRIFUGES
DRUM DRYERS
FLUIDIZED BED DRYERS
LOOP DRYERS
VACUUM DRYERS

CENTRIFUGAL DRYING
UF SPIN DRYING
BT DRYING
RT CENTRIFUGAL FORCE
DIELECTRIC DRYING
FREEZE DRYING
HOT AIR DRYING
INFRARED DRYING
LINE DRYING (EASY CARE GARMENTS)
TUMBLE DRYING
VACUUM DRYING

CENTRIFUGAL EXTRACTORS
USE CENTRIFUGAL DRYERS

CENTRIFUGAL FORCE
UF CENTRIPETAL FORCE
BT FORCE
RT ACCELERATION (MECHANICAL)
ANGULAR VELOCITY
ACCELEROMETERS
BALLOON DYNAMICS
CENTRIFUGAL DRYERS
CENTRIFUGAL DRYING
CENTRIFUGAL PUMPS
CENTRIFUGES
CORIOLIS ACCELERATION
CORIOLIS FORCE
FRICTION
FRICTIONAL FORCE
INERTIA
NORMAL FORCE
ROTATION
SEPARATION (SOLUTION)
SPINNING DYNAMICS
TRANSVERSE FORCE
TRAVELERS
ULTRACENTRIFUGES
VELOCITY

CENTRIFUGAL PUMPS
BT PUMPS
RT CENTRIFUGAL FORCE

CENTRIFUGAL SPINNING
USE POT SPINNING

CENTRIFUGES
RT DRYERS
CENTRIFUGAL DRYERS
CENTRIFUGAL FORCE
CENTRIFUGAL PUMPS
EXTRACTION
SCOURING TRAINS
SEPARATION (SOLUTION)
ULTRACENTRIFUGES

CENTRIPETAL FORCE
USE CENTRIFUGAL FORCE

CERAMIC GUIDES
BT GUIDES
RT CERAMICS
THREAD GUIDES

CERAMIC FIBERS
UF REFRACTORY FIBERS
VITREOUS FIBERS
NT ALUMINUM SILICATE FIBERS
BORON NITRIDE FIBERS
SILICON CARBIDE FIBERS
BT HIGH TEMPERATURE FIBERS
INORGANIC FIBERS (MAN MADE)
MAN MADE FIBERS
RT CERAMICS
FIRING (REFRACTORY)
GLASS FIBERS
METALLIC FIBERS
METALLIC YARNS
MINERAL FIBERS

CERAMICS
RT ABRASIVES
CERAMIC FIBERS
CERAMIC GUIDES
CLAYS
CORUNDUM
EMERY
FIRING (REFRACTORY)
GLASS
GRIT
MATERIALS

CEREMONIAL CLOTHING
NT CLERICAL CLOTHES
COPES
GOWNS
HOODS
MITRES
ROBES
SOUTANES
UNIFORMS
VESTMENTS
BT GARMENTS
RT EMBROIDERY

CERIUM COMPOUNDS
NT CERIUM SULFATE
RT CATALYSTS
GRAFTING

CERIUM SULFATE
BT CERIUM SULFATE
SULFATES

CESIUM COMPOUNDS
NT CESIUM HYDROXIDE

CESIUM HYDROXIDE
BT CESIUM COMPOUNDS
HYDROXIDES

CETYL ALCOHOL
BT FATTY ALCOHOLS

CHAFE MARKS
UF ABRASION MARKS
BRUISED PLACES
BT FABRIC DEFECTS
RT ABRASION
DAMAGE

CHAFER FABRICS
RT TIRE FABRICS

CHAIN DRAFT
RT DESIGN
DRAFT (PATTERN)
DRAFTING (PATTERN)
FABRIC DESIGN
HARNESS DRAFT
PATTERN (FABRICS)

CHAIN DRIVES
BT DRIVES
RT BELT DRIVES
FLUID DRIVES
GEAR DRIVES
HYDRAULIC DRIVES
MECHANICAL DRIVES
VARIABLE SPEED DRIVES

CHAIN FOLDING
RT CRYSTALLITES
CRYSTALLIZATION
LINEAR CHAIN MOLECULES
LONG CHAIN MOLECULES

CHAIN MERCERIZERS
BT MERCERIZING MACHINES
RT CHAINLESS MERCERIZERS
MERCERIZING

CHAIN MOLECULES
USE LONG CHAIN MOLECULES

CHAIN SEWING
BT GARMENT MANUFACTURE
SEWING
RT ANGULAR SEAMING
BACK TACKING
CURVED SEAMING
GATHERING
HAND SEWING
HEMMING
LOOP CATCHING
MACHINE SEWING
MALI PROCESSES (TN)
NEEDLE PENETRATION
PINCHING
POSITIONING
PUCKERING
RUFFLING
SEAMING
SEWABILITY
SEWING CYCLES
SEWING MACHINES
SEWING NEEDLES
SHIRRING
STITCHES (SEWING)
STITCHING
TACKING
TAILORING
TAKE UP (SEWING)
TUCKING (GARMENT MANUFACTURE)

CHAIN STITCHES
NT SINGLE THREAD CHAIN STITCHES
BT STITCHES (SEWING)
RT LOCKSTITCHES
STITCH BONDING

CHAINLESS MERCERIZERS
BT MERCERIZING MACHINES
RT CHAIN MERCERIZERS
MERCERIZING
WET FINISHING

CHAINS
RT ACCESSORIES
(CONT.)

RT BELTS
DEVICES
MECHANISMS

CHAMBRAY
BT FLAT WOVEN FABRICS
WOVEN FABRICS
RT PLAIN WEAVES

CHANTILLY (LACE)
BT HAND MADE LACES
PILLOW LACES (BOBBIN LACES)

CHARACTER RECOGNITION
RT FILM READING
INFORMATION RETRIEVAL
INFORMATION SYSTEMS
OPTICAL SCANNING

CHARACTERISTICS
RT AESTHETIC PROPERTIES
CHEMICAL PROPERTIES
MECHANICAL PROPERTIES
PERFORMANCE
PHYSICAL PROPERTIES (EXCLUDING MECHANICAL)
PROPERTIES
QUALITY IMPROVEMENT

CHARGE (ELECTRICAL)
RT CAPACITANCE (ELECTRICAL)
DIPOLES
POLARITY

CHARGING
RT CONVEYORS
DELIVERY RATE
FEED RATE
FEED ROLLS
FEEDERS
HOPPER FEED
INPUT
LOADING (PROCESS)
THROUGHPUT

CHARMANETTE SATIN
BT FLAT WOVEN FABRICS
WOVEN FABRICS
RT TWILL WEAVES

CHASE
USE TRAVERSE

CHECK SHIRTING
BT FLAT WOVEN FABRICS
WOVEN FABRICS

CHECK SPRINGS
RT SEWING CYCLES
SEWING MACHINES
TAKE UP MECHANISMS
TENSION DEVICES
THREAD SLACK

CHECK STRAPS
USE PICKER STRAPS

CHECKER BOARD WEAVES
BT WEAVE (TYPES)
RT FLAT WOVEN FABRICS

CHEESE CLOTH
BT FLAT WOVEN FABRICS
WOVEN FABRICS
RT PLAIN WEAVES

CHEESE DYEING
BT PACKAGE DYEING
YARN DYEING
RT BEAM DYEING (YARN)
BOBBIN DYEING
CAKE DYEING
CHEESE DYEING MACHINES
CONE DYEING
COP DYEING
HANK DYEING
MUFF DYEING
PRESSURE DYEING

CHEESE DYEING MACHINES
BT PACKAGE DYEING MACHINES
YARN DYEING MACHINES
RT BEAM DYEING MACHINES (YARN)
BOBBIN DYEING MACHINES
CAKE DYEING MACHINES
CHEESE DYEING
CONE DYEING MACHINES
COP DYEING MACHINES
HANK DYEING MACHINES
WARP DYEING MACHINES

CHEESE WINDERS
BT WINDERS
RT CHEESES
CLOSE WIND
COMBINATION WIND
CONERS
CONES
DRUM WINDERS
FILLING WIND
OPEN WIND
PACKAGES
PRECISION WINDERS
QUILLERS
QUILLS
SLUB CATCHERS
STOP MOTIONS
TENSION CONTROL
TENSION DEVICES
THREAD GUIDES
WARP WIND
WARPERS (MACHINE)

CHEESE WINDING
BT WINDING
RT AUTOMATIC KNOTTING
CHEESES
CONES
DOFFING (PACKAGE)
PACKAGES
PACKAGING (OPERATION)
QUILLING
SLUB CATCHING
TENSION CONTROL

CHEESES
BT PACKAGES
SELF SUPPORTING PACKAGES
RT BEAMS
BOBBINS
CAKES
CHEESE WINDERS
CHEESE WINDING
COMBINATION WIND
CONES
CREELS
JACKSPOOLS
LARGE PACKAGES
PACKAGE DYEING
PACKAGING (OPERATION)
PRECISION WINDERS
QUILLS
SPINDLES
WARP WIND
WARPERS (MACHINE)
WINDING

CHELATES
RT COBALT COMPOUNDS
NICKEL COMPOUNDS
ORGANOMETALLIC COMPOUNDS (EXCLUDING SILICONES)

CHELATING
UF CHELATION
SEQUESTERING
BT REACTIONS (CHEMICAL)
RT BLEACHING
CALGON (TN)
COMPLEXING
DYEING
EDTA
HARD WATER
METALLIZED DYES

CHELATING AGENTS
UF SEQUESTERING AGENTS
RT BUILDERS (DETERGENTS)
CALGON (TN)
CITRIC ACID
COMPLEXING
EDTA
METALLIZED DYES
PYROPHOSPHATES
TETRAPOTASSIUM PYROPHOSPHATE
TSPP

CHELATION
USE CHELATING

CHEMICAL ANALYSIS
NT CHROMATOGRAPHY
COLORIMETRIC ANALYSIS
CONDUCTOMETRY
GRAVIMETRIC ANALYSIS
MASS SPECTROMETRY
POLAROGRAPHY
VOLUMETRIC ANALYSIS
RT ANALYZING
CHEMICAL COMPOSITION
DETERMINATION
FIBER IDENTIFICATION
FRACTIONATION
IODINE NUMBER
PHYSICAL ANALYSIS
QUALITATIVE ANALYSIS
QUANTITATIVE ANALYSIS
REAGENTS
SPECTROPHOTOMETERS
SPECTROSCOPY
SPECTRUM ANALYSIS
STATISTICAL ANALYSIS
STRESS ANALYSIS
STRUCTURAL ANALYSIS
THERMAL ANALYSIS
X RAY ANALYSIS

CHEMICAL ANALYSIS (FIBER IDENTIFICATION)
BT FIBER IDENTIFICATION
RT CHEMICAL COMPOSITION
MICROSCOPIC ANALYSIS (FIBER IDENTIFICATION)
REAGENT TESTING (FIBER IDENTIFICATION)
SPECIFIC GRAVITY TESTING (FIBER IDENTIFICATION)
STAINING TESTING (FIBER IDENTIFICATION)
SWELLING TESTING (FIBER IDENTIFICATION)
THERMAL TESTING (FIBER IDENTIFICATION)

CHEMICAL BONDS
NT COVALENT BONDS
DISULFIDE BONDS
ELECTROSTATIC BONDS
HYDROGEN BONDS
HYDROPHOBIC BONDS
IONIC BONDS
METALLIC BONDS
RT BOND DISSOCIATION ENERGY
BOND ENERGY
BONDING
VAN DER WAALS FORCES

CHEMICAL COMPOSITION
RT CHEMICAL ANALYSIS
CHEMICAL ANALYSIS (FIBER IDENTIFICATION)

CHEMICAL ENGINEERING
RT COLLOID CHEMISTRY
COLLOIDS
CONTROL SYSTEMS
DIFFUSION
DISTILLATION
EXTRACTION
INSTRUMENTATION
OSMOSIS
PERMEABILITY
PROCESS CONTROL
PROCESS DESIGN
SEPARATION (SOLUTION)
SORPTION
SURFACE CHEMISTRY
UNIT OPERATIONS

CHEMICAL MODIFICATION (FIBERS)
BT REACTIONS (CHEMICAL)
RT ACETYLATED COTTON
ACETYLATION
ACRYLAMIDE
ACRYLONITRILE
BENZYLATION
CARBAMOYLETHYLATED COTTON
CARBOXYMETHYLATION
CELLULOSE ESTERS
CELLULOSE ETHERS
CHEMICALLY MODIFIED CELLULOSIC FIBERS
CHEMICALLY MODIFIED COTTON
CHEMICALLY MODIFIED PROTEIN FIBERS
COPOLYMERIZATION
CROSSLINKING
CYANOETHYLATED COTTON
CYANOETHYLATION
CYANOETHYLCELLULOSE
CYSTINE
DECRYSTALLIZED COTTON
DIALDEHYDE CELLULOSE
DIAZOMETHANE
DIFFERENTIAL DISTRIBUTION (CROSSLINKING)
DS (CELLULOSE)
ESTERIFICATION
ETHERIFICATION
ETHYLENE OXIDE
FIBER REACTIVITY
FINISHING PROCESS (GENERAL)
GRAFTING
METHYLATION
POLYMER DEPOSITION
REACTANTS
VINYL COMPOUNDS
WASH WEAR TREATMENTS

CHEMICAL PROPERTIES
NT ACID BINDING
ACID SOLUBILITY
ALKALI SOLUBILITY
BIODEGRADABILITY
CHEMICAL STABILITY
COPPER NUMBER
CUPRAMMONIUM FLUIDITY
DYEING PROPERTIES (GENERAL)
FINISH (PROPERTY)
ISOELECTRIC POINT
ISOIONIC POINT
SOLUBILITY
UREA BISULFITE SOLUBILITY
RT ACID DEGRADATION
AESTHETIC PROPERTIES

RT AGING (MATERIALS)
ALKALI DEGRADATION
BIOCHEMICAL PROPERTIES
CHARACTERISTICS
CHLORINE DAMAGE
DEGRADATION
DURABLE PRESS
DYEABILITY
END USE PROPERTIES
FABRIC PROPERTIES
FABRIC PROPERTIES (PHYSICAL EXCLUDING MECHANICAL)
FIBER PROPERTIES
FIRE RETARDANCY
HYDROPHILIC FIBERS
HYDROPHOBIC FIBERS
LAUNDRY DEGRADATION
MECHANICAL PROPERTIES
MERCERIZED FINISH
MICROBIOLOGICAL DEGRADATION
PHOTOCHEMICAL DEGRADATION
PHYSICAL CHEMICAL PROPERTIES
PHYSICAL PROPERTIES (EXCLUDING MECHANICAL)
PROPERTIES
REACTIVITY
SORPTION OF GASES
SORPTION OF LIQUIDS
SORPTION OF WATER
SURFACE AREA
SWELLING
THERMAL DEGRADATION
VISCOSITY
WEATHERING

CHEMICAL REACTIONS
USE REACTIONS (CHEMICAL)

CHEMICAL RESISTANT LEATHER
BT LEATHER
RT CHEMICAL STABILITY

CHEMICAL STABILITY
UF STABILITY TO CHEMICALS
BT CHEMICAL PROPERTIES
RT ACID HYDROLYSIS
ALKALI HYDROLYSIS
CHEMICAL RESISTANT LEATHER
CHLORINE FASTNESS (OF FINISH)
COLORFASTNESS
DECOMPOSITION
DEGRADATION
DEGRADATION PROPERTIES
HYDROLYSIS
OXIDATION
STABILITY
STABILIZERS (AGENTS)
THERMAL STABILITY

CHEMICAL TESTING
BT TESTING

CHEMICALLY MODIFIED CELLULOSIC FIBERS
NT CHEMICALLY MODIFIED COTTON
BT CELLULOSIC FIBERS
RT CELLULOSE DERIVATIVES
CHEMICAL MODIFICATION (FIBERS)
CHEMICALLY MODIFIED PROTEIN FIBERS

CHEMICALLY MODIFIED COTTON
NT ACETYLATED COTTON
AMINIZED COTTON
BENZOYLATED COTTON
CARBAMOYLETHYLATED COTTON
CARBOXYMETHYLATED COTTON
CYANOETHYLATED COTTON
DIAZOBENZOYLATED COTTON
VINYLATED COTTON
BT CHEMICALLY MODIFIED CELLULOSIC FIBERS
RT BENZYLATION
CELLULOSE ESTERS
CELLULOSE ETHERS
CHEMICAL MODIFICATION (FIBERS)
COTTON
CYANOETHYLATION
CYANOETHYLCELLULOSE
DECRYSTALLIZED COTTON
DS (CELLULOSE)
POLYMER DEPOSITION
REACTANTS
VINYLATION
WASHFASTNESS (OF FINISH)

CHEMICALLY MODIFIED PROTEIN FIBERS
NT CHLORINATED WOOL
BT PROTEIN FIBERS
RT CHEMICAL MODIFICATION (FIBERS)
CHEMICALLY MODIFIED CELLULOSIC FIBERS

CHEMICKING
USE CHLORINE BLEACHING AGENTS

CHEMILUMINESCENCE
USE LUMINESCENCE

CHEMISORPTION
BT SORPTION
RT PHYSICAL CHEMICAL PROPERTIES
SORPTION OF DYES
SORPTION OF GASES
SORPTION OF LIQUIDS
SORPTION OF WATER

CHENILLE (YARN)
USE CHENILLE YARNS

CHENILLE AXMINSTER CARPETS
USE CHENILLE CARPETS

CHENILLE CARPETS
UF CHENILLE AXMINSTER CARPETS
BT CARPETS
PILE FABRICS (WOVEN)
WOVEN CARPETS
WOVEN FABRICS
RT AXMINSTER CARPETS
CHENILLE YARNS
FLOCKED CARPETS
NONWOVEN CARPETS
TAPESTRIES
TUFTED CARPETS
VELVET CARPETS
WILTON CARPETS

CHENILLE FABRICS (TUFTED FABRIC)
BT TUFTED FABRICS
RT TUFTED CARPETS

CHENILLE YARNS
UF CHENILLE (YARN)
RT CHENILLE CARPETS

CHEVIOT TWEED
BT WOVEN FABRICS
RT BANNOCK BURN

CHEVIOT WOOL
BT MEDIUM WOOL
WOOL

CHEVRON TWILLS
BT TWILLS
RT FANCY TWILLS
HERRINGBONE TWILLS
LOW TWILLS
REGULAR TWILLS
S TWILLS
STEEP TWILLS
TWILL ANGLES
TWILL WEAVES
Z TWILLS

CHI SQUARED DISTRIBUTION
RT BINOMIAL DISTRIBUTION
DISTRIBUTION
F DISTRIBUTION
NORMAL DISTRIBUTION
T DISTRIBUTION

CHIFFON
BT FLAT WOVEN FABRICS
WOVEN FABRICS
RT PLAIN WEAVES

CHILDRENS WEAR
RT INFANTS CLOTHING
PLAYSUITS
SPORTSWEAR

CHINCHILLA MACHINES
RT DRY FINISHING
NAPPING MACHINES
PILE FABRICS
SUEDING MACHINES

CHINTZ
BT FLAT WOVEN FABRICS
WOVEN FABRICS
RT COTTON
PLAIN WEAVES

CHIPPING
RT BREAKING
COMMINUTION
CRACK PROPAGATION
CUTTING
DELAMINATING
FLAKING
FRACTURING
GRINDING (COMMINUTION)
GRINDING (MATERIAL REMOVAL)
SHARPENING
SPLITTING

CHITIN
BT NATURAL POLYMERS
RT CARBOHYDRATES
DEACETYLATED CHITIN
POLYMERS

CHLORAL
BT ALDEHYDES

CHLORATES
NT SODIUM CHLORATE
RT CHLORINE COMPOUNDS
OXIDIZING AGENTS
PERCHLORATES

CHLORIDES
NT BERYLLIUM CHLORIDE
BISMUTH CHLORIDE
BORON TRICHLORIDE
CADMIUM CHLORIDE
CALCIUM CHLORIDE
CUPRIC CHLORIDE
CUPROUS CHLORIDE
FERRIC CHLORIDE
GOLD CHLORIDE
LITHIUM CHLORIDE
MAGNESIUM CHLORIDE
MERCURIC CHLORIDE
POTASSIUM CHLORIDE
SODIUM CHLORIDE
THPC
ZINC CHLORIDE
BT HALIDES
RT BROMIDES
CHLORINE
CHLORINE COMPOUNDS
FLUORIDES
HYDROCHLORIC ACID
IODIDES

CHLORINATED HYDROCARBONS
NT CARBON TETRACHLORIDE
CHLOROFORM
DICHLOROBENZENE
METHYLENE CHLORIDE
PERCHLOROETHYLENE
PERCLENE (TN)
TRICHLOROBENZENE
TRICHLOROETHYLENE
BT HALOGEN COMPOUNDS
RT CARRIERS (DYEING)
CHLORINATED SOLVENTS
CHLORINE
SOLVENTS

CHLORINATED SOLVENTS
BT SOLVENTS
RT CARBON TETRACHLORIDE
CHLORINATED HYDROCARBONS
CHLORINE
CHLOROFORM
DICHLOROBENZENE
DRY CLEANING
DRY CLEANING SOLVENTS
METHYLENE CHLORIDE
PERCLENE (TN)
PERCHLOROETHYLENE
TRICHLOROBENZENE
TRICHLOROETHYLENE

CHLORINATED WOOL
BT CHEMICALLY MODIFIED PROTEIN FIBERS
RT ANTIFELTING TREATMENTS
CHLORINATION (SHRINK PROOFING)
CHLORINE DAMAGE
SHRINK PROOFING
SHRINK RESISTANCE
SHRINK RESISTANCE FINISHES
WOOL

CHLORINATION (CHEMICAL REACTION)
BT HALOGENATION
REACTIONS (CHEMICAL)
RT CHLORINATION (SHRINK PROOFING)
CHLORINE
CHLORINE COMPOUNDS

CHLORINATION (SHRINK PROOFING)
BT SHRINK PROOFING
RT ANTIFELTING TREATMENTS
CHLORINATED WOOL
CHLORINATION (CHEMICAL REACTION)
CHLORINE
OXIDATION
PERMANGANATES
WOOL

CHLORINE
RT BROMINE
CHLORIDES
CHLORINATED HYDROCARBONS
CHLORINATED SOLVENTS
CHLORINATION (CHEMICAL REACTION)
CHLORINATION (SHRINK PROOFING)
CHLORINE BLEACHING AGENTS
CHLORINE COMPOUNDS
CHLORINE DAMAGE
CHLORINE RETENTION
IODINE

CHLORINE BLEACHING AGENTS
UF CHEMICKING
BT BLEACHING AGENTS
RT BLEACHING
(CONT.)

RT CHLORINE DAMAGE
CHLORINE DIOXIDE
CHLORINE FASTNESS (COLOR)
CHLORINE FASTNESS (OF FINISH)
CHLORITES
CHLOROCYANURIC ACID
HYPOBROMITES
HYPOCHLORITES
LITHIUM HYPOCHLORITE
OXIDIZING AGENTS
SODIUM HYPOCHLORITE

CHLORINE COMPOUNDS
NT ALLYL CHLORIDE
BENZYL CHLORIDE
CHLORINATED HYDROCARBONS
CHLOROACETIC ACID
CHLOROCYANURIC ACID
CHLOROHYDRINS
CYANURIC CHLORIDE
METHALLYL CHLORIDE
VINYL CHLORIDE
BT HALOGEN COMPOUNDS
RT ACID CHLORIDES
CHLORATES
CHLORIDES
CHLORINATION (CHEMICAL REACTION)
CHLORINE
CHLORINE DIOXIDE
CHLORITES
HALIDES
HYDROCHLORIC ACID
HYPOCHLORITES
PERCHLORATES
PERCHLORIC ACID

CHLORINE DAMAGE
UF CHLORINE RETENTION DAMAGE
BT DEGRADATION
RT BLEACHING
BLEACHING AGENTS
CHEMICAL PROPERTIES
CHLORINATED WOOL
CHLORINE
CHLORINE BLEACHING AGENTS
CHLORINE DIOXIDE
CHLORINE FASTNESS (COLOR)
CHLORINE FASTNESS (OF FINISH)
CHLORINE RETENTION
CHLORITES
DECOMPOSITION
HYPOCHLORITES
LAUNDRY DEGRADATION
OXIDATION
SCORCHING

CHLORINE DIOXIDE
BT OXIDES
RT BLEACHING
CHLORINE BLEACHING AGENTS
CHLORINE COMPOUNDS
CHLORINE DAMAGE
CHLORITES
HYPOCHLORITES
OXIDATION
OXIDIZING AGENTS
PEROXIDES

CHLORINE FASTNESS (COLOR)
BT COLORFASTNESS
DEGRADATION PROPERTIES
END USE PROPERTIES
RT BLEACHING
CHLORINE BLEACHING AGENTS
CHLORINE DAMAGE
CHLORINE FASTNESS (OF FINISH)
DURABLE FINISH (GENERAL)
DYEING
LIGHTFASTNESS (COLOR)
WASHFASTNESS (COLOR)

CHLORINE FASTNESS (OF FINISH)
BT DEGRADATION PROPERTIES
END USE PROPERTIES
RT BLEACHING
CHLORINE BLEACHING AGENTS
CHLORINE DAMAGE
CHLORINE FASTNESS (COLOR)
CHLORINE RETENTION
DURABILITY
DURABLE FINISH (GENERAL)
LAUNDERABILITY

CHLORINE RETENTION
RT CHLORINE
CHLORINE DAMAGE
CHLORINE FASTNESS (OF FINISH)
WASH WEAR FINISHES
WHITENESS RETENTION

CHLORINE RETENTION DAMAGE
USE CHLORINE DAMAGE

CHLORITES
RT BLEACHING
CHLORINE BLEACHING AGENTS
CHLORINE COMPOUNDS
CHLORINE DAMAGE
CHLORINE DIOXIDE
HYPOCHLORITES
OXIDATION
OXIDIZING AGENTS
PEROXIDES

CHLOROACETIC ACID
UF TRICHLOROACETIC ACID
BT CARBOXYLIC ACIDS
CHLORINE COMPOUNDS
ORGANIC ACIDS

CHLOROACETYLAMINO DYES
BT REACTIVE DYES

CHLOROBENZOTHIAZOLE DYES
BT REACTIVE DYES

CHLOROCYANURIC ACID
UF DICHLOROCYANURIC ACID
TRICHLOROCYANURIC ACID
BT CHLORINE COMPOUNDS
TRIAZINES
RT ANTIFELTING AGENTS
CHLORINE BLEACHING AGENTS
CYANURIC CHLORIDE
SHRINK RESISTANCE FINISHES

CHLOROETHANOL
USE ETHYLENE CHLOROHYDRIN

CHLOROFORM
BT CHLORINATED HYDROCARBONS
RT CHLORINATED SOLVENTS

CHLOROHYDRINS
NT CHLOROPROPANOL
DICHLOROHYDRIN
ETHYLENE CHLOROHYDRIN
GLYCERIN DICHLOROHYDRIN
GLYCERIN MONOCHLOROHYDRIN
BT ALCOHOLS
CHLORINE COMPOUNDS
HALOHYDRINS
REACTANTS
RT EPICHLOROHYDRIN
RESIN FINISHES
SOLVENTS

CHLOROPROPANOL
BT CHLOROHYDRINS

CHLOROPYRIMIDINE DYES
UF DICHLOROPYRIMIDINE DYES
DRIMARENE (TN)
REACTONE (TN)
TRICHLOROPYRIMIDINE DYES
BT REACTIVE DYES

CHLOROTRIAZINE DYES
UF CIBACRON (TN)
DICHLOROTRIAZINE DYES
MONOCHLOROTRIAZINE DYES
PROCION (TN)
PROCION H (TN)
BT REACTIVE DYES
RT TRIAZINES

CHLOROUS ACID
BT INORGANIC ACIDS

CHOPPING
USE CUTTING

CHROMA
USE SATURATION (COLOR)

CHROMATICITY DIAGRAM
BT CIE SYSTEM
RT COLOR
COLORIMETRY
SPECTROPHOTOMETRY
TRICHROMATIC COEFFICIENTS
TRISTIMULUS VALUES

CHROMATOGRAPHY
BT CHEMICAL ANALYSIS
RT ABSORBENCY (MATERIAL)
ABSORPTION (MATERIAL)
ADSORPTION
PARTITION COEFFICIENT
SORPTION
WICKING

CHROME
USE CHROMIUM COMPOUNDS

CHROME DYEING
USE MORDANT DYEING

CHROME DYES
USE MORDANT DYES

CHROME PRINTING COLORS
USE MORDANT DYES

CHROMIUM COMPOUNDS
UF CHROME
NT CHROMIUM FLUORIDE
DICHROMATES
SODIUM DICHROMATE
RT MORDANT DYES
TANNING AGENTS

CHROMIUM FLUORIDE
BT CHROMIUM COMPOUNDS
FLUORIDES
RT MORDANT DYEING

CHROMOTROPISM
RT FADING
METACHROMASY

CHRYSOTILE
USE ASBESTOS

CIBACRON (TN)
USE CHLOROTRIAZINE DYES

CIBAPHASOL (TN)
USE COACERVATE SYSTEMS

CIE SYSTEM
UF ICI SYSTEM
NT CHROMATICITY DIAGRAM
TRICHROMATIC COEFFICIENTS
TRISTIMULUS VALUES
BT COLOR DESIGNATION
RT COLOR
COLOR MATCHING
COLORIMETRY
JUDD NBS COLOR SYSTEM
MUNSELL COLOR SYSTEM

CIGARETTE FILTERS
RT FILTER CLOTH
FILTER PAPER
FILTERS (FLUID)
FILTRATION
TOW

CIRCULAR COMBING
USE NOBLE COMBING

CINNAMIC ACID
BT CARBOXYLIC ACIDS
ORGANIC ACIDS

CIRCULAR COMBS
USE NOBLE COMBS

CIRCULAR HOSIERY
BT HOSIERY
RT BOARDING
DENIER
FOOTWEAR
FULLY FASHIONED HOSIERY
HALF HOSE
HOSIERY MACHINES
KNITTING
KNITTING MACHINES
LADDERING
MOCK FASHIONING MARKS
MOCK SEAMS
PRESSING
RUN RESISTANCE
RUN RESISTANT HOSIERY
SEAMLESS HOSIERY
SHAPING
SHEERNESS
SOCKS
STOCKINGS

CIRCULAR KNITTED FABRICS
UF TUBULAR KNITTED FABRICS
BT FILLING KNITTED FABRICS
KNITTED FABRICS
TUBULAR FABRICS
RT DOUBLE KNIT FABRICS
FULL CARDIGAN FABRICS
HALF CARDIGAN FABRICS
KNITTING
MACHINE KNITTING
PLAIN KNITTED FABRICS
RIB KNITTED FABRICS

CIRCULAR KNITTING
BT FILLING KNITTING
KNITTING
RT BOARDING
CARDIGANS
FILLING KNITTING MACHINES
FLAT KNITTING
FULLY FASHIONED KNITTING
JACQUARD KNITTING
JERKINS
JUMPERS
MOCK FASHIONING MARKS
PLAIN KNITTED FABRICS
PULL OVERS
SEAMLESS HOSIERY
SLIP ONS
SWEATERS
TORSIONAL STABILITY

CIRCULAR KNITTING MACHINES
NT SEAMLESS HOSIERY MACHINES
BT FILLING KNITTING MACHINES
KNITTING MACHINES
RT CYLINDER NEEDLES
DIAL NEEDLES
FILLING KNITTING
FLAT KNITTING MACHINES
FULLY FASHIONED KNITTING MACHINES
HOSIERY MACHINES
KNITTING
KNITTING NEEDLES
LINKS LINKS MACHINES
MOCK SEAMS
NEEDLE CAMS
PATTERN WHEELS
POSITIVE FEED
SEAMLESS HOSIERY
SINKERS
SPIRALITY (KNITTED FABRICS)
STITCH CAMS
WARP KNITTING MACHINES
WEBHOLDERS
YARN CHANGING UNIT (KNITTING)

CIRCULAR LOOMS
BT LOOMS
RT WEAVING

CIRCULAR TRAVELERS
BT SPINNING TRAVELERS
TRAVELERS
RT ELLIPTICAL TRAVELERS
NYLON TRAVELERS
PLYING TRAVELERS
RING SPINNING
RINGS
SPINNING
STEEL TRAVELERS
TRAVELER CHATTER
TWIST
TWISTING
YARN TENSION

CIRCULARITY
RT DIAMETER

CIRCULATION
RT CIRCULATION PUMPS
CURRENTS
DELIVERY
DISPERSING
DISSOLVING
MIXING
SEPARATION
STIRRING

CIRCULATION PUMPS
RT CIRCULATION
PACKAGE DYEING
YARN DYEING

CIRCUMFERENCE
RT CROSS SECTIONAL AREA
DIAMETER
PERIMETER

CITRIC ACID
BT CARBOXYLIC ACIDS
ORGANIC ACIDS
TRIBASIC ACIDS
RT CHELATING AGENTS

CLASS (FIBERS)
NT COTTON CLASS
WOOL CLASS
RT CLASSING
COEFFICIENT OF FINENESS VARIATION
COLOR GRADING
FIBER FINENESS
FIBER LENGTH
FIBERS
GRADE (FIBERS)
GRADING
IMMATURITY (FIBER)
LINEAR DENSITY

CLASSING
RT CLASS (FIBERS)
COLOR GRADING
COTTON CLASS
FIBER TENDERNESS
FINENESS TESTING
GRADING
TESTING
WOOL CLASS

CLAYS
UF KAOLIN
NT MONTMORILLONITE
RT ARTIFICIAL SOIL
CERAMICS
SILICATES
SOIL
SYNTHETIC SOIL

CLEANERS
RT BLOWERS
TRAVELING BLOWERS
VACUUM CLEANERS

CLEANING
NT BACKWASHING
DEGREASING
DESIZING
DEWAXING (COTTON)
DRY CLEANING
GINNING
LAUNDERING
SCOURING
SOIL REMOVAL
STAIN REMOVAL
RT ABRASION
ANTIREDEPOSITION AGENTS
BATHS
BLEACHING
BLEACHING AGENTS
CARBONIZING
CLEANLINESS
DEODORIZING
DETERGENTS
DRY CLEANING MACHINES
JET SCOURING
PICKING (OPENING)
POLISHING
SCRUBBING
ULTRASONICS

CLEANLINESS
UF CLEANNESS
CLEARNESS
RT AESTHETIC PROPERTIES
BLEACHING
BURR PICKING
CARDING
CLEANING
CROSS ROLLS
DRY CLEANING
FUZZINESS
IMPURITIES
LAUNDERING
OIL STAINS
OPENING
PERALTA ROLLS
PICKING (OPENING)
SCOURING
SHEDDING (FIBER)
STAIN RESISTANCE AGENTS
TRASH
WASTE

CLEANNESS
USE CLEANLINESS

CLEARANCES
RT ADJUSTMENTS
ALIGNMENT
ECCENTRICITY
FITTING
SEPARATION

CLEARERS (CARD)
UF CARD CLEARERS
RT BURR BEATERS
BURR CRUSHERS
BURRS
CARD CYLINDERS
CARD WASTE
CARDING
CARDS
CLEARERS (LINT)
COTTON CARDS
MOTE KNIVES
NEPS
TRASH
WOOLEN CARDING
WOOLEN CARDS

CLEARERS (LINT)
RT CLEARERS (CARD)
DRAWFRAMES
LAP UP
LINT (FIBER)
LINT (FLY)
ROVING FRAMES
SPINNING FRAMES
TOP ROLLS

CLEARERS (YARN)
USE SLUB CATCHERS

CLEARING (YARN)
USE SLUB CATCHING

CLEARING LEASE
BT LEASE
RT SILK SYSTEM WARPING

CLEARNESS
USE CLEANLINESS

CLEMSON STRENGTH TESTING
UF CLEMSON STRENGTH TESTS
BT TESTING
RT FIBER BUNDLE STRENGTH
FIBER STRENGTH
PRESSLEY TEST

CLEMSON STRENGTH TESTS
USE CLEMSON STRENGTH TESTING

CLERICAL CLOTHES
BT CEREMONIAL CLOTHING

CLICKERS
BT CUTTING MACHINES

CLIP TENTERS
BT TENTER FRAMES
RT DRY FINISHING
TENTERING
WET FINISHING

CLIPPINGS
RT CUTTING (TAILORING)
FLY (WASTE)
GARNETTING
NOILS
TAILORING
TRASH
WASTE
WASTE MACHINES

CLIPSPOT WEAVES
BT WEAVE (TYPES)
RT FLAT WOVEN FABRICS

CLOAKS
USE COATS

CLOCKWISE
RT ANTICLOCKWISE
SENSE (DIRECTION)

CLOSE FACE FINISH
USE FACE FINISH

CLOSE WIND
RT CHEESE WINDERS
CLOSE WINDING
COMBINATION WIND
CONING
DRUM WINDERS
DRUM WINDING
FILLING WIND
LAYER LOCKING
NOSE BUNCHING
OPEN WIND
OPEN WINDING
PACKAGES
PRECISION WINDERS
PRECISION WINDING
QUILLING
RIBBON BREAKING
ROVING WIND
WARP WIND
WARPING
WIND
WINDING
WINDING ANGLE

CLOSE WINDING
BT WINDING
RT BUILDER MOTIONS
CLOSE WIND
CONING
CROSS WINDING
DRUM WINDING
GAIN (WINDING)
OPEN WIND
OPEN WINDING
PACKAGES
PACKAGING (OPERATION)
PARALLEL WINDING
PRECISION WINDING
RIBBON BREAKING
TRAVERSE

CLOSED LOOP CONTROL
RT CONTROL SYSTEMS
CONTROL THEORY

CLOSED SHED
BT SHED (WEAVING)
RT HARNESSES
LOOMS
NEGATIVE SHEDDING
OPEN SHED
PICKING (WEAVING)
POSITIVE SHEDDING
SHEDDING
SHUTTLE SMASH
SHUTTLES
SPLIT SHED
WEAVING

CLOSED STITCHES
BT KNITTED STITCHES
RT OPEN STITCHES
TRICOT STITCHES

CLOSURES
USE FASTENERS

CLOSURES (FOOTWEAR)
BT FOOTWEAR COMPONENTS

CLOTH
USE FABRICS

CLOTH ANALYSIS
USE FABRIC ANALYSIS

CLOTH FELL
UF FELL (CLOTH)
RT BEATING UP
BOUNCE (WEAVING)
CRIMP INTERCHANGE
DENTS
JAMMING
LOOM TAKE UP
LOOMS
PICKING (WEAVING)
REEDS
SHEDDING
STARTING MARKS
TEMPLES
WARP ENDS
WEAVING
WOVEN FABRICS

CLOTH GEOMETRY
USE FABRIC GEOMETRY

CLOTH GUIDES
UF FABRIC GUIDES
GUIDERS (CLOTH)
NT CENTER GUIDES
EDGE GUIDES
TEMPLES
TENTER CLIPS
TENTER PINS
BT GUIDES
RT COMPENSATORS (CLOTH)

CLOTH INSPECTION
USE FABRIC INSPECTION

CLOTH ROLLS
RT LOOM TAKE UP
LOOMS
PIECE DYEING
WEAVING

CLOTH STRUCTURE
USE FABRIC STRUCTURE

CLOTHING
USE GARMENTS

CLUNY (LACE)
BT CROCHETED LACES
HAND MADE LACES

CLUTCHES
RT ACCESSORIES
BRAKES (FOR ARRESTING MOTION)
MECHANISMS

CMC
USE CRITICAL MICELLE CONCENTRATION

COACERVATE SYSTEMS
UF CIBAPHASOL (TN)
RT DYEING

COAGULATING
RT AGGREGATION
BATHS
COAGULATING BATHS
DEPOSITION
GELLING
NEUTRALIZING (FIBER EXTRUSION)
PRECIPITATION
SOLIDIFICATION
THICKENING
THICKENING AGENTS
WET SPINNING
ZETA POTENTIAL

COAGULATING BATHS
UF SPINNING BATHS
RT FILAMENT YARNS
MAN MADE FIBERS
MONOFILAMENT YARNS
NEUTRALIZING (FIBER EXTRUSION)
REGENERATED FIBERS (EXCLUDING CELLULOSIC AND PROTEIN)
SOLVENT SPINNING
SPINNERETS
SPINNING (EXTRUSION)
SYNTHETIC FIBERS
THICKENING AGENTS
WET SPINNING

COAL TAR
RT CARBON

COARSE YARNS
BT YARNS
RT FINE YARNS
HEAVY WEIGHT
HIGH TWIST
LIGHT WEIGHT
LOW TWIST
MEDIUM WEIGHT
PLIED YARNS
SINGLES YARNS

COARSENESS
USE FINENESS

COATED FABRICS
UF FABRICS (COATED)
NT BUNA N COATED FABRICS
BUTYL COATED FABRICS
LACQUER COATED FABRICS
NEOPRENE COATED FABRICS
RUBBER COATED FABRICS
SBR COATED FABRICS
VINYL COATED FABRICS
BT COMPOSITES
FABRICS (ACCORDING TO STRUCTURE)
RT BOMBAZINE
CALENDER COATING
CAST COATING
COATING (PROCESS)
COATINGS (SUBSTANCES)
CORFAM (TN)
DIP COATING
DELAMINATING
EXTRUSION COATING
FOAM BACKED FABRICS
HEAT SEALING
INDIANA CLOTH
KNIFE COATING
LAMINATED FABRICS
LAWN
LEATHER CLOTH
NIP COATING
PLASTICS
PRINT CLOTH
QUILTING
RESINS
ROLLER COATING
SCUFFING
SYNTHETIC LEATHER
UPHOLSTERY FABRICS
WOVEN FABRICS

COATING (PROCESS)
UF BACK COATING
NT CALENDER COATING
CAST COATING
DIP COATING
ELECTROPLATING
EXTRUSION COATING
FLEXIBLE FILM LAMINATING
KISS COATING
KNIFE COATING
NIP COATING
ROLLER COATING
VACUUM COATING
RT BACK FILLING MACHINES
BATHS
COATED FABRICS
COATINGS (SUBSTANCES)
DEPOSITION
DOCTOR BLADES
ELECTRODEPOSITION
IMPREGNATION
ORGANOSOLS
PAPER MAKING
RESINS
SEALING
SUBSTRATES
WATERPROOFING

COATING MACHINES
RT BACK FILLING MACHINES
DOCTOR BLADES
IMPREGNATING MACHINES

COATING RESINS
BT RESINS

COATINGS (SUBSTANCES)
RT BUNA N COATED FABRICS
BUTYL COATED FABRICS
COATED FABRICS
COATING (PROCESS)
COTTON WAX
DEPOSITION
FINISH (SUBSTANCE ADDED)
LACQUER COATED FABRICS
LACQUERS
NATURAL POLYMERS
NEOPRENE COATED FABRICS
PAINTS
POLYMERS
RUBBER
SBR COATED FABRICS
SYNTHETIC LEATHER
SYNTHETIC RUBBER
VINYL COATED FABRICS

COBALT COMPOUNDS
RT CHELATES
COBALT CHLORIDE METHOD
DYE FIXING AGENTS
METALLIZED DYES
OLEFIN DYEING

WATERPROOFING AGENTS
WAXES

COATS
UF CLOAKS
BT GARMENTS
OUTERWEAR
RT DRESSCLOTHING
DRESSES
FROCKS
GLOVES
GOWNS
HOSIERY
JACKETS
MILITARY CLOTHING
OVERCOATS
PONCHOS
RAINWEAR
SCARVES
SUITS
WORK CLOTHING

COBALT CHLORIDE METHOD
RT COBALT COMPOUNDS
KARL FISCHER TITRATION METHOD
MOISTURE CONTENT TESTING
OVEN METHOD (MOISTURE TESTING)

COCKLE (DEFECT)
BT FABRIC DEFECTS
RT BARRE
COCKLED BAR
RELAXATION SHRINKAGE

COCKLE (TAILORING)
USE PUCKERS (TAILORING)

COCKLE (YARN)
UF COCKLE YARNS
KNUCKLE YARNS
BT NOVELTY YARNS
RT STRETCH

COCKLE YARNS
USE COCKLE (YARN)

COCKLED BAR
RT COCKLE (DEFECT)

COCKLED SELVEDGES
USE SLACK SELVEDGES

COCKLING
RT BUCKLING
GATHERING
PINCHING
PUCKERING
RIPPLES
RUFFLING
SEAM PUCKER
SEAMING
SEWING
SHIRRING
STITCHING
STRETCH PUCKERS

COCONUT DIETHANOLAMIDE
BT DIETHANOLAMIDE SURFACTANTS

COCOONS
RT SILK

COEFFICIENT OF CROSS VISCOSITY
BT STRESS STRAIN PROPERTIES
RT FIRMOVISCOSITY
RELAXATION SPECTRUM
RELAXATION TIME (MECHANICAL)
RHEOLOGY
YIELD POINT

COEFFICIENT OF EXPANSION
RT GLASS TRANSITION TEMPERATURE
SECONDARY TRANSITION TEMPERATURES

COEFFICIENT OF FINENESS VARIATION
BT FIBER PROPERTIES
RT CLASS (FIBERS)
FIBER FINENESS
GRADE (FIBERS)
IMMATURITY (FIBER)
MICRONAIRE
STATISTICAL MEASURES

COEFFICIENT OF FRICTION
UF FRICTION COEFFICIENT
BT MECHANICAL PROPERTIES
SURFACE PROPERTIES (MECHANICAL)
RT ABRASION
APPARENT CONTACT AREA

RT ASPERITIES
BELT SLIPPAGE
COEFFICIENT OF VISCOSITY
COHESION
DYNAMIC FRICTION
FIBER FIBER FRICTION
FIBER FRICTION
FIBER SLIPPAGE
FIBER TO METAL FRICTION
FRICTION
FRICTIONAL CHARACTERISTICS
FRICTIONAL FORCE
FRICTION THEORY
HOT SPOTS
HYSTERESIS (MECHANICAL)
INTERNAL FRICTION
NORMAL FORCE
PLASTIC FLOW
PLOUGHING
PROCESS VARIABLES
ROUGHNESS
ROUGHNESS COEFFICIENT
TRUE CONTACT AREA
WELDING
YARN TO METAL FRICTION
YARN TO YARN FRICTION
RHEOLOGY
STATIC FRICTION
STICK SLIP FRICTION

COEFFICIENT OF LENGTH VARIATION
BT FIBER PROPERTIES
RT COEFFICIENT OF VARIATION
COTTON CLASS
DIGITAL FIBROGRAPH
FIBER ARRAY
FIBER DIAGRAM
FIBER LENGTH
FIBER LENGTH DISTRIBUTION
FIBROGRAPH
FIBROGRAPH MEAN LENGTH
MEAN FIBER LENGTH
SERVO FIBROGRAPH
STAPLE FIBERS
STAPLING
UPPER HALF MEAN LENGTH
UPPER QUARTILE LENGTH
VARIANCE

COEFFICIENT OF PARTITION
USE PARTITION COEFFICIENT

COEFFICIENT OF VARIATION
BT STATISTICAL MEASURES
RT COEFFICIENT OF LENGTH VARIATION
CORRELATION COEFFICIENT
DEGREES OF FREEDOM
DISPERSION (STATISTICAL)
DISTRIBUTION
IRREGULARITY
MEAN
MEDIAN
SAMPLING
STANDARD DEVIATION
STANDARD ERROR
STATISTICAL ANALYSIS
STATISTICAL INFERENCE
STATISTICS
VARIANCE

COEFFICIENT OF VISCOSITY
BT STRESS STRAIN PROPERTIES
RT COEFFICIENT OF FRICTION
HYSTERESIS (MECHANICAL)
RHEOLOGY

COHESION
BT MECHANICAL PROPERTIES
RT ADHESION
ADHESIVES
BONDING
BONDING STRENGTH
CEMENTING
COEFFICIENT OF FRICTION
DRAG
ENTANGLEMENTS
FRICTION
FUSION (MELTING)
GLUING
JOINING
LAMINATING
MOLECULAR ATTRACTION
PEELING
PHYSICAL PROPERTIES (EXCLUDING **MECHANICAL)**
SEALING
SEAMING
SELF BONDING FIBERS
THERMOPLASTICITY
WELDING

COILER CANS
USE SLIVER CANS

COILERS
RT CARD SLIVERS
CARDS
COMBED SLIVERS
COMBS
DRAWFRAMES
GILL BOXES
PIN DRAFTERS
SLIVER CANS
SLIVERS
TRUMPETS

COIR
BT CELLULOSIC FIBERS
FRUIT FIBERS

COLD CLIMATE FOOTWEAR
BT FOOTWEAR

COLD DRAWING
BT DRAWING (FILAMENT)
RT DEGREE OF ORIENTATION
DRAW RATIO
DRAWTWISTING
ELASTIC STRAIN
EXTRUSION
HOT DRAWING
MOLECULAR ORIENTATION
NATURAL DRAW RATIO
NECKING (FILAMENT)
PLASTIC FLOW
PLASTIC STRAIN
PREORIENTATION
RHEOLOGY
STRAIN
STRESS
STRESS STRAIN CURVES
STRESS STRAIN PROPERTIES (TENSILE)
STRETCH BREAKING
VISCOELASTICITY
YIELD (MECHANICAL)
YIELD POINT

COLD DYEING
USE LOW TEMPERATURE DYEING

COLD SETTING (SYNTHETICS)
BT WET FINISHING
RT HAIR SETTING
HEAT SETTING (SYNTHETICS)
LOW TEMPERATURE
SETTING (EXCEPT SYNTHETICS)
SWELLING AGENTS

COLLAGEN
NT GUT (COLLAGEN)
BT PROTEINS
RT COLLAGEN FIBERS
SUTURING

COLLAGEN FIBERS
BT AZLON (REGENERATED PROTEIN FIBERS)
PROTEIN FIBERS
RT COLLAGEN
GUT (COLLAGEN)
HIDES
LEATHER
MEDICAL TEXTILES
SUTURING

COLLAPSED BALLOON SPINNING
BT SPINNING
RT BALLOON COLLAPSE
BALLOON STABILITY
RING SPINNING

COLLAR STIFFENERS
BT GARMENT COMPONENTS
RT COLLARS
STIFFENERS (AGENTS)

COLLARS
BT GARMENT COMPONENTS
RT COLLAR STIFFENERS
TAILORING

COLLECTING
USE COLLECTION

COLLECTION
UF COLLECTING
RT ACCUMULATION
ACQUISITION
AGGREGATION
ASSEMBLIES
INTEGRATION
SAMPLING

COLLOID CHEMISTRY
RT CHEMICAL ENGINEERING
ZETA POTENTIAL

COLLOIDAL ALUMINA
BT ALUMINA
RT COLLOIDAL SILICA
SOIL RESISTANCE FINISHES

COLLOIDAL SILICA
NT LUDOX (TN)
SYTON (TN)
BT SILICA
RT ANTISLIP AGENTS
COLLOIDAL ALUMINA

COLLOIDS
NT GELS
SOLS
BT DISPERSE SYSTEMS
RT CHEMICAL ENGINEERING
DISPERSING AGENTS
ELECTROPHORESIS
EMULSIONS
FOAMS (FROTH)
PARTICLE SIZE
SOLUTIONS (LIQUID)
SUSPENSIONS
ZETA POTENTIAL

COLOR
BT OPTICAL PROPERTIES
RT ABSORPTION (RADIATION)
AESTHETIC APPEAL
AESTHETIC PROPERTIES
APPEARANCE
BLOOM
BRIGHTNESS
CHROMATICITY DIAGRAM
CIE SYSTEM
COLOR DESIGNATION
COLOR YIELD
COLORFASTNESS
COLORIMETERS
COLORIMETRY
DAYLIGHT
DESIGN
DOMINANT WAVELENGTH
DRABNESS
DULLNESS
FABRIC DESIGN
FABRIC PROPERTIES
FABRIC PROPERTIES (AESTHETIC)
FABRIC PROPERTIES (PHYSICAL EXCLUDING MECHANICAL)
FLUORESCENCE
HUE
ILLUMINATION
INKS
JUDD NBS COLOR SYSTEM
LIGHT
LUMINESCENCE
LUSTER
LUXURIOUSNESS
METAMERISM
MULTICOLORS
MUNSELL COLOR SYSTEM
PATTERN DEFINITION
PHOTOTROPY
RICHNESS
SATURATION (COLOR)
SHADE
SHEEN
SOLID COLORS
SPARKLE
SPECTROPHOTOMETRY
TOP MATCHING
TRICHROMATIC COEFFICIENTS
TRISTIMULUS VALUES
WAVE LENGTH

COLOR BLEEDING
USE BLEEDING

COLOR BLINDNESS
USE COLOR VISION

COLOR CONTROL
RT AUTOMATIC CONTROL
COLOR STANDARDS

COLOR DESIGNATION
NT CIE SYSTEM
JUDD NBS COLOR SYSTEM
MUNSELL COLOR SYSTEM
RT COLOR
COLOR MATCHING
COLOR STANDARDS
COLOR VISION
COLORIMETRY

COLOR FADING
USE FADING

COLOR GRADE
USE COLOR GRADING

COLOR GRADING
UF COLOR GRADE
BT GRADING
RT CLASS (FIBERS)
CLASSING
COLOR STANDARDS
GRADE (FIBERS)

COLOR MATCHING
NT INSTRUMENTAL COLOR MATCHING
RT CIE SYSTEM
COLOR DESIGNATION
(CONT.)

RT COLOR STANDARDS
COLOR VISION
COLORIMETERS
COLORIMETRY
DAYLIGHT
DOMINANT WAVELENGTH
DYEING
FABRIC ANALYSIS
ILLUMINATION
JUDD NBS COLOR SYSTEM
KUBELKA-MUNK EQUATION
METAMERISM
MUNSELL COLOR SYSTEM
SHADE
SHADING (TAILORING)
SPECTRUM ANALYSIS
SPECTROPHOTOMETRY
TOP MATCHING
TRICHROMATIC COEFFICIENTS
TRISTIMULUS VALUES

COLOR MEASUREMENT
USE COLORIMETRY

COLOR MEASUREMENT INSTRUMENTS
USE COLORIMETERS

COLOR STANDARDS
RT COLOR CONTROL
COLOR GRADING
COLOR MATCHING
COLOR VISION
COLORIMETRY
STANDARDS
FADING

COLOR TEMPERATURE
BT THERMAL PROPERTIES
RT BLACKBODY
EMISSIVITY

COLOR TESTING
USE COLORIMETRY

COLOR TROUGH
RT PRINTING
PRINTING ROLLS
SCREEN PRINTING

COLOR VISION
UF COLOR BLINDNESS
RT COLOR DESIGNATION
COLOR MATCHING
COLOR STANDARDS
MUNSELL COLOR SYSTEM
SHADING (TAILORING)

COLOR YIELD
RT COLOR
DYES
YIELD (MECHANICAL)

COLORANTS
NT DYES
PIGMENTS
RT YELLOWING

COLORAY (TN)
BT CELLULOSIC FIBERS
RAYON (REGENERATED CELLULOSE FIBERS)
RT DOPE DYED YARNS

COLORFASTNESS
UF DYEFASTNESS
FASTNESS (COLOR)
NT CHLORINE FASTNESS (COLOR)
CROCKFASTNESS
DRY CLEANING FASTNESS (COLOR)
GAS FUME FASTNESS
IRONING FASTNESS
LIGHTFASTNESS (COLOR)
RESISTANCE TO BLEEDING
SEAWATER FASTNESS (COLOR)
WASHFASTNESS (COLOR)
BT DEGRADATION PROPERTIES
END USE PROPERTIES
RT BLEACHING
CHEMICAL STABILITY
COLOR
DAYLIGHT
DYE FIXING
DYEING
DYEING DEFECTS
FABRIC DEFECTS
FABRIC PROPERTIES
FABRIC PROPERTIES (PHYSICAL EXCLUDING MECHANICAL)
FADING
LAUNDERABILITY
LEVELNESS (DYEING)
LIGHT
N FADING
NITROGEN OXIDES
O FADING
PERSPIRATION
PERSPIRATION RESISTANCE
SKITTERINESS
RT SPECKINESS
STREAKINESS
SUNLIGHT
THERMAL STABILITY
TIPPINESS
WEATHER RESISTANCE

COLORFASTNESS TESTING
NT BLEEDING TESTING
CROCK TESTING
GAS FUME FASTNESS TESTING
LIGHTFASTNESS TESTING
WASHFASTNESS TESTING
WEATHERING TESTING
BT TESTING
RT FADING
GREY SCALE
XENOTEST (TN)

COLORFASTNESS TO HEAT
USE IRONING FASTNESS

COLORFASTNESS TO HOT PRESSING
USE IRONING FASTNESS

COLORFASTNESS TO LIGHT TESTING
USE LIGHTFASTNESS TESTING

COLORIMETERS
UF COLOR MEASUREMENT INSTRUMENTS
BT MEASURING INSTRUMENTS
OPTICAL INSTRUMENTS
RT COLOR
COLOR MATCHING
COLORIMETRY
OPTICAL PROPERTIES
PHOTOMETERS
REFLECTOMETERS
SPECTROPHOTOMETERS

COLORIMETRIC ANALYSIS
BT CHEMICAL ANALYSIS
RT COLORIMETRY

COLORIMETRY
UF COLOR MEASUREMENT
COLOR TESTING
RT ABSORPTION (RADIATION)
CHROMATICITY DIAGRAM
CIE SYSTEM
COLOR
COLOR DESIGNATION
COLOR MATCHING
COLOR STANDARDS
COLORIMETERS
COLORIMETRIC ANALYSIS
DAYLIGHT
DOMINANT WAVELENGTH
SPECTROPHOTOMETRY
TRICHROMATIC COEFFICIENTS
TRISTIMULUS VALUES

COLORING
USE DYEING

COMB BRUSHES
RT DETACHING ROLLS
HALF LAP
RECTILINEAR COMBS
TOP COMBS

COMB CIRCLES
NT LARGE CIRCLE (COMB)
SMALL CIRCLE (COMB)
RT COMBING
DABBING BRUSHES
NOBLE COMBS
NOIL KNIVES
SETOVER

COMB CYLINDERS
RT DETACHING ROLLS
HALF LAP
LAP PLATES
NIPPER JAWS
NIPPER KNIVES
NIPPER PLATES
RECTILINEAR COMBS
TOP COMBS

COMB LEAD
RT LACE MACHINES

COMB NOILS
USE NOILS

COMB PINS
RT DABBING BRUSHES
DETACHING ROLLS
HALF LAP
LAP PLATES
NIPPER JAWS
NIPPER KNIVES
NIPPER PLATES
NOBLE COMBS
RECTILINEAR COMBS
TOP COMBS

COMB SEGMENTS
RT DETACHING ROLLS
HALF LAP
LAP PLATES
NIPPER JAWS
RECTILINEAR COMBS
TOP COMBS

COMB SPINNING
BT OPEN END SPINNING
SPINNING
RT AIR SPINNING
FLUID SPINNING

COMB TEAR
UF COMB TEARAGE
RT COMBED SLIVERS
COMBING
NOBLE COMBING
NOIL PER CENT
NOIL REMOVAL
NOILS
SETOVER
SLIVER NOIL RATIO
TOP NOIL RATIO
TOPS
YIELD (RETURN)

COMB TEARAGE
USE COMB TEAR

COMB WASTE
USE NOILS

COMBAT SUITS
BT MILITARY CLOTHING

COMBED SLIVERS
UF COMBING SLIVERS
BT FIBER ASSEMBLIES
SLIVERS
RT AUTOLEVELLERS
BALLING HEADS
CARD SLIVERS
COILERS
COMB TEAR
COMBER LAP MACHINES
COMBER LAPS
COMBING
COMBS
DETACHING ROLLS
FIBER ORIENTATION
HALF LAP
NOIL REMOVAL
NOILS
RECTILINEAR COMBS
SLIVER CANS
SLIVER KNITTING
SLIVER TENDERNESS
SLIVER TESTERS
TOP COMBS
TOPS

COMBED TOPS
USE TOPS

COMBED YARNS
BT SPUN YARNS
YARNS
RT CARDED YARNS
COMBING
COTTON SPUN YARNS
WOOLEN SPUN YARNS
WORSTED SPUN YARNS

COMBER LAP MACHINES
RT BALLING HEADS
COMBED SLIVERS
COMBER LAPS
COMBING
COMBS
COTTON SYSTEM PROCESSING
LAP ARBOURS
LAP RODS
LAP WASTE
LAPS
RIBBON LAP MACHINES

COMBER LAPS
BT LAPS
RT CARD SLIVERS
COMBED SLIVERS
COMBER LAP MACHINES
COMBING
COMBS
COTTON COMBING
COTTON SYSTEM PROCESSING
LAP ARBOURS
LAP RODS
LAP WASTE
LAP WINDING
NOIL REMOVAL
NOILS
PICKER LAPS
RIBBON LAPS
SLIVER NOIL RATIO

COMBERS
USE COMBS

COMBINATION STITCHES
BT STITCHES (SEWING)
RT SEWING

COMBINATION WEAVES
BT WEAVE (TYPES)
RT FLAT WOVEN FABRICS

COMBINATION WIND
BT WIND
RT CHEESE WINDERS
CHEESES
CLOSE WIND
CONERS
CONES
CREELING
DRUM WINDERS
FILLING WIND
OPEN WIND
PACKAGES
PRECISION WINDERS
QUILLERS
QUILLS
ROVING WIND
TRAVERSE
WARP WIND
WINDERS

COMBING
NT COTTON COMBING
HAND COMBING
NOBLE COMBING
RECTILINEAR COMBING
WORSTED COMBING
RT BACKWASHING
BALLING HEADS
BLENDING
BRADFORD SYSTEM PROCESSING
CARDING
COMB TEAR
COMBED SLIVERS
COMBED YARNS
COMBER LAP MACHINES
COMBER LAPS
COMBING EFFICIENCY
COMBING WOOL
COMBS
CONTINENTAL SYSTEM PROCESSING
COTTON SYSTEM PROCESSING
DABBING (COMBING)
DETACHING
DETACHING EFFICIENCY
DETACHING ROLLS
FIBER BREAKAGE
FIBER HOOKS
FIBER LENGTH DISTRIBUTION
FIBER ORIENTATION
HALF LAP
HOOK REMOVAL
HOOKED FIBERS
LAP PLATES
NEW BRADFORD SYSTEM PROCESSING
NIPPER JAWS
NIPPER KNIVES
NIPPER PLATES
NIPPING
NOBLE COMBS
NOIL PER CENT
NOIL REMOVAL
NOILS
PARALLELIZATION
PUNCH BALLS (COMBING)
RECOMBING
RECTILINEAR COMBS
RIBBON LAP MACHINES
SETOVER
SLIVER NOIL RATIO
SLIVERS
TOP COMBING (COMB CYCLE)
TOP COMBS
TOP MATCHING
TOP NOIL RATIO
TOPS
WORSTED SYSTEM PROCESSING
YIELD (RETURN)

COMBING EFFICIENCY
NT DETACHING EFFICIENCY
BT PROCESS EFFICIENCY
RT COMBING
SLIVER NOIL RATIO

COMBING SLIVERS
USE COMBED SLIVERS

COMBING WOOL
BT WOOL CLASS
RT COMBING

COMBS
UF COMBERS
NT NOBLE COMBS
RECTILINEAR COMBS
RT COILERS
COMBED SLIVERS
COMBER LAP MACHINES
COMBER LAPS
COMBING
HALF LAP
LAP PLATES
PACKAGE HOLDERS
TOP COMBS

COMBS (LEASING)
RT LEASING
WARPERS (MACHINE)

COMFORT
BT AESTHETIC PROPERTIES
END USE PROPERTIES
FABRIC PROPERTIES
FABRIC PROPERTIES (AESTHETIC)
FABRIC PROPERTIES (PHYSICAL EXCLUDING MECHANICAL)
RT BAGGINESS
COMPRESSIBILITY
FINISH (PROPERTY)
FIT
FOUNDATION GARMENTS
GARMENTS
HAIRINESS
HAND
HARSHNESS
LOOSENESS
MOISTURE VAPOR TRANSMISSION
POWER (FABRIC)
PRICKLINESS
ROUGHNESS
SMOOTHNESS
STRETCH FABRICS
SURFACE CONTOUR
SURFACE FRICTION
SURFACE PROPERTIES (MECHANICAL)
TAILORING
TEXTURE
THERMAL PROPERTIES
TIGHTNESS
TRIMNESS
WARMTH

COMMINUTION
NT GRINDING (COMMINUTION)
RT ATOMIZING
CHIPPING
CUTTING
POWDERING

COMMON (WOOL GRADE)
BT GRADE (FIBERS)
WOOL GRADE

COMPACTING
BT COMPRESSIVE SHRINKING
DRY FINISHING
RT DIMENSIONAL STABILITY
FABRIC SHRINKAGE
FELTING SHRINKAGE
FUSION CONTRACTION SHRINKAGE
LONDON SHRINKING
PRESHRINKING
PRESSURE ROLLS
RELAXATION SHRINKAGE
RESIDUAL SHRINKAGE
SANFORIZING (TN)
SHRINK RESISTANCE
SHRINK RESISTANCE FINISHES
SHRINKAGE
SHRINKING
WARP SHRINKAGE

COMPACTOR
BT COMPRESSIVE SHRINKAGE MACHINES
RT FABRIC SHRINKAGE
MANGLES
OVERFED PIN TENTERS
PADDERS
PRESS SECTION
ROLLING MILLS
SANDWICH ROLLING MILLS
SANFORIZING MACHINES

COMPARISON
RT ADJUSTMENTS
CALIBRATION
COMPENSATION
CORRECTION
CORRELATION
DIFFERENCES
ERRORS

COMPATIBLE DYES
RT DYEING
DYES

COMPENSATION
RT ACCURACY
ADAPTATION
ADJUSTMENTS
BALANCE
CALIBRATION
COMPARISON
CONTROLLING
CORRECTION
DIFFERENCES
DRIFT
ERRORS
ESTIMATION
FLUCTUATIONS
LEVELLING
LINEARITY

COMPENSATORS (CLOTH)
RT CLOTH GUIDES
GUIDES
J BOXES

COMPILATIONS
RT HISTORY
LITERATURE SURVEYS
REVIEWS
SURVEYS
YEARS COVERAGE (1)
YEARS COVERAGE (5)
YEARS COVERAGE (10)
YEARS COVERAGE (25)
YEARS COVERAGE (50)
YEARS COVERAGE (75)
YEARS COVERAGE (100)

COMPLETELY AUTOMATED SPINNING
USE AUTOMATED SPINNING (STAPLE FIBER)

COMPLEXING
BT REACTIONS (CHEMICAL)
RT CHELATING
CHELATING AGENTS
EDTA
METALLIZED DYES

COMPLIANCE
UF TEXTILE COMPLIANCE
BT STRESS STRAIN PROPERTIES
STRESS STRAIN PROPERTIES (TENSILE)
RT COMPRESSIBILITY
COMPRESSIVE MODULUS
CREPING
DYNAMIC MODULUS
ELASTIC MODULUS (TENSILE)
ELASTIC STRAIN
EXTENSIBILITY
HARDNESS
MODULUS
MODULUS OF CROSS ELASTICITY
PAPERINESS
PLASTIC STRAIN
SECANT MODULUS
SHEAR MODULUS
STIFFNESS
STRAIN
STRESS
STRESS STRAIN CURVES
TANGENT MODULUS
VISCOELASTICITY
YIELD POINT

COMPONENTS
RT ACCESSORIES
APPARATUS
ASSEMBLIES
DEVICES
EQUIPMENT
INSTRUMENTATION
MACHINERY
MECHANISMS
TESTING EQUIPMENT
TEXTILE MACHINERY (GENERAL)

COMPOSITES
UF FABRICS (COMPOSITES)
NT BUNA N COATED FABRICS
BUTYL COATED FABRICS
COATED FABRICS
FABRIC REINFORCED COMPOSITES
FABRIC TO FABRIC LAMINATES
FIBER REINFORCED COMPOSITES
FILAMENT WOUND COMPOSITES
FOAM BACKED FABRICS
HONEYCOMB LAMINATES
LACQUER COATED FABRICS
LAMINATED FABRICS
LAMINATES
NEOPRENE COATED FABRICS
POST FORMED LAMINATES
REINFORCED COMPOSITES
SANDWICH LAMINATES
SBR COATED FABRICS
VINYL COATED FABRICS
YARN REINFORCED COMPOSITES
RT ANISOTROPIC MATERIALS
BONDED YARN FABRICS
CROSS LAID YARN FABRICS
FABRICS
FABRICS (ACCORDING TO STRUCTURE)
FILLER
FOAMS
NEEDLED FABRICS
NONWOVEN FABRICS
(CONT.)

RT RESINS
STITCH BONDED FABRICS
STITCH REINFORCED FABRICS
STITCHED PILE FABRICS
WADDINGS

COMPOUND NEEDLES
BT KNITTING NEEDLES
NEEDLES
RT CASTING OFF
GUIDE BARS
LAPPING (WARP KNITTING)
LATCH NEEDLES
MILANESE KNITTING MACHINES
OVERLAP
PRESSER BARS
RASCHEL KNITTING MACHINES
SPRING NEEDLES
TRICOT KNITTING
TRICOT KNITTING MACHINES
TRICOT STITCHES
UNDERLAP
WARP KNITTING MACHINES

COMPOUNDING
RT BLENDING
MIXING
STIRRING

COMPRESSIBILITY
BT AESTHETIC PROPERTIES
STRESS STRAIN PROPERTIES
STRESS STRAIN PROPERTIES (COMPRESSIVE)
RT BENDING RIGIDITY
BULK MODULUS
COMFORT
COMPLIANCE
COMPRESSIVE MODULUS
CONTACT AREA
DENSITY
EXTENSIBILITY
FABRIC PROPERTIES
FABRIC PROPERTIES (MECHANICAL)
HAIRINESS
HAND
PACKAGE DENSITY
PACKAGE HARDNESS
PACKING
PADDING
RESILIENCE
ROUGHNESS
SANDWICH ROLLING MILLS
SOFTNESS
STUFFING
WADDINGS

COMPRESSION
RT BIAXIAL STRESS
BUCKLING
COMPRESSIVE MODULUS
COMPRESSIVE STRENGTH
CONTACT
MATTING
PRESSURE
TENSION
UNIAXIAL STRESS
YARN BUCKLING

COMPRESSION MOLDING
BT MOLDING
RT CONTACT MOLDING
FLEXIBLE PLUNGER MOLDING
INJECTION MOLDING
MATCHED DIE MOLDING
PRESSURE BAG MOLDING
REINFORCED COMPOSITES
VACUUM BAG MOLDING

COMPRESSION TESTING
BT STRENGTH TESTING
TESTING
RT RESILIENCE

COMPRESSIVE BUCKLING
UF EULER BUCKLING
BT BUCKLING
RT DIMENSIONAL STABILITY
TENSILE BUCKLING
TORSIONAL BUCKLING

COMPRESSIVE MODULUS
UF ELASTIC MODULUS (COMPRESSIVE)
BT MODULUS
STRESS STRAIN PROPERTIES
STRESS STRAIN PROPERTIES (COMPRESSIVE)
RT BULK MODULUS
COMPLIANCE
COMPRESSIBILITY
COMPRESSION
DYNAMIC MODULUS
ELASTIC MODULUS (TENSILE)
HARDNESS
MODULUS OF CROSS ELASTICITY
SANDWICH ROLLING MILLS
SHEAR MODULUS
TENSION

COMPRESSIVE SHRINKAGE
BT FABRIC SHRINKAGE
SHRINKAGE
RT BLANKET (FINISHING)
DRY FINISHING
RELAXATION SHRINKAGE
ROLLING

COMPRESSIVE SHRINKAGE MACHINES
UF RUBBER SHRINKING BLANKETS
NT COMPACTOR
SANFORIZING MACHINES
RT COMPRESSIVE SHRINKING
DIMENSIONAL STABILITY
DRY FINISHING
FELTING MACHINES
FILLING STRAIGHTENING MACHINES
OVERFED PIN TENTERS

COMPRESSIVE SHRINKING
NT COMPACTING
SANFORIZING (TN)
BT DRY FINISHING
SHRINKING
RT COMPRESSIVE SHRINKAGE MACHINES
DIMENSIONAL STABILITY
FELTING SHRINKAGE
FILLING SHRINKAGE
FUSION CONTRACTION SHRINKAGE
LONDON SHRINKING
MOLDING SHRINKAGE
PAPER MAKING
PRESHRINKING
RELAXATION SHRINKAGE
RESIDUAL SHRINKAGE
ROLLING
SHRINK PROOFING
SHRINK RESISTANCE
SHRINK RESISTANCE FINISHES
SHRINKAGE
WARP SHRINKAGE

COMPRESSIVE STRENGTH
BT STRENGTH
STRESS STRAIN PROPERTIES
STRESS STRAIN PROPERTIES (COMPRESSIVE)
RT BEARING STRENGTH
BREAKING STRENGTH
BRITTLENESS
COMPRESSION
COMPRESSIVE MODULUS
CONTACT AREA
CREEP RUPTURE STRENGTH
ELASTIC MODULUS (TENSILE)
FABRIC STRENGTH
FIBER STRENGTH
HARDNESS
IMPACT STRENGTH
KNOT EFFICIENCY
KNOT STRENGTH
LOOP EFFICIENCY
LOOP STRENGTH
PERALTA ROLLS
RUPTURE
SHEAR STRENGTH
STRAIN
STRENGTH ELONGATION TESTING
STRENGTH OF MATERIALS
STRENGTH TESTING
STRESS
STRESS STRAIN CURVES
TEAR STRENGTH
TESTING EQUIPMENT
WET STRENGTH
YARN STRENGTH

COMPRESSIVE STRESS
BT STRESS
RT SHEAR STRESS
STRESS STRAIN PROPERTIES (COMPRESSIVE)
TENSILE STRESS
UNIAXIAL STRESS

COMPUTATIONS
UF CALCULATING
CALCULATIONS
COMPUTING
NT ADDITION
DIVISION
MULTIPLICATION
SUBTRACTION
RT ARITHMETIC
COMPUTERS
COUNTING
DATA REDUCTION
DESIGN
DIFFERENTIATION
EXTRAPOLATION
FORMULAS
GEOMETRY
GRAPHS
INTEGRATION
INTERPOLATION
ROOTS (MATHEMATICAL)
SLIDE RULES
SOLUTIONS (MATHEMATICAL)
TRIAL AND ERROR SOLUTIONS

COMPUTERS
RT ANALOGUES
AUTOMATIC CONTROL
AUTOMATION
COMPUTATIONS
COUNTERS
ELECTRONIC INSTRUMENTS
INSTRUMENTAL COLOR MATCHING
MAGNETIC TAPE
MODELS (MATHEMATICAL)
PROCESS CONTROL
PUNCHED CARDS
TAPES (DATA MEDIA)
TRIAL AND ERROR SOLUTIONS

COMPUTING
USE COMPUTATIONS

CONCENTRATION
RT CONCENTRATION GRADIENT
DIFFUSION
DILUTION
DYEING
EXTRACTION
FILTRATION
MASS TRANSFER
OSMOSIS
SATURATED SOLUTIONS
SATURATION (MATERIAL)
SOLUBILITY

CONCENTRATION GRADIENT
RT CONCENTRATION
DIFFUSION
DILUTION
DYEING
MASS TRANSFER
OSMOSIS
SATURATED SOLUTIONS

CONDENSATION (CHEMICAL)
BT REACTIONS (CHEMICAL)
RT CONDENSATION POLYMERIZATION
CONDENSATION POLYMERS

CONDENSATION POLYMERIZATION
BT POLYMERIZATION
RT ADDITION POLYMERIZATION
ADDITION POLYMERS
CONDENSATION POLYMERS
CONDENSATION (CHEMICAL)
COPOLYMERIZATION
CROSSLINKING
INTERFACIAL POLYMERIZATION

CONDENSATION POLYMERS
RT BLOCK POLYMERS
CONDENSATION (CHEMICAL)
CONDENSATION POLYMERIZATION
COPOLYMERS
POLYAMIDES
POLYESTERS
POLYMERS
POLYUREAS
POLYURETHANES

CONDENSED COUNT
RT CONDENSERS
SLUBBINGS

CONDENSED NAPHTHALENE SULFONATES
USE NAPHTHALENE FORMALDEHYDE SULFONATES

CONDENSED YARNS
UF CONDENSER SPUN YARNS
NT ANGOLA (YARN)
BT SPUN YARNS
YARNS
RT CARDED YARNS
CARDING
CONDENSER TAPES
CONDENSERS
CONDENSING (YARN MANUFACTURING)
HOOKED FIBERS
JACKSPOOLS
SLUBBINGS
SPINNING
WOOL
WOOLEN DRAFTING
WOOLEN SPINNING FRAMES
WOOLEN SPUN YARNS
WOOLEN SYSTEM PROCESSING

CONDENSER BOBBINS
USE JACKSPOOLS

CONDENSER SCREENS
RT AIR BRIDGES
CURLATORS (TN)
RANDO FEEDER (TN)
RANDO WEBBER (TN)
RANDO WEBS (TN)
TUFT FORMERS
WEBS

CONDENSER SPUN YARNS
  USE CONDENSED YARNS

CONDENSER TAPES
  BT TAPE
  RT CONDENSED YARNS
    CONDENSERS
    CONDENSING (YARN MANUFACTURING)
    TAPE CONDENSERS

CONDENSERS
  NT RING CONDENSERS
    TAPE CONDENSERS
  RT CARD WEBS
    CARDS
    CARDING
    CONDENSED COUNT
    CONDENSED YARNS
    CONDENSER TAPES
    DOFFERS (CARD)
    DOFFING (OF A CARD)
    FANCY (NOUN)
    FINISHER PART
    JACKSPOOLS
    RUBBING LEATHERS
    SLIVER CANS
    SLUBBINGS
    WOOLEN CARDING
    WOOLEN CARDS

CONDENSING (YARN MANUFACTURING)
  BT WOOLEN SYSTEM PROCESSING
  RT CARDING
    CONDENSED YARNS
    CONDENSER TAPES
    HOOK FORMATION
    HOOKED FIBERS
    SLUBBINGS
    WOOL
    WOOLEN CARDS

CONDENSING FUNNELS
  USE TRUMPETS

CONDITION (TEXTILES)
  RT CONDITIONING
    EQUILIBRIUM MOISTURE CONTENT
    HUMIDITY
    HYGROSCOPICITY
    REGAIN

CONDITIONING
  NT MECHANICAL CONDITIONING
  RT ABSORBENCY (MATERIAL)
    CONDITION (TEXTILES)
    CRABBING
    DRY RELAXATION
    EQUILIBRIUM MOISTURE CONTENT
    FABRIC RELAXATION
    FABRIC SHRINKAGE
    LONDON SHRINKING
    MOISTURE CONTENT TESTING
    REGAIN
    RELAXATION TIME (MECHANICAL)
    STRESS RELAXATION
    WET RELAXATION

CONDUCTION
  RT CONDUCTIVITY (THERMAL)
    CONVECTION
    DRYING
    HEAT
    HEAT TRANSFER
    HEATING
    RADIATION

CONDUCTIVITY (ELECTRICAL)
  USE RESISTIVITY

CONDUCTIVITY (THERMAL)
  UF HEAT CONDUCTIVITY
    THERMAL CONDUCTIVITY
    THERMAL RESISTANCE
  BT PHYSICAL PROPERTIES (EXCLUDING MECHANICAL)
    TRANSFER PROPERTIES
    THERMAL PROPERTIES
  RT CONDUCTION
    EMISSIVITY
    HEAT TRANSFER
    REFLECTANCE
    SPECIFIC HEAT
    THERMAL TESTING
    THERMAL TRANSMITTANCE TESTING
    WARMTH

CONDUCTOMETRY
  BT CHEMICAL ANALYSIS
  RT ELECTROLYTES
    RESISTIVITY

CONDUCTOR BOXES
  RT NOBLE COMBS

CONE DRAWING
  RT BRADFORD OPEN DRAWING
    DRAFTING (STAPLE FIBER)
    DRAWFRAMES

CONE DYEING
  BT PACKAGE DYEING
    YARN DYEING
  RT BEAM DYEING (YARN)
    BOBBIN DYEING
    CAKE DYEING
    CHEESE DYEING
    CONE DYEING MACHINES
    COP DYEING
    HANK DYEING
    MUFF DYEING
    PRESSURE DYEING

CONE DYEING MACHINES
  BT PACKAGE DYEING MACHINES
    YARN DYEING MACHINES
  RT BEAM DYEING MACHINES (YARN)
    BOBBIN DYEING MACHINES
    CAKE DYEING MACHINES
    CHEESE DYEING MACHINES
    CONE DYEING
    COP DYEING MACHINES
    HANK DYEING MACHINES
    WARP DYEING MACHINES

CONE WINDING
  USE CONING

CONERS
  NT UNICONER (TN)
  BT WINDERS
  RT CHEESE WINDERS
    COMBINATION WIND
    CONING
    OPEN WIND
    PRECISION WINDERS
    WARPERS (MACHINE)

CONES
  BT SELF SUPPORTING PACKAGES
    PACKAGES
  RT BEAMS
    BOBBINS
    CAKES
    CHEESE WINDERS
    CHEESE WINDING
    CHEESES
    COMBINATION WIND
    CONING
    CREELS
    JACKSPOOLS
    LARGE PACKAGES
    PACKAGE DYEING
    PACKAGING (OPERATION)
    PRECISION WINDERS
    QUILLS
    RIBBON BREAKERS
    SPINDLES
    TAPER (BOBBIN)
    UNICONER (TN)
    WARP WIND
    WARPERS (MACHINE)
    WINDERS
    WINDING
    YARN CARRIERS

CONFIDENCE COEFFICIENT
  BT STATISTICAL MEASURES
  RT CORRELATION COEFFICIENT
    DEGREES OF FREEDOM
    LEVEL OF SIGNIFICANCE
    STANDARD DEVIATION

CONFIDENCE INTERVAL
  USE CONFIDENCE LIMITS

CONFIDENCE LIMITS
  UF CONFIDENCE INTERVAL
  BT STATISTICAL MEASURES
  RT ACCURACY
    ANALYZING
    CORRELATION COEFFICIENT
    DEGREES OF FREEDOM
    DISTRIBUTION
    ERRORS
    ESTIMATION
    EXTRAPOLATION
    LEVEL OF SIGNIFICANCE
    LOWER CONTROL LIMIT
    PROBABILITY
    QUALITY CONTROL
    STANDARD DEVIATION
    STATISTICAL ANALYSIS
    UPPER CONTROL LIMIT
    SHAPE
    SYMMETRY

CONFIGURATION
  RT ASSEMBLIES
    CONVENTIONAL PRACTICE
    DESIGN
    DIMENSIONS
    ENTANGLEMENTS
    INVERSION
    MODIFICATION
    ORIENTATION
    PARALLELIZATION
    SHAPE
  RT SHAPE FACTOR
    STRUCTURES
    SYMMETRY
    TOPOLOGY

CONGRUENCE
  RT SHAPE
    SYMMETRY

CONICAL BEARINGS
  BT BEARINGS
  RT AIR BEARINGS
    BALL BEARINGS
    JOURNAL BEARINGS
    NEEDLE BEARINGS
    NYLON BEARINGS
    ROLLER BEARINGS
    SLEEVE BEARINGS
    THRUST BEARINGS

CONING
  UF CONE WINDING
  BT WINDING
  RT AUTOMATIC KNOTTING
    CLOSE WIND
    CLOSE WINDING
    CONERS
    CONES
    DRUM WINDING
    OPEN WINDING
    PRECISION WINDING
    QUILLING
    RIBBON BREAKERS
    RIBBON BREAKING
    RIBBONING
    SLUB CATCHING
    TENSION CONTROL
    WARPING
    WIND
    WINDING ANGLE

CONJUGATE SPINNING
  USE BICOMPONENT SPINNING

CONSTANT RATE PERIOD
  RT CRITICAL MOISTURE CONTENT
    DRYING
    DRYING TIME
    FALLING RATE PERIOD

CONSTANT SPEED
  BT VELOCITY
  RT ACCELERATION (MECHANICAL)
    CONTROL SYSTEMS
    DRIVES
    HIGH SPEED
    MOMENTUM
    MOTION
    MOTORS
    ROTATION
    SPEED CONTROL
    SPINDLE SPEED
    STARTING TIME
    VARIABLE SPEED

CONSUMPTION (MATERIAL)
  UF MATERIAL CONSUMPTION
  RT FLOW (MATERIALS)
    MANAGEMENT
    PROCESS EFFICIENCY
    PRODUCTION
    WASTE

CONTACT
  RT COMPRESSION
    CONTACT AREA
    FRICTION
    HAND

CONTACT ANGLES
  RT CAPILLARITY
    HYDROPHILIC PROPERTY
    HYDROPHOBIC PROPERTY
    INTERFACIAL TENSION
    ORIENTATION
    SURFACE CHEMISTRY
    SURFACE PROPERTIES (PHYSICAL CHEMICAL)
    SURFACE TENSION
    WATER REPELLENCY
    WATERPROOFING
    WETTING

CONTACT AREA
  UF AREA OF CONTACT
  NT APPARENT CONTACT AREA
    TRUE CONTACT AREA
  RT COMPRESSIBILITY
    COMPRESSIVE STRENGTH
    CONTACT
    FRICTION
    HAND
    HEAT TRANSFER
    MANGLING
    NIP
    NIP PRESSURE
    SURFACE CONTOUR

CONTACT MOLDING
BT MOLDING
RT COMPRESSION MOLDING
FLEXIBLE PLUNGER MOLDING
INJECTION MOLDING
MATCHED DIE MOLDING
PRESSURE BAG MOLDING
REINFORCED COMPOSITES
VACUUM BAG MOLDING

CONTAMINATION
RT AIR POLLUTION
IMPURITIES
POLLUTION
POLYMERS
WASTE
WATER POLLUTION

CONTINENTAL COMBS
USE RECTILINEAR COMBS

CONTINENTAL SYSTEM PROCESSING
UF FRENCH SYSTEM PROCESSING
BT WORSTED SYSTEM PROCESSING
RT AMERICAN SYSTEM PROCESSING
BACKWASHING
BRADFORD OPEN DRAWING
BRADFORD SYSTEM PROCESSING
CAP SPINNING
CARBONIZING
CARDING
COMBING
COTTON SYSTEM PROCESSING
GILLING
NEW BRADFORD SYSTEM PROCESSING
PICKING (OPENING)
PIN DRAFTING
SCOURING
SPINNING
TOP MAKING
TWISTING
WOOL
WOOLEN SYSTEM PROCESSING
WORSTED CARDING
WORSTED COMBING
WORSTED SYSTEM SPINNING

CONTINUOUS BLEACHING
BT BLEACHING
CONTINUOUS FINISHING
WET FINISHING
RT BLEACHING AGENTS
BLEACHING RANGES
CONTINUOUS DRYING
CONTINUOUS DYEING
CONTINUOUS SCOURING

CONTINUOUS CONTROL
RT AUTOMATIC CONTROL
CONTROL SYSTEMS
CONTROL THEORY
DIFFERENTIAL CONTROL
INTEGRAL CONTROL
ON OFF CONTROL

CONTINUOUS DRYING
BT CONTINUOUS FINISHING
DRYING
RT BATCH DRYING
CONTINUOUS BLEACHING
CONTINUOUS DYEING
CONTINUOUS FINISHING
CONTINUOUS SCOURING
FREEZE DRYING
TUMBLE DRYING
VACUUM DRYING

CONTINUOUS DYEING
NT CONTINUOUS TOP DYEING
THERMOSOL PROCESS (TN)
BT DYEING (BY PROCESS FLOW)
RT ACID SHOCK METHOD
BATCH DYEING
CONTINUOUS BLEACHING
CONTINUOUS DRYING
CONTINUOUS DYEING RANGE
CONTINUOUS FINISHING
CONTINUOUS SCOURING
HIGH TEMPERATURE DYEING
OPEN WIDTH DYEING
PAD DYEING
PAD STEAM DYEING
PIECE DYEING
SEMI CONTINUOUS DYEING
SIMULTANEOUS DYEING AND FINISHING
THERMOFIXING (DYEING)
UNIT OPERATIONS
VAPOCOL PROCESS (TN)

CONTINUOUS DYEING RANGE
BT DYEING MACHINES
DYEING MACHINES (BY PROCESS FLOW)
RT BATCH DYEING MACHINES
BOOSTER BOXES
CONTINUOUS DYEING
CONTINUOUS SCOURING
DYEING
DYEING (BY PROCESS FLOW)
COOLING CAN
FLUIDIZED BEDS
J BOXES
MOLTEN METAL BATHS
OIL BATHS
OXIDIZING COMPARTMENT (DYEING)
PIECE DYEING MACHINES
PADDERS
PAD FIX MACHINES
REDUCTION PADDERS
SEMI CONTINUOUS DYEING RANGE
STEAMERS
WILLIAMS UNIT (TN)

CONTINUOUS FILAMENT YARNS
USE FILAMENT YARNS

CONTINUOUS FINISHING
NT CONTINUOUS BLEACHING
CONTINUOUS DRYING
CONTINUOUS SCOURING
JET SCOURING
BT WET FINISHING
RT BATCH FINISHING
CONTINUOUS DYEING
CURING RANGES
FINISHING MACHINERY (GENERAL)
PAD DRY CURE PROCESS
PAD FINISHING
PAD STEAM FINISHING
SIMULTANEOUS DYEING AND FINISHING
UNIT OPERATIONS

CONTINUOUS SCOURING
NT JET SCOURING
BT CONTINUOUS FINISHING
SCOURING
RT BATCH SCOURING
BLEACHING
CONTINUOUS BLEACHING
CONTINUOUS DYEING
CONTINUOUS DYEING RANGE
DESIZING
DETERGENTS
INTEGRATED SYSTEMS
J BOXES
KIER SCOURING
OPEN END SPINNING
RINSING
SCOURING BATHS

CONTINUOUS SPINNING (STAPLE FIBER)
BT SPINNING
RT AIR SPINNING
AUTOMATED SPINNING (STAPLE FIBER)
FLUID SPINNING

CONTINUOUS TOP DYEING
BT CONTINUOUS DYEING
TOP DYEING
RT WOOL DYEING

CONTOURS
USE SHAPE

CONTRACTION
RT DIFFERENTIAL SHRINKAGE
DIMENSIONAL STABILITY
FILLING SHRINKAGE
SHRINKAGE
SUPERCONTRACTION
TWIST CONTRACTION
WARP SHRINKAGE

CONTRAST EFFECT
RT RESERVE EFFECT
TONE IN TONE DYEING
TWO COLOR DYEING

CONTROL
USE CONTROL SYSTEMS

CONTROL APRONS
USE APRONS

CONTROL CHARACTERISTICS
RT AUTOMATIC CONTROL
CONTROLLING
FREQUENCY RESPONSE
HUNTING
RESPONSE TIME

CONTROL CHARTS
RT CONTROL SYSTEMS
LOWER CONTROL LIMIT
PROCESS CONTROL
QUALITY CONTROL
UPPER CONTROL LIMIT

CONTROL EQUIPMENT
UF CONTROL INSTRUMENTS
CONTROL PANELS
NT ELECTRIC CONTROLLERS
PNEUMATIC INSTRUMENTS
RT AUTOMATIC CONTROL
AUTOMATION
CONTROL SYSTEMS
ELECTROMAGNETIC DEVICES
ELECTRONIC INSTRUMENTS
EQUIPMENT
HYDRAULICS
SCHUPP REGULATOR

CONTROL INSTRUMENTS
USE CONTROL EQUIPMENT

CONTROL PANELS
USE CONTROL EQUIPMENT

CONTROL POINTS
RT AUTOMATIC CONTROL
CONTROL SYSTEMS
PROCESS CONTROL

CONTROL SURFACES
UF DRAFT CONTROL SURFACES
NT APRON CONTROL
APRONS
RT ACTUAL DRAFT
AERODYNAMICS
AMBLER SUPERDRAFT
APRON DRAFTING
AUTOMATIC CONTROL
BACK ROLLS
CAPS (SPINNING)
CASABLANCA SYSTEM
CONTROL SYSTEMS
DRAFT
DRAFTING (STAPLE FIBER)
DRAFTING ROLLS
DRAFTING THEORY
DRAFTING WAVES
DRAFTING ZONE
FIBER CONTROL
FIBER SLIPPAGE
FLOATING FIBERS
GILLING
HIGH DRAFT
HIGH DRAFT SPINNING
IRREGULARITY CONTROL
PROCESS CONTROL
RAPER AUTOLEVELLER (TN)
RATCH
ROLLS
SINGLE APRONS
SPINNING
TWIST CONTROL

CONTROL **SYSTEMS**
UF **CONTROL**
CONTROLLERS
REGULATION
RT **AUTOCOUNT (TN)**
AUTOLEVELLERS
AUTOMATIC CONTROL
AUTOMATION
CENTRALIZED CONTROL
CHEMICAL ENGINEERING
CLOSED LOOP CONTROL
CONSTANT SPEED
CONTINUOUS CONTROL
CONTROL CHARTS
CONTROL POINTS
CONTROL SURFACES
CONTROL THEORY
CONTROL VALVES
CONTROLLING
COST CONTROL
DOUBLE SYSTEMS
FEEDBACK CONTROL
FLOW CONTROL
FLUCTUATIONS
FLUIDICS
HUMIDITY CONTROL
HYDRAULIC CONTROL
IRREGULARITY CONTROL
MACHINE DESIGN
MAN MACHINE CONTROL SYSTEMS
MANUAL CONTROL
MEMORY WHEELS
ON OFF CONTROL
OPEN LOOP CONTROL
PNEUMATIC CONTROL
PROCESS CONTROL
REMOTE CONTROL
RESPONSE TIME
STARTING TIME
SYSTEMS ENGINEERING
TRANSDUCERS
VOLTAGE REGULATION
WEIGHT CONTROL

CONTROL THEORY
RT AUTOMATIC CONTROL
AUTOMATION
CLOSED LOOP CONTROL
CONTINUOUS CONTROL
CONTROL SYSTEMS
CONTROLLING
DRAFTING THEORY
FEEDBACK CONTROL
FLUIDICS

RT MATHEMATICAL ANALYSIS
ON OFF CONTROL
OPEN LOOP CONTROL

CONTROL VALVES
RT CONTROL SYSTEMS

CONTROLLERS
USE CONTROL SYSTEMS

CONTROLLING
RT ADJUSTMENTS
ALIGNMENT
COMPENSATION
CONTROL CHARACTERISTICS
CONTROL SYSTEMS
CONTROL THEORY
CORRECTION
POSITIONING

CONVECTION
RT AIR FLOW
CONDUCTION
DRYING
HEAT
HEAT TRANSFER
HEATING
MASS TRANSFER
RADIATION

CONVENTIONAL PRACTICE
RT ABSENCE
CONFIGURATION
INVERSION
METHODS
PRESENCE
TECHNIQUES

CONVERTED TICKING
RT TICKING

CONVERTERS (TOW)
UF TOW CONVERTERS
NT COURTAULDS CONVERTER
GREENFIELD CONVERTER
HALLE SEYDEL STRETCH BREAKER
PACIFIC CONVERTER
PERLOK SYSTEM
RIETER CONVERTER
SACO LOWELL DIRECT SPINNER (TN)
STAINS DIRECT SPINNER
TOHO DIRECT SPINNER (TN)
TOW TO TOP MACHINES
TOW TO YARN MACHINES
TURBOSTAPLER
RT STRETCH BREAKING
TOW
TOW CONVERSION
TOW TO TOP CONVERSION
TOW TO YARN CONVERSION

CONVEYOR BELTS
BT BELTS
INDUSTRIAL FABRICS
RT BELT DRIVES
PAPERMAKERS FELTS

CONVEYOR BUCKETS
USE CONVEYORS

CONVEYORS
UF CONVEYOR BUCKETS
RT CHARGING
DELIVERY RATE
FEED LATTICES
FEED RATE
FEED ROLLS
FEEDERS
FLOW (MATERIALS)
HOPPER FEED
INPUT
MATERIALS HANDLING
THROUGHPUT
TRANSFERRING (MATERIAL)

CONVOLUTIONS
BT FIBER GEOMETRY
RT COTTON
CRIMP
FIBER TWIST
FIBERS
FIBRIL REVERSALS
FIBRILS
IMMATURE FIBERS
LUMEN
NATURAL CRIMP
PRIMARY WALLS
SECONDARY WALLS
SEED HAIR FIBERS (GENERAL)

COOLING
RT AIR CONDITIONING
DRYING
FREEZING
HEATING
MELTING
REFRIGERATION

COOLING CAN
RT CONTINUOUS DYEING RANGE

COP BUILD
USE FILLING WIND

COP DYEING
BT PACKAGE DYEING
YARN DYEING
RT CAKE DYEING
CHEESE DYEING
COP DYEING MACHINES
HANK DYEING
MUFF DYEING

COP DYEING MACHINES
BT PACKAGE DYEING MACHINES
YARN DYEING MACHINES
RT BEAM DYEING MACHINES (YARN)
BOBBIN DYEING MACHINES
CAKE DYEING MACHINES
CHEESE DYEING MACHINES
CONE DYEING MACHINES
HANK DYEING MACHINES
WARP DYEING MACHINES

COP END EFFECT
RT COPS

COP WIND
USE FILLING WIND

COPES
BT CEREMONIAL CLOTHING

COPOLYMERIZATION
NT GRAFTING
BT REACTIONS (CHEMICAL)
RT ADDITION POLYMERIZATION
BUTADIENE
CHEMICAL MODIFICATION (FIBERS)
CONDENSATION POLYMERIZATION
COPOLYMERS
CROSSLINKING
MOLECULAR WEIGHT
MONOMERS
POLYMERIZATION
PREPARATION (CHEMICAL)
VINYL COMPOUNDS

COPOLYMERS
NT BLOCK POLYMERS
GRAFT POLYMERS
OLEFIN COPOLYMERS
RT ADDITION POLYMERS
CONDENSATION POLYMERS
COPOLYMERIZATION
CROSSLINKING
ELASTOMERS
HOMOPOLYMERS
MAN MADE FIBERS
PLASTICS
POLYMERS
RESINS
SYNTHETIC FIBERS
TERPOLYMERS

COPPER
BT METALS
RT COPPER COMPOUNDS

COPPER COMPOUNDS
NT COPPER FORMATE
CUPRIC COMPOUNDS
CUPROUS COMPOUNDS
RT COPPER

COPPER FORMATE
BT COPPER COMPOUNDS
RT MILDEW RESISTANCE FINISHES

COPPER NAPHTHENATE
BT CUPRIC COMPOUNDS
RT MILDEW RESISTANCE FINISHES

COPPER NUMBER
BT CHEMICAL PROPERTIES
DEGRADATION PROPERTIES
RT ACID DEGRADATION
CELLULOSE
HYDROCELLULOSE
OXYCELLULOSE

COPPER PENTACHLOROPHENATE
BT CUPRIC COMPOUNDS
PHENOLS
RT MILDEW RESISTANCE FINISHES

COPPER 8-HYDROXYQUINOLATE
BT CUPRIC COMPOUNDS
RT MILDEW RESISTANCE FINISHES

COPS
BT PACKAGES
SELF SUPPORTING PACKAGES
RT CARRIAGES (SPINNING MULE)
COP END EFFECT
CORES (CENTER)
LARGE PACKAGES
LOOMS
MULE SPINNING
MULES
QUADRANTS (MULE)
QUILLING
QUILLS
SHUTTLES
TUBES

CORDAGE
NT CORDS
HAWSERS
ROPES
TIRE CORDS
BT FIBER ASSEMBLIES
RT BAST FIBERS
BRAIDS
CABLES
CABLING
CORDAGE MACHINES
FISHING LINES
HELIX ANGLE
HEMP
JUTE
KNOT EFFICIENCY
KNOT STRENGTH
LOOP EFFICIENCY
LOOP STRENGTH
MANILA
PLIED YARNS
PLYING
ROPES
SAILCLOTH
SEWING THREADS
SINGLES YARNS
SISAL
STRANDS
STRING
TWINE
TWIST
YARN GEOMETRY
YARN STRUCTURE

CORDAGE MACHINES
RT CORDAGE
ROPE MACHINES
ROPES
TWISTERS

CORDURA (TN)
UF FIBER G (TN)
BT CELLULOSIC FIBERS
HIGH TENACITY VISCOSE RAYON
HIGH WET MODULUS RAYON

CORDUROY
BT CUT PILE FABRICS
PILE FABRICS (WOVEN)
WOVEN FABRICS
RT VELVET

CORDS
NT PARACHUTE CORDS
BT CORDAGE
RT FISHING LINES

CORE (FIBER)
BT FINE STRUCTURE (FIBERS)
RT FIBER CROSS SECTIONS
SKIN (FIBER)

CORE BRAIDS
NT PARACHUTE CORDS
BT BRAIDS
REGULAR BRAIDS

CORE SPINNING
BT SPINNING
RT BLENDED YARNS (FILAMENT)
BLENDED YARNS (STAPLE)
CORE SPUN YARNS
COVERED YARNS
RING SPINNING
WRAPPING MACHINES

CORE SPUN YARNS
UF CORED YARNS
BT YARNS
RT BICOMPONENT FIBER YARNS (FILAMENT)
BICOMPONENT FIBER YARNS (STAPLE)
BLENDED YARNS (FILAMENT)
BLENDED YARNS (STAPLE)
CORE SPINNING
EXTENSIBILITY
FIBER MIGRATION
NOVELTY YARNS
PLIED YARNS
POWER (FABRIC)
RUBBER
SINGLES YARNS
SPANDEX FIBERS
STRETCH FABRICS
STRETCH YARNS (FILAMENT)
YARN GEOMETRY
YARN STRUCTURE

CORED YARNS
USE CORE SPUN YARNS

CORES (CENTER)
RT ABSENCE
BOBBINS
COPS
CORING
HEADS (TOP)
HOLLOWNESS
PACKAGES
ROUNDNESS
SPACE

CORFAM (TN)
BT SYNTHETIC LEATHER
RT COATED FABRICS
COMPOSITES
HIDES
LEATHER
NONWOVEN FABRICS

CORING
UF CORING DEVICES
RT ANALYZING
CORES (CENTER)
QUALITY CONTROL
SAMPLING
TESTING

CORING DEVICES
USE CORING

CORIOLIS ACCELERATION
BT ACCELERATION (MECHANICAL)
RT ACCELEROMETERS
ANGULAR VELOCITY
CENTRIFUGAL FORCE
CORIOLIS FORCE
FORCE
GRAVITATION
INERTIA
ROTATION
VELOCITY

CORIOLIS FORCE
BT FORCE
RT CENTRIFUGAL FORCE
CORIOLIS ACCELERATION
DALEMBERTS FORCE
INERTIAL REFERENCE
ROTATION

CORK
RT ROLL COVERINGS
ROLLS

CORN STARCH
BT STARCH
RT POTATO STARCH
PRINTING PASTES
THICKENING AGENTS

CORONA
RT IONIZATION
STATIC ELECTRICITY

CORONIZING (TN)
BT HEAT TREATMENT
RT FIRING (REFRACTORY)
HEAT SETTING (DRY)
HEAT SETTING (SYNTHETICS)

CORRECTION
RT ACCURACY
ADJUSTMENTS
ALIGNMENT
CALIBRATION
COMPARISON
COMPENSATION
CONTROLLING
DIFFERENCES
DRIFT
ERRORS
ESTIMATION
FLUCTUATIONS
FOCUSING
INCREASE
MINIMIZATION
OPTIMIZATION
POSITIONING
QUALITY
TOLERANCES

CORRELATION
RT AUTOCORRELATION
COMPARISON
CORRELATION COEFFICIENT
DIFFERENCES
REGRESSION
REGRESSION COEFFICIENT
STANDARD DEVIATION
STANDARD ERROR
STATISTICAL METHODS

CORRELATION COEFFICIENT
BT STATISTICAL MEASURES
RT COEFFICIENT OF VARIATION
CONFIDENCE COEFFICIENT
CONFIDENCE LIMITS
CORRELATION
DEGREES OF FREEDOM
DISTRIBUTION
EXTRAPOLATION
FREQUENCY
KURTOSIS
LEVEL OF SIGNIFICANCE
MEAN
MEDIAN
MODE
RANGE
REGRESSION COEFFICIENT
STANDARD DEVIATION
STANDARD ERROR
STATISTICS
VARIANCE

CORROSION
UF RUSTING
BT DEGRADATION
RT CORROSION INHIBITORS
CORROSION RESISTANCE
CORROSION TESTING
ELECTROLYSIS
FATIGUE
FATIGUE RESISTANCE
OXIDATION
REACTIONS (CHEMICAL)
SCALE (CORROSION)
SURFACE PROPERTIES (PHYSICAL CHEMICAL)

CORROSION INHIBITORS
RT CORROSION
CORROSION RESISTANCE
INHIBITION
SCALE (CORROSION)

CORROSION RESISTANCE
UF RUST RESISTANCE
RT CORROSION
CORROSION INHIBITORS
CORROSION TESTING
FATIGUE
FATIGUE RESISTANCE
HYDRAZIDES

CORROSION TESTING
BT TESTING
RT CORROSION
CORROSION RESISTANCE
FATIGUE TESTING

CORSET BATISTE
BT WOVEN FABRICS
RT BATISTE
POPLIN

CORSETS
BT FOUNDATION GARMENTS

CORTEX
BT FINE STRUCTURE (FIBERS)
RT CUTICLE
EPICUTICLE
EXOCUTICLE
MEDULLA
MICROFIBRILS
ORTHOCORTEX
PARACORTEX
WOOL

CORUNDUM
NT EMERY
BT ABRASIVES
RT ABRASION
CERAMICS
GRIT
PUMICE

COST CONTROL
RT CONTROL SYSTEMS
COSTS
INVENTORIES
INVENTORY CONTROL
MATERIALS HANDLING

COSTS
RT COST CONTROL
ECONOMICS
FACILITIES
INVENTORIES
INVESTMENTS
JOB ANALYSIS
MACHINERY
MARKETING
MERCHANDISING
PLANT
PROCESS CONTROL
PROCESS EFFICIENCY
PRODUCTION
PRODUCTION CONTROL
RESEARCH
SALES
TEXTILE MACHINERY (GENERAL)
WAGES

COTS (ROLLERS)
RT BEARINGS
DRAFTING ROLLS
FLUTED ROLLS
ROLLER SHAFTS
ROLLS

COTSWOLD WOOL
BT LONG WOOL
WOOL

COTTON
UF COTTON FIBERS
NT AMERICAN EGYPTIAN COTTON
AMERICAN UPLAND COTTON
CALIFORNIA COTTON
CHEMICALLY MODIFIED COTTON
EGYPTIAN COTTON
INDIAN COTTON
LONG STAPLE COTTON
PIMA COTTON
SEA ISLAND COTTON
SHORT STAPLE COTTON
TEXAS COTTON
BT CELLULOSIC FIBERS
FIBERS
NATURAL FIBERS
RT BENZOYLATED COTTON
BOLLS
CAUSTICAIRE SCALE
CAUSTICAIRE VALUE
CELLULOSE
CELLULOSE DYES
CHEMICALLY MODIFIED COTTON
CONVOLUTIONS
COTTON FABRICS
COTTON GINS
COTTON GRADE
COTTON LINTERS
COTTON SEEDS
COTTON SYSTEM PROCESSING
COTTON WAX
DECRYSTALLIZED COTTON
DEWAXING (COTTON)
DIAZOBENZOYLATED COTTON
FIBER FINENESS
FIBER LENGTH
FIBRIL REVERSALS
FIBRILS
FIBROGRAPH
GINNING
HEMICELLULOSE
IMMATURE FIBERS
IMMATURITY (FIBER)
IMMATURITY TESTING (COTTON FIBER)
LINT (FIBER)
LINT COLOR
LINT CONTENT
LUMEN
MATURITY INDEX
MERCERIZED COTTON
MICRONAIRE
NEPS
PICKER LAPS
PRESSLEY STRENGTH
PRIMARY WALLS
RAYON (REGENERATED CELLULOSE FIBERS)
SECONDARY WALLS
SEED HAIR FIBERS (GENERAL)
SEEDS (TRASH)
SRRL OPENER
STAPLE FIBERS
VINYLATED COTTON

COTTON BALES
BT BALES
RT COTTON
BALE BREAKERS
BALING
COTTON GINS
WOOL BALES

COTTON BLENDS
BT BLENDS (FIBERS)
CELLULOSE BLENDS
RT LONG STAPLE COTTON
POLYESTER FIBER BLENDS

COTTON BOLLS
USE BOLLS

COTTON CARDING
BT CARDING
COTTON SYSTEM PROCESSING
RT AERODYNAMIC CARDING
CARD CYLINDERS
CARD FLATS
CARD SLIVERS
CARD WASTE
CARD WEBS
CARDING ACTION
CARDING EFFICIENCY
CARDS
DOFFER COMBS
DOFFERS (CARD)
DOFFING (OF A CARD)

RT FEED ROLLS
FETTLING
FLAT SETTINGS (CARD)
LICKER IN
MOTE KNIVES
NEP COUNT
NEPS
SLIVER CANS
SLIVERS
STRIPPING ACTION
TRUMPETS
WOOLEN CARDING
WORSTED CARDING

COTTON CARDS
NT DUO CARDS
BT CARDS
RT CARD CLOTHING
CARD CYLINDERS
CARD FLATS
CARD GRINDING
CARD SLIVERS
CARD WEBS
CARDING
CARDING EFFICIENCY
CLEARERS (CARD)
COTTON SYSTEM PROCESSING
CROSROL WEB PURIFIER (TN)
CYLINDER SCREENS
DOFFER COMBS
DOFFERS (CARD)
FEED LATTICES
FEED PLATES
FEED ROLLS
FLAT CHAINS (CARD)
FLAT STRIPPING COMBS
HACKLE COMBS
KNIFE PLATES
LAP ARBOURS
LAP GLIDES
LEADING HOOKS
LICKER IN
LICKER IN SCREENS
MOTE KNIVES
MOTES
NEP COUNT
NEPS
PERCENTAGE PLATES
PICKER LAPS
RANDO WEBBER (TN)
SLIVER CANS
SRRL CARD
STRIPPERS
TRAILING HOOKS
TRUMPETS
WASTE CONTROL KNIVES

COTTON CLASS
BT CLASS (FIBERS)
RT CLASSING
COEFFICIENT OF LENGTH VARIATION
COTTON CLASSING
COTTON GRADE
COTTON QUALITY
FIBER LENGTH
FIBROGRAPH
GRADE (FIBERS)
MEAN FIBER LENGTH
UPPER QUARTILE LENGTH

COTTON CLASSING
RT COTTON CLASS
COTTON QUALITY
EFFECTIVE LENGTH (FIBER)
FIBER ARRAY
FIBER DIAGRAM
FIBER LENGTH DETERMINATION
FIBER LENGTH DISTRIBUTION
FIBROGRAPH
GRADING
LENGTH UNIFORMITY RATIO (FIBER)
MEAN FIBER LENGTH
SHORT FIBERS INDEX
SORTING
SPAN LENGTH
STAPLING
SUTER WEBB TESTING
UPPER HALF MEAN LENGTH

COTTON COMBING
BT COMBING
COTTON SYSTEM PROCESSING
RT BLENDING
COMBED SLIVERS
COMBED YARNS
COMBER LAP MACHINES
COMBER LAPS
COMBING EFFICIENCY
COTTON SYSTEM PROCESSING
DETACHING
DETACHING EFFICIENCY
FIBER BREAKAGE
FIBER LENGTH DISTRIBUTION
FIBER ORIENTATION
HOOKED FIBERS
HOOK REMOVAL

RT NIPPING
NOBLE COMBING
NOIL PER CENT
NOILS
PARALLELIZATION
SLIVER NOIL RATIO
TOP COMBING (COMB CYCLE)
RECTILINEAR COMBS
WORSTED COMBING

COTTON COMBS
USE RECTILINEAR COMBS

COTTON COUNT
BT COUNT
INDIRECT COUNT
RT COUNT TESTING
DENIER

COTTON DYEING
BT CELLULOSE DYEING
DYEING (BY FIBER CLASSES)
RT ACETATE DYEING
ANIMALIZING
BLEND DYEING
RAYON DYEING
SILK DYEING
SYNTHETIC DYEING
WOOL DYEING

COTTON DYES
USE CELLULOSE DYES

COTTON FABRICS
BT FABRICS (ACCORDING TO FIBER)
RT COTTON
COTTON SPUN YARNS
WOOL FABRICS

COTTON FIBERS
USE COTTON

COTTON GINNING
USE GINNING

COTTON GINS
UF GINS (COTTON)
RT BALING
BOLLS
COTTON
COTTON BALES
COTTON GRADE
COTTON SEEDS
COTTON SYSTEM PROCESSING
FIBER LENGTH
GIN CUT
GIN FALL
GINNING
KNIFE BLADE GINNING
LINT COLOR
LINT CONTENT
ROLLER GINNING
SAW BLADE GINNING
SEED HAIR FIBERS (GENERAL)
WASTE

COTTON GRADE
NT GOOD MIDDLING (COTTON GRADE)
GOOD ORDINARY (COTTON GRADE)
LOW MIDDLING (COTTON GRADE)
MIDDLING (COTTON GRADE)
MIDDLING FAIR (COTTON GRADE)
STRICT GOOD MIDDLING (COTTON GRADE)
STRICT GOOD ORDINARY (COTTON GRADE)
STRICT LOW MIDDLING (COTTON GRADE)
STRICT MIDDLING (COTTON GRADE)
BT GRADE (FIBERS)
RT COTTON
COTTON CLASS
COTTON GINS
COTTON QUALITY
FIBER FINENESS
GRADING
LINT CONTENT
WOOL GRADE

COTTON GRADING
BT GRADING
RT COTTON CLASSING
COTTON GRADE

COTTON LASTING
BT LASTING
RT LININGS

COTTON LINT
USE LINT (FIBER)

COTTON LINTERS
RT COTTON
GINNING
LINT (FIBER)
LINT CONTENT

COTTON PROCESSING
USE COTTON SYSTEM PROCESSING

COTTON QUALITY
RT COTTON CLASS
COTTON CLASSING
COTTON GRADE
EFFECTIVE LENGTH (FIBER)
FIBER DIAGRAM
FIBER FINENESS
FIBER LENGTH DETERMINATION
FIBER LENGTH DISTRIBUTION
FIBROGRAPH
GRADING
LENGTH UNIFORMITY RATIO (FIBER)
MEAN FIBER LENGTH
SHORT FIBERS INDEX
SORTING
SPAN LENGTH
STAPLING
SUTER WEBB TESTING
UPPER HALF MEAN LENGTH

COTTON SEEDS
RT COTTON
COTTON GINS
GINNING

COTTON SPINNING
USE COTTON SYSTEM SPINNING

COTTON SPUN YARNS
UF COTTON TYPE YARNS
BT SPUN YARNS
YARNS
RT CARDED YARNS
COMBED YARNS
COTTON FABRICS
COTTON SYSTEM SPINNING
WOOLEN SPUN YARNS
WORSTED SPUN YARNS

COTTON SYSTEM PROCESSING
UF COTTON PROCESSING
NT BALE BREAKING
COTTON CARDING
COTTON COMBING
COTTON SYSTEM SPINNING
GINNING
MERCERIZING
OPENING
PICKING (OPENING)
ROLLER DRAFTING
RT AMERICAN SYSTEM PROCESSING
AUTOMATED SPINNING (STAPLE FIBER)
BALE BREAKERS
BALE PLUCKERS
BALING
BLENDING
BRADFORD SYSTEM PROCESSING
CARDING
CASABLANCA SYSTEM
COMBER LAPS
COMBER LAP MACHINES
COMBING
CONTINENTAL SYSTEM PROCESSING
COTTON
COTTON CARDS
COTTON GINS
DRAFTING (STAPLE FIBER)
FLAX SYSTEM PROCESSING
HIGH DRAFT SPINNING
MERCERIZED COTTON
PICKER LAPS
PICKING (HARVESTING)
PLYING
RECTILINEAR COMBS
RIBBON LAPS
RING FRAMES
SHORT STAPLE SPINNING
SILK SYSTEM PROCESSING
SLASHING
SLIVER TO YARN SPINNING
SLIVERS
SPINNING
SPINNING FRAMES
TWISTING
WARPING
WINDING
WOOLEN SYSTEM PROCESSING
WORSTED SYSTEM PROCESSING

COTTON SYSTEM SPINNING
UF COTTON SPINNING
BT COTTON SYSTEM PROCESSING
SPINNING
RT COTTON SPUN YARNS
SPINNING FRAMES
WOOLEN SYSTEM SPINNING
WORSTED SYSTEM SPINNING

COTTON SYSTEM WARPING
BT WARPING
RT SECTIONAL WARPING
SILK SYSTEM WARPING

COTTON THREADS
BT SEWING THREADS
THREADS
RT BOBBIN THREADS (SEWING)
LINEN THREADS
LONG STAPLE COTTON
NEEDLE THREADS
NYLON THREADS
SEAMING
SEWING
STITCHING
THREAD BREAKAGES

COTTON TYPE YARNS
USE COTTON SPUN YARNS

COTTON WAX
BT WAXES
RT BLEACHING
COATINGS (SUBSTANCES)
COTTON
DEWAXING (COTTON)
IMPURITIES
SCOURING

COULIER MOTION (KNITTING)
USE DRAW MECHANISM (KNITTING)

COUNT
UF YARN COUNT
NT COTTON COUNT
CUT SYSTEM
DENIER
DIRECT COUNT
EQUIVALENT COUNT
GREX
HANK NUMBER
INDIRECT COUNT
LEA COUNT
RUN SYSTEM
TEX
TYPP COUNT
WORSTED COUNT
YORKSHIRE SKEIN WOOLEN COUNT
BT YARN PROPERTIES
RT COUNT STRENGTH PRODUCT
COUNT TESTERS
COUNT TESTING
DENSITY
DIAMETER
FIBER DENIER
FIBER FINENESS
LINEAR DENSITY
MICRONAIRE
SINGLES YARNS
TWIST MULTIPLIER
YARN CROSS SECTIONS
YARN DIAMETER
YARN GEOMETRY
YARN STRUCTURE
YARNS

COUNT CROWN AND GAUGE
RT CARD CLOTHING
CARD CLOTHING FOUNDATION
CARD GRINDING
CARD NAILING
CARD WIRES
FILLET CLOTHING
KNEE
METALLIC CARD CLOTHING
NOGG
RIB SET
SHEET CLOTHING
TWILL SET

COUNT STRENGTH PRODUCT
BT BREAKING STRENGTH
STRENGTH
STRESS STRAIN PROPERTIES (TENSILE)
YARN STRENGTH
YARN PROPERTIES
RT BREAKING ELONGATION
BREAKING LENGTH
BUNDLE STRENGTH
COUNT
DENIER
END BREAKAGES
IMPACT STRENGTH
LEA STRENGTH PRODUCT
LOOP EFFICIENCY
LOOP STRENGTH
PRESSLEY TEST
STRENGTH EFFICIENCY
STRENGTH ELONGATION TESTING
TEAR STRENGTH
TENACITY
THREAD BREAKAGES
WEAK SPOTS

COUNT TESTERS
NT VIBRASCOPE
BT TESTING EQUIPMENT
RT COUNT

COUNT TESTING
BT TESTING
RT COTTON COUNT
COUNT
DENIER
GREX
LINEAR DENSITY
MEASURING INSTRUMENTS
TEX
VIBRASCOPE

COUNTERCLOCKWISE
USE ANTICLOCKWISE

COUNTERFALLERS
RT CARRIAGES (SPINNING MULE)
HEADSTOCKS
MULE SPINNING
MULES
QUADRANTS (MULE)
TWISTING AT THE HEAD

COUNTERFLOW PROCESSES
RT AIR FLOW
DRYING
FLOW (MATERIALS)
FLUID FLOW
HEAT TRANSFER
SCOURING
SCOURING MACHINES
SCOURING TRAINS

COUNTERS
NT NEP COUNTERS
PHOTOELECTRIC COUNTERS (NEPS)
PICK COUNTERS
RT COMPUTERS
COUNTING

COUNTING
RT ANALYZING
COMPUTATIONS
COUNTERS
DATA REDUCTION
DETECTION
DETERMINATION
OPTICAL SCANNING
PHOTOELECTRIC SCANNING
RADIATION COUNTERS
SAMPLING

COUPLER CURVES
RT KINEMATICS

COUPLER POINTS
RT KINEMATICS

COUPLES
RT ANGULAR DEFLECTION
ANGULAR VELOCITY
BENDING
BENDING RIGIDITY
CURVATURE
MOMENTS
SENSE (DIRECTION)
TORSION (GEOMETRIC)
TORSIONAL RIGIDITY
TWIST
TWISTING MOMENTS

COUPLING (CHEMICAL)
BT REACTIONS (CHEMICAL)
RT AZOIC BASES
AZOIC DYEING
AZOIC DYES
DIAZOBENZOYLATED COTTON
DIAZO COMPOUNDS
DIAZOTIZATION
NAPHTHOLS

COUPLINGS (MECHANICAL)
RT ACCESSORIES
COMPONENTS
DEVICES
JOINTS
MECHANISMS

COURSE SPACING
USE COURSES PER INCH

COURSES
RT COURSES PER INCH
COVER
HAND KNITTING
KNITTED FABRICS
KNITTING
LOCKING COURSES
LOOP DISTORTION
MACHINE KNITTING
SLACK COURSE
STITCH CLARITY
STITCH DENSITY
STITCH LENGTH
WALES
WALES PER INCH

COURSES PER INCH
UF COURSE SPACING
BT FABRIC PROPERTIES
FABRIC PROPERTIES (STRUCTURAL)
RT COURSES
COVER FACTOR
DESIGN
ENDS PER INCH
FABRIC DESIGN
FABRIC GEOMETRY
FABRIC STRUCTURE
KNITTED FABRICS
KNITTED STITCHES
KNITTING
LOOP DISTORTION
PICKS PER INCH
RUNNER RATIO
STITCH DENSITY
STITCH LENGTH
TEXTURE
TIGHTNESS
WALES
WALES PER INCH

COURTAULDS CONVERTER
BT CONVERTERS (TOW)
TOW TO TOP MACHINES
RT TOW TO TOP CONVERSION

COUTIL
BT FLAT WOVEN FABRICS
WOVEN FABRICS
RT COTTON
TWILL WEAVES

COVALENT BONDS
BT CHEMICAL BONDS
RT HYDROGEN BONDS
IONIC BONDS
METALLIC BONDS

COVER
NT BOTTOM COVER
FABRIC COVER
TOP COVER
YARN COVER
BT FABRIC PROPERTIES
FABRIC PROPERTIES (STRUCTURA
OPTICAL PROPERTIES
RT AESTHETIC PROPERTIES
APPEARANCE
BACK (FABRIC)
BULK
COURSES
COVER FACTOR
DESIGN
DRAPE
FABRIC GEOMETRY
FACE (FABRIC)
FILLING EFFECT (TWILL)
FINISH (PROPERTY)
HAIRINESS
HAND
KNITTED FABRICS
PACKING FACTOR
PATTERN (FABRICS)
PATTERN DEFINITION
POROSITY
STITCH CLARITY
STITCH DENSITY
TEXTURE
TRANSLUCENCY
TRANSPARENCY

COVER FACTOR
BT FABRIC PROPERTIES
FABRIC PROPERTIES (STRUCTURA
RT AIR PERMEABILITY
AIR RESISTANCE
COURSES PER INCH
COVER
CRIMP INTERCHANGE
DENSITY
DOUBLE JAMMING
ENDS PER INCH
FABRIC ANALYSIS
FABRIC COVER
FABRIC GEOMETRY
FABRIC STRUCTURE
JAMMING
PACKING FACTOR
PICKS PER INCH
POROSITY
STITCH DENSITY
TIGHTNESS
WALES PER INCH
YARN COVER

COVER PLATES
RT NOBLE COMBS
SETOVER

COVER STITCHES
BT STITCHES (SEWING)

COVERALLS
BT WORK CLOTHING
RT FATIGUES

COVERED ROLLS
BT DRAFTING ROLLS
ROLLS
RT FLUTED ROLLS

COVERED SEAMS
BT SEAMS
RT BUTTED SEAMS
BUTTERFLY SEAMS
FELLED SEAMS
FELLING
FRENCH SEAMS
FULL SEAM WIDTH
HEM SEAMS
JOINTS
LAPPED SEAMS
PINKED SEAMS
PIPED SEAMS
SANDWICH SEAMS
SEAM ABRASIVE STRENGTH
SEAM ALLOWANCE
SEAM EFFICIENCY
SEAM GRIN
SEAM HEADING
SEAM LET OUT
SEAM QUALITY
SEAM SIZE
SEAM SLIPPAGE
SEAM SLIPPAGE STRENGTH
SEAM SPECIFICATIONS
SEAM STRENGTH
SEAM TENSILE STRENGTH
SEAM THICKNESS
SEAM WIDTH
SEWING
STITCH ELASTICITY
STITCH SEQUENCE
STITCH SIZE (SEWING)
STITCH TENSION
STITCHES (SEWING)
STITCHING
STRETCHING (SEAMING)
SUPERPOSED SEAMS
TAILORED SEAMS
YARN SEVERANCE

COVERED YARNS
UF WRAPPED YARNS
RT BLENDED YARNS (FILAMENT)
BLENDED YARNS (STAPLE)
CORE SPINNING
CORE SPUN YARNS
FOUNDATION GARMENTS
POWER (FABRIC)
RUBBER
SPANDEX FIBERS
STRETCH FABRICS
YARN GEOMETRY
YARN STRUCTURE

COVERT
BT FLAT WOVEN FABRICS
WOVEN FABRICS
RT TWILL WEAVES

COW HAIR
BT HAIR
KERATIN FIBERS
PROTEIN FIBERS
RT COWS

COWHIDE
BT LEATHER
RT COWS
SYNTHETIC LEATHER

COWS
BT ANIMALS
RT ANIMAL HUSBANDRY
COW HAIR
COWHIDE

CRABBING
BT WET FINISHING
RT CONDITIONING
FABRIC RELAXATION
FABRIC SHRINKAGE
HEAT SETTING (SYNTHETICS)
LONDON SHRINKING
RELAXATION
RELAXATION SHRINKAGE
RELAXATION TIME (MECHANICAL)
RESIDUAL STRESS
SPONGING
STRAIN
STRESS
STRESS RELAXATION
WET RELAXATION

CRABBING MACHINES
RT CRABBING
WET RELAXATION

CRACK PROPAGATION
RT BREAKING
BRITTLE FRACTURE
BRITTLENESS
CHIPPING
CRACKS (MECHANICAL)
FATIGUE
FRACTURING
IMPACT
IMPACT STRENGTH
RT MECHANICAL DETERIORATION PROPERTIES
RUPTURE
SHOCK RESISTANCE
SLITTING
SPLITTING
STRESS PROPAGATION

CRACKED SELVEDGES
BT FABRIC DEFECTS
RT CENTER SELVEDGES
DOG LEGGED SELVEDGES
LENO SELVEDGES
LOOP SELVEDGES
PULLED IN SELVEDGES
SEALED SELVEDGES
SELVEDGE WARPS
SELVEDGES
SLACK SELVEDGES
TIGHT SELVEDGES
WEAVING
WOVEN FABRICS

CRACKS (FABRIC)
BT DEFECTS
DYEING DEFECTS
FABRIC DEFECTS
RT STREAKINESS

CRACKS (MECHANICAL)
BT DEFECTS
RT BREAKS
CRACK PROPAGATION
GRAIN BOUNDARIES
SLIP PLANES

CRADLES
RT DRAFTING ROLLS

CRANKS
RT ACCESSORIES
DEVICES
MECHANISMS

CRASH TOWELING
BT FLAT WOVEN FABRICS
WOVEN FABRICS
RT ABSORBENCY (MATERIAL)
PLAIN WEAVES
TERRYCLOTH
TOWELS

CRATCH
RT CREELS
NARROW FABRICS

CRAVATS
BT GARMENTS
RT NECKTIES
SCARVES

CREASE ACCEPTANCE
BT END USE PROPERTIES
FABRIC PROPERTIES
FABRIC PROPERTIES (MECHANICAL)
FINISH (PROPERTY)
RT CREASES
CREASE RETENTION
CREASING

CREASE MARKS
BT DYEING DEFECTS
FABRIC DEFECTS
RT ROPE MARKS

CREASE RECOVERY
USE WRINKLE RECOVERY

CREASE RESISTANCE
USE WRINKLE RESISTANCE

CREASE RESISTANT FABRICS
USE EASY CARE FABRICS

CREASE RETENTION
BT END USE PROPERTIES
FABRIC PROPERTIES
FABRIC PROPERTIES (MECHANICAL)
FINISH (PROPERTY)
RT BENDING
BENDING RIGIDITY
CREASE ACCEPTANCE
CREASES
CREASING
DEFORMATION
DIMENSIONAL STABILITY
DMEU
DURABLE PRESS
DURABLE PRESS FINISHES
DURABLE PRESS TREATMENTS
EASY CARE FABRICS
PERMANENT DEFORMATION
PERMANENT PLEATING
PERMANENT SET
PRESSING
RESILIENCE
WRINKLE RECOVERY
WRINKLE RESISTANCE
WRINKLING

CREASES
BT FABRIC DEFECTS
RT CREASE ACCEPTANCE
CREASE RETENTION
CREASING
GARMENT COMPONENTS
PLEATS
WRINKLE RECOVERY
WRINKLES
WRINKLING

CREASING
RT CREASE ACCEPTANCE
CREASES
BENDING
BENDING RIGIDITY
CREASE RETENTION
DEFORMATION
DISTORTION
EASY CARE FABRICS
HEAT SETTING (SYNTHETICS)
IRONING
PERMANENT DEFORMATION
PERMANENT PLEATING
PERMANENT SET
PLEATING
PRESSING
RESILIENCE
WRINKLING

CREELING
BT WARP PREPARATION
RT BEAM DYEING (YARN)
BEAMS
CREELS
PACKAGES
SECTIONAL WARPING
STOP MOTIONS
TENSION DEVICES
UNWINDING
WARP ENDS
WARPERS (MACHINE)
WARPING
WINDERS
WINDING

CREELS
RT CHEESES
CONES
CRATCH
CREELING
PACKAGE HOLDERS
PACKAGES
WARPERS (MACHINE)

CREEP
NT DELAYED RECOVERY
PERMANENT SET
BT MECHANICAL DETERIORATION PROPERTIES
STRESS STRAIN PROPERTIES
STRESS STRAIN PROPERTIES (TENSILE)
RT BAGGINESS
CREEP RECOVERY
CREEP RUPTURE
CREEP STRENGTH
CREEP TESTING
CYCLIC STRESS
DIMENSIONAL STABILITY
DRAWING (FILAMENT)
DYNAMIC MODULUS
ELASTIC RECOVERY
ELASTICITY
FATIGUE
FATIGUE RESISTANCE
FLOW (PLASTIC)
GROWTH (FABRIC)
LINEAR VISCOELASTICITY
NONLINEAR VISCOELASTICITY
PLASTIC STRAIN
RELAXATION TIME (MECHANICAL)
RHEOLOGY
STRAIN
UNIAXIAL STRESS
VISCOELASTICITY
VISCOUS FLOW
WRINKLE RECOVERY

CREEP RECOVERY
BT RECOVERY (SELF RESTORATION TO ORIGINAL CONDITION)
STRESS STRAIN PROPERTIES
STRESS STRAIN PROPERTIES (TENSILE)
RT BENDING RECOVERY
CRIMP RECOVERY
DELAYED RECOVERY
ELASTIC RECOVERY
LINEAR VISCOELASTICITY
NONLINEAR VISCOELASTICITY
MAXWELL MODEL
PERMANENT SET
RELAXATION
RELAXATION SPECTRUM
RELAXATION TIME (MECHANICAL)
RESILIENCE
RHEOLOGY
(CONT.)

RT STRESS RELAXATION
VISCOELASTICITY
VOIGT MODEL
WORK RECOVERY
WRINKLE RECOVERY

CREEP RESISTANCE
USE CREEP STRENGTH

CREEP RUPTURE
RT CREEP
CREEP RUPTURE STRENGTH
CREEP TESTING
FATIGUE RUPTURE
RUPTURE

CREEP RUPTURE STRENGTH
BT STRENGTH
STRESS STRAIN PROPERTIES (TENSILE)
RT BUNDLE STRENGTH
COMPRESSIVE STRENGTH
CREEP RUPTURE
CREEP STRENGTH
CREEP TESTING
STRENGTH ELONGATION TESTING

CREEP STRENGTH
UF CREEP RESISTANCE
BT STRENGTH
STRESS STRAIN PROPERTIES (TENSILE)
RT CREEP
CREEP RECOVERY
CREEP RUPTURE STRENGTH
CREEP TESTING
DIMENSIONAL STABILITY
DURABILITY

CREEP TESTING
UF CREEP TESTS
BT TESTING
RT ABRASION TESTING
CREEP
CREEP RUPTURE
CREEP RUPTURE STRENGTH
CREEP STRENGTH
CUTTING TESTING
DELAYED RECOVERY
FATIGUE TESTING
IMPACT PENETRATION TESTING
IMPACT TESTING
PERMANENT SET
PILLING TESTING
SNAG TESTING
TEAR TESTING
WEAR TESTING

CREEP TESTS
USE CREEP TESTING

CREPE
UF CREPE FABRICS
NT MAROCAIN
BT FLAT WOVEN FABRICS
WOVEN FABRICS
RT CREPE EMBOSSING
CREPE WEAVES
CREPE YARNS
HIGH TWIST
PLAIN WEAVES
PUCKERS (FOR EFFECT)
TORSIONAL BUCKLING

CREPE CORDS
BT NARROW FABRICS

CREPE EMBOSSING
RT CREPE
EMBOSSING
PUCKERING

CREPE FABRICS
USE CREPE

CREPE SUZETTE
USE CREPON GEORGETTE

CREPE WEAVES
RT CREPE
PLAIN WEAVES

CREPE YARNS
RT CREPE
GEORGETTE
HIGH TWIST
S TWIST YARNS
TORSIONAL BUCKLING
Z TWIST YARNS

CREPING
RT BUCKLING
CREPE
CREPE YARNS
CURL
DIMENSIONAL STABILITY
PAPER MAKING
PUCKERING

RT PUCKERS (FOR EFFECT)
STRETCH YARNS (FILAMENT)
TORSIONAL BUCKLING
TWIST
YARN BUCKLING

CREPON GEORGETTE
UF CREPE SUZETTE
BT FLAT WOVEN FABRICS
GEORGETTE
WOVEN FABRICS

CREPSET (TN)
BT NYLON (POLYAMIDE FIBERS)
NYLON 6

CRESLAN (TN)
BT ACRYLIC FIBERS

CRETONNE
BT FLAT WOVEN FABRICS
WOVEN FABRICS
RT PLAIN WEAVES
SHEETING (FABRIC)

CRIMP
UF CRINKLE
NT FABRIC CRIMP
FIBER CRIMP
FOLLOW THE LEADER CRIMP
HELICAL CRIMP
LATENT CRIMP
NATURAL CRIMP
YARN CRIMP
Z CRIMP
BT PHYSICAL PROPERTIES (EXCLUDING MECHANICAL)
STRUCTURAL PROPERTIES
RT BICOMPONENT FIBERS
BICOMPONENT FIBER YARNS (FILAMENT)
BICOMPONENT FIBER YARNS (STAPLE)
BULK
CONVOLUTIONS
CRIMP AMPLITUDE (FIBER)
CRIMP BALANCE
CRIMP FREQUENCY
CRIMP INDEX
CRIMP INTERCHANGE
CRIMP PERCENT
CRIMP RADIUS
CRIMP REMOVAL
CRIMPED YARNS
CRIMPING
CRIMPING MACHINES
DIFFERENTIAL SHRINKAGE
EDGE CRIMPING
FABRIC ANALYSIS
FABRIC GEOMETRY
FABRIC STRUCTURE
FALSE TWIST YARNS
FIBER PROPERTIES
FIBRIL REVERSALS
SELF CRIMPING YARNS
STRETCH
STRETCH YARNS (FILAMENT)
TEXTURING
WAVINESS
WOOL
YARN GEOMETRY
YARN PROPERTIES
YARN STRUCTURE

CRIMP AMPLITUDE (FIBER)
BT FIBER PROPERTIES
RT AMPLITUDE
CRIMP
CRIMP FREQUENCY
CRIMP INDEX
CRIMP RADIUS
FIBER CRIMP

CRIMP BALANCE
RT CRIMP
CRIMP INTERCHANGE
FABRIC CRIMP
FABRIC GEOMETRY
FABRIC STRUCTURE
YARN CRIMP

CRIMP FIXING
BT CRIMPING
RT CURING
HEAT SETTING (SYNTHETICS)
HEAT TREATMENT
REACTANTS
RESIN FINISHES
WET FINISHING

CRIMP FIXING AGENTS
USE REACTANTS

CRIMP FREQUENCY
BT FIBER PROPERTIES
RT AMPLITUDE
CRIMP
CRIMP AMPLITUDE (FIBER)

RT CRIMP INDEX
CRIMP RADIUS
CRIMPED YARNS
DIFFERENTIAL SHRINKAGE
FIBER CRIMP
HELICAL CRIMP
LATENT CRIMP
NATURAL CRIMP
SELF CRIMPING YARNS
TEXTURED YARNS (FILAMENT)
YARN CRIMP
YARN GEOMETRY
YARN STRUCTURE

CRIMP INDEX
BT FIBER PROPERTIES
RT CRIMP
CRIMP AMPLITUDE (FIBER)
CRIMP FREQUENCY
CRIMP RADIUS
CRIMP REMOVAL
CRIMPED YARNS
FIBER CRIMP
HELICAL CRIMP
LATENT CRIMP
NATURAL CRIMP
SELF CRIMPING YARNS
TEXTURED YARNS (FILAMENT)
YARN CRIMP
YARN GEOMETRY
YARN STRUCTURE
Z CRIMP

CRIMP INTERCHANGE
RT BEATING UP
BIAXIAL STRESS
CLOTH FELL
COVER FACTOR
CRIMP
CRIMP BALANCE
CRIMP PERCENT
CRIMP REMOVAL
ENDS PER INCH
FABRIC ANALYSIS
FABRIC CRIMP
FABRIC GEOMETRY
JAMMING
JAMMING POINT
LOOM TAKE UP
PICKS PER INCH
SELF CRIMPING YARNS
SHEDDING
STRESS STRAIN CURVES
TEMPLES
WARP TENSION
WEAVING
WEAVING DYNAMICS
YARN CRIMP
Z CRIMP

CRIMP PERCENT
UF PERCENTAGE CRIMP
CRIMP RATIO
RT CRIMP
CRIMP INTERCHANGE
CRIMP REMOVAL
FABRIC GEOMETRY
FABRIC STRUCTURE
YARN CRIMP
YARN GEOMETRY
YARN STRUCTURE

CRIMP RATIO
USE CRIMP PERCENT

CRIMP RADIUS
RT CRIMP
CRIMP AMPLITUDE (FIBER)
CRIMP FREQUENCY
CRIMP INDEX
CRIMP REMOVAL
FIBER CRIMP
HELICAL CRIMP
NATURAL CRIMP
YARN GEOMETRY
YARN STRUCTURE

CRIMP RECOVERY
BT RECOVERY (SELF RESTORATION TO ORIGINAL CONDITION)
RT BENDING RECOVERY
CREEP RECOVERY
CRIMP RETENTION
DELAYED RECOVERY
ELASTIC RECOVERY
FIBER CRIMP
WORK RECOVERY
YARN CRIMP

CRIMP REMOVAL
RT CRIMP
CRIMP INDEX
CRIMP INTERCHANGE
CRIMP PERCENT
CRIMP RADIUS
CRIMP RETENTION
CRIMP TENDENCY

CRIMP RETENTION
RT CRIMP RECOVERY
CRIMP REMOVAL
DIMENSIONAL STABILITY
FIBER CRIMP
PERMANENT SET

CRIMP TENDENCY
UF CRIMPING POTENTIAL
CRIMPING TENDENCY
RT BICOMPONENT FIBER YARNS (FILAMENT)
BICOMPONENT FIBERS
CRIMP REMOVAL
DIFFERENTIAL SHRINKAGE
FIBER CRIMP
SELF CRIMPING YARNS
YARN CRIMP
Z CRIMP

CRIMP TESTERS
BT TESTING EQUIPMENT
RT CRIMP
FIBER CRIMP
YARN CRIMP

CRIMPED YARNS
NT AGILON (TN)
BANLON (TN)
BICOMPONENT FIBER YARNS (FILAMENT)
BICOMPONENT FIBER YARNS (STAPLE)
CELACRIMP (TN)
EDGE CRIMPED YARNS
FALSE TWIST YARNS
NYLOFT (TN)
SELF CRIMPING YARNS
SPUNIZE (TN)
STUFFER BOX CRIMPED YARNS
BT FILAMENT YARNS
TEXTURED YARNS (FILAMENT)
YARNS
RT BULKED YARNS
BULKING
CRIMP
CRIMP FREQUENCY
CRIMP INDEX
CRIMPING
CRIMPING MACHINES
CRINKLE TYPE YARNS
EDGE CRIMPING
FALSE TWIST
FALSE TWISTING (TEXTURING)
LOOP YARNS
STRETCH FABRICS
STRETCH KNITTED FABRICS
STRETCH WOVEN FABRICS
STRETCH YARNS (FILAMENT)
STUFFER BOX CRIMPING
TEXTRALIZED YARNS (TN)
TEXTURING
YARN CRIMP
Z CRIMP

CRIMPERS
USE CRIMPING MACHINES

CRIMPING
NT AIR JET TEXTURING
BELT CRIMPING
CRIMP FIXING
EDGE CRIMPING
GEAR CRIMPING
STUFFER BOX CRIMPING
TOW CRIMPING
BT TEXTURING
RT AIR JET TEXTURING MACHINES
AIR JET TEXTURED YARNS
BANLON (TN)
BICOMPONENT FIBER YARNS (FILAMENT)
BICOMPONENT FIBER YARNS (STAPLE)
BULKED YARNS
BULKING
CREPING
CRIMP
CRIMPED YARNS
EDGE CRIMPED YARNS
FALSE TWISTING (TEXTURING)
HEAT SETTING (SYNTHETICS)
LATENT CRIMP
SANDWICH ROLLING MILLS
SELF CRIMPING YARNS
STUFFER BOXES
TEXTURED YARNS (FILAMENT)
THROWING
YARN CRIMP
Z CRIMP

CRIMPING MACHINES
UF CRIMPERS
NT AIR JET TEXTURING MACHINES
EDGE CRIMPING MACHINES
STUFFER BOXES
BT THROWING MACHINES
RT BICOMPONENT FIBER YARNS (FILAMENT)
BICOMPONENT FIBER YARNS (STAPLE)
CRIMP
CRIMPED YARNS
FALSE TWISTERS
GEAR CRIMPING
TEXTURING

CRIMPING POTENTIAL
USE CRIMP TENDENCY

CRIMPING TENDENCY
USE CRIMP TENDENCY

CRINKLE
USE CRIMP

CRINKLE TEXTURED YARNS
USE CRINKLE TYPE YARNS

CRINKLE TYPE YARNS
UF CRINKLE TEXTURED YARNS
KNIT DEKNIT YARNS
BT BULKED YARNS
STRETCH YARNS (FILAMENT)
TEXTURED YARNS (FILAMENT)
YARNS
RT CRIMPED YARNS
HEAT SETTING (SYNTHETICS)
KNITTING
UNRAVELLING

CRINOLINE
BT WOVEN FABRICS
RT SHEETING (FABRIC)
STARCH
TOBACCO CLOTH
THROWING

CRITICAL MICELLE CONCENTRATION
UF CMC
RT MICELLES
SURFACTANTS

CRITICAL MOISTURE CONTENT
BT MOISTURE CONTENT
RT ABSORBENCY (MATERIAL)
CONSTANT RATE PERIOD
DEW POINT
DIFFUSIVITY
DRY BULB TEMPERATURE
DRYING
EQUILIBRIUM MOISTURE CONTENT
FABRIC PROPERTIES
FALLING RATE PERIOD
HUMIDITY
MOISTURE STRENGTH CURVES
PSEUDO WET BULB TEMPERATURE
PSYCHROMETRY
REGAIN
RELATIVE HUMIDITY
SPECIFIC HUMIDITY
VAPOR PRESSURE
WET BULB TEMPERATURE

CRITICAL VELOCITY
RT IMPACT
IMPACT TESTING
IMPACT PENETRATION TESTING
SHOCK RESISTANCE

CROCHET NEEDLES
BT NEEDLES
RT CROCHETING
CROCHETING MACHINES
KNITTING NEEDLES

CROCHETED LACES
NT CLUNY (LACE)
IRISH (LACE)
BT HAND MADE LACES
LACES
RT CROCHETING MACHINES

CROCHETING
RT CROCHET NEEDLES
CROCHETING MACHINES
HAND SEWING

CROCHETING MACHINES
RT CROCHET NEEDLES
CROCHETED LACES
CROCHETING
LACE MACHINES

CROCK TESTING
UF CROCK TESTS
BT COLORFASTNESS TESTING
TESTING
RT CROCKFASTNESS
CROCKING

CROCK TESTS
USE CROCK TESTING

CROCKFASTNESS
UF CROCKING RESISTANCE
RUBFASTNESS
BT COLORFASTNESS
END USE PROPERTIES
RT CROCK TESTING
CROCKING
DYEING
SOAPING
SOIL RESISTANCE
STAIN RESISTANCE
WASHFASTNESS (COLOR)
WET CROCKING

CROCKING
NT WET CROCKING
RT BLEEDING
CROCK TESTING
CROCKFASTNESS
DYEING
FADING
RUBBING
SCRUBBING
SOILING
STAINING

CROCKING RESISTANCE
USE CROCKFASTNESS

CROPPING
USE SHEARING (FINISHING)

CROSROL WEB CLEANERS
USE CROSROL WEB PURIFIER (TN)

CROSROL WEB PURIFIER (TN)
UF CROSROL WEB CLEANERS
BT WEB PURIFIERS
RT CARD SLIVERS
CARDS
COTTON CARDS
CROSS ROLLS
HIGH PRODUCTION CARDS
PERALTA ROLLS
TRUMPETS
WEBS

CROSS DYEING
NT TONE IN TONE DYEING
TWO COLOR DYEING
TWO TONE DYEING
BT DYEING (FOR EFFECT)
RT BLEND DYEING
MULTICOLORS
UNION DYEING

CROSS LAID WEBS
RT BONDED FIBER FABRICS
CARD WEBS
NONWOVEN FABRICS
RANDO WEBS (TN)
WEB ORIENTATION

CROSS LAID YARN FABRICS
NT MALIMO FABRIC (TN)
BT FABRICS (ACCORDING TO STRUCTURE)
RT ARACHNE SYSTEMS (TN)
BONDED YARN FABRICS
COMPOSITES
CROSS LAID YARN FABRICS
MALI PROCESSES (TN)
NEEDLED FABRICS
NONWOVEN FABRICS
STITCH BONDED FABRICS
STITCH REINFORCED FABRICS
STITCHED PILE FABRICS
WOVEN FABRICS

CROSS PLAITING
RT KNITTING

CROSS ROLLS
UF CRUSH ROLLS
BT WEB PURIFIERS
RT BURR CRUSHERS
CARD WEBS
CARDS
CLEANLINESS
CROSROL WEB PURIFIER (TN)
NIP PRESSURE
PERALTA ROLLS

CROSS SECTIONS
RT CROSS SECTIONAL AREA
DIAMETER
FIBER CROSS SECTIONS
FINENESS
NEUTRAL AXIS
PERIMETER
ROUNDNESS
SHAPE

CROSS SECTIONAL AREA
RT AREA
CIRCUMFERENCE
CROSS SECTIONS
DIAMETER
FINENESS
PERIMETER

CROSS SECTIONAL SHAPE METHOD (COTTON IMMATURITY)
BT IMMATURITY TESTING (COTTON FIBER)

CROSS STITCHES
BT STITCHES (SEWING)

CROSS WINDING
BT WINDING
RT CLOSE WINDING
OPEN WINDING
PARALLEL WINDING

CROSSBRED WOOL
BT WOOL
RT GRADE (FIBERS)
LAMBSWOOL
LONG WOOL
MEDIUM WOOL
MERINO WOOL
MOUNTAIN WOOL
WOOL GRADE

CROSSLINK DENSITY
RT CROSSLINKING
CURING
ELASTIC MODULUS (TENSILE)
POLYMERIZATION
REACTANTS
RUBBER ELASTICITY

CROSSLINKING
BT REACTIONS (CHEMICAL)
RT ACETAL RESINS
ADIPALDEHYDE
AZIDES
CARBODIIMIDES
CHEMICAL MODIFICATION (FIBERS)
CONDENSATION POLYMERIZATION
COPOLYMERIZATION
COPOLYMERS
CROSSLINK DENSITY
CURING

CYSTINE
DIALDEHYDE STARCH
DIFFERENTIAL DISTRIBUTION (CROSSLINKING)
DISULFIDE BONDS
DISULFIDES (ORGANIC)
DMEU
EPICHLOROHYDRIN
FORMALDEHYDE
GELLING
GELS
GRAFTING
HALOHYDRINS
IMBIBITION
IONIZING RADIATION
IRRADIATION
ISOCYANATES
MOLECULAR STRUCTURE
PHASE BOUNDARY CROSSLINKING
POLYMERIZATION
REACTANTS
WATER IMBIBITION
WATER RETENTION

CROSSLINKING AGENTS (FIBERS)
USE REACTANTS

CROWNS
RT FABRIC GEOMETRY
FABRIC STRUCTURE

CROWNS (ROLL)
UF ROLL CROWNS
RT BOWING
CALENDER ROLLS
ROLL BOSSES
ROLL ECCENTRICITY
ROLLS

CRUSH RESISTANCE
BT END USE PROPERTIES
FABRIC PROPERTIES (MECHANICAL)
RT MATTING
STRESS STRAIN PROPERTIES
STRESS STRAIN PROPERTIES (COMPRESSIVE)

CRUSH ROLLS
USE CROSS ROLLS

CRUSHING
USE GRINDING (COMMINUTION)

CRYSTAL DEFECTS
RT CRYSTALLIZATION
DISLOCATIONS
SLIP PLANES

CRYSTAL LATTICE
RT CELLULOSE I
CELLULOSE II
CELLULOSE III
CELLULOSE IV
CHAIN FOLDING
CRYSTALLINE REGION
CRYSTALLITES
DEGREE OF CRYSTALLINITY
MELT CRYSTALLIZATION
SPHERULITES

CRYSTALLINE ORIENTATION
RT COLD DRAWING
CRYSTALLINE POLYMERS
CRYSTALLINE REGION
FINE STRUCTURE (FIBERS)

CRYSTALLINE POLYMERS
RT AMORPHOUS POLYMERS
CRYSTALLINE REGION
CRYSTALLITES
CRYSTALLIZATION
DEGREE OF POLYMERIZATION
FIBRILS
FINE STRUCTURE (FIBERS)
FRINGED MICELLES
LONG CHAIN MOLECULES
POLYMERS

CRYSTALLINE REGION
BT FINE STRUCTURE (FIBERS)
RT AMORPHOUS REGION
CRYSTAL LATTICE
CRYSTALLITES
CRYSTALLIZATION
DEGREE OF CRYSTALLINITY
FIBRILLATION
FIBRILS
FRINGED MICELLES
GRAIN BOUNDARIES
LONG CHAIN MOLECULES
MELT CRYSTALLIZATION
MOLECULAR STRUCTURE
SPHERULITES
SPHERULITIC FIBRILS

CRYSTALLINITY
RT ACCESSIBILITY (INTERNAL)
ALPHA BETA TRANSFORMATION
CRYSTALLIZATION
DECRYSTALLIZATION
DEGREE OF ORIENTATION
DISCOLORATION
FINE STRUCTURE (FIBERS)
MOLECULAR STRUCTURE
RECRYSTALLIZATION
SPINNING (EXTRUSION)
STRUCTURE OF MATERIALS
X RAY ANALYSIS

CRYSTALLITES
BT FINE STRUCTURE (FIBERS)
RT CRYSTAL LATTICE
CRYSTALLINE POLYMERS
CRYSTALLINE REGION
CRYSTALLIZATION
FIBRIL REVERSALS
FIBRILS
MICROFIBRILS

CRYSTALLIZATION
NT SPHERULITIC CRYSTALLIZATION
RT ANNEALING
CHAIN FOLDING
CRYSTAL DEFECTS
CRYSTALLINE POLYMERS
CRYSTALLINE REGION
CRYSTALLINITY
CRYSTALLITES
DECRYSTALLIZATION
DEGREE OF CRYSTALLINITY
DENSITY
DRAWING (FILAMENT)
ELECTRON MICROSCOPY
FINE STRUCTURE (FIBERS)
FREEZING
HEAT SETTING (SYNTHETICS)
MELT CRYSTALLIZATION
MELT SPINNING
METAL FIBERS
MICELLES
ORIENTATION
QUENCHING (FILAMENT)
RECRYSTALLIZATION
SATURATED SOLUTIONS
SOLIDIFICATION
SHRINKING
SOLUBILITY
SPHERULITES
STRESS RELAXATION
THERMOPLASTICITY
X RAY DIFFRACTION

CUEN
USE CUPRIETHYLENEDIAMINE HYDROXIDE

CUFFS
BT GARMENT COMPONENTS

CUMULOFT (TN)
BT NYLON (POLYAMIDE FIBERS)
NYLON 66
TEXTURED YARNS (FILAMENT)

CUOXAM
USE CUPRAMMONIUM HYDROXIDE

CUPRAMMONIUM FLUIDITY
BT CHEMICAL PROPERTIES
DEGRADATION PROPERTIES
RT DEGREE OF POLYMERIZATION
MOLECULAR WEIGHT
MOLECULAR WEIGHT CONTROL
VISCOSITY

CUPRAMMONIUM HYDROXIDE
UF CUOXAM
BT CUPRIC COMPOUNDS
RT CADOXEN
CELLULOSE
CUPRIETHYLENEDIAMINE HYDROXIDE
FERRIC TARTRATE
MARSCHALL SOLUTION
REGENERATED CELLULOSE
SOLVENTS
VISCOSITY

CUPRAMMONIUM RAYON
UF RAYON (CUPRAMMONIUM)
BEMBERG FIBERS
BT CELLULOSIC FIBERS
MAN MADE FIBERS
RAYON (REGENERATED CELLULOSE FIBERS)
RT ACETATE FIBERS
CELLULOSE ESTER FIBERS
FORTISAN (TN)
TRIACETATE FIBERS
VISCOSE RAYON

CUPREL (TN)
BT TEXTURED YARNS (FILAMENT)

CUPRIC CHLORIDE
BT CHLORIDES
CUPRIC COMPOUNDS
RT DYE FIXING AGENTS
MORDANT DYEING

CUPRIC CHROMATE
BT CUPRIC COMPOUNDS

CUPRIC COMPOUNDS
UF CUPRIC SULFITE
NT COPPER NAPHTHENATE
COPPER PENTACHLOROPHENATE
COPPER 8-HYDROXYQUINOLATE
CUPRAMMONIUM HYDROXIDE
CUPRIC CHLORIDE
CUPRIC CHROMATE
CUPRIC SULFATE
CUPRIETHYLENEDIAMINE HYDROXIDE
BT COPPER COMPOUNDS
RT CUPROUS COMPOUNDS
FERRIC COMPOUNDS
SODIUM CUPRATE

CUPRIC SULFATE
BT CUPRIC COMPOUNDS
SULFATES
RT DYE FIXING AGENTS
WASHFASTNESS (COLOR)

CUPRIC SULFITE
USE CUPRIC COMPOUNDS

CUPRIETHYLENEDIAMINE HYDROXIDE
UF CUEN
BT CUPRIC COMPOUNDS
RT CADOXEN
CELLULOSE
CUPRAMMONIUM HYDROXIDE
FERRIC TARTRATE
MARSCHALL SOLUTION
SOLVENTS
VISCOSITY

CUPROUS CHLORIDE
BT CHLORIDES
CUPROUS COMPOUNDS
RT CATALYSTS

CUPROUS COMPOUNDS
NT CUPROUS CHLORIDE
BT COPPER COMPOUNDS
RT CUPRIC COMPOUNDS
FERROUS COMPOUNDS

CURING
UF THERMOSETTING
NT DRY CURING
WET CURING
BT DRY FINISHING
WET FINISHING
RT BONDED FIBER FABRICS
CATALYSTS
CRIMP FIXING
CROSSLINK DENSITY
CROSSLINKING

CURING OVENS
CURING RANGES
DURABLE PRESS

RT DEFERRED CURE METHOD
DOUBLE CURE METHOD
DURABLE PRESS TREATMENTS
DYEING
HEAT
HEAT SETTING (SYNTHETICS)
HEAT TREATMENT
HEATING
HOT PRESS METHOD (DURABLE PRESS)
IRONING
KORATRON (TN)
OVENS
PAD BATCH CURE PROCESS
PAD DRY CURE PROCESS
PERMANENT PLEATING
PRE-CURE METHOD
REACTANTS
RESINS
STEAMING
THERMOPLASTICITY
THERMOSETTING POWDERS
THERMOSETTING RESINS
THERMOSOL PROCESS (TN)
WASH WEAR TREATMENTS

CURING AGENTS
USE REACTANTS

CURING OVENS
UF RESIN CURING MACHINES
BT CURING RANGES
RT CURING
DRY CURING
HEATING EQUIPMENT
OVENS
PAD DRY CURE PROCESS

CURING RANGES
NT CURING OVENS
RT CONTINUOUS FINISHING
CURING
DRY CURING
PAD DRY CURE PROCESS

CURL
BT MODES OF DEFORMATION
RT BENDING
BENDING ENERGY
DIMENSIONAL STABILITY
PLAIN KNITTED FABRICS
STABILITY
TORSIONAL BUCKLING

CURL PILE FABRICS
BT PILE FABRICS
WOVEN FABRICS
RT CUT PILE FABRICS
LOOP PILE FABRICS
PILE DENSITY
PILE HEIGHT
SCULPTURED PILE FABRICS

CURL TENDENCY
USE CURLING TENDENCY

CURLATORS (TN)
RT AIR BRIDGES
CONDENSER SCREENS
RANDO WEBBER (TN)
RANDO WEBS (TN)
TUFT FORMERS

CURLING
BT FABRIC DEFECTS
RT DISTORTION

CURLING TENDENCY
UF CURL TENDENCY
BT FABRIC PROPERTIES
FABRIC PROPERTIES (MECHANICAL)
RT MECHANICAL PROPERTIES

CURRENT (ELECTRICAL)
UF ELECTRIC CURRENT
RT CAPACITANCE (ELECTRICAL)
VOLTAGE

CURRENTS
RT AERODYNAMICS
AIR DRAG
AIR FLOW
AIR STREAMS
CIRCULATION
EDDIES
FLUID FLOW
FLUID SPINNING

CURTAIN MACHINES
UF LACE FURNISHING MACHINES
BT LACE MACHINES
RT BARMEN MACHINES
BAR WARP MACHINES
BOBBINET MACHINES
DOUBLE LOCKER MACHINES
GO THROUGH MACHINES
LEAVERS MACHINES
MECHLIN MACHINES
ROLLING LOCKER MACHINES
SIVAL MACHINES
STRING WARP MACHINES
WARP LACE MACHINES

CURTAINS
USE DRAPES

CURVATURE
RT BENDING
BENDING RIGIDITY
CAMBER
COUPLES
DIFFERENTIAL GEOMETRY
MOMENT CURVATURE CURVES
RADIUS OF CURVATURE
SHAPE
STRAIGHTENING
TORSION (GEOMETRIC)

CURVED SEAMING
BT GARMENT MANUFACTURE
SEAMING
RT ANGULAR SEAMING
BACK TACKING
BASTING
CHAIN SEWING
GATHERING
HAND SEWING
HEMMING
MACHINE SEWING
SEAM BINDING
SEAMSTRESSES
SEWABILITY
SEWING
SEWING MACHINES
TACKING
TAILORING
TUCKING (GARMENT MANUFACTURE)

CURVILINEAR SCALE
RT CAUSTICAIRE SCALE
CAUSTICAIRE VALUE
COTTON
FIBER FINENESS
FINENESS TESTING
MICRONAIRE

CUSHIONS
RT FURNISHING FABRICS
FURNISHINGS
HOUSEHOLD FABRICS
UPHOLSTERY
UPHOLSTERY FABRICS

CUT SYSTEM
UF AMERICAN CUT
BT COUNT
INDIRECT COUNT
RT RUN SYSTEM

CUT MARKS (SLASHING)
UF KEEL (SLASHING)
RT SLASHERS
WARP PREPARATION

CUT PILE FABRICS
NT VELVET
VELVET CARPETS
VELVETEEN
BT PILE FABRICS (WOVEN)
WOVEN FABRICS
RT CURL PILE FABRICS
LOOP PILE FABRICS
PILE DENSITY
PILE HEIGHT
SCULPTURED PILE FABRICS

CUT PILE TUFTING MACHINES
BT TUFTING MACHINES
RT LOOP PILE TUFTING MACHINES

CUT RUCHE
BT RUCHE
RT PIPING

CUTICLE
BT FINE STRUCTURE (FIBERS)
RT CORTEX
EPICUTICLE
MEDULLA
ORTHOCORTEX
PARACORTEX
PRIMARY WALLS
SCALES (WOOL FIBERS)
SECONDARY WALLS
WOOL

CUTTERS
NT STAPLE CUTTERS
THREAD CUTTERS
RT CUTTING MACHINES
KNIVES
SCISSORS

CUTTING
UF CHOPPING
RT ABRASION
CHIPPING
COMMINUTION
CUTTING RATIO
CUTTING RESISTANCE
DRAFT CUT
EMBEDDING MEDIUM
ENGRAVING
FIBER BREAKAGE
FIBER CUTTING
FLAKING
FRACTURING
GALLING
GRINDING (MATERIAL REMOVAL)
KNIVES
MICROTOMES
PAPER MAKING
PLOUGHING
POLISHING
SCISSORS
SCORING
SCRATCHING
SCUFFING
SHARPENING
SPLITTING
STAPLE CUTTERS
TEARING (TAILORING)
YARN SEVERANCE

CUTTING (TAILORING)
UF KNIFING
SHEARING (TAILORING)
BT GARMENT MANUFACTURE
RT APPAREL DESIGN
BUNDLING (TAILORING)
CLIPPINGS
CUTTING DRILLS
CUTTING MACHINES
CUTTING SHEARS
DRESSMAKING
GARMENTS
KNIVES
MARKING
PATTERN (APPAREL)
PINKING (TAILORING)
PUNCHING
RIPPING
SEWING
SHADING (TAILORING)
SHARPENING
SLITTING
SLOPING
SPREADING (TAILORING)
TAILORING
TEARING (TAILORING)
TRIMMING (OPERATION)

CUTTING DRILLS
NT AWL CUTTING DRILLS
HYPODERMIC CUTTING DRILLS
BT CUTTING MACHINES
GARMENT MANUFACTURING MACHINES
RT CUTTING (TAILORING)
GARMENT MANUFACTURE
KNIVES
TAILORING

CUTTING MACHINES
NT BAND KNIFE CUTTING MACHINES
CLICKERS
CUTTING DRILLS
CUTTING SHEARS
DIE CUTTERS
RECIPROCATING CUTTING MACHINES
ROTARY CUTTING MACHINES
SCISSORS
SLITTERS (CLOTH)
BT GARMENT MANUFACTURING MACHINES
RT CUTTERS
CUTTING (TAILORING)
GARMENT MANUFACTURE
KNIVES
TAILORING

CUTTING MARKERS
RT MARKING

CUTTING RATIO
RT CUTTING
FIBER CUTTING
FIBER LENGTH

CUTTING RESISTANCE
BT MECHANICAL DETERIORATION PROPERTIES
MECHANICAL PROPERTIES
STRESS STRAIN PROPERTIES
STRESS STRAIN PROPERTIES (SHEAR)
RT CUTTING
FABRIC PROPERTIES
FABRIC PROPERTIES (MECHANICAL)
FIBER CUTTING
SHEAR RESISTANCE
STRENGTH
TEAR STRENGTH

CUTTING SHEARS
BT CUTTING MACHINES
(CONT.)

RT CUTTING (TAILORING)
KNIVES
SCISSORS

CUTTING TESTING
BT TESTING
RT ABRASION TESTING
CUTTING RESISTANCE
TEAR TESTING

CYANATES
NT POTASSIUM CYANATE

CYANIDES
NT POTASSIUM CYANIDE
SODIUM CYANIDE

CYANO COMPOUNDS
USE NITRILES

CYANOETHYLCELLULOSE
BT CELLULOSE ETHERS
RT ACETATE FIBERS
ALKALI CELLULOSE
CARBOXYMETHYLCELLULOSE
CELLULOSE XANTHATE
CHEMICAL MODIFICATION (FIBERS)
CHEMICALLY MODIFIED COTTON
CYANOETHYLATED COTTON
RAYON (REGENERATED CELLULOSE FIBERS)
REGENERATED CELLULOSE

CYANOETHYLATED COTTON
BT CHEMICALLY MODIFIED COTTON
RT ACETYLATED COTTON
ACRYLONITRILE
BENZOYLATED COTTON
CARBAMOYLETHYLATED COTTON
CHEMICAL MODIFICATION (FIBERS)
CYANOETHYLATION
CYANOETHYLCELLULOSE
DS (CELLULOSE)
ETHERIFICATION
REACTANTS

CYANOETHYLATION
BT ETHERIFICATION
RT ACRYLONITRILE
CHEMICAL MODIFICATION (FIBERS)
CHEMICALLY MODIFIED COTTON
CYANOETHYLATED COTTON
NITRILES

CYANURIC ACID
BT TRIAZINES
RT CHLOROCYANURIC ACID
CYANURIC CHLORIDE

CYANURIC CHLORIDE
BT CHLORINE COMPOUNDS
BT TRIAZINES
RT CHLOROCYANURIC ACID
CYANURIC ACID
MELAMINE

CYBERNETICS
RT AUTOMATIC CONTROL
AUTOMATION

CYCLIC COMPRESSION
USE CYCLIC STRESS

CYCLIC STRESS
UF CYCLIC COMPRESSION
CYCLIC TENSION
RT CREEP
FATIGUE
HYSTERESIS (MECHANICAL)

CYCLIC TENSION
USE CYCLIC STRESS

CYCLOHEXANE, 1,4-BIS (METHYLAMINE)-
BT AMINES
RT MONOMERS
NYLON (POLYAMIDE FIBERS)
POLYAMIDES

CYCLOSET YARNS (TN)
BT FILAMENT YARNS
RT ACETATE FIBERS
FALSE TWIST
INTERLACING (FILAMENT IN YARN)
PRODUCERS TWIST
PRODUCERS YARNS
REAL TWIST
ROTOSET YARNS (TN)
TEXTURED YARNS (FILAMENT)
ZERO TWIST

CYLINDER DRYERS
USE DRUM DRYERS

CYLINDER MACHINES
BT PAPER MACHINES
RT FOURDRINIER MACHINES
INVERFORM MACHINES
KAMYR MACHINES
ROTAFORMERS
VERTAFORMERS
WET MACHINES (PAPERMAKING)

CYLINDER NEEDLES
BT KNITTING NEEDLES
NEEDLES
RT CIRCULAR KNITTING MACHINES
DIAL NEEDLES
HOSIERY MACHINES
KNITTING
KNITTING MACHINES
LATCH NEEDLES
SEAMLESS HOSIERY MACHINES

CYLINDER PRINTING
USE ROLLER PRINTING

CYLINDER SCREENS
RT CARD FLATS
COTTON CARDS
LICKER IN SCREENS
MOTES
PERCENTAGE PLATES
WASTE

CYLINDER STRIPS
BT CARD WASTE
RT FLAT STRIPS
IMPURITIES

CYLINDERS
NT ENGRAVED CYLINDERS
PERFORATED CYLINDERS
RT ELLIPTICITY
GEOMETRY
GROOVES
HYDRAULICS
ROUNDNESS
SPHERICITY

CYLINDERS (CARD)
USE CARD CYLINDERS

CYSTEINE
BT AMINO ACIDS
MERCAPTANS
RT CYSTINE
DISULFIDE INTERCHANGE
KERATIN
PROTEINS
REDUCTION
WOOL

CYSTINE
BT AMINO ACIDS
DISULFIDES (ORGANIC)
RT CHEMICAL MODIFICATION (FIBERS)
CROSSLINKING
CYSTEINE
DISULFIDE BONDS
DISULFIDE INTERCHANGE
DURABLE PRESS
KERATIN
PROTEINS
REDUCTION
THIOETHERS
WOOL

CYSTINE BRIDGE
USE DISULFIDE BONDS

DABBING (COMBING)
RT COMBING
DABBING BRUSHES
NOBLE COMBING

DABBING BRUSHES
RT COMB PINS
DABBING (COMBING)
NOBLE COMBS
SETOVER

DACRON (TN)
UF AMILAR (TN)
FIBER V (TN)
BT POLYESTER FIBERS
RT POLY (ETHYLENE GLYCOL TEREPHTHALATE)

DALEMBERTS FORCE
BT FORCE
RT CORIOLIS FORCE
INERTIA

DAMAGE
UF MECHANICAL DAMAGE
NT SEAM DAMAGE
SEWING DAMAGE
YARN DAMAGE
RT CHAFE MARKS
DECOMPOSITION
DEFORMATION
DEGRADATION
DURABILITY
FATIGUE
FLAKING
FRACTURING
LARVAE
LAUNDRY DEGRADATION
MECHANICAL HISTORY
MECHANICAL PROPERTIES
TENDERING
WEAR

DAMASK
NT TABLE DAMASK
BT FIGURED FABRICS
FLAT WOVEN FABRICS
WOVEN FABRICS
RT BROCADES
BROCATELLES
JACQUARD WEAVES
SATIN WEAVES

DAMPENING
UF DAMPING (MOISTURE)
DEWING
BT WET FINISHING
RT IRONING
PRESSING

DAMPERS
NT VIBRATION DAMPERS
RT DAMPING
DASHPOTS
ENERGY ABSORPTION
IMPACT
IMPULSES
RESONANT FREQUENCY
SHOCK
SHOCK ABSORBERS
SHOCK RESISTANCE
VIBRATION
VIBRATION ISOLATORS
VISCOUS DAMPING

DAMPING
NT MECHANICAL DAMPING
VISCOUS DAMPING
RT DAMPERS
DISSIPATION
DYNAMIC CHARACTERISTICS
DYNAMIC RESPONSE
DYNAMICS
ENERGY ABSORPTION
EXHAUSTION (MECHANICAL)
HYSTERESIS (MECHANICAL)
INTERNAL FRICTION
SHOCK ABSORBERS
VIBRATION
VIBRATION DAMPERS
VIBRATION ISOLATORS
VISCOELASTICITY

DAMPING (MOISTURE)
USE DAMPENING

DARLAN (TN)
USE DARVAN (TN)

DARNED LACES
BT HAND MADE LACES
LACES

DARNING
RT HAND SEWING
STITCHING

DARNING NEEDLES
BT NEEDLES
RT BURLING
DARNING
KNITTING NEEDLES
SEWING NEEDLES

DARTS
BT GARMENT COMPONENTS
RT GARMENT MANUFACTURE
SEWING
STITCHING
TAILORING

DARVAN (TN)
UF DARLAN (TN)
BT NYTRIL FIBERS

DASHPOTS
RT DAMPERS
ENERGY ABSORPTION
MAXWELL MODEL
MECHANISMS
SHOCK ABSORBERS
SPRINGS
VISCOELASTICITY
VISCOUS DAMPING
VOIGT MODEL

DATA REDUCTION
RT ARITHMETIC
COMPUTATIONS
COUNTING
STATISTICAL METHODS

DAWSON WHEELS
USE PATTERN WHEELS

DAYLIGHT
NT ARTIFICIAL DAYLIGHT
SUNLIGHT
BT LIGHT
RT BRIGHTNESS
COLOR
COLOR MATCHING
COLORFASTNESS
COLORIMETRY
FADING
FLUORESCENT LIGHT
ILLUMINATION
INCANDESCENT LIGHT
LIGHTFASTNESS (COLOR)
LIGHTFASTNESS (OF FINISH)
MONOCHROMATIC LIGHT
POLARIZED LIGHT

DEACETYLATED CHITIN
BT SIZES (SLASHING)
RT CARBOHYDRATES
CHITIN
POLYMERS

DEAD WEIGHT LOADING
BT LOADING
RT DRAWFRAMES
LET OFF MOTIONS
SPRING LOADING
TOP ROLLS
WEIGHT

DEAFNESS
RT OCCUPATIONAL HAZARDS
SAFETY

DECATING
UF DECATIZING
DRY DECATING
WET DECATING
RT DECATING MACHINES
HEAT SETTING (SYNTHETICS)
LUSTER
NAPPING
PRESSING
SHRINK RESISTANCE
SHRINKING
TAILORING

DECATING MACHINES
BT GARMENT MANUFACTURING MACHINES
RT DECATING
STEAMERS

DECATIZING
USE DECATING

DECELERATION
USE ACCELERATION (MECHANICAL)

DECERESOL OT (TN)
BT DIOCTYL SULFOSUCCINATE

DECOLORIZING
RT BLEACHING
COLORFASTNESS
DYEING
SCOURING
STRIPPING (COLOR)
WASHFASTNESS (COLOR)

DECOLORIZING CARBON
USE ACTIVATED CARBON

DECOMPOSITION
RT CHEMICAL STABILITY
CHLORINE DAMAGE
DAMAGE
DEGRADATION
DEPOLYMERIZATION
HYDROLYSIS
OXIDATION

DECORATIVE FABRICS
BT FABRICS (BY END USES)
RT BUNTING
CARPETS
DRAPES
END USES
FABRICS (ACCORDING TO STRUCTURE)
FLAGS
FURNISHING FABRICS
TAPESTRIES

DECORTICATING
RT BAST FIBERS
FLAX
LINEN
SEPARATION

DECRYSTALLIZATION
RT ANNEALING
CRYSTALLINITY
CRYSTALLIZATION
DRAWING (FILAMENT)
HEAT SETTING (SYNTHETICS)
MELT SPINNING
RECRYSTALLIZATION
SWELLING

DECRYSTALLIZED COTTON
RT AMINES
CHEMICAL MODIFICATION (FIBERS)
CHEMICALLY MODIFIED COTTON
COTTON
SWELLING

DEFECTS
UF BLEMISHES
FAULTS
FLAWS
IMPERFECTIONS
NT BREAKS
CRACKS (FABRIC)
CRACKS (MECHANICAL)
FABRIC DEFECTS
FROSTING
KNOTS
NEPS
NUBS
PILLS
SLUBS
SOILING
STAINS
STREAKINESS
THICK SPOTS
YARN DEFECTS
RT BURLING
DETECTION
DRAFTING WAVES
END BREAKAGES
ENTANGLEMENTS
FABRIC INSPECTION
IMPURITIES
INSPECTION
IRREGULARITY
MENDING
NEP POTENTIAL
QUALITY CONTROL
REDYEING
SPECKING
WEAK SPOTS

DEFERRED CURE METHOD
UF DELAYED CURE
POST CURE
BT DURABLE PRESS TREATMENTS
RT DOUBLE CURE METHOD
DRY CURING
DURABLE PRESS FINISHES
HOT PRESS METHOD (DURABLE PRESS)
PRE-CURE METHOD
RESIN FINISHES
WET CURING

DEFLECTION
RT BENDING
BENDING LENGTH
BENDING MOMENTS
BENDING RIGIDITY
CAMBER
DEFORMATION
DISTORTION
DRAPE
EXTENSIBILITY
LOAD DEFLECTION CURVES
PERMANENT DEFORMATION
(CONT.)

RT RESILIENCE
SAG
SPRING CONSTANT
STRAIN

DEFOAMERS
USE ANTIFOAM AGENTS

DEFORMATION
UF DEFORMING
NT PERMANENT DEFORMATION
RT BENDING
BENDING RIGIDITY
BOWING
BUCKLING
CAMBER
CREASE RETENTION
CREASING
DAMAGE
DEFLECTION
DIMENSIONAL STABILITY
DISTORTION
DRAPE
DRAWING (FILAMENT)
ELASTIC STRAIN
EXTENSIBILITY
MECHANICAL PROPERTIES
PLASTIC STRAIN
RESILIENCE
SAG
SLIP PLANES
SPRING CONSTANT
STRAIN
VISCOELASTICITY
YIELD (MECHANICAL)
YIELD POINT

DEFORMING
USE DEFORMATION

DEGRADATION
UF DETERIORATION
NT ALKALI DEGRADATION
ACID DEGRADATION
BACTERIAL DEGRADATION
CHLORINE DAMAGE
ENZYMATIC DEGRADATION
LAUNDRY DEGRADATION
MICROBIOLOGICAL DEGRADATION
PHOTOCHEMICAL DEGRADATION
THERMAL DEGRADATION
RT ACID SOLUBILITY
AGING (MATERIALS)
ALKALI SOLUBILITY
CARBOXYL CONTENT
CHEMICAL PROPERTIES
CHEMICAL STABILITY
CREEP
DAMAGE
DECOMPOSITION
DEPOLYMERIZATION
DISCOLORATION
DURABILITY
FABRIC PROPERTIES
(DEGRADATION)
FADING
FATIGUE
FUNGUS
HYDROLYSIS
INSECT RESISTANCE
INSECT RESISTANCE TREATMENTS
IONIZING RADIATION
MILDEW
ODORS
OXIDATION
RADIATION
REACTIONS (CHEMICAL)
STABILITY
STABILIZERS (AGENTS)
TENDERING
UREA BISULFITE SOLUBILITY
WEAR
YELLOWING

DEGRADATION PROPERTIES
UF FABRIC PROPERTIES
(DEGRADATION)
NT ACID SOLUBILITY
ALKALI SOLUBILITY
BACTERIAL INHIBITION
BIODEGRADABILITY
COLORFASTNESS
CHLORINE FASTNESS (COLOR)
CHLORINE FASTNESS (OF FINISH)
COPPER NUMBER
CUPRAMMONIUM FLUIDITY
DRY CLEANING FASTNESS (COLOR)
DRY CLEANING FASTNESS (OF
FINISH)
GAS FUME FASTNESS
INSECT RESISTANCE
LIGHTFASTNESS (COLOR)
LIGHTFASTNESS (OF FINISH)
MILDEW RESISTANCE
PERSPIRATION RESISTANCE
UREA BISULFITE SOLUBILITY
WASHFASTNESS (COLOR)
WASHFASTNESS (OF FINISH)
NT WEATHER RESISTANCE
BT END USE PROPERTIES
RT ACID DEGRADATION
AGING (MATERIALS)
AGING (STORAGE)
BIOCHEMICAL PROPERTIES
CHEMICAL PROPERTIES
CHEMICAL STABILITY
DURABLE FINISH (GENERAL)
FABRIC PROPERTIES
FIBER PROPERTIES
MECHANICAL PROPERTIES
MICROBIOLOGICAL DEGRADATION
PHYSICAL CHEMICAL PROPERTIES
PHYSICAL PROPERTIES (EXCLUDING
MECHANICAL)
PROPERTIES

DEGRADED CELLULOSE
NT HYDROCELLULOSE
OXYCELLULOSE
BT CELLULOSE
CELLULOSE DERIVATIVES

DEGREASING
BT CLEANING
RT DESIZING
DETERGENTS
DEWAXING (COTTON)
LAUNDERING
LUBRICATION
SOAPS
SOIL REMOVAL

DEGREE OF CRYSTALLINITY
RT AMORPHOUS REGION
CRYSTAL LATTICE
CRYSTALLINE REGION
CRYSTALLIZATION
DEGREE OF POLYMERIZATION

DEGREE OF ORIENTATION
RT ANISOTROPIC MATERIALS
ANISOTROPY (STRESS STRAIN)
BIREFRINGENCE
COLD DRAWING
CRYSTALLINITY
DRAFTING (STAPLE FIBER)
DRAWING (FILAMENT)
FIBER ORIENTATION
HOT DRAWING
ISOTROPIC MATERIALS
LEADING HOOKS
MOLECULAR ORIENTATION
ORIENTATION
ORIENTATION BIREFRINGENCE
PARALLELIZATION
PREORIENTATION
TRAILING HOOKS
WEB ORIENTATION
X RAY DIFFRACTION

DEGREE OF POLYMERIZATION
UF DP
RT ACID DEGRADATION
ACID END GROUPS
AMINE END GROUPS
AMORPHOUS POLYMERS
CARBOXYL END GROUPS
CRYSTALLINE POLYMERS
CUPRAMMONIUM FLUIDITY
DEGREE OF CRYSTALLINITY
END GROUPS
LONG CHAIN MOLECULES
MOLECULAR STRUCTURE
MOLECULAR WEIGHT
MOLECULAR WEIGHT CONTROL
MOLECULAR WEIGHT DISTRIBUTION
MONOMERS
POLYMERIZATION
POLYMERS
SULFONIC END GROUPS
ULTRACENTRIFUGES

DEGREE OF SUBSTITUTION (CELLULOSE)
USE DS (CELLULOSE)

DEGREES OF FREEDOM
BT STATISTICAL MEASURES
RT ANALYSIS OF VARIANCE
COEFFICIENT OF VARIATION
CONFIDENCE COEFFICIENT
CONFIDENCE LIMITS
CORRELATION COEFFICIENT
DISTRIBUTION
LEVEL OF SIGNIFICANCE
MEAN
MEDIAN
MODE
MODES OF DEFORMATION
RANGE
REGRESSION COEFFICIENT
STANDARD DEVIATION
STANDARD ERROR
STATISTICS
VARIANCE

DEGUMMING
RT BOILING OFF
FIBROIN (SILK)
SERICIN
SILK

DEHAIRING
RT CASHMERE
GUARD HAIRS
HAIR
HAIR PROCESSING MACHINES
HAIRINESS

DEHYDRATION
RT DRYING
WATER

DELAINE (WOOL CLASS)
BT WOOL CLASS

DELAMINATING
RT ADHESION
CHIPPING
COATED FABRICS
FLAKING
LAMINATED FABRICS
LAMINATING
PEELING
SPLITTING

DELAYED CURE
USE DEFERRED CURE METHOD

DELAYED ELASTIC RECOVERY
USE DELAYED RECOVERY

DELAYED RECOVERY
UF DELAYED ELASTIC RECOVERY
BT RECOVERY (SELF RESTORATION TO
ORIGINAL CONDITION)
STRESS STRAIN PROPERTIES
STRESS STRAIN PROPERTIES
(TENSILE)
RT BENDING RECOVERY
CREEP
CREEP RECOVERY
CRIMP RECOVERY
DYNAMIC MODULUS
ELASTIC RECOVERY
ELASTIC STRAIN
ELASTICITY
LINEAR VISCOELASTICITY
LOADING
MAXWELL MODEL
NONLINEAR VISCOELASTICITY
PERMANENT SET
PLASTIC FLOW
PLASTIC STRAIN
PRIMARY CREEP
RELAXATION
RELAXATION SPECTRUM
RESILIENCE
RHEOLOGY
STRAIN
STRESS
STRESS RELAXATION
STRESS STRAIN CURVES
VISCOELASTICITY
WORK RECOVERY
WRINKLE RESISTANCE

DELIQUESCENT AGENTS
BT TEXTILE CHEMICALS
RT CALCIUM CHLORIDE
GLYCEROL
MAGNESIUM CHLORIDE
MOISTURE
MOISTURE CONTENT
SIZES (SLASHING)
SOFTENERS

DELIVERY
RT ACQUISITION
CIRCULATION
DELIVERY RATE
DELIVERY ROLLS
FEED RATE
FLOW (MATERIALS)
MARKETING
MATERIALS HANDLING
PACKAGING (MATERIAL)
PACKAGING (OPERATION)
PRODUCTION
TRANSFERRING (MATERIAL)

DELIVERY RATE
BT RATE
RT CHARGING
CONVEYORS
DELIVERY
DELIVERY ROLLS
FEED RATE
FEED ROLLS
FEEDERS
FLOW (MATERIALS)
INPUT
OUTPUT
PRODUCTION
THROUGHPUT

DELIVERY ROLLS
BT ROLLS
RT CARRIAGES (SPINNING MULE)
DELIVERY RATE
FEED ROLLS
FLOW (MATERIALS)
FRONT ROLLS

DELUSTERANTS
USE DELUSTERING AGENTS

DELUSTERING
BT WET FINISHING
RT BRIGHTNESS
HOLES
LUSTER

DELUSTERING AGENTS
UF DELUSTERANTS
BT FINISH (SUBSTANCE ADDED)
RT BRIGHTNESS
LUSTER
TITANIUM DIOXIDE

DEMINERALIZING
BT SEPARATION (SOLUTION)
RT CALGON (TN)
ION EXCHANGE
WATER TREATMENT

DENIER
UF YARN DENIER
BT COUNT
DIRECT COUNT
YARN PROPERTIES
RT COTTON COUNT
COUNT STRENGTH PRODUCT
COUNT TESTING
FIBER DENIER
FILAMENTS
GREX
INDIRECT COUNT
LINEAR DENSITY
TENACITY
TEX
WORSTED COUNT
YARN CROSS SECTIONS

DENIMS
BT FLAT WOVEN FABRICS
INDUSTRIAL CLOTHING
WORK CLOTHING
WOVEN FABRICS
RT DUNGAREES
OVERALLS
TWILL WEAVES

DENSITY
UF MASS DENSITY
NT BULK DENSITY
LINEAR DENSITY
BT STRUCTURAL PROPERTIES
RT BULK
COMPRESSIBILITY
COUNT
COVER FACTOR
CRYSTALLIZATION
FABRIC COVER
FABRIC PROPERTIES
FABRIC PROPERTIES (PHYSICAL EXCLUDING MECHANICAL)
JAMMING POINT
MASS
PACKAGE DENSITY
PACKING FACTOR
PICKS PER INCH
POROSITY
SPACE
SPECIFIC GRAVITY
STITCH DENSITY
TENACITY
TIGHTNESS

DENTS
UF REED SPACING
RT CLOTH FELL
LIGHT DENT
LOOMS
REED DRAFT
REED WIDTH
REED WIRES
REEDING
REEDS
SWOLLEN DENT
WARPING

DEODORANTS
RT ACTIVATED CARBON
ANTIMICROBIAL FINISHES
DEODORIZING
ODOR CONTROL
PERSPIRATION
PERSPIRATION RESISTANCE
TEXTILE CHEMICALS

DEODORIZING
BT WET FINISHING
RT ANTIMICROBIAL FINISHES
CLEANING
DEODORANTS
DRY CLEANING
LAUNDERING
ODOR CONTROL

DEODORIZING CARBON
USE ACTIVATED CARBON

DEPOLYMERIZATION
RT DEGRADATION
DEGREE OF POLYMERIZATION
POLYMERIZATION
POLYMERS

DEPOSITION
RT ANTIREDEPOSITION AGENTS
COAGULATING
COATING (PROCESS)
COATINGS (SUBSTANCES)
GRAFTING
POLYMER DEPOSITION
POLYMERIZATION
POLYMERS
SCALE MASKING
SOIL REDEPOSITION

DERBY RIB
USE RIB WEAVES

DERMATITIS
RT BIOCHEMICAL PROPERTIES
OCCUPATIONAL HAZARDS
SAFETY
SKIN IRRITATION
TOXICITY

DESIGN
UF DESIGNING
NT APPAREL DESIGN
DRESS DESIGN
FABRIC DESIGN
MACHINE DESIGN
PRODUCT DESIGN
RT AESTHETIC APPEAL
AESTHETIC PROPERTIES
APPEARANCE
CHAIN DRAFT
COLOR
COMPUTATIONS
CONFIGURATION
COVER
DEVELOPMENTS
DRAFTING (PATTERN)
DRESSMAKING
ENGRAVING
FABRIC ANALYSIS
FABRIC STRUCTURE
FLOAT DESIGN
HARNESS DRAFT
IMPROVEMENTS
INTERLACINGS (YARN IN FABRIC)
LUXURIOUSNESS
MODIFICATION
MULTICOLORS
OPTICAL PROPERTIES
PATTERN (APPAREL)
PATTERN (FABRICS)
PATTERN DEFINITION
PATTERN REPEATS
PATTERN WHEELS
PRINTING
PRINTING ROLLS
REPEATS
RICHNESS
SHADE
SHAPE
STITCH CLARITY
STRUCTURAL ANALYSIS
SYSTEMS ENGINEERING
WEAVE

DESIGNING
USE DESIGN

DESIZING
BT CLEANING
WET FINISHING
RT BATHS
BOILING OFF
CONTINUOUS SCOURING
DEGREASING
DESIZING AGENTS
DESIZING MACHINES
DEWAXING (COTTON)
ENZYMES
GELATINE
LAUNDERING
PROTEOLYTIC ENZYMES
SCOURING
SCOURING BATHS
SIMULTANEOUS FINISHING PROCESSES
SIZE BOXES
SIZES (SLASHING)
SLASHING
SODIUM BROMATE
SODIUM BROMITE
SODIUM HYDROXIDE
SQUEEZING
STRETCH ROLLS

DESIZING AGENTS
BT TEXTILE CHEMICALS
RT DESIZING
ENZYMES
SIZES (SLASHING)
SODIUM BROMATE
SODIUM BROMITE
SODIUM HYDROXIDE

DESIZING MACHINES
RT DESIZING
SCOURING MACHINES
SIZES (SLASHING)
SLASHING
WARPING

DESORPTION
BT SORPTION
RT ABSORPTION (MATERIAL)
ADSORPTION
SORPTION OF DYES
SORPTION OF GASES
SORPTION OF LIQUIDS
SORPTION OF WATER

DETACHING
RT COMBING
COTTON COMBING
DETACHING EFFICIENCY
NIPPING
TOP COMBING (COMB CYCLE)

DETACHING EFFICIENCY
BT COMBING EFFICIENCY
RT COMBING
COTTON COMBING
DETACHING
SLIVER NOIL RATIO

DETACHING ROLLS
RT COMB BRUSHES
COMB CYLINDERS
COMB PINS
COMB SEGMENTS
COMBED SLIVERS
COMBING
COMBS
COTTON SYSTEM PROCESSING
HALF LAP
LAP PLATES
NIPPER JAWS
NIPPER KNIVES
NIPPER PLATES
NOILS
RECTILINEAR COMBS
RIBBON LAPS
SLIVERS
TOP COMBS
TRUMPETS

DETECTION
RT ACCURACY
ANALYZING
COUNTING
DEFECTS
DETERMINATION
DIFFERENCES
ERRORS
ESTIMATION
EVALUATION
EXAMINATION
FABRIC DEFECTS
OPTICAL SCANNING
PHOTOELECTRIC SCANNING
QUALITY CONTROL
SEAM DETECTORS
YARN DEFECTS

DETERGENCY
RT ANTIREDEPOSITION AGENTS
BUILDERS (DETERGENTS)
DETERGENTS
DRY CLEANING
LAUNDERING
RESIDUAL FAT
SCOURING
SEBUM
SOILING
SURFACTANTS
WETTING
WETTING AGENTS

DETERGENTS
BT TEXTILE CHEMICALS
RT AMPHOTERIC SURFACTANTS
ANIONIC SURFACTANTS
ANTIREDEPOSITION AGENTS
BIODEGRADABILITY
BLEACHING AGENTS
BUILDERS (DETERGENTS)
CATIONIC SURFACTANTS
CLEANING
CONTINUOUS SCOURING
DEGREASING
(CONT.)

RT DETERGENCY
DYEING AUXILIARIES
EMULSIFYING AGENTS
JET SCOURING
LAUNDERING
LEVELLING AGENTS
NONIONIC SURFACTANTS
SCOURING
SCRUBBING
SOAP SODA SCOURING
SOAPS
SOIL REMOVAL
SURFACTANT APPLICATIONS
SURFACTANTS
SYNTHETIC DETERGENTS
WETTING
WETTING AGENTS

DETERIORATION
USE DEGRADATION

DETERMINATION
RT ANALYZING
CHEMICAL ANALYSIS
COUNTING
DETECTION
DIFFERENCES
MEASURING
MEASURING INSTRUMENTS
RECORDING INSTRUMENTS

DEUTERIUM
RT HYDROGEN

DEVELOPED DYES
USE AZOIC DYES

DEVELOPERS (DYEING)
BT DYEING AUXILIARIES
RT AZOIC DYEING

DEVELOPMENTS
RT DESIGN
IMPROVEMENTS
MACHINE DESIGN
MODIFICATION
OPTIMIZATION
PERFORMANCE

DEVICES
NT ELECTROMAGNETIC DEVICES
ELECTROMECHANICAL DEVICES
GUARDING DEVICES
MECHANICAL DEVICES
SELF THREADING DEVICES
TENSION DEVICES
RT ACCESSORIES
APPARATUS
BELTS
CHAINS
COMPONENTS
FLUID DRIVES
GADGETS
GEAR DRIVES
GRIPPERS (YARN)
INVENTIONS
LINKAGES
MECHANISMS
SPRINGS
STANDS

DEWAXING (COTTON)
BT CLEANING
RT COTTON
COTTON WAX
DEGREASING
DESIZING
IMPURITIES
LAUNDERING
LUBRICATION
SCOURING
SOIL REMOVAL

DEWING
USE DAMPENING

DEW POINT
RT CRITICAL MOISTURE CONTENT
DRY BULB TEMPERATURE
DRYING
EQUILIBRIUM MOISTURE CONTENT
HUMIDITY
HYGROMETERS
MOISTURE CONTENT
PSEUDO WET BULB TEMPERATURE
PSYCHROMETRIC CHARTS
PSYCHROMETRY
REGAIN
RELATIVE HUMIDITY
SATURATED VAPOR
SPECIFIC HUMIDITY
VAPOR PRESSURE
WET BULB TEMPERATURE

DEXTRAN
BT POLYSACCHARIDES
RT DEXTRINS
GUMS
STARCH

DEXTRINS
BT STARCH
RT DEXTRAN
SIZES (SLASHING)
STARCH GUMS

DEXTROSE
USE GLUCOSE

DHEU
UF DIHYDROXYETHYLENE UREA
BT UREA DERIVATIVES
RT METHYLOL UREAS

DIACETALS
USE ACETALS

DIACETONE ACRYLAMIDE
USE ACRYLAMIDE

DIACIDS
USE DIBASIC ACIDS

DIAGONAL STITCHES
BT STITCHES (SEWING)
RT SEWING
STITCHING

DIAL NEEDLES
BT KNITTING NEEDLES
NEEDLES
RT CIRCULAR KNITTING MACHINES
CYLINDER NEEDLES
HOSIERY MACHINES
KNITTING MACHINES
LATCH NEEDLES
SEAMLESS HOSIERY MACHINES

DIALDEHYDE CELLULOSE
BT ALDEHYDES
CELLULOSE DERIVATIVES
RT CHEMICAL MODIFICATION (FIBERS)
DIALDEHYDE STARCH
POLYACROLEIN

DIALDEHYDE STARCH
BT ALDEHYDES
MODIFIED STARCHES
RT CROSSLINKING
DIALDEHYDE CELLULOSE
POLYACROLEIN
OXIDIZED STARCHES
REACTANTS

DIALDEHYDES
NT ADIPALDEHYDE
GLUTARALDEHYDE
GLYOXAL
BT ALDEHYDES
REACTANTS

DIALYSIS
NT ELECTRODIALYSIS
RT DIAPHRAGMS
DIFFUSION
MASS TRANSFER
MEMBRANES
OSMOSIS
PENETRATION
PERMEABILITY
PURIFICATION
SEPARATION (SOLUTION)

DIAMETER
RT CIRCULARITY
CIRCUMFERENCE
COUNT
CROSS SECTIONAL AREA
CROSS SECTIONS
ELLIPTICITY
FIBER DIAMETER
HEIGHT
MAGNITUDE
NECKING (FILAMENT)
PERIMETER
RADIUS
ROUNDNESS
SPHERICITY
THICKNESS
WIDTH
YARN DIAMETER

DIAMINES
USE AMINES

DIAMOND BARRING
BT BARRE
FABRIC DEFECTS

DIAMOND WEAVES
BT WEAVE (TYPES)
RT FLAT WOVEN FABRICS

DIAPERS
UF NAPPIES
BT INFANTS CLOTHING

DIAPHRAGM BURST TESTING
UF DIAPHRAGM BURST TESTS
BT BURST TESTING
TESTING
RT BALL BURST TESTING
BIAXIAL STRESS
BREAKING STRENGTH
BURSTING STRENGTH
MECHANICAL PROPERTIES
MULLEN BURST TESTING

DIAPHRAGM BURST TESTS
USE DIAPHRAGM BURST TESTING

DIAPHRAGMS
RT DIALYSIS
DIFFUSION
INDUSTRIAL FABRICS
MEMBRANES
OSMOSIS

DIAZO COMPOUNDS
RT AZOIC BASES
AZOIC DYEING
AZOIC DYES
COUPLING (CHEMICAL)
DIAZOBENZOYLATED COTTON

DIAZOBENZOYLATED COTTON
BT CHEMICALLY MODIFIED COTTON
RT AZOIC BASES
COTTON
COUPLING (CHEMICAL)
DIAZO COMPOUNDS
DIAZOTIZATION

DIAZOMETHANE
BT REACTANTS
RT ALKYLATION
CHEMICAL MODIFICATION (FIBERS)
METHYLATION

DIAZOTIZATION
BT REACTIONS (CHEMICAL)
RT AZOIC BASES
AZOIC DYEING
AZOIC DYES
COUPLING (CHEMICAL)
DIAZOBENZOYLATED COTTON
NITRITES
POTASSIUM NITRITE

DIBASIC ACIDS
UF DIACIDS
NT ADIPIC ACID
ISOPHTHALIC ACID
ITACONIC ACID
OXALIC ACID
PHTHALIC ACID
SUCCINIC ACID
SULFURIC ACID
TEREPHTHALIC ACID
BT ACIDS
RT REACTANTS
INORGANIC ACIDS
ORGANIC ACIDS

DIBORANE
BT BORON COMPOUNDS
BORON HYDRIDES

DICHLOROBENZENE
BT CHLORINATED HYDROCARBONS
RT CARRIERS (DYEING)
CHLORINATED SOLVENTS
INSECT RESISTANCE FINISHES
TRICHLOROBENZENE

DICHLOROCYANURIC ACID
USE CHLOROCYANURIC ACID

DICHLORCHYDRIN
BT CHLOROHYDRINS

DIHYDROXYETHYLENE UREA
USE DHEU

DICHLOROPROPANOL
USE GLYCERIN DICHLOROHYDRIN

DICHLOROPYRIMIDINE DYES
USE CHLOROPYRIMIDINE DYES

DICHLOROQUINOXALINE DYES
UF LEVAFIX E (TN)
BT REACTIVE DYES

DICHLOROTRIAZINE DYES
USE CHLOROTRIAZINE DYES

DICHROIC RATIO
BT DICHROISM
RT MOLECULAR STRUCTURE
ORIENTATION

DICHROISM
NT DICHROIC RATIO
BT OPTICAL PROPERTIES

RT BIREFRINGENCE
PHOTOELASTICITY
REFRACTANCE

DICHROMATES
NT SODIUM DICHROMATE
BT CHROMIUM COMPOUNDS

DIE CUTTERS
BT CUTTING MACHINES
RT CLICKERS
PRESSES
RECIPROCATING CUTTING MACHINES
ROTARY CUTTING MACHINES

DIELDRIN
RT INSECT RESISTANCE FINISHES

DIELECTRIC CONSTANT
BT DIELECTRIC PROPERTIES
RT CAPACITANCE (ELECTRICAL)
DIELECTRICS
DIPOLE MOMENT
DISCHARGE (ELECTRIC)

DIELECTRIC DRYERS
BT DRYERS
RT CENTRIFUGAL DRYERS
DIELECTRICS
DRUM DRYERS
HEAT TRANSFER
HOT AIR DRYERS
INFRARED DRYERS
LOOP DRYERS

DIELECTRIC DRYING
BT DRYING
RT FREEZE DRYING
HOT AIR DRYING
INFRARED DRYING
VACUUM DRYING

DIELECTRIC HEAT SEALING
BT HEAT SEALING
SEALING

DIELECTRIC HEATING
BT HEATING
RT DRYING
ELECTRIC HEATING (RESISTANCE)
GAS HEATING (GAS FUEL)
RADIANT HEATING
OIL HEATING (OIL FUEL)

DIELECTRIC PROPERTIES
NT DIELECTRIC CONSTANT
BT ELECTRICAL PROPERTIES
RT CAPACITANCE (ELECTRICAL)
DIELECTRICS
DIPOLES
POLARITY

DIELECTRIC STRENGTH
BT ELECTRICAL PROPERTIES
RT CAPACITANCE (ELECTRICAL)
IMPEDANCE
INDUCTANCE
INSULATION (ELECTRICAL)
RESISTANCE (ELECTRICAL)
RESISTIVITY

DIELECTRICS
RT CAPACITANCE (ELECTRICAL)
DIELECTRIC CONSTANT
DIELECTRIC DRYERS
DIELECTRIC PROPERTIES
DISCHARGE (ELECTRIC)
LEAKAGE RESISTANCE
(ELECTRICAL)

DIENES
NT BUTADIENE
BT ALKENES

DIEPOXIDES
USE OXIRANE COMPOUNDS

DIETHANOLAMIDE SURFACTANTS
UF KRITCHEVSKY PRODUCTS
NT CAPRYL DIETHANOLAMIDE
COCONUT DIETHANOLAMIDE
LAURYL DIETHANOLAMIDE
BT NONIONIC SURFACTANTS
RT ALKANOLAMIDES
DIETHANOLAMINE SOAPS
SARCOSIDE SURFACTANTS

DIETHANOLAMIDES
BT ALKANOLAMIDES

DIETHANOLAMINE
BT ALKANOLAMINES
SECONDARY AMINES
RT DIETHYLAMINE

DIETHANOLAMINE SOAPS
BT ALKANOLAMINE SOAPS
RT DIETHANOLAMIDE SURFACTANTS

DIETHYLAMINE
BT SECONDARY AMINES
RT DIETHANOLAMINE

DIETHYLENEGLYCOL ESTER SURFACTANTS
USE DIGLYCOL ESTER SURFACTANTS

DIFFERENCES
RT COMPARISON
COMPENSATION
CORRECTION
CORRELATION
DETECTION
DETERMINATION
ERRORS
NUMERICAL ANALYSIS

DIFFERENTIAL CONTROL
BT AUTOMATIC CONTROL
RT CONTINUOUS CONTROL
FEEDBACK CONTROL
INTEGRAL CONTROL
ON OFF CONTROL
PROCESS CONTROL

DIFFERENTIAL DISTRIBUTION
(CROSSLINKING)
RT CHEMICAL MODIFICATION (FIBERS)
CROSSLINKING
DISTRIBUTION
REACTANTS

DIFFERENTIAL DYEING METHOD (COTTON
IMMATURITY)
BT IMMATURITY TESTING (COTTON
FIBER)
RT CROSS SECTIONAL SHAPE METHOD
(COTTON IMMATURITY)
POLARIZED LIGHT METHOD (COTTON
IMMATURITY)

DIFFERENTIAL EQUATIONS
NT LINEAR DIFFERENTIAL EQUATIONS
NONLINEAR DIFFERENTIAL
EQUATIONS
ORDINARY DIFFERENTIAL
EQUATIONS
PARTIAL DIFFERENTIAL EQUATIONS
BT EQUATIONS
RT GENERAL SOLUTIONS
INTEGRATION
MATHEMATICAL ANALYSIS
PARTICULAR SOLUTIONS

DIFFERENTIAL FRICTION
BT FRICTION
RT ANTIFELTING TREATMENTS
FELTING
FELTS
FIBER FIBER FRICTION
FRICTION
SCALES (WOOL FIBERS)
WOOL

DIFFERENTIAL GEOMETRY
RT CURVATURE
FABRIC GEOMETRY
GEOMETRY
HELICAL CRIMP
HELIX ANGLE
STRUCTURAL MECHANICS
TENSORS
TORSION (GEOMETRIC)
VECTORS
YARN GEOMETRY
YARN STRUCTURE

DIFFERENTIAL SHRINKAGE
BT SHRINKAGE
RT BICOMPONENT FIBER YARNS
(FILAMENT)
BICOMPONENT FIBER YARNS
(STAPLE)
BICOMPONENT FIBERS
BULKED YARNS
CONTRACTION
CRIMP
CRIMP FREQUENCY
CRIMP TENDENCY
LATENT CRIMP

DIFFERENTIAL THERMAL ANALYSIS
BT THERMAL ANALYSIS
RT CALORIMETRY
MEASURING INSTRUMENTS
THERMAL TESTING

DIFFERENTIATION
RT COMPUTATIONS
DIFFERENTIAL EQUATIONS
INTEGRATION
SOLUTIONS (MATHEMATICAL)

DIFFRACTION
RT BIREFRINGENCE
DIFFRACTION PATTERNS
REFRACTANCE
X RAY DIFFRACTION

DIFFRACTION PATTERNS
RT BIREFRINGENCE
DIFFRACTION
INTERFERENCE FRINGES
MOIRE PATTERNS
X RAY ANALYSIS

DIFFUSION
NT FICKIAN DIFFUSION
RT ABSORPTION (MATERIAL)
ACCESSIBILITY (INTERNAL)
ADSORPTION
CHEMICAL ENGINEERING
CONCENTRATION
CONCENTRATION GRADIENT
CRITICAL MOISTURE CONTENT
DEW POINT
DIALYSIS
DIAPHRAGMS
DIFFUSION COEFFICIENT
DIFFUSIVITY
DILUTION
DISSIPATION
DISSOLVING
DRYING
DYE MIGRATION
DYE PENETRATION
DYEING PROPERTIES (GENERAL)
DYEING RATE
DYEING THEORY
EQUILIBRIUM (CHEMICAL)
EQUILIBRIUM (DYEING)
EQUILIBRIUM MOISTURE CONTENT
EVAPORATION
EXHAUSTION (MECHANICAL)
EXTRACTION
MASS TRANSFER
MEMBRANES
MIGRATION (SUBSTANCE)
OSMOSIS
PERMEABILITY
PRIMARY WALLS
PSYCHROMETRY
REGAIN
RELATIVE HUMIDITY
SATURATED SOLUTIONS
SEPARATION (SOLUTION)
SORPTION
SPECIFIC HUMIDITY
VAPOR PRESSURE

DIFFUSION COEFFICIENT
UF DIFFUSION CONSTANT
BT PHYSICAL CHEMICAL PROPERTIES
RT DIFFUSION
DIFFUSIVITY
DRYING
DYEING RATE
DYEING THEORY

DIFFUSION CONSTANT
USE DIFFUSION COEFFICIENT

DIFFUSIVITY
BT TRANSFER PROPERTIES
RT CRITICAL MOISTURE CONTENT
DIFFUSION
DIFFUSION COEFFICIENT
EQUILIBRIUM MOISTURE CONTENT
HUMIDITY
MEMBRANES
MOISTURE CONTENT
OSMOSIS
PENETRATION
PERMEABILITY
RELATIVE HUMIDITY
SEPARATION (SOLUTION)
SPECIFIC HUMIDITY
SWELLING
VAPOR PRESSURE

DIGITAL FIBROGRAPH
BT FIBROGRAPH
RT COEFFICIENT OF LENGTH
VARIATION
FIBER ARRAY
FIBER DIAGRAM
FIBER LENGTH
FIBER LENGTH DISTRIBUTION
FIBROGRAPH MEAN LENGTH
MEAN FIBER LENGTH
SERVO FIBROGRAPH
STAPLING
UPPER HALF MEAN LENGTH
UPPER QUARTILE LENGTH

DIGLYCIDYL ETHERS
BT OXIRANE COMPOUNDS
RT EPOXY RESINS

DIGLYCOL ESTER SURFACTANTS
UF DIETHYLENEGLYCOL ESTER
SURFACTANTS
BT GLYCOL ESTER SURFACTANTS

DIISOCYANATES
USE ISOCYANATES

DILATANCY
BT STRESS STRAIN PROPERTIES
RT BULK MODULUS
FREE VOLUME
RHEOLOGY

DILUTION
RT CONCENTRATION
CONCENTRATION GRADIENT
DIFFUSION
DISSIPATION
DISSOLVING
OSMOSIS
SOLUBILITY
SOLUTES
SOLUTIONS (LIQUID)
SOLVENTS
THINNERS

DIMENSIONAL ANALYSIS
RT ANALYZING
DIMENSIONS
FLUID FLOW
FLUID MECHANICS
HEAT TRANSFER
MASS TRANSFER

DIMENSIONAL STABILITY
BT END USE PROPERTIES
FINISH (PROPERTY)
MECHANICAL PROPERTIES
RT BAGGINESS
BELFAST (TN)
BOARDING
COMPACTING
COMPRESSIVE SHRINKAGE MACHINES
COMPRESSIVE SHRINKING
CONTRACTION
CREASE RETENTION
CREEP
CREPING
CRIMP RETENTION
CURL
DEFORMATION
DISTORTION
ELASTIC STRAIN
ENTANGLEMENTS
EXTENSIBILITY
FABRIC PROPERTIES
FABRIC PROPERTIES (MECHANICAL)
FABRIC RELAXATION
FABRIC SHRINKAGE
FATIGUE RESISTANCE
FELTING RESISTANCE
FELTING SHRINKAGE
FILLING SHRINKAGE
FULLING
FUSION CONTRACTION SHRINKAGE
GROWTH (FABRIC)
HEAT SETTING (SYNTHETICS)
LIVELINESS
LOOP DISTORTION
MOLDING SHRINKAGE
PERMANENT SET
PLASTIC STRAIN
PRESHRINKING
PUCKER (DEFECT)
REACTANTS
RELAXATION SHRINKAGE
RESIDUAL SHRINKAGE
SAG
SANFORIZING (TN)
SEAM GRIN
SEAM PUCKER
SEAM SLIPPAGE
SHRINK RESISTANCE
SHRINKAGE
SHRINKING
SLIP RESISTANCE
SNARLING
SNARLING TENDENCY
STABILITY
STRETCH
STRETCH FABRICS
SWELLING
THREAD SLIPPAGE
TORSIONAL BUCKLING
TWIST LIVELINESS
UNRAVELLING
WARP SHRINKAGE
WEIGHT
WILDNESS
WRINKLE RESISTANCE
YARN SLIPPAGE

DIMENSIONS
UF SIZE (MAGNITUDE)
NT LENGTH
RADIUS
THICKNESS
WIDTH
BT STRUCTURAL PROPERTIES
RT AMPLITUDE
AREA
CONFIGURATION
DIMENSIONAL ANALYSIS
HEADS (TOP)
HEIGHT
MAGNITUDE
SHAPE
VOLUME

DIMETHYL ACETAL
BT ACETALS

DIMETHYL ACETAMIDE
BT AMIDES
RT SOLVENTS

DIMETHYL FORMAMIDE
BT AMIDES
RT SOLVENTS

DIMETHYL SULFOXIDE
RT SOLVENTS
SWELLING AGENTS

DIMETHYLOL ACETAMIDE
BT METHYLOL AMIDES

DIMETHYLOL DIHYDROXYETHYLENE UREA
USE DMDHEU

DIMETHYLOL ETHYL CARBAMATE
USE DMEC

DIMETHYLOL ETHYL TRIAZONE
USE DMET

DIMETHYLOL ETHYLENE UREA
USE DMEU

DIMETHYLOL FORMAMIDE
BT METHYLOL AMIDES

DIMETHYLOL HYDROXYETHYLENE UREA
USE DMHEU

DIMETHYLOL HYDROXYPROPYLENE UREA
USE DMHPU

DIMETHYLOL ISOPROPYL CARBAMATE
BT METHYLOL CARBAMATES

DIMETHYLOL ITACONAMIDE
BT METHYLOL AMIDES
RT ITACONIC ACID

DIMETHYLOL METHYL CARBAMATE
BT METHYLOL CARBAMATES

DIMETHYLOL METHYL TRIAZONE
BT METHYLOL TRIAZONES

DIMETHYLOL PROPYLENE UREA
USE DMPU

DIMETHYLOL UREA
USE DMU

DIMETHYLOL URON
USE DMUR

DIMITY
BT FLAT WOVEN FABRICS
WOVEN FABRICS
RT COTTON
PLAIN WEAVES

DIOCTYL SULFOSUCCINATE
NT DECERESOL OT (TN)
BT SULFOSUCCINIC ESTERS

DIOLS
USE GLYCOLS

DIP BONDING
BT BONDING
RT BONDED YARN FABRICS
DRY POWDER BONDING
FLAME BONDING
NONWOVEN FABRICS
SPRAY BONDING

DIP COATING
BT COATING (PROCESS)
RT CALENDER COATING
CAST COATING
COATED FABRICS
EXTRUSION COATING
FLEXIBLE FILM LAMINATING
KNIFE COATING
NIP COATING
PAPER MAKING
ROLLER COATING

DIPOLE MOMENT
BT PHYSICAL CHEMICAL PROPERTIES
RT DIELECTRIC CONSTANT
DIPOLES
DISPERSION FORCES
MOLECULAR ATTRACTION
POLAR COMPOUNDS
POLARITY
VAN DER WAALS FORCES

DIPOLES
RT CHARGE (ELECTRICAL)
DIELECTRIC PROPERTIES
DIPOLE MOMENT
POLAR COMPOUNDS
POLARITY

DIPPING
USE PADDING

DIRECT COUNT
NT DENIER
GREX
TEX
BT COUNT
RT INDIRECT COUNT
LINEAR DENSITY

DIRECT DYEING
BT DYEING (BY DYE CLASSES)
RT ACID DYEING
AZOIC DYEING
BASIC DYEING
DIRECT DYES
DISPERSE DYEING
METALLIZED DYEING
MORDANT DYEING
NEUTRAL DYEING
PIGMENT DYEING
REACTIVE DYEING
SULFUR DYEING
VAT DYEING

DIRECT DYES
BT ANIONIC DYES
DYES (BY CHEMICAL CLASSES)
RT AZOIC DYES
CELLULOSE DYES
DIRECT DYEING
DYEING (BY DYE CLASSES)
DYES (BY FIBER CLASSES)
METALLIZED DYES
MORDANT DYES
REACTIVE DYES
SULFUR DYES
VAT DYES

DIRECT MOLDED SOLE (FOOTWEAR)
BT FOOTWEAR COMPONENTS

DIRECT PRINTING
USE ROLLER PRINTING

DIRECT SPINNING
USE TOW TO YARN CONVERSION

DIRECT SPINNING MACHINES
USE TOW TO YARN MACHINES

DIRECT SPUN YARNS
BT SPUN YARNS
YARNS
RT CARDED YARNS
COMBED YARNS
CONDENSED YARNS
STAINS DIRECT SPINNER
TOW TO YARN CONVERSION
TOW TO YARN MACHINES

DIRECT TENSION DEVICES
BT TENSION DEVICES
RT HYDRAULIC TENSION DEVICES
INDIRECT TENSION DEVICES
MAGNETIC TENSION DEVICES

DIRECTION (TWILL)
NT S TWILLS
Z TWILLS
RT BALANCED YARNS
DIRECTION (TWIST)
S TWIST
SENSE (DIRECTION)
TWIST
Z TWIST

DIRT
RT ARTIFICIAL SOIL
FABRIC INSPECTION
FOGGY YARNS
FOGMARKING
SOIL
SOIL RESISTANCE
SOILING
STAINING
STAINS

DIRT REMOVAL
USE SOIL REMOVAL

DISC TENSION
RT BALL AND SOCKET TENSION
BELT TENSION
END BREAKAGES
FINGER TENSION
FRICTION
GATE TENSION
KNITTING TENSION
STOP MOTIONS

RT TAKE DOWN TENSION
TENSION
TENSION CONTROL
TENSION DEVICES
TIGHTNESS
TWISTING TENSION
WARP TENSION
WINDING TENSION
WRAP
YARN TENSION

DISCHARGE (ELECTRIC)
RT CAPACITANCE (ELECTRICAL)
DIELECTRIC CONSTANT
DIELECTRICS
DISSIPATION
RESISTIVITY
STATIC ELECTRICITY
TRIBOELECTRICITY

DISCHARGE PRINTING
UF EXTRACT PRINTING
BT PRINTING
RT AZOIC DYEING
BLOCK PRINTING
DYEING
DYES
MELANGE PRINTING
RESIST PRINTING
SCREEN PRINTING
THROUGH PRINTING
VAT DYEING

DISCOLORATION
RT COLORFASTNESS
DEGRADATION
FADING
LIGHTFASTNESS (OF FINISH)
SOILING
SPOTTING
STAINING
STAINING (DYEING EFFECTS)
WHITENESS RETENTION
YELLOWING

DISCRETE DISTRIBUTION
RT BINOMIAL DISTRIBUTION
CHI SQUARED DISTRIBUTION
DISTRIBUTION
F DISTRIBUTION
NORMAL DISTRIBUTION
T DISTRIBUTION

DISINFECTANTS
USE ANTIMICROBIAL FINISHES

DISINFECTION
USE ANTIMICROBIAL TREATMENTS

DISLOCATIONS
RT CRACKS (MECHANICAL)
CRYSTAL DEFECTS
CRYSTALLINITY
GRAIN BOUNDARIES
SLIP PLANES

DISORIENTATION
USE ORIENTATION

DISPERSANTS
USE DISPERSING AGENTS

DISPERSE DYEING
BT DYEING (BY DYE CLASSES)
RT ACID DYEING
AZOIC DYEING
BASIC DYEING
DIRECT DYEING
DISPERSE DYES
DISPERSE SYSTEMS
DISPERSING
DISPERSING AGENTS
METALLIZED DYEING
MORDANT DYEING
NEUTRAL DYEING
PARTITION COEFFICIENT
PIGMENT DYEING
REACTIVE DYEING
SULFUR DYEING
TRIACETATE DYEING
VAT DYEING

DISPERSE DYES
BT DYES (BY CHEMICAL CLASSES)
RT ACETATE DYES
DISPERSE DYEING
DYEING (BY DYE CLASSES)
DYES (BY FIBER CLASSES)

DISPERSE SYSTEMS
UF DISPERSIONS
NT COLLOIDS
EMULSIONS
RT DISPERSE DYEING
DISPERSING
DISPERSING AGENTS
PIGMENT DISPERSIONS

DISPERSING
RT CALGON (TM)
CIRCULATION
DISPERSE DYEING
DISPERSE SYSTEMS
DISPERSING AGENTS
DISPERSION FORCES
DISSIPATION
DISSOLVING
EMULSIFYING
EXHAUSTION (MECHANICAL)
GRINDING (COMMINUTION)
MIXING
SEPARATION
STIRRING
SUSPENSIONS
SYNTHETIC DYEING

DISPERSING AGENTS
UF DISPERSANTS
BT SURFACTANT APPLICATIONS
RT COLLOIDS
DISPERSE DYEING
DISPERSE SYSTEMS
DISPERSING
DISPERSION FORCES
DYEING AUXILIARIES
EMULSIFYING AGENTS
FOAMS (FROTH)
LIGNIN SULFONATES
NAPHTHALENE FORMALDEHYDE SULFONATES
PARTICLE SIZE
PLASTICIZERS
SURFACTANTS
SUSPENDING AGENTS
SUSPENSIONS
WETTING AGENTS

DISPERSION (STATISTICAL)
RT COEFFICIENT OF VARIATION
DISTRIBUTION

DISPERSION FORCES
UF LONDON FORCES
LONDON-HEITLER FORCES
BT FORCE
RT DIPOLE MOMENT
DISPERSING
DISPERSING AGENTS
MOLECULAR ATTRACTION
SOLUBILITY PARAMETER
VAN DER WAALS FORCES

DISPLACEMENT
RT DISTORTION
DYNAMICS
MOTION
POSITION
STRAIN
VELOCITY
VIBRATION

DISSIPATION
RT CAPACITANCE (ELECTRICAL)
DAMPING
DIFFUSION
DILUTION
DISCHARGE (ELECTRIC)
DISPERSING
EXHAUSTION (MECHANICAL)
RESISTIVITY
STATIC ELECTRICITY

DISSOLVING
RT CIRCULATION
DIFFUSION
DILUTION
DISPERSING
EXTRACTION
SOLIDIFICATION
SOLUBILITY
SOLUTES
SOLUTIONS (LIQUID)
SOLVENTS

DISPERSIONS
USE DISPERSE SYSTEMS

DISTILLATION
RT CHEMICAL ENGINEERING
SUBLIMATION

DISTILLED WATER
BT WATER
RT HARD WATER

DISTORTION
RT BENDING
BOW
BOWING
BUCKLING
CAMBER
CREASING
CURLING
DEFLECTION
DEFORMATION
DIMENSIONAL STABILITY
DISPLACEMENT
SEAM GRIN
SEAM SLIPPAGE
SKEW
SNAGGING
SNAGS
SWELLING
TWISTING

DISTRIBUTION
UF FREQUENCY DISTRIBUTION
BT STATISTICAL MEASURES
RT ANALYSIS OF VARIANCE
BINOMIAL DISTRIBUTION
CAUCHY DISTRIBUTION
CHI SQUARED DISTRIBUTION
COEFFICIENT OF VARIATION
CONFIDENCE LIMITS
CORRELATION COEFFICIENT
DEGREES OF FREEDOM
DIFFERENTIAL DISTRIBUTION (CROSSLINKING)
DISCRETE DISTRIBUTION
DISPERSION (STATISTICAL)
DISTRIBUTION FUNCTION
EXPONENTIAL DISTRIBUTION
F DISTRIBUTION
GAMMA DISTRIBUTION
KURTOSIS
MAXWELLIAN DISTRIBUTION
MEAN
MODE
NORMAL DISTRIBUTION
OPERATIONS RESEARCH
POISSON DISTRIBUTION
POPULATION
PROBABILITY
PROBABILITY DENSITY FUNCTION
RADIAL DISTRIBUTION
RANDOM DISTRIBUTION
RANGE
SAMPLING
STANDARD DEVIATION
STANDARD ERROR
STATISTICAL DISTRIBUTION
STATISTICAL INFERENCE
STATISTICAL METHODS
STATISTICS
STOCHASTIC PROCESSES
T DISTRIBUTION
TIME SERIES
VARIANCE

DISTRIBUTION FUNCTION
RT DISTRIBUTION
PROBABILITY

DISULFIDE BONDS
UF CYSTINE BRIDGE
BT CHEMICAL BONDS
RT CROSSLINKING
CYSTINE
DISULFIDE INTERCHANGE
DISULFIDES (ORGANIC)
DURABLE PRESS
HAIR
KERATIN
SULFIDES (INORGANIC)
SULFUR DYES
THIOETHERS
WASH WEAR FINISHES
WOOL

DISULFIDE INTERCHANGE
RT CYSTEINE
CYSTINE
DISULFIDE BONDS
KERATIN
SUPERCONTRACTION
WOOL

DISULFIDES (INORGANIC)
USE SULFIDES (INORGANIC)

DISULFIDES (ORGANIC)
NT CYSTINE
BT SULFUR COMPOUNDS
RT CROSSLINKING
DISULFIDE BONDS
DURABLE PRESS
HAIR
KERATIN
MERCAPTANS
REDUCING AGENTS
SILK
SULFUR DYES
THIOETHERS
WOOL

DITHIOCARBAMATES
BT ESTERS
SULFUR COMPOUNDS

DITHIONITES
UF HYDROSULFITES
NT SODIUM DITHIONITE
BT SULFUR COMPOUNDS
RT REDUCING AGENTS
VAT DYEING

DIVIDER ROLLS
    USE DIVIDERS (CARD ROLLS)

DIVIDERS (CARD ROLLS)
    UF DIVIDER ROLLS
    BT CARD ROLLS
    RT WORSTED CARDING
       WORSTED CARDS

DIVIDERS (KNITTING)
    RT FILLING KNITTING
       FILLING KNITTING MACHINES
       FULLY FASHIONED KNITTING MACHINES
       KINKING (KNITTING)
       NEEDLE BARS
       SINKERS

DIVIDING (KNITTING)
    RT FILLING KNITTING
       FILLING KNITTING MACHINES
       KNITTING
       LANDING LOOPS
       LAYING IN (KNITTING)
       SINKING (KNITTING)

DIVINYL SULFONE
    BT REACTANTS
       SULFONES
    RT BIS (HYDROXYETHYL) SULFONE
       SULFONIUM REACTANTS
       WASH WEAR FINISHES

DIVISION
    BT COMPUTATIONS
    RT ADDITION
       MULTIPLICATION
       SUBTRACTION

DMDHEU
    UF DIMETHYLOL DIHYDROXYETHYLENE UREA
    BT METHYLOL UREAS

DMEC
    UF DIMETHYLOL ETHYL CARBAMATE
    BT METHYLOL CARBAMATES

DMET
    UF DIMETHYLOL ETHYL TRIAZONE
    BT METHYLOL TRIAZONES

DMEU
    UF DIMETHYLOL ETHYLENE UREA
       ETHYLENE UREA FORMALDEHYDE CONDENSATES
    BT METHYLOL UREAS
    RT CREASE RETENTION
       CROSSLINKING
       DMPU
       DURABLE PRESS FINISHES
       FORMALDEHYDE
       UREA-FORMALDEHYDE RESINS
       WASH WEAR FINISHES
       WASHFASTNESS (OF FINISH)

DMHEU
    UF DIMETHYLOL HYDROXYETHYLENE UREA
    BT METHYLOL UREAS
    RT DMHPU

DMHPU
    UF DIMETHYLOL HYDROXYPROPYLENE UREA
    BT METHYLOL UREAS
    RT DMHEU
       DMPU

DMPU
    UF DIMETHYLOL PROPYLENE UREA
    BT METHYLOL UREAS
    RT DMEU
       DMHPU
       WASH WEAR FINISHES

DMU
    UF DIMETHYLOL UREA
    BT METHYLOL UREAS

DMUR
    UF DIMETHYLOL URON
    BT METHYLOL URONS

DOBBIES
    USE DOBBY HEADS

DOBBY HEADS
    UF DOBBIES
    RT LOOMS

DOBBY LOOMS
    BT LOOMS
    RT AUTOMATIC LOOMS
       BOX LOOMS
       CAM LOOMS
       HARNESSES
       JACQUARD LOOMS
       LENO LOOMS

DOBBY WEAVING
    BT WEAVING
    RT JACQUARD WEAVING

DOCTOR BLADES
    RT COATING (PROCESS)
       COATING MACHINES
       KNIFE COATING
       PRINTING
       PRINTING MACHINES
       SQUEEGEES

DOCUMENTATION
    RT INFORMATION RETRIEVAL
       INFORMATION SYSTEMS
       THESAURI

DODECYL SULFATE
    BT ALKYL SULFATES

DOESKIN
    BT FLAT WOVEN FABRICS
       WOVEN FABRICS
    RT SATIN WEAVES
       TWILL WEAVES

DOFF
    USE DOFFING (PACKAGE)

DOFFER BRUSHES
    RT BRUSHING
       BRUSHING ACTION
       DOFFING (OF A CARD)
       DOFFING (OPENER)
       OPENERS
       SRRL OPENER

DOFFER COMBS
    UF FLY COMBS
    RT BREAKERS
       CARD WEBS
       CARDING
       CARDS
       COTTON CARDING
       COTTON CARDS
       DOFFERS (CARD)
       DOFFING (OF A CARD)
       FINISHER PART

DOFFER STRIPS
    BT CARD WASTE
    RT CARD STRIPS
       CARDING
       DOFFERS (CARD)
       FETTLING
       FLAT STRIPS
       FLY (WASTE)
       IMPURITIES

DOFFERS (CARD)
    UF CARD DOFFERS
    BT CARD ROLLS
    RT AERODYNAMIC CARDING
       AUTOCOUNT (TN)
       BREAKERS
       CARD CLOTHING
       CARD CYLINDERS
       CARD GRINDING
       CARD SLIVERS
       CARD STRIPS
       CARD WEBS
       CARDING
       CARDING ACTION
       CARDING EFFICIENCY
       CARDS
       CONDENSERS
       COTTON CARDING
       COTTON CARDS
       DOFFER COMBS
       DOFFER STRIPS
       DOFFING (OF A CARD)
       FANCY (NOUN)
       FINISHER PART
       NEP COUNT
       NEPS
       SLIVERS
       STRIPPERS
       STRIPPING ACTION
       WEBS
       WOOLEN CARDING
       WORKERS
       WORSTED CARDING

DOFFERS (PACKAGE)
    RT RING FRAMES
       RING SPINNING
       TWISTERS
       TWISTING

DOFFING (OF A CARD)
    RT AERODYNAMIC CARDING
       AUTOCOUNT (TN)
       BREAKERS
       CARD CLOTHING
       CARD CYLINDERS
       CARD FLATS
       CARD SLIVERS
       CARD WEBS
       CARDING
       CARDING ACTION
       CARDING EFFICIENCY
       CARDS
       CONDENSERS
       COTTON CARDING
       DOFFER BRUSHES
       DOFFER COMBS
       DOFFERS (CARD)
       DOFFING (OPENER)
       FANCY (NOUN)
       FETTLING
       SCRIBBLING
       SLIVERS
       STRIPPERS
       STRIPPING ACTION
       WOOLEN CARDING
       WORSTED CARDING

DOFFING (OPENER)
    RT DOFFER BRUSHES
       DOFFING (OF A CARD)
       OPENERS
       SRRL OPENER

DOFFING (PACKAGE)
    UF DOFF
    NT AUTOMATIC DOFFING
    RT BOBBINS
       CAP SPINNING FRAMES
       CAPS (SPINNING)
       CHEESE WINDING
       DOWN TIME
       FLYER SPINNING FRAMES
       FULL PACKAGES
       MATERIALS HANDLING
       MULE SPINNING
       PACKAGES
       PACKAGING (OPERATION)
       POT SPINNING FRAMES
       RING FRAMES
       SPINNING
       SPINNING EFFICIENCY
       SPINNING FRAMES
       TWISTERS
       TWO FOR ONE TWISTING
       UPTWISTERS
       UPTWISTING
       WINDERS
       WINDING

DOG LEGGED SELVEDGES
    BT FABRIC DEFECTS
    RT CENTER SELVEDGES
       COP END EFFECT
       CRACKED SELVEDGES
       LENO SELVEDGES
       LOOMS
       LOOP SELVEDGES
       PULLED IN SELVEDGES
       SEALED SELVEDGES
       SELVEDGE WARPS
       SELVEDGES
       SLACK SELVEDGES
       TEMPLES
       TIGHT SELVEDGES
       WEAVING
       WOVEN FABRICS

DOGBONE CROSS SECTION
    BT FIBER CROSS SECTIONS

DOGSTOOTH WEAVES
    BT WEAVE (TYPES)
    RT FLAT WOVEN FABRICS

DOLLY
    USE ROPE SCOURING MACHINES

DOMESTIC SEWING MACHINES
    BT SEWING MACHINES
    RT INDUSTRIAL SEWING MACHINES

DOMINANT WAVELENGTH
    RT COLOR
       COLOR MATCHING
       COLORIMETRY

DONNAN EQUILIBRIUM
    RT DONNAN THEORY
       EQUILIBRIUM (DYEING)
       DYEING THEORY
       MEMBRANES
       THERMODYNAMICS

DONNAN THEORY
    BT DYEING THEORY
    RT DONNAN EQUILIBRIUM
       GILBERT-RIDEAL THEORY

DOPE (POLYMER)
    UF VISCOSE SOLUTION
    RT DOPE DYED YARNS
       EXTRUSION
       DOPE DYEING
       DOPE DYEING EQUIPMENT
       POLYMERS
       SPINNERETS
       SPINNING (EXTRUSION)

DOPE DYED YARNS
BT YARNS
RT COLORAY (TN)
DOPE (POLYMER)
DOPE DYEING
DOPE DYEING EQUIPMENT

DOPE DYEING
UF MASS DYEING
SOLUTION DYEING
SPIN DYEING
BT DYEING (BY MATERIAL ASSEMBLY)
RT ACETATE DYEING
DELUSTERING
DOPE (POLYMER)
DOPE DYED YARNS
DOPE DYEING EQUIPMENT
JETSPUN (TN)
PIECE DYED FABRICS
PIECE DYEING
RAYON DYEING
STOCK DYEING
SYNTHETIC DYEING
TOP DYEING
TOW DYEING
YARN DYED FABRICS
YARN DYEING

DOPE DYEING EQUIPMENT
BT DYEING MACHINES (BY MATERIAL ASSEMBLY)
RT DOPE (POLYMER)
DOPE DYED YARNS
DOPE DYEING
DYEING
PACKAGE DYEING MACHINES
PIECE DYEING MACHINES
PROPORTIONATING FEEDERS
STOCK DYEING MACHINES
TOP DYEING MACHINES
TOW DYEING MACHINES
YARN DYEING MACHINES

DORLON (TN)
BT POLYURETHANE FIBERS

DORSET WOOL
BT MEDIUM WOOL
WOOL

DOT DASH STITCHES
BT STITCHES (SEWING)

DOTTED MUSLIN
BT FLAT WOVEN FABRICS
WOVEN FABRICS

DOUBLE APRONS
BT APRONS
RT CASABLANCA SYSTEM
DRAFTING (STAPLE FIBER)

DOUBLE BOW
USE BOW

DOUBLE BRAIDS
BT BRAIDS
REGULAR BRAIDS

DOUBLE CLOTH
RT FLAT WOVEN FABRICS
MULTIPLE LAYER FABRICS
PILLOW TUBING

DOUBLE CURE METHOD
BT DURABLE PRESS TREATMENTS
RT DEFERRED CURE METHOD
DRY CURING
DURABLE PRESS FINISHES
HOT PRESS METHOD (DURABLE PRESS)
PRE-CURE METHOD
RESIN FINISHES
WET CURING

DOUBLE DIAGONAL STITCHES
BT STITCHES (SEWING)

DOUBLE FACED FABRICS (KNITTED)
UF BACK TO BACK FABRICS (KNITTED)
BT KNITTED FABRICS
RT KNITTING

DOUBLE HEADED NEEDLES
BT KNITTING NEEDLES
RT LATCH NEEDLES
LINKS LINKS MACHINES

DOUBLE JAMMING
BT JAMMING
RT COVER FACTOR
TIGHTNESS

DOUBLE JERSEY
USE INTERLOCK (KNITTED FABRICS)

DOUBLE KNIT FABRICS
BT FILLING KNITTED FABRICS
BT KNITTED FABRICS
RT BACK (FABRIC)
CIRCULAR KNITTED FABRICS
COVER
FACE (FABRIC)
FASHIONING
FULLY FASHIONED FABRICS
HAND KNITTING
HOSIERY
KNITTED STITCHES
KNITTING
KNITTING MACHINES
KNITTING YARNS
LOOP DISTORTION
MACHINE KNITTING
NEEDLE LOOPS
PLAIN KNITTED FABRICS
PURL KNITTED FABRICS
RIB KNITTED FABRICS
SINKER LOOPS
STITCH DENSITY
TRICOT KNITTED FABRICS
WARP KNITTED FABRICS

DOUBLE LOCKER MACHINES
BT LACE MACHINES
RT BAR WARP MACHINES
BARMEN MACHINES
BOBBINET MACHINES
CURTAIN MACHINES
GO THROUGH MACHINES
LEAVERS MACHINES
MECHLIN MACHINES
ROLLING LOCKER MACHINES
SIVAL MACHINES
STRING WARP MACHINES
WARP LACE MACHINES

DOUBLE PICKS
BT FABRIC DEFECTS
RT FABRIC INSPECTION
LOOMS
PICKS
WEAVING

DOUBLE PIQUE
NT FRENCH DOUBLE PIQUE
SWISS DOUBLE PIQUE
BT KNITTED FABRICS
RT RIB KNITTING

DOUBLE RING SYSTEM
RT BALLOON CONTROL
BALLOON CONTROL RINGS
DOUBLE SYSTEMS
RING FRAMES
RING SPINNING

DOUBLE ROVINGS
RT RING SPINNING
ROVINGS
SPINNING

DOUBLE ROW STITCHES
BT STITCHES (SEWING)

DOUBLE SYSTEMS
RT CONTROL SYSTEMS
DOUBLE RING SYSTEM
DOUBLE TRUMPETS
MACHINE DESIGN
MULTIPLES
SYSTEMS ENGINEERING
TWO FOR ONE TWISTERS

DOUBLE TRUMPETS
RT CARD SLIVERS
CARDING
DRAFTING (STAPLE FIBER)
DOUBLE SYSTEMS
DRAWFRAMES
ROVING FRAMES
SPINNING FRAMES
TRUMPETS

DOUBLE TWIST SPINDLES
BT SPINDLES
RT TWO FOR ONE TWISTERS

DOUBLED YARNS
USE PLIED YARNS

DOUBLING (DRAFTING)
RT BLENDING
CARD SLIVERS
DRAFT
DRAFTING (STAPLE FIBER)
DRAFTING THEORY
DRAFTING WAVES
DRAWFRAMES
IRREGULARITY
PROCESS VARIABLES

DOUBLING (TWISTING)
USE PLYING

DOUP
RT BACK CROSSING HEDDLES
DOUP ENDS
FRONT CROSSING HEDDLES
GROUND HEDDLES
HARNESSES
LENO WEAVES
LENO WEAVING
STANDARD HEDDLES

DOUP ENDS
RT DOUP
LENO WEAVES
LENO WEAVING
SLIVERS
VARIANCE
WEAK SPOTS

DOUBLING TRAVELERS
USE PLYING TRAVELERS

DOUP WEAVES
USE LENO WEAVES

DOUP WEAVING
USE LENO WEAVING

DOWN TIME
RT DOFFING (PACKAGE)
END BREAKAGES
IMPROVEMENTS
MODIFICATION
PERFORMANCE
PROCESS EFFICIENCY
SET UP TIME
SPINNING EFFICIENCY
STOPS
WEAVING EFFICIENCY
WINDING EFFICIENCY

DOWNTWISTERS
BT TWISTERS
RT THROWSTERS
TWO FOR ONE TWISTERS
UPTWISTERS

DOWNTWISTING
BT TWISTING
RT BALLOON CONTROL
BALLOON CONTROL RINGS
BALLOON DYNAMICS
BALLOON SEPARATORS
BALLOON TENSION
BALLOONS
CAP SPINNING
FALSE TWISTING (EXCEPT TEXTURING)
FLYER SPINNING
PACKAGING (OPERATION)
PLYING
POT SPINNING
REAL TWIST
RING SPINNING
RING TWISTING
SPINNING
SPINNING BALLOONS
THROWING
TWIST
TWO FOR ONE TWISTING
UPTWISTING

DP
USE DEGREE OF POLYMERIZATION

DRABNESS
BT AESTHETIC PROPERTIES
FABRIC PROPERTIES
FABRIC PROPERTIES (AESTHETIC)
OPTICAL PROPERTIES
RT APPEARANCE
BLOOM
BRIGHTNESS
COLOR
DULLNESS
LUSTER
LUXURIOUSNESS
RICHNESS
SHADE
SPARKLE

DRAFT
UF DRAFT RATIO
DRAW
MECHANICAL DRAFT
NT ACTUAL DRAFT
BREAK DRAFT
FALSE DRAFT
RT AMBLER SUPERDRAFT
APRON CONTROL
CONTROL SURFACES
DOUBLING (DRAFTING)
DRAFT CONSTANT
DRAFT CUT
DRAFTING (STAPLE FIBER)
DRAFTING FORCE
DRAFTING MACHINES
DRAFTING TWIST
DRAFTING WAVES
DRAFTING ZONE
DRAW RATIO
(CONT.)

RT FIBER ORIENTATION
FLOATING FIBERS
FRONT ROLLS
GILLING
HIGH DRAFT
HIGH DRAFT SPINNING
HOOK REMOVAL
IRREGULARITY
LEADING HOOKS
MULE DRAFT
PROCESS VARIABLES
ROLLER DRAFT
ROVINGS
SLIVERS
SPINNING
TRAILING HOOKS
VELOCITY CHANGE POINT
WOOLEN DRAFTING

DRAFT (PATTERN)
RT CHAIN DRAFT
DRAFTING (PATTERN)
HARNESS DRAFT
PATTERN (FABRICS)
WEAVING

DRAFT CONSTANT
RT ACTUAL DRAFT
DRAFT
DRAFT CUT
DRAFTING (STAPLE FIBER)
DRAFTING FORCE
DRAFTING ZONE
FIBER WITHDRAWAL FORCE

DRAFT CONTROL
RT AUTOLEVELLERS
AUTOMATIC CONTROL
DRAFTING (STAPLE FIBER)
DRAFTING FORCE
DRAFTING FORCE TESTERS
DRAFTING ZONE
IRREGULARITY CONTROL
PROCESS CONTROL
RAPER AUTOLEVELLER (TN)
SERVO DRAFT
WEIGHT CONTROL

DRAFT CONTROL SURFACES
USE CONTROL SURFACES

DRAFT CUT
RT ABRASION
CUTTING
DRAFT
DRAFT CONSTANT
DRAFTING (STAPLE FIBER)
DRAFTING WAVES
FIBER BREAKAGE
FIBER DIAGRAM
FIBER SHUFFLING
GREENFIELD CONVERTER
HALLE SEYDEL STRETCH BREAKER
PACIFIC CONVERTER
PERLOK SYSTEM
STAINS DIRECT SPINNER
STRETCH BREAKING
TOW
TOW CONVERSION
TOW TO YARN CONVERSION
TURBOSTAPLER

DRAFT RATIO
USE DRAFT

DRAFTING (STAPLE FIBER)
UF DRAFTING SYSTEMS
DRAWING (STAPLE FIBER)
REDUCING (WORSTED PROCESS)
NT APRON DRAFTING
GILLING
MULE DRAFT
PIN DRAFTING
ROLLER DRAFTING
WOOLEN DRAFTING
RT ACTUAL DRAFT
AMBLER SUPERDRAFT
APRON CONTROL
AUTOLEVELLERS
BACK ROLLS
BLEND DRAFTING
BRADFORD SYSTEM PROCESSING
CARD SLIVERS
CASABLANCA SYSTEM
CONTROL SURFACES
COTTON SYSTEM PROCESSING
DEGREE OF ORIENTATION
DOUBLE TRUMPETS
DOUBLING (DRAFTING)
DRAFT
DRAFT CONSTANT
DRAFT CONTROL
DRAFT CUT
DRAFTING DYNAMICS
DRAFTING FORCE TESTERS
DRAFTING ROLLS
DRAFTING THEORY
DRAFTING TWIST

RT DRAFTING WAVES
DRAFTING ZONE
DRAG
DRAWFRAMES
DUO ROTH SYSTEM
EQUALIZER (DRAFTING)
EQUALIZING (DRAFTING)
FALSE TWIST TUBES
FIBER BREAKAGE
FIBER CONTROL
FIBER HOOKS
FIBER ORIENTATION
FIBER SHUFFLING
FIBER SLIPPAGE
FIBER WITHDRAWAL FORCE
FINISHER GILLING
FLOATING FIBERS
FRONT ROLLS
GILL BOXES
HIGH DRAFT SPINNING
HOOK DIRECTION
HOOK FORMATION
HOOK REMOVAL
IRREGULARITY
IRREGULARITY (PERIODIC)
IRREGULARITY CONTROL
LAP UP
LEADING HOOKS
LINDSLEY RATIOS
MULE SPINNING
MULES
NEW BRADFORD SYSTEM PROCESSING
OPEN END SPINNING
OVERLAP (FIBERS)
PARALLELIZATION
RAPER AUTOLEVELLER (TN)
RATCH
RING SPINNING
ROLL WEIGHTING
ROLLER DRAFT
ROLLER LAP
ROLLER SLIPPAGE
ROVING (OPERATION)
ROVINGS
SERVO DRAFT
SHAW SYSTEM
SLIPPAGE
SLIVER CANS
SLIVER REVERSALS
SLIVERS
SPINNING
SPINNING FRAMES
SURGING
TOPS
TRAILING HOOKS
TWIST CONTROL
TWISTING AT THE HEAD
VELOCITY CHANGE POINT
WEIGHT CONTROL
WORSTED SYSTEM PROCESSING
YARNS

DRAFTING (PATTERN)
NT FLOAT DESIGN
RT CHAIN DRAFT
DRAFT (PATTERN)
FABRIC ANALYSIS
FABRIC DESIGN
HARNESS DRAFT
HEDDLES
PATTERN (FABRICS)
PATTERN REPEATS
WEAVE
WEAVING

DRAFTING DYNAMICS
BT PROCESS DYNAMICS
RT CARDING DYNAMICS
DRAFTING (STAPLE FIBER)
DRAFTING FORCE TESTERS
DRAFTING THEORY
DRAFTING TWIST
DRAFTING WAVES
FIBER CONTROL
FIBER WITHDRAWAL FORCE
KNITTING DYNAMICS
MECHANISM (FUNDAMENTAL)
SPINNING DYNAMICS
TWISTING DYNAMICS
WEAVING DYNAMICS
WINDING DYNAMICS
YARN DYNAMICS

DRAFTING FORCE
UF DRAFTING RESISTANCE
BT FORCE
RT DRAFT
DRAFT CONSTANT
DRAFTING FORCE TESTERS
DRAFTING THEORY
DRAFTING ZONE
DRAWING TENSION (FILAMENT)
FIBER CONTROL
FIBER WITHDRAWAL FORCE
ROLLER SLIPPAGE

DRAFTING FORCE TESTERS
UF WEST POINT COHESION TESTER

BT TESTING EQUIPMENT
RT DRAFT CONTROL
DRAFTING (STAPLE FIBER)
DRAFTING DYNAMICS
DRAFTING FORCE
DRAFTING THEORY
FIBER FIBER FRICTION

DRAFTING MACHINES
NT DRAWFRAMES
GILL BOXES
PIN DRAFTERS
RT AMBLER SUPERDRAFT
AUTOLEVELLERS
CASABLANCA SYSTEM
DRAFT
DRAFTING ROLLS
PORCUPINE DRAWING
RAPER AUTOLEVELLER (TN)
ROVINGS
SLIVERS
SPINNING FRAMES

DRAFTING RESISTANCE
USE DRAFTING FORCE

DRAFTING ROLLS
NT BACK ROLLS
BOTTOM ROLLS
COVERED ROLLS
DRIVEN ROLLS
DRIVING ROLLS
FLUTED ROLLS
FRONT ROLLS
INTERMEDIATE ROLLS
PLAIN ROLLS
TOP ROLLS
TUMBLERS (ROLLS)
BT ROLLS
RT ACTUAL DRAFT
APRONS
BOWING
BREAK DRAFT
CAP SPINNING FRAMES
CAPS (SPINNING)
CONTROL SURFACES
COTS (ROLLERS)
CRADLES
DRAFTING (STAPLE FIBER)
DRAFTING MACHINES
DRAFTING ZONE
MAGNADRAFT (TN)
NIP
NIP PRESSURE
PNEUMAFIL (TN)
PRESSURE ROLLS
RATCH
RING FRAMES
ROLL BOSSES
ROLL CLEARERS
ROLL COVERINGS
ROLL ECCENTRICITY
ROLL WEIGHTING
ROLLER DRAFTING
ROLLER LAP
ROVINGS
SLIVERS
SPINNING FRAMES

DRAFTING SYSTEMS
USE DRAFTING (STAPLE FIBER)

DRAFTING THEORY
RT ACTUAL DRAFT
APRONS
AUTOCORRELATION
AUTOLEVELLING
BACK ROLLS
BREAK DRAFT
CONTROL SURFACES
CONTROL THEORY
DOUBLING (DRAFTING)
DRAFTING (STAPLE FIBER)
DRAFTING DYNAMICS
DRAFTING FORCE
DRAFTING FORCE TESTERS
DRAFTING TWIST
DRAFTING WAVES
FIBER CONTROL
FIBER HOOKS
FIBER ORIENTATION
FIBER SLIPPAGE
FIBER WITHDRAWAL FORCE
FLOATING FIBERS
FRONT ROLLS
HOOK REMOVAL
HOOKED FIBERS
IRREGULARITY
IRREGULARITY (PERIODIC)
LEADING HOOKS
MARTINDALE LIMIT
MECHANISM (FUNDAMENTAL)
RATCH
SLIVER REVERSALS
TRAILING HOOKS
TWIST DISTRIBUTION
VARIANCE
VELOCITY CHANGE POINT

DRAFTING TWIST
RT DRAFT
DRAFTING (STAPLE FIBER)
DRAFTING DYNAMICS
DRAFTING THEORY
FALSE TWIST TUBES
MULE SPINNING
TWIST

DRAFTING WAVES
RT AMPLITUDE
BACK ROLLS
CONTROL SURFACES
DEFECTS
DOUBLING (DRAFTING)
DRAFT
DRAFT CUT
DRAFTING (STAPLE FIBER)
DRAFTING DYNAMICS
DRAFTING THEORY
DRAFTING ZONE
DRAWFRAMES
FIBER CONTROL
FIBER LENGTH
FIBER ORIENTATION
FIBER SLIPPAGE
FLOATING FIBERS
FREQUENCY
HIGH DRAFT SPINNING
IRREGULARITY
IRREGULARITY (PERIODIC)
IRREGULARITY (SHORT TERM)
PERIODICITY
RATCH
ROVINGS
SLIVERS
SURGING
VELOCITY CHANGE POINT
WEAK SPOTS

DRAFTING ZONE
RT ACTUAL DRAFT
CONTROL SURFACES
DRAFT
DRAFT CONSTANT
DRAFT CONTROL
DRAFTING (STAPLE FIBER)
DRAFTING DYNAMICS
DRAFTING FORCE
DRAFTING ROLLS
DRAFTING WAVES
HOOK REMOVAL
ROLLER DRAFTING
ROVINGS
SLIVERS

DRAG
RT AERODYNAMIC CONFIGURATIONS
AERODYNAMICS
AIR DRAG
AIR RESISTANCE
BALLOON DYNAMICS
BOUNDARY LAYER
COHESION
DRAFTING (STAPLE FIBER)
FLUID FLOW
FLUID FRICTION
FRICTION
LAMINAR FLOW
LIFT (FLUID)

DRAPE
UF DYNAMIC DRAPE
BT AESTHETIC PROPERTIES
END USE PROPERTIES
FABRIC PROPERTIES
FABRIC PROPERTIES (AESTHETIC)
FABRIC PROPERTIES (MECHANICAL)
RT AESTHETIC APPEAL
APPEARANCE
BENDING
BENDING LENGTH
BENDING RIGIDITY
BOWING
BUCKLING
COVER
CREPING
DEFLECTION
DEFORMATION
DRAPEOMETER
FRL DRAPE TESTER
HAND
HEART LOOP
LUXURIOUSNESS
MECHANICAL PROPERTIES
OPTICAL PROPERTIES
PAPERINESS
SHEAR RESISTANCE
STIFFNESS
WAVINESS

DRAPEOMETER
RT DRAPE
FRL DRAPE TESTER

DRAPER SHUTTLELESS LOOMS (TN)
BT SHUTTLELESS LOOMS
RT AIR JET LOOMS
JET LOOMS
OUTSIDE FILLING SUPPLY
RAPIER LOOMS
SHUTTLELESS WEAVING
SULZER LOOMS (TN)
WATER JET LOOMS

DRAPES
UF CURTAINS
RT DECORATIVE FABRICS
FURNISHING FABRICS
FURNISHINGS
HOUSEHOLD FABRICS
TAPESTRIES
UPHOLSTERY

DRAVES TEST
BT TESTING
RT HYDROPHILIC PROPERTY
WATER ABSORPTION TESTING
WATER REPELLENCY
WETTING
WETTING AGENTS

DRAW
USE DRAFT

DRAW MECHANISM (KNITTING)
UF COULIER MOTION (KNITTING)
RT KNITTING

DRAW RATIO
RT COLD DRAWING
DRAFT
DRAWTWISTING
DRAWING (FILAMENT)
DRAWING TENSION (FILAMENT)
HOT DRAWING
NATURAL DRAW RATIO
NECKING (FILAMENT)

DRAWERS
BT UNDERWEAR
RT BOXER SHORTS
BRIEFS
SHORTS
TRUNKS

DRAWFRAMES
UF DRAWING FRAMES
BT DRAFTING MACHINES
RT AUTOLEVELLING
BRADFORD OPEN DRAWING
CASABLANCA SYSTEM
CLEARERS (LINT)
COILERS
CONE DRAWING
DEAD WEIGHT LOADING
DOUBLE TRUMPETS
DOUBLING (DRAFTING)
DRAFTING (STAPLE FIBER)
DRAFTING WAVES
DUO ROTH SYSTEM
EQUALIZER (DRAFTING)
FALLER SCREWS
FLYER SPINNING FRAMES
FRONT ROLLS
GILL BOXES
GILLING
INTERSECTING GILL BOXES
PORCUPINE DRAWING
ROLL WEIGHTING
ROLLER SHAFTS
ROVING FRAMES
ROVINGS
ROVOMATIC (TM)
SHAW SYSTEM
SLIVER CANS
SLIVER REVERSALS
SLIVERS
SPINNING FRAMES
SPRING LOADING
TUMBLERS (ROLLS)

DRAWING (FILAMENT)
NT COLD DRAWING
DRAWTWISTING
HOT DRAWING
RT ANNEALING
CREEP
CRYSTALLIZATION
DECRYSTALLIZATION
DEFORMATION
DEGREE OF ORIENTATION
DRAW RATIO
DRAWING TENSION (FILAMENT)
ELASTICITY
EXTRUSION
FIBER STRENGTH
FILAMENT YARNS
FILAMENTS
FLOW (PLASTIC)
LOADING
MAN MADE FIBERS
MELT SPINNING
MODULUS
MOLECULAR ORIENTATION
RT MONOFILAMENT YARNS
MULTIFILAMENT YARNS
NATURAL DRAW RATIO
NECKING (FILAMENT)
PLASTIC FLOW
PLASTICITY
RECOVERY (SELF RESTORATION TO ORIGINAL CONDITION)
RHEOLOGY
SPACE DRAWING
SPINNING (EXTRUSION)
STRAIN
STRESS
STRESS RELAXATION
STRESS STRAIN CURVES
STRESS STRAIN PROPERTIES (TENSILE)
TENACITY
VISCOELASTICITY
YIELD (MECHANICAL)

DRAWING (STAPLE FIBER)
USE DRAFTING (STAPLE FIBER)

DRAWING (WIRE)
USE WIRE DRAWING

DRAWING FRAMES
USE DRAWFRAMES

DRAWING IN
NT REEDING
RT AUTOMATIC KNOTTING
ENDS PER INCH
HARNESS DRAFT
HARNESSES
HEDDLES
LEASE RODS
LEASING
LIGHT DENT
LOOMS
REED DRAFT
REED HOOKS
REED WIRES
REEDS
SWOLLEN DENT
THREADING
TYING IN
WARP BEAMS
WARP ENDS
WARP PREPARATION
WARPING
WEAVE
WEAVING

DRAWING IN MACHINES
RT KNOTTERS
LOOMS
REED HOOKS
TYING IN MACHINES
WARPING

DRAWING TENSION (FILAMENT)
BT TENSION
RT DRAFTING FORCE
DRAW RATIO
DRAWTWISTING
DRAWING (FILAMENT)
NECKING (FILAMENT)
PLASTIC FLOW
SPINNING TENSION (EXTRUSION)
STRESS STRAIN PROPERTIES (TENSILE)
VISCOELASTICITY
TWISTING TENSION
WINDING TENSION
YARN TENSION

DRAWN PIECE
USE BOWED FABRICS

DRAWN PILE FINISH
RT DRESS FACE FINISH
NAP
NAPPING
TEAZLE GIG

DRAWTWISTING
BT DRAWING (FILAMENT)
RT COLD DRAWING
DRAW RATIO
DRAWING TENSION (FILAMENT)
FILAMENT YARNS
FILAMENTS
HOT DRAWING
MONOFILAMENT YARNS
NECKING (FILAMENT)
PLASTIC FLOW
SPACE DRAWING
TWISTING
WIRE DRAWING
YIELD (MECHANICAL)

DRESDEN POINT
BT HAND MADE LACES
PILLOW LACES (BOBBIN LACES)

DRESS BANDINGS
BT LININGS
RT POCKET LININGS

DRESS DESIGN
BT APPAREL DESIGN
DESIGN
RT DRESSMAKING
FABRIC DESIGN
GARMENT MANUFACTURE
GARMENTS
MANNEQUINS
PATTERN (APPAREL)
STYLING (APPAREL)
TAILORING

DRESS FACE FINISH
RT DRAWN PILE FINISH
NAPPING
SHEARING (FINISHING)

DRESSCLOTHING
UF DRESSWEAR
BT GARMENTS
OUTERWEAR
RT COATS
DRESSES
FROCKS
GOWNS
HOSIERY
JACKETS
NECKTIES
OVERCOATS
SUITS

DRESSGOODS
USE APPAREL FABRICS

DRESSES
NT SACKS (DRESSES)
BT GARMENTS
OUTERWEAR
RT BLOUSES
COATS
DRESSCLOTHING
FROCKS
GOWNS
HOSIERY
JACKETS
SKIRTS
SUITS

DRESSING (WARP)
USE WARPING

DRESSING SELVEDGES (LACE)
BT SELVEDGES
RT CENTER SELVEDGES
LACES
LENO SELVEDGES
LOOP SELVEDGES
SEALED SELVEDGES

DRESSMAKING
RT BASTING
CUTTING (TAILORING)
DESIGN
DRESS DESIGN
MARKING
PATTERN (APPAREL)
PATTERN (FABRICS)
PATTERN REPEATS
PRESSING
PRODUCT DESIGN
SEAMING
SEWING
STITCHING
TAILORING

DRESSWEAR
USE DRESSCLOTHING

DRIFT
RT ACCURACY
ADJUSTMENTS
CALIBRATION
COMPENSATION
CORRECTION
DYNAMIC CHARACTERISTICS
ERRORS
ESTIMATION
EXTRAPOLATION
FLUCTUATIONS
HUNTING
STABILITY
TOLERANCES

DRILLS (FABRICS)
BT FLAT WOVEN FABRICS
WOVEN FABRICS

DRIMARENE (TN)
USE CHLOROPYRIMIDINE DYES

DRIP DRY FABRICS
USE EASY CARE FABRICS

DRIP DRYING
USE LINE DRYING (EASY CARE GARMENTS)

DRIVEN ROLLS
BT DRAFTING ROLLS
ROLLS
RT DRIVING ROLLS
FLUTED ROLLS

DRIVES
NT BELT DRIVES
CHAIN DRIVES
FLUID DRIVES
GEAR DRIVES
HYDRAULIC DRIVES
MECHANICAL DRIVES
TAPE DRIVES
VARIABLE SPEED DRIVES
RT CONSTANT SPEED
GEARS
MOTION
MOTORS
POWER CONSUMPTION
STARTING TIME
VARIABLE SPEED

DRIVING BANDS
RT CARRIAGES (SPINNING MULE)
HEADSTOCKS
MULE DRAFT
MULE SPINNING
MULES
SLUBBINGS

DRIVING ROLLS
BT DRAFTING ROLLS
ROLLS
RT DRIVEN ROLLS
FLUTED ROLLS

DROP BOX LOOMS
USE BOX LOOMS

DROP PENETRATION TESTING
UF DROP PENETRATION TESTS
BT WATER RESISTANCE TESTING
RT PERMEABILITY TESTING
RAIN PENETRATION TESTING
WATER REPELLENCY TESTING

DROP PENETRATION TESTS
USE DROP PENETRATION TESTING

DROP WIRES
BT STOP MOTIONS
RT BANGING OFF
FEELER MECHANISMS
LOOMS
PROCESS CONTROL
WARP TENSION
WEAVING
WINDING

DROPPED STITCHES
BT FABRIC DEFECTS
RT MISPICKS

DRUM (DYEING)
RT HOSIERY DYEING MACHINES

DRUM DRYERS
UF CYLINDER DRYERS
BT DRYERS
RT ATMOSPHERIC DRYERS
CENTRIFUGAL DRYERS
DIELECTRIC DRYERS
DRUM SCOURING MACHINES
FLUIDIZED BED DRYERS
HOT AIR DRYERS
INFRARED DRYERS
LOOP DRYERS
VACUUM DRYERS

DRUM SCOURING MACHINES
BT SCOURING MACHINES
RT CASCADE WASHERS
DRUM DRYERS
SCOURING
SCOURING BATHS
SCOURING TRAINS

DRUM WINDERS
NT SPLIT DRUM WINDERS
UNICONER (TN)
BT WINDERS
RT BUILDER MOTIONS
CHEESE WINDERS
CLOSE WIND
COMBINATION WIND
CONING
DRUM WINDING
GAIN (WINDING)
GROOVES
LAYER LOCKING
OPEN WIND
PACKAGES
PRECISION WINDERS
RIBBON BREAKERS
SLUB CATCHERS
TENSION CONTROL
WIND
WINDING

DRUM WINDING
BT WINDING
RT AUTOMATIC WINDING
BUILDER MOTIONS
CLOSE WIND
CLOSE WINDING
CONING
CREELING
DRUM WINDERS
KNOTS
KNOTTING
LAYER LOCKING
OPEN WIND
OPEN WINDING
PACKAGES
PACKAGING (OPERATION)
PRECISION WINDING
QUILLING
RIBBON BREAKERS
RIBBON BREAKING
SLOUGHING
SLUB CATCHING
TENSION CONTROL
TRAVERSE
UNWINDING
WARPING
WIND
WINDERS
WINDING ANGLE

DRY BULB TEMPERATURE
RT CRITICAL MOISTURE CONTENT
DEW POINT
DIFFUSION
DIFFUSIVITY
DRYING
EQUILIBRIUM MOISTURE CONTENT
HYGROMETERS
PSEUDO WET BULB TEMPERATURE
PSYCHROMETRIC CHARTS
PSYCHROMETRY
REGAIN
RELATIVE HUMIDITY
SPECIFIC HUMIDITY
WET BULB TEMPERATURE

DRY CLEANING
BT CLEANING
RT ARTIFICIAL SOIL
CARBON TETRACHLORIDE
CHLORINATED SOLVENTS
CLEANLINESS
DEODORIZING
DETERGENCY
DRY CLEANING FASTNESS (COLOR)
DRY CLEANING FASTNESS (OF FINISH)
DRY CLEANING MACHINES
DRY CLEANING SOLVENTS
GARMENT DYEING
GARMENT DYEING MACHINES
LAUNDERING
PERCHLOROETHYLENE
PERCLENE (TN)
SOIL DEPOSITION
SOIL REMOVAL
SOIL RESISTANCE
SOLVENTS
SPOTTING
STAIN REMOVAL
STODDARD SOLVENT
SURFACTANT APPLICATIONS
WEAR TESTING

DRY CLEANING FASTNESS (COLOR)
BT COLORFASTNESS
DEGRADATION PROPERTIES
END USE PROPERTIES
RT DRY CLEANING
DRY CLEANING FASTNESS (OF FINISH)
DRY CLEANING SOLVENTS
DYEING
DYES
PERCHLOROETHYLENE
PERCLENE (TN)

DRY CLEANING FASTNESS (OF FINISH)
BT DEGRADATION PROPERTIES
END USE PROPERTIES
RT DRY CLEANING
DRY CLEANING FASTNESS (COLOR)
DRY CLEANING SOLVENTS
DRY FINISHING
DURABLE FINISH (GENERAL)
DURABILITY
FINISH (PROPERTY)
FINISH (SUBSTANCE ADDED)
PERCHLOROETHYLENE
PERCLENE (TN)
WET FINISHING

DRY CLEANING MACHINES
RT CLEANING
DRY CLEANING
LAUNDERING
SOLVENTS
WASHING MACHINES

DRY CLEANING SOLVENTS
BT SOLVENTS
RT CARBON TETRACHLORIDE
CHLORINATED SOLVENTS
DRY CLEANING
DRY CLEANING FASTNESS (COLOR)
DRY CLEANING FASTNESS (OF FINISH)
NAPHTHA
PERCLENE (TN)
STODDARD SOLVENT
TRICHLOROETHYLENE

DRY CREASE RECOVERY
USE DRY WRINKLE RECOVERY

DRY CURING
UF BAKING
BT CURING
DRY FINISHING
HEAT TREATMENT
RT CURING OVENS
CURING RANGES
DEFERRED CURE METHOD
DOUBLE CURE METHOD
DURABLE PRESS TREATMENTS
HOT PRESS METHOD (DURABLE PRESS)
PAD DRY CURE PROCESS
PRE-CURE METHOD
WASH WEAR TREATMENTS
WET CURING

DRY DECATING
USE DECATING

DRY FINISHING
UF MECHANICAL FINISHING
NT BEETLING
BLOWN FINISH
BOARDING
BRUSHING
BURLING
BUTTON BREAKING
CALENDERING
COMPACTING
COMPRESSIVE SHRINKING
CURING
DRY CURING
DRY RELAXATION
DRYING
FABRIC RELAXATION
FOLDING (OF CLOTH)
GLAZING
HEAT SETTING (DRY)
HEAT SETTING (SYNTHETICS)
IRONING
MANGLING
MENDING
NAPPING
NEEDLING
PERCHING
PILL RESISTANCE TREATMENTS
PLEATING
PRESHRINKING
PRESSING
RELAXATION
SANFORIZING (TN)
SCHREINERING
SCUTCHING (FINISHING)
SETTING (EXCEPT SYNTHETICS)
SHEARING (FINISHING)
SINGEING
SPECKING
SPONGING
STRESS RELAXATION
SUEDING
TENTERING
BT FINISHING PROCESS (GENERAL)
RT AFTERTREATMENTS (GENERAL)
DRY CLEANING FASTNESS (OF FINISH)
EMBOSSING CALENDERS
FACE FINISH
FINISHED FABRICS
HEAT TREATMENT
INSPECTION
PRETREATMENTS (GENERAL)
PROCESS DESIGN
SETTING TREATMENTS (GENERAL)
TEXTILE MACHINERY (GENERAL)
TEXTILE PROCESSES (GENERAL)
WET FINISHING

DRY IN BLEACHING
BT BLEACHING
RT DRYING

DRY POWDER BONDING
BT BONDING
RT ADHESIVES
BONDED FIBER FABRICS
DIP BONDING
FLAME BONDING
NONWOVEN FABRICS
SPRAY BONDING

DRY RELAXATION
UF DRY RELAXING
BT DRY FINISHING
FABRIC RELAXATION
RELAXATION
RT CONDITIONING
FABRIC SHRINKAGE
HEAT SETTING (SYNTHETICS)
RELAXATION SHRINKAGE
RELAXATION TIME (MECHANICAL)
RESIDUAL STRESS
STRAIN
STRESS
STRESS RELAXATION
WET RELAXATION

DRY RELAXING
USE DRY RELAXATION

DRY SPINNING
USE SOLVENT SPINNING

DRY STRENGTH
BT STRENGTH
RT FABRIC STRENGTH
STRENGTH ELONGATION TESTING
WET STRENGTH

DRY WEIGHT
BT WEIGHT
RT HUMIDITY
HUMIDITY CONTROL
MOISTURE CONTENT
MOISTURE CONTENT CONTROL

DRY WRINKLE RECOVERY
UF DRY CREASE RECOVERY
BT WRINKLE RECOVERY
RT DURABLE PRESS
WET WRINKLE RECOVERY
WRINKLE RECOVERY ANGLE
WRINKLE RECOVERY TESTING
WRINKLE RESISTANCE

DRYERS
UF DRYING MACHINES
NT ATMOSPHERIC DRYERS
CENTRIFUGAL DRYERS
DIELECTRIC DRYERS
DRUM DRYERS
DRYING CANS
FLUIDIZED BED DRYERS
HOT AIR DRYERS
INFRARED DRYERS
LOOP DRYERS
VACUUM DRYERS
RT AIR FLOW
CENTRIFUGES
CURING OVENS
HEAT
HEAT EXCHANGERS
HEAT TRANSFER
HEATING EQUIPMENT
LAUNDERING
MANGLES
MASS TRANSFER
OVENS
PAPER MACHINES
TUMBLE DRYING

DRYING
NT BATCH DRYING
CENTRIFUGAL DRYING
CONTINUOUS DRYING
DIELECTRIC DRYING
FREEZE DRYING
HOT AIR DRYING
INFRARED DRYING
LINE DRYING (EASY CARE GARMENTS)
TUMBLE DRYING
VACUUM DRYING
BT DRY FINISHING
RT AIR
AIR FLOW
AMBIENT TEMPERATURE
BOILING
BOUND WATER
BOUNDARY LAYER
CAPILLARITY
CAPILLARY WATER
CONDUCTION
CONSTANT RATE PERIOD
CONVECTION
COUNTERFLOW PROCESSES
CRITICAL MOISTURE CONTENT
DEHYDRATION
DEW POINT
DIELECTRIC HEATING
DIFFUSION
DIFFUSION COEFFICIENT
DRY BULB TEMPERATURE
DRY IN BLEACHING
DRYERS
DRYING CANS
DRYING TIME
ELECTRIC HEATING (RESISTANCE)
FALLING RATE PERIOD
GAS HEATING (GAS FUEL)
HEAT
HEAT TRANSFER
HEATING
HUMIDITY
HUMIDITY CONTROL
HYGROMETERS
IMBIBITION
MANGLES
MANGLING
MASS TRANSFER
MELTING
MIGRATION (SUBSTANCE)
MOISTURE CONTENT
MOISTURE CONTENT TESTING
OVENS
PAPER MAKING
PARTIAL PRESSURE
PRESS SECTION
PRESSING
PSEUDO WET BULB TEMPERATURE
PSYCHROMETRIC CHARTS
PSYCHROMETRY
RADIATION
RADIANT HEATING
REGAIN
RELATIVE HUMIDITY
SATURATED VAPOR
SCORCHING MARKS
SPECIFIC HUMIDITY
SQUEEZING
SUPERHEATED STEAM
TENTERING
THERMAL DEGRADATION
THERMAL PROPERTIES
VACUUM EXTRACTION
WET BULB TEMPERATURE

DRYING CANS
UF CANS (DRYING)
BT DRYERS
RT ATMOSPHERIC DRYERS
CENTRIFUGAL DRYERS
DRYING
FLUIDIZED BED DRYERS
HEAT TRANSFER
HOT AIR DRYERS
INFRARED DRYERS
MASS TRANSFER
TUMBLE DRYING

DRYING MACHINES
USE DRYERS

DRYING TIME
RT CONSTANT RATE PERIOD
DRYING
EQUILIBRIUM MOISTURE CONTENT
FALLING RATE PERIOD
HEAT TRANSFER
MASS TRANSFER
TIME

DS (CELLULOSE)
UF DEGREE OF SUBSTITUTION (CELLULOSE)
RT ACETYLATED COTTON
CARBAMOYLETHYLATED COTTON
CELLULOSE DERIVATIVES
CHEMICAL MODIFICATION (FIBERS)
CHEMICALLY MODIFIED COTTON
CYANOETHYLATED COTTON

DUCHESS (LACE)
BT HAND MADE LACES
LACES

DUCK
UF CANVAS
NT NUMBERED DUCK
BT FLAT WOVEN FABRICS
INDUSTRIAL FABRICS
WOVEN FABRICS
RT AWNING CLOTH
EQUIPAGE
PLAIN WEAVES
TARPAULINS
TENTS

DUCTILITY
BT STRESS STRAIN PROPERTIES
STRESS STRAIN PROPERTIES (TENSILE)
RT BENDING
BRITTLENESS
EXTENSIBILITY
NECKING (FILAMENT)
PENETRATION
PLASTICITY
SEEPAGE (RHEOLOGY)
STRAIN
TACK
THIXOTROPY
TOUGHNESS

DUFFEL
UF DUFFLE
RT FLAT WOVEN FABRICS
NAPPING
WOOLEN FABRICS

DUFFLE
USE DUFFEL

DULLNESS
BT AESTHETIC PROPERTIES
FABRIC PROPERTIES
FABRIC PROPERTIES (AESTHETIC)
OPTICAL PROPERTIES
RT APPEARANCE
BLOOM
BRIGHTNESS
COLOR
DRABNESS
LUSTER
OPACITY
RICHNESS
SHADE
SMOOTHNESS
SPARKLE
SPECTROPHOTOMETRY
TRANSPARENCY

DUMMY NEEDLES
UF BLANK NEEDLES
RT KNITTING
KNITTING NEEDLES

DUMMY SLIDERS
UF LATCH GUARDS
RT KNITTING

DUNGAREES
BT GARMENTS
INDUSTRIAL CLOTHING
WORK CLOTHING
RT DENIMS
JEANS
OVERALLS
SAFETY CLOTHING
SPORTSWEAR

DUNNAGE BAGS
RT INDUSTRIAL FABRICS

DUOMEENS (TN)
BT POLYALKYLAMINE SURFACTANTS

DUO CARDS
BT CARDS
COTTON CARDS
RT AERODYNAMIC CARDS
HIGH PRODUCTION CARDS
WOOLEN CARDS

DUO ROTH SYSTEM
RT ACTUAL DRAFT
CASABLANCA SYSTEM
DRAFTING (STAPLE FIBER)
DRAWFRAMES
HIGH DRAFT
ROVING FRAMES
SHAW SYSTEM
SPINNING FRAMES

DUPLEX PRINTING
UF DUPLICATE PRINTING
REGISTERED PRINTING
BT PRINTING
RT DUPLEX PRINTING MACHINES
PRINTING MACHINES
SCREEN PRINTING

DUPLEX PRINTING MACHINES
BT PRINTING MACHINES
RT BLOCK PRINTING EQUIPMENT
DUPLEX PRINTING
MELANGE PRINTING EQUIPMENT
SCREEN PRINTING MACHINES

DUPLICATE PRINTING
USE DUPLEX PRINTING

DURABILITY
BT END USE PROPERTIES
MECHANICAL PROPERTIES
RT ABRASION
ABRASION RESISTANCE
ABRASION RESISTANCE FINISHES
CHLORINE FASTNESS (OF FINISH)
CUTTING RESISTANCE
DAMAGE
DEGRADATION
DURABLE FINISH (GENERAL)
FATIGUE RESISTANCE
FRICTION
LAUNDERABILITY
LIGHTFASTNESS (OF FINISH)
SNAG RESISTANCE
STABILITY
TOUGHNESS
WASHFASTNESS (COLOR)
WASHFASTNESS (OF FINISH)
WEAR
WEAR RESISTANCE

DURABLE FINISH (GENERAL)
UF PERMANENT FINISHES (GENERAL)
RT CHLORINE FASTNESS (OF FINISH)
DEGRADATION PROPERTIES
DRY CLEANING FASTNESS (OF FINISH)
DURABILITY
FINISH (PROPERTY)
FINISH (SUBSTANCE ADDED)
FINISHING PROCESS (GENERAL)
LIGHTFASTNESS (OF FINISH)
WASHFASTNESS (OF FINISH)

DURABLE PRESS
UF PERMANENT CREASING
PERMANENT PRESS
BT END USE PROPERTIES
FINISH (PROPERTY)
RT CATALYSTS
CHEMICAL PROPERTIES
CREASE RETENTION
CURING
CURING OVENS
CYSTINE
DISULFIDE BONDS
DISULFIDES (ORGANIC)
DRY WRINKLE RECOVERY
DURABLE PRESS FINISHES
DURABLE PRESS TREATMENTS
EASY CARE FABRICS
IRONING
KORATRON (TN)
LAUNDERABILITY
METHYLOL CARBAMATES
PERMANENT PLEATING
POLYESTER FIBER BLENDS
PRESSING
WRINKLE RECOVERY
WRINKLE RESISTANCE

DURABLE PRESS FINISHES
BT FINISH (SUBSTANCE ADDED)
RT AMINO RESINS
APO
CREASE RETENTION
DEFERRED CURE METHOD
DMEU
DOUBLE CURE METHOD
DURABLE PRESS
DURABLE PRESS TREATMENTS
HOT PRESS METHOD (DURABLE PRESS)
METHYLOL COMPOUNDS
PRE-CURE METHOD
REACTANTS

DURABLE PRESS TREATMENTS
UF PERMANENT SETTING
NT DEFERRED CURE METHOD
DOUBLE CURE METHOD
HOT PRESS METHOD (DURABLE PRESS)
IWS FINISH 6 (TN)
KORATRON (TN)
PRE-CURE METHOD
WILLIAMSON DICKIE (TN)
BT WET FINISHING
RT CREASE RETENTION
CURING
DRY CURING
DURABLE PRESS
DURABLE PRESS FINISHES
PAD BATCH CURE PROCESS
PAD DRY CURE PROCESS
SETTING (EXCEPT SYNTHETICS)
WET CURING

DURAFIL (TN)
BT CELLULOSIC FIBERS
HIGH TENACITY VISCOSE RAYON
RAYON (REGENERATED CELLULOSE FIBERS)
VISCOSE RAYON

DURASPAN (TN)
BT SPANDEX FIBERS

DUSTING MACHINES
NT WILLOWS
RT FEARNOUGHTS

DYE ABSORPTION
USE SORPTION OF DYES

DYE ADSORPTION
USE SORPTION OF DYES

DYE AFFINITY
UF AFFINITY (DYES)
BT FREE ENERGY
RT ABSORPTION (MATERIAL)
ADSORPTION
ADSORPTION ISOTHERM
ANIMALIZING
DYE MIGRATION
DYEABILITY
DYEING RATE
ENTHALPY OF DYEING
RT EQUILIBRIUM (DYEING)
EXHAUSTION (DYEING)
EXHAUSTION (WET PROCESS)
FABRIC PROPERTIES
FABRIC PROPERTIES (PHYSICAL EXCLUDING MECHANICAL)
FREE ENERGY OF DYEING
PREFERENTIAL ADSORPTION
SORPTION OF DYES
SUBSTANTIVITY

DYE BATHS
UF DYE LIQUORS
LIQUOR (DYEING)
BT BATHS
RT BLEACHING
DYE SOLUTIONS
DYEING
DYEING MACHINES
DYEING RATE
FINISHING PROCESS (GENERAL)
PADDING
SCOURING
SCOURING BATHS

DYED FIBER TRACER TECHNIQUES
USE TRACER TECHNIQUES

DYED FIBER TRACERS
USE TRACER FIBERS

DYE FIXING
UF FIXING (DYES)
NT ACID SHOCK METHOD
THERMOFIXING (DYEING)
BT WET FINISHING
RT AGING (STEAMING)
COLORFASTNESS
DYEING
MOLTEN METAL BATHS
NICKEL COMPOUNDS
OIL BATHS
PRINTING
STANDFAST (TN)

DYE FIXING AGENTS
BT DYEING AUXILIARIES
RT COBALT COMPOUNDS
CUPRIC SULFATE
TANNIC ACID

DYE HOUSE
RT BLEACHERIES
DYEING
FINISHING PROCESS (GENERAL)
PRINTING

DYE KETTLES
USE KETTLES (DYEING)

DYE LIQUORS
USE DYE BATHS

DYE MIGRATION
BT MIGRATION (SUBSTANCE)
RT ABSORPTION (MATERIAL)
ADSORPTION
DIFFUSION
DYE AFFINITY
DYE PENETRATION
DYEABILITY
DYEING
DYEING THEORY
EQUILIBRIUM (DYEING)
EXHAUSTION (DYEING)
LEVELLING AGENTS
LEVELNESS (DYEING)
SORPTION OF DYES

DYE PENETRATION
BT PENETRATION
RT BEAM DYEING (YARN)
DIFFUSION
DYE MIGRATION
EQUILIBRIUM (DYEING)
EXHAUSTION (DYEING)
PERMEABILITY
SORPTION OF DYES

DYE SPECKS
BT FABRIC DEFECTS
RT FABRIC INSPECTION

DYE SOLUTIONS
BT SOLUTIONS (LIQUID)
RT ACID SOLUTIONS
ALKALINE SOLUTIONS

DYE BATHS
NONAQUEOUS SOLUTIONS
WATER SOLUTIONS

DYE UNIFORMITY
RT BEAM DYEING (YARN)
LEVELNESS (DYEING)

DYEABILITY
UF SUBSTANTIVITY (DYES)
RT ANIONIC SITES
CATIONIC SITES
CHEMICAL PROPERTIES

RT DYE AFFINITY
DYE MIGRATION
DYEING
DYEING RATE
DYES
MIGRATION (SUBSTANCE)
POLY (VINYL PYRROLIDONE)
SITES (DYEING)
SORPTION OF DYES

DYED FABRICS
NT PIECE DYED FABRICS
YARN DYED FABRICS
RT FABRICS
FINISHED FABRICS
PRINTED FABRICS

DYEFASTNESS
USE COLORFASTNESS

DYEING
UF COLORING
NT DYEING (BY DYE CLASSES)
DYEING (BY ENVIRONMENTAL CONDITIONS)
DYEING (BY FIBER CLASSES)
DYEING (BY MATERIAL ASSEMBLY)
DYEING (BY PROCESS FLOW)
DYEING (FOR EFFECT)
REDYEING
RT AFTERTREATMENTS (GENERAL)
AGERS
ANIMALIZING
BLEACHING
BLEEDING
BLOTCHES
BOILING
BOTTOMING
CHELATING
CHLORINE FASTNESS (COLOR)
COACERVATE SYSTEMS
COLOR MATCHING
COLORFASTNESS
COMPATIBLE DYES
CURING
DRY CLEANING FASTNESS (COLOR)
DRYING
DYEING AUXILIARIES
DYEING DEFECTS
DYEING MACHINES
DYEING MACHINES (BY ENVIRONMENTAL CONDITIONS)
DYEING THEORY
DYES
DYE HOUSE
DYE MIGRATION
DYEABILITY
EQUILIBRIUM (CHEMICAL)
EQUILIBRIUM (DYEING)
EXHAUSTION (DYEING)
FINISHING PROCESS (GENERAL)
FORMULATIONS
IMPREGNATION
MIGRATION (SUBSTANCE)
NATURAL DYES
PAPER MAKING
PIGMENTS
PRETREATMENTS (GENERAL)
PRINTING
SHADE
SITES (DYEING)
SPECTROPHOTOMETRY
SQUEEZING
STAINING
STAINS
STREAKINESS
STRIPPING (COLOR)
THERMOFIXING (DYEING)
VAPOR PHASE
WET FINISHING

DYEING (BY DYE CLASSES)
NT ACID DYEING
AZOIC DYEING
BASIC DYEING
DIRECT DYEING
DISPERSE DYEING
METALLIZED DYEING
MORDANT DYEING
NEUTRAL DYEING
PIGMENT DYEING
REACTIVE DYEING
SULFUR DYEING
VAT DYEING
BT DYEING

DYEING (BY ENVIRONMENTAL CONDITIONS)
NT CARRIER DYEING
HIGH TEMPERATURE DYEING
LOW TEMPERATURE DYEING
PRESSURE DYEING
SOLVENT ASSISTED DYEING
SOLVENT DYEING
SPECK DYEING
VACUUM DYEING
VAPOR PHASE DYEING
BT DYEING
RT DYEING MACHINES (BY ENVIRONMENTAL CONDITIONS)

DYEING (BY FIBER CLASSES)
NT ACETATE DYEING
ACRYLIC DYEING
BLEND DYEING
CELLULOSE DYEING
COTTON DYEING
HAIR DYEING (HUMAN HAIR)
NYLON DYEING
OLEFIN DYEING
POLYESTER DYEING
RAYON DYEING
SILK DYEING
SYNTHETIC DYEING
TRIACETATE DYEING
WOOL DYEING
BT DYEING

DYEING (BY MATERIAL ASSEMBLY)
NT DOPE DYEING
GARMENT DYEING
PIECE DYEING
STOCK DYEING
STRIP DYEING
TOP DYEING
TOW DYEING
YARN DYEING
BT DYEING
RT DOPE DYEING EQUIPMENT
GARMENT DYEING MACHINES
HOSIERY DYEING MACHINES
PACKAGE DYEING MACHINES
PIECE DYEING MACHINES
STOCK DYEING MACHINES
TOP DYEING MACHINES
TOW DYEING MACHINES
YARN DYEING MACHINES

DYEING (BY PROCESS FLOW)
NT BATCH DYEING
CONTINUOUS DYEING
PAD DYEING
SEMI CONTINUOUS DYEING
SIMULTANEOUS DYEING AND FINISHING
BT DYEING
RT DYEING MACHINES (BY PROCESS FLOW)
ONE BATH PROCESS
TWO BATH PROCESS

DYEING (FOR EFFECT)
UF EFFECT DYEING
NT BATIK DYEING
CROSS DYEING
SPACE DYEING
UNION DYEING
BT DYEING
RT SPECKLED FABRIC

DYEING ASSISTANTS
USE DYEING AUXILIARIES

DYEING AUXILIARIES
UF DYEING ASSISTANTS
NT ACCELERANTS (DYEING)
CARRIERS (DYEING)
DEVELOPERS (DYEING)
DYE FIXING AGENTS
LEVELLING AGENTS
RETARDING AGENTS
STRIPPING AGENTS
BT TEXTILE CHEMICALS
RT ACETALDEHYDE SULFOXYLATES
ADDITIVES (CHEMICAL)
AMMONIUM SULFATE
BISULFATES
CARRIER DYEING
DETERGENTS
DISPERSING AGENTS
EMULSIFYING AGENTS
FORMALDEHYDE SULFOXYLATES
GLAUBERS SALT
HYDROXYLAMINE
POTASSIUM CARBONATE
PRINTING PASTES
SODIUM SULFIDE
SOLVENT ASSISTED DYEING
SODIUM BICARBONATE
SODIUM BISULFATE
SODIUM CARBONATE
SODIUM CHLORIDE
SODIUM SESQUICARBONATE
SODIUM SULFATE
SULFURIC ACID
SURFACTANT APPLICATIONS
SURFACTANTS
TSPP
WETTING
WETTING AGENTS

DYEING DEFECTS
UF PRINTING DEFECTS
NT BACK AND FACE EFFECTS
BARRE
BRONZING
CRACKS (FABRIC)
CREASE MARKS
OFF SHADE
ROPE MARKS
SHADE BARS
SKITTERINESS
SPECKINESS
STAINING (DYEING DEFECTS)
STREAKINESS
TIPPINESS
WEFT STREAKS
BT FABRIC DEFECTS
RT COLORFASTNESS
LEVELNESS (DYEING)

DYEING MACHINES
NT BATCH DYEING MACHINES
CONTINUOUS DYEING RANGE
DOPE DYEING EQUIPMENT
DYEING MACHINES (BY ENVIRONMENTAL CONDITIONS)
DYEING MACHINES (BY MATERIAL ASSEMBLY)
DYEING MACHINES (BY PROCESS FLOW)
GARMENT DYEING MACHINES
PACKAGE DYEING MACHINES
PIECE DYEING MACHINES
PRESSURE DYEING MACHINES
SEMI CONTINUOUS DYEING RANGE
STOCK DYEING MACHINES
TOP DYEING MACHINES
TOW DYEING MACHINES
YARN DYEING MACHINES
RT BECKS
CIRCULATION PUMPS
DYE BATHS
DYEING
FINISHING MACHINERY (GENERAL)
JIGS (DYEING)
KETTLES (DYEING)
PADDERS
PROPELLERS
REEL MACHINE (DYEING)
SCOURING MACHINES
STEAMERS
TEXTILE PROCESSES (GENERAL)

DYEING MACHINES (BY ENVIRONMENTAL CONDITIONS)
NT PRESSURE DYEING MACHINES
BT DYEING MACHINES
RT DYEING
DYEING (BY ENVIRONMENTAL CONDITIONS)
PRESSURE KETTLES

DYEING MACHINES (BY MATERIAL ASSEMBLY)
NT DOPE DYEING EQUIPMENT
GARMENT DYEING MACHINES
HOSIERY DYEING MACHINES
PACKAGE DYEING MACHINES
PIECE DYEING MACHINES
ROTARY DYEING MACHINES
STOCK DYEING MACHINES
TOP DYEING MACHINES
TOW DYEING MACHINES
YARN DYEING MACHINES
BT DYEING MACHINES
RT BATCH DYEING MACHINES
DYEING
DYEING (BY MATERIAL ASSEMBLY)

DYEING MACHINES (BY PROCESS FLOW)
NT BATCH DYEING MACHINES
CONTINUOUS DYEING RANGE
SEMI CONTINUOUS DYEING RANGE
BT DYEING MACHINES
RT DYEING
DYEING (BY PROCESS FLOW)
PADDERS
STEAMERS

DYEING PROPERTIES (GENERAL)
NT COLORFASTNESS
DYE AFFINITY
DYEABILITY
DYEING RATE
LEVELNESS (DYEING)
BT CHEMICAL PROPERTIES
RT DIFFUSION
EXHAUSTION (DYEING)
SORPTION OF LIQUIDS
SORPTION OF WATER
SURFACE AREA
SWELLING

DYEING RATE
UF EXHAUSTION RATE (DYEING
BT RATE
RT ABSORPTION (MATERIAL)
ADSORPTION
DIFFUSION
DIFFUSION COEFFICIENT
DYE AFFINITY
DYE BATHS
DYEABILITY
DYEING THEORY
KINETICS (CHEMICAL)
REACTION RATE

DYEING THEORY
NT DONNAN THEORY
GILBERT-RIDEAL THEORY
RT ABSORPTION (MATERIAL)
ADSORPTION
ADSORPTION ISOTHERM
ANIONIC SITES
CATIONIC SITES
DONNAN EQUILIBRIUM
DYE MIGRATION
DYEING RATE
DIFFUSION
DIFFUSION COEFFICIENT
ENTHALPY OF DYEING
EQUILIBRIUM (CHEMICAL)
EQUILIBRIUM (DYEING)
FREUNDLICH ISOTHERM
HYDROPHOBIC BONDS
ION EXCHANGE RESINS
LANGMUIR ISOTHERM
MECHANISM (FUNDAMENTAL)
OVERDYEING
PARTITION COEFFICIENT
SITES (DYEING)

DYES
UF DYESTUFFS
NT DYES (BY CHEMICAL CLASSES)
DYES (BY FIBER CLASSES)
FOOD DYES
BT COLORANTS
TEXTILE CHEMICALS
RT AGGREGATION
CARRIERS (DYEING)
COLOR YIELD
COLORFASTNESS
COMPATIBLE DYES
DRY CLEANING FASTNESS (COLOR)
DYEABILITY
DYEING
DYEING AUXILIARIES
DYEING MACHINES
FLUORESCENT BRIGHTENERS
INKS
PAINTS
PHOTOCHROMY
PIGMENTS
PRINTING
SOLVENTS

DYES (BY CHEMICAL CLASSES)
NT ACID DYES
ANILINE BLACK DYES
ANIONIC DYES
AZOIC DYES
BASIC DYES
DIRECT DYES
DISPERSE DYES
FLUORESCENT DYES
INGRAIN DYES
METALLIZED DYES
MORDANT DYES
NATURAL DYES
NEUTRAL DYES
OXIDATION DYES
POLYMERIC DYES
REACTIVE DYES
SOLVENT DYES
SULFUR DYES
VAT DYES
VINYLATED DYES

BT DYES
RT DYEING
DYEING MACHINES
DYES (BY FIBER CLASSES)
PHTHALOCYANINE COMPOUNDS
PIGMENTS

DYES (BY FIBER CLASSES)
NT ACETATE DYES
ACRYLIC DYES
CELLULOSE DYES
NYLON DYES
POLYESTER DYES
SILK DYES
WOOL DYES
BT DYES
RT DYEING
DYEING MACHINES
DYES (BY CHEMICAL CLASSES)
FOOD DYES

DYESTUFFS
USE DYES

DYNALOFT (TN)
BT FALSE TWIST YARNS
STRETCH YARNS (FILAMENT)
RT BULKED YARNS
FALSE TWISTING (TEXTURING)
FLUFLON (TN)
HELANCA (TN)
SAABA (TN)
SUPERLOFT (TN)

DYNAMIC ABSORPTION (WATER)
BT ABSORPTION (MATERIAL)
WATER ABSORPTION
RT DYNAMIC ABSORPTION TESTING
STATIC ABSORPTION (WATER)
STATIC ABSORPTION TESTING
WETTING

DYNAMIC ABSORPTION TESTING
UF DYNAMIC ABSORPTION TESTS
BT WATER ABSORPTION TESTING
RT DYNAMIC ABSORPTION (WATER)
IMMERSION TESTING
PERMEABILITY TESTING
STATIC ABSORPTION TESTING
WATER REPELLENCY
WATER RESISTANCE TESTING
WATERPROOFING

DYNAMIC ABSORPTION TESTS
USE DYNAMIC ABSORPTION TESTING

DYNAMIC BALANCING
BT BALANCING
RT ACCELEROMETERS
DYNAMIC BALANCING MACHINES
DYNAMIC CHARACTERISTICS
DYNAMIC RESPONSE
ECCENTRICITY
LONGITUDINAL VIBRATIONS
RESONANT FREQUENCY
ROTATION
SHAFTS
SPINDLE VIBRATION
SPINDLES
VIBRATION
VIBRATION DAMPERS

DYNAMIC BALANCING MACHINES
RT DYNAMIC BALANCING

DYNAMIC BUCKLING
BT BUCKLING
RT SNAP BACK

DYNAMIC CHARACTERISTICS
RT AUTOMATIC CONTROL
DAMPING
DRIFT
DYNAMIC BALANCING
DYNAMIC MODULUS
DYNAMIC RESPONSE
FREQUENCY RESPONSE
HYSTERESIS (MECHANICAL)
RESONANT FREQUENCY
RESPONSE TIME
STABILITY
TIME CONSTANT
TRANSIENT RESPONSE

DYNAMIC DRAPE
USE DRAPE

DYNAMIC FRICTION
UF KINETIC FRICTION
BT FRICTION
RT ABRASION
BOUNDARY LUBRICATION
COEFFICIENT OF FRICTION
FABRIC TO FABRIC FRICTION
FIBER FIBER FRICTION
FRICTIONAL FORCE
HYDRODYNAMIC LUBRICATION
INTERNAL FRICTION
ROLLING
ROLLING FRICTION
RUBBING
SLIDING
SLIDING FRICTION
RT SLIPPAGE
STATIC FRICTION
STICK SLIP FRICTION
WEAR
YARN TO METAL FRICTION
YARN TO YARN FRICTION

DYNAMIC MODULUS
NT IMAGINARY COMPONENT (DYNAMIC MODULUS)
LOSS TANGENT
REAL COMPONENT (DYNAMIC MODULUS)
BT MODULUS
STRESS STRAIN PROPERTIES
STRESS STRAIN PROPERTIES (TENSILE)
RT COMPLIANCE
COMPRESSIVE MODULUS
CREEP
DELAYED RECOVERY
DYNAMIC CHARACTERISTICS
DYNAMIC MODULUS TESTERS
ELASTIC MODULUS (TENSILE)
ELASTIC RECOVERY
ELASTICITY
FIBER PROPERTIES
LINEAR VISCOELASTICITY
MODULUS OF CROSS ELASTICITY
NONLINEAR VISCOELASTICITY
PULSE PROPAGATION METER
SONIC VELOCITY
SOUND
STRAIN RATE
STRESS PROPAGATION
STRESS STRAIN CURVES
TIRE CORDS
UNIAXIAL STRESS
YARN PROPERTIES
YARN STRENGTH

DYNAMIC MODULUS TESTERS
RT DYNAMIC MODULUS
IMPACT TESTERS
IMPULSES
PULSE PROPAGATION METER
SONIC VELOCITY

DYNAMIC RECOVERY
USE RECOVERY (SELF RESTORATION TO ORIGINAL CONDITION)

DYNAMIC RESPONSE
NT TRANSIENT RESPONSE
RT AMPLITUDE
DAMPING
DYNAMIC BALANCING
DYNAMIC CHARACTERISTICS
FREQUENCY
FREQUENCY RESPONSE
HIGH FREQUENCY
IMPULSES
RELAXATION TIME (MECHANICAL)
RESONANT FREQUENCY
STABILITY
TIME CONSTANT

DYNAMICS
NT AERODYNAMICS
PROCESS DYNAMICS
BT MECHANICS
RT DAMPING
DISPLACEMENT
DYNAMIC CHARACTERISTICS
DYNAMIC RESPONSE
KINEMATICS
KINETICS
MOTION
STATICS
STRUCTURAL MECHANICS

DYNAMOMETERS
BT MEASURING INSTRUMENTS
RT ACCELERATION (MECHANICAL)
ACCELEROMETERS
FORCE
MECHANICAL PROPERTIES
MECHANICAL SHOCK
TENSION

DYNEL (TN)
BT MODACRYLIC FIBERS

EAR MUFFS
BT HEADGEAR
RT CAPS (HEADGEAR)
GARMENTS
HATS
HELMETS
HOODS

EASY CARE FABRICS
UF CREASE RESISTANT FABRICS
DRIP DRY FABRICS
SELF SMOOTHING FABRICS
WASH WEAR FABRICS
RT AMINO RESINS
APO
BELFAST (TN)
BIS (HYDROXYETHYL) SULFONE
CATALYSTS
CREASE RETENTION
CREASING
DURABLE PRESS
DURABLE PRESS TREATMENTS
END USE PROPERTIES
FINISHING PROCESS (GENERAL)
HALOHYDRINS
LAUNDERING
OIL REPELLENTS
OIL REPELLENT TREATMENTS
PERMANENT PLEATING
REACTANTS
STAIN RESISTANCE AGENTS
WASH WEAR FINISHES
WASH WEAR PROPERTIES
WASH WEAR TREATMENTS
WRINKLE RECOVERY
WRINKLE RESISTANCE

EASY CARE FINISHES
USE WASH WEAR FINISHES

EASY CARE TREATMENTS
USE WASH WEAR TREATMENTS

ECCENTRICITY
UF RUN OUT
NT ROLL ECCENTRICITY
SPINDLE ECCENTRICITY
RT CLEARANCES
DAMPING
DRAFTING ROLLS
DYNAMIC BALANCING
DYNAMICS
ELLIPTICITY
ROLLS
SPINDLES
TOLERANCES
VIBRATION

ECONOMICS
RT COSTS
EDUCATION
INVESTMENTS
MANAGEMENT
MARKETING
POWER CONSUMPTION
PROCESS EFFICIENCY
PRODUCTION

EDDIES
RT AIR FLOW
CURRENTS
FLUID FLOW
FLUID MECHANICS

EDGE CRIMPED YARNS
NT AGILON (TN)
BT BULKED YARNS
CRIMPED YARNS
STRETCH YARNS (FILAMENT)
TEXTURED YARNS (FILAMENT)
RT AIR JET TEXTURED YARNS
BICOMPONENT FIBER YARNS (FILAMENT)
BICOMPONENT FIBER YARNS (STAPLE)
BULKING
CRIMPING
EDGE CRIMPING
EDGE CRIMPING MACHINES
FALSE TWIST YARNS
FALSE TWISTING (TEXTURING)
GEAR CRIMPED YARNS
HEAT SETTING (SYNTHETICS)
HELICAL CRIMP
STUFFER BOX CRIMPED YARNS
STUFFER BOX CRIMPING

EDGE CRIMPING
BT CRIMPING
TEXTURING
RT AGILON (TN)
AIR JET TEXTURING
BULKED YARNS
BULKING
CRIMP
CRIMPED YARNS
EDGE CRIMPED YARNS
EDGE CRIMPING MACHINES
RT FALSE TWISTING (TEXTURING)
GEAR CRIMPING
HEAT SETTING (SYNTHETICS)
HELICAL CRIMP
STRETCH
STUFFER BOX CRIMPING
TEXTURED YARNS (FILAMENT)

EDGE CRIMPING MACHINES
BT CRIMPING MACHINES
RT EDGE CRIMPED YARNS
EDGE CRIMPING

EDGE GUIDES
BT CLOTH GUIDES
GUIDES
RT CENTER GUIDES
SELVEDGES

EDGEING (LACE)
BT LACES

EDTA
UF ETHYLENEDIAMINE TETRAACETIC ACID
BT CARBOXYLIC ACIDS
ORGANIC ACIDS
RT CALGON (TN)
CHELATING
CHELATING AGENTS
COMPLEXING

EDUCATION
RT ECONOMICS
MANAGEMENT
RESEARCH
TRAINING

EFFECT DYEING
USE DYEING (FOR EFFECT)

EFFECT OF WEATHER
USE WEATHERING

EFFECTIVE LENGTH (FIBER)
RT COTTON CLASSING
COTTON QUALITY
FIBER ARRAY
FIBER DIAGRAM
FIBER LENGTH DETERMINATION
FIBER LENGTH DISTRIBUTION
FIBROGRAPH
GRADING
LENGTH UNIFORMITY RATIO (FIBER)
MEAN FIBER LENGTH
SHORT FIBERS INDEX
SORTING
SPAN LENGTH
STAPLING
SUTER WEBB TESTING
UPPER HALF MEAN LENGTH

EFFECTIVE POROSITY
USE POROSITY

EFFICIENCY (STRUCTURAL)
UF STRUCTURAL EFFICIENCY
NT BUNDLE EFFICIENCY
KNOT EFFICIENCY
LOOP EFFICIENCY
SEAM EFFICIENCY
STRENGTH EFFICIENCY
BT MECHANICAL PROPERTIES
STRUCTURAL PROPERTIES
RT PROCESS EFFICIENCY
WEAK SPOTS
YARN BUNDLE COHESION
YARN STRENGTH
YARN STRENGTH EFFICIENCY

EFFICIENCY (PROCESS)
USE PROCESS EFFICIENCY

EFFLUENT TREATMENT
USE WASTE TREATMENT

EFFLUENTS
RT EJECTION
EMPTYING
POLLUTION
WASTE
WASTE DISPOSAL
WASTE TREATMENT

EGYPTIAN COTTON
BT COTTON
RT LONG STAPLE COTTON

EIDERDOWNS
RT BEDSPREADS
BLANKETS
HOUSEHOLD FABRICS
PILLOWS

EIGHTLOCK (KNITTED FABRICS)
BT KNITTED FABRICS
RT INTERLOCK (KNITTED FABRICS)
KNITTING
RIB KNITTED FABRICS

EJECTION
RT EFFLUENTS
EMPTYING
EXHAUSTION (MECHANICAL)
QUILL CHANGING
REMOVAL
SHUTTLE CHANGING
UNLOADING (PROCESS)
UNROLLING
UNWINDING

ELASTIC DEFORMATION
USE ELASTIC STRAIN

ELASTIC FABRICS
USE STRETCH FABRICS

ELASTIC FOAMS
BT FOAMS
RT FOAM RUBBER
RIGID FOAMS

ELASTIC LIMIT
USE YIELD POINT

ELASTIC LIQUIDS
BT LIQUIDS
RT FLUID FLOW
FLUID MECHANICS
PHENOMENOLOGICAL RHEOLOGY
RHEOLOGY

ELASTIC MODULUS (COMPRESSIVE)
USE COMPRESSIVE MODULUS

ELASTIC MODULUS (TENSILE)
UF MODULUS OF ELASTICITY
STATIC MODULUS
TENSION MODULUS
YOUNGS MODULUS
BT MODULUS
STRESS STRAIN PROPERTIES
STRESS STRAIN PROPERTIES (TENSILE)
RT BENDING
BENDING RIGIDITY
BUCKLING
COMPLIANCE
COMPRESSIVE MODULUS
COMPRESSIVE STRENGTH
CROSSLINK DENSITY
DYNAMIC MODULUS
ELASTIC PERFORMANCE
ELASTIC STRAIN
ELASTICITY
GLASS TRANSITION TEMPERATURE
HOOKES LAW
INITIAL MODULUS
LINEAR VISCOELASTICITY
MODULUS OF CROSS ELASTICITY
NONLINEAR VISCOELASTICITY
SECANT MODULUS
SECONDARY TRANSITION TEMPERATURES
SHEAR MODULUS
STRAIN
STRESS
STRESS STRAIN CURVES
TANGENT MODULUS
TIRE CORDS
UNIAXIAL STRESS
YIELD POINT

ELASTIC PERFORMANCE
RT COMPLIANCE
ELASTIC MODULUS (TENSILE)
ELASTIC RECOVERY
ELASTICITY
ENERGY OF RETRACTION
STRESS STRAIN CURVES
WORK RECOVERY

ELASTIC RECOVERY
UF IMMEDIATE ELASTIC RECOVERY
BT RECOVERY (SELF RESTORATION TO ORIGINAL CONDITION)
STRESS STRAIN PROPERTIES
STRESS STRAIN PROPERTIES (TENSILE)
RT BENDING RECOVERY
CREEP
CREEP RECOVERY
CRIMP RECOVERY
DELAYED RECOVERY
DYNAMIC MODULUS
ELASTIC PERFORMANCE
ELASTIC STRAIN
ENERGY OF RETRACTION
RESILIENCE
STRESS STRAIN CURVES
VISCOELASTICITY
WORK RECOVERY
WRINKLE RECOVERY

ELASTIC REGION
USE ELASTIC STRAIN

ELASTIC STRAIN
- UF ELASTIC DEFORMATION
  - ELASTIC REGION
  - HOOKEAN REGION
  - HOOKEAN STRAIN
- BT STRAIN
  - STRESS STRAIN PROPERTIES
  - STRESS STRAIN PROPERTIES (TENSILE)
- RT BIAXIAL STRESS
  - BREAKING ELONGATION
  - COLD DRAWING
  - COMPLIANCE
  - DEFORMATION
  - DELAYED RECOVERY
  - DIMENSIONAL STABILITY
  - ELASTIC MODULUS (TENSILE)
  - ELASTIC RECOVERY
  - ELASTICITY
  - EXTENSIBILITY
  - FILLINGWISE STRETCH
  - GUT THREAD
  - HOOKES LAW
  - INITIAL MODULUS
  - NECKING (FILAMENT)
  - PERMANENT SET
  - PLASTIC FLOW
  - PLASTIC STRAIN
  - SECANT MODULUS
  - STRESS
  - STRESS STRAIN CURVES
  - STRETCH
  - TANGENT MODULUS
  - UNIAXIAL STRESS
  - VISCOUS FLOW
  - WARPWISE STRETCH
  - YIELD (MECHANICAL)
  - YIELD POINT

ELASTICITY
- NT APPLIED ELASTICITY
  - KINETIC ELASTICITY
  - RUBBER ELASTICITY
- BT STRESS STRAIN PROPERTIES
  - STRESS STRAIN PROPERTIES (TENSILE)
- RT BIAXIAL STRESS
  - BREAKING ELONGATION
  - BREAKING ENERGY
  - BREAKING STRENGTH
  - BRITTLENESS
  - COMPRESSIBILITY
  - CREEP
  - DELAYED RECOVERY
  - DRAWING (FILAMENT)
  - DYNAMIC MODULUS
  - ELASTIC LIQUIDS
  - ELASTIC MODULUS (TENSILE)
  - ELASTIC PERFORMANCE
  - ELASTIC STRAIN
  - EXTENSIBILITY
  - FILLINGWISE STRETCH
  - HOOKES LAW
  - INITIAL MODULUS
  - LINEAR VISCOELASTICITY
  - NECKING (FILAMENT)
  - NONLINEAR VISCOELASTICITY
  - PERMANENT SET
  - PLASTIC STRAIN
  - PLASTICITY
  - PLASTICOVISCOELASTICITY
  - PLASTICOVISCOSITY
  - POWER (FABRIC)
  - RHEOLOGY
  - RUBBER
  - STITCH ELASTICITY
  - STRAIN
  - STRENGTH
  - STRENGTH OF MATERIALS
  - STRESS STRAIN CURVES
  - STRETCH
  - STRETCH KNITTED FABRICS
  - STRETCH WOVEN FABRICS
  - TENACITY
  - UNIAXIAL STRESS
  - VISCOELASTICITY
  - VISCOSITY
  - WARPWISE STRETCH
  - WET MODULUS
  - YARN PROPERTIES
  - YIELD POINT

ELASTIQUE
- BT FLAT WOVEN FABRICS
  - WOVEN FABRICS
- RT TWILL WEAVES

ELASTOMER FIBERS
- NT RUBBER FIBER (NATURAL)
  - SPANDEX FIBERS
  - SYNTHETIC RUBBER FIBERS
- RT ELASTOMERS
  - POLYMERS
  - POLYURETHANES
  - RUBBER

ELASTOMERS
- RT COPOLYMERS
  - ELASTOMER FIBERS
  - NATURAL RUBBER
  - POLYMERS
  - POLYURETHANES
  - RUBBER
  - SILICONE RUBBER
  - SYNTHETIC RUBBER

ELECTRIC BLANKETS
- BT BLANKETS
- RT HEATING
  - RESISTANCE (ELECTRICAL)

ELECTRIC BRAKES
- BT BRAKES (FOR ARRESTING MOTION)
- RT ELECTROMAGNETIC DEVICES
  - MAGNETIC BRAKES
  - MECHANICAL BRAKES

ELECTRIC CONTROLLERS
- BT CONTROL EQUIPMENT
- RT AUTOMATIC CONTROL
  - MEASURING INSTRUMENTS

ELECTRIC CURRENT
- USE CURRENT (ELECTRICAL)

ELECTRIC EQUIPMENT
- RT ELECTROMAGNETIC DEVICES
  - ELECTRONIC INSTRUMENTS
  - ELECTRONICS
  - EQUIPMENT
  - MOTORS
  - TRANSDUCERS

ELECTRIC HEATING (RESISTANCE)
- BT HEATING
- RT DIELECTRIC HEATING
  - DRYING
  - GAS HEATING (GAS FUEL)
  - INFRARED DRYERS
  - OIL HEATING (OIL FUEL)
  - RADIANT HEATING

ELECTRICAL CONDUCTIVITY
- USE RESISTIVITY

ELECTRICAL INPUT TRANSDUCERS
- BT TRANSDUCERS

ELECTRICAL OUTPUT TRANSDUCERS
- BT TRANSDUCERS

ELECTRICAL PROPERTIES
- NT CAPACITANCE (ELECTRICAL)
  - DIELECTRIC PROPERTIES
  - DIELECTRIC STRENGTH
  - IMPEDANCE
  - INDUCTANCE
  - POTENTIAL (ELECTRICAL)
  - REACTANCE (ELECTRICAL)
  - REDOX POTENTIAL
  - RESISTANCE (ELECTRICAL)
  - RESISTIVITY
  - ZETA POTENTIAL
- BT PHYSICAL PROPERTIES (EXCLUDING MECHANICAL)
- RT ACOUSTIC PROPERTIES
  - AESTHETIC PROPERTIES
  - BIOCHEMICAL PROPERTIES
  - CHEMICAL PROPERTIES
  - ELECTRIC EQUIPMENT
  - ELECTRONICS
  - FABRIC PROPERTIES
  - FABRIC PROPERTIES (PHYSICAL EXCLUDING MECHANICAL)
  - FIBER PROPERTIES
  - MECHANICAL PROPERTIES
  - OPTICAL PROPERTIES
  - PHYSICAL CHEMICAL PROPERTIES
  - STRESS OPTICAL PROPERTIES
  - STRUCTURAL PROPERTIES
  - THERMAL PROPERTIES
  - TRANSFER PROPERTIES

ELECTRICAL RESISTIVITY
- USE RESISTIVITY

ELECTRO-OSMOSIS
- BT OSMOSIS
- RT ELECTRODIALYSIS
  - ELECTROPHORESIS

ELECTRODEPOSITION
- RT COATING (PROCESS)
  - ELECTROLYSIS
  - ELECTROLYTES
  - ELECTROPHORESIS
  - ELECTROPLATING

ELECTRODIALYSIS
- BT DIALYSIS
- RT ELECTRO-OSMOSIS
  - MEMBRANES

ELECTROKINETIC POTENTIAL
- USE ZETA POTENTIAL

ELECTROLYSIS
- RT CORROSION
  - ELECTRODEPOSITION
  - ELECTROLYTES
  - ELECTROPLATING
  - OXIDATION
  - REDUCTION

ELECTROLYTES
- NT POLYELECTROLYTES
- RT ACIDS
  - BASES
  - CATIONS
  - CONDUCTOMETRY
  - ELECTRODEPOSITION
  - ELECTROLYSIS
  - ELECTROPLATING
  - IONS
  - SALTS (GENERAL)

ELECTROMAGNETIC DEVICES
- BT DEVICES
- RT AUTOMATIC CONTROL
  - CONTROL EQUIPMENT
  - ELECTRIC BRAKES
  - ELECTRIC EQUIPMENT
  - ELECTROMECHANICAL DEVICES
  - ELECTRONIC INSTRUMENTS
  - MECHANICAL DEVICES

ELECTROMECHANICAL DEVICES
- BT DEVICES
- RT ELECTROMAGNETIC DEVICES
  - MECHANISMS
  - TENSION DEVICES

ELECTROMETERS
- BT ELECTRONIC INSTRUMENTS

ELECTRON MICROSCOPE
- USE ELECTRON MICROSCOPY

ELECTRON MICROSCOPY
- UF ELECTRON MICROSCOPE
- NT STEREOSCAN MICROSCOPY
- BT MICROSCOPY
- RT CRYSTALLIZATION
  - FIBRILS
  - FINE STRUCTURE (FIBERS)
  - INFRARED MICROSCOPY
  - INTERFERENCE MICROSCOPY
  - LIGHT MICROSCOPY (OPTICAL)
  - OPTICS
  - ULTRAVIOLET MICROSCOPY
  - X RAY MICROSCOPY

ELECTRON SPIN RESONANCE
- BT PHYSICAL PROPERTIES (EXCLUDING MECHANICAL)
- RT NUCLEAR MAGNETIC RESONANCE
  - SPECTROSCOPY

ELECTRONIC INSTRUMENTS
- NT AMPLIFIERS
  - ELECTROMETERS
  - GEIGER COUNTERS
  - OSCILLATORS
  - OSCILLOSCOPES
  - PHOTOELECTRIC CELLS
  - STROBOSCOPES
  - THERMOPILES
  - TRANSDUCERS
- BT INSTRUMENTATION
- RT AUTOMATIC CONTROL
  - CAPACITANCE BRIDGES
  - COMPUTERS
  - CONTROL EQUIPMENT
  - ELECTRIC EQUIPMENT
  - ELECTROMAGNETIC DEVICES
  - ELECTRON MICROSCOPY
  - ELECTRONICS
  - IMPEDANCE BRIDGES
  - MEASURING INSTRUMENTS
  - NULL METHODS
  - OPTICAL INSTRUMENTS
  - PYROMETERS

ELECTRONIC MULES
- USE MULES

ELECTRONIC SLUB CATCHERS
- BT SLUB CATCHERS
- RT MECHANICAL SLUB CATCHERS
  - QUALITY CONTROL
  - SLUBS
  - SLUB CATCHING
  - UNICONER (TN)
  - WINDING
  - YARN DEFECTS

ELECTRONICS
- RT ELECTRIC EQUIPMENT
  - ELECTRICAL PROPERTIES
  - ELECTRONIC INSTRUMENTS

ELECTROPHORESIS
- UF CATAPHORESIS
- RT COLLOIDS

RT ELECTRO-OSMOSIS
ELECTRODEPOSITION
ELECTROPLATING

ELECTROPLATING
BT COATING (PROCESS)
RT ELECTRODEPOSITION
ELECTROLYSIS
ELECTROLYTES
ELECTROPHORESIS

ELECTROSTATIC BONDS
BT CHEMICAL BONDS
RT BOND ENERGY
COVALENT BONDS
HYDROGEN BONDS
IONIC BONDS
METALLIC BONDS
VAN DER WAALS FORCES

ELECTROSTATIC FLOCKING
BT FLOCKING
RT FLOCK
FLOCK COATING
FLOCK PRINTING
FLOCKED CARPETS
FLOCKED FABRICS
MECHANICAL FLOCKING
STATIC ELECTRICITY

ELECTROSTATICS
USE STATIC ELECTRICITY

ELLIPTICAL TRAVELERS
BT SPINNING TRAVELERS
TRAVELERS
RT CIRCULAR TRAVELERS
NYLON TRAVELERS
PLYING TRAVELERS
RING SPINNING
RINGS
SPINNING
STEEL TRAVELERS
TWISTING

ELLIPTICITY
RT CYLINDERS
DIAMETER
ECCENTRICITY
ROUNDNESS
SHAPE
SHAPE FACTOR
SPHERICITY

ELMENDORF TEAR STRENGTH
BT FABRIC PROPERTIES
FABRIC PROPERTIES (MECHANICAL)
TEAR STRENGTH
RT TONGUE TEAR STRENGTH
TRAPEZOID TEAR STRENGTH

ELMENDORF TEAR TESTING
UF ELMENDORF TEAR TESTS
BT TEAR TESTING
RT TEAR STRENGTH
TONGUE TEAR TESTING
TRAPEZOID TEAR TESTING

ELMENDORF TEAR TESTS
USE ELMENDORF TEAR TESTING

ELONGATION
USE STRAIN

ELONGATION AT BREAK
USE BREAKING ELONGATION

ELONGATION PERCENT
USE STRAIN

EMBEDDING MEDIUM
RT CROSS SECTIONS
CUTTING
FIBER CROSS SECTIONS
MICROSCOPY
MICROTOMES
RESINOGRAPHY

EMBOSSED CREPE
BT FLAT WOVEN FABRICS
WOVEN FABRICS
RT EMBOSSED FABRICS
PLAIN WEAVES
PRINT CLOTH

EMBOSSED FABRICS
RT ARMURE
COATED FABRICS
EMBOSSED CREPE
EMBOSSING
EMBOSSING CALENDERS

EMBOSSING
RT BLANKET (FINISHING)
BLANKET MARKS
CALENDERING
CREPE EMBOSSING
EMBOSSED FABRICS
EMBOSSING CALENDERS
FRICTION CALENDERING
GRINDING (MATERIAL REMOVAL)
POLISHING
SCHREINERING
SHEARING (FINISHING)

EMBOSSING CALENDERS
BT CALENDER ROLLS
CALENDERING MACHINES
RT CALENDERING
DRY FINISHING
EMBOSSED FABRICS
EMBOSSING
ENGRAVING
SCHREINERING

EMBROIDERY
NT EYELET EMBROIDERY
SCHIFFLE EMBROIDERY
RT CEREMONIAL CLOTHING
EMBROIDERY FLOSS
EMBROIDERY MACHINES
HAND SEWING
KNITTING
LOOP STITCHES
NEEDLEWORK
SEWING
STITCHING

EMBROIDERY FLOSS
RT EMBROIDERY

EMBROIDERY MACHINES
RT EMBROIDERY

EMERY
BT ABRASIVES
CORUNDUM
RT ABRASION
CERAMICS
GRINDING (MATERIAL REMOVAL)
GRIT
PUMICE
WEAR

EMISSIVITY
BT PHYSICAL PROPERTIES (EXCLUDING MECHANICAL)
THERMAL PROPERTIES
TRANSFER PROPERTIES
RT ABSORPTIVITY (RADIATION)
BLACKBODY
COLOR TEMPERATURE
CONDUCTIVITY (THERMAL)
HEAT TRANSFER
RADIATION
REFLECTANCE
THERMAL TESTING
TRANSMISSIVITY

EMPIRICAL ANALYSIS
RT ANALYZING
EXAMINATION
EXPERIMENTAL ANALYSIS
EXPERIMENTAL STRESS ANALYSIS
EXPERIMENTATION
FABRIC ANALYSIS
MATHEMATICAL ANALYSIS
QUALITATIVE ANALYSIS
QUANTITATIVE ANALYSIS
STATISTICAL ANALYSIS
THEORETICAL ANALYSIS

EMPTY PACKAGES
BT PACKAGES
RT BOBBINS
BUILDER MOTIONS
FULL PACKAGES
GROOVES
NOSE BUNCHING
QUILLS
SPINNING
WINDING
WINDING ANGLE
WINDING TENSION

EMPTYING
RT EFFLUENTS
EJECTION
EXHAUSTION (MECHANICAL)
REMOVAL
UNLOADING (PROCESS)

EMULSIFYING
RT DISPERSING
EMULSIFYING AGENTS
EMULSIONS
HOMOGENIZERS
MIXING
STIRRING

EMULSIFYING AGENTS
UF EMULSION STABILIZERS
PROTECTIVE COLLOIDS
BT SURFACTANT APPLICATIONS
RT CARRIER DYEING
CARRIERS (DYEING)
DETERGENTS
DISPERSING AGENTS
DYEING AUXILIARIES
EMULSIFYING
EMULSIONS
EMULSION POLYMERS
EMULSION PRINTING
FOAMS (FROTH)
HYDROXYALKYLCELLULOSE
LATICES
LEVELLING AGENTS
OIL IN WATER EMULSIONS
MINERAL OIL
PECTINS
SLURRY
SOAPS
SOLVENT ASSISTED DYEING
SOLVENT EMULSIONS
SURFACTANTS
SUSPENDING AGENTS
WATER IN OIL EMULSIONS
WETTING AGENTS

•EMULSION BINDERS (NONWOVENS)
BT BINDERS (NONWOVENS)
RT BONDED FIBER FABRICS
NONWOVEN FABRICS

EMULSION BINDERS (PIGMENTS)
BT BINDERS (PIGMENTS)
RT EMULSIONS
OIL IN WATER EMULSIONS
WATER IN OIL EMULSIONS

EMULSION DYEING
USE CARRIER DYEING

EMULSION POLYMERS
RT EMULSIFYING AGENTS
EMULSION POLYMERIZATION
EMULSIONS
LATICES
OIL IN WATER EMULSIONS
POLYMERS
SOLVENT EMULSIONS
SUSPENSION POLYMERS
WATER IN OIL EMULSIONS

EMULSION PRINTING
UF NEOFATF PROCESS (TN)
SAICATEX PROCESS (TN)
BT PRINTING
RT EMULSIFYING AGENTS
EMULSIONS

EMULSION POLYMERIZATION
BT POLYMERIZATION
RT BULK POLYMERIZATION
EMULSION POLYMERS
SUSPENSION POLYMERIZATION

EMULSION STABILIZERS
USE EMULSIFYING AGENTS

EMULSIONS
NT OIL IN WATER EMULSIONS
LATICES
SOLVENT EMULSIONS
WATER IN OIL EMULSIONS
BT DISPERSE SYSTEMS
RT COLLOIDS
EMULSIFYING
EMULSIFYING AGENTS
EMULSION BINDERS (PIGMENTS)
EMULSION BINDERS (NONWOVEN
EMULSION POLYMERS
EMULSION PRINTING
FINISH (SUBSTANCE ADDED)
OPACITY
PRINTING
STAIN RESISTANCE AGENTS
WATER REPELLENTS

END BREAKAGES
UF ENDS DOWN
YARN BREAKAGES
RT ALLOTMENT
AUTOMATIC KNOTTING
AUTOMATIC STOP MOTIONS
BALLOON COLLAPSE
BALLOON CONTROL RINGS
BALLOON TENSION
BREAKING
BREAKING ELONGATION
BREAKING LENGTH
BREAKING STRENGTH
BREAKS
BROKEN END COLLECTORS
CATCHING
COUNT STRENGTH PRODUCT
DEFECTS
DOWN TIME
IMPACT STRENGTH
IRREGULARITY
KNOTS
LASHING END RADIUS
LEA STRENGTH PRODUCT
LOOM STOPS
MECHANICAL SLUB CATCHERS
(CONT.)

RT PATROLLING
PERFORMANCE
PIECING
PNEUMAFIL (TN)
PROCESS EFFICIENCY
RING SPINNING
RUNS (KNITTING)
SHUTTLE SMASH
SLIVER TENDERNESS
SLUB CATCHERS
SLUB CATCHING
SNAGS
SPINNING EFFICIENCY
STOP MOTIONS
STOPS
STRENGTH
TENSION
TENSION CONTROL
THREAD BREAKAGES
WARP ENDS
WARP TENSION
WEAK SPOTS
WEAVING
WEAVING EFFICIENCY
WINDING EFFICIENCY
WINDING TENSION
YARN DEFECTS
YARN DYNAMICS
YARN STRENGTH
YARN TENSION

END FENT
USE LEADER CLOTH

END GROUPS
NT ACID END GROUPS
AMINE END GROUPS
CARBOXYL END GROUPS
SULFONIC END GROUPS
RT ALDEHYDE GROUPS
ALKOXYL GROUPS
AMINO GROUPS
CARBONYL GROUPS
CARBOXYL GROUPS
DEGREE OF POLYMERIZATION
MOLECULAR WEIGHT
MOLECULAR WEIGHT CONTROL
NUMBER AVERAGE MOLECULAR WEIGHT
POLYMERIZATION
POLYMERS
WEIGHT AVERAGE MOLECULAR WEIGHT

END USE PROPERTIES
UF PERFORMANCE PROPERTIES
NT ABRASION RESISTANCE
ABSORBENCY (MATERIAL)
ANTISTATIC BEHAVIOR
APPEARANCE
BACTERIAL INHIBITION
BREAKING STRENGTH
CHLORINE FASTNESS (COLOR)
CHLORINE FASTNESS (OF FINISH)
COLORFASTNESS
COMFORT
CREASE ACCEPTANCE
CREASE RETENTION
CROCKFASTNESS
CRUSH RESISTANCE
DIMENSIONAL STABILITY
DRY CLEANING FASTNESS (COLOR)
DRY CLEANING FASTNESS (OF FINISH)
DURABLE PRESS
DURABILITY
DRAPE
FATIGUE RESISTANCE
FELTING RESISTANCE
FINISH (PROPERTY)
FIRE RESISTANCE
FIRE RETARDANCY
GAS FUME FASTNESS
INSECT RESISTANCE
IRONING FASTNESS
LIGHTFASTNESS (COLOR)
LIGHTFASTNESS (OF FINISH)
LIVELINESS
MILDEW RESISTANCE
OIL REPELLENCY
PERSPIRATION RESISTANCE
RESILIENCE
ROUGHNESS
RUN RESISTANCE
SEWABILITY
SHEEN
SHRINK RESISTANCE
SLICKNESS
SLIP RESISTANCE
SMOOTHNESS
SNAG RESISTANCE
SOFTNESS
SOIL RESISTANCE
STAIN RESISTANCE
STIFFNESS
TEAR STRENGTH
TOUGHNESS
TRANSLUCENCY
NT TRANSPARENCY
WARMTH
WASHFASTNESS (COLOR)
WASHFASTNESS (OF FINISH)
WATER REPELLENCY
WATER RESISTANCE
WEAR RESISTANCE
WEATHER RESISTANCE
WRINKLE RECOVERY
WRINKLE RESISTANCE
RT AESTHETIC PROPERTIES
APPAREL FABRICS
BIOCHEMICAL PROPERTIES
CHEMICAL PROPERTIES
EASY CARE FABRICS
FABRIC PROPERTIES
FABRIC PROPERTIES (AESTHETIC)
FURNISHING FABRICS
HOUSEHOLD FABRICS
MECHANICAL PROPERTIES
OPTICAL PROPERTIES
PERFORMANCE
PHYSICAL CHEMICAL PROPERTIES
PHYSICAL PROPERTIES (EXCLUDING MECHANICAL)
PROPERTIES

END USES
RT AGRICULTURAL FABRICS
APPAREL FABRICS
APPLICATIONS
ARRESTERS
AUTOMOTIVE FABRICS
DECORATIVE FABRICS
EQUIPAGE
FABRICS (BY END USES)
FURNISHING FABRICS
FURNISHINGS
GARMENTS
HOUSEHOLD FABRICS
INDUSTRIAL FABRICS
INFLATABLE FABRICS
MEDICAL TEXTILES
PARACHUTES
PRODUCTS
RADOMES
SHELTERS
TENTS
UPHOLSTERY
UPHOLSTERY FABRICS
WEARING (OF APPAREL)

ENDOCUTICLE
BT FINE STRUCTURE (FIBERS)
RT CORTEX
EPICUTICLE
EXOCUTICLE
FIBRILS
MEDULLA
MICROFIBRILS
ORTHOCORTEX
PARACORTEX
SCALES (WOOL FIBERS)
WOOL

ENDS (WEAVING)
USE WARP ENDS

ENDS DOWN
USE END BREAKAGES

ENDS OUT
BT FABRIC DEFECTS

END SPACING
USE ENDS PER INCH

ENDS PER INCH
UF END SPACING
BT FABRIC PROPERTIES
FABRIC PROPERTIES (STRUCTURAL)
RT BALANCED WEAVES
COURSES PER INCH
COVER FACTOR
CRIMP INTERCHANGE
DENTS
DESIGN
DRAWING IN
FABRIC ANALYSIS
FABRIC DESIGN
FABRIC GEOMETRY
FABRIC STRUCTURE
JAMMING
PICK COUNTERS
PICK GLASS
PICKS PER INCH
SET (WOVEN FABRIC)
WALES PER INCH
WARP ENDS
WARP PREPARATION

ENERGY
NT ACTIVATION ENERGY
BENDING ENERGY
BOND DISSOCIATION ENERGY
BOND ENERGY
BREAKING ENERGY
NT ENERGY OF RETRACTION
INTERNAL ENERGY
KINETIC ENERGY
POTENTIAL ENERGY
STRAIN ENERGY
TORSIONAL ENERGY
RT ENERGY ABSORPTION
SURFACE TENSION

ENERGY ABSORPTION
BT MECHANICAL PROPERTIES
STRESS STRAIN PROPERTIES
STRESS STRAIN PROPERTIES (TENSILE)
RT ARRESTERS
BRITTLE FRACTURE
CREPING
DAMPERS
DAMPING
DUCTILITY
ENERGY
FREQUENCY RESPONSE
HYSTERESIS (MECHANICAL)
IMPACT
IMPULSES
RESONANT FREQUENCY
SHOCK
SHOCK ABSORBERS
SHOCK RESISTANCE
SOUND
TOUGHNESS
VIBRATION
VIBRATION DAMPERS
VIBRATION ISOLATORS
VISCOUS DAMPING
WORK RECOVERY

ENERGY OF DEFORMATION
USE STRAIN ENERGY

ENERGY OF RETRACTION
BT ENERGY
STRESS STRAIN PROPERTIES (TENSILE)
RT ELASTIC PERFORMANCE
ELASTIC RECOVERY
HYSTERESIS (MECHANICAL)
RESILIENCE
STRESS STRAIN CURVES
TOUGHNESS
WORK RECOVERY

ENGRAVED CYLINDERS
BT CYLINDERS
RT ENGRAVING
PERFORATED CYLINDERS
PRINTING ROLLS
ROLLER PRINTING MACHINES

ENGRAVING
RT CUTTING
DESIGN
EMBOSSING CALENDERS
ENGRAVED CYLINDERS
PATTERN (FABRICS)
POLISHING
PRINTING ROLLS
SCORING

ENKALOFT (TN)
BT NYLON (POLYAMIDE FIBERS)
NYLON 6

ENKALON (TN)
BT NYLON (POLYAMIDE FIBERS)
NYLON 6

ENKALURE (TN)
BT NYLON (POLYAMIDE FIBERS)
NYLON 6

ENKATRON (TN)
BT NYLON (POLYAMIDE FIBERS)
NYLON 6

ENTANGLEMENTS
UF FIBER ENTANGLEMENTS
RT BUCKLING
CARDING
CARDING ACTION
CARDING DYNAMICS
COHESION
CONFIGURATION
DEFECTS
DIMENSIONAL STABILITY
FIBER HOOKS
FIBER WITHDRAWAL FORCE
FUZZ RESISTANCE
FUZZING
KNOTS
LOOPS
NEP SIZE
NEPS
NEPTEX (TN)
NEPTOMETER (TN)
OPENING
PICKING (OPENING)
PILLING

RT PILLS
SLENDERNESS RATIO
SLOUGHING
SLUBS
SNAGGING
SNAGS
SNAP BACK
SNARLING
SPINNING LIMIT
TORSIONAL BUCKLING
WILDNESS
YARN DEFECTS

ENTERING LEASE
BT LEASE

ENTHALPY
NT ENTHALPY OF DYEING
ENTHALPY OF HYDRATION
ENTHALPY OF REACTION
ENTHALPY OF SORPTION
ENTHALPY OF SOLUTION
ENTHALPY OF SWELLING
ENTHALPY OF WETTING
BT PHYSICAL PROPERTIES (EXCLUDING MECHANICAL)
THERMAL PROPERTIES
RT BOILING POINT
ENTROPY
EQUILIBRIUM (CHEMICAL)
FREE ENERGY
HEAT CONTENT
LATENT HEAT
SATURATED VAPOR
SPECIFIC HEAT
TEMPERATURE
THERMODYNAMICS

ENTHALPY OF DYEING
UF HEAT OF DYEING
BT ENTHALPY
RT DYE AFFINITY
DYEING THEORY
ENTHALPY OF REACTION
ENTROPY OF DYEING

ENTHALPY OF HYDRATION
UF HEAT OF HYDRATION
BT ENTHALPY
RT ENTHALPY OF REACTION
ENTHALPY OF WETTING
THERMODYNAMICS

ENTHALPY OF REACTION
UF HEAT OF REACTION
BT ENTHALPY
RT ENTHALPY OF DYEING
ENTHALPY OF HYDRATION
ENTHALPY OF SOLUTION
ENTHALPY OF SWELLING
ENTHALPY OF WETTING
ENTROPY OF REACTION
FREE ENERGY
REACTIONS (CHEMICAL)
THERMAL ANALYSIS
THERMODYNAMICS

ENTHALPY OF SOLUTION
UF HEAT OF SOLUTION
BT ENTHALPY
RT ENTHALPY OF REACTION
ENTHALPY OF WETTING

ENTHALPY OF SORPTION
BT ENTHALPY
RT SORPTION

ENTHALPY OF SWELLING
UF HEAT OF SWELLING
BT ENTHALPY
RT ENTHALPY OF REACTION
ENTHALPY OF WETTING
SWELLING
THERMODYNAMICS

ENTHALPY OF WETTING
UF HEAT OF WETTING
BT ENTHALPY
RT ENTHALPY OF HYDRATION
ENTHALPY OF REACTION
ENTHALPY OF SOLUTION
ENTHALPY OF SWELLING
THERMODYNAMICS
WETTING

ENTROPY
NT ENTROPY OF DYEING
ENTROPY OF REACTION
BT PHYSICAL PROPERTIES (EXCLUDING MECHANICAL)
THERMAL PROPERTIES
RT BOILING POINT
ENTHALPY
EQUILIBRIUM (CHEMICAL)
FREE ENERGY
HEAT CONTENT
LATENT HEAT
SPECIFIC HEAT
TEMPERATURE
THERMODYNAMICS

ENTROPY OF DYEING
BT ENTROPY
RT ENTHALPY OF DYEING

ENTROPY OF REACTION
BT ENTROPY
RT ENTHALPY OF REACTION

ENZYMATIC DEGRADATION
BT DEGRADATION
RT BACTERIAL DEGRADATION
ENZYMES
MICROBIOLOGICAL DEGRADATION

ENZYMES
NT CELLULASE
PROTEOLYTIC ENZYMES
RT CATALYSTS
DESIZING
DESIZING AGENTS
ENZYMATIC DEGRADATION
YEAST

EPICHLOROHYDRIN
BT EPIHALOGENOHYDRIN
OXIRANE COMPOUNDS
RT CHLOROHYDRINS
CROSSLINKING
EPOXY RESINS
FINISH (SUBSTANCE ADDED)
WET WRINKLE RECOVERY
WRINKLE RECOVERY

EPIHALOGENOHYDRIN
NT EPICHLOROHYDRIN
BT OXIRANE COMPOUNDS

EPICUTICLE
BT FINE STRUCTURE (FIBERS)
RT CORTEX
CUTICLE
ENDOCUTICLE
MEDULLA
MICROFIBRILS
ORTHOCORTEX
PARACORTEX
PRIMARY WALLS
SCALES (WOOL FIBERS)
SECONDARY WALLS
WOOL

EPOXIDES
USE OXIRANE COMPOUNDS

EPOXY COMPOUNDS
USE OXIRANE COMPOUNDS

EPOXY RESINS
BT RESINS
RT DIGLYCIDYL ETHERS
EPICHLOROHYDRIN
HALOHYDRINS
POLYMERS

EQUALIZER (DRAFTING)
RT DRAFTING (STAPLE FIBER)
DRAWFRAMES
EQUALIZING (DRAFTING)
EVENERS
IRREGULARITY CONTROL
PROCESS CONTROL

EQUALIZING (DRAFTING)
RT AUTOLEVELLING
DRAFTING (STAPLE FIBER)
EQUALIZER (DRAFTING)
IRREGULARITY CONTROL

EQUATIONS
NT DIFFERENTIAL EQUATIONS
LINEAR DIFFERENTIAL EQUATIONS
NONLINEAR DIFFERENTIAL EQUATIONS
ORDINARY DIFFERENTIAL EQUATIONS
PARTIAL DIFFERENTIAL EQUATIONS
RT FORMULAS
GENERAL SOLUTIONS
MATHEMATICAL ANALYSIS
NUMERICAL ANALYSIS
PARTICULAR SOLUTIONS
ROOTS (MATHEMATICAL)
SOLUTIONS (MATHEMATICAL)

EQUILIBRIUM (CHEMICAL)
NT SORPTION EQUILIBRIUM
RT ADSORPTION ISOTHERM
DIFFUSION
DYEING
DYEING THEORY
ENTHALPY
ENTROPY
EQUILIBRIUM (DYEING)
EQUILIBRIUM (PHYSICAL)
EQUILIBRIUM MOISTURE CONTENT
FREE ENERGY
KINETICS (CHEMICAL)
LANGMUIR ISOTHERM
MIGRATION (SUBSTANCE)
MOISTURE CONTENT
REACTION KINETICS
REACTIONS (CHEMICAL)
REGAIN
SORPTION RATE
THERMODYNAMICS

EQUILIBRIUM (DYEING)
RT ADSORPTION ISOTHERM
DIFFUSION
DONNAN EQUILIBRIUM
DYE AFFINITY
DYE MIGRATION
DYE PENETRATION
DYEING
DYEING THEORY
EQUILIBRIUM (CHEMICAL)
EQUILIBRIUM (PHYSICAL)
EXHAUSTION (DYEING)
OVERDYEING
SORPTION EQUILIBRIUM
SORPTION OF DYES

EQUILIBRIUM (PHYSICAL)
RT EQUILIBRIUM (CHEMICAL)
EQUILIBRIUM (DYEING)
EQUILIBRIUM MOISTURE CONTENT
MECHANICAL CONDITIONING
MECHANICAL HISTORY
PRESSURE
PRODUCTION
STRESS
STRESS HISTORY
TEMPERATURE

EQUILIBRIUM MOISTURE CONTENT
BT MOISTURE CONTENT
RT ABSORBENCY (MATERIAL)
BOILING POINT
CONDITION (TEXTILES)
CONDITIONING
DEW POINT
DIFFUSION
DIFFUSIVITY
DRY BULB TEMPERATURE
DRYING
DRYING TIME
EQUILIBRIUM (CHEMICAL)
EQUILIBRIUM (PHYSICAL)
FREEZING POINT
HUMIDITY
MOISTURE
MOISTURE CONTENT TESTING
MOISTURE STRENGTH CURVES
PSYCHROMETRY
REGAIN
RELATIVE HUMIDITY
SPECIFIC HUMIDITY
VAPOR PRESSURE
WET BULB TEMPERATURE
WET STRENGTH

EQUIPAGE
BT MILITARY PRODUCTS
RT DUCK
END USES
TENTS

EQUIPMENT
RT APPARATUS
ASSEMBLIES
COMPONENTS
CONTROL EQUIPMENT
ELECTRIC EQUIPMENT
FACILITIES
HEATING EQUIPMENT
INSTRUMENTATION
INVENTORIES
INVESTMENTS
MACHINERY
MATERIALS HANDLING
MOTORS
PLANT
PUNCHED CARD EQUIPMENT
TESTING EQUIPMENT
TEXTILE MACHINERY (GENERAL)

EQUIVALENT COUNT
BT COUNT

ERRORS
NT STANDARD ERROR
RT ACCURACY
ADJUSTMENTS
APPROXIMATIONS
CALIBRATION
COMPARISON
COMPENSATION
CONFIDENCE LIMITS
CORRECTION
DETECTION
DIFFERENCES
DRIFT
ESTIMATION
EXPERIMENTAL DESIGN
EXPERIMENTATION
FEEDBACK CONTROL
(CONT.)

RT FLUCTUATIONS
HYSTERESIS (MECHANICAL)
LINEARITY
OPEN LOOP CONTROL
POSITIONING
STANDARD DEVIATION
STATISTICS
TOLERANCES

ESTERIFICATION
BT REACTIONS (CHEMICAL)
RT ACETYLATED COTTON
ACETYLATION
ACYLATION
ALKYLATION
CHEMICAL MODIFICATION (FIBERS)
ESTERS
HYDROLYSIS

ESTERS
NT ACETATE ESTERS
ACETYLTRIETHYL CITRATE
ACRYLIC ESTERS
ALKYLENE CARBONATES
BENZOATE ESTERS
CELLULOSE ESTERS
DITHIOCARBAMATES
ISOCYANATES
ISOTHIOCYANATES
LACTONES
NITRILES
PHTHALATE ESTERS
POLY (VINYL ACETATE)
POLYACRYLATES
POLYMETHACRYLATES
STEARATE ESTERS
RT ALCOHOLS
ESTERIFICATION
ORGANIC ACIDS
POLYESTERS
SALTS (GENERAL)
SOLVENTS
SULTONES

ESTIMATION
RT ANALYZING
APPROXIMATIONS
COMPENSATION
CONFIDENCE LIMITS
CORRECTION
DETECTION
DRIFT
ERRORS
EVALUATION
EXAMINATION
EXTRAPOLATION
FLUCTUATIONS
FORECASTING
INSPECTION
STANDARD DEVIATION
STANDARD ERROR
STATISTICS

ESTRON (TN)
BT ACETATE FIBERS
CELLULOSE ESTER FIBERS

ETHANOL
UF ETHYL ALCOHOL
BT ALCOHOLS

ETHERALCOHOL SULFATES
USE ALKYL ETHOXY SULFATES

ETHERIFICATION
NT BENZYLATION
CYANOETHYLATION
VINYLATION
BT REACTIONS (CHEMICAL)
RT ALKYLATION
CARBAMOYLETHYLATED COTTON
CHEMICAL MODIFICATION (FIBERS)
CYANOETHYLATED COTTON
ETHERS
ETHOXYLATION
ETHYLENE OXIDE

ETHERS
NT CELLULOSE ETHERS
VINYL ETHERS
RT ACETALS
ALCOHOLS
ALKOXYL GROUPS
ETHERIFICATION
POLYETHERS
SOLVENTS
THIOETHERS

ETHODUOMEENS (TN)
USE POLYETHOXY ALKYLAMINE SURFACTANTS

ETHOFATS (TN)
BT POLYETHOXY ESTERS

ETHOMEENS (TN)
BT POLYETHOXY ALKYLAMINE SURFACTANTS

ETHOMIDS (TN)
BT POLYETHOXY AMIDES

ETHOXY PROPOXY BLOCK POLYMERS
USE POLYETHOXY POLYPROPOXY SURFACTANTS

ETHOXYLATED ALKYL PHENOLS
USE POLYETHOXY ALKYLARYL ETHERS

ETHOXYLATION
RT ALKYLATION
ETHERIFICATION
ETHYLENE OXIDE
GRAFTING
NONIONIC SURFACTANTS

ETHYL ACETATE
BT ACETATE ESTERS
RT SOLVENTS

ETHYL ACRYLATE
BT ACRYLIC COMPOUNDS
ACRYLIC ESTERS
RT BUTYL ACRYLATE
METHYL ACRYLATE
POLY (ETHYL ACRYLATE)

ETHYL ALCOHOL
USE ETHANOL

ETHYL MERCAPTAN
BT MERCAPTANS

ETHYL TRIAZONE
BT TRIAZONES
RT METHYLOL TRIAZONES

ETHYLATION
BT ALKYLATION
RT METHYLATION

ETHYLCELLULOSE
BT CELLULOSE ETHERS
RT HYDROXYALKYLCELLULOSE
METHYLCELLULOSE

ETHYLENE CARBONATE
BT ALKYLENE CARBONATES
RT PROPYLENE CARBONATE

ETHYLENE CHLOROHYDRIN
UF CHLOROETHANOL
BT CHLOROHYDRINS

ETHYLENE DIBROMIDE
BT BROMINE COMPOUNDS
RT VINYLATED COTTON

ETHYLENE GLYCOL
BT GLYCOLS
RT GLYCEROL
POLY (ETHYLENE GLYCOL)

ETHYLENE IMINE
BT AZIRIDINE COMPOUNDS
RT APO
GRAFTING
PROPYLENE IMINE

ETHYLENE OXIDE
BT OXIRANE COMPOUNDS
RT CHEMICAL MODIFICATION (FIBERS)
ETHERIFICATION
ETHOXYLATION
GRAFTING
POLY (ETHYLENE GLYCOL)

ETHYLENE UREA
USE UREA DERIVATIVES

ETHYLENE UREA FORMALDEHYDE CONDENSATES
USE DMEU

ETHYLENEDIAMINE
BT AMINES

ETHYLENEDIAMINE TETRAACETIC ACID
USE EDTA

EULER BUCKLING
USE COMPRESSIVE BUCKLING

EVALUATION
RT ANALYZING
DETECTION
ESTIMATION
EXAMINATION
EXPERIMENTATION
JOB ANALYSIS
REVIEWS
TESTING
TRAINING

EVAPORATION
RT CONSTANT RATE PERIOD
CONVECTION
CRYSTALLIZATION
DIFFUSION
DRYING
FALLING RATE PERIOD
MOISTURE CONTENT TESTING
PRECIPITATION
RELATIVE HUMIDITY
SUBLIMATION
WET BULB TEMPERATURE

EVENERS
RT AUTOLEVELLERS
EQUALIZER (DRAFTING)
EVEREVEN (DRAFTING)
IRREGULARITY
IRREGULARITY CONTROL
PICKER EVENERS
PROCESS CONTROL

EVENESS
USE IRREGULARITY

EVEREVEN (DRAFTING)
RT DRAFT CONTROL
DRAFTING (STAPLE FIBER)
DRAFTING WAVES
EVENERS
IRREGULARITY CONTROL
PROCESS CONTROL

EVERLASTING (LACE)
BT HAND MADE LACES
LACES

EWNN
USE FERRIC TARTRATE

EXAMINATION
RT ACCURACY
ANALYZING
DETECTION
EMPIRICAL ANALYSIS
ESTIMATION
EVALUATION
EXPERIMENTATION

EXHAUSTION (DYEING)
BT EXHAUSTION (WET PROCESS)
RT DYE AFFINITY
DYE MIGRATION
DYE PENETRATION
DYEING
DYEING PROPERTIES (GENERAL)
EQUILIBRIUM (DYEING)
LEVELLING AGENTS
LEVELNESS (DYEING)

EXHAUSTION (MECHANICAL)
RT DAMPING
DIFFUSION
DISPERSING
DISSIPATION
EJECTION
EMPTYING
REMOVAL

EXHAUSTION (WET PROCESS)
NT EXHAUSTION (DYEING)
RT DYE AFFINITY
MIGRATION (SUBSTANCE)
PREFERENTIAL ADSORPTION
SUBSTANTIVITY

EXHAUSTION RATE (DYEING)
USE DYEING RATE

EXOCUTICLE
BT FINE STRUCTURE (FIBERS)
RT CORTEX
ENDOCUTICLE
EPICUTICLE
MEDULLA
MICROFIBRILS
ORTHOCORTEX
PARACORTEX
PRIMARY WALLS
SCALES (WOOL FIBERS)
SECONDARY WALLS
WOOL

EXPANDED PLASTICS
RT FOAMS

EXPANDERS
RT BOW STRAIGHTENERS
SKEW STRAIGHTENERS
TENTER FRAMES

EXPANSION TANKS
RT PRESSURE DYEING MACHINES

EXPERIMENTAL ANALYSIS
NT EXPERIMENTAL STRESS ANALYSIS
RT ANALYZING
EMPIRICAL ANALYSIS
EXPERIMENTATION
INTERFERENCE FRINGES
MATHEMATICAL ANALYSIS
PREPARATION (CHEMICAL)

RT QUALITATIVE ANALYSIS
QUANTITATIVE ANALYSIS
STATISTICAL ANALYSIS
THEORETICAL ANALYSIS
THEORIES
TRIAL AND ERROR SOLUTIONS

EXPERIMENTAL DESIGN
RT ACCURACY
ANALYSIS OF VARIANCE
DEGREES OF FREEDOM
ERRORS
EXPERIMENTATION
OPERATIONS RESEARCH
PROCESS VARIABLES
STOCHASTIC PROCESSES
STATISTICAL METHODS
STATISTICS
VARIABLES

EXPERIMENTAL STRESS ANALYSIS
BT EXPERIMENTAL ANALYSIS
STRESS ANALYSIS
RT ACCURACY
ANALYZING
BIREFRINGENCE
EMPIRICAL ANALYSIS
ERRORS
EVALUATION
EXAMINATION
EXPERIMENTATION
PHOTOELASTICITY
STRESS
STRESS CONCENTRATION
STRESS OPTICAL PROPERTIES

EXPERIMENTATION
UF EXPERIMENTS
RT ANALOGUES
EXPERIMENTAL ANALYSIS
EXPERIMENTAL DESIGN
EXPERIMENTAL STRESS ANALYSIS
INSPECTION
MATHEMATICAL ANALYSIS
TESTING EQUIPMENT
THEORIES
TRIAL AND ERROR SOLUTIONS

EXPERIMENTS
USE EXPERIMENTATION

EXPONENTIAL DISTRIBUTION
RT BINOMIAL DISTRIBUTION
CHI SQUARED DISTRIBUTION
DISTRIBUTION
F DISTRIBUTION
NORMAL DISTRIBUTION
T DISTRIBUTION

EXTENSIBILITY
RT BENDING RIGIDITY
BIAXIAL STRESS
BREAKING ELONGATION
COMPLIANCE
COMPRESSIBILITY
DEFLECTION
DEFORMATION
DIMENSIONAL STABILITY
DUCTILITY
ELASTIC STRAIN
ELASTICITY
FILLINGWISE STRETCH
GUT THREAD
INITIAL MODULUS
NECKING (FILAMENT)
PLASTIC STRAIN
POWER (FABRIC)
STRAIN
STRESS STRAIN CURVES
STRETCH
STRETCH KNITTED FABRICS
STRETCH WOVEN FABRICS
STRETCH YARNS (FILAMENT)
UNIAXIAL STRESS
WARPWISE STRETCH

EXTENSION
USE STRAIN

EXTENSION AT BREAK
USE BREAKING ELONGATION

EXTRACT PRINTING
USE DISCHARGE PRINTING

EXTRACTION
NT VACUUM EXTRACTION
BT SEPARATION (SOLUTION)
RT ABSORPTION (MATERIAL)
ACQUISITION
ADSORPTION
CENTRIFUGES
CHEMICAL ENGINEERING
CONCENTRATION
DIFFUSION
DISSOLVING
FILTRATION
MEMBRANES
OSMOSIS
PARTITION COEFFICIENT
SOLUBILITY
SOLUTIONS (LIQUID)
SOLVENTS
SORPTION

EXTRAPOLATION
RT COMPUTATIONS
CONFIDENCE LIMITS
CORRELATION COEFFICIENT
DRIFT
ESTIMATION
FORECASTING
GRAPHICAL SOLUTIONS
INTERPOLATION
NUMERICAL ANALYSIS

EXTRUDERS
RT EXTRUSION
EXTRUSION COATING
EXTRUSION PUMPS
SPINNERETS
SPINNING MACHINES (EXTRUSION)

EXTRUSION
NT MELT SPINNING
SPINNING (EXTRUSION)
SOLVENT SPINNING
WET SPINNING
RT CAKES
COLD DRAWING
DOPE (POLYMER)
DRAWING (FILAMENT)
EXTRUDERS
EXTRUSION COATING
EXTRUSION PUMPS
FIBERS
FILAMENT YARNS
FILAMENTS
GLASS FIBERS
HOT DRAWING
INORGANIC FIBERS (MAN MADE)
MAN MADE FIBERS
MELT VISCOSITY
METALLIC FIBERS
MONOFILAMENTS
SPINNERETS
SPINNING ASSEMBLIES (EXTRUSION)
SYNTHETIC FIBERS
TOW

EXTRUSION COATING
BT COATING (PROCESS)
RT CALENDER COATING
CAST COATING
COATED FABRICS
EXTRUDERS
EXTRUSION
FLEXIBLE FILM LAMINATING
KNIFE COATING
NIP COATING
ROLLER COATING

EXTRUSION PUMPS
BT PUMPS
RT EXTRUDERS
EXTRUSION
SPINNING (EXTRUSION)

EYELET EMBROIDERY
BT EMBROIDERY

EYELETS (GUIDES)
USE THREAD GUIDES

F DISTRIBUTION
RT BINOMIAL DISTRIBUTION
CHI SQUARED DISTRIBUTION
DISTRIBUTION
GAMMA DISTRIBUTION
NORMAL DISTRIBUTION
T DISTRIBUTION

FABRIC ANALYSIS
UF CLOTH ANALYSIS
RT ANALYZING
COLOR MATCHING
COURSES PER INCH
COVER FACTOR
CRIMP
CRIMP INTERCHANGE
DESIGN
DRAFTING (PATTERN)
EMPIRICAL ANALYSIS
ENDS PER INCH
FABRIC CRIMP
FABRIC CROSS SECTIONS
FABRIC DESIGN
FABRIC GEOMETRY
FABRIC INSPECTION
FABRIC PROPERTIES (STRUCTURAL)
FABRIC STRUCTURE
INTERLACINGS (YARN IN FABRIC)
PATTERN (FABRICS)
PICKS PER INCH
STITCH LENGTH
WALES PER INCH
WEAVE

FABRIC BOLTS
BT PACKAGES

FABRIC COVER
BT COVER
RT BOTTOM COVER
COVER FACTOR
DENSITY
TOP COVER
TRANSLUCENCY
TRANSPARENCY
YARN COVER

FABRIC CRIMP
BT CRIMP
FABRIC PROPERTIES (STRUCTURAL)
RT CRIMP BALANCE
CRIMP INTERCHANGE
FABRIC ANALYSIS
FABRIC CROSS SECTIONS
FABRIC GEOMETRY
FIBER CRIMP
JAMMING POINT
YARN CRIMP

FABRIC CROSS SECTIONS
RT CROSS SECTIONS
FABRIC ANALYSIS
FABRIC CRIMP
FABRIC GEOMETRY
FABRIC PROPERTIES (STRUCTURAL)
FABRIC STRUCTURE
FIBER CROSS SECTIONS
YARN CROSS SECTIONS

FABRIC DEFECTS
NT BACK AND FACE EFFECTS
BARRE
BLEEDING
BOW
BRIGHT SPECKS
BROKEN SELVEDGES
BROKEN THREADS
BRONZING
BOWED FABRICS
CENTER LOOPS
CHAFE MARKS
COCKLE (DEFECT)
CRACKED SELVEDGES
CRACKS (FABRIC)
CREASE MARKS
CREASES
CURLING
DIAMOND BARRING
DOG LEGGED SELVEDGES
DOUBLE PICKS
DROPPED STITCHES
DYE SPECKS
DYEING DEFECTS
ENDS CUT
FILLING BANDS
FILLING BARS
FILLING STREAKS
FOGMARKING
FROSTING
HARNESS SKIPS
HOLES (FABRIC DEFECTS)
KINKY FILLING
KNOTS
LASHED IN WEFT
MISPICKS
MISREED
OFF SHADE
PILLS
PIRN BARRE
PUCKER (DEFECT)
PULLED IN SELVEDGES
REED MARKS
REED RAKE
ROPE MARKS
RUNS (KNITTING)
SHADE BARS
SHINERS
SHUTTLE MARKS
SINGLINGS
SKEW
SKITTERINESS
SLACK COURSE
SLACK SELVEDGES
SLACK THREADS
SLUBS
SPECKINESS
SPECKS
SPIRALITY (KNITTED FABRICS)
SPOTS
SPREAD LOOPS
STAINING (DYEING DEFECTS)
STAINS
STARTING MARKS
STREAKINESS
SWOLLEN DENT
SWOLLEN HEDDLE
TEMPLE MARKS
TIGHT SELVEDGES
TIPPINESS
WARP STREAKS
WATER SPOTS
WRONG DRAW
BT DEFECTS
RT APPEARANCE
COLORFASTNESS
END BREAKAGES
FABRIC INSPECTION
FABRIC PROPERTIES
FABRIC PROPERTIES (AESTHETIC)
FABRIC PROPERTIES (STRUCTURAL)
FUZZ RESISTANCE
IRREGULARITY
INSPECTION
LEVELNESS (DYEING)
MENDING
NEPS
NUBS
PERCHING
PILL RESISTANCE
PILL WEAR OFF
PILLING
REEDY FABRICS
SECONDS (FABRICS)
SHEDDING (FIBER)
SKEW STRAIGHTENERS
SOILING
SPECKING

FABRIC DESIGN
BT DESIGN
RT APPAREL DESIGN
CHAIN DRAFT
COLOR
COURSES PER INCH
DRAFTING (PATTERN)
DRESS DESIGN
ENDS PER INCH
FABRIC ANALYSIS
FABRIC PROPERTIES
FABRIC PROPERTIES (AESTHETIC)
FLOAT DESIGN
HARNESS DRAFT
INTERLACINGS (YARN IN FABRIC)
PATTERN REPEATS
PATTERN (FABRICS)
PICKS PER INCH
SHADE
STRIPES
STRUCTURAL ANALYSIS
WALES PER INCH
WEAVE

FABRIC DISTORTION (TAILORING)
NT GATHERING
PINCHING
PUCKERING
RUFFLING
SHIRRING
STRETCHING (SEAMING)
RT DISTORTION
SEAMING
SEAMS
SEWING
SHEAR (MODE OF DEFORMATION)
STITCH DISTORTION
STITCHING

FABRIC GEOMETRY
UF CLOTH GEOMETRY
BT FABRIC PROPERTIES (STRUCTURAL)
GEOMETRY
RT COURSES PER INCH
COVER
COVER FACTOR
CRIMP
CRIMP BALANCE
CRIMP INTERCHANGE
CRIMP PERCENT
CROWNS
DIFFERENTIAL GEOMETRY
DRY RELAXATION
ENDS PER INCH
FABRIC ANALYSIS
FABRIC CRIMP
FABRIC CROSS SECTIONS
FABRIC INTERSTICES
FABRIC STRUCTURE
FLOAT LENGTH
HEAT SETTING (SYNTHETICS)
INTERLACINGS (YARN IN FABRIC)
JAMMING
JAMMING POINT
LOOP DISTORTION
MATRIX FREEDOM
PACKING FACTOR
PICKS PER INCH
POROSITY
STITCH DENSITY
STITCH LENGTH
STRUCTURAL MECHANICS
STRUCTURAL PROPERTIES
STRUCTURES
TIGHTNESS
TRELLIS MODEL
TWILL ANGLES
TWILLS
WALES PER INCH
WET RELAXATION
YARN BUCKLING
YARN FLATTENING
YARN GEOMETRY
YARN SLIPPAGE
YARN STRUCTURE

FABRIC GUIDES
USE CLOTH GUIDES

FABRIC INSPECTION
UF CLOTH INSPECTION
RT BARRE
BROKEN SELVEDGES
BROKEN THREADS
BURLING
BURLING MACHINES
CENTER LOOPS
COCKLE (DEFECT)
CREASES
DEFECTS
DIAMOND BARRING
DIRT
DYE SPECKS
END BREAKAGES
FABRIC ANALYSIS
FABRIC DEFECTS
FILLING BARS
FILLING STREAKS
FINISHING PROCESS (GENERAL)
HARNESS SKIPS
HOLES (FABRIC DEFECTS)
IRREGULARITY
KNOTS
LINT (WASTE)
MENDING
MISPICKS
MISREED
NEPS
PICK GLASS
PULLED IN SELVEDGES
QUALITY CONTROL
REED MARKS
SECONDS (FABRICS)
SEWING
SHINERS
SINGLINGS
SLACK THREADS
SLUBS
SPECK DYEING
SPECKING
SPECKS
SPOTS
STAINS
STARTING MARKS
STREAKINESS
TEMPLE MARKS
TIGHT SELVEDGES
WARP STREAKS

FABRIC INTERSTICES
BT PORES
RT AIR RESISTANCE
FABRIC PROPERTIES (STRUCTURAL)
WATER PERMEABILITY
WATER REPELLENCY
WATER RESISTANCE

FABRIC LAMINATES
USE LAMINATED FABRICS

FABRIC PROPERTIES
NT AESTHETIC APPEAL
APPEARANCE
BAGGINESS
BIAXIAL STRENGTH
BLOOM
BOW

NT BRIGHTNESS
BURSTING STRENGTH
COMFORT
COURSES PER INCH
COVER
COVER FACTOR
CREASE ACCEPTANCE
CREASE RETENTION
CURLING TENDENCY
DRABNESS
DRAPE
DULLNESS
ELMENDORF TEAR STRENGTH
ENDS PER INCH
FABRIC GEOMETRY
FABRIC PROPERTIES (AESTHETIC)
FABRIC PROPERTIES (MECHANICAL)
FABRIC PROPERTIES (PHYSICAL EXCLUDING MECHANICAL)
FABRIC PROPERTIES (STRUCTURAL)
FABRIC WEIGHT
FELTING RESISTANCE
FINISH (PROPERTY)
GRAB STRENGTH
HAND
HARSHNESS
LIVELINESS
LUXURIOUSNESS
PATTERN (FABRICS)
PATTERN DEFINITION
PICKS PER INCH
PILL RESISTANCE
PUNCTURE RESISTANCE
RAVEL STRIP STRENGTH
RICHNESS
SCROOP
SET (WOVEN FABRIC)
SEWABILITY
SHEEN
SHRINK RESISTANCE
SLIP RESISTANCE
SNAG RESISTANCE
STITCH CLARITY
STITCH DENSITY
TEAR STRENGTH
TEXTURE
TONGUE TEAR STRENGTH
TRAPEZOID TEAR STRENGTH
WALES PER INCH
WATER REPELLENCY
WATER RESISTANCE
WEAVE
WRINKLE RECOVERY
WRINKLE RESISTANCE
RT ABRASION RESISTANCE
ABSORBENCY (MATERIAL)
ACOUSTIC PROPERTIES
AESTHETIC PROPERTIES
BENDING RECOVERY
BENDING RIGIDITY
BIOCHEMICAL PROPERTIES
BREAKING STRENGTH
BULK
CHEMICAL PROPERTIES
COLOR
COLORFASTNESS
COMPRESSIBILITY
CRITICAL MOISTURE CONTENT
CUTTING RESISTANCE
DEGRADATION PROPERTIES
DENSITY
DIMENSIONAL STABILITY
DYE AFFINITY
ELECTRICAL PROPERTIES
END USE PROPERTIES
FABRIC DEFECTS
FABRIC DESIGN
FABRICS
FLUORESCENCE
GEOMETRY
GLOSS
HAIRINESS
IMPACT STRENGTH
LUSTER
MECHANICAL DETERIORATION PROPERTIES
MECHANICAL PROPERTIES
MILDEW RESISTANCE
OPTICAL PROPERTIES
PACKING FACTOR
PERMEABILITY
PHYSICAL CHEMICAL PROPERTIES
PHYSICAL PROPERTIES (EXCLUDING MECHANICAL)
POROSITY
REFLECTANCE
REFRACTIVE INDEX
RESILIENCE
RESISTIVITY
ROUGHNESS
SHADE
SHAPE
SKEW
SMOOTHNESS
SOFTNESS
SOIL RESISTANCE
SPARKLE
STAIN RESISTANCE
RT STATIC RESISTANCE
STIFFNESS
STRESS OPTICAL PROPERTIES
STRESS STRAIN PROPERTIES
STRETCH
STRUCTURAL PROPERTIES
SURFACE CONTOUR
SURFACE FRICTION
SURFACE PROPERTIES (MECHANICAL)
SURFACE PROPERTIES (PHYSICAL CHEMICAL)
THERMAL PROPERTIES
THICKNESS
TRANSFER PROPERTIES
TRANSLUCENCY
TRANSPARENCY
TRELLIS MODEL
UNIAXIAL STRENGTH
WATER REPELLENCY
WEAR RESISTANCE
WEATHER RESISTANCE
YARN PROPERTIES

FABRIC PROPERTIES (AESTHETIC)
UF SUBJECTIVE MECHANICAL PROPERTIES
NT AESTHETIC APPEAL
APPEARANCE
BAGGINESS
BLOOM
BRIGHTNESS
COMFORT
DRABNESS
DRAPE
DULLNESS
FINISH (PROPERTY)
HAND
HARSHNESS
LUXURIOUSNESS
PATTERN (FABRICS)
PATTERN DEFINITION
PRICKLINESS
RICHNESS
SCROOP
SHEEN
STITCH CLARITY
TEXTURE
BT FABRIC PROPERTIES
RT AESTHETIC PROPERTIES
COLOR
FABRIC DEFECTS
FABRIC DESIGN
FLUORESCENCE
GLOSS
HAIRINESS
LUSTER
MECHANICAL PROPERTIES
OPTICAL PROPERTIES
ROUGHNESS
SHADE
SHAPE

SMOOTHNESS
SOFTNESS
SPARKLE
SURFACE PROPERTIES (MECHANICAL)
SURFACE PROPERTIES (PHYSICAL CHEMICAL)

FABRIC PROPERTIES (DEGRADATION)
USE DEGRADATION PROPERTIES

FABRIC PROPERTIES (MECHANICAL)
NT BAGGINESS
BIAXIAL STRENGTH
BURSTING STRENGTH
CREASE ACCEPTANCE
CREASE RETENTION
CRUSH RESISTANCE
CURLING TENDENCY
DRAPE
ELMENDORF TEAR STRENGTH
FABRIC RELAXATION
FABRIC SHRINKAGE
FABRIC STRENGTH
FABRIC TO FABRIC FRICTION
FINISH (PROPERTY)
GRAB STRENGTH
LIVELINESS
PILL RESISTANCE
PUNCTURE RESISTANCE
RAVEL STRIP STRENGTH
SCROOP
SEAM EFFICIENCY
SEAM STRENGTH
SEWABILITY
SHRINK RESISTANCE
SLIP RESISTANCE
SNAG RESISTANCE
TEAR STRENGTH
TONGUE TEAR STRENGTH
TRAPEZOID TEAR STRENGTH
WRINKLE RECOVERY
WRINKLE RESISTANCE
BT FABRIC PROPERTIES
RT ABRASION RESISTANCE
RT BENDING RECOVERY
BENDING RIGIDITY
BREAKING STRENGTH
COMPRESSIBILITY
CUTTING RESISTANCE
DIMENSIONAL STABILITY
HAIRINESS
IMPACT STRENGTH
MECHANICAL DETERIORATION PROPERTIES
MECHANICAL PROPERTIES
PHYSICAL PROPERTIES (EXCLUDING MECHANICAL)
RESILIENCE
ROUGHNESS
SMOOTHNESS
SOFTNESS
STIFFNESS
STRESS OPTICAL PROPERTIES
STRESS STRAIN PROPERTIES
STRETCH
SURFACE FRICTION
SURFACE PROPERTIES (MECHANICAL)
SURFACE PROPERTIES (PHYSICAL CHEMICAL)
TRANSFER PROPERTIES
UNIAXIAL STRENGTH
WEAR RESISTANCE

FABRIC PROPERTIES (PHYSICAL EXCLUDING MECHANICAL)
NT BRIGHTNESS
COMFORT
FABRIC WEIGHT
FINISH (PROPERTY)
WATER REPELLENCY
WATER RESISTANCE
BT FABRIC PROPERTIES
RT ABSORBENCY (MATERIAL)
ACOUSTIC PROPERTIES
CHEMICAL PROPERTIES
COLOR
COLORFASTNESS
CRITICAL MOISTURE CONTENT
DENSITY
DYE AFFINITY
ELECTRICAL PROPERTIES
FLUORESCENCE
MECHANICAL PROPERTIES
OPTICAL PROPERTIES
PERMEABILITY
PHYSICAL CHEMICAL PROPERTIES
PHYSICAL PROPERTIES (EXCLUDING MECHANICAL)
POROSITY
REFLECTANCE
RESISTIVITY
SHADE
SOIL RESISTANCE
STAIN RESISTANCE
STATIC RESISTANCE
STRUCTURAL PROPERTIES
SURFACE CONTOUR
SURFACE FRICTION
SURFACE PROPERTIES (MECHANICAL)
SURFACE PROPERTIES (PHYSICAL CHEMICAL)
THERMAL CONDUCTIVITY
THERMAL PROPERTIES
TRANSFER PROPERTIES
TRANSLUCENCY
TRANSPARENCY
WATER REPELLENCY

FABRIC PROPERTIES (STRUCTURAL)
NT COURSES PER INCH
COVER
COVER FACTOR
ENDS PER INCH
FABRIC CRIMP
FABRIC GEOMETRY
FLOAT LENGTH
PATTERN (FABRICS)
PATTERN DEFINITION
PICKS PER INCH
SET (WOVEN FABRIC)
STITCH CLARITY
STITCH DENSITY
TEXTURE
WALES PER INCH
WEAVE
BT FABRIC PROPERTIES
RT AESTHETIC PROPERTIES
FABRIC ANALYSIS
FABRIC CROSS SECTIONS
FABRIC DEFECTS
FABRIC INTERSTICES
FABRIC STRUCTURE
GEOMETRY
MECHANICAL PROPERTIES
PACKING FACTOR
POROSITY
SHAPE
SKEW
STRUCTURAL PROPERTIES
SURFACE PROPERTIES (MECHANICAL)
THICKNESS

FABRIC REINFORCED COMPOSITES
BT COMPOSITES
LAMINATES
REINFORCED COMPOSITES
RT HONEYCOMB LAMINATES
POST FORMED LAMINATES

FABRIC RELAXATION
NT DRY RELAXATION
WET RELAXATION
BT DRY FINISHING
FABRIC PROPERTIES (MECHANICAL)
RELAXATION
WET FINISHING
RT BOARDING
CONDITIONING
CRABBING
DIMENSIONAL STABILITY
DRAPE
FABRIC SHRINKAGE
HAND
HEAT SETTING (SYNTHETICS)
LONDON SHRINKING
RELAXATION SHRINKAGE
RELAXATION TIME (MECHANICAL)
RESIDUAL STRESS
SHAPING
SPONGING
STRAIN
STRESS
STRESS RELAXATION

FABRIC SHRINKAGE
NT COMPRESSIVE SHRINKAGE
FELTING SHRINKAGE
RELAXATION SHRINKAGE
BT FABRIC PROPERTIES (MECHANICAL)
SHRINKAGE
RT COMPACTING
COMPACTOR
CONDITIONING
CRABBING
DIMENSIONAL STABILITY
DRY RELAXATION
FABRIC RELAXATION
LONDON SHRINKING
PRESHRINKING
RELAXATION
RELAXATION TIME (MECHANICAL)
RESIDUAL STRESS
RIGMEL PROCESS
SANFORIZING (TN)
SHRINK RESISTANCE
SHRINKING
STRESS RELAXATION
WET RELAXATION

FABRIC STRENGTH
BT BREAKING STRENGTH
FABRIC PROPERTIES
FABRIC PROPERTIES (MECHANICAL)
STRENGTH
RT BURSTING STRENGTH
COMPRESSIVE STRENGTH
DRY STRENGTH
FABRICS
FIBER STRENGTH
GRAB STRENGTH
PUNCTURE RESISTANCE
RAVEL STRIP STRENGTH
STRENGTH ELONGATION TESTING
STRENGTH TESTING
TEAR STRENGTH
WET STRENGTH
YARN STRENGTH

FABRIC STIFFENERS
USE STIFFENERS (AGENTS)

FABRIC STRUCTURE
UF CLOTH STRUCTURE
NT KNITTED FABRIC STRUCTURE (FILLING KNIT)
KNITTED FABRIC STRUCTURE (WARP KNIT)
NONWOVEN FABRIC STRUCTURE
WOVEN FABRIC STRUCTURE
RT BACK (FABRIC)
BACKED FABRICS
COUNT
COURSES PER INCH
COVER FACTOR
CRIMP
CRIMP BALANCE
CRIMP PERCENT
CROWNS
DESIGN
ENDS PER INCH
FABRIC ANALYSIS
FABRIC CROSS SECTIONS
FABRIC DESIGN
FABRIC GEOMETRY
FABRIC PROPERTIES (STRUCTURAL)
FABRICS (ACCORDING TO STRUCTURE)
FACE (FABRIC)
FILLING EFFECT (TWILL)
FILLING FACE
RT FILLING FLOATS
FLOAT DESIGN
FLOAT LENGTH
FLOATS
JAMMING
JAMMING POINT
KNITTED FABRICS
LAMINATED FABRICS
MALI FABRICS (TN)
MULTIPLE LAYER FABRICS
NEEDLED FABRICS
NONWOVEN FABRICS
PICKS PER INCH

PILE FABRICS (WOVEN)
POROSITY
SELVEDGES
SET (WOVEN FABRIC)
STITCH DENSITY
STRUCTURAL MECHANICS
STRUCTURAL PROPERTIES
STRUCTURES
TUFTED FABRICS
TWIST
WALES PER INCH
WARP EFFECT (TWILL)
WARP FACE
WEAVE
WOVEN FABRICS
YARN GEOMETRY
YARN STRUCTURE
Z TWILLS

FABRIC TO FABRIC FRICTION
BT FABRIC PROPERTIES (MECHANICAL)
FRICTION
RT ABRASION
DYNAMIC FRICTION
FIBER FIBER FRICTION
FIBER FRICTION
FRICTIONAL FORCE
FUZZING
NORMAL FORCE
PILLING
RUBBING
STATIC FRICTION
YARN TO METAL FRICTION
YARN TO YARN FRICTION

FABRIC TO FABRIC LAMINATES
BT COMPOSITES
LAMINATED FABRICS
LAMINATES
RT FOAM BACKED FABRICS

FABRIC UPPERS (FOOTWEAR)
BT FOOTWEAR COMPONENTS

FABRIC WEIGHT
BT FABRIC PROPERTIES
FABRIC PROPERTIES (PHYSICAL EXCLUDING MECHANICAL)
RT COVER
DENSITY
HEAVY WEIGHT
LIGHT WEIGHT
MEDIUM WEIGHT
PACKING FACTOR
THICKNESS
WEIGHT

FABRICS
UF CLOTH
NT ALBERT TWILL
ALHAMBRA QUILT
ALL OVER (LACE)
ALPACA FABRICS
ANGOLA (FABRIC)
APPAREL FABRICS
ARGYLE GIMP
ARMURE
ASTRAKHAN
AUTOMOTIVE FABRICS
AWNING CLOTH
BACKED FABRICS
BAIZE
BANNOCK BURN
BARATHEA
BATISTE
BEADING (LACE)
BEAVER CLOTH
BEDFORD CORD
BILLIARD CLOTH
BLUETTE
BOLTON SHEETING
BOMBAZINE
BONDED FIBER FABRICS
BRACE WEB
BRAIDS
BRASSIER CLOTH
BRATTICE CLOTH
BREAKER FABRICS
BRILLIANT SPOT
BROADCLOTH
BUCKRAM
BUILDER FABRICS
BUNTING
CALICO
NT CAMBRIC
CANDLEWICK FABRICS (TUFTED FABRIC)
CARPETS
CASEMENT
CAVALRY TWILL
CHAMBRAY
CHARMANETTE SATIN
CHECK SHIRTING
CHEESE CLOTH
CHENILLE FABRICS (TUFTED FABRIC)
CHIFFON
CHINTZ
CORSET BATISTE
COUTIL
COVERT
CRASH TOWELING
CREPE
CREPON GEORGETTE
CRETONNE
DAMASK
DENIMS
DIMITY
DOESKIN
DOTTED MUSLIN
DOUBLE KNIT FABRICS
DOUBLE PIQUE
DRILLS (FABRICS)
DUCK
DUFFEL
EDGEING (LACE)
ELASTIQUE
EMBOSSED CREPE
FAILLE
FANCY LENO
FELTS
FILLING KNITTED FABRICS
FLANNEL
FLAT BRAIDS
FLAT WOVEN FABRICS
FLOCKED CARPETS
FLOCKED FABRICS
FLOUNCING (LACE)
FOULARD (FABRIC)
FULL CARDIGAN FABRICS
GABARDINE
GALLOON
GAUZE
GENOA VELVET
GEORGETTE
GINGHAM
HALF CARDIGAN FABRICS
HAND MADE LACES
HARVARD
HESSIAN
HIMALAYA CLOTH
HONEYCOMB LAMINATES
HOPSACK
HOUSEHOLD FABRICS
IMITATION GINGHAM
INDIA LINON
INDIANA CLOTH
INDUSTRIAL FABRICS
INSERTION (LACE)
INTERLOCK (KNITTED FABRICS)
ITALIAN CLOTH
JACONET
JEANS
KERSEY
KNITTED FABRICS
KNITTED NETS
KNOTTED NETS
KNITTED PILE FABRICS
LACE NETS
LACES
LAMINATED FABRICS
LAMINATED JERSEY (FOAM)
LASTING
LAWN
LEATHER CLOTH
LINGETTE
LOCKKNIT
LONGCLOTH
MACHINE MADE LACES
MAINSOOK CHECK
MALI FABRICS (TN)
MARQUISETTE
MELTON
MESALINE
MOCK LENO
MOTIFS (LACE)
MULTIPLE LAYER FABRICS
MUSLIN
NEEDLED FABRICS
NETS
NETTING
NONWOVEN FABRICS
OPALINE
OSNABURG
PAJAMA CHECK
PERCALE
PERCALINE
PILE FABRICS
PILLOW TUBING
PIQUE
PLAIN KNITTED FABRICS
PLISSE

NT PONGEE
POPLIN
PRINT CLOTH
PURL KNITTED FABRICS
REGULAR BRAIDS
REPP
RIB KNITTED FABRICS
RIBBON
SCREENING
SCRIM
SEERSUCKER
SERGE
SHADOW STRIPE
SHARKSKIN
SHEETING (FABRIC)
SHIRTING
SILESIA
SKIP DENT
SOLID BRAIDS
SPECKLED FABRIC
SPLASH VOILE
SPORT SATIN
STRETCH FABRICS
STRETCH KNITTED FABRICS
STRETCH WOVEN FABRICS
TAFFETA
TAPE
TICKING
TIRE FABRICS
TOBACCO CLOTH
TRICOT KNITTED FABRICS
TROPICAL WORSTED
TUBULAR FABRICS
TUFTED FABRICS
TWILL CHECKBOARD
UMBRELLA CLOTH
UPHOLSTERY FABRICS
VEILING (LACE)
VELOUR
VOILE
WARP KNITTED FABRICS
WARP REPP
WEBBING
WHIPCORD
WIGAN
WINDOW HOLLAND
WOVEN FABRICS
WOVEN NETS
BT FIBER ASSEMBLIES
RT COATED FABRICS
COMPOSITES
DYED FABRICS
END USES
FABRIC PROPERTIES
FABRIC STRUCTURE
FABRICS (ACCORDING TO FIBER)
FABRICS (ACCORDING TO STRUCTURE)
FIBERS
FILMS
FINISHED FABRICS
FLOCK
FOAMS
GARMENTS
HEAVY WEIGHT
KNITTED OUTERWEAR
LAMINATED FABRICS
LIGHT WEIGHT
LININGS
LININGS (FOOTWEAR)
MEDIUM WEIGHT
PAPER
PIECE DYED FABRICS
PIECE DYEING
PRINTED FABRICS
SAMPLES
SUBSTRATES
TEXTILE MATERIALS
WEAVE
WEAVE (TYPES)
WEBS
YARN DYED FABRICS
YARNS

FABRICS (ACCORDING TO FIBER)
NT COTTON FABRICS
WOOL FABRICS
RT FABRICS
FABRICS (ACCORDING TO STRUCTURE)

FABRICS (ACCORDING TO STRUCTURE)
NT BONDED FIBER FABRICS
BONDED YARN FABRICS
BRAIDS
COATED FABRICS
CROSS LAID YARN FABRICS
FABRIC REINFORCED COMPOSITES
FLAT WOVEN FABRICS
FLOCKED FABRICS
KNITTED FABRICS
LACES
LAMINATED FABRICS
LAMINATED JERSEY (FOAM)
MULTIPLE LAYER FABRICS
NEEDLED FABRICS
NETS
NONWOVEN FABRICS
NT PILE FABRICS
PRESSED FELTS
STITCH BONDED FABRICS
STITCH REINFORCED FABRICS
STITCHED PILE FABRICS
TUFTED FABRICS
WOVEN FABRICS
BT FIBER ASSEMBLIES
RT AGRICULTURAL FABRICS
APPAREL FABRICS
ARACHNE SYSTEMS (TN)
AUTOMOTIVE FABRICS
COMPOSITES
DECORATIVE FABRICS
FABRIC STRUCTURE
FABRICS
FABRICS (ACCORDING TO FIBER)
FABRICS (BY END USES)
FURNISHING FABRICS
HOUSEHOLD FABRICS
INDUSTRIAL FABRICS
MALI PROCESSES (TN)
MEDICAL TEXTILES
REVERSIBLE FABRICS
UPHOLSTERY FABRICS

FABRICS (BRAIDS)
USE BRAIDS

FABRICS (BY END USES)
NT AGRICULTURAL FABRICS
APPAREL FABRICS
AUTOMOTIVE FABRICS
DECORATIVE FABRICS
FURNISHING FABRICS
HOUSEHOLD FABRICS
INDUSTRIAL FABRICS
UPHOLSTERY FABRICS
RT APPLICATIONS
END USES
FABRICS (ACCORDING TO STRUCTURE)
FIBER ASSEMBLIES
PRODUCTS

FABRICS (COATED)
USE COATED FABRICS

FABRICS (COMPOSITES)
USE COMPOSITES

FABRICS (FLAT WOVEN)
USE FLAT WOVEN FABRICS

FABRICS (FLOCKED)
USE FLOCKED FABRICS

FABRICS (KNITTED)
USE KNITTED FABRICS

FABRICS (LACE)
USE LACES

FABRICS (MALI)
USE MALI FABRICS (TN)

FABRICS (MULTIPLE LAYER)
USE MULTIPLE LAYER FABRICS

FABRICS (NET)
USE NETS

FABRICS (NONWOVEN)
USE NONWOVEN FABRICS

FABRICS (PILE)
USE PILE FABRICS

FABRICS (TUFTED)
USE TUFTED FABRICS

FABRICS (WOVEN)
USE WOVEN FABRICS

FACE (FABRIC)
RT BACK (FABRIC)
COURSES
COVER
FABRIC STRUCTURE
FABRICS
FACE FINISH
FACE GOODS
FACE YARNS
FACING
FILLING FACE
WARP FACE

FACE FINISH
UF CLOSE FACE FINISH
FULL FACE FINISH
RT BRUSHING
FACE (FABRIC)
FACE GOODS
FINISHING PROCESS (GENERAL)
NAPPING
PRESSING
SHEARING (FINISHING)

FACE FINISHED WOOLENS
USE FACE GOODS

FACE GOODS
UF FACE FINISHED WOOLENS
RT BEAVER CLOTH
DOESKIN
FACE (FABRIC)
FACE FINISH
KERSEY
MELTON
VELOUR

FACE YARNS
RT BACK FILLING
BACK WARP
FACE (FABRIC)
MULTIPLE LAYER FABRICS
PLAITING (KNITTING)

FACILITIES
RT COSTS
EQUIPMENT
INVENTORIES
INVESTMENTS
MATERIALS HANDLING
PLANT

FACING
BT GARMENT COMPONENTS
RT FACE (FABRIC)
GARMENTS
LAPELS
TAILORING

FACTORS (PHYSICAL)
USE VARIABLES

FADEOMETER (TN)
BT LIGHTFASTNESS TESTERS
RT COLORFASTNESS TESTING
LIGHTFASTNESS (COLOR)
LIGHTFASTNESS TESTING
XENOTEST (TN)

FADING
UF COLOR FADING
NT FUME FADING
RT BLEACHING
CHROMOTROPISM
COLOR STANDARDS
COLORFASTNESS
COLORFASTNESS TESTING
CROCKING
DAYLIGHT
DEGRADATION
LIGHT
LIGHTFASTNESS TESTING
METACHROMASY
OFF SHADE
PHOTOCHEMICAL REACTIONS
SUNLIGHT
WET CROCKING
YELLOWING

FAILLE
BT WOVEN FABRICS
RT PLAIN WEAVES
POULT
TAFFETA

FALL PLATE FABRICS
BT RASCHEL FABRICS
WARP KNITTED FABRICS
RT TUCK STITCHES

FALLER BARS
UF FALLERS (GILL BOXES)
RT ACTUAL DRAFT
BREAK DRAFT
FALLER SCREWS
FALLER SPEED
FINISHER GILLING
GILL BOXES
GILLING
INTERSECTING GILL BOXES

FALLER SCREWS
RT BACK ROLLS
DRAWFRAMES
FALLER BARS
FALLER SPEED
FRONT ROLLS
GILL BOXES
GILLING
INTERSECTING GILL BOXES

FALLER SPEED
RT FALLER BARS
FALLER SCREWS
GILL BOXES
GILLING
INTERSECTING GILL BOXES
PROCESS VARIABLES

FALLERS (GILL BOXES)
USE FALLER BARS

FALLERS (MULE)
RT CARRIAGES (SPINNING MULE)
HEADSTOCKS
MULE SPINNING
MULES
QUADRANTS (MULE)
TWISTING AT THE HEAD

FALLING RATE PERIOD
RT CONSTANT RATE PERIOD
CRITICAL MOISTURE CONTENT
DRYING
DRYING TIME

FALSE DRAFT
BT DRAFT
RT ACTUAL DRAFT
BREAK DRAFT

FALSE TWIST
BT TWIST
RT ALTERNATING TWIST
CRIMPED YARNS
CYCLOSET YARNS (TN)
FALSE TWIST SPINDLES
FALSE TWIST TUBES
FALSE TWIST YARNS
FALSE TWISTING (EXCEPT TEXTURING)
FLUID TWISTING
HEAT SETTING (SYNTHETICS)
HIGH TWIST
INTERLACING (FILAMENT IN YARN)
LOW TWIST
REAL TWIST
ROTOSET YARNS (TN)
S TWIST
SLIVER CANS
STRETCH YARNS (FILAMENT)
TEXTURED YARNS (FILAMENT)
TWIST CONTROL
TWISTERS
TWISTING
WOOLEN DRAFTING
Z TWIST
ZERO TWIST

FALSE TWIST SPINDLES
BT SPINDLES
RT AIR BEARINGS
AXLES
FALSE TWIST
FALSE TWIST TUBES
FALSE TWIST YARNS
FALSE TWISTERS
FALSE TWISTING (EXCEPT TEXTURING)
HEAT SETTING (SYNTHETICS)
SPINNING
STRETCH YARNS (FILAMENT)
TEXTURED YARNS (FILAMENT)
TEXTURING
TWIST
TWISTERS
TWISTING

FALSE TWIST TUBES
RT DRAFTING (STAPLE FIBER)
DRAFTING TWIST
FALSE TWIST
FALSE TWIST SPINDLES
FALSE TWISTING (EXCEPT TEXTURING)
RING FRAMES
TWIST
TWIST CONTROL
TWISTING
WOOLEN DRAFTING
WOOLEN SYSTEM PROCESSING

FALSE TWIST YARNS
NT DYNALOFT (TN)
FLUFLON (TN)
HELANCA (TN)
SUPERLOFT (TN)
BT CRIMPED YARNS
STRETCH YARNS (FILAMENT)
TEXTURED YARNS (FILAMENT)
YARNS
RT AGILON (TN)
AIR JET TEXTURED YARNS
BICOMPONENT FIBER YARNS (FILAMENT)
BICOMPONENT FIBER YARNS (STAPLE)
BULK
BULKED YARNS
CRIMP
EDGE CRIMPED YARNS
FALSE TWIST
FALSE TWIST SPINDLES
FALSE TWISTERS
FALSE TWISTING (TEXTURING)
GEAR CRIMPED YARNS
HEAT SETTING (SYNTHETICS)
ROTOSET YARNS (TN)
STRETCH
STUFFER BOX CRIMPED YARNS
RT TEXTURE
TWIST
TWISTING

FALSE TWISTERS
BT THROWING MACHINES
TWISTERS
RT FALSE TWIST SPINDLES
FALSE TWIST TUBES
FALSE TWIST YARNS
THROWING
UPTWISTERS

FALSE TWISTING (TEXTURING)
BT TEXTURING
RT BULKED YARNS
BULKING
CRIMPED YARNS
CRIMPING
DYNALOFT (TN)
EDGE CRIMPED YARNS
EDGE CRIMPING
FALSE TWIST
FALSE TWIST SPINDLES
FALSE TWIST YARNS
FLUFLON (TN)
GEAR CRIMPING
HEAT SETTING (SYNTHETICS)
HELANCA (TN)
INTERLACING (FILAMENT IN YARN)
STRETCH YARNS (FILAMENT)
STRETCH YARNS (SPUN)
STUFFER BOX CRIMPING
SUPERLOFT (TN)
TEXTURED YARNS (FILAMENT)
THROWING

FALSE TWISTING (EXCEPT TEXTURING)
BT TWISTING
RT DOWNTWISTING
FALSE TWIST
FALSE TWIST SPINDLES
FALSE TWIST TUBES
REAL TWIST
SPINNING
TWIST
WOOLEN DRAFTING
WOOLEN SPINNING FRAMES
WOOLEN SYSTEM PROCESSING

FANCY (NOUN)
BT CARD ROLLS
RT BREAKERS
BRUSHING ACTION
CARD CLOTHING
CARD CYLINDERS
CARDING
CARDING ACTION
CARDS
CONDENSERS
DOFFERS (CARD)
DOFFING (OF A CARD)
FINISHER PART
FLY (WASTE)
SCRIBBLING
STRIPPERS
STRIPPING ACTION
WOOLEN CARDING
WOOLEN CARDS
WORKERS
WORSTED CARDING

FANCY KNITTING
USE JACQUARD KNITTING

FANCY LENO
BT FLAT WOVEN FABRICS
WOVEN FABRICS
RT LENO WEAVES

FANCY TWILLS
BT TWILLS
RT CHEVRON TWILLS
HERRINGBONE TWILLS
LOW TWILLS
REGULAR TWILLS
S TWILLS
STEEP TWILLS
Z TWILLS

FANCY YARNS
USE NOVELTY YARNS

FASHION (APPAREL)
RT APPAREL DESIGN
DRESS DESIGN
MANNEQUINS
STYLING (APPAREL)

FASHION MODELS
USE MANNEQUINS

FASHIONING
NT NARROWING (KNITTING)
WIDENING (KNITTING)
RT BOARDING
DOUBLE KNIT FABRICS
FASHIONING MARKS
RT FILLING KNITTING
FILLING KNITTING MACHINES
FLAT KNITTING
FULLY FASHIONED HOSIERY
FULLY FASHIONED KNITTING
FULLY FASHIONED KNITTING MACHINES
HAND KNITTING
HOSIERY
KNITTED FABRICS
KNITTING
LANDING LOOPS
LAYING IN (KNITTING)
LINKING (KNITTING)
MACHINE KNITTING
MOCK FASHIONING MARKS
NARROWING COMBS
SEAMLESS HOSIERY
SOCKS
STOCKINGS
STRETCH HOSIERY
TUCKING (KNITTING)
WELTING

FASHIONING MARKS
RT FASHIONING
FILLING KNITTING
FILLING KNITTING MACHINES
FULLY FASHIONED HOSIERY
FULLY FASHIONED KNITTING
HOSIERY
KNITTED FABRICS
KNITTING
LOOP TRANSFER
MOCK FASHIONING MARKS

FASTENERS
UF CLOSURES
NT BUCKLES
BUTTONS
HOOK CLOSURES
RIVETS
SNAP FASTENERS
VELCRO (TN)
ZIPPERS
BT GARMENT COMPONENTS
RT JOINTS
LINKAGES

FASTENING
USE JOINING

FASTNESS (COLOR)
USE COLORFASTNESS

FATIGUE
NT FLEXURAL FATIGUE
RT ABRASION
BENDING
BREAKING
CORROSION
CORROSION RESISTANCE
CRACK PROPAGATION
CREEP RUPTURE
CYCLIC STRESS
DAMAGE
FATIGUE RESISTANCE
FATIGUE RUPTURE
FATIGUE TESTING
FIBRILLATION
FRACTURING
MECHANICAL CONDITIONING
SHOCK RESISTANCE

FATIGUE RESISTANCE
BT END USE PROPERTIES
MECHANICAL DETERIORATION PROPERTIES
MECHANICAL PROPERTIES
RT ABRASION
ABRASION RESISTANCE
ABRASION RESISTANCE FINISHES
CORROSION
CORROSION RESISTANCE
CREEP
DIMENSIONAL STABILITY
DURABILITY
FATIGUE
FRICTION
SNAG RESISTANCE
TIRE CORDS
TOUGHNESS
WEAR RESISTANCE
WEAVING DYNAMICS

FATIGUE RUPTURE
RT CREEP RUPTURE
FATIGUE
FATIGUE TESTING
RUPTURE

FATIGUE TESTING
BT TESTING
RT ABRASION TESTING
CORROSION TESTING
CREEP TESTING
CUTTING TESTING
FATIGUE

RT FATIGUE RUPTURE
IMPACT PENETRATION TESTING
IMPACT TESTING
PILLING TESTING
SNAG TESTING
WEAR TESTING

FATIGUES
BT MILITARY CLOTHING
RT COVERALLS
WORK CLOTHING

FATS
UF LIPIDS
NT ANIMAL FATS
RESIDUAL FAT
VEGETABLE FATS
RT FATTY ACIDS
FATTY ALCOHOLS
GLYCERYL ESTER SURFACTANTS
IODINE NUMBER
LUBRICANTS
OILS
OLEIC ACID
SEBUM
WAXES

FATTY ACIDS
NT LAURIC ACID
OLEIC ACID
STEARIC ACID
BT ALIPHATIC COMPOUNDS
CARBOXYLIC ACIDS
ORGANIC ACIDS
RT ANIMAL FATS
ANIONIC SURFACTANTS
FATS
FATTY ALCOHOLS
FATTY KETONES
LUBRICANTS
OILS
OLEATE SALTS
RADICALS (CHEMICAL)
ROSIN
SOAPS

FATTY ALCOHOL SULFATES
USE ALKYL SULFATES

FATTY ALCOHOLS
NT CETYL ALCOHOL
LAURYL ALCOHOL
MYRISTYL ALCOHOL
OLEYL ALCOHOL
STEARYL ALCOHOL
BT ALCOHOLS
ALIPHATIC COMPOUNDS
RT FATS
FATTY ACIDS
FATTY KETONES
RADICALS (CHEMICAL)

FATTY DIGLYCERIDES
BT GLYCERYL ESTER SURFACTANTS

FATTY KETONES
BT KETONES
RT FATTY ACIDS
FATTY ALCOHOLS

FATTY MONOGLYCERIDES
BT GLYCERYL ESTER SURFACTANTS

FAULTS
USE DEFECTS

FEARNOUGHTS
RT DUSTING MACHINES
GARNETTING MACHINES
RECOVERED WOOL
WILLOWS

FEED BOXES
UF FEEDER BOXES
RT BALLING HEADS
FEEDERS
NOBLE COMBS

FEED DOGS
RT PRESSER FOOT
SEWING
SEWING MACHINES
THROAT PLATES

FEED LATTICES
RT CARDS
CONVEYORS
COTTON CARDS
FEED ROLLS
FEEDERS
HOPPER FEED
LICKER IN
OPENERS
PICKERS (OPENING)
WOOLEN CARDS
WORSTED CARDS

FEED PLATES
RT CARDING
COTTON CARDS
NIPPER PLATES

FEED RATE
BT RATE
RT CHARGING
CONVEYORS
DELIVERY
DELIVERY RATE
FEED ROLLS
FEEDERS
FLOW (MATERIALS)
HOPPER FEED
INPUT
LOADING (PROCESS)
PROCESS VARIABLES
PRODUCTION
THROUGHPUT

FEED ROLLS
BT CARD ROLLS
ROLLS
RT BREAKERS
CARDING
CARDS
CHARGING
CONVEYORS
COTTON CARDING
COTTON CARDS
DELIVERY RATE
DELIVERY ROLLS
FEED LATTICES
FEED RATE
FEEDERS
FINISHER PART
FLOW (MATERIALS)
HOPPER FEED
HYDROSTATIC LOADING
INPUT
LICKER IN
LOADING (PROCESS)
OPENING
PICKING (OPENING)
PRESSURE ROLLS
ROLL COVERINGS
SCOTCH FEED
THROUGHPUT
WOOLEN CARDING
WORSTED CARDING

FEEDBACK CONTROL
BT AUTOMATIC CONTROL
RT AUTOLEVELLING
CONTROL SYSTEMS
CONTROL THEORY
DIFFERENTIAL CONTROL
ERRORS
INTEGRAL CONTROL
ON OFF CONTROL
OPEN LOOP CONTROL
PICKER EVENERS

FEEDER BOXES
USE FEED BOXES

FEEDERS
RT CARRIER RODS
CHARGING
CONVEYORS
CREELS
DELIVERY RATE
FEED BOXES
FEED LATTICES
FEED RATE
FEED ROLLS
FLOW (MATERIALS)
HOPPER FEED
INPUT
LOADING (PROCESS)
RANDO FEEDER (TN)
THROUGHPUT
YARN CARRIERS

FEEL
USE HAND

FEELER MECHANISMS
RT BANGING OFF
DROP WIRES
LOOMS
STOP MOTIONS
WEFT FORKS

FELL (CLOTH)
USE CLOTH FELL

FELLED SEAMS
BT SEAMS
RT BUTTED SEAMS
BUTTERFLY SEAMS
COVERED SEAMS
FELLING
LAPPED SEAMS
PINKED SEAMS
SEAMING
SEWING

RT STITCHING
TAILORED SEAMS
TAILORING
TOP STITCHED SEAMS

FELLING
RT BUTTERFLY SEAMS
COVERED SEAMS
FELLED SEAMS
LAPPED SEAMS
SEAMING
SEAMS
SEWING
STITCHING

FELT BOOTS
BT BOOTS
FOOTWEAR

FELT PROOFING
USE ANTIFELTING TREATMENTS

FELTED FABRICS
USE FELTS

FELTING
BT WET FINISHING
RT ANTIFELTING AGENTS
ANTIFELTING TREATMENTS
BRUSHING
DIFFERENTIAL SHRINKAGE
FELTS
FELTING SHRINKAGE
FIBER MIGRATION
FULLING
HARDENING (FELTS)
HARDENERS (FELTS)
NAPPING
NONWOVEN FABRICS
PILLS
PLATEN HARDENERS
ROLLER HARDENERS
SCALE MASKING
SCALES (WOOL FIBERS)
SHRINK PROOFING
SHRINKING
WEAVING
WOVEN FELTS

FELTING MACHINES
RT COMPRESSIVE SHRINKAGE MACHINES
FULLING MACHINES
HARDENERS (FELTS)
HARDENING (FELTS)
NEEDLE LOOMS (NEEDLING)
PLATEN HARDENERS
ROLLER HARDENERS

FELTING RESISTANCE
BT FABRIC PROPERTIES
FINISH (PROPERTY)
RT ANTIFELTING TREATMENTS
DIMENSIONAL STABILITY
FELTING SHRINKAGE
SHRINK RESISTANCE

FELTING SHRINKAGE
BT END USE PROPERTIES
FABRIC SHRINKAGE
SHRINKAGE
RT ANTIFELTING TREATMENTS
COMPACTING
COMPRESSIVE SHRINKING
DIMENSIONAL STABILITY
FELTS
FILLING SHRINKAGE
FUSION CONTRACTION SHRINKAGE
MOLDING SHRINKAGE
PRESHRINKING
RELAXATION SHRINKAGE
RESIDUAL SHRINKAGE
SCOURING SHRINKAGE
SHRINK RESISTANCE
WARP SHRINKAGE

FELTS
UF FELTED FABRICS
NT NEEDLE PUNCHED FELTS
PAPERMAKERS FELTS
PRESSED FELTS
SYNTHETIC FELTS
WOVEN FELTS
BT FIBER ASSEMBLIES
RT ARACHNE FABRIC (TN)
BAIZE
BONDED YARN FABRICS
CARD WEBS
DIFFERENTIAL FRICTION
FELTING
FELTING SHRINKAGE
FULLING
HARDENERS (FELTS)
HARDENING (FELTS)
HATS
NEEDLED FABRICS
NONWOVEN FABRICS
PAPER
PLATEN HARDENERS
ROLLER HARDENERS
SCALES (WOOL FIBERS)
WEBS

FERRIC CHLORIDE
BT CHLORIDES
FERRIC COMPOUNDS
IRON COMPOUNDS
RT CATALYSTS
FERRIC TARTRATE

FERRIC COMPOUNDS
NT FERRIC CHLORIDE
FERRIC TARTRATE
BT IRON COMPOUNDS
RT CUPRIC COMPOUNDS
FERROUS COMPOUNDS

FERRIC TARTRATE
UF EWNN
BT FERRIC COMPOUNDS
IRON COMPOUNDS
RT CADOXEN
CELLULOSE
CUPRAMMONIUM HYDROXIDE
CUPRIETHYLENEDIAMINE HYDROXIDE
FERRIC CHLORIDE
MARSCHALL SOLUTION
SOLVENTS
VISCOSITY

FERROUS COMPOUNDS
BT IRON COMPOUNDS
RT CUPROUS COMPOUNDS
FERRIC COMPOUNDS
REDOX CATALYSIS

FETTLING
UF CARD STRIPPING
STRIPPING (CARDING)
RT CARD CLOTHING
CARD CYLINDERS
CARD FLATS
CARD ROLLS
CARD STRIPS
CARD WASTE
CARD WIRES
CARDING
CARDS
COTTON CARDING
DOFFING (OF A CARD)
FLAT STRIPPING COMBS
FLAT STRIPS
LICKER IN
STRIPPERS
STRIPPING ACTION
WOOLEN CARDING
WORKERS
WORSTED CARDING

FETTLING COMBS
USE FLAT STRIPPING COMBS

FETTLING WASTE
BT CARD WASTE
RT FLAT STRIPS
STRIPPING ACTION

FIBER (TRACER)
USE TRACER FIBERS

FIBER ANALYSIS
USE FIBER IDENTIFICATION

FIBER ARRANGEMENT
RT BUNCHING COEFFICIENT
FIBER ASSEMBLIES
FIBER CONFIGURATION
FIBER CRIMP
FIBER GEOMETRY
FIBER HOOKS
FIBER MIGRATION
FIBER ORIENTATION
FIBER PROPERTIES
LINDSLEY RATIOS
ORIENTATION UNIFORMITY
OVERLAP (FIBERS)
TUFTS (FIBER)
WEB ORIENTATION

FIBER ARRAY
UF STAPLE ARRAY
BT FIBER LENGTH DETERMINATION
FIBER PROPERTIES
RT STAPLING
COEFFICIENT OF LENGTH
VARIATION
COTTON QUALITY
DIGITAL FIBROGRAPH
EFFECTIVE LENGTH (FIBER)
FIBER CONFIGURATION
FIBER DIAGRAM
FIBER ENDS
FIBER LENGTH
FIBER LENGTH DISTRIBUTION
FIBERS
FIBROGRAPH
FIBROGRAPH MEAN LENGTH
MEAN FIBER LENGTH
OVERLAP (FIBERS)
SERVO FIBROGRAPH
SHORT FIBERS INDEX
RT SORTING
SPAN LENGTH
SUTER WEBB TESTING
UPPER QUARTILE LENGTH

FIBER ASSEMBLIES
NT BRAIDS
BUNDLES
CABLES
CARD SLIVERS
COMBED SLIVERS
CORDS
FABRICS
FABRICS (ACCORDING TO
STRUCTURE)
FIBERS
FELTS
FLOCK
LAPS
NONWOVEN FABRICS
PAPER
PLIED YARNS
ROPES
ROVINGS
SLIVERS
SLUBBINGS
STOCK (FIBER)
STRANDS
TOPS
TOW
TUFTS (FIBER)
TWINE
WEBS
YARNS
RT ASSEMBLIES
BLENDED YARNS
FABRICS (BY END USES)
FIBER ARRANGEMENT
FIBERS
SKEINS
STRUCTURAL PROPERTIES
STRUCTURES

FIBER BLENDS
USE BLENDS (FIBERS)

FIBER BREAKAGE
RT BREAKING
BREAKING ELONGATION
BREAKING LENGTH
BREAKING STRENGTH
BROKEN FIBERS
CARDING
CARDING ACTION
COMBING
CUTTING
DRAFT CUT
DRAFTING (STAPLE FIBER)
FIBER DIAGRAM
FIBER LENGTH
FIBER LENGTH DISTRIBUTION
FIBER SLIPPAGE
FIBER STRENGTH
FIBER TENDERNESS
FIBERS
GINNING
MEAN FIBER LENGTH
NEPS
PICKING (OPENING)
SLIPPAGE
SNAP BACK
STRESS CONCENTRATION
STRESS STRAIN CURVES
STRETCH BREAKING
TOW CONVERSION
WEAK SPOTS
YARN STRENGTH

FIBER BUNDLE STRENGTH
NT PRESSLEY STRENGTH
BT FIBER PROPERTIES
RT BREAKING ELONGATION
BREAKING STRENGTH
CLEMSON STRENGTH TESTING
FIBER STRENGTH
FIBER TENDERNESS

FIBER BUNDLE TESTING
NT PRESSLEY TEST
BT TESTING
RT CLEMSON STRENGTH TESTING
FIBER LENGTH DETERMINATION
PRESSLEY STRENGTH

FIBER COARSENESS
USE FIBER FINENESS

FIBER CONFIGURATION
RT FIBER ARRANGEMENT
FIBER ARRAY
FIBER CRIMP
FIBER GEOMETRY
FIBER HOOKS
FIBER ORIENTATION
FIBER PROPERTIES
TUFTS (FIBER)

FIBER CONTENT
BT YARN PROPERTIES
RT BLENDED FIBER RATIO
BLENDED YARNS
BLENDED YARNS (FILAMENT)
BLENDED YARNS (STAPLE)
BLENDING

FIBER CONTROL
RT APRON CONTROL
CONTROL SURFACES
DRAFTING (STAPLE FIBER)
DRAFTING DYNAMICS
DRAFTING FORCE
DRAFTING THEORY
DRAFTING WAVES
FIBER SLIPPAGE
FLOATING FIBERS
HOOK REMOVAL
IRREGULARITY CONTROL
SINGLE APRONS
VELOCITY CHANGE POINT

FIBER COUNT
USE FIBER DENIER

FIBER CRIMP
NT NATURAL CRIMP
BT CRIMP
FIBER GEOMETRY
FIBER PROPERTIES
RT CRIMP AMPLITUDE (FIBER)
CRIMP FREQUENCY
CRIMP INDEX
CRIMP RADIUS
FABRIC CRIMP
FIBER ARRANGEMENT
FIBER CONFIGURATION
FIBER ENDS
FOLLOW THE LEADER CRIMP
HELICAL CRIMP
LATENT CRIMP
YARN CRIMP
Z CRIMP

FIBER CROSS SECTIONS
NT DOGBONE CROSS SECTION
FLAT CROSS SECTION
IRREGULAR CROSS SECTION
ROUND CROSS SECTION
SERRATED CROSS SECTION
TRILOBAL CROSS SECTION
BT FIBER GEOMETRY
FIBER PROPERTIES
RT CORE (FIBER)
CROSS SECTIONS
FIBER DIAMETER
FIBER GEOMETRY
FIBER SURFACE
FILAMENTS
FINENESS
HOLLOW FILAMENT YARNS
IMMATURE FIBERS
IMMATURITY (FIBER)
IMMATURITY TESTING (COTTON
FIBER)
SHAPE
SHAPE FACTOR
SKIN (FIBER)
SPARKLE
SPINNERETS
YARN CROSS SECTIONS

FIBER CUTTING
RT ABRASION
CUTTING
CUTTING RATIO
CUTTING RESISTANCE
FIBER BREAKAGE
FIBER SHUFFLING
FIBRILLATION
STAPLE CUTTERS
STRETCH BREAKING
TOW
TOW CONVERSION
TOW TO TOP CONVERSION
TOW TO YARN CONVERSION

FIBER DENIER
UF FIBER COUNT
FILAMENT COUNT
FILAMENT DENIER
BT FIBER PROPERTIES
RT COUNT
DENIER
FIBER DIAMETER
FIBER FINENESS
GREX
SPINNING LIMIT
TEX

FIBER DIAGRAM
UF STAPLE DIAGRAM
BT FIBER LENGTH DETERMINATION
FIBER PROPERTIES
RT COEFFICIENT OF LENGTH
VARIATION
COTTON CLASSING

RT COTTON QUALITY
DIGITAL FIBROGRAPH
DRAFT CUT
EFFECTIVE LENGTH (FIBER)
FIBER ARRAY
FIBER BREAKAGE
FIBER LENGTH
FIBER LENGTH DISTRIBUTION
FIBERS
FIBROGRAPH
FIBROGRAPH MEAN LENGTH
MEAN FIBER LENGTH
SERVO FIBROGRAPH
SHORT FIBERS INDEX
SORTING
SPAN LENGTH
STAPLE FIBERS
STAPLING
SUTER WEBB TESTING
UPPER HALF MEAN LENGTH
UPPER QUARTILE LENGTH

FIBER DIAMETER
BT FIBER PROPERTIES
RT DIAMETER
FIBER CROSS SECTIONS
FIBER DENIER
FIBER FINENESS
FIBER LENGTH
GAMMA DISTRIBUTION
MICRONAIRE

FIBER ENDS
RT FIBER ARRAY
FIBER CRIMP
FIBER DIAMETER
FIBER EXTENT
FIBER GEOMETRY
FIBER LENGTH
FIBER LENGTH DISTRIBUTION
FIBER SURFACE

FIBER ENTANGLEMENTS
USE ENTANGLEMENTS

FIBER EXTENT
BT FIBER PROPERTIES
RT FIBER ENDS
FIBER ORIENTATION
FIBERS
HOOKED FIBERS
ORIENTATION
PARALLELIZATION

FIBER FIBER FRICTION
BT FRICTION
RT BENDING RIGIDITY
COEFFICIENT OF FRICTION
DIFFERENTIAL FRICTION
DRAFTING FORCE TESTERS
DYNAMIC FRICTION
FABRIC TO FABRIC FRICTION
FIBER FRICTION
FIBER SLIPPAGE
FIBER WITHDRAWAL FORCE
FORCE
FRICTIONAL FORCE
INTERNAL FRICTION
MATRIX FREEDOM
NORMAL FORCE
PILL RESISTANCE
PILLING
RUBBING
SHEDDING (FIBER)
SLICKNESS
SPINNING LIMIT
STATIC FRICTION
STICK SLIP FRICTION
SYTON (TN)
TEFLON FIBER (TN)
VELOCITY (RELATIVE)
WRINKLE RECOVERY
YARN STRENGTH
YARN TO METAL FRICTION
YARN TO YARN FRICTION

FIBER FINENESS
UF FIBER COARSENESS
BT FIBER PROPERTIES
RT CAUSTICAIRE SCALE
CAUSTICAIRE VALUE
CLASS (FIBERS)
COEFFICIENT OF FINENESS VARIATION
COTTON
COTTON GRADE
COTTON QUALITY
COUNT
CURVILINEAR SCALE
FIBER DENIER
FIBER DIAMETER
FIBERS
FINENESS
FINENESS TESTING
GRADE (FIBERS)
GRADING
IMMATURE FIBERS
LINEAR DENSITY

RT MATURITY INDEX
MICRONAIRE
NEP POTENTIAL
NEPS
SPINNING LIMIT
VIBRASCOPE
WEDGE SCALE
WOOL GRADE
WOOL SORTING

FIBER FLOCK
USE FLOCK

FIBER FRICTION
BT FIBER PROPERTIES
FRICTION
RT COEFFICIENT OF FRICTION
FABRIC TO FABRIC FRICTION
FIBER FIBER FRICTION
FIBER TO METAL FRICTION
LUBRICATION
SLICKNESS

FIBER G (TN)
USE CORDURA (TN)

FIBER GEOMETRY
UF FIBER PROPERTIES (GEOMETRIC)
NT CONVOLUTIONS
FIBER CRIMP
FIBER CROSS SECTIONS
FIBER HOOKS
FIBER SURFACE
BT FIBER PROPERTIES
GEOMETRY
RT FIBER ARRANGEMENT
FIBER CONFIGURATION
FIBER DIAMETER
IMMATURITY (FIBER)
SCALES (WOOL FIBERS)
SPINNING LIMIT

FIBER HOOKS
UF HOOKS (FIBER)
NT LEADING HOOKS
TRAILING HOOKS
BT FIBER GEOMETRY
RT CARD WEBS
CARDING
COMBING
DRAFTING (STAPLE FIBER)
DRAFTING THEORY
ENTANGLEMENTS
FIBER ARRANGEMENT
FIBER CONFIGURATION
FIBER LENGTH
FIBER ORIENTATION
FIBERS
HIGH DRAFT SPINNING
HOOK DIRECTION
HOOK FORMATION
HOOK REMOVAL
HOOKED FIBERS
LINDSLEY RATIOS
NEPS
PARALLELIZATION
PROCESSING DIRECTION
SLIVER REVERSALS
SLIVERS
SPINNING

FIBER IDENTIFICATION
UF FIBER ANALYSIS
IDENTIFICATION (FIBERS)
NT CHEMICAL ANALYSIS (FIBER IDENTIFICATION)
MICROSCOPIC ANALYSIS (FIBER IDENTIFICATION)
REAGENT TESTING (FIBER IDENTIFICATION)
SPECIFIC GRAVITY TESTING (FIBER IDENTIFICATION)
STAINING TESTING (FIBER IDENTIFICATION)
SWELLING TESTING (FIBER IDENTIFICATION)
THERMAL TESTING (FIBER IDENTIFICATION)
BT TESTING
RT ANALYZING
LEAD COMPOUNDS
TINTING

FIBER IMMATURITY
USE IMMATURITY (FIBER)

FIBER LENGTH
UF STAPLE LENGTH
BT FIBER PROPERTIES
RT CLASS (FIBERS)
COEFFICIENT OF LENGTH VARIATION
COTTON CLASS
CUTTING RATIO
DIGITAL FIBROGRAPH
DRAFTING WAVES
FIBER ARRAY
FIBER BREAKAGE

RT FIBER DIAGRAM
FIBER DIAMETER
FIBER ENDS
FIBER LENGTH DETERMINATION
FIBER LENGTH DISTRIBUTION
FIBERS
FIBROGRAPH
GRADING
LENGTH
MEAN FIBER LENGTH
PROCESS VARIABLES
RATCH
SERVO FIBROGRAPH
SPINNING LIMIT
STAPLE FIBERS
STAPLING
UPPER HALF MEAN LENGTH
UPPER QUARTILE LENGTH
WOOL CLASS
WOOL SORTING

FIBER LENGTH DETERMINATION
NT FIBER ARRAY
FIBER DIAGRAM
FIBER LENGTH DISTRIBUTION
FIBROGRAPH
STAPLING
SUTER WEBB TESTING
BT TESTING
RT COTTON CLASSING
COTTON QUALITY
EFFECTIVE LENGTH (FIBER)
FIBER BUNDLE TESTING
FIBER LENGTH
FINENESS TESTING
GRADING
IMMATURITY TESTING (COTTON FIBER)
MEAN FIBER LENGTH
SHORT FIBERS INDEX
SORTING
SPAN LENGTH
TWO POINT FIVE PERCENT SPAN LENGTH
UPPER HALF MEAN LENGTH
UPPER QUARTILE LENGTH

FIBER LENGTH DISTRIBUTION
UF FIBER LENGTH DISTRIBUTION TESTING
FREQUENCY DISTRIBUTION (FIBER LENGTH)
LENGTH UNIFORMITY
LENGTH VARIATION (FIBER)
BT FIBER LENGTH DETERMINATION
FIBER PROPERTIES
RT COEFFICIENT OF LENGTH VARIATION
COMBING
COTTON CLASSING
COTTON QUALITY
DIGITAL FIBROGRAPH
DRAFTING (STAPLE FIBER)
EFFECTIVE LENGTH (FIBER)
FIBER BREAKAGE
FIBER DIAGRAM
FIBER ENDS
FIBER LENGTH
FIBERS
FIBROGRAPH
FIBROGRAPH MEAN LENGTH
FLOATING FIBERS
MEAN FIBER LENGTH
SERVO FIBROGRAPH
SHORT FIBERS INDEX
SPAN LENGTH
SPINNING LIMIT
STAPLE FIBERS
STAPLING
TESTING
TWO POINT FIVE PERCENT SPAN LENGTH
UNIFORMITY RATIO
UPPER HALF MEAN LENGTH
UPPER QUARTILE LENGTH

FIBER LENGTH DISTRIBUTION TESTING
USE FIBER LENGTH DISTRIBUTION

FIBER MATURITY
USE IMMATURITY (FIBER)

FIBER MIGRATION
UF MIGRATION (FIBER)
RT CORE SPUN YARNS
FELTING
FIBER ARRANGEMENT
FILAMENT YARNS
FUZZ RESISTANCE
HAIRINESS
MIGRATION (SUBSTANCE)
MIGRATION PERIOD
NOVELTY YARNS
PILL RESISTANCE
PILL WEAR OFF
PILLING
PILLS
RADIAL DISTRIBUTION
(CONT.)

FIBER MIGRATION
RT SHEDDING (FIBER)
SPINNING LIMIT
SPUN YARNS
TRACER FIBERS
TWIST
TWIST TRIANGLE
TWISTING
YARN GEOMETRY
YARN STRUCTURE

FIBER OPTICS
BT OPTICS
RT BIREFRINGENCE
INTERNAL REFLECTION
MONOFILAMENT YARNS
OPTICAL PROPERTIES

FIBER ORIENTATION
RT BUNCHING COEFFICIENT
CARD SLIVERS
CARD WEBS
COMBED SLIVERS
COMBING
DEGREE OF ORIENTATION
DRAFT
DRAFTING (STAPLE FIBER)
DRAFTING THEORY
DRAFTING WAVES
FIBER ARRANGEMENT
FIBER CONFIGURATION
FIBER EXTENT
FIBER HOOKS
FIBERS
FLOATING FIBERS
GILLING
HOOK REMOVAL
LEADING HOOKS
NONWOVEN FABRIC STRUCTURE
NONWOVEN FABRICS
ORIENTATION
ORIENTATION UNIFORMITY
PARALLELIZATION
SLIVER REVERSALS
TOPS
TRAILING HOOKS
WEBS

FIBER PROPERTIES
NT COEFFICIENT OF FINENESS VARIATION
COEFFICIENT OF LENGTH VARIATION
CRIMP AMPLITUDE (FIBER)
CRIMP FREQUENCY
CRIMP INDEX
FIBER ARRAY
FIBER BUNDLE STRENGTH
FIBER CRIMP
FIBER CROSS SECTIONS
FIBER DENIER
FIBER DIAGRAM
FIBER DIAMETER
FIBER EXTENT
FIBER FINENESS
FIBER FRICTION
FIBER GEOMETRY
FIBER LENGTH
FIBER LENGTH DISTRIBUTION
FIBER STRENGTH
FIBER TENDERNESS
FIBER TWIST
FINE STRUCTURE (FIBERS)
GRADE (FIBERS)
IMMATURITY (FIBER)
MEAN FIBER LENGTH
MICRONAIRE FINENESS
NATURAL DRAW RATIO
PRESSLEY STRENGTH
RT ACOUSTIC PROPERTIES
AESTHETIC PROPERTIES
BIOCHEMICAL PROPERTIES
BLENDED FIBER RATIO
CHEMICAL PROPERTIES
DEGRADATION PROPERTIES
ELECTRICAL PROPERTIES
FABRIC PROPERTIES
FIBER ARRANGEMENT
FIBER CONFIGURATION
MECHANICAL PROPERTIES
MECHANICAL DETERIORATION PROPERTIES
OPTICAL PROPERTIES
PHYSICAL CHEMICAL PROPERTIES
PHYSICAL PROPERTIES (EXCLUDING MECHANICAL)
PROPERTIES
SPINNING LIMIT
STRESS OPTICAL PROPERTIES
STRESS STRAIN PROPERTIES
STRESS STRAIN PROPERTIES (COMPRESSIVE)
STRESS STRAIN PROPERTIES (SHEAR)
STRESS STRAIN PROPERTIES (TENSILE)
STRUCTURAL PROPERTIES
SURFACE PROPERTIES (MECHANICAL)
RT SURFACE PROPERTIES (PHYSICAL CHEMICAL)
THERMAL PROPERTIES
TRANSFER PROPERTIES
YARN PROPERTIES

FIBER PROPERTIES (GEOMETRIC)
USE FIBER GEOMETRY

FIBER QUALITY
USE GRADE (FIBERS)

FIBER REACTIVE COMPOUNDS
USE REACTANTS

FIBER REACTIVITY
RT CHEMICAL MODIFICATION (FIBERS)
REACTANTS
REACTIVITY

FIBER REINFORCED COMPOSITES
UF FIBER REINFORCED PLASTICS
BT COMPOSITES
REINFORCED COMPOSITES
RT COATED FABRICS
FILAMENT WOUND COMPOSITES
GLASS MAT
LAMINATES
PLASTICS
YARN REINFORCED ELASTOMERS

FIBER REINFORCED PLASTICS
USE FIBER REINFORCED COMPOSITES

FIBER SHUFFLING
RT DRAFT CUT
DRAFTING (STAPLE FIBER)
FIBER CUTTING
FIBER SLIPPAGE
TOW CONVERSION
TOW TO TOP CONVERSION
TOW TO YARN CONVERSION

FIBER SLIPPAGE
BT SLIPPAGE
RT BREAKING STRENGTH
COEFFICIENT OF FRICTION
CONTROL SURFACES
DRAFTING (STAPLE FIBER)
DRAFTING THEORY
DRAFTING WAVES
FIBER BREAKAGE
FIBER CONTROL
FIBER FIBER FRICTION
FIBER SHUFFLING
FIBER WITHDRAWAL FORCE
FIBERS
FRICTION
NORMAL FORCE
ROLLER SLIPPAGE
SLIVERS
SPUN YARNS
STICK SLIP FRICTION
STRENGTH EFFICIENCY
TWIST CONTROL
VELOCITY (RELATIVE)
YARN STRENGTH

FIBER SOUNDNESS
USE FIBER TENDERNESS

FIBER SPINNING
USE SPINNING (EXTRUSION)

FIBER STRENGTH
UF STAPLE STRENGTH
BT BREAKING ELONGATION
FIBER PROPERTIES
STRENGTH
RT BREAKING LENGTH
BREAKING STRENGTH
BUNDLE STRENGTH
CLEMSON STRENGTH TESTING
COMPRESSIVE STRENGTH
DRAWING (FILAMENT)
FABRIC STRENGTH
FIBER BREAKAGE
FIBER TENDERNESS
LOOP EFFICIENCY
LOOP STRENGTH
PRESSLEY TEST
SNAP BACK
STELOMETER STRENGTH TESTING
STRENGTH ELONGATION TESTING
STRESS
STRESS STRAIN CURVES
STRETCH BREAKING
TENACITY
WEAK SPOTS
YARN STRENGTH
YARN STRENGTH EFFICIENCY

FIBER STRUCTURE
USE FINE STRUCTURE (FIBERS)

FIBER SURFACE
BT FIBER GEOMETRY
RT CAUSTICAIRE SCALE
CAUSTICAIRE VALUE
FIBER FINENESS
FIBER LENGTH
MICRONAIRE
SKIN (FIBER)
SURFACE FRICTION
SURFACE PROPERTIES (MECHANICAL)
SURFACE PROPERTIES (PHYSICAL CHEMICAL)
SURFACES

FIBER TENDERNESS
UF FIBER SOUNDNESS
FIBER WEAKNESS
BT FIBER PROPERTIES
RT CLASSING
FIBER BREAKAGE
FIBER STRENGTH
FIBERS
WEAK SPOTS

FIBER TO METAL FRICTION
BT FRICTION
RT COEFFICIENT OF FRICTION
FIBER FIBER FRICTION
FIBER FRICTION

FIBER TWIST
BT FIBER PROPERTIES
TWIST
RT CONVOLUTIONS

FIBER V (TN)
USE DACRON (TN)

FIBER WEAKNESS
USE FIBER TENDERNESS

FIBER WITHDRAWAL FORCE
RT ACTUAL DRAFT
BENDING RIGIDITY
CARDING ACTION
CARDING DYNAMICS
DRAFT CONSTANT
DRAFTING (STAPLE FIBER)
DRAFTING DYNAMICS
DRAFTING FORCE
DRAFTING THEORY
ENTANGLEMENTS
FIBER FIBER FRICTION
FIBER SLIPPAGE
FLOATING FIBERS

FIBER WOVEN (TN)
BT NEEDLED FABRICS
NONWOVEN FABRICS
RT ARACHNE FABRIC (TN)
MALIWATT FABRIC (TN)
NEEDLE LOOMS (NEEDLING)
NEEDLE PUNCHED FELTS
RANDO WEBBER (TN)
SYNTHETIC FELTS

FIBERGLAS (TN)
BT GLASS FIBERS
HIGH TEMPERATURE FIBERS
INORGANIC FIBERS (MAN MADE)

FIBERS
NT ANIMAL FIBERS (GENERAL)
ARTIFICIAL SILK (ARCHAIC)
AZLON (REGENERATED PROTEIN FIBERS)
BAST FIBERS
CELLULOSE ESTER FIBERS
CELLULOSIC FIBERS
COTTON
FLUOROCARBON FIBERS
FRUIT FIBERS
HAIR
HIGH TEMPERATURE FIBERS
INORGANIC FIBERS (MAN MADE)
KERATIN FIBERS
LEAF FIBERS
MAN MADE FIBERS
MINERAL FIBERS
NATURAL FIBERS
NYLON (POLYAMIDE FIBERS)
OLEFIN FIBERS
POLYAMIDOESTER FIBERS
POLYESTER FIBERS
POLYURETHANE FIBERS
PROTEIN FIBERS
RAYON (REGENERATED CELLULOSE FIBERS)
REGENERATED FIBERS (EXCLUDING CELLULOSIC AND PROTEIN)
SEED HAIR FIBERS (GENERAL)
SILK
SPANDEX FIBERS
SYNTHETIC FIBERS
SYNTHETIC RUBBER FIBERS
VEGETABLE FIBERS (GENERAL)
VINYL FIBERS (GENERAL)
WOOD FIBERS
WOOL
RT CLASS (FIBERS)

RT CONVOLUTIONS
COUNT
DENIER
FABRICS
FIBER ASSEMBLIES
FIBER BREAKAGE
FIBER HOOKS
FIBER LENGTH DETERMINATION
FIBER ORIENTATION
FIBER PROPERTIES
FIBER SLIPPAGE
FIBRILS
FIBROGRAPH
FILAMENTS
FILMS
FINE STRUCTURE (FIBERS)
FOAMS
GRADE (FIBERS)
MONOFILAMENTS
POLYMER TYPES (GENERAL)
POLYMERS
RHEOLOGY
SAMPLES
STAPLE FIBERS
SYNTHETIC LEATHER
TEXTILE MATERIALS
TINSEL
WEBS
YARNS

FIBERS (BICOMPONENT)
USE BICOMPONENT FIBERS

FIBRIL REVERSALS
BT FINE STRUCTURE (FIBERS)
RT CONVOLUTIONS
COTTON
CRIMP
CRYSTALLITES
FIBRILS
HELIX REVERSALS
LUMEN
MICELLES
PRIMARY WALLS
SECONDARY WALLS
SPHERULITES

FIBRILLAR STRUCTURE
BT FINE STRUCTURE (FIBERS)
RT FIBRILLATION
FIBRILS
SPHERULITIC FIBRILS
STRUCTURE OF MATERIALS

FIBRILLATION
RT CRYSTALLINE POLYMERS
CRYSTALLINE REGION
FATIGUE
FIBER CUTTING
FIBRILLAR STRUCTURE
FIBRILS
FINE STRUCTURE (FIBERS)
FROSTING
LONG CHAIN MOLECULES
MECHANICAL DETERIORATION
PROPERTIES
SPHERULITIC FIBRILS

FIBRILS
NT MACROFIBRILS
MICROFIBRILS
SPHERULITIC FIBRILS
BT FINE STRUCTURE (FIBERS)
RT CELLULOSE
CONVOLUTIONS
COTTON
CRYSTALLINE POLYMERS
CRYSTALLINE REGION
CRYSTALLITES
ELECTRON MICROSCOPY
ENDOCUTICLE
FIBERS
FIBRIL REVERSALS
FIBRILLATION
FRINGED FIBRILS
FRINGED MICELLES
LUMEN
MATRIX (FINE STRUCTURE)
MEDULLA
MICELLES
ORTHOCORTEX
PRIMARY WALLS
SECONDARY WALLS
SPHERULITIC CRYSTALLIZATION
SYNTHETIC FIBERS
VEGETABLE FIBERS (GENERAL)
WOOD FIBRILS
WOOL

FIBRO (TN)
BT CELLULOSIC FIBERS
RAYON (REGENERATED CELLULOSE FIBERS)
VISCOSE RAYON
RT STAPLE FIBERS

FIBROCETA (TN)
BT ACETATE FIBERS
CELLULOSE ESTER FIBERS

FIBROGRAPH
NT DIGITAL FIBROGRAPH
SERVO FIBROGRAPH

BT FIBER LENGTH DETERMINATION
RT COEFFICIENT OF LENGTH VARIATION
COTTON
COTTON CLASS
COTTON CLASSING
COTTON QUALITY
EFFECTIVE LENGTH (FIBER)
FIBER ARRAY
FIBER DIAGRAM
FIBER LENGTH
FIBERS
FIBROGRAPH MEAN LENGTH
GRADING
MEAN FIBER LENGTH
SHORT FIBERS INDEX
SORTING
SPAN LENGTH
STAPLE FIBERS
STAPLING
SUTER WEBB TESTING
TWO POINT FIVE PERCENT SPAN LENGTH
UNIFORMITY RATIO
UPPER HALF MEAN LENGTH
UPPER QUARTILE LENGTH

FIBROGRAPH MEAN LENGTH
RT COEFFICIENT OF LENGTH VARIATION
DIGITAL FIBROGRAPH
FIBER ARRAY
FIBER DIAGRAM
FIBER LENGTH DISTRIBUTION
FIBROGRAPH
MEAN FIBER LENGTH
SERVO FIBROGRAPH
STAPLING
UPPER HALF MEAN LENGTH
UPPER QUARTILE LENGTH

FIBROIN (SILK)
RT DEGUMMING
SERICIN
SILK

FIBROLANE (TN)
BT AZLON (REGENERATED PROTEIN FIBERS)
PROTEIN FIBERS

FIBROUS MATERIALS
USE TEXTILE MATERIALS

FICKIAN DIFFUSION
BT DIFFUSION

FIELD EMISSION MICROSCOPY
BT MICROSCOPY
RT ELECTRON MICROSCOPY
INFRARED MICROSCOPY
INTERFERENCE MICROSCOPY
LIGHT MICROSCOPY (OPTICAL)
ULTRAVIOLET MICROSCOPY
X RAY MICROSCOPY

FIELDEN WALKER IRREGULARITY TESTER (TN)
BT CAPACITANCE TYPE IRREGULARITY TESTERS
IRREGULARITY TESTERS
RT CAPACITANCE (ELECTRICAL)
IRREGULARITY
IRREGULARITY TESTING
PNEUMATIC IRREGULARITY TESTERS
ROVINGS
SACO LOWELL IRREGULARITY TESTER (TN)
SLIVERS
USTER IRREGULARITY TESTER (TN)
YARNS
ZELLWEGER TESTER (TN)

FIGURED FABRICS
NT BROCADES
BROCATELLES
DAMASK
FIGURED REPP
GRECIAN ALHAMERA
RT DOBBY WEAVING

FIGURED REPP
BT FIGURED FABRICS

FILAMENT COUNT
USE FIBER DENIER

FILAMENT DENIER
USE FIBER DENIER

FILAMENT WOUND COMPOSITES
BT COMPOSITES
REINFORCED COMPOSITES
YARN REINFORCED COMPOSITES

BT FIBER REINFORCED COMPOSITES
RT HONEYCOMB LAMINATES
YARN REINFORCED COMPOSITES
YARN REINFORCED ELASTOMERS

FILAMENT YARNS
UF YARNS (CONTINUOUS FILAMENT)
NT AIR JET TEXTURED YARNS
BICOMPONENT FIBER YARNS (FILAMENT)
BULKED YARNS
CRIMPED YARNS
CYCLOSET YARNS (TN)
HOLLOW FILAMENT YARNS
MONOFILAMENT YARNS
MULTIFILAMENT YARNS
ROTOSET YARNS (TN)
STRETCH YARNS (FILAMENT)
STUFFER BOX CRIMPED YARNS
TEXTURED YARNS (FILAMENT)
BT YARNS
RT COAGULATING BATHS
DRAWTWISTING
DRAWING (FILAMENT)
EXTRUSION
FIBER MIGRATION
FILAMENTS
MAN MADE FIBERS
MELT SPINNING
MONOFILAMENTS
NEUTRALIZING (FIBER EXTRUSION)
PRODUCERS TWIST
SILK
SINGLES YARNS
SOLVENT SPINNING
SPIN FINISHES
SPINNERETS
SPINNING (EXTRUSION)
SPUN YARNS
SYNTHETIC FIBERS
TEXTURING
THROWING
TOW
UNCONVENTIONAL YARNS
WET SPINNING
YARN GEOMETRY
YARN STRUCTURE
Z TWIST YARNS

FILAMENTS
NT MONOFILAMENTS
RT DENIER
DRAWTWISTING
DRAWING (FILAMENT)
EXTRUSION
FIBER CROSS SECTIONS
FIBER DIAMETER
FIBER SURFACE
FIBERS
FILAMENT YARNS
GLASS FIBERS
MAN MADE FIBERS
METALLIC FIBERS
SILK
SPINNERETS
SPINNING (EXTRUSION)
SYNTHETIC FIBERS
THICK SPOTS
TOW
YARNS

FILET (LACE)
BT HAND MADE LACES

FILLER
RT BATTING
COMPOSITES
PACKING
PADDING (MATERIAL)
WADDINGS
WEBS

FILLET CLOTHING
UF WIRE FILLET CLOTHING
BT CARD CLOTHING
RT CARD CLOTHING FOUNDATION
CARD CROWN
CARD GAUGE
CARD GRINDING
CARD NAILING
CARD WIRES
COUNT CROWN AND GAUGE
KNEE
METALLIC CARD CLOTHING
NOGG
RIB SET
SHEET CLOTHING
TWILL SET

FILLING
USE FILLING YARNS

FILLING BANDS
BT FABRIC DEFECTS
RT FILLING BARS
FILLING STREAKS
PIRN BARRE

FILLING BARS
BT FABRIC DEFECTS
RT FILLING BANDS
FILLING STREAKS
LOOMS
PIRN BARRE

FILLING EFFECT (TWILL)
RT COVER
FABRIC STRUCTURE
FILLING FACE
TWILLS
WARP EFFECT (TWILL)
WEAVE (TYPES)

FILLING FACE
RT FABRIC STRUCTURE
FACE (FABRIC)
FILLING EFFECT (TWILL)
WARP FACE

FILLING FLOATS
RT FABRIC STRUCTURE
FILLING EFFECT (TWILL)
FILLING FACE
LOOMS

FILLING KNITTED FABRICS
UF WEFT KNITTED FABRICS
NT ACCORDION FABRICS
ASTRAKHAN
CIRCULAR KNITTED FABRICS
DOUBLE KNIT FABRICS
FLAT KNITTED FABRICS
FULL CARDIGAN FABRICS
FULLY FASHIONED FABRICS
HALF CARDIGAN FABRICS
INTERLOCK (KNITTED FABRICS)
JACQUARD KNITTED FABRICS
KNITTED PILE FABRICS
PLAIN KNITTED FABRICS
RIB KNITTED FABRICS
BT KNITTED FABRICS
RT FILLING KNITTING
HOSIERY
JACQUARD KNITTING
KNITTED OUTERWEAR
KNITTED UNDERWEAR
KNITTING
LAYING IN (KNITTING)
MACHINE KNITTING
SWEATERS
TRICOT KNITTED FABRICS

FILLING KNITTING
NT CIRCULAR KNITTING
FULLY FASHIONED KNITTING
JACQUARD KNITTING
BT KNITTING
RT CIRCULAR KNITTING MACHINES
DIVIDERS (KNITTING)
DIVIDING (KNITTING)
FASHIONING
FASHIONING MARKS
FILLING KNITTED FABRICS
FILLING YARNS
FULLY FASHIONED KNITTING MACHINES
HOSIERY
KNITTED FABRICS
LANDING LOOPS
LAYING IN (KNITTING)
MALI PROCESSES (TN)
MOCK FASHIONING MARKS
NEEDLE BARS
NEEDLES
PATTERN WHEELS
PRESSER BARS
PRESSING (BEARDED NEEDLES)
RACKING
SINKERS
SINKING (KNITTING)
TUCKING (KNITTING)
WEAVING
WELTING

FILLING KNITTING MACHINES
UF WEFT KNITTING MACHINERY
NT CIRCULAR KNITTING MACHINES
FULLY FASHIONED KNITTING MACHINES
BT KNITTING MACHINES
RT CASTING OFF
CIRCULAR KNITTING
DIVIDERS (KNITTING)
DIVIDING (KNITTING)
FASHIONING
FASHIONING MARKS
FILLING KNITTING
FULLY FASHIONED KNITTING
HOSIERY
JACQUARD KNITTING
KNITTED FABRICS
KNITTING
LANDING LOOPS
LAYING IN (KNITTING)
MOCK FASHIONING MARKS
NEEDLE BARS
RT NEEDLES
PATTERN WHEELS
PRESSER BARS
PRESSING (BEARDED NEEDLES)
RACKING
SINKERS
SINKING (KNITTING)
TUCKING (KNITTING)
WARP KNITTING MACHINES
WELTING

FILLING SHRINKAGE
BT SHRINKAGE
RT COMPACTING
COMPRESSIVE SHRINKING
CONTRACTION
DIMENSIONAL STABILITY
FELTING SHRINKAGE
FULLING
FUSION CONTRACTION SHRINKAGE
MOLDING SHRINKAGE
PRESHRINKING
RELAXATION SHRINKAGE
RESIDUAL SHRINKAGE
SHRINK RESISTANCE
SHRINKING
TENTERING
WARP SHRINKAGE

FILLING STRAIGHTENING MACHINES
UF WEFT STRAIGHTENERS
RT BOW STRAIGHTENERS
COMPRESSIVE SHRINKAGE MACHINES
SELVEDGE UNCURLERS
SKEW STRAIGHTENERS

FILLING STREAKS
BT FABRIC DEFECTS
RT FILLING BANDS
FILLING BARS
PIRN BARRE
WARP STREAKS

FILLING TENSION
BT TENSION
RT STOP MOTIONS
WARP TENSION
WEAVING
WEAVING DYNAMICS
YARN TENSION

FILLING WARP BLENDS
USE ORTHOBLENDS

FILLING WIND
UF COP BUILD
COP WIND
BT WIND
RT CLOSE WIND
COMBINATION WIND
OPEN WIND
PACKAGES
QUILLING
ROVING WIND
TRAVERSE
WARP WIND
WINDERS
WINDING

FILLING YARNS
UF FILLING
WEFT
WEFT YARNS
NT BACK FILLING
BT YARNS
RT BEATING UP
FILLING KNITTING
JAMMING
JAMMING POINT
KNITTING YARNS
LOOMS
MISPICKS
PICK AND PICK
PICKER STICKS
PICKING (WEAVING)
PICKS
PICKS PER INCH
QUILLERS
QUILLING
QUILLS
ROTARY MAGAZINES
SHEDDING
SHUTTLE BOXES
SHUTTLES
THREADING
WARP ENDS
WEFT FORKS
WINDING

FILLINGWISE STRETCH
BT STRETCH
RT ELASTIC STRAIN
ELASTICITY
EXTENSIBILITY
RECOVERY (SELF RESTORATION TO ORIGINAL CONDITION)
STRESS STRAIN CURVES
STRETCH FABRICS
RT STRETCH WOVEN FABRICS
STRETCH YARNS (FILAMENT)
TEXTURED YARNS (FILAMENT)
UNIAXIAL STRESS
WARPWISE STRETCH

FILM READING
RT CHARACTER RECOGNITION
OPTICAL SCANNING

FILMS
NT CELLOPHANE
MYLAR (TN)
RT FABRICS
FIBERS
FOAMS
LACQUERS
PACKAGING (MATERIAL)
WEBS
YARNS

FILTER CAKES
RT ACCUMULATION
FILTER CLOTH
FILTER PAPER
FILTRATION
PRESSURE DROP

FILTER CLOTH
UF FILTER FABRICS
BT INDUSTRIAL FABRICS
RT CIGARETTE FILTERS
FILTER PAPER
FILTERS (FLUID)
FILTRATION

FILTER FABRICS
USE FILTER CLOTH

FILTER PACKS
NT SPINNING FILTERS
RT FILTER PAPER
FILTERS (FLUID)
SPINNING (EXTRUSION)
SPINNING ASSEMBLIES (EXTRUSION)
FILTRATION

FILTER PAPER
RT CIGARETTE FILTERS
FILTER CAKES
FILTER CLOTH
FILTER PACKS
FILTERS (FLUID)
FILTRATION

FILTERS (ELECTRICAL)
RT CAPACITANCE (ELECTRICAL)

FILTERS (FLUID)
RT CIGARETTE FILTERS
FILTER CLOTH
FILTER PACKS
FILTER PAPER
FILTRATION
INDUSTRIAL FABRICS
POROUS MATERIALS
SCREENING

FILTRATION
RT ABSORBENCY (MATERIAL)
ACCUMULATION
ACQUISITION
ADSORPTION
CIGARETTE FILTERS
CONCENTRATION
DIFFUSION
EXTRACTION
FILTER CAKES
FILTER CLOTH
FILTER PACKS
FILTER PAPER
FILTERS (FLUID)
MASS TRANSFER
POROUS MATERIALS
PRESSURE DROP
SORPTION

FINDINGS
BT GARMENT COMPONENTS
RT GARMENT MANUFACTURE
PIPING
TAILORING

FINE (WOOL GRADE)
BT GRADE (FIBERS)
WOOL GRADE

FINE MEDIUM (WOOL GRADE)
BT GRADE (FIBERS)
WOOL GRADE

FINE STRUCTURE (FIBERS)
UF FIBER STRUCTURE
MORPHOLOGY
NT ACCESSIBILITY (INTERNAL)
AMORPHOUS REGION
CORE (FIBER)

NT CORTEX
CRYSTALLINE REGION
CRYSTALLITES
CUTICLE
ENDOCUTICLE
EPICUTICLE
EXOCUTICLE
FIBRIL REVERSALS
FIBRILLAR STRUCTURE
FIBRILS
FRINGED FIBRILS
FRINGED MICELLES
LUMEN
MACROFIBRILS
MEDULLA
MICELLES
MICROFIBRILS
ORTHOCORTEX
PARACORTEX
PARACRYSTALLINE REGION
PRIMARY WALLS
SCALES (WOOL FIBERS)
SECONDARY WALLS
SKIN (FIBER)
SPHERULITES
SPHERULITIC FIBRILS
BT FIBER PROPERTIES
RT AMORPHOUS POLYMERS
CHAIN MOLECULES
CRYSTALLINE POLYMERS
CRYSTALLINITY
CRYSTALLIZATION
ELECTRON MICROSCOPY
FIBERS
FIBRILLATION
GRAIN BOUNDARIES
LONG CHAIN MOLECULES
MATRIX (FINE STRUCTURE)
MOLECULAR ORIENTATION
MOLECULAR STRUCTURE
MOLECULES
POLYMERIZATION
POLYMERS
STRUCTURAL PROPERTIES
STRUCTURE OF MATERIALS
STRUCTURES
X RAY ANALYSIS
X RAY DIFFRACTION

FINE WOOL
USE MERINO WOOL

FINE YARNS
BT YARNS
RT COARSE YARNS
HEAVY WEIGHT
HIGH TWIST
LIGHT WEIGHT
LOW TWIST
MEDIUM WEIGHT
PLIED YARNS
SINGLES YARNS

FINENESS
UF COARSENESS
RT AREA
CROSS SECTIONS
CROSS SECTIONAL AREA
FIBER CROSS SECTIONS
FIBER FINENESS
FINENESS TESTING
PARTICLE SIZE
SURFACE AREA
TEXTURE

FINENESS TESTERS
NT MICRONAIRE
BT TESTING EQUIPMENT
RT FIBER FINENESS
FINENESS

FINENESS TESTING
BT TESTING
RT CAUSTICAIRE SCALE
CLASSING
CURVILINEAR SCALE
FIBER LENGTH DETERMINATION
FINENESS
FINENESS TESTERS
GRADING
IMMATURITY (FIBER)
MICRONAIRE FINENESS
WEDGE SCALE

FINGER TENSION
RT ANGLE OF WRAP
BALL AND SOCKET TENSION
BELT TENSION
DISC TENSION
GATE TENSION
TENSION
TENSION CONTROL
WINDING
WRAP
YARN TENSION

FINGERING YARNS
UF HAND KNITTING YARNS
BT KNITTING YARNS
YARNS
RT HAND KNITTING

FINISH (AGENTS)
USE FINISH (SUBSTANCE ADDED)

FINISH (PROCESS)
USE FINISHING PROCESS (GENERAL)

FINISH (PROPERTY)
NT ABRASION RESISTANCE
ANTISTATIC BEHAVIOR
BACTERIAL INHIBITION
CREASE ACCEPTANCE
CREASE RETENTION
DIMENSIONAL STABILITY
DURABLE PRESS
FELTING RESISTANCE
FIRE RESISTANCE
FIRE RETARDANCY
INSECT RESISTANCE
LUSTER
MERCERIZED FINISH
MILDEW RESISTANCE
OIL REPELLENCY
ROUGHNESS
SEWABILITY
SHEEN
SHRINK RESISTANCE
SLICKNESS
SLIP RESISTANCE
SMOOTHNESS
SOFTNESS
SOIL RESISTANCE
STAIN RESISTANCE
WATER REPELLENCY
WATER RESISTANCE
WEATHER RESISTANCE
WRINKLE RECOVERY
WRINKLE RESISTANCE
BT END USE PROPERTIES
FABRIC PROPERTIES
FABRIC PROPERTIES (AESTHETIC)
FABRIC PROPERTIES (MECHANICAL)
FABRIC PROPERTIES (PHYSICAL EXCLUDING MECHANICAL)
RT AESTHETIC PROPERTIES
APPEARANCE
CHEMICAL PROPERTIES
COMFORT
COVER
DRY CLEANING FASTNESS (OF FINISH)
DURABLE FINISH (GENERAL)
FINISHING PROCESS (GENERAL)
HAIRINESS
HAND
LIVELINESS
OPTICAL PROPERTIES
PATTERN (FABRICS)
PATTERN DEFINITION
RICHNESS
SPARKLE
STITCH CLARITY
SURFACES
TRANSLUCENCY
TRANSPARENCY

FINISH (SUBSTANCE ADDED)
UF FINISH (AGENTS)
FINISHING AGENTS
NT ABRASION RESISTANCE FINISHES
ANTIBACTERIAL FINISHES
ANTIFELTING AGENTS
ANTIMICROBIAL FINISHES
ANTISLIP AGENTS
ANTISTATIC AGENTS
BLUEING AGENTS
DELUSTERING AGENTS
DURABLE PRESS FINISHES
FIRE PROOFING AGENTS
FIRE RETARDANCY AGENTS
INSECT RESISTANCE FINISHES
METHYLOL CARBAMATES
METHYLOL MELAMINES
METHYLOL UREAS
MILDEW RESISTANCE FINISHES
OIL REPELLENTS
PILL RESISTANCE FINISHES
PRESERVATIVES
REACTANTS
RESIN FINISHES
SHRINK RESISTANCE FINISHES
SOFTENERS
SOIL RELEASING FINISHES
SOIL RESISTANCE FINISHES
SPIN FINISHES
STAIN RESISTANCE AGENTS
STARCH
STIFFENERS (AGENTS)
WASH WEAR FINISHES
WATER REPELLENTS
WATERPROOFING AGENTS
YARN FINISH
BT TEXTILE CHEMICALS
RT ACETAL RESINS
ACETALS
ADD ON
ADDITIVES (CHEMICAL)
ALDEHYDES
AMINO RESINS
BLEACHING AGENTS
CATALYSTS
COATINGS (SUBSTANCES)

DRY CLEANING FASTNESS (OF FINISH)
DURABLE FINISH (GENERAL)
DURABLE PRESS TREATMENTS
EMULSIONS
EPICHLOROHYDRIN
FINISHING PROCESS (GENERAL)
FLUOROCARBON POLYMERS
FLUOROCARBONS
HALOHYDRINS
INSECT RESISTANCE TREATMENTS
LUBRICANTS
MERCERIZED COTTON
MILDEW RESISTANCE TREATMENTS
RESINS
SETTING AGENTS
WAXES

FINISHED FABRICS
RT DRY FINISHING
DYED FABRICS
FABRICS
FINISHING PROCESS (GENERAL)
GREY GOODS
PRINTED FABRICS
WET FINISHING

FINISHER GILLING
BT GILLING
RT ACTUAL DRAFT
BRADFORD SYSTEM PROCESSING
BREAK DRAFT
DRAFTING (STAPLE FIBER)
FALLER BARS
GILL BOXES
INTERSECTING GILL BOXES
LONG STAPLE SPINNING
PIN DRAFTING
SLIVERS

FINISHER PART
UF CARDER PART
RT ANGLE STRIPPERS
AUTOCOUNT (TN)
BLAMIRE FEED
BREAKERS
CARD CLOTHING
CARD CYLINDERS
CARD SLIVERS
CARD WEBS
CARDING
CARDING EFFICIENCY
CARDS
CONDENSERS
DOFFER COMBS
DOFFERS (CARD)
FANCY (NOUN)
FEED ROLLS
INTERMEDIATE PART
LICKER IN
SCOTCH FEED
SCRIBBLING
SLIVER CANS
SLIVERS
SLUBBINGS
STRIPPERS
WOOLEN CARDING
WOOLEN CARDS
WORKERS
WORSTED CARDING

FINISHING AGENTS
USE FINISH (SUBSTANCE ADDED)

FINISHING MACHINERY (GENERAL)
NT BLEACHING MACHINES
BLEACHING RANGES
BOIL OFF MACHINES
BRUSHING MACHINES
BURLING MACHINES
CALENDERING MACHINES
CARBONIZERS
COMPRESSIVE SHRINKAGE MACHINES
CRABBING MACHINES
CURING OVENS
CURING RANGES
DECATING MACHINES
DESIZING MACHINES
DRY CLEANING MACHINES
DRYERS
FELTING MACHINES
FOLDING MACHINES (FOR CLOTH)
FULLING MACHINES
HEAT SETTING MACHINES
IMPREGNATING MACHINES
MANGLES
MERCERIZING MACHINES
NAPPING MACHINES
OVERFED PIN TENTERS
PAD FIX MACHINES
PAD ROLL MACHINES
(CONT.)

NT PAD STEAM MACHINES
PADDERS
PRINTING MACHINES
ROPE SCOURING MACHINES
SANFORIZING MACHINES
SCOURING MACHINES
SHEARING MACHINES
SLASHERS
TENTER FRAMES
RT BATCH FINISHING
CONTINUOUS FINISHING
DYEING MACHINES

INVESTMENTS
LEADER CLOTH
MACHINERY
PAD BATCH CURE PROCESS
PAD DRY CURE PROCESS
PLANT LAYOUT
PROCESSING MACHINERY
ROLLS

FINISHING PROCESS (GENERAL)
UF FINISH (PROCESS)
NT ANTIFELTING TREATMENTS
ANTIMICROBIAL TREATMENTS
BATCH FINISHING
BAULK FINISHING
BLEACHING
BLOWN FINISH
BOARDING
BURLING
BUTTON BREAKING
CALENDERING
CARBONIZING
CARROTING
COLD SETTING (SYNTHETICS)
COMPACTING
COMPRESSIVE SHRINKING
CONTINUOUS BLEACHING
CONTINUOUS FINISHING
CONTINUOUS SCOURING
CORONIZING (TN)
CRABBING
CURING
DAMPING
DELUSTERING
DEODORIZING
DESIZING
DRY CURING
DRY FINISHING
DRY RELAXATION
DRYING
DURABLE PRESS TREATMENTS
DYE FIXING
FABRIC RELAXATION
FELTING
FIRE PROOFING TREATMENTS
FIRE RETARDANCY TREATMENTS
FOLDING (OF CLOTH)
FULLING
GLAZING
HEAT SETTING (DRY)
HEAT SETTING (MOIST)
HEAT SETTING (SYNTHETICS)
INSECT RESISTANCE TREATMENTS
IRONING
LONDON SHRINKING
MANGLING
MENDING
MERCERIZING
MILDEW RESISTANCE TREATMENTS
NAPPING
NEEDLING
OIL REPELLENT TREATMENTS
PAD BATCH CURE PROCESS
PAD DRY CURE PROCESS
PAD FINISHING
PADDING
PERCHING
PHASE BOUNDARY CROSSLINKING
PILL RESISTANCE TREATMENTS
PLEATING
PRESHRINKING
PRESSING
PRINTING
PROCESSES
RELAXATION
RINSING
SANFORIZING (TN)
SCHREINERING
SCOURING
SCUTCHING (FINISHING)
SETTING (EXCEPT SYNTHETICS)
SETTING TREATMENTS (GENERAL)
SHEARING (FINISHING)
SHRINK PROOFING
SIMULTANEOUS DYEING AND FINISHING
SINGEING
SLASHING
SOFTENING
SOLVENT TREATMENTS
SPECKING
STARCHING
STEAMING
STIFFENING TREATMENTS
STRESS RELAXATION
SUEDING
NT **TENTERING**
VAPOR PHASE TREATMENTS
WASH WEAR TREATMENTS
WATER REPELLENCY TREATMENTS
WATERPROOFING
WET CURING
WET FINISHING
WET RELAXATION
RT **AFTERTREATMENTS (GENERAL)**
CHEMICAL MODIFICATION (FIBERS)
DYE HOUSE
DYEING
DURABLE FINISH (GENERAL)
EASY CARE FABRICS
FABRIC INSPECTION
FACE FINISH
FINISH (PROPERTY)
FINISH (SUBSTANCE ADDED)
FINISHED FABRICS
FINISHING MACHINERY (GENERAL)
IMPREGNATION
LEADER CLOTH
PRETREATMENTS (GENERAL)
PRINTING ROLLS
TEXTILE MACHINERY (GENERAL)
TEXTILE PROCESSES (GENERAL)
UNIT OPERATIONS

FIRE PROOFING TREATMENTS
NT PROBAN PROCESS (TN)
BT WET FINISHING
RT FIRE PROOFING AGENTS
FIRE RESISTANCE
FIRE RETARDANCY
FIRE RETARDANCY TREATMENTS
HEAT RESISTANCE

FIRE PROOFING AGENTS
UF FLAME PROOFING AGENTS
BT FINISH (SUBSTANCE ADDED)
RT APO
FIRE PROOFING TREATMENTS
FIRE RETARDANCY
FIRE RETARDANCY AGENTS
FIRE RETARDANCY TREATMENTS
PHOSPHORUS COMPOUNDS

FIRE RESISTANCE
UF FLAME RESISTANCE
FLAMEPROOF (PROPERTY)
FLAMMABILITY
INFLAMMABILITY
BT END USE PROPERTIES
FINISH (PROPERTY)
RT BURNING
FIRE PROOFING TREATMENTS
FIRE RESISTANCE TESTERS
SCORCH TESTERS
SCORCH TESTING
SCORCHING
SCORCHING MARKS
FIRE RETARDANCY
FLASH POINT
HEAT RESISTANCE

FIRE RESISTANCE TESTERS
UF FLAMMABILITY TESTERS
BT TESTING EQUIPMENT
RT FIRE RESISTANCE
FIRE RETARDANCY
SCORCH TESTERS

FIRE RESISTANCE TESTING
BT TESTING
RT THERMAL TESTING

FIRE RETARDANCY
BT END USE PROPERTIES
FINISH (PROPERTY)
RT BURNING
CHEMICAL PROPERTIES
FIRE PROOFING AGENTS
FIRE PROOFING TREATMENTS
FIRE RESISTANCE
FIRE RESISTANCE TESTERS
FIRE RETARDANCY AGENTS
FIRE RETARDANCY TREATMENTS

FIRE RETARDANCY AGENTS
UF FLAME RETARDANTS
BT FINISH (SUBSTANCE ADDED)
RT APO
FIRE PROOFING AGENTS
FIRE RESISTANCE
FIRE RETARDANCY
FIRE RETARDANCY TREATMENTS
MAPO
THPC
PHOSPHORUS COMPOUNDS

FIRE RETARDANCY TREATMENTS
NT BANCROFT PROCESS (TN)
BT WET FINISHING
RT FIRE PROOFING AGENTS
FIRE PROOFING TREATMENTS
FIRE RETARDANCY
FIRE RETARDANCY AGENTS

FIRING (REFRACTORY)
RT CERAMIC FIBERS
CERAMICS
CORONIZING (TN)
HEAT
HEAT TREATMENT
HEATING
HIGH TEMPERATURE

FIRMNESS
USE HARDNESS

FIRMOVISCOSITY
BT STRESS STRAIN PROPERTIES
RT COEFFICIENT OF CROSS VISCOSITY
ELASTICITY
PHENOMENOLOGICAL RHEOLOGY
PLASTICITY
PLASTICOVISCOSITY
RELAXATION SPECTRUM
RELAXATION TIME (MECHANICAL)
VISCOELASTICITY
VISCOSITY
YIELD POINT

FIRST ORDER TRANSITION TEMPERATURE
USE MELTING POINT

FISHING LINES
RT CABLING
CORDAGE
CORDS
HAWSERS
INDUSTRIAL FABRICS
MONOFILAMENTS
ROPES
STRANDS
STRING
TWINE

FIT
RT ANTHROPOMETRIC KINEMATICS
BAGGINESS
COMFORT
FITTERS (GARMENTS)
FITTING
FOUNDATION GARMENTS
GARMENT MANUFACTURE
GARMENTS
LOOSENESS
POWER (FABRIC)
STRETCH
STRETCH FABRICS
STRETCH KNITTED FABRICS
STRETCH WOVEN FABRICS
TAILORING
TIGHTNESS

FITTERS (GARMENTS)
RT FIT
FITTING
GARMENT MANUFACTURE

FITTING
BT GARMENT MANUFACTURE
RT ADJUSTMENTS
ALTERATIONS
CLEARANCES
COMFORT
FIT
FITTERS (GARMENTS)
SETTING (ADJUSTMENTS)
TAILORING

FIXED CAPACITORS
RT CAPACITANCE (ELECTRICAL)
VARIABLE CAPACITORS

FIXED REEDS
BT REEDS
RT LOOMS
SLEYS (LOOM)
WEAVING

FIXING (DYES)
USE DYE FIXING

FLAGS
RT BUNTING
DECORATIVE FABRICS

FLAKE YARNS
BT NOVELTY YARNS
RT BEAD YARNS
BOUCLES
FRILLS (NOVELTY YARNS)
KNICKERBOCKER YARNS
LOOPS (NOVELTY YARNS)
SEEDS (NOVELTY YARNS)
SPIRALS (NOVELTY YARNS)
SPLASHES

FLAKING
RT CHIPPING
CUTTING
DAMAGE
DELAMINATING
FRACTURING

RT PEELING
SEPARATION
SPLITTING

FLAME BACKING
USE FLAME BONDING

FLAME BONDING
UF FLAME BACKING
FLAME LAMINATING
BT BONDING
LAMINATING
RT BONDED FIBER FABRICS
DIP BONDING
DRY POWDER BONDING
FOAM BONDING
HEAT BONDING
HEAT SEALING
LAMINATED FABRICS
SATURATION BONDING
SOLVENT BONDING
SPRAY BONDING
THERMOPLASTICITY

FLAME LAMINATING
USE FLAME BONDING

FLAME PROOFING AGENTS
USE FIRE PROOFING AGENTS

FLAME RESISTANCE
USE FIRE RESISTANCE

FLAME RETARDANTS
USE FIRE RETARDANCY AGENTS

FLAMEPROOF (PROPERTY)
USE FIRE RESISTANCE

FLAMES (NOVELTY YARNS)
BT NOVELTY YARNS
RT BEAD YARNS
BOUCLES
FRILLS (NOVELTY YARNS)
KNICKERBOCKER YARNS
LOOPS (NOVELTY YARNS)
SEEDS (NOVELTY YARNS)
SPIRALS (NOVELTY YARNS)
SPLASHES

FLAMMABILITY
USE FIRE RESISTANCE

FLAMMABILITY TESTERS
USE FIRE RESISTANCE TESTERS

FLAMMABILITY TESTING
RT FIRE RESISTANCE TESTING

FLANGES
RT BEAMS
BOBBINS
PACKAGES
RINGS

FLANNEL
UF BABY FLANNEL
BT FLAT WOVEN FABRICS
WOVEN FABRICS
RT PLAIN WEAVES

FLASH AGING
BT AGING (STEAMING)
RT ACID AGING
AGERS
PRINTING
TWO STAGE PRINTING

FLASH POINT
BT PHYSICAL CHEMICAL PROPERTIES
PHYSICAL PROPERTIES (EXCLUDING MECHANICAL)
THERMAL PROPERTIES
RT BOILING POINT
FIRE RESISTANCE
TEMPERATURE
VAPOR PRESSURE

FLAT BRAIDS
BT BRAIDS
RT NARROW FABRICS
RIBBON
TAPE

FLAT CHAINS (CARD)
RT CARD FLATS
CARDS
COTTON CARDS

FLAT CLOTHING
USE CARD CLOTHING

FLAT CROSS SECTION
UF RIBBON CROSS SECTION
BT FIBER CROSS SECTIONS
RT ROVANA (TN)

FLAT KNITTED FABRICS
BT FILLING KNITTED FABRICS
KNITTED FABRICS
RT FULL CARDIGAN FABRICS
HALF CARDIGAN FABRICS
KNITTING
PLAIN KNITTED FABRICS
RIB KNITTED FABRICS

FLAT KNITTING
BT KNITTING
RT CIRCULAR KNITTING
FASHIONING
FLAT KNITTING MACHINES
FULLY FASHIONED KNITTING
KNITTED FABRICS
PLAIN KNITTED FABRICS
RACKING
TRICOT KNITTED FABRICS
WARP KNITTING

FLAT KNITTING MACHINES
NT FULLY FASHIONED KNITTING MACHINES
BT KNITTING MACHINES
RT CIRCULAR KNITTING MACHINES
FLAT KNITTING

FLAT PRESSING
BT GARMENT MANUFACTURE
PRESSING
RT TAILORING

FLAT SCREEN PRINTING
BT PRINTING
SCREEN PRINTING
RT ROTARY SCREEN PRINTING
SCREEN PRINTING MACHINES

FLAT SETTINGS (CARD)
RT ADJUSTMENTS
CARD CLOTHING
CARD FLATS
COTTON CARDING
FLAT STRIPS
PROCESS VARIABLES
SETTING (ADJUSTMENTS)

FLAT SPOTTING (TIRES)
RT THERMAL PROPERTIES
TIRE CORDS
TIRE FABRICS
TIRES

FLAT STRIPPING BRUSHES
USE FLAT STRIPPING COMBS

FLAT STRIPPING COMBS
UF FETTLING COMBS
FLAT STRIPPING BRUSHES
RT CARD WASTE
COTTON CARDS
FETTLING
FLAT STRIPS
HACKLE COMBS
WASTE

FLAT STRIPS
BT CARD STRIPS
CARD WASTE
RT CARD FLATS
CYLINDER STRIPS
DOFFER STRIPS
FETTLING
FLAT SETTINGS (CARD)
FLAT STRIPPING COMBS
HACKLE COMBS
IMPURITIES

FLAT TOP CARDS
USE REVOLVING FLAT CARDS

FLAT WOVEN FABRICS
UF FABRICS (FLAT WOVEN)
NT ALBERT TWILL
ALPACA FABRICS
AWNING CLOTH
BARATHEA
BATISTE
BEDFORD CORD
BOMBAZINE
BROCADES
BROCATELLES
BRASSIER CLOTH
BRILLIANT SPOT
BROADCLOTH
BUCKRAM
BUNTING
CALICO
CAMBRIC
CASEMENT
CAVALRY TWILL
CHAMBRAY
CHARMANETTE SATIN
CHECK SHIRTING
CHEESE CLOTH
CHIFFON
CHINTZ

NT **COUTIL**
COVERT
CRASH TOWELING
CREPE
CREPON GEORGETTE
CRETONNE
DAMASK
DENIMS
DIMITY
DOESKIN
DOTTED MUSLIN
DRILLS (FABRICS)
DUCK
ELASTIQUE
EMBOSSED CREPE
FANCY LENO
FLANNEL
GABARDINE
GAUZE
GINGHAM
HIMALAYA CLOTH
HOPSACK
IMITATION GINGHAM
INDIA LINON
JEANS
KERSEY
LAWN
LINGETTE
LONGCLOTH
MAINSOOK CHECK
MARQUISETTE
MELTON
MOCK LENO
MUSLIN
NETTING
ORGANDIE
OSNABURG
PAJAMA CHECK
PERCALE
PILLOW TUBING
PIQUE
PLISSE
PONGEE
POPLIN
PRINT CLOTH
RAYON STRIPE
REPP
RIBBON
SATIN
SATEEN
SCREENING
SCRIM
SEERSUCKER
SERGE
SHADOW STRIPE
SHARKSKIN
SHEETING (FABRIC)
SHIRTING
SILESIA
SKIP DENT
SPLASH VOILE
SPORT SATIN
SWISS
TABLE DAMASK
TAFFETA
TAPE
TICKING
TIRE FABRICS
TOBACCO CLOTH
TROPICAL WORSTED
TWILL CHECKBOARD
TWILL WEAVES
UMBRELLA CLOTH
VELOUR
VICTORIA LAWN
VOILE
WARP REPP
WEBBING
WHIPCORD
WINDOW HOLLAND
BT **FABRICS**
FABRICS (ACCORDING TO STRUCTURE)
WOVEN FABRICS
RT **BACKED FABRICS**
BASKET WEAVES
CELLULAR FABRICS
CHECKER BOARD WEAVES
CLIPSPOT WEAVES
COMBINATION WEAVES
DIAMOND WEAVES
DOGSTOOTH WEAVES
GRANITE WEAVES
HONEYCOMB WEAVES
HUCKABACK WEAVES
JACQUARD WEAVES
LAPPET WEAVES
LENO WEAVES
MULTIPLE LAYER FABRICS
OXFORD WEAVES
PILE FABRICS (WOVEN)
PIQUE WEAVES
PLAIN WEAVES
RIB WEAVES
SATIN WEAVES
SPOT WEAVES
STITCH REINFORCED FABRICS
SWIVEL WEAVES
(CONT.)

RT TABBY WEAVES
TAFFETA WEAVES
TWILL WEAVES
TWILLS
WEAVE (TYPES)

FLATLOCK STITCHES
BT STITCHES (SEWING)
RT SEWING

FLATS
USE CARD FLATS

FLAWS
USE DEFECTS

FLAX
BT BAST FIBERS
CELLULOSIC FIBERS
RT DECORTICATING
FLAX SYSTEM PROCESSING
HEMP
JUTE
LINEN
RAMIE
RETTING

FLAX SYSTEM PROCESSING
RT COTTON SYSTEM PROCESSING
FLAX
SILK SYSTEM PROCESSING
WOOLEN SYSTEM PROCESSING
WORSTED SYSTEM PROCESSING

FLEECES
UF WOOL FLEECES
RT PULLED WOOL
SHEEP
SORTING
WOOL
WOOL SHEARING

FLEXIBILITY
USE BENDING RIGIDITY

FLEXIBLE FILM LAMINATING
BT COATING (PROCESS)
LAMINATING
RT CALENDER COATING
CAST COATING
DIP COATING
EXTRUSION COATING
KNIFE COATING
LAMINATED FABRICS
NIP COATING
ROLLER COATING

FLEXIBLE PLUNGER MOLDING
BT MOLDING
RT CONTACT MOLDING
COMPRESSION MOLDING
INJECTION MOLDING
MATCHED DIE MOLDING
PRESSURE BAG MOLDING
REINFORCED COMPOSITES
VACUUM BAG MOLDING

FLEXING
USE BENDING

FLEXURAL FATIGUE
BT FATIGUE
RT BENDING RIGIDITY
FLEXURAL STRENGTH

FLEXURAL RIGIDITY
USE BENDING RIGIDITY

FLEXURAL STRENGTH
UF BENDING STRENGTH
BT MECHANICAL PROPERTIES
RT BENDING RIGIDITY
FLEXURAL FATIGUE

FLIPPER FABRICS
USE TIRE FABRICS

FLOAT (OPERATION)
USE FLOATING (KNITTING)

FLOAT DESIGN
BT DRAFTING (PATTERN)
RT DESIGN
FABRIC DESIGN
FABRIC STRUCTURE
FLOATS
LAID IN YARNS

FLOAT LENGTH
BT FABRIC PROPERTIES
RT FABRIC GEOMETRY
FABRIC STRUCTURE
FLOATS

FLOAT STITCHES (KNITTING)
BT KNITTED STITCHES
RT PELERINE STITCHES
PLAIN STITCHES (KNITTING)
RACKED STITCHES
TUCK STITCHES

FLOATING (KNITTING)
UF FLOAT (OPERATION)
RT KNITTING
WELTING

FLOATING FIBERS
RT BACK ROLLS
CONTROL SURFACES
DRAFT
DRAFTING (STAPLE FIBER)
DRAFTING THEORY
DRAFTING WAVES
FIBER CONTROL
FIBER LENGTH DISTRIBUTION
FIBER ORIENTATION
FIBER WITHDRAWAL FORCE
FRONT ROLLS
IRREGULARITY
VELOCITY CHANGE POINT

FLOATS
RT FABRIC STRUCTURE
FLOAT DESIGN
FLOAT LENGTH
INTERLACINGS (YARN IN FABRIC)
LAID IN YARNS

FLOCK
UF FIBER FLOCK
BT FIBER ASSEMBLIES
RT ELECTROSTATIC FLOCKING
FABRICS
FLOCK COATING
FLOCK DOTS
FLOCK POWDERS
FLOCK PRINTING
FLOCKED CARPETS
FLOCKED FABRICS
FLOCKING
MECHANICAL FLOCKING
PILE
PILE FABRICS (FLOCKED)
STUFFING

FLOCK COATING
RT COATING (PROCESS)
ELECTROSTATIC FLOCKING
FLOCK
FLOCK PRINTING
FLOCKING
MECHANICAL FLOCKING

FLOCK DOTS
RT FLOCK
FLOCK PRINTING
FLOCKING

FLOCK POWDERS
RT ADHESIVES
FLOCK

FLOCK PRINTING
RT ELECTROSTATIC FLOCKING
FLOCK
FLOCK COATING
FLOCKING
MECHANICAL FLOCKING
PRINTING

FLOCKED CARPETS
BT CARPETS
FLOCKED FABRICS
PILE FABRICS (FLOCKED)
RT AXMINSTER CARPETS
CHENILLE CARPETS
ELECTROSTATIC FLOCKING
FABRICS
FLOCK
FLOCKING
MECHANICAL FLOCKING
WILTON CARPETS

FLOCKED FABRICS
UF FABRICS (FLOCKED)
NT FLOCKED CARPETS
BT FABRICS
FABRICS (ACCORDING TO STRUCTURE)
PILE FABRICS (FLOCKED)
RT BONDED YARN FABRICS
ELECTROSTATIC FLOCKING
FLOCK
FLOCKING
MECHANICAL FLOCKING
NONWOVEN FABRICS
STITCHED PILE FABRICS
SYNTHETIC LEATHER
TUFTED FABRICS

FLOCKING
NT ELECTROSTATIC FLOCKING
MECHANICAL FLOCKING
RT FLOCK
FLOCK COATING
FLOCK PRINTING

FLOUNCING (LACE)
BT LACES

FLOW (MATERIALS)
UF MATERIALS FLOW
RT ALLOTMENT
CARRIER RODS
CONVEYORS
DELIVERY
DELIVERY RATE
DELIVERY ROLLS
CONSUMPTION (MATERIAL)
COUNTERFLOW PROCESSES
FEED RATE
FEED ROLLS
FEEDERS
FLOW CONTROL
INPUT
INVENTORIES
LOADING (PROCESS)
MANAGEMENT
MANUFACTURING
OUTPUT
PROCESS CONTROL
PRODUCTION
PRODUCTION CONTROL
SURGING
THROUGHPUT
TRANSFERRING (MATERIAL)
UNLOADING (PROCESS)
YARN CARRIERS

FLOW (PLASTIC)
USE PLASTIC FLOW

FLOW (FLUIDS)
USE FLUID FLOW

FLOW CONTROL
UF ROTAMETERS
RT AUTOMATIC CONTROL
CONTROL SYSTEMS
DELIVERY RATE
FEED RATE
FLOW (MATERIALS)
FLUID FLOW
FLUIDICS
PRESSURE CONTROL
PROCESS CONTROL
THROUGHPUT

FLUCTUATIONS
RT ADJUSTMENTS
COMPENSATION
CONTROL SYSTEMS
CONTROLLING
CORRECTION
DRIFT
ERRORS
ESTIMATION
MOTION
OSCILLATIONS
QUALITY CONTROL
SETTING (ADJUSTMENTS)
STANDARD DEVIATION
VARIABLES
VARIATION

FLUFLON (TN)
BT FALSE TWIST YARNS
STRETCH YARNS (FILAMENT)
TEXTURED YARNS (FILAMENT)
RT BULKED YARNS
DYNALOFT (TN)
FALSE TWISTING (TEXTURING)
HELANCA (TN)
SAABA (TN)
SUPERLOFT (TN)

FLUID DRIVES
BT DRIVES
RT BELT DRIVES
CHAIN DRIVES
DEVICES
GEAR DRIVES
MECHANICAL DEVICES
MECHANISMS

FLUID FLOW
UF FLOW (FLUIDS)
NT LAMINAR FLOW
TURBULENT FLOW

RT AERODYNAMICS
AIR DRAG
AIR FLOW
AIR RESISTANCE
AIR STREAMS
BOUNDARY LAYER
BOUNDARY LAYER CONTROL
BOUNDARY LUBRICATION
COUNTERFLOW PROCESSES
CURRENTS
DRAG
EDDIES
FLOW CONTROL
FLUID DRIVES
FLUID FRICTION
FLUID MECHANICS
FLUID SPINNING
FLUIDICS

RT HYPERSONIC FLOW
LIFT (FLUID)
LIQUIDS
MASS TRANSFER
PISTONS
PRESSURE DROP
SHEAR (MODE OF DEFORMATION)
SURGING
TURBULENCE
VELOCITY
VISCOSITY
WATER

FLUID FRICTION
BT FRICTION
RT AERODYNAMICS
AIR DRAG
AIR FLOW
AIR RESISTANCE
BOUNDARY LAYER
BOUNDARY LUBRICATION
DRAG
FLUID FLOW
HYDRODYNAMIC LUBRICATION
LAMINAR FLOW
LIFT (FLUID)
PRESSURE DROP
SHEAR (MODE OF DEFORMATION)
VISCOSITY

FLUID MECHANICS
BT MECHANICS
RT AIR FLOW
DIMENSIONAL ANALYSIS
EDDIES
FLUID FLOW
FLUID TWISTING
HEAT TRANSFER
HYDRAULICS
MASS TRANSFER
OPEN END SPINNING

FLUID SPINNING
NT AIR SPINNING
BT SPINNING
RT AERODYNAMICS
AIR FLOW
COMB SPINNING
CONTINUOUS SPINNING (STAPLE FIBER)
FLUID FLOW
FLUID TWISTING
OPEN END SPINNING
RING SPINNING

FLUID TWISTING
UF AIR TWISTING
BT TWISTING
RT AIR FLOW
AIR JET TEXTURING
AIR SPINNING
ALTERNATING TWIST
FALSE TWIST
FLUID MECHANICS
FLUID SPINNING
OPEN END SPINNING
REAL TWIST
ZERO TWIST

FLUIDICS
RT CONTROL SYSTEMS
CONTROL THEORY
FLOW CONTROL
FLUID FLOW
HYDRAULIC CONTROL

FLUIDITY
USE VISCOSITY

FLUIDIZED BED DRYERS
BT DRYERS
FLUIDIZED BEDS
RT ATMOSPHERIC DRYERS
CENTRIFUGAL DRYERS
DRUM DRYERS
DRYING CANS
HOT AIR DRYERS
INFRARED DRYERS
LOOP DRYERS
VACUUM DRYERS

FLUIDIZED BEDS
NT FLUIDIZED BED DRYERS
RT AIR FLOW
CONTINUOUS DYEING RANGE
POROUS MATERIALS

FLUORESCENCE
BT OPTICAL PROPERTIES
RT COLOR
FABRIC PROPERTIES
FABRIC PROPERTIES (AESTHETIC)
FABRIC PROPERTIES (PHYSICAL EXCLUDING MECHANICAL)
LUMINESCENCE
ULTRAVIOLET MICROSCOPY

FLUORESCENCE MICROSCOPY
BT MICROSCOPY
RT ELECTRON MICROSCOPY
INFRARED MICROSCOPY
INTERFERENCE MICROSCOPY
LIGHT MICROSCOPY (OPTICAL)
ULTRAVIOLET MICROSCOPY
X RAY MICROSCOPY

FLUORESCENT BRIGHTENERS
UF OPTICAL BLEACHES
OPTICAL BLEACHING
OPTICAL BRIGHTENING
OPTICAL WHITENERS
NT BLANCOPHOR (TN)
RT BLEACHING
BLEACHING AGENTS
BLUEING AGENTS
BRIGHTNESS
DYES
FLUORESCENT DYES
LAUNDERING
LIGHTFASTNESS (COLOR)
WHITENESS

FLUORESCENT DYES
BT DYES (BY CHEMICAL CLASSES)
RT FLUORESCENT BRIGHTENERS

FLUORESCENT LIGHT
BT LIGHT
RT ARTIFICIAL DAYLIGHT
DAYLIGHT
ILLUMINATION
INCANDESCENT LIGHT
POLARIZED LIGHT
SUNLIGHT

FLUORIDES
NT CHROMIUM FLUORIDE
POTASSIUM FLUORIDE
SILICOFLUORIDES
SODIUM FLUORIDE
STANNOUS FLUORIDE
BT HALIDES
RT BROMIDES
CHLORIDES
FLUORINE COMPOUNDS
IODIDES

FLUORINE COMPOUNDS
NT FLUOROCARBONS
BT HALOGEN COMPOUNDS
RT FLUORIDES

FLUOROCARBON COMPOUNDS
USE FLUOROCARBONS

FLUOROCARBON CONTAINING POLYMERS
USE FLUOROCARBON POLYMERS

FLUOROCARBON FIBERS
NT FTORLON (TN)
POLYTETRAFLUOROETHYLENE FIBERS
POLYTRIFLUOROCHLOROETHYLENE FIBERS
TEFLON FIBER (TN)
BT FIBERS
MAN MADE FIBERS
SYNTHETIC FIBERS
RT FLUOROCARBON POLYMERS
FLUOROCARBONS

FLUOROCARBON POLYMERS
UF FLUOROCARBON CONTAINING POLYMERS
NT POLYTETRAFLUOROETHYLENE
TEFLON (TN)
BT FLUOROCARBONS
RT FINISH (SUBSTANCE ADDED)
FLUOROCARBON FIBERS
LUBRICANTS
OIL REPELLENCY
STAIN RESISTANCE
WATER REPELLENCY

FLUOROCARBONS
UF FLUOROCARBON COMPOUNDS
PERFLUOROALKYL COMPOUNDS
PERFLUOROCARBONS
NT FLUOROCARBON POLYMERS
BT FLUORINE COMPOUNDS
RT FINISH (SUBSTANCE ADDED)
FLUOROCARBON FIBERS
LUBRICANTS
OIL REPELLENCY
PROPELLANTS (AEROSOLS)
SILICOFLUORIDES
STAIN RESISTANCE
WATER REPELLENCY

FLUTED ROLLS
BT DRAFTING ROLLS
ROLLS
RT COTS (ROLLERS)
COVERED ROLLS
DRAWFRAMES
DRIVEN ROLLS
DRIVING ROLLS
FLUTES
PLAIN ROLLS
ROLLER NIP
SPINNING FRAMES

FLUTES
RT DRAFTING ROLLS
FLUTED ROLLS
ROLLS

FLY (GARMENT)
BT GARMENT COMPONENTS

FLY (WASTE)
BT CARD WASTE
RT CARD STRIPS
CARDS
CLIPPINGS
FANCY (NOUN)
IRREGULARITY
SLUBS
NEPS
TRASH
WASTE
YIELD (RETURN)

FLY BEAMS (LOOM)
USE SLEYS (LOOM)

FLY COMBS
USE DOFFER COMBS

FLY SHUTTLES
BT SHUTTLES
RT BLANKET SHUTTLES
BOAT SHUTTLES
PICKER STICKS
STICK SHUTTLES
WEAVING

FLYER FRAMES
USE FLYER SPINNING FRAMES

FLYER LEAD
RT BOBBIN LEAD
FLYER LEG
FLYER SPINNING FRAMES
PRESSURE FINGERS

FLYER LEG
RT BOBBINS
FLYER LEAD
FLYER SPINNING
FLYER SPINNING FRAMES
FLYERS
ROVING FRAMES

FLYER SPINNING
BT SPINNING
RT BALLOONS
CAP SPINNING
DOWNTWISTING
FLYER LEG
MULE SPINNING
PRESSURE FINGERS
RING SPINNING

FLYER SPINNING FRAMES
UF FLYER FRAMES
THROSTLE SPINNING FRAMES
BT SPINNING FRAMES
RT BOBBIN LEAD
BOBBINS
BUILDER MOTIONS
CAP SPINNING FRAMES
CREELING
DOFFING (PACKAGE)
DRAWFRAMES
FLYER LEAD
FLYER LEG
FLYERS
LIFTER PLATES
MULES
PACKAGES
POT SPINNING FRAMES
PRESSURE FINGERS
RING FRAMES
SPINDLES
SPINNING
THREAD GUIDES
TWISTING

FLYERS
RT BOBBINS
CAPS (SPINNING)
FLYER LEG
FLYER SPINNING FRAMES
PRESSURE FINGERS
RINGS
ROVINGS
SPINDLES
SPINNING
SPINNING FRAMES
TRAVELERS
TWISTING

FLYSHOT LOOMS
BT LOOMS
NARROW FABRIC LOOMS
RT NARROW FABRICS
WEAVING

FOAM BACKED FABRICS
UF FOAM LAMINATES
NT LAMINATED JERSEY (FOAM)
BT COMPOSITES
LAMINATED FABRICS
LAMINATES
RT ADHESION
BACKING (MATERIAL)
BACKING FABRICS
BONDED FIBER FABRICS
CARPETS
COATED FABRICS
FABRIC TO FABRIC LAMINATES
FOAM BONDING
FOAM RUBBER
FUSION (MELTING)
HEAT SEALING
NONWOVEN FABRICS
POST FORMED LAMINATES

FOAM BACKING
BT BACKING (MATERIAL)
RT BACKED FABRICS
BACKING COMPOUNDS
BACKING FABRICS
FOAM BONDING
FOAMS

FOAM BONDING
BT BONDING
RT BONDED FIBER FABRICS
DIP BONDING
DRY POWDER BONDING
FLAME BONDING
FOAM BACKED FABRICS
FOAM BACKING
FOAMS
HEAT BONDING
SPRAY BONDING

FOAM LAMINATES
USE FOAM BACKED FABRICS

FOAM RUBBER
BT FOAMS
RT BACKING COMPOUNDS
CARPETS
ELASTIC FOAMS
FOAM BACKED FABRICS
FOAM BACKING
POLYMERS
URETHANE FOAMS
VINYL FOAMS

FOAM YARNS
USE CELLULAR YARNS

FOAMING
RT FOAMS
FOAMS (FROTH)

FOAMING AGENTS
BT SURFACTANT APPLICATIONS
RT ANTIFOAM AGENTS
BUILDERS (DETERGENTS)
FOAMS
FOAMS (FROTH)
SURFACTANTS

FOAMS
UF CELLULAR PLASTICS
NT ELASTIC FOAMS
FOAM RUBBER
RIGID FOAMS
STYROFOAM (TN)
URETHANE FOAMS
VINYL FOAMS
BT BACKING COMPOUNDS
RT CELLULAR YARNS
COMPOSITES
EXPANDED PLASTICS
FABRICS
FIBERS
FILMS
FOAM BACKED FABRICS
FOAM BACKING
FOAM BONDING
FOAMING
FOAMING AGENTS
INSULATION (THERMAL)
LAMINATES
LAMINATING RESINS
PACKAGING (MATERIAL)
POROUS MATERIALS
WEBS
YARNS

FOAMS (FROTH)
RT AEROSOLS
ANTIFOAM AGENTS
COLLOIDS
DISPERSING AGENTS
EMULSIFYING AGENTS
FOAMING
FOAMING AGENTS
LAUNDERING
SCOURING
SOAPS
SURFACTANTS
SUSPENSIONS
WETTING
WETTING AGENTS

FOCUSING
RT ADJUSTMENTS
CORRECTION
MICROSCOPY
OPTICAL INSTRUMENTS
OPTICS

FOGGY YARNS
BT YARN DEFECTS
RT DIRT

FOGMARKING
BT FABRIC DEFECTS
RT DIRT
STATIC ELECTRICITY
STREAKINESS

FOIL
NT ALUMINUM FOIL
BT HIGH TEMPERATURE FIBERS
INORGANIC FIBERS (MAN MADE)
METALLIC FIBERS
RT METAL FIBERS
METALLIC YARNS
MINERAL FIBERS
TINSEL
WHISKER FIBERS (METAL)

FOLDED YARNS
USE PLIED YARNS

FOLDING (OF CLOTH)
UF PLAITING (FOLDING)
BT DRY FINISHING
RT PACKAGING (OPERATION)

FOLDING (TWISTING)
USE PLYING

FOLDING MACHINES (FOR CLOTH)
UF PLAITING MACHINES (FOLDING)
RT PILERS

FOLK WEAVES
BT WEAVE (TYPES)
RT HAND WOVEN FABRICS

FOLLICLE
RT ANIMAL FIBERS (GENERAL)
HAIR
SKIN
WOOL

FOLLOW THE LEADER CRIMP
BT CRIMP
FIBER CRIMP
RT BICOMPONENT FIBER YARNS (FILAMENT)
BICOMPONENT FIBER YARNS (STAPLE)
FIBER GEOMETRY
FIBER PROPERTIES
LATENT CRIMP
SELF CRIMPING YARNS
YARN CRIMP
Z CRIMP

FOOD DYES
BT DYES
RT DYES (BY FIBER CLASSES)

FOOTSTEP BEARINGS
BT BEARINGS
RT PLAIN BEARINGS
ROVING FRAMES
SLEEVE BEARINGS
SPINDLES
SPINNING FRAMES
THRUST BEARINGS

FOOTWEAR
NT BOOTS
COLD CLIMATE FOOTWEAR
FELT BOOTS
GALOSHES
HEELS (FOOTWEAR)
INSULATED BOOTS
LEATHER BOOTS
LEGGINGS
ORTHOPEDIC FOOTWEAR
PUMPS (FOOTWEAR)
RUBBER FOOTWEAR
SAFETY FOOTWEAR
SHOEPAC
SHOES
SLIPPERS
SNEAKERS
NT TROPICAL FOOTWEAR
WATER RESISTANT FOOTWEAR
RT BINDINGS (FOOTWEAR)
CIRCULAR HOSIERY
FULLY FASHIONED HOSIERY
GARMENTS
HALF HOSE
HOSIERY
INSOLES (FOOTWEAR)
LACES (SHOES)
LASTS
LEATHER
LININGS (FOOTWEAR)
RAWHIDE
SOCKS
SOLES
SPATS
STITCHING
STOCKINGS
SURGICAL STOCKINGS
SYNTHETIC LEATHER
TONGUES
UPPERS
VINYL COATED FABRICS

FOOTWEAR COMPONENTS
NT ADHESIVES (FOOTWEAR)
BOTTOM FILLER (FOOTWEAR)
BOX TOES
CLOSURES (FOOTWEAR)
DIRECT MOLDED SOLE (FOOTWEAR)
FABRIC UPPERS (FOOTWEAR)
GUSSETS (FOOTWEAR)
HEELS (FOOTWEAR)
INSOLES (FOOTWEAR)
LACES (SHOES)
LASTS
SOLES
TONGUES
UPPERS
VAMPS
VENTILATING INSOLES
WELTS (FOOTWEAR)
RT GARMENT COMPONENTS

FORCE
NT AIR DRAG
BALLOON TENSION
CENTRIFUGAL FORCE
CORIOLIS FORCE
DALEMBERTS FORCE
DISPERSION FORCES
DRAFTING FORCE
FRICTIONAL FORCE
GRAVITATION
NORMAL FORCE
SHEAR FORCE
TANGENTIAL FORCE
TENSION
TRANSVERSE FORCE
VAN DER WAALS FORCES
RT ACCELERATION (MECHANICAL)
ACCELEROMETERS
BRAKING
CORIOLIS ACCELERATION
DYNAMOMETERS
FIBER FIBER FRICTION
IMPACT
INERTIA
LOADING
MASS
MECHANICAL SHOCK
MOMENTUM
MOMENTS
STRENGTH
STRESS
TENACITY
WEDGING

FORECASTING
RT ANALYZING
ESTIMATION
EXTRAPOLATION

FORCED CONVECTION DRYERS
USE HOT AIR DRYERS

FORM PERSUASIVE GARMENTS
BT GARMENTS
UNDERWEAR
RT FOUNDATION GARMENTS
SPANDEX FIBERS
STRETCH FABRICS
STRETCH YARNS (FILAMENT)
STRETCH YARNS (SPUN)
TWO WAY STRETCH
WARPWISE STRETCH

FORMALDEHYDE
NT PARAFORMALDEHYDE
BT ALDEHYDES
RT CROSSLINKING
DMEU
METHYLOL COMPOUNDS
METHYLOLATION
ODORS
PRESERVATIVES
RESIN FINISHES
WASH WEAR FINISHES

FORMALDEHYDE PRECONDENSATES
    USE METHYLOL COMPOUNDS

FORMALDEHYDE SULFOXYLATES
    UF RONGALITE (TN)
        SODIUM SULFOXYLATE FORMALDEHYDE
    NT ZINC FORMALDEHYDE SULFOXYLATE
    RT ACETALDEHYDE SULFOXYLATES
        DYEING AUXILIARIES
        PRINTING
        REDUCING AGENTS
        VAT DYEING
        VAT PRINTING

FORMALS
    BT ACETALS
    RT HEMIACETALS
        METHYLOLATION

FORMATION
    USE FORMING

FORMIC ACID
    BT CARBOXYLIC ACIDS
        ORGANIC ACIDS
    RT MARSCHALL SOLUTION
        PERFORMIC ACID

FORMING
    UF FORMATION
    RT BOARDING
        MOLDING
        PRESSING
        SHAPING

FORMULAS
    RT COMPUTATIONS
        EQUATIONS
        MATHEMATICAL ANALYSIS

FORMULATIONS
    RT DYEING
        FINISH (SUBSTANCE ADDED)
        MIXTURES
        WET FINISHING

FORTE MOISTURE METER
    BT TESTING EQUIPMENT
    RT MOISTURE CONTENT TESTING

FORTISAN (TN)
    BT CELLULOSIC FIBERS
        RAYON (REGENERATED CELLULOSE FIBERS)
    RT CUPRAMMONIUM RAYON
        VISCOSE RAYON

FORTREL (TN)
    BT POLYESTER FIBERS

FOULARD (DYEING)
    USE PADDERS

FOULARD (FABRIC)
    BT WOVEN FABRICS
    RT TWILL WEAVES

FOUNDATION GARMENTS
    NT BRASSIERES
        CORSETS
        GIRDLES
    BT GARMENTS
        UNDERWEAR
    RT BRIEFS
        COMFORT
        FABRICS
        FIT
        FORM PERSUASIVE GARMENTS
        LINGERIE
        LYCRA (TN)
        PANTIES
        POWER (FABRIC)
        SLIPS
        SPANDEX
        STRETCH FABRICS
        STRETCH KNITTED FABRICS
        SUPPORT HOSIERY
        WARPWISE STRETCH

FOUR NEEDLE COVER STITCHES
    BT STITCHES (SEWING)

FOURDRINIER MACHINES
    BT PAPER MACHINES
    RT CYLINDER MACHINES
        INVERFORM MACHINES
        KAMYR MACHINES
        ROTAFORMERS
        WET MACHINES (PAPERMAKING)

FRACTIONATION
    RT CHEMICAL ANALYSIS
        MOLECULAR WEIGHT
        POLYMERS
        RHEOLOGY

FRACTURE
    USE FRACTURING

FRACTURING
    UF FRACTURE
    NT BRITTLE FRACTURE
    RT BREAKING
        BREAKING ELONGATION
        BRITTLENESS
        CHIPPING
        CRACK PROPAGATION
        CUTTING
        DAMAGE
        FLAKING
        IMPACT
        IMPACT STRENGTH
        MECHANICAL PROPERTIES
        RUPTURE
        SEPARATION
        SHOCK RESISTANCE
        SPLITTING
        STRENGTH

FRAMES (MACHINES)
    USE TENTER FRAMES

FRAMING
    USE TENTERING

FRAZIER PERMEOMETER
    BT PERMEABILITY TESTERS
        TESTING EQUIPMENT
    RT AIR PERMEABILITY
        AIR PERMEABILITY TESTING
        GURLEY PERMEOMETER
        PERMEABILITY TESTING

FREE ENERGY
    NT FREE ENERGY OF DYEING
        FREE ENERGY OF REACTION
        FREE ENERGY OF SOLUTION
        FREE ENERGY OF SWELLING
        FREE ENERGY OF WETTING
    RT ENTHALPY
        ENTHALPY OF REACTION
        ENTROPY
        EQUILIBRIUM (CHEMICAL)
        THERMODYNAMICS

FREE ENERGY OF DYEING
    BT FREE ENERGY
    RT DYE AFFINITY

FREE ENERGY OF REACTION
    BT FREE ENERGY

FREE ENERGY OF SOLUTION
    BT FREE ENERGY

FREE ENERGY OF SWELLING
    BT FREE ENERGY

FREE ENERGY OF WETTING
    BT FREE ENERGY

FREE RADICALS
    BT RADICALS (CHEMICAL)
    RT ADDITION POLYMERIZATION
        ANIONS
        AZIDES
        CATIONS
        GRAFTING
        INITIATORS
        IONS
        OXIDATION
        REACTIONS (CHEMICAL)

FREE VOLUME
    RT ACTIVATION ENERGY
        BULK MODULUS
        DILATANCY
        METARHEOLOGY
        RHEOLOGY
        STRESS STRAIN PROPERTIES

FREEZE DRYING
    BT DRYING
    RT BATCH DRYING
        CENTRIFUGAL DRYING
        CONTINUOUS DRYING
        DIELECTRIC DRYING
        FREEZING
        HOT AIR DRYING
        INFRARED DRYING
        LINE DRYING (EASY CARE GARMENTS)
        TUMBLE DRYING
        VACUUM DRYING

FREEZING
    RT COOLING
        CRYSTALLIZATION
        FREEZE DRYING
        MELTING
        REFRIGERATION
        SOLIDIFICATION

FREEZING POINT
    BT PHYSICAL CHEMICAL PROPERTIES
        PHYSICAL PROPERTIES (EXCLUDING MECHANICAL)
    RT BOILING POINT
        DEW POINT
        MELTING POINT
        REFRIGERATION

FRENCH COMBING
    USE RECTILINEAR COMBING

FRENCH COMBING WOOL
    BT WOOL CLASS

FRENCH COMBS
    USE RECTILINEAR COMBS

FRENCH CREPE CORDS
    BT CABLE CORDS
    RT LACING CORDS

FRENCH DOUBLE PIQUE
    BT DOUBLE PIQUE
    RT KNITTED FABRICS
        RIB KNITTED FABRICS
        SWISS DOUBLE PIQUE

FRENCH SEAMS
    BT SEAMS
    RT BUTTED SEAMS
        BUTTERFLY SEAMS
        COVERED SEAMS
        IMITATION FRENCH PIPING
        PIPED SEAMS
        SANDWICH SEAMS
        SEAMING
        SEWING
        STITCHING
        STRIPPED SEAMS
        TAILORED SEAMS

FRENCH SYSTEM PROCESSING
    USE CONTINENTAL SYSTEM PROCESSING

FREQUENCY
    NT HIGH FREQUENCY
    RT AMPLITUDE
        CORRELATION COEFFICIENT
        DRAFTING WAVES
        DYNAMIC RESPONSE
        FREQUENCY ANALYSERS
        FREQUENCY DENSITY
        FREQUENCY RESPONSE
        IRREGULARITY (PERIODIC)
        MULTIPLES
        OSCILLATIONS
        PERIODICITY
        REPEATS
        RESONANT FREQUENCY
        STATISTICAL MEASURES
        STATISTICS
        SURGING
        TRANSVERSE VIBRATIONS
        VIBRASCOPE
        VIBRATION

FREQUENCY ANALYSERS
    RT FREQUENCY
        FREQUENCY RESPONSE

FREQUENCY DENSITY
    BT STATISTICAL MEASURES
    RT FREQUENCY

FREQUENCY DISTRIBUTION
    USE DISTRIBUTION

FREQUENCY DISTRIBUTION (FIBER LENGTH)
    USE FIBER LENGTH DISTRIBUTION

FREQUENCY RESPONSE
    RT CONTROL CHARACTERISTICS
        DYNAMIC CHARACTERISTICS
        DYNAMIC RESPONSE
        ENERGY ABSORPTION
        FREQUENCY
        FREQUENCY ANALYSERS
        HIGH FREQUENCY
        HUNTING
        OSCILLATIONS
        RESONANT FREQUENCY
        VIBRATION
        VIBRATION DAMPERS
        VIBRATION ISOLATORS

FREUNDLICH ISOTHERM
    BT ADSORPTION ISOTHERM
    RT BET ISOTHERM
        DYEING THEORY
        LANGMUIR ISOTHERM
        PARTITION COEFFICIENT

FRICTION
    NT DIFFERENTIAL FRICTION
        DYNAMIC FRICTION
        FABRIC TO FABRIC FRICTION
        FIBER FIBER FRICTION
        (CONT.)

NT FIBER FRICTION
FIBER TO METAL FRICTION
FLUID FRICTION
INTERNAL FRICTION
ROLLING FRICTION
SLIDING FRICTION
SURFACE FRICTION
STATIC FRICTION
STICK SLIP FRICTION
YARN TO METAL FRICTION
YARN TO YARN FRICTION
RT ABRASION
ABRASION RESISTANCE
ABRASION RESISTANCE FINISHES
ABRASIVES
ANGLE OF WRAP
APPARENT CONTACT AREA
ASPERITIES
BELT SLIPPAGE
BOUNDARY LUBRICATION
BRAKES (FOR ARRESTING MOTION)
BRAKING
CENTRIFUGAL FORCE
COEFFICIENT OF FRICTION
COHESION
DISC TENSION
DRAG
DURABILITY
FATIGUE RESISTANCE
FIBER SLIPPAGE
FRICTION THEORY
FRICTIONAL CHARACTERISTICS
FRICTIONAL FORCE
GRINDING (MATERIAL REMOVAL)
HOT SPOTS
HYDRODYNAMIC LUBRICATION
INTERNAL FRICTION
LUBRICATION
MAGNETIC BRAKES
MECHANICAL BRAKES
NORMAL FORCE
PLASTIC FLOW
PLOUGHING
PRESSURE DROP
ROLLER SLIPPAGE
ROLLING
ROUGHNESS
ROUGHNESS COEFFICIENT
RUBBING
SCUFFING
SKIN IRRITATION
SLIDING
SLIPPAGE
SNAG RESISTANCE
SYTON (TN)
TEFLON BEARINGS
TEFLON FIBER (TN)
TRUE CONTACT AREA
UNDERLAY (CARPETS)
VELOCITY (RELATIVE)
VOLUME WEAR
WEAR
WEAR RESISTANCE
WEAVING DYNAMICS
WEDGING
WELDING
WRAP
WRINKLE RECOVERY

FRICTION CALENDERING
BT CALENDERING
RT EMBOSSING
POLISHING
PRESSING
SCHREINERING

FRICTION COEFFICIENT
USE COEFFICIENT OF FRICTION

FRICTION LET OFF
BT LET OFF
RT LET OFF BRACKETS
LET OFF CHAINS
LET OFF SPOOLS
LET OFF WEIGHT ARM
LET OFF WEIGHTS
LOOMS
TAKE UP
WARP BEAMS
WARP TENSION
WEAVING

FRICTION THEORY
RT APPARENT CONTACT AREA
ASPERITIES
COEFFICIENT OF FRICTION
FRICTION
HOT SPOTS
MECHANISM (FUNDAMENTAL)
NORMAL FORCE
PLASTIC FLOW
PLOUGHING
THEORIES
TRUE CONTACT AREA
WELDING

FRICTIONAL CHARACTERISTICS
BT MECHANICAL PROPERTIES
RT COEFFICIENT OF FRICTION
DIFFERENTIAL FRICTION
FRICTION
FRICTIONAL FORCE
STICK SLIP FRICTION

FRICTIONAL FORCE
BT FORCE
RT ABRASION
CENTRIFUGAL FORCE
COEFFICIENT OF FRICTION
DYNAMIC FRICTION
FABRIC TO FABRIC FRICTION
FIBER FIBER FRICTION
FIBER TO METAL FRICTION
FRICTION
FRICTIONAL CHARACTERISTICS
INTERNAL FRICTION
NORMAL FORCE
RUBBING
STATIC FRICTION
TANGENTIAL FORCE
YARN TO METAL FRICTION
YARN TO YARN FRICTION

FRILLS (NOVELTY YARNS)
BT NOVELTY YARNS
RT BEAD YARNS
BOUCLES
BUG YARNS
FLAKE YARNS
FLAMES (NOVELTY YARNS)
KNICKERBOCKER YARNS
LOOPS (NOVELTY YARNS)
NOVELTY TWISTERS
NUB YARNS
RATINES
SEEDS (NOVELTY YARNS)
SLUB YARNS
SPIRALS (NOVELTY YARNS)
SPLASHES

FRINGE (NARROW FABRIC)
UF HEADING (NARROW FABRIC)
BT NARROW FABRICS
RT SELVEDGES

FRINGED FIBRILS
BT FINE STRUCTURE (FIBERS)
RT FIBRILS
MICROFIBRILS

FRINGED MICELLES
BT FINE STRUCTURE (FIBERS)
MICELLES
RT AMORPHOUS REGION
CRYSTALLINE REGION
CHAIN FOLDING
CRYSTAL LATTICE
FIBRILS
MICROFIBRILS
SPHERULITIC FIBRILS

FRL DRAPE TESTER
BT TESTING EQUIPMENT
RT BENDING RIGIDITY
BENDING TESTING
DRAPE
DRAPEOMETER

FROCKS
BT GARMENTS
OUTERWEAR
RT COATS
DRESSCLOTHING
DRESSES
JACKETS
SACKS (DRESSES)
SUITS

FRONT CROSSING HEDDLES
RT BACK CROSSING HEDDLES
BENZYLCELLULOSE
DOUP
GROUND HEDDLES
LENO WEAVES
LENO WEAVING
STANDARD HEDDLES

FRONT ROLLERS
USE FRONT ROLLS

FRONT ROLLS
UF FRONT ROLLERS
BT DRAFTING ROLLS
ROLLS
RT BACK ROLLS
DELIVERY ROLLS
DRAFT
DRAFTING (STAPLE FIBER)
DRAFTING THEORY
DRAWFRAMES
FALLER SCREWS
FLOATING FIBERS
GILL BOXES
INTERSECTING GILL BOXES
RATCH
RING SPINNING
RT ROLLER NIP
SPINNING FRAMES
VELOCITY CHANGE POINT

FRONTS
BT GARMENT COMPONENTS
RT FACING
INTERLININGS
LAPELS

FROSTING
UF NON-FROSTING
BT DEFECTS
FABRIC DEFECTS
RT ABRASION
FIBRILLATION

FRUIT FIBERS
NT COIR
BT CELLULOSIC FIBERS
FIBERS
NATURAL FIBERS
RT BAST FIBERS
LEAF FIBERS
SEED HAIR FIBERS (GENERAL)

FTORLON (TN)
BT FLUOROCARBON FIBERS
HIGH TEMPERATURE FIBERS
POLYTRIFLUOROCHLOROETHYLENE FIBERS
RT POLYTETRAFLUOROETHYLENE
TEFLON FIBER (TN)

FUGITIVE DYES
UF FUGITIVE TINTS
RT BACKWASHING
BLEACHING
BLEND DYEING
TINTING

FUGITIVE TINTS
USE FUGITIVE DYES

FULL CARDIGAN FABRICS
BT FILLING KNITTED FABRICS
KNITTED FABRICS
RT CIRCULAR KNITTED FABRICS
FLAT KNITTED FABRICS
FULLY FASHIONED FABRICS
HALF CARDIGAN FABRICS

FULL FACE FINISH
USE FACE FINISH

FULL PACKAGES
BT PACKAGES
RT EMPTY PACKAGES
BOBBINS
WINDING
WINDING TENSION
WINDING ANGLE
BUILDER MOTIONS
SLOUGHING
SPINNING
QUILLS
DOFFING (PACKAGE)

FULL SEAM WIDTH
BT SEAM SIZE
SEAM SPECIFICATIONS
RT BUTTED SEAMS
BUTTERFLY SEAMS
COVERED SEAMS
SEAM THICKNESS
SEAM WIDTH
SEAMING
SEAMS
SEWING
STITCHING

FULL SET
RT GUIDE BARS
HALF SET
OVERLAP
RASCHEL KNITTING MACHINES
RUNNER RATIO
SET (WARP KNITTING)
TRICOT KNITTING MACHINES
WARP KNITTING MACHINES

FULLING
UF MILLING
BT WET FINISHING
RT BAULK FINISHING
BEETLING
BRUSHING
DIMENSIONAL STABILITY
FELTING
FELTS
FILLING SHRINKAGE
GATTERWALKING (FULLING MACHINES)
HARDENING (FELTS)
NONWOVEN FABRICS
OPEN WIDTH FINISHERS (FULLING)
PATTERN DEFINITION
PILGRIMSTEP (FULLING MACHINES)

RT PLANKING (FULLING MACHINES)
REFULLING MACHINES
SHRINKAGE
SIMULTANEOUS FINISHING PROCESSES

FULLING MACHINES
NT GATTERWALKING (FULLING MACHINES)
MILLS (FULLING ROTARY)
MILLS (KICKER)
OPEN WIDTH FINISHERS (FULLING)
PILGRIMSTEP (FULLING MACHINES)
PLANKING (FULLING MACHINES)
REFULLING MACHINES
RT FELTING MACHINES

FULLY FASHIONED FABRICS
BT FILLING KNITTED FABRICS
KNITTED FABRICS
RT DOUBLE KNIT FABRICS
FULL CARDIGAN FABRICS
HALF CARDIGAN FABRICS
KNITTING
MACHINE KNITTING
PLAIN KNITTED FABRICS
RIB KNITTED FABRICS

FULLY FASHIONED HOSIERY
UF FULLY FASHIONED STOCKINGS
BT HOSIERY
RT CIRCULAR HOSIERY
DENIER
FASHIONING
FASHIONING MARKS
FOOTWEAR
FULLY FASHIONED KNITTING MACHINES
HALF HOSE
HOSIERY MACHINES
KNITTING
KNITTING MACHINES
LADDERING
MOCK FASHIONING MARKS
MOCK SEAMS
NARROWING (KNITTING)
PRESSING
RUN RESISTANCE
RUN RESISTANT HOSIERY
SEAMLESS HOSIERY
SHAPING
SHEERNESS
SOCKS
STITCH TRANSFER
STOCKINGS
STRETCH HOSIERY
WIDENING (KNITTING)

FULLY FASHIONED HOSIERY MACHINES
BT FULLY FASHIONED KNITTING MACHINES
HOSIERY MACHINES
RT FULLY FASHIONED HOSIERY
MOCK SEAMS
POSITIVE FEED
SEAMLESS HOSIERY MACHINES

FULLY FASHIONED KNITTING
BT FILLING KNITTING
KNITTING
RT CARDIGANS
CIRCULAR KNITTING
FASHIONING
FASHIONING MARKS
FILLING KNITTING MACHINES
FLAT KNITTING
FULLY FASHIONED KNITTING MACHINES
HOSIERY
JACQUARD KNITTING
JERKINS
JUMPERS
KNITTED FABRICS
LANDING LOOPS
NARROWING (KNITTING)
NEEDLE BARS
PULLOVERS
SLIP ONS
SWEATERS
WIDENING (KNITTING)
SOCKS

FULLY FASHIONED KNITTING MACHINES
NT FULLY FASHIONED HOSIERY MACHINES
BT FILLING KNITTING MACHINES
FLAT KNITTING MACHINES
KNITTING MACHINES
RT CARRIER RODS
CASTING OFF
CIRCULAR KNITTING MACHINES
DIVIDERS (KNITTING)
FASHIONING
FILLING KNITTING
FULLY FASHIONED HOSIERY
FULLY FASHIONED KNITTING
HOSIERY
KNITTING STATIONS
KNOCKOVER
LANDING LOOPS
LINKS LINKS MACHINES
LOOP TRANSFER
NARROWING COMBS
NEEDLE BARS
NEEDLES
PRESSER BARS
PRESSING (BEARDED NEEDLES)
RACKING
RIB TRANSFER
SINKERS
SINKING (KNITTING)
SLURCOCKS
WEBHOLDERS
YARN CARRIERS

FULLY FASHIONED STOCKINGS
USE FULLY FASHIONED HOSIERY

FUME FADING
NT N FADING
O FADING
BT FADING
RT GAS FUME FASTNESS

FUNGICIDES
UF FUNGISTATS
BT ANTIMICROBIAL FINISHES
PRESERVATIVES
RT SOIL BURIAL TEST

FUNGISTATS
USE FUNGICIDES

FUNGUS
NT YEAST
RT DEGRADATION
DEGRADATION PROPERTIES
FABRIC PROPERTIES (DEGRADATION)
MICROBIOLOGICAL DEGRADATION
MILDEW RESISTANCE

FUNGUS RESISTANCE
USE MILDEW RESISTANCE

FUR
BT NATURAL FIBERS
PROTEIN FIBERS
RT ARTIFICIAL FURS
CARROTING
GUARD HAIRS
HAIR
SILK
VEGETABLE FIBERS (GENERAL)
WOOL

FURNISHING FABRICS
BT FABRICS (BY END USES)
WOVEN FABRICS
RT AGRICULTURAL FABRICS
AUTOMOTIVE FABRICS
CARPETS
DECORATIVE FABRICS
DRAPES
END USE PROPERTIES
END USES
FABRICS (ACCORDING TO STRUCTURE)
FURNISHINGS
HOUSEHOLD FABRICS
INDUSTRIAL FABRICS
TAPESTRIES
UNDERLAY (CARPETS)
UPHOLSTERY
UPHOLSTERY FABRICS

FURNISHINGS
UF FURNITURE
RT AUTOMOTIVE FABRICS
CARPETS
CUSHIONS
DRAPES
END USES
FURNISHING FABRICS
LAMPSHADES
MATS
TAPESTRIES
UPHOLSTERY
UPHOLSTERY FABRICS

FURNITURE
USE FURNISHINGS

FUSING (MELTING)
USE FUSION (MELTING)

FUSION (MELTING)
UF FUSING (MELTING)
RT ADHESION
BASTING
BONDED FIBER FABRICS
BONDING
CEMENTING
COHESION
FOAM BACKED FABRICS
GLUING
HEAT BONDING
HEAT OF FUSION
HEAT SEALING
JOINING
LAMINATED FABRICS
MELTING
POST FORMED LAMINATES
SEALING
SEAMING
SEAMS
TAPING
THERMOPLASTICITY
YARN SEVERANCE

FUSION CONTRACTION SHRINKAGE
BT SHRINKAGE
RT COMPACTING
COMPRESSIVE SHRINKING
DIMENSIONAL STABILITY
FELTING SHRINKAGE
FILLING SHRINKAGE
LONDON SHRINKING
MOLDING SHRINKAGE
PRESHRINKING
RELAXATION SHRINKAGE
RESIDUAL SHRINKAGE
SHRINK GRADING
SHRINK RESISTANCE
SHRINKING
WARP SHRINKAGE

FUZZ
BT FABRIC DEFECTS
RT BOTTOM COVER
FUZZ RESISTANCE
HAIRINESS
HAND
LINT (WASTE)
NAP
PILLS
TOP COVER

FUZZ RESISTANCE
RT ENTANGLEMENTS
FABRIC DEFECTS
FIBER MIGRATION
FUZZ
KNITTED FABRICS
NEP POTENTIAL
PILL RESISTANCE
PILL WEAR OFF
PILLING
PILLS

FUZZINESS
RT CLEANLINESS
HAIRINESS
HAND
SHEDDING (FIBER)
SMOOTHNESS

FUZZING
RT ENTANGLEMENTS
FIBER MIGRATION
PILLING
RUBBING

FUZZY MOTES
RT MOTE KNIVES

GABARDINE
BT FLAT WOVEN FABRICS
WOVEN FABRICS
RT TWILL WEAVES

GADGETS
RT ACCESSORIES
DEVICES
MECHANISMS

GAIN (WINDING)
RT CLOSE WINDING
DRUM WINDERS
LAYER LOCKING
OPEN WIND
OPEN WINDING
PRECISION WINDERS
PRECISION WINDING
PROCESS VARIABLES
TRAVERSE
WIND
WINDING ANGLE

GALLING
RT ABRASION
CUTTING
FRICTION
GRINDING (MATERIAL REMOVAL)
PLOUGHING
POLISHING
SCORING
SCOURING
SCRATCHING
WEAR
WEDGING

GALLOON
BT LACES

GALOSHES
BT FOOTWEAR
RAINWEAR
RT BOOTS
LEGGINGS
RUBBER
RUBBER FOOTWEAR
SHOES
SNEAKERS
SOCKS
SPATS

GAMMA DISTRIBUTION
RT BINOMIAL DISTRIBUTION
DISTRIBUTION
F DISTRIBUTION
FIBER DIAMETER
NORMAL DISTRIBUTION

GAMMA RADIATION
USE GAMMA RAYS

GAMMA RAYS
UF GAMMA RADIATION
BT IONIZING RADIATION
RADIATION
RT GRAFTING
IRRADIATION
PHOTONS
X RAYS

GARDINOL (TN)
BT ALKYL SULFATES

GARMENT COMPONENTS
NT BELTS (APPAREL)
BODICES
BUSTS
BUTTONHOLES
BUTTONS
COLLAR STIFFENERS
COLLARS
CUFFS
DARTS
FACING
FASTENERS
FINDINGS
FLY (GARMENT)
FRONTS
HEMS
HOOK CLOSURES
INTERLININGS
LAPELS
LININGS
PLEATS
POCKETS
REVERES
RUFFLES
SLEEVES
SNAP FASTENERS
STIFFENINGS (GARMENT)
STITCHES (SEWING)
TRIMMINGS
VELCRO (TN)
ZIPPERS
RT CREASES
FOOTWEAR COMPONENTS
GARMENT MANUFACTURE
GARMENT MANUFACTURING MACHINES
TAILORING

GARMENT DESIGN
USE APPAREL DESIGN

GARMENT DYEING
NT HOSIERY DYEING
BT DYEING (BY MATERIAL ASSEMBLY)
RT DRY CLEANING
GARMENT DYEING MACHINES
GARMENTS
PIECE DYEING
STRIP DYEING

GARMENT DYEING MACHINES
NT HOSIERY DYEING MACHINES
BT DYEING MACHINES
DYEING MACHINES (BY MATERIAL ASSEMBLY)
RT BATCH DYEING MACHINES
DRY CLEANING
DYEING
DYEING (BY MATERIAL ASSEMBLY)
GARMENT DYEING
PIECE DYEING MACHINES
ROTARY DYEING MACHINES

GARMENT MANUFACTURE
UF MAKING UP
NT ALTERATIONS
APPAREL DESIGN
BACK TACKING
BASTING
BLIND STITCHING
BOARDING
BUNDLING (TAILORING)
CHAIN SEWING
CURVED SEAMING
CUTTING (TAILORING)
DESIGN
FITTING
FLAT PRESSING
HEMMING
JOINING
LABELLING
LETTING OUT (GARMENTS)
LOOPING
MARKING
MOLDING (TAILORING)
PINKING (TAILORING)
PLEATING
PRESSING
RIPPING
SEAMING
SEWING
SHADING (TAILORING)
SHAPING
SHRINKING
SLOPING
SPONGING
SPREADING (TAILORING)
STAPLING (WIRE)
STITCHING
STYLING (APPAREL)
TAKING IN (GARMENTS)
TAILORING
TEARING (TAILORING)
TRIMMING (OPERATION)
TUCKING (GARMENT MANUFACTURE)
TUMBLING
RT CUTTING DRILLS
CUTTING MACHINES
DRESS DESIGN
DYEING
FIT
FITTERS (GARMENTS)
GARMENT COMPONENTS
GARMENT MANUFACTURING MACHINES
GARMENTS
KNITTING
LOOP CATCHING
MEASURING
MODELLING
PUCKER
RIVETS
SEAM ALLOWANCE
SEAM BASTING
SEAM BINDING
SEAM PUCKER
SEAMS
SEWING MACHINES
SEWING NEEDLES
SEWING THREADS
TAILORED SEAMS

GARMENT MANUFACTURING MACHINES
UF TAILORING MACHINES
NT CUTTING DRILLS
CUTTING MACHINES
DECATING MACHINES
LABELLING MACHINES
MARKERS
MEASURING MACHINES
PINKING MACHINES
PLEATING MACHINES
SEWING MACHINES
SLITTERS (CLOTH)
SPREADERS (CLOTH)
STEAMERS
RT GARMENT COMPONENTS
GARMENT MANUFACTURE

GARMENTS
UF APPAREL
CLOTHING
NT ATHLETIC CLOTHING
BELTS (APPAREL)
BRASSIERES
BRIEFS
CARDIGANS
CEREMONIAL CLOTHING
COATS
CRAVATS
DRESSCLOTHING
DRESSES
DUNGAREES
FORM PERSUASIVE GARMENTS
FOUNDATION GARMENTS
FROCKS
GARTER BELTS
GARTERS
GIRDLES
INDUSTRIAL CLOTHING
INFANTS CLOTHING
JACKETS
JERKINS
JUMPERS
KNITTED OUTERWEAR
KNITTED UNDERWEAR
LEOTARDS
LINGERIE
LODEN COATS
MILITARY CLOTHING
NECKTIES
NIGHTWEAR
OUTERWEAR
OVERALLS
OVERCOATS
PANTIES
PLAYSUITS
PONCHOS
PULL OVERS
RAINWEAR
REVERSIBLE GARMENTS
SACKS (DRESSES)
SAFETY CLOTHING
SCARVES
SHIRTS
SHORTS
SKI PANTS
SKIRTS
SLACKS
SLIP ONS
SLIPS
SMOCKS
SPORTSWEAR
SUITS
SUSPENDERS
SWEATERS
SWEATSHIRTS
TROUSERS
UNDERWEAR
UNIFORMS
VESTS
WORK CLOTHING
RT ALTERATIONS
APPAREL DESIGN
APPAREL FABRICS
BAGGINESS
CAPS (HEADGEAR)
COMFORT
CUTTING (TAILORING)
DRESS DESIGN
EAR MUFFS
END USES
FABRICS
FIT
GARMENT DYEING
GARMENT MANUFACTURE
GLOVE LININGS
HATS
HELMETS
HOODS
HOSIERY
HOUSEHOLD FABRICS
LININGS
POCKET LININGS
PRODUCT DESIGN
PUCKER
RIVETS
TAILORING

GARNETT WIRES
USE METALLIC CARD CLOTHING

GARNETTING
RT CARDING
CLIPPINGS
METALLIC CARD CLOTHING
MUNGO
OPENING
PICKING (OPENING)
RANDO FEEDER (TN)
RECOVERED WOOL
SHODDY
WASTE

GARNETTING MACHINES
BT NONWOVEN FABRIC MACHINES
RT CARDS
FEARNOUGHTS
METALLIC CARD CLOTHING

RT MUNGO
OPENERS
RAG PULLING MACHINES
RANDO WEBBER (TN)
SHODDY
WASTE MACHINES

GARTER BANDS
USE AFTERWELT

GARTER BELTS
BT BELTS (APPAREL)
GARMENTS
UNDERWEAR
RT BRASSIERES
BRIEFS
GARTERS
LINGERIE
SUSPENDERS

GARTERS
BT GARMENTS
UNDERWEAR
RT BRASSIERES
BRIEFS
GARTER BELTS
SUSPENDERS

GAS (FUEL)
RT GAS HEATERS
GAS HEATING (GAS FUEL)

GAS FUME FASTNESS
BT COLORFASTNESS
DEGRADATION PROPERTIES
END USE PROPERTIES
RT LIGHTFASTNESS (COLOR)
FUME FADING
WASHFASTNESS (COLOR)

GAS FUME FASTNESS TESTING
BT COLORFASTNESS TESTING

GAS HEATERS
BT HEATING EQUIPMENT
RT GAS (FUEL)
GAS HEATING (GAS FUEL)

GAS HEATING (GAS FUEL)
BT HEATING
RT DIELECTRIC HEATING
DRYING
ELECTRIC HEATING (RESISTANCE)
GAS (FUEL)
GAS HEATERS
HEATING EQUIPMENT
OIL HEATING (OIL FUEL)
RADIANT HEATING

GAS PHASE
RT VAPOR PHASE

GASKETS
RT INDUSTRIAL FABRICS
MECHANISMS
SEALS
WASHERS

GATE TENSION
RT BALL AND SOCKET TENSION
DISC TENSION
FINGER TENSION
HYDRAULIC TENSION DEVICES
MAGNETIC TENSION DEVICES
TENSION
TENSION CONTROL
TENSION DEVICES
TENSION DISCS
TENSION SPRINGS
WASHERS
WINDING
WRAP

GATHERING
BT FABRIC DISTORTION (TAILORING)
RT BACK TACKING
BASTING
BLIND STITCHING
CHAIN SEWING
COCKLING
CURVED SEAMING
GATHERS (TAILORING)
HEMMING
PINCHING
PUCKERING
RIPPLES
RUFFLING
SEAM PUCKER
SEAMING
SEAMS
SERGING
SEWING
SHIRRING
STITCH QUALITY
STITCHES (SEWING)
STITCHING
TAILORING
TRIMMING (OPERATION)
YARN SEVERANCE

GATHERS (TAILORING)
RT GATHERING
PLEATS
PUCKERS (TAILORING)
STITCHES (SEWING)

GATTERWALKING (FULLING MACHINES)
BT FULLING MACHINES
RT FULLING
MILLS (FULLING ROTARY)
MILLS (KICKER)
OPEN WIDTH FINISHERS (FULLING)
PILGRIMSTEP (FULLING MACHINES)
PLANKING (FULLING MACHINES)
REFULLING MACHINES

GAUGE (WIRE)
USE CARD GAUGE

GAUGE LENGTH
UF SPECIMEN LENGTH
RT BREAKING ELONGATION
RATCH
STRAIN
STRAIN RATE
STRENGTH ELONGATION TESTING
STRESS STRAIN CURVES

GAUSSIAN DISTRIBUTION
USE NORMAL DISTRIBUTION

GAUZE
BT FLAT WOVEN FABRICS
MEDICAL TEXTILES
WOVEN FABRICS
RT BANDAGES
GAUZE WEAVES
LENO WEAVES
PLAIN WEAVES
SANITARY NAPKINS

GAUZE WEAVES
RT GAUZE
LENO WEAVES
MARQUISETTE

GEAR CRIMPED YARNS
BT TEXTURED YARNS (FILAMENT)
RT AIR JET TEXTURED YARNS
BICOMPONENT FIBER YARNS (FILAMENT)
BICOMPONENT FIBER YARNS (STAPLE)
EDGE CRIMPED YARNS
FALSE TWIST YARNS
STUFFER BOX CRIMPED YARNS

GEAR CRIMPING
BT CRIMPING
TEXTURING
RT AIR JET TEXTURING
BULKED YARNS
CRIMPED YARNS
CRIMPING MACHINES
EDGE CRIMPING
FALSE TWISTING (TEXTURING)
STUFFER BOX CRIMPING
TOW CRIMPING
Z CRIMP

GEAR DRIVES
BT DRIVES
RT BELT DRIVES
CHAIN DRIVES
DEVICES
FLUID DRIVES
GEARS
HYDRAULIC DRIVES
MECHANICAL DEVICES
MECHANICAL DRIVES
TWIST GEARS
VARIABLE SPEED DRIVES

GEARS
NT TWIST GEARS
RT ACCESSORIES
BELT DRIVES
CAMS
DEVICES
DRIVES
GEAR DRIVES
LINKAGES
MECHANISMS

GEIGER COUNTERS
BT ELECTRONIC INSTRUMENTS
MEASURING INSTRUMENTS

GELATINE
BT PROTEINS
RT DESIZING
GLUES
LUBRICANTS
PRESERVATIVES
RESINS
SIZES (SLASHING)
SLASHING
STARCH

GELLATION
USE GELLING

GELLING
UF GELLATION
RT COAGULATING
CROSSLINKING
GELS
POLYMERS
THICKENING

GELS
BT COLLOIDS
RT CROSSLINKING
GELLING
POLYMERS
SOLS
SOLUTIONS (LIQUID)
THERMAL STABILITY

GENERAL SOLUTIONS
BT SOLUTIONS (MATHEMATICAL)
RT DIFFERENTIAL EQUATIONS
EQUATIONS
GRAPHICAL SOLUTIONS
INTEGRATION
LINEAR DIFFERENTIAL EQUATIONS
MATHEMATICAL ANALYSIS
NUMERICAL SOLUTIONS
PARTICULAR SOLUTIONS
TRIAL AND ERROR SOLUTIONS

GENOA VELVET
BT VELVET
WOVEN FABRICS

GEOMETRIC POROSITY
USE POROSITY

GEOMETRICAL PROPERTIES
USE STRUCTURAL PROPERTIES

GEOMETRY
NT FABRIC GEOMETRY
FIBER GEOMETRY
YARN GEOMETRY
RT COMPUTATIONS
CYLINDERS
DIFFERENTIAL GEOMETRY
FABRIC PROPERTIES
FABRIC PROPERTIES (STRUCTURAL)
GROOVES
HOLLOWNESS
LENGTH
SHAPE
SURFACES
TRELLIS MODEL

GEORGETTE
NT CREPON GEORGETTE
BT WOVEN FABRICS
RT CREPE YARNS
PLAIN WEAVES

GERMICIDES
USE ANTIMICROBIAL FINISHES

GIANT SADDLE STITCHES
BT STITCHES (SEWING)
RT SEWING

GIGGING
USE NAPPING

GILBERT-RIDEAL THEORY
BT DYEING THEORY
RT DONNAN THEORY

GILL BOXES
NT INTERSECTING GILL BOXES
BT DRAFTING MACHINES
RT ACTUAL DRAFT
BACK ROLLS
BRADFORD OPEN DRAWING
BREAK DRAFT
DRAFTING (STAPLE FIBER)
DRAWFRAMES
FALLER BARS
FALLER SCREWS
FINISHER GILLING
FRONT ROLLS
GILLING
PIN DRAFTERS
RATCH
WORSTED SYSTEM PROCESSING
WORSTED SYSTEM SPINNING

GILL DRAFTING
USE GILLING

GILL PREPARING
USE GILLING

GILLING
UF GILL DRAFTING
GILL PREPARING
NT FINISHER GILLING
BT DRAFTING (STAPLE FIBER)
(CONT.)

BT WORSTED SYSTEM PROCESSING
RT ACTUAL DRAFT
APRONS
BRADFORD OPEN DRAWING
BRADFORD SYSTEM PROCESSING
BREAK DRAFT
CONTROL SURFACES
DRAFT
DRAFTING (STAPLE FIBER)
DRAWFRAMES
FALLER BARS
FALLER SCREWS
FALLER SPEED
FIBER ORIENTATION
GILL BOXES
HIGH DRAFT SPINNING
INTERSECTING GILL BOXES
LONG STAPLE SPINNING
NEW BRADFORD SYSTEM PROCESSING
PIN DRAFTING
SLIVERS
WORSTED SYSTEM SPINNING

GIN CUT
RT COTTON GINS
GINNING

GIN FALL
RT COTTON GINS
GINNING

GINGHAM
BT FLAT WOVEN FABRICS
WOVEN FABRICS
RT PLAIN WEAVES

GINNING
UF COTTON GINNING
NT KNIFE BLADE GINNING
ROLLER GINNING
SAW BLADE GINNING
BT CLEANING
COTTON SYSTEM PROCESSING
RT BALES
BALING
BOLLS
CLEANING
COTTON
COTTON GINS
COTTON SEEDS
FIBER BREAKAGE
GIN CUT
GIN FALL
LINT (FIBER)
LINT COLOR
LINT CONTENT
NEPS
PICKING (HARVESTING)
TRASH
WASTE

GINS (COTTON)
USE COTTON GINS

GIRDLES
BT FOUNDATION GARMENTS
UNDERWEAR
RT BRASSIERES
BRIEFS
LINGERIE
PANTIES
POWER (FABRIC)
SLIPS
SPANDEX FIBERS

GLASS
RT ASBESTOS
BRITTLENESS
CERAMICS
GLASS CLOTH
GLASS FIBERS
GLASS MAT
INORGANIC FIBERS (MAN MADE)
MATERIALS
MINERAL FIBERS

GLASS CLOTH
RT GLASS
GLASS FIBERS
LAMINATED FABRICS
REINFORCED COMPOSITES

GLASS FIBERS
UF GLASS FILAMENTS
NT FIBERGLAS (TN)
BT HIGH TEMPERATURE FIBERS
INORGANIC FIBERS (MAN MADE)
MAN MADE FIBERS
RT CERAMIC FIBERS
EXTRUSION
FILAMENTS
GLASS
GLASS MAT
MELT SPINNING
METALLIC FIBERS
MINERAL FIBERS
SYNTHETIC FIBERS
YARN REINFORCED COMPOSITES

GLASS FILAMENTS
USE GLASS FIBERS

GLASS MAT
RT FIBER REINFORCED COMPOSITES
GLASS
GLASS FIBERS
LAMINATED FABRICS
LAMINATES
REINFORCED COMPOSITES
YARN REINFORCED COMPOSITES

GLASS RUBBER TRANSITION
BT TRANSITIONS (POLYMERS)
RT GLASS TRANSITION TEMPERATURE
MOISTURE CONTENT
PRESSURE

GLASS TRANSITION TEMPERATURE
UF PRIMARY TRANSITION
TEMPERATURES
BT TRANSITION TEMPERATURES
RT BIREFRINGENCE
COEFFICIENT OF EXPANSION
DIFFUSION COEFFICIENT
ELASTIC MODULUS (TENSILE)
GLASS RUBBER TRANSITION
HEAT CAPACITY
HEAT OF FUSION
MECHANICAL DAMPING
MELTING
MELTING POINT
MODULUS
MOLECULAR CHAIN MOVEMENT
NUCLEAR MAGNETIC RESONANCE
SECOND ORDER TRANSITION
TEMPERATURE
SECONDARY TRANSITION
TEMPERATURES
SOFTENING POINT
STRESS OPTICAL COEFFICIENT
SUBGROUP MOVEMENT (MOLECULAR)
TEMPERATURE
TRANSITIONS
VOLUME TEMPERATURE
MEASUREMENTS
X RAY ANALYSIS

GLAUBERS SALT
BT SODIUM SULFATE
SULFATES
RT DYEING AUXILIARIES
LEVELLING AGENTS
RETARDING AGENTS

GLAZE
RT CALENDERING
GLOSS
LUSTER
OPACITY
PRESSING
SINGEING

GLAZING
BT DRY FINISHING
WET FINISHING

GLOSS
RT BLOOM
FABRIC PROPERTIES
FABRIC PROPERTIES (AESTHETIC)
GLAZE
LUSTER
OPACITY
SMOOTHNESS
SPARKLE

GLOSPAN (TN)
BT SPANDEX FIBERS

GLOVE LININGS
BT LININGS
RT BOX LININGS
CASKET LININGS
FABRICS
GARMENTS
GLOVES
INTERLININGS
POCKET LININGS
SHOE LININGS

GLOVES
BT ATHLETIC CLOTHING
INDUSTRIAL CLOTHING
OUTERWEAR
RT COATS
GLOVE LININGS
WORK CLOTHING

GLUCOSE
UF DEXTROSE
BT SUGARS
RT SUCROSE

GLUES
RT ADHESIVES
BONDING STRENGTH
CEMENTS
GLUING
GUMS
PASTES
SEALERS

GLUING
BT JOINING
RT ADHESION
BONDING
CEMENTING
COHESION
FUSION (MELTING)
LAMINATING
PEELING
SEALING
SEAMING
TAPING

GLUTARALDEHYDE
BT DIALDEHYDES
RT ADIPALDEHYDE

GLYCERIN
USE GLYCEROL

GLYCERIN DICHLOROHYDRIN
UF DICHLOROPROPANOL
BT CHLOROHYDRINS

GLYCERIN MONOCHLOROHYDRIN
UF MONOCHLOROHYDRIN
BT CHLOROHYDRINS

GLYCEROL
UF GLYCERIN
BT POLYOLS
RT DELIQUESCENT AGENTS
ETHYLENE GLYCOL
GLYCOLS
HUMECTANTS
PENTAERYTHRITOL

GLYCEROPHOSPHATES
UF LECITHIN
PHOSPHATED GLYCERIDES
BT PHOSPHATE SURFACTANTS

GLYCERYL ESTER SURFACTANTS
NT FATTY DIGLYCERIDES
FATTY MONOGLYCERIDES
POLYGLYCERYL ESTERS
BT NONIONIC SURFACTANTS
RT FATS
TRIGLYCERIDES

GLYCIDYL (PREFIX)
USE OXIRANE COMPOUNDS

GLYCIDYL COMPOUNDS
USE OXIRANE COMPOUNDS

GLYCOL ESTER SURFACTANTS
NT DIGLYCOL ESTER SURFACTANTS
PROPANEDIOL ESTER SURFACTANTS
BT NONIONIC SURFACTANTS

GLYCOLIC ACID
USE HYDROXYACETIC ACID

GLYCOLS
UF DIOLS
NT ETHYLENE GLYCOL
POLY (ETHYLENE GLYCOL)
POLY (PROPYLENE GLYCOL)
BT POLYOLS
RT GLYCEROL
HUMECTANTS

GLYOXAL
BT DIALDEHYDES
RT METHYLOL ACETYLENE DIUREA

GOAT BREEDING
BT ANIMAL HUSBANDRY
RT GOAT HAIR
GOATS
SHEEP BREEDING

GO THROUGH MACHINES
BT LACE MACHINES
RT BAR WARP MACHINES
BARMEN MACHINES
BOBBINET MACHINES
CURTAIN MACHINES
DOUBLE LOCKER MACHINES
LEAVERS MACHINES
MECHLIN MACHINES
ROLLING LOCKER MACHINES
SIVAL MACHINES
STRING WARP MACHINES
WARP LACE MACHINES

GOAT HAIR
BT HAIR
KERATIN FIBERS
PROTEIN FIBERS
RT GOAT BREEDING
GOATS

GOATS
BT ANIMALS
RT ANIMAL HUSBANDRY
GOAT BREEDING
GOAT HAIR

GOBELIN TAPESTRIES
BT TAPESTRIES

GODET ROLLS
BT ROLLS
RT SOLVENT SPINNING
SPINNING ASSEMBLIES (EXTRUSION)
WET SPINNING

GOING PART (LOOM)
USE SLEYS (LOOM)

GOLD CHLORIDE
BT CHLORIDES
GOLD COMPOUNDS

GOLD COMPOUNDS
NT GOLD CHLORIDE
RT MOLYBDENUM COMPOUNDS
SILVER COMPOUNDS
TUNGSTEN COMPOUNDS

GOLDEN CAPROLAN (TN)
BT NYLON (POLYAMIDE FIBERS)
NYLON 6

GOOD MIDDLING (COTTON GRADE)
BT COTTON GRADE
GRADE (FIBERS)
RT MIDDLING (COTTON GRADE)

GOOD ORDINARY (COTTON GRADE)
BT COTTON GRADE
GRADE (FIBERS)
RT MIDDLING (COTTON GRADE)
STRICT GOOD ORDINARY (COTTON GRADE)

GOTHIC TAPESTRIES
BT TAPESTRIES

GOWNS
BT CEREMONIAL CLOTHING
NIGHTWEAR
OUTERWEAR
RT COATS
DRESSCLOTHING
DRESSES

GRAB STRENGTH
BT FABRIC PROPERTIES
FABRIC PROPERTIES (MECHANICAL)
STRENGTH
RT BREAKING ELONGATION
BREAKING LENGTH
BREAKING STRENGTH
FABRIC STRENGTH
RAVEL STRIP STRENGTH
STRENGTH ELONGATION TESTING
UNIAXIAL STRESS

GRADE (FIBERS)
UF FIBER QUALITY
GRADE STANDARDS
NT BRAID (WOOL GRADE)
COMMON (WOOL GRADE)
COTTON GRADE
FINE (WOOL GRADE)
FINE MEDIUM (WOOL GRADE)
GOOD MIDDLING (COTTON GRADE)
GOOD ORDINARY (COTTON GRADE)
LOW MIDDLING (COTTON GRADE)
LOW QUARTER BLOOD (WOOL GRADE)
MEDIUM (WOOL GRADE)
MEDIUM FINE (WOOL GRADE)
MIDDLING (COTTON GRADE)
MIDDLING FAIR (COTTON GRADE)
ONE HALF BLOOD (WOOL GRADE)
QUARTER BLOOD (WOOL GRADE)
STRICT GOOD MIDDLING (COTTON GRADE)
STRICT GOOD ORDINARY (COTTON GRADE)
STRICT LOW MIDDLING (COTTON GRADE)
STRICT MIDDLING (COTTON GRADE)
THREE EIGHTHS BLOOD (WOOL GRADE)
WOOL GRADE
BT FIBER PROPERTIES
RT CAUSTICAIRE VALUE
CLASS (FIBERS)
COEFFICIENT OF FINENESS VARIATION
COLOR GRADING
COTTON CLASS
CROSSBRED WOOL
FIBER FINENESS
FIBERS
GRADING
IMMATURE FIBERS
IMMATURITY (FIBER)
LINEAR DENSITY
MATURITY INDEX
MERINO WOOL
MICRONAIRE

GRADE STANDARDS
USE GRADE (FIBERS)

GRADING
NT COLOR GRADING
COTTON GRADING
WOOL GRADE
RT CLASS (FIBERS)
CLASSING
COTTON CLASSING
COTTON GRADE
COTTON QUALITY
EFFECTIVE LENGTH (FIBER)
FIBER FINENESS
FIBER LENGTH
FIBER LENGTH DETERMINATION
FIBROGRAPH
FINENESS TESTING
GRADE (FIBERS)
LAMBSWOOL
SORTING
SPAN LENGTH
STAPLING
SUTER WEBB TESTING
TESTING
TOP MATCHING
WOOL
WOOL SORTING

GRAFT COPOLYMERS
USE GRAFT POLYMERS

GRAFT POLYMERIZATION
USE GRAFTING

GRAFT POLYMERS
UF GRAFT COPOLYMERS
BT COPOLYMERS
RT BLOCK POLYMERS
GRAFTING
RADIATION GRAFTING

GRAFTING
UF GRAFT POLYMERIZATION
NT RADIATION GRAFTING
BT COPOLYMERIZATION
RT CERIUM COMPOUNDS
CHEMICAL MODIFICATION (FIBERS)
CROSSLINKING
DEPOSITION
ETHOXYLATION
ETHYLENE IMINE
ETHYLENE OXIDE
FREE RADICALS
GAMMA RAYS
GRAFT POLYMERS
INITIATION
INITIATORS
IONIZING RADIATION
IRRADIATION
LACTONES
STYRENE
VINYL COMPOUNDS

GRAIN BOUNDARIES
RT AMORPHOUS REGION
CRACKS (MECHANICAL)
CRYSTALLINE REGION
DISLOCATIONS
FINE STRUCTURE (FIBERS)

GRAMS PER DENIER
USE TENACITY

GRANITE WEAVES
BT WEAVE (TYPES)
RT FLAT WOVEN FABRICS

GRAPHICAL SOLUTIONS
BT SOLUTIONS (MATHEMATICAL)
RT EXTRAPOLATION
GENERAL SOLUTIONS
INTERPOLATION
NOMOGRAMS
NUMERICAL SOLUTIONS
PARTICULAR SOLUTIONS
TRIAL AND ERROR SOLUTIONS

GRAPHITE
BT CARBON
RT GRAPHITE FIBERS
LUBRICANTS

GRAPHITE FIBERS
BT HIGH TEMPERATURE FIBERS
INORGANIC FIBERS (MAN MADE)
MAN MADE FIBERS
RT CARBON
GRAPHITE

GRAPHS
RT COMPUTATIONS
NOMOGRAMS

GRASS BLEACHING
BT BLEACHING
WET FINISHING
RT CONTINUOUS BLEACHING

GRAVIMETRIC ANALYSIS
BT CHEMICAL ANALYSIS
RT QUANTITATIVE ANALYSIS
VOLUMETRIC ANALYSIS

GRAVITATION
BT FORCE
RT ACCELERATION (MECHANICAL)
ACCELEROMETERS
CORIOLIS ACCELERATION
INERTIA
MASS
MOMENTUM
VELOCITY
WEIGHT

GREASE
RT GREASE RECOVERY
LANOLIN
LUBRICANTS
LUBRICATION
SCOURING
SUINT
WOOL GREASE
WOOL WAX

GREASE RECOVERY
RT GREASE
LANOLIN
LUBRICATION
RECOVERY (WASTE)
SCOURING
SCOURING WASTE
SOLVENT SCOURING
SUINT
WOOL GREASE
WOOL WAX

GREASY WOOL
BT WOOL
RT PULLED WOOL
SCOURED WOOL
SCOURING WASTE
SCOURING YIELD
WOOL CLASS
WOOL GREASE
WOOL WAX

GRECIAN ALHAMBRA
BT FIGURED FABRICS
RT GRECIAN WEAVES

GRECIAN WEAVES
BT WEAVE (TYPES)
RT GRECIAN ALHAMBRA

GREENFIELD CONVERTER
BT CONVERTERS (TOW)
RT DRAFT CUT
STRETCH BREAKING
TOW
TOW CONVERSION

GREX
BT COUNT
DIRECT COUNT
RT COUNT TESTING
DENIER
FIBER DENIER
INDIRECT COUNT
TEX

GREY GOODS
RT FINISHED FABRICS
WEAVING

GREY SCALE
RT COLORFASTNESS TESTING
LIGHTFASTNESS TESTING

GRID BARS
RT BEATERS (OPENING)
OPENERS
PICKERS (OPENING)
PICKING (OPENING)

GRILENE (TN)
BT POLYESTER FIBERS

GRINDING (COMMINUTION)
UF CRUSHING
PULVERIZING
BT COMMINUTION
RT ATOMIZING
CHIPPING
DISPERSING
MILLS (COLLOID)
MIXING
PIGMENTS
POWDERING

GRINDING (MATERIAL REMOVAL)
RT ABRASION
(CONT.)

RT ABRASIVES
CHIPPING
CUTTING
EMBOSSING
EMERY
FRICTION
GALLING
GRIT
PLOUGHING
POLISHING
RUBBING
SCORING
SCRATCHING
SHARPENING
SPLITTING

GRINDING ROLLS (CARDS)
UF CARD GRINDERS
RT CARD GRINDING
ROLLS

GRIPPERS (YARN)
RT DEVICES
LOOMS
MECHANISMS
SHUTTLE CHANGING

GRIT
RT ABRASIVES
CERAMICS
CORUNDUM
EMERY
GRINDING (MATERIAL REMOVAL)
PUMICE
SCORING
SCRATCHING

GROOVES
RT CYLINDERS
DRUM WINDERS
EMPTY PACKAGES
BOBBINS
GEOMETRY
QUILLS
ROUGHNESS
SHAPE
SURFACE FRICTION

GROUND HEDDLES
RT BACK CROSSING HEDDLES
DOUP
FRONT CROSSING HEDDLES
HARNESSES
LENO WEAVES
LENO WEAVING
STANDARD HEDDLES

GROWTH (FABRIC)
RT BAGGINESS
CREEP
DIMENSIONAL STABILITY
STRETCH
STRETCH FABRICS
TIRES

GUARD HAIRS
BT HAIR
KERATIN FIBERS
PROTEIN FIBERS
RT ARTIFICIAL FURS
DEHAIRING
FUR

GUARDING DEVICES
BT DEVICES
RT AUTOMATIC CONTROL

GUARDS (ATHLETIC CLOTHING)
BT ATHLETIC CLOTHING

GUIDE BARS
RT COMPOUND NEEDLES
FULL SET
GUIDES
HALF SET
LAPPING (WARP KNITTING)
MILANESE KNITTING MACHINES
NEEDLE BARS
OVERLAP
PATTERN CHAINS
RASCHEL KNITTING MACHINES
RUNNER RATIO
RUNNERS
SET (WARP KNITTING)
TRICOT KNITTING
TRICOT KNITTING MACHINES
UNDERLAP
WARP KNITTING MACHINES

GUIDE EYES
USE THREAD GUIDES

GUIDERS (CLOTH)
USE CLOTH GUIDES

GUIDES
NT CENTER GUIDES
CERAMIC GUIDES
CLOTH GUIDES
EDGE GUIDES
LAP GUIDES
THREAD GUIDES
RT COMPENSATORS (CLOTH)
GUIDE BARS
GUIDERS (CLOTH)
TEMPLES
TENTER CLIPS
TENTER FRAMES

GUIPURE
BT HAND MADE LACES
LACES

GUMS
RT ADHESIVES
DEXTRAN
GLUES
HYDROXYALKYLCELLULOSE
HYDROXYETHYLCELLULOSE
METHYLCELLULOSE
NATURAL POLYMERS
PASTES
PECTINS

GUNNY SACKS
RT AGRICULTURAL FABRICS
BAGS
JUTE

GURLEY PERMEMETER
BT PERMEABILITY TESTERS
TESTING EQUIPMENT
RT AIR PERMEABILITY
AIR PERMEABILITY TESTING
FRAZIER PERMEOMETER
PERMEABILITY TESTING

GUSSETS (FOOTWEAR)
BT FOOTWEAR COMPONENTS

GUT (COLLAGEN)
BT COLLAGEN
RT AZLON (REGENERATED PROTEIN FIBERS)
COLLAGEN FIBERS
MEDICAL TEXTILES
SUTURING

GUT THREAD
RT DIMENSIONAL STABILITY
ELASTIC STRAIN
EXTENSIBILITY
KNITTING
STIFFNESS
WEAVING

GYMP
RT LACES (SHOES)
RAWHIDE
THREADS

HACK STANDS
RT PATTERN REEDS
REEDS
WARPERS (MACHINE)

HACKLE COMBS
RT CARD WASTE
CARDS
COTTON CARDS
FLAT STRIPPING COMBS
FLAT STRIPS

HAIR
NT ALPACA
CAMEL HAIR
CASHMERE
COW HAIR
GOAT HAIR
GUARD HAIRS
HORSE HAIR
HUMAN HAIR
MOHAIR
RABBIT HAIR
VICUNA
BT FIBERS
KERATIN FIBERS
NATURAL FIBERS
PROTEIN FIBERS
RT DISULFIDE BONDS
DISULFIDES (ORGANIC)
FOLLICLE
HAIRINESS

HAIR DYEING (HUMAN HAIR)
BT DYEING (BY FIBER CLASSES)
RT HAIR SETTING
HUMAN HAIR
WOOL DYEING

HAIR PROCESSING MACHINES
RT DEHAIRING

HAIRINESS
BT AESTHETIC PROPERTIES
SURFACE PROPERTIES
(MECHANICAL)
YARN PROPERTIES
RT APPEARANCE
BALDNESS
COMFORT
COMPRESSIBILITY
COVER
DEHAIRING
FABRIC PROPERTIES
FABRIC PROPERTIES (AESTHETIC)
FABRIC PROPERTIES (MECHANICAL)
FIBER MIGRATION
FINISH (PROPERTY)
FUZZ
FUZZINESS
HAIR
HAND
OPTICAL PROPERTIES
PATTERN DEFINITION
PILLING
PILLS
PRICKLINESS
ROUGHNESS
SHEARING (FINISHING)
SHEDDING (FIBER)
SINGEING
SIZES (SLASHING)
SMOOTHNESS
SURFACE FRICTION
TEXTURE
YARN COVER
YARN DIAMETER
YARN GEOMETRY
YARN STRUCTURE

HAIR SETTING
UF HUMAN HAIR SETTING
RT COLD SETTING (SYNTHETICS)
HAIR DYEING (HUMAN HAIR)
HEAT SETTING (DRY)
HEAT SETTING (MOIST)
HEAT SETTING (SYNTHETICS)
HUMAN HAIR
LINT (WASTE)
SETTING (EXCEPT SYNTHETICS)
SETTING AGENTS
SETTING TREATMENTS (GENERAL)
STARCHING

HALCHING
RT LEASING
WINDING

HALF CARDIGAN FABRICS
BT FILLING KNITTED FABRICS
RT CIRCULAR KNITTED FABRICS
FLAT KNITTED FABRICS
FULL CARDIGAN FABRICS
FULLY FASHIONED FABRICS

HALF HOSE
UF HALF HOSIERY
BT HOSIERY
RT BOARDING
CIRCULAR HOSIERY
FOOTWEAR
FULLY FASHIONED HOSIERY
KNITTED OUTERWEAR
KNITTING
LEGGINGS
RUN RESISTANT HOSIERY
SEAMLESS HOSIERY
SOCKS
STOCKINGS
STRETCH HOSIERY
SURGICAL STOCKINGS

HALF HOSIERY
USE HALF HOSE

HALF LAP
RT COMB BRUSHES
COMB CYLINDERS
COMB PINS
COMB SEGMENTS
COMBED SLIVERS
COMBING
COMBS
COTTON SYSTEM PROCESSING
DETACHING ROLLS
LAP PLATES
NIPPER JAWS
NIPPER KNIVES
NIPPER PLATES
NOILS
RECTILINEAR COMBS
RIBBON LAPS
SLIVERS
TOP COMBS
TRUMPETS
WORSTED SYSTEM PROCESSING

HALF SET
RT FULL SET
GUIDE BARS
LAPPING (WARP KNITTING)
MILANESE KNITTING MACHINES
OVERLAP
RASCHEL KNITTING MACHINES
RUNNER RATIO
RUNNERS
SET (WARP KNITTING)
TRICOT KNITTING
TRICOT KNITTING MACHINES
UNDERLAP
WARP KNITTING MACHINES

HALIDES
NT BROMIDES
CHLORIDES
FLUORIDES
IODIDES
RT BROMINE COMPOUNDS
CHLORINE COMPOUNDS
HALOGEN COMPOUNDS

HALLE SEYDEL STRETCH BREAKER
UF SEYDEL MACHINE
BT CONVERTERS (TOW)
TOW TO TOP MACHINES
RT DRAFT CUT
TOW CONVERSION
TOW TO TOP CONVERSION

HALOGEN COMPOUNDS
UF ORGANIC HALOGEN COMPOUNDS
NT BROMINE COMPOUNDS
CHLORINE COMPOUNDS
FLUORINE COMPOUNDS
IODINE COMPOUNDS
RT HALIDES

HALOGENATION
NT CHLORINATION (CHEMICAL
REACTION)
BT REACTIONS (CHEMICAL)

HALOHYDRINS
NT CHLOROHYDRINS
BT ALCOHOLS
REACTANTS
RT CROSSLINKING
EASY CARE FABRICS
EPOXY RESINS
FINISH (SUBSTANCE ADDED)
OXIRANE COMPOUNDS
WASH WEAR FINISHES

HALTERS
BT SPORTSWEAR

HAMBURG (LACE)
BT HAND MADE LACES
LACES

HAMPSHIRE WOOL
BT MEDIUM WOOL
WOOL
RT WOOL GRADE

HAND
UF FEEL
HANDLE (TEXTILE)
TOUCH
BT AESTHETIC PROPERTIES
FABRIC PROPERTIES
FABRIC PROPERTIES (AESTHETIC)
SURFACE PROPERTIES
(MECHANICAL)
RT BENDING RIGIDITY
BOTTOM COVER
BULK
COMFORT
COMPRESSIBILITY
CONTACT
CONTACT AREA
COVER
DRAPE
FINISH (PROPERTY)
FUZZ
HAIRINESS
HARDNESS
HARSHNESS
LIVELINESS
LUXURIOUSNESS
MECHANICAL PROPERTIES
NAP
PACKAGE HARDNESS
PAPERINESS
PRICKLINESS
RESILIENCE
ROUGHNESS
SKIN
SLICKNESS
SMOOTHNESS
SOFTNESS
SPECIFIC HEAT
STIFFNESS
SURFACE CONTOUR
SURFACE FRICTION
TEXTURE
TOP COVER
WARMTH

HAND COMBING
BT COMBING

HAND KNITTING
BT KNITTING
RT COURSES
DOUBLE KNIT FABRICS
FINGERING YARNS
HAND SEWING
KNITTED FABRICS
MACHINE KNITTING

HAND KNITTING YARNS
USE FINGERING YARNS

HAND LOOMS
BT LOOMS
RT AUTOMATIC LOOMS
BOX LOOMS
DOBBY LOOMS
HAND WEAVING
HAND WOVEN FABRICS

HAND MADE LACES
NT ALENCON
ALEO
ANTIQUE (LACE)
ARMENIAN (LACE)
BATTENBERG
BINCHE
BLONDE (LACE)
BOHEMIAN (LACE)
BOURDON
BRABANT
CARRICKMACROSS (LACE)
CHANTILLY (LACE)
CLUNY (LACE)
CROCHETED LACES
DARNED LACES
DRESDEN POINT
DUCHESS (LACE)
EVERLASTING (LACE)
FILET (LACE)
GUIPURE
HAMBURG (LACE)
IRISH (LACE)
LAZIES
LILLE (LACE)
MALINES
MALTESE (LACE)
MILAN (LACE)
MIRECOURT
NEEDLE POINT LACES
NORMANDIE
PARAGUAY (LACE)
PILLOW LACES (BOBBIN LACES)
POINT DE PARIS
POINT DESPRIT
RENAISSANCE (LACE)
ROSE POINT
RUSSIAN POINT
SPANISH (LACE)
TATTING LACES
TORCHON
(CONT.)

NT TUCK LACES
VALENCIENNES
VENICE (LACE)
YAK
BT LACES

HAND PICK STITCHES
BT STITCHES (SEWING)

HAND SADDLE STITCHES
BT STITCHES (SEWING)

HAND SEWING
BT SEWING
RT BACK TACKING
BASTING
CHAIN SEWING
CROCHETING
CURVED SEAMING
DARNING
EMBROIDERY
HAND KNITTING
LOOP CATCHING
MACHINE SEWING
SEAM STRENGTH
SEAMING
SEWING NEEDLES
SEWING THREADS
STITCHING

HAND WEAVING
BT WEAVING
RT HAND LOOMS
HAND WOVEN FABRICS
TAPESTRIES
WEAVE
WEAVE (TYPES)
WOVEN FABRICS

HAND WOVEN FABRICS
BT WOVEN FABRICS
RT FOLK WEAVES
HAND LOOMS
HAND WEAVING

HANDLE (TEXTILE)
USE HAND

HANK DYEING
BT DYEING (BY MATERIAL ASSEMBLY)
YARN DYEING
RT HANK DYEING MACHINES
HANKS
MUFF DYEING
PACKAGE DYEING

HANK DYEING MACHINES
BT PACKAGE DYEING MACHINES
YARN DYEING MACHINES
RT BEAM DYEING MACHINES (YARN)
BOBBIN DYEING MACHINES
CAKE DYEING MACHINES
CHEESE DYEING MACHINES
CONE DYEING MACHINES
COP DYEING MACHINES
HANK DYEING
HANKS
WARP DYEING MACHINES

HANK NUMBER
BT COUNT
INDIRECT COUNT
RT HANKS
ROVINGS
SKEINS

HANK SIZING
BT SLASHING
RT HANKS

HANKS
RT BUNDLES
HANK DYEING
HANK DYEING MACHINES
HANK NUMBER
HANK SIZING
INDIRECT COUNT
PACKAGES
ROVINGS
SKEINS
YARNS

HARD WATER
BT WATER
RT CARBONATES
CHELATING
DISTILLED WATER
ION EXCHANGE
WATER TREATMENT

HARDENERS (FELTS)
NT ROLLER HARDENERS
RT FELTING MACHINES
FELTING
FELTS
HARDENING (FELTS)
PLATEN HARDENERS

HARDENING (FELTS)
RT FELTING
FELTING MACHINES
FELTS
FULLING
HARDENERS (FELTS)

HARDNESS
UF FIRMNESS
BT MECHANICAL PROPERTIES
STRESS STRAIN PROPERTIES
STRESS STRAIN PROPERTIES (COMPRESSIVE)
RT BEARING STRENGTH
BRITTLENESS
COMPLIANCE
COMPRESSIVE MODULUS
COMPRESSIVE STRENGTH
HAND
LAYER LOCKING
PACKAGE DENSITY
PACKAGE HARDNESS
PAPERINESS
PIRN DENSITY
PLASTICITY
SOFTNESS
STIFFNESS

HARES
USE RABBITS

HARNESS CHAINS
RT HARNESSES
LOOMS

HARNESS CORDS
RT HARNESSES
LOOMS

HARNESS DRAFT
RT CHAIN DRAFT
DESIGN
DRAFT (PATTERN)
DRAFTING (PATTERN)
DRAWING IN
FABRIC DESIGN
HARNESSES
HEDDLES
PATTERN (FABRICS)
PATTERN REPEATS
WARP ENDS
WEAVE
WEAVING

HARNESS DROPS
RT HARNESSES
LOOMS

HARNESS LEVERS
RT HARNESSES
LOOMS

HARNESS SKIPS
BT FABRIC DEFECTS
RT FABRIC INSPECTION
HARNESSES
LOOMS
WEAVING
WOVEN FABRICS

HARNESS STRAPS
RT HARNESSES
LOOMS

HARNESSES
NT JACQUARD HARNESSES
RT BACK CROSSING HEDDLES
CLOSED SHED
DOBBY LOOMS
DOUP
DRAWING IN
FRONT CROSSING HEDDLES
GROUND HEDDLES
HARNESS CHAINS
HARNESS CORDS
HARNESS DRAFT
HARNESS DROPS
HARNESS LEVERS
HARNESS SKIPS
HARNESS STRAPS
HEDDLES
JACQUARD LOOMS
LET OFF
LINGOE
LOOMS
MAIL (JACQUARD LOOM)
NEGATIVE SHEDDING
OPEN SHED
PICKING (WEAVING)
POSITIVE SHEDDING
REEDS
SHED (WEAVING)
SHEDDING
SPLIT SHED
STANDARD HEDDLES
WARP ENDS
WEAVING

HARROW FORKS (SCOURING)
RT SCOURING BATHS
SCOURING TRAINS

HARSHNESS
BT FABRIC PROPERTIES
FABRIC PROPERTIES (AESTHETIC)
RT COMFORT
HAND
ROUGHNESS
SMOOTHNESS
SOFTNESS

HART MOISTURE METER
BT TESTING EQUIPMENT
RT MOISTURE CONTENT
MOISTURE CONTENT CONTROL
MOISTURE CONTENT TESTING

HARVARD
RT TWILL WEAVES

HATS
BT HEADGEAR
RT CAPS (HEADGEAR)
FELTS
GARMENTS
RAINWEAR

HAWSERS
BT CORDAGE
RT CABLING
CORDS
FISHING LINES
PLIED YARNS
PLYING
ROPES
SAILCLOTH
STRANDS

HEADGEAR
NT CAPS (HEADGEAR)
EAR MUFFS
HATS
HELMETS
HOODS
MITRES
NIGHTCAPS (TEXTILE)

HEADING (NARROW FABRIC)
USE FRINGE (NARROW FABRIC)

HEADS (TOP)
RT BOBBINS
BUILDER MOTIONS
CORES (CENTER)
DIMENSIONS
HOLLOWNESS
INVERSION
PACKAGES
SHAPE

HEADSTOCKS
RT CARRIAGES (SPINNING MULE)
COUNTERFALLERS
DELIVERY ROLLS
DRIVING BANDS
FALLERS (MULE)
MULE CARRIAGES
MULE DRAFT
MULE SPINNING
MULES
QUADRANTS (MULE)
SPINDLES
SURFACE DRUMS

HEALDS
USE HEDDLES

HEART LOOP
RT BENDING LENGTH
BENDING RIGIDITY
DRAPE
HAND

HEART VALVES (FABRIC)
BT MEDICAL TEXTILES
RT ARTERIAL REPLACEMENTS

HEAT
RT CONDUCTION
CONVECTION
CURING
DRYERS
DRYING
FIRING (REFRACTORY)
HEAT EXCHANGERS
HEAT TREATMENT
HEATING EQUIPMENT
HOT AIR DRYERS
HUMIDITY
HUMIDITY CONTROL
OVENS
TEMPERATURE
WASTE HEAT RECOVERY

HEAT BONDING
BT BONDING
RT FLAME BONDING
FOAM BONDING
FUSION (MELTING)
HEAT SEALING
LAMINATED FABRICS

HEAT CAPACITY
RT GLASS TRANSITION TEMPERATURE
SECONDARY TRANSITION
TEMPERATURES

HEAT CONDUCTIVITY
USE CONDUCTIVITY (THERMAL)

HEAT CONTENT
BT PHYSICAL PROPERTIES (EXCLUDING
MECHANICAL)
THERMAL PROPERTIES
RT BOILING POINT
ENTHALPY
ENTROPY
LATENT HEAT
SPECIFIC HEAT
TEMPERATURE
THERMODYNAMICS

HEAT EXCHANGERS
NT MOLTEN METAL BATHS
OIL BATHS
RT DRYERS
HEAT
HEAT TRANSFER
HEATING EQUIPMENT
OVENS
WASTE HEAT RECOVERY

HEAT OF DYEING
USE ENTHALPY OF DYEING

HEAT OF FUSION
BT THERMAL PROPERTIES
RT FUSION (MELTING)
GLASS TRANSITION TEMPERATURE
LATENT HEAT
MELTING
MELTING POINT
THERMOPLASTICITY

HEAT OF HYDRATION
USE ENTHALPY OF HYDRATION

HEAT OF REACTION
USE ENTHALPY OF REACTION

HEAT OF SOLUTION
USE ENTHALPY OF SOLUTION

HEAT OF SWELLING
USE ENTHALPY OF SWELLING

HEAT OF WETTING
USE ENTHALPY OF WETTING

HEAT RESISTANCE
**UF HEAT RESISTANCE (HIGH
TEMPERATURE)**
RESISTANCE (HEAT)
**BT PHYSICAL PROPERTIES (EXCLUDING
MECHANICAL)**
THERMAL PROPERTIES
RT FIRE PROOFING TREATMENTS
FIRE RESISTANCE
MELTING POINT
NOMEX (TN)
POLYBENZIMIDAZOLES
SCORCHING
SHRINKAGE
TEMPERATURE
THERMAL DEGRADATION

HEAT RESISTANCE (HIGH TEMPERATURE)
USE HEAT RESISTANCE

HEAT SEALING
NT DIELECTRIC HEAT SEALING
BT SEALING
RT ADHESION
BACKING (MATERIAL)
BACKING FABRICS
BONDED FIBER FABRICS
COATED FABRICS
CURING AGENTS
DECRYSTALLIZATION
FALSE TWISTING (TEXTURING)
FLAME BONDING
FOAM BACKED FABRICS
FUSION (MELTING)
HEAT BONDING
LAMINATED FABRICS
LAMINATING
NONWOVEN FABRICS
PLASTICS
POST FORMED LAMINATES
STITCHING
TAILORING
TAPING
THERMOPLASTICITY

HEAT SETTING (DRY)
BT DRY FINISHING
HEAT SETTING (SYNTHETICS)
HEAT TREATMENT
RT COLD SETTING (SYNTHETICS)
CORONIZING (TN)
HAIR SETTING
HEAT SETTING (MOIST)
SETTING (EXCEPT SYNTHETICS)

HEAT SETTING (MOIST)
UF HYDROSETTING
BT HEAT SETTING (SYNTHETICS)
HEAT TREATMENT
STEAMING
WET FINISHING
RT COLD SETTING (SYNTHETICS)
HAIR SETTING
HEAT SETTING (DRY)
SETTING (EXCEPT SYNTHETICS)

HEAT SETTING (SYNTHETICS)
NT HEAT SETTING (DRY)
HEAT SETTING (MOIST)
BT DRY FINISHING
HEAT TREATMENT
RT ANNEALING
BOARDING
BULKING
COLD SETTING (SYNTHETICS)
CORONIZING (TN)
CRABBING
CREASING
CRIMP FIXING
CRIMPING
CRINKLE TYPE YARNS
CRYSTALLIZATION
CURING
CURING OVENS
DECATING
DIMENSIONAL STABILITY
DRY RELAXATION
EDGE CRIMPED YARNS
EDGE CRIMPING
FABRIC RELAXATION
FALSE TWIST
FALSE TWIST SPINDLES
FALSE TWIST YARNS
HAIR SETTING
HEAT SETTING MACHINES
IRONING
PERMANENT DEFORMATION
PERMANENT PLEATING
PRESSING
RECOVERY (SELF RESTORATION TO
ORIGINAL CONDITION)
RELAXATION
RELAXATION TIME (MECHANICAL)
RESIDUAL STRESS
RESINS
SCORCHING MARKS
SEAMLESS HOSIERY
SETTING (EXCEPT SYNTHETICS)
SHRINKING
SOCKS
STEAMING
STRESS RELAXATION
STRETCH HOSIERY
STRETCH YARNS (FILAMENT)
STRETCHING (YARNS)
STUFFER BOX CRIMPING
STUFFER BOXES
TEMPERATURE CONTROL
TENTERING
TEXTURING
THERMAL DEGRADATION
THERMOFIXING (DYEING)
THERMOPLASTICITY
THERMOSOL PROCESS (TN)

HEAT SETTING MACHINES
RT HEAT SETTING (SYNTHETICS)
TENTER FRAMES

HEAT STABILIZERS
UF STABILIZERS AGAINST THERMAL
DEGRADATION
BT STABILIZERS (AGENTS)
RT ANTIOXIDANTS
STABILITY
STABILIZATION
THERMAL DEGRADATION
THERMAL STABILITY
ULTRAVIOLET STABILIZERS

HEAT TRANSFER
RT AIR FLOW
CONDUCTION
CONDUCTIVITY (THERMAL)
COUNTERFLOW PROCESSES
CONVECTION
DIELECTRIC DRYERS
DIMENSIONAL ANALYSIS
DRYERS
DRYING
DRYING CANS
DRYING TIME
EMISSIVITY

RT HEAT EXCHANGERS
HEATING
HOT AIR DRYERS
HOT AIR HEATING
INFRARED DRYERS
MASS TRANSFER
MOLTEN METAL BATHS
MOMENTUM TRANSFER
OIL BATHS
OVENS
RADIATION
TRANSFER PROPERTIES
WARMTH
WASTE HEAT RECOVERY

HEAT TREATMENT
NT CORONIZING (TN)
DRY CURING
HEAT SETTING (DRY)
HEAT SETTING (MOIST)
HEAT SETTING (SYNTHETICS)
RT CRIMP FIXING
CURING
DRY FINISHING
FIRING (REFRACTORY)
HEAT
HEATING
STRETCHING (YARNS)
TEMPERATURE
WET FINISHING

HEATING
NT DIELECTRIC HEATING
ELECTRIC HEATING (RESISTANCE)
GAS HEATING (GAS FUEL)
HOT AIR HEATING
INFRARED HEATING
OIL HEATING (OIL FUEL)
RADIANT HEATING
STEAM HEATING
RT AERODYNAMIC HEATING
AIR CONDITIONING
AIR FLOW
CONDUCTION
CONVECTION
COOLING
CURING
DRYING
ELECTRIC BLANKETS
FIRING (REFRACTORY)
HEAT TRANSFER
HEAT TREATMENT
HIGH TEMPERATURE
HUMIDITY
HUMIDITY CONTROL
OVENS
PYROMETERS
RADIATION
SUPERHEATED STEAM
TEMPERATURE CONTROL
THERMAL DEGRADATION
THERMAL PROPERTIES
THERMAL STABILITY
TRAVELER BURNOUT

HEATING EQUIPMENT
NT GAS HEATERS
RT CURING OVENS
DRYERS
EQUIPMENT
GAS HEATING (GAS FUEL)
HEAT
HEAT EXCHANGERS
HOT AIR HEATING
OVENS
STEAM
TENTER FRAMES

HEAVY METAL SOAPS
BT METALLIC SOAPS

HEAVY WEIGHT
BT WEIGHT
RT COARSE YARNS
FABRIC WEIGHT
FABRICS
FINE YARNS

HEDDLE BARS
RT HARNESSES
HEDDLES
LOOMS

HEDDLE EYES
RT HARNESSES
HEDDLES
LOOMS

HEDDLE HOOKS
RT HARNESSES
HEDDLES

HEELS (HOSIERY)
USE HEELS (STOCKINGS)

HEELS (STOCKINGS)
UF HEELS (HOSIERY)
(CONT.)

RT HOSIERY
REINFORCED KNITTED FABRICS
STOCKINGS
TOES
WELTS (KNITTED)

HEDDLE SMASH
RT HARNESSES
HEDDLES
LOOMS
WEAVING

HEDDLES
UF HEALDS
RT DRAFTING (PATTERN)
DRAWING IN
HARNESS DRAFT
HARNESSES
HEDDLE BARS
HEDDLE EYES
HEDDLE HOOKS
HEDDLE SMASH
LOOMS
REEDS
SHED (WEAVING)
SHEDDING
SLEYS (LOOM)
WARP ENDS
WARP PREPARATION
WEAVING

HEELS (FOOTWEAR)
BT FOOTWEAR
FOOTWEAR COMPONENTS

HEIGHT
RT BALLOON HEIGHT
DIAMETER
DIMENSIONS
LENGTH
RADIUS
SHAPE
THICKNESS
WIDTH

HEILMANN COMBS
USE RECTILINEAR COMBS

HELANCA (TN)
BT FALSE TWIST YARNS
STRETCH YARNS (FILAMENT)
TEXTURED YARNS (FILAMENT)
RT BULKED YARNS
DYNALOFT (TN)
FALSE TWISTING (TEXTURING)
FLUFLON (TN)
SAABA (TN)
SUPERLOFT (TN)

HELICAL CRIMP
BT CRIMP
RT BICOMPONENT FIBERS
CRIMP FREQUENCY
CRIMP INDEX
CRIMP RADIUS
DIFFERENTIAL GEOMETRY
EDGE CRIMPED YARNS
FIBER CRIMP
FOLLOW THE LEADER CRIMP
HELIX ANGLE
HELIX REVERSALS
NATURAL CRIMP
STRETCH YARNS (FILAMENT)
Z CRIMP

HELIX ANGLE
UF ANGLE OF TWIST
RT ANGLE OF INCIDENCE
ANGULAR DEFLECTION
CORDS
CRIMP INDEX
DIFFERENTIAL GEOMETRY
HELICAL CRIMP
REAL TWIST
ROPES
S TWIST
TORSION (GEOMETRIC)
TORSIONAL ENERGY
TORSIONAL RIGIDITY
TWIST
TWIST MULTIPLIER
TWIST TESTING
TWISTING
TWISTING MOMENTS
YARN CROSS SECTIONS
YARN GEOMETRY
YARN STRUCTURE

HELIX REVERSALS
RT BICOMPONENT FIBERS
EDGE CRIMPED YARNS
FALSE TWIST YARNS
FIBRIL REVERSALS
HELICAL CRIMP
TWIST

HELMETS
BT ATHLETIC CLOTHING
HEADGEAR
SAFETY CLOTHING
RT GARMENTS

HEM SEAMS
BT SEAMS
RT BUTTED SEAMS
BUTTERFLY SEAMS
COVERED SEAMS
SEAMING
SEWING
STITCHING

HEMIACETALS
BT ACETALS
RT FORMALS

HEMICELLULOSE
BT CARBOHYDRATES
RT CELLULOSE
CELLULOSE DERIVATIVES
COTTON
HYDROCELLULOSE

HEMMING
BT GARMENT MANUFACTURE
RT BACK TACKING
BASTING
BLIND STITCHING
CHAIN SEWING
CURVED SEAMING
GATHERING
HEMS
PINCHING
PUCKERING
RUFFLING
SEAMING
SERGING
SEWING
SHIRRING
STITCHES (SEWING)
STITCHING
TAILORING
TRIMMING (OPERATION)

HEMP
BT BAST FIBERS
CELLULOSIC FIBERS
VEGETABLE FIBERS (GENERAL)
RT CORDAGE
FLAX
JUTE
RAMIE
ROPES

HEMS
BT GARMENT COMPONENTS
RT HEMMING
SEAMS

HERCULON (TN)
BT OLEFIN FIBERS
POLYPROPYLENE FIBERS

HERRINGBONE TWILLS
BT TWILLS
RT CHEVRON TWILLS
FANCY TWILLS
LOW TWILLS
REGULAR TWILLS
S TWILLS
STEEP TWILLS
Z TWILLS

HESSIAN
BT WOVEN FABRICS
RT BAGS
BAST FIBERS
JUTE
PLAIN WEAVES

HETEROCYCLIC COMPOUNDS
NT PYRIDINE
PYRIDINE DERIVATIVES
RHODANINE
TRIAZINES
TRIAZONES
VINYL MORPHOLINE
VINYL PYRROLIDONE
RT ALICYCLIC COMPOUNDS
ALIPHATIC COMPOUNDS
AROMATIC COMPOUNDS

HEXAMETHYLENEDIAMINE
BT AMINES
RT MONOMERS
NYLON 610
NYLON 66
POLY (HEXAMETHYLENE ADIPAMIDE)
POLY (HEXAMETHYLENE SEBACAMIDE)
POLY (HEXAMETHYLENE TEREPHTHALAMIDE)

HEXAMETHYLENE DIISOCYANATE
BT ISOCYANATES

HEXAMETHYLOL MELAMINE
BT METHYLOL MELAMINES

HIBISCUS
UF KENAF
BT BAST FIBERS
CELLULOSIC FIBERS
RT CORDAGE
JUTE
TWINE

HIDES
RT COLLAGEN FIBERS
CORFAM (TN)
LEATHER
SYNTHETIC LEATHER
TANNING (LEATHER)

HIGH DRAFT
RT ACTUAL DRAFT
AMBLER SUPERDRAFT
AUTOLEVELLING
CASABLANCA SYSTEM
CONTROL SURFACES
DRAFT
DUO ROTH SYSTEM
HIGH DRAFT SPINNING
ROTH SYSTEM
SHAW SYSTEM
SLIVERS
SPINNING

HIGH DRAFT SPINNING
BT SPINNING
RT ACTUAL DRAFT
AMBLER SUPERDRAFT
APRONS
CASABLANCA SYSTEM
CONTROL SURFACES
COTTON SYSTEM PROCESSING
DRAFT
DRAFTING (STAPLE FIBER)
DRAFTING WAVES
FIBER HOOKS
GILLING
HIGH DRAFT
LONG STAPLE SPINNING
TWIST CONTROL
WORSTED SYSTEM PROCESSING

HIGH FREQUENCY
BT FREQUENCY
RT VIBRATION
DYNAMIC RESPONSE
FREQUENCY RESPONSE
OSCILLATIONS
TRANSVERSE VIBRATIONS

HIGH PRODUCTION CARDS
BT CARDS
RT CROSROL WEB PURIFIER (TN)
DUO CARDS
PRODUCTION

HIGH SPEED
BT VELOCITY
RT ACCELERATION (MECHANICAL)
CONSTANT SPEED
TERMINAL VELOCITY
VARIABLE SPEED
VIBRATION

HIGH TEMPERATURE
BT TEMPERATURE
RT FIRING (REFRACTORY)
HEATING
LOW TEMPERATURE
ROOM TEMPERATURE
THERMOMETERS
THERMOPILES
SUPERHEATED STEAM
WARMTH

HIGH TEMPERATURE DYEING
NT PRESSURE DYEING
THERMOSOL PROCESS (TN)
BT DYEING (BY ENVIRONMENTAL CONDITIONS)
RT CARRIER DYEING
CONTINUOUS DYEING RANGE
DYEING MACHINES
JIG DYEING
LOW TEMPERATURE DYEING
PRESSURE DYEING MACHINES
SOLVENT ASSISTED DYEING
SOLVENT DYEING
THERMOFIXING (DYEING)
VACUUM DYEING

HIGH TEMPERATURE FIBERS
NT ALUMINUM FOIL
ALUMINUM SILICATE FIBERS
ASBESTOS
BORON NITRIDE FIBERS
CERAMIC FIBERS
FIBERGLAS (TN)
FOIL
FTORLON (TN)

NT GLASS FIBERS
GRAPHITE FIBERS
INORGANIC FIBERS (MAN MADE)
LAME (TN)
LUREX (TN)
METAL FIBERS
METALLIC FIBERS
METLON (TN)
MINERAL FIBERS
NOMEX (TN)
POLYTETRAFLUOROETHYLENE FIBERS
POLYTRIFLUOROCHLOROETHYLENE FIBERS
SILICON CARBIDE FIBERS
TEFLON FIBER (TN)
TINSEL
WHISKER FIBERS (METAL)
BT FIBERS
RT SYNTHETIC FIBERS

HIGH TENACITY VISCOSE RAYON
NT CORDURA (TN)
DURAFIL (TN)
TENAX (TN)
BT CELLULOSIC FIBERS
RAYON (REGENERATED CELLULOSE FIBERS)
MAN MADE FIBERS
VISCOSE RAYON
RT HIGH WET MODULUS RAYON

HIGH TWIST
BT TWIST
RT ALTERNATING TWIST
COARSE YARNS
CREPE
CREPE YARNS
FALSE TWIST
FINE YARNS
LOW TWIST
PLY TWIST
PRODUCERS TWIST
REAL TWIST
S TWIST
SINGLES TWIST
SNARLING
TORSIONAL BUCKLING
TURNS PER INCH
TWIST MULTIPLIER
TWISTING
YARNS
Z TWIST
ZERO TWIST

HIGH WET MODULUS RAYON
NT AVRIL (TN)
CORDURA (TN)
POLYNOSIC FIBERS
BT CELLULOSIC FIBERS
MAN MADE FIBERS
RAYON (REGENERATED CELLULOSE FIBERS)
VISCOSE RAYON
RT HIGH TENACITY VISCOSE RAYON

HILOW BULKED YARNS
BT TEXTURED SPUN YARNS
RT BICOMPONENT FIBER YARNS (STAPLE)
BULKED YARNS

HIMALAYA CLOTH
BT FLAT WOVEN FABRICS
WOVEN FABRICS
RT PLAIN WEAVES
SHANTUNG

HISTORY
RT BIBLIOGRAPHY
COMPILATIONS
LITERATURE SURVEYS
MECHANICAL HISTORY
REVIEWS
STRESS HISTORY
SUMMARIES
SURVEYS
THEORIES
YEARS COVERAGE (1)
YEARS COVERAGE (5)
YEARS COVERAGE (10)
YEARS COVERAGE (25)
YEARS COVERAGE (50)
YEARS COVERAGE (75)
YEARS COVERAGE (100)

HOLDERS (PACKAGE)
USE PACKAGE HOLDERS

HOLES
UF CAVITIES
RT ABSENCE
ASPERITIES
DELUSTERING
HOLES (FABRIC DEFECTS)
HOLLOWNESS
ORIFICES
PACKING FACTOR
PENETRATION
PERFORATED BEAMS
PERFORATED BELTS
PERFORATED CAGES
PERFORATED CYLINDERS
PERFORATED TUBES (DYEING)
PERMEABILITY
PERMEABILITY TESTING
PORES
ROUGHNESS
ROUNDNESS
SPINNERETS

HOLES (FABRIC DEFECTS)
BT FABRIC DEFECTS
RT HOLES
PICKING (SNAGGING)
RUNS (KNITTING)
SNAGGING

HOLLOW BRAIDS
BT BRAIDS
REGULAR BRAIDS

HOLLOW FILAMENT YARNS
UF MACARONI YARNS
BT FILAMENT YARNS
RT FIBER CROSS SECTIONS

HOLLOWNESS
RT ABSENCE
CORES (CENTER)
GEOMETRY
HEADS (TOP)
HOLES
MATRIX FREEDOM
PACKING FACTOR
INVERSION
PRESENCE
ROUNDNESS
SHAPE
SPACE
SYMMETRY
VACUUM
VOLUME

HOMOGENEOUS MATERIALS
RT ANISOTROPIC MATERIALS
ISOTROPIC MATERIALS
MACRORHEOLOGY
QUASI HOMOGENEOUS MATERIALS

HOMOGENIZERS
RT BLENDERS
BLENDING
EMULSIFYING
MASS TRANSFER
MILLS (COLLOID)
MIXING
STIRRING

HOMOPOLYMERS
RT BLOCK POLYMERS
COPOLYMERS
POLYMERS

HONEYCOMB LAMINATES
BT COMPOSITES
REINFORCED COMPOSITES
RT FABRIC REINFORCED COMPOSITES
LAMINATED FABRICS
SANDWICH LAMINATES

HONEYCOMB WEAVES
BT WEAVE (TYPES)
RT CELLULAR FABRICS
FLAT WOVEN FABRICS

HOODS
BT CEREMONIAL CLOTHING
HEADGEAR
RT GARMENTS

HOOK CLOSURES
BT FASTENERS
GARMENT COMPONENTS
RT SNAP FASTENERS
VELCRO (TN)
ZIPPERS

HOOK DIRECTION
NT LEADING HOOKS
TRAILING HOOKS
RT CARDING
DRAFTING (STAPLE FIBER)
FIBER HOOKS
HOOK FORMATION
HOOK REMOVAL
HOOKED FIBERS
LINDSLEY RATIOS
PROCESSING DIRECTION

HOOK FORMATION
RT CARDING
CONDENSING (YARN MANUFACTURING)
DRAFTING (STAPLE FIBER)
FIBER HOOKS
HOOK DIRECTION
HOOK REMOVAL
HOOKED FIBERS
LEADING HOOKS
TRAILING HOOKS

HOOK REMOVAL
RT CARDING
COMBING
DRAFT
DRAFTING (STAPLE FIBER)
DRAFTING THEORY
DRAFTING ZONE
FIBER CONTROL
FIBER HOOKS
FIBER ORIENTATION
HOOK DIRECTION
HOOK FORMATION
HOOKED FIBERS
LEADING HOOKS
SLIVER REVERSALS
SLIVERS
TRAILING HOOKS

HOOK SHAFTS
RT LOOP CATCHERS
SEWING BOBBINS
SEWING MACHINES

HOOKEAN REGION
USE ELASTIC STRAIN

HOOKEAN STRAIN
USE ELASTIC STRAIN

HOOKED FIBERS
RT CARDING
COMBING
CONDENSED YARNS
CONDENSING (YARN MANUFACTURING)
DRAFTING THEORY
FIBER EXTENT
FIBER HOOKS
HOOK DIRECTION
HOOK FORMATION
HOOK REMOVAL
LEADING HOOKS
LINDSLEY RATIOS
SLIVER REVERSALS
SLIVERS
SPINNING
TRACER FIBERS
TRAILING HOOKS

HOOKES LAW
RT ELASTIC MODULUS (TENSILE)
ELASTIC STRAIN
INITIAL MODULUS
STRESS STRAIN CURVES

HOOKS (FIBER)
USE FIBER HOOKS

HOOP STRESS
BT STRESS
RT BIAXIAL STRESS
UNIAXIAL STRESS

HOPPER FEED
RT BLAMIRE FEED
BREAKERS
CARDING
CARDS
CHARGING
CONVEYORS
FEED LATTICES
FEED RATE
FEED ROLLS
FEEDERS
HOPPERS
INPUT
IRREGULARITY CONTROL
LOADING (PROCESS)
OPENERS
PICKERS (OPENING)
PICKING (OPENING)
PROCESS CONTROL
SCOTCH FEED
STOCK (FIBER)
THROUGHPUT
WEIGHT CONTROL
WOOLEN CARDING
WOOLEN CARDS
WORSTED CARDING
WORSTED CARDS

HOPPERS
RT FEED LATTICES
HOPPER FEED

HOPSACK
BT FLAT WOVEN FABRICS
WOVEN FABRICS
RT PLAIN WEAVES

HORSE HAIR
BT HAIR
KERATIN FIBERS
PROTEIN FIBERS
RT HORSEHIDE

HORSEHIDE
BT LEATHER
RT HORSE HAIR

HOSE
USE HOSIERY

HOSES (PIPE)
USE PIPES

HOSIERY
UF HOSE
NT CIRCULAR HOSIERY
FULLY FASHIONED HOSIERY
HALF HOSE
RUN RESISTANT HOSIERY
SEAMLESS HOSIERY
SOCKS
STOCKINGS
STRETCH HOSIERY
SUPPORT HOSIERY
RT BOARDING
DENIER
DOUBLE KNIT FABRICS
FASHIONING
FASHIONING MARKS
FILLING KNITTED FABRICS
FILLING KNITTING
FILLING KNITTING MACHINES
FOOTWEAR
FULLY FASHIONED KNITTING
FULLY FASHIONED KNITTING MACHINES
GARMENTS
HEELS (STOCKINGS)
HOSIERY DYEING MACHINES
HOSIERY MACHINES
KNITTED OUTERWEAR
KNITTING
KNITTING MACHINES
LISLE THREADS
MACHINE KNITTING
MESH (KNITTING)
MICROMESH (KNITTING)
MOCK FASHIONING MARKS
MOCK SEAMS
MONOFILAMENT YARNS
OUTERWEAR
PICOT
PRESSING
REINFORCED KNITTED FABRICS
RUN RESISTANCE
SHAPE
SHEERNESS
SNAG TESTING
WELTING
WORK CLOTHING

HOSIERY DYEING
BT GARMENT DYEING
RT HOSIERY DYEING MACHINES
PIECE DYEING

HOSIERY DYEING MACHINES
BT DYEING MACHINES (BY MATERIAL ASSEMBLY)
RT DRUM (DYEING)
GARMENT DYEING
GARMENT DYEING MACHINES
HOSIERY
HOSIERY DYEING
PERFORATED CAGES
ROTARY DYEING MACHINES

HOSIERY MACHINES
NT FULLY FASHIONED HOSIERY MACHINES
SEAMLESS HOSIERY MACHINES
BT KNITTING MACHINES
RT BOARDING
CIRCULAR HOSIERY
CIRCULAR KNITTING MACHINES
CYLINDER NEEDLES
DIAL NEEDLES
FULLY FASHIONED HOSIERY
HOSIERY
MOCK FASHIONING MARKS
MOCK SEAMS
NEEDLES
POSITIVE FEED
SINKERS
STITCH CAMS
TAKE DOWN
WEBHOLDERS

HOT AIR DRYERS
UF FORCED CONVECTION DRYERS
NATURAL CONVECTION DRYERS
BT DRYERS
RT ATMOSPHERIC DRYERS
DRUM DRYERS
FLUIDIZED BED DRYERS
HEAT TRANSFER
HOT AIR HEATING
LOOP DRYERS
VACUUM DRYERS
TENTER FRAMES

HOT AIR DRYING
BT DRYING
RT AIR FLOW
CENTRIFUGAL DRYING
DIELECTRIC DRYING
FREEZE DRYING
HEAT
INFRARED DRYING
VACUUM DRYING

HOT AIR HEATING
BT HEATING
RT AIR FLOW
HEAT
HEAT TRANSFER
HEATING EQUIPMENT
HOT AIR DRYERS

HOT DRAWING
BT DRAWING (FILAMENT)
RT COLD DRAWING
DRAWTWISTING
DEGREE OF ORIENTATION
DRAW RATIO
EXTRUSION
LOADING
MOLECULAR ORIENTATION
NATURAL DRAW RATIO
NECKING (FILAMENT)
PLASTIC FLOW
PLASTIC STRAIN
RHEOLOGY
STRAIN
STRESS
YIELD (MECHANICAL)
YIELD POINT

HOT PRESS METHOD (DURABLE PRESS)
BT DURABLE PRESS TREATMENTS
RT DEFERRED CURE METHOD
DOUBLE CURE METHOD
DRY CURING
DURABLE PRESS FINISHES
PRE-CURE METHOD
RESIN FINISHES
WET CURING

HOT SPOTS
RT COEFFICIENT OF FRICTION
FRICTION
FRICTION THEORY

HOUSEHOLD FABRICS
UF LINENS
BT FABRICS
FABRICS (BY END USES)
RT AGRICULTURAL FABRICS
APPAREL FABRICS
AUTOMOTIVE FABRICS
BEDSPREADS
BLANKETS
CARPETS
CUSHIONS
DRAPES
EIDERDOWNS
END USE PROPERTIES
END USES
FURNISHING FABRICS
GARMENTS
INDUSTRIAL FABRICS
MATS
MATTRESSES
NAPKINS
PILLOWCASES
PILLOWS
QUILTS
SHEETING (FABRIC)
TABLECLOTHS
TERRYCLOTH
TOWELS
UPHOLSTERY
UPHOLSTERY FABRICS
WINDOW HOLLAND
WOVEN CARPETS
WOVEN FABRICS

HUCKABACK WEAVES
BT WEAVE (TYPES)
RT FLAT WOVEN FABRICS

HUE
BT **MUNSELL COLOR SYSTEM**
RT **ABSORPTION (RADIATION)**
COLOR
SATURATION (COLOR)
SHADE
SPECTROPHOTOMETRY
VALUE (MUNSELL)

HULLS
RT GINNING
MOTE KNIVES
TRASH

HUMAN HAIR
BT HAIR
KERATIN FIBERS
PROTEIN FIBERS
RT HAIR DYEING (HUMAN HAIR)
HAIR SETTING

HUMAN HAIR SETTING
USE HAIR SETTING

HUMECTANTS
RT GLYCEROL
GLYCOLS
SALTS (GENERAL)
SODIUM ACETATE

HUMIDITY
NT RELATIVE HUMIDITY
SPECIFIC HUMIDITY
RT CONDITION (TEXTILES)
CRITICAL MOISTURE CONTENT
DIFFUSION
DIFFUSIVITY
DRY BULB TEMPERATURE
DRY WEIGHT
DRYING
EQUILIBRIUM MOISTURE CONTENT
HEAT
HEATING
HYGROMETERS
HYGROSCOPICITY
IONIZATION
MOISTURE CONTENT
PSEUDO WET BULB TEMPERATURE
PSYCHROMETRY
REGAIN
VAPOR PRESSURE
WATER
WET BULB TEMPERATURE

HUMIDITY CONTROL
RT AIR CONDITIONING
AUTOMATIC CONTROL
CONTROL SYSTEMS
DRY WEIGHT
DRYING
HEATING
MOISTURE CONTENT CONTROL
REGAIN
TEMPERATURE CONTROL

HUNTING
RT CONTROL CHARACTERISTICS
DRIFT
FREQUENCY RESPONSE

HUSKS
RT SILK

HYDRAULIC BRAKES
BT BRAKES (FOR ARRESTING MOTION)
RT HYDRAULIC CYLINDERS
MAGNETIC BRAKES
MECHANICAL BRAKES

HYDRAULIC CONTROL
RT CONTROL SYSTEMS
FLUIDICS
MANUAL CONTROL
PNEUMATIC CONTROL

HYDRAULIC CYLINDERS
RT BRAKES (FOR ARRESTING MOTION)
HYDRAULIC BRAKES
HYDRAULIC CONTROL

HYDRAULIC TENSION DEVICES
BT TENSION DEVICES
RT DIRECT TENSION DEVICES
GATE TENSION
IMPACT TESTERS
INDIRECT TENSION DEVICES
MAGNETIC TENSION DEVICES
PNEUMATIC CONTROL
PRESSURE
PRESSURE CONTROL
TENSION DISCS
TENSION SPRINGS
VALVES
WASHERS

HYDRAULIC DRIVES
BT DRIVES
RT BELT DRIVES
CHAIN DRIVES
GEAR DRIVES
MECHANICAL DRIVES
VARIABLE SPEED DRIVES

HYDRAULICS
RT CONTROL EQUIPMENT
CYLINDERS
FLUID MECHANICS
PRESSURE

HYDRAZIDES
RT AMIDES
CORROSION RESISTANCE

HYDROCARBON SULFONATES
USE ALKYL SULFONATES

HYDROCARBONS
NT ALICYCLIC HYDROCARBONS
ALIPHATIC HYDROCARBONS
AROMATIC HYDROCARBONS
UNSATURATED HYDROCARBONS

RT MINERAL OIL
STODDARD SOLVENT

HYDROCELLULOSE
BT CELLULOSE DERIVATIVES
DEGRADED CELLULOSE
RT ACID DEGRADATION
COPPER NUMBER
HEMICELLULOSE
OXYCELLULOSE
YELLOWING

HYDROCHLORIC ACID
BT INORGANIC ACIDS
RT CHLORIDES
CHLORINE COMPOUNDS

HYDRODYNAMIC LUBRICATION
BT LUBRICATION
PHYSICAL CHEMICAL PROPERTIES
SURFACE PROPERTIES (PHYSICAL CHEMICAL)
RT ABSORBENCY (MATERIAL)
BOUNDARY LAYER
BOUNDARY LUBRICATION
CONTACT ANGLES
DYNAMIC FRICTION
FLUID FRICTION
FRICTION
LUBRICANTS
ROLLING
SHEAR (MODE OF DEFROMATION)
SLIDING
VISCOSITY
WEAR
YARN FINISH

HYDROGEN
RT DEUTERIUM
PH

HYDROGEN BONDS
BT CHEMICAL BONDS
RT BOND ENERGY
COVALENT BONDS
ELECTROSTATIC BONDS
HYDROPHOBIC BONDS
IONIC BONDS
METALLIC BONDS
POLAR COMPOUNDS
SWELLING
VAN DER WAALS FORCES
WET STRENGTH

HYDROGEN IODIDE
BT INORGANIC ACIDS
RT IODIDES

HYDROGEN ION CONCENTRATION
USE PH

HYDROGEN PEROXIDE
BT PEROXIDES
RT BLEACHING
OXIDATION
PERACETIC ACID
PERBORATES
PERFORMIC ACID
PEROXIDE BLEACHING AGENTS
SILICATES
SODIUM HYPOCHLORITE
SODIUM PEROXIDE
STRIPPING (COLOR)

HYDROLYSIS
NT ACID HYDROLYSIS
ALKALI HYDROLYSIS
BT REACTIONS (CHEMICAL)
RT CELLULASE
CHEMICAL STABILITY
DECOMPOSITION
DEGRADATION
ESTERIFICATION
YELLOWING

HYDROMETERS
BT MEASURING INSTRUMENTS
RT INSTRUMENTATION
SPECIFIC GRAVITY

HYDROPHILIC FIBERS
RT CHEMICAL PROPERTIES
HYDROPHILIC PROPERTY
HYDROPHOBIC FIBERS
HYGROSCOPICITY
NATURAL FIBERS
SURFACE PROPERTIES (PHYSICAL CHEMICAL)
SURFACE TENSION
SWELLING
SYNTHETIC FIBERS
WETTING

HYDROPHILIC PROPERTY
BT PHYSICAL CHEMICAL PROPERTIES
SURFACE PROPERTIES (PHYSICAL CHEMICAL)
RT ABSORBENCY (MATERIAL)
CONTACT ANGLES
DRAVES TEST
HYDROPHILIC FIBERS
HYDROPHOBIC PROPERTY
HYGROSCOPICITY
LIPOPHILIC PROPERTY
ORIENTATION (SURFACE)
REGAIN
SURFACE CHEMISTRY
SURFACE TENSION
SWELLING
WATER REPELLENCY
WATER REPELLENCY TREATMENTS
WATERPROOFING
WETTING
WETTING AGENTS
WETTING TIME

HYDROPHOBIC BONDS
BT CHEMICAL BONDS
RT DYEING THEORY
HYDROGEN BONDS
VAN DER WAALS FORCES

HYDROPHOBIC FIBERS
RT CHEMICAL PROPERTIES
HYDROPHILIC FIBERS
HYDROPHOBIC PROPERTY
SURFACE PROPERTIES (PHYSICAL CHEMICAL)
SURFACE TENSION
SWELLING
SYNTHETIC FIBERS
WETTING

HYDROPHOBIC PROPERTY
RT HYDROPHILIC PROPERTY
HYGROSCOPICITY
LIPOPHILIC PROPERTY
ORIENTATION (SURFACE)
RADICALS (CHEMICAL)
REGAIN
SURFACE CHEMISTRY
SURFACE TENSION
SWELLING
WATER REPELLENCY
WATER REPELLENCY TREATMENTS
WATERPROOFING
WETTING

HYDROSETTING
USE HEAT SETTING (MOIST)

HYDROSOLS
BT SOLS
RT AEROSOLS
ORGANOSOLS

HYDROSTATIC HEAD TESTING
USE HYDROSTATIC PRESSURE TESTING

HYDROSTATIC LOADING
RT CALENDER ROLLS
FEED ROLLS
HYDROSTATIC PRESSURE
NIP
PERALTA ROLLS
PRESS SECTION
PRESSING
PRESSURE ROLLS
ROLLS
SQUEEZE ROLLS
SUCTION PRESSES

HYDROSTATIC PRESSURE
BT PRESSURE
RT HYDROSTATIC LOADING
NORMAL FORCE
PARTIAL PRESSURE

HYDROSTATIC PRESSURE TESTING
UF HYDROSTATIC HEAD TESTING
HYDROSTATIC PRESSURE TESTS
BT TESTING
WATER RESISTANCE TESTING
RT PERMEABILITY TESTING
WATERPROOFING
WATER REPELLENCY

HYDROSTATIC PRESSURE TESTS
USE HYDROSTATIC PRESSURE TESTING

HYDROSTATIC STRESS
BT STRESS
RT BIAXIAL STRESS
SHEAR STRESS
UNIAXIAL STRESS

HYDROSULFITES
USE DITHIONITES

HYDROXIDES
NT BARIUM HYDROXIDE
BERYLLIUM HYDROXIDE
CALCIUM HYDROXIDE
CESIUM HYDROXIDE
LITHIUM HYDROXIDE
MAGNESIUM HYDROXIDE
POTASSIUM HYDROXIDE
RUBIDIUM HYDROXIDE
SODIUM HYDROXIDE
BT ALKALIES
BASES
RT ALCOHOLS

HYDROXYACETIC ACID
UF GLYCOLIC ACID
BT CARBOXYLIC ACIDS
ORGANIC ACIDS
RT ACETIC ACID
LACTIC ACID

HYDROXYACETONE
BT ALCOHOLS
KETONES
RT ACETONE
REDUCING AGENTS
VAT DYEING
VAT PRINTING

HYDROXYALKYLCELLULOSE
NT HYDROXYETHYLCELLULOSE
BT CELLULOSE ETHERS
RT EMULSIFYING AGENTS
ETHYLCELLULOSE
GUMS
STIFFENERS (AGENTS)

HYDROXYETHYLCELLULOSE
BT CELLULOSE ETHERS
HYDROXYALKYLCELLULOSE
RT GUMS
THICKENING AGENTS

HYDROXYLAMINE
RT ACID DYEING
AMINES
AMMONIA
REDUCING AGENTS

HYDROXYMETHYL COMPOUNDS
USE METHYLOL COMPOUNDS

HYGROMETERS
BT MEASURING INSTRUMENTS
RT DEW POINT
DRY BULB TEMPERATURE
DRYING
HUMIDITY
INSTRUMENTATION
RELATIVE HUMIDITY
WET BULB TEMPERATURE

HYGROSCOPICITY
BT PHYSICAL CHEMICAL PROPERTIES
RT ABSORBENCY (MATERIAL)
ABSORPTION (MATERIAL)
CAPILLARITY
CONDITION (TEXTILES)
HUMIDITY
HYDROPHILIC FIBERS
HYDROPHILIC PROPERTY
HYDROPHOBIC PROPERTY
MOISTURE CONTENT CONTROL
MOISTURE CONTENT TESTING
REGAIN

HYPOCHLORITES
NT LITHIUM HYPOCHLORITE
SODIUM HYPOCHLORITE
RT BLEACHING
CHLORINE BLEACHING AGENTS
CHLORINE COMPOUNDS
CHLORINE DAMAGE
CHLORINE DIOXIDE
CHLORITES
OXIDATION
OXIDIZING AGENTS
PEROXIDES

HYPERSONIC FLOW
RT AERODYNAMICS
AIR FLOW
FLUID FLOW

HYPOBROMITES
RT BLEACHING AGENTS
CHLORINE BLEACHING AGENTS
OXIDIZING AGENTS
SODIUM HYPOCHLORITE

HYPODERMIC CUTTING DRILLS
BT CUTTING DRILLS

HYSTERESIS (MECHANICAL)
BT MECHANICAL DETERIORATION PROPERTIES
STRESS STRAIN PROPERTIES
RT ACCURACY
COEFFICIENT OF FRICTION
COEFFICIENT OF VISCOSITY
CREEP
CYCLIC STRESS
DAMPING
DYNAMIC CHARACTERISTICS
ENERGY ABSORPTION
(CONT.)

HYSTERESIS (MECHANICAL)

RT ENERGY OF RETRACTION
ERRORS
INTERNAL FRICTION
LINEAR VISCOELASTICITY
LOSS TANGENT
MECHANICAL CONDITIONING
MECHANICAL PROPERTIES
MOMENT CURVATURE CURVES
NONLINEAR VISCOELASTICITY
STRESS RELAXATION
VIBRATION DAMPERS
VISCOUS DAMPING
WORK RECOVERY

ICI SYSTEM
USE CIE SYSTEM

IDENTIFICATION (FIBERS)
USE FIBER IDENTIFICATION

IGEPON A (TN)
BT ACYL ESTER SULFONATES

IGEPON T (TN)
BT METHYLTAURIDE SURFACTANTS

ILLUMINATION
UF LIGHTING
RT ARTIFICIAL DAYLIGHT
BRIGHTNESS
COLOR
COLOR MATCHING
DAYLIGHT
FLUORESCENT LIGHT
INCANDESCENT LIGHT
LIGHT
PHOTONS
SHADING (TAILORING)
SUNLIGHT

IMAGINARY COMPONENT (DYNAMIC MODULUS)
BT DYNAMIC MODULUS
STRESS STRAIN PROPERTIES
RT LOSS TANGENT
REAL COMPONENT (DYNAMIC MODULUS)

IMBIBITION
RT ABSORBENCY (MATERIAL)
CROSSLINKING
DRYING
MOISTURE ABSORPTION

IMIDAZOLIDINONES
USE METHYLOL UREAS

IMIDAZOLINE SURFACTANTS
BT AMINE SURFACTANTS
RT ALKYLAMINE SURFACTANTS
AMIDOAMINE SURFACTANTS
OXAZOLINE SURFACTANTS

IMIDES
RT AMIDES
CARBODIIMIDES

IMINES
NT AZIRIDINE COMPOUNDS
RT AMINES

IMITATION FRENCH PIPING
RT FRENCH SEAMS

IMITATION GINGHAM
BT FLAT WOVEN FABRICS
WOVEN FABRICS
RT PLAIN WEAVES

IMMATURE FIBERS
RT CAUSTICAIRE VALUE
CONVOLUTIONS
COTTON
FIBER CROSS SECTIONS
FIBER FINENESS
FIBER GEOMETRY
GRADE (FIBERS)
IMMATURITY (FIBER)
LUMEN
MATURITY INDEX
MICRONAIRE
NEPS
PRIMARY WALLS
SECONDARY WALLS
SEED HAIR FIBERS (GENERAL)
VEGETABLE FIBERS (GENERAL)

IMMATURITY (FIBER)
UF FIBER IMMATURITY
FIBER MATURITY
MATURITY (FIBER)
BT FIBER PROPERTIES
RT CLASS (FIBERS)
COTTON
COEFFICIENT OF FINENESS VARIATION
FIBER CROSS SECTIONS
FIBER GEOMETRY
FINENESS TESTING
GRADE (FIBERS)
IMMATURITY TESTING (COTTON FIBER)
LUMEN
MATURITY INDEX
NEP COUNT
NEP POTENTIAL
NEP SIZE
NEPPINESS INDEX
NEPS
NEPTOMETER GRADE
PRIMARY WALLS
SECONDARY WALLS

IMMATURITY TESTING (COTTON FIBER)
NT CROSS SECTIONAL SHAPE METHOD (COTTON IMMATURITY)
DIFFERENTIAL DYEING METHOD (COTTON IMMATURITY)
POLARIZED LIGHT METHOD (COTTON IMMATURITY)
SWELLING METHOD (COTTON IMMATURITY)
BT TESTING
RT COTTON
FIBER CROSS SECTIONS
FIBER LENGTH DETERMINATION
IMMATURITY (FIBER)
LUMEN
PRIMARY WALLS
SECONDARY WALLS

IMMEDIATE ELASTIC RECOVERY
USE ELASTIC RECOVERY

IMMERSION
RT IMMERSION TESTING
LIQUIDS
WETTING
WETTING TIME

IMMERSION ROLLS
BT ROLLS
RT PADDING

IMMERSION TESTING
BT TESTING
RT DYNAMIC ABSORPTION TESTING
IMMERSION
STATIC ABSORPTION TESTING
WATER REPELLENCY
WATER RESISTANCE TESTING

IMPACT
RT ACCELERATION (MECHANICAL)
ACCELEROMETERS
ANGLE OF INCIDENCE
BRITTLE FRACTURE
CRACK PROPAGATION
CRITICAL VELOCITY
DAMPERS
ENERGY ABSORPTION
FORCE
FRACTURING
IMPACT STRENGTH
IMPULSES
MECHANICAL PROPERTIES
MECHANICAL SHOCK
OSCILLATIONS
SHOCK
SHOCK RESISTANCE
STRENGTH OF MATERIALS
VIBRATION
VIBRATION DAMPERS
VIBRATION ISOLATORS
WEAVING DYNAMICS

IMPACT PENETRATION TESTING
UF IMPACT PENETRATION TESTS
BT TESTING
WATER RESISTANCE TESTING
RT CRITICAL VELOCITY
PENETRATION
PERMEABILITY TESTING
STRESS CONCENTRATION
TRANSVERSE VIBRATIONS
WATER REPELLENCY
WATERPROOFING

IMPACT PENETRATION TESTS
USE IMPACT PENETRATION TESTING

IMPACT RESISTANCE
USE IMPACT STRENGTH

IMPACT STRENGTH
UF IMPACT RESISTANCE
BT BREAKING STRENGTH
STRENGTH
STRESS STRAIN PROPERTIES
RT ARRESTERS
BALLISTIC FABRICS
BALLISTIC TESTING
BEARING STRENGTH
BREAKING ELONGATION
BREAKING ENERGY
BREAKING LENGTH
BRITTLENESS
BULLET PROOF CLOTHING
BUNDLE STRENGTH
COMPRESSIVE STRENGTH
COUNT STRENGTH PRODUCT
CRACK PROPAGATION
DUCTILITY
END BREAKAGES
FABRIC PROPERTIES
FABRIC PROPERTIES (MECHANICAL)
FRACTURING
IMPACT
IMPACT TESTING
LEA STRENGTH PRODUCT
LOOP EFFICIENCY
SHOCK RESISTANCE
STRAIN RATE
STRENGTH ELONGATION TESTING
TEAR STRENGTH
YIELD (MECHANICAL)

IMPACT TESTERS
BT TESTING EQUIPMENT
RT DYNAMIC MODULUS TESTERS
HYDRAULIC CYLINDERS
IMPACT TESTING
WEAVING DYNAMICS

IMPACT TESTING
UF IMPACT TESTS
BT STRENGTH TESTING
TESTING
RT ABRASION TESTING
BALLISTIC LIMIT
BALLISTIC TESTING
BENDING TESTING
BURST TESTING
CRITICAL VELOCITY
IMPACT STRENGTH
IMPACT TESTERS

IMPACT TESTS
USE IMPACT TESTING

IMPEDANCE
BT ELECTRICAL PROPERTIES
RT CAPACITANCE (ELECTRICAL)
DIELECTRIC STRENGTH
IMPEDANCE BRIDGES
INDUCTANCE
MEASURING INSTRUMENTS
REACTANCE (ELECTRICAL)
RESISTANCE (ELECTRICAL)

IMPEDANCE BRIDGES
RT CAPACITANCE (ELECTRICAL)
CAPACITANCE BRIDGES
ELECTRONIC INSTRUMENTS
IMPEDANCE

IMPELLERS
UF BLADES (ROLLER)
RT BEATERS (OPENING)
OPENERS
SAW TOOTHED ROLLS

IMPERFECTIONS
USE DEFECTS

IMPERIAL SATEEN
BT SATEEN
WEAVE (TYPES)

IMPREGNATING MACHINES
RT BACK FILLING MACHINES
COATING MACHINES
IMPREGNATION
PADDERS
SPRAYING MACHINES

IMPREGNATION
RT ADD ON
COATING (PROCESS)
DYEING
FINISHING PROCESS (GENERAL)
IMPREGNATING MACHINES
LAUNDERING
PADDING

IMPROVEMENTS
RT ADAPTATION
ADVANTAGES
DESIGN
DEVELOPMENTS
DOWN TIME
INCREASE
INVENTIONS
MACHINE DESIGN
MINIMIZATION
MODIFICATION
OPTIMIZATION
PERFORMANCE
PROCESS EFFICIENCY
PRODUCTION
QUALITY
QUALITY IMPROVEMENT
REMODELLING
REPAIRS

IMPULSES
RT ACCELERATION (MECHANICAL)
DAMPERS
DYNAMIC RESPONSE
ENERGY ABSORPTION
IMPACT
MOMENTUM
PULSE PROPAGATION METER
SURGING
VIBRATION
VIBRATION DAMPERS
VIBRATION ISOLATORS

IMPURITIES
NT TRASH
RT BACKWASHING
BURRS
CARDING
CLEANLINESS
CONTAMINATION
COTTON WAX
CYLINDER STRIPS
DEFECTS
DEWAXING (COTTON)
DOFFER STRIPS
FLAT STRIPS
MOTES
NEPS
OPENING
PICKING (OPENING)
SCOURING
SEBUM
SEEDS (TRASH)
SLUBS
SOIL
SOILING
SUINT
VEGETABLE MATTER
WASTE

INCANDESCENT LIGHT
BT LIGHT
RT ARTIFICIAL DAYLIGHT
DAYLIGHT
FLUORESCENT LIGHT
ILLUMINATION
POLARIZED LIGHT
SUNLIGHT

INCREASE
RT ADJUSTMENTS
ADVANTAGES
ALTERATIONS
CORRECTION
IMPROVEMENTS
MINIMIZATION
MODIFICATION
OPTIMIZATION
QUALITY
QUALITY IMPROVEMENT

INDIA LINON
BT FLAT WOVEN FABRICS
WOVEN FABRICS
RT LAWN
LAWN CLOTH
PLAIN WEAVES

INDIAN COTTON
BT COTTON

INDIANA CLOTH
BT WOVEN FABRICS
RT COATED FABRICS
LAWN

INDICATORS (INSTRUMENTATION)
RT AUTOMATIC CONTROL
INSTRUMENTATION
MEASURING INSTRUMENTS
RECORDING INSTRUMENTS

INDIGOID DYES
USE VAT DYES

INDIGOSOLS
USE LEUCO ESTER DYES

INDIRECT COUNT
NT COTTON COUNT
CUT SYSTEM
HANK NUMBER
LEA COUNT
METRIC COUNT
RUN SYSTEM
TYPP COUNT
WORSTED COUNT
YORKSHIRE SKEIN WOOLEN COUNT
BT COUNT
RT DENIER
DIRECT COUNT
GREX
HANKS
LINEAR DENSITY
ROVINGS
SKEINS
TEX
YARNS

INDIRECT TENSION DEVICES
BT TENSION DEVICES
RT DIRECT TENSION DEVICES
HYDRAULIC TENSION DEVICES
MAGNETIC TENSION DEVICES

INDUCTANCE
BT ELECTRICAL PROPERTIES
RT CAPACITANCE (ELECTRICAL)
DIELECTRIC STRENGTH
IMPEDANCE
REACTANCE (ELECTRICAL)
RESISTANCE (ELECTRICAL)

INDUSTRIAL CLOTHING
NT APRONS (CLOTHING)
DENIMS
DUNGAREES
GLOVES
JEANS
OVERALLS
RAINWEAR
SAFETY CLOTHING
SWEATSHIRTS
UNIFORMS
BT GARMENTS
WORK CLOTHING

INDUSTRIAL FABRICS
NT ACOUSTIC BAFFLING
BELTS
BOOKBINDING
BRATTICE CLOTH
BREAKER FABRICS
BUILDER FABRICS
DUCK
FILTER CLOTH
INFLATABLE FABRICS
LEATHER CLOTH
PAPERMAKERS FELTS
PRINTERS BLANKETS
SAILCLOTH
TARPAULINS
WIPING CLOTHS
BT FABRICS
FABRICS (BY END USES)
RT AGRICULTURAL FABRICS
APPAREL FABRICS
AUTOMOTIVE FABRICS
AWNINGS
BAGS
BUFFING WHEELS
CANOPIES
CASKET LININGS
CONVEYOR BELTS
DIAPHRAGMS
DUNNAGE BAGS
FABRICS (ACCORDING TO STRUCTURE)
FILTERS (FLUID)
FISHING LINES
FURNISHING FABRICS
END USES
GASKETS
HOUSEHOLD FABRICS
INSULATION (THERMAL)
LAUNDRY NETS
LIFE RAFTS
PACKAGING (MATERIAL)
PIPE COVERINGS
PRESSURE SUITS
RADOMES
SEALS
STUFFING
TAPE
TIRES
TYPEWRITER RIBBONS

INDUSTRIAL SEWING MACHINES
BT SEWING MACHINES
RT DOMESTIC SEWING MACHINES

INERTIA
RT ACCELERATION (MECHANICAL)
ACCELEROMETERS
ANGULAR VELOCITY
BRAKING
CENTRIFUGAL FORCE
CORIOLIS ACCELERATION
FORCE
GRAVITATION
MASS
MECHANICAL SHOCK
MOMENTUM

INERTIAL REFERENCE
RT CORIOLIS FORCE

INFANTS CLOTHING
NT DIAPERS
ROMPERS
BT GARMENTS
RT CHILDRENS WEAR

INFLAMMABILITY
USE FIRE RESISTANCE

INFLATABLE FABRIC STRUCTURES
USE INFLATABLE FABRICS

INFLATABLE FABRICS
UF INFLATABLE FABRIC STRUCTURES
BT INDUSTRIAL FABRICS
RT END USES
MILITARY PRODUCTS
PRESSURE SUITS
RADOMES
TIRES

INFORMATION RETRIEVAL
RT CHARACTER RECOGNITION
DOCUMENTATION
INFORMATION SYSTEMS
THESAURI

INFORMATION SYSTEMS
RT AUTOMATION
CHARACTER RECOGNITION
DOCUMENTATION
INFORMATION RETRIEVAL
OPTICAL SCANNING
THESAURI

INFRARED ABSORBERS
BT STABILIZERS (AGENTS)
RT INFRARED RADIATION
LIGHTFASTNESS (COLOR)
LIGHTFASTNESS (OF FINISH)
PHOTOCHEMICAL DEGRADATION
ULTRAVIOLET ABSORBERS

INFRARED DRYERS
BT DRYERS
RT ATMOSPHERIC DRYERS
ELECTRIC HEATING (RESISTANCE)
CENTRIFUGAL DRYERS
DIELECTRIC DRYERS
DRUM DRYERS
FLUIDIZED BED DRYERS
HEAT TRANSFER
INFRARED DRYING
INFRARED HEATING
LOOP DRYERS
RADIANT HEATING
VACUUM DRYERS

INFRARED DRYING
BT DRYING
RT CENTRIFUGAL DRYING
DIELECTRIC DRYING
FREEZE DRYING
HEATING
HOT AIR DRYING
INFRARED DRYERS
INFRARED HEATING
INFRARED RADIATION
VACUUM DRYING

INFRARED HEATING
BT HEATING
RT INFRARED DRYERS
INFRARED DRYING
INFRARED RADIATION

INFRARED MICROSCOPY
BT MICROSCOPY
RT ELECTRON MICROSCOPY
INTERFERENCE MICROSCOPY
LIGHT MICROSCOPY (OPTICAL)
ULTRAVIOLET MICROSCOPY
X RAY MICROSCOPY

INFRARED RADIATION
BT RADIATION
RT INFRARED ABSORBERS
INFRARED DRYING
INFRARED HEATING
INFRARED SPECTROSCOPY
IRRADIATION
LIGHT

INFRARED SPECTRA
BT OPTICAL PROPERTIES
SPECTRA
RT INFRARED SPECTROPHOTOMETERS
INFRARED SPECTROSCOPY
SPECTRUM ANALYSIS

INFRARED SPECTROPHOTOMETERS
BT SPECTROPHOTOMETERS
RT INFRARED SPECTRA
INFRARED SPECTROSCOPY

INFRARED SPECTROSCOPY
BT SPECTROSCOPY
RT INFRARED RADIATION
INFRARED SPECTRA
INFRARED SPECTROPHOTOMETERS
SPECTRA
SPECTROPHOTOMETERS

INGRAIN CARPETS
UF KIDDERMINSTER CARPETS
SCOTCH CARPETS
BT CARPETS
PILE FABRICS
WOVEN FABRICS

INGRAIN DYES
NT ALCIAN BLUE (TN)
BT DYES (BY CHEMICAL CLASSES)
RT DYEING (BY DYE CLASSES)
DYES (BY FIBER CLASSES)

INHIBITION
RT AGING (MATERIALS)
ANTIOXIDANTS
BACTERIAL INHIBITION
CATALYSIS
CORROSION INHIBITORS
OXIDATION
STABILIZATION

INITIAL MODULUS
BT MODULUS
STRESS STRAIN PROPERTIES
STRESS STRAIN PROPERTIES (TENSILE)
RT ELASTIC MODULUS (TENSILE)
ELASTICITY
EXTENSIBILITY
PAPERINESS
SECANT MODULUS
STIFFNESS
TANGENT MODULUS

INITIATION
RT CATALYSIS
GRAFTING
POLYMERIZATION

INITIATORS
BT CATALYSTS
RT ADDITION POLYMERIZATION
FREE RADICALS
GRAFTING
PEROXYGEN COMPOUNDS

INJECTION MOLDING
BT MOLDING
RT COMPRESSION MOLDING
CONTACT MOLDING
FLEXIBLE PLUNGER MOLDING
MATCHED DIE MOLDING
PRESSURE BAG MOLDING
REINFORCED COMPOSITES
VACUUM BAG MOLDING

INKS
RT COLOR
DYES
MARKING
PIGMENTS
PRINTING PASTES

INORGANIC ACIDS
NT CHLOROUS ACID
HYDROCHLORIC ACID
HYDROGEN IODIDE
NITRIC ACID
NITROUS ACID
PERCHLORIC ACID
PHOSPHORIC ACID
SULFURIC ACID
BT ACIDS
RT DIBASIC ACIDS
TRIBASIC ACIDS

INORGANIC COMPOUNDS (GENERAL)
NT ALUMINUM COMPOUNDS
AMMONIUM COMPOUNDS
BARIUM COMPOUNDS
BERYLLIUM COMPOUNDS
BISMUTH COMPOUNDS
BORON COMPOUNDS
CADMIUM COMPOUNDS
CALCIUM COMPOUNDS
CERIUM COMPOUNDS
CESIUM COMPOUNDS
CHROMIUM COMPOUNDS
COBALT COMPOUNDS
COPPER COMPOUNDS
GOLD COMPOUNDS
INORGANIC ACIDS
IRON COMPOUNDS
LEAD COMPOUNDS
LITHIUM COMPOUNDS
MAGNESIUM COMPOUNDS
MANGANESE COMPOUNDS
MERCURY COMPOUNDS
NICKEL COMPOUNDS
PHOSPHORUS COMPOUNDS
POTASSIUM COMPOUNDS (GENERAL)
RUBIDIUM COMPOUNDS
SILICON COMPOUNDS
SILVER COMPOUNDS
SODIUM COMPOUNDS (GENERAL)
SULFUR COMPOUNDS
TIN COMPOUNDS
TITANIUM COMPOUNDS
ZINC COMPOUNDS
ZIRCONIUM COMPOUNDS
RT ORGANIC COMPOUNDS (GENERAL)
ORGANOMETALLIC COMPOUNDS (EXCLUDING SILICONES)
OXIDES
SALTS (GENERAL)

INORGANIC FIBER BLENDS
BT BLENDS (FIBERS)

INORGANIC FIBERS (MAN MADE)
NT ALUMINUM FOIL
ALUMINUM SILICATE FIBERS
BORON NITRIDE FIBERS
CERAMIC FIBERS
FIBERGLAS (TN)
FOIL
GLASS FIBERS
GRAPHITE FIBERS
LAME (TN)
LUREX (TN)
METAL FIBERS
METALLIC FIBERS
METLON (TN)
SILICON CARBIDE FIBERS
TINSEL
WHISKER FIBERS (METAL)
BT FIBERS
HIGH TEMPERATURE FIBERS
MAN MADE FIBERS
RT ASBESTOS
METALLIC YARNS
MINERAL FIBERS
NATURAL FIBERS
RAYON (REGENERATED CELLULOSE FIBERS)
SYNTHETIC FIBERS

INPUT
RT CHARGING
CONVEYORS
DELIVERY RATE
FEED RATE
FEED ROLLS
FEEDERS
FLOW (MATERIALS)
HOPPER FEED
LOADING (PROCESS)
OUTPUT
PRODUCTION
THROUGHPUT

INSECT DEGRADATION
USE INSECT RESISTANCE

INSECT RESISTANCE
UF INSECT DEGRADATION
MOSQUITO RESISTANCE
MOTH RESISTANCE
BT DEGRADATION PROPERTIES
END USE PROPERTIES
FINISH (PROPERTY)
RT DEGRADATION
DEGRADATION PROPERTIES
INSECT RESISTANCE FINISHES
INSECT RESISTANCE TREATMENTS
INSECTS
LARVAE
MOTHS
PENETRATION

INSECT RESISTANCE FINISHES
UF LARVICIDES
MOTH RESISTANCE FINISHES
BT FINISH (SUBSTANCE ADDED)
PRESERVATIVES
RT ANTIMICROBIAL FINISHES
DIELDRIN
INSECT RESISTANCE
LARVAE

INSECT RESISTANCE TREATMENTS
UF MOTH PROOFING
BT WET FINISHING
RT DEGRADATION
FABRIC PROPERTIES (DEGRADATION)
FINISH (SUBSTANCE ADDED)
INSECT RESISTANCE
INSECT RESISTANCE FINISHES

INSECTS
NT MOTHS
RT INSECT RESISTANCE
LARVAE
SILKWORMS

INSERTION
RT ADAPTATION
ADJUSTMENTS
ALIGNMENT
INSTALLATION
POSITIONING

INSERTION (LACE)
BT LACES

INSOLES (FOOTWEAR)
NT VENTILATING INSOLES
BT FOOTWEAR COMPONENTS

INSOLUBLE AZO DYES
USE AZOIC DYES

INSPECTION
NT FABRIC INSPECTION
RT ANALYZING
DEFECTS
DRY FINISHING
ESTIMATION
EXPERIMENTATION
FABRIC DEFECTS
INSPECTION MACHINES
IRREGULARITY
PERCHING
PROCESS CONTROL
QUALITY CONTROL
SLOTS
SLUB CATCHERS
SLUB CATCHING

INSPECTION MACHINES
RT INSPECTION
PERCHING
SLUB CATCHERS

INSTALLATION
RT ADAPTATION
ADJUSTMENTS
ALIGNMENT
INSERTION
POSITIONING

INSTRON TENSILE TESTER (TN)
BT TENSILE TESTERS
RT CAMBRIDGE EXTENSOMETER (TN)
SCOTT TENSILE TESTER (TN)

INSTRUMENTAL COLOR MATCHING
BT COLOR MATCHING
RT COMPUTERS

INSTRUMENTATION
UF INSTRUMENTS
NT ELECTRONIC INSTRUMENTS
MEASURING INSTRUMENTS
OPTICAL INSTRUMENTS
PNEUMATIC INSTRUMENTS
RECORDING INSTRUMENTS
RT AMPLIFIERS
APPARATUS
AUTOMATIC CONTROL
AUTOMATION
CHEMICAL ENGINEERING
COMPONENTS
EQUIPMENT
HYDROMETERS
HYGROMETERS
INDICATORS (INSTRUMENTATION)
NULL METHODS
PROCESS CONTROL
PULSE PROPAGATION METER
TESTING EQUIPMENT
TRANSDUCERS
ULTRACENTRIFUGES

INSTRUMENTS
USE INSTRUMENTATION

INSULATED BOOTS
BT FOOTWEAR
BOOTS

INSULATION (ACOUSTICAL)
USE ACOUSTIC INSULATION

INSULATION (ELECTRICAL)
RT DIELECTRIC STRENGTH
LEAKAGE RESISTANCE (ELECTRICAL)
RESISTANCE (ELECTRICAL)
RESISTIVITY

INSULATION (THERMAL)
RT ASBESTOS
BATTING
BULK DENSITY
COMFORT
GLASS FIBERS
FOAMS
INDUSTRIAL FABRICS
LINING FELT
LININGS
WARMTH

INTEGRAL CONTROL
BT AUTOMATIC CONTROL
RT CONTINUOUS CONTROL
DIFFERENTIAL CONTROL
FEEDBACK CONTROL
ON OFF CONTROL
PROCESS CONTROL

INTEGRATED SYSTEMS
RT AUTOMATED SPINNING (STAPLE FIBER)
AUTOMATIC CONTROL
CONTINUOUS SPINNING (STAPLE FIBER)
SYSTEMS ENGINEERING

INTEGRATION
RT ACCUMULATION
COLLECTION
COMPUTATIONS
DIFFERENTIAL EQUATIONS
DIFFERENTIATION
GENERAL SOLUTIONS
PARTICULAR SOLUTIONS
SOLUTIONS (MATHEMATICAL)

INTENSITY
RT ABSENCE
BRIGHTNESS
MAGNITUDE

INTERFACES
RT INTERFACIAL POLYMERIZATION
SURFACES

INTERFACIAL POLYMERIZATION
BT POLYMERIZATION
RT CONDENSATION POLYMERIZATION
INTERFACES
PHASE BOUNDARY CROSSLINKING
POLYMERS
WURLAN (TN)

INTERFACIAL TENSION
RT CONTACT ANGLES
SURFACE CHEMISTRY
SURFACE PROPERTIES (PHYSICAL CHEMICAL)
SURFACE TENSION
WATER REPELLENCY

INTERFERENCE FRINGES
UF INTERFERENCE PATTERNS
RT BIREFRINGENCE
DIFFRACTION PATTERNS
EXPERIMENTAL STRESS ANALYSIS
INTERFEROMETRY
MOIRE PATTERNS
POLARIZED LIGHT

INTERFERENCE MICROSCOPY
BT MICROSCOPY
RT ELECTRON MICROSCOPY
INFRARED MICROSCOPY
LIGHT MICROSCOPY (OPTICAL)
ULTRAVIOLET MICROSCOPY
X RAY MICROSCOPY

INTERFERENCE PATTERNS
USE INTERFERENCE FRINGES

INTERFEROMETRY
RT ELECTRON MICROSCOPY
INFRARED MICROSCOPY
INTERFERENCE FRINGES
INTERFERENCE MICROSCOPY
LIGHT MICROSCOPY (OPTICAL)
MICROSCOPY
OPTICAL INSTRUMENTS
ULTRAVIOLET MICROSCOPY
X RAY MICROSCOPY

INTERLACING (FILAMENT IN YARN)
RT AIR JET TEXTURING
ALTERNATING TWIST
CYCLOSET YARNS (TN)
FALSE TWIST
FALSE TWISTING (TEXTURING)
OPEN END SPINNING
REAL TWIST
ROTOSET YARNS (TN)
SPINNING
TWISTING
ZERO TWIST

INTERLACINGS (YARN IN FABRIC)
RT DESIGN
FABRIC ANALYSIS
FABRIC DESIGN
FABRIC GEOMETRY
FABRIC INTERSTICES
FABRICS
FLOATS
LOOPS
WEAVING
PORES

INTERLININGS
BT GARMENT COMPONENTS
LININGS
RT BOX LININGS
CASKET LININGS
GLOVE LININGS
POCKET LININGS
SHOE LININGS

INTERLOCK (KNITTED FABRICS)
UF DOUBLE JERSEY
INTERLOCK FABRICS
INTERLOCK STITCH
BT FILLING KNITTED FABRICS
KNITTED FABRICS
RT EIGHTLOCK (KNITTED FABRICS)
KNITTING
RIB KNITTING

INTERLOCK FABRICS
USE INTERLOCK (KNITTED FABRICS)

INTERLOCK STITCH
USE INTERLOCK (KNITTED FABRICS)

INTERMEDIATE FEED
NT BLAMIRE FEED
SCOTCH FEED
RT BREAKERS
INTERMEDIATE PART
WOOLEN CARDING

INTERMEDIATE FRAMES
USE ROVING FRAMES

INTERMEDIATE PART
RT BLAMIRE FEED
BREAKERS
CARD CYLINDERS
CARDING
FINISHER PART
SCOTCH FEED
WOOLEN CARDING
WORKERS
WORSTED CARDING

INTERMEDIATE ROLLS
BT DRAFTING ROLLS
ROLLS

INTERNAL ENERGY
BT ENERGY
RT KINETIC ENERGY
POTENTIAL ENERGY
STRAIN ENERGY
TORSIONAL ENERGY

INTERNAL FRICTION
BT FRICTION
RT BENDING RIGIDITY
COEFFICIENT OF FRICTION
DAMPING
DYNAMIC FRICTION
FIBER FIBER FRICTION
FRICTIONAL FORCE
HYSTERESIS (MECHANICAL)
MATRIX FREEDOM
RHEOLOGY
STATIC FRICTION
SURFACE FRICTION
VISCOELASTICITY

INTERNAL REFLECTION
BT OPTICAL PROPERTIES
RT FIBER OPTICS
REFLECTANCE

INTERPOLATION
RT COMPUTATIONS
EXTRAPOLATION
GRAPHICAL SOLUTIONS

INTERSECTING GILL BOXES
BT GILL BOXES
RT ACTUAL DRAFT
BACK ROLLS
BREAK DRAFT
DRAWFRAMES
FALLER BARS
FALLER SCREWS
FALLER SPEED
FINISHER GILLING
FRONT ROLLS
GILLING
RATCH

INTERSPAN (TN)
BT SPANDEX FIBERS

INTERSTICES
USE PORES

INTIMATE BLENDS
RT BLEND DRAFTING
BLENDED FABRICS
BLENDED YARNS
BLENDED YARNS (STAPLE)
BLENDING
BLENDS (FIBERS)
ORTHOBLENDS

INTRINSIC VISCOSITY
BT VISCOSITY
RT MELT VISCOSITY
MOLECULAR WEIGHT
RELATIVE VISCOSITY
VISCOSIMETERS

INTURNED WELTS
BT WELTS (KNITTED)
RT KNITTED FABRICS
KNITTING

INVENTIONS
RT ACCESSORIES
DEVICES
GADGETS
IMPROVEMENTS
INVESTMENTS
MECHANISMS

INVENTORIES
RT ALLOTMENT
COST CONTROL
COSTS
EQUIPMENT
FACILITIES
FLOW (MATERIALS)
INVENTORY CONTROL
INVESTMENTS
MACHINERY
MANAGEMENT
MANUFACTURING
MARKETING
MATERIALS HANDLING
PLANT
PROCESS CONTROL
PROCESS EFFICIENCY
PRODUCTION
PRODUCTION CONTROL
TEXTILE MACHINERY (GENERAL)

INVENTORY CONTROL
RT COST CONTROL
INVENTORIES
PROCESS CONTROL
PRODUCTION CONTROL

INVERFORM MACHINES
BT PAPER MACHINES
RT CYLINDER MACHINES
FOURDRINIER MACHINES
KAMYR MACHINES
ROTAFORMERS
VERTAFORMERS
WET MACHINES (PAPERMAKING)

INVERSION
RT ABSENCE
CONVENTIONAL PRACTICE
CONFIGURATION
HEADS (TOP)
HOLLOWNESS
MODIFICATION
PRESENCE

INVESTMENTS
RT COSTS
ECONOMICS
EQUIPMENT
FACILITIES
FINISHING MACHINERY (GENERAL)
INVENTIONS
INVENTORIES
JOB ANALYSIS
MACHINERY
MANAGEMENT
PLANT
TEXTILE MACHINERY (GENERAL)

INVISCID FLOW
BT FLUID FLOW
RT LAMINAR FLOW
TURBULENT FLOW
VISCOUS FLOW

IODIDES
NT POTASSIUM IODIDE
SODIUM IODIDE
BT HALIDES
RT BROMIDES
CHLORIDES
FLUORIDES
HYDROGEN IODIDE
IODINE
IODINE COMPOUNDS

IODINE
RT BROMINE
CHLORINE
IODIDES
IODINE COMPOUNDS
IODINE NUMBER

IODINE ABSORPTION
RT ABSORPTION (MATERIAL)
IODINE NUMBER

IODINE COMPOUNDS
BT HALOGEN COMPOUNDS
RT IODIDES
IODINE
IODONIUM COMPOUNDS
PERIODATES

IODINE NUMBER
RT CHEMICAL ANALYSIS
FATS
IODINE
IODINE ABSORPTION
OILS
OXIDATION
UNSATURATED COMPOUNDS

IODONIUM COMPOUNDS
BT ONIUM COMPOUNDS
RT IODINE COMPOUNDS

ION EXCHANGE
UF ANION EXCHANGE
CATION EXCHANGE
RT ANION EXCHANGE RESINS
ANIONIC SITES
CATION EXCHANGE RESINS
CATIONIC SITES
DEMINERALIZING
HARD WATER
ION EXCHANGE RESINS
SITES (DYEING)
WATER TREATMENT

ION EXCHANGE RESINS
  NT ANION EXCHANGE RESINS
     CATION EXCHANGE RESINS
  BT RESINS
  RT DYEING THEORY
     ION EXCHANGE
     PERMUTITE (TN)
     POLYELECTROLYTES

IONIC BONDS
  BT CHEMICAL BONDS
  RT BOND ENERGY
     COVALENT BONDS
     ELECTROSTATIC BONDS
     HYDROGEN BONDS
     METALLIC BONDS

IONIZATION
  RT CORONA
     HUMIDITY
     IONIZING RADIATION
     IONS
     PH
     REACTIONS (CHEMICAL)

IONIZING RADIATION
  NT BETA RAYS
     GAMMA RAYS
     X RAYS
  BT RADIATION
  RT CROSSLINKING
     DEGRADATION
     GRAFTING
     IONIZATION
     IRRADIATION

IONS
  NT ANIONS
     CATIONS
  RT ELECTROLYTES
     FREE RADICALS
     IONIZATION
     PH
     SALTS (GENERAL)

IRISH (LACE)
  BT CROCHETED LACES
     HAND MADE LACES

IRON
  BT METALS

IRON COMPOUNDS
  NT FERRIC CHLORIDE
     FERRIC COMPOUNDS
     FERRIC TARTRATE
     FERROUS COMPOUNDS

IRONING
  BT DRY FINISHING
  RT BOARDING
     CREASING
     CURING
     DURABLE PRESS
     HEAT SETTING (SYNTHETICS)
     IRONING FASTNESS
     LAUNDERING
     PRESSING
     SCORCHING MARKS
     TENTERING
     THERMOPLASTICITY

IRONING FASTNESS
  UF COLORFASTNESS TO HEAT
     COLORFASTNESS TO HOT PRESSING
  BT COLORFASTNESS
     END USE PROPERTIES

IRONS (BURLING MACHINE)
  USE BURLING MACHINES

IRRADIATION
  UF RADIATION TREATMENT
  RT CROSSLINKING
     GAMMA RAYS
     GRAFTING
     INFRARED RADIATION
     IONIZING RADIATION
     PHOTOCHEMICAL REACTIONS
     RADIATION
     ULTRAVIOLET RADIATION
     X RAYS

IRREGULAR CROSS SECTION
  BT FIBER CROSS SECTIONS

IRREGULARITY
  UF EVENESS
     LEVELNESS (YARN)
     NONUNIFORMITY
     REGULARITY
     UNEVENNESS
     UNIFORMITY
     VARIABILITY
     WEIGHT UNIFORMITY
  NT IRREGULARITY (LONG TERM)
     IRREGULARITY (PERIODIC)
     IRREGULARITY (SHORT TERM)
  RT AMPLITUDE
     AUTOCORRELATION
     AUTOCOUNT (TN)
     AUTOLEVELLING
     CAPACITANCE TYPE IRREGULARITY
      TESTERS
     CARD SLIVERS
     COEFFICIENT OF VARIATION
     DEFECTS
     DOUBLING (DRAFTING)
     DRAFT
     DRAFTING (STAPLE FIBER)
     DRAFTING THEORY
     DRAFTING WAVES
     END BREAKAGES
     FABRIC DEFECTS
     FABRIC INSPECTION
     FIELDEN WALKER IRREGULARITY
      TESTER (TN)
     FLOATING FIBERS
     FLY (WASTE)
     INSPECTION
     IRREGULARITY CONTROL
     IRREGULARITY TESTERS
     IRREGULARITY TESTING
     KNOTS
     LINEAR DENSITY
     MASS UNIFORMITY
     MECHANICAL SLUB CATCHERS
     NEPS
     ORIENTATION UNIFORMITY
     PATCHINESS
     PERIODICITY
     PNEUMATIC IRREGULARITY TESTERS
     SACO LOWELL IRREGULARITY
      TESTER (TN)
     SERVO DRAFT
     SLIVERS
     SLUB CATCHERS
     SLUBS
     SPINNING
     SPOTTINESS
     STRUCTURAL PROPERTIES
     SURGING
     THICK SPOTS
     TOPS
     USTER IRREGULARITY TESTER (TN)
     VARIANCE LENGTH CURVES
     VARIATION
     WEAK SPOTS
     WEAVING EFFICIENCY
     WEB UNIFORMITY
     WINDING
     YARN DEFECTS
     YARN PROPERTIES
     YARN STRENGTH

IRREGULARITY (LONG TERM)
  BT IRREGULARITY
  RT IRREGULARITY (PERIODIC)
     IRREGULARITY (SHORT TERM)
     IRREGULARITY CONTROL
     MASS UNIFORMITY

IRREGULARITY (PERIODIC)
  BT IRREGULARITY
  RT DRAFTING (STAPLE FIBER)
     DRAFTING THEORY
     DRAFTING WAVES
     FREQUENCY
     IRREGULARITY (LONG TERM)
     IRREGULARITY (SHORT TERM)
     IRREGULARITY CONTROL
     MASS UNIFORMITY
     PERIODICITY

IRREGULARITY (SHORT TERM)
  BT IRREGULARITY
  RT DRAFTING WAVES
     IRREGULARITY (LONG TERM)
     IRREGULARITY (PERIODIC)
     IRREGULARITY CONTROL
     MASS UNIFORMITY
     THICK SPOTS
     WEAK SPOTS

IRREGULARITY CONTROL
  RT AUTODRAFTER
     AUTOLEVELLERS
     AUTOMATIC CONTROL
     CONTROL SURFACES
     CONTROL SYSTEMS
     DRAFT CONTROL
     DRAFTING (STAPLE FIBER)
     EQUALIZER (DRAFTING)
     EQUALIZING (DRAFTING)
     EVEREVEN (DRAFTING)
     FIBER CONTROL
     HOPPER FEED
     IRREGULARITY
     IRREGULARITY (LONG TERM)
     IRREGULARITY (PERIODIC)
     IRREGULARITY (SHORT TERM)
     PIANO MOTIONS
     PICKER EVENERS
     PNEUMATIC LAP WEIGHT CONTROL
     PROCESS CONTROL
     QUALITY CONTROL
     RAPER AUTOLEVELLER (TN)
     SERVO DRAFT
     SUPERLEVELLER (DRAFTING)
     SWING DOOR MECHANISMS
     WEIGHT CONTROL

IRREGULARITY TESTERS
  NT CAPACITANCE TYPE IRREGULARITY
      TESTERS
     FIELDEN WALKER IRREGULARITY
      TESTER (TN)
     PNEUMATIC IRREGULARITY TESTERS
     SACO LOWELL IRREGULARITY
      TESTER (TN)
     USTER IRREGULARITY TESTER (TN)
     WIRA WEB LEVELNESS TESTER
     ZELLWEGER TESTER (TN)
  BT TESTING EQUIPMENT
  RT AUTOLEVELLING
     CAPACITANCE (ELECTRICAL)
     IRREGULARITY
     IRREGULARITY TESTING
     OPTICAL SCANNERS
     QUALITY CONTROL
     ROVINGS
     SCANNING
     SLIVERS
     VARIANCE LENGTH CURVES
     YARNS

IRREGULARITY TESTING
  BT TESTING
  RT FIELDEN WALKER IRREGULARITY
      TESTER (TN)
     IRREGULARITY
     IRREGULARITY TESTERS
     USTER IRREGULARITY TESTER (TN)
     VARIANCE LENGTH CURVES

IRRITATION (SKIN)
  USE SKIN IRRITATION

ISOCYANATE POLYMERS
  USE POLYURETHANES

ISOCYANATES
  UF DIISOCYANATES
  NT HEXAMETHYLENE DIISOCYANATE
     TOLLENE ISOCYANATES
  BT ESTERS
     REACTANTS
  RT CARBODIIMIDES
     CROSSLINKING
     ISOTHIOCYANATES
     POLYURETHANES
     SPANDEX FIBERS
     UREA
     URETHANE FOAMS
     URETHANE RUBBERS
     URETHANES

ISOELECTRIC POINT
  BT CHEMICAL PROPERTIES
     PHYSICAL CHEMICAL PROPERTIES
  RT ISOIONIC POINT

ISOIONIC POINT
  BT CHEMICAL PROPERTIES
     PHYSICAL CHEMICAL PROPERTIES
  RT ISOELECTRIC POINT

ISOMERS (MOLECULAR)
  RT MOLECULAR STRUCTURE

ISOPHTHALIC ACID
  BT CARBOXYLIC ACIDS
     DIBASIC ACIDS
     ORGANIC ACIDS
  RT PHTHALIC ACID
     TEREPHTHALIC ACID

ISOPROPYL ALCOHOL
  BT ALCOHOLS

ISOTACTIC POLYMERS
  BT TACTIC POLYMERS

ISOTHERMAL PROCESSES
  RT MACRORHEOLOGY
     THERMODYNAMICS

ISOTHIOCYANATES
  BT ESTERS
     SULFUR COMPOUNDS
  RT ISOCYANATES
     POLYURETHANES
     THIOCYANATES
     URETHANE FOAMS
     URETHANES

ISOTOPES
  RT RADIOACTIVE COMPOUNDS
     RADIOACTIVE TRACERS

ISOTROPIC MATERIALS
  RT ANISOTROPIC MATERIALS
     DEGREE OF ORIENTATION
     HOMOGENEOUS MATERIALS
      (CONT.)

ISOTROPIC MATERIALS

RT NONWOVEN FABRIC STRUCTURE
ORIENTATION
ORIENTED WEBS
RANDO WEBS (TN)
STRUCTURAL ANALYSIS
STRUCTURAL MECHANICS
STRUCTURAL PROPERTIES

ITACONIC ACID
BT CARBOXYLIC ACIDS
DIBASIC ACIDS
ORGANIC ACIDS
RT DIMETHYLOL ITACONAMIDE

ITALIAN CLOTH
BT WOVEN FABRICS
RT TWILL WEAVES
TWILLS

ITALIAN SILK
BT SILK

IWS FINISH 6 (TN)
BT DURABLE PRESS TREATMENTS

J BOXES
RT BLEACHING RANGES
COMPENSATORS (CLOTH)
CONTINUOUS DYEING RANGE
CONTINUOUS SCOURING
DYEING
SEMI CONTINUOUS DYEING RANGE

JACKETS
BT GARMENTS
OUTERWEAR
SPORTSWEAR
RT ATHLETIC CLOTHING
COATS
DRESSCLOTHING
DRESSES
FROCKS
SUITS
WORK CLOTHING

JACKS (KNITTING)
RT KNITTING MACHINES
KNITTING NEEDLES
PATTERN WHEELS

JACKSPOOLS
UF CONDENSER BOBBINS
BT PACKAGES
RT BEAMS
BOBBINS
CARRIAGES (SPINNING MULE)
CHEESES
CONDENSED YARNS
CONDENSERS
CONES
MULES
SLUBBINGS
SURFACE DRUMS
WARPING

JACONET
BT WOVEN FABRICS
RT MUSLIN
PLAIN WEAVES

JACQUARD HARNESSES
BT HARNESSES
RT JACQUARD LOOMS

JACQUARD HEADS
RT JACQUARD LOOMS
LOOMS

JACQUARD KNITTED FABRICS
BT FILLING KNITTED FABRICS
RT JACQUARD KNITTING

JACQUARD KNITTING
UF FANCY KNITTING
BT FILLING KNITTING
KNITTING
RT CARD CYLINDERS (JACQUARD KNITTING)
CIRCULAR KNITTING
FILLING KNITTED FABRICS
FILLING KNITTING MACHINES
FULLY FASHIONED KNITTING
JACQUARD KNITTED FABRICS
JACQUARD KNITTING MACHINES
KNITTED FABRICS
PATTERN WHEEL KNITTING

JACQUARD KNITTING MACHINES
BT KNITTING MACHINES
RT JACQUARD KNITTING

JACQUARD LOOMS
BT LOOMS
RT AUTOMATIC LOOMS
BOX LOOMS
CAM LOOMS
DOBBY LOOMS
HARNESSES

JACQUARD PUNCHED CARDS
RT JACQUARD LOOMS
LOOMS

JACQUARD WEAVES
BT WEAVE (TYPES)
RT FLAT WOVEN FABRICS

JACQUARD WEAVING
BT WEAVING
RT DOBBY WEAVING

JAMMING
NT DOUBLE JAMMING
RT BEATING UP
BOUNCE (WEAVING)
CLOTH FELL
COVER FACTOR
CRIMP INTERCHANGE
ENDS PER INCH
FABRIC GEOMETRY
FABRIC STRUCTURE
FILLING YARNS
JAMMING POINT
RT PICKING (WEAVING)
PICKS PER INCH
WARP ENDS
WEAVING
WEAVING DYNAMICS

JAMMING POINT
RT CRIMP INTERCHANGE
DENSITY
FABRIC CRIMP
FABRIC GEOMETRY
FABRIC STRUCTURE
FILLING YARNS
JAMMING
PICKING (WEAVING)
PICKS PER INCH
TIGHTNESS

JEANS
BT FLAT WOVEN FABRICS
INDUSTRIAL CLOTHING
WORK CLOTHING
WOVEN FABRICS
RT DUNGAREES
OVERALLS
TWILL WEAVES

JERKINS
BT GARMENTS
OUTERWEAR
SPORTSWEAR
RT CARDIGANS
CIRCULAR KNITTING
FULLY FASHIONED KNITTING
JUMPERS
KNITTED FABRICS
KNITTING
PULL OVERS
SLIP ONS
SWEATERS

JERSEY (FILLING KNITTING)
USE PLAIN KNITTED FABRICS

JERSEY (WARP KNITTING)
USE TRICOT KNITTED FABRICS

JET BULKING
USE AIR JET TEXTURING

JET CRIMPING
USE AIR JET TEXTURING

JET LOOMS
NT AIR JET LOOMS
WATER JET LOOMS
BT LOOMS
SHUTTLELESS LOOMS
RT DRAPER SHUTTLELESS LOOMS (TN)
JETS
NOZZLES
OUTSIDE FILLING SUPPLY
RAPIER LOOMS
SHUTTLELESS WEAVING
SULZER LOOMS (TN)

JET LOOPING
USE AIR JET TEXTURING

JET QUENCHING
RT AERODYNAMICS
AIR JET TEXTURED YARNS
AIR JET TEXTURING
ANNEALING
JETS
NOZZLES
SPINNING (EXTRUSION)
STEAM JET TEXTURING

JET SCOURING
BT CONTINUOUS SCOURING
SCOURING
RT CLEANING
DETERGENTS
JETS
NOZZLES

JETS
RT AIR JET TEXTURING
JET LOOMS
JET QUENCHING
JET SCOURING
NOZZLES
SPRAYING
STEAM JET TEXTURING

JETSPUN (TN)
RT DOPE DYEING
VISCOSE RAYON

JIG DYEING
BT PIECE DYEING
RT BATCH DYEING
BECK DYEING
HIGH TEMPERATURE DYEING
OPEN WIDTH DYEING
PAD JIG DYEING
PRESSURE DYEING

JIGGERS
USE JIGS (DYEING)

JIGS (DYEING)
UF JIGGERS
RT SEMI CONTINUOUS DYEING RANGE

JOB ANALYSIS
RT ANALYZING
COSTS
EVALUATION
INVESTMENTS
MANAGEMENT
PATROLLING
PLANT
PERSONNEL
PROCESS CONTROL
PROCESS EFFICIENCY
PRODUCTION
TIME AND MOTION STUDIES
TRAINING
WAGES

JOINING
UF FASTENING
NT BONDING
CEMENTING
GLUING
LAMINATING
LOOPING
SEAMING
STITCHING
TACKING
TAPING
BT GARMENT MANUFACTURE
RT ADHESION
COHESION
FUSION (MELTING)
JOINTS
KNOTS
PEELING
PIECING
SEALING
STITCH QUALITY
THERMOPLASTICITY

JOINTS
RT BUCKLES
BUTTED SEAMS
BUTTERFLY SEAMS
BUTTONS
COUPLINGS (MECHANICAL)
COVERED SEAMS
FASTENERS
JOINING
LINKAGES
SEAMS
SNAP FASTENERS
ZIPPERS

JOURNAL BEARINGS
BT BEARINGS
RT AIR BEARINGS
BALL BEARINGS
CONICAL BEARINGS
MAGNETIC BEARINGS
NEEDLE BEARINGS
NYLON BEARINGS
PACKING
ROLLER BEARINGS
SLEEVE BEARINGS
TEFLON BEARINGS
THRUST BEARINGS

JUDD NBS COLOR SYSTEM
BT COLOR DESIGNATION
RT CIE SYSTEM
COLOR
COLOR MATCHING
MUNSELL COLOR SYSTEM
SHADE
SPECTROPHOTOMETRY

JUMPERS
BT GARMENTS
KNITTED OUTERWEAR
OUTERWEAR
SPORTSWEAR
RT CARDIGANS
CIRCULAR KNITTING
FULLY FASHIONED KNITTING
JERKINS
KNITTED FABRICS
KNITTING
PULL OVERS
SLIP ONS
SWEATERS

JUTE
BT BAST FIBERS
CELLULOSIC FIBERS
RT CORDAGE
FLAX
HEMP
HESSIAN
RAMIE
RETTING
TWINE

KAMYR MACHINES
BT PAPER MACHINES
RT CYLINDER MACHINES
FOURDRINIER MACHINES
INVERFORM MACHINES
ROTAFORMERS
VERTAFORMERS
WET MACHINES (PAPERMAKING)

KAOLIN
USE CLAYS

KAPOK
BT CELLULOSIC FIBERS
RT COTTON
MATTRESSES
PILLOWS

KAPRON (TN)
BT NYLON (POLYAMIDE FIBERS)
NYLON 6

KARL FISCHER TITRATION METHOD
RT COBALT CHLORIDE METHOD
MOISTURE CONTENT TESTING
OVEN METHOD (MOISTURE TESTING)

KEEL (SLASHING)
USE CUT MARKS (SLASHING)

KENAF
USE HIBISCUS

KERATIN
UF ALPHA KERATIN
BETA KERATIN
BT PROTEINS
RT ACID SOLUBILITY
ALKALI SOLUBILITY
ALPHA BETA TRANSFORMATION
AMINO ACIDS
ANIMAL FIBERS (GENERAL)
AZLON (REGENERATED PROTEIN FIBERS)
CYSTEINE
CYSTINE
DISULFIDE BONDS
DISULFIDE INTERCHANGE
DISULFIDES (ORGANIC)
KERATIN FIBERS
LANTHIONINE
LYSINE
MATRIX (FINE STRUCTURE)
SUPERCONTRACTION
THIOETHERS
UREA BISULFITE SOLUBILITY
WOOL

KERATIN FIBERS
NT ALPACA
CAMEL HAIR
CASHMERE
COW HAIR
GOAT HAIR
GUARD HAIRS
HAIR
HORSE HAIR
HUMAN HAIR
LAMBSWOOL
MOHAIR
OXFORD WOOL
RABBIT HAIR
SHORT WOOL
VICUNA
WOOL
BT FIBERS
NATURAL FIBERS
RT AMINO GROUPS
KERATIN

KERSEY
BT FLAT WOVEN FABRICS
WOVEN FABRICS
RT FACE GOODS

KETONES
NT ACETONE
FATTY KETONES
HYDROXYACETONE
POLY (METHYL VINYL KETONE)
PYRUVIC ACID
RT ALDEHYDES
CARBONYL GROUPS
POLYKETONES
SOLVENTS

KETTLES (DYEING)
UF DYE KETTLES
BT PIECE DYEING MACHINES
VESSELS
RT AUTOCLAVES
DYEING MACHINES
KETTLES (HEATING)
PACKAGE DYEING MACHINES
PRESSURE KETTLES
SEMI CONTINUOUS DYEING RANGE
STOCK DYEING MACHINES
TRANSPORTING VESSELS
YARN DYEING MACHINES

KETTLES (HEATING)
BT VESSELS
RT AUTOCLAVES
KETTLES (DYEING)
PRESSURE KETTLES
SIZES (SLASHING)
SLASHING
TRANSPORTING VESSELS

KID LEATHER
BT LEATHER
RT SUEDE
SYNTHETIC LEATHER

KIDDERMINSTER CARPETS
USE INGRAIN CARPETS

KIER BOILING
USE KIER SCOURING

KIER SCOURING
UF KIER BOILING
BT SCOURING
RT BATCH SCOURING
BLEACHING
BOILING
CONTINUOUS SCOURING
KIERS

KIERS
BT VESSELS
RT BLEACHING MACHINES
KIER SCOURING
SCOURING MACHINES

KILOTEX
USE TEX

KINEMATIC ANALYSIS
BT KINEMATICS
RT KINEMATIC SYNTHESIS
MOTION

KINEMATIC SYNTHESIS
BT KINEMATICS
RT KINEMATIC ANALYSIS
MOTION

KINEMATICS
NT KINEMATIC ANALYSIS
KINEMATIC SYNTHESIS
RT COUPLER CURVES
COUPLER POINTS
DYNAMICS
KINETICS
LINKAGES
MACHINE DESIGN
MECHANICS
MOTION
STRUCTURAL MECHANICS

KINETIC ELASTICITY
BT ELASTICITY
RT METARHEOLOGY

KINETIC ENERGY
BT ENERGY
RT BENDING ENERGY
INTERNAL ENERGY
MOMENTUM
POTENTIAL ENERGY
STRAIN ENERGY
TORSIONAL ENERGY
TRANSLATION
VELOCITY

KINETIC FRICTION
USE DYNAMIC FRICTION

KINETICS
BT MECHANICS
RT DYNAMIC FRICTION
DYNAMICS
KINEMATICS
KINETICS (CHEMICAL)
RATE
TIME

KINETICS (CHEMICAL)
NT REACTION KINETICS
RT EQUILIBRIUM (CHEMICAL)
DYEING RATE
KINETICS
MECHANISM (FUNDAMENTAL)
RATE
REACTION RATE
STATISTICAL MECHANICS
TIME

KINKING (KNITTING)
RT DIVIDERS (KNITTING)
KNITTING
LOOPING

KINKY FILLING
BT FABRIC DEFECTS

KISS COATING
BT COATING (PROCESS)
ROLLER COATING
RT COATING MACHINES
KISS ROLLS

KISS ROLLS
BT ROLLS
RT KISS COATING

KNEE
RT **CARD CLOTHING**
CARD CROWN
CARD WIRES
COUNT CROWN AND GAUGE
FILLET CLOTHING
NOGG
RIB SET
SHEET CLOTHING
TWILL SET

KNICKERBOCKER YARNS
BT NOVELTY YARNS
RT BEAD YARNS
BOUCLES
BUG YARNS
FLAKE YARNS
FLAMES (NOVELTY YARNS)
FRILLS (NOVELTY YARNS)
LOOPS (NOVELTY YARNS)
NOVELTY TWISTERS
NUB YARNS
RATINES
SEEDS (NOVELTY YARNS)
SLUB YARNS
SPIRALS (NOVELTY YARNS)
SPLASHES

KNICKERS
BT ATHLETIC CLOTHING
SPORTSWEAR

KNIFE BLADE GINNING
BT GINNING
RT COTTON GINS
ROLLER GINNING
SAW BLADE GINNING

KNIFE COATING
BT COATING (PROCESS)
RT BACK FILLING MACHINES
CALENDER COATING
CAST COATING
COATED FABRICS
DIP COATING
DOCTOR BLADES
EXTRUSION COATING
FLEXIBLE FILM LAMINATING
NIP COATING
PAPER MAKING
ROLLER COATING

KNIFE PLATES
RT CARD WASTE
CARDS
CLEARERS (CARD)
COTTON CARDS

KNIFE TUCKING
BT TUCKING (GARMENT MANUFACTURE)

KNIFING
USE CUTTING

KNIT DEKNIT YARNS
USE CRINKLE TYPE YARNS

KNITTED CARPETS
BT CARPETS
RT TUFTED CARPETS
WOVEN CARPETS

KNITTED FABRIC STRUCTURE (FILLING KNIT)
BT FABRIC STRUCTURE
RT KNITTED FABRIC STRUCTURE (WARP KNIT)
NONWOVEN FABRIC STRUCTURE
WOVEN FABRIC STRUCTURE
YARN STRUCTURE

KNITTED FABRIC STRUCTURE (WARP KNIT)
BT FABRIC STRUCTURE
RT KNITTED FABRIC STRUCTURE (FILLING KNIT)
NONWOVEN FABRIC STRUCTURE
WOVEN FABRIC STRUCTURE
YARN STRUCTURE

KNITTED FABRICS
UF FABRICS (KNITTED)
KNITWEAR
NT ACCORDION FABRICS
CIRCULAR KNITTED FABRICS
DOUBLE FACED FABRICS (KNITTED)
DOUBLE KNIT FABRICS
DOUBLE PIQUE
EIGHTLOCK (KNITTED FABRICS)

NT FILLING KNITTED FABRICS
FLAT KNITTED FABRICS
FULL CARDIGAN FABRICS
FULLY FASHIONED FABRICS
INTERLOCK (KNITTED FABRICS)
KNITTED LACE
LOCKKNIT
MILANESE FABRICS
PLAIN KNITTED FABRICS
PURL KNITTED FABRICS
RASCHEL FABRICS
REINFORCED KNITTED FABRICS
RIB KNITTED FABRICS
SIMPLEX FABRICS
STRETCH KNITTED FABRICS
TRICOT KNITTED FABRICS
WARP KNITTED FABRICS
BT FABRICS
FABRICS (ACCORDING TO STRUCTURE)
RT BACK (FABRIC)
BOARDING
BRIEFS
CARDIGANS
COURSES
COURSES PER INCH
COVER
FABRIC STRUCTURE
FACE (FABRIC)
FALL PLATE FABRICS
FASHIONING
FASHIONING MARKS
FILLING KNITTING
FILLING KNITTING MACHINES
FLAT KNITTING
FULLY FASHIONED KNITTING
FUZZ RESISTANCE
HAND KNITTING
INTURNED WELTS
JACQUARD KNITTING
JERKINS
JUMPERS
KNITTED NETS
KNITTED OUTERWEAR
KNITTED UNDERWEAR
KNITTING MACHINES
KNITTING YARNS
LACES
LANDING LOOPS
LAYING IN (KNITTING)
LINGERIE
LOOP DISTORTION
MACHINE KNITTING
MALI FABRICS (TN)
MALIPOL FABRIC (TN)
MESH (KNITTING)
MICROMESH (KNITTING)
MOCK FASHIONING MARKS
NEEDLE LOOPS
NONWOVEN FABRICS
PICOT
PILL RESISTANCE
PILL WEAR OFF
PILLING
PILLS
PULL OVERS
SELVEDGES
SHORTS
SINKER LOOPS
SLACK COURSE
SLIP ONS
SNAGS
SPIRALITY (KNITTED FABRICS)
SPREAD LOOPS
STARTING MARKS
STITCH BONDED FABRICS
STITCH CLARITY
STITCH DENSITY
STITCH DISTORTION
STITCH LENGTH
STRETCH
STRIP DYEING
SWEATERS
TRICOT STITCHES
TUFTED FABRICS
UNDERWEAR
VESTS
WALES
WALES PER INCH
WOVEN FABRICS

KNITTED LACE
BT KNITTED FABRICS
LACES
RT MACHINE KNITTING
RASCHEL FABRICS
WARP KNITTING

KNITTED LOOPS
USE KNITTED STITCHES

KNITTED NETS
BT NETS
RT KNITTED FABRICS
KNOTTED NETS
RASCHEL FABRICS

KNITTED OUTERWEAR
NT CARDIGANS
JUMPERS
SWEATERS
BT GARMENTS
OUTERWEAR
RT DOUBLE KNIT FABRICS
FABRICS
FILLING KNITTED FABRICS
HALF HOSE
HOSIERY
KNITTED FABRICS
KNITTED UNDERWEAR
KNITTING
MACHINE KNITTING
SOCKS
STOCKINGS

KNITTED PILE FABRICS
NT PLUSH
BT FILLING KNITTED FABRICS
PILE FABRICS
RT PILE

KNITTED STITCHES
UF KNITTED LOOPS
KNITTING STITCHES
STITCHES (KNITTING)
NT CLOSED STITCHES
FLOAT STITCHES (KNITTING)
KNOTTED STITCHES (KNITTING)
OPEN STITCHES
PELERINE STITCHES
PLAIN STITCHES (KNITTING)
PURL STITCHES
RACKED STITCHES
TUCK STITCHES
RT COURSES PER INCH
LOOP DISTORTION
LOOP STITCHES
LOOPER CLIPS
LOOPER POINTS
LOOPERS
LOOPING
LOOPS
MESH (KNITTING)
MICROMESH (KNITTING)
NEEDLE LOOPS
PICOT
SINKER LOOPS

KNITTED UNDERWEAR
BT GARMENTS
UNDERWEAR
RT DOUBLE KNIT FABRICS
FILLING KNITTED FABRICS
KNITTED FABRICS
KNITTED OUTERWEAR
KNITTING
MACHINE KNITTING

KNITTING
NT CIRCULAR KNITTING
FILLING KNITTING
FLAT KNITTING
FULLY FASHIONED KNITTING
HAND KNITTING
JACQUARD KNITTING
MACHINE KNITTING
PATTERN WHEEL KNITTING
PEG BOARD KNITTING
PLAIN KNITTING
PURL KNITTING
RIB KNITTING
TRICOT KNITTING
WARP KNITTING
RT AFTERWELT
ARACHNE PROCESS (TN)
AUTOMATIC TRANSFER (KNITTING)
BALLET TOES (KNITTING)
BOARDING
BOLT CAMS (KNITTING)
BRAIDING
CARDIGANS
CARRIER RODS
CASTING OFF
CIRCULAR HOSIERY
CIRCULAR KNITTED FABRICS
CIRCULAR KNITTING MACHINES
COURSES
COURSES PER INCH
CRINKLE TYPE YARNS
CROSS PLAITING
CYLINDER NEEDLES
DIVIDING (KNITTING)
DOUBLE FACED FABRICS (KNITTED)
DOUBLE KNIT FABRICS
DRAW MECHANISM (KNITTING)
DUMMY NEEDLES
DUMMY SLIDERS
EIGHTLOCK (KNITTED FABRICS)
EMBROIDERY
FASHIONING
FASHIONING MARKS
FILLING KNITTED FABRICS
FILLING KNITTING MACHINES
FLAT KNITTED FABRICS
FLOATING (KNITTING)
RT FULLY FASHIONED FABRICS
FULLY FASHIONED HOSIERY
GARMENT MANUFACTURE
HALF HOSE
HOSIERY
INTERLOCK (KNITTED FABRICS)
INTURNED WELTS
JERKINS
JUMPERS
KINKING (KNITTING)
KNITTED OUTERWEAR
KNITTED UNDERWEAR
KNITTING DYNAMICS
KNITTING MACHINES
KNITTING NEEDLES
KNITTING TENSION
KNITTING YARNS
KNOCKOVER
LANDING LOOPS
LAPPING (WARP KNITTING)
LAYING IN (KNITTING)
LINKING (KNITTING)
LINKS LINKS MACHINES
LISLE THREADS
LOCKING COURSES
LOOP DISTORTION
LOOP STITCHES
LOOP TRANSFER
LOOPER CLIPS
LOOPER POINTS
LOOPERS
LOOPING
LOOPS
MALI PROCESSES (TN)
MALIMO PROCESS (TN)
MALIPOL PROCESS (TN)
MALIWATT PROCESS (TN)
MOCK FASHIONING MARKS
NEEDLE BARS
NEEDLE LOOPS
NEEDLES
NEEDLING
PATTERN WHEELS
PLAITING (KNITTING)
PRESSING (BEARDED NEEDLES)
PRESSING OFF
PULL OVERS
RACKING
RELAXATION
REVERSE PLAITING
REVERSE WELTING
RIB TRANSFER
RUN RESISTANT HOSIERY
SEAMLESS HOSIERY
SHOGGING
SINKER LOOPS
SINKERS
SINKING (KNITTING)
SLACK FEEDERS (KNITTING)
SLIP ONS
SLIVER KNITTING
SLURGALLING
SOCKS
SPLICING (KNITTING)
STARTING MARKS
STITCH CAMS
STITCH DENSITY
STITCH LENGTH
STITCH QUALITY
STITCHING
STOCKINGS
STOP MOTIONS
STRETCH HOSIERY
SUTURING
SWEATERS
TAKE DOWN
TAKE DOWN TENSION
TRANSFER TAILS
TUCKING (KNITTING)
TUFTING
WALES
WALES PER INCH
WARP KNITTING MACHINES
WEAVING
WEBHOLDERS
WELTING
WELTS (KNITTED)
YARN CARRIERS
YARN CHANGING UNIT (KNITTING)
YARN DYNAMICS
YARN TENSION

KNITTING DYNAMICS
BT PROCESS DYNAMICS
RT CARDING DYNAMICS
DRAFTING DYNAMICS
KNITTING
MECHANISM (FUNDAMENTAL)
SPINNING DYNAMICS
TWISTING DYNAMICS
WEAVING DYNAMICS
WINDING DYNAMICS
YARN DYNAMICS

KNITTING MACHINERY
USE KNITTING MACHINES

KNITTING MACHINES
UF KNITTING MACHINERY
NT CIRCULAR KNITTING MACHINES
FILLING KNITTING MACHINES
FLAT KNITTING MACHINES
FULLY FASHIONED KNITTING MACHINES
HOSIERY MACHINES
JACQUARD KNITTING MACHINES
LINKS LINKS MACHINES
MILANESE KNITTING MACHINES
RASCHEL KNITTING MACHINES
SEAMLESS HOSIERY MACHINES
SIMPLEX KNITTING MACHINES
TRICOT KNITTING MACHINES
WARP KNITTING MACHINES
RT AUXILIARY CAMS
BOLT CAMS (KNITTING)
CARRIER RODS
CYLINDER NEEDLES
DIAL NEEDLES
DOUBLE KNIT FABRICS
FULLY FASHIONED HOSIERY
HOSIERY
JACKS (KNITTING)
KNITTED FABRICS
KNITTING
KNITTING NEEDLES
LOOMS
LOOP DISTORTION
MACHINE KNITTING
MOCK SEAMS
PATTERN WHEELS
POSITIVE FEED
SINKERS
SOCKS
STITCH CAMS
STOCKINGS
TRANSFER TAILS
WEBHOLDERS
YARN CARRIERS

KNITTING NEEDLES
NT COMPOUND NEEDLES
CYLINDER NEEDLES
DIAL NEEDLES
DOUBLE HEADED NEEDLES
LATCH NEEDLES
SPRING NEEDLES
BT NEEDLES
RT CAM ANGLES
CARRIER RODS
CIRCULAR KNITTING MACHINES
CROCHET NEEDLES
DUMMY NEEDLES
JACKS (KNITTING)
KNITTING
KNITTING MACHINES
NEEDLE BARS
NEEDLE BEARDS
NEEDLE BUTTS
NEEDLE CAMS
NEEDLE LATCHES
NEEDLE LOOPS
NEEDLE TRICKS
PATTERN WHEELS
PRESSER BARS
SEWING NEEDLES
STITCH CAMS
TUFTING NEEDLES
WELTING
YARN CARRIERS

KNITTING STATIONS
RT CIRCULAR KNITTING MACHINES
FULLY FASHIONED KNITTING MACHINES
KNITTING MACHINES

KNITTING STITCHES
USE KNITTED STITCHES

KNITTING TENSION
BT TENSION
RT DISC TENSION
KNITTING
STITCH LENGTH
TAKE DOWN TENSION
YARN TENSION

KNITTING YARNS
NT FINGERING YARNS
LISLE THREADS
BT YARNS
RT DOUBLE KNIT FABRICS
FILLING YARNS
KNITTED FABRICS
KNITTING
LOOP DISTORTION
MACHINE KNITTING
WARP ENDS

KNITWEAR
USE KNITTED FABRICS

KNIVES
RT CUTTERS
CUTTING
CUTTING (TAILORING)
CUTTING DRILLS
CUTTING MACHINES
CUTTING SHEARS
SCISSORS
SHARPENING
SHEARING (FINISHING)

KNOCK OFF
RT BANGING OFF
LOOM STOPS
LOOMS
STOP MOTIONS

KNOCKING OVER
USE KNOCKOVER

KNOCKOVER
UF KNOCKING OVER
RT FULLY FASHIONED KNITTING MACHINES
KNITTING
KNOCKOVER BITS
RASCHEL KNITTING MACHINES
TAKE DOWN

KNOCKOVER BITS
RT KNOCKOVER
WEBHOLDERS

KNOP YARNS
USE SEEDS (NOVELTY YARNS)

KNOPPED YARNS
USE SEEDS (NOVELTY YARNS)

KNOPT YARNS
USE SEEDS (NOVELTY YARNS)

KNOT EFFICIENCY
BT EFFICIENCY (STRUCTURAL)
MECHANICAL PROPERTIES
RT BEARING STRENGTH
COMPRESSIVE STRENGTH
CORDAGE
FRICTION
KNOT STRENGTH
KNOTS
KNOTTING
LOOP EFFICIENCY
LOOP STRENGTH
STRENGTH
STRENGTH EFFICIENCY
STRENGTH ELONGATION TESTING
STRENGTH OF MATERIALS
STRESS STRAIN CURVES

KNOT STRENGTH
BT STRENGTH
STRESS STRAIN PROPERTIES
RT BEARING STRENGTH
COMPRESSIVE STRENGTH
CORDAGE
KNOT EFFICIENCY
KNOTS
KNOTTING
LOOP EFFICIENCY
LOOP STRENGTH
STRENGTH ELONGATION TESTING
STRESS CONCENTRATION

KNOTS
NT WEAVERS KNOTS
BT DEFECTS
FABRIC DEFECTS
YARN DEFECTS
RT AUTOMATIC KNOTTING
DRUM WINDING
END BREAKAGES
ENTANGLEMENTS
FABRIC INSPECTION
IRREGULARITY
JOINING
KNOT EFFICIENCY
KNOT STRENGTH
KNOTTERS
KNOTTING
LOOPS
MECHANICAL SLUB CATCHERS
NEPS
PIECING
PILLS
SLOTS
SLUB CATCHERS
SNARLING
WINDING

KNOTTED NETS
BT NETS
RT KNITTED NETS

KNOTTED STITCHES (KNITTING)
BT KNITTED STITCHES
RT MESH (KNITTING)
TUCK STITCHES

KNOTTERS
RT AUTOMATIC KNOTTING
DRAWING IN MACHINES
KNOTTING
SLUB CATCHERS
TYING IN MACHINES
WINDERS

KNOTTING
UF TYING
NT AUTOMATIC KNOTTING
RT DRUM WINDING
END BREAKAGES
KNOT EFFICIENCY
KNOT STRENGTH
KNOTS
KNOTTERS
MALI PROCESSES (TN)
MECHANICAL SLUB CATCHERS
PIECING
QUILLING
SLUB CATCHERS
SLUBS
SPINNING EFFICIENCY
TYING IN
WEAVING
WINDING

KNUCKLE YARNS
USE COCKLE (YARN)

KODEL (TN)
BT POLYESTER FIBERS
RT VESTAN (TN)
POLY (ETHYLENE GLYCOL TEREPHTHALATE)

KORATRON (TN)
BT DURABLE PRESS TREATMENTS
RT DURABLE PRESS
WILLIAMSON DICKIE (TN)

KRITCHEVSKY PRODUCTS
USE DIETHANOLAMIDE SURFACTANTS

KUBELKA-MUNK EQUATION
RT COLOR MATCHING
REFLECTANCE

KURTOSIS
BT STATISTICAL MEASURES
STATISTICS
RT CORRELATION COEFFICIENT
DISTRIBUTION
STANDARD DEVIATION
VARIANCE

LABELLING
BT GARMENT MANUFACTURE
RT CUTTING (TAILORING)
GARMENTS
JOINING
SEAMING
SEWING
STITCHING
TAILORING

LABELLING MACHINES
BT GARMENT MANUFACTURING MACHINES
RT GARMENT MANUFACTURE
GARMENTS

LABOR
USE PERSONNEL

LABORATORY APPARATUS
BT APPARATUS
RT INDICATORS (INSTRUMENTATION)
INSTRUMENTATION
TESTING
TESTING EQUIPMENT

LABOR COST
USE WAGES

LACE FURNISHING MACHINES
BT LACE MACHINES
RT BAR WARP MACHINES
BOBBINET MACHINES
CURTAIN MACHINES
DOUBLE LOCKER MACHINES
GO THROUGH MACHINES
LEAVERS MACHINES
MECHLIN MACHINES
ROLLING LOCKER MACHINES
SIVAL MACHINES
STRING WARP MACHINES
WARP LACE MACHINES

LACE MACHINES
NT BARMEN MACHINES
BAR WARP MACHINES
BOBBINET MACHINES
CURTAIN MACHINES
DOUBLE LOCKER MACHINES
GO THROUGH MACHINES
MECHLIN MACHINES
ROLLING LOCKER MACHINES
SIVAL MACHINES
STRING WARP MACHINES
WARP LACE MACHINES
RT CATCH BARS
COMB LEAD
CROCHETING MACHINES
LANDING BARS
RASCHEL KNITTING MACHINES
TRICOT KNITTING MACHINES
WARP KNITTING MACHINES

LACE MAKING
RT KNITTING
LACE MACHINES
LACES
MALI PROCESSES (TN)
TRICOT KNITTING
WARP KNITTING
WEAVING

LACE NETS
BT NETS

LACES
UF FABRICS (LACE)
NT ALL OVER (LACE)
ARMENIAN (LACE)
BEADING (LACE)
BINCHE
BLONDE (LACE)
BOBBINETTE LACE
BOHEMIAN (LACE)
BOURDON
BRABANT
CARRICKMACROSS (LACE)
CROCHETED LACES
DARNED LACES
DUCHESS (LACE)
EDGEING (LACE)
EVERLASTING (LACE)
FLOUNCING (LACE)
GALLOON
GUIPURE
HAMBURG (LACE)
HAND MADE LACES
INSERTION (LACE)
KNITTED LACE
LAZIES
LEAVERS LACE
MACHINE MADE LACES
MALINES
MALTESE (LACE)
MIRECOURT
MOTIFS (LACE)
NEEDLE POINT LACES
NORMANDIE
NOTTINGHAM LACE
ORIENTAL LACE
PARAGUAY (LACE)
PILLOW LACES (BOBBIN LACES)
POINT DE PARIS
POINT DESPRIT
RUSSIAN POINT
SCHIFFLI LACE
SHADOW LACE
SPANISH (LACE)
TATTING LACES
TORCHON
TUCK LACES
VEILING (LACE)
YAK
BT FABRICS
FABRICS (ACCORDING TO STRUCTURE)
RT DRESSING SELVEDGES (LACE)
LACES (SHOES)
KNITTED FABRICS
NONWOVEN FABRICS
RASCHEL KNITTING MACHINES
TUFTED FABRICS
WOVEN FABRICS

LACES (SHOES)
BT BINDINGS (FOOTWEAR)
FOOTWEAR COMPONENTS
RT BOOTS
FOOTWEAR
GYMP
PUMPS (FOOTWEAR)
RAWHIDE
SHOES
SLIPPERS
SNEAKERS
SOLES
THREADS
TONGUES
UPPERS

LACING CORDS
BT CABLE CORDS
RT FRENCH CREPE CORDS

LACQUER COATED FABRICS
BT COATED FABRICS
COMPOSITES
RT BUNA N COATED FABRICS
BUTYL COATED FABRICS
COATINGS (SUBSTANCES)
NEOPRENE COATED FABRICS
RUBBER COATED FABRICS
SBR COATED FABRICS
VINYL COATED FABRICS

LACQUERS
RT CELLULOSE ACETATE BUTYRATE
COATINGS (SUBSTANCES)
FILMS
PAINTS
PIGMENTS

LACTAMS
NT CAPROLACTAM
BT MONOMERS
RT AMIDES
AMINO ACIDS
POLYAMIDES

LACTIC ACID
BT CARBOXYLIC ACIDS
ORGANIC ACIDS
RT HYDROXYACETIC ACID

LACTONES
NT BUTYROLACTONE
PROPIOLACTONE
BT ESTERS
RT GRAFTING

LADDER STITCHES
BT STITCHES (SEWING)

LADDER WEBBING
BT NARROW FABRICS
WEBBING

LADDERING
RT CIRCULAR HOSIERY
FULLY FASHIONED HOSIERY
PLAIN KNITTED FABRICS
RUN RESISTANCE
SNAGGING
SNAG RESISTANCE

LADDERS (HOSIERY)
USE RUNS (KNITTING)

LAID IN YARNS
RT FABRIC DESIGN
FLOAT DESIGN
FLOATING (KNITTING)
FLOATS
KNITTING
LAYING IN (KNITTING)

LAMBS
BT SHEEP
RT LAMBSWOOL

LAMBSWOOL
BT KERATIN FIBERS
PROTEIN FIBERS
WOOL
RT CASHMERE
CLASSING
CRIMP
CROSSBRED WOOL
FIBER LENGTH
GRADING
HAIR
LAMBS
MERINO WOOL
RECOVERED WOOL
SHEEP
SUINT
VIRGIN WOOL
WOOL CLASS
WOOL GREASE
WOOL SORTING
WOOL WAX
WOOLEN FABRICS
WOOLEN SPUN YARNS
WOOLEN SYSTEM PROCESSING
WORSTED FABRICS
WORSTED SPUN YARNS
WORSTED SYSTEM PROCESSING

LAME (TN)
BT HIGH TEMPERATURE FIBERS
INORGANIC FIBERS (MAN MADE)
METALLIC FIBERS
METALLIC YARNS
RT LUREX (TN)
METLON (TN)
TINSEL

LAMEPONS
UF PROTEIN FATTY ACID CONDENSATES
BT CARBOXYLATE SURFACTANTS

LAMINAR FLOW
BT FLUID FLOW
RT AERODYNAMICS
AIR FLOW
AIR STREAMS
BOUNDARY LAYER
DRAG
FLUID FRICTION
INVISCID FLOW
TURBULENCE
TURBULENT FLOW
VISCOUS FLOW

LAMINARIN
USE CARBOHYDRATES

LAMINATED FABRICS
UF BONDED FABRICS (LAMINATED)
FABRIC LAMINATES
NT FABRIC TO FABRIC LAMINATES
FOAM BACKED FABRICS
LAMINATED JERSEY (FOAM)
BT COMPOSITES
FABRICS (ACCORDING TO STRUCTURE)
LAMINATES
RT ADHESION
BACKED FABRICS
BACKING (MATERIAL)
BACKING FABRICS
BONDED FIBER FABRICS
COATED FABRICS
FABRIC STRUCTURE
FABRICS
FLAME BONDING
FLEXIBLE FILM LAMINATING
FUSION (MELTING)
GLASS CLOTH
GLASS MAT
HEAT BONDING
HEAT SEALING
HONEYCOMB LAMINATES
MALIPOL FABRIC (TN)
NONWOVEN FABRICS
PLASTICS
POST FORMED LAMINATES
QUILTING
SANDWICH LAMINATES
TUFTED FABRICS
UPHOLSTERY FABRICS
YARN REINFORCED COMPOSITES

LAMINATED JERSEY (FOAM)
BT FOAM BACKED FABRICS
LAMINATED FABRICS
LAMINATES

LAMINATED YARNS
BT UNCONVENTIONAL YARNS
YARNS
RT BONDED YARNS
PAPER YARNS
SLIT FILM YARNS

LAMINATES
    NT FABRIC TO FABRIC LAMINATES
        FOAM BACKED FABRICS
        HONEYCOMB LAMINATES
        LAMINATED FABRICS
        LAMINATED JERSEY (FOAM)
        POST FORMED LAMINATES
        SANDWICH LAMINATES
    BT COMPOSITES
    RT BACKING (MATERIAL)
        BACKING COMPOUNDS
        BACKING FABRICS
        BONDED YARN FABRICS
        CELLOPHANE
        FOAMS
        GLASS MAT
        ORGANOSOLS
        QUILTING
        RESINS
        SYNTHETIC LEATHER
LAMINATING
    UF LAMINATION
    NT FLAME BONDING
        FLEXIBLE FILM LAMINATING
    BT JOINING
    RT ADHESION
        BONDING
        CEMENTING
        COHESION
        DELAMINATING
        FABRIC REINFORCED COMPOSITES
        GLUING
        PEELING
        POST FORMING (LAMINATES)
        QUILTING
        VINYL COATED FABRICS
LAMINATING RESINS
    BT RESINS
    RT ADHESIVES
        BACKING COMPOUNDS
        BINDERS (NONWOVENS)
        FOAMS

LAMINATION
    USE LAMINATING

LAMPSHADES
    RT FURNISHINGS
        WINDOW HOLLAND

LANDING (LOOPS)
    USE LANDING LOOPS

LANDING BARS
    RT LACE MACHINES
        LEAVERS MACHINES

LANDING LOOPS
    UF LANDING (LOOPS)
    RT CASTING OFF
        CIRCULAR KNITTING
        CIRCULAR KNITTING MACHINES
        DIVIDING (KNITTING)
        FASHIONING
        FILLING KNITTING
        FILLING KNITTING MACHINES
        FULLY FASHIONED KNITTING
        FULLY FASHIONED KNITTING
         MACHINES
        KNITTED FABRICS
        KNITTING
        LAYING IN (KNITTING)
        NEEDLE BARS
        NEEDLES
        PRESSER BARS
        RACKING
        SINKING (KNITTING)
        TUCKING (KNITTING)
        WELTING

LANGMUIR ISOTHERM
    BT ADSORPTION ISOTHERM
    RT ADSORPTION
        BET ISOTHERM
        DYEING THEORY
        EQUILIBRIUM (CHEMICAL)
        FREUNDLICH ISOTHERM
        PARTITION COEFFICIENT

LANITAL (TN)
    BT AZLON (REGENERATED PROTEIN
         FIBERS)
        PROTEIN FIBERS

LANOLIN
    UF LANOLINE
    RT GREASE
        GREASE RECOVERY
        SCOURING
        SCOURING BATHS
        SCOURING TRAINS
        SCOURING WASTE
        SCOURING YIELD
        SOAP SODA SCOURING
        SOLVENT SCOURING
        SUINT
        WOOL GREASE

LANOLINE
    USE LANOLIN

LANOLISED NYLON
    RT NYLON (POLYAMIDE FIBERS)

LANTHIONINE
    BT AMINO ACIDS
    RT ANIMAL FIBERS (GENERAL)
        KERATIN
        PROTEINS

LAP ARBOURS
    UF LAP ROLLS
    RT COMBER LAP MACHINES
        COMBER LAPS
        COTTON CARDS
        LAP RODS
        LAPS
        PACKAGES
        PICKER LAPS
        RIBBON LAP MACHINES
        RIBBON LAPS
        ROLLER LAP

LAP GUIDES
    BT GUIDES
    RT CARDS
        COTTON CARDS
        LAP RODS
        LAPS
        PICKER LAPS

LAP PLATES
    RT COMB CYLINDERS
        COMB PINS
        COMB SEGMENTS
        COMBING
        COMBS
        COTTON SYSTEM PROCESSING
        DETACHING ROLLS
        HALF LAP
        NIPPER JAWS
        NIPPER KNIVES
        NIPPER PLATES
        RECTILINEAR COMBS
        RIBBON LAPS
        TOP COMBS

LAP RODS
    UF LAP STICKS
    RT COMBER LAP MACHINES
        COMBER LAPS
        LAP ARBOURS
        LAP GUIDES
        LAPS
        LAP WASTE
        PICKER LAPS
        RIBBON LAP MACHINES
        RIBBON LAPS

LAP ROLLS
    USE LAP ARBOURS

LAP STICKS
    USE LAP RODS

LAP UP
    UF LAPPING (OF ROLLS)
        LICKING (YARNS)
        WRAP UP
    RT CLEARERS (LINT)
        DRAFTING (STAPLE FIBER)
        LINT (FIBER)
        PROCESS EFFICIENCY
        SPINNING
        WINDING

LAP WASTE
    RT COMBER LAP MACHINES
        COMBER LAPS
        LAP RODS
        LAPS
        PICKER LAPS
        RIBBON LAP MACHINES
        RIBBON LAPS
        WASTE

LAP WEIGHT CONTROL
    RT AUTOMATIC CONTROL
        PICKER EVENERS
        PROCESS CONTROL
        WEIGHT CONTROL

LAP WINDING
    RT COMBER LAPS
        COMBING

LAPELS
    BT GARMENT COMPONENTS
    RT COLLARS
        FACING
        FRONTS

LAPPED SEAMS
    BT SEAMS
    RT BUTTED SEAMS
        BUTTERFLY SEAMS
        COVERED SEAMS
        FELLED SEAMS
        SEAMING
        SEWING
        STITCHING

LAPPERS
    NT RIBBON LAP MACHINES
        SLIVER LAP MACHINES

LAPPET LOOMS
    BT LOOMS
    RT AUTOMATIC LOOMS
        DOBBY LOOMS

LAPPET WEAVES
    BT WEAVE (TYPES)
    RT FLAT WOVEN FABRICS

LAPPING (OF ROLLS)
    USE LAP UP

LAPPING (WARP KNITTING)
    RT COMPOUND NEEDLES
        GUIDE BARS
        HALF SET
        KNITTING
        OVERLAP
        PATTERN CHAINS
        RASCHEL KNITTING MACHINES
        RUNNER RATIO
        SET (WARP KNITTING)
        SHOGGING
        TRICOT KNITTED FABRICS
        TRICOT KNITTING MACHINES
        TRICOT STITCHES
        UNDERLAP
        WARP KNITTED FABRICS
        WARP KNITTING
        WARP KNITTING MACHINES

LAPS
    NT COMBER LAPS
        PICKER LAPS
        RIBBON LAPS
    BT FIBER ASSEMBLIES
    RT COMBER LAP MACHINES
        LAP ARBOURS
        LAP GUIDES
        LAP RODS
        LAP WASTE
        NEPS
        RIBBON LAP MACHINES
        ROLLER LAP
        ROVINGS
        SLIVERS
        STOCK (FIBER)
        STRANDS
        TOPS
        TOW
        TUFTS (FIBER)
        WEBS

LARGE CIRCLE (COMB)
    RT COMB CIRCLES
        COMB PINS
        CONDUCTOR BOXES
        COVER PLATES
        DABBING BRUSHES
        NOBLE COMBS
        SETOVER
        SMALL CIRCLE (COMB)

LARGE PACKAGES
    BT PACKAGES
    RT BALLOON HEIGHT
        BALLOON TENSION
        BOBBINS
        CAKES
        CHEESES
        CONES
        COPS
        LIFT (WINDING)
        QUILLS
        SPINNING
        TAPER (BOBBIN)
        TRAVERSE
        VIBRATION
        WINDING

LARVAE
    RT DAMAGE
        INSECT RESISTANCE
        INSECT RESISTANCE FINISHES
        INSECTS
        MOTHS

LARVICIDES
    USE INSECT RESISTANCE FINISHES

LAS
    USE LINEAR ALKYLBENZENE SULFONATES

LASHED IN WEFT
    UF LASHING IN
    BT FABRIC DEFECTS

LASHING END RADIUS
RT END BREAKAGES
RING SPINNING
SPINNING
TWISTING
WINDING

LASHING IN
USE LASHED IN WEFT

LASTING
NT COTTON LASTING
WORSTED LASTING
BT WOVEN FABRICS
RT LEATHER
SATIN WEAVES
SHOE LININGS
SHOES
TWILL WEAVES

LASTS
UF LASTS (FOOTWEAR)
BT FOOTWEAR COMPONENTS
RT BOOTS
FOOTWEAR
SHOES
SHOE SIZES
SOLES
UPPERS

LASTS (FOOTWEAR)
USE LASTS

LATCH GUARDS
USE DUMMY SLIDERS

LATCH NEEDLES
BT KNITTING NEEDLES
NEEDLES
RT COMPOUND NEEDLES
CYLINDER NEEDLES
DIAL NEEDLES
DOUBLE HEADED NEEDLES
KNITTING MACHINES
SPRING NEEDLES

LATCHES
USE NEEDLE LATCHES

LATENT CRIMP
BT CRIMP
RT BICOMPONENT FIBER YARNS (FILAMENT)
BICOMPONENT FIBER YARNS (STAPLE)
BULKING
CRIMP FREQUENCY
CRIMP INDEX
CRIMPING
DIFFERENTIAL SHRINKAGE
FIBER CRIMP
FOLLOW THE LEADER CRIMP
NATURAL CRIMP
SELF CRIMPING YARNS
TEXTURING
WAVINESS
YARN CRIMP
Z CRIMP

LATENT HEAT
BT PHYSICAL CHEMICAL PROPERTIES
PHYSICAL PROPERTIES (EXCLUDING MECHANICAL)
THERMAL PROPERTIES
RT BOILING POINT
ENTHALPY
ENTROPY
HEAT CONTENT
HEAT OF FUSION
MELTING
MELTING POINT
SPECIFIC HEAT
TEMPERATURE
THERMODYNAMICS

LATEXES
USE LATICES

LATHES (LOOM)
USE SLEYS (LOOM)

LATICES
UF LATEXES
BT EMULSIONS
RT EMULSIFYING AGENTS
EMULSION POLYMERS
OIL IN WATER EMULSIONS
SOLVENT EMULSIONS
WATER IN OIL EMULSIONS

LAUNDERABILITY
UF WASHABILITY
RT BLEACHING
BLEEDING
CHLORINE FASTNESS (OF FINISH)
DETERGENTS
LAUNDERABILITY TESTERS
LAUNDERABILITY TESTING
LAUNDERING
LAUNDEROMETER
LAUNDRY DEGRADATION
SCOURING
SHRINK RESISTANCE
WASHFASTNESS (COLOR)
WASHFASTNESS (OF FINISH)

LAUNDERABILITY TESTERS
NT LAUNDEROMETER
BT TESTING EQUIPMENT
RT LAUNDERABILITY
LAUNDERABILITY TESTING
LAUNDERING
SHRINKAGE TESTERS
WASHFASTNESS TESTERS
WASHING MACHINES

LAUNDERABILITY TESTING
NT SHRINKAGE TESTING
WASHFASTNESS TESTING
BT TESTING
RT LAUNDERABILITY
LAUNDERABILITY TESTERS
LAUNDERING
LAUNDEROMETER
LAUNDRY DEGRADATION
SCOURING
SHRINK RESISTANCE
WASH WEAR RATING

LAUNDERING
UF WASHING (LAUNDERING)
BT CLEANING
RT ANTIFOAM AGENTS
ANTIREDEPOSITION AGENTS
BATHS
BLEACHING
BLEACHING AGENTS
BLEACHING MACHINES
BLEEDING
CALGON (TN)
CARBOXYMETHYLCELLULOSE
CENTRIFUGAL DRYING
CLEANLINESS
DEGREASING
DEODORIZING
DESIZING
DETERGENCY
DETERGENTS
DEWAXING (COTTON)
DRY CLEANING
DRY CLEANING MACHINES
DRYERS
EASY CARE FABRICS
FLUORESCENT BRIGHTENERS
FOAMS (FROTH)
IMPREGNATION
IRONING
LAUNDERABILITY
LAUNDERABILITY TESTING
LAUNDRY NETS
LINT (WASTE)
LINT CATCHERS
NEUTRALIZING
PILLING
PUCKER (DEFECT)
RINSING
SCOURING
SCRUBBING
SOAPS
SOIL REMOVAL
SOIL RESISTANCE
SQUEEZING
STARCH
STARCHING
SURFACTANT APPLICATIONS
TUMBLE DRYING
WASHFASTNESS (COLOR)
WASHFASTNESS (OF FINISH)
WASHING MACHINES
WEAR TESTING
WETTING AGENTS
WHITENESS RETENTION

LAUNDEROMETER
BT LAUNDERABILITY TESTERS
RT LAUNDERABILITY
LAUNDERABILITY TESTING
WASHFASTNESS TESTING

LAUNDRY DEGRADATION
BT DEGRADATION
RT CHEMICAL PROPERTIES
CHLORINE DAMAGE
DAMAGE
LAUNDERABILITY
WASHFASTNESS (COLOR)
WASHFASTNESS (OF FINISH)

LAUNDRY NETS
BT NETS
RT INDUSTRIAL FABRICS
KNITTED NETS
KNOTTED NETS
LAUNDERING
WASHING MACHINES

LAURIC ACID
BT FATTY ACIDS

LAURYL ALCOHOL
BT FATTY ALCOHOLS

LAURYL DIETHANOLAMIDE
BT DIETHANOLAMIDE SURFACTANTS

LAWN
BT FLAT WOVEN FABRICS
WOVEN FABRICS
RT BATISTE
COATED FABRICS
INDIA LINON
INDIANA CLOTH
OPALINE
ORGANDIE
PLAIN WEAVES
VICTORIA LAWN

LAWN CLOTH
RT INDIA LINON

LAWN FINISH
RT LAWN
ORGANDIE
STARCH

LAY (ROPE)
RT CABLES
CORDAGE
ROPES
STRANDS
WIRE ROPES

LAY (WIND)
RT PACKAGES
RING RAILS
WINDING

LAYER LOCKING
RT BUILDER MOTIONS
CLOSE WIND
DRUM WINDERS
DRUM WINDING
GAIN (WINDING)
HARDNESS
NOSE BUNCHING
PACKAGE DENSITY
PACKAGES
PACKAGING (OPERATION)
PIRN DENSITY
QUILLERS
QUILLING
SLOUGHING
TRAVERSE
WINDING

LAYING IN (KNITTING)
RT CASTING OFF
DIVIDING (KNITTING)
FASHIONING
FILLING KNITTED FABRICS
FILLING KNITTING
FILLING KNITTING MACHINES
KNITTED FABRICS
KNITTING
LAID IN YARNS
LANDING LOOPS
RACKING
SINKING (KNITTING)
TUCKING (KNITTING)
WARP KNITTING
WELTING

LAYS (LOOM)
USE SLEYS (LOOM)

LAZIES
BT HAND MADE LACES
LACES

LEA COUNT
BT COUNT
INDIRECT COUNT
RT LEA STRENGTH PRODUCT
LINEN THREADS
SEWING THREADS

LEA STRENGTH PRODUCT
BT BREAKING STRENGTH
STRENGTH
STRESS STRAIN PROPERTIES (TENSILE)
YARN STRENGTH
RT BOBBIN THREADS (SEWING)
BREAKING ELONGATION
BREAKING LENGTH
BUNDLE STRENGTH
COUNT STRENGTH PRODUCT
END BREAKAGES
IMPACT STRENGTH
LEA COUNT
LINEN THREADS
LOOP EFFICIENCY
LOOP STRENGTH
STRENGTH EFFICIENCY
(CONT.)

RT STRENGTH ELONGATION TESTING
TENACITY
WEAK SPOTS

LEAD COMPOUNDS
RT FIBER IDENTIFICATION

LEADER CLOTH
UF END FENT
RT FINISHING MACHINERY (GENERAL)
FINISHING PROCESS (GENERAL)

LEADING HOOKS
BT FIBER HOOKS
HOOK DIRECTION
RT CARD SLIVERS
CARD WEBS
CARDING
COTTON CARDS
DEGREE OF ORIENTATION
DRAFT
DRAFTING (STAPLE FIBER)
DRAFTING THEORY
FIBER ORIENTATION
HOOK FORMATION
HOOK REMOVAL
HOOKED FIBERS
LINDSLEY RATIOS
NEPS
PROCESSING DIRECTION
SLIVER CANS
SLIVER REVERSALS
SLIVERS
TOPS
TRAILING HOOKS
WORSTED CARDING

LEAF FIBERS
NT MANILA
RAFFIA
SISAL
BT CELLULOSIC FIBERS
FIBERS
NATURAL FIBERS
RT BAST FIBERS
CELLULOSE
FRUIT FIBERS
SEED HAIR FIBERS (GENERAL)

LEAKAGE RESISTANCE (ELECTRICAL)
RT CAPACITANCE (ELECTRICAL)
DIELECTRIC STRENGTH
INSULATION (ELECTRICAL)
RESISTANCE (ELECTRICAL)

LEASE
NT CLEARING LEASE
ENTERING LEASE
SECTION LEASE
WARPING LEASE
WEAVING LEASE
RT LEASE RODS
LEASING
LOOMS
WEAVING

LEASE RODS
RT DRAWING IN
LEASE
LEASING
LET OFF
LOOMS
SPLIT RODS
WARP BEAMS
WARP ENDS
WARP PREPARATION
WARPING
WEAVING

LEASING
RT COMBS (LEASING)
DRAWING IN
HALCHING
LEASE
LEASE RODS
TYING IN
WARP ENDS
WARP PREPARATION
WEAVING

LEATHER
NT ALLIGATOR
CALFSKIN
CHEMICAL RESISTANT LEATHER
COWHIDE
HORSEHIDE
KID LEATHER
MILDEW RESISTANT LEATHER
PIGSKIN
SYNTHETIC LEATHER
WATER RESISTANT LEATHER
RT BINDINGS (FOOTWEAR)
BOOTS
COLLAGEN FIBERS
CORFAM (TN)
FOOTWEAR
HIDES
LASTING
LEGGINGS
LININGS (FOOTWEAR)
PUMPS (FOOTWEAR)
SHOES
SLIPPERS
SNEAKERS
SOLES
SPATS
STITCH QUALITY
SUEDE
SYNTHETIC LEATHER
TANNING (LEATHER)
TANNING AGENTS
TONGUES
UPHOLSTERY FABRICS
UPPERS
VINYL COATED FABRICS

LEATHER BOOTS
BT FOOTWEAR
BOOTS

LEATHER CLOTH
BT INDUSTRIAL FABRICS
RT COATED FABRICS
LEATHER
MELTON
SYNTHETIC LEATHER
UPHOLSTERY FABRICS

LEAVERS LACE
BT LACES
MACHINE MADE LACES
RT LEAVERS MACHINES

LEAVERS MACHINES
BT LACE MACHINES
RT BARMEN MACHINES
BAR WARP MACHINES
BOBBINET MACHINES
CURTAIN MACHINES
DOUBLE LOCKER MACHINES
GO THROUGH MACHINES
LANDING BARS
LEAVERS LACE
ROLLING LOCKER MACHINES
SIVAL MACHINES
STRING WARP MACHINES
WARP LACE MACHINES

LEAVES
BT WASTE
RT BURRS
MOTE KNIVES
MOTES
SEEDS (TRASH)
TRASH

LECITHIN
USE GLYCEROPHOSPHATES

LECITHIN SULFATE
NT ALKYL GLYCEROSULFATES

LEFT HAND COORDINATE SYSTEMS
RT LEFT HAND RULE
RIGHT HAND COORDINATE SYSTEMS
RIGHT HAND RULE
SENSE (DIRECTION)

LEFT HAND RULE
RT LEFT HAND COORDINATE SYSTEMS
RIGHT HAND COORDINATE SYSTEMS
RIGHT HAND RULE
SENSE (DIRECTION)

LEFT HAND TWIST
USE Z TWIST

LEGGINGS
BT FOOTWEAR
MILITARY CLOTHING
RT GALOSHES
HALF HOSE
LEATHER
SHOES
SOLES
SPATS
STOCKINGS
SYNTHETIC LEATHER

LEICESTER WOOL
BT KERATIN FIBERS
WOOL

LENGTH
BT DIMENSIONS
RT AMPLITUDE
AREA
BENDING LENGTH
BREAKING LENGTH
FIBER LENGTH
GEOMETRY
HEADS (TOP)
HEIGHT
HOLLOWNESS
LENGTH MEASURING DEVICES
MAGNITUDE
MEAN FIBER LENGTH
THICKNESS
WIDTH

LENGTH DIAMETER RATIO
USE SLENDERNESS RATIO

LENGTH MEASURING DEVICES
RT LENGTH
MEASURING
MEASURING INSTRUMENTS
MEASURING MACHINES
REELING
REELING MACHINES (WINDING)

LENGTH PER UNIT WEIGHT
USE LINEAR DENSITY

LENGTH UNIFORMITY
USE FIBER LENGTH DISTRIBUTION

LENGTH UNIFORMITY RATIO (FIBER)
RT COTTON CLASSING
COTTON QUALITY
EFFECTIVE LENGTH (FIBER)
SHORT FIBERS INDEX
SPAN LENGTH
STAPLING

LENGTH VARIATION (FIBER)
USE FIBER LENGTH DISTRIBUTION

LENO
USE LENO WEAVES

LENO CELLULAR FABRICS
USE CELLULAR FABRICS

LENO LOOMS
BT LOOMS
RT AUTOMATIC LOOMS
BOX LOOMS
DOBBY LOOMS
LENO WEAVING

LENO SELVEDGES
BT SELVEDGES
RT CENTER SELVEDGES
CRACKED SELVEDGES
DOG LEGGED SELVEDGES
DRESSING SELVEDGES (LACE)
LOOMS
LOOP SELVEDGES
PULLED IN SELVEDGES
SEALED SELVEDGES
SELVEDGE WARPS
SLACK SELVEDGES
TAPE SELVEDGES
TEMPLES
TIGHT SELVEDGES
WEAVING
WOVEN FABRICS

LENO WEAVES
UF DOUP WEAVES
LENO
BT WEAVE (TYPES)
RT BACK CROSSING HEDDLES
CELLULAR FABRICS
DOUP
DOUP ENDS
FANCY LENO
FLAT WOVEN FABRICS
FRONT CROSSING HEDDLES
GAUZE
GROUND HEDDLES
LENO WEAVING
MARQUISETTE
MOCK LENO
MOCK LENO WEAVES
NETTING
SCREENING
STANDARD HEDDLES

LENO WEAVING
UF DOUP WEAVING
BT WEAVING
RT BACK CROSSING HEDDLES
DOUP
DOUP ENDS
FRONT CROSSING HEDDLES
GROUND HEDDLES
LENO LOOMS
LENO WEAVES
STANDARD HEDDLES

LEOTARDS
BT ATHLETIC CLOTHING
GARMENTS
RT SKI PANTS
SLACKS
SPORTSWEAR
STRETCH WOVEN FABRICS

LET OFF
UF LOOM LET OFF
NT FRICTION LET OFF
RT BEATING UP

RT HARNESSES
LEASE RODS
LET OFF BRACKETS
LET OFF CHAINS
LET OFF SPOOLS
LET OFF WEIGHT ARM
LET OFF WEIGHTS
LOOM TAKE UP
LOOMS
SHEDDING
TAKE UP
WARP BEAMS
WARP ENDS
WARP TENSION
WEAVING

LET OFF BRACKETS
RT BEATING UP
FRICTION LET OFF
LET OFF
LET OFF CHAINS
LET OFF SPOOLS
LET OFF WEIGHT ARM
LET OFF WEIGHTS
LOOMS
TAKE UP
WARP BEAMS
WARP ENDS
WARP TENSION
WEAVING

LET OFF CHAINS
RT FRICTION LET OFF
LET OFF
LET OFF BRACKETS
LET OFF SPOOLS
LET OFF WEIGHT ARM
LET OFF WEIGHTS
LOOMS
WARP BEAMS
WARP ENDS
WEAVING

LET OFF MOTIONS
NT DEAD WEIGHT LOADING
RT BEAMS
BRAKES (FOR ARRESTING MOTION)
LOOMS
MECHANICAL BRAKES
PICKS PER INCH
SPRING LOADING
TAKE UP
TENSION DEVICES
TENSION SPRINGS
WARP ENDS
WEAVING
WEAVING DYNAMICS

LET OFF SPOOLS
BT BOBBINS
RT FRICTION LET OFF
LET OFF
LET OFF BRACKETS
LET OFF CHAINS
LET OFF WEIGHT ARM
LET OFF WEIGHTS
LOOMS
WARP BEAMS
WARP ENDS
WEAVING

LET OFF WEIGHT ARM
RT FRICTION LET OFF
LET OFF
LET OFF BRACKETS
LET OFF CHAINS
LET OFF SPOOLS
LET OFF WEIGHTS
LOOMS
WEAVING

LET OFF WEIGHTS
RT FRICTION LET OFF
LET OFF
LET OFF BRACKETS
LET OFF CHAINS
LET OFF SPOOLS
LET OFF WEIGHT ARM
LOOMS
WARP BEAMS
WARP ENDS
WEAVING

LETTING OUT (GARMENTS)
BT ALTERATIONS
GARMENT MANUFACTURE
RT TAILORING
TAKING IN (GARMENTS)

LEUCO ESTER DYES
UF INDIGOSOLS
LEUCO VAT ESTER DYES
SOLUBLE VAT DYES
VAT ESTER DYES
BT VAT DYES
RT OXIDIZING COMPARTMENT (DYEING)
VAT DYEING

LEUCO VAT ESTER DYES
USE LEUCO ESTER DYES

LEVAFIX (TN)
USE SULFONAMIDE DYES

LEVAFIX E (TN)
USE DICHLOROQUINOXALINE DYES

LEVEL DYEING
USE LEVELNESS (DYEING)

LEVEL OF SIGNIFICANCE
BT STATISTICAL MEASURES
STATISTICS
RT CONFIDENCE COEFFICIENT
CONFIDENCE LIMITS
CORRELATION COEFFICIENT
DEGREES OF FREEDOM
STANDARD DEVIATION
VARIANCE

LEVELLING
RT ADJUSTMENTS
BALANCE
COMPENSATION
SETTING (ADJUSTMENTS)

LEVELLING AGENTS
NT RETARDING AGENTS
BT DYEING AUXILIARIES
RT AMPHOTERIC SURFACTANTS
ANIONIC SURFACTANTS
CARRIER DYEING
CARRIERS (DYEING)
CATIONIC SURFACTANTS
DETERGENTS
DYE MIGRATION
DYES
EMULSIFYING AGENTS
EXHAUSTION (DYEING)
FOAMS (FROTH)
GLAUBERS SALT
LEVELNESS (DYEING)
LIGNIN
NONIONIC SURFACTANTS
POLY (VINYL PYRROLIDONE)
SOAPS
SODIUM CHLORIDE
SODIUM SULFATE
SOLVENT ASSISTED DYEING
SURFACTANT APPLICATIONS
SURFACTANTS
UNION DYEING
WETTING
WETTING AGENTS

LEVELNESS (DYEING)
UF LEVEL DYEING
UNLEVEL DYEING
RT COLORFASTNESS
DYE MIGRATION
DYE UNIFORMITY
DYEING DEFECTS
EXHAUSTION (DYEING)
FABRIC DEFECTS
SKITTERINESS
SPECKINESS
TIPPINESS

LEVELNESS (YARN)
USE IRREGULARITY

LICKER IN
UF TAKER IN
BT CARD ROLLS
RT ANGLE STRIPPERS
BREAKERS
CARD CLOTHING
CARD CYLINDERS
CARDING
CARDING EFFICIENCY
CARDS
COTTON CARDING
COTTON CARDS
FEED LATTICES
FEED ROLLS
FETTLING
FINISHER PART
LICKER IN SCREENS
MOTE KNIVES
SCRIBBLING
STRIPPERS
STRIPPING ACTION
WOOLEN CARDING
WORSTED CARDING

LICKER IN SCREENS
RT AERODYNAMICS
AIR FLOW
CARD WASTE
CARDS
COTTON CARDS
CYLINDER SCREENS
LICKER IN
MOTE KNIVES
MOTES
TRASH

LICKING (YARNS)
USE LAP UP

LIFE RAFTS
BT MILITARY PRODUCTS
RT COATED FABRICS
INDUSTRIAL FABRICS
INFLATABLE FABRICS
KAPOK
PRESSURE SUITS
RADOMES
TIRES

LIFT (FLUID)
RT AERODYNAMICS
AIR DRAG
DRAG
FLUID FLOW
FLUID FRICTION
WIND TUNNELS

LIFT (WINDING)
RT BALLOONS
BALLOON HEIGHT
BOBBINS
BUILDER MOTIONS
LARGE PACKAGES
PACKAGES
RING RAILS
SPINDLE RAILS
TRAVERSE
SPINNING
WINDING

LIFTER PLATES
RT FLYER SPINNING FRAMES
TRAVERSE

LIGHT
NT ARTIFICIAL DAYLIGHT
DAYLIGHT
FLUORESCENT LIGHT
INCANDESCENT LIGHT
MONOCHROMATIC LIGHT
POLARIZED LIGHT
SUNLIGHT
BT RADIATION
RT ANGLE OF INCIDENCE
BRIGHTNESS
COLOR
COLORFASTNESS
FADING
ILLUMINATION
INFRARED RADIATION
LIGHTFASTNESS (COLOR)
LIGHTFASTNESS (OF FINISH)
PHOTOCHROMY
PHOTONS
PHOTOTROPY
ULTRAVIOLET RADIATION

LIGHT DENT
RT DENTS
DRAWING IN
REED MARKS
REED WIRES
REEDING
REEDS
SKIP DENT
SWOLLEN DENT
WARP ENDS
WARP PREPARATION

LIGHT METERS
USE PHOTOMETERS

LIGHT MICROSCOPY (OPTICAL)
BT MICROSCOPY
RT ELECTRON MICROSCOPY
INFRARED MICROSCOPY
INTERFERENCE MICROSCOPY
ULTRAVIOLET MICROSCOPY
X RAY MICROSCOPY

LIGHT TRANSMISSION
USE TRANSMITTANCE

LIGHT WEIGHT
BT WEIGHT
RT COARSE YARNS
FABRIC WEIGHT
FABRICS
FINE YARNS

LIGHTFASTNESS (COLOR)
BT COLORFASTNESS
DEGRADATION PROPERTIES
END USE PROPERTIES
RT CHLORINE FASTNESS (COLOR)
CROCKFASTNESS
DAYLIGHT
DYEING
FADEOMETER (TN)
FLUORESCENT BRIGHTENERS
GAS FUME FASTNESS
INFRARED ABSORBERS
LIGHT
LIGHTFASTNESS TESTERS
(CONT.)

RT LIGHTFASTNESS TESTING
OZONE
PHOTOCHEMICAL REACTIONS
PHOTOCHROMY
RESISTANCE TO BLEEDING
SUNLIGHT
ULTRAVIOLET ABSORBERS
ULTRAVIOLET RADIATION
ULTRAVIOLET STABILIZERS
WASHFASTNESS (COLOR)
XENOTEST (TN)

LIGHTFASTNESS (OF FINISH)
BT DEGRADATION PROPERTIES
END USE PROPERTIES
RT DAYLIGHT
DURABILITY
INFRARED ABSORBERS
LIGHT
PHOTOCHEMICAL DEGRADATION
PHOTOCHEMICAL REACTIONS
SUNLIGHT
ULTRAVIOLET ABSORBERS
ULTRAVIOLET RADIATION
ULTRAVIOLET STABILIZERS
YELLOWING

LIGHTFASTNESS TESTERS
NT FADEOMETER (TN)
XENOTEST (TN)
BT TESTING EQUIPMENT
RT LIGHTFASTNESS (COLOR)
LIGHTFASTNESS TESTING

LIGHTFASTNESS TESTING
UF COLORFASTNESS TO LIGHT TESTING
SUNLIGHT TESTS
BT COLORFASTNESS TESTING
TESTING
RT BLEEDING TESTING
CROCK TESTING
FADEOMETER (TN)
GAS FUME FASTNESS TESTING
GREY SCALE
LIGHTFASTNESS (COLOR)
LIGHTFASTNESS TESTERS
REFLECTANCE
WASHFASTNESS TESTING
WEATHERING TESTING
XENOTEST (TN)

LIGHTING
USE ILLUMINATION

LIGHTNESS
RT ABSORPTION (RADIATION)

LIGNIN
BT POLYMERS
RT LEVELLING AGENTS
WOOD

LIGNIN SULFONATES
BT SULFONATE SURFACTANTS
RT DISPERSING AGENTS

LILION (TN)
BT NYLON (POLYAMIDE FIBERS)
NYLON 6

LILLE (LACE)
BT HAND MADE LACES
PILLOW LACES (BOBBIN LACES)

LILLE TAPESTRIES
BT TAPESTRIES

LIME SOAP
RT CALCIUM SOAPS
CALCIUM STEARATE
MAGNESIUM COMPOUNDS
METALLIC SOAPS
SOAPS

LIMITING FRICTION
USE STATIC FRICTION

LINCOLN WOOL
BT LONG WOOL
WOOL

LINDSLEY RATIOS
RT CARD SLIVERS
CARDING
DRAFTING (STAPLE FIBER)
FIBER ARRANGEMENT
FIBER HOOKS
HOOK DIRECTION
HOOKED FIBERS
LEADING HOOKS
ROVINGS
TRAILING HOOKS

LINE DRYING (EASY CARE GARMENTS)
UF DRIP DRYING
BT DRYING
RT CENTRIFUGAL DRYING
FREEZE DRYING
TUMBLE DRYING
VACUUM DRYING

LINEAR ALKYLBENZENE SULFONATES
UF LAS
BT ALKYLBENZENE SULFONATES

LINEAR CHAIN MOLECULES
BT LONG CHAIN MOLECULES
MOLECULES
RT AMORPHOUS POLYMERS
BRANCHED CHAIN MOLECULES
CRYSTALLINE POLYMERS
MOLECULAR STRUCTURE
MOLECULAR WEIGHT

LINEAR DENSITY
UF LENGTH PER UNIT WEIGHT
MASS PER UNIT LENGTH
WEIGHT PER UNIT LENGTH
BT DENSITY
RT BULK DENSITY
CLASS (FIBERS)
COUNT
COUNT TESTING
DENIER
DIRECT COUNT
FIBER FINENESS
GRADE (FIBERS)
INDIRECT COUNT
IRREGULARITY
TEX
VIBRASCOPE
WEIGHT

LINEAR DIFFERENTIAL EQUATIONS
BT DIFFERENTIAL EQUATIONS
EQUATIONS
RT GENERAL SOLUTIONS
NONLINEAR DIFFERENTIAL EQUATIONS
ORDINARY DIFFERENTIAL EQUATIONS
PARTIAL DIFFERENTIAL EQUATIONS
PARTICULAR SOLUTIONS
SOLUTIONS (MATHEMATICAL)

LINEAR VISCOELASTICITY
BT VISCOELASTICITY
RT CREEP
CREEP RECOVERY
DELAYED RECOVERY
DYNAMIC MODULUS
ELASTIC MODULUS (TENSILE)
ELASTICITY
FIRMOVISCOSITY
HYSTERESIS (MECHANICAL)
MAXWELL MODEL
PLASTICITY
PLASTICOVISCOSITY
RELAXATION TIME (MECHANICAL)
STRESS RELAXATION
STRESS STRAIN PROPERTIES
VISCOSITY
VOIGT MODEL

LINEARITY
RT ACCURACY
COMPENSATION
ERRORS
MATHEMATICAL ANALYSIS

LINEN
RT DECORTICATING
FLAX
LINEN THREADS

LINEN THREADS
BT SEWING THREADS
THREADS
RT BOBBIN THREADS (SEWING)
COTTON THREADS
LEA COUNT
LEA STRENGTH PRODUCT
NEEDLE THREADS
NYLON THREADS
SEAMING
SEWING
STITCHING

LINENS
USE HOUSEHOLD FABRICS

LINGERIE
NT BRIEFS
NEGLIGEES
PANTIES
SLIPS
BT GARMENTS
UNDERWEAR
RT BRASSIERES
BRIEFS
FABRICS
FOUNDATION GARMENTS
GARTER BELTS
GIRDLES
KNITTED FABRICS

LINGETTE
BT FLAT WOVEN FABRICS
WOVEN FABRICS

LINGOE
RT HARNESSES
JACQUARD LOOMS
SHEDDING

LINING FABRICS
USE LININGS

LINING FELT
BT LININGS
RT BOX LININGS
CASKET LININGS
INSULATION (THERMAL)
SHOE LININGS

LININGS
UF LINING FABRICS
NT BOX LININGS
CASKET LININGS
DRESS BANDINGS
GLOVE LININGS
INTERLININGS
LINING FELT
LININGS (FOOTWEAR)
POCKET LININGS
SHOE LININGS
BT GARMENT COMPONENTS
WOVEN FABRICS
RT FABRICS
GARMENTS
MESALINE
PERCALINE
PRINT CLOTH
SATEEN
SCHREINERING
SHEETING (FABRIC)
TAFFETA
WIGAN

LININGS (FOOTWEAR)
NT SHOE LININGS
BT LININGS
RT BINDINGS (FOOTWEAR)
BOOTS
FOOTWEAR
LEATHER
PUMPS (FOOTWEAR)
SHOES
SLIPPERS
SNEAKERS
SOLES
TONGUES
UPPERS

LINKAGES
RT ACCESSORIES
AXLES
DEVICES
FASTENERS
JOINTS
KINEMATICS
MECHANISMS

LINKING (KNITTING)
RT FASHIONING
KNITTING
SEWING

LINKS LINKS FABRICS
USE PURL KNITTED FABRICS

LINKS LINKS MACHINES
BT KNITTING MACHINES
RT CABLE STITCH
CIRCULAR KNITTING MACHINES
DOUBLE HEADED NEEDLES
FULLY FASHIONED KNITTING MACHINES
KNITTING
NEEDLE JACKS
WARP KNITTING MACHINES

LINT (FIBER)
UF COTTON LINT
LINTERS
WORKABLE FIBERS
BT CELLULOSIC FIBERS
RT BOLLS
CLEARERS (LINT)
COTTON
GINNING
LAP UP
LINT COLOR
LINT CONTENT

LINT (FLY)
USE LINT (WASTE)

LINT (FUZZ)
USE LINT (WASTE)

LINT (WASTE)
UF LINT (FLY)
LINT (FUZZ)
RT CARD STRIPS
CLEARERS (LINT)
FABRIC INSPECTION
FUZZ
HAIRINESS

RT LAUNDERING
LINT BLADES
LINT CATCHERS
NEPS
PILLING
SHEDDING (FIBER)
SPECKING
SPECKS
TRASH

LINT BLADES
RT LINT (WASTE)
LINT CATCHERS
PRINTING ROLLS
SPECKS

LINT CATCHERS
RT LAUNDERING
LINT (WASTE)
LINT BLADES
PILLING
SPECKS
WASHING MACHINES

LINT COLOR
RT COTTON
COTTON GINS
GINNING
LINT (FIBER)
LINT CONTENT

LINT CONTENT
UF NON LINT CONTENT
RT COTTON
COTTON GINS
COTTON GRADE
COTTON LINTERS
GINNING
LINT (FIBER)
SEED HAIR FIBERS (GENERAL)

LINTERS
USE LINT (FIBER)

LIPIDS
USE FATS

LIPOPHILIC PROPERTY
RT HYDROPHILIC PROPERTY
HYDROPHOBIC PROPERTY
STAIN RESISTANCE

LIQUIDS
NT ELASTIC LIQUIDS
RT FLUID FLOW
IMMERSION
OILS
SOLIDS
SOLUTIONS (LIQUID)
WATER

LIQUOR (DYEING)
USE DYE BATHS

LIQUOR RATIO
USE BATH RATIO

LISLE THREADS
BT KNITTING YARNS
RT HOSIERY
KNITTING

LISTINGS
USE SPINDLE TAPES

LITERATURE SURVEYS
RT BIBLIOGRAPHY
COMPILATIONS
HISTORY
REVIEWS
SURVEYS
YEARS COVERAGE (1)
YEARS COVERAGE (5)
YEARS COVERAGE (10)
YEARS COVERAGE (25)
YEARS COVERAGE (50)
YEARS COVERAGE (75)
YEARS COVERAGE (100)

LITHIUM BROMIDE
BT BROMIDES
LITHIUM COMPOUNDS
RT LITHIUM CHLORIDE

LITHIUM CHLORIDE
BT LITHIUM COMPOUNDS
CHLORIDES
RT LITHIUM BROMIDE

LITHIUM COMPOUNDS
NT LITHIUM BROMIDE
LITHIUM CHLORIDE
LITHIUM HYDROXIDE
LITHIUM HYPOCHLORITE
LITHIUM STEARATE

LITHIUM HYDROXIDE
BT HYDROXIDES
LITHIUM COMPOUNDS

LITHIUM HYPOCHLORITE
BT HYPOCHLORITES
LITHIUM COMPOUNDS
RT BLEACHING
CHLORINE BLEACHING AGENTS

LITHIUM STEARATE
BT LITHIUM COMPOUNDS
METALLIC SOAPS

LIVELINESS
NT TWIST LIVELINESS
BT AESTHETIC PROPERTIES
END USE PROPERTIES
FABRIC PROPERTIES
FABRIC PROPERTIES (MECHANICAL)
MECHANICAL PROPERTIES
RT AESTHETIC APPEAL
DIMENSIONAL STABILITY
FINISH (PROPERTY)
HAND
RECOVERY (SELF RESTORATION TO ORIGINAL CONDITION)
RESILIENCE
SPONGINESS
WORK RECOVERY

LOAD BEARING CAPACITY
RT BEARING STRENGTH
BEARINGS
BREAKING STRENGTH
CAPACITY (GENERAL)
COMPRESSIVE STRENGTH

LOAD DEFLECTION CURVES
RT BENDING MOMENT DIAGRAMS
BENDING MOMENTS
DEFLECTION
LOADING
MOMENT CURVATURE CURVES
SHEAR FORCE
SHEAR FORCE DIAGRAMS
STRESS STRAIN CURVES

LOAD ELONGATION CURVES
USE STRESS STRAIN CURVES

LOAD EXTENSION CURVES
USE STRESS STRAIN CURVES

LOADING
NT SPRING LOADING
RT BREAKING STRENGTH
COLD DRAWING
CREEP
DEAD WEIGHT LOADING
DELAYED RECOVERY
DRAWING (FILAMENT)
FORCE
HOT DRAWING
LOAD DEFLECTION CURVES
PERMANENT SET
STRAIN
STRENGTH ELONGATION TESTING
STRESS
STRESS CONCENTRATION
WEIGHT
YIELD POINT

LOADING (PROCESS)
RT CHARGING
FEED RATE
FEED ROLLS
FEEDERS
FLOW (MATERIALS)
HOPPER FEED
INPUT
OUTPUT
THROUGHPUT
UNLOADING (PROCESS)

LOCKING COURSES
RT COURSES
KNITTING
PURL KNITTING
RIB KNITTING
RUN RESISTANT STITCHES

LOCKKNIT
BT KNITTED FABRICS
RT WARP KNITTING

LOCKSTITCH BLINDSTITCHES
BT LOCKSTITCHES
STITCHES (SEWING)
RT STITCHING

LOCKSTITCHES
NT LOCKSTITCH BLINDSTITCHES
SINGLE NEEDLE LOCKSTITCHES
BT STITCHES (SEWING)
RT CHAIN STITCHES
STITCHING

LODEN COATS
BT GARMENTS
RT OVERCOATS
PARKAS

LOFT
USE BULK

LOFTURA (TN)
BT ACETATE FIBERS
CELLULOSE ESTER FIBERS
TEXTURED YARNS (FILAMENT)

LOGWOOD
BT NATURAL DYES

LONDON FORCES
USE DISPERSION FORCES

LONDON SHRINKING
BT SHRINKING
WET FINISHING
WET RELAXATION
RT COMPACTING
COMPRESSIVE SHRINKING
CONDITIONING
CRABBING
FABRIC RELAXATION
FABRIC SHRINKAGE
FUSION CONTRACTION SHRINKAGE
PRESHRINKING
RELAXATION
RELAXATION SHRINKAGE
RELAXATION TIME (MECHANICAL)
RESIDUAL STRESS
SHRINK RESISTANCE
SHRINKAGE
SPONGING
STRAIN
STRESS RELAXATION
TAILORING

LONDON-HEITLER FORCES
USE DISPERSION FORCES

LONG CHAIN MOLECULES
NT BRANCHED CHAIN MOLECULES
LINEAR CHAIN MOLECULES
BT MOLECULES
RT AMORPHOUS POLYMERS
AMORPHOUS REGION
CHAIN FOLDING
CRYSTALLINE POLYMERS
CRYSTALLINE REGION
DEGREE OF POLYMERIZATION
FINE STRUCTURE (FIBERS)

LONG STAPLE COTTON
BT COTTON
RT AMERICAN UPLAND COTTON
COTTON THREADS
EGYPTIAN COTTON

LONG STAPLE SPINNING
RT BRADFORD SYSTEM PROCESSING
FINISHER GILLING
GILLING
HIGH DRAFT SPINNING
NEW BRADFORD SYSTEM PROCESSING
SHORT STAPLE SPINNING
WORSTED SPUN YARNS
WORSTED SYSTEM PROCESSING

LONG WOOL
NT COTSWOLD WOOL
LEICESTER WOOL
LINCOLN WOOL
ROMNEY MARSH WOOL
BT WOOL
RT CROSSBRED WOOL
MEDIUM WOOL
MERINO WOOL
MOUNTAIN WOOL
SHEEP BREEDS
SHORT WOOL
WOOL GRADE

LONGCLOTH
UF MAINSOOK
BT FLAT WOVEN FABRICS
WOVEN FABRICS
RT PLAIN WEAVES

LONGITUDINAL VIBRATIONS
BT VIBRATION
RT DYNAMIC BUCKLING
DYNAMIC MODULUS
ENERGY ABSORPTION
IMPULSES
PULSE PROPAGATION METER
SNAP BACK
SONIC VELOCITY
SOUND
STRAIN WAVES
STRESS PROPAGATION
TRANSVERSE VIBRATIONS
WAVE LENGTH
YARN DYNAMICS

LOOM LET OFF
USE LET OFF

LOOM STOPS
BT STOPS
RT END BREAKAGES
BANGING OFF
KNOCK OFF
LOOMS
STOP MOTIONS
WEAVING
WEAVING EFFICIENCY

LOOM TAKE UP
RT BEATING UP
CLOTH FELL
CLOTH ROLLS
CRIMP INTERCHANGE
LET OFF
LOOMS
PICKS PER INCH
SHEDDING
TEMPLES
WARP TENSION
WEAVING
WEAVING DYNAMICS

LOOMS
UF **POWER LOOMS**
NT **AIR JET LOOMS**
AUTOMATIC LOOMS
AXMINSTER LOOMS
BOX LOOMS
CAM LOOMS
CIRCULAR LOOMS
DOBBY LOOMS
DRAPER SHUTTLELESS LOOMS (TN)
FLYSHOT LOOMS
HAND LOOMS
JACQUARD LOOMS
JET LOOMS
LAPPET LOOMS
LENO LOOMS
NARROW FABRIC LOOMS
NEEDLE LOOMS (NARROW FABRICS)
PILE FABRIC LOOMS
RAPIER LOOMS
SHUTTLELESS LOOMS
SULZER LOOMS (TN)
SWIVEL LOOMS
WATER JET LOOMS
WILTON LOOMS
RT ALTERNATE PICKS
BANGING OFF
BEAMS
BEATING UP
BLANKET SHUTTLES
BOAT SHUTTLES
BOUNCE (WEAVING)
BREAST BEAMS
CLOSED SHED
CLOTH FELL
CLOTH ROLLS
COPS
DENTS
DOBBY HEADS
DOUBLE PICKS
DRAWING IN
DROP WIRES
FEELER MECHANISMS
FILLING BARS
FILLING FLOATS
FILLING YARNS
FIXED REEDS
FRICTION LET OFF
GRIPPERS (YARN)
HARNESS SKIPS
HARNESSES
HEDDLE BARS
HEDDLE EYES
HEDDLE SMASH
HEDDLES
JACQUARD HEADS
JACQUARD PUNCHED CARDS
KNITTING MACHINES
KNOCK OFF
LEASE RODS
LET OFF
LET OFF BRACKETS
LET OFF CHAINS
LET OFF MOTIONS
LET OFF SPOOLS
LET OFF WEIGHTS
LOOM STOPS
LOOM TAKE UP
LOOSE REEDS
MARIONETTE
NEGATIVE SHEDDING
OPEN SHED
OUTSIDE FILLING SUPPLY
PATTERN CHAINS
PICK AND PICK
PICKER SHOES
PICKER STICKS
PICKER STRAPS
PICKERS (LOOM)
PICKING (WEAVING)
POSITIVE SHEDDING
QUILLS
RACE BOARDS
RAPIERS (LOOMS)
RT REED MARKS
REED WIDTH
REED WIRES
REEDING
REEDS
ROTARY MAGAZINES
SELVEDGE WARPS
SELVEDGES
SHED (WEAVING)
SHEDDING
SHUTTLE BINDERS
SHUTTLE BOXES
SHUTTLE CHECKING
SHUTTLE DEFLECTORS
SHUTTLE FEELERS
SHUTTLE GUARDS
SHUTTLE MARKS
SHUTTLE SMASH
SHUTTLE SMASH PROTECTORS
SHUTTLELESS WEAVING
SHUTTLES
SLASHING
SLEYS (LOOM)
SPLIT SHED
STICK SHUTTLES
STOP MOTIONS
TAKE UP
TAKE UP ROLLS
TEMPLE MARKS
TEMPLES
THREAD CUTTERS
TRANSFER TAILS
TREACLES
TYING IN
UNIFIL (TN)
WARP ENDS
WARPING
WARP PREPARATION
WARP TENSION
WEAVERS KNOTS
WEAVING
WEAVING LEASE
WEFT FORKS
WHIP ROLLS
WOVEN FABRICS

LOOP CATCHERS
RT HOOK SHAFTS
LOOP CATCHING
NEEDLE PENETRATION
SEWING
SEWING MACHINES

LOOP CATCHING
RT CHAIN SEWING
GARMENT MANUFACTURE
HAND SEWING
LOOP CATCHERS
MACHINE SEWING
NEEDLE PENETRATION
SEAMING
SEWING
SEWING CYCLES
SEWING MACHINES
SEWING NEEDLES
STITCHES (SEWING)
STITCHING

LOOP DISTORTION
RT COURSES
COURSES PER INCH
DIMENSIONAL STABILITY
DOUBLE KNIT FABRICS
FABRIC GEOMETRY
KNITTED FABRICS
KNITTED STITCHES
KNITTING
KNITTING MACHINES
KNITTING YARNS
NEEDLE LOOPS
SINKER LOOPS
STITCH CLARITY
STITCH DENSITY
STITCH LENGTH
STRETCH
WALES
WALES PER INCH

LOOP DRYERS
BT DRYERS
RT ATMOSPHERIC DRYERS
CENTRIFUGAL DRYERS
DIELECTRIC DRYERS
DRUM DRYERS
FLUIDIZED BED DRYERS
HOT AIR DRYERS
INFRARED DRYERS
VACUUM DRYERS

LOOP EFFICIENCY
BT STRESS STRAIN PROPERTIES
EFFICIENCY (STRUCTURAL)
MECHANICAL PROPERTIES
RT BEARING STRENGTH
BREAKING STRENGTH
BUNDLE STRENGTH
COMPRESSIVE STRENGTH
CORDAGE
RT COUNT STRENGTH PRODUCT
FIBER STRENGTH
IMPACT STRENGTH
KNOT EFFICIENCY
KNOT STRENGTH
LEA STRENGTH PRODUCT
LOOP STRENGTH
PILL RESISTANCE
PILL WEAR OFF
PILLING
SHEAR STRENGTH
STRAIN
STRENGTH
STRENGTH EFFICIENCY
STRENGTH ELONGATION TESTING
STRENGTH OF MATERIALS
STRENGTH TESTING
STRESS
STRESS STRAIN CURVES
STRESS STRAIN PROPERTIES (SHEAR)
TEAR STRENGTH
WET STRENGTH
YARN STRENGTH

LOOP LENGTH
USE STITCH LENGTH

LOOP PILE FABRICS
BT PILE FABRICS
WOVEN FABRICS
RT PILE DENSITY
PILE HEIGHT

LOOP PILE TUFTING MACHINES
BT TUFTING MACHINES
RT CUT PILE TUFTING MACHINES
TUFTING MACHINES

LOOP SELVEDGES
BT SELVEDGES
RT CENTER SELVEDGES
CRACKED SELVEDGES
DOG LEGGED SELVEDGES
DRESSING SELVEDGES (LACE)
LENO SELVEDGES
PULLED IN SELVEDGES
SEALED SELVEDGES
SELVEDGE WARPS
SLACK SELVEDGES
TAPE SELVEDGES
TIGHT SELVEDGES

LOOP STITCHES
RT EMBROIDERY
KNITTED STITCHES
KNITTING
LOOP YARNS
LOOPER CLIPS
LOOPER POINTS
LOOPERS
LOOPING
LOOPS

LOOP STRENGTH
BT MECHANICAL PROPERTIES
STRENGTH
STRESS STRAIN PROPERTIES
RT BEARING STRENGTH
BREAKING STRENGTH
BUNDLE STRENGTH
BURSTING STRENGTH
COMPRESSIVE STRENGTH
CORDAGE
COUNT STRENGTH PRODUCT
FIBER STRENGTH
KNOT EFFICIENCY
KNOT STRENGTH
LEA STRENGTH PRODUCT
LOOP EFFICIENCY
PILL WEAR OFF
PRESSLEY STRENGTH
RUPTURE
SHEAR STRENGTH
STRAIN
STRENGTH EFFICIENCY
STRENGTH ELONGATION TESTING
STRENGTH TESTING
STRESS
STRESS CONCENTRATION
STRESS STRAIN CURVES
TEAR STRENGTH
YARN STRENGTH

LOOP TRANSFER
UF TRANSFERRING (LOOPS)
RT AUTOMATIC TRANSFER (KNITTING)
FASHIONING
FASHIONING MARKS
FULLY FASHIONED KNITTING MACHINES
KNITTING

LOOP TYPE YARNS
USE LOOP YARNS

LOOP YARNS
UF LOOPED YARNS

UF LOOP TYPE YARNS
NT TASLAN (TN)
BT AIR JET TEXTURED YARNS
BULKED YARNS
TEXTURED YARNS (FILAMENT)
RT CRIMPED YARNS
LOOPS (NOVELTY YARNS)

LOOPED YARNS
USE LOOP YARNS

LOOPER CLIPS
RT KNITTED STITCHES
KNITTING
LOOP STITCHES
LOOPER POINTS
LOOPERS
LOOPING
LOOPS
SOCKS
STOCKINGS

LOOPER POINTS
RT KNITTED STITCHES
KNITTING
LOOP STITCHES
LOOPER CLIPS
LOOPERS
LOOPING
LOOPS

LOOPERS
RT KNITTED STITCHES
KNITTING
LOOP STITCHES
LOOPER CLIPS
LOOPER POINTS
LOOPING
LOOPS

LOOPING
BT GARMENT MANUFACTURE
JOINING
RT BENDING
KINKING (KNITTING)
KNITTED STITCHES
KNITTING
LOOP STITCHES
LOOPER CLIPS
LOOPER POINTS
LOOPERS
LOOPS
RESIDUAL TORQUE
SNARLING TENDENCY

LOOPS
RT ENTANGLEMENTS
FLOATS
INTERLACINGS (YARN IN FABRIC)
KNITTED STITCHES
KNITTING
KNOTS
LOOP STITCHES
LOOPER CLIPS
LOOPER POINTS
LOOPERS
LOOPING
TORSIONAL BUCKLING

LOOPS (NOVELTY YARNS)
BT NOVELTY YARNS
RT BEAD YARNS
BOUCLES
BUG YARNS
FLAKE YARNS
FLAMES (NOVELTY YARNS)
FRILLS (NOVELTY YARNS)
KNICKERBOCKER YARNS
LOOP YARNS
NOVELTY TWISTERS
NUB YARNS
PICOT
RATINES
RING TWISTERS
SEEDS (NOVELTY YARNS)
SLUB YARNS
SPIRALS (NOVELTY YARNS)
SPLASHES
TWISTERS

LOOSE REEDS
BT REEDS
RT LOOMS
SLEYS (LOOM)
WEAVING

LOOSE STOCK DYEING
USE STOCK DYEING

LOOSE STOCK DYEING MACHINES
USE STOCK DYEING MACHINES

LOOSENESS
RT BAGGINESS
COMFORT
FIT
TIGHTNESS

LOSS TANGENT
BT DYNAMIC MODULUS
RT HYSTERESIS (MECHANICAL)
IMAGINARY COMPONENT (DYNAMIC MODULUS)
REAL COMPONENT (DYNAMIC MODULUS)
RHEOLOGY
VISCOELASTICITY

LOW MIDDLING (COTTON GRADE)
BT COTTON GRADE
GRADE (FIBERS)
RT COTTON

LOW QUARTER BLOOD (WOOL GRADE)
BT GRADE (FIBERS)
WOOL GRADE
RT WOOL

LOW TEMPERATURE
BT TEMPERATURE
RT COLD SETTING (SYNTHETICS)
HIGH TEMPERATURE
REFRIGERATION
ROOM TEMPERATURE
THERMOMETERS

LOW TEMPERATURE DYEING
UF COLD DYEING
BT DYEING (BY ENVIRONMENTAL CONDITIONS)
RT CARRIER DYEING
HIGH TEMPERATURE DYEING
SOLVENT ASSISTED DYEING
SOLVENT DYEING
VACUUM DYEING

LOW TWILLS
BT TWILLS
RT CHEVRON TWILLS
FANCY TWILLS
HERRINGBONE TWILLS
REGULAR TWILLS
S TWILLS
STEEP TWILLS
Z TWILLS

LOW TWIST
BT TWIST
RT ALTERNATING TWIST
COARSE YARNS
FALSE TWIST
FINE YARNS
HIGH TWIST
PLY TWIST
PRODUCERS TWIST
S TWIST
SINGLES TWIST
TURNS PER INCH
TWIST MULTIPLIER
TWISTING
YARNS
Z TWIST
ZERO TWIST

LOWER CONTROL LIMIT
RT CONFIDENCE LIMITS
CONTROL CHARTS
STATISTICAL MEASURES
UPPER CONTROL LIMIT

LUBRICANTS
UF SLICKENERS
NT CARDING OIL
SPINNING LUBRICANTS
BT TEXTILE CHEMICALS
RT ABRASION RESISTANCE FINISHES
CATIONIC SOFTENERS
EMULSIONS
FATS
FATTY ACIDS
FINISH (SUBSTANCE ADDED)
FLUOROCARBON POLYMERS
FLUOROCARBONS
GELATINE
GRAPHITE
GREASE
HYDRODYNAMIC LUBRICATION
LUBRICATION
LUBRICITY
MINERAL OIL
NEEDLE HEATING
OILING
OILS
OLEATE SALTS
OLEIC ACID
PARAFFIN WAXES
RESIDUAL FAT
SEWABILITY
SILICONES
SIZES (SLASHING)
SLICKNESS
SOFTENERS
YARN FINISH
WAXES

LUBRICATED RINGS
UF WICK LUBRICATED RINGS
BT RINGS
RT ANTIWEDGE RINGS
FRICTION
LUBRICATION
PLYING
RING SPINNING
SINTERED METAL RINGS
SPINNING
TRAVELERS
TWISTING
UNLUBRICATED RINGS
WEAR

LUBRICATING
USE LUBRICATION

LUBRICATION
UF LUBRICATING
NT BOUNDARY LUBRICATION
HYDRODYNAMIC LUBRICATION
RT AIR BEARINGS
BEARINGS
BLENDING
BUSHINGS
CARDING OIL
DEGREASING
DEWAXING (COTTON)
FRICTION
GREASE
GREASE RECOVERY
LUBRICATED RINGS
LUBRICANTS
LUBRICITY
MAINTENANCE
OILING
PLOUGHING
ROLLER SLIPPAGE
ROLLING
SLICKNESS
SLIDING
SLIPPAGE
SPINNING LUBRICANTS
STICK SLIP FRICTION
SUINT
UNLUBRICATED RINGS
WEAR
YARN FINISH

LUBRICITY
RT ABRASION RESISTANCE
LUBRICANTS
LUBRICATION
SLICKNESS
YARN FINISH

LUDOX (TN)
BT COLLOIDAL SILICA
SILICA
RT ANTISLIP AGENTS
SYTON (TN)

LUG STRAPS
RT LOOMS
PICKER STICKS

LUGGAGE
RT END USES
INDUSTRIAL FABRICS
SYNTHETIC LEATHER
UPHOLSTERY FABRICS
VINYL COATED FABRICS

LUMEN
BT FINE STRUCTURE (FIBERS)
RT CONVOLUTIONS
COTTON
FIBRIL REVERSALS
FIBRILS
IMMATURE FIBERS
IMMATURITY (FIBER)
IMMATURITY TESTING (COTTON FIBER)
MATURITY INDEX
MEDULLA
MERCERIZED COTTON
PRIMARY WALLS
SECONDARY WALLS

LUMINESCENCE
UF CHEMILUMINESCENCE
BT OPTICAL PROPERTIES
RT APPEARANCE
COLOR
FLUORESCENCE

LUREX (TN)
BT HIGH TEMPERATURE FIBERS
INORGANIC FIBERS (MAN MADE)
METALLIC FIBERS
METALLIC YARNS
RT LAME (TN)
METLON (TN)

LUSTER
NT YARN LUSTER
BT AESTHETIC PROPERTIES
(CONT.)

BT FINISH (PROPERTY)
OPTICAL PROPERTIES
RT ABSORPTION (RADIATION)
AESTHETIC APPEAL
APPEARANCE
BLOOM
BRIGHTNESS
COLOR
DELUSTERING AGENTS
DRABNESS
DULLNESS
FABRIC PROPERTIES
FABRIC PROPERTIES (AESTHETIC)
GLAZE
GLOSS
LUSTERING MACHINES
LUXURIOUSNESS
PATTERN (FABRICS)
PATTERN DEFINITION
PRESSING
REFLECTANCE
RICHNESS
RT ROUGHNESS
SHADE
SHEEN
SMOOTHNESS
SPARKLE
TEXTURE
TINSEL
TITANIUM DIOXIDE
TRANSLUCENCY
TRANSPARENCY

LUSTERING MACHINES
RT CALENDERING MACHINES
LUSTER
SCHREINERING

LUXURIOUSNESS
BT AESTHETIC PROPERTIES
FABRIC PROPERTIES
FABRIC PROPERTIES (AESTHETIC)
RT APPEARANCE
BLOOM
RT COLOR
DESIGN
DRABNESS
DRAPE
HAND
LUSTER
PATTERN (FABRICS)
RICHNESS
ROUGHNESS
SHADE
SMOOTHNESS
TEXTURE
WARMTH

LYCRA (TN)
BT SPANDEX FIBERS
RT FOUNDATION GARMENTS
STRETCH FABRICS

LYSINE
BT AMINO ACIDS
RT KERATIN
PROTEINS
WOOL

MACARONI YARNS
USE HOLLOW FILAMENT YARNS

MACHINE DESIGN
BT DESIGN
RT CONTROL SYSTEMS
DEVELOPMENTS
DOUBLE SYSTEMS
IMPROVEMENTS
KINEMATICS
MACHINERY
MAN MACHINE CONTROL SYSTEMS
MODIFICATION
PERFORMANCE
STRENGTH OF MATERIALS
TEXTILE MACHINERY (GENERAL)

MACHINE INTERFERENCE
RT MANAGEMENT
MANUFACTURING
PATROLLING
PROCESS EFFICIENCY
PRODUCTIVITY
QUALITY CONTROL
WORKLOADS

MACHINE KNITTING
BT KNITTING
RT BOARDING
CIRCULAR KNITTED FABRICS
COURSES
DOUBLE KNIT FABRICS
FASHIONING
FILLING KNITTED FABRICS
FULLY FASHIONED FABRICS
HAND KNITTING
HOSIERY
KNITTED FABRICS
KNITTED LACE
KNITTED OUTERWEAR
KNITTED UNDERWEAR
KNITTING MACHINES
KNITTING YARNS
PLAIN KNITTED FABRICS
PURL KNITTED FABRICS
RIB KNITTED FABRICS
TRICOT KNITTED FABRICS
WALES
WARP KNITTED FABRICS

MACHINE MADE LACES
NT BOBBINETTE LACE
BRETON
LEAVERS LACE
NOTTINGHAM LACE
SCHIFFLI LACE
BT LACES
RT BATTENBERG
ORIENTAL LACE
PRINCESS LACE
RASCHEL KNITTING MACHINES
SHADOW LACE

MACHINE SEWING
BT SEWING
RT ANGULAR SEAMING
BASTING
BOBBIN THREADS (SEWING)
CHAIN SEWING
CURVED SEAMING
HAND SEWING
LOOP CATCHING
NEEDLE THREADS
SEAM STRENGTH
SEAMING
SEWABILITY
SEWING MACHINES
SEWING THREADS
STITCHING

MACHINERY
UF MACHINES
RT ACCESSORIES
COMPONENTS
COSTS
FINISHING MACHINERY (GENERAL)
INVENTORIES
INVESTMENTS
MACHINE DESIGN
MAINTENANCE
MECHANISMS
MOTORS
PLANT
PLANT LAYOUT
POWER CONSUMPTION
TEXTILE MACHINERY (GENERAL)

MACHINES
USE MACHINERY

MACROFIBRILS
BT FIBRILS
FINE STRUCTURE (FIBERS)
RT FIBRIL REVERSALS
FIBRILLATION
FRINGED FIBRILS
MICROFIBRILS
WOOD FIBRILS

MACRORHEOLOGY
BT RHEOLOGY
RT HOMOGENEOUS MATERIALS
ISOTHERMAL PROCESSES
METARHEOLOGY
MICRORHEOLOGY
PHENOMENOLOGICAL RHEOLOGY
QUASI HOMOGENEOUS MATERIALS

MAGNADRAFT (TN)
RT DRAFTING (STAPLE FIBER)
DRAFTING ROLLS
DRIVEN ROLLS
DRIVING ROLLS
FLUTED ROLLS
PRESSURE ROLLS
ROLL BOSSES
SPINNING FRAMES

MAGNESIUM CHLORIDE
BT CHLORIDES
MAGNESIUM COMPOUNDS
RT CATALYSTS
DELIQUESCENT AGENTS
MAGNESIUM NITRATE
ZINC NITRATE

MAGNESIUM COMPOUNDS
NT MAGNESIUM CHLORIDE
MAGNESIUM HYDROXIDE
MAGNESIUM NITRATE
MAGNESIUM SULFATE
RT LIME SOAP

MAGNESIUM HYDROXIDE
BT HYDROXIDES
MAGNESIUM COMPOUNDS

MAGNESIUM NITRATE
BT MAGNESIUM COMPOUNDS
NITRATES
RT CATALYSTS
MAGNESIUM CHLORIDE
ZINC CHLORIDE
ZINC NITRATE

MAGNESIUM SULFATE
BT MAGNESIUM COMPOUNDS
SULFATES

MAGNETIC BEARINGS
BT BEARINGS
RT AIR BEARINGS
JOURNAL BEARINGS
MAGNETIC BRAKES

MAGNETIC BRAKES
BT BRAKES (FOR ARRESTING MOTION)
RT AIR BRAKES
AUTOMATIC STOP MOTIONS
BRAKE DISCS
BRAKE DRUMS
BRAKE LININGS
BRAKING
ELECTRIC BRAKES
FRICTION
HYDRAULIC BRAKES
MAGNETIC BEARINGS
MECHANICAL BRAKES
POWER BRAKES
RETARDERS
WATER BRAKES

MAGNETIC RINGS
BT RINGS
RT RING SPINNING

MAGNETIC TAPE
BT TAPES (DATA MEDIA)
RT COMPUTERS
PUNCHED TAPE

MAGNETIC TENSION DEVICES
BT TENSION DEVICES
RT DIRECT TENSION DEVICES
GATE TENSION
HYDRAULIC TENSION DEVICES
INDIRECT TENSION DEVICES
TENSION DISCS
TENSION SPRINGS
WASHERS

MAGNITUDE
RT AMPLITUDE
DIAMETER
DIMENSIONS
INTENSITY
LENGTH
NUMBERS
SENSE (DIRECTION)

MAIL (JACQUARD LOOM)
RT HARNESSES
JACQUARD LOOMS

MAINSOCK
USE LONGCLOTH

MAINSOOK CHECK
BT FLAT WOVEN FABRICS
WOVEN FABRICS

MAINTENANCE
UF OVERHAULING
PATCHING
PREVENTIVE MAINTENANCE
REPAIRS (MACHINERY)
UPKEEP
RT ADJUSTMENTS
ALIGNMENT
LUBRICATION
MACHINERY
MANAGEMENT
REPAIRS
SETTING (ADJUSTMENTS)
TRAINING

MAKING UP
USE GARMENT MANUFACTURE

MALEIC ACID
USE MALEIC ANHYDRIDE

MALEIC ANHYDRIDE
UF MALEIC ACID
BT ACID ANHYDRIDES
RT PHTHALIC ANHYDRIDE

MALI FABRICS (TN)
UF FABRICS (MALI)
NT MALIMO FABRIC (TN)
MALIPOL FABRIC (TN)
MALIWATT FABRIC (TN)
RT FABRIC STRUCTURE
KNITTED FABRICS
MALI PROCESSES (TN)
MALIMO PROCESS (TN)
MALIPOL PROCESS (TN)
MALIWATT PROCESS (TN)
NONWOVEN FABRICS
WOVEN FABRICS

MALI PROCESSES (TN)
NT MALIMO PROCESS (TN)
MALIPOL PROCESS (TN)
MALIWATT PROCESS (TN)
RT ARACHNE PROCESS (TN)
ARACHNE SYSTEMS (TN)
BONDING
BRAIDING
CHAIN SEWING
CROSS LAID YARN FABRICS
FABRICS (ACCORDING TO STRUCTURE)
FILLING KNITTING
KNITTING
KNOTTING
LACE MAKING
MALI FABRICS (TN)
NEEDLES
NEEDLING
NONWOVEN FABRIC MACHINES
NONWOVEN FABRICS
SEWING
STITCH BONDED FABRICS
STITCH BONDING MACHINES
STITCH REINFORCED FABRICS
STITCHED PILE FABRICS
STITCHING
TUFTING
WARP KNITTING
WEAVING
WEBS

MALIMO FABRIC (TN)
BT CROSS LAID YARN FABRICS
MALI FABRICS (TN)
STITCH BONDED FABRICS (TN)
RT MALIPOL FABRIC (TN)
MALIWATT FABRIC (TN)

MALIMO PROCESS (TN)
BT MALI PROCESSES (TN)
STITCH BONDING
RT ARACHNE PROCESS (TN)
BEAMS
KNITTING
MALI FABRICS (TN)
MALIPOL PROCESS (TN)
MALIWATT PROCESS (TN)
STITCHING
WARPING
WEAVING
WEBS

MALINES
BT HAND MADE LACES
LACES

MALIPOL FABRIC (TN)
BT MALI FABRICS (TN)
NONWOVEN FABRICS
STITCH BONDED FABRICS
STITCHED PILE FABRICS
TUFTED FABRICS

(CONT.)

RT ARACHNE FABRIC (TN)
KNITTED FABRICS
LAMINATED FABRICS
MALIMO FABRIC (TN)
MALIPOL PROCESS (TN)
MALIWATT FABRIC (TN)
NEEDLING
TUFTED CARPETS
TUFTING
WOVEN FABRICS

MALIPOL PROCESS (TN)
BT MALI PROCESSES (TN)
STITCH BONDING
RT ARACHNE PROCESS (TN)
KNITTING
MALI FABRICS (TN)
MALIMO PROCESS (TN)
MALIPOL FABRIC (TN)
MALIWATT PROCESS (TN)
NEEDLING
NONWOVEN FABRICS
SEWING
STITCH BONDED FABRICS
STITCHED PILE FABRICS
STITCHING
TUFTING
WEAVING
WEBS

MALIWATT FABRIC (TN)
BT MALI FABRICS
NONWOVEN FABRICS
STITCH BONDED FABRICS
STITCH REINFORCED FABRICS

MALIWATT PROCESS (TN)
BT MALI PROCESSES (TN)
STITCH BONDING
RT ARACHNE FABRIC (TN)
ARACHNE PROCESS (TN)
FIBER WOVEN (TN)
KNITTING
MALI FABRICS (TN)
MALIMO FABRIC (TN)
MALIMO PROCESS (TN)
MALIPOL FABRIC (TN)
MALIPOL PROCESS (TN)
MALIWATT FABRIC (TN)
MALIWATT PROCESS (TN)
NEEDLE PUNCHED FELTS
NEEDLED FABRICS
NEEDLING
RANDO WEBBER (TN)
SEWING
STITCHING
SYNTHETIC FELTS
TUFTED FABRICS
WEAVING
WEBS

MALTESE (LACE)
BT HAND MADE LACES
LACES

MAN MACHINE CONTROL SYSTEMS
RT CONTROL SYSTEMS
MACHINE DESIGN

MAN MADE FIBER BLENDS
USE BLENDS (FIBERS)

MAN MADE FIBERS
NT ACETATE FIBERS
ACRYLIC FIBERS
ALGINATE FIBERS
ARTIFICIAL SILK (ARCHAIC)
AZLON (REGENERATED PROTEIN FIBERS)
BUTADIENE RUBBER FIBER
CELLULOSE ESTER FIBERS
CERAMIC FIBERS
CUPRAMMONIUM RAYON
FLUOROCARBON FIBERS
GLASS FIBERS
HIGH TENACITY VISCOSE RAYON
HIGH WET MODULUS RAYON
INORGANIC FIBERS (MAN MADE)
METAL FIBERS
METALLIC FIBERS
MODACRYLIC FIBERS
NITRILE RUBBER FIBER
NYLON (POLYAMIDE FIBERS)
NYLON 11
NYLON 6
NYLON 610
NYLON 66
NYLON 7
NYTRIL FIBERS
OLEFIN FIBERS
POLYAMIDOESTER FIBERS
POLYESTER FIBERS
POLYETHYLENE FIBERS
POLYNOSIC FIBERS
POLYPROPYLENE FIBERS
POLYTETRAFLUOROETHYLENE FIBERS
POLYTRIFLUOROCHLOROETHYLENE FIBERS
NT POLYURETHANE FIBERS
RAYON (REGENERATED CELLULOSE FIBERS)
REGENERATED FIBERS (EXCLUDING CELLULOSIC AND PROTEIN)
RUBBER FIBER (NATURAL)
SARAN POLY (VINYLIDENE CHLORIDE) FIBERS
SPANDEX FIBERS
SYNTHETIC FIBERS
SYNTHETIC RUBBER FIBERS
TRIACETATE FIBERS
VINAL POLY (VINYL ALCOHOL) FIBERS
VINYL FIBERS (GENERAL)
VINYON POLY (VINYL CHLORIDE) FIBERS
VISCOSE RAYON
BT FIBERS
RT ANIMAL FIBERS (GENERAL)
BICOMPONENT FIBERS
BICONSTITUENT FIBERS
BLENDS (FIBERS)
COAGULATING BATHS
COPOLYMERS
DRAWING (FILAMENT)
EXTRUSION
FILAMENT YARNS
FILAMENTS
MELT SPINNING
MINERAL FIBERS
MONOFILAMENT YARNS
MULTIFILAMENT YARNS
NATURAL FIBERS
NEUTRALIZING (FIBER EXTRUSION)
POLYMERS
PROTEIN FIBERS
RUBBER
SOLVENT SPINNING
SPINNERETS
SPINNING (EXTRUSION)
STAPLE FIBERS
SYNTHETIC DYEING
SYNTHETIC FELTS
TEXTURED YARNS (FILAMENT)
TEXTURING
TOW
VEGETABLE FIBERS (GENERAL)
WET SPINNING

MAN POWER
USE PERSONNEL

MANAGEMENT
RT ALLOTMENT
CONSUMPTION (MATERIAL)
DOWN TIME
ECONOMICS
FLOW (MATERIALS)
INVENTORIES
INVESTMENTS
JOB ANALYSIS
MACHINE INTERFERENCE
MAINTENANCE
MANUFACTURING
MARKETING
MATERIAL PROCESSING
MATERIALS HANDLING
MERCHANDISING
OCCUPATIONAL HAZARDS
PATROLLING
PERSONNEL
PLANNING
PLANT
PLANT LAYOUT
PROCESS CONTROL
PRODUCTION
RESEARCH
SAFETY
SALES
SET UP TIME
TIME AND MOTION STUDIES
TRAINING
WORKLOADS
WAGES
YIELD (RETURN)

MANDRELS
RT MECHANISMS
PACKAGE HOLDERS
PICKER LAPS
PICKERS (OPENING)
PRINTING MACHINES
PRINTING ROLLS

MANGANESE COMPOUNDS
NT PERMANGANATES

MANGLE EXPRESSION
USE WET PICKUP

MANGLES
RT CENTRIFUGES
DRYERS
PADDERS
PAPER MACHINES
PRESSURE ROLLS
ROLLS
RT SQUEEZE ROLLS
SUCTION PRESSES
WET FINISHING

MANGLING
BT DRY FINISHING
WET FINISHING
RT ADD ON
CONTACT AREA
DRYING
NIP
NIP PRESSURE
NIP ROLLS
PAPER MAKING
RATCH
SQUEEZING
SUCTION PRESSES
WET PICKUP

MANILA
BT CELLULOSIC FIBERS
LEAF FIBERS
RT CORDAGE
SISAL

MANNEQUINS
UF FASHION MODELS
BT MODELS (FASHION)
RT APPAREL DESIGN
DRESS DESIGN
FASHION (APPAREL)
MODELLING
STYLING (APPAREL)

MANUAL CONTROL
BT CONTROL SYSTEMS
RT AUTOMATIC CONTROL
CONTINUOUS CONTROL
CONTROL SYSTEMS
FEEDBACK CONTROL
ON OFF CONTROL
PROCESS CONTROL
REMOTE CONTROL

MANUFACTURE
USE MANUFACTURING

MANUFACTURING
UF MANUFACTURE
RT FLOW (MATERIALS)
GARMENT MANUFACTURE
INVENTORIES
MACHINE INTERFERENCE
MANAGEMENT
MARKETING
MATERIAL PROCESSING
MATERIALS HANDLING
MERCHANDISING
PATROLLING
PLANT
PREPARATION (CHEMICAL)
PROCESS CONTROL
PROCESS EFFICIENCY
PROCESSES
PRODUCTION
RESEARCH
SALES

MAPO
UF TRIS (METHYLAZIRIDINYL) PHOSPHINE OXIDE
BT AZIRIDINE COMPOUNDS
PHOSPHORUS COMPOUNDS
RT APO
FIRE RETARDANCY AGENTS
PROPYLENE IMINE

MARIONETTE
RT LOOMS
WEAVING

MARK OFF
USE STAINING

MARKER MAKING
USE MARKING

MARKER STENCILS
RT MARKERS
MARKING
MARKING MACHINES

MARKERS
BT GARMENT MANUFACTURING MACHINES
RT CUTTING MARKERS
MARKING
MARKING MACHINES
MARKING PAPER
TAILORING

MARKETING
RT COSTS
DELIVERY
ECONOMICS
INVENTORIES
MANAGEMENT
MANUFACTURING
PRODUCT DESIGN

RT PRODUCTION
RESEARCH
STYLING (APPAREL)

MARKING
UF MARKER MAKING
POSITIONING MARKING
BT GARMENT MANUFACTURE
RT CUTTING (TAILORING)
CUTTING MARKERS
DRESSMAKING
INKS
MARKER STENCILS
MARKERS
MARKING MACHINES
MARKING PAPER
PATTERN (APPAREL)
SHADING (TAILORING)
SPREADING (TAILORING)
TAILORING

MARKING MACHINES
RT MARKER STENCILS
MARKERS
MARKING

MARKING PAPER
RT MARKER STENCILS
MARKERS
MARKING

MAROCAIN
BT CREPE
RT RIB WEAVES

MARKOV CHAIN
RT MATHEMATICAL ANALYSIS
PROBABILITY
STATISTICAL ANALYSIS
STATISTICAL INFERENCE

MARQUISETTE
BT FLAT WOVEN FABRICS
WOVEN FABRICS
RT GAUZE WEAVES
LENO WEAVES
LENO WEAVING
NETTING
PLAIN WEAVES
SCREENING
SCRIM

MARRIED YARNS
RT END BREAKAGES
SPINNING
YARN DEFECTS

MARSCHALL SOLUTION
RT CADOXEN
CELLULOSE
CUPRAMMONIUM HYDROXIDE
CUPRIETHYLENEDIAMINE HYDROXIDE
FERRIC TARTRATE
FORMIC ACID
SOLVENTS
VISCOSITY
ZINC CHLORIDE

MARTINDALE LIMIT
RT DRAFTING (STAPLE FIBER)
DRAFTING THEORY
IRREGULARITY
SPINNING

MASS
RT CENTER OF GRAVITY
DENSITY
FORCE
GRAVITATION
INERTIA
MASS TRANSFER
MASS UNIFORMITY
MOMENTUM
SPRING CONSTANT
STICK SLIP FRICTION
WEIGHT

MASS DENSITY
USE DENSITY

MASS DYEING
USE DOPE DYEING

MASS MOMENT OF INERTIA
USE MOMENT OF INERTIA

MASS PER UNIT LENGTH
USE LINEAR DENSITY

MASS SPECTROMETRY
BT CHEMICAL ANALYSIS
RT SPECTROGRAPHY
SPECTROSCOPY
SPECTRUM ANALYSIS

MASS TRANSFER
RT AIR FLOW
CONCENTRATION
CONCENTRATION GRADIENT
CONVECTION
DIALYSIS
DIFFUSION
DRYERS
DRYING
FILTRATION
FLUID FLOW
HEAT TRANSFER
MASS
MEMBRANES
MOMENTUM TRANSFER
PERMEABILITY
POROSITY
POROUS MATERIALS
SATURATED SOLUTIONS
TURBULENCE
TURBULENT FLOW

MASS UNIFORMITY
RT IRREGULARITY
IRREGULARITY (LONG TERM)
IRREGULARITY (PERIODIC)
IRREGULARITY (SHORT TERM)
MASS
ORIENTATION UNIFORMITY
PATCHINESS
SPOTTINESS
WEB UNIFORMITY

MATCHED DIE MOLDING
BT MOLDING
RT COMPRESSION MOLDING
CONTACT MOLDING
FLEXIBLE PLUNGER MOLDING
INJECTION MOLDING
PRESSURE BAG MOLDING
REINFORCED COMPOSITES
VACUUM BAG MOLDING

MATCHINGS
RT SORTING
WOOL

MATERIAL CONSUMPTION
USE CONSUMPTION (MATERIAL)

MATERIAL PROCESSING
BT MATERIALS ENGINEERING
RT MANAGEMENT
MANUFACTURING
MATERIALS HANDLING
MATERIALS SHAPING
PROCESS EFFICIENCY
PRODUCTION
TRANSFERRING (MATERIAL)

MATERIALS
RT CERAMICS
GLASS
MECHANICAL PROPERTIES
METALS
POLYMERS
RUBBER
TEXTILE MATERIALS

MATERIALS ENGINEERING
NT MATERIAL PROCESSING
RT CHEMICAL ENGINEERING
MATERIALS SHAPING
MECHANICAL PROPERTIES
PHYSICAL PROPERTIES (EXCLUDING MECHANICAL)
PROPERTIES

MATERIALS FLOW
USE FLOW (MATERIALS)

MATERIALS HANDLING
RT CONVEYORS
COST CONTROL
DELIVERY
DISTRIBUTION
DOFFING (PACKAGE)
EQUIPMENT
FACILITIES
FEED LATTICES
HOPPERS
INVENTORIES
MANAGEMENT
MANUFACTURING
MATERIAL PROCESSING
PLANT
PLANT LAYOUT
PROCESS CONTROL
PROCESS EFFICIENCY
PRODUCTION
THROUGHPUT
TRANSFERRING (MATERIAL)

MATERIALS SHAPING
RT MATERIALS ENGINEERING
MATERIAL PROCESSING

MATHEMATICAL ANALYSIS
UF ANALYSIS (MATHEMATICAL)
RT ANALYZING
CONTROL THEORY
DIFFERENTIAL EQUATIONS
EQUATIONS
EMPIRICAL ANALYSIS
EXPERIMENTAL ANALYSIS
EXPERIMENTATION
GENERAL SOLUTIONS
LINEARITY
MARKOV CHAIN
MODELS (MATHEMATICAL)
MATRICES
NUMERICAL SOLUTIONS
ROOTS (MATHEMATICAL)
SOLUTIONS (MATHEMATICAL)
STATISTICAL ANALYSIS
THEORETICAL ANALYSIS
THEORIES

MATHEMATICAL MODEL
USE MODELS (MATHEMATICAL)

MATRICES
RT MATHEMATICAL ANALYSIS
TENSORS

MATRIX (FINE STRUCTURE)
RT AMORPHOUS REGION
FINE STRUCTURE (FIBERS)
FIBRILS
KERATIN
WOOL

MATRIX FREEDOM
BT STRUCTURAL PROPERTIES
RT BATTING
BENDING RIGIDITY
FABRIC GEOMETRY
FIBER FIBER FRICTION
HOLLOWNESS
INTERLACINGS (YARN IN FABRIC)
NEEDLE PENETRATION
PACKING FACTOR
PILLING
POROSITY
SHEDDING (FIBER)
SPACE
TIGHTNESS
WRINKLE RECOVERY
YARN CROSS SECTIONS
YARN DIAMETER
YARN FLATTENING
YARN GEOMETRY
YARN SLIPPAGE

MATS
RT BATTING
CARPETS
FOAM RUBBER
FOAMS
FURNISHINGS
HOUSEHOLD FABRICS
KAPOK

MATTING
RT COMPRESSION
CRUSH RESISTANCE
PILLING
RECOVERY (SELF RESTORATION TO ORIGINAL CONDITION)
SHEDDING

MATTRESSES
RT HOUSEHOLD FABRICS
KAPOK
TICKING

MATURITY (FIBER)
USE IMMATURITY (FIBER)

MATURITY INDEX
RT CAUSTICAIRE VALUE
COTTON
FIBER FINENESS
GRADE (FIBERS)
IMMATURE FIBERS
IMMATURITY (FIBER)
LUMEN
MICRONAIRE
NEP COUNT

MAXWELL MODEL
RT CREEP RECOVERY
DASHPOTS
DELAYED RECOVERY
FLOW (PLASTIC)
LINEAR VISCOELASTICITY
NONLINEAR VISCOELASTICITY
PERMANENT SET
RECOVERY (SELF RESTORATION TO ORIGINAL CONDITION)
RELAXATION
RELAXATION SPECTRUM
RELAXATION TIME (MECHANICAL)
RHEOLOGY
SPRINGS
VISCOELASTICITY
VOIGT MODEL

MAXWELLIAN DISTRIBUTION
RT BINOMIAL DISTRIBUTION
CHI SQUARED DISTRIBUTION
DISTRIBUTION
F DISTRIBUTION
NORMAL DISTRIBUTION
T DISTRIBUTION

MEAN
RT COEFFICIENT OF VARIATION
CORRELATION COEFFICIENT
DEGREES OF FREEDOM
DISTRIBUTION
MEAN FIBER LENGTH
MEDIAN
MODE
STANDARD DEVIATION
STANDARD ERROR

MEAN FIBER LENGTH
UF MEAN STAPLE LENGTH
BT FIBER PROPERTIES
RT COEFFICIENT OF LENGTH VARIATION
COTTON CLASS
COTTON CLASSING
COTTON QUALITY
DIGITAL FIBROGRAPH
EFFECTIVE LENGTH (FIBER)
FIBER BREAKAGE
FIBER DIAGRAM
FIBER LENGTH
FIBER LENGTH DETERMINATION
FIBER LENGTH DISTRIBUTION
FIBROGRAPH
FIBROGRAPH MEAN LENGTH
LENGTH
MEAN
SERVO FIBROGRAPH
SHORT FIBERS INDEX
SPAN LENGTH
STAPLE FIBERS
STAPLING
SUTER WEBB TESTING
UPPER HALF MEAN LENGTH
UPPER QUARTILE LENGTH

MEAN STAPLE LENGTH
USE MEAN FIBER LENGTH

MEANS
USE METHODS

MEASUREMENTS
USE MEASURING

MEASURING
UF MEASUREMENTS
MENSURATION
RT ANALYZING
ANTHROPOMETRIC MEASUREMENTS
DETERMINATION
GARMENT MANUFACTURE
LENGTH MEASURING DEVICES
MEASURING INSTRUMENTS
RECORDING INSTRUMENTS

MEASURING INSTRUMENTS
UF METERS (INSTRUMENTS)
NT ACCELEROMETERS
COLORIMETERS
DYNAMOMETERS
ELECTROMETERS
GEIGER COUNTERS
HYDROMETERS
HYGROMETERS
OSCILLOSCOPES
PH METERS
PHOTOMETERS
PYROMETERS
REFLECTOMETERS
SPECTROPHOTOMETERS
STRAIN GAUGES
STRAINOMETER
TENSIOMETERS
THERMOMETERS
THERMOPILES
VISCOSIMETERS
BT APPARATUS
INSTRUMENTATION
RT AUTOMATIC CONTROL
CALORIMETRY
CAPACITANCE BRIDGES
COUNT TESTING
COUNTERS
DETERMINATION
DIFFERENTIAL THERMAL ANALYSIS
ELECTRONIC INSTRUMENTS
ELECTRIC CONTROLLERS
IMPEDANCE BRIDGES
INDICATORS (INSTRUMENTATION)
LABORATORY APPARATUS
LENGTH MEASURING DEVICES
MEASURING
METALLOGRAPHY
MICROSCOPY
NULL METHODS
PROXIMITY TESTERS
RT RECORDING INSTRUMENTS
SPECTROSCOPY
TESTING
TESTING EQUIPMENT
THERMAL ANALYSIS

MEASURING MACHINES
BT GARMENT MANUFACTURING MACHINES
RT LENGTH MEASURING DEVICES

MECHANICAL BONDING
BT BONDING
RT BONDED FIBER FABRICS
NEEDLING
SEWING
STITCH BONDING
STITCHING

MECHANICAL BRAKES
BT BRAKES (FOR ARRESTING MOTION)
RT AIR BRAKES
AUTOMATIC STOP MOTIONS
BRAKE DISCS
BRAKE DRUMS
BRAKE LININGS
BRAKING
ELECTRIC BRAKES
FRICTION
HYDRAULIC BRAKES
LET OFF MOTIONS
MAGNETIC BRAKES
POWER BRAKES
RETARDERS
WATER BRAKES

MECHANICAL CONDITIONING
BT CONDITIONING
RT CYCLIC STRESS
DELAYED RECOVERY
EQUILIBRIUM (PHYSICAL)
FATIGUE
HYSTERESIS (MECHANICAL)
LINEAR VISCOELASTICITY
MECHANICAL HISTORY
MECHANICAL PROPERTIES
PERMANENT SET
VISCOELASTICITY

MECHANICAL DAMAGE
USE DAMAGE

MECHANICAL DAMPING
BT DAMPING
RT GLASS TRANSITION TEMPERATURE
HYSTERESIS (MECHANICAL)
MECHANICAL PROPERTIES
SECONDARY TRANSITION TEMPERATURES
VISCOUS DAMPING

MECHANICAL DETERIORATION PROPERTIES
NT ABRASION RESISTANCE
CREEP
CUTTING RESISTANCE
FATIGUE RESISTANCE
HYSTERESIS (MECHANICAL)
PERMANENT SET
PUNCTURE RESISTANCE
RESISTANCE TO DELAMINATION
SNAG RESISTANCE
WEAR RESISTANCE
BT MECHANICAL PROPERTIES
RT AESTHETIC PROPERTIES
BIOCHEMICAL PROPERTIES
CHEMICAL PROPERTIES
CRACK PROPAGATION
FABRIC PROPERTIES
FABRIC PROPERTIES (MECHANICAL)
FIBER PROPERTIES
FIBRILLATION
MECHANICAL HISTORY
PHYSICAL CHEMICAL PROPERTIES
PHYSICAL PROPERTIES (EXCLUDING MECHANICAL)
PLASTIC FLOW
STRESS OPTICAL PROPERTIES
STRESS STRAIN PROPERTIES
SURFACE PROPERTIES (MECHANICAL)
TRANSFER PROPERTIES

MECHANICAL DEVICES
BT DEVICES
RT ELECTROMAGNETIC DEVICES
FLUID DRIVES
GEAR DRIVES
MECHANISMS
TENSION DEVICES

MECHANICAL DRAFT
USE DRAFT

MECHANICAL DRIVES
BT DRIVES
RT BELT DRIVES
CHAIN DRIVES
GEAR DRIVES
HYDRAULIC DRIVES
VARIABLE SPEED DRIVES

MECHANICAL FINISHING
USE DRY FINISHING

MECHANICAL FLOCKING
BT FLOCKING
RT ELECTROSTATIC FLOCKING
FLOCK
FLOCK COATING
FLOCK PRINTING
FLOCKED CARPETS
FLOCKED FABRICS

MECHANICAL HISTORY
RT CYCLIC STRESS
DAMAGE
HISTORY
EQUILIBRIUM (PHYSICAL)
MECHANICAL CONDITIONING
MECHANICAL DETERIORATION PROPERTIES
STRESS HISTORY

MECHANICAL MIXERS
RT MASS TRANSFER
MILLS (COLLOID)

MECHANICAL PROPERTIES
UF PHYSICAL PROPERTIES (MECHANICAL)
NT ABRASION RESISTANCE
ADHESION
BEARING STRENGTH
BENDING RECOVERY
BENDING RIGIDITY
COEFFICIENT OF FRICTION
COHESION
CRIMP
CUTTING RESISTANCE
DIMENSIONAL STABILITY
DURABILITY
EFFICIENCY (STRUCTURAL)
FATIGUE RESISTANCE
FLEXURAL STRENGTH
FRICTIONAL CHARACTERISTICS
HARDNESS
KNOT EFFICIENCY
LIVELINESS
LOOP EFFICIENCY
LOOP STRENGTH
MECHANICAL DETERIORATION PROPERTIES
MODULUS
PAPERINESS
PERMEABILITY
PILL RESISTANCE
PUNCTURE RESISTANCE
RESISTANCE TO DELAMINATION
SHEAR RESISTANCE
SHOCK RESISTANCE
SHRINK RESISTANCE
SNAG RESISTANCE
SNARLING TENDENCY
SOFTNESS
STRESS STRAIN PROPERTIES
SURFACE PROPERTIES (MECHANICAL)
TACK
TEAR STRENGTH
TEXTURE
TORSIONAL RIGIDITY
TRANSFER PROPERTIES
WEAR RESISTANCE
RT ABRASION
AESTHETIC PROPERTIES
BIOCHEMICAL PROPERTIES
CHARACTERISTICS
CHEMICAL PROPERTIES
CURLING TENDENCY
DAMAGE
DEFORMATION
DRAPE
DYNAMOMETERS
END USE PROPERTIES
FABRIC PROPERTIES
FABRIC PROPERTIES (AESTHETIC)
FABRIC PROPERTIES (MECHANICAL)
FABRIC PROPERTIES (PHYSICAL EXCLUDING MECHANICAL)
FABRIC PROPERTIES (STRUCTURAL)
FIBER PROPERTIES
FRACTURING
HAND
HYSTERESIS (MECHANICAL)
IMPACT
MATERIALS
MATERIALS ENGINEERING
MECHANICAL CONDITIONING
PHYSICAL CHEMICAL PROPERTIES
PHYSICAL PROPERTIES (EXCLUDING MECHANICAL)
PROPERTIES
RESIDUAL TORQUE
STRENGTH ELONGATION TESTING
UNIAXIAL STRENGTH
WRINKLE RECOVERY
WRINKLE RESISTANCE
WILDNESS
YARN PROPERTIES

MECHANICAL SHOCK
RT ACCELERATION (MECHANICAL)
ACCELEROMETERS
BRAKING
DYNAMOMETERS
FATIGUE
FORCE
FRACTURING
IMPACT
IMPACT STRENGTH
INERTIA
RUPTURE
SHOCK RESISTANCE
VIBRATION

MECHANICAL SLUB CATCHERS
RT ELECTRONIC SLUB CATCHERS
END BREAKAGES
IRREGULARITY
KNOTS
KNOTTING
QUALITY CONTROL
SLUBS
WINDING

MECHANICS
NT AERODYNAMICS
DYNAMICS
FLUID MECHANICS
KINEMATICS
KINETICS
STATICS
STRUCTURAL MECHANICS
RT MECHANISMS
MOTION
STATISTICAL MECHANICS
STRESS ANALYSIS

MECHANISM (FUNDAMENTAL)
RT CARDING DYNAMICS
DRAFTING DYNAMICS
DRAFTING THEORY
DYEING THEORY
FRICTION THEORY
KINETICS (CHEMICAL)
KNITTING DYNAMICS
MECHANISMS
PROCESS DYNAMICS
SPINNING DYNAMICS
THEORETICAL ANALYSIS
THEORIES
TWISTING DYNAMICS
WEAVING DYNAMICS
WINDING DYNAMICS
YARN DYNAMICS

MECHANISMS
RT ACCESSORIES
BEARINGS
BELTS
BRAKES (FOR ARRESTING MOTION)
CAMS
CHAINS
CLUTCHES
COMPONENTS
COUPLINGS (MECHANICAL)
CRANKS
DASHPOTS
DEVICES
ELECTROMECHANICAL DEVICES
FLUID DRIVES
GADGETS
GASKETS
GEARS
GRIPPERS (YARN)
KINEMATICS
LINKAGES
MANDRELS
MECHANICS
MECHANICAL DEVICES
MECHANISM (FUNDAMENTAL)
MOTORS
PISTONS
SEALS
SHAFTS
SPRINGS
VALVES

MECHLIN MACHINES
BT LACE MACHINES
RT BAR WARP MACHINES
BARMEN MACHINES
BOBBINET MACHINES
CURTAIN MACHINES
DOUBLE LOCKER MACHINES
GO THROUGH MACHINES
LEAVERS MACHINES
ROLLING LOCKER MACHINES
SIVAL MACHINES
STRING WARP MACHINES
WARP LACE MACHINES

MEDIAN
BT STATISTICAL MEASURES
RT COEFFICIENT OF VARIATION
CORRELATION COEFFICIENT
DEGREES OF FREEDOM
MEAN
MODE

MEDICAL TEXTILES
NT ARTERIAL REPLACEMENTS
BANDAGES
GAUZE
HEART VALVES (FABRIC)
SUPPORT BANDAGES
SURGICAL STOCKINGS
RT BATTING
COLLAGEN FIBERS
END USES
FABRICS (ACCORDING TO STRUCTURE)
SANITARY NAPKINS
TAPE

MEDIUM (WOOL GRADE)
BT GRADE (FIBERS)
WOOL GRADE

MEDIUM FINE (WOOL GRADE)
BT GRADE (FIBERS)
WOOL GRADE

MEDIUM WEIGHT
BT WEIGHT
RT COARSE YARNS
FABRIC WEIGHT
FABRICS
FINE YARNS

MEDIUM WOOL
NT CHEVIOT WOOL
DORSET WOOL
HAMPSHIRE WOOL
OXFORD WOOL
RYELAND WOOL
SHROPSHIRE WOOL
SOUTHDOWN WOOL
SUFFOLK WOOL
BT WOOL
RT CROSSBRED WOOL
LONG WOOL
MERINO WOOL
MOUNTAIN WOOL
SHEEP BREEDS
SHORT WOOL

MEDULLA
BT FINE STRUCTURE (FIBERS)
RT ANIMAL FIBERS (GENERAL)
CORTEX
CRIMP
CUTICLE
ENDOCUTICLE
EPICUTICLE
EXOCUTICLE
FIBER LENGTH
FIBRILS
LUMEN
MICROFIBRILS
ORTHOCORTEX
PARACORTEX
SCALES (WOOL FIBERS)
WOOL

MELAMINE
BT AMINES
TRIAZINES
RT CYANURIC CHLORIDE
METHYLOL MELAMINES

MELAMINE RESINS
BT RESINS
THERMOSETTING RESINS
RT EPOXY RESINS
PHENOLIC RESINS

MELAMINE FORMALDEHYDE PRECONDENSATES
USE METHYLOL MELAMINES

MELAMINE FORMALDEHYDE RESINS
BT AMINO RESINS
RT UREA-FORMALDEHYDE RESINS

MELANGE PRINTING
UF VIGOUREUX PRINTING
BT PRINTING
RT BLOCK PRINTING
DISCHARGE PRINTING
DYEING
MELANGE PRINTING EQUIPMENT
RESIST PRINTING
ROLLER PRINTING
SCREEN PRINTING
SLUBBINGS
THROUGH PRINTING

MELANGE PRINTING EQUIPMENT
BT PRINTING MACHINES
RT DUPLEX PRINTING MACHINES
MELANGE PRINTING
SCREEN PRINTING MACHINES

MELT CRYSTALLIZATION
BT CRYSTALLIZATION
RT CRYSTAL LATTICE
CRYSTALLINE POLYMERS
CRYSTALLINE REGION
CRYSTALLITES
SPHERULITES
SPHERULITIC FIBRILS

MELT SPINNING
BT EXTRUSION
SPINNING (EXTRUSION)
RT ANNEALING
CRYSTALLIZATION
DECRYSTALLIZATION
DRAWING (FILAMENT)
FILAMENT YARNS
GLASS FIBERS
MAN MADE FIBERS
MELT VISCOSITY
MOLECULAR ORIENTATION
MONOFILAMENT YARNS
MULTIFILAMENT YARNS
NEUTRALIZING (FIBER EXTRUSION)
NYLON (POLYAMIDE FIBERS)
ORIENTATION BIREFRINGENCE
POLYAMIDES
POLYESTER FIBERS
POLYESTERS
POLYMERS
QUENCHING (FILAMENT)
SOLVENT SPINNING
SPINNERETS
SYNTHETIC FIBERS
WET SPINNING

MELT VISCOSITY
BT VISCOSITY
RT EXTRUSION
INTRINSIC VISCOSITY
MELT SPINNING
MELTING
MELTING POINT
MOLECULAR WEIGHT
POLYMERS
RELATIVE VISCOSITY
RHEOLOGY

MELTING
RT COOLING
DRYING
FREEZING
FUSION (MELTING)
HEAT OF FUSION
LATENT HEAT
MELT VISCOSITY
MELTING POINT
SELF BONDING FIBERS
SPECIFIC HEAT
THERMOPLASTICITY

MELTING POINT
UF FIRST ORDER TRANSITION TEMPERATURE
BT PHYSICAL CHEMICAL PROPERTIES
PHYSICAL PROPERTIES (EXCLUDING MECHANICAL)
THERMAL PROPERTIES
TRANSITION TEMPERATURES
RT FUSION (MELTING)
GLASS RUBBER TRANSITION
GLASS TRANSITION TEMPERATURE
HEAT OF FUSION
HEAT RESISTANCE
LATENT HEAT
MELT VISCOSITY
MELTING
SECONDARY TRANSITION TEMPERATURES
SINGEING
SOFTENING POINT
SPECIFIC HEAT
TRANSITIONS (POLYMERS)

MELTON
BT FLAT WOVEN FABRICS
WOVEN FABRICS
RT FACE GOODS

MEMBRANES
RT DIALYSIS
DIAPHRAGMS
DIFFUSION
DIFFUSIVITY
DONNAN EQUILIBRIUM
ELECTRODIALYSIS
EXTRACTION
MASS TRANSFER
OSMOSIS
PENETRATION
PERMEABILITY
PURIFICATION
SEMIPERMEABLE MEMBRANES
SEPARATION (SOLUTION)

MEMORY WHEELS
RT AUTOMATIC CONTROL
CONTROL SYSTEMS

MENDING
BT DRY FINISHING
RT BURLING
(CONT.)

RT DEFECTS
FABRIC DEFECTS
FABRIC INSPECTION
SPECKING

MENSURATION
USE MEASURING

MERAKLON (TN)
BT OLEFIN FIBERS
POLYPROPYLENE FIBERS

MERCAPTANS
UF THIOLS
NT CYSTEINE
ETHYL MERCAPTAN
THIOGLYCOLIC ACID
BT SULFUR COMPOUNDS
RT DISULFIDES (ORGANIC)
THIOETHERS

MERCERIZATION
USE MERCERIZING

MERCERIZED COTTON
RT CELLULOSE
COTTON
COTTON SYSTEM PROCESSING
DYEING
FINISH (SUBSTANCE ADDED)
LUMEN
LUSTER
MERCERIZING
PRIMARY WALLS
SECONDARY WALLS
SWELLING

MERCERIZED FINISH
BT FINISH (PROPERTY)
RT ABSORBENCY (MATERIAL)
BARIUM ACTIVITY NUMBER
CHEMICAL PROPERTIES
MERCERIZING
PHYSICAL PROPERTIES (EXCLUDING MECHANICAL)

MERCERIZERS
USE MERCERIZING MACHINES

MERCERIZING
UF MERCERIZATION
NT SLACK MERCERIZING
BT COTTON SYSTEM PROCESSING
WET FINISHING
RT BARIUM ACTIVITY NUMBER
CHAIN MERCERIZERS
CHAINLESS MERCERIZERS
MERCERIZED COTTON
MERCERIZED FINISH
SCHREINERING
SODIUM HYDROXIDE

MERCERIZING MACHINES
UF MERCERIZERS
NT CHAIN MERCERIZERS
CHAINLESS MERCERIZERS

MERCHANDISING
RT COSTS
MANAGEMENT
MANUFACTURING
MARKETING
PRODUCT DESIGN
PRODUCTION
SALES
SAMPLES
STYLING (APPAREL)

MERCURIC ACETATE
BT ACETATE SALTS
MERCURY COMPOUNDS
RT ANTIMICROBIAL FINISHES

MERCURIC CHLORIDE
BT CHLORIDES
MERCURY COMPOUNDS

MERCURY COMPOUNDS
NT MERCURIC ACETATE
MERCURIC CHLORIDE
ORGANOMERCURY COMPOUNDS
PHENYLMERCURIC ACETATE
PHENYLMERCURIC DIOCTYLSULFOSUCCINATE
PHENYLMERCURIC SUCCINATE
RT CARROTING
SILVER COMPOUNDS

MERINO WOOL
UF FINE WOOL
NT AMERICAN MERINO WOOL
AUSTRALIAN MERINO WOOL
RAMBOUILLET WOOL
SAXONY WOOL
SILESIAN WOOL
SOUTH AFRICAN MERINO WOOL
SOUTH AMERICAN MERINO WOOL
BT WOOL
RT BOTANY WOOL
CROSSBRED WOOL
GRADE (FIBERS)
LAMBSWOOL
LONG WOOL
MEDIUM WOOL
MOUNTAIN WOOL
SHEEP BREEDS

MERROW STITCHES
BT STITCHES (SEWING)

MESALINE
BT WOVEN FABRICS
RT LININGS
SATEEN

MESH (KNITTING)
NT MICROMESH (KNITTING)
RT HOSIERY
KNITTED FABRICS
KNITTED STITCHES
KNOTTED STITCHES (KNITTING)
RUN RESISTANT HOSIERY
STOCKINGS

METACHROMASY
RT FADING

METACHROME DYEING
USE MORDANT DYEING

METACHROME DYES
USE MORDANT DYES

METAL CHELATING DYES
USE METALLIZED DYES

METAL FIBERS
UF METAL FILAMENTS
NT WHISKER FIBERS (METAL)
BT HIGH TEMPERATURE FIBERS
INORGANIC FIBERS (MAN MADE)
MAN MADE FIBERS
METALLIC FIBERS
RT CRYSTALLIZATION
FOIL
METALLIC YARNS
MINERAL FIBERS

METAL FILAMENTS
USE METAL FIBERS

METALLIC ALLOYS
NT STAINLESS STEEL
BT ALLOYS
RT POLYMER ALLOYS

METALLIC BONDS
BT CHEMICAL BONDS
RT BOND ENERGY
COVALENT BONDS
ELECTROSTATIC BONDS
HYDROGEN BONDS
IONIC BONDS
VAN DER WAALS FORCES

METALLIC CARD CLOTHING
UF GARNETT WIRES
SAW TOOTH CLOTHING
BT CARD CLOTHING

METALLIC FIBERS
UF METALLIZED FIBERS
NT ALUMINUM FOIL
FOIL
LAME (TN)
LUREX (TN)
METLON (TN)
METAL FIBERS
TINSEL
WHISKER FIBERS (METAL)
BT HIGH TEMPERATURE FIBERS
INORGANIC FIBERS (MAN MADE)
MAN MADE FIBERS
RT CERAMIC FIBERS
GLASS FIBERS
VACUUM COATING

METALLIC SOAPS
NT ALUMINUM ABIETATE
ALUMINUM SOAPS
ALUMINUM STEARATE
CALCIUM SOAPS
CALCIUM STEARATE
HEAVY METAL SOAPS
LITHIUM STEARATE
ZINC SOAPS
ZINC STEARATE
BT SOAPS
RT ALKALI SOAPS
LIME SOAP
OLEATE SALTS
WATER REPELLENTS

METALLIC YARNS
NT LAME (TN)
LUREX (TN)
METLON (TN)
TINSEL
RT CERAMIC FIBERS
FOIL
INORGANIC FIBERS (MAN MADE)
METAL FIBERS
METALLIC FIBERS
METALS
MINERAL FIBERS
WHISKER FIBERS (METAL)

METALLIZED DYEING
BT DYEING (BY DYE CLASSES)
RT ACID DYEING
AZOIC DYEING
BASIC DYEING
DIRECT DYEING
DISPERSE DYEING
METALLIZED DYES
MORDANT DYEING
NEUTRAL DYEING
PIGMENT DYEING
REACTIVE DYEING
SULFUR DYEING
VAT DYEING

METALLIZED DYES
UF METAL CHELATING DYES
PREMETALLIZED DYES
BT DYES
DYES (BY CHEMICAL CLASSES)
RT CHELATING
CHELATING AGENTS
COBALT COMPOUNDS
COMPLEXING
DIRECT DYES
DYEING (BY DYE CLASSES)
DYES (BY FIBER CLASSES)
METALLIZED DYEING
MORDANT DYES
NICKEL COMPOUNDS
ORGANOMETALLIC COMPOUNDS (EXCLUDING SILICONES)
REACTIVE DYES

METALLIZED FIBERS
USE METALLIC FIBERS

METALLOGRAPHY
BT MICROSCOPY
RT ELECTRON MICROSCOPY
INFRARED MICROSCOPY
INTERFERENCE MICROSCOPY
LIGHT MICROSCOPY (OPTICAL)
MEASURING INSTRUMENTS
ULTRAVIOLET MICROSCOPY
X RAY MICROSCOPY

METALORGANIC COMPOUNDS
USE ORGANOMETALLIC COMPOUNDS (EXCLUDING SILICONES)

METALS
NT ALUMINUM
COPPER
IRON
NICKEL
PLATINUM
TIN
RT ALLOYS
INORGANIC FIBERS (MAN MADE)
MATERIALS
METALLIC FIBERS
METALLIC YARNS

METAMERISM
RT COLOR
COLOR MATCHING

METARHEOLOGY
BT RHEOLOGY
RT ACTIVATION ENERGY
FREE VOLUME
KINETIC ELASTICITY
MACRORHEOLOGY
MICRORHEOLOGY
PHENOMENOLOGICAL RHEOLOGY
RATE PROCESSES
SURFACE TENSION

METER PUMPS
BT PUMPS
RT EXTRUSION
FILTER PACKS
SPINNERETS
SPINNING ASSEMBLIES (EXTRUSION)
SPINNING FILTERS

METERS (INSTRUMENTS)
USE MEASURING INSTRUMENTS

METHACROLEIN
USE ACROLEIN

METHACRYLAMIDE
BT ACRYLIC COMPOUNDS
AMIDES
RT ACRYLAMIDE
METHYLOL METHACRYLAMIDE

METHACRYLIC ACID
BT ACRYLIC COMPOUNDS
CARBOXYLIC ACIDS
ORGANIC ACIDS
RT ACRYLIC ACID
METHACRYLIC ESTERS
POLYMETHACRYLATES
POLYMETHACRYLIC ACID

METHACRYLIC ESTERS
NT METHYL METHACRYLATE
BT ACRYLIC COMPOUNDS
RT ACRYLIC ESTERS
METHACRYLIC ACID
POLYMETHACRYLATES

METHACRYLONITRILE
BT ACRYLIC COMPOUNDS
NITRILES

METHALLYL CHLORIDE
BT ALLYL COMPOUNDS
CHLORINE COMPOUNDS
RT ALLYL CHLORIDE

METHANOL
BT ALCOHOLS

METHAPHOSPHATES
USE CALGON (TN)

METHODS
UF MEANS
MODES
NT NULL METHODS
RT CONVENTIONAL PRACTICE
PERFORMANCE
SOLUTIONS (MATHEMATICAL)
TECHNIQUES

METHOXYMETHYL COMPOUNDS
USE METHYLOL COMPOUNDS

METHOXYMETHYL MELAMINE
BT METHYLOL MELAMINES

METHYL ACRYLATE
BT ACRYLIC COMPOUNDS
ACRYLIC ESTERS
RT BUTYL ACRYLATE
ETHYL ACRYLATE
METHYL METHACRYLATE
POLY (METHYL ACRYLATE)

METHYL METHACRYLATE
BT ACRYLIC COMPOUNDS
METHACRYLIC ESTERS
RT METHYL ACRYLATE
POLY (METHYL METHACRYLATE)

METHYL TRIAZONE
BT TRIAZONES
RT METHYLOL TRIAZONES

METHYLATED METHYLOL MELAMINE
BT METHYLOL MELAMINES

METHYLATED METHYLOL UREA
BT METHYLOL UREAS

METHYLATION
BT ALKYLATION
RT CHEMICAL MODIFICATION (FIBERS)
DIAZOMETHANE
ETHYLATION

METHYLCELLULOSE
BT CELLULOSE ETHERS
RT ETHYLCELLULOSE
GUMS
THICKENING AGENTS

METHYLENE BLUE
BT BASIC DYES
BLUEING AGENTS
RT METHYLENE BLUE NUMBER

METHYLENE BLUE NUMBER
RT ACID DEGRADATION
CELLULOSE
METHYLENE BLUE

METHYLENE CHLORIDE
BT CHLORINATED HYDROCARBONS
RT CHLORINATED SOLVENTS

METHYLOL ACETAMIDE
BT METHYLOL AMIDES

METHYLOL ACETYLENE DIUREA
UF ACETYLENE DIUREA FORMALDEHYDE
BT METHYLOL UREAS
RT GLYOXAL
UREA

METHYLOL ACRYLAMIDE
BT ACRYLIC COMPOUNDS
METHYLOL AMIDES
RT ACRYLAMIDE
METHYLOL METHACRYLAMIDE

METHYLOL ALKYL CARBAMATE
USE METHYLOL CARBAMATES

METHYLOL AMIDES
NT DIMETHYLOL ACETAMIDE
DIMETHYLOL FORMAMIDE
DIMETHYLOL ITACONAMIDE
METHYLOL ACETAMIDE
METHYLOL ACRYLAMIDE
METHYLOL FORMAMIDE
METHYLOL METHACRYLAMIDE
TMCEA
BT METHYLOL COMPOUNDS
RT AMIDES
METHYLOL UREAS

METHYLOL CARBAMATES
UF ALKYL CARBAMATE FORMALDEHYDE
METHYLOL ALKYL CARBAMATE
MONOALKYL CARBAMATE
FORMALDEHYDE
NT DIMETHYLOL ISOPROPYL CARBAMATE
DIMETHYLOL METHYL CARBAMATE
DMEC
METHYLOL ETHYL CARBAMATE
METHYLOL ISOPROPYL CARBAMATE
METHYLOL METHYL CARBAMATE
MONOMETHYLOL ETHYL CARBAMATE
MONOMETHYLOL METHYL CARBAMATE
BT METHYLOL COMPOUNDS
RT DURABLE PRESS
URETHANES
WASH WEAR FINISHES

METHYLOL COMPOUNDS
UF FORMALDEHYDE PRECONDENSATES
HYDROXYMETHYL COMPOUNDS
METHOXYMETHYL COMPOUNDS
N-METHYLOL COMPOUNDS
NT METHYLOL AMIDES
METHYLOL CARBAMATES
METHYLOL MELAMINES
METHYLOL TRIAZINES
METHYLOL UREAS
METHYLOL URONS
BT REACTANTS
RT AMINO RESINS
DURABLE PRESS FINISHES
FORMALDEHYDE
METHYLOLATION
PRECONDENSATES
WASH WEAR FINISHES

METHYLOL DYES
BT REACTIVE DYES

METHYLOL ETHYL CARBAMATE
BT METHYLOL CARBAMATES

METHYLOL ETHYL TRIAZONE
BT METHYLOL TRIAZONES

METHYLOL FORMAMIDE
BT METHYLOL AMIDES

METHYLOL ISOPROPYL CARBAMATE
BT METHYLOL CARBAMATES

METHYLOL MELAMINES
UF MELAMINE FORMALDEHYDE
PRECONDENSATES
METHYLOL TRIAZINES
NT HEXAMETHYLOL MELAMINE
METHOXYMETHYL MELAMINE
METHYLATED METHYLOL MELAMINE
TRIMETHYLOL MELAMINE
BT METHYLOL COMPOUNDS
RT MELAMINE

METHYLOL METHACRYLAMIDE
BT ACRYLIC COMPOUNDS
METHYLOL AMIDES
RT METHACRYLAMIDE
METHYLOL ACRYLAMIDE

METHYLOL METHYL CARBAMATE
BT METHYLOL CARBAMATES

METHYLOL METHYL TRIAZONE
BT METHYLOL TRIAZONES

METHYLOL TRIAZINES
USE METHYLOL MELAMINES

METHYLOL TRIAZONES
NT DIMETHYLOL METHYL TRIAZONE
DMET
METHYLOL ETHYL TRIAZONE
METHYLOL METHYL TRIAZONE
BT METHYLOL COMPOUNDS
TRIAZONES
RT ETHYL TRIAZONE
METHYL TRIAZONE

METHYLOL UREAS
UF IMIDAZOLIDINONES
UREA-FORMALDEHYDE
UREA-FORMALDEHYDE
PRECONDENSATES
NT DMDHEU
DMEU
DMHEL
DMHPU
DMPU
DMU
METHYLATED METHYLOL UREA
METHYLOL ACETYLENE DIUREA
MONOMETHYLOL UREA
BT METHYLOL COMPOUNDS
UREA DERIVATIVES
RT DHEU
METHYLOL AMIDES
UREA
UREA-FORMALDEHYDE RESINS

METHYLOL URONS
UF URON-FORMALDEHYDE CONDENSATES
NT DMUR
BT METHYLOL COMPOUNDS

METHYLOLATION
BT REACTIONS (CHEMICAL)
RT ACETALS
FORMALDEHYDE
FORMALS
METHYLOL COMPOUNDS

METHYLTAURIDE SURFACTANTS
NT IGEPON T (TN)
BT ALKYLAMIDE SULFONATES

METLON (TN)
BT HIGH TEMPERATURE FIBERS
INORGANIC FIBERS (MAN MADE)
METALLIC FIBERS
METALLIC YARNS
RT LAME (TN)
LUREX (TN)

METRIC COUNT
BT COUNT
INDIRECT COUNT
RT GREX
TEX

MICELLES
NT FRINGED MICELLES
BT FINE STRUCTURE (FIBERS)
RT AMORPHOUS REGION
CRITICAL MICELLE CONCENTRATION
CRYSTALLINE REGION
CRYSTALLITES
CRYSTALLIZATION
FIBRIL REVERSALS
FIBRILS
MICROFIBRILS

MICROBIAL DEGRADATION
USE MICROBIOLOGICAL DEGRADATION

MICROBIOLOGICAL DEGRADATION
UF MICROBIAL DEGRADATION
BT DEGRADATION
RT ANTIMICROBIAL FINISHES
BACTERIAL INHIBITION
CHEMICAL PROPERTIES
DEGRADATION PROPERTIES
ENZYMATIC DEGRADATION
FABRIC PROPERTIES
(DEGRADATION)
FUNGUS
MILDEW RESISTANCE
PERSPIRATION
PERSPIRATION RESISTANCE

MICROFIBRILS
BT FIBRILS
FINE STRUCTURE (FIBERS)
RT CORTEX
CRYSTALLITES
ENDOCUTICLE
EPICUTICLE
EXOCUTICLE
FIBRIL REVERSALS
FIBRILLATION
FRINGED FIBRILS
FRINGED MICELLES
MACROFIBRILS
MEDULLA
MICELLES
ORTHOCORTEX
PARACORTEX
SPHERULITES
WOOD FIBRILS

MICROMESH (KNITTING)
BT MESH (KNITTING)
RT HOSIERY
KNITTED FABRICS
KNITTED STITCHES

MICRONAIRE
BT FINENESS TESTERS
RT AIR FLOW
AIR RESISTANCE
CAUSTICAIRE SCALE
CAUSTICAIRE VALUE
COEFFICIENT OF FINENESS
VARIATION
(CONT.)

RT COTTON
COUNT
CURVILINEAR SCALE
FIBER DIAMETER
FIBER FINENESS
FIBER SURFACE
GRADE (FIBERS)
IMMATURE FIBERS
MATURITY INDEX
MICRONAIRE FINENESS
NEP COUNT
NEP POTENTIAL

MICRONAIRE FINENESS
BT FIBER PROPERTIES
RT CAUSTICAIRE SCALE
CAUSTICAIRE VALUE
FINENESS TESTING
MICRONAIRE

MICRORADIOGRAPHY
BT MICROSCOPY
RT ELECTRON MICROSCOPY
INFRARED MICROSCOPY
INTERFERENCE MICROSCOPY
LIGHT MICROSCOPY (OPTICAL)
OPTICAL INSTRUMENTS
ULTRAVIOLET MICROSCOPY
X RAY MICROSCOPY

MICRORHEOLOGY
BT RHEOLOGY
RT MACRORHEOLOGY
METARHEOLOGY
PHENOMENOLOGICAL RHEOLOGY

MICROSCOPIC ANALYSIS (FIBER IDENTIFICATION)
BT FIBER IDENTIFICATION
RT CHEMICAL ANALYSIS (FIBER IDENTIFICATION)
MICROSCOPY
MICROTOMES
REAGENT TESTING (FIBER IDENTIFICATION)
SPECIFIC GRAVITY TESTING (FIBER IDENTIFICATION)
STAINING TESTING (FIBER IDENTIFICATION)
SWELLING TESTING (FIBER IDENTIFICATION)
THERMAL TESTING (FIBER IDENTIFICATION)

MICROSCOPY
NT ELECTRON MICROSCOPY
FIELD EMISSION MICROSCOPY
FLUORESCENCE MICROSCOPY
INFRARED MICROSCOPY
INTERFERENCE MICROSCOPY
LIGHT MICROSCOPY (OPTICAL)
METALLOGRAPHY
MICRORADIOGRAPHY
PHASE MICROSCOPY
PROJECTION MICROSCOPY
RESINOGRAPHY
STEREOSCAN MICROSCOPY
STEREOSCOPIC MICROSCOPY
ULTRAMICROSCOPY
ULTRAVIOLET MICROSCOPY
X RAY MICROSCOPY
RT EMBEDDING MEDIUM
FOCUSING
INTERFEROMETRY
MEASURING INSTRUMENTS
MICROSCOPIC ANALYSIS (FIBER IDENTIFICATION)
MICROTOMES
OPTICAL INSTRUMENTS
PHOTOGRAPHY
PHOTOMICROGRAPHS
POLARIZING MICROSCOPE
REFRACTOMETRY

MICROTOMES
RT CUTTING
EMBEDDING MEDIUM
MICROSCOPIC ANALYSIS (FIBER IDENTIFICATION)
MICROSCOPY

MIDDLING (COTTON GRADE)
BT COTTON GRADE
GRADE (FIBERS)
RT GOOD MIDDLING (COTTON GRADE)

MIDDLING FAIR (COTTON GRADE)
BT COTTON GRADE
GRADE (FIBERS)

MIGRATION (FIBER)
USE FIBER MIGRATION

MIGRATION (SUBSTANCE)
NT BINDER MIGRATION
DYE MIGRATION
RT DIFFUSION
DRYING
RT DYEABILITY
DYEING
EQUILIBRIUM (CHEMICAL)
EXHAUSTION (WET PROCESS)
FIBER MIGRATION
RADIAL DISTRIBUTION

MIGRATION PERIOD
RT FIBER MIGRATION
SPINNING
TWISTING

MILAN (LACE)
BT HAND MADE LACES
NEEDLE POINT LACES

MILANESE FABRICS
BT KNITTED FABRICS
WARP KNITTED FABRICS
RT MILANESE KNITTING MACHINES

MILANESE KNITTING MACHINES
BT KNITTING MACHINES
WARP KNITTING MACHINES
RT COMPOUND NEEDLES
GUIDE BARS
HALF SET
LATCH NEEDLES
MILANESE FABRICS
NEEDLE BARS
OVERLAP
PICKER POINTS
PRESSER BARS
RASCHEL KNITTING MACHINES
SET (WARP KNITTING)
SILK LAP (WARP KNITTING)
SIMPLEX KNITTING MACHINES
SINKER BARS
SINKERS
SPRING NEEDLES
THREAD PRESSERS
TRICOT KNITTING MACHINES
UNDERLAP
WARP KNITTING

MILDEW
RT DEGRADATION
DEGRADATION PROPERTIES
FABRIC PROPERTIES (DEGRADATION)
FUNGUS
MICROBIOLOGICAL DEGRADATION
MILDEW RESISTANCE
MILDEW RESISTANCE FINISHES
MILDEW RESISTANCE TREATMENTS
MILDEW RESISTANT LEATHER

MILDEW RESISTANCE
UF FUNGUS RESISTANCE
ROT RESISTANCE
ROTTING
BT DEGRADATION PROPERTIES
END USE PROPERTIES
FINISH (PROPERTY)
RT ANTIMICROBIAL FINISHES
BACTERIAL INHIBITION
FABRIC PROPERTIES
FUNGUS
INSECT RESISTANCE
MILDEW
MILDEW RESISTANCE FINISHES
MILDEW RESISTANCE TREATMENTS
MILDEW RESISTANT LEATHER
PHOTOCHEMICAL DEGRADATION

MILDEW RESISTANCE FINISHES
UF ROT RESISTANCE FINISHES
BT ANTIMICROBIAL FINISHES
FINISH (SUBSTANCE ADDED)
PRESERVATIVES
RT CADMIUM SELENIDE
COPPER FORMATE
COPPER NAPHTHENATE
COPPER PENTACHLOROPHENATE
COPPER 8-HYDROXYQUINOLATE
INSECT RESISTANCE FINISHES
INSECT RESISTANCE TREATMENTS
MILDEW
MILDEW RESISTANCE
MILDEW RESISTANCE TREATMENTS
MILDEW RESISTANT LEATHER
SOIL BURIAL TEST

MILDEW RESISTANCE TREATMENTS
UF ROT RESISTANCE TREATMENTS
BT ANTIMICROBIAL TREATMENTS
WET FINISHING
RT ANTIBACTERIAL TREATMENTS
FINISH (SUBSTANCE ADDED)
INSECT RESISTANCE TREATMENTS
MILDEW
MILDEW RESISTANCE
MILDEW RESISTANCE FINISHES
MILDEW RESISTANT LEATHER

MILDEW RESISTANT LEATHER
BT LEATHER
RT MILDEW
RT MILDEW RESISTANCE
MILDEW RESISTANCE FINISHES
MILDEW RESISTANCE TREATMENTS

MILITARY CLOTHING
NT BATTLEDRESS
BULLET PROOF CLOTHING
COMBAT SUITS
FATIGUES
LEGGINGS
RAINWEAR
UNIFORMS
BT GARMENTS
RT BALLISTIC FABRICS
COATS
MUFTI
PARKAS

MILITARY PRODUCTS
NT ARRESTERS
CAMOUFLAGES
EQUIPAGE
LIFE RAFTS
PARACHUTES
PRESSURE SUITS
RADOMES
RETARDERS
SHELTERS
TENTS
RT DUCK
INFLATABLE FABRICS
PRODUCTS

MILLING
USE FULLING

MILLING ACID DYES
USE MILLING DYES

MILLING DYES
UF ACID MILLING DYES
MILLING ACID DYES
BT ACID DYES

MILLS (COLLOID)
RT GRINDING (COMMINUTION)
HOMOGENIZERS
POWDERING

MILLS (FULLING ROTARY)
BT FULLING MACHINES
RT FULLING
GATTERWALKING (FULLING MACHINES)
OPEN WIDTH FINISHERS (FULLING)
PILGRIMSTEP (FULLING MACHINES)
PLANKING (FULLING MACHINES)
REFULLING MACHINES

MILLS (KICKER)
BT FULLING MACHINES
RT GATTERWALKING (FULLING MACHINES)
OPEN WIDTH FINISHERS (FULLING)
PILGRIMSTEP (FULLING MACHINES)
PLANKING (FULLING MACHINES)
REFULLING MACHINES

MINERAL FIBERS
NT ASBESTOS
BT FIBERS
HIGH TEMPERATURE FIBERS
NATURAL FIBERS
RT CERAMIC FIBERS
FOIL
GLASS FIBERS
INORGANIC FIBERS (MAN MADE)
MAN MADE FIBERS
METAL FIBERS
METALLIC FIBERS
METALLIC YARNS
TINSEL
VEGETABLE FIBERS (GENERAL)
WHISKER FIBERS (METAL)

MINERAL OIL
RT EMULSIFYING AGENTS
HYDROCARBONS
LUBRICANTS
OIL STAINS
OILING
OILS

MINIMIZATION
RT ADJUSTMENTS
ADVANTAGES
ALTERATIONS
CORRECTION
IMPROVEMENTS
INCREASE
MODIFICATION
OPTIMIZATION
PERFORMANCE

MIRECOURT
BT HAND MADE LACES
LACES

MISPICKS
BT FABRIC DEFECTS
RT DROPPED STITCHES
FABRIC INSPECTION
FILLING YARNS
MISREED
PICKER STICKS
PICKING (WEAVING)
STREAKINESS
WEAVING
WOVEN FABRICS

MISREED
BT FABRIC DEFECTS
RT MISPICKS
REED MARKS

MITRES
BT CEREMONIAL CLOTHING
HEADGEAR

MIXING
NT BLENDING
RT CIRCULATION
COMPOUNDING
DISPERSING
EMULSIFYING
GRINDING (COMMINUTION)
HOMOGENIZERS
MASS TRANSFER
MIXTURES
POWDERING
STIRRERS
STIRRING

MIXTURES
RT FORMULATIONS
MIXING
SOLUTIONS (LIQUID)

MOCK FASHIONING MARKS
RT CIRCULAR HOSIERY
CIRCULAR KNITTING
FASHIONING
FASHIONING MARKS
FILLING KNITTING
FILLING KNITTING MACHINES
FULLY FASHIONED HOSIERY
HOSIERY
HOSIERY MACHINES
KNITTED FABRICS
KNITTING
MOCK SEAMS
SEAMLESS HOSIERY MACHINES

MOCK LENO
BT FLAT WOVEN FABRICS
WOVEN FABRICS
RT LENO WEAVES

MOCK LENO WEAVES
BT WEAVE (TYPES)
RT BALANCED WEAVES
CELLULAR FABRICS
LENO WEAVES

MOCK SEAMS
RT CIRCULAR HOSIERY
CIRCULAR KNITTING MACHINES
CYLINDER NEEDLES
FULLY FASHIONED HOSIERY
FULLY FASHIONED HOSIERY
MACHINES
HOSIERY
HOSIERY MACHINES
KNITTING MACHINES
MOCK FASHIONING MARKS
NEEDLES
SEAMLESS HOSIERY
SEAMLESS HOSIERY MACHINES
SINKERS
STITCH CAMS
STOCKINGS
TAKE DOWN
WEBHOLDERS

MODACRYLIC FIBERS
UF MODACRYLICS
NT AERESS (TN)
DYNEL (TN)
VEREL (TN)
BT MAN MADE FIBERS
SYNTHETIC FIBERS
RT ACRYLIC DYES
ACRYLIC FIBERS
OLEFIN FIBERS
POLYACRYLATES
POLYACRYLONITRILE
POLYVINYLS

MODACRYLICS
USE MODACRYLIC FIBERS

MODE
BT STATISTICAL MEASURES
RT CORRELATION COEFFICIENT
DEGREES OF FREEDOM
DISTRIBUTION
MEAN
MEDIAN
RANGE
STATISTICS

MODELLING
RT APPAREL DESIGN
GARMENT MANUFACTURE
MANNEQUINS
MODELS (FASHION)

MODELS (FASHION)
NT MANNEQUINS
RT MODELLING

MODELS (MATHEMATICAL)
UF MATHEMATICAL MODEL
RT ANALOGUES
COMPUTERS
MATHEMATICAL ANALYSIS
OPERATIONS RESEARCH
STATISTICAL ANALYSIS
TRELLIS MODEL

MODES
USE METHODS

MODES OF DEFORMATION
NT BENDING
BUCKLING
CURL
STRAIGHTENING
STRAIN
STRETCH
TWIST
RT COMPRESSION
DEGREES OF FREEDOM
DIMENSIONAL STABILITY
TENSION
TORSION (GEOMETRIC)
VIBRATION

MODIFICATION
RT ADAPTATION
ADVANTAGES
CONFIGURATION
DESIGN
DEVELOPMENTS
DOWN TIME
IMPROVEMENTS
INCREASE
MACHINE DESIGN
MINIMIZATION
OPTIMIZATION
PROCESS EFFICIENCY
PRODUCTION
QUALITY

MODIFIED CONTINUOUS FILAMENT YARNS
USE TEXTURED YARNS (FILAMENT)

MODIFIED STARCHES
NT DIALDEHYDE STARCH
OXIDIZED STARCHES
STARCH ETHERS
STARCH PHOSPHATE
BT STARCH

MODULUS
UF TEXTILE MODULUS
NT BULK MODULUS
COMPRESSIVE MODULUS
DYNAMIC MODULUS
ELASTIC MODULUS (TENSILE)
INITIAL MODULUS
MODULUS OF CROSS ELASTICITY
SECANT MODULUS
SHEAR MODULUS
TANGENT MODULUS
WET MODULUS
BT STRESS STRAIN PROPERTIES
STRESS STRAIN PROPERTIES
(TENSILE)
RT COMPLIANCE
DRAWING (FILAMENT)
FIBER PROPERTIES
GLASS TRANSITION TEMPERATURE
RELAXATION TIME (MECHANICAL)
SECONDARY TRANSITION
TEMPERATURES
YARN PROPERTIES
YIELD POINT

MODULUS OF CROSS ELASTICITY
BT MODULUS
STRESS STRAIN PROPERTIES
RT COMPLIANCE
COMPRESSIVE MODULUS
DYNAMIC MODULUS
ELASTIC MODULUS (TENSILE)
MODULUS
SHEAR MODULUS

MODULUS OF ELASTICITY
USE ELASTIC MODULUS (TENSILE)

MOHAIR
UF ANGORA (GOAT)
BT HAIR
KERATIN FIBERS
PROTEIN FIBERS

MOHRS CIRCLE
RT BIAXIAL STRESS
ELASTIC STRAIN
STRAIN
STRENGTH OF MATERIALS
STRESS

MOIRE PATTERNS
RT DIFFRACTION PATTERNS
INTERFERENCE FRINGES

MOISTURE
RT ACCESSIBILITY (INTERNAL)
BOUND WATER
CAPILLARY WATER
DELIQUESCENT AGENTS
HUMIDITY
MOISTURE CONTENT
MOISTURE STRENGTH CURVES
RELATIVE HUMIDITY

MOISTURE ABSORPTION
BT ABSORPTION (MATERIAL)
RT IMBIBITION
MOISTURE CONTENT
MOISTURE CONTENT TESTERS
MOISTURE CONTENT TESTING
MOISTURE STRENGTH CURVES
WATER ABSORPTION
WETTING

MOISTURE CONTENT
NT CRITICAL MOISTURE CONTENT
EQUILIBRIUM MOISTURE CONTENT
RT BOUND WATER
DELIQUESCENT AGENTS
DEW POINT
DIFFUSION
DIFFUSIVITY
DRY BULB TEMPERATURE
DRY WEIGHT
DRYING
EQUILIBRIUM (CHEMICAL)
HART MOISTURE METER
HUMIDITY
MOISTURE ABSORPTION
MOISTURE CONTENT TESTERS
MOISTURE CONTENT TESTING
PSYCHROMETRY
REGAIN
RELATIVE HUMIDITY
SATURATED VAPOR
SOAKING
SPECIFIC HUMIDITY
SQUEEZE ROLLS
SQUEEZING
STATIC ABSORPTION (WATER)
STATIC ABSORPTION TESTING
STEAM
SUCTION PRESSES
VAPOR PRESSURE
VACUUM EXTRACTION
WATER
WATER RETENTION
WET BULB TEMPERATURE
WET MODULUS
WET PICKUP
WET RELAXATION
WET STRENGTH
WET WRINKLE RECOVERY
WETTING
WETTING AGENTS

MOISTURE CONTENT CONTROL
RT AUTOMATIC CONTROL
DRY WEIGHT
HART MOISTURE METER
HUMIDITY CONTROL
HYGROMETERS
HYGROSCOPICITY

MOISTURE CONTENT TESTERS
NT FORTE MOISTURE METER
HART MOISTURE METER
SHIRLEY MOISTURE METER
RT MOISTURE ABSORPTION
MOISTURE CONTENT
MOISTURE CONTENT TESTING

MOISTURE CONTENT TESTING
BT TESTING
RT COBALT CHLORIDE METHOD
CONDITIONING
DRYING
EQUILIBRIUM MOISTURE CONTENT
EVAPORATION
FORTE MOISTURE METER
HART MOISTURE METER
HYGROMETERS
KARL FISCHER TITRATION METHOD
MOISTURE ABSORPTION
MOISTURE CONTENT
MOISTURE CONTENT TESTERS
MOISTURE STRENGTH CURVES
(CONT.)

RT OVEN METHOD (MOISTURE TESTING)
REGAIN
SHIRLEY MOISTURE METER

MOISTURE REGAIN
USE REGAIN

MOISTURE STRENGTH CURVES
RT BUNDLE STRENGTH
CRITICAL MOISTURE CONTENT
EQUILIBRIUM MOISTURE CONTENT
HYGROSCOPICITY
MOISTURE
MOISTURE ABSORPTION
MOISTURE CONTENT TESTING
REGAIN
STRENGTH
STRENGTH ELONGATION TESTING
STRESS STRAIN CURVES
WET STRENGTH

MOISTURE VAPOR TRANSMISSION
USE WATER VAPOR TRANSMISSION

MOLDING
NT COMPRESSION MOLDING
CONTACT MOLDING
FLEXIBLE PLUNGER MOLDING
INJECTION MOLDING
MATCHED DIE MOLDING
PRESSURE BAG MOLDING
VACUUM BAG MOLDING
RT FORMING
REINFORCED COMPOSITES
SHAPING

MOLDING (TAILORING)
BT GARMENT MANUFACTURE
RT BOARDING
TAILORING

MOLDING SHRINKAGE
BT SHRINKAGE
RT COMPRESSIVE SHRINKING
DIMENSIONAL STABILITY
FELTING SHRINKAGE
FILLING SHRINKAGE
FUSION CONTRACTION SHRINKAGE
PRESHRINKING
RELAXATION SHRINKAGE
RESIDUAL SHRINKAGE
SHRINK RESISTANCE
SHRINK RESISTANCE FINISHES
SHRINKING
THERMOPLASTICITY
WARP SHRINKAGE

MOLECULAR ATTRACTION
BT PHYSICAL CHEMICAL PROPERTIES
RT COHESION
DIPOLE MOMENT
DISPERSION FORCES
MOLECULES
VAN DER WAALS FORCES

MOLECULAR CHAIN MOVEMENT
RT BIREFRINGENCE
GLASS TRANSITION TEMPERATURE
MELTING POINT
MOLECULES
SECONDARY TRANSITION
TEMPERATURES
STRESS OPTICAL COEFFICIENT
SUBGROUP MOVEMENT (MOLECULAR)
TRANSITIONS (POLYMERS)
X RAY ANALYSIS

MOLECULAR ORIENTATION
RT BIREFRINGENCE
COLD DRAWING
CRYSTALLINITY
DEGREE OF ORIENTATION
DRAWING (FILAMENT)
FIBRILS
FINE STRUCTURE (FIBERS)
HOT DRAWING
LONG CHAIN MOLECULES
MELT SPINNING
MOLECULAR STRUCTURE
MOLECULES
ORIENTATION
PLASTIC STRAIN
POLYMERS
TEXTURING

MOLECULAR STRUCTURE
RT AMORPHOUS REGION
CHAIN MOLECULES
CROSSLINKING
CRYSTALLINE REGION
CRYSTALLINITY
CRYSTALLITES
DEGREE OF POLYMERIZATION
DICHROIC RATIO
FIBRILS
FINE STRUCTURE (FIBERS)
ISOMERS (MOLECULAR)
LINEAR CHAIN MOLECULES
RT LONG CHAIN MOLECULES
MICELLES
MOLECULAR ORIENTATION
MOLECULAR WEIGHT
MOLECULES
NUCLEAR MAGNETIC RESONANCE
POLYMERS
SIDE CHAINS (MOLECULAR)
SPHERULITIC FIBRILS
STRUCTURAL PROPERTIES
STRUCTURES
X RAY DIFFRACTION

MOLECULAR WEIGHT
NT MOLECULAR WEIGHT CONTROL
NUMBER AVERAGE MOLECULAR
WEIGHT
WEIGHT AVERAGE MOLECULAR
WEIGHT
RT CHAIN MOLECULES
COPOLYMERIZATION
CUPRAMMONIUM FLUIDITY
DEGREE OF POLYMERIZATION
END GROUPS
FRACTIONATION
INTRINSIC VISCOSITY
LINEAR CHAIN MOLECULES
LONG CHAIN MOLECULES
MELT VISCOSITY
MOLECULAR STRUCTURE
MOLECULAR WEIGHT DISTRIBUTION
MOLECULES
POLYMERIZATION
POLYMERS
ULTRACENTRIFUGES
VISCOSIMETERS

MOLECULAR WEIGHT CONTROL
BT MOLECULAR WEIGHT
RT ACID END GROUPS
AMINE END GROUPS
CUPRAMMONIUM FLUIDITY
DEGREE OF POLYMERIZATION
END GROUPS
MOLECULAR WEIGHT DISTRIBUTION
MOLECULES
VISCOSITY

MOLECULAR WEIGHT DISTRIBUTION
RT DEGREE OF POLYMERIZATION
MOLECULAR WEIGHT
MOLECULAR WEIGHT CONTROL
MOLECULES
NUMBER AVERAGE MOLECULAR
WEIGHT
ULTRACENTRIFUGES
WEIGHT AVERAGE MOLECULAR
WEIGHT

MOLECULES
NT BRANCHED CHAIN MOLECULES
LINEAR CHAIN MOLECULES
LONG CHAIN MOLECULES
RT FINE STRUCTURE (FIBERS)
MOLECULAR ATTRACTION
MOLECULAR CHAIN MOVEMENT
MOLECULAR ORIENTATION
MOLECULAR STRUCTURE
MOLECULAR WEIGHT
MOLECULAR WEIGHT CONTROL
MOLECULAR WEIGHT DISTRIBUTION

MOLTEN METAL BATHS
BT HEAT EXCHANGERS
RT CONTINUOUS DYEING RANGE
DYE FIXING
DYEING
HEAT TRANSFER
OIL BATHS
STANDFAST (TN)

MOLYBDENUM COMPOUNDS
RT GOLD COMPOUNDS
SILVER COMPOUNDS
TUNGSTEN COMPOUNDS

MOMENT CURVATURE CURVES
RT BENDING MOMENT DIAGRAMS
BENDING MOMENTS
BENDING RECOVERY
BENDING RIGIDITY
CURVATURE
HYSTERESIS (MECHANICAL)
LOAD DEFLECTION CURVES
MOMENTS
RADIUS OF CURVATURE

MOMENT OF INERTIA
UF AREA MOMENT OF INERTIA
MASS MOMENT OF INERTIA
SECTIONAL MOMENT OF INERTIA
RT BENDING
BENDING RIGIDITY
NEUTRAL AXIS
PAPERINESS
SHAPE FACTOR
STIFFNESS
TORSIONAL RIGIDITY

MOMENTS
NT BENDING MOMENTS
TWISTING MOMENTS
RT COUPLES
FORCE
MOMENT CURVATURE CURVES

MOMENTUM
RT ACCELERATION (MECHANICAL)
ANGULAR VELOCITY
CONSTANT SPEED
FORCE
GRAVITATION
IMPULSES
INERTIA
KINETIC ENERGY
MASS
MOMENTUM TRANSFER
TRANSLATION
VARIABLE SPEED
VELOCITY
VELOCITY (RELATIVE)

MOMENTUM TRANSFER
RT HEAT TRANSFER
MASS TRANSFER
MOMENTUM
TURBULENCE

MONFORTS REACTOR
RT PRESSURE DYEING MACHINES

MONOALKYL CARBAMATE FORMALDEHYDE
USE METHYLOL CARBAMATES

MONOCHLOROHYDRIN
USE GLYCERIN MONOCHLOROHYDRIN

MONOCHLOROTRIAZINE DYES
USE CHLOROTRIAZINE DYES

MONOCHROMATIC LIGHT
BT LIGHT
RT DAYLIGHT
LIGHT MICROSCOPY (OPTICAL)
MICROSCOPY
OPTICAL INSTRUMENTS
OPTICS
PHOTOGRAPHY
POLARIZED LIGHT
SPECTROPHOTOMETRY
SUNLIGHT

MONOETHANOLAMIDES
BT ALKANOLAMIDES

MONOETHANOLAMINE
BT ALKANOLAMINES
RT MONOETHANOLAMINE CARBONATE

MONOETHANOLAMINE BISULFITE
USE BISULFITES

MONOETHANOLAMINE CARBONATE
BT CARBONATES
RT MONOETHANOLAMINE

MONOETHANOLAMINE SULFITE
USE SULFITES

MONOFILAMENT YARNS
UF SINGLE FILAMENT YARNS
BT FILAMENT YARNS
YARNS
RT COAGULATING BATHS
CONTINUOUS FILAMENT YARNS
DRAWTWISTING
DRAWING (FILAMENT)
FIBER OPTICS
HOSIERY
MAN MADE FIBERS
MELT SPINNING
MULTIFILAMENT YARNS
NEUTRALIZING (FIBER EXTRUSION)
PRODUCERS YARNS
SOLVENT SPINNING
SPINNERETS
SPUN YARNS
SYNTHETIC FIBERS
WET SPINNING

MONOFILAMENTS
BT FILAMENTS
RT COUNT
DENIER
EXTRUSION
FIBERS
FILAMENT YARNS
FISHING LINES
MAN MADE FIBERS
MONOFILAMENT YARNS
SILK
SPINNERETS
SPINNING (EXTRUSION)
SYNTHETIC FIBERS

MONOGLYCERIDE SULFATES
USE ALKYL GLYCEROSULFATES

MONOMERS
NT ACRYLIC COMPOUNDS
ALLYL COMPOUNDS
LACTAMS
VINYL COMPOUNDS
VINYLIDENE COMPOUNDS
RT ADIPIC ACID
BUTADIENE
COPOLYMERIZATION
CYCLOHEXANE, 1,4-BIS (METHYLAMINE)-
DEGREE OF POLYMERIZATION
HEXAMETHYLENEDIAMINE
OLIGOMERS
POLYMERIZATION
POLYMERS
TEREPHTHALIC ACID

MONOMETHYLOL ETHYL CARBAMATE
BT METHYLOL CARBAMATES

MONOMETHYLOL METHYL CARBAMATE
BT METHYLOL CARBAMATES

MONOMETHYLOL UREA
BT METHYLOL UREAS

MONOSACCHARIDES
USE SUGARS

MONOSHEER (TN)
BT NYLON (POLYAMIDE FIBERS)
NYLON 6

MONOUNSATURATED COMPOUNDS
BT OLEFINIC COMPOUNDS

MONSANTO CREASE RECOVERY
USE WRINKLE RECOVERY TESTERS

MONTMORILLONITE
BT CLAYS
RT ANTISTATIC AGENTS
SILICATES

MORDANT DYEING
UF AFTERCHROME DYEING
CHROME DYEING
METACHROME DYEING
BT DYEING (BY DYE CLASSES)
RT ACID DYEING
AZOIC DYEING
BASIC DYEING
CHROMIUM FLUORIDE
CUPRIC CHLORIDE
DIRECT DYEING
DISPERSE DYEING
METALLIZED DYEING
MORDANT DYES
NEUTRAL DYEING
PIGMENT DYEING
REACTIVE DYEING
SODIUM DICHROMATE
SULFUR DYEING
TANNIC ACID
VAT DYEING

MORDANT DYES
UF AFTERCHROME DYES
CHROME DYES
CHROME PRINTING COLORS
METACHROME DYES
BT DYES (BY CHEMICAL CLASSES)
RT CHROMIUM COMPOUNDS
DIRECT DYES
DYEING (BY DYE CLASSES)
DYES (BY FIBER CLASSES)
METALLIZED DYES
MORDANT DYEING
REACTIVE DYES

MORPHOLINE SOAPS
BT AMINE SOAPS

MORPHOLINIUM COMPOUNDS
BT ONIUM COMPOUNDS

MORPHOLOGY
USE FINE STRUCTURE (FIBERS)

MOSQUITO RESISTANCE
USE INSECT RESISTANCE

MOTE KNIVES
BT WASTE CONTROL KNIVES
RT BEARDED MOTES
BOLLS
BREAKERS
BURR BEATERS
BURR CRUSHERS
CARD WASTE
CARDING
CARDS
CLEARERS (CARD)
COTTON CARDING
COTTON CARDS
FUZZY MOTES
HULLS
LEAVES
LICKER IN
LICKER IN SCREENS
MOTES
NEPS
SEED COAT NEPS
SEEDS (TRASH)
TRASH
WOOLEN CARDING

MOTES
BT CARD WASTE
TRASH
WASTE
RT CARDING
COTTON
COTTON CARDS
CYLINDER SCREENS
IMPURITIES
LICKER IN SCREENS
MOTE KNIVES
NEPS
SEEDS (TRASH)

MOTH PROOFING
USE INSECT RESISTANCE TREATMENTS

MOTH RESISTANCE
USE INSECT RESISTANCE

MOTH RESISTANCE FINISHES
USE INSECT RESISTANCE FINISHES

MOTHS
BT INSECTS
RT INSECT RESISTANCE
LARVAE

MOTIFS (LACE)
BT LACES

MOTION
UF MOVEMENT
NT AXIAL MOTION
PARALLEL MOTION
PLANE MOTION
RT ACCELERATION (MECHANICAL)
ANGULAR VELOCITY
BRAKES (FOR ARRESTING MOTION)
BRAKING
CONSTANT SPEED
DISPLACEMENT
DRIVES
DYNAMICS
FLUCTUATIONS
FORCE
INERTIA
KINEMATICS
MECHANICS
MOMENTUM
POSITION
REVERSAL
VARIABLE SPEED
VELOCITY
VIBRATION
VISCOSITY

MOTORS
RT ACCESSORIES
CONSTANT SPEED
DRIVES
ELECTRIC EQUIPMENT
EQUIPMENT
MACHINERY
MECHANISMS
POWER CONSUMPTION
STARTING TIME
TEXTILE MACHINERY (GENERAL)
VARIABLE SPEED
VELOCITY

MOUNTAIN WOOL
NT SCOTCH BLACKFACE WOOL
WELSH MOUNTAIN WOOL
BT WOOL
RT CROSSBRED WOOL
LAMBSWOOL
LONG WOOL
MEDIUM WOOL
MERINO WOOL
SHEEP BREEDS

MOVEMENT
USE MOTION

MUFF DYEING
BT PACKAGE DYEING
YARN DYEING
RT BOBBIN DYEING
CAKE DYEING
CHEESE DYEING
CONE DYEING
COP DYEING

MUFTI
RT MILITARY CLOTHING

MULE CARRIAGES
RT DRAFT
HEADSTOCKS
MULE DRAFT
MULES
SPINDLES

MULE DRAFT
UF MULE DRAFTING
MULE DRAW
SPINDLE DRAFTING
BT DRAFTING (STAPLE FIBER)
RT CARRIAGES (SPINNING MULE)
DRAFT
DRIVING BANDS
HEADSTOCKS
MULE CARRIAGES
MULE SPINNING
MULES
PAYING OUT
SLUBBINGS
TWISTING AT THE HEAD
WOOLEN DRAFTING

MULE DRAFTING
USE MULE DRAFT

MULE DRAW
USE MULE DRAFT

MULE FRAMES
USE MULES

MULE SPINNING
BT SPINNING
RT CAP SPINNING
CARRIAGES (SPINNING MULE)
COPS
COUNTERFALLERS
DOFFING (PACKAGE)
DRAFTING (STAPLE FIBER)
DRIVING BANDS
END BREAKAGES
FALLERS (MULE)
FLYER SPINNING
HEADSTOCKS
MULE DRAFT
MULES
PAYING OUT
RING SPINNING
SLUBBINGS
TWIST CONTROL
TWISTING
TWISTING AT THE HEAD
WINDING
YARNS

MULE SPINNING FRAMES
USE MULES

MULES
UF ELECTRONIC MULES
MULE FRAMES
MULE SPINNING FRAMES
SELF ACTOR MULES
SPINNING MULES
BT SPINNING FRAMES
RT CAP SPINNING FRAMES
CARRIAGES (SPINNING MULE)
COPS
COUNTERFALLERS
DELIVERY ROLLS
DRIVING BANDS
END BREAKAGES
FALLERS (MULE)
FLYER SPINNING FRAMES
HEADSTOCKS
JACKSPOOLS
MULE CARRIAGES
MULE DRAFT
MULE SPINNING
POT SPINNING FRAMES
QUADRANTS (MULE)
RING FRAMES
SLUBBINGS
SPINDLES
SURFACE DRUMS
TWISTING AT THE HEAD

MULLEN BURST TESTING
BT BURST TESTING
RT BALL BURST TESTING
BURSTING STRENGTH
DIAPHRAGM BURST TESTING

MULTICOLORS
RT COLOR
CROSS DYEING
DESIGN
PATTERN (APPAREL)
PATTERN (FABRICS)
SOLID COLORS
YARN DYEING

MULTIFILAMENT YARNS
BT FILAMENT YARNS
YARNS
RT COAGULATING BATHS
(CONT.)

RT DRAWING (FILAMENT)
FILAMENT YARNS
MAN MADE FIBERS
MELT SPINNING
MONOFILAMENT YARNS
NEUTRALIZING (FIBER EXTRUSION)
PRODUCERS YARNS
SOLVENT SPINNING
SPINNERETS
SYNTHETIC FIBERS
TEXTURED YARNS (FILAMENT)
WET SPINNING

MULTIFOLD YARNS
USE MULTIPLIED YARNS

MULTIPLE LAYER FABRICS
UF FABRICS (MULTIPLE LAYER)
NT TRILOCK (TN)
WOVEN TUBES
BT FABRICS (ACCORDING TO STRUCTURE)
WOVEN FABRICS
RT BACK FILLING
BACK WARP
BACKED FABRICS
CARPETS
DOUBLE CLOTH
FABRIC STRUCTURE
FACE YARNS
FLAT WOVEN FABRICS
PILE FABRICS (WOVEN)
QUILTING

MULTIPLES
RT ADDITION
DOUBLE SYSTEMS
FREQUENCY
MULTIPLICATION
NUMBERS

MULTIPLICATION
BT COMPUTATIONS
RT ADDITION
ARITHMETIC
DIVISION
MULTIPLES
SUBTRACTION

MULTIPLIED YARNS
UF MULTIFOLD YARNS
BT PLIED YARNS
RT BALANCED YARNS
PLY TWIST
PLYING
S TWIST
SINGLES YARNS
TWIST LIVELINESS
TWISTING
TWO PLY YARNS
Z TWIST

MUNGO
RT GARNETTING
GARNETTING MACHINES
RECOVERED WOOL
SHODDY

MUNSELL COLOR SYSTEM
NT BRIGHTNESS
HUE
SATURATION (COLOR)
VALUE (MUNSELL)
BT COLOR DESIGNATION
RT CIE SYSTEM
COLOR
COLOR MATCHING
COLOR VISION
JUDD NBS COLOR SYSTEM
SHADE
SPECTROPHOTOMETRY

MUSLIN
BT FLAT WOVEN FABRICS
WOVEN FABRICS
RT PLAIN WEAVES

MUSS RESISTANCE
USE WRINKLE RESISTANCE

MYLAR (TN)
BT FILMS
POLYESTERS
RT POLY (ETHYLENE GLYCOL TEREPHTHALATE)
DACRON (TN)

MYRISTYL ALCOHOL
BT FATTY ALCOHOLS

N FADING
BT FUME FADING
RT COLORFASTNESS
NITROGEN OXIDES

N-METHYLOL COMPOUNDS
USE METHYLOL COMPOUNDS

NACCONOL NR (TN)
BT ALKYLBENZENE SULFONATES

NAP
RT BOTTOM COVER
COVER
DRAWN PILE FINISH
FUZZ
HAIRINESS
HAND
NAPPING
NAPPING MACHINES
PATTERN DEFINITION
PILLS
STITCH CLARITY
TEAZLE GIG
TEAZLES
TOP COVER

NAPHTHA
RT DRY CLEANING SOLVENTS
PERCLENE (TN)
TRICHLOROETHYLENE

NAPHTHALENE FORMALDEHYDE SULFONATES
UF CONDENSED NAPHTHALENE SULFONATES
BT SULFONATE SURFACTANTS
RT ALKYLARYL SULFONATES
ALKYLNAPHTHALENE SULFONATES
DISPERSING AGENTS
TETRAHYDRONAPHTHALENE SULFONATES

NAPHTHOL DYES
USE AZOIC DYES

NAPHTHOLS
RT AZOIC BASES
AZOIC DYES
COUPLING (CHEMICAL)
PHENOLS

NAPKINS
RT END USES
HOUSEHOLD FABRICS
NONWOVEN FABRICS
PAPER

NAPPIES
USE DIAPERS

NAPPING
UF GIGGING
RAISING
BT DRY FINISHING
RT BRUSHING
BULKING
BUTTON BREAKING
CHINCHILLA MACHINES
FACE FINISH
FULLING
FUZZ
NAP
NAPPING MACHINES
PATTERN DEFINITION
SHEARING (FINISHING)
SUEDING
TEAZLE GIG
TEAZLES

NAPPING MACHINES
RT BUTTON BREAKERS
CHINCHILLA MACHINES
NAP
NAPPING
SUEDING MACHINES

NARROW FABRIC LOOMS
NT FLYSHOT LOOMS
NEEDLE LOOMS (NARROW FABRICS)
BT LOOMS
RT AUTOMATIC LOOMS
DOBBY LOOMS
NARROW FABRICS
TAPE
WEBBING

NARROW FABRICS
NT ARGYLE GIMP
BRACE WEB
CRATCH
CREPE CORDS
FRINGE (NARROW FABRIC)
LADDER WEBBING
RIBBON
TAPE
WEBBING
RT BRAIDS
CORDAGE
NARROW FABRIC LOOMS

NARROWING
RT BAGGINESS
DIMENSIONAL STABILITY
SKEW
WIDENING

NARROWING (KNITTING)
BT FASHIONING
RT FASHIONING MARKS
FULLY FASHIONED HOSIERY
FULLY FASHIONED KNITTING
NARROWING COMBS
WIDENING (KNITTING)

NARROWING COMBS
RT FASHIONING
FULLY FASHIONED KNITTING MACHINES
NARROWING

NASMITH COMBS
USE RECTILINEAR COMBS

NATURAL CONVECTION DRYERS
USE HOT AIR DRYERS

NATURAL CRIMP
BT CRIMP
FIBER CRIMP
RT BICOMPONENT FIBERS
CONVOLUTIONS
CRIMP FREQUENCY
CRIMP INDEX
CRIMP RADIUS
HELICAL CRIMP
LATENT CRIMP

NATURAL DRAW RATIO
BT FIBER PROPERTIES
RT DRAW RATIO
DRAWING (FILAMENT)
MOLECULAR ORIENTATION
NECKING (FILAMENT)

NATURAL DYES
NT LOGWOOD
BT DYES (BY CHEMICAL CLASSES)
RT DYEING

NATURAL FIBER BLENDS
USE BLENDS (FIBERS)

NATURAL FIBERS
NT ANIMAL FIBERS (GENERAL)
BAST FIBERS
COTTON
FRUIT FIBERS
HAIR
KERATIN FIBERS
LEAF FIBERS
MINERAL FIBERS
SEED HAIR FIBERS (GENERAL)
SILK
VEGETABLE FIBERS (GENERAL)
WOOD FIBERS
WOOL
BT FIBERS
RT CELLULOSIC FIBERS
HYDROPHILIC FIBERS
INORGANIC FIBERS (MAN MADE)
MAN MADE FIBERS
NATURAL POLYMERS
PROTEIN FIBERS
STAPLE FIBERS
SYNTHETIC FIBERS

NATURAL FREQUENCY
USE RESONANT FREQUENCY

NATURAL POLYMERS
UF POLYMERS (NATURAL)
NT CELLULOSE
CHITIN
NATURAL RUBBER
PECTINS
POLYSACCHARIDES
PROTEINS
STARCH
RT CARBOHYDRATES
GUMS
NATURAL FIBERS
ORGANIC COMPOUNDS (GENERAL)
POLYMER TYPES (GENERAL)
POLYMERS

NATURAL RUBBER
BT NATURAL POLYMERS
RUBBER
RT ELASTOMERS
RUBBER ELASTICITY
RUBBER FIBER (NATURAL)
RUBBER FOOTWEAR
SYNTHETIC RUBBER

NATURAL SOIL
BT SOIL

NATURAL STARCH
USE STARCH

NECK FORMATION
USE NECKING (FILAMENT)

NECKING (FILAMENT)
UF NECK FORMATION
RT BREAKING ELONGATION
COLD DRAWING
DIAMETER
DRAW RATIO
DRAWING TENSION (FILAMENT)
DRAWTWISTING
DRAWING (FILAMENT)
DUCTILITY
ELASTIC STRAIN
ELASTICITY
EXTENSIBILITY
HOT DRAWING
NATURAL DRAW RATIO
PLASTIC FLOW
PLASTIC STRAIN
PLASTICITY
RHEOLOGY
SPINNING (EXTRUSION)
STRAIN
STRESS STRAIN CURVES
UNIAXIAL STRESS
YIELD POINT

NECKTIES
BT GARMENTS
OUTERWEAR
RT CRAVATS
DRESSCLOTHING
HOSIERY
SCARVES

NEEDLE BARS
RT DIVIDERS (KNITTING)
FILLING KNITTING
FILLING KNITTING MACHINES
FULLY FASHIONED KNITTING
FULLY FASHIONED KNITTING MACHINES
GUIDE BARS
KNITTING
KNITTING NEEDLES
MILANESE KNITTING MACHINES
NEEDLE LEADS
NEEDLES
PRESSER BARS
RASCHEL KNITTING MACHINES
SINKERS
WARP KNITTING MACHINES

NEEDLE BEARDS
RT KNITTING NEEDLES
NEEDLE EYE
NEEDLES
SPRING NEEDLES

NEEDLE BEARINGS
BT BEARINGS
RT AIR BEARINGS
BALL BEARINGS
CONICAL BEARINGS
JOURNAL BEARINGS
NYLON BEARINGS
ROLLER BEARINGS
SLEEVE BEARINGS
THRUST BEARINGS

NEEDLE BEDS
RT KNITTING MACHINES
KNITTING NEEDLES
NEEDLE BARS
NEEDLE LEADS
NEEDLE TRICKS
NEEDLES

NEEDLE BUTTS
UF BUTTS (NEEDLES)
RT CAM ANGLES
KNITTING MACHINES
KNITTING NEEDLES
NEEDLE CAMS
NEEDLES

NEEDLE CAMS
BT CAMS
RT CIRCULAR KNITTING MACHINES
KNITTING MACHINES
KNITTING NEEDLES
NEEDLES
PATTERN WHEELS
STITCH CAMS

NEEDLE EYE
RT NEEDLE HOOK
NEEDLE GROOVE
NEEDLE POINT
NEEDLE SCARF
NEEDLE SHAFT (SEWING)
NEEDLE SHANK (SEWING)
NEEDLES
SEWING
(CONT.)

RT SEWING MACHINES
SEWING NEEDLES
STITCHES (SEWING)
STITCHING
THREADING

NEEDLE GROOVE
RT NEEDLE EYE
NEEDLE POINT
NEEDLE SCARF
NEEDLE SHAFT (SEWING)
NEEDLE SHANK (SEWING)
NEEDLES
SEWING
SEWING MACHINES
SEWING NEEDLES
STITCHING

NEEDLE HEATING
RT CUTTING RESISTANCE
FRICTION
LUBRICANTS
SEAM DAMAGE
SEAM SLIPPAGE STRENGTH
SEAM STRENGTH
SEAMING
SEWABILITY
SEWING
SEWING THREADS
SOFTENERS
STITCHING
THREAD BREAKAGES
TUFTING
TUFTING NEEDLES
YARN DAMAGE
YARN SEVERANCE

NEEDLE HOOK
RT NEEDLE EYE
NEEDLES

NEEDLE JACKS
RT KNITTING MACHINES
KNITTING NEEDLES
LINKS LINKS MACHINES

NEEDLE LATCHES
UF LATCHES
RT KNITTING MACHINES
KNITTING NEEDLES
LATCH NEEDLES
NEEDLES

NEEDLE LEADS
RT NEEDLE BARS
NEEDLE BEDS

NEEDLE LOOMS (NARROW FABRICS)
BT LOOMS
NARROW FABRIC LOOMS
RT AUTOMATIC LOOMS
RAPIER LOOMS

NEEDLE LOOMS (NEEDLING)
UF NEEDLE PUNCHING MACHINES
BT NONWOVEN FABRIC MACHINES
RT ARACHNE FABRIC (TN)
FELTING MACHINES
FIBER WOVEN (TN)
NEEDLED FABRICS

NEEDLE LOOPS
RT BOBBIN LOOPS
DOUBLE KNIT FABRICS
KNITTED FABRICS
KNITTED STITCHES
KNITTING
KNITTING NEEDLES
LOOP DISTORTION
NEEDLES
ROBBING BACK
SINKER LOOPS
STITCH LENGTH

NEEDLE PENETRATION
RT CHAIN SEWING
LOOP CATCHERS
LOOP CATCHING
MATRIX FREEDOM
NEEDLE POINT
NEEDLING
SEAM DAMAGE
SEAM PUCKER
SEAM QUALITY
SEAMING
SEWABILITY
SEWING
SEWING CYCLES
SEWING MACHINES
SEWING NEEDLES
STITCHING
TUFTING
TUFTING MACHINES
TUFTING NEEDLES
YARN DAMAGE

NEEDLE POINT
RT NEEDLE EYE
NEEDLE GROOVE
NEEDLE PENETRATION
NEEDLE SCARF
NEEDLE SHAFT (SEWING)
NEEDLE SHANK (SEWING)
NEEDLES
SEAM DAMAGE
SEWING
SEWING MACHINES
SEWING NEEDLES
STITCHING
TUFTING
TUFTING NEEDLES
YARN DAMAGE

NEEDLE POINT LACES
NT ALENCON
MILAN (LACE)
VENICE (LACE)
BT HAND MADE LACES
LACES

NEEDLE PRESSER BARS
USE PRESSER BARS

NEEDLE PUNCHED FELTS
BT FELTS
NEEDLED FABRICS
RT ARACHNE FABRIC (TN)
FIBER WOVEN (TN)
MALIWATT FABRIC (TN)
NEEDLE LOOMS (NEEDLING)
NEEDLES
NEEDLING
NONWOVEN FABRICS

NEEDLE PUNCHING
USE NEEDLING

NEEDLE PUNCHING MACHINES
USE NEEDLE LOOMS (NEEDLING)

NEEDLE SCARF
RT NEEDLE EYE
NEEDLE GROOVE
NEEDLE POINT
NEEDLE SHAFT (SEWING)
NEEDLE SHANK (SEWING)
NEEDLES
SEWING
SEWING MACHINES
SEWING NEEDLES

NEEDLE SHAFT (SEWING)
RT NEEDLE EYE
NEEDLE GROOVE
NEEDLE POINT
NEEDLE SCARF
NEEDLE SHANK (SEWING)
NEEDLES
SEWING
SEWING MACHINES
SEWING NEEDLES
STITCHING

NEEDLE SHANK (SEWING)
RT KNITTING NEEDLES
NEEDLE EYE
NEEDLE GROOVE
NEEDLE POINT
NEEDLE SCARF
NEEDLE SHAFT (SEWING)
NEEDLES
SEWING
SEWING MACHINES
SEWING NEEDLES
STITCHING
TUFTING
TUFTING NEEDLES

NEEDLE THREADS
BT SEWING THREADS
THREADS
RT BOBBIN THREADS (SEWING)
COTTON THREADS
LINEN THREADS
MACHINE SEWING
NEEDLES
THREADING
THREADS

NEEDLE TRANSFER
RT FASHIONING
NEEDLES
RACKING
STITCH TRANSFER

NEEDLE TRICKS
RT KNITTING NEEDLES
NEEDLES

NEEDLED FABRICS
NT ARACHNE FABRIC (TN)
FIBER WOVEN (TN)
NEEDLE PUNCHED FELTS
BT FABRICS (ACCORDING TO STRUCTURE)
RT BONDED FIBER FABRICS
BONDED YARN FABRICS
CARPETS
COMPOSITES
CROSS LAID YARN FABRICS
FABRIC STRUCTURE
FELTS
MALI FABRICS (TN)
MALIWATT PROCESS (TN)
NEEDLE LOOMS (NEEDLING)
NEEDLING
NONWOVEN FABRICS
RANDO WEBBER (TN)
STITCH BONDED FABRICS
STITCH REINFORCED FABRICS
STITCHED PILE FABRICS
TUFTED FABRICS

NEEDLES
NT COMPOUND NEEDLES
CROCHET NEEDLES
CYLINDER NEEDLES
DARNING NEEDLES
DIAL NEEDLES
KNITTING NEEDLES
LATCH NEEDLES
SEWING NEEDLES
SPRING NEEDLES
TUFTING NEEDLES
RT FILLING KNITTING
FILLING KNITTING MACHINES
FULLY FASHIONED KNITTING MACHINES
HOSIERY MACHINES
KNITTING
LANDING LOOPS
MALI PROCESSES (TN)
NEEDLE BARS
NEEDLE BEARDS
NEEDLE BUTTS
NEEDLE CAMS
NEEDLE EYE
NEEDLE GROOVE
NEEDLE HOOK
NEEDLE LATCHES
NEEDLE LOOPS
NEEDLE POINT
NEEDLE PUNCHED FELTS
NEEDLE SHAFT (SEWING)
NEEDLE SHANK (SEWING)
NEEDLE THREADS
NEEDLE TRANSFER
NEEDLE TRICKS
NEEDLEWORK
NEEDLING
RIVETS
SEAMLESS HOSIERY MACHINES
SEWING
SEWING MACHINES
SHANKS
TUFTING

NEEDLEWORK
RT EMBROIDERY
NEEDLES
SEWING

NEEDLING
UF NEEDLE PUNCHING
BT DRY FINISHING
RT ARACHNE PROCESS (TN)
KNITTING
MALI PROCESSES (TN)
MALIMO PROCESS (TN)
MALIPOL FABRIC (TN)
MALIPOL PROCESS (TN)
MALIWATT PROCESS (TN)
NEEDLE PENETRATION
NEEDLE PUNCHED FELTS
NEEDLED FABRICS
NEEDLES
SEWING
STITCH BONDING
STITCHING
TUFTED FABRICS
TUFTING
TUFTING MACHINES
TUFTING NEEDLES
WEAVING
WEBS

NEGATIVE SHEDDING
BT SHEDDING
RT CLOSED SHED
HARNESSES
LOOMS
OPEN SHED
PICKING (WEAVING)
POSITIVE SHEDDING
SHUTTLE SMASH
SHUTTLES
SPLIT SHED
WEAVING

NEGLIGEES
BT LINGERIE
NIGHTWEAR

NEKAL A (TN)
 BT ALKYLNAPHTHALENE SULFONATES
 RT NEKAL NS (TN)
 WETTING AGENTS

NEKAL B (TN)
 BT ALKYLNAPHTHALENE SULFONATES
 RT NEKAL NS (TN)
 WETTING AGENTS

NEKAL NS (TN)
 BT TRIHEXYL SULFOCARBALLYLATE
 RT NEKAL A (TN)
 NEKAL B (TN)
 WETTING AGENTS

NEOPATE PROCESS (TN)
 USE EMULSION PRINTING

NEOPRENE COATED FABRICS
 BT COATED FABRICS
 COMPOSITES
 RT BUNA N COATED FABRICS
 BUTYL COATED FABRICS
 COATINGS (SUBSTANCES)
 LACQUER COATED FABRICS
 RUBBER COATED FABRICS
 SBR COATED FABRICS
 VINYL COATED FABRICS

NEP COUNT
 UF NEPPINESS
 RT CARD WEBS
 CARDING
 CARDING ACTION
 COTTON CARDING
 COTTON CARDS
 DEFECTS
 DOFFERS (CARD)
 IMMATURITY (FIBER)
 MATURITY INDEX
 MICRONAIRE
 NEP COUNTERS
 NEP POTENTIAL
 NEP SIZE
 NEP TESTERS
 NEPPINESS INDEX
 NEPS
 NEPTEX (TN)
 NEPTOMETER (TN)
 NEPTOMETER GRADE
 PHOTOELECTRIC COUNTERS (NEPS)
 SLUBBINGS
 TRASH CONTENT
 WOOLEN CARDING
 WORSTED CARDING

NEP COUNTERS
 NT PHOTOELECTRIC COUNTERS (NEPS)
 BT COUNTERS
 RT NEP COUNT
 NEP TESTERS
 OPTICAL SCANNERS
 OPTICAL SCANNING

NEP FORMATION
 USE NEPS

NEP POTENTIAL
 RT DEFECTS
 FIBER FINENESS
 FUZZ RESISTANCE
 IMMATURITY (FIBER)
 MICRONAIRE
 NEP COUNT
 NEP TESTERS
 NEPPINESS INDEX
 NEPS
 NEPTOMETER (TN)
 NEPTOMETER GRADE
 PILL RESISTANCE
 PILLING

NEP SIZE
 RT ENTANGLEMENTS
 IMMATURITY (FIBER)
 NEP COUNT
 NEPS
 NEPTEX (TN)
 NEPTOMETER (TN)

NEP TESTERS
 NT NEPTEX (TN)
 NEPTOMETER (TN)
 BT TESTING EQUIPMENT
 RT NEP COUNT
 NEP POTENTIAL
 NEPS

NEP YARNS
 BT NOVELTY YARNS
 YARNS
 RT NEPS
 SLUB YARNS
 SLUBS
 THICK AND THIN YARNS

NEPPINESS
 USE NEP COUNT

NEPPINESS INDEX
 RT DEFECTS
 IMMATURITY (FIBER)
 NEP COUNT
 NEP POTENTIAL
 NEPS
 NEPTOMETER (TN)
 NEPTOMETER GRADE
 WEBS

NEPS
 UF NEP FORMATION
 BT DEFECTS
 YARN DEFECTS
 RT BURR CRUSHERS
 CARD SLIVERS
 CARD WASTE
 CARD WEBS
 CARDING
 CARDING ACTION
 CARDS
 CLEARERS (CARD)
 COTTON
 COTTON CARDING
 COTTON CARDS
 DOFFERS (CARD)
 ENTANGLEMENTS
 FABRIC DEFECTS
 FABRIC INSPECTION
 FIBER BREAKAGE
 FIBER FINENESS
 FIBER HOOKS
 GINNING
 IMMATURITY (FIBER)
 IMPURITIES
 IRREGULARITY
 KNOTS
 LAPS
 LEADING HOOKS
 LINT (WASTE)
 MOTE KNIVES
 MOTES
 NEP COUNT
 NEP POTENTIAL
 NEP SIZE
 NEP TESTERS
 NEP YARNS
 NEPPINESS INDEX
 NEPTEX (TN)
 NEPTOMETER (TN)
 NEPTOMETER GRADE
 NOVELTY YARNS
 PATCHINESS
 PERALTA ROLLS
 PICKING (OPENING)
 PILLING
 PILLS
 ROVINGS
 SLENDERNESS RATIO
 SLIVERS
 SLUB CATCHERS
 SLUB YARNS
 SLUBBINGS
 SLUBS
 SNAGS
 SPOTTINESS
 THICK SPOTS
 TRAILING HOOKS
 TUFTS (FIBER)
 WEB UNIFORMITY
 WEBS
 WOOLEN CARDING
 WORSTED CARDING
 YARNS

NEPTEX (TN)
 BT NEP TESTERS
 RT ENTANGLEMENTS
 NEP COUNT
 NEP POTENTIAL
 NEP SIZE
 NEPS
 NEPTOMETER (TN)

NEPTOMETER (TN)
 BT NEP TESTERS
 RT ENTANGLEMENTS
 NEP COUNT
 NEP POTENTIAL
 NEP SIZE
 NEPPINESS INDEX
 NEPS
 NEPTEX (TN)
 NEPTOMETER GRADE

NEPTOMETER GRADE
 RT DEFECTS
 IMMATURITY (FIBER)
 NEP COUNT
 NEP POTENTIAL
 NEPPINESS INDEX
 NEPS
 NEPTOMETER (TN)

NETS
 UF FABRICS (NET)
 NT KNITTED NETS
 KNOTTED NETS
 LACE NETS
 LAUNDRY NETS
 WOVEN NETS
 BT FABRICS
 FABRICS (ACCORDING TO STRUCTURE)
 RT NETTING
 NONWOVEN FABRICS
 TUFTED FABRICS

NETT SILK
 BT SILK

NETTING
 BT FLAT WOVEN FABRICS
 WOVEN FABRICS
 RT LENO WEAVES
 MARQUISETTE
 NETS
 PLAIN WEAVES
 SCREENING
 SCRIM

NEUTRAL AXIS
 RT BENDING
 BENDING RECOVERY
 BENDING RIGIDITY
 CROSS SECTIONS
 MOMENT OF INERTIA
 RADIUS OF CURVATURE

NEUTRAL DYEING
 BT DYEING (BY DYE CLASSES)
 RT ACID DYEING
 AZOIC DYEING
 BASIC DYEING
 DIRECT DYEING
 DISPERSE DYEING
 METALLIZED DYEING
 MORDANT DYEING
 NEUTRAL DYES
 PIGMENT DYEING
 REACTIVE DYEING
 SULFUR DYEING
 VAT DYEING

NEUTRAL DYES
 BT DYES (BY CHEMICAL CLASSES)
 RT ACID DYES
 BASIC DYES
 DYEING (BY DYE CLASSES)
 DYES (BY FIBER CLASSES)
 NEUTRAL DYEING

NEUTRALIZING
 RT LAUNDERING
 SCOURING
 SCOURING TRAINS
 SOAP SODA SCOURING

NEUTRALIZING (FIBER EXTRUSION)
 RT BATHS
 CARBONIZING
 COAGULATING
 COAGULATING BATHS
 FILAMENT YARNS
 MAN MADE FIBERS
 MELT SPINNING
 MONOFILAMENT YARNS
 SOLVENT SPINNING
 SPINNERETS
 SYNTHETIC FIBERS
 WET SPINNING

NEW BRADFORD SYSTEM PROCESSING
 BT WORSTED SYSTEM PROCESSING
 RT AMBLER SUPERDRAFT
 AMERICAN SYSTEM PROCESSING
 BACKWASHING
 BLENDING
 BRADFORD OPEN DRAWING
 BRADFORD SYSTEM PROCESSING
 CAP SPINNING
 CARDING
 COMBING
 CONTINENTAL SYSTEM PROCESSING
 CONTROL SURFACES
 DRAFTING (STAPLE FIBER)
 GILLING
 LONG STAPLE SPINNING
 NOBLE COMBS
 OILING
 PIN DRAFTING
 RAPER AUTOLEVELLER (TN)
 SCOURING
 SPINNING
 TOP MAKING
 TWISTING
 WOOL
 WOOLEN SYSTEM PROCESSING
 WORSTED CARDING
 WORSTED COMBING
 WORSTED FABRICS
 WORSTED SPUN YARNS
 WORSTED SYSTEM PROCESSING
 WORSTED SYSTEM SPINNING

NEWTONIAN LIQUIDS
RT PHENOMENOLOGICAL RHEOLOGY

NICKEL
BT METALS
RT NICKEL COMPOUNDS

NICKEL COMPOUNDS
RT CHELATES
DYE FIXING AGENTS
METALLIZED DYES
NICKEL
OLEFIN DYEING

NIGHTCAPS (TEXTILE)
BT HEADGEAR
NIGHTWEAR

NIGHTSHIRTS
BT NIGHTWEAR
RT PAJAMAS

NIGHTWEAR
NT BEDJACKETS
BEDSOCKS
GOWNS
NEGLIGEES
NIGHTCAPS (TEXTILE)
NIGHTSHIRTS
PAJAMAS
ROBES
BT GARMENTS

NIP
RT CONTACT AREA
DRAFTING ROLLS
HYDROSTATIC LOADING
MANGLING
NIP PRESSURE
NIP ROLLS
PAPER MACHINES
PRESS SECTION
PRESSING
PRESSURE ROLLS
RATCH
ROLL COVERINGS
ROLLS
SUCTION PRESSES
WET PICKUP

NIP COATING
BT COATING (PROCESS)
RT BACK FILLING MACHINES
CALENDER COATING
CAST COATING
COATED FABRICS
EXTRUSION COATING
FLEXIBLE FILM LAMINATING
KNIFE COATING
NIP PRESSURE
ROLLER COATING

NIP PRESSURE
RT BURR CRUSHERS
CALENDER ROLLS
CONTACT AREA
CROSS ROLLS
DRAFTING ROLLS
MANGLING
NIP
NIP COATING
PERALTA ROLLS
PRESS SECTION
PRESSURE ROLLS
SQUEEZE ROLLS
WET PICKUP

NIP ROLLS
BT PRESSURE ROLLS
ROLLS
RT MANGLING
NIP
PADDING
PRESSING
SQUEEZING
WET PICKUP

NIPPER JAWS
RT COMB CYLINDERS
COMB PINS
COMB SEGMENTS
COMBING
DETACHING ROLLS
HALF LAP
LAP PLATES
NIPPER KNIVES
NIPPER PLATES
RECTILINEAR COMBS
TOP COMBS

NIPPER KNIVES
RT COMB CYLINDERS
COMB PINS
COMBING
DETACHING ROLLS
HALF LAP
LAP PLATES
RT NIPPER JAWS
NIPPER PLATES
RECTILINEAR COMBS
TOP COMBS

NIPPER PLATES
RT COMB CYLINDERS
COMB PINS
COMBING
DETACHING ROLLS
HALF LAP
LAP PLATES
NIPPER JAWS
NIPPER KNIVES
RECTILINEAR COMBS
TOP COMBS

NIPPING
RT COMBING
COTTON COMBING
DETACHING
TOP COMBING (COMB CYCLE)

NITRATES
NT MAGNESIUM NITRATE
SODIUM NITRATE
ZINC NITRATE
RT CELLULOSE NITRATE
NITRIC ACID
NITRITES

NITRIC ACID
BT INORGANIC ACIDS
RT CARROTING
NITRATES
NITRITES

NITRILE RUBBER FIBER
BT MAN MADE FIBERS
SYNTHETIC FIBERS
SYNTHETIC RUBBER FIBERS
RT RUBBER FIBER (NATURAL)

NITRILES
UF CYANO COMPOUNDS
NT ACETONITRILE
ACRYLONITRILE
METHACRYLONITRILE
POLYACRYLONITRILE
BT ESTERS
RT CYANOETHYLATION

NITRITES
NT POTASSIUM NITRITE
SODIUM NITRITE
RT AZOIC DYES
DIAZOTIZATION
NITRATES
NITRIC ACID
NITROUS ACID

NITRO COMPOUNDS
RT CELLULOSE NITRATE

NITROCELLULOSE
USE CELLULOSE NITRATE

NITROGEN OXIDES
RT COLORFASTNESS
N FADING

NITRON (TN)
BT ACRYLIC FIBERS

NITROUS ACID
BT INORGANIC ACIDS
RT NITRITES

NMR
USE NUCLEAR MAGNETIC RESONANCE

NO CORE BRAIDS
BT BRAIDS
REGULAR BRAIDS

NO TORQUE YARNS
RT BALANCED YARNS
PLIED YARNS
PLYING
ZERO TWIST

NOBLE COMBING
UF CIRCULAR COMBING
OIL COMBING
BT COMBING
RT BRADFORD OPEN DRAWING
BRADFORD SYSTEM PROCESSING
COMB TEAR
COMBED YARNS
COTTON COMBING
DABBING (COMBING)
NOBLE COMBS
NOIL PER CENT
OIL COMBED TOPS
PUNCH BALLS (COMBING)
RECTILINEAR COMBING
SETOVER
SLIVER NOIL RATIO
RT TOP MAKING
TOP NOIL RATIO
WORSTED COMBING
WORSTED SYSTEM PROCESSING
YIELD (RETURN)

NOBLE COMBS
UF CIRCULAR COMBS
BT COMBS
RT BALLING HEADS
BRADFORD SYSTEM PROCESSING
COMB CIRCLES
COMB PINS
COMBING
CONDUCTOR BOXES
COVER PLATES
DABBING BRUSHES
LARGE CIRCLE (COMB)
NEW BRADFORD SYSTEM PROCESSING
NOBLE COMBING
NOIL CHUTES
NOIL KNIVES
NOILS
PLOUGH KNIVES
PRESSER KNIVES
PUNCH BALLS (COMBING)
RECTILINEAR COMBS
SETOVER
SLIVER CANS
SLIVERS
SMALL CIRCLE (COMB)
STAR WHEELS
TOP NOIL RATIO
TOPS
WORSTED SYSTEM PROCESSING

NOGG
RT CARD CLOTHING
CARD CLOTHING FOUNDATION
CARD WIRES
COUNT CROWN AND GAUGE
FILLET CLOTHING
KNEE
METALLIC CARD CLOTHING
NOGGS PER INCH
RIB SET
SHEET CLOTHING
TWILL SET

NOGGS PER INCH
RT CARD CLOTHING
CARD WIRES
NOGG
TWILL SET

NOIL CHUTES
RT COMBING
NOBLE COMBS
NOIL KNIVES
NOIL REMOVAL
NOILS

NOIL KNIVES
RT COMBING
NOBLE COMBS
NOIL CHUTES
NOIL REMOVAL
NOILS

NOIL PER CENT
RT COMB TEAR
COMBING
NOBLE COMBING
NOILS
RECTILINEAR COMBS
SLIVER NOIL RATIO
TOP NOIL RATIO
YIELD (RETURN)

NOIL REMOVAL
RT COMB TEAR
COMBED SLIVERS
COMBER LAPS
COMBING
NOIL CHUTES
NOIL KNIVES
NOILS
WASTE

NOILS
UF COMB NOILS
COMB WASTE
NT CARBONIZED NOILS
SILK NOILS
RT CLIPPINGS
COMB TEAR
COMBED SLIVERS
COMBER LAPS
COMBING
DETACHING ROLLS
HALF LAP
NOBLE COMBS
NOIL CHUTES
NOIL KNIVES
NOIL PER CENT
NOIL REMOVAL
RECTILINEAR COMBS
SLIVER NOIL RATIO

RT SLIVERS
TOP COMBS
TOP NOIL RATIO
TOPS
WASTE
WOOL
WOOL CLASS

NOISE
RT DEAFNESS
MANAGEMENT
OCCUPATIONAL HAZARDS
PRODUCTION
SAFETY

NOMENCLATURE
UF TERMINOLOGY
RT THESAURI

NOMEX (TN)
BT HIGH TEMPERATURE FIBERS
NYLON (POLYAMIDE FIBERS)
RT HEAT RESISTANCE
NYLON 66
TEFLON FIBER (TN)

NOMOGRAMS
UF ALIGNMENT CHARTS
NOMOGRAPHS
RT GRAPHICAL SOLUTIONS
GRAPHS
NUMBERS
SLIDE RULES

NOMOGRAPHS
USE NOMOGRAMS

NON-FROSTING
USE FROSTING

NON-SOAP DETERGENTS
USE SYNTHETIC DETERGENTS

NON LINT CONTENT
USE LINT CONTENT

NONAQUEOUS SOLUTIONS
BT SOLUTIONS (LIQUID)
RT DYE SOLUTIONS
WATER SOLUTIONS

NONIONIC SOFTENERS
BT SOFTENERS
RT ANIONIC SOFTENERS
CATIONIC SOFTENERS
NONIONIC SURFACTANTS
POLYETHOXY ESTERS
POLYETHOXY ETHERS
SILICONES

NONIONIC SURFACTANTS
NT DIETHANOLAMIDE SURFACTANTS
GLYCERYL ESTER SURFACTANTS
GLYCOL ESTER SURFACTANTS
POLYETHOXY ALKYLARYL ETHERS
POLYETHOXY AMIDES
POLYETHOXY ESTERS
POLYETHOXY ETHERS
POLYETHOXY GLYCERIDES
POLYETHOXY POLYPROPOXY
SURFACTANTS
POLYETHOXY THIOETHERS
SORBITAN ESTER SURFACTANTS
SUCROSE ESTER SURFACTANTS
BT SURFACTANTS
RT AMPHOTERIC SURFACTANTS
ANIONIC SURFACTANTS
CATIONIC SURFACTANTS
DYEING AUXILIARIES
EMULSIFYING AGENTS
ETHOXYLATION
LEVELLING AGENTS
NONIONIC SOFTENERS
POLYOXYETHYLENE
RETARDING AGENTS
WETTING
WETTING AGENTS

NONLINEAR DIFFERENTIAL EQUATIONS
BT DIFFERENTIAL EQUATIONS
EQUATIONS
RT LINEAR DIFFERENTIAL EQUATIONS
ORDINARY DIFFERENTIAL
EQUATIONS
PARTIAL DIFFERENTIAL EQUATIONS

NONLINEAR VISCOELASTICITY
BT VISCOELASTICITY
RT LINEAR VISCOELASTICITY
RHEOLOGY
STRESS STRAIN PROPERTIES

NONUNIFORMITY
USE IRREGULARITY

NONWOVEN CARPETS
BT CARPETS
RT AXMINSTER CARPETS
CHENILLE CARPETS
NONWOVEN FABRICS
TUFTED CARPETS
WILTON CARPETS
WOVEN CARPETS

NONWOVEN FABRIC MACHINES
NT NEEDLE LOOMS (NEEDLING)
RANDO WEBBER (TN)
RODNEY HUNT SATURATOR
WALDRON SATURATOR
RT ARACHNE PROCESS (TN)
BATT MAKING MACHINES
CARDS
GARNETTING MACHINES
MALI PROCESSES (TN)
PAPER MACHINES
WEB FORMING MACHINES

NONWOVEN FABRIC STRUCTURE
BT FABRIC STRUCTURE
RT ANISOTROPIC MATERIALS
BINDERS (NONWOVENS)
BONDED FIBER FABRICS
BUNCHING COEFFICIENT
DRAPE
FABRICS
FIBER ORIENTATION
IRREGULARITY
ISOTROPIC MATERIALS
KNITTED FABRIC STRUCTURE
(FILLING KNIT)
KNITTED FABRIC STRUCTURE (WARP
KNIT)
NONWOVEN FABRICS
ORTHOTROPIC MATERIALS
PAPERINESS
PATCHINESS
SHEAR RESISTANCE
STRUCTURES
UNIT CELLS (NONWOVEN FABRICS)
WOVEN FABRIC STRUCTURE
YARN STRUCTURE

NONWOVEN FABRICS
UF FABRICS (NONWOVEN)
NONWOVENS
NT ADHESIVE BONDED (NONWOVENS)
ARABEVA FABRIC (TN)
ARALCOP FABRIC (TN)
BONDED FIBER FABRICS
FIBER WOVEN (TN)
MALIPOL FABRIC (TN)
MALIWATT FABRIC (TN)
NONWOVEN SCRIMS
PRESSED FELTS
SELF BONDED FABRICS
(NONWOVENS)
SPUNBONDED (NONWOVENS)
BT FABRICS
FABRICS (ACCORDING TO
STRUCTURE)
FIBER ASSEMBLIES
RT ADHESION
ADHESIVES
ARACHNE SYSTEMS (TN)
BINDERS (NONWOVENS)
BONDED YARN FABRICS
BONDING
BRAIDS
BUNCHING COEFFICIENT
CARD WEBS
COMPOSITES
CREPING
CROSS LAID WEBS
CROSS LAID YARN FABRICS
DIP BONDING
EMULSION BINDERS (NONWOVENS)
FABRIC STRUCTURE
FELTING
FELTS
FIBER ORIENTATION
FLOCKED FABRICS
FOAM BACKED FABRICS
FULLING
HEAT SEALING
KNITTED FABRICS
LACES
LAMINATED FABRICS
MALI PROCESSES (TN)
MALIMO PROCESS (TN)
MALIPOL PROCESS (TN)
MALIWATT PROCESS (TN)
NAPKINS
NEEDLE PUNCHED FELTS
NEEDLED FABRICS
NETS
NONWOVEN FABRIC STRUCTURE
PACKAGING (MATERIAL)
PAPER
PAPERINESS
PILE FABRICS
RANDO WEBBER (TN)
RANDO WEBS (TN)
SANITARY NAPKINS
STITCH BONDED FABRICS
STITCH REINFORCED FABRICS
STITCHED PILE FABRICS
SUBSTRATES
SYNTHETIC LEATHER
THERMOPLASTIC FIBERS
TUFTED FABRICS
WEBS
WET PROCESS (WEB)
WOVEN FABRICS

NONWOVEN SCRIMS
BT NONWOVEN FABRICS
SCRIM
RT BONDED FIBER FABRICS
BONDED YARN FABRICS

NONWOVENS
USE NONWOVEN FABRICS

NONYLBENZENE SULFONATES
BT ALKYLBENZENE SULFONATES

NONYLNAPHTHALENE SULFONATES
BT ALKYLNAPHTHALENE SULFONATES

NORMAL ALIPHATIC HYDROCARBONS
BT ALIPHATIC HYDROCARBONS
RT BRANCHED ALIPHATIC
HYDROCARBONS

NORMAL DISTRIBUTION
UF GAUSSIAN DISTRIBUTION
RT BINOMIAL DISTRIBUTION
CHI SQUARED DISTRIBUTION
DISTRIBUTION
F DISTRIBUTION
GAMMA DISTRIBUTION
T DISTRIBUTION

NORMAL FORCE
BT FORCE
RT CENTRIFUGAL FORCE
COEFFICIENT OF FRICTION
CONTACT AREA
FABRIC TO FABRIC FRICTION
FIBER FIBER FRICTION
FIBER SLIPPAGE
FRICTION
FRICTIONAL FORCE
GRAVITATION
PRESSURE
RESOLUTION (OF FORCES)
RUBBING
SHEAR FORCE
TANGENTIAL FORCE
TENSION
TRANSVERSE FORCE
YARN TO METAL FRICTION
YARN TO YARN FRICTION

NORMANDIE
BT HAND MADE LACES
LACES

NOSE BUNCHING
RT BUILDER MOTIONS
CLOSE WIND
EMPTY PACKAGES
LAYER LOCKING
PACKAGES
PACKAGING (OPERATION)
QUILLERS
QUILLING
QUILLS
TRAVERSE
UNWINDING
WINDING

NOTTINGHAM LACE
BT LACES
MACHINE MADE LACES

NOVELTY TWISTERS
BT TWISTERS
RT BEAD YARNS
BOUCLES
FRILLS (NOVELTY YARNS)
KNICKERBOCKER YARNS
LOOPS (NOVELTY YARNS)
NOVELTY YARNS
PLIED YARNS
PLY TWISTERS
PLYING
RING SPINNING
RING TWISTERS
RINGS
SEEDS (NOVELTY YARNS)
SPINDLES
SPIRALS (NOVELTY YARNS)
TRAVELERS
TWIST
TWO FOR ONE TWISTERS

NOVELTY YARNS
UF FANCY YARNS
SPECIALTY YARNS
NT BATINES
BEAD YARNS
BOUCLES
BUG YARNS
(CONT.)

NT COCKLE (YARN)
FLAKE YARNS
FLAMES (NOVELTY YARNS)
FRILLS (NOVELTY YARNS)
KNICKERBOCKER YARNS
LOOPS (NOVELTY YARNS)
NEP YARNS
NUB YARNS
RATINES
SEEDS (NOVELTY YARNS)
SLUB YARNS
SPIRALS (NOVELTY YARNS)
SPLASHES
THICK AND THIN YARNS
BT SPUN YARNS
YARNS
RT CORE SPUN YARNS
NEPS
NOVELTY TWISTERS
PLIED YARNS
PLY TWIST
RING TWISTERS
SINGLES YARNS
SLUBS
TEXTURED SPUN YARNS
YARN GEOMETRY
YARN STRUCTURE

NOZZLES
RT AIR JET TEXTURING
AIR JET TEXTURING MACHINES
JET LOOMS
JET QUENCHING
JET SCOURING
JETS
SPRAYING MACHINES

NUB YARNS
BT NOVELTY YARNS
YARNS
RT BEAD YARNS
BOUCLES
BUG YARNS
FRILLS (NOVELTY YARNS)
KNICKERBOCKER YARNS
LOOPS (NOVELTY YARNS)
NUBS
SEEDS (NOVELTY YARNS)
SLUB YARNS
SPIRALS (NOVELTY YARNS)
SPLASHES
THICK AND THIN YARNS

NUBS
BT DEFECTS
FABRIC DEFECTS
YARN DEFECTS
RT NUB YARNS

NUCLEAR MAGNETIC RESONANCE
UF NMR
BT PHYSICAL PROPERTIES (EXCLUDING MECHANICAL)
RT ELECTRON SPIN RESONANCE
GLASS TRANSITION TEMPERATURE
MOLECULAR STRUCTURE
SECONDARY TRANSITION TEMPERATURES
SPECTROSCOPY

NULL METHODS
BT METHODS
RT ELECTRONIC INSTRUMENTS
INSTRUMENTATION
MEASURING INSTRUMENTS
RECORDING INSTRUMENTS

NUMA (TN)
BT SPANDEX FIBERS

NUMBER AVERAGE MOLECULAR WEIGHT
BT MOLECULAR WEIGHT
RT END GROUPS
MOLECULAR WEIGHT DISTRIBUTION
WEIGHT AVERAGE MOLECULAR WEIGHT

NUMBERED DUCK
BT DUCK
FLAT WOVEN FABRICS

NUMBERS
RT COMPUTERS
MAGNITUDE
MULTIPLES
NOMOGRAMS
NUMERICAL ANALYSIS
NUMERICAL INTEGRATION
NUMERICAL SOLUTIONS

NUMERICAL ANALYSIS
RT ACCURACY
ANALYZING
DIFFERENCES
EQUATIONS
EXTRAPOLATION
MATHEMATICAL ANALYSIS
NUMBERS
NUMERICAL SOLUTIONS

NUMERICAL INTEGRATION
BT INTEGRATION
RT NUMBERS
NUMERICAL ANALYSIS
NUMERICAL SOLUTIONS

NUMERICAL SOLUTIONS
BT SOLUTIONS (MATHEMATICAL)
RT GENERAL SOLUTIONS
GRAPHICAL SOLUTIONS
MATHEMATICAL ANALYSIS
NOMOGRAMS
NUMBERS
NUMERICAL ANALYSIS
NUMERICAL INTEGRATION
PARTICULAR SOLUTIONS
TRIAL AND ERROR SOLUTIONS

NYLEX (TN)
BT NYLON (POLYAMIDE FIBERS)
NYLON 66

NYLOFT (TN)
BT CRIMPED YARNS
NYLON (POLYAMIDE FIBERS)
NYLON 6
RT BULKED YARNS
STUFFER BOX CRIMPING

NYLON
USE NYLON (POLYAMIDE FIBERS)

NYLON (POLYAMIDE FIBERS)
UF NYLON
NT ANTRON (TN)
BLANC DE BLANCS (TN)
BLUE C NYLON (TN)
BRI-NYLON (TN)
CADON (TN)
CANTRECE (TN)
CAPROLAN (TN)
CELON (TN)
CREPSET (TN)
CUMULOFT (TN)
ENKALOFT (TN)
ENKALON (TN)
ENKALURE (TN)
ENKATRON (TN)
GOLDEN CAPROLAN (TN)
KAPRON (TN)
LILION (TN)
MONOSHEER (TN)
NOMEX (TN)
NYLEX (TN)
NYLOFT (TN)
NYLON 11
NYLON 6
NYLON 610
NYLON 66
NYLON 7
NYLTEST (TN)
NYPEL (TN)
PERLON (TN)
POLIAFIL (TN)
RILSAN (TN)
SPECKELON (TN)
STEELON (TN)
TYNEX (TN)
VECTRA NYLON (TN)
BT FIBERS
MAN MADE FIBERS
SYNTHETIC FIBERS
RT CYCLOHEXANE, 1,4-BIS (METHYLAMINE)-
LANOLISED NYLON
MELT SPINNING
NYLON DYEING
NYLON DYES
NYLON THREADS
POLYAMIDES
POLYAMIDOESTER FIBERS
SILK
WOOL

NYLON BEARINGS
BT BEARINGS
RT AIR BEARINGS
ANTIFRICTION BEARINGS
BALL BEARINGS
BUSHINGS
CONICAL BEARINGS
JOURNAL BEARINGS
NEEDLE BEARINGS
ROLLER BEARINGS
SLEEVE BEARINGS
TEFLON BEARINGS
THRUST BEARINGS

NYLON BLENDS
BT BLENDS (FIBERS)

NYLON DYEING
BT DYEING (BY FIBER CLASSES)
SYNTHETIC DYEING
RT NYLON (POLYAMIDE FIBERS)
NYLON DYES

NYLON DYES
UF POLYAMIDE DYES
BT DYES (BY FIBER CLASSES)
RT ACETATE DYES
CELLULOSE DYES
DYEING
DYES (BY CHEMICAL CLASSES)
NYLON (POLYAMIDE FIBERS)
NYLON DYEING
POLYESTER DYES
SILK DYES
WOOL DYES

NYLON THREADS
BT SEWING THREADS
THREADS
RT BOBBIN THREADS (SEWING)
COTTON THREADS
LINEN THREADS
NYLON (POLYAMIDE FIBERS)
SEAMING
SEWING
STITCHING

NYLON TRAVELERS
BT TRAVELERS
RT CIRCULAR TRAVELERS
ELLIPTICAL TRAVELERS
PLYING TRAVELERS
POLYAMIDES
RING SPINNING
RINGS
SPINNING
SPINNING TRAVELERS
STEEL TRAVELERS
TWISTING

NYLON 11
NT RILSAN (TN)
BT MAN MADE FIBERS
NYLON (POLYAMIDE FIBERS)
SYNTHETIC FIBERS

NYLON 6
NT BLANC DE BLANCS (TN)
CAPROLAN (TN)
CELON (TN)
CREPSET (TN)
ENKALOFT (TN)
ENKALON (TN)
ENKALURE (TN)
ENKATRON (TN)
GOLDEN CAPROLAN (TN)
KAPRON (TN)
LILION (TN)
MONOSHEER (TN)
NYLOFT (TN)
PERLON (TN)
STEELON (TN)
BT MAN MADE FIBERS
NYLON (POLYAMIDE FIBERS)
SYNTHETIC FIBERS
RT AMINO ACIDS
CANTRECE (TN)
CAPROLACTAM
NYLON 66
POLYCAPROLACTAM
VECTRA NYLON (TN)

NYLON 610
BT MAN MADE FIBERS
NYLON (POLYAMIDE FIBERS)
SYNTHETIC FIBERS
RT HEXAMETHYLENEDIAMINE
POLY (HEXAMETHYLENE SEBACAMIDE)
VECTRA NYLON (TN)

NYLON 66
NT ANTRON (TN)
BLUE C NYLON (TN)
BRI-NYLON (TN)
CADON (TN)
CUMULOFT (TN)
NYLEX (TN)
NYLTEST (TN)
NYPEL (TN)
POLIAFIL (TN)
SPECKELON (TN)
TYNEX (TN)
BT MAN MADE FIBERS
NYLON (POLYAMIDE FIBERS)
SYNTHETIC FIBERS
RT ADIPIC ACID
CANTRECE (TN)
HEXAMETHYLENEDIAMINE
NOMEX (TN)
NYLON 6
POLY (HEXAMETHYLENE ADIPAMIDE)

NYLON 7
BT MAN MADE FIBERS
NYLON (POLYAMIDE FIBERS)
SYNTHETIC FIBERS

NYLTEST (TN)
BT NYLON (POLYAMIDE FIBERS)
NYLON 66

NYPEL (TN)
    BT NYLON (POLYAMIDE FIBERS)
        NYLON 66

NYTRIL FIBERS
    UF POLY (VINYLIDENE CYANIDE) FIBERS
    NT DARVAN (TN)
    BT MAN MADE FIBERS
        SYNTHETIC FIBERS
    RT ACRYLIC FIBERS
        VINYLIDENE COMPOUNDS

O FADING
UF OZONE FADING
BT FUME FADING
RT COLORFASTNESS
OZONE

OCCUPATIONAL HAZARDS
RT ACCIDENTS
DEAFNESS
MANAGEMENT
NOISE
PRODUCTION
SAFETY
TOXICITY

OCTADECYL ETHYLENE UREA
BT AZIRIDINE COMPOUNDS
UREA DERIVATIVES
RT SOFTENERS
WATER REPELLENTS

OCTYL SULFATE
BT ALKYL SULFATES

OCTYLBENZENE SULFONATES
BT ALKYLBENZENE SULFONATES

ODOR CONTROL
RT BACTERIAL INHIBITION
DEODORANTS
DEODORIZING
ODORS

ODORS
RT BACTERIAL DEGRADATION
DEGRADATION
FORMALDEHYDE
ODOR CONTROL
PERSPIRATION
PERSPIRATION RESISTANCE
REACTANTS
WASH WEAR FINISHES

OFF SHADE
BT DYEING DEFECTS
FABRIC DEFECTS
RT FADING
SHADE BARS

OIL BATHS
BT HEAT EXCHANGERS
RT CONTINUOUS DYEING RANGE
DYE FIXING
HEAT TRANSFER
MOLTEN METAL BATHS

OIL COMBED TOPS
BT TOPS
RT BRADFORD OPEN DRAWING
NOBLE COMBING

OIL COMBING
USE NOBLE COMBING

OIL HEATING (OIL FUEL)
BT HEATING
RT ELECTRIC HEATING (RESISTANCE)
GAS HEATING (GAS FUEL)
RADIANT HEATING

OIL IN WATER EMULSIONS
BT EMULSIONS
RT BINDERS (PIGMENTS)
EMULSIFYING AGENTS
EMULSION POLYMERS
LATICES
SOLVENT EMULSIONS
WATER IN OIL EMULSIONS

OIL PROOFING
RT SEALING

OIL REPELLENCY
BT END USE PROPERTIES
FINISH (PROPERTY)
RT FLUOROCARBON POLYMERS
FLUOROCARBONS
STAIN RESISTANCE
SURFACE CHEMISTRY

OIL REPELLENT FINISHES
USE OIL REPELLENTS

OIL REPELLENT TREATMENTS
BT WET FINISHING
RT EASY CARE FABRICS
OIL REPELLENTS
STAIN RESISTANCE

OIL REPELLENTS
UF OIL REPELLENT FINISHES
BT FINISH (SUBSTANCE ADDED)
RT EASY CARE FABRICS
STAIN RESISTANCE

OIL STAINS
BT STAINS
RT CLEANLINESS
RT DEFECTS
FABRIC DEFECTS
MINERAL OIL
SOILING
STAIN REMOVAL
STAIN RESISTANCE
WATER SPOTS
WIPING CLOTHS

OILING
RT BACKWASHING
BLENDING
BRADFORD SYSTEM PROCESSING
CARBONIZING
CARDING OIL
LUBRICANTS
LUBRICATION
MINERAL OIL
NEW BRADFORD SYSTEM PROCESSING
SCOURING
SOAP SODA SCOURING
WOOLEN SYSTEM PROCESSING
WORSTED SYSTEM PROCESSING
YARN FINISH

OILS
RT FATS
FATTY ACIDS
IODINE NUMBER
LIQUIDS
LUBRICANTS
MINERAL OIL
RESIDUAL FAT
SEBUM
STAINING
STAINS
WAXES
WIPING CLOTHS

OLEATE SALTS
NT POTASSIUM OLEATE
SODIUM OLEATE
BT SOAPS
RT ALKALI SOAPS
ANIONIC SURFACTANTS
FATTY ACIDS
LUBRICANTS
METALLIC SOAPS
OLEIC ACID

OLEFIN COPOLYMERS
BT COPOLYMERS
RT OLEFIN FIBERS
POLYOLEFINS

OLEFIN DYEING
BT DYEING (BY FIBER CLASSES)
SYNTHETIC DYEING
RT OLEFIN FIBERS

OLEFIN FIBER BLENDS
BT BLENDS (FIBERS)

OLEFIN FIBERS
UF OLEFINS
NT HERCULON (TN)
MERAKLON (TN)
POLYETHYLENE FIBERS
POLYPROPYLENE FIBERS
VECTRA POLYPROPYLENE (TN)
BT FIBERS
MAN MADE FIBERS
SYNTHETIC FIBERS
RT ACRYLIC FIBERS
MODACRYLIC FIBERS
OLEFIN COPOLYMERS
OLEFIN DYEING
POLYETHYLENE
POLYOLEFINS
POLYPROPYLENE
POLYTETRAFLUOROETHYLENE FIBERS
VINYL FIBERS (GENERAL)

OLEFINS
USE OLEFIN FIBERS

OLEFINIC COMPOUNDS
NT ALKENES
MONOUNSATURATED COMPOUNDS
POLYUNSATURATED COMPOUNDS
BT UNSATURATED COMPOUNDS

OLEFINIC HYDROCARBONS
USE ALKENES

OLEIC ACID
BT FATTY ACIDS
RT ANIONIC SURFACTANTS
FATS
LUBRICANTS
OLEATE SALTS
OLEYL ALCOHOL
POTASSIUM OLEATE
SOAPS
SODIUM OLEATE

OLEYL ALCOHOL
BT FATTY ALCOHOLS
RT OLEIC ACID

OLIGOMERS
RT MONOMERS
POLYMERS

ON OFF CONTROL
BT AUTOMATIC CONTROL
RT CONTINUOUS CONTROL
CONTROL SYSTEMS
CONTROL THEORY
DIFFERENTIAL CONTROL
INTEGRAL CONTROL

ONE BATH PROCESS
BT WET FINISHING
RT DYEING (BY PROCESS FLOW)
TWO BATH PROCESS

ONE HALF BLOOD (WOOL GRADE)
BT GRADE (FIBERS)
WOOL GRADE

ONIUM COMPOUNDS
NT ARSONIUM COMPOUNDS
IODONIUM COMPOUNDS
MORPHOLINIUM COMPOUNDS
PHOSPHONIUM COMPOUNDS
PYRIDINIUM COMPOUNDS
QUATERNARY AMMONIUM COMPOUNDS
SULFONIUM COMPOUNDS

ONIUM SURFACTANTS
NT PHOSPHONIUM SURFACTANTS
QUATERNARY AMMONIUM
SURFACTANTS
SULFONIUM SURFACTANTS
BT CATIONIC SURFACTANTS
RT AMINE SURFACTANTS

OPACITY
BT OPTICAL PROPERTIES
RT DULLNESS
EMULSIONS
GLAZE
GLOSS
REFLECTANCE
TRANSLUCENCY
TRANSPARENCY
TURBIDITY
YARN LUSTER

OPALINE
BT WOVEN FABRICS
RT LAWN

OPEN DRAWING
USE BRADFORD OPEN DRAWING

OPEN END SPINNING
UF BREAK SPINNING
NT AIR SPINNING
COMB SPINNING
FLUID SPINNING
BT SPINNING
RT AUTOMATED SPINNING (STAPLE FIBER)
CONTINUOUS SPINNING (STAPLE
FIBER)
DRAFTING (STAPLE FIBER)
FLUID FLOW
FLUID MECHANICS
FLUID TWISTING
INTERLACING (FILAMENT IN YARN)

OPEN LOOP CONTROL
BT AUTOMATIC CONTROL
RT CONTROL SYSTEMS
CONTROL THEORY
ERRORS
FEEDBACK CONTROL

OPEN SHED
BT SHED (WEAVING)
RT CLOSED SHED
HARNESSES
LOOMS
NEGATIVE SHEDDING
PICKING (WEAVING)
PICKS
POSITIVE SHEDDING
REEDS
SHEDDING
SHUTTLE SMASH
SHUTTLES
SPLIT SHED
WARP ENDS
WEAVING

OPEN STITCHES
BT KNITTED STITCHES
RT CLOSED STITCHES
PELERINE STITCHES
TRICOT STITCHES

OPEN WIDTH BLEACHING
BT BLEACHING
OPEN WIDTH PROCESSING
WET FINISHING
RT OPEN WIDTH DYEING
OPEN WIDTH SCOURING
ROPE BLEACHING

OPEN WIDTH DYEING
NT JIG DYEING
BT OPEN WIDTH PROCESSING
PIECE DYEING
RT BATIK DYEING
CARPET DYEING
CONTINUOUS DYEING
OPEN WIDTH BLEACHING
OPEN WIDTH SCOURING
ROPE DYEING
SEMI CONTINUOUS DYEING

OPEN WIDTH FINISHERS (FULLING)
BT FULLING MACHINES
RT FULLING
MILLS (FULLING ROTARY)
MILLS (KICKER)
REFULLING MACHINES

OPEN WIDTH PROCESSING
NT OPEN WIDTH BLEACHING
OPEN WIDTH DYEING
OPEN WIDTH SCOURING
RT ROPE PROCESSING
WET FINISHING

OPEN WIDTH SCOURING
BT OPEN WIDTH PROCESSING
SCOURING
RT OPEN WIDTH BLEACHING
OPEN WIDTH DYEING
ROPE SCOURING

OPEN WIND
BT WIND
RT CHEESE WINDERS
CLOSE WIND
CLOSE WINDING
COMBINATION WIND
CONERS
DRUM WINDERS
DRUM WINDING
FILLING WIND
GAIN (WINDING)
OPEN WINDING
PACKAGES
PRECISION WINDERS
PRECISION WINDING
RIBBONING
ROVING WIND
WARP WIND
WINDERS
WINDING ANGLE

OPEN WINDING
BT WINDING
RT BUILDER MOTIONS
CLOSE WIND
CLOSE WINDING
CONING
CROSS WINDING
DRUM WINDING
GAIN (WINDING)
OPEN WIND
PACKAGES
PACKAGING (OPERATION)
PARALLEL WINDING
PRECISION WINDING
RIBBON BREAKERS
RIBBON BREAKING
TRAVERSE

OPENERS
UF OPENING MACHINES
NT SRRL OPENER
RT BEATERS (OPENING)
BLENDERS
DOFFER BRUSHES
DOFFING (OPENER)
FEED LATTICES
GARNETTING MACHINES
GRID BARS
HOPPER FEED
IMPELLERS
OPENING
PICKERS (OPENING)
SAW TOOTHED ROLLS

OPENING
BT COTTON SYSTEM PROCESSING
RT CLEANLINESS
ENTANGLEMENTS
FEED ROLLS
GARNETTING
IMPURITIES
OPENERS
PICKER LAPS
PICKING (OPENING)
STOCK (FIBER)
TUFTS (FIBER)

OPENING MACHINES
USE OPENERS

OPERATIONS RESEARCH
RT DISTRIBUTION
EXPERIMENTAL DESIGN
MODELS (MATHEMATICAL)
PROBABILITY
STATISTICAL ANALYSIS

OPTICAL BLEACHES
USE FLUORESCENT BRIGHTENERS

OPTICAL BLEACHING
USE FLUORESCENT BRIGHTENERS

OPTICAL BRIGHTENING
USE FLUORESCENT BRIGHTENERS

OPTICAL DENSITY
RT ABSORPTION (RADIATION)

OPTICAL INSTRUMENTS
NT COLORIMETERS
OPTICAL SCANNERS
PHOTOMETERS
REFLECTOMETERS
SPECTROPHOTOMETERS
BT INSTRUMENTATION
RT ELECTRONIC INSTRUMENTS
FOCUSING
INTERFEROMETRY
MICRORADIOGRAPHY
MICROSCOPY
REFRACTOMETRY
SPECTROGRAPHY
SPECTROPHOTOMETRY
SPECTROSCOPY

OPTICAL PROPERTIES
NT ABSORPTIVITY (RADIATION)
BIREFRINGENCE
BRIGHTNESS
COLOR
COVER
DICHROISM
FLUORESCENCE
INFRARED SPECTRA
LUMINESCENCE
LUSTER
OPACITY
REFLECTIVITY
REFRACTIVE INDEX
SHADE
SPARKLE
STRESS OPTICAL PROPERTIES
TRANSLUCENCY
TRANSMISSIVITY
TRANSPARENCY
TURBIDITY
BT PHYSICAL PROPERTIES (EXCLUDING MECHANICAL)
RT ABSORPTION (RADIATION)
ACOUSTIC PROPERTIES
AESTHETIC PROPERTIES
APPEARANCE
BIOCHEMICAL PROPERTIES
BLOOM
CHEMICAL PROPERTIES
COLORIMETERS
DESIGN
DRABNESS
DRAPE
DULLNESS
ELECTRICAL PROPERTIES
END USE PROPERTIES
FABRIC PROPERTIES (AESTHETIC)
FABRIC PROPERTIES (PHYSICAL EXCLUDING MECHANICAL)
FIBER OPTICS
FIBER PROPERTIES
FINISH (PROPERTY)
HAIRINESS
MECHANICAL PROPERTIES
PATTERN (FABRICS)
PATTERN DEFINITION
PHOTOELASTICITY
PHYSICAL CHEMICAL PROPERTIES
REFLECTANCE
REFLECTOMETERS
RICHNESS
SPECTROPHOTOMETERS
STITCH CLARITY
STRUCTURAL PROPERTIES
TEXTURE
THERMAL PROPERTIES
TRANSFER PROPERTIES
TRANSMITTANCE
WEAVE

OPTICAL SCANNERS
NT PHOTOELECTRIC COUNTERS (NEPS)
BT OPTICAL INSTRUMENTS
RT AUTOCOUNT (TN)
AUTOMATIC CONTROL
BOW STRAIGHTENERS
INFORMATION SYSTEMS
IRREGULARITY TESTERS
NEP COUNTERS
SCANNING
SEAM DETECTORS
SKEW STRAIGHTENERS
SLUB CATCHERS
STOP MOTIONS

OPTICAL SCANNING
NT PHOTOELECTRIC SCANNING
BT SCANNING
RT AUTOCOUNT (TN)
AUTOMATIC CONTROL
CARD WEBS
CHARACTER RECOGNITION
COUNTING
FILM READING
NEP COUNTERS
SEAM DETECTORS
SKEW STRAIGHTENERS
SLUB CATCHERS

OPTICAL WHITENERS
USE FLUORESCENT BRIGHTENERS

OPTICS
NT **FIBER** OPTICS
RT **BIREFRINGENCE**
ELECTRON MICROSCOPY
FOCUSING
MICROSCOPY
MONOCHROMATIC LIGHT
OPTICAL INSTRUMENTS
OPTICAL SCANNING
REFRACTIVE INDEX
SPECTROPHOTOMETRY
SPECTROSCOPY
STRESS OPTICAL PROPERTIES

OPTIMIZATION
RT ADJUSTMENTS
ADVANTAGES
ALTERATIONS
CORRECTION
DEVELOPMENTS
IMPROVEMENTS
INCREASE
MINIMIZATION
MODIFICATION
PERFORMANCE
QUALITY
QUALITY IMPROVEMENT

ORDINARY DIFFERENTIAL EQUATIONS
BT DIFFERENTIAL EQUATIONS
EQUATIONS
RT GENERAL SOLUTIONS
NONLINEAR DIFFERENTIAL EQUATIONS
PARTIAL DIFFERENTIAL EQUATIONS
PARTICULAR SOLUTIONS

ORGANDIE
BT FLAT WOVEN FABRICS
WOVEN FABRICS
RT LAWN
LAWN FINISH
STARCH

ORGANIC ACIDS
NT ACETIC ACID
ACRYLIC ACID
ADIPIC ACID
ALGINIC ACID
AMINO ACIDS
BENZOIC ACID
CARBOXYLIC ACIDS
CHLOROACETIC ACID
CINNAMIC ACID
CITRIC ACID
FATTY ACIDS
FORMIC ACID
HYDROXYACETIC ACID
ISOPHTHALIC ACID
ITACONIC ACID
LACTIC ACID
METHACRYLIC ACID
OXALIC ACID
PHOSPHONIC ACIDS
PHTHALIC ACID
POLYACRYLIC ACID
POLYMETHACRYLIC ACID
PYRUVIC ACID
SUCCINIC ACID
SULFINIC ACIDS
SULFONIC ACIDS
TANNIC ACID
TEREPHTHALIC ACID
THIOGLYCOLIC ACID
THIOSULFINIC ACIDS
BT ACIDS
RT ACID END GROUPS
DIBASIC ACIDS
ESTERS
THIOSULFONATES
TRIBASIC ACIDS

ORGANIC COMPOUNDS (GENERAL)
NT ALICYCLIC COMPOUNDS
ALIPHATIC COMPOUNDS
AROMATIC COMPOUNDS
DIAZO COMPOUNDS
HALOGEN COMPOUNDS
HETEROCYCLIC COMPOUNDS
HYDROCARBONS
NITRO COMPOUNDS
ORGANIC ACIDS
ORGANOMETALLIC COMPOUNDS (EXCLUDING SILICONES)
PHOSPHORUS COMPOUNDS
SILICONES
RT INORGANIC COMPOUNDS (GENERAL)
NATURAL POLYMERS
(CONT.)

RT POLAR COMPOUNDS
POLYMERS
RADICALS (CHEMICAL)

ORGANIC HALOGEN COMPOUNDS
USE HALOGEN COMPOUNDS

ORGANOBISMUTH COMPOUNDS
BT BISMUTH COMPOUNDS
ORGANOMETALLIC COMPOUNDS (EXCLUDING SILICONES)
RT ANTIMICROBIAL FINISHES
ORGANOMERCURY COMPOUNDS
ORGANOTIN COMPOUNDS

ORGANOMERCURY COMPOUNDS
NT PHENYLMERCURIC ACETATE
PHENYLMERCURIC DIOCTYLSULFOSUCCINATE
PHENYLMERCURIC SUCCINATE
BT MERCURY COMPOUNDS
ORGANOMETALLIC COMPOUNDS (EXCLUDING SILICONES)
RT ANTIMICROBIAL FINISHES
CARROTING
ORGANOBISMUTH COMPOUNDS
ORGANOTIN COMPOUNDS
PRESERVATIVES

ORGANOMETALLIC COMPOUNDS (EXCLUDING SILICONES)
UF METALORGANIC COMPOUNDS
NT ORGANOBISMUTH COMPOUNDS
ORGANOMERCURY COMPOUNDS
ORGANOTIN COMPOUNDS
RT CHELATES
INORGANIC COMPOUNDS (GENERAL)
METALLIZED DYES
ORGANIC COMPOUNDS (GENERAL)
SILICONES

ORGANOSILICON COMPOUNDS
USE SILICONES

ORGANOSOLS
BT DISPERSE SYSTEMS
RT COATING (PROCESS)
HYDROSOLS
LAMINATES
RESINS

ORGANOTIN COMPOUNDS
BT ORGANOMETALLIC COMPOUNDS (EXCLUDING SILICONES)
TIN COMPOUNDS
RT ANTIMICROBIAL FINISHES
CATALYSTS
ORGANOBISMUTH COMPOUNDS
ORGANOMERCURY COMPOUNDS
WATER REPELLENTS

ORGANZINE
RT SILK
TRAM
YARNS

ORIENTAL LACE
BT LACES
RT MACHINE MADE LACES

ORIENTATION
UF DISORIENTATION
NT PARALLELIZATION
RT ALIGNMENT
ANISOTROPIC MATERIALS
ANISOTROPY (STRESS STRAIN)
BUNCHING COEFFICIENT
CONFIGURATION
CRYSTALLIZATION
DEGREE OF ORIENTATION
DICHROIC RATIO
FIBER EXTENT
FIBER ORIENTATION
ISOTROPIC MATERIALS
MOLECULAR ORIENTATION
ORIENTATION BIREFRINGENCE
PREORIENTATION
QUENCHING (FILAMENT)
SPINNING (EXTRUSION)
WEBS
X RAY ANALYSIS

ORIENTATION (SURFACE)
BT STRUCTURAL PROPERTIES
RT CONTACT ANGLES
HYDROPHILIC PROPERTY
HYDROPHOBIC PROPERTY
SURFACE TENSION
WATERPROOFING

ORIENTATION BIREFRINGENCE
BT BIREFRINGENCE
RT DEGREE OF ORIENTATION
ORIENTATION
PREORIENTATION
STRESS OPTICAL PROPERTIES

ORIENTATION UNIFORMITY
RT BLOTCHINESS
BUNCHING COEFFICIENT
CARD WEBS
CARDING
FIBER ARRANGEMENT
FIBER ORIENTATION
IRREGULARITY
MASS UNIFORMITY
PATCHINESS
SPOTTINESS
WEB ORIENTATION
WEB UNIFORMITY
WEBS
WIRA WEB LEVELNESS TESTER

ORIENTED WEBS
RT ISOTROPIC MATERIALS
NONWOVEN FABRICS
ORTHOTROPIC MATERIALS
RANDO WEBS (TN)

ORIFICES
RT AERODYNAMICS
FLUID FLOW
HOLES
PORES
SPINNERETS
SPINNING (EXTRUSION)
VISCOUS FLOW

ORLON (TN)
NT ORLON SAYELLE (TN)
BT ACRYLIC FIBERS
RT POLYACRYLATES
POLYACRYLONITRILE
POLYMETHACRYLATES

ORLON SAYELLE (TN)
BT ACRYLIC FIBERS
BICOMPONENT FIBERS
ORLON (TN)
RT BICOMPONENT FIBER YARNS (FILAMENT)
BICOMPONENT FIBER YARNS (STAPLE)
TEXTURED YARNS (FILAMENT)

ORCFIL (TN)
BT SPANDEX FIBERS

ORTHOBLENDS
UF FILLING WARP BLENDS
WARP FILLING BLENDS
RT BLENDED FABRICS
BLENDING
BLENDS (FIBERS)
INTIMATE BLENDS

ORTHOCORTEX
BT FINE STRUCTURE (FIBERS)
RT BICOMPONENT FIBERS
ENDOCUTICLE
EPICUTICLE
EXOCUTICLE
FIBRILS
MEDULLA
MICROFIBRILS
PARACORTEX
SCALES (WOOL FIBERS)
WOOL

ORTHOPEDIC FOOTWEAR
BT FOOTWEAR
RT MEDICAL TEXTILES

ORTHOTROPIC MATERIALS
RT ANISOTROPIC MATERIALS
ISOTROPIC MATERIALS
NONWOVEN FABRIC STRUCTURE
ORIENTED WEBS
RANDO WEBS (TN)
WEB ORIENTATION

OSCILLATIONS
RT AMPLITUDE
FLUCTUATIONS
FREQUENCY
FREQUENCY RESPONSE
HIGH FREQUENCY
IMPACT
IMPULSES
SURGING
VARIABLES
VIBRATION

OSCILLATORS
BT ELECTRONIC INSTRUMENTS
RT AMPLIFIERS
TRANSDUCERS

OSCILLOSCOPES
BT ELECTRONIC INSTRUMENTS
MEASURING INSTRUMENTS

OSMOSIS
NT ELECTRO-OSMOSIS
RT ABSORBENCY (MATERIAL)
CHEMICAL ENGINEERING
CONCENTRATION
CONCENTRATION GRADIENT
DIALYSIS
DIAPHRAGMS
DIFFUSION
DIFFUSIVITY
DILUTION
EXTRACTION
MEMBRANES
PERMEABILITY
SEPARATION (SOLUTION)
SWELLING

OSNABURG
BT FLAT WOVEN FABRICS
WOVEN FABRICS
RT PLAIN WEAVES

OUTERWEAR
NT CARDIGANS
COATS
DRESSCLOTHING
DRESSES
FROCKS
GLOVES
GOWNS
JACKETS
JERKINS
JUMPERS
KNITTED OUTERWEAR
NECKTIES
OVERCOATS
PONCHOS
PULL OVERS
RAINWEAR
SCARVES
SHORTS
SLIP ONS
SUITS
SWEATERS
TRUNKS
VESTS
BT GARMENTS
RT HOSIERY

OUTLINE
USE SHAPE

OUTPUT
RT BYPRODUCTS
CAPACITY (GENERAL)
CARD YIELD
DELIVERY
DELIVERY RATE
FEEDERS
FLOW (MATERIALS)
INPUT
LOADING (PROCESS)
PROCESS EFFICIENCY
PRODUCTION
THROUGHPUT

OUTSIDE FILLING SUPPLY
RT AIR JET LOOMS
DRAPER SHUTTLELESS LOOMS (TN)
JET LOOMS
LOOMS
RAPIER LOOMS
SHUTTLELESS LOOMS
SHUTTLELESS WEAVING
SULZER LOOMS (TN)
WATER JET LOOMS
WEAVING

OVEN METHOD (MOISTURE TESTING)
BT MOISTURE CONTENT TESTING
RT COBALT CHLORIDE METHOD
HYGROMETERS
KARL FISCHER TITRATION METHOD

OVENS
RT CURING
CURING OVENS
DRYERS
DRYING
HEAT TRANSFER
HEATING
HEATING EQUIPMENT
HEAT EXCHANGERS
PAPER MACHINES
TENTER FRAMES
TENTERING
TWIST SETTERS

OVERALLS
BT GARMENTS
INDUSTRIAL CLOTHING
SAFETY CLOTHING
WORK CLOTHING
RT APRONS (CLOTHING)
DENIMS
DUNGAREES
JEANS

OVERCAST STITCHES
BT STITCHES (SEWING)

OVERCOATS
    BT GARMENTS
       OUTERWEAR
    RT COATS
       DRESSCLOTHING
       LODEN COATS
       PARKAS
       PONCHOS
       RAINWEAR

OVERDYEING
    RT ACID DYEING
       AMINE END GROUPS
       DYEING THEORY
       EQUILIBRIUM (DYEING)
       SITES (DYEING)

OVEREDGE STITCHES
    BT STITCHES (SEWING)
    RT UNRAVELLING
       WELTS (SEWN)

OVEREND UNWINDING
    BT UNWINDING
    RT BALLOONS
       TWIST
       TWIST CONTROL
       TWIST DISTRIBUTION
       TWIST SENSE
       TWO FOR ONE TWISTING
       UNROLLING
       UNWINDING ACCELERATORS
       UNWINDING BALLOONS
       UPTWISTERS
       UPTWISTING
       WINDING

OVERFED PIN TENTERS
    BT PIN TENTERS
    RT COMPACTOR
       COMPRESSIVE SHRINKAGE MACHINES
       SANFORIZING MACHINES

OVERHAULING
    USE MAINTENANCE

OVERLAP
    RT COMPOUND NEEDLES
       FULL SET
       GUIDE BARS
       HALF SET
       LAPPING (WARP KNITTING)
       MILANESE KNITTING MACHINES
       PROCESS VARIABLES
       RASCHEL KNITTING MACHINES
       RUNNER RATIO
       RUNNERS
       SET (WARP KNITTING)
       TRICOT KNITTED FABRICS
       TRICOT KNITTING
       TRICOT KNITTING MACHINES
       TRICOT STITCHES
       UNDERLAP
       WARP KNITTING MACHINES

OVERLAP (FIBERS)
    RT DRAFTING (STAPLE FIBER)
       FIBER ARRANGEMENT
       FIBER ASSEMBLIES
       ROVINGS
       SLIVERS

OVERLOCK STITCHES
    BT STITCHES (SEWING)

OVERSEAM STITCHES
    BT STITCHES (SEWING)

OXALIC ACID
    BT CARBOXYLIC ACIDS
       DIBASIC ACIDS
       ORGANIC ACIDS
    RT SUCCINIC ACID

OXAZOLINE SURFACTANTS
    BT AMINE SURFACTANTS
    RT ALKYLAMINE SURFACTANTS
       IMIDAZOLINE SURFACTANTS

OXFORD WEAVES
    BT WEAVE (TYPES)
    RT FLAT WOVEN FABRICS
       SHARKSKIN

OXFORD WOOL
    BT MEDIUM WOOL
       WOOL

OXIDANTS
    USE OXIDIZING AGENTS

OXIDATION
    BT REACTIONS (CHEMICAL)
    RT ANTIOXIDANTS
       BLEACHING
       CHEMICAL STABILITY
       CHLORINATION (SHRINK PROOFING)
       CHLORINE DAMAGE
       CHLORINE DIOXIDE
       CHLORITES
       CORROSION
       DECOMPOSITION
       DEGRADATION
       ELECTROLYSIS
       FREE RADICALS
       HYDROGEN PEROXIDE
       HYPOCHLORITES
       INHIBITION
       IODINE NUMBER
       OXIDIZING AGENTS
       OXIDIZING COMPARTMENT (DYEING)
       OXYCELLULOSE
       OXYGEN
       OZONE
       REDOX CATALYSIS
       REDOX REACTIONS
       REDUCING AGENTS
       REDUCTION
       SODIUM HYPOCHLORITE
       THERMAL STABILITY

OXIDATION DYES
    BT DYES (BY CHEMICAL CLASSES)
    RT DYEING (BY DYE CLASSES)
       DYES (BY FIBER CLASSES)

OXIDES
    NT ALUMINA
       CHLORINE DIOXIDE
       SILICA
       SULFUR DIOXIDE
       TITANIUM DIOXIDE
       ZINC OXIDE
    RT INORGANIC COMPOUNDS (GENERAL)
       PEROXYGEN COMPOUNDS
       SOIL RESISTANCE FINISHES

OXIDIZED STARCHES
    BT MODIFIED STARCHES
       STARCH
    RT DIALDEHYDE STARCH

OXIDIZING AGENTS
    UF OXIDANTS
    RT AIR
       CHLORATES
       CHLORINE BLEACHING AGENTS
       CHLORINE DIOXIDE
       CHLORITES
       HYPOBROMITES
       HYPOCHLORITES
       OXIDATION
       OXIDIZING COMPARTMENT (DYEING)
       PERACETIC ACID
       PERBORATES
       PERCHLORATES
       PERFORMIC ACID
       PERIODATES
       PERMANGANATES
       PEROXIDE BLEACHING AGENTS
       PEROXIDES
       PEROXYGEN COMPOUNDS
       PERSULFATES
       REDOX CATALYSIS
       REDUCING AGENTS
       SODIUM BISULFITE
       SODIUM CHLORATE
       SODIUM HYPOCHLORITE

OXIDIZING COMPARTMENT (DYEING)
    RT CONTINUOUS DYEING RANGE
       LEUCO ESTER DYES
       OXIDATION
       OXIDIZING AGENTS
       VAT DYEING
       VAT DYES

OXIRANE COMPOUNDS
    UF DIEPOXIDES
       EPOXIDES
       EPOXY COMPOUNDS
       GLYCIDYL (PREFIX)
       GLYCIDYL COMPOUNDS
    NT ALLYL GLYCIDYL ETHER
       BUTADIENE DIEPOXIDE
       DIGLYCIDYL ETHERS
       EPICHLOROHYDRIN
       EPIHALOGENOHYDRIN
       ETHYLENE OXIDE
    BT REACTANTS
    RT HALOHYDRINS
       WASH WEAR FINISHES

OXIRANE DYES
    BT REACTIVE DYES

OXYCELLULOSE
    BT CELLULOSE DERIVATIVES
       DEGRADED CELLULOSE
    RT BLEACHING AGENTS
       COPPER NUMBER
       HYDROCELLULOSE
       OXIDATION
       PERIODATES
       YELLOWING

OXYGEN
    RT AIR
       BIOCHEMICAL OXYGEN DEMAND
       OXIDATION

OZONE
    RT LIGHTFASTNESS (COLOR)
       O FADING
       OXIDATION

OZONE FADING
    USE O FADING

PACIFIC CONVERTER
BT CONVERTERS (TOW)
TOW TO TOP MACHINES
RT DRAFT CUT
TOW
TOW CONVERSION
TOW TO TOP CONVERSION

PACKAGE DENSITY
NT PIRN DENSITY
RT COMPRESSIBILITY
DENSITY
HARDNESS
LAYER LOCKING
PACKAGE HARDNESS
PACKAGES
PACKING FACTOR
RIBBONING
SLOUGHING
SPINNING
WINDING
WINDING ANGLE
WINDING DYNAMICS
WINDING EFFICIENCY
WINDING TENSION

PACKAGE DYEING
NT BEAM DYEING (YARN)
BOBBIN DYEING
CAKE DYEING
CHEESE DYEING
CONE DYEING
COP DYEING
MUFF DYEING
BT YARN DYEING
RT CHEESES
CONES
CONTINUOUS DYEING
HANK DYEING
PACKAGE DYEING MACHINES
PACKAGE SCOURING
PACKAGES
PIECE DYEING
PRESSURE DYEING
SKEIN DYEING
STOCK DYEING
TOP DYEING
TOW DYEING

PACKAGE DYEING MACHINES
NT BEAM DYEING MACHINES (YARN)
BOBBIN DYEING MACHINES
CAKE DYEING MACHINES
CHEESE DYEING MACHINES
CONE DYEING MACHINES
COP DYEING MACHINES
BT DYEING MACHINES (BY MATERIAL ASSEMBLY)
YARN DYEING MACHINES
RT BATCH DYEING MACHINES
CIRCULATION PUMPS
DOPE DYEING EQUIPMENT
DYEING
DYEING (BY MATERIAL ASSEMBLY)
HANK DYEING MACHINES
KETTLES (DYEING)
PACKAGE DYEING
PACKAGES
PERFORATED BEAMS
PERFORATED CAGES
PERFORATED CYLINDERS
PERFORATED PLATES
PERFORATED TUBES (DYEING)
PIECE DYEING MACHINES
PRESSURE DYEING MACHINES
STOCK DYEING MACHINES
TOP DYEING MACHINES
TOW DYEING MACHINES

PACKAGE HARDNESS
RT COMPRESSIBILITY
HAND
HARDNESS
PACKAGE DENSITY
PACKING FACTOR
PIRN DENSITY
SOFTNESS
TAPER (BOBBIN)
WINDING
WINDING ANGLE
WINDING DYNAMICS
WINDING EFFICIENCY
WINDING TENSION

PACKAGE HOLDERS
UF HOLDERS (PACKAGE)
RT BOBBINS
COMBS
CREELS
MANDRELS
PACKAGES

PACKAGE SCOURING
BT BATCH SCOURING
SCOURING
RT PACKAGES
ROPE SCOURING

PACKAGES
NT BAGS
BALES
BALL WARPS
BEAMS
BOBBINS
BOTTLE BOBBINS
BOXES
BRAIDER BOBBINS
CAKES
CARTONS
CHEESES
CONES
COPS
EMPTY PACKAGES
FABRIC BOLTS
FULL PACKAGES
JACKSPOOLS
LARGE PACKAGES
PUNCH BALLS (COMBING)
QUILLS
SELF SUPPORTING PACKAGES
SEWING BOBBINS
SPINNING BOBBINS
TWISTING BOBBINS
RT BAGGING MACHINES
BUILDER MOTIONS
CAP SPINNING FRAMES
CHEESE WINDERS
CHEESE WINDING
CLOSE WIND
CLOSE WINDING
COMBINATION WIND
CORES (CENTER)
CREELING
CREELS
DOFFING (PACKAGE)
DRUM WINDERS
DRUM WINDING
FILLING WIND
FLANGES
FLYER SPINNING FRAMES
HANKS
HEADS (TOP)
LAP ARBOURS
LAY (WIND)
LAYER LOCKING
LIFT (WINDING)
NOSE BUNCHING
OPEN WIND
OPEN WINDING
PACKAGE DENSITY
PACKAGE DYEING
PACKAGE DYEING MACHINES
PACKAGE HARDNESS
PACKAGE HOLDERS
PACKAGE SCOURING
PACKAGING (OPERATION)
PIRN DENSITY
POT SPINNING FRAMES
PRECISION WIND
PRECISION WINDERS
PRECISION WINDING
PRESSURE CONTAINERS
RIBBON BREAKERS
RING FRAMES
RING RAILS
RING SPINNING
SKEINS
SLIVER CANS
SLOUGHING
SPINDLES
SPINNING FRAMES
STRAPPING
STRAPPING MACHINES
TAPER (BOBBIN)
TRAVERSE
TWISTERS
UPTWISTERS
WARP BEAMS
WARP ENDS
WARP PREPARATION
WARP WIND
WARPERS (MACHINE)
WARPING
WIND
WINDERS
WINDING
WINDING ANGLE

PACKAGING (MATERIAL)
RT CELLOPHANE
DELIVERY
FILMS
FOAMS
INDUSTRIAL FABRICS
NONWOVEN FABRICS
PAPER
PAPERBOARD

PACKAGING (OPERATION)
RT BAGGING MACHINES
BALING
CHEESE WINDING
CHEESES
CLOSE WINDING
CONES
CREELING
DOFFING (PACKAGE)
DOWNTWISTING
DRUM WINDING
FINISHING PROCESS (GENERAL)
FOLDING (OF CLOTH)
LAYER LOCKING
NOSE BUNCHING
OPEN WINDING
PACKAGES
PACKAGING EQUIPMENT
PAPER MAKING
PRECISION WINDING
QUILLING
QUILLS
SPINNING
THROWING
TRAVERSE
TWISTING
TWO FOR ONE TWISTERS
UPTWISTING
WARP PREPARATION
WARPING
WINDING
WRAPPING MACHINES
YARNS

PACKAGING EQUIPMENT
RT BAGGING MACHINES
BALING MACHINES
PACKAGING (OPERATION)
WRAPPING MACHINES

PACKING
RT BATTING
BEARINGS
COMPRESSIBILITY
FILLER
JOURNAL BEARINGS
STUFFING
WADDINGS

PACKING DENSITY
USE PACKING FACTOR

PACKING FACTOR
UF PACKING DENSITY
BT PHYSICAL PROPERTIES (EXCLUDING MECHANICAL)
STRUCTURAL PROPERTIES
RT AIR PERMEABILITY
BULK
BULK DENSITY
COVER
COVER FACTOR
DENSITY
FABRIC GEOMETRY
FABRIC PROPERTIES
FABRIC PROPERTIES (STRUCTURAL)
HOLES
HOLLOWNESS
MATRIX FREEDOM
PACKAGE DENSITY
PORES
POROSITY
SPACE
SPECIFIC GRAVITY
TEXTURE
TIGHTNESS
TWIST
TWIST MULTIPLIER
VOLUME
YARN CROSS SECTIONS
YARN DIAMETER
YARN FLATTENING
YARN GEOMETRY
YARN STRUCTURE

PAD BATCH CURE PROCESS
BT WET FINISHING
RT BATCH FINISHING
CURING
DURABLE PRESS TREATMENTS
PAD DRY CURE PROCESS
PAD ROLL FINISHING
REACTANTS
WASH WEAR TREATMENTS
WET CURING

PAD DRY CURE PROCESS
UF PDC PROCESS
BT WET FINISHING
RT CONTINUOUS FINISHING
CURING
CURING OVENS
CURING RANGES
DRY CURING
DURABLE PRESS TREATMENTS
PAD BATCH CURE PROCESS
PAD FINISHING
REACTANTS
WASH WEAR TREATMENTS

PAD DYEING
NT PAD JIG DYEING
PAD ROLL DYEING
PAD STEAM DYEING
BT DYEING (BY PROCESS FLOW)
RT BATCH DYEING

RT CONTINUOUS DYEING
PAD FINISHING
PADDING
SEMI CONTINUOUS DYEING
SIMULTANEOUS DYEING AND FINISHING
WET PICKUP

PAD FINISHING
NT PAD JIG FINISHING
PAD ROLL FINISHING
PAD STEAM FINISHING
BT WET FINISHING
RT BATCH FINISHING
PAD BATCH CURE PROCESS
PAD DRY CURE PROCESS
PAD DYEING
PADDING

PAD FIX MACHINES
RT CONTINUOUS DYEING RANGE
PAD ROLL MACHINES
PAD STEAM MACHINES

PAD JIG DYEING
BT PAD DYEING
RT BATCH DYEING
SEMI CONTINUOUS DYEING
JIG DYEING
PAD JIG FINISHING
PAD ROLL DYEING
PAD STEAM DYEING
PADDING

PAD JIG FINISHING
BT PAD FINISHING
WET FINISHING
RT BATCH FINISHING
PAD JIG DYEING
PAD ROLL FINISHING
PAD STEAM FINISHING
PADDING

PAD ROLL DYEING
BT PAD DYEING
RT BATCH DYEING
PAD JIG DYEING
PAD ROLL FINISHING
PAD ROLL MACHINES
PAD STEAM DYEING
PADDING
SEMI CONTINUOUS DYEING

PAD ROLL FINISHING
BT PAD FINISHING
WET FINISHING
RT BATCH FINISHING
PAD BATCH CURE PROCESS
PAD JIG FINISHING
PAD ROLL DYEING
PAD ROLL MACHINES
PAD STEAM FINISHING
PADDING

PAD ROLL MACHINES
RT PAD FIX MACHINES
PAD ROLL DYEING
PAD ROLL FINISHING
PADDERS
SEMI CONTINUOUS DYEING RANGE

PAD STEAM DYEING
BT PAD DYEING
RT BATCH DYEING
CONTINUOUS DYEING
PAD JIG DYEING
PAD ROLL DYEING
PAD STEAM FINISHING
PAD STEAM MACHINES
PADDERS
PADDING
SEMI CONTINUOUS DYEING

PAD STEAM FINISHING
BT PAD FINISHING
WET FINISHING
RT BATCH FINISHING
CONTINUOUS FINISHING
PAD JIG FINISHING
PAD ROLL FINISHING
PAD STEAM DYEING
PAD STEAM MACHINES
PADDERS
PADDING
STEAMING

PAD STEAM MACHINES
RT PAD FIX MACHINES
PAD STEAM DYEING
PAD STEAM FINISHING
PADDERS
SEMI CONTINUOUS DYEING RANGE

PADDERS
UF FOULARD (DYEING)
PADDING MANGLES
RT CONTINUOUS DYEING RANGE
DYEING MACHINES
DYEING MACHINES (BY PROCESS FLOW)
IMPREGNATING MACHINES
MANGLES
PAD FIX MACHINES
RT PAD ROLL MACHINES
PAD STEAM DYEING
PAD STEAM FINISHING
PAD STEAM MACHINES
PADDING
PIECE DYEING MACHINES
REDUCTION PADDERS
WET PICKUP

PADDING
UF DIPPING
BT WET FINISHING
RT ADD ON
COMPRESSIBILITY
DYE BATHS
IMMERSION ROLLS
IMPREGNATION
NIP ROLLS
PAD DYEING
PAD FINISHING
PAD JIG DYEING
PAD JIG FINISHING
PAD ROLL DYEING
PAD ROLL FINISHING
PAD STEAM DYEING
PAD STEAM FINISHING
PADDERS
WARMTH
WET PICKUP

PADDING (MATERIAL)
RT BATTING
FILLER
PACKING
STUFFING
WADDINGS

PADDING (SLEEVES)
USE PADS

PADDING MANGLES
USE PADDERS

PADS
UF PADDING (SLEEVES)
RT GARMENTS
TAILORING

PAINTS
RT CELLULOSE ACETATE BUTYRATE
COATINGS (SUBSTANCES)
DYES
LACQUERS
PIGMENTS

PAJAMA CHECK
BT FLAT WOVEN FABRICS
WOVEN FABRICS

PAJAMAS
BT NIGHTWEAR
RT BEDJACKETS
NIGHTSHIRTS

PANTIES
BT GARMENTS
LINGERIE
UNDERWEAR
RT BRASSIERES
BRIEFS
FOUNDATION GARMENTS
GIRDLES
SHORTS
SLIPS

PAPER
UF PAPERBOARD
BT BONDED FIBER FABRICS
FIBER ASSEMBLIES
RT ABRASION TESTING
BENDING TESTING
BONDED YARN FABRICS
BURST TESTING
CHEMICAL TESTING
FABRICS
FELTS
KNITTED FABRICS
MOISTURE CONTENT TESTING
NAPKINS
NONWOVEN FABRICS
PACKAGING (MATERIAL)
PAPER MACHINES
PAPER MAKING
PAPERINESS
PERMEABILITY TESTING
PROPERTIES
PULP
STRENGTH TESTING
TEAR TESTING
TENSILE TESTING
WATER RESISTANCE TESTING
WET PROCESS (WEB)
WOOD FIBERS
WOVEN FABRICS

PAPER MACHINES
NT CYLINDER MACHINES
FOURDRINIER MACHINES
NT INVERFORM MACHINES
KAMYR MACHINES
ROTAFORMERS
VERTAFORMERS
WET MACHINES (PAPERMAKING)
RT DRYERS
MANGLES
NIP
NONWOVEN FABRIC MACHINES
OVENS
PAPER
PAPER MAKING
PAPERMAKERS FELTS
PRESS SECTION
ROLLS
SUCTION PRESSES
TENSION CONTROL
WEB FORMING MACHINES
WINDERS

PAPER MAKING
RT BLEACHING
BLENDING
CALENDER COATING
CALENDERING
COATING (PROCESS)
COMPRESSIVE SHRINKING
CREPING
CUTTING
DIP COATING
DRYING
DYEING
GLAZING
KNIFE COATING
MANGLING
MIXING
MOISTURE CONTENT
PACKAGING (OPERATION)
PAPER
PAPER MACHINES
PRESSING
PULPING
RESINS
ROLLER COATING
SIZES (SLASHING)
SIZING (PAPERMAKING)
SLASHING
SQUEEZING
SUCTION PRESSES
WET PROCESS (WEB)
WINDING

PAPER YARNS
BT YARNS
RT LAMINATED YARNS
PAPER
SLIT FILM YARNS
SPLITTING
TWISTING
UNCONVENTIONAL YARNS

PAPERBOARD
RT PACKAGING (MATERIAL)
PAPER

PAPERINESS
BT MECHANICAL PROPERTIES
RT BENDING RIGIDITY
COMPLIANCE
DRAPE
HAND
HARDNESS
INITIAL MODULUS
MOMENT OF INERTIA
NONWOVEN FABRICS
PAPER
SOFTNESS
SPRING CONSTANT
STARCH
STIFFENERS (AGENTS)
STIFFNESS
WRINKLE RESISTANCE

PAPERMAKERS FELTS
BT FELTS
INDUSTRIAL FABRICS
RT BAGGINESS
BELTS
COMPRESSIBILITY
CONVEYOR BELTS
DIMENSIONAL STABILITY
MANGLING
MASS TRANSFER
NEEDLED FABRICS
PAPER MACHINES
PAPER MAKING
PRESS SECTION
PRESSING
SKEW
TUBULAR FABRICS
WATER ABSORPTION
WATER PERMEABILITY
WIDENING

PARACHUTE CORDS
BT BRAIDS
CORDS
CORE BRAIDS
(CONT.)

BT REGULAR BRAIDS
RT ARRESTERS
PARACHUTES

PARACHUTES
BT MILITARY PRODUCTS
RT AERODYNAMICS
AIR DRAG
AIR PERMEABILITY
AIR RESISTANCE
ARRESTERS
END USES
PARACHUTE CORDS
RECOVERY SYSTEMS (AERODYNAMIC)
SHOCK ABSORBERS
STABILIZATION
WIND TUNNELS

PARACORTEX
BT FINE STRUCTURE (FIBERS)
RT BICOMPONENT FIBERS
CORTEX
CUTICLE
ENDOCUTICLE
EPICUTICLE
EXOCUTICLE
MEDULLA
MICROFIBRILS
ORTHOCORTEX
WOOL

PARACRYSTALLINE REGION
BT FINE STRUCTURE (FIBERS)
RT AMORPHOUS REGION
CRYSTALLINE REGION
FRINGED MICELLES
MICELLES

PARAFFIN WAXES
BT WAXES
RT LUBRICANTS

PARAFORMALDEHYDE
BT FORMALDEHYDE

PARAGUAY (LACE)
BT HAND MADE LACES
LACES

PARALLEL MOTION
BT MOTION
RT AXIAL MOTION
PLANE MOTION

PARALLEL WINDING
BT WINDING
RT CLOSE WINDING
CROSS WINDING
OPEN WINDING

PARALLELIZATION
BT ORIENTATION
RT ANISOTROPIC MATERIALS
CARDING
COMBING
CONFIGURATION
DEGREE OF ORIENTATION
DRAFTING (STAPLE FIBER)
FIBER EXTENT
FIBER HOOKS
FIBER ORIENTATION
STRUCTURAL PROPERTIES

PARKAS
BT SPORTSWEAR
RT LODEN COATS
MILITARY CLOTHING
OVERCOATS

PARTIAL DIFFERENTIAL EQUATIONS
BT DIFFERENTIAL EQUATIONS
EQUATIONS
RT LINEAR DIFFERENTIAL EQUATIONS
NONLINEAR DIFFERENTIAL EQUATIONS
ORDINARY DIFFERENTIAL EQUATIONS

PARTIAL PRESSURE
RT DEW POINT
DIFFUSION
DRYING
HUMIDITY
HYDROSTATIC PRESSURE
PRESSURE
RELATIVE HUMIDITY
SATURATED VAPOR
SPECIFIC HUMIDITY
TEMPERATURE
VAPOR PRESSURE

PARTICLE SIZE
RT ABSORPTION (MATERIAL)
ADSORPTION
AMPLITUDE
AREA
COLLOIDS
DISPERSING AGENTS
RT FINENESS
PENETRATION
PORES

PARTICULAR SOLUTIONS
BT SOLUTIONS (MATHEMATICAL)
RT DIFFERENTIAL EQUATIONS
EQUATIONS
GENERAL SOLUTIONS
GRAPHICAL SOLUTIONS
INTEGRATION
LINEAR DIFFERENTIAL EQUATIONS
NUMERICAL SOLUTIONS
TRIAL AND ERROR SOLUTIONS

PARTITION COEFFICIENT
UF COEFFICIENT OF PARTITION
RT ADSORPTION ISOTHERM
CHROMATOGRAPHY
DISPERSE DYEING
DYEING THEORY
EXTRACTION
FREUNDLICH ISOTHERM
LANGMUIR ISOTHERM
SOLUBILITY

PASTES
RT ADHESION
ADHESIVES
BONDING STRENGTH
CEMENTS
COHESION
GLUES
GUMS
PRINTING PASTES
SIZES (SLASHING)
STARCH

PATCHINESS
RT APPEARANCE
BLOTCHINESS
BUNCHING COEFFICIENT
CARD WEBS
IRREGULARITY
MASS UNIFORMITY
NEPS
NONWOVEN FABRICS
ORIENTATION UNIFORMITY
SPOTTINESS
WEB ORIENTATION
WEB UNIFORMITY
WEBS
WIRA WEB LEVELNESS TESTER

PATCHING
USE MAINTENANCE

PATROLLING
UF SUPERVISING
RT ALLOTMENT
DOWN TIME
END BREAKAGES
JOB ANALYSIS
MACHINE INTERFERENCE
MANAGEMENT
MANUFACTURING
PERSONNEL
PROCESS EFFICIENCY
PRODUCTION
SPINNING
SPINNING EFFICIENCY
STOPS
TRAINING
WEAVING EFFICIENCY
WINDING
WINDING EFFICIENCY
WORKLOADS

PATTERN (APPAREL)
RT APPAREL DESIGN
BUNDLING (TAILORING)
CUTTING (TAILORING)
DRESS DESIGN
DRESSMAKING
MARKING
MULTICOLORS
PATTERN GRADING
PATTERN MAKING
SLOPING
TAILORING

PATTERN (FABRICS)
BT FABRIC PROPERTIES
FABRIC PROPERTIES (AESTHETIC)
FABRIC PROPERTIES (STRUCTURAL)
RT AESTHETIC PROPERTIES
APPEARANCE
CHAIN DRAFT
COVER
DESIGN
DRAFT (PATTERN)
DRAFTING (PATTERN)
ENGRAVING
FABRIC ANALYSIS
FABRIC DESIGN
FINISH (PROPERTY)
HARNESS DRAFT
MULTICOLORS
RT OPTICAL PROPERTIES
PATTERN CHAINS
PATTERN DEFINITION
PATTERN REPEATS
PATTERN WHEELS
PRINTING
PRINTING ROLLS
REEDS
REPEATS
RICHNESS
SCREEN PRINTING
STITCH CLARITY
TWILL ANGLES
WEAVE

PATTERN AREA
RT PATTERN REPEATS
PATTERN WHEELS

PATTERN CHAINS
RT GUIDE BARS
LAPPING (WARP KNITTING)
LOOMS
PATTERN (FABRICS)
WARP KNITTING MACHINES

PATTERN DEFINITION
BT AESTHETIC PROPERTIES
FABRIC PROPERTIES
FABRIC PROPERTIES (AESTHETIC)
FABRIC PROPERTIES (STRUCTURAL)
RT AESTHETIC PROPERTIES
APPEARANCE
BOTTOM COVER
BRUSHING
COLOR
COVER
DESIGN
FINISH (PROPERTY)
FULLING
HAIRINESS
LUSTER
NAP
NAPPING
OPTICAL PROPERTIES
PATTERN (FABRICS)
SHADE
SINGEING
STITCH CLARITY
TEXTURE
TOP COVER
WEAVE (TYPES)

PATTERN GRADING
RT APPAREL DESIGN
CUTTING (TAILORING)
GARMENT MANUFACTURE
PATTERN MAKING
TAILORING

PATTERN MAKING
RT APPAREL DESIGN
GARMENT MANUFACTURE
PATTERN GRADING

PATTERN REEDS
RT HACK STANDS
WARP PREPARATION

PATTERN REPEATS
RT DESIGN
DRAFTING (PATTERN)
DRESSMAKING
FABRIC DESIGN
HARNESS DRAFT
LACE MAKING
PATTERN (FABRICS)
PATTERN WHEELS

PATTERN WHEEL JACKS
RT KNITTING
KNITTING MACHINES
KNITTING NEEDLES
PATTERN WHEELS

PATTERN WHEEL KNITTING
BT KNITTING
RT JACQUARD KNITTING
PATTERN WHEELS

PATTERN WHEELS
UF DAWSON WHEELS
RT CIRCULAR KNITTING MACHINES
DESIGN
FILLING KNITTING
FILLING KNITTING MACHINES
JACKS (KNITTING)
KNITTING
KNITTING MACHINES
KNITTING NEEDLES
NEEDLE CAMS
PATTERN (FABRICS)
PATTERN AREA
PATTERN REPEATS
PATTERN WHEEL JACKS
PATTERN WHEEL KNITTING
RAISING CAMS
STITCH CAMS

PAYING OUT
    RT DRAFTING (STAPLE FIBER)
        MULE CRAFT
        MULE SPINNING
        MULES
        TWISTING AT THE HEAD

PDC PROCESS
    USE PAD DRY CURE PROCESS

PECTINS
    BT CARBOHYDRATES
        NATURAL POLYMERS
    RT EMULSIFYING AGENTS
        GUMS
        POLYSACCHARIDES

PEDESTAL STANDS
    USE STANDS

PEELING
    RT ADHESION
        BONDING
        CEMENTING
        COHESION
        DELAMINATING
        FLAKING
        GLUING
        JOINING
        LAMINATING
        SCUFFING
        SEALING
        SEAMING
        SEPARATION
        TACK
        TAPING
        VISCOUS FLOW

PEG BOARD KNITTING
    BT KNITTING
    RT HAND KNITTING
        MACHINE KNITTING

PELERINE STITCHES
    BT KNITTED STITCHES
    RT CLOSED STITCHES
        FLOAT STITCHES (KNITTING)
        OPEN STITCHES
        PURL STITCHES

PENETRANTS
    USE WETTING AGENTS

PENETRATING
    USE PENETRATION

PENETRATION
    UF PENETRATING
    NT DYE PENETRATION
    RT ABSORPTION (MATERIAL)
        ACCESSIBILITY (INTERNAL)
        DIALYSIS
        DIFFUSION
        DIFFUSIVITY
        DUCTILITY
        HOLES
        IMPACT PENETRATION TESTING
        INSECT RESISTANCE
        MEMBRANES
        PARTICLE SIZE
        PERMEABILITY
        PORES
        PUNCTURE RESISTANCE
        PUNCTURING
        RAIN PENETRATION TESTING
        RUN RESISTANCE
        SEEPAGE (RHEOLOGY)
        SNAGGING
        SNAGS
        STRESS CONCENTRATION
        STRESS STRAIN PROPERTIES
        TACK
        THIXOTROPY
        WATER PERMEABILITY
        WATER REPELLENCY
        WATERPROOFING

PENTACHLOROPHENYL LAURATE
    BT PHENOLS
    RT ANTIMICROBIAL FINISHES

PENTAERYTHRITOL
    BT POLYOLS
    RT GLYCEROL

PEPTIDES
    USE POLYPEPTIDES

PERACETIC ACID
    BT PEROXYGEN COMPOUNDS
    RT ACETIC ACID
        HYDROGEN PEROXIDE
        OXIDIZING AGENTS
        PERBORATES
        PERFORMIC ACID
        PEROXIDE BLEACHING AGENTS

PERALTA ROLLS
    BT CARD ROLLS
        ROLLS
        WEB PURIFIERS
    RT BEARING STRENGTH
        BREAKERS
        BURR BEATERS
        BURR CRUSHERS
        CALENDER ROLLS
        CARD CYLINDERS
        CARD WASTE
        CARD WEBS
        CARDING
        CARDS
        CLEANLINESS
        COMPRESSIVE STRENGTH
        CROSROL WEB PURIFIER (TN)
        CROSS ROLLS
        HYDROSTATIC LOADING
        NEPS
        NIP PRESSURE
        PRESS SECTION
        PRESSURE ROLLS
        ROLL COVERINGS
        SUCTION PRESSES
        WOOLEN CARDING
        WOOLEN CARDS
        WORSTED CARDING

PERBORATES
    NT SODIUM PERBORATE
    BT PEROXYGEN COMPOUNDS
    RT HYDROGEN PEROXIDE
        OXIDIZING AGENTS
        PERACETIC ACID
        PEROXIDE BLEACHING AGENTS

PERCALE
    BT FLAT WOVEN FABRICS
        WOVEN FABRICS
    RT PLAIN WEAVES

PERCALINE
    BT WOVEN FABRICS
    RT LININGS
        TAFFETA

PERCENTAGE CRIMP
    USE CRIMP PERCENT

PERCENTAGE PLATES
    RT CARD CYLINDERS
        CARD FLATS
        CARDS
        COTTON CARDS
        CYLINDER SCREENS
        FIBER LENGTH DISTRIBUTION
        FLAT STRIPS
        FLY (WASTE)

PERCENTAGE STRAIN
    USE STRAIN

PERCHING
    BT DRY FINISHING
    RT BURLING
        FABRIC DEFECTS
        INSPECTION
        INSPECTION MACHINES

PERCHLORATES
    RT CHLORATES
        CHLORINE COMPOUNDS
        OXIDIZING AGENTS
        PERCHLORIC ACID

PERCHLORIC ACID
    BT INORGANIC ACIDS
    RT CHLORINE COMPOUNDS
        PERCHLORATES

PERCHLOROETHYLENE
    NT PERCLENE (TN)
    BT CHLORINATED HYDROCARBONS
    RT CHLORINATED SOLVENTS
        DRY CLEANING
        DRY CLEANING FASTNESS (COLOR)
        DRY CLEANING FASTNESS (OF FINISH)
        TRICHLOROETHYLENE

PERCLENE (TN)
    BT CHLORINATED HYDROCARBONS
        PERCHLOROETHYLENE
    RT CARBON TETRACHLORIDE
        CHLORINATED SOLVENTS
        DRY CLEANING
        DRY CLEANING FASTNESS (COLOR)
        DRY CLEANING FASTNESS (OF FINISH)
        DRY CLEANING SOLVENTS
        NAPHTHA
        TRICHLOROETHYLENE

PERFLUOROALKYL COMPOUNDS
    USE FLUOROCARBONS

PERFLUOROCARBONS
    USE FLUOROCARBONS

PERFORATED BEAMS
    RT HOLES
        PENETRATION
        PERFORATED BELTS
        PERFORATED CAGES
        PERFORATED CYLINDERS
        PERFORATED PLATES
        PERFORATED TUBES (DYEING)

PERFORATED BELTS
    RT HOLES
        PENETRATION
        PERFORATED BEAMS
        PERFORATED CAGES
        PERFORATED CYLINDERS
        PERFORATED PLATES
        PERFORATED TUBES (DYEING)

PERFORATED CAGES
    RT HOLES
        HOSIERY DYEING MACHINES
        PENETRATION
        PERFORATED BEAMS
        PERFORATED BELTS
        PERFORATED CYLINDERS
        PERFORATED PLATES
        PERFORATED TUBES (DYEING)

PERFORATED CYLINDERS
    BT CYLINDERS
    RT ENGRAVED CYLINDERS
        HOLES
        PENETRATION
        PERFORATED BEAMS
        PERFORATED BELTS
        PERFORATED CAGES
        PERFORATED PLATES
        PERFORATED TUBES (DYEING)
        SUCTION PRESSES

PERFORATED PLATES
    RT HOLES
        PENETRATION
        PERFORATED BEAMS
        PERFORATED BELTS
        PERFORATED CAGES
        PERFORATED CYLINDERS
        PERFORATED TUBES (DYEING)

PERFORATED TUBES (DYEING)
    RT HOLES
        PACKAGE DYEING MACHINES
        PENETRATION
        PERFORATED BEAMS
        PERFORATED BELTS
        PERFORATED CAGES
        PERFORATED CYLINDERS
        PERFORATED PLATES
        YARN DYEING MACHINES

PERFORMANCE
    RT ADVANTAGES
        CHARACTERISTICS
        DESIGN
        DEVELOPMENTS
        DOWN TIME
        END BREAKAGES
        END USE PROPERTIES
        IMPROVEMENTS
        MACHINE DESIGN
        METHODS
        MINIMIZATION
        OPTIMIZATION
        PROCESS EFFICIENCY
        PRODUCTION
        QUALITY
        QUALITY IMPROVEMENT
        SPINNING EFFICIENCY
        TECHNIQUES
        WEAR RESISTANCE
        WEAVING EFFICIENCY
        WINDING EFFICIENCY
        WRINKLE RESISTANCE

PERFORMANCE PROPERTIES
    USE END USE PROPERTIES

PERFORMIC ACID
    BT PEROXYGEN COMPOUNDS
    RT FORMIC ACID
        HYDROGEN PEROXIDE
        OXIDIZING AGENTS
        PERACETIC ACID
        PEROXIDE BLEACHING AGENTS

PERIMETER
    RT AREA
        CIRCUMFERENCE
        CROSS SECTIONAL AREA
        CROSS SECTIONS
        DIAMETER
        GEOMETRY
        SHAPE
        SURFACE AREA

PERIODATES
RT IODINE COMPOUNDS
OXIDIZING AGENTS
OXYCELLULOSE

PERIODICITY
RT AMPLITUDE
AUTOCORRELATION
BARRE
DRAFTING WAVES
FREQUENCY
IRREGULARITY
IRREGULARITY (PERIODIC)
REPEATS
VARIATION
VIBRATION

PERLOK SYSTEM
BT CONVERTERS (TOW)
TOW TO TOP MACHINES
RT DRAFT CUT
STRETCH BREAKING
TOW
TOW CONVERSION

PERLON (TN)
BT NYLON (POLYAMIDE FIBERS)
NYLON 6

PERLON U (TN)
BT POLYURETHANE FIBERS

PERMANENT CREASING
USE DURABLE PRESS

PERMANENT DEFORMATION
BT DEFORMATION
RT CREASE RETENTION
CREASING
CREEP
DEFLECTION
DELAYED RECOVERY
ELASTIC STRAIN
GROWTH (FABRIC)
HEAT SETTING (SYNTHETICS)
PERMANENT PLEATING
PERMANENT SET
PLASTIC STRAIN
SAG
SEAM GRIN
SEAM SLIPPAGE
STRAIN
STRESS RELAXATION
STRETCH
TWIST SETTERS
WRINKLE RESISTANCE
WRINKLING
YARN SLIPPAGE

PERMANENT FINISHES (GENERAL)
USE DURABLE FINISH (GENERAL)

PERMANENT PLEATING
BT WET FINISHING
RT CREASE RETENTION
CREASING
CURING
DURABLE PRESS
EASY CARE FABRICS
GARMENT MANUFACTURE
HEAT SETTING (SYNTHETICS)
PERMANENT DEFORMATION
PERMANENT SET
PLEATING
PRESSING

PERMANENT PRESS
USE DURABLE PRESS

PERMANENT SET
UF SECONDARY CREEP
BT CREEP
MECHANICAL DETERIORATION
PROPERTIES
STRESS STRAIN PROPERTIES
(TENSILE)
RT BENDING RECOVERY
CREASING
CREEP RECOVERY
CRIMP RETENTION
DELAYED RECOVERY
DIMENSIONAL STABILITY
ELASTIC STRAIN
ELASTICITY
LOADING
MAXWELL MODEL
PLASTIC FLOW
PLASTIC STRAIN
RECOVERY (SELF RESTORATION TO
ORIGINAL CONDITION)
RELAXATION
RELAXATION SPECTRUM
RELAXATION TIME (MECHANICAL)
RESILIENCE
RHEOLOGY
STRAIN
STRESS
STRESS RELAXATION
RT STRESS STRAIN CURVES
VISCOELASTICITY
VOIGT MODEL
WORK RECOVERY
WRINKLING
YIELD POINT

PERMANENT SETTING
USE DURABLE PRESS TREATMENTS

PERMANGANATES
BT MANGANESE COMPOUNDS
RT ANTIFELTING AGENTS
ANTIFELTING TREATMENTS
CHLORINATION (SHRINK PROOFING)
OXIDIZING AGENTS
SHRINK PROOFING

PERMEABILITY
NT AIR PERMEABILITY
WATER PERMEABILITY
BT MECHANICAL PROPERTIES
RT ABSORPTION (MATERIAL)
ACCESSIBILITY (INTERNAL)
AERODYNAMICS
CHEMICAL ENGINEERING
DIALYSIS
DIFFUSION
DIFFUSIVITY
DYE PENETRATION
FABRIC INTERSTICES
FABRIC PROPERTIES
FABRIC PROPERTIES (PHYSICAL
EXCLUDING MECHANICAL)
FABRIC PROPERTIES (STRUCTURAL)
HOLES
MASS TRANSFER
MEMBRANES
OSMOSIS
PENETRATION
PERMEABILITY TESTING
PORES
POROSITY
SEEPAGE (RHEOLOGY)
SEMIPERMEABLE MEMBRANES
WATER REPELLENCY
WATERPROOFING

PERMEABILITY TESTERS
UF PERMEOMETERS
NT FRAZIER PERMEOMETER
GURLEY PERMEOMETER
BT TESTING EQUIPMENT
RT PERMEABILITY
PERMEABILITY TESTING

PERMEABILITY TESTING
UF PERMEABILITY TESTS
NT AIR PERMEABILITY TESTING
WATER RESISTANCE TESTING
BT TESTING
RT AIR PERMEABILITY
BURST TESTING
DROP PENETRATION TESTING
DYNAMIC ABSORPTION TESTING
FRAZIER PERMEOMETER
GURLEY PERMEOMETER
HOLES
HYDROSTATIC PRESSURE TESTING
IMPACT PENETRATION TESTING
PERMEABILITY
PERMEABILITY TESTERS
PORES
RAIN PENETRATION TESTING
SPRAY TESTING
STATIC ABSORPTION TESTING
WATER PERMEABILITY
WATER REPELLENCY

PERMEABILITY TESTS
USE PERMEABILITY TESTING

PERMEOMETERS
USE PERMEABILITY TESTERS

PERMUTITE (TN)
RT ION EXCHANGE RESINS
WATER TREATMENT

PEROXIDE BLEACHES
USE PEROXIDE BLEACHING AGENTS

PEROXIDE BLEACHING AGENTS
UF PEROXIDE BLEACHES
BT BLEACHING AGENTS
RT HYDROGEN PEROXIDE
PERACETIC ACID
PERBORATES
PERFORMIC ACID
OXIDIZING AGENTS
PEROXYGEN COMPOUNDS
PERSULFATES
SODIUM PERBORATE
SODIUM PEROXIDE

PEROXIDES
NT HYDROGEN PEROXIDE
SODIUM PEROXIDE
BT PEROXYGEN COMPOUNDS
RT BLEACHING
CHLORINE DIOXIDE
CHLORITES
HYPOCHLORITES
OXIDIZING AGENTS
STRIPPING (COLOR)

PEROXYGEN COMPOUNDS
NT PERACETIC ACID
PERBORATES
PERFORMIC ACID
PEROXIDES
PERSULFATES
RT BLEACHING
INITIATORS
OXIDES
OXIDIZING AGENTS
PEROXIDE BLEACHING AGENTS

PERSONNEL
UF LABOR
MAN POWER
RT JOB ANALYSIS
MANAGEMENT
PATROLLING
SAFETY
TIME AND MOTION STUDIES
TRAINING
WAGES
WORKLOADS

PERSPIRATION
RT BACTERIAL DEGRADATION
BACTERIAL INHIBITION
COLORFASTNESS
DEODORANTS
MICROBIOLOGICAL DEGRADATION
ODORS
PERSPIRATION RESISTANCE
STAIN RESISTANCE
STAINS

PERSPIRATION RESISTANCE
BT DEGRADATION PROPERTIES
END USE PROPERTIES
RT BIOCHEMICAL PROPERTIES
BACTERIAL DEGRADATION
BACTERIAL INHIBITION
COLORFASTNESS
DEODORANTS
MICROBIOLOGICAL DEGRADATION
ODORS
PERSPIRATION
STAIN RESISTANCE
STAINS

PERSULFATES
UF PERSULFURIC ACID
NT POTASSIUM PERSULFATE
SODIUM PERSULFATE
BT PEROXYGEN COMPOUNDS
SULFUR COMPOUNDS
RT ACRYLIC DYES
CATALYSTS
OXIDIZING AGENTS
PEROXIDE BLEACHING AGENTS
SULFATES
SULFIDES (INORGANIC)
SULFITES
THIOSULFATES

PERSULFURIC ACID
USE PERSULFATES

PETROLEUM SULFONATES
USE ALKYL SULFONATES

PH
UF HYDROGEN ION CONCENTRATION
RT ACID SOLUTIONS
ACIDITY
ALKALINE SOLUTIONS
ALKALINITY
HYDROGEN
IONIZATION
IONS
PH CONTROL
PH METERS

PH CONTROL
RT AUTOMATIC CONTROL
DYEING
PH
PH METERS
POTASSIUM CARBONATE
PROCESS CONTROL
SODIUM BICARBONATE
SODIUM CARBONATE
SODIUM SESQUICARBONATE
TSPP
WATER TREATMENT

PH METERS
BT MEASURING INSTRUMENTS
RT PH
PH CONTROL

PHASE BOUNDARY CROSSLINKING
BT WET FINISHING
RT CROSSLINKING
INTERFACIAL POLYMERIZATION

PHASE MICROSCOPY
BT MICROSCOPY
RT ELECTRON MICROSCOPY
INFRARED MICROSCOPY
INTERFERENCE MICROSCOPY
LIGHT MICROSCOPY (OPTICAL)
STEREOSCAN MICROSCOPY
ULTRAVIOLET MICROSCOPY
X RAY MICROSCOPY

PHENOL-FORMALDEHYDE RESINS
BT PHENOLIC RESINS
THERMOSETTING RESINS
RT PHENOLIC PLASTICS

PHENOLIC PLASTICS
BT PLASTICS
RT PHENOL-FORMALDEHYDE RESINS
PHENOLIC RESINS
POLYMERS

PHENOLIC RESINS
NT PHENOL-FORMALDEHYDE RESINS
BT RESINS
THERMOSETTING RESINS
RT EPOXY RESINS
MELAMINE RESINS
PHENOLIC PLASTICS
POLYESTER RESINS
SILICONE RESINS

PHENOLS
NT COPPER PENTACHLOROPHENATE
PENTACHLOROPHENYL LAURATE
SALICYLANILIDE
SALICYLATE SALTS
SALICYLIC ACID
RT ALCOHOLS
AROMATIC COMPOUNDS
CARRIERS (DYEING)
NAPHTHOLS
PRESERVATIVES
SOLVENTS
TANNIC ACID

PHENOMENOLOGICAL RHEOLOGY
BT RHEOLOGY
RT ELASTIC LIQUIDS
FIRMOVISCOSITY
MACRORHEOLOGY
METARHEOLOGY
MICRORHEOLOGY
NEWTONIAN LIQUIDS
PLASTICOELASTICITY
VISCOELASTICITY

PHENYLMERCURIC ACETATE
BT MERCURY COMPOUNDS
ORGANOMERCURY COMPOUNDS
RT ANTIMICROBIAL FINISHES
PHENYLMERCURIC
DIOCTYLSULFOSUCCINATE

PHENYLMERCURIC DIOCTYLSULFOSUCCINATE
BT MERCURY COMPOUNDS
ORGANOMERCURY COMPOUNDS
RT ANTIMICROBIAL FINISHES
PHENYLMERCURIC ACETATE

PHENYLMERCURIC SUCCINATE
BT MERCURY COMPOUNDS
ORGANOMERCURY COMPOUNDS
RT ANTIMICROBIAL FINISHES

PHOSPHATE SURFACTANTS
NT ALKYL ETHOXY PHOSPHATES
ALKYL PHOSPHATE SURFACTANTS
ALKYLARYL ETHOXY PHOSPHATES
GLYCIROPHOSPHATES
BT ANIONIC SURFACTANTS
RT PHOSPHONIC ACIDS
PHOSPHORUS COMPOUNDS

PHOSPHATED GLYCERIDES
USE GLYCEROPHOSPHATES

PHOSPHATES
NT CALGON (TN)
PYROPHOSPHATES
TETRAPOTASSIUM PYROPHOSPHATE
TRISODIUM PHOSPHATE
TSPP
BT PHOSPHORUS COMPOUNDS
RT BUILDERS (DETERGENTS)
CELLULOSE PHOSPHATE
PHOSPHONIC ACIDS
PHOSPHORIC ACID

PHOSPHITES
BT PHOSPHORUS COMPOUNDS
RT PHOSPHATES

PHOSPHONIC ACIDS
BT ORGANIC ACIDS
PHOSPHORUS COMPOUNDS
RT PHOSPHATE SURFACTANTS
PHOSPHATES
PHOSPHONIUM COMPOUNDS

PHOSPHONIUM COMPOUNDS
NT THPC
BT ONIUM COMPOUNDS
PHOSPHORUS COMPOUNDS
RT PHOSPHONIC ACIDS

PHOSPHONIUM SURFACTANTS
BT ONIUM SURFACTANTS
RT PHOSPHORUS COMPOUNDS
QUATERNARY AMMONIUM
SURFACTANTS
SULFONIUM SURFACTANTS

PHOSPHORIC ACID
BT INORGANIC ACIDS
PHOSPHORUS COMPOUNDS
TRIBASIC ACIDS
RT PHOSPHATES

PHOSPHORUS COMPOUNDS
NT APO
MAPO
PHOSPHATES
PHOSPHITES
PHOSPHONIC ACIDS
PHOSPHONIUM COMPOUNDS
PHOSPHORIC ACID
THPC
RT FIRE PROOFING AGENTS
FIRE RETARDANCY AGENTS
PHOSPHATE SURFACTANTS
PHOSPHONIUM SURFACTANTS

PHOTOCHEMICAL DEGRADATION
UF ACTINIC DEGRADATION
BT DEGRADATION
RT ANTIOXIDANTS
BACTERIAL DEGRADATION
DEGRADATION PROPERTIES
FABRIC PROPERTIES
(DEGRADATION)
INFRARED ABSORBERS
MICROBIOLOGICAL DEGRADATION
PHOTOCHEMICAL REACTIONS
PHYSICAL CHEMICAL PROPERTIES
LIGHTFASTNESS (OF FINISH)
PHOTOCHROMY
PHOTOTROPY
RADIATION
STABILIZERS (AGENTS)
THERMAL DEGRADATION
ULTRAVIOLET ABSORBERS
ULTRAVIOLET STABILIZERS

PHOTOCHEMICAL REACTIONS
BT REACTIONS (CHEMICAL)
RT FADING
IRRADIATION
LIGHTFASTNESS (COLOR)
LIGHTFASTNESS (OF FINISH)
PHOTOCHEMICAL DEGRADATION

PHOTOCHROMY
RT DYES
LIGHT
LIGHTFASTNESS (COLOR)
PHOTOCHEMICAL DEGRADATION
PHOTOTROPY

PHOTOELASTICITY
RT BIREFRINGENCE
DICHROISM
EXPERIMENTAL STRESS ANALYSIS
OPTICAL PROPERTIES
REFRACTIVE INDEX
STRAIN
STRESS
STRESS OPTICAL PROPERTIES

PHOTOELECTRIC CELLS
BT ELECTRONIC INSTRUMENTS
RT PHOTOELECTRIC COUNTERS (NEPS)
PHOTOELECTRIC SCANNING
PHOTOMETERS

PHOTOELECTRIC COUNTERS (NEPS)
BT COUNTERS
NEP COUNTERS
OPTICAL SCANNERS
RT NEP COUNT
PHOTOELECTRIC CELLS
PHOTOELECTRIC SCANNING

PHOTOELECTRIC SCANNING
BT OPTICAL SCANNING
SCANNING
RT AUTOCOUNT (TM)
COUNTING
FABRIC DEFECTS
IRREGULARITY
IRREGULARITY CONTROL
IRREGULARITY TESTING

RT PHOTOELECTRIC CELLS
PHOTOELECTRIC COUNTERS (NEPS)
SLUB CATCHING
WEB UNIFORMITY
YARN DEFECTS

PHOTOGRAPHY
RT MICROSCOPY
OPTICAL INSTRUMENTS
PHOTOELASTICITY
PHOTOMICROGRAPHS
RECORDING INSTRUMENTS
STROBOSCOPES

PHOTOMETERS
UF LIGHT METERS
NT SPECTROPHOTOMETERS
BT MEASURING INSTRUMENTS
OPTICAL INSTRUMENTS
RT COLORIMETERS
PHOTOELECTRIC CELLS
REFLECTOMETERS

PHOTOMICROGRAPHS
RT MICROSCOPY
PHOTOGRAPHY

PHOTONS
RT GAMMA RAYS
ILLUMINATION
LIGHT
RADIATION

PHOTOTROPY
RT COLOR
LIGHT
PHOTOCHEMICAL DEGRADATION
PHOTOCHROMY

PHTHALATE ESTERS
BT ESTERS
RT PHTHALIC ACID
PHTHALIC ANHYDRIDE
PLASTICIZERS

PHTHALIC ACID
BT CARBOXYLIC ACIDS
DIBASIC ACIDS
ORGANIC ACIDS
RT ISOPHTHALIC ACID
PHTHALATE ESTERS
PHTHALIC ANHYDRIDE
TEREPHTHALIC ACID

PHTHALIC ANHYDRIDE
BT ACID ANHYDRIDES
RT MALEIC ANHYDRIDE
PHTHALATE ESTERS
PHTHALIC ACID

PHTHALOCYANINE COMPOUNDS
RT DYES (BY CHEMICAL CLASSES)
PIGMENTS
PRINTING

PHYSICAL ANALYSIS
RT ANALYZING
CHEMICAL ANALYSIS
MATHEMATICAL ANALYSIS
QUANTITATIVE ANALYSIS
SPECTRUM ANALYSIS
STATISTICAL ANALYSIS
STRESS ANALYSIS
STRUCTURAL ANALYSIS
THERMAL ANALYSIS
X RAY ANALYSIS

PHYSICAL CHEMICAL PROPERTIES
NT ABSORBENCY (MATERIAL)
ACTIVATION ENERGY
BOILING POINT
DIFFUSION COEFFICIENT
FIRE RETARDANCY
FLASH POINT
FREEZING POINT
GLASS TRANSITION TEMPERATURE
HYDROPHILIC PROPERTY
HYDROPHOBIC PROPERTY
HYGROSCOPICITY
ISOELECTRIC POINT
ISOIONIC POINT
LATENT HEAT
MELTING POINT
MOLECULAR ATTRACTION
SECONDARY TRANSITION
TEMPERATURES
SOIL RESISTANCE
SOLUBILITY
SPECIFIC HEAT
SURFACE PROPERTIES (PHYSICAL
CHEMICAL)
TRANSFER PROPERTIES
RT ABSORPTION (MATERIAL)
AESTHETIC PROPERTIES
BIOCHEMICAL PROPERTIES
CHEMICAL PROPERTIES
CHEMISORPTION
DIPOLE MOMENT
(CONT.)

RT END USE PROPERTIES
FABRIC PROPERTIES
FABRIC PROPERTIES (PHYSICAL EXCLUDING MECHANICAL)
FIBER PROPERTIES
MECHANICAL PROPERTIES
PHYSICAL PROPERTIES (EXCLUDING MECHANICAL)
PHOTOCHEMICAL DEGRADATION
PROPERTIES
SORPTION OF GASES
SORPTION OF LIQUIDS
SORPTION OF WATER

PHYSICAL PROPERTIES (EXCLUDING MECHANICAL)
NT ACOUSTIC PROPERTIES
BULK
BULK DENSITY
CRIMP
ELECTRICAL PROPERTIES
ELECTRON SPIN RESONANCE
ENTHALPY
ENTROPY
FLASH POINT
FREEZING POINT
HEAT RESISTANCE
LATENT HEAT
MELTING POINT
NUCLEAR MAGNETIC RESONANCE
OPTICAL PROPERTIES
PACKING FACTOR
SOFTENING POINT
STRESS OPTICAL PROPERTIES
STRUCTURAL PROPERTIES
TEMPERATURE
THERMAL PROPERTIES
WARMTH
RT AESTHETIC PROPERTIES
CHARACTERISTICS
CHEMICAL PROPERTIES
COHESION
END USE PROPERTIES
FABRIC PROPERTIES
FABRIC PROPERTIES (PHYSICAL EXCLUDING MECHANICAL)
FIBER PROPERTIES
MATERIALS ENGINEERING
MECHANICAL PROPERTIES
PROPERTIES
YARN PROPERTIES

PHYSICAL PROPERTIES (MECHANICAL)
USE MECHANICAL PROPERTIES

PIANO MOTIONS
RT IRREGULARITY CONTROL
PICKER EVENERS
PICKER LAPS
PICKING (OPENING)
PROCESS CONTROL

PICK AND PICK
RT FILLING YARNS
LOOMS
PICKING (WEAVING)
SHUTTLE BOXES
WEAVING

PICK COUNTERS
BT COUNTERS
RT ENDS PER INCH
PICK GLASS
PICKS
PICKS PER INCH

PICK GLASS
RT ENDS PER INCH
FABRIC INSPECTION
PICK COUNTERS
PICKS PER INCH

PICK INSERTION
USE PICKING (WEAVING)

PICK SPACING
USE PICKS PER INCH

PICK STITCHES
BT STITCHES (SEWING)
RT PLAIN STITCHES (KNITTING)
SADDLE STITCHES

PICKER EVENERS
RT AUTOMATIC CONTROL
EVENERS
FEEDBACK CONTROL
IRREGULARITY CONTROL
PIANO MOTIONS
PICKER LAPS
PICKING (OPENING)
PROCESS CONTROL

PICKER LAPS
UF SCUTCHER LAPS
BT LAPS
RT COMBER LAPS
COTTON
COTTON CARDS
COTTON SYSTEM PROCESSING
LAP ARBOURS
LAP GUIDES
LAP RODS
LAP WASTE
MANDRELS
OPENING
PICKERS (OPENING)
PICKING (OPENING)
RIBBON LAPS

PICKER POINTS
RT MILANESE KNITTING MACHINES

PICKER SHOES
RT LOOMS
PICKERS (LOOM)
SHUTTLE BINDERS
SHUTTLE BOXES
SHUTTLES
STICK SHUTTLES

PICKER STICKS
RT BEATING UP
FILLING YARNS
LOOMS
LUG STRAPS
MISPICKS
PICKER STRAPS
PICKERS (LOOM)
PICKING (WEAVING)
PICKS
PICKS PER INCH
SHED (WEAVING)
SHEDDING
SHUTTLE BINDERS
SHUTTLE BOXES
SHUTTLES
STICK SHUTTLES
WEAVING

PICKER STRAPS
UF CHECK STRAPS
RT LOOMS
PICKER STICKS
PICKING (WEAVING)

PICKERS (OPENING)
RT AUTOMATIC CONTROL
BEATERS (OPENING)
CARDS
FEED LATTICES
GRID BARS
HOPPER FEED
MANDRELS
OPENERS
PIANO MOTIONS
PICKER EVENERS
PICKER LAPS
PICKING (OPENING)

PICKERS (LOOM)
RT LOOMS
PICKER SHOES
PICKER STICKS
PICKING (WEAVING)
WEAVING

PICKING (HARVESTING)
RT BOLLS
COTTON SYSTEM PROCESSING
GINNING

PICKING (OPENING)
UF SCUTCHING (PICKING)
BT COTTON SYSTEM PROCESSING
RT AMERICAN SYSTEM PROCESSING
BLENDING
CARDING
CLEANING
CONTINENTAL SYSTEM PROCESSING
ENTANGLEMENTS
FEED ROLLS
FIBER BREAKAGE
GARNETTING
GRID BARS
HOPPER FEED
IMPURITIES
NEPS
OPENING
PICKER LAPS
PICKERS (OPENING)
TRASH
WASTE
WOOLEN SYSTEM PROCESSING

PICKING (SNAGGING)
RT CATCHING
PICKING RESISTANCE
PUNCTURING
RIPPING
RUN RESISTANCE
RUNS (KNITTING)
SNAG RESISTANCE
SNAG TESTERS
SNAGGING
SNAGS
STRESS CONCENTRATION

PICKING (WEAVING)
UF PICK INSERTION
PICKING MOTIONS
WEFT INSERTION
RT BEATING UP
BLANKET SHUTTLES
BOAT SHUTTLES
BOUNCE (WEAVING)
CLEANLINESS
CLOSED SHED
CLOTH FELL
FILLING YARNS
HARNESSES
JAMMING
JAMMING POINT
LOOMS
MISPICKS
NEGATIVE SHEDDING
OPEN SHED
PICK AND PICK
PICKER STICKS
PICKER STRAPS
PICKS
PICKS PER INCH
POSITIVE SHEDDING
RACE BOARDS
REEDS
ROTARY MAGAZINES
SHED (WEAVING)
SHEDDING
SHUTTLE BINDERS
SHUTTLE BOXES
SHUTTLE DEFLECTORS
SHUTTLE MARKS
SHUTTLE SMASH PROTECTORS
SHUTTLELESS LOOMS
SHUTTLES
SLEYS (LOOM)
SPLIT SHED
STICK SHUTTLES
TAKE UP
THREAD CUTTERS
WEAVE
WEAVING

PICKING MOTIONS
USE PICKING (WEAVING)

PICKING RESISTANCE
RT DURABILITY
PICKING (SNAGGING)
PUNCTURE RESISTANCE
RUN RESISTANCE
SNAG RESISTANCE
SNAGGING

PICKING TONGS
USE BURLING MACHINES

PICKS
RT BEATING UP
DOUBLE PICKS
FILLING YARNS
OPEN SHED
PICKER STICKS
PICKING (WEAVING)
PICKS PER INCH
QUILLS
SHUTTLES
WARP ENDS

PICKS PER INCH
UF PICK SPACING
BT FABRIC PROPERTIES
FABRIC PROPERTIES (STRUCTURAL)
RT BALANCED WEAVES
BEATING UP
COURSES PER INCH
COVER FACTOR
CRIMP INTERCHANGE
DENSITY
DESIGN
ENDS PER INCH
FABRIC ANALYSIS
FABRIC DESIGN
FABRIC GEOMETRY
FABRIC STRUCTURE
FILLING YARNS
JAMMING
JAMMING POINT
LET OFF MOTIONS
LOOM TAKE UP
PICK COUNTERS
PICK GLASS
PICKER STICKS
PICKING (WEAVING)
PICKS
PROCESS VARIABLES
SET (WOVEN FABRIC)
SHUTTLES
TAKE UP
TIGHTNESS

PICOETTA FAGOTING (STITCH)
BT STITCHES (SEWING)

PIECE DYED FABRICS
BT DYED FABRICS

RT DOPE DYEING
FABRICS
PIECE DYEING
PRINTED FABRICS
STOCK DYEING
YARN DYED FABRICS
YARN DYEING

PICOT
RT HOSIERY
LOOPS
KNITTED FABRICS
KNITTED STITCHES

PIECE DYEING
NT BECK DYEING
CARPET DYEING
HOSIERY DYEING
JIG DYEING
OPEN WIDTH DYEING
ROPE DYEING
BT DYEING (BY MATERIAL ASSEMBLY)
RT BATCH DYEING
CONTINUOUS DYEING
CROSS DYEING
DOPE DYEING
FABRICS
GARMENT DYEING
PIECE DYED FABRICS
PIECE DYEING MACHINES
SOLID COLORS
STOCK DYEING
TONE IN TONE DYEING
TOP DYEING
TOW DYEING
UNION DYEING
YARN DYED FABRICS
YARN DYEING

PIECE DYEING MACHINES
NT BAROTOR MACHINES
BEAM DYEING MACHINES (FABRIC)
BECKS
JIGS (DYEING)
KETTLES (DYEING)
BT DYEING MACHINES (BY MATERIAL ASSEMBLY)
DYEING MACHINES
RT BATCH DYEING MACHINES
CONTINUOUS DYEING RANGE
DOPE DYEING EQUIPMENT
DYEING
DYEING (BY MATERIAL ASSEMBLY)
FABRICS
GARMENT DYEING MACHINES
PACKAGE DYEING MACHINES
PADDERS
PIECE DYEING
PRESSURE DYEING MACHINES
REEL MACHINE (DYEING)
STEAMERS
STOCK DYEING MACHINES
TOP DYEING MACHINES
TOW DYEING MACHINES
YARN DYEING MACHINES

PIECED UP SECTIONS
USE PIECING

PIECENINGS
USE PIECING

PIECING
UF PIECED UP SECTIONS
PIECENINGS
RT AUTOMATIC KNOTTING
BREAKS
END BREAKAGES
JOINING
KNOTS
KNOTTING
SPINNING
SPLICING (KNITTING)
WARP ENDS
WINDING

PIGMENT DISPERSIONS
RT DISPERSE SYSTEMS
PIGMENT DYEING
PIGMENTS

PIGMENT DYEING
BT DYEING (BY DYE CLASSES)
RT ACID DYEING
AZOIC DYEING
BASIC DYEING
BINDERS (PIGMENTS)
DIRECT DYEING
DISPERSE DYEING
METALLIZED DYEING
MORDANT DYEING
NEUTRAL DYES
PIGMENT DISPERSIONS
PIGMENT PRINTING
PIGMENTS
REACTIVE DYEING
SULFUR DYEING
THINNERS
VAT DYEING

PIGMENT PRINTING
BT PRINTING
RT BINDERS (PIGMENTS)
EMULSION PRINTING
PIGMENT DYEING
PIGMENTS
REACTIVE PRINTING
VAT PRINTING
ZINC OXIDE

PIGMENTS
NT ULTRAMARINE BLUE
BT COLORANTS
RT ALUMINA
DYEING
DYES
GRINDING (COMMINUTION)
INKS
LACQUERS
PAINTS
PHTHALOCYANINE COMPOUNDS
PIGMENT DISPERSIONS
PIGMENT DYEING
PIGMENT PRINTING
POWDERING
TITANIUM COMPOUNDS
TITANIUM DIOXIDE
ZINC OXIDE

PIGSKIN
BT LEATHER
RT SKIN
SYNTHETIC LEATHER

PIGTAILS (GUIDES)
USE THREAD GUIDES

PILE
RT FLOCK
KNITTED PILE FABRICS
PILE DENSITY
PILE FABRICS (FLOCKED)
PILE FABRICS (TUFTED)
PILE FABRICS (WOVEN)
PILE HEIGHT

PILE DENSITY
RT CUT PILE FABRICS
LOOP PILE FABRICS
PILE
PILE FABRICS
PILE FABRICS (WOVEN)
PILE HEIGHT

PILE FABRIC LOOMS
BT LOOMS
RT AUTOMATIC LOOMS
BOX LOOMS
DOBBY LOOMS
HAND LOOMS
JACQUARD LOOMS
LAPPET LOOMS
PILE FABRICS
PILE FABRICS (WOVEN)
SHUTTLELESS LOOMS

PILE FABRICS
NT CURL PILE FABRICS
CUT PILE FABRICS
KNITTED PILE FABRICS
LOOP PILE FABRICS
PILE FABRICS (FLOCKED)
PILE FABRICS (TUFTED)
PILE FABRICS (WOVEN)
SCULPTURED PILE FABRICS
STITCHED PILE FABRICS
BT FABRICS
FABRICS (ACCORDING TO STRUCTURE)
RT NONWOVEN FABRICS
PILE
PILE DENSITY
PILE HEIGHT

PILE FABRICS (FLOCKED)
NT FLOCKED CARPETS
FLOCKED FABRICS
BT PILE FABRICS
RT FLOCK
FLOCKING
PILE

PILE FABRICS (KNITTED)
USE KNITTED PILE FABRICS

PILE FABRICS (TUFTED)
NT TUFTED CARPETS
TUFTED FABRICS
BT PILE FABRICS
RT PILE
PILE DENSITY
STITCHED PILE FABRICS
TUFTING
TUFTING MACHINES

PILE FABRICS (WOVEN)
UF FABRICS (PILE)
NT ARTIFICIAL FURS
AXMINSTER CARPETS
CARPETS
CHENILLE CARPETS
CORDUROY
CURL PILE FABRICS
CUT PILE FABRICS
INGRAIN CARPETS
LOOP PILE FABRICS
PLUSH
TERRYCLOTH
TUFTED CARPETS
TUFTED FABRICS
UNCUT PILE FABRICS
VELVET
VELVETEEN
WILTON CARPETS
WOVEN CARPETS
BT PILE FABRICS
WOVEN FABRICS
RT BACKED FABRICS
CHINCHILLA MACHINES
FABRIC STRUCTURE
FILLING KNITTED FABRICS
FLAT WOVEN FABRICS
FLOCK
MULTIPLE LAYER FABRICS
PILE
PILE DENSITY
PILE HEIGHT
SLIVER KNITTING
TUFTING
TUFTING MACHINES
VELOUR

PILE HEIGHT
UF PILE LENGTH
RT CUT PILE FABRICS
LOOP PILE FABRICS
PILE
PILE DENSITY
PILE FABRICS
PILE FABRICS (WOVEN)

PILE LENGTH
USE PILE HEIGHT

PILERS
RT FOLDING MACHINES (FOR CLOTH)
MATERIALS HANDLING

PILGRIMSTEP (FULLING MACHINES)
BT FULLING MACHINES
RT FULLING
GATTERWALKING (FULLING MACHINES)
MILLS (FULLING ROTARY)
MILLS (KICKER)
OPEN WIDTH FINISHERS (FULLING)
PLANKING (FULLING MACHINES)
REFULLING MACHINES

PILL RESISTANCE
BT FABRIC PROPERTIES
FABRIC PROPERTIES (MECHANICAL)
MECHANICAL PROPERTIES
SURFACE PROPERTIES (MECHANICAL)
RT BENDING RIGIDITY
FABRIC DEFECTS
FIBER FIBER FRICTION
FIBER MIGRATION
FUZZ RESISTANCE
KNITTED FABRICS
LOOP EFFICIENCY
LOOP STRENGTH
NEP POTENTIAL
PILL RESISTANCE FINISHES
PILL RESISTANCE TREATMENTS
PILL WEAR OFF
PILLING
PILLING TESTING
PILLS
SHEDDING (FIBER)
TENACITY
TORSIONAL RIGIDITY

PILL RESISTANCE FINISHES
BT FINISH (SUBSTANCE ADDED)
RT PILL RESISTANCE
PILL RESISTANCE TREATMENTS
PILL WEAR OFF
PILLING
PILLING TESTING
PILLS

PILL RESISTANCE TREATMENTS
BT DRY FINISHING
WET FINISHING
RT PILL RESISTANCE
PILL RESISTANCE FINISHES
PILLING
PILLING TESTING

PILL WEAR OFF
RT ABRASION RESISTANCE
FUZZ RESISTANCE
LOOP STRENGTH
SHEDDING (FIBER)

PILLING
RT BENDING RIGIDITY
ENTANGLEMENTS
FABRIC DEFECTS
FABRIC TO FABRIC FRICTION
FIBER FIBER FRICTION
FIBER MIGRATION
FUZZ RESISTANCE
FUZZING
HAIRINESS
KNITTED FABRICS
LAUNDERING
LINT (WASTE)
LINT CATCHERS
LOOP EFFICIENCY
LOOP STRENGTH
MATRIX FREEDOM
MATTING
NEP POTENTIAL
NEPS
PILL RESISTANCE
PILL RESISTANCE FINISHES
PILL RESISTANCE TREATMENTS
PILL WEAR OFF
PILLING TESTING
PILLS
RUBBING
SHEDDING (FIBER)
SNARLING
SNARLING TENDENCY
TENACITY
TORSIONAL RIGIDITY

PILLING TESTING
BT TESTING
RT CREEP TESTING
CUTTING TESTING
IMPACT PENETRATION TESTING
IMPACT TESTING
PILL RESISTANCE
PILL RESISTANCE FINISHES
PILL RESISTANCE TREATMENTS
PILLING
SNAG TESTING
TEAR TESTING
WEAR TESTING

PILLOW LACES (BOBBIN LACES)
NT ALEC
ANTIQUE (LACE)
BATTENBERG
CHANTILLY (LACE)
DRESDEN POINT
LILLE (LACE)
RENAISSANCE (LACE)
ROSE POINT
VALENCIENNES
BT HAND MADE LACES
LACES

PILLOW TUBING
BT FLAT WOVEN FABRICS
TUBULAR FABRICS
WOVEN FABRICS
RT COTTON
DOUBLE CLOTH
PILLOWCASES
PILLOWS
PLAIN WEAVES

PILLOWCASES
RT HOUSEHOLD FABRICS
PILLOW TUBING
PILLOWS

PILLOWS
RT BEDSPREADS
BLANKETS
EIDERDOWNS
HOUSEHOLD FABRICS
KAPOK
PILLOW TUBING
PILLOWCASES

PILLS
BT DEFECTS
FABRIC DEFECTS
RT ABRASION
ENTANGLEMENTS
FELTING
FIBER MIGRATION
FUZZ
FUZZ RESISTANCE
HAIRINESS
KNITTED FABRICS
NEPS
PILL RESISTANCE
PILL WEAR OFF
PILLING
SLENDERNESS RATIO
SNARLING
SNARLING TENDENCY

PIMA COTTON
BT COTTON

PIN DRAFTERS
RT DRAFTING MACHINES
DRAFTING (STAPLE FIBER)
GILL BOXES
PIN DRAFTING

PIN DRAFTING
BT DRAFTING (STAPLE FIBER)
WORSTED SYSTEM PROCESSING
RT BRADFORD SYSTEM PROCESSING
FINISHER GILLING
GILLING
PIN DRAFTERS

PIN POINT SADDLE STITCHES
BT STITCHES (SEWING)

PIN TACKS
RT TACKING

PIN TENTERS
NT OVERFED PIN TENTERS
BT TENTER FRAMES
RT DRY FINISHING
TENTER CLIPS
TENTER PINS
TENTERING
WET FINISHING

PINCHING
BT FABRIC DISTORTION (TAILORING)
RT BACK TACKING
BASTING
CHAIN SEWING
COCKLING
CURVED SEAMING
GATHERING
HEMMING
PUCKERING
RIPPLES
RUFFLING
SEAM PUCKER
SEAMING
SEAMS
SERGING
SEWING
SHIRRING
STITCHES (SEWING)
STITCHING
TAILORING
TRIMMING (OPERATION)
YARN SEVERANCE

PINKED SEAMS
BT SEAMS
RT BUTTED SEAMS
BUTTERFLY SEAMS
COVERED SEAMS
FELLED SEAMS
PINKING (TAILORING)
SEWING

PINKING (TAILORING)
BT GARMENT MANUFACTURE
RT CUTTING (TAILORING)
PINKED SEAMS
PINKING MACHINES
PINKING SHEARS
SEAMING
SEAMS
SEWING
STITCHES (SEWING)
TAILORING

PINKING MACHINES
BT GARMENT MANUFACTURING MACHINES
RT GARMENT MANUFACTURE
PINKING (TAILORING)

PINKING SHEARS
RT PINKING (TAILORING)

PIPE COVERINGS
RT INDUSTRIAL FABRICS
INSULATION (THERMAL)

PIPED SEAMS
BT SEAMS
RT BUTTED SEAMS
BUTTERFLY SEAMS
COVERED SEAMS
FRENCH SEAMS
SEWING

PIPES
UF HOSES (PIPE)
RT REINFORCED HOSES

PIPING
RT CUT RUCHE
FINDINGS
NARROW FABRICS
RUCHE
TRIMMINGS

PIQUE
BT FLAT WOVEN FABRICS
WOVEN FABRICS
RT BEDFORD CORD
BIRDS EYE WEAVES

PIQUE WEAVES
BT WEAVE (TYPES)
RT FLAT WOVEN FABRICS

PIRN BARRE
BT BARRE
FABRIC DEFECTS
RT SHINERS

PIRN DENSITY
BT PACKAGE DENSITY
RT HARDNESS
LAYER LOCKING
PACKAGES
SLOUGHING

PIRN WINDERS
USE QUILLERS

PIRN WINDING
USE QUILLING

PIRNS
USE QUILLS

PISTONS
RT ACCESSORIES
FLUID FLOW
MECHANISMS
PRESSURE

PLAIN BEARINGS
BT BEARINGS
RT AIR BEARINGS
BALL BEARINGS
CONICAL BEARINGS
FOOTSTEP BEARINGS
JOURNAL BEARINGS
NEEDLE BEARINGS
NYLON BEARINGS
ROLLER BEARINGS
SLEEVE BEARINGS
TEFLON BEARINGS
THRUST BEARINGS

PLAIN KNITTED FABRICS
UF BALBRIGGAN
JERSEY (FILLING KNITTING)
STOCKINETTE
BT FILLING KNITTED FABRICS
KNITTED FABRICS
RT CIRCULAR KNITTED FABRICS
CIRCULAR KNITTING
CURL
DOUBLE KNIT FABRICS
FLAT KNITTED FABRICS
FLAT KNITTING
FULLY FASHIONED FABRICS
LADDERING
MACHINE KNITTING
PLAIN KNITTING
PURL KNITTED FABRICS
RIB KNITTED FABRICS
RUN RESISTANCE
SPIRALITY (KNITTED FABRICS)

PLAIN KNITTING
BT KNITTING
RT INTERLOCK (KNITTED FABRICS)
PLAIN KNITTED FABRICS
PURL KNITTING
RIB KNITTING

PLAIN NET MACHINES
USE BOBBINET MACHINES

PLAIN ROLLS
BT DRAFTING ROLLS
ROLLS
RT FLUTED ROLLS

PLAIN STITCHES (KNITTING)
BT KNITTED STITCHES
RT FLOAT STITCHES (KNITTING)
PURL STITCHES

PLAIN WEAVES
BT WEAVE (TYPES)
RT ANGOLA (FABRIC)
AWNING CLOTH
BALANCED WEAVES
BEDFORD CORD
BROADCLOTH
BUCKRAM
BUNTING
CALICO
CAMBRIC
CASEMENT
CHAMBRAY
CHEESE CLOTH
CHIFFON
CHINTZ
CRASH TOWELING
CREPE
CRETONNE
DIMITY
DUCK
EMBOSSED CREPE

RT FAILLE
FLANNEL
FLAT WOVEN FABRICS
GAUZE
GEORGETTE
GINGHAM
HESSIAN
HIMALAYA CLOTH
HOPSACK
IMITATION GINGHAM
INDIA LINON
LAWN
LONGCLOTH
MARQUISETTE
MUSLIN
NETTING
OSNABURG
PERCALE
PILLOW TUBING
PONGEE
POPLIN
PRINT CLOTH
REPP
RIBBON
SCREENING
SCRIM
SEERSUCKER
SERGE
SHADOW STRIPE
SHARKSKIN
SHEETING (FABRIC)
SILESIA
SPLASH VOILE
TICKING
TOBACCO CLOTH
TROPICAL WORSTED
UMBRELLA CLOTH
VELOUR
VOILE
WARP REPP

PLAITING (FOLDING)
USE FOLDING (OF CLOTH)

PLAITING (BRAIDING)
USE BRAIDING

PLAITING (KNITTING)
RT FACE YARNS
KNITTING
REVERSE PLAITING

PLAITING MACHINES (BRAIDING)
USE BRAIDERS

PLAITING MACHINES (FOLDING)
USE FOLDING MACHINES (FOR CLOTH)

PLANE MOTION
BT MOTION
RT AXIAL MOTION
PARALLEL MOTION

PLANKING (FULLING MACHINES)
BT FULLING MACHINES
RT FULLING
GATTERWALKING (FULLING MACHINES)
MILLS (FULLING ROTARY)
MILLS (KICKER)
OPEN WIDTH FINISHERS (FULLING)
PILGRIMSTEP (FULLING MACHINES)
REFULLING MACHINES

PLANNING
RT CAPACITY (GENERAL)
MANAGEMENT
PRODUCTION
SALES

PLANT
RT COSTS
EQUIPMENT
FACILITIES
INVENTORIES
INVESTMENTS
JOB ANALYSIS
MACHINERY
MANAGEMENT
MANUFACTURING
MATERIALS HANDLING
PLANT LAYOUT
PROCESS DESIGN
PROCESSES
PRODUCTION
TEXTILE MACHINERY (GENERAL)
TEXTILE PROCESSES (GENERAL)

PLANT LAYOUT
RT FINISHING MACHINERY (GENERAL)
MACHINERY
MANAGEMENT
MATERIALS HANDLING
PLANT
TEXTILE MACHINERY (GENERAL)

PLASTIC DEFORMATION
USE PLASTIC STRAIN

PLASTIC FLOW
UF FLOW (PLASTIC)
BT STRESS STRAIN PROPERTIES
STRESS STRAIN PROPERTIES (TENSILE)
RT COEFFICIENT OF FRICTION
COLD DRAWING
CREEP
DELAYED RECOVERY
DRAWTWISTING
DRAWING (FILAMENT)
FRICTION
HOT DRAWING
MECHANICAL DETERIORATION PROPERTIES
NECKING (FILAMENT)
PERMANENT DEFORMATION
PERMANENT SET
RELAXATION TIME (MECHANICAL)
RHEOLOGY
SOFTENING POINT
STRESS STRAIN CURVES
VISCOELASTICITY
VISCOUS FLOW
YIELD (MECHANICAL)
YIELD POINT

PLASTIC REGION
USE PLASTIC STRAIN

PLASTIC STRAIN
UF PLASTIC DEFORMATION
PLASTIC REGION
BT STRAIN
RT BREAKING ELONGATION
COLD DRAWING
CREEP
DEFORMATION
DELAYED RECOVERY
DIMENSIONAL STABILITY
ELASTIC STRAIN
ELASTICITY
EXTENSIBILITY
HOT DRAWING
MOLECULAR ORIENTATION
NECKING (FILAMENT)
PERMANENT SET
STRESS STRAIN CURVES
STRETCH
UNIAXIAL STRESS
YIELD (MECHANICAL)
YIELD POINT

PLASTICITY
BT STRESS STRAIN PROPERTIES
STRESS STRAIN PROPERTIES (TENSILE)
RT BRITTLENESS
DRAWING (FILAMENT)
DUCTILITY
ELASTICITY
FIRMOVISCOSITY
HARDNESS
LINEAR VISCOELASTICITY
NECKING (FILAMENT)
NONLINEAR VISCOELASTICITY
PLASTICOVISCOSITY
PLASTICOVISCOELASTICITY
SOFTENING POINT
STRENGTH
STRENGTH OF MATERIALS
THERMOPLASTICITY

PLASTICIZERS
RT ACETYLTRIETHYL CITRATE
COATINGS (SUBSTANCES)
DISPERSING AGENTS
PHTHALATE ESTERS
SOLVENTS
SURFACTANTS

PLASTICOELASTICITY
BT STRESS STRAIN PROPERTIES
RT PHENOMENOLOGICAL RHEOLOGY
VISCOELASTICITY
VISCOSITY

PLASTICOVISCOELASTICITY
BT STRESS STRAIN PROPERTIES
RT ELASTICITY
PLASTICITY
PLASTICOVISCOSITY
VISCOELASTICITY
VISCOSITY

PLASTICOVISCOSITY
UF BINGHAM BODY
BT STRESS STRAIN PROPERTIES
RT ELASTICITY
FIRMOVISCOSITY
LINEAR VISCOELASTICITY
PLASTICITY
RHEOLOGY
VISCOELASTICITY
VISCOSITY

PLASTICS
NT PHENOLIC PLASTICS
RT COATED FABRICS
COPOLYMERS
FIBER REINFORCED COMPOSITES
FILAMENT WOUND COMPOSITES
HEAT SEALING
LAMINATED FABRICS
POLYMERS
POLYSTYRENE
POST FORMED LAMINATES
REINFORCED COMPOSITES
RESINS

PLATEN HARDENERS
BT HARDENERS (FELTS)
RT FELTING
FELTING MACHINES
FELTS
ROLLER HARDENERS

PLATINUM
BT METALS

PLAYSUITS
BT GARMENTS
SPORTSWEAR
RT CHILDRENS WEAR

PLEATING
BT DRY FINISHING
GARMENT MANUFACTURE
RT BENDING
CREASING
PERMANENT PLEATING
PRESSING
TAILORING
WRINKLING

PLEATING MACHINES
BT GARMENT MANUFACTURING MACHINES
RT PLEATING
PLEATS

PLEATS
BT GARMENT COMPONENTS
RT CREASES
GATHERS (TAILORING)
PLEATING
PLEATING MACHINES
WRINKLES

PLIED YARNS
UF CABLED YARNS
DOUBLED YARNS
FOLDED YARNS
NT MULTIPLIED YARNS
TWO PLY YARNS
BT YARNS
RT BALANCED YARNS
BRAIDING
CABLES
CABLING
COARSE YARNS
CORE SPUN YARNS
CORDS
FINE YARNS
HAWSERS
NO TORQUE YARNS
NOVELTY TWISTERS
NOVELTY YARNS
PLY TWIST
PLYING
RING TWISTERS
ROPES
S TWIST
S TWIST YARNS
SINGLES YARNS
SPIRALS (NOVELTY YARNS)
TIRE CORDS
TWIST LIVELINESS
TWISTERS
TWISTING
WIRE ROPES
YARN GEOMETRY
YARN STRUCTURE
Z TWIST
Z TWIST YARNS

PLISSE
BT FLAT WOVEN FABRICS
WOVEN FABRICS

PLOUGH KNIVES
RT NOBLE COMBS

PLOUGHING
RT ABRASION
COEFFICIENT OF FRICTION
CUTTING
FRICTION
FRICTION THEORY
GRINDING (MATERIAL REMOVAL)
LUBRICATION
SLICING
WEAR
WEDGING

PLUNGER CAMS
USE BOLT CAMS (KNITTING)

PLURONICS (TN)
BT POLYETHOXY POLYPROPOXY SURFACTANTS
RT BLOCK POLYMERS
POLY (ETHYLENE GLYCOL)
POLY (PROPYLENE GLYCOL)
POLYOXYALKYL COMPOUNDS

PLUSH
BT KNITTED PILE FABRICS
PILE FABRICS (WOVEN)
WOVEN FABRICS
RT PILE
PILE DENSITY
PILE HEIGHT

PLY TWIST
BT TWIST
RT BALANCED YARNS
HIGH TWIST
LOW TWIST
MULTIPLIED YARNS
NOVELTY YARNS
PLIED YARNS
PLYING
REAL TWIST
S TWIST
SINGLES TWIST
TWIST LIVELINESS
TWISTING
TWO PLY YARNS
YARNS
Z TWIST

PLY TWISTERS
BT TWISTERS
RT NOVELTY TWISTERS

PLYING
UF DOUBLING (TWISTING)
FOLDING (TWISTING)
RT BALANCED YARNS
BRAIDING
CABLES
CABLING
CORDAGE
CORDS
COTTON SYSTEM PROCESSING
DOWNTWISTING
HAWSERS
LUBRICATED RINGS
MULTIPLIED YARNS
NO TORQUE YARNS
NOVELTY TWISTERS
PLIED YARNS
PLY TWIST
RING TWISTERS
RINGS
S TWIST
SINGLES YARNS
SINTERED METAL RINGS
SPINNING
THROWING
TWIST
TWIST LIVELINESS
TWISTERS
TWISTING
TWO PLY YARNS
UNLUBRICATED RINGS
Z TWIST

PLYING TRAVELERS
UF DOUBLING TRAVELERS
BT TRAVELERS
RT CIRCULAR TRAVELERS
ELLIPTICAL TRAVELERS
NYLON TRAVELERS
RING SPINNING
RINGS
SPINNING
SPINNING TRAVELERS
STEEL TRAVELERS
TWISTING

PNEUMAFIL (TN)
BT BROKEN END COLLECTORS
RT AERODYNAMICS
DRAFTING ROLLS
END BREAKAGES
VACUUM END COLLECTORS

PNEUMATIC BROKEN END COLLECTORS
USE VACUUM END COLLECTORS

PNEUMATIC CONTROL
RT AIR FLOW
AUTOMATIC CONTROL
CONTROL SYSTEMS
HYDRAULIC CYLINDERS
PNEUMATIC INSTRUMENTS
PNEUMATIC LAP WEIGHT CONTROL
PRESSURE
PROCESS CONTROL
SERVOMECHANISMS

PNEUMATIC INSTRUMENTS
BT CONTROL EQUIPMENT
RT AIR
AIR FLOW
AUTOMATIC CONTROL
INSTRUMENTATION
MEASURING INSTRUMENTS
PNEUMATIC IRREGULARITY TESTERS
PRESSURE
PROCESS CONTROL
RECORDING INSTRUMENTS
REMOTE CONTROL

PNEUMATIC IRREGULARITY TESTERS
BT INSTRUMENTATION
IRREGULARITY TESTERS
RT AIR FLOW
CAPACITANCE TYPE IRREGULARITY TESTERS
FIELDEN WALKER IRREGULARITY TESTER (TN)
IRREGULARITY
QUALITY CONTROL
PNEUMATIC INSTRUMENTS
PRESSURE
ROVINGS
SLIVERS
USTER IRREGULARITY TESTER (TN)
VARIANCE LENGTH CURVES
YARNS
ZELLWEGER TESTER (TN)

PNEUMATIC LAP WEIGHT CONTROL
RT IRREGULARITY CONTROL
PNEUMATIC CONTROL
PNEUMATIC INSTRUMENTS

POCKET LININGS
BT LININGS
RT BOX LININGS
CASKET LININGS
DRESS BANDINGS
FABRICS
GARMENTS
GLOVE LININGS
INTERLININGS
POCKETS
SHOE LININGS

POCKETS
BT GARMENT COMPONENTS
RT TAILORING

POINT DE PARIS
BT HAND MADE LACES
LACES

POINT DESPRIT
BT HAND MADE LACES
LACES

POISONOUS
USE TOXICITY

POISSON DISTRIBUTION
RT BINOMIAL DISTRIBUTION
CHI SQUARED DISTRIBUTION
DISTRIBUTION
F DISTRIBUTION
NORMAL DISTRIBUTION
T DISTRIBUTION

POISSONS RATIO
RT BIAXIAL STRESS
COMPRESSIVE MODULUS
ELASTIC MODULUS (TENSILE)
ELASTIC STRAIN
MODULUS
PLASTIC STRAIN
SHEAR MODULUS
TRELLIS MODEL
UNIAXIAL STRESS

POLAR COMPOUNDS
RT DIPOLE MOMENT
DIPOLES
HYDROGEN BONDS
ORGANIC COMPOUNDS (GENERAL)
POLARITY

POLAR MOMENT OF INERTIA
RT MOMENT OF INERTIA
TORSION (GEOMETRIC)
TORSIONAL ENERGY
TORSIONAL RIGIDITY
TWIST
TWISTING MOMENTS

POLARITY
RT CHARGE (ELECTRICAL)
DIELECTRIC PROPERTIES
DIPOLE MOMENT
DIPOLES
POLAR COMPOUNDS

POLARIZED LIGHT
BT LIGHT
RT ARTIFICIAL DAYLIGHT
BIREFRINGENCE
DAYLIGHT
FLUORESCENT LIGHT
INCANDESCENT LIGHT
INTERFERENCE FRINGES
MONOCHROMATIC LIGHT
OPTICAL INSTRUMENTS
OPTICS
REFRACTIVE INDEX
SPECTROPHOTOMETRY
STRESS OPTICAL PROPERTIES
SUNLIGHT

POLARIZED LIGHT METHOD (COTTON IMMATURITY)
BT IMMATURITY TESTING (COTTON FIBER)
RT CROSS SECTIONAL SHAPE METHOD (COTTON IMMATURITY)
DIFFERENTIAL DYEING METHOD (COTTON IMMATURITY)
SWELLING METHOD (COTTON IMMATURITY)

POLARIZING MICROSCOPE
RT ELECTRON MICROSCOPY
INFRARED MICROSCOPY
INTERFERENCE MICROSCOPY
LIGHT MICROSCOPY (OPTICAL)
MICROSCOPY
STEREOSCAN MICROSCOPY
ULTRAVIOLET MICROSCOPY
X RAY MICROSCOPY

POLAROGRAPHY
BT CHEMICAL ANALYSIS

POLIAFIL (TN)
BT NYLON (POLYAMIDE FIBERS)
NYLON 66

POLISHING
RT ABRASION
ABRASIVES
BRUSHING
CALENDERING
CLEANING
CUTTING
EMBOSSING
ENGRAVING
FRICTION CALENDERING
GALLING
GRINDING (MATERIAL REMOVAL)
PUMICE
SCORING
SCRATCHING
SHARPENING

POLLUTION
NT AIR POLLUTION
WATER POLLUTION
RT CONTAMINATION
EFFLUENTS
WASTE DISPOSAL
WASTE TREATMENT

POLY (ETHYL ACRYLATE)
BT POLYACRYLATES
POLYVINYLS
RT ACRYLIC COMPOUNDS
ETHYL ACRYLATE
POLY (METHYL ACRYLATE)
POLY (METHYL METHACRYLATE)

POLY (ETHYLENE GLYCOL TEREPHTHALATE)
BT POLYESTERS
RT DACRON (TN)
MYLAR (TN)
TEREPHTHALIC ACID
TERON (TN)

POLY (ETHYLENE GLYCOL)
BT GLYCOLS
RT ETHYLENE GLYCOL
ETHYLENE OXIDE
PLURONICS (TN)
POLY (ETHYLENE OXIDE)
POLYETHERS
POLYOXYETHYLENE

POLY (ETHYLENE OXIDE)
BT POLYETHERS
RT ETHYLENE OXIDE
POLY (ETHYLENE GLYCOL)

POLY (HEXAMETHYLENE ADIPAMIDE)
BT POLYAMIDES
RT ADIPIC ACID
HEXAMETHYLENEDIAMINE
NYLON 66

POLY (HEXAMETHYLENE SEBACAMIDE)
BT POLYAMIDES
RT HEXAMETHYLENEDIAMINE
NYLON 610

POLY (HEXAMETHYLENE TEREPHTHALAMIDE)
BT POLYAMIDES
RT HEXAMETHYLENEDIAMINE
TEREPHTHALIC ACID

POLY (M-XYLYLENE ADIPAMIDE)
BT POLYAMIDES
RT ADIPIC ACID

POLY (METHYL ACRYLATE)
BT POLYACRYLATES
POLYVINYLS
RT METHYL ACRYLATE
POLY (ETHYL ACRYLATE)
POLY (METHYL METHACRYLATE)

POLY (METHYL METHACRYLATE)
BT POLYMETHACRYLATES
POLYVINYLS
RT METHYL METHACRYLATE
POLY (ETHYL ACRYLATE)
POLY (METHYL ACRYLATE)

POLY (METHYL VINYL KETONE)
BT POLYKETONES
POLYVINYLS

POLY (PROPYLENE GLYCOL)
BT GLYCOLS
RT PLURONICS (TN)
POLY (PROPYLENE OXIDE)
POLYETHERS

POLY (PROPYLENE OXIDE)
BT POLYETHERS
RT POLY (PROPYLENE GLYCOL)

POLY (VINYL ACETAL)
NT POLY (VINYL BUTYRAL)
BT ACETALS
POLYVINYLS
RT POLY (VINYL ALCOHOL)

POLY (VINYL ACETATE)
BT ESTERS
POLYVINYLS
RT VINYL ACETATE

POLY (VINYL ALCOHOL)
BT ALCOHOLS
POLYOLS
POLYVINYLS
RT POLY (VINYL ACETAL)
VINAL POLY (VINYL ALCOHOL) FIBERS

POLY (VINYL BUTYRAL)
BT ACETALS
POLY (VINYL ACETAL)
POLYVINYLS

POLY (VINYL CHLORIDE)
UF PVC
BT POLYVINYLS
RT VINYL CHLORIDE
VINYON POLY (VINYL CHLORIDE) FIBERS

POLY (VINYL OXAZOLIDINONE)
BT POLYVINYLS

POLY (VINYL PYRIDINE)
BT POLYVINYLS
RT POLY (VINYL PYRROLIDINE)
POLYELECTROLYTES
VINYL PYRIDINE

POLY (VINYL PYRROLIDINE)
BT POLYVINYLS
RT POLY (VINYL PYRIDINE)
POLY (VINYL PYRROLIDONE)
POLYELECTROLYTES

POLY (VINYL PYRROLIDONE)
BT POLYVINYLS
RT ANTIREDEPOSITION AGENTS
DYEABILITY
LEVELLING AGENTS
POLY (VINYL PYRROLIDINE)
STRIPPING AGENTS

POLY (VINYLIDENE CHLORIDE)
BT POLYVINYLS
RT SARAN POLY (VINYLIDENE CHLORIDE) FIBERS
VINYLIDENE COMPOUNDS

POLY (VINYLIDENE CYANIDE) FIBERS
USE NYTRIL FIBERS

POLYACROLEIN
BT ALDEHYDES
POLYVINYLS
RT ACROLEIN
DIALDEHYDE CELLULOSE
DIALDEHYDE STARCH

POLYACRYLAMIDE
BT AMIDES
POLYVINYLS
RT ACRYLAMIDE

POLYACRYLATES
NT POLY (ETHYL ACRYLATE)
POLY (METHYL ACRYLATE)
BT ESTERS
POLYVINYLS
RT ACRYLIC DYES
ACRYLIC ESTERS
ACRYLIC FIBERS
ACRYLIC SALTS
MODACRYLIC FIBERS
ORLON (TN)
POLYACRYLIC ACID
POLYACRYLONITRILE
POLYMETHACRYLATES

POLYACRYLIC ACID
BT CARBOXYLIC ACIDS
ORGANIC ACIDS
POLYVINYLS
RT ACRYLIC ACID
POLYACRYLATES
POLYELECTROLYTES

POLYACRYLONITRILE
BT NITRILES
POLYVINYLS
RT ACRYLIC FIBERS
ACRYLONITRILE
MODACRYLIC FIBERS
ORLON (TN)
POLYACRYLATES
POLYMETHACRYLATES

POLYALKYLAMINE SURFACTANTS
NT DUOMEENS (TN)
BT AMINE SURFACTANTS
RT ALKYLAMINE SURFACTANTS
AMIDOAMINE SURFACTANTS
POLYETHOXY ALKYLAMINE SURFACTANTS

POLYALKYLENEOXIDE BLOCK COPOLYMERS
USE POLYETHOXY POLYPROPOXY SURFACTANTS

POLYAMIDE DYES
USE NYLON DYES

POLYAMIDES
NT POLY (HEXAMETHYLENE ADIPAMIDE)
POLY (HEXAMETHYLENE SEBACAMIDE)
POLY (HEXAMETHYLENE TEREPHTHALAMIDE)
POLY (M-XYLYLENE ADIPAMIDE)
POLYCAPROLACTAM
BT POLYMERS
RT AMIDES
AMINO RESINS
CARBOXYL CONTENT
CONDENSATION POLYMERS
CYCLOHEXANE, 1,4-BIS (METHYLAMINE)-
LACTAMS
NYLON (POLYAMIDE FIBERS)
POLYBENZIMIDAZOLES
POLYPEPTIDES
POLYUREAS
POLYURETHANES
PROTEINS
SYNTHETIC FIBERS

POLYAMIDOESTER FIBERS
BT FIBERS
MAN MADE FIBERS
SYNTHETIC FIBERS
RT NYLON (POLYAMIDE FIBERS)
POLYESTER FIBERS

POLYAMINES
BT POLYMERS
RT AMINES
AMINO GROUPS
ANTISTATIC AGENTS
POLYELECTROLYTES
POLYETHYLENEIMINE

POLYBENZIMIDAZOLES
BT POLYMERS
RT HEAT RESISTANCE
POLYAMIDES

POLYBUTADIENE
BT POLYOLEFINS
RT BUTADIENE
BUTADIENE RUBBER FIBER

POLYCAPROAMIDE
USE POLYCAPROLACTAM

POLYCAPROLACTAM
UF POLYCAPROAMIDE
BT POLYAMIDES
RT CAPROLACTAM
NYLON 6

POLYCARBONATES
BT POLYMERS
RT POLYESTERS
THERMOPLASTIC RESINS

POLYELECTROLYTES
BT ELECTROLYTES
RT CARBOXYMETHYLCELLULOSE
CATIONIC SITES
ION EXCHANGE RESINS
POLY (VINYL PYRIDINE)
POLY (VINYL PYRROLIDINE)
POLYACRYLIC ACID
POLYAMINES
POLYMERS
POLYMETHACRYLIC ACID

POLYENES
BT ALKENES

POLYESTER DYEING
BT DYEING (BY FIBER CLASSES)
SYNTHETIC DYEING

POLYESTER DYES
BT DYES (BY FIBER CLASSES)
RT ACETATE DYES
ACRYLIC DYES
CELLULOSE DYES
DYEING
DYES (BY CHEMICAL CLASSES)
NYLON DYES
POLYESTERS
SILK DYES
WOOL DYES

POLYESTER FIBER BLENDS
BT BLENDS (FIBERS)
RT COTTON
COTTON BLENDS
DURABLE PRESS
LONG STAPLE COTTON

POLYESTER FIBERS
NT BLUE C POLYESTER (TN)
DACRON (TN)
FORTREL (TN)
GRILENE (TN)
KODEL (TN)
TERGAL (TN)
TERON (TN)
TERYLENE (TN)
VECTRA POLYESTER (TN)
VESTAN (TN)
VYCRON (TN)
BT FIBERS
MAN MADE FIBERS
SYNTHETIC FIBERS
RT MELT SPINNING
POLYAMIDOESTER FIBERS
POLYESTERS

POLYESTER RESINS
UF ALKYD RESINS
BT RESINS
THERMOPLASTIC RESINS
RT ACRYLIC RESINS
EPOXY RESINS
MELAMINE RESINS
PHENOLIC RESINS
POLYESTERS

POLYESTERS
NT POLY (ETHYLENE GLYCOL TEREPHTHALATE)
BT POLYMERS
RT CARBOXYL CONTENT
CONDENSATION POLYMERS
ESTERS
POLYCARBONATES
POLYESTER FIBERS

POLYETHERS
NT POLY (ETHYLENE OXIDE)
POLY (PROPYLENE OXIDE)
POLYOXYALKYL COMPOUNDS
POLYOXYETHYLENE
POLYOXYETHYLENE COMPOUNDS
POLYOXYMETHYLENE
BT POLYMERS
RT ACETALS
ETHERS
POLY (ETHYLENE GLYCOL)
POLY (PROPYLENE GLYCOL)

POLYETHOXY ALCOHOLS
USE POLYETHOXY ETHERS

POLYETHOXY ALKYL PHENOLS
USE POLYETHOXY ALKYLARYL ETHERS

POLYETHOXY ALKYL POLYAMINES
USE POLYETHOXY ALKYLAMINE SURFACTANTS

POLYETHOXY ALKYLAMINE SURFACTANTS
UF ETHODUOMEENS (TN)
ETHOMEENS (TN)
POLYETHOXY ALKYL POLYAMINES
POLYETHOXY AMINE SURFACTANTS
POLYOXYETHYLENE ALKYLAMINES
BT AMINE SURFACTANTS
RT ALKYLAMINE SURFACTANTS
POLYALKYLAMINE SURFACTANTS
POLYETHOXY AMIDES

POLYETHOXY ALKYLARYL ETHERS
UF ALKYLARYL POLYETHOXY ETHERS
ETHOXYLATED ALKYL PHENOLS
POLYETHOXY ALKYL PHENOLS
POLYOXYETHYLENE ALKYLPHENOLS
BT NONIONIC SURFACTANTS
RT POLYETHOXY ETHERS

POLYETHOXY AMIDES
UF POLYOXYETHYLENE ALKYLAMIDES
NT ETHOMIDS (TN)
BT NONIONIC SURFACTANTS
RT POLYETHOXY ALKYLAMINE SURFACTANTS

POLYETHOXY AMINE SURFACTANTS
USE POLYETHOXY ALKYLAMINE SURFACTANTS

POLYETHOXY ANHYDROSORBITOL ESTERS
USE SORBITAN ESTER SURFACTANTS

POLYETHOXY ESTERS
UF POLYETHOXY FATTY ACIDS
POLYETHOXY POLYOL ESTERS
POLYETHYLENE GLYCOL ESTERS
POLYGLYCOL ESTERS
POLYOXYETHYLENE ESTERS
NT ETHOFATS (TN)
BT NONIONIC SURFACTANTS
RT NONIONIC SOFTENERS
POLYETHOXY ETHERS
POLYETHOXY GLYCERIDES
POLYETHOXYLATED CASTOR OIL

POLYETHOXY ETHERS
UF POLYETHOXY ALCOHOLS
POLYETHYLENE GLYCOL ETHERS
POLYOXYETHYLENE ALCOHOLS
BT NONIONIC SURFACTANTS
RT NONIONIC SOFTENERS
POLYETHOXY ALKYLARYL ETHERS
POLYETHOXY ESTERS
POLYETHOXY GLYCERIDES
POLYETHOXY POLYPROPOXY SURFACTANTS
POLYETHOXYLATED CASTOR OIL

POLYETHOXY FATTY ACIDS
USE POLYETHOXY ESTERS

POLYETHOXY GLYCERIDES
NT POLYETHOXYLATED CASTOR OIL
BT NONIONIC SURFACTANTS
RT POLYETHOXY ESTERS
POLYETHOXY ETHERS

POLYETHOXY POLYOL ESTERS
USE POLYETHOXY ESTERS

POLYETHOXY POLYPROPOXY ETHYLENEDIAMINE
UF TETRONICS (TN)
BT AMINE SURFACTANTS
BLOCK POLYMERS
RT ALKYLAMINE SURFACTANTS

POLYETHOXY POLYPROPOXY SURFACTANTS
UF ETHOXY PROPOXY BLOCK POLYMERS
POLYALKYLENEOXIDE BLOCK COPOLYMERS
NT PLURONICS (TN)
BT NONIONIC SURFACTANTS
RT POLYETHOXY ETHERS

POLYETHOXY THIOETHERS
UF POLYOXYETHYLENE THIOETHERS
BT NONIONIC SURFACTANTS

POLYETHOXYLATED CASTOR OIL
BT POLYETHOXY GLYCERIDES
RT POLYETHOXY ESTERS
POLYETHOXY ETHERS

POLYETHYLENE
UF POLYTHENE
NT POLYETHYLENE (HIGH DENSITY)
BT POLYOLEFINS
RT SOFTENERS
OLEFIN FIBERS
POLYETHYLENE FIBERS

POLYETHYLENE (HIGH DENSITY)
BT POLYETHYLENE

POLYETHYLENE FIBERS
BT MAN MADE FIBERS
OLEFIN FIBERS
SYNTHETIC FIBERS
RT POLYETHYLENE

POLYETHYLENE GLYCOL ESTERS
USE POLYETHOXY ESTERS

POLYETHYLENE GLYCOL ETHERS
USE POLYETHOXY ETHERS

POLYETHYLENEIMINE
BT POLYMERS
RT POLYAMINES

POLYGLYCERYL ESTERS
BT GLYCERYL ESTER SURFACTANTS

POLYGLYCOL ESTERS
USE POLYETHOXY ESTERS

POLYKETONES
NT POLY (METHYL VINYL KETONE)
BT POLYMERS
RT KETONES

POLYMER ALLOYS
BT ALLOYS
RT METALLIC ALLOYS

POLYMER DEPOSITION
RT CHEMICAL MODIFICATION (FIBERS)
CHEMICALLY MODIFIED COTTON
DEPOSITION
POLYMERIZATION

POLYMER TYPES (GENERAL)
NT ADDITION POLYMERS
AMORPHOUS POLYMERS
ATACTIC POLYMERS
BLENDS (POLYMERS)
BLOCK POLYMERS
CONDENSATION POLYMERS
COPOLYMERS
EMULSION POLYMERS
HOMOPOLYMERS
TACTIC POLYMERS
RT FIBERS
NATURAL POLYMERS
POLYMERS

POLYMERIC DYES
BT DYES (BY CHEMICAL CLASSES)
RT POLYMERS
REACTIVE DYES

POLYMERIZATION
NT ADDITION POLYMERIZATION
BULK POLYMERIZATION
CONDENSATION POLYMERIZATION
EMULSION POLYMERIZATION
INTERFACIAL POLYMERIZATION
SUSPENSION POLYMERIZATION
BT REACTIONS (CHEMICAL)
RT ACID END GROUPS
ADDITION POLYMERS
AMINE END GROUPS
AZIDES
BUTADIENE
CATALYSTS
CONDENSATION POLYMERS
COPOLYMERIZATION
CROSSLINK DENSITY
CROSSLINKING
CRYSTALLIZATION
DEGREE OF POLYMERIZATION
DEPOSITION
END GROUPS
FINE STRUCTURE (FIBERS)
INITIATION
INITIATORS
MOLECULAR WEIGHT
MONOMERS
PLASTICIZERS
POLYMER DEPOSITION
POLYMERS
PREPARATION (CHEMICAL)
REDOX CATALYSIS

POLYMER TRANSITIONS
USE TRANSITIONS (POLYMERS)

POLYMERS
UF SYNTHETIC POLYMERS
NT POLYAMIDES
POLYAMINES
POLYBENZIMIDAZOLES
POLYCARBONATES
POLYESTERS
POLYETHERS
POLYETHYLENEIMINE
POLYKETONES
POLYOLEFINS
POLYSULFONES
POLYUREAS
POLYURETHANES
POLYVINYLS
RT ACID END GROUPS
ADDITION POLYMERS
AMORPHOUS POLYMERS
ATACTIC POLYMERS
BLENDS (POLYMERS)
BLOCK POLYMERS
CARBOXYL END GROUPS
CELLULOSE DERIVATIVES
CHITIN
CONDENSATION POLYMERS
CONTAMINATION
COPOLYMERS
CRYSTALLINE POLYMERS
DEACETYLATED CHITIN
DEGREE OF POLYMERIZATION
DEPOSITION
DOPE (POLYMER)
ELASTOMER FIBERS
ELASTOMERS
EMULSION POLYMERS
END GROUPS
FIBERS
FOAM RUBBER
FRACTIONATION
GELLING
GELS
HOMOPOLYMERS
MAN MADE FIBERS
MATERIALS
MELT VISCOSITY
MONOMERS
NATURAL POLYMERS
OLIGOMERS
ORGANIC COMPOUNDS (GENERAL)
PHENOLIC PLASTICS
PLASTICS
POLYELECTROLYTES
POLYMERIC DYES
POLYSACCHARIDES
REGENERATED CELLULOSE
RESINS
RUBBER
SULFONIC END GROUPS
SYNTHETIC FIBERS
SYNTHETIC LEATHER
TACTIC POLYMERS
TERPOLYMERS
VINYLIDENE COMPOUNDS
VISCOSIMETERS

POLYMERS (NATURAL)
USE NATURAL POLYMERS

POLYMETHACRYLATES
NT POLY (METHYL METHACRYLATE)
BT ESTERS
POLYVINYLS
RT ACRYLIC FIBERS
METHACRYLIC ACID
METHACRYLIC ESTERS
ORLON (TN)
POLYACRYLATES
POLYACRYLONITRILE
POLYMETHACRYLIC ACID

POLYMETHACRYLIC ACID
BT CARBOXYLIC ACIDS
ORGANIC ACIDS
POLYVINYLS
RT METHACRYLIC ACID
POLYELECTROLYTES
POLYMETHACRYLATES

POLYNOSIC FIBERS
BT CELLULOSIC FIBERS
HIGH WET MODULUS RAYON
MAN MADE FIBERS
RAYON (REGENERATED CELLULOSE FIBERS)
VISCOSE RAYON

POLYOLEFINS
NT POLYBUTADIENE
POLYETHYLENE
POLYPROPYLENE
BT POLYMERS
RT OLEFIN COPOLYMERS
OLEFIN FIBERS
POLYTETRAFLUOROETHYLENE

POLYOLS
NT GLYCEROL
GLYCOLS
PENTAERYTHRITOL
POLY (VINYL ALCOHOL)
BT ALCOHOLS

POLYOXYALKYL COMPOUNDS
BT POLYETHERS
RT PLURONICS (TN)

POLYOXYETHYLENE
BT POLYETHERS
RT NONIONIC SURFACTANTS
POLY (ETHYLENE GLYCOL)
POLYOXYETHYLENE COMPOUNDS
POLYOXYMETHYLENE

POLYOXYETHYLENE ALCOHOLS
USE POLYETHOXY ETHERS

POLYOXYETHYLENE ALKYLAMIDES
USE POLYETHOXY AMIDES

POLYOXYETHYLENE ALKYLAMINES
USE POLYETHOXY ALKYLAMINE SURFACTANTS

POLYOXYETHYLENE ALKYLPHENOLS
USE POLYETHOXY ALKYLARYL ETHERS

POLYOXYETHYLENE COMPOUNDS
BT POLYETHERS
RT POLYOXYETHYLENE

POLYOXYETHYLENE ESTERS
    USE POLYETHOXY ESTERS

POLYOXYETHYLENE THIOETHERS
    USE POLYETHOXY THIOETHERS

POLYOXYMETHYLENE
    BT POLYETHERS
    RT POLYOXYETHYLENE

POLYPEPTIDES
    UF PEPTIDES
    NT PROTEINS
    RT AMINO ACIDS
       ANIMAL FIBERS (GENERAL)
       AZLON (REGENERATED PROTEIN FIBERS)
       POLYAMIDES
       WOOL

POLYPROPYLENE
    BT POLYOLEFINS
    RT OLEFIN FIBERS
       POLYPROPYLENE FIBERS

POLYPROPYLENE FIBERS
    NT HERCULON (TN)
       MERAKLON (TN)
       VECTRA POLYPROPYLENE (TN)
    BT MAN MADE FIBERS
       OLEFIN FIBERS
       SYNTHETIC FIBERS
    RT POLYPROPYLENE

POLYSACCHARIDES
    UF CELLOTETRAOSE
    NT ALGINIC ACID
       CELLULOSE
       DEXTRAN
       STARCH
    BT CARBOHYDRATES
       NATURAL POLYMERS
    RT PECTINS
       POLYMERS

POLYSILOXANES
    USE SILICONES

POLYSTYRENE
    BT POLYVINYLS
    RT PLASTICS
       STYRENE

POLYSULFIDES (INORGANIC)
    USE SULFIDES (INORGANIC)

POLYSULFONES
    BT POLYMERS
    RT SULFONES

POLYTETRAFLUOROETHYLENE
    BT FLUOROCARBON POLYMERS
       POLYVINYLS
    RT POLYOLEFINS
       POLYTETRAFLUOROETHYLENE FIBERS
       TEFLON (TN)
       TEFLON FIBER (TN)

POLYTETRAFLUOROETHYLENE FIBER BLENDS
    BT BLENDS (FIBERS)

POLYTETRAFLUOROETHYLENE FIBERS
    NT TEFLON FIBER (TN)
    BT FLUOROCARBON FIBERS
       HIGH TEMPERATURE FIBERS
       MAN MADE FIBERS
       SYNTHETIC FIBERS
    RT FTORLON (TN)
       OLEFIN FIBERS
       POLYTETRAFLUOROETHYLENE
       POLYTRIFLUOROCHLOROETHYLENE FIBERS
       RAYON (REGENERATED CELLULOSE FIBERS)
       VINYL FIBERS (GENERAL)

POLYTHANE (TN)
    BT SPANDEX FIBERS

POLYTHENE
    USE POLYETHYLENE

POLYTRIFLUOROCHLOROETHYLENE FIBERS
    NT FTORLON (TN)
    BT FLUOROCARBON FIBERS
       HIGH TEMPERATURE FIBERS
       MAN MADE FIBERS
       SYNTHETIC FIBERS
    RT POLYTETRAFLUOROETHYLENE FIBERS
       TEFLON FIBER (TN)

POLYUNSATURATED COMPOUNDS
    BT OLEFINIC COMPOUNDS

POLYUREAS
    BT POLYMERS
    RT CONDENSATION POLYMERS
       POLYAMIDES
       POLYURETHANES

POLYURETHANE FIBERS
    NT DORLON (TN)
       PERLON U (TN)
    BT FIBERS
       MAN MADE FIBERS
       SYNTHETIC FIBERS
    RT POLYURETHANES

POLYURETHANES
    UF ISOCYANATE POLYMERS
    BT POLYMERS
    RT CONDENSATION POLYMERS
       ELASTOMER FIBERS
       ELASTOMERS
       ISOCYANATES
       ISOTHIOCYANATES
       POLYAMIDES
       POLYUREAS
       POLYURETHANE FIBERS
       RUBBER
       SPANDEX FIBERS
       SYNTHETIC RUBBER FIBERS
       TOLUENE ISOCYANATES
       URETHANES

POLYVINYL DERIVATIVES
    USE POLYVINYLS

POLYVINYLS
    UF POLYVINYL DERIVATIVES
    NT POLY (ETHYL ACRYLATE)
       POLY (METHYL ACRYLATE)
       POLY (METHYL METHACRYLATE)
       POLY (METHYL VINYL KETONE)
       POLY (VINYL ACETAL)
       POLY (VINYL ACETATE)
       POLY (VINYL ALCOHOL)
       POLY (VINYL BUTYRAL)
       POLY (VINYL CHLORIDE)
       POLY (VINYL OXAZOLIDINONE)
       POLY (VINYL PYRIDINE)
       POLY (VINYL PYRROLIDINE)
       POLY (VINYL PYRROLIDONE)
       POLY (VINYLIDENE CHLORIDE)
       POLYACROLEIN
       POLYACRYLAMIDE
       POLYACRYLATES
       POLYACRYLIC ACID
       POLYACRYLONITRILE
       POLYMETHACRYLATES
       POLYMETHACRYLIC ACID
       POLYSTYRENE
       POLYTETRAFLUOROETHYLENE
    BT POLYMERS
    RT ACRYLIC FIBERS
       MODACRYLIC FIBERS
       VINYL COMPOUNDS
       VINYL ETHERS
       VINYL FIBERS (GENERAL)

PONCHOS
    BT GARMENTS
       RAINWEAR
       OUTERWEAR
    RT COATS
       OVERCOATS

PONGEE
    BT FLAT WOVEN FABRICS
       WOVEN FABRICS
    RT PLAIN WEAVES

POPLIN
    BT FLAT WOVEN FABRICS
       WOVEN FABRICS
    RT BATISTE
       CORSET BATISTE
       PLAIN WEAVES

POPULATION
    RT DISPERSION (STATISTICAL)
       DISTRIBUTION
       PROBABILITY
       PROBABILITY DENSITY FUNCTION
       SAMPLING
       STATISTICAL MEASURES
       STATISTICS

PORCUPINE DRAWING
    RT CONTINENTAL SYSTEM PROCESSING
       DRAFTING MACHINES
       DRAWFRAMES
       GILLING
       PORCUPINE ROLLS

PORCUPINE ROLLS
    BT ROLLS
    RT DRAFTING ROLLS

PORES
    UF INTERSTICES
    NT FABRIC INTERSTICES
    RT AIR PERMEABILITY
       AIR RESISTANCE
       BULK DENSITY
       CAPILLARITY
       INTERLACINGS (YARN IN FABRIC)
       ORIFICES
    RT PACKING FACTOR
       PARTICLE SIZE
       PENETRATION
       PERMEABILITY TESTING
       POROSITY
       ROUNDNESS
       WATER REPELLENCY
       WATER RESISTANCE

POROSITY
    UF EFFECTIVE POROSITY
       GEOMETRIC POROSITY
    BT STRUCTURAL PROPERTIES
    RT ABSORBENCY (MATERIAL)
       ABSORPTION (MATERIAL)
       ADSORPTION
       AERODYNAMICS
       AIR RESISTANCE
       BULK
       BULK DENSITY
       CAPILLARITY
       COVER
       COVER FACTOR
       DENSITY
       FABRIC GEOMETRY
       FABRIC PROPERTIES
       FABRIC PROPERTIES (PHYSICAL EXCLUDING MECHANICAL)
       FABRIC PROPERTIES (STRUCTURAL)
       FABRIC STRUCTURE
       INTERLACINGS (YARN IN FABRIC)
       MASS TRANSFER
       MATRIX FREEDOM
       PACKING FACTOR
       PERMEABILITY
       PORES
       POROUS MATERIALS
       SPACE

POROUS MATERIALS
    UF POROUS MEDIA
    RT AIR RESISTANCE
       BULK DENSITY
       BUSHINGS
       CELLULAR YARNS
       FILTERS (FLUID)
       FILTRATION
       FLUIDIZED BEDS
       FOAMS
       MASS TRANSFER
       PERMEABILITY
       PORES
       POROSITY
       PRESSURE DROP

POROUS MEDIA
    USE POROUS MATERIALS

PORTER
    USE BEER

POSITION
    RT ACCELERATION (MECHANICAL)
       DISPLACEMENT
       MOTION
       REVERSAL
       STRAIN
       VELOCITY

POSITIONING
    RT ADJUSTMENTS
       ALIGNMENT
       CHAIN SEWING
       CONTROLLING
       CORRECTION
       ERRORS
       INSERTION
       INSTALLATION
       SETTING (ADJUSTMENTS)
       SEWING

POSITIONING MARKING
    USE MARKING

POSITIVE DISPLACEMENT PUMPS
    BT PUMPS

POSITIVE FEED
    UF POSITIVE FEED MECHANISMS
    RT BARRE
       CIRCULAR KNITTING MACHINES
       FULLY FASHIONED HOSIERY MACHINES
       HOSIERY MACHINES
       KNITTING
       KNITTING MACHINES
       SEAMLESS HOSIERY MACHINES
       STITCH LENGTH
       STITCH LENGTH CONTROL
       TENSION CONTROL

POSITIVE FEED MECHANISMS
    USE POSITIVE FEED

POSITIVE SHEDDING
    BT SHEDDING
    RT CLOSED SHED
       HARNESSES
       (CONT.)

RT LOOMS
NEGATIVE SHEDDING
OPEN SHED
PICKING (WEAVING)
SHUTTLE SMASH
SHUTTLES
SPLIT SHED
WEAVING

POST CURE
USE DEFERRED CURE METHOD

POST FORMED LAMINATES
BT COMPOSITES
LAMINATES
REINFORCED COMPOSITES
RT FABRIC REINFORCED COMPOSITES
FIBER REINFORCED COMPOSITES
FILAMENT WOUND COMPOSITES
FOAM BACKED FABRICS
FUSION (MELTING)
HEAT SEALING
HONEYCOMB LAMINATES
LAMINATED FABRICS
PLASTICS
POST FORMING (LAMINATES)
SANDWICH LAMINATES

POST FORMING (LAMINATES)
RT DURABLE PRESS
LAMINATING
POST FORMED LAMINATES

POT SPINNING
UF BUCKET SPINNING
CENTRIFUGAL SPINNING
BT SPINNING
RT CAKES
CAP SPINNING
DOWNTWISTING
FLYER SPINNING
RING SPINNING
SOLVENT SPINNING
WET SPINNING

POT SPINNING FRAMES
BT SPINNING FRAMES
RT BUILDER MOTIONS
CAKES
CAP SPINNING FRAMES

DOFFING (PACKAGE)
FLYER SPINNING FRAMES
MULES
PACKAGES
RING FRAMES
SPINNING
SPINNING (EXTRUSION)
SPINNING POTS
TWISTING

POTASH
USE POTASSIUM CARBONATE

POTASSIUM ACETATE
BT ACETATE SALTS
RT SODIUM ACETATE

POTASSIUM BICARBONATE
BT BICARBONATES

POTASSIUM BROMIDE
BT BROMIDES

POTASSIUM CARBONATE
UF POTASH
BT CARBONATES
RT DYEING AUXILIARIES
PH CONTROL

POTASSIUM CHLORIDE
BT CHLORIDES

POTASSIUM COMPOUNDS (GENERAL)
NT POTASSIUM ACETATE
POTASSIUM BICARBONATE
POTASSIUM BROMIDE
POTASSIUM CARBONATE
POTASSIUM CHLORIDE
POTASSIUM CYANATE
POTASSIUM CYANIDE
POTASSIUM FLUORIDE
POTASSIUM HYDROXIDE
POTASSIUM IODIDE
POTASSIUM NITRITE
POTASSIUM OLEATE
POTASSIUM PERSULFATE
POTASSIUM SOAPS
POTASSIUM STEARATE
TETRAPOTASSIUM PYROPHOSPHATE

POTASSIUM CYANATE
BT CYANATES

POTASSIUM CYANIDE
BT CYANIDES

POTASSIUM FLUORIDE
BT FLUORIDES

POTASSIUM HYDROXIDE
UF CAUSTIC POTASH
BT HYDROXIDES

POTASSIUM IODIDE
BT IODIDES

POTASSIUM NITRITE
BT NITRITES
RT AZOIC DYES
DIAZOTIZATION

POTASSIUM OLEATE
BT ALKALI SOAPS
OLEATE SALTS
RT OLEIC ACID

POTASSIUM PERSULFATE
BT PERSULFATES

POTASSIUM SOAPS
BT ALKALI SOAPS

POTASSIUM STEARATE
BT ALKALI SOAPS

POTATO STARCH
BT STARCH
RT CORN STARCH
PRINTING PASTES
THICKENING AGENTS

POTENTIAL (ELECTRICAL)
NT REDOX POTENTIAL
BT ELECTRICAL PROPERTIES
RT PHYSICAL ANALYSIS
SURFACE PROPERTIES (PHYSICAL CHEMICAL)

POTENTIAL ENERGY
BT ENERGY
RT BENDING ENERGY
KINETIC ENERGY
STRAIN ENERGY
TORSIONAL ENERGY

POTS (SPINNING)
USE SPINNING POTS

POULT
RT FAILLE
PLAIN WEAVES
TAFFETA

POWDERING
RT ATOMIZING
COMMINUTION
GRINDING (COMMINUTION)
MILLS (COLLOID)
MIXING
PIGMENTS

POWER (FABRIC)
RT COMFORT
CORE SPUN YARNS
COVERED YARNS
ELASTICITY
EXTENSIBILITY
FIT
FOUNDATION GARMENTS
GIRDLES
SPANDEX FIBERS
STRETCH
STRETCH FABRICS
STRETCH KNITTED FABRICS
STRETCH WOVEN FABRICS
STRETCH YARNS (FILAMENT)

POWER BRAKES
BT BRAKES (FOR ARRESTING MOTION)
RT ARRESTERS
FRICTION LET OFF
MAGNETIC BRAKES
MECHANICAL BRAKES
STOP MOTIONS

POWER CONSUMPTION
RT DRIVES
MACHINERY
MOTORS
PROCESS DESIGN
PROCESS EFFICIENCY
PRODUCTION
TEXTILE MACHINERY (GENERAL)
TEXTILE PROCESSES (GENERAL)

POWER LOOMS
USE LOOMS

PRE-CURE METHOD
UF RE-CURE
BT CURING
DURABLE PRESS TREATMENTS
RT DEFERRED CURE METHOD
DOUBLE CURE METHOD
DRY CURING
DURABLE PRESS FINISHES
HOT PRESS METHOD (DURABLE PRESS)
RESIN FINISHES
WET CURING

PREBOARDING
USE BOARDING

PRECIPITATION
RT COAGULATING
CRYSTALLIZATION
DIFFUSION
DISSOLVING
EMULSIONS
EVAPORATION
FILTRATION
RELATIVE HUMIDITY
SETTLING TANKS
SOLUBILITY
SOLUTIONS (LIQUID)

PRECISION WIND
RT CLOSE WIND
OPEN WIND
PACKAGES
PRECISION WINDERS
PRECISION WINDING
WINDING

PRECISION WINDERS
BT WINDERS
RT BALLOONS
CHEESE WINDERS
CHEESES
CLOSE WIND
COMBINATION WIND
CONERS
CONES
CREELING
DRUM WINDERS
GAIN (WINDING)
OPEN WIND
PACKAGES
QUILLERS
QUILLS
SLUB CATCHERS
STOP MOTIONS
TENSION CONTROL
TENSION DEVICES
THREAD GUIDES
UNWINDING
UNWINDING ACCELERATORS
WARPERS (MACHINE)
WINDING ANGLE

PRECISION WINDING
BT WINDING
RT AUTOMATIC WINDING
BUILDER MOTIONS
CLOSE WIND
CLOSE WINDING
CONING
DRUM WINDING
GAIN (WINDING)
OPEN WIND
OPEN WINDING
PACKAGES
PACKAGING (OPERATION)
RIBBON BREAKING
TRAVERSE
WIND
WINDING ANGLE

PRECONDENSATES
UF RESIN PRECONDENSATES
RT AMINO RESINS
METHYLOL COMPOUNDS
UREA-FORMALDEHYDE RESINS

PREFERENTIAL ADSORPTION
BT ADSORPTION
SORPTION
RT ABSORPTION (MATERIAL)
DIFFUSION
DYE AFFINITY
EXHAUSTION (WET PROCESS)
SUBSTANTIVITY

PREFERENTIAL DRAFT
RT DRAFT
DRAFTING (STAPLE FIBER)
TWIST CONTROL
WOOLEN DRAFTING

PREMETALLIZED DYES
USE METALLIZED DYES

PREORIENTATION
RT BIREFRINGENCE
COLD DRAWING
DEGREE OF ORIENTATION
DRAWING (FILAMENT)
ORIENTATION
ORIENTATION BIREFRINGENCE
POLYMERS

PREPARATION (CHEMICAL)
    UF SYNTHESIS
    RT ANALYZING
        COPOLYMERIZATION
        MANUFACTURING
        POLYMERIZATION
        REACTIONS (CHEMICAL)

PREPARING (WORSTED)
    RT BACKWASHING
        BRADFORD SYSTEM PROCESSING
        CARDING
        GILLING
        PREPARING WOOL

PREPARING WOOL
    BT WOOL CLASS
    RT CARDING WOOL
        PREPARING (WORSTED)
        WORSTED SPUN YARNS

PRESENCE
    RT ABSENCE
        APPEARANCE
        HOLLOWNESS
        VACUUM

PRESERVATIVES
    NT ANTIBACTERIAL FINISHES
        ANTIMICROBIAL FINISHES
        FUNGICIDES
        INSECT RESISTANCE FINISHES
        MILDEW RESISTANCE FINISHES
    BT FINISH (SUBSTANCE ADDED)
    RT BACTERIAL DEGRADATION
        BENZOATE SALTS
        BENZOIC ACID
        FORMALDEHYDE
        GELATINE
        ORGANOMERCURY COMPOUNDS
        PHENOLS
        SALICYLANILIDE
        SALICYLATE SALTS
        SALICYLIC ACID
        SIZES (SLASHING)
        STABILIZERS (AGENTS)

PRESHRINKING
    BT DRY FINISHING
        SHRINKING
        WET FINISHING
    RT COMPACTING
        COMPRESSIVE SHRINKING
        DIMENSIONAL STABILITY
        FABRIC SHRINKAGE
        FELTING SHRINKAGE
        FILLING SHRINKAGE
        FUSION CONTRACTION SHRINKAGE
        LONDON SHRINKING
        MOLDING SHRINKAGE
        RELAXATION SHRINKAGE
        RESIDUAL SHRINKAGE
        SANFORIZING (TN)
        SHRINK GRADING
        SHRINK RESISTANCE
        SHRINK RESISTANCE FINISHES
        SHRINKAGE
        WARP SHRINKAGE

PRESS FINISHING
    USE BOARDING

PRESS SECTION
    BT ROLLS
    RT CALENDER ROLLS
        COMPACTING
        DRAFTING ROLLS
        FEED ROLLS
        HYDROSTATIC LOADING
        MOISTURE CONTENT
        NIP
        NIP PRESSURE
        PAPER MACHINES
        PAPER MAKING
        PERALTA ROLLS
        PRESSING
        PRESSURE ROLLS
        ROLL COVERINGS
        SQUEEZE ROLLS
        SQUEEZERS (CLOTH)
        SQUEEZING
        SUCTION PRESSES
        WET PICKUP

PRESSED FELTS
    BT FABRICS (ACCORDING TO
         STRUCTURE)
        FELTS
        NONWOVEN FABRICS
    RT NEEDLE PUNCHED FELTS
        PAPERMAKERS FELTS
        WOVEN FELTS

PRESSER BARS
    UF NEEDLE PRESSER BARS
    RT COMPOUND NEEDLES
        FILLING KNITTING
        FILLING KNITTING MACHINES
        FULLY FASHIONED KNITTING
         MACHINES
        KNITTING NEEDLES
        LANDING LOOPS
        MILANESE KNITTING MACHINES
        NEEDLE BARS
        SPRING NEEDLES
        TRICOT KNITTING
        TRICOT KNITTING MACHINES
        WARP KNITTING MACHINES

PRESSER FOOT
    RT FEED DOGS
        SEWING MACHINES
        SEWING NEEDLES
        SHIRRING

PRESSER KNIVES
    RT COMB CIRCLES
        COMBING
        NOBLE COMBS
        NOIL KNIVES
        SETOVER

PRESSES
    UF PRESSING MACHINES
    BT GARMENT MANUFACTURING MACHINES
    RT CREASING
        DURABLE PRESS
        FLAT PRESSING
        GARMENT MANUFACTURE
        PLEATING
        SUCTION PRESSES

PRESSING
    NT FLAT PRESSING
    BT DRY FINISHING
        GARMENT MANUFACTURE
    RT BOARDING
        CREASE RETENTION
        CREASING
        DAMPENING
        DECATING
        DRESSMAKING
        DRYING
        DURABLE PRESS
        FACE FINISH
        FORMING
        FRICTION CALENDERING
        HEAT SETTING (SYNTHETICS)
        HOSIERY
        HYDROSTATIC LOADING
        IRONING
        KNITTING
        LUSTER
        NIP
        NIP ROLLS
        PAPER MAKING
        PLEATING
        PRESS SECTION
        PRESSURE ROLLS
        SCORCHING
        SCORCHING MARKS
        SETTING (EXCEPT SYNTHETICS)
        SOCKS
        STOCKINGS
        STRETCH HOSIERY
        SUCTION PRESSES
        TAILORING
        WRINKLE RESISTANCE

PRESSING (BEARDED NEEDLES)
    RT DIVIDING (KNITTING)
        FILLING KNITTING
        NEEDLE BARS
        PRESSER BARS
        SINKING (KNITTING)
        SPRING NEEDLES
        WARP KNITTING

PRESSING MACHINES
    USE PRESSES

PRESSING OFF
    RT CASTING OFF
        KNITTING

PRESSLEY INDEX
    USE PRESSLEY STRENGTH

PRESSLEY STRENGTH
    UF PRESSLEY INDEX
    BT BUNDLE STRENGTH
        FIBER BUNDLE STRENGTH
        FIBER PROPERTIES
        STRENGTH
    RT COTTON
        FIBER FINENESS
        LOOP STRENGTH
        MICRONAIRE
        STRENGTH ELONGATION TESTING

PRESSLEY TEST
    BT FIBER BUNDLE TESTING
    RT BREAKING LENGTH
        BREAKING STRENGTH
        CLEMSON STRENGTH TESTING
        COTTON
        COUNT STRENGTH PRODUCT
        END BREAKAGES
        FIBER STRENGTH
        MECHANICAL PROPERTIES
        STELOMETER STRENGTH TESTING
        STRENGTH OF MATERIALS
        TENACITY

PRESSURE
    NT HYDROSTATIC PRESSURE
        VACUUM
    RT CAPILLARY PRESSURE
        COMPRESSION
        COMPRESSIVE MODULUS
        EQUILIBRIUM (PHYSICAL)
        HYDRAULICS
        NORMAL FORCE
        HYDRAULIC CYLINDERS
        PARTIAL PRESSURE
        PRESSURE CONTAINERS
        PRESSURE CONTROL
        PRESSURE DISTRIBUTION
        PRESSURE DROP
        STRESS
        TEMPERATURE
        VAPOR PRESSURE

PRESSURE BAG MOLDING
    BT MOLDING
    RT COMPRESSION MOLDING
        CONTACT MOLDING
        FLEXIBLE PLUNGER MOLDING
        INJECTION MOLDING
        MATCHED DIE MOLDING
        REINFORCED COMPOSITES
        VACUUM BAG MOLDING

PRESSURE BECKS
    RT PRESSURE DYEING MACHINES

PRESSURE CONTAINERS
    RT BURSTING STRENGTH
        INFLATABLE FABRICS
        LIFE RAFTS
        PACKAGES
        PRESSURE
        PRESSURE CONTROL
        PRESSURE SUITS
        RADOMES

PRESSURE CONTROL
    RT AUTOMATIC CONTROL
        CONTROL SYSTEMS
        FLOW CONTROL
        HYDRAULIC CYLINDERS
        PH CONTROL
        PNEUMATIC CONTROL
        PRESSURE
        PRESSURE CONTAINERS
        PRESSURE DISTRIBUTION
        PRESSURE DROP
        PROCESS CONTROL
        TEMPERATURE CONTROL
        VALVES
        WEIGHT CONTROL

PRESSURE DISTRIBUTION
    RT DISTRIBUTION
        PRESSURE
        PRESSURE CONTROL
        STRESS DISTRIBUTION
        TEMPERATURE DISTRIBUTION

PRESSURE DROP
    RT AERODYNAMICS
        AIR DRAG
        DRAG
        FILTRATION
        FLUID FLOW
        FLUID FRICTION
        FRICTION
        POROUS MATERIALS
        PRESSURE
        PRESSURE CONTROL
        PRESSURE DISTRIBUTION
        SUCTION PRESSES
        TEMPERATURE GRADIENT

PRESSURE DYEING
    BT DYEING (BY ENVIRONMENTAL
         CONDITIONS)
        HIGH TEMPERATURE DYEING
    RT JIG DYEING
        LOW TEMPERATURE DYEING
        PACKAGE DYEING
        PRESSURE
        PRESSURE CONTROL
        PRESSURE DYEING MACHINES
        THERMOSOL PROCESS (TN)
        VACUUM DYEING

PRESSURE DYEING MACHINES
    NT BARCTOR MACHINES
        BEAM DYEING MACHINES (FABRIC)
    BT DYEING MACHINES (BY
         ENVIRONMENTAL CONDITIONS)
    RT AUTOCLAVES
        EXPANSION TANKS
         (CONT.)

RT HIGH TEMPERATURE DYEING
MONFORTS REACTOR
PACKAGE DYEING MACHINES
PIECE DYEING MACHINES
PRESSURE
PRESSURE BECKS
PRESSURE CONTROL
PRESSURE DYEING
PRESSURE JET DYEING MACHINES
PRESSURE JIGS (DYEING)
PRESSURE KETTLES
STOCK DYEING MACHINES
YARN DYEING MACHINES

PRESSURE FINGERS
RT BOBBIN LEAD
FLYER LEAD
FLYER SPINNING
FLYER SPINNING FRAMES
FLYERS

PRESSURE JET DYEING MACHINES
RT DYEING
PRESSURE
PRESSURE CONTROL
PRESSURE DYEING
PRESSURE DYEING MACHINES

PRESSURE JIGS (DYEING)
RT PRESSURE
PRESSURE CONTROL
PRESSURE DYEING
PRESSURE DYEING MACHINES

PRESSURE KETTLES
BT VESSELS
RT AUTOCLAVES
DYEING
DYEING MACHINES (BY ENVIRONMENTAL CONDITIONS)
KETTLES (HEATING)
PRESSURE
PRESSURE CONTROL
PRESSURE DYEING
PRESSURE DYEING MACHINES

PRESSURE ROLLS
NT NIP ROLLS
BT DRAFTING ROLLS
ROLLS
RT CALENDER ROLLS
COMPACTING
FEED ROLLS
HYDROSTATIC LOADING
NIP
NIP PRESSURE
PERALTA ROLLS
PRESS SECTION
PRESSING
PRESSURE
ROLL COVERINGS
ROLL WEIGHTING
SCHREINERING
SQUEEZE ROLLS
SUCTION PRESSES

PRESSURE SUITS
BT MILITARY PRODUCTS
RT INDUSTRIAL FABRICS
INFLATABLE FABRICS
PRESSURE
PRESSURE CONTROL

PRESSURE VESSELS
USE AUTOCLAVES

PRETREATMENTS (GENERAL)
NT BLEACHING
CARBONIZING
DESIZING
HEAT SETTING (SYNTHETICS)
SCOURING
SINGEING
SLASHING
RT AFTERTREATMENTS (GENERAL)
DRY FINISHING
DYEING
FINISHING PROCESS (GENERAL)
PRINTING
PROCESSES
WET FINISHING

PREVENTIVE MAINTENANCE
USE MAINTENANCE

PRICKLINESS
BT FABRIC PROPERTIES (AESTHETIC)
RT COMFORT
HAIRINESS
HAND
ROUGHNESS
SMOOTHNESS

PRIMARY AMINE SURFACTANTS
USE ALKYLAMINE SURFACTANTS

PRIMARY AMINES
BT AMINES

PRIMARY CREEP
USE DELAYED RECOVERY

PRIMARY TRANSITION TEMPERATURES
USE GLASS TRANSITION TEMPERATURE

PRIMARY TRANSITIONS
BT TRANSITIONS (POLYMERS)
RT GLASS TRANSITION TEMPERATURE
SECONDARY TRANSITIONS

PRIMARY WALLS
BT FINE STRUCTURE (FIBERS)
RT CONVOLUTIONS
COTTON
CUTICLE
DIFFUSION
EPICUTICLE
EXOCUTICLE
FIBRIL REVERSALS
FIBRILS
IMMATURE FIBERS
IMMATURITY (FIBER)
IMMATURITY TESTING (COTTON FIBER)
LUMEN
MERCERIZED COTTON
MERCERIZING
SECONDARY WALLS
SHRINKAGE
SWELLING
WOOD FIBERS

PRIMAZIN (TN)
USE ACRYLAMIDO DYES

PRINCESS LACE
RT MACHINE MADE LACES

PRINT CLOTH
BT FLAT WOVEN FABRICS
WOVEN FABRICS
RT BOMBAZINE
COATED FABRICS
LININGS
PLAIN WEAVES
PRINTED FABRICS
PRINTING
PRINTING ROLLS
SCREEN PRINTING
SHEETING (FABRIC)
WIGAN

PRINT ROLLS
USE PRINTING ROLLS

PRINTED FABRICS
UF PRINTS
RT DYED FABRICS
FABRICS
FINISHED FABRICS
PIECE DYED FABRICS
PRINT CLOTH
PRINTING
PRINTING ROLLS
SCREEN PRINTING

PRINTERS BLANKETS
BT INDUSTRIAL FABRICS
RT PRINTING
PRINTING ROLLS
SCREEN PRINTING

PRINTING
NT BLOCK PRINTING
DISCHARGE PRINTING
DUPLEX PRINTING
EMULSION PRINTING
FLAT SCREEN PRINTING
MELANGE PRINTING
PIGMENT PRINTING
REACTIVE PRINTING
RESIST PRINTING
ROLLER PRINTING
ROTARY SCREEN PRINTING
SCREEN PRINTING
THROUGH PRINTING
TWO STAGE PRINTING
VAT PRINTING
BT WET FINISHING
RT ACETALDEHYDE SULFOXYLATES
ACID AGING
AFTERTREATMENTS (GENERAL)
AGERS
AGING (STEAMING)
BACK GRAY
CARBOXYMETHYLCELLULOSE
CARRIER DYEING
COLOR TROUGH
DESIGN
DOCTOR BLADES
DYE HOUSE
DYEING
FLASH AGING
FLOCK PRINTING
FORMALDEHYDE SULFOXYLATES
OIL IN WATER EMULSIONS
PATTERN (FABRICS)
PHTHALOCYANINE COMPOUNDS
PRETREATMENTS (GENERAL)

RT PRINT CLOTH
PRINTED FABRICS
PRINTERS BLANKETS
PRINTING MACHINES
PRINTING PASTES
PRINTING ROLLS
THICKENING AGENTS
WATER IN OIL EMULSIONS

PRINTING DEFECTS
USE DYEING DEFECTS

PRINTING MACHINES
NT BLOCK PRINTING EQUIPMENT
DUPLEX PRINTING MACHINES
ROLLER PRINTING MACHINES
SCREEN PRINTING MACHINES
RT BACK GRAY
BLANKET (FINISHING)
BLOCK PRINTING
COLOR TROUGH
DISCHARGE PRINTING
DOCTOR BLADES
DUPLEX PRINTING
LINT BLADES
MANDRELS
PRINTERS BLANKETS
PRINTING
PRINTING PASTES
PRINTING ROLLS
RESIST PRINTING
ROLLER PRINTING
SCREEN PRINTING

PRINTING PASTES
RT ALGINIC ACID
CORN STARCH
DYEING AUXILIARIES
INKS
POTATO STARCH
PRINTING
PRINTING MACHINES
PRINTING ROLLS
SCREEN PRINTING
STARCH GUMS
THICKENING AGENTS

PRINTING ROLLS
UF PRINT ROLLS
PRINTING SHELLS
BT ROLLER PRINTING MACHINES
ROLLS
RT DESIGN
ENGRAVED CYLINDERS
ENGRAVING
LINT BLADES
MANDRELS
PATTERN (FABRICS)
PRINT CLOTH
PRINTED FABRICS
PRINTERS BLANKETS
PRINTING
PRINTING MACHINES
PRINTING PASTES
PRINTING SCREENS
ROLLER PRINTING
ROLLER PRINTING MACHINES

PRINTING SCREENS
UF SCREENS (PRINTING)
BT SCREEN PRINTING MACHINES
RT PRINTING ROLLS
SCREEN PRINTING

PRINTING SHELLS
USE PRINTING ROLLS

PRINTS
USE PRINTED FABRICS

PROBABILITY
RT CONFIDENCE LIMITS
DISTRIBUTION
DISTRIBUTION FUNCTION
MARKOV CHAIN
MEAN
OPERATIONS RESEARCH
POPULATION
PROBABILITY DENSITY FUNCTION
STANDARD DEVIATION
STANDARD ERROR
STATISTICAL INFERENCE
STATISTICAL MECHANICS
STATISTICS
VARIANCE

PROBABILITY DENSITY FUNCTION
RT DISTRIBUTION
DISTRIBUTION FUNCTION
POPULATION
PROBABILITY

PROBAN PROCESS (TN)
BT FIRE PROOFING TREATMENTS

PROCESS CONTROL
RT ALLOTMENT
AUTOCOUNT (TN)

RT AUTODRAFTER
AUTOLEVELLERS
AUTOMATIC CONTROL
AUTOMATION
BOW STRAIGHTENERS
CHEMICAL ENGINEERING
CLOTH GUIDES
COMPUTERS
CONTROL CHARTS
CONTROL SURFACES
CONTROL SYSTEMS
COSTS
DIFFERENTIAL CONTROL
DRAFT CONTROL
DROP WIRES
EQUALIZER (DRAFTING)
EVENERS
EVEREVEN (DRAFTING)
HOPPER FEED
INSTRUMENTATION
INTEGRAL CONTROL
INVENTORIES
IRREGULARITY CONTROL
JOB ANALYSIS
MANAGEMENT
MANUFACTURING
MATERIALS HANDLING
LAP WEIGHT CONTROL
PIANO MOTIONS
PICKER EVENERS
PROCESSES
PROCESSING DIRECTION
QUALITY CONTROL
RAPER AUTOLEVELLER (TN)
SERVO DRAFT
SLUB CATCHERS
STITCH LENGTH CONTROL
STOP MOTIONS
SUPERLEVELLER (DRAFTING)
SWING DOOR MECHANISMS
SYSTEMS ENGINEERING
TENSION CONTROL
WEIGHT CONTROL

PROCESS DESIGN
RT CHEMICAL ENGINEERING
DRY FINISHING
PLANT
POWER CONSUMPTION
PROCESS DYNAMICS
PROCESS EFFICIENCY
PROCESS VARIABLES
PROCESSING MACHINERY
PROCESSES
RESEARCH
TEXTILE MACHINERY (GENERAL)
TEXTILE PROCESSES (GENERAL)
WET FINISHING

PROCESS DYNAMICS
BT DYNAMICS
NT CARDING DYNAMICS
DRAFTING DYNAMICS
KNITTING DYNAMICS
SPINNING DYNAMICS
TWISTING DYNAMICS
WEAVING DYNAMICS
WINDING DYNAMICS
YARN DYNAMICS
RT MECHANISM (FUNDAMENTAL)
PROCESS DESIGN
PROCESS EFFICIENCY
PROCESSING DIRECTION

PROCESS EFFICIENCY
UF EFFICIENCY (PROCESS)
NT CARDING EFFICIENCY
COMBING EFFICIENCY
SEWING EFFICIENCY
SPINNING EFFICIENCY
WINDING EFFICIENCY
WEAVING EFFICIENCY
RT ALLOTMENT
BYPRODUCTS
CARD YIELD
CONSUMPTION (MATERIAL)
COSTS
DOWN TIME
ECONOMICS
EFFICIENCY (STRUCTURAL)
END BREAKAGES
IMPROVEMENTS
INVENTORIES
JOB ANALYSIS
LAP UP
MACHINE INTERFERENCE
MANUFACTURING
MATERIAL PROCESSING
MATERIALS HANDLING
MODIFICATION
OUTPUT
PERFORMANCE
POWER CONSUMPTION
PROCESS DESIGN
PROCESS DYNAMICS
PROCESSES
SAFETY
SET UP TIME

RT THREAD BREAKAGES
TIME AND MOTION STUDIES
TRAINING
WORKLOADS

PROCESS VARIABLES
RT ACTUAL DRAFT
ADJUSTMENTS
BREAK DRAFT
COEFFICIENT OF FRICTION
DOUBLING (DRAFTING)
DRAFT
EXPERIMENTAL DESIGN
FALLER SPEED
FEED RATE
FIBER LENGTH
FLAT SETTINGS (CARD)
GAIN (WINDING)
OVERLAP
PICKS PER INCH
PROCESS DESIGN
PROCESS DYNAMICS
PROCESS EFFICIENCY
PROCESSES
PROCESSING DIRECTION
RATCH
RUNNER RATIO
SEPARATOR SPACING
SETOVER
SPINNING LIMIT
STITCH LENGTH
TAKE DOWN TENSION
THROUGHPUT
TWIST
UNDERLAP
VELOCITY
WARP TENSION
WIND
WINDING ANGLE
YARN TENSION

PROCESSES
UF PROCESSING
RT AFTERTREATMENTS (GENERAL)
FINISHING PROCESS (GENERAL)
MANUFACTURING
PLANT
PRETREATMENTS (GENERAL)
PROCESS CONTROL
PROCESS DESIGN
PROCESS DYNAMICS
PROCESS EFFICIENCY
PROCESS VARIABLES
PROCESSING DIRECTION
PRODUCTION
TEXTILE MACHINERY (GENERAL)
TEXTILE PROCESSES (GENERAL)
UNIT OPERATIONS

PROCESSING
USE PROCESSES

PROCESSING DIRECTION
RT FIBER HOOKS
HOOK DIRECTION
LEADING HOOKS
PROCESS CONTROL
PROCESS DYNAMICS
PROCESS VARIABLES
TRAILING HOOKS

PROCESSING MACHINERY
RT FINISHING MACHINERY (GENERAL)
PROCESS DESIGN
PROCESS VARIABLES
PROCESSES
REVERSAL
TEXTILE MACHINERY (GENERAL)

PROCILAN (TN)
USE ACRYLAMIDO DYES

PROCION (TN)
USE CHLOROTRIAZINE DYES

PROCION H (TN)
USE CHLOROTRIAZINE DYES

PRODUCERS TWIST
BT TWIST
RT CYCLOSET YARNS (TN)
FILAMENT YARNS
HIGH TWIST
LOW TWIST
PRODUCERS YARNS
ROTOSET YARNS (TN)
THROWING
ZERO TWIST

PRODUCERS YARNS
RT AIR JET TEXTURED YARNS
BICOMPONENT FIBER YARNS
(FILAMENT)
CYCLOSET YARNS (TN)
FILAMENT YARNS
MONOFILAMENT YARNS
MULTIFILAMENT YARNS
PRODUCERS TWIST
ROTOSET YARNS (TN)
TEXTURED YARNS (FILAMENT)

PRODUCT DESIGN
BT DESIGN
RT APPAREL DESIGN
DRESS DESIGN
DRESSMAKING
FABRIC DESIGN
GARMENTS
MACHINE DESIGN
MARKETING
PRODUCTION
PRODUCTION CONTROL
PRODUCTS

PRODUCTION
UF PRODUCTION RATE
PRODUCTIVITY
RT ALLOTMENT
CONSUMPTION (MATERIAL)
COSTS
DELIVERY
DELIVERY RATE
DOWN TIME
ECONOMICS
EQUILIBRIUM (PHYSICAL)
FEED RATE
FLOW (MATERIALS)
HIGH PRODUCTION CARDS
IMPROVEMENTS
INPUT
INVENTORIES
JOB ANALYSIS
MACHINE INTERFERENCE
MANAGEMENT
MANUFACTURING
MARKETING
MATERIAL PROCESSING
MATERIALS HANDLING
MERCHANDISING
MODIFICATION
OCCUPATIONAL HAZARDS
OUTPUT
PATROLLING
PERFORMANCE
PLANT
POWER CONSUMPTION
PROCESS EFFICIENCY
PROCESSES
PRODUCT DESIGN
PRODUCTION CONTROL
PRODUCTS
SAFETY
SALES
SET UP TIME
SPECIFICATIONS
STANDARDS
STARTING TIME
STOPS
TRAINING
THROUGHPUT
WORKLOADS
YIELD (MECHANICAL)

PRODUCTION CONTROL
RT ANALYZING
AUTOMATIC CONTROL
AUTOMATION
CAPACITY (GENERAL)
COST CONTROL
COSTS
FLOW (MATERIALS)
INVENTORIES
INVENTORY CONTROL
PROCESS CONTROL
PRODUCT DESIGN
PRODUCTS

PRODUCTION RATE
USE PRODUCTION

PRODUCTIVITY
USE PRODUCTION

PRODUCTS
RT APPLICATIONS
END USES
FABRICS
FABRICS (BY END USES)
GARMENTS
MILITARY PRODUCTS
PRODUCT DESIGN
PRODUCTION
QUALITY CONTROL
SAMPLES

PROJECTION MICROSCOPY
BT MICROSCOPY
RT ELECTRON MICROSCOPY
INFRARED MICROSCOPY
INTERFERENCE MICROSCOPY
LIGHT MICROSCOPY (OPTICAL)
STEREOSCAN MICROSCOPY
ULTRAVIOLET MICROSCOPY
X RAY MICROSCOPY

PROPANEDIOL ESTER SURFACTANTS
BT GLYCOL ESTER SURFACTANTS

PROPELLANTS (AEROSOLS)
RT FLUOROCARBONS

PROPELLERS
RT PUMPS
STOCK DYEING MACHINES

PROPERTIES
RT AESTHETIC PROPERTIES
BIOCHEMICAL PROPERTIES
CHARACTERISTICS
CHEMICAL PROPERTIES
DEGRADATION PROPERTIES
END USE PROPERTIES
FABRIC PROPERTIES
FIBER PROPERTIES
MATERIALS ENGINEERING
MECHANICAL PROPERTIES
PHYSICAL CHEMICAL PROPERTIES
PHYSICAL PROPERTIES (EXCLUDING MECHANICAL)
QUALITY IMPROVEMENT
SURFACE PROPERTIES (MECHANICAL)
SURFACE PROPERTIES (PHYSICAL CHEMICAL)
TRANSFER PROPERTIES
YARN PROPERTIES

PROPIOLACTONE
BT LACTONES

PROPIONIC ANHYDRIDE
BT ACID ANHYDRIDES

PROPORTIONATING FEEDERS
RT DOPE DYEING EQUIPMENT

PROPYLENE CARBONATE
BT ALKYLENE CARBONATES
RT ETHYLENE CARBONATE

PROPYLENE IMINE
BT AZIRIDINE COMPOUNDS
RT APO
ETHYLENE IMINE
MAPO

PROTECTIVE COLLOIDS
USE EMULSIFYING AGENTS

PROTEIN FATTY ACID CONDENSATES
USE LAMEPONS

PROTEIN FIBERS
NT ALPACA
ANIMAL FIBERS (GENERAL)
ARALAC (TN)
ARDIL (TN)
AZLON (REGENERATED PROTEIN FIBERS)
CASHMERE
CHEMICALLY MODIFIED PROTEIN FIBERS
COLLAGEN FIBERS
COW HAIR
FIBROLANE (TN)
FUR
GOAT HAIR
GUARD HAIRS
HAIR
HORSE HAIR
HUMAN HAIR
LAMBSWOOL
LANITAL (TN)
MOHAIR
RABBIT HAIR
SILK
VICARA (TN)
VICUNA
WOOL
BT FIBERS
RT MAN MADE FIBERS
NATURAL FIBERS
PROTEINS
PYRUVIC ACID

PROTEINS
NT ALBUMIN
CASEIN
COLLAGEN
GELATINE
KERATIN
ZEIN
BT NATURAL POLYMERS
POLYPEPTIDES
RT AMINO ACIDS
AMINO GROUPS
AMINO RESINS
CARBOXYL CONTENT
CARBOXYL GROUPS
CYSTEINE
CYSTINE
LANTHIONINE
LYSINE
POLYAMIDES
PROTEIN FIBERS
PROTEOLYTIC ENZYMES
RT PYRUVIC ACID
SILK
WOOL

PROTEOLYTIC ENZYMES
BT ENZYMES
RT DESIZING
PROTEINS

PROXIMITY TESTERS
RT CAPACITANCE (ELECTRICAL)
INSTRUMENTATION
IRREGULARITY TESTERS
MEASURING INSTRUMENTS

PSEUDO WET BULB TEMPERATURE
RT DIFFUSION
DIFFUSIVITY
DRY BULB TEMPERATURE
DRYING
MOISTURE CONTENT
PSYCHROMETRY
REGAIN
WET BULB TEMPERATURE

PSYCHROMETRIC CHARTS
RT DEW POINT
DRY BULB TEMPERATURE
DRYING
HUMIDITY
REGAIN
RELATIVE HUMIDITY
WET BULB TEMPERATURE

PSYCHROMETRY
RT BOILING POINT
CRITICAL MOISTURE CONTENT
DEW POINT
DIFFUSION
DIFFUSIVITY
DRY BULB TEMPERATURE
DRYING
ENTHALPY
EQUILIBRIUM MOISTURE CONTENT
FREEZING POINT
HEAT TRANSFER
MOISTURE CONTENT
PSEUDO WET BULB TEMPERATURE
REGAIN
RELATIVE HUMIDITY
SATURATED VAPOR
SPECIFIC HUMIDITY
THERMODYNAMICS
VAPOR PRESSURE
WET BULB TEMPERATURE

PUCKER (DEFECT)
BT FABRIC DEFECTS
RT BUCKLING
DIFFERENTIAL SHRINKAGE
DIMENSIONAL STABILITY
GARMENT MANUFACTURE
GARMENTS
LAUNDERING
SEAM PUCKER
SEAMS
SEWING
STITCHING
THREAD SLACK
THREAD SLIPPAGE
THREAD TENSION
WRINKLING
YARN SEVERANCE
YARN SLIPPAGE
YARN TENSION

PUCKERED SELVEDGES
USE SLACK SELVEDGES

PUCKERING
BT FABRIC DISTORTION (TAILORING)
RT BACK TACKING
BASTING
BUTTED SEAMS
CHAIN SEWING
COCKLING
CREPE EMBOSSING
CURVED SEAMING
GATHERING
HEMMING
LOOP CATCHING
PINCHING
PUCKERS (FOR EFFECT)
PUCKERS (TAILORING)
RIPPLES
RUFFLING
SEAM PUCKER
SEAMING
SEAMS
SERGING
SEWING
SHIRRING
STITCH SIZE (SEWING)
STITCHES (SEWING)
STITCHING
STRETCH PUCKERS
TAILORING
TRIMMING (OPERATION)
YARN SEVERANCE

PUCKERS (FOR EFFECT)
RT BUCKLING
CREPE
CREPING
PUCKERING

PUCKERS (TAILORING)
UF COCKLE (TAILORING)
RT GATHERS (TAILORING)
PUCKERING
STITCHES (SEWING)

PULL OVERS
BT GARMENTS
OUTERWEAR
SPORTSWEAR
RT CARDIGANS
CIRCULAR KNITTING
FULLY FASHIONED KNITTING
JERKINS
JUMPERS
KNITTED FABRICS
KNITTING
SLIP ONS
SWEATERS

PULLED IN SELVEDGES
UF UNEVEN SELVEDGES
BT FABRIC DEFECTS
RT CENTER SELVEDGES
CRACKED SELVEDGES
DOG LEGGED SELVEDGES
LENO SELVEDGES
LOOMS
LOOP SELVEDGES
SEALED SELVEDGES
SELVEDGES
SLACK SELVEDGES
TEMPLES
TIGHT SELVEDGES
WEAVING
WOVEN FABRICS

PULLED WOOL
UF SKIN WOOL
SLIPE (WOOL)
BT WOOL
RT FLEECES
GREASY WOOL
RECOVERED WOOL
SCOURED WOOL
SUINT
VIRGIN WOOL

PULP
UF CELLULOSE PULP
WOOD PULP
BT CELLULOSIC FIBERS
RT PAPER
PAPER MAKING
PULPING
SLURRY
WOOD
WOOD FIBERS
WOOD FIBRILS

PULPING
RT PAPER
PAPER MAKING
PULP
SLURRY
WET LAY RANDOM WEBS
WET PROCESS (WEB)

PULSE PROPAGATION METER
BT TESTING EQUIPMENT
RT DYNAMIC MODULUS
DYNAMIC MODULUS TESTERS
IMPULSES
INSTRUMENTATION
SONIC VELOCITY
SOUND

PULVERIZING
USE GRINDING (COMMINUTION)

PUMICE
RT ABRASIVES
CORUNDUM
EMERY
GRIT
POLISHING

PUMPS
NT CENTRIFUGAL PUMPS
EXTRUSION PUMPS
METER PUMPS
POSITIVE DISPLACEMENT PUMPS
VACUUM PUMPS
RT PROPELLERS

PUMPS (FOOTWEAR)
BT FOOTWEAR
RT BINDINGS (FOOTWEAR)
BOOTS
LACES (SHOES)
LEATHER
LININGS (FOOTWEAR)
SHOES
SLIPPERS

RT SNEAKERS
SOLES
STITCHING
TONGUES
UPPERS

PUNCH BALLS (COMBING)
BT PACKAGES
RT BALLING HEADS
COMBING
NOBLE COMBING
NOBLE COMBS

PUNCHED CARD EQUIPMENT
RT AUTOMATIC CONTROL
COMPUTERS
EQUIPMENT
JACQUARD LOOMS
PUNCHED CARDS
PUNCHED TAPE
TAPES (DATA MEDIA)

PUNCHED CARDS
RT COMPUTERS
PUNCHED CARD EQUIPMENT
PUNCHED TAPE
TAPES (DATA MEDIA)

PUNCHED TAPE
BT TAPES (DATA MEDIA)
RT CONTROL SYSTEMS
MAGNETIC TAPE
PUNCHED CARD EQUIPMENT
PUNCHED CARDS

PUNCHING
RT CUTTING (TAILORING)
PUNCTURING
SEWING
STITCHING
TUFTING

PUNCTURE RESISTANCE
BT FABRIC PROPERTIES
FABRIC PROPERTIES (MECHANICAL)
MECHANICAL DETERIORATION PROPERTIES
MECHANICAL PROPERTIES
RT BALLISTIC TESTING
BEARING STRENGTH
BULLET PROOF CLOTHING
BURSTING STRENGTH
CATCHING
COMPRESSIVE STRENGTH
CUTTING RESISTANCE
FABRIC STRENGTH
IMPACT PENETRATION TESTING

RT PENETRATION
PUNCTURING
RUN RESISTANCE
RUNS (KNITTING)
SNAG RESISTANCE
SNAG TESTERS
SNAGGING
SNAGS
STRESS CONCENTRATION
TEAR STRENGTH

PUNCTURING
RT CATCHING
CUTTING
NEEDLING
PENETRATION
PICKING (SNAGGING)
PUNCHING
PUNCTURE RESISTANCE
RUN RESISTANCE
RUNS (KNITTING)
SEWING
SNAG RESISTANCE
SNAG TESTERS
SNAGGING
SNAGS
STITCHING
STRENGTH
TEAR STRENGTH
TUFTING

PURIFICATION
RT DIALYSIS
MEMBRANES
WASTE TREATMENT
WATER TREATMENT

PURL FABRICS
USE PURL KNITTED FABRICS

PURL KNIT FABRICS
USE PURL KNITTED FABRICS

PURL KNITTED FABRICS
UF LINKS LINKS FABRICS
PURL FABRICS
PURL KNIT FABRICS
BT KNITTED FABRICS
RT DOUBLE KNIT FABRICS
MACHINE KNITTING
PLAIN KNITTED FABRICS
PURL KNITTING

PURL KNITTING
BT KNITTING
RT LINKS LINKS MACHINES
LOCKING COURSES
PLAIN KNITTING
PURL KNITTED FABRICS
RIB KNITTING

PURL STITCHES
BT KNITTED STITCHES
RT PELERINE STITCHES
PLAIN STITCHES (KNITTING)
RACKED STITCHES
TUCK STITCHES

PVC
USE POLY (VINYL CHLORIDE)

PYRIDINE
BT HETEROCYCLIC COMPOUNDS
RT PYRIDINE DERIVATIVES

PYRIDINE DERIVATIVES
NT PYRIDINIUM COMPOUNDS
VINYL PYRIDINE
BT HETEROCYCLIC COMPOUNDS
RT AMINES
AMMONIUM COMPOUNDS
PYRIDINE
WATER REPELLENTS

PYRIDINIUM COMPOUNDS
BT ONIUM COMPOUNDS
PYRIDINE DERIVATIVES

PYROLYSIS
USE THERMAL DEGRADATION

PYROMETERS
BT MEASURING INSTRUMENTS
RT ELECTRONIC INSTRUMENTS
HEATING
RECORDING INSTRUMENTS
THERMAL TESTING
THERMOMETERS

PYROPHOSPHATES
NT TETRAPOTASSIUM PYROPHOSPHATE
TSPP
BT PHOSPHATES
RT CHELATING AGENTS

PYROSULFITES
BT SULFUR COMPOUNDS
RT BISULFITES
REDUCING AGENTS

PYRUVIC ACID
BT CARBOXYLIC ACIDS
KETONES
ORGANIC ACIDS
RT PROTEIN FIBERS
PROTEINS

QUADRANTS (MULE)
RT CARRIAGES (SPINNING MULE)
COPS
COUNTERFALLERS
FALLERS (MULE)
HEADSTOCKS
MULE SPINNING
MULES

QUALITATIVE ANALYSIS
RT CHEMICAL ANALYSIS
EMPIRICAL ANALYSIS
EXPERIMENTAL ANALYSIS
EXPERIMENTAL STRESS ANALYSIS
FABRIC ANALYSIS
FIBER IDENTIFICATION
PHYSICAL ANALYSIS
QUANTITATIVE ANALYSIS
SPECTRUM ANALYSIS
STATISTICAL ANALYSIS
STRESS ANALYSIS
STRUCTURAL ANALYSIS
THEORETICAL ANALYSIS
THERMAL ANALYSIS
X RAY ANALYSIS

QUALITY
RT ADVANTAGES
CORRECTION
IMPROVEMENTS
INCREASE
MODIFICATION
OPTIMIZATION
PERFORMANCE
QUALITY CONTROL
QUALITY IMPROVEMENT

QUALITY CONTROL
RT AUTOMATIC CONTROL
CAPACITANCE TYPE IRREGULARITY TESTERS
CONFIDENCE LIMITS
CONTROL CHARTS
CONTROL POINTS
CORING
DEFECTS
FABRIC INSPECTION
FLUCTUATIONS
INSPECTION
IRREGULARITY CONTROL
IRREGULARITY TESTERS
PNEUMATIC IRREGULARITY TESTERS
PROCESS CONTROL
PRODUCTION CONTROL
PROPERTIES
QUALITY
QUALITY IMPROVEMENT
SPECIFICATIONS
STANDARDS
TESTING
UPPER CONTROL LIMIT
USTER IRREGULARITY TESTER (TN)
WORKLOADS

QUALITY IMPROVEMENT
RT CHARACTERISTICS
INCREASE
OPTIMIZATION
PERFORMANCE
PROPERTIES
QUALITY
QUALITY CONTROL

QUANTITATIVE ANALYSIS
RT CHEMICAL ANALYSIS
EMPIRICAL ANALYSIS
EXPERIMENTAL ANALYSIS
EXPERIMENTAL STRESS ANALYSIS
FABRIC ANALYSIS
FIBER IDENTIFICATION
GRAVIMETRIC ANALYSIS
PHYSICAL ANALYSIS
QUALITATIVE ANALYSIS
SPECTRUM ANALYSIS
STATISTICAL ANALYSIS
STRESS ANALYSIS
STRUCTURAL ANALYSIS
THEORETICAL ANALYSIS
THERMAL ANALYSIS
X RAY ANALYSIS

QUARTER BLOOD (WOOL GRADE)
BT GRADE (FIBERS)
WOOL GRADE

QUASI HOMOGENEOUS MATERIALS
RT HOMOGENEOUS MATERIALS
MACRORHEOLOGY

QUATERNARY AMMONIUM COMPOUNDS
BT AMINES
AMMONIUM COMPOUNDS
ONIUM COMPOUNDS
RT ANTIBACTERIAL FINISHES
QUATERNARY AMMONIUM SURFACTANTS
SOFTENERS
SULFONIUM COMPOUNDS

QUATERNARY AMMONIUM SURFACTANTS
BT ONIUM SURFACTANTS
RT AMINE SURFACTANTS
CATIONIC SOFTENERS
PHOSPHONIUM SURFACTANTS
QUATERNARY AMMONIUM COMPOUNDS
SULFONIUM SURFACTANTS

QUENCHING (FILAMENT)
RT BATHS
BIREFRINGENCE
CRYSTALLIZATION
MELT SPINNING
ORIENTATION
SHRINKING
SPINNING (EXTRUSION)
TEXTURING

QUILL CHANGING
RT EJECTION
QUILLS
SHUTTLE CHANGING
WEAVING

QUILL STRIPPERS
USE BOBBIN STRIPPING MACHINES

QUILLERS
UF PIRN WINDERS
QUILLING MACHINES
NT UNIFIL (TN)
BT WINDERS
RT CHEESE WINDERS
COMBINATION WIND
FILLING YARNS
LAYER LOCKING
NOSE BUNCHING
PRECISION WINDERS
QUILLING
QUILLS
WARPERS (MACHINE)

QUILLING
UF PIRN WINDING
BT WINDING
RT CHEESE WINDING
CLOSE WIND
CONING
COPS
DRUM WINDING
FILLING WIND
FILLING YARNS
KNOTTING
LAYER LOCKING
NOSE BUNCHING
PACKAGING (OPERATION)
QUILLERS
QUILLS
SHUTTLES
STOP MOTIONS
TENSION CONTROL
TRAVERSE
UNIFIL (TN)

QUILLING MACHINES
USE QUILLERS

QUILLS
UF BASTARD COPS
PIRNS
WEFT PIRNS
BT PACKAGES
SELF SUPPORTING PACKAGES
RT BEAMS
BLANKET SHUTTLES
BOAT SHUTTLES
BOBBINS
CAKES
CHEESES
COMBINATION WIND
CONES
COPS
EMPTY PACKAGES
FILLING YARNS
FULL PACKAGES
GROOVES
LARGE PACKAGES
LOOMS
NOSE BUNCHING
PACKAGING (OPERATION)
PICKS
PRECISION WINDERS
QUILL CHANGING
QUILLERS
QUILLING
SHUTTLE ACCESSORIES
SHUTTLES
SPINDLES
STICK SHUTTLES
TUBES
UNIFIL (TN)
WARP WIND
WEAVING
WINDERS
WINDING

QUILTING
RT COATED FABRICS
LAMINATED FABRICS
LAMINATES
LAMINATING
MULTIPLE LAYER FABRICS
SANDWICH LAMINATES
SEWING
STITCHING

QUILTS
RT BEDSPREADS
BLANKETS
HOUSEHOLD FABRICS

QUINONE
UF BENZOQUINONE

R C CIRCUITS
RT CAPACITANCE (ELECTRICAL)
IMPEDANCE
INDUCTANCE
RESISTANCE (ELECTRICAL)
RESONANT FREQUENCY

RABBIT HAIR
UF ANGORA (RABBIT)
BT HAIR
KERATIN FIBERS
PROTEIN FIBERS
RT ALPACA
CAMEL HAIR
COW HAIR
FELTS
GOAT HAIR
HATS
MOHAIR
RABBITS
VICUNA

RABBITS
UF HARES
BT ANIMALS
RT RABBIT HAIR

RACE (LOOM)
USE RACE BOARDS

RACE BOARDS
UF RACE (LOOM)
RACE PLATES
RACEWAYS
RT BLANKET SHUTTLES
BOAT SHUTTLES
LOOMS
PICKING (WEAVING)
REEDS
SHUTTLE BINDERS
SHUTTLE DEFLECTORS
SHUTTLE SMASH PROTECTORS
SHUTTLES
SLEYS (LOOM)
STICK SHUTTLES
WEAVING

RACE PLATES
USE RACE BOARDS

RACEWAYS
USE RACE BOARDS

RACKED STITCHES
BT KNITTED STITCHES
RT FLOAT STITCHES (KNITTING)
KNITTING
PURL STITCHES
RACKING

RACKING
RT FILLING KNITTING
FLAT KNITTING
FULLY FASHIONED KNITTING MACHINES
KNITTING
LANDING LOOPS
LAYING IN (KNITTING)
RACKED STITCHES
WELTING

RADIAL BEARINGS
BT BEARINGS
RT AIR BEARINGS
BALL BEARINGS
CONICAL BEARINGS
JOURNAL BEARINGS
THRUST BEARINGS

RADIAL DISTRIBUTION
RT BLEND DRAFTING
BLENDED YARNS
BLENDED YARNS (FILAMENT)
BLENDED YARNS (STAPLE)
BLENDS (FIBERS)
DISTRIBUTION
FIBER MIGRATION
MIGRATION (SUBSTANCE)
SKIN (FIBER)
SPINNING
TRACER FIBERS
TWIST TRIANGLE
TWISTING
YARN STRUCTURE

RADIANT HEATING
BT HEATING
RT DIELECTRIC HEATING
DRYING
ELECTRIC HEATING (RESISTANCE)
GAS HEATING (GAS FUEL)
INFRARED DRYERS
OIL HEATING (OIL FUEL)

RADIATION
NT BETA RAYS
GAMMA RAYS
INFRARED RADIATION
IONIZING RADIATION
LIGHT
ULTRAVIOLET RADIATION
X RAYS
RT ABSORPTIVITY (RADIATION)
ALPHA PARTICLES
BETA PARTICLES
CONDUCTION
CONVECTION
DEGRADATION
DRYING
EMISSIVITY
HEAT TRANSFER
HEATING
IRRADIATION
PHOTOCHEMICAL DEGRADATION
PHOTONS
RADIATION COUNTERS
RADIATION GRAFTING
RADIOACTIVE COMPOUNDS
SUNLIGHT

RADIATION COUNTERS
RT COUNTING
RADIATION
RADIOACTIVE COMPOUNDS
TRACER FIBERS
TRACER TECHNIQUES

RADIATION GRAFTING
BT GRAFTING
RT GRAFT POLYMERS
RADIATION

RADIATION TREATMENT
USE IRRADIATION

RADICALS (CHEMICAL)
NT FREE RADICALS
RT FATTY ACIDS
FATTY ALCOHOLS
HYDROPHOBIC PROPERTY
ORGANIC COMPOUNDS (GENERAL)

RADIOACTIVE COMPOUNDS
RT ISOTOPES
RADIATION
RADIATION COUNTERS
RADIOACTIVE TRACERS
TRACER TECHNIQUES

RADIOACTIVE STATIC ELIMINATORS
USE STATIC ELIMINATORS

RADIOACTIVE TRACER TECHNIQUES
USE TRACER TECHNIQUES

RADIOACTIVE TRACERS
BT TRACERS
RT COUNTING
ISOTOPES
RADIATION COUNTERS
RADIOACTIVE COMPOUNDS
TRACER FIBERS
TRACER TECHNIQUES

RADIUS
RT AMPLITUDE
AREA
CROSS SECTIONAL AREA
DIAMETER
GEOMETRY
HEIGHT
MOMENT OF INERTIA
PERIMETER
POLAR MOMENT OF INERTIA
RADIUS OF CURVATURE
SHAPE

RADIUS OF CURVATURE
RT BENDING
BENDING RIGIDITY
CURVATURE
MOMENT CURVATURE CURVES
MOMENT OF INERTIA
NEUTRAL AXIS
RADIUS
STRAIN
TORSION (GEOMETRIC)

RADOMES
BT MILITARY PRODUCTS
RT END USES
INDUSTRIAL FABRICS
INFLATABLE FABRICS

RAFFIA
BT CELLULOSIC FIBERS
LEAF FIBERS
RT BAST FIBERS

RAG PULLING MACHINES
RT GARNETTING MACHINES
MUNGO
RECOVERED WOOL
SHODDY
WASTE MACHINES

RAIN PENETRATION TESTING
BT WATER RESISTANCE TESTING
RT DROP PENETRATION TESTING
HYDROSTATIC PRESSURE TESTING
IMPACT PENETRATION TESTING
PENETRATION
PERMEABILITY TESTING
WATER REPELLENCY
WATER RESISTANCE
WATER VAPOR TRANSMISSION
WATERPROOFING

RAINCOATS
USE RAINWEAR

RAINWEAR
UF RAINCOATS
NT GALOSHES
PONCHOS
RUBBER FOOTWEAR
BT GARMENTS
INDUSTRIAL CLOTHING
OUTERWEAR
RT CAPS (HEADGEAR)
COATS
HATS
OVERCOATS
UMBRELLA CLOTH
UMBRELLAS
VINYL COATED FABRICS
WATER REPELLENCY
WATERPROOFING
WORK CLOTHING

RAISING
USE NAPPING

RAISING CAMS
BT CAMS
RT KNITTING
KNITTING MACHINES
PATTERN WHEELS
STITCH CAMS

RAMBOUILLET WOOL
BT MERINO WOOL

RAMIE
BT BAST FIBERS
CELLULOSIC FIBERS
RT CORDAGE
FLAX
HEMP
JUTE
RETTING

RAMS
BT SHEEP
RT LAMBS
WOOL

RANDO FEEDER (TN)
RT AIR BRIDGES
CARDING
CONDENSER SCREENS
FEEDERS
GARNETTING
RANDO WEBBER (TN)
RANDO WEBS (TN)
TUFT FORMERS

RANDO WEBBER (TN)
BT NONWOVEN FABRIC MACHINES
RT AIR BRIDGES
ARACHNE FABRIC (TN)
BONDED FIBER FABRICS
CARDING
CARDS
CONDENSER SCREENS
COTTON CARDS
CURLATORS (TN)
FIBER WOVEN (TN)
GARNETTING MACHINES
MALIWATT FABRIC (TN)
NEEDLED FABRICS
NONWOVEN FABRICS
PAPER MACHINES
RANDO FEEDER (TN)
RANDO WEBS (TN)
TUFT FORMERS
WEBS
WET PROCESS (WEB)
WORSTED CARDING

RANDO WEBS (TN)
UF AIR LAY RANDOM WEB
RANDOM LAID WEBS
BT WEBS
RT AIR BRIDGES
ANISOTROPIC MATERIALS
CARD WEBS
CONDENSER SCREENS
CROSS LAID WEBS
CURLATORS (TN)
ISOTROPIC MATERIALS
NONWOVEN FABRICS
ORIENTED WEBS
(CONT.)

RANCO WEBS (TN)
RT ORTHOTROPIC MATERIALS
RANCO FEEDER (TN)
RANDO WEBBER (TN)
WEBS
WET LAY RANDOM WEBS
WET PROCESS (WEB)

RANDOM DISTRIBUTION
RT BINOMIAL DISTRIBUTION
CHI SQUARED DISTRIBUTION
DISTRIBUTION
F DISTRIBUTION
NORMAL DISTRIBUTION
T DISTRIBUTION

RANDOM DYEING
USE SPACE DYEING

RANDOM LAID WEBS
USE RANDO WEBS (TN)

RANGE
BT STATISTICAL MEASURES
RT CONFIDENCE LIMITS
CORRELATION COEFFICIENT
DEGREES OF FREEDOM
DISTRIBUTION
MEAN
MEDIAN
MODE
STANDARD DEVIATION
STANDARD ERROR
VARIANCE

RAPER AUTOLEVELLER (TN)
BT AUTOLEVELLERS
RT AUTOLEVELLING
AUTOMATIC CONTROL
CONTROL SURFACES
DRAFTING (STAPLE FIBER)
DRAFTING MACHINES
IRREGULARITY CONTROL
NEW BRADFORD SYSTEM PROCESSING
PROCESS CONTROL
WEIGHT CONTROL
WORSTED SYSTEM PROCESSING

RAPIER LOOMS
BT LOOMS
SHUTTLELESS LOOMS
RT AIR JET LOOMS
DRAPER SHUTTLELESS LOOMS (TN)
JET LOOMS
NEEDLE LOOMS (NARROW FABRICS)
OUTSIDE FILLING SUPPLY
RAPIERS (LOOMS)
SHUTTLELESS WEAVING
SULZER LOOMS (TN)
WATER JET LOOMS
WEAVING

RAPIERS (LOOMS)
RT LOOMS
RAPIER LOOMS
SHUTTLELESS LOOMS
SHUTTLES

RASCHEL FABRICS
NT FALL PLATE FABRICS
BT KNITTED FABRICS
WARP KNITTED FABRICS
RT LACES
LAUNDRY NETS
NETS

RASCHEL KNITTING MACHINES
UF RASCHEL LOOMS
BT KNITTING MACHINES
WARP KNITTING MACHINES
RT COMPOUND NEEDLES
FULL SET
GUIDE BARS
HALF SET
KNOCKOVER
LACE MACHINES
LACES
LAPPING (WARP KNITTING)
LATCH NEEDLES
MACHINE MADE LACES
MILANESE KNITTING MACHINES
NEEDLE BARS
NEEDLE BEDS
OVERLAP
PATTERN CHAINS
SET (WARP KNITTING)
SHOGGING
SIMPLEX KNITTING MACHINES
TAKE DOWN
TRICOT KNITTING MACHINES
UNDERLAP
WARP KNITTED FABRICS

RASCHEL LOOMS
USE RASCHEL KNITTING MACHINES

RATCH
RT BACK ROLLS
CONTROL SURFACES
DRAFTING (STAPLE FIBER)
DRAFTING ROLLS
DRAFTING THEORY
DRAFTING WAVES
FIBER LENGTH
FRONT ROLLS
GAUGE LENGTH
GILL BOXES
INTERSECTING GILL BOXES
NIP
PROCESS VARIABLES
ROLLS
VELOCITY CHANGE POINT

RATE PROCESSES
RT METARHEOLOGY
REACTION KINETICS
RHEOLOGY
STRAIN RATE

RATE
NT DELIVERY RATE
DYEING RATE
FEED RATE
REACTION RATE
SORPTION RATE
STRAIN RATE
RT KINETICS
KINETICS (CHEMICAL)
TIME
VELOCITY

RATINES
BT NOVELTY YARNS
RT BEAD YARNS
BOUCLES
FRILLS (NOVELTY YARNS)
KNICKERBOCKER YARNS
LOOPS (NOVELTY YARNS)
SEEDS (NOVELTY YARNS)
SPIRALS (NOVELTY YARNS)
SPLASHES

RAVEL STRIP STRENGTH
BT FABRIC PROPERTIES
FABRIC PROPERTIES (MECHANICAL)
STRENGTH
RT BREAKING LENGTH
BREAKING STRENGTH
FABRIC STRENGTH
GRAB STRENGTH
TEAR STRENGTH
UNIAXIAL STRESS

RAW STOCK DYEING
USE STOCK DYEING

RAW STOCK DYEING MACHINES
USE STOCK DYEING MACHINES

RAWHIDE
RT BOOTS
FOOTWEAR
GYMP
LACES (SHOES)
LEATHER
STITCH QUALITY
THREADS

RAYON
USE RAYON (REGENERATED CELLULOSE FIBERS)

RAYON (CUPRAMMONIUM)
USE CUPRAMMONIUM RAYON

RAYON (REGENERATED CELLULOSE FIBERS)
UF RAYON
REGENERATED CELLULOSE FIBERS
NT AVRIL (TN)
COLORAY (TN)
CORDURA (TN)
CUPRAMMONIUM RAYON
DURAFIL (TN)
FIBRO (TN)
FORTISAN (TN)
HIGH TENACITY VISCOSE RAYON
HIGH WET MODULUS RAYON
POLYNOSIC FIBERS
TENASCO (TN)
TENAX (TN)
VISCOSE RAYON
BT CELLULOSIC FIBERS
FIBERS
MAN MADE FIBERS
RT ALKALI CELLULOSE
AZLON (REGENERATED PROTEIN FIBERS)
CELLULOSE
CELLULOSE DERIVATIVES
CELLULOSE ESTER FIBERS
CELLULOSE ESTERS
CELLULOSE NITRATE
CYANOETHYLCELLULOSE
INORGANIC FIBERS (MAN MADE)
POLYTETRAFLUOROETHYLENE FIBERS
RAYON BLENDS
REGENERATED FIBERS (EXCLUDING CELLULOSIC AND PROTEIN)
RT SYNTHETIC FIBERS
SYNTHETIC RUBBER FIBERS
VINYL FIBERS (GENERAL)
WOOD
WOOD FIBERS

RAYON BLENDS
BT BLENDS (FIBERS)
CELLULOSE BLENDS
RT RAYON (REGENERATED CELLULOSE FIBERS)

RAYON DYEING
BT CELLULOSE DYEING
DYEING (BY FIBER CLASSES)
RT ACETATE DYEING
BLEND DYEING
CELLULOSE DYES
COTTON DYEING
DOPE DYEING
SILK DYEING
SYNTHETIC DYEING
TOW DYEING
WOOL DYEING

RAYON DYES
USE CELLULOSE DYES

RAYON STRIPE
BT FLAT WOVEN FABRICS
WOVEN FABRICS

REACTANCE (ELECTRICAL)
RT CAPACITANCE (ELECTRICAL)
IMPEDANCE
INDUCTANCE
RESISTANCE (ELECTRICAL)

RE-CURE
USE PRE-CURE METHOD

REACTANTS
UF CELLULOSE REACTANTS
CRIMP FIXING AGENTS
CROSSLINKING AGENTS (FIBERS)
FIBER REACTIVE COMPOUNDS
NT ACETALS
ALDEHYDES
AZIRIDINE COMPOUNDS
BIS (HYDROXYETHYL) SULFONE
CARBODIIMIDES
CHLOROHYDRINS
DIALDEHYDES
DIAZOMETHANE
DIVINYL SULFONE
HALOHYDRINS
ISOCYANATES
METHYLOL COMPOUNDS
OXIRANE COMPOUNDS
SULFONES
SULFONIUM REACTANTS
THPC
TMCEA
BT FINISH (SUBSTANCE ADDED)
RT ACRYLAMIDE
CARBAMOYLETHYLATED COTTON
CATALYSIS
CHEMICAL MODIFICATION (FIBERS)
CHEMICALLY MODIFIED COTTON
CRIMP FIXING
CROSSLINK DENSITY
CROSSLINKING

CYANOETHYLATED COTTON
DIALDEHYDE STARCH
DIBASIC ACIDS
DIFFERENTIAL DISTRIBUTION (CROSSLINKING)
DURABLE PRESS FINISHES
EASY CARE FABRICS
FIBER REACTIVITY
ODORS
PAD BATCH CURE PROCESS
PAD DRY CURE PROCESS
REACTIONS (CHEMICAL)
REAGENTS
SHRINK RESISTANCE FINISHES
SULTONES
WASH WEAR PROPERTIES
WASH WEAR FINISHES
WASHFASTNESS (OF FINISH)

REACTION KINETICS
BT KINETICS (CHEMICAL)
RT EQUILIBRIUM (CHEMICAL)
RATE PROCESSES
REACTION RATE
REACTIONS (CHEMICAL)

REACTION RATE
BT RATE
RT DYEING RATE
REACTION KINETICS
REACTIONS (CHEMICAL)

REACTIONS (CHEMICAL)
UF CHEMICAL REACTIONS
NT ACETYLATION

NT ACYLATION
ALKYLATION
CARBOXYMETHYLATION
CHELATING
CHEMICAL MODIFICATION (FIBERS)
CHLORINATION (CHEMICAL REACTION)
COMPLEXING
CONDENSATION (CHEMICAL)
COPOLYMERIZATION
COUPLING (CHEMICAL)
CROSSLINKING
DIAZOTIZATION
ESTERIFICATION
ETHERIFICATION
HALOGENATION
HYDROLYSIS
METHYLOLATION
OXIDATION
PHOTOCHEMICAL REACTIONS
POLYMERIZATION
REDOX REACTIONS
REDUCTION
RT ACID DEGRADATION
CATALYSIS
CORROSION
DEGRADATION
ENTHALPY OF REACTION
EQUILIBRIUM (CHEMICAL)
FREE RADICALS
IONIZATION
PREPARATION (CHEMICAL)
REACTANTS
REACTION KINETICS
REACTION RATE
REACTIVITY
VAPOR PHASE

REACTIVE DYEING
BT DYEING (BY DYE CLASSES)
RT ACID DYEING
AZOIC DYEING
BASIC DYEING
DIRECT DYEING
DISPERSE DYEING
METALLIZED DYEING
MORDANT DYEING
NEUTRAL DYEING
PIGMENT DYEING
REACTIVE DYES
REACTIVE PRINTING
SIMULTANEOUS DYEING AND FINISHING
SULFUR DYEING
VAT DYEING

REACTIVE DYES
NT ACRYLAMIDO DYES
AZIDOSULFONYL DYES
CHLOROACETYLAMINO DYES
CHLOROBENZOTHIAZOLE DYES
CHLOROPYRIMIDINE DYES
DICHLOROQUINOXALINE DYES
METHYLOL DYES
OXIRANE DYES
SULFONAMIDE DYES
VINYLSULFONE DYES
BT DYES (BY CHEMICAL CLASSES)
RT ALCIAN BLUE (TN)
DYEING (BY DYE CLASSES)
DYES (BY FIBER CLASSES)
DIRECT DYES
METALLIZED DYES
MORDANT DYES
POLYMERIC DYES
REACTIVE DYEING
REACTIVE PRINTING
VINYLATED DYES

REACTIVE PRINTING
BT PRINTING
RT AGERS
AGING (STEAMING)
PIGMENT PRINTING
REACTIVE DYEING
REACTIVE DYES
TWO STAGE PRINTING
VAT PRINTING

REACTIVITY
BT CHEMICAL PROPERTIES
RT FIBER REACTIVITY
REACTIONS (CHEMICAL)
REACTIVE DYEING

REACTONE (TN)
USE CHLOROPYRIMIDINE DYES

REAGENT TESTING (FIBER IDENTIFICATION)
BT FIBER IDENTIFICATION
RT CHEMICAL ANALYSIS (FIBER IDENTIFICATION)
MICROSCOPIC ANALYSIS (FIBER IDENTIFICATION)
REAGENTS
SPECIFIC GRAVITY TESTING (FIBER IDENTIFICATION)
STAINING TESTING (FIBER IDENTIFICATION)
SWELLING TESTING (FIBER IDENTIFICATION)
THERMAL TESTING (FIBER IDENTIFICATION)

REAGENTS
RT CATALYSTS
CHEMICAL ANALYSIS
REACTANTS
REAGENT TESTING (FIBER IDENTIFICATION)

REAL COMPONENT (DYNAMIC MODULUS)
BT DYNAMIC MODULUS
STRESS STRAIN PROPERTIES
RT ELASTIC MODULUS (TENSILE)
IMAGINARY COMPONENT (DYNAMIC MODULUS)
LOSS TANGENT

REAL TWIST
UF TRUE TWIST
BT TWIST
RT ALTERNATING TWIST
BALANCED YARNS
CYCLOSET YARNS (TN)
DOWNTWISTING
FALSE TWIST
FALSE TWISTING (EXCEPT TEXTURING)
FALSE TWISTING (TEXTURING)
FLUID TWISTING
HELIX ANGLE
HIGH TWIST
INTERLACING (FILAMENT IN YARN)
PLY TWIST
ROTOSET YARNS (TN)
S TWIST
SINGLES TWIST
SPINNING FRAMES
TORSION (GEOMETRIC)
TWIST CONTRACTION
TWIST DISTRIBUTION
TWIST MULTIPLIER
TWIST TRIANGLE
TWISTERS
TWISTING
TWISTING MOMENTS
TWO FOR ONE TWISTING
UPTWISTING
Z TWIST
ZERO TWIST

RECIPROCATING CUTTING MACHINES
BT CUTTING MACHINES
RT BAND KNIFE CUTTING MACHINES
CLICKERS
CUTTING (TAILORING)
CUTTING SHEARS
DIE CUTTERS
ROTARY CUTTING MACHINES
SCISSORS

RECIPROCATING TRAVERSE
BT TRAVERSE
RT PACKAGES
ROTARY TRAVERSE
WIND
WINDING

RECOMBED TOPS
BT TOPS
RT COMBING
DYEING

RECOMBING
RT COMBING
DYEING
TOP DYEING

RECORDING INSTRUMENTS
BT INSTRUMENTATION
RT AMPLIFIERS
AUTOMATIC CONTROL
DETERMINATION
MEASURING INSTRUMENTS
NULL METHODS
PHOTOGRAPHY
TRANSDUCERS

RECOVERABILITY
USE RECOVERY (SELF RESTORATION TO ORIGINAL CONDITION)

RECOVERED WOOL
BT WOOL
RT FEARNOUGHTS
GARNETTING
LAMBSWOOL
MUNGO
PULLED WOOL
SHODDY
VIRGIN WOOL

RECOVERING
USE RECOVERY (SELF RESTORATION TO ORIGINAL CONDITION)

RECOVERY (CHEMICAL)
NT SOLVENT RECOVERY
RT RECOVERY (WASTE)
WASTE HEAT RECOVERY
WASTE TREATMENT

RECOVERY (SELF RESTORATION TO ORIGINAL CONDITION)
UF DYNAMIC RECOVERY
RECOVERABILITY
RECOVERING
NT BENDING RECOVERY
CREEP RECOVERY
CRIMP RECOVERY
DELAYED RECOVERY
ELASTIC RECOVERY
WORK RECOVERY
WRINKLE RECOVERY
BT STRESS STRAIN PROPERTIES
STRESS STRAIN PROPERTIES (TENSILE)
RT CREASE RESISTANCE
DRAWING (FILAMENT)
FILLINGWISE STRETCH
HEAT SETTING (SYNTHETICS)
LIVELINESS
MATTING
MAXWELL MODEL
PERMANENT SET
RELAXATION
RELAXATION SPECTRUM
RELAXATION TIME (MECHANICAL)
RESIDUAL STRAIN
RESIDUAL STRESS
RESILIENCE
SAG
STRAIN
STRESS RELAXATION
STRETCH
UNIAXIAL STRESS
VOIGT MODEL
WARPWISE STRETCH
WRINKLE RESISTANCE
YIELD POINT

RECOVERY (WASTE)
RT GREASE RECOVERY
RECOVERY (CHEMICAL)
WASTE
WASTE TREATMENT

RECOVERY SYSTEMS (AERODYNAMIC)
RT AERODYNAMICS
AIR DRAG
AIR RESISTANCE
ARRESTERS
MILITARY PRODUCTS
PARACHUTES
SHOCK ABSORBERS
STABILIZATION
WIND TUNNELS

RECRYSTALLIZATION
RT CRYSTALLINITY
CRYSTALLIZATION
DECRYSTALLIZATION

RECTILINEAR COMBS
UF CONTINENTAL COMBS
COTTON COMBS
FRENCH COMBS
HEILMANN COMBS
NASMITH COMBS
BT COMBS
RT COMB BRUSHES
COMB CYLINDERS
COMB PINS
COMB SEGMENTS
COMBED SLIVERS
COMBING
COTTON COMBING
COTTON SYSTEM PROCESSING
DETACHING ROLLS
HALF LAP
LAP PLATES
NIPPER JAWS
NIPPER KNIVES
NIPPER PLATES
NOBLE COMBS
NOIL PER CENT
NOILS
RIBBON LAPS
SLIVER NOIL RATIO
SLIVERS
TOP COMBS
TRUMPETS
WORSTED SYSTEM PROCESSING

RECTILINEAR COMBING
UF FRENCH COMBING
BT COMBING
RT NOBLE COMBING
WORSTED COMBING

REDOX CATALYSIS
BT CATALYSIS
RT FERROUS COMPOUNDS
OXIDATION
(CONT.)

RT OXIDIZING AGENTS
POLYMERIZATION
REDOX POTENTIAL
REDOX REACTIONS
REDUCING AGENTS
REDUCTION

REDOX POTENTIAL
BT POTENTIAL (ELECTRICAL)
ELECTRICAL PROPERTIES
RT REDOX CATALYSIS
REDOX REACTIONS

REDOX REACTIONS
BT REACTIONS (CHEMICAL)
RT OXIDATION
REDOX CATALYSIS
REDOX POTENTIAL
REDUCING AGENTS
REDUCTION

REDUCING (WORSTED PROCESS)
USE DRAFTING (STAPLE FIBER)

REDUCING AGENTS
RT ACETALDEHYDE SULFOXYLATES
ALDEHYDES
BISULFITES
BORON HYDRIDES
DISULFIDES (ORGANIC)
DITHIONITES
FORMALDEHYDE SULFOXYLATES
HYDROXYACETONE
HYDROXYLAMINE
OXIDATION
OXIDIZING AGENTS
PYROSULFITES
REDOX CATALYSIS
REDOX REACTIONS
REDUCTION
SODIUM BISULFITE
SODIUM BOROHYDRIDE
SODIUM DITHIONITE
SODIUM SULFIDE
STRIPPING (COLOR)
SULFINIC ACIDS
THIOUREA DIOXIDE
VAT DYES
ZINC FORMALDEHYDE SULFOXYLATE

REDUCTION
BT REACTIONS (CHEMICAL)
RT OXIDATION
BISULFITES
CYSTEINE
CYSTINE
ELECTROLYSIS
REDOX CATALYSIS
REDOX REACTIONS
REDUCING AGENTS
SODIUM DITHIONITE

REDUCTION PADDERS
RT CONTINUOUS DYEING RANGE
PADDERS

REDYEING
BT DYEING
RT DEFECTS
STRIPPING (COLOR)

REED DRAFT
UF REED PLAN
RT DENTS
DRAWING IN
REED WIDTH
REEDING
REEDS
WARP ENDS
WARPING

REED HOOKS
RT DRAWING IN
DRAWING IN MACHINES
HEDDLES
REEDING
REEDS
WARP ENDS
WARPING

REED MARKS
UF REED OMBRE
BT FABRIC DEFECTS
RT BEATING UP
BOUNCE (WEAVING)
COVER
FABRIC INSPECTION
INTERLACINGS (YARN IN FABRIC)
LIGHT DENT
LOOMS
PATTERN DEFINITION
REED RAKE
REED WIRES
REEDING
REEDS
REEDY FABRICS
STREAKINESS
SWOLLEN DENT

RT VIBRATION
WARP STREAKS
WEAVE (TYPES)
WOVEN FABRICS

REED OMBRE
USE REED MARKS

REED PLAN
USE REED DRAFT

REED RAKE
BT FABRIC DEFECTS
RT REED MARKS
REEDS

REED SPACING
USE DENTS

REED WIDTH
RT DENTS
LOOMS
REED DRAFT
REEDING
REEDS
WARP ENDS
WARPERS (MACHINE)
WEAVING

REED WIRES
RT DENTS
DRAWING IN
LIGHT DENT
LOOMS
REED MARKS
REEDING
REEDS
SKIP DENT
SWOLLEN DENT
WARP ENDS
WARP PREPARATION
WARPERS (MACHINE)

REEDING
UF SLEYING
BT DRAWING IN
WARP PREPARATION
RT BEER
DENTS
LIGHT DENT
LOOMS
REED DRAFT
REED HOOKS
REED MARKS
REED WIDTH
REED WIRES
REEDS
SKIP DENT
SWOLLEN DENT
WARP ENDS
WARPING

REEDS
NT FIXED REEDS
LOOSE REEDS
RT ABRASION
BEATING UP
BEER
BOUNCE (WEAVING)
CLOTH FELL
DENTS
DRAWING IN
HACK STANDS
HARNESSES
HEDDLES
LIGHT DENT
LOOMS
OPEN SHED
PATTERN (FABRICS)
PICKING (WEAVING)
RACE BOARDS
REED DRAFT
REED HOOKS
REED MARKS
REED RAKE
REED WIDTH
REED WIRES
REEDING
REEDY FABRICS
SHEDDING
SHUTTLE DEFLECTORS
SHUTTLE MARKS
SHUTTLE SMASH PROTECTORS
SHUTTLES
SLEYS (LOOM)
STICK SHUTTLES
SWOLLEN DENT
WARP ENDS
WARP PREPARATION
WARP TENSION

REEDY FABRICS
RT APPEARANCE
COVER
FABRIC DEFECTS
FABRICS
REED MARKS
REEDS
TRANSPARENCY

REEL MACHINE (DYEING)
BT BATIK DYEING MACHINES
PIECE DYEING MACHINES
RT DYEING

REELING
RT LENGTH MEASURING DEVICES
REELING MACHINES (WINDING)
SKEINS
YARNS

REELING MACHINES (WINDING)
RT LENGTH MEASURING DEVICES
REELING
SKEINS
YARNS

REELS
USE BOBBINS

REFLECTANCE
RT ABSORPTIVITY (RADIATION)
ANGLE OF INCIDENCE
CONDUCTIVITY (THERMAL)
EMISSIVITY
FABRIC PROPERTIES
FABRIC PROPERTIES (PHYSICAL EXCLUDING MECHANICAL)
INTERNAL REFLECTION
KUBELKA-MUNK EQUATION
LIGHTFASTNESS TESTING
LUSTER
OPACITY
OPTICAL PROPERTIES
REFLECTIVITY
REFLECTOMETERS
SCHREINERING
THERMAL PROPERTIES
THERMAL TESTING
TRANSFER PROPERTIES
TRANSLUCENCY
TRANSPARENCY

REFLECTIVITY
BT OPTICAL PROPERTIES
THERMAL PROPERTIES
RT ABSORPTION (RADIATION)
ABSORPTIVITY (RADIATION)
CONDUCTIVITY (THERMAL)
EMISSIVITY
REFLECTANCE
TRANSMISSIVITY

REFLECTOMETERS
BT MEASURING INSTRUMENTS
OPTICAL INSTRUMENTS
RT COLORIMETERS
OPTICAL PROPERTIES
PHOTOMETERS
REFLECTANCE
SPECTROPHOTOMETERS

REFRACTANCE
BT OPTICAL PROPERTIES
RT DICHROISM
DIFFRACTION
REFRACTIVE INDEX

REFRACTIVE INDEX
BT OPTICAL PROPERTIES
RT BIREFRINGENCE
FABRIC PROPERTIES
OPTICS
ORIENTATION BIREFRINGENCE
REFRACTOMETRY
STRESS OPTICAL PROPERTIES
TRANSFER PROPERTIES

REFRACTOMETRY
RT ELECTRON MICROSCOPY
INFRARED MICROSCOPY
INTERFERENCE MICROSCOPY
LIGHT MICROSCOPY (OPTICAL)
MICROSCOPY
OPTICAL INSTRUMENTS
ULTRAVIOLET MICROSCOPY
X RAY MICROSCOPY

REFRACTORY FIBERS
USE CERAMIC FIBERS

REFRIGERATION
RT AIR CONDITIONING
COOLING
FREEZING
FREEZING POINT
LOW TEMPERATURE

REFULLING MACHINES
BT FULLING MACHINES
RT FULLING
GATTERWALKING (FULLING MACHINES)
MILLS (FULLING ROTARY)
MILLS (KICKER)
OPEN WIDTH FINISHERS (FULLING)
PILGRIMSTEP (FULLING MACHINES)
PLANKING (FULLING MACHINES)

REGAIN
UF MOISTURE REGAIN
RT ABSORBENCY (MATERIAL)
BOILING POINT
BOUND WATER
CONDITION (TEXTILES)
CONDITIONING
CRITICAL MOISTURE CONTENT
DEW POINT
DIFFUSION
DIFFUSIVITY
DRY BULB TEMPERATURE
DRYING
EQUILIBRIUM (CHEMICAL)
EQUILIBRIUM MOISTURE CONTENT
FREEZING POINT
HUMIDITY
HUMIDITY CONTROL
HYDROPHILIC PROPERTY
HYDROPHOBIC PROPERTY
HYGROSCOPICITY
MOISTURE CONTENT
MOISTURE CONTENT TESTING
MOISTURE STRENGTH CURVES
PSEUDO WET BULB TEMPERATURE
PSYCHROMETRY
RELATIVE HUMIDITY
SPECIFIC HUMIDITY
VAPOR PRESSURE
WET BULB TEMPERATURE

REGENERATED CELLULOSE
NT CELLOPHANE
BT POLYMERS
RT ALKALI CELLULOSE
CELLULOSE
CELLULOSE DERIVATIVES
CELLULOSE XANTHATE
CUPRAMMONIUM HYDROXIDE
CYANOETHYLCELLULOSE
POLYMERS
RAYON (REGENERATED CELLULOSE FIBERS)
VISCOSE RAYON

REGENERATED CELLULOSE FIBERS
USE RAYON (REGENERATED CELLULOSE FIBERS)

REGENERATED FIBERS (EXCLUDING CELLULOSIC AND PROTEIN)
NT ALGINATE FIBERS
RUBBER FIBER (NATURAL)
BT FIBERS
MAN MADE FIBERS
RT AZLON (REGENERATED PROTEIN FIBERS)
RAYON (REGENERATED CELLULOSE FIBERS)
RUBBER
SYNTHETIC FIBERS

REGENERATED PROTEIN FIBERS
USE AZLON (REGENERATED PROTEIN FIBERS)

REGISTERED PRINTING
USE DUPLEX PRINTING

REGRESSION
RT AUTOCORRELATION
CORRELATION
REGRESSION ANALYSIS
REGRESSION COEFFICIENT
STATISTICAL ANALYSIS
STATISTICS

REGRESSION ANALYSIS
RT OPERATIONS RESEARCH
REGRESSION
REGRESSION COEFFICIENT
STATISTICAL ANALYSIS

REGRESSION COEFFICIENT
BT STATISTICAL MEASURES
RT CORRELATION
CORRELATION COEFFICIENT
DEGREES OF FREEDOM
REGRESSION
REGRESSION ANALYSIS

REGULAR BRAIDS
NT CORE BRAIDS
DOUBLE BRAIDS
HOLLOW BRAIDS
NO CORE BRAIDS
PARACHUTE CORDS
BT BRAIDS

REGULAR SADDLE STITCHES
BT STITCHES (SEWING)

REGULAR TWILLS
BT TWILLS
RT CHEVRON TWILLS
FANCY TWILLS
HERRINGBONE TWILLS
LOW TWILLS
S TWILLS
STEEP TWILLS
Z TWILLS

REGULARITY
USE IRREGULARITY

REGULATION
USE CONTROL SYSTEMS

REGULATORS
RT AUTOLEVELLERS
AUTOMATIC CONTROL
AUTOMATION
CONTROL EQUIPMENT
CONTROL SYSTEMS
CONTROL VALVES

REINFORCED COMPOSITES
UF REINFORCED PLASTICS
NT FABRIC REINFORCED COMPOSITES
FIBER REINFORCED COMPOSITES
FILAMENT WOUND COMPOSITES
HONEYCOMB LAMINATES
POST FORMED LAMINATES
YARN REINFORCED COMPOSITES
BT COMPOSITES
RT COATED FABRICS
COMPRESSION MOLDING
CONTACT MOLDING
FLEXIBLE PLUNGER MOLDING
GLASS CLOTH
GLASS MAT
INJECTION MOLDING
LAMINATES
MATCHED DIE MOLDING
MOLDING
PLASTICS
PRESSURE BAG MOLDING
VACUUM BAG MOLDING

REINFORCED HOSES
BT YARN REINFORCED ELASTOMERS
RT FIBER REINFORCED COMPOSITES
FILAMENT WOUND COMPOSITES
INFLATABLE FABRICS
PIPES
PRESSURE SUITS
TIRES
V BELTS

REINFORCED KNITTED FABRICS
BT KNITTED FABRICS
RT HEELS (STOCKINGS)
HOSIERY
SOCKS
TOES

REINFORCED PLASTICS
USE REINFORCED COMPOSITES

RELATIVE DENSITY
USE SPECIFIC GRAVITY

RELATIVE HUMIDITY
BT HUMIDITY
RT BOILING POINT
CRITICAL MOISTURE CONTENT
DEW POINT
DIFFUSION
DIFFUSIVITY
DRY BULB TEMPERATURE
DRYING
EQUILIBRIUM MOISTURE CONTENT
EVAPORATION
FREEZING POINT
HUMIDITY
HYGROMETERS
MOISTURE CONTENT
PSEUDO WET BULB TEMPERATURE
PSYCHROMETRY
REGAIN
ROOM TEMPERATURE
SATURATED VAPOR
SPECIFIC HUMIDITY
VAPOR PRESSURE
WET BULB TEMPERATURE

RELATIVE VISCOSITY
BT VISCOSITY
RT INTRINSIC VISCOSITY
MELT VISCOSITY
SPINNING (EXTRUSION)
VISCOSIMETERS

RELAXATION
UF RELAXING
NT DRY RELAXATION
FABRIC RELAXATION
STRESS RELAXATION
WET RELAXATION
BT DRY FINISHING
WET FINISHING
RT CRABBING
CREEP RECOVERY
DELAYED RECOVERY
FABRIC SHRINKAGE
HEAT SETTING (SYNTHETICS)
KNITTING
LONDON SHRINKING
MAXWELL MODEL
PERMANENT SET
RECOVERY (SELF RESTORATION TO ORIGINAL CONDITION)
RELAXATION SHRINKAGE
RELAXATION SPECTRUM
RELAXATION TIME (MECHANICAL)
RESIDUAL STRESS
RHEOLOGY
SHRINKING
VISCOELASTICITY
VOIGT MODEL
WRINKLE RECOVERY

RELAXATION SHRINKAGE
BT FABRIC SHRINKAGE
SHRINKAGE
RT COCKLE (DEFECT)
COMPACTING
COMPRESSIVE SHRINKAGE
COMPRESSIVE SHRINKING
CRABBING
DIMENSIONAL STABILITY
DRY RELAXATION
FABRIC RELAXATION
FELTING SHRINKAGE
FILLING SHRINKAGE
FUSION CONTRACTION SHRINKAGE
LONDON SHRINKING
MOLDING SHRINKAGE
PRESHRINKING
RELAXATION
RESIDUAL SHRINKAGE
SHRINK RESISTANCE
STRESS RELAXATION
WARP SHRINKAGE
WET RELAXATION

RELAXATION SPECTRUM
RT COEFFICIENT OF CROSS VISCOSITY
CREEP RECOVERY
DELAYED RECOVERY
MAXWELL MODEL
PERMANENT SET
RECOVERY (SELF RESTORATION TO ORIGINAL CONDITION)
RELAXATION
RELAXATION TIME (MECHANICAL)
RHEOLOGY
STRESS RELAXATION
STRESS STRAIN PROPERTIES
VISCOELASTICITY
VOIGT MODEL

RELAXATION TIME (MECHANICAL)
BT STRESS STRAIN PROPERTIES
RT CONDITIONING
CRABBING
CREEP
CREEP RECOVERY
COEFFICIENT OF CROSS VISCOSITY
DELAYED RECOVERY
DRY RELAXATION
DYNAMIC RESPONSE
FABRIC RELAXATION
FABRIC SHRINKAGE
FIRMOVISCOSITY
HEAT SETTING (SYNTHETICS)
LINEAR VISCOELASTICITY
LONDON SHRINKING
MAXWELL MODEL
MODULUS
NONLINEAR VISCOELASTICITY
RELAXATION
PERMANENT SET
PLASTIC FLOW
RECOVERY (SELF RESTORATION TO ORIGINAL CONDITION)
RELAXATION SPECTRUM
RESIDUAL STRESS
RHEOLOGY
STRAIN
STRESS
STRESS RELAXATION
TIME
VISCOELASTICITY
VISCOSITY
VOIGT MODEL
WET RELAXATION
YIELD POINT

RELAXING
USE RELAXATION

REMAZOL (TN)
USE VINYLSULFONE DYES

REMODELLING
UF RESTYLING
RT ALTERATIONS
GARMENT MANUFACTURE
IMPROVEMENTS
LETTING OUT (GARMENTS)
REPAIRS
REVISIONS
TAILORING
TAKING IN (GARMENTS)

REMOTE CONTROL
RT AUTOMATIC CONTROL
AUTOMATION
CONTINUOUS CONTROL
CONTROL SYSTEMS
DIFFERENTIAL CONTROL
FEEDBACK CONTROL
INTEGRAL CONTROL
MANUAL CONTROL
ON OFF CONTROL

REMOVAL
RT EJECTION
EMPTYING
EXHAUSTION (MECHANICAL)
SOIL REMOVAL
STAIN REMOVAL
UNLOADING (PROCESS)
UNROLLING
UNWINDING

RENAISSANCE (LACE)
BT HAND MADE LACES
PILLOW LACES (BOBBIN LACES)

REPAIRS
RT ALTERATIONS
GARMENT MANUFACTURE
IMPROVEMENTS
INSPECTION
LETTING OUT (GARMENTS)
MAINTENANCE
MENDING
PERCHING
REMODELLING
TAILORING
TAKING IN (GARMENTS)

REPAIRS (MACHINERY)
USE MAINTENANCE

REPEATS
RT AMPLITUDE
AUTOCORRELATION
BARRE
DESIGN
FREQUENCY
PATTERN (FABRICS)
PATTERN REPEATS
PERIODICITY
VARIATION
WEAVE

REPP
BT FLAT WOVEN FABRICS
WOVEN FABRICS
RT PLAIN WEAVES

RESEARCH
RT COSTS
MANAGEMENT
MANUFACTURING
MARKETING
OPERATIONS RESEARCH
PROCESS DESIGN
PRODUCT DESIGN
PRODUCTION

RESERVE EFFECT
RT CONTRAST EFFECT
DYEING (FOR EFFECT)
TONE IN TONE DYEING
TWO COLOR DYEING

RESERVE PRINTING
USE RESIST PRINTING

RESIDUAL FAT
BT FATS
RT DETERGENCY
LUBRICANTS
OILS
SCOURED WOOL
SCOURING
WOOL

RESIDUAL SHRINKAGE
BT SHRINKAGE
RT COMPACTING
COMPRESSIVE SHRINKING
DIMENSIONAL STABILITY
FELTING SHRINKAGE
FILLING SHRINKAGE
FUSION CONTRACTION SHRINKAGE
MOLDING SHRINKAGE
PRESHRINKING
RELAXATION SHRINKAGE
SANFORIZING (TN)
SHRINK GRADING
SHRINK RESISTANCE
SHRINK RESISTANCE FINISHES
SHRINKING
WARP SHRINKAGE

RESIDUAL STRAIN
RT BENDING RECOVERY
CREASE RETENTION
PLASTIC STRAIN
RECOVERY (SELF RESTORATION TO ORIGINAL CONDITION)
RESIDUAL SHRINKAGE
RESIDUAL STRESS
STRAIN
TEXTURING

RESIDUAL STRESS
RT BENDING RECOVERY
CRABBING
DRY RELAXATION
FABRIC RELAXATION
FABRIC SHRINKAGE
HEAT SETTING (SYNTHETICS)
LONDON SHRINKING
PLASTIC STRAIN
RECOVERY (SELF RESTORATION TO ORIGINAL CONDITION)
RELAXATION
RELAXATION TIME (MECHANICAL)
RESIDUAL STRAIN
RESIDUAL TORQUE
STRESS
STRESS RELAXATION
TEXTURING
WET RELAXATION

RESIDUAL TORQUE
RT BALANCED YARNS
BUCKLING
CREPE
CURL
LOOPING
MECHANICAL PROPERTIES
RESIDUAL STRESS
SNARLING TENDENCY
TEXTURING
TORSIONAL BUCKLING
TORSIONAL ENERGY
TWIST LIVELINESS
TWISTING MOMENTS
WILDNESS

RESILIENCE
UF SPRINGINESS
BT END USE PROPERTIES
STRESS STRAIN PROPERTIES
STRESS STRAIN PROPERTIES (TENSILE)
RT AESTHETIC PROPERTIES
BENDING
BULK
BULK DENSITY
COMPRESSIBILITY
COMPRESSION TESTING
CREASE RETENTION
CREASING
CREEP RECOVERY
DEFLECTION
DEFORMATION
DELAYED RECOVERY
ELASTIC RECOVERY
ENERGY OF RETRACTION
FABRIC PROPERTIES
FABRIC PROPERTIES (MECHANICAL)
HAND
LIVELINESS
PERMANENT SET
RECOVERY (SELF RESTORATION TO ORIGINAL CONDITION)
STRESS STRAIN CURVES
WORK RECOVERY

RESIN CURING MACHINES
USE CURING OVENS

RESIN FINISHES
BT FINISH (SUBSTANCE ADDED)
RT ABRASION RESISTANCE FINISHES
ACETALS
ANTIFELTING AGENTS
ANTISLIP AGENTS
CATALYSIS
CHLOROHYDRINS
CRIMP FIXING
DEFERRED CURE METHOD
DIMENSIONAL STABILITY
DOUBLE CURE METHOD
DURABLE PRESS FINISHES
FORMALDEHYDE
HOT PRESS METHOD (DURABLE PRESS)
PRE-CURE METHOD
RESINS
SHRINK RESISTANCE FINISHES
STIFFENERS (AGENTS)
WATERPROOFING
WASH WEAR FINISHES

RESIN PRECONDENSATES
USE PRECONDENSATES

RESINOGRAPHY
BT MICROSCOPY
RT ELECTRON MICROSCOPY
EMBEDDING MEDIUM
INFRARED MICROSCOPY
INTERFERENCE MICROSCOPY
LIGHT MICROSCOPY (OPTICAL)
ULTRAVIOLET MICROSCOPY
X RAY MICROSCOPY

RESINS
NT ACETAL RESINS
ACRYLIC RESINS
AMINO RESINS
COATING RESINS
EPOXY RESINS
ION EXCHANGE RESINS
LAMINATING RESINS
MELAMINE RESINS
PHENOLIC RESINS
POLYESTER RESINS
SILICONE RESINS
THERMOPLASTIC RESINS
THERMOSETTING RESINS
BT POLYMERS
RT ADHESION
ADHESIVES
COATED FABRICS
COATING (PROCESS)
COMPOSITES
COPOLYMERS
CURING

CURING OVENS

FINISH (SUBSTANCE ADDED)
GELATINE
GLUES
HEAT SETTING (SYNTHETICS)
LAMINATES
ORGANOSOLS
PAPER MAKING
PLASTICS
POLYMERS
RESIN FINISHES
ROSIN
SIZES (SLASHING)
SYNTHETIC LEATHER
VINYL COATED FABRICS

RESIST PRINTING
UF RESERVE PRINTING
BT PRINTING
RT BATIK DYEING
BLOCK PRINTING
DISCHARGE PRINTING
DUPLEX PRINTING
MELANGE PRINTING
ROLLER PRINTING
SCREEN PRINTING
THROUGH PRINTING

RESISTANCE (ELECTRICAL)
BT ELECTRICAL PROPERTIES
RT CAPACITANCE (ELECTRICAL)
DIELECTRIC STRENGTH
IMPEDANCE
INDUCTANCE
INSULATION (ELECTRICAL)
LEAKAGE RESISTANCE (ELECTRICAL)
REACTANCE (ELECTRICAL)

RESISTANCE (HEAT)
USE HEAT RESISTANCE

RESISTANCE TO BLEEDING
BT COLORFASTNESS
END USE PROPERTIES
RT BLEEDING
BLEEDING TESTING
CROCKFASTNESS
LIGHTFASTNESS (COLOR)
WASHFASTNESS (COLOR)

RESISTANCE TO DELAMINATION
BT MECHANICAL DETERIORATION PROPERTIES
MECHANICAL PROPERTIES
RT COMPOSITES
DELAMINATING
LAMINATED FABRICS
LAMINATES
PEELING

RESISTIVITY
UF CONDUCTIVITY (ELECTRICAL)
ELECTRICAL CONDUCTIVITY
ELECTRICAL RESISTIVITY
RT CONDUCTOMETRY
DIELECTRIC STRENGTH
DISCHARGE (ELECTRIC)
DISSIPATION
FABRIC PROPERTIES
FABRIC PROPERTIES (PHYSICAL EXCLUDING MECHANICAL)
INSULATION (ELECTRICAL)

RESOLUTION (OF FORCES)
RT FORCE
NORMAL FORCE
TANGENTIAL FORCE

RESONANCE
USE RESONANT FREQUENCY

RESONANT FREQUENCY
UF NATURAL FREQUENCY
RESONANCE
RT CAPACITANCE (ELECTRICAL)
DAMPERS
DYNAMIC BALANCING
DYNAMIC CHARACTERISTICS
DYNAMIC RESPONSE
ENERGY ABSORPTION
FREQUENCY
FREQUENCY RESPONSE
SPRING CONSTANT
TRANSVERSE VIBRATIONS
VIBRASCOPE
VIBRATION ISOLATORS

RESPONSE TIME
RT AUTOMATIC CONTROL
CONTROL CHARACTERISTICS
CONTROL SYSTEMS
DYNAMIC CHARACTERISTICS

RESTRAINING AGENTS
USE RETARDING AGENTS

RESTYLING
USE REMODELLING

RESULTS
RT COMPUTATIONS
EXPERIMENTATION
MATHEMATICAL ANALYSIS
PRODUCTION
RESEARCH
ROOTS (MATHEMATICAL)
SOLUTIONS (MATHEMATICAL)

RETARDATION
USE ACCELERATION (MECHANICAL)

RETARDERS
BT MILITARY PRODUCTS
RT ARRESTERS
BRAKES (FOR ARRESTING MOTION)
MAGNETIC BRAKES
MECHANICAL BRAKES
PARACHUTES

RETARDING AGENTS
UF RESTRAINING AGENTS
BT DYEING AUXILIARIES
LEVELLING AGENTS
RT ACCELERANTS (DYEING)
ANIONIC SURFACTANTS
CATALYSTS
CATIONIC SURFACTANTS
GLAUBERS SALT
NONIONIC SURFACTANTS
SODIUM CHLORIDE
SODIUM SULFATE
UNION DYEING

RETTING
RT BAST FIBERS
FLAX
HEMP
JUTE
SCUTCHING (FLAX)

REVERES
BT GARMENT COMPONENTS
RT TAILORING

REVERSAL
RT ACCELERATION (MECHANICAL)
ADJUSTMENTS
MOTION
POSITION
PROCESSING DIRECTION
VELOCITY

REVERSE PLAITING
RT KNITTING
PLAITING (KNITTING)
REVERSE WELTING

REVERSE WELTING
RT KNITTING
REVERSE PLAITING
WELTING

REVERSIBLE FABRICS
RT FABRICS (ACCORDING TO STRUCTURE)
REVERSIBLE GARMENTS
WOVEN FABRICS

REVERSIBLE GARMENTS
BT GARMENTS
RT REVERSIBLE FABRICS

REVIEWING
USE REVIEWS

REVIEWS
UF REVIEWING
RT ANALYZING
BIBLIOGRAPHY
COMPILATIONS
EVALUATION
HISTORY
LITERATURE SURVEYS
SURVEYS
THEORIES
YEARS COVERAGE (1)
YEARS COVERAGE (5)
YEARS COVERAGE (10)
YEARS COVERAGE (25)
YEARS COVERAGE (50)
YEARS COVERAGE (75)
YEARS COVERAGE (100)

REVISIONS
RT ADJUSTMENTS
ALTERATIONS
REMODELLING

REVOLVING FLAT CARDS
UF FLAT TOP CARDS
BT CARDS
RT COTTON CARDS
ROLLER TOP CARDS
SRRL CARD

REWETTING AGENTS
BT SURFACTANT APPLICATIONS
RT SURFACTANTS
WETTING AGENTS

RHEOLOGICAL PROPERTIES
USE STRESS STRAIN PROPERTIES

RHEOLOGY
NT MACRORHEOLOGY
METARHEOLOGY
MICRORHEOLOGY
PHENOMENOLOGICAL RHEOLOGY
RT BULGING (FILAMENT)
COEFFICIENT OF CROSS VISCOSITY
COEFFICIENT OF FRICTION
COEFFICIENT OF VISCOSITY
COLD DRAWING
CREEP
CREEP RECOVERY
DELAYED RECOVERY
DILATANCY
DRAWING (FILAMENT)
ELASTIC STRAIN
ELASTICITY
FIBERS
FIRMOVISCOSITY
FRACTIONATION
HOT DRAWING
INTERNAL FRICTION
LINEAR VISCOELASTICITY
MAXWELL MODEL
MELT VISCOSITY
NECKING (FILAMENT)
NONLINEAR VISCOELASTICITY
PERMANENT SET
PLASTIC FLOW
PLASTICITY
PLASTICOVISCOSITY
POLYMERS
RELAXATION
RELAXATION SPECTRUM
RELAXATION TIME (MECHANICAL)
STRENGTH OF MATERIALS
THIXOTROPY
VISCOELASTICITY
VISCOSITY
VISCOUS FLOW
VOIGT MODEL
YIELD (MECHANICAL)

RHODANINE
BT HETEROCYCLIC COMPOUNDS
SULFUR COMPOUNDS
RT ULTRAVIOLET STABILIZERS

RIB KNITTED FABRICS
BT FILLING KNITTED FABRICS
KNITTED FABRICS
RT CIRCULAR KNITTED FABRICS
DOUBLE KNIT FABRICS
EIGHTLOCK (KNITTED FABRICS)
FLAT KNITTED FABRICS
FULLY FASHIONED FABRICS
MACHINE KNITTING
PLAIN KNITTED FABRICS
RIB KNITTING

RIB KNITTING
BT KNITTING
RT DOUBLE PIQUE
INTERLOCK (KNITTED FABRICS)
LOCKING COURSES
PLAIN KNITTING
PURL KNITTING
RIB KNITTED FABRICS

RIB SET
RT CARD CLOTHING
CARD CLOTHING FOUNDATION
CARD WIRES
COUNT CROWN AND GAUGE
FILLET CLOTHING
KNEE
METALLIC CARD CLOTHING
NOGG
SHEET CLOTHING
TWILL SET

RIB TRANSFER
RT FULLY FASHIONED KNITTING MACHINES
KNITTING

RIB WEAVES
UF DERBY RIB
BT WEAVE (TYPES)
RT FLAT WOVEN FABRICS
MAROCAIN
STRIPES

RIBBON
NT TYPEWRITER RIBBONS
BT FLAT WOVEN FABRICS
NARROW FABRICS
WOVEN FABRICS
RT PLAIN WEAVES
TAPE

RIBBON BREAKERS
RT BUILDER MOTIONS
CONES
CONING
DRUM WINDERS
DRUM WINDING
ENTANGLEMENTS
OPEN WINDING
PACKAGES
RIBBON BREAKING
RIBBONING
SLOUGHING
UNWINDING
WINDING

RIBBON BREAKING
RT CLOSE WIND
CLOSE WINDING
CONING
DRUM WINDING
OPEN WINDING
PRECISION WINDING
RIBBON BREAKERS
RIBBONING
WIND
WINDING

RIBBON CROSS SECTION
USE FLAT CROSS SECTION

RIBBON LAP MACHINES
BT LAPPERS
RT COMBER LAP MACHINES
COMBING
LAP ARBOURS
LAP RODS
LAP WASTE
LAPS
RIBBON LAPS

RIBBON LAPS
BT LAPS
RT COMBER LAPS
COTTON SYSTEM PROCESSING
DETACHING ROLLS
HALF LAP
LAP ARBOURS
LAP PLATES
LAP RODS
LAP WASTE
PICKER LAPS
RECTILINEAR COMBS
RIBBON LAP MACHINES

RIBBONING
RT CONING
ENTANGLEMENTS
OPEN WIND
PACKAGE DENSITY
RIBBON BREAKERS
RIBBON BREAKING
SLOUGHING
UNWINDING
WIND

RICHNESS
BT AESTHETIC PROPERTIES
FABRIC PROPERTIES
FABRIC PROPERTIES (AESTHETIC)
RT AESTHETIC APPEAL
APPEARANCE
BLOOM
BRIGHTNESS
COLOR
DESIGN
DRABNESS
DULLNESS
(CONT.)

RT FINISH (PROPERTY)
LUSTER
LUXURIOUSNESS
OPTICAL PROPERTIES
PATTERN (FABRICS)
SHADE
SPARKLE
WEAVE

RIETER CONVERTER
BT CONVERTERS (TOW)
TOW TO TOP MACHINES
RT TOW TO TOP CONVERSION

RIGHT HAND COORDINATE SYSTEMS
RT LEFT HAND COORDINATE SYSTEMS
LEFT HAND RULE
RIGHT HAND RULE
SENSE (DIRECTION)

RIGHT HAND RULE
RT LEFT HAND COORDINATE SYSTEMS
LEFT HAND RULE
RIGHT HAND COORDINATE SYSTEMS
SENSE (DIRECTION)

RIGHT HAND TWILLS
USE S TWILLS

RIGHT HAND TWIST
USE S TWIST

RIGID FOAMS
BT FOAMS
RT ELASTIC FOAMS
FOAM RUBBER
URETHANE FOAMS
VINYL FOAMS

RIGIDITY
USE STIFFNESS

RIGMEL PROCESS
RT FABRIC SHRINKAGE
SHRINK PROOFING
SHRINKING

RILSAN (TN)
BT NYLON (POLYAMIDE FIBERS)
NYLON 11

RING CONDENSERS
BT CONDENSERS
RT TAPE CONDENSERS
WOOLEN SYSTEM PROCESSING

RING FRAME SPINNING
USE RING SPINNING

RING FRAMES
UF RING SPINNING FRAMES
BT SPINNING FRAMES
RT AMBLER SUPERDRAFT
BALLOON CONTROL RINGS
BALLOON SEPARATORS
BALLOONS
BOBBINS
BUILDER MOTIONS
CAP SPINNING FRAMES
CASABLANCA SYSTEM
CONTROL SURFACES
COTTON SYSTEM PROCESSING
CREELING
DOFFERS (PACKAGE)
DOFFING (PACKAGE)
DOUBLE RING SYSTEM
DRAFTING ROLLS
FALSE TWIST TUBES
FLYER SPINNING FRAMES
MULES
PACKAGES
POT SPINNING FRAMES
RING RAILS
RING SPINNING
RINGS
SPINDLE RAILS
SPINDLES
SPINNING
THREAD GUIDES
TRAVELERS
WOOLEN SYSTEM PROCESSING
WORSTED SYSTEM PROCESSING

RING RAILS
RT BALLOON SEPARATORS
BUILDER MOTIONS
LAY (WIND)
LIFT (WINDING)
PACKAGES
RING FRAMES
RING SPINNING
RINGS
SPINDLE RAILS
SPINNING
SPINNING FRAMES
TRAVERSE
TWISTERS

RING SPINNING
UF RING FRAME SPINNING
BT SPINNING
RT AIR SPINNING
ANTIWEDGE RINGS
BALLOON COLLAPSE
BALLOON CONTROL
BALLOON CONTROL RINGS
BALLOON DYNAMICS
BALLOON SEPARATORS
BALLOON TENSION
BALLOONS
CAP SPINNING
CIRCULAR TRAVELERS
COLLAPSED BALLOON SPINNING
CORE SPINNING
CORE SPUN YARNS
DOFFERS (PACKAGE)
DOUBLE RING SYSTEM
DOUBLE ROVINGS
DOWNTWISTING
DRAFTING (STAPLE FIBER)
ELLIPTICAL TRAVELERS
END BREAKAGES
FIBER MIGRATION
FLUID SPINNING
FLYER SPINNING
FRONT ROLLS
HAIRINESS
LASHING END RADIUS
LUBRICATED RINGS
MAGNETIC RINGS
MIGRATION PERIOD
MULE SPINNING
NOVELTY TWISTERS
NYLON TRAVELERS
PACKAGES
PLYING TRAVELERS
POT SPINNING
RING FRAMES
RING RAILS
RING TWISTERS
RING TWISTING
RINGS
SINTERED METAL RINGS
SPINDLES
SPINNING (EXTRUSION)
SPINNING BALLOONS
SPINNING TRAVELERS
STEEL TRAVELERS
TRAVELER CHATTER
TRAVELERS
TWIST
TWIST TRIANGLE
TWISTING
TWO FOR ONE TWISTING
UNLUBRICATED RINGS
WINDING
WORSTED SPUN YARNS
YARN GEOMETRY

RING SPINNING FRAMES
USE RING FRAMES

RING TRAVELERS
USE TRAVELERS

RING TWISTERS
BT TWISTERS
RT LOOPS (NOVELTY YARNS)
NOVELTY TWISTERS
NOVELTY YARNS
PLIED YARNS
PLYING
RING RAILS
RING SPINNING
RING TWISTING
RINGS
SPINDLES
TRAVELERS

RING TWISTING
BT TWISTING
RT DOWNTWISTING
FIBER MIGRATION
MIGRATION PERIOD
RING RAILS
RING SPINNING
RING TWISTERS
RINGS
TWIST TRIANGLE

RINGS
UF SPINNING RINGS
NT ANTIWEDGE RINGS
LUBRICATED RINGS
MAGNETIC RINGS
SINTERED METAL RINGS
UNLUBRICATED RINGS
RT BALLOON DYNAMICS
BALLOON SEPARATORS
BALLOON TENSION
BALLOONS
CAPS (SPINNING)
CIRCULAR TRAVELERS
ELLIPTICAL TRAVELERS
FLANGES
FLYERS

RT NOVELTY TWISTERS
NYLON TRAVELERS
PLYING
PLYING TRAVELERS
RING FRAMES
RING RAILS
RING SPINNING
RING TWISTERS
SPINNING
SPINNING FRAMES
SPINNING TRAVELERS
STEEL TRAVELERS
TRAVELER CHATTER
TRAVELERS
TWISTERS
TWISTING
WEDGING

RINGS (BOBBINS)
USE BOBBIN RINGS

RINSING
BT WET FINISHING
RT CARBONIZING
CONTINUOUS SCOURING
LAUNDERING
MANGLING
NEUTRALIZING
PADDING
SCOURING
SCOURING BATHS
SCOURING TRAINS
SOAP SODA SCOURING
SPRAYING

RINSING BATHS (SCOURING)
USE SCOURING BATHS

RIPPING
BT GARMENT MANUFACTURE
RT ALTERATIONS
CUTTING (TAILORING)
PICKING (SNAGGING)
SEAMS
SEWING
SNAGGING
STITCHES (SEWING)
TAILORING
TEARING (TAILORING)

RIPPLES
RT BUCKLING
COCKLING
GATHERING
PINCHING
PUCKERING
RUFFLING
SEAM PUCKER
SEAMING
SEWING
SHIRRING
STITCHING
STRETCH PUCKERS
THREAD TENSION
WRINKLING
YARN SLIPPAGE

RIVETS
BT FASTENERS
RT BUCKLES
BUTTONS
GARMENT MANUFACTURE
GARMENTS
HOOK CLOSURES
SNAP FASTENERS
ZIPPERS

ROBBING BACK
RT KNITTING
NEEDLE LOOPS
SINKER LOOPS
STITCH LENGTH
YARN TENSION

ROBES
BT ATHLETIC CLOTHING
CEREMONIAL CLOTHING
NIGHTWEAR

RODNEY HUNT SATURATOR
BT NONWOVEN FABRIC MACHINES
SATURATORS
RT BINDERS (NONWOVENS)
BONDED FIBER FABRICS
CARD WEBS
CURING OVENS
CURLATORS (TN)
PADDERS
WALDRON SATURATOR

ROLL BOSSES
RT CROWNS (ROLL)
DRAFTING ROLLS
DRIVEN ROLLS
DRIVING ROLLS
MAGNADRAFT (TN)
PRESSURE ROLLS
ROLL WEIGHTING
ROLLER NIP
ROLLS

ROLL CLEARERS
RT DRAFTING ROLLS
LINT BLADES
ROLLS
VACUUM END COLLECTORS

ROLL COVERINGS
UF ROLLER LEATHERS
RT BELTS
CALENDER ROLLS
CORK
DRAFTING ROLLS
FEED ROLLS
NIP
PAPERMAKERS FELTS
PERALTA ROLLS
PRESS SECTION
PRESSURE ROLLS
ROLLER NIP
ROLLS
RUBBER
SQUEEZE ROLLS
SUCTION PRESSES
TEFLON (TN)

ROLL CROWNS
USE CROWNS (ROLL)

ROLL ECCENTRICITY
RT BOWING
CROWNS (ROLL)
DRAFTING ROLLS
ECCENTRICITY

ROLL WEIGHTING
RT DRAFTING (STAPLE FIBER)
DRAFTING ROLLS
DRAWFRAMES
PRESSURE ROLLS
ROLL BOSSES
WEIGHT

ROLLER BEARINGS
NT TAPERED ROLLER BEARINGS
BT BEARINGS
RT AIR BEARINGS
BALL BEARINGS
CONICAL BEARINGS
JOURNAL BEARINGS
NEEDLE BEARINGS
NYLON BEARINGS
SLEEVE BEARINGS
THRUST BEARINGS

ROLLER COATING
NT KISS COATING
BT COATING (PROCESS)
RT BACK FILLING MACHINES
CALENDER COATING
CAST COATING
COATED FABRICS
DIP COATING
EXTRUSION COATING
FLEXIBLE FILM LAMINATING
KNIFE COATING
NIP COATING
PAPER MAKING

ROLLER DRAFT
RT ACTUAL DRAFT
BREAK DRAFT
DRAFT
DRAFTING (STAPLE FIBER)
MULE DRAFT

ROLLER DRAFTING
BT COTTON SYSTEM PROCESSING
DRAFTING (STAPLE FIBER)
RT ACTUAL DRAFT
APRON DRAFTING
DRAFT
DRAFTING ROLLS
DRAFTING ZONE
DRAWFRAMES
MULE DRAFT
ROLLER DRAFT

ROLLER GINNING
BT GINNING
RT COTTON GINS
GIN CUT
GIN FALL
KNIFE BLADE GINNING
NEPS
SAW BLADE GINNING

ROLLER HARDENERS
BT HARDENERS (FELTS)
RT FELTING
FELTING MACHINES
FELTS
PLATEN HARDENERS

ROLLER LAP
RT DRAFTING (STAPLE FIBER)
DRAFTING ROLLS
LAP ARBOURS
LAP RODS
RT LAPS
ROLLS
SPINNING

ROLLER LEATHERS
USE ROLL COVERINGS

ROLLER NIP
RT COMPRESSIBILITY
CONTACT AREA
FIBER SLIPPAGE
FLUTED ROLLS
FRONT ROLLS
PAPER MACHINES
PRESSURE ROLLS
RATCH
ROLL COVERINGS
ROLL ECCENTRICITY
ROLLER SLIPPAGE
ROLLS

ROLLER PRINTING
UF CALENDER PRINTING
CYLINDER PRINTING
DIRECT PRINTING
BT PRINTING
RT BLOCK PRINTING
DISCHARGE PRINTING
DUPLEX PRINTING
MELANGE PRINTING
PRINTING MACHINES
PRINTING ROLLS
RESIST PRINTING
SCREEN PRINTING
TWO STAGE PRINTING

ROLLER PRINTING MACHINES
BT PRINTING MACHINES
RT BLOCK PRINTING EQUIPMENT
DUPLEX PRINTING MACHINES
ENGRAVED CYLINDERS
PRINTING
PRINTING ROLLS
SCREEN PRINTING MACHINES

ROLLER SHAFTS
RT COTS (ROLLERS)
DRAFTING (STAPLE FIBER)
DRAWFRAMES
ROLL BOSSES
ROLL ECCENTRICITY
ROLLS

ROLLER SLIP
USE ROLLER SLIPPAGE

ROLLER SLIPPAGE
UF ROLLER SLIP
BT SLIPPAGE
RT ALLOWANCE FACTOR (WARP PREPARATION)
CALENDERING
CONTACT AREA
CROWNS (ROLL)
DRAFTING (STAPLE FIBER)
DRAFTING FORCE
FIBER SLIPPAGE
FRICTION
LUBRICATION
ROLL ECCENTRICITY
ROLLER NIP
ROLLING
ROLLING FRICTION
ROLLS
SHEAR (MODE OF DEFORMATION)
SLIDING
STICK SLIP FRICTION

ROLLER TOP CARDS
BT CARDS
RT REVOLVING FLAT CARDS
WOOLEN CARDS
WORSTED CARDS

ROLLERS
USE ROLLS

ROLLING
RT BOUNDARY LUBRICATION
COMPACTING
COMPRESSIVE SHRINKING
DRAWING (FILAMENT)
DYNAMIC FRICTION
FRICTION
HYDRODYNAMIC LUBRICATION
LUBRICATION
ROLLER SLIPPAGE
ROLLING FRICTION
ROLLING MILLS
SANDWICH ROLLING MILLS
SLIDING
WEAR
WIRE DRAWING

ROLLING FRICTION
BT FRICTION
RT COEFFICIENT OF FRICTION
DYNAMIC FRICTION
RT ROLLER SLIPPAGE
ROLLING
SLIDING FRICTION
STATIC FRICTION

ROLLING LOCKER MACHINES
BT LACE MACHINES
RT BAR WARP MACHINES
BARMEN MACHINES
BOBBINET MACHINES
CURTAIN MACHINES
DOUBLE LOCKER MACHINES
GO THROUGH MACHINES
LEAVERS MACHINES
MECHLIN MACHINES
SIVAL MACHINES
STRING WARP MACHINES
WARP LACE MACHINES

ROLLING MILLS
RT COMPACTOR
MANGLES
PADDERS
PAPER MACHINES
ROLLING
SANDWICH ROLLING MILLS
SANFORIZING MACHINES
WIRE DRAWING

ROLLS
UF BOWLS
ROLLERS
NT BACK ROLLS
BOTTOM ROLLS
CALENDER ROLLS
CARD ROLLS
COVERED ROLLS
DELIVERY ROLLS
DRAFTING ROLLS
DRIVEN ROLLS
DRIVING ROLLS
FEED ROLLS
FLUTED ROLLS
FRONT ROLLS
GODET ROLLS
IMMERSION ROLLS
INTERMEDIATE ROLLS
KISS ROLLS
NIP ROLLS
PERALTA ROLLS
PLAIN ROLLS
PORCUPINE ROLLS
PRESS SECTION
PRESSURE ROLLS
PRINTING ROLLS
SAW TOOTHED ROLLS
SQUEEZE ROLLS
SUCTION PRESSES
TOP ROLLS
TUMBLERS (ROLLS)
RT APRONS
CONTROL SURFACES
CORK
COTS (ROLLERS)
CROWNS (ROLL)
FINISHING MACHINERY (GENERAL)
GRINDING ROLLS (CARDS)
HYDROSTATIC LOADING
MANGLES
NIP
PAPER MACHINES
RATCH
ROLL COVERINGS
ROLLER LAP
ROLLER NIP
ROLLER SHAFTS
ROLLER SLIPPAGE
RUBBER

ROMNEY MARSH WOOL
BT LONG WOOL
WOOL

ROMPERS
BT INFANTS CLOTHING

RONGALITE (TN)
USE FORMALDEHYDE SULFOXYLATES

ROOTS (MATHEMATICAL)
RT COMPUTATIONS
EQUATIONS
MATHEMATICAL ANALYSIS
RESULTS
SOLUTIONS (MATHEMATICAL)

ROOM TEMPERATURE
BT TEMPERATURE
RT AMBIENT TEMPERATURE
HIGH TEMPERATURE
LOW TEMPERATURE
RELATIVE HUMIDITY

ROPE BLEACHING
BT BLEACHING
ROPE PROCESSING
WET FINISHING
RT OPEN WIDTH BLEACHING
ROPE DYEING
ROPE SCOURING

ROPE DYEING
BT PIECE DYEING
ROPE PROCESSING
RT BECK DYEING
CORDAGE MACHINES
OPEN WIDTH DYEING
ROPE BLEACHING
ROPE MARKS
ROPE SCOURING

ROPE FORM PROCESSING
USE ROPE PROCESSING

ROPE MACHINES
RT CORDAGE
CORDAGE MACHINES
ROPES
TWISTERS

ROPE MARKS
BT DYEING DEFECTS
FABRIC DEFECTS
RT CREASE MARKS
ROPE DYEING
WET FINISHING

ROPE PROCESSING
UF ROPE FORM PROCESSING
NT ROPE BLEACHING
ROPE DYEING
ROPE SCOURING
RT OPEN WIDTH PROCESSING
WET FINISHING

ROPE SCOURING
BT ROPE PROCESSING
SCOURING
RT CONTINUOUS SCOURING

KIER SCOURING
OPEN WIDTH SCOURING
PACKAGE SCOURING
ROPE BLEACHING
ROPE SCOURING MACHINES

ROPE SCOURING MACHINES
UF DOLLY
BT SCOURING MACHINES
RT ROPE SCOURING

ROPES
NT WIRE ROPES
BT CORDAGE
FIBER ASSEMBLIES
RT BRAIDS
CABLES
CABLING
CORDAGE MACHINES
CORDS
FISHING LINES
HAWSERS
HELIX ANGLE
HEMP
LAY (ROPE)
PLIED YARNS
SAILCLOTH
SISAL
STRANDS
TIRE CORDS
TWIST
YARNS

ROSE POINT
BT HAND MADE LACES
PILLOW LACES (BOBBIN LACES)

ROSIN
UF ROSIN DERIVATIVES
RT ALUMINUM ABIETATE
FATTY ACIDS
RESINS
SIZES (SLASHING)
SIZING (PAPERMAKING)
SLASHING
SODIUM ROSINATE

ROSIN DERIVATIVES
USE ROSIN

ROT RESISTANCE
USE MILDEW RESISTANCE

ROT RESISTANCE FINISHES
USE MILDEW RESISTANCE FINISHES

ROT RESISTANCE TREATMENTS
USE MILDEW RESISTANCE TREATMENTS

ROTAFORMERS
BT PAPER MACHINES
RT CYLINDER MACHINES
FOURDRINIER MACHINES
INVERFORM MACHINES
KAMYR MACHINES
VERTAFORMERS
WET MACHINES (PAPERMAKING)

ROTAMETERS
USE FLOW CONTROL

ROTARY BATTERIES
USE ROTARY MAGAZINES

ROTARY CUTTING MACHINES
BT CUTTING MACHINES
RT BAND KNIFE CUTTING MACHINES
CLICKERS
CUTTING (TAILORING)
RECIPROCATING CUTTING MACHINES
SCISSORS

ROTARY DYEING MACHINES
BT DYEING MACHINES (BY MATERIAL ASSEMBLY)
RT GARMENT DYEING MACHINES
HOSIERY DYEING MACHINES

ROTARY MAGAZINES
UF ROTARY BATTERIES
RT FILLING YARNS
LOOMS
PICKING (WEAVING)
QUILLS
SHUTTLE BOXES
SHUTTLES
UNIFIL (TN)
WEAVING

ROTARY SCREEN PRINTING
BT PRINTING
SCREEN PRINTING
RT FLAT SCREEN PRINTING
SCREEN PRINTING MACHINES

ROTARY TRAVERSE
BT TRAVERSE
RT RECIPROCATING TRAVERSE
WINDERS
WINDING

ROTATION
RT ANGULAR VELOCITY
AXIAL MOTION
CENTRIFUGAL FORCE
CONSTANT SPEED
CORIOLIS ACCELERATION
CORIOLIS FORCE
DYNAMIC BALANCING
FORCE
SPINDLE SPEED
TRANSLATION
VARIABLE SPEED

ROTATIONAL VELOCITY
USE ANGULAR VELOCITY

ROTH SYSTEM
RT ACTUAL DRAFT
CASABLANCA SYSTEM
DRAFTING (STAPLE FIBER)
DUO ROTH SYSTEM
HIGH DRAFT
SHAW SYSTEM
SPINNING FRAMES

ROTOSET YARNS (TN)
BT FILAMENT YARNS
RT AIR JET TEXTURED YARNS
CYCLOSET YARNS (TN)
FALSE TWIST
FALSE TWIST YARNS
INTERLACING (FILAMENT IN YARN)
PRODUCERS TWIST
PRODUCERS YARNS
REAL TWIST
ZERO TWIST

ROTTING
USE MILDEW RESISTANCE

ROUGHNESS
BT AESTHETIC PROPERTIES
STRUCTURAL PROPERTIES
SURFACE PROPERTIES (MECHANICAL)
RT APPEARANCE
COEFFICIENT OF FRICTION
COMFORT
COMPRESSIBILITY
FABRIC PROPERTIES
FABRIC PROPERTIES (AESTHETIC)
FABRIC PROPERTIES (MECHANICAL)
FINISH (PROPERTY)
FRICTION
GROOVES
HAIRINESS
HAND
HARSHNESS
HOLES
LUSTER
LUXURIOUSNESS
PRICKLINESS
ROUGHNESS COEFFICIENT
SKIN IRRITATION
SLIDING

RT SLIP RESISTANCE
SMOOTHNESS
SOFTNESS
SURFACE CONTOUR
SURFACE FRICTION
TEXTURE

ROUGHNESS COEFFICIENT
BT SURFACE PROPERTIES (MECHANICAL)
RT COEFFICIENT OF FRICTION
FRICTION
HAIRINESS
ROUGHNESS
SMOOTHNESS
SURFACE CONTOUR

ROUND CROSS SECTION
BT FIBER CROSS SECTIONS

ROUNDNESS
BT SHAPE
RT CORES (CENTER)
CROSS SECTIONS
CYLINDERS
DIAMETER
ELLIPTICITY
HOLES
HOLLOWNESS
PACKING FACTOR
PORES
SHAPE FACTOR
SPHERICITY
SYMMETRY
YARN DIAMETER
YARN PROPERTIES

ROVANA (TN)
BT SARAN POLY (VINYLIDENE CHLORIDE) FIBERS
RT FLAT CROSS SECTION

ROVING (OPERATION)
RT DRAFTING (STAPLE FIBER)
ROVING FRAMES
ROVING WIND
ROVINGS
SPINNING
TWIST
TWISTING

ROVING FRAMES
UF INTERMEDIATE FRAMES
ROVING MACHINES
SLUBBING FRAMES
SPEED FRAMES
NT ROVOMATIC (TN)
RT CLEARERS (LINT)
DRAWFRAMES
DOUBLE TRUMPETS
DUO ROTH SYSTEM
FLYER LEG
FOOTSTEP BEARINGS
ROVING (OPERATION)
ROVING WIND
ROVINGS
SPINNING FRAMES

ROVING MACHINES
USE ROVING FRAMES

ROVING WIND
BT WIND
RT CLOSE WIND
COMBINATION WIND
FILLING WIND
OPEN WIND
ROVING (OPERATION)
ROVING FRAMES
WARP WIND
WINDING

ROVINGS
BT FIBER ASSEMBLIES
RT CAP SPINNING FRAMES
CAPACITANCE TYPE IRREGULARITY TESTERS
CARD SLIVERS
DOUBLE ROVINGS
DRAFT
DRAFTING (STAPLE FIBER)
DRAFTING MACHINES
DRAFTING ROLLS
DRAFTING WAVES
DRAFTING ZONE
DRAWFRAMES
FIELDEN WALKER IRREGULARITY TESTER (TN)
FLYERS
HANK NUMBER
HANKS
INDIRECT COUNT
IRREGULARITY TESTERS
LAPS
LINDSLEY RATIOS
NEPS
OVERLAP (FIBERS)
PNEUMATIC IRREGULARITY TESTEI

RT ROVING (OPERATION)
ROVING FRAMES
SKEINS
SLIVERS
SLUBBINGS
STRANDS
THICK SPOTS
TOPS
TOW
TWIST
TWISTING
USTER IRREGULARITY TESTER (TN)
WEBS
YARNS

ROVOMATIC (TN)
BT ROVING FRAMES
RT DRAWFRAMES

RUBBER
NT NATURAL RUBBER
SYNTHETIC RUBBER
RT ELASTICITY
ELASTOMER FIBERS
ELASTOMERS
GALOSHES
MAN MADE FIBERS
MATERIALS
POLYMERS
POLYURETHANES
REGENERATED FIBERS (EXCLUDING CELLULOSIC AND PROTEIN)
ROLL COVERINGS
ROLLS
RUBBER COATED FABRICS
RUBBER ELASTICITY
RUBBER FIBER (NATURAL)
SNEAKERS
SPANDEX FIBERS
SYNTHETIC RUBBER FIBERS
VISCOELASTICITY

RUBBER COATED FABRICS
BT COATED FABRICS
RT BUNA N COATED FABRICS
LACQUER COATED FABRICS
NEOPRENE COATED FABRICS
RUBBER
SBR COATED FABRICS
VINYL COATED FABRICS

RUBBER ELASTICITY
BT ELASTICITY
RT CROSSLINK DENSITY
NATURAL RUBBER
RUBBER
VISCOELASTICITY

RUBBER FIBER (NATURAL)
BT ELASTOMER FIBERS
MAN MADE FIBERS
REGENERATED FIBERS (EXCLUDING CELLULOSIC AND PROTEIN)
RT BUTADIENE RUBBER FIBER
NATURAL RUBBER
NITRILE RUBBER FIBER
RUBBER
SPANDEX FIBERS
SYNTHETIC RUBBER FIBERS

RUBBER FOOTWEAR
BT FOOTWEAR
RAINWEAR
RT BOOTS
COATED FABRICS
GALOSHES
NATURAL RUBBER
RUBBER
SHOES
SNEAKERS
VINYL COATED FABRICS

RUBBER SHRINKING BLANKETS
USE COMPRESSIVE SHRINKAGE MACHINES

RUBBING
RT ABRASION
COEFFICIENT OF FRICTION
CROCKING
DYNAMIC FRICTION
FABRIC TO FABRIC FRICTION
FIBER FIBER FRICTION
FRICTION
FRICTIONAL FORCE
FUZZING
GRINDING (MATERIAL REMOVAL)
NORMAL FORCE
PILLING
PLOUGHING
ROLLING
SCRUBBING
SCUFFING
SHEAR (MODE OF DEFORMATION)
SHEDDING (FIBER)
SLIDING
SLIDING FRICTION
SLIPPAGE
STATIC FRICTION
STICK SLIP FRICTION
WEAR
WELDING
WET CROCKING
YARN TO METAL FRICTION
YARN TO YARN FRICTION

RUBBING LEATHERS
RT CONDENSERS
WOOLEN CARDS

RUBFASTNESS
USE CROCKFASTNESS

RUBIDIUM COMPOUNDS
NT RUBIDIUM HYDROXIDE

RUBIDIUM HYDROXIDE
BT HYDROXIDES
RUBIDIUM COMPOUNDS

RUCHE
NT CUT RUCHE
RT PIPING
UPHOLSTERY

RUFFLES
BT GARMENT COMPONENTS
RT GARMENT MANUFACTURE
LACES
NETS
PLEATS
REVERES
RIPPLES
RUFFLING

RUFFLING
BT FABRIC DISTORTION (TAILORING)
RT BACK TACKING
BASTING
CHAIN SEWING
COCKLING
CURVED SEAMING
GATHERING
HEMMING
PINCHING
PUCKERING
RIPPLES
RUFFLES
SEAM PUCKER
SEAMING
SEAMS
SERGING
SEWING
SHIRRING
STITCHES (SEWING)
STITCHING
TAILORING
TRIMMING (OPERATION)
YARN SEVERANCE

RUGS
USE CARPETS

RUN OUT
USE ECCENTRICITY

RUN RESISTANCE
BT END USE PROPERTIES
RT CATCHING
CIRCULAR HOSIERY
FULLY FASHIONED HOSIERY
HOSIERY
LADDERING
PENETRATION
PICKING (SNAGGING)
PICKING RESISTANCE
PLAIN KNITTED FABRICS
PUNCTURE RESISTANCE
PUNCTURING
RUN RESISTANT HOSIERY
RUN RESISTANT STITCHES
RUNS (KNITTING)
SNAG RESISTANCE
SNAG TESTERS
SNAGGING
SNAGS
STOCKINGS

RUN RESISTANT HOSIERY
BT HOSIERY
RT CIRCULAR HOSIERY
FULLY FASHIONED HOSIERY
HALF HOSE
KNITTING
MESH (KNITTING)
RUN RESISTANCE
RUN RESISTANT STITCHES
RUNS (KNITTING)
SEAMLESS HOSIERY
SNAG RESISTANCE
SNAGS
SOCKS
STOCKINGS

RUN RESISTANT STITCHES
RT LOCKING COURSES
PELERINE STITCHES
RUN RESISTANCE
RUN RESISTANT HOSIERY
RUNS (KNITTING)
SNAGS

RUN SYSTEM
UF AMERICAN RUN
BT COUNT
INDIRECT COUNT
RT CUT SYSTEM

RUNNER RATIO
RT COURSES PER INCH
FULL SET
GUIDE BARS
HALF SET
LAPPING (WARP KNITTING)
OVERLAP
PROCESS VARIABLES
RUNNERS
SET (WARP KNITTING)
TRICOT KNITTED FABRICS
TRICOT KNITTING
TRICOT KNITTING MACHINES
UNDERLAP
WARP KNITTING MACHINES

RUNNERS
RT GUIDE BARS
HALF SET
OVERLAP
RUNNER RATIO
TRICOT KNITTING MACHINES
UNDERLAP
WARP KNITTING MACHINES

RUNS (KNITTING)
UF LADDERS (HOSIERY)
BT FABRIC DEFECTS
RT CATCHING
COURSES
PICKING (SNAGGING)
PUNCTURE RESISTANCE
PUNCTURING
RUN RESISTANCE
RUN RESISTANT HOSIERY
RUN RESISTANT STITCHES
SNAG RESISTANCE
SNAG TESTERS
SNAGGING
SNAGS
WALES

RUPTURE
UF RUPTURING
RT BREAKING
BRITTLE FRACTURE
BUNDLE STRENGTH
COMPRESSIVE STRENGTH
CRACK PROPAGATION
CREEP RUPTURE
FATIGUE RUPTURE
LOOP STRENGTH
SHOCK
SHOCK RESISTANCE
STRENGTH
STRENGTH ELONGATION TESTING
STRESS CONCENTRATION

RUPTURE STRENGTH
USE BREAKING STRENGTH

RUPTURE TENACITY
USE BREAKING STRENGTH

RUPTURING
USE RUPTURE

RUSSIAN POINT
BT HAND MADE LACES
LACES

RUST RESISTANCE
USE CORROSION RESISTANCE

RUSTING
USE CORROSION

RYELAND WOOL
BT MEDIUM WOOL
WOOL

S TWILLS
  UF RIGHT HAND TWILLS
  BT DIRECTION (TWILL)
    TWILLS
  RT CHEVRON TWILLS
    FANCY TWILLS
    HERRINGBONE TWILLS
    LOW TWILLS
    REGULAR TWILLS
    SENSE (DIRECTION)
    TWISTING MOMENTS
    Z TWILLS

S TWIST
  UF RIGHT HAND TWIST
  BT TWIST
  RT ALTERNATING TWIST
    BALANCED YARNS
    DIRECTION (TWILL)
    FALSE TWIST
    HELIX ANGLE
    HIGH TWIST
    LOW TWIST
    MULTIPLIED YARNS
    PLIED YARNS
    PLY TWIST
    PLYING
    REAL TWIST
    SENSE (DIRECTION)
    SINGLES YARNS
    TWIST LIVELINESS
    TWIST SENSE
    TWISTING
    TWO PLY YARNS
    Z TWIST
    ZERO TWIST

S TWIST YARNS
  BT YARNS
  RT CREPE YARNS
    PLIED YARNS
    SINGLES YARNS
    Z TWIST YARNS

SAABA (TN)
  BT TEXTURED YARNS (FILAMENT)
  RT BULKED YARNS
    DYNALOFT (TN)
    FLUFLON (TN)
    HELANCA (TN)
    SUPERLOFT (TN)

SACKCLOTH
  RT BARRAS
    FLAT WOVEN FABRICS
    MUSLIN

SACKS (BAGS)
  USE BAGS

SACKS (DRESSES)
  BT DRESSES
    GARMENTS
  RT FROCKS
    SUITS

SACO LOWELL DIRECT SPINNER (TN)
  BT CONVERTERS (TOW)
    TOW TO YARN MACHINES
  RT TOW
    TOW TO YARN CONVERSION

SACO LOWELL IRREGULARITY TESTER (TN)
  BT IRREGULARITY TESTERS
  RT CAPACITANCE TYPE IRREGULARITY TESTERS
    FIELDEN WALKER IRREGULARITY TESTER (TN)
    IRREGULARITY
    USTER IRREGULARITY TESTER (TN)
    ZELLWEGER TESTER (TN)

SADDLE STITCHES
  BT STITCHES (SEWING)

SAFETY
  RT ACCIDENTS
    DEAFNESS
    MANAGEMENT
    NOISE
    OCCUPATIONAL HAZARDS
    PERSONNEL
    PROCESS EFFICIENCY
    PRODUCTION
    SAFETY BELTS
    SAFETY CLOTHING
    TOXICITY

SAFETY BELTS
  RT BELTS
    BELTS (APPAREL)
    INDUSTRIAL CLOTHING
    INDUSTRIAL FABRICS
    MILITARY PRODUCTS
    NARROW FABRICS
    SAFETY
    SAFETY CLOTHING
    SAFETY FOOTWEAR

SAFETY CLOTHING
  NT HELMETS
    OVERALLS
  BT GARMENTS
    INDUSTRIAL CLOTHING
    WORK CLOTHING
  RT APRONS (CLOTHING)
    DUNGAREES
    SAFETY
    SAFETY BELTS
    SAFETY FOOTWEAR

SAFETY FOOTWEAR
  BT FOOTWEAR
  RT SAFETY
    SAFETY BELTS
    SAFETY CLOTHING

SAG
  UF SAGGING
  RT BENDING
    BOWING
    CREEP
    DEFLECTION
    DEFORMATION
    DIMENSIONAL STABILITY
    DISTORTION
    PERMANENT DEFORMATION
    RECOVERY (SELF RESTORATION TO ORIGINAL CONDITION)
    STRETCH

SAGGING
  USE SAG

SAICATEX PROCESS (TN)
  USE EMULSION PRINTING

SAILCLOTH
  BT INDUSTRIAL FABRICS
  RT CORDAGE
    FLAT WOVEN FABRICS
    HAWSERS
    MILITARY PRODUCTS
    ROPES

SALES
  RT CAPACITY (GENERAL)
    COSTS
    MANAGEMENT
    MANUFACTURING
    MARKETING
    MERCHANDISING
    PRODUCTION

SALICYLANILIDE
  BT PHENOLS
  RT ANTIMICROBIAL FINISHES
    PRESERVATIVES
    SALICYLIC ACID

SALICYLATE SALTS
  BT PHENOLS
  RT PRESERVATIVES
    SALICYLIC ACID

SALICYLIC ACID
  BT CARBOXYLIC ACIDS
    PHENOLS
  RT PRESERVATIVES
    SALICYLANILIDE
    SALICYLATE SALTS

SALT
  USE SODIUM CHLORIDE

SALTS (GENERAL)
  NT ACETATE SALTS
    ACRYLIC SALTS
    BENZOATE SALTS
    BICARBONATES
    BISULFATES
    BISULFITES
    CARBONATES
    CHLORATES
    CYANATES
    CYANIDES
    DITHIONITES
    HALIDES
    HYPOCHLORITES
    NITRATES
    NITRITES
    OLEATE SALTS
    PERBORATES
    PERCHLORATES
    PERIODATES
    PERMANGANATES
    PERSULFATES
    PHOSPHATES
    PYROPHOSPHATES
    SALICYLATE SALTS
    SILICATES
    SULFATES
    SULFIDES (INORGANIC)
    SULFITES
    THIOCYANATES
    THIOSULFONATES
  RT ACIDS
    ANIONS
    BASES
    CATIONS
    ELECTROLYTES
    ESTERS
    HUMECTANTS
    INORGANIC COMPOUNDS (GENERAL)
    IONS

SAMPLES
  RT FABRICS
    FIBERS
    MERCHANDISING
    PRODUCTS
    SAMPLING
    SPECIMEN PREPARATION
    TESTING
    YARNS

SAMPLING
  RT ANALYZING
    COEFFICIENT OF VARIATION
    COLLECTION
    CORING
    COUNTING
    DISTRIBUTION
    POPULATION
    QUALITY CONTROL
    SAMPLES
    SPECIMEN PREPARATION
    STATISTICAL ANALYSIS
    STATISTICS
    TESTING
    WEAR TESTING

SANDWICH BLENDING
  BT BLENDING
  RT BLENDS (FIBERS)
    INTIMATE BLENDS
    ORTHOBLENDS

SANDWICH LAMINATES
  BT COMPOSITES
    LAMINATES
  RT FIBER REINFORCED COMPOSITES
    FOAM BACKED FABRICS
    HONEYCOMB LAMINATES
    LAMINATED FABRICS
    POST FORMED LAMINATES

SANDWICH ROLLING MILLS
  RT COMPRESSIVE SHRINKAGE
    COMPRESSIVE SHRINKAGE MACHINES
    COMPRESSIVE SHRINKING
    CRIMPING
    MANGLES
    PADDERS
    PAPER MACHINES
    PRESS SECTION
    ROLLING MILLS
    SANFORIZING MACHINES
    SOFTENING

SANDWICH SEAMS
  BT SEAMS
  RT BUTTED SEAMS
    BUTTERFLY SEAMS
    COVERED SEAMS
    FRENCH SEAMS
    SEAMING
    SEWING

SANFORIZING (TN)
  BT COMPRESSIVE SHRINKING
    DRY FINISHING
  RT COMPACTING
    DIMENSIONAL STABILITY
    FABRIC SHRINKAGE
    PRESHRINKING
    RESIDUAL SHRINKAGE
    SANFORIZING MACHINES
    SHRINK PROOFING
    SHRINK RESISTANCE
    SHRINKING

SANFORIZING MACHINES
  BT COMPRESSIVE SHRINKAGE MACHINES
  RT COMPACTOR
    MANGLES
    OVERFED PIN TENTERS
    PADDERS
    ROLLING MILLS
    SANDWICH ROLLING MILLS
    SANFORIZING (TN)

SANITARY NAPKINS
  UF SANITARY TOWELS
  RT ABSORPTION (MATERIAL)
    CARBOXYMETHYLCELLULOSE
    GAUZE
    MEDICAL TEXTILES
    NONWOVEN FABRICS

SANITARY TOWELS
  USE SANITARY NAPKINS

SANITIZING AGENTS
    USE ANTIMICROBIAL FINISHES

SAPONIFICATION
    USE ALKALI HYDROLYSIS

SARAN
    USE SARAN POLY (VINYLIDENE CHLORIDE) FIBERS

SARAN POLY (VINYLIDENE CHLORIDE) FIBERS
    UF SARAN
    NT ROVANA (TN)
       VECTRA SARAN (TN)
       VELON (TN)
    BT MAN MADE FIBERS
       SYNTHETIC FIBERS
    RT POLY (VINYLIDENE CHLORIDE)
       VINAL POLY (VINYL ALCOHOL) FIBERS
       VINYLIDENE COMPOUNDS
       VINYON POLY (VINYL CHLORIDE) FIBERS

SARCOSIDE SURFACTANTS
    BT CARBOXYLATE SURFACTANTS
    RT DIETHANOLAMIDE SURFACTANTS
       SOAPS

SATEEN
    NT IMPERIAL SATEEN
    BT FLAT WOVEN FABRICS
       WOVEN FABRICS
    RT LININGS
       MESALINE

SATIN
    NT SPORT SATIN
    BT FLAT WOVEN FABRICS
       WOVEN FABRICS
    RT SATIN WEAVES
       TICKING

SATIN WEAVES
    BT WEAVE (TYPES)
    RT DAMASK
       DOESKIN
       FLAT WOVEN FABRICS
       LASTING
       SATIN
       SPORT SATIN
       TICKING
       UMBRELLA CLOTH

SATURATED ALIPHATIC COMPOUNDS
    USE SATURATED COMPOUNDS

SATURATED ALIPHATIC HYDROCARBONS
    USE ALKANES

SATURATED COMPOUNDS
    UF SATURATED ALIPHATIC COMPOUNDS
    NT ALKANES
    BT ALIPHATIC COMPOUNDS
    RT UNSATURATED COMPOUNDS

SATURATED HYDROCARBONS
    USE ALKANES

SATURATED SOLUTIONS
    BT SOLUTIONS (LIQUID)
    RT CONCENTRATION
       CONCENTRATION GRADIENT
       CRYSTALLIZATION
       DIFFUSION
       MASS TRANSFER
       SATURATED VAPOR
       SATURATION (MATERIAL)
       SOLUBILITY

SATURATED VAPOR
    RT DEW POINT
       DRYING
       ENTHALPY
       MOISTURE CONTENT
       PARTIAL PRESSURE
       PSYCHROMETRY
       RELATIVE HUMIDITY
       SATURATED SOLUTIONS
       SATURATION (MATERIAL)
       SOLUBILITY
       SPECIFIC HUMIDITY
       THERMODYNAMICS
       VAPOR PRESSURE

SATURATION (COLOR)
    UF CHROMA
    BT MUNSELL COLOR SYSTEM
    RT COLOR
       HUE
       SHADE
       SPECTROPHOTOMETRY
       VALUE (MUNSELL)

SATURATION (MATERIAL)
    RT CONCENTRATION
       CRYSTALLIZATION
       DIFFUSION
       DRYING
       MOISTURE CONTENT
       PSYCHROMETRY
       REGAIN
       RELATIVE HUMIDITY
       SATURATED SOLUTIONS
       SATURATED VAPOR
       SOLUBILITY

SATURATION BONDING
    BT BONDING
    RT BONDED FIBER FABRICS
       DIP BONDING
       DRY POWDER BONDING
       FLAME BONDING
       FOAM BONDING
       NONWOVEN FABRICS
       SPRAY BONDING

SATURATORS
    NT RODNEY HUNT SATURATOR
       WALDRON SATURATOR
    BT NONWOVEN FABRIC MACHINES
    RT BINDERS (NONWOVENS)
       BONDED FIBER FABRICS
       PADDERS

SAVONNERIE TAPESTRIES
    BT TAPESTRIES

SAW BLADE GINNING
    BT GINNING
    RT COTTON GINS
       GIN CUT
       GIN FALL
       KNIFE BLADE GINNING
       LINT (FIBER)
       LINT CONTENT
       NEPS
       ROLLER GINNING

SAW TOOTH CLOTHING
    USE METALLIC CARD CLOTHING

SAW TOOTH CRIMP
    USE Z CRIMP

SAW TOOTHED ROLLS
    BT ROLLS
    RT BEATERS (OPENING)
       IMPELLERS
       OPENERS

SAXONY WOOL
    BT MERINO WOOL
       WOOL

SBR COATED FABRICS
    BT COATED FABRICS
       COMPOSITES
    RT BUNA N COATED FABRICS
       BUTYL COATED FABRICS
       COATINGS (SUBSTANCES)
       LACQUER COATED FABRICS
       NEOPRENE COATED FABRICS
       RUBBER COATED FABRICS
       VINYL COATED FABRICS

SCALE (CORROSION)
    RT CORROSION
       CORROSION INHIBITORS

SCALE MASKING
    RT ANTIFELTING TREATMENTS
       DEPOSITION
       FELTING
       SCALES (WOOL FIBERS)
       WOOL

SCALES (WOOL FIBERS)
    UF WOOL SCALES
    BT FINE STRUCTURE (FIBERS)
    RT COEFFICIENT OF FRICTION
       CUTICLE
       DIFFERENTIAL FRICTION
       ENDOCUTICLE
       EPICUTICLE
       EXOCUTICLE
       FELTING
       FELTS
       FIBER CRIMP
       FRICTION
       MEDULLA
       ORTHOCORTEX
       SCALE MASKING
       WOOL

SCANNING
    NT OPTICAL SCANNING
       PHOTOELECTRIC SCANNING
    RT FABRIC INSPECTION
       IRREGULARITY TESTERS
       OPTICAL SCANNERS
       PERCHING
       PHOTOELECTRIC COUNTERS (NEPS)
       SEAM DETECTORS
       SLUB CATCHING

SCANNING ELECTRON MICROSCOPY
    USE STEREOSCAN MICROSCOPY

SCARVES
    BT GARMENTS
       OUTERWEAR
    RT COATS
       CRAVATS
       NECKTIES

SCHAPPE SILK
    BT SILK

SCHAPPE SYSTEM PROCESSING
    USE SILK SYSTEM PROCESSING

SCHIEFER ABRASION TESTING
    BT ABRASION TESTING
    RT STOLL QM ABRASION TESTING
       TABER ABRASION TESTING

SCHIFFLE EMBROIDERY
    BT EMBROIDERY

SCHIFFLI LACE
    BT LACES
       MACHINE MADE LACES

SCHLAFORST UNWINDING ACCELERATOR
    BT UNWINDING ACCELERATORS
    RT UNWINDING
       WINDING

SCHREINERING
    UF SCHREINERIZING
    BT DRY FINISHING
    RT BLOOM
       CALENDERING
       EMBOSSING
       EMBOSSING CALENDERS
       FRICTION CALENDERING
       LININGS
       LUSTERING MACHINES
       MERCERIZING
       PRESSURE ROLLS
       REFLECTANCE
       TEXTILE MACHINERY (GENERAL)

SCHREINERIZING
    USE SCHREINERING

SCHUPP REGULATOR
    RT AUTOMATIC CONTROL
       CONTROL EQUIPMENT
       WEIGHT CONTROL

SCISSORS
    RT CUTTERS
       CUTTING
       CUTTING MACHINES
       CUTTING SHEARS
       KNIVES
       SHARPENING
       SLITTERS (CLOTH)

SCORCH TESTERS
    BT TESTING EQUIPMENT
    RT FIRE RESISTANCE
       FIRE RESISTANCE TESTERS
       SCORCH TESTING
       SCORCHING

SCORCH TESTING
    BT TESTING
    RT FIRE RESISTANCE
       SCORCH TESTERS
       SCORCHING

SCORCHING
    RT BURNING
       CARBONIZING
       CHLORINE DAMAGE
       DRYING
       FIRE RESISTANCE
       HEAT RESISTANCE
       PRESSING
       SCORCH TESTERS
       SCORCH TESTING
       SCORCHING MARKS

SCORCHING MARKS
    RT BURNING
       CARBONIZING
       DRYING
       FIRE RESISTANCE
       HEAT SETTING (SYNTHETICS)
       IRONING
       PRESSING
       SCORCHING

SCORING
    RT ABRASION
       ABRASIVES
       BREAKING
       CUTTING
       ENGRAVING
       GALLING
       GRINDING (MATERIAL REMOVAL)
       (CONT.)

RT GRIT
POLISHING
SCRATCHING
SCUFFING
SNAGGING
WEAR
WEDGING

SCOTCH BLACKFACE WOOL
BT MOUNTAIN WOOL

SCOTCH CARPETS
USE INGRAIN CARPETS

SCOTCH FEED
BT INTERMEDIATE FEED
RT BLAMIRE FEED
CARDING
CARDS
FEED ROLLS
FINISHER PART
HOPPER FEED
INTERMEDIATE PART
SCRIBBLING
SLIVERS
WEBS
WOOLEN CARDING
WOOLEN CARDS

SCOTT TENSILE TESTER (TN)
BT TENSILE TESTERS
RT CAMBRIDGE EXTENSOMETER (TN)
INSTRON TENSILE TESTER (TN)

SCOURED WOOL
BT WOOL
RT GREASY WOOL
PULLED WOOL
RESIDUAL FAT
SCOURING WASTE
SCOURING YIELD
SUINT
WOOL CLASS
WOOL GREASE
WOOL WAX

SCOURING
UF WASHING (PROCESSING)
NT AQUEOUS SCOURING
BATCH SCOURING
CAUSTIC SCOURING
CONTINUOUS SCOURING
JET SCOURING
KIER SCOURING
OPEN WIDTH SCOURING
PACKAGE SCOURING
ROPE SCOURING
SOAP SODA SCOURING
SOLVENT SCOURING
BT CLEANING
WET FINISHING
RT AMERICAN SYSTEM PROCESSING
ANTIFOAM AGENTS
BACKWASHING
BATHS
BLEACHING
BLEACHING AGENTS
BLEACHING MACHINES
BOILING OFF
BOTTOMING
BRADFORD SYSTEM PROCESSING
CARBONIZING
CASCADE WASHERS
CLEANLINESS
CONTINENTAL SYSTEM PROCESSING
COTTON WAX
COUNTERFLOW PROCESSES
DECOLORIZING
DESIZING
DETERGENCY
DETERGENTS
DEWAXING (COTTON)

DRUM SCOURING MACHINES
DRYING
DYE BATHS
FOAMS (FROTH)
GREASE
GREASE RECOVERY
IMPURITIES
LANOLIN
LAUNDERABILITY
LAUNDERABILITY TESTING
LAUNDERING
NEUTRALIZING
NEW BRADFORD SYSTEM PROCESSING
OILING
RESIDUAL FAT
RINSING
SCOURING BATHS
SCOURING MACHINES
SCOURING SHRINKAGE
SCOURING TRAINS
SCOURING WASTE
SCOURING YIELD
SCRUBBING
SIMULTANEOUS FINISHING PROCESSES

RT **SOAPING**
SOAPS
SOIL REMOVAL
SQUEEZING
STOCK (FIBER)
STRIPPING (COLOR)
SUINT
SURFACTANT APPLICATIONS
VEGETABLE MATTER
WETTING
WETTING AGENTS
WOOL GREASE
WOOL WAX
WOOLEN SYSTEM PROCESSING
WORSTED SYSTEM PROCESSING

SCOURING BATHS
UF RINSING BATHS (SCOURING)
SCOURING BOWLS
BT BATHS
SCOURING MACHINES
RT BATH RATIO
CARBONIZING
CONTINUOUS SCOURING
DESIZING
DRUM SCOURING MACHINES
DYE BATHS
HARROW FORKS (SCOURING)
LANOLIN
RINSING
SCOURING
SCOURING TRAINS
SCOURING WASTE
SCOURING YIELD
SETTLING TANKS
SOAP SODA SCOURING
SOLVENT SCOURING

SCOURING BOWLS
USE SCOURING BATHS

SCOURING MACHINES
UF WASHING MACHINES (PROCESSING)
NT DRUM SCOURING MACHINES
BACKWASHING MACHINES
CASCADE WASHERS
ROPE SCOURING MACHINES
SCOURING BATHS
SCOURING TRAINS
RT BLEACHING MACHINES
BOIL OFF MACHINES
COUNTERFLOW PROCESSES
DESIZING MACHINES
DYEING MACHINES
SCOURING

SCOURING SETS
USE SCOURING TRAINS

SCOURING SHRINKAGE
BT SHRINKAGE
RT CARBONIZING
FELTING SHRINKAGE
SCOURING
SCOURING TRAINS
SCOURING YIELD
SOAP SODA SCOURING
SOLVENT SCOURING

SCOURING TRAINS
UF SCOURING SETS
BT SCOURING MACHINES
RT CARBONIZING
CENTRIFUGES
COUNTERFLOW PROCESSES
DRUM SCOURING MACHINES
HARROW FORKS (SCOURING)
LANOLIN
NEUTRALIZING
RINSING
SCOURING
SCOURING BATHS
SCOURING SHRINKAGE
SCOURING WASTE
SCOURING YIELD
SETTLING TANKS
SOAP SODA SCOURING
SQUEEZING
SUINT
WOOL GREASE

SCOURING WASTE
RT CARBONIZING
GREASE RECOVERY
GREASY WOOL
LANOLIN
SCOURED WOOL
SCOURING
SCOURING BATHS
SCOURING TRAINS
SCOURING YIELD
SOAP SODA SCOURING
SOLVENT SCOURING
SUINT
VEGETABLE MATTER
WASTE
WOOL GREASE
WOOL WAX

SCOURING YIELD
RT CARBONIZING
GREASY WOOL
LANOLIN
SCOURED WOOL
SCOURING
SCOURING BATHS
SCOURING SHRINKAGE
SCOURING TRAINS
SCOURING WASTE
SOAP SODA SCOURING
SOLVENT SCOURING
SUINT
VEGETABLE MATTER
WOOL GREASE
YIELD (MECHANICAL)

SCRATCHING
RT ABRASION
ABRASIVES
BREAKING
CUTTING
FRICTION
GALLING
GRINDING (MATERIAL REMOVAL)
GRIT
PICKING (SNAGGING)
POLISHING
SCORING
SCUFFING
SLITTING
SNAGGING
WEAR
WEDGING

SCREEN PRINTING
NT FLAT SCREEN PRINTING
ROTARY SCREEN PRINTING
BT PRINTING
RT COLOR TROUGH
DESIGN
DISCHARGE PRINTING
DUPLEX PRINTING
DYEING
MELANGE PRINTING
PATTERN (FABRICS)
PRINT CLOTH
PRINTED FABRICS
PRINTERS BLANKETS
PRINTING MACHINES
PRINTING PASTES
PRINTING SCREENS
RESIST PRINTING
ROLLER PRINTING
SQUEEGEES
TOPPING
TWO STAGE PRINTING

SCREEN PRINTING MACHINES
NT PRINTING SCREENS
BT PRINTING MACHINES
RT BLOCK PRINTING EQUIPMENT
DUPLEX PRINTING MACHINES
FLAT SCREEN PRINTING
ROLLER PRINTING MACHINES
ROTARY SCREEN PRINTING
SQUEEGEES

SCREENING
BT FLAT WOVEN FABRICS
WOVEN FABRICS
RT FILTERS (FLUID)
LENO WEAVES
MARQUISETTE
NETTING
PLAIN WEAVES
SCRIM

SCREENS (PRINTING)
USE PRINTING SCREENS

SCRIBBLER PART
USE BREAKERS

SCRIBBLING
BT WOOLEN CARDING
RT BREAKERS
CARDING
CARDING ACTION
DOFFING (OF A CARD)
FANCY (NOUN)
FINISHER PART
LICKER IN
SCOTCH FEED
STRIPPING ACTION

SCRIM
NT NONWOVEN SCRIMS
BT FLAT WOVEN FABRICS
WOVEN FABRICS
RT MARQUISETTE
NETTING
PLAIN WEAVES
SCREENING

SCROOP
BT FABRIC PROPERTIES
FABRIC PROPERTIES (AESTHETIC)

BT FABRIC PROPERTIES (MECHANICAL)
RT ACOUSTIC PROPERTIES
HARSHNESS
SILK
SOFTNESS

SCRUBBING
RT ABRASION
CLEANING
CROCKING
DETERGENTS
LAUNDERING
RUBBING
SCOURING
SCUFFING
SLIDING
WET CROCKING

SCUFFING
RT ABRASION
COATED FABRICS
CUTTING
FRICTION
PEELING
RUBBING
SCORING
SCRATCHING
SCRUBBING
SHOES
WEAR
WEDGING

SCULPTURED CARPETS
UF CARVED PILE FABRICS
BT WOVEN CARPETS
RT SCULPTURED PILE FABRICS
TUFTED CARPETS

SCULPTURED PILE FABRICS
BT PILE FABRICS
RT CUT PILE FABRICS
SCULPTURED CARPETS

SCUTCHER LAPS
USE PICKER LAPS

SCUTCHING (FINISHING)
BT DRY FINISHING
WET FINISHING
RT DRYING
SELVEDGE UNCURLERS
SPREADERS (CLOTH)
SPREADING (TAILORING)
SPREADING MACHINES (TAILORING)
TENTERING

SCUTCHING (FLAX)
RT BAST FIBERS
FLAX
GINNING
LINEN
RETTING

SCUTCHING (PICKING)
USE PICKING (OPENING)

SEA ISLAND COTTON
BT COTTON

SEALED SELVEDGES
BT SELVEDGES
RT CENTER SELVEDGES
CRACKED SELVEDGES
DOG LEGGED SELVEDGES
DRESSING SELVEDGES (LACE)
KNITTED FABRICS
LENO SELVEDGES
LOOMS
LOOP SELVEDGES
PULLED IN SELVEDGES
SELVEDGE WARPS
SLACK SELVEDGES
TAPE SELVEDGES
TEMPLES
TIGHT SELVEDGES
WEAVING
WOVEN FABRICS

SEALERS
RT ADHESIVES
BONDING STRENGTH
CEMENTS
GLUES
PASTES
SEALING
SIZES (SLASHING)
TAPE

SEALING
NT DIELECTRIC HEAT SEALING
HEAT SEALING
RT ADHESION
BONDING
CEMENTING
COATING (PROCESS)
COHESION
FUSION (MELTING)
GLUING
JOINING
OIL PROOFING
PEELING
SEALERS
SEAMING
SPRAYING
TAPING
THERMOPLASTICITY
WATERPROOFING

SEALS
RT ACCESSORIES
GASKETS
INDUSTRIAL FABRICS
MECHANISMS
VALVES
WASHERS

SEAM ABRASIVE STRENGTH
BT SEAM SPECIFICATIONS
SEAM STRENGTH
RT ABRASION
ABRASION RESISTANCE
SEAM QUALITY
SEAM SLIPPAGE STRENGTH
SEAM TENSILE STRENGTH
SEAMS
SEWING

SEAM ALLOWANCE
RT BUTTERFLY SEAMS
GARMENT MANUFACTURE
SEAM QUALITY
SEAM SLIPPAGE STRENGTH
SEAM TENSILE STRENGTH
SEAM WIDTH
SEAMS
SEWING
TAILORING

SEAM BASTING
RT ANGULAR SEAMING
BACK TACKING
CURVED SEAMING
GARMENT MANUFACTURE
SEAMING

SEAM BINDING
UF BINDING (SEAMS)
SEAM PIPING
SEAM TAPING
RT BACK TACKING
BASTING
CURVED SEAMING
GARMENT MANUFACTURE
OVEREDGE STITCHES
OVERSEAM STITCHES
SEAMING
SEWING

SEAM DAMAGE
BT DAMAGE
SEAM SPECIFICATIONS
RT ABRASION
COMPRESSIBILITY
CUTTING RESISTANCE
FUSION (MELTING)
NEEDLE HEATING
NEEDLE PENETRATION
NEEDLE POINT
RIPPLES
SEAM ABRASIVE STRENGTH
SEAM GRIN
SEAM PUCKER
SEAM QUALITY
SEAM SLIPPAGE
SEAM STRENGTH
SEAM TENSILE STRENGTH
SEWABILITY
SEWING
SEWING DAMAGE
SEWING EFFICIENCY
STITCHING
THREAD BREAKAGES
THROAT PLATES
YARN SEVERANCE
YARN SLIPPAGE

SEAM DETECTORS
RT AUTOMATIC CONTROL
DETECTION
OPTICAL SCANNERS
OPTICAL SCANNING
PROCESS CONTROL
SCANNING
SHEARING (FINISHING)

SEAM EFFICIENCY
UF SEAM TENSILE STRENGTH EFFICIENCY
BT EFFICIENCY (STRUCTURAL)
FABRIC PROPERTIES (MECHANICAL)
SEAM SPECIFICATIONS
RT NEEDLE HEATING
NEEDLE PENETRATION
SEAM DAMAGE
SEAM GRIN
SEAM SLIPPAGE
SEAM SLIPPAGE STRENGTH
SEAM STRENGTH
SEAM TENSILE STRENGTH
SEAMING
SEAMS
SEWABILITY
SEWING
SEWING DAMAGE
SEWING EFFICIENCY
STITCH QUALITY
STITCH SIZE (SEWING)
STRENGTH EFFICIENCY
YARN DAMAGE
YARN SEVERANCE
YARN SLIPPAGE

SEAM ELASTICITY
USE STITCH ELASTICITY

SEAM ELONGATION
USE STITCH ELASTICITY

SEAM GRIN
UF SEAM GRINNING
RT COVER FACTOR
DIMENSIONAL STABILITY
NEEDLE PENETRATION
PERMANENT DEFORMATION
RIPPLES
SEAM PUCKER
SEAM SLIPPAGE
SEAMS
SEWING
STITCH SIZE (SEWING)
STITCH TENSION
THREAD SLIPPAGE
THREAD TENSION
TIGHTNESS
YARN SLIPPAGE

SEAM GRINNING
USE SEAM GRIN

SEAM HEADING
RT PUCKERING
RIPPLES
RUFFLING
SEAM QUALITY
SEAMING
SEAMS
SEWING
SHIRRING
STITCH TENSION
STITCHING
STRETCH PUCKERS
THREAD TENSION
TIGHTNESS
YARN BUCKLING
YARN SLIPPAGE

SEAM LET OUT
RT ALTERATIONS
LETTING OUT (GARMENTS)
RECOVERY (SELF RESTORATION TO ORIGINAL CONDITION)
SEAM QUALITY
SEAMS
SEWING
TAILORING
TAKING IN (GARMENTS)

SEAM PIPING
USE SEAM BINDING

SEAM PUCKER
RT BUCKLING
COCKLING
COMPRESSIBILITY
COVER FACTOR
DIMENSIONAL STABILITY
GARMENT MANUFACTURE
GATHERING
MATRIX FREEDOM
NEEDLE PENETRATION
NEEDLE POINT
PINCHING
PUCKER (DEFECT)

SEAM QUALITY
BT SEAM SPECIFICATIONS
RT NEEDLE PENETRATION
SEAM ABRASIVE STRENGTH
SEAM ALLOWANCE
SEAM HEADING
SEAM LET OUT
SEAM SIZE
SEAM SLIPPAGE STRENGTH
SEAM TENSILE STRENGTH
SEAM THICKNESS
SEAMS
SEWING
STITCH SIZE (SEWING)

SEAM SIZE
NT FULL SEAM WIDTH
SEAM THICKNESS
SEAM WIDTH
BT SEAM SPECIFICATIONS
(CONT.)

RT SEAM QUALITY
SEAMS
SEWING

SEAM SLIPPAGE
BT SLIPPAGE
RT COMPRESSIBILITY
DIMENSIONAL STABILITY
MATRIX FREEDOM
PERMANENT DEFORMATION
SEAM DAMAGE
SEAM EFFICIENCY
SEAM GRIN
SEAM SLIPPAGE STRENGTH
SEAM STRENGTH
SEAMS
SEWING
STITCH SIZE (SEWING)
STITCH TENSION
THREAD SLIPPAGE
THREAD TENSION
TIGHTNESS
YARN BUCKLING
YARN SLIPPAGE

SEAM SLIPPAGE STRENGTH
BT SEAM SPECIFICATIONS
SEAM STRENGTH
RT NEEDLE HEATING
SEAM ABRASIVE STRENGTH
SEAM EFFICIENCY
SEAM QUALITY
SEAM SLIPPAGE
SEAM TENSILE STRENGTH
SEAMING
SEAMS
SEWABILITY
SEWING
SEWING THREADS
STITCHING

SEAM SPECIFICATIONS
NT FULL SEAM WIDTH
SEAM ABRASIVE STRENGTH
SEAM DAMAGE
SEAM EFFICIENCY
SEAM QUALITY
SEAM SIZE
SEAM SLIPPAGE STRENGTH
SEAM STRENGTH
SEAM TENSILE STRENGTH
SEAM THICKNESS
SEAM WIDTH
STITCH SIZE (SEWING)
RT SEAMING
SEAMS

SEAM STRENGTH
NT SEAM ABRASIVE STRENGTH
SEAM SLIPPAGE STRENGTH
SEAM TENSILE STRENGTH
BT FABRIC PROPERTIES (MECHANICAL)
SEAM SPECIFICATIONS
RT BREAKING STRENGTH
GRAB STRENGTH
MACHINE SEWING
NEEDLE HEATING
SEAM DAMAGE
SEAM EFFICIENCY
SEAM SLIPPAGE
SEAMING
SEAMS
SEWABILITY
SEWING
STITCH SIZE (SEWING)
STITCHES (SEWING)
STITCHING
TEAR STRENGTH
THREAD BREAKAGES
THREAD TENSION
YARN DAMAGE
YARN SEVERANCE

SEAM TAPING
USE SEAM BINDING

SEAM TENSILE STRENGTH
BT SEAM SPECIFICATIONS
SEAM STRENGTH
RT MATRIX FREEDOM
NEEDLE PENETRATION
SEAM ABRASIVE STRENGTH
SEAM DAMAGE
SEAM EFFICIENCY
SEAM QUALITY
SEAM SLIPPAGE STRENGTH
SEAMS
SEWING
SEWING DAMAGE
STITCH ELASTICITY
THREAD BREAKAGES
THREAD TENSION
YARN DAMAGE

SEAM TENSILE STRENGTH EFFICIENCY
USE SEAM EFFICIENCY

SEAM THICKNESS
BT SEAM SIZE
SEAM SPECIFICATIONS
RT COMPRESSIBILITY
FULL SEAM WIDTH
SEAM QUALITY
SEAM WIDTH
SEAMS
SEWING
THICKNESS
THREAD TENSION

SEAM WIDTH
BT SEAM SIZE
SEAM SPECIFICATIONS
RT FULL SEAM WIDTH
SEAM ALLOWANCE
SEAM THICKNESS
SEAMS
SEWING

SEAMING
NT ANGULAR SEAMING
CURVED SEAMING
BT GARMENT MANUFACTURE
JOINING
RT ADHESION
BACK TACKING
BASTING
BLIND STITCHING
BONDING
CEMENTING
CHAIN SEWING
COCKLING
COTTON THREADS
DRESSMAKING
FABRIC DISTORTION (TAILORING)
FELLED SEAMS
FELLING
FRENCH SEAMS
FULL SEAM WIDTH
FUSION (MELTING)
GATHERING
GLUING
HAND SEWING
HEM SEAMS
HEMMING
LAPPED SEAMS
LINEN THREADS
LOOP CATCHING
MACHINE SEWING
NEEDLE HEATING
NEEDLE PENETRATION
NYLON THREADS
PINCHING
PINKING (TAILORING)
PUCKERING
RIPPLES
RUFFLING
SANDWICH SEAMS
SEALING
SEAM BASTING
SEAM BINDING
SEAM EFFICIENCY
SEAM PUCKER
SEAM SLIPPAGE STRENGTH
SEAM SPECIFICATIONS
SEAM STRENGTH
SEAMS
SEAMSTRESSES
SERGING
SEWABILITY
SEWING
SEWING MACHINES
SEWING THREADS
SHIRRING
STAPLING (WIRE)
STITCH ELASTICITY
STITCH QUALITY
STITCH SEQUENCE
STITCH SIZE (SEWING)
STITCHES (SEWING)
STITCHING
STRETCHING (SEAMING)
SUPERPOSED SEAMS
TACKING
TAILORED SEAMS
TAILORING
TAPING
THREADS
TRIMMING (OPERATION)
TUCKING (GARMENT MANUFACTURE)

SEAMLESS HOSIERY
UF SEAMLESS STOCKINGS
BT HOSIERY
RT BOARDING
CIRCULAR HOSIERY
CIRCULAR KNITTING
CIRCULAR KNITTING MACHINES
FASHIONING
FULLY FASHIONED HOSIERY
HALF HOSE
HEAT SETTING (SYNTHETICS)
KNITTING
MOCK SEAMS
RUN RESISTANT HOSIERY
SEAMS
SOCKS
STOCKINGS

SEAMLESS HOSIERY MACHINES
BT CIRCULAR KNITTING MACHINES
HOSIERY MACHINES
KNITTING MACHINES
RT BOARDING
CYLINDER NEEDLES
DIAL NEEDLES
FULLY FASHIONED HOSIERY MACHINES
MOCK FASHIONING MARKS
MOCK SEAMS
NEEDLES
POSITIVE FEED
SINKERS
STITCH CAMS
TAKE DOWN
WEBHOLDERS

SEAMLESS STOCKINGS
USE SEAMLESS HOSIERY

SEAMS
NT FELLED SEAMS
FRENCH SEAMS
HEM SEAMS
LAPPED SEAMS
PINKED SEAMS
PIPED SEAMS
SANDWICH SEAMS
STRIPPED SEAMS
SUPERPOSED SEAMS
TAILORED SEAMS
TOP STITCHED SEAMS
RT ADHESIVES
ANGULAR SEAMING
BASTING
BONDING STRENGTH
FABRIC DISTORTION (TAILORING)
FELLING
FULL SEAM WIDTH
FUSION (MELTING)
GARMENT MANUFACTURE
GATHERING
JOINTS
PINCHING
PINKING (TAILORING)
PUCKERING
RIPPING
RUFFLING
SEAM ABRASIVE STRENGTH
SEAM ALLOWANCE
SEAM EFFICIENCY
SEAM GRIN
SEAM HEADING
SEAM LET OUT
SEAM QUALITY
SEAM SIZE
SEAM SLIPPAGE
SEAM SLIPPAGE STRENGTH
SEAM SPECIFICATIONS
SEAM STRENGTH
SEAM TENSILE STRENGTH
SEAM THICKNESS
SEAM WIDTH
SEAMING
SEAMLESS HOSIERY
SEAMSTRESSES
SEWABILITY
SEWING
SHIRRING
STITCH ELASTICITY
STITCH SEQUENCE
STITCH SIZE (SEWING)
STITCH TENSION
STITCHES (SEWING)
STITCHING
STRETCHING (SEAMING)
YARN SEVERANCE

SEAMSTRESSES
RT GARMENT MANUFACTURE
SEAMING
SEAMS
SEWING
STITCHING
TAILORING

SEAWATER FASTNESS (COLOR)
BT COLORFASTNESS
RT WASHFASTNESS (COLOR)

SEBUM
RT DETERGENCY
FATS
IMPURITIES
OILS

SECANT MODULUS
BT MODULUS
STRESS STRAIN PROPERTIES
STRESS STRAIN PROPERTIES (TENSILE)
RT COMPLIANCE
ELASTIC MODULUS (TENSILE)
ELASTIC STRAIN
INITIAL MODULUS
STRESS STRAIN CURVES
TANGENT MODULUS
YIELD POINT

SECOND ORDER TRANSITION TEMPERATURE
BT TRANSITION TEMPERATURES
RT GLASS TRANSITION TEMPERATURE
MELTING POINT

SECONDARY AMINE SURFACTANTS
USE ALKYLAMINE SURFACTANTS

SECONDARY AMINES
NT DIETHANOLAMINE
DIETHYLAMINE
BT AMINES

SECONDARY TRANSITION TEMPERATURE (AMORPHOUS PHASE)
BT SECONDARY TRANSITION TEMPERATURES
TRANSITION TEMPERATURES

SECONDARY TRANSITION TEMPERATURE (CRYSTALLINE PHASE)
BT SECONDARY TRANSITION TEMPERATURES
TRANSITION TEMPERATURES

SECONDARY TRANSITION TEMPERATURES
NT SECONDARY TRANSITION TEMPERATURE (AMORPHOUS PHASE)
SECONDARY TRANSITION TEMPERATURE (CRYSTALLINE PHASE)
BT TRANSITION TEMPERATURES
RT BIREFRINGENCE
COEFFICIENT OF EXPANSION
ELASTIC MODULUS (TENSILE)
GLASS TRANSITION TEMPERATURE
HEAT CAPACITY
MECHANICAL DAMPING
MELTING POINT
MODULUS
MOLECULAR CHAIN MOVEMENT
NUCLEAR MAGNETIC RESONANCE
STRESS OPTICAL COEFFICIENT
SUBGROUP MOVEMENT (MOLECULAR)
TRANSITIONS (POLYMERS)
VOLUME TEMPERATURE MEASUREMENTS
X RAY ANALYSIS

SECONDARY TRANSITIONS
BT TRANSITIONS (POLYMERS)
RT PRIMARY TRANSITIONS
SECONDARY TRANSITION TEMPERATURES

SECONDARY CREEP
USE PERMANENT SET

SECONDARY WALLS
BT FINE STRUCTURE (FIBERS)
RT CONVOLUTIONS
CORTEX
COTTON
CUTICLE
EPICUTICLE
EXOCUTICLE
FIBRIL REVERSALS
FIBRILS
IMMATURE FIBERS
IMMATURITY (FIBER)
IMMATURITY TESTING (COTTON FIBER)
LUMEN
MERCERIZED COTTON
PARACORTEX
PRIMARY WALLS
SEED HAIR FIBERS (GENERAL)
WOOD FIBERS

SECONDS (FABRICS)
RT BURLING
FABRIC DEFECTS
FABRIC INSPECTION
MENDING
PERCHING
QUALITY CONTROL

SECTION LEASE
BT LEASE
RT WARPING

SECTIONAL BEAMING
USE SECTIONAL WARPING

SECTIONAL DRESSING
USE SECTIONAL WARPING

SECTIONAL MOMENT OF INERTIA
USE MOMENT OF INERTIA

SECTIONAL WARPING
UF SECTIONAL BEAMING
SECTIONAL DRESSING
BT WARP PREPARATION
RT BEAMS
SILK SYSTEM WARPING
SLASHING
WARP ENDS
WARPERS (MACHINE)
WARPING
WINDING

SEED COAT NEPS
RT COTTON SEEDS
GINNING
KNIFE BLADE GINNING
MOTE KNIVES
NEPS
ROLLER GINNING
SAW BLADE GINNING

SEED HAIR FIBERS (GENERAL)
NT COTTON
KAPOK
LINT (FIBER)
BT CELLULOSIC FIBERS
FIBERS
NATURAL FIBERS
RT BAST FIBERS
BOLLS
CONVOLUTIONS
COTTON GINS
FRUIT FIBERS
IMMATURE FIBERS
LEAF FIBERS
LINT CONTENT
SEEDS (TRASH)
STAPLE FIBERS

SEEDS (NOVELTY YARNS)
UF KNOP YARNS
KNOPPED YARNS
KNOPT YARNS
RT BEAD YARNS
BOUCLES
BUG YARNS
FLAKE YARNS
FLAMES (NOVELTY YARNS)
FRILLS (NOVELTY YARNS)
KNICKERBOCKER YARNS
LOOPS (NOVELTY YARNS)
NOVELTY TWISTERS
NUB YARNS
RATINES
SLUB YARNS
SPIRALS (NOVELTY YARNS)
SPLASHES

SEEDS (TRASH)
BT WASTE
TRASH
VEGETABLE MATTER
RT BOLLS
COTTON
IMPURITIES
MOTE KNIVES
MOTES
SEED HAIR FIBERS (GENERAL)

SEEPAGE (RHEOLOGY)
BT STRESS STRAIN PROPERTIES
RT DUCTILITY
PENETRATION
PERMEABILITY
TACK
THIXOTROPY

SEERSUCKER
BT FLAT WOVEN FABRICS
WOVEN FABRICS
RT COTTON
PLAIN WEAVES

SELF ACTOR MULES
USE MULES

SELF ADJUSTING ACTION
USE ADJUSTMENTS

SELF BONDED FABRICS (NONWOVENS)
BT BONDED FIBER FABRICS
NONWOVEN FABRICS
RT BINDERS (NONWOVENS)
BONDED YARN FABRICS
BONDING
BONDING STRENGTH
INTIMATE BLENDS
SELF BONDING FIBERS
SPUNBONDED (NONWOVENS)

SELF BONDING FIBERS
RT BONDED FIBER FABRICS
COHESION
FUSION (MELTING)
MELTING
MELTING POINT
NONWOVEN FABRICS
SELF BONDED FABRICS (NONWOVENS)
SPUNBONDED (NONWOVENS)
THERMOPLASTICITY

SELF CRIMPING YARNS
NT BICOMPONENT FIBER YARNS (FILAMENT)
BICOMPONENT FIBER YARNS (STAPLE)
BT CRIMPED YARNS
TEXTURED YARNS (FILAMENT)
RT BICOMPONENT FIBERS
BULKED YARNS
BULKING
CRIMP
CRIMP FREQUENCY
CRIMP INDEX
CRIMP INTERCHANGE
CRIMP TENDENCY
CRIMPING
FOLLOW THE LEADER CRIMP
LATENT CRIMP
QUENCHING (FILAMENT)
STRETCH YARNS (FILAMENT)
TEXTURING
WAVINESS
YARN CRIMP
Z CRIMP

SELF SMOOTHING FABRICS
USE EASY CARE FABRICS

SELF SUPPORTING PACKAGES
NT CHEESES
CONES
COPS
QUILLS
BT PACKAGES

SELF THREADING DEVICES
BT DEVICES
RT THREADING
THREADS
TYING IN MACHINES
YARNS

SELVAGES
USE SELVEDGES

SELVEDGE UNCURLERS
RT CURL
FILLING STRAIGHTENING MACHINES
SCUTCHING (FINISHING)
SPREADERS (CLOTH)

SELVEDGE WARPS
RT CENTER SELVEDGES
CRACKED SELVEDGES
DOG LEGGED SELVEDGES
EDGE GUIDES
LENO SELVEDGES
LOOMS
LOOP SELVEDGES
SEALED SELVEDGES
SELVEDGES
SLACK SELVEDGES
TAPE SELVEDGES
TEMPLES
TIGHT SELVEDGES
WARP ENDS
WEAVING
WOVEN FABRICS

SELVEDGES
UF SELVAGES
NT CENTER SELVEDGES
DRESSING SELVEDGES (LACE)
LENO SELVEDGES
LOOP SELVEDGES
SEALED SELVEDGES
TAPE SELVEDGES
RT CRACKED SELVEDGES
DOG LEGGED SELVEDGES
FABRIC STRUCTURE
KNITTED FABRICS
LOOMS
PULLED IN SELVEDGES
SELVEDGE WARPS
SLACK SELVEDGES
TEMPLES
TIGHT SELVEDGES
WEAVING
WOVEN FABRICS

SEMI CONTINUOUS DYEING
BT DYEING (BY PROCESS FLOW)
RT BATCH DYEING
BECK DYEING
CONTINUOUS DYEING
OPEN WIDTH DYEING
PAD DYEING
PAD JIG DYEING
PAD ROLL DYEING
PAD STEAM DYEING
SEMI CONTINUOUS DYEING RANGE
SIMULTANEOUS DYEING AND FINISHING

SEMI CONTINUOUS DYEING RANGE
BT DYEING MACHINES
DYEING MACHINES (BY PROCESS FLOW)
RT BATCH DYEING MACHINES
BECKS
CONTINUOUS DYEING RANGE
DYEING
DYEING (BY PROCESS FLOW)
(CONT.)

RT J BOXES
JIGS (DYEING)
KETTLES (DYEING)
PAD FIX MACHINES
PAD ROLL DYEING
PAD STEAM DYEING
PADDERS
SEMI CONTINUOUS DYEING
STEAMERS
WARP DYEING MACHINES
WILLIAMS UNIT (TN)

SEMIPERMEABLE MEMBRANES
BT MEMBRANES
RT DIFFUSION
OSMOSIS
PERMEABILITY

SENSE (DIRECTION)
RT ANTICLOCKWISE
CLOCKWISE
COUPLES
DIRECTION (TWILL)
LEFT HAND COORDINATE SYSTEMS
LEFT HAND RULE
MAGNITUDE
RIGHT HAND COORDINATE SYSTEMS
RIGHT HAND RULE
S TWILLS
S TWIST
TENSORS
TWIST
TWIST SENSE
TWISTING MOMENTS
VECTORS
Z TWILLS
Z TWIST

SEPARATING
USE SEPARATION

SEPARATION
UF SEPARATING
RT CLEARANCES
DECORTICATING
DELAMINATING
DISPERSING
FLAKING
FRACTURING
PEELING
RUPTURE
SPLITTING

SEPARATION (SOLUTION)
NT DEMINERALIZING
EXTRACTION
RT CENTRIFUGAL FORCE
CENTRIFUGES
CHEMICAL ENGINEERING
CIRCULATION
DIALYSIS
DIFFUSION
DIFFUSIVITY
MEMBRANES
OSMOSIS

SEPARATOR SPACING
RT BALLOON DYNAMICS
BALLOON SEPARATORS
PROCESS VARIABLES
SEPARATORS
SPINNING
SPINNING BALLOONS
TWISTING

SEPARATORS
USE BALLOON SEPARATORS

SEQUESTERING
USE CHELATING

SEQUESTERING AGENTS
USE CHELATING AGENTS

SERGE
BT FLAT WOVEN FABRICS
WOVEN FABRICS
RT PLAIN WEAVES
TWILL WEAVES

SERGING
RT BLIND STITCHING
GARMENT MANUFACTURE
GATHERING
HEMMING
OVEREDGE STITCHES
PINCHING
PUCKERING
RUFFLING
SEAMING
SERGING STITCHES
SEWING
SHIRRING
STITCH QUALITY
STITCHES (SEWING)
STITCHING
TAILORING
TRIMMING (OPERATION)

SERGING STITCHES
BT STITCHES (SEWING)
RT OVEREDGE STITCHES
SERGING

SERICIN
RT BOILING OFF
DEGUMMING
FIBROIN (SILK)
SILK

SERRATED CROSS SECTION
BT FIBER CROSS SECTIONS

SERVICEABILITY
RT ABRASION RESISTANCE
DIMENSIONAL STABILITY
DURABILITY
END USE PROPERTIES
FATIGUE RESISTANCE
FELTING RESISTANCE
LAUNDERABILITY
MILDEW RESISTANCE
INSECT RESISTANCE
ROT RESISTANCE
RUN RESISTANCE
SHRINK RESISTANCE
SOIL RESISTANCE
WEAR RESISTANCE
WEATHER RESISTANCE

SERVO DRAFT
RT AUTOLEVELLING
AUTOMATIC CONTROL
DRAFT CONTROL
DRAFTING (STAPLE FIBER)
IRREGULARITY
IRREGULARITY CONTROL
PROCESS CONTROL

SERVO FIBROGRAPH
BT FIBROGRAPH
RT COEFFICIENT OF LENGTH VARIATION
DIGITAL FIBROGRAPH
FIBER ARRAY
FIBER DIAGRAM
FIBER LENGTH
FIBER LENGTH DISTRIBUTION
FIBROGRAPH MEAN LENGTH
MEAN FIBER LENGTH
STAPLING
UPPER HALF MEAN LENGTH
UPPER QUARTILE LENGTH

SERVOMECHANISMS
RT AUTOMATIC CONTROL
CONTROL SYSTEMS
PNEUMATIC CONTROL
PROCESS CONTROL
REMOTE CONTROL

SET (WARP KNITTING)
RT FULL SET
GUIDE BARS
HALF SET
LAPPING (WARP KNITTING)
MILANESE KNITTING MACHINES
OVERLAP
RASCHEL KNITTING MACHINES
RUNNER RATIO
TRICOT KNITTED FABRICS
TRICOT KNITTING
TRICOT KNITTING MACHINES
UNDERLAP
WARP KNITTING MACHINES

SET (WOVEN FABRIC)
BT FABRIC PROPERTIES
FABRIC PROPERTIES (STRUCTURAL)
RT BEER
COVER FACTOR
ENDS PER INCH
FABRIC STRUCTURE
PICKS PER INCH
STITCH DENSITY
TEXTURE
TIGHTNESS

SET UP TIME
RT DOWN TIME
MANAGEMENT
PROCESS EFFICIENCY
PRODUCTION

SETOVER
UF THROWOVER
RT COMB TEAR
COMBING
COVER PLATES
DABBING BRUSHES
LARGE CIRCLE (COMB)
NOBLE COMBING
NOBLE COMBS
PROCESS VARIABLES
SMALL CIRCLE (COMB)

SETTING (ADJUSTMENTS)
RT ADJUSTMENTS
CALIBRATION
FITTING
FLAT SETTINGS (CARD)
FLUCTUATIONS
LEVELLING
MAINTENANCE
POSITIONING

SETTING (EXCEPT SYNTHETICS)
BT DRY FINISHING
WET FINISHING
RT COLD SETTING (SYNTHETICS)
DURABLE PRESS TREATMENTS
HAIR SETTING
HEAT SETTING (DRY)
HEAT SETTING (MOIST)
HEAT SETTING (SYNTHETICS)
PRESSING
WOOL
WOOL FABRICS

SETTING AGENTS
RT FINISH (SUBSTANCE ADDED)
HAIR SETTING
RESINS
STIFFENERS (AGENTS)

SETTING TIME
RT ANNEALING
BONDING STRENGTH
POLYMERIZATION
RELAXATION TIME (MECHANICAL)
STRESS RELAXATION
TIME

SETTING TREATMENTS (GENERAL)
NT COLD SETTING (SYNTHETICS)
HEAT SETTING (SYNTHETICS)
SETTING (EXCEPT SYNTHETICS)
RT DRY FINISHING
HAIR SETTING
WET FINISHING

SETTLING TANKS
RT COAGULATING BATHS
FILTRATION
GREASE RECOVERY
PRECIPITATION
SCOURING BATHS
SCOURING TRAINS
SEPARATION (SOLUTION)
SEWING EFFICIENCY

SEWABILITY
BT FABRIC PROPERTIES
FABRIC PROPERTIES (MECHANICAL)
FINISH (PROPERTY)
RT BOBBIN THREADS (SEWING)
COMPRESSIBILITY
LUBRICANTS
MACHINE SEWING
MATRIX FREEDOM
NEEDLE HEATING
NEEDLE PENETRATION
SEAM DAMAGE
SEAM EFFICIENCY
SEAM SLIPPAGE STRENGTH
SEAM STRENGTH
SEAMING
SEAMS
SEWING
SEWING THREADS
SOFTENERS
STITCH ELASTICITY
STITCHING
THREAD BREAKAGES
THREAD TENSION
YARN DAMAGE

SEWING
NT CHAIN SEWING
HAND SEWING
MACHINE SEWING
TACKING
BT GARMENT MANUFACTURE
RT ANGULAR SEAMING
BACK TACKING
BASTING
BLIND STITCHING
BOBBIN THREADS (SEWING)
BUNDLING (TAILORING)
BUTTED SEAMS
BUTTERFLY SEAMS
COCKLING
COTTON THREADS
COVERED SEAMS
CURVED SEAMING
CUTTING (TAILORING)
DRESSMAKING
EMBROIDERY
FABRIC DISTORTION (TAILORING)
FABRIC INSPECTION
FELLED SEAMS
FELLING
FRENCH SEAMS
FULL SEAM WIDTH

RT GATHERING
HEM SEAMS
HEMMING
LAPPED SEAMS
LINEN THREADS
LINKING (KNITTING)
LOOP CATCHING
MALI PROCESSES (TN)
NEEDLE EYE
NEEDLE GROOVE
NEEDLE HEATING
NEEDLE PENETRATION
NEEDLE POINT
NEEDLE SCARF
NEEDLE SHAFT (SEWING)
NEEDLE SHANK (SEWING)
NEEDLES
NEEDLING
NYLON THREADS
PINCHING
PINKED SEAMS
PINKING (TAILORING)
PIPED SEAMS
POSITIONING
PUCKER (DEFECT)
PUCKERING
QUILTING
RIPPING
RIPPLES
RUFFLING
SANDWICH SEAMS
SEAM ABRASIVE STRENGTH
SEAM ALLOWANCE
SEAM BINDING
SEAM DAMAGE
SEAM EFFICIENCY
SEAM GRIN
SEAM HEADING
SEAM LET OUT
SEAM PUCKER
SEAM QUALITY
SEAM SIZE
SEAM SLIPPAGE
SEAM SLIPPAGE STRENGTH
SEAM STRENGTH
SEAM TENSILE STRENGTH
SEAM THICKNESS
SEAM WIDTH
SEAMING
SEAMS
SEAMSTRESSES
SERGING
SEWABILITY
SEWING BOBBINS
SEWING CYCLES
SEWING EFFICIENCY
SEWING MACHINES
SEWING NEEDLES
SEWING THREADS
SHADING (TAILORING)
SHIRRING
SLOPING
STITCH ELASTICITY
STITCH QUALITY
STITCH SEQUENCE
STITCH SIZE (SEWING)
STITCH TENSION
STITCHES (SEWING)
STITCHING
STRETCHING (SEAMING)
SUPERPOSED SEAMS
TAILORED SEAMS
TAILORING
TAKE UP (SEWING)
TAPING
TENSION SPRINGS
THREAD TENSION
THREADLINE
THREADS
TRIMMING (OPERATION)
TUCKING (GARMENT MANUFACTURE)
TUFTING
YARN SEVERANCE
WELTS (SEWN)

SEWING BOBBINS
BT BOBBINS
PACKAGES
RT BOBBIN CASES
BOBBIN HOOKS
BOBBIN THREADS (SEWING)
SEWING
SEWING MACHINES
SEWING THREADS

SEWING CYCLES
RT CHAIN SEWING
CHECK SPRINGS
LOOP CATCHING
NEEDLE PENETRATION
SEWING
SEWING MACHINES
TAKE UP (SEWING)
THREAD BREAKAGES
THREAD TENSION

SEWING DAMAGE
BT DAMAGE
RT FUSION (MELTING)
SEAM DAMAGE
SEAM EFFICIENCY
SEAM GRIN
SEAM PUCKER
SEAM TENSILE STRENGTH
STITCHING
THREAD BREAKAGES
TUFTING
YARN DAMAGE
YARN SEVERANCE

SEWING EFFICIENCY
BT PROCESS EFFICIENCY
RT SEAM DAMAGE
SEAM EFFICIENCY
SEWABILITY
SEWING
THREAD BREAKAGES

SEWING MACHINES
NT DOMESTIC SEWING MACHINES
INDUSTRIAL SEWING MACHINES
ZIG ZAG MACHINES
BT GARMENT MANUFACTURING MACHINES
RT BACK TACKING
BASTING
BOBBIN LOOPS
BOBBIN THREADS (SEWING)
BOBBINS
CHAIN SEWING
CHECK SPRINGS
FEED DOGS
GARMENT MANUFACTURE
HOOK SHAFTS
LOOP CATCHERS
LOOP CATCHING
MACHINE SEWING
NEEDLE EYE
NEEDLE GROOVE
NEEDLE HEATING
NEEDLE PENETRATION
NEEDLE POINT
NEEDLE SCARF
NEEDLE SHAFT (SEWING)
NEEDLE SHANK (SEWING)
NEEDLES
PRESSER FOOT
SEAMING
SEWING
SEWING BOBBINS
SEWING CYCLES
SEWING NEEDLES
SEWING THREADS
STITCHES (SEWING)
STITCHING
TAKE UP (SEWING)
TAKE UP MECHANISMS
TENSION DEVICES
THROAT PLATES

SEWING NEEDLES
BT NEEDLES
RT CHAIN SEWING
GARMENT MANUFACTURE
HAND SEWING
KNITTING NEEDLES
LOOP CATCHING
MACHINE SEWING
NEEDLE EYE
NEEDLE GROOVE
NEEDLE HEATING
NEEDLE PENETRATION
NEEDLE POINT
NEEDLE SCARF
NEEDLE SHAFT (SEWING)
NEEDLE SHANK (SEWING)
SEWING
SEWING MACHINES
STITCHES (SEWING)
STITCHING
THREADING

SEWING THREADS
UF SEWING YARNS
NT BOBBIN THREADS (SEWING)
COTTON THREADS
LINEN THREADS
NEEDLE THREADS
NYLON THREADS
BT THREADS
RT CORDAGE
GARMENT MANUFACTURE
HAND SEWING
LEA COUNT
MACHINE SEWING
NEEDLE HEATING
SEAM SLIPPAGE STRENGTH
SEAMING
SEWABILITY
SEWING
SEWING BOBBINS
SEWING MACHINES
STITCHING

SEWING YARNS
USE SEWING THREADS

SEWN WELTS
USE WELTS (SEWN)

SEYDEL MACHINE
USE HALLE SEYDEL STRETCH BREAKER

SHADE
BT OPTICAL PROPERTIES
RT AESTHETIC PROPERTIES
APPEARANCE
BLOOM
BRIGHTNESS
COLOR
COLOR MATCHING
DESIGN
DRABNESS
DULLNESS
DYEING
FABRIC DESIGN
FABRIC PROPERTIES
FABRIC PROPERTIES (AESTHETIC)
FABRIC PROPERTIES (PHYSICAL EXCLUDING MECHANICAL)
FABRICS
HUE
JUDD NBS COLOR SYSTEM
LUSTER
LUXURIOUSNESS
MUNSELL COLOR SYSTEM
PATTERN DEFINITION
RICHNESS
SATURATION (COLOR)
SPARKLE
SPECTROPHOTOMETRY
VALUE (MUNSELL)

SHADE BARS
BT DYEING DEFECTS
FABRIC DEFECTS
RT BARRE
OFF SHADE

SHADING (TAILORING)
UF SIDE TO SIDE SHADING
BT GARMENT MANUFACTURE
RT COLOR MATCHING
COLOR VISION
CUTTING (TAILORING)
ILLUMINATION
MARKING
SEWING
TAILORING

SHADOW LACE
BT LACES
RT MACHINE MADE LACES

SHADOW STRIPE
BT FLAT WOVEN FABRICS
WOVEN FABRICS
RT PLAIN WEAVES

SHADOW WELT
USE AFTERWELT

SHAFTS
RT ACCESSORIES
AXLES
BEARINGS
BUSHINGS
DYNAMIC BALANCING
MECHANISMS
SPINDLES
VIBRATION

SHANKS
RT NEEDLES

SHANTUNG
RT HIMALAYA CLOTH
NOVELTY YARNS
PLAIN WEAVES
SILK

SHAPE
UF CONTOURS
OUTLINE
NT ROUNDNESS
SPHERICITY
RT CONGRUENCE
CROSS SECTIONS
CURVATURE
DESIGN
DIMENSIONS
ELLIPTICITY
FABRIC PROPERTIES
FABRIC PROPERTIES (AESTHETIC)
FABRIC PROPERTIES (STRUCTURAL)
FIBER CROSS SECTIONS
GEOMETRY
GROOVES
HEADS (TOP)
HEIGHT
HOLLOWNESS
SURFACE PROPERTIES (MECHANICAL)
SYMMETRY
TAPER (BOBBIN)
TOPOLOGY

SHAPE FACTOR
RT CONFIGURATION
ELLIPTICITY
FIBER CROSS SECTIONS
MOMENT OF INERTIA
ROUNDNESS

SHAPING
NT BOARDING
BT GARMENT MANUFACTURE
RT ANNEALING
CIRCULAR HOSIERY
CONFIGURATION
FABRIC RELAXATION
FASHIONING
FORMING
FULLY FASHIONED HOSIERY
HEAT SETTING (SYNTHETICS)
MOLDING
PRESSING
SHAPE

SHARKSKIN
BT FLAT WOVEN FABRICS
WOVEN FABRICS
RT OXFORD WEAVES
PLAIN WEAVES
TWILL WEAVES

SHARPENING
RT CHIPPING
CUTTING
GRINDING (MATERIAL REMOVAL)
KNIVES
POLISHING
SCISSORS

SHAW SYSTEM
RT ACTUAL DRAFT
CASABLANCA SYSTEM
DRAFTING (STAPLE FIBER)
DUO ROTH SYSTEM
HIGH DRAFT
ROTH SYSTEM
SPINNING FRAMES

SHEAR (MODE OF DEFORMATION)
UF SHEAR STRAIN
SHEARING (MODE OF DEFORMATION)
BT STRESS STRAIN PROPERTIES (SHEAR)
RT BOUNDARY LAYER
COMPRESSION
CUTTING
DEFORMATION
DISTORTION
DRAPE
FLUID FLOW
FLUID FRICTION
HYDRODYNAMIC LUBRICATION
ROLLER SLIPPAGE
RUBBING
SHEAR FORCE
SHEAR FORCE DIAGRAMS
SHEAR STRESS
SLIDING
SLIPPAGE
TENSION

SHEAR FORCE
BT FORCE
RT BENDING MOMENTS
LOAD DEFLECTION CURVES
NORMAL FORCE
SHEAR (MODE OF DEFORMATION)
SHEAR FORCE DIAGRAMS
SHEAR STRENGTH
SHEAR STRESS
TRANSVERSE FORCE
TWISTING MOMENTS

SHEAR FORCE DIAGRAMS
RT BENDING MOMENT DIAGRAMS
LOAD DEFLECTION CURVES
SHEAR FORCE

SHEAR MODULUS
BT MODULUS
STRESS STRAIN PROPERTIES
STRESS STRAIN PROPERTIES (SHEAR)
RT COMPLIANCE
COMPRESSIVE MODULUS
DYNAMIC MODULUS
ELASTIC MODULUS (TENSILE)
MODULUS OF CROSS ELASTICITY
SHEAR (MODE OF DEFORMATION)
SHEAR FORCE
TORSIONAL RIGIDITY

SHEAR RESISTANCE
BT MECHANICAL PROPERTIES
STRESS STRAIN PROPERTIES
RT CUTTING RESISTANCE
DRAPE
PAPERINESS
SHEAR (MODE OF DEFORMATION)
SHEAR FORCE

SHEAR STRAIN
USE SHEAR (MODE OF DEFORMATION)

SHEAR STRENGTH
BT STRENGTH
STRESS STRAIN PROPERTIES
STRESS STRAIN PROPERTIES (SHEAR)
RT BEARING STRENGTH
COMPRESSIVE STRENGTH
CUTTING RESISTANCE
LOOP EFFICIENCY
LOOP STRENGTH
SHEAR (MODE OF DEFORMATION)
SHEAR FORCE
SHEAR STRESS
STRENGTH ELONGATION TESTING

SHEAR STRESS
BT STRESS
STRESS STRAIN PROPERTIES
STRESS STRAIN PROPERTIES (SHEAR)
RT BIAXIAL STRESS
COMPRESSIVE STRESS
HYDROSTATIC STRESS
SHEAR (MODE OF DEFORMATION)
SHEAR FORCE
SHEAR STRENGTH
TENSILE STRESS
TWISTING MOMENTS
UNIAXIAL STRESS

SHEARING (FINISHING)
UF CROPPING
BT DRY FINISHING
RT BRUSHING
FUZZ
HAIRINESS
KNIVES
KNOTS
NAPPING
PILLS
SHEAR STRENGTH
SHEARING MACHINES
SINGEING
SUEDING

SHEARING (MODE OF DEFORMATION)
USE SHEAR (MODE OF DEFORMATION)

SHEARING (TAILORING)
USE CUTTING (TAILORING)

SHEARING (WOOL)
USE WOOL SHEARING

SHEARING MACHINES
RT SHEARING (FINISHING)

SHEATH (FIBER)
USE SKIN (FIBER)

SHED (WEAVING)
UF SHED OPENING
NT BALANCED SHED
CLOSED SHED
OPEN SHED
SPLIT SHED
UNBALANCED SHED
RT BEATING UP
HARNESSES
HEDDLES
LOOMS
PICKER STICKS
PICKING (WEAVING)
SHEDDING
SHUTTLES
WARP ENDS
WEAVING

SHED OPENING
USE SHED (WEAVING)

SHEDDING
NT NEGATIVE SHEDDING
POSITIVE SHEDDING
RT BALANCED SHED
BEATING UP
BLANKET SHUTTLES
BOAT SHUTTLES
CLOSED SHED
CLOTH FELL
CRIMP INTERCHANGE
FILLING YARNS
HARNESSES
HEDDLES
LET OFF
LOOM TAKE UP
LOOMS
MATTING
OPEN SHED
PICKER STICKS
PICKING (WEAVING)
REEDS
SHED (WEAVING)
SHUTTLE BOXES
SHUTTLE DEFLECTORS

RT SHUTTLE MARKS
SHUTTLE SMASH PROTECTORS
SHUTTLES
SPLIT SHED
STICK SHUTTLES
TAKE UP
TREADLES
UNBALANCED SHED
WARP ENDS
WARP PREPARATION
WARP TENSION
WEAVING

SHEDDING (FIBER)
RT CLEANLINESS
FABRIC DEFECTS
FIBER FIBER FRICTION
FIBER MIGRATION
FUZZING
HAIRINESS
LINT (WASTE)
MATRIX FREEDOM
PILL RESISTANCE
PILL WEAR OFF
PILLING
RUBBING

SHEEN
UF SHINE
BT FABRIC PROPERTIES
FABRIC PROPERTIES (AESTHETIC)
FINISH (PROPERTY)
RT AESTHETIC PROPERTIES
BLOOM
BRIGHTNESS
COLOR
DULLNESS
LUSTER
OPTICAL PROPERTIES
SMOOTHNESS
SPARKLE

SHEEP
NT LAMBS
RAMS
BT ANIMALS
RT ANIMAL FIBERS (GENERAL)
ANIMAL HUSBANDRY
FLEECES
GRADE (FIBERS)
LAMBSWOOL
PULLED WOOL
SHEEP BREEDS
VIRGIN WOOL
WOOL
WOOL GRADE
WOOL SHEARING

SHEEP BREEDING
BT ANIMAL HUSBANDRY
RT GOAT BREEDING

SHEEP BREEDS
RT ANIMAL HUSBANDRY
LONG WOOL
MEDIUM WOOL
MERINO WOOL
MOUNTAIN WOOL
SHEEP
WOOL
WOOL GRADE

SHEERNESS
RT AESTHETIC PROPERTIES
APPEARANCE
CIRCULAR HOSIERY
COVER
FABRIC PROPERTIES (AESTHETIC)
FULLY FASHIONED HOSIERY
HOSIERY
LINGERIE
STOCKINGS
TRANSLUCENCY
TRANSPARENCY

SHEET CLOTHING
BT CARD CLOTHING
RT CARD CLOTHING FOUNDATION
CARD GRINDING
CARD NAILING
CARD WIRES
CARDS
COUNT CROWN AND GAUGE
FILLET CLOTHING
KNEE
METALLIC CARD CLOTHING
NOGG
RIB SET
TWILL SET

SHEETING (FABRIC)
BT FLAT WOVEN FABRICS
WOVEN FABRICS
RT CRETONNE
CRINOLINE
HOUSEHOLD FABRICS
PLAIN WEAVES
TOBACCO CLOTH
WIGAN

SHELL OFF
USE SLOUGHING

SHELTERS
BT MILITARY PRODUCTS
RT END USES
INDUSTRIAL FABRICS
INFLATABLE FABRICS
RADOMES
TARPAULINS

SHINE
USE SHEEN

SHINERS
UF BRIGHT PICKS
BT FABRIC DEFECTS
RT FABRIC INSPECTION
PIRN BARRE
QUILLING
STREAKINESS
WINDING
YARN CRIMP
YARN TENSION

SHINGUARDS
BT ATHLETIC CLOTHING

SHIRLEY ANALYZER
RT COTTON SYSTEM PROCESSING
LINT CONTENT
QUALITY CONTROL
TRASH

SHIRLEY MOISTURE METER
BT MOISTURE CONTENT TESTERS
RT FORTE MOISTURE METER
HART MOISTURE METER
KARL FISCHER TITRATION METHOD
MOISTURE CONTENT
MOISTURE CONTENT TESTING

SHIRRER BLADES
RT SHIRRING

SHIRRING
BT FABRIC DISTORTION (TAILORING)
RT BACK TACKING
BASTING
CHAIN SEWING
COCKLING
CURVED SEAMING
GATHERING
HEMMING
PINCHING
PRESSER FOOT
PUCKERING
RIPPLES
RUFFLING
SEAM PUCKER
SEAMING
SEAMS
SERGING
SEWING
SHIRRER BLADES
STITCHES (SEWING)
STITCHING
TAILORING
TRIMMING (OPERATION)

SHIRTING
BT FLAT WOVEN FABRICS
WOVEN FABRICS

SHIRTS
BT ATHLETIC CLOTHING
GARMENTS
WORK CLOTHING
RT NECKTIES
SWEATSHIRTS

SHOCK
RT BRITTLE FRACTURE
CRACK PROPAGATION
DAMPERS
ENERGY ABSORPTION
FRACTURING
IMPACT
RUPTURE
SHOCK ABSORBERS
SHOCK RESISTANCE
SNAGGING
VIBRATION
VIBRATION DAMPERS
VIBRATION ISOLATORS

SHOCK ABSORBERS
RT DAMPERS
DAMPING
ENERGY ABSORPTION
PARACHUTES
RECOVERY SYSTEMS (AERODYNAMIC)
SHOCK
SHOCK RESISTANCE
VIBRATION
VIBRATION DAMPERS
VIBRATION ISOLATORS

SHOCK RESISTANCE
BT MECHANICAL PROPERTIES
STRENGTH
STRESS STRAIN PROPERTIES
RT BREAKING LENGTH
BREAKING STRENGTH
CRACK PROPAGATION
CRITICAL VELOCITY
DAMPERS
ENERGY ABSORPTION
FATIGUE
FRACTURING
IMPACT
IMPACT STRENGTH
RUPTURE
SHOCK
SHOCK ABSORBERS
TEAR STRENGTH
VISCOELASTICITY

SHODDY
RT GARNETTING
GARNETTING MACHINES
MUNGO
RECOVERED WOOL
WASTE MACHINES
WOOL

SHOE LININGS
BT LININGS
LININGS (FOOTWEAR)
RT BOOTS
BOX LININGS
CASKET LININGS
FABRICS
GLOVE LININGS
INTERLININGS
LASTING
LINING FELT
POCKET LININGS
SHOES

SHOE SIZES
RT ANTHROPOMETRIC MEASUREMENTS
COMFORT
FIT
LASTS
SHOES

SHOEPAC
BT BOOTS
FOOTWEAR
RT SHOES

SHOES
BT FOOTWEAR
RT BINDINGS (FOOTWEAR)
BOOTS
COMFORT
CORFAM (TN)
FIT
GALOSHES
LACES (SHOES)
LASTING
LASTS
LEATHER
LEGGINGS
LININGS (FOOTWEAR)
PUMPS (FOOTWEAR)
RUBBER FOOTWEAR
SCUFFING
SHOE LININGS
SHOEPAC
SLIPPERS
SNEAKERS
SOLES
SPATS
STITCH QUALITY
STITCHING
SYNTHETIC LEATHER
TONGUES
UPPERS
VINYL COATED FABRICS

SHOGGING
RT GUIDE BARS
KNITTING
LAPPING (WARP KNITTING)
RACKING
RASCHEL KNITTING MACHINES
TRICOT KNITTING MACHINES
WARP KNITTING

SHORN WOOL
BT WOOL
RT PULLED WOOL
VIRGIN WOOL
WOOL CLASS
WOOL GRADE
WOOL SHEARING

SHORT FIBERS INDEX
RT COTTON CLASSING
COTTON QUALITY
EFFECTIVE LENGTH (FIBER)
FIBER ARRAY
FIBER DIAGRAM
FIBER LENGTH DETERMINATION
FIBER LENGTH DISTRIBUTION
FIBROGRAPH
LENGTH UNIFORMITY RATIO (FIBER)
MEAN FIBER LENGTH
SORTING
SPAN LENGTH
STAPLING
SUTER WEBB TESTING
UPPER HALF MEAN LENGTH

SHORT LONG STITCHES
BT STITCHES (SEWING)

SHORT STAPLE COTTON
BT COTTON

SHORT STAPLE SPINNING
RT COTTON SYSTEM PROCESSING
LONG STAPLE SPINNING

SHORT WOOL
BT WOOL
RT LONG WOOL
MEDIUM WOOL
WOOL GRADE

SHORTS
BT ATHLETIC CLOTHING
GARMENTS
OUTERWEAR
SPORTSWEAR
UNDERWEAR
RT BRASSIERES
BRIEFS
DRAWERS
KNITTED FABRICS
PANTIES
TRUNKS
VESTS

SHRINK GRADING
RT FUSION CONTRACTION SHRINKAGE
PRESHRINKING
RESIDUAL SHRINKAGE
SHRINKAGE
WARP SHRINKAGE

SHRINK PROOFING
NT CHLORINATION (SHRINK PROOFING)
WURLAN (TN)
BT WET FINISHING
RT ANTIFELTING TREATMENTS
COMPACTING
COMPRESSIVE SHRINKING
FELTING
PERMANGANATES
RIGMEL PROCESS
SANFORIZING (TN)
SHRINK RESISTANCE
SHRINK RESISTANCE FINISHES

SHRINK RESISTANCE
UF SHRINKAGE RESISTANCE
BT END USE PROPERTIES
FABRIC PROPERTIES
FABRIC PROPERTIES (MECHANICAL)
FINISH (PROPERTY)
MECHANICAL PROPERTIES
RT CHLORINATED WOOL
COMPACTING
COMPRESSIVE SHRINKING
DIMENSIONAL STABILITY
FABRIC SHRINKAGE
FELTING SHRINKAGE
FILLING SHRINKAGE
FUSION CONTRACTION SHRINKAGE
LAUNDERABILITY TESTING
LONDON SHRINKING
MOLDING SHRINKAGE
PRESHRINKING
RELAXATION SHRINKAGE
RESIDUAL SHRINKAGE
SANFORIZING (TN)
SHRINK PROOFING
SHRINK RESISTANCE FINISHES
SHRINKAGE
SHRINKING
WARP SHRINKAGE
WRINKLE RESISTANCE

SHRINK RESISTANCE FINISHES
UF ANTISHRINK COMPOUNDS
NT WURLAN (TN)
BT FINISH (SUBSTANCE ADDED)
RT CHLORINATED WOOL
CHLOROCYANURIC ACID
COMPACTING
COMPRESSIVE SHRINKING
MOLDING SHRINKAGE
PRESHRINKING
REACTANTS
RESIDUAL SHRINKAGE
SANFORIZING (TN)
SHRINK PROOFING
SHRINK RESISTANCE
SHRINKAGE
SHRINKING
WASH WEAR FINISHES

SHRINKAGE
NT COMPRESSIVE SHRINKAGE
DIFFERENTIAL SHRINKAGE
FABRIC SHRINKAGE
FELTING SHRINKAGE
FILLING SHRINKAGE
FUSION CONTRACTION SHRINKAGE
MOLDING SHRINKAGE
RELAXATION SHRINKAGE
RESIDUAL SHRINKAGE
SCOURING SHRINKAGE
WARP SHRINKAGE
RT BUCKLING
COMPACTING
COMPRESSIVE SHRINKING
CONTRACTION
CREPING
CURL
DIMENSIONAL STABILITY
FULLING
HEAT RESISTANCE
LONDON SHRINKING
PRESHRINKING
PUCKERING
SHRINK GRADING
SHRINK RESISTANCE
SHRINK RESISTANCE FINISHES
SHRINKING
SWELLING
THERMAL STABILITY
YARN SLIPPAGE

SHRINKAGE RESISTANCE
USE SHRINK RESISTANCE

SHRINKAGE TESTERS
BT TESTING EQUIPMENT
RT LAUNDERABILITY TESTERS
LAUNDEROMETER
WASHFASTNESS TESTERS
WASHING MACHINES

SHRINKAGE TESTING
BT LAUNDERABILITY TESTING
TESTING
RT WASHFASTNESS TESTING

SHRINKING
NT COMPRESSIVE SHRINKING
LONDON SHRINKING
PRESHRINKING
BT GARMENT MANUFACTURE
WET FINISHING
RT BOARDING
COMPACTING
DECATING
DIMENSIONAL STABILITY
FABRIC SHRINKAGE
FELTING
FILLING SHRINKAGE
FULLING
FUSION CONTRACTION SHRINKAGE
HEAT SETTING (SYNTHETICS)
MOLDING SHRINKAGE
NAPPING
RELAXATION
RESIDUAL SHRINKAGE
RIGMEL PROCESS
SANFORIZING (TN)
SHRINK PROOFING
SHRINK RESISTANCE
SHRINK RESISTANCE FINISHES
SHRINKAGE
TAILORING
WARP SHRINKAGE

SHROPSHIRE WOOL
BT MEDIUM WOOL
WOOL

SHUTTLE ACCESSORIES
RT BOBBINS
QUILLS
SHUTTLE BINDERS
SHUTTLE BOXES
SHUTTLE DEFLECTORS
SHUTTLE FEELERS
SHUTTLE GUARDS
SHUTTLE SMASH PROTECTORS
SHUTTLES

SHUTTLE BINDERS
RT LOOMS
PICKER SHOES
PICKER STICKS
PICKING (WEAVING)
RACE BOARDS
SHUTTLE ACCESSORIES
SHUTTLE BOXES
SHUTTLE CHECKING
SHUTTLE SMASH PROTECTORS
SHUTTLES
SLEYS (LOOM)
WEAVING

SHUTTLE BOXES
RT AUTOMATIC LOOMS
FILLING YARNS
LOOMS
PICK AND PICK
PICKER SHOES
PICKER STICKS
PICKING (WEAVING)
ROTARY MAGAZINES
SHEDDING
SHUTTLE ACCESSORIES
SHUTTLE BINDERS
SHUTTLE CHANGING
SHUTTLE CHECKING
SHUTTLE SMASH PROTECTORS
SHUTTLES
SLEYS (LOOM)

SHUTTLE BREAKING
USE SHUTTLE CHECKING

SHUTTLE CHANGING
RT AUTOMATIC LOOMS
EJECTION
GRIPPERS (YARN)
QUILL CHANGING
SHUTTLE BOXES
SHUTTLES

SHUTTLE CHECKING
UF SHUTTLE BREAKING
RT LOOMS
SHUTTLE BINDERS
SHUTTLE BOXES
SHUTTLES
SLEYS (LOOM)
WEAVING

SHUTTLE DEFLECTORS
RT LOOMS
PICKING (WEAVING)
RACE BOARDS
REEDS
SHEDDING
SHUTTLE ACCESSORIES
SHUTTLE SMASH PROTECTORS
SHUTTLES
SLEYS (LOOM)
WEAVING

SHUTTLE FEELERS
RT AUTOMATIC CONTROL
BANGING OFF
LOOMS
NOSE BUNCHING
QUILLS
SHUTTLE ACCESSORIES
SHUTTLES
STOP MOTIONS

SHUTTLE GUARDS
RT LOOMS
SHUTTLE ACCESSORIES
SHUTTLES

SHUTTLE MARKS
BT FABRIC DEFECTS
RT LOOMS
PICKING (WEAVING)
REEDS
SHEDDING
SHUTTLE SMASH
SHUTTLE SMASH PROTECTORS
SHUTTLES
SLEYS (LOOM)
WEAVING

SHUTTLE SMASH
UF SMASH
RT CLOSED SHED
END BREAKAGES
LOOMS
NEGATIVE SHEDDING
OPEN SHED
POSITIVE SHEDDING
SHUTTLE MARKS
SHUTTLE SMASH PROTECTORS
SHUTTLES
SPLIT SHED

SHUTTLE SMASH **PROTECTORS**
RT **LOOMS**
PICKING (WEAVING)
RACE BOARDS
REEDS
SHEDDING
SHUTTLE ACCESSORIES
SHUTTLE BINDERS
SHUTTLE BOXES
SHUTTLE DEFLECTORS
SHUTTLE MARKS
SHUTTLE SMASH
SHUTTLES
SLEYS (LOOM)
STOP MOTIONS
WEAVING

SHUTTLE WEAVING
USE WEAVING

SHUTTLELESS LOOMS
NT DRAPER SHUTTLELESS LOOMS (TN)
JET LOOMS
RAPIER LOOMS
SULZER LOOMS (TN)
WATER JET LOOMS
BT LOOMS
RT AUTOMATIC LOOMS
BOX LOOMS
DOBBY LOOMS
HAND LOOMS
JACQUARD LOOMS
LAPPET LOOMS
OUTSIDE FILLING SUPPLY
PICKING (WEAVING)
PILE FABRIC LOOMS
RAPIERS (LOOMS)
SHUTTLE BINDERS

SHUTTLELESS WEAVING
BT WEAVING
RT AIR JET LOOMS
DRAPER SHUTTLELESS LOOMS (TN)
JET LOOMS
LOOMS
OUTSIDE FILLING SUPPLY
RAPIER LOOMS
SULZER LOOMS (TN)
WATER JET LOOMS

SHUTTLES
NT BLANKET SHUTTLES
BOAT SHUTTLES
FLY SHUTTLES
STICK SHUTTLES
RT CLOSED SHED
COPS
FILLING YARNS
LOOMS
NEGATIVE SHEDDING
OPEN SHED
PICKER SHOES
PICKER STICKS
PICKING (WEAVING)
PICKS
PICKS PER INCH
POSITIVE SHEDDING
QUILLING
QUILLS
RACE BOARDS
REEDS
ROTARY MAGAZINES
SHED (WEAVING)
SHEDDING
SHUTTLE ACCESSORIES
SHUTTLE BINDERS
SHUTTLE BOXES
SHUTTLE CHANGING
SHUTTLE CHECKING
SHUTTLE DEFLECTORS
SHUTTLE MARKS
SHUTTLE SMASH
SHUTTLE SMASH PROTECTORS
SLEYS (LOOM)
SPLIT SHED
THREADING
WEAVING

SIDE CHAINS (MOLECULAR)
RT CHAIN MOLECULES
CROSSLINKING
CRYSTALLINITY
DEGREE OF POLYMERIZATION
LONG CHAIN MOLECULES
MOLECULAR ORIENTATION
MOLECULAR STRUCTURE
POLYMERS

SIDE TO SIDE SHADING
USE SHADING (TAILORING)

SILANES
USE SILICONES

SILESIA
BT FLAT WOVEN FABRICS
WOVEN FABRICS
RT PLAIN WEAVES

SILESIAN WOOL
BT MERINO WOOL

SILICA
NT COLLOIDAL SILICA
LUDOX (TN)
SYTON (TN)
BT OXIDES
SILICON COMPOUNDS

SILICATES
NT SODIUM SILICATE
BT SILICON COMPOUNDS
RT BLEACHING
CLAYS
HYDROGEN PEROXIDE
MONTMORILLONITE

SILICOFLUORIDES
BT FLUORIDES
SILICON COMPOUNDS
RT FLUOROCARBONS

SILICON CARBIDE FIBERS
BT CERAMIC FIBERS
HIGH TEMPERATURE FIBERS
INORGANIC FIBERS (MAN MADE)
RT ALUMINUM SILICATE FIBERS
BORON NITRIDE FIBERS

SILICON COMPOUNDS
NT SILICA
SILICATES
SILICOFLUORIDES
SILICONES

SILICONE FLUIDS
USE SILICONES

SILICONE RESINS
BT RESINS
SILICONES

SILICONE RUBBER
BT SILICONES
SYNTHETIC RUBBER
RT ELASTOMERS

SILICONES
UF ORGANOSILICON COMPOUNDS
POLYSILOXANES
SILANES
SILICONE FLUIDS
SILOXANES
NT SILICONE RESINS
SILICONE RUBBER
BT SILICON COMPOUNDS
RT ANTIFOAM AGENTS
LUBRICANTS
NONIONIC SOFTENERS
ORGANOMETALLIC COMPOUNDS
(EXCLUDING SILICONES)
SOFTENERS
WATER REPELLENTS

SILK
NT CANTON SILK
ITALIAN SILK
NETT SILK
SCHAPPE SILK
SPUN SILK
TSATLEE SILK
TUSSAH SILK
WHITE CHINA SILK
WHITE JAPAN SILK
YELLOW CHINA SILK
YELLOW JAPAN SILK
BT FIBERS
NATURAL FIBERS
PROTEIN FIBERS
RT AMINO ACIDS
ARTIFICIAL SILK (ARCHAIC)
BAVE (SILK)
BOOKS (SILK)
BRINS (SILK)
COCOONS
DEGUMMING
DENIER
DISULFIDES (ORGANIC)
FIBROIN (SILK)
FILAMENT YARNS
FILAMENTS
FUR
HUSKS
MONOFILAMENTS
NYLON (POLYAMIDE FIBERS)
ORGANZINE
PROTEINS
SCROOP
SERICIN
SILK BLENDS
SILK DYEING
SILK DYES
SILK NOILS
SILK SYSTEM PROCESSING
SILK THROWING
SILKWORMS
SKEINS
TRAM

SILK BLENDS
BT BLENDS (FIBERS)
RT SILK
SILK DYEING
SILK DYES

SILK DYEING
BT DYEING (BY FIBER CLASSES)
RT ACETATE DYEING
BLEND DYEING
CELLULOSE DYEING
COTTON DYEING
RAYON DYEING
SILK
SILK BLENDS
SILK DYES
SYNTHETIC DYEING
WOOL DYEING

SILK DYES
BT DYES (BY FIBER CLASSES)
RT ACETATE DYES
ACRYLIC DYES
CELLULOSE DYES
DYEING
DYES (BY CHEMICAL CLASSES)
NYLON DYES
POLYESTER DYES
SILK
SILK BLENDS
SILK DYEING
WOOL DYES

SILK LAP (WARP KNITTING)
RT LAPPING (WARP KNITTING)
MILANESE KNITTING MACHINES
OVERLAP
UNDERLAP
WARP KNITTING

SILK NOILS
BT NOILS
RT SILK

SILK SYSTEM PROCESSING
UF SCHAPPE SYSTEM PROCESSING
RT COTTON SYSTEM PROCESSING
FLAX SYSTEM PROCESSING
SILK
SILK BLENDS
SILK NOILS
WOOLEN SYSTEM PROCESSING
WORSTED SYSTEM PROCESSING

SILK SYSTEM WARPING
RT CLEARING LEASE
SECTIONAL WARPING
SILK
SILK SYSTEM PROCESSING
WARPING
WARPING LEASE

SILK THROWING
RT SILK
SILK SYSTEM PROCESSING
THROWING
THROWSTERS

SILKWORMS
RT ANIMALS
INSECTS
SILK

SILOXANES
USE SILICONES

SILVER COMPOUNDS
RT ANTIMICROBIAL FINISHES
GOLD COMPOUNDS
MERCURY COMPOUNDS
MOLYBDENUM COMPOUNDS
TUNGSTEN COMPOUNDS

SIMPLEX FABRICS
BT KNITTED FABRICS
WARP KNITTED FABRICS
RT MILANESE FABRICS
RASCHEL FABRICS
TRICOT KNITTED FABRICS

SIMPLEX KNITTING MACHINES
BT KNITTING MACHINES
WARP KNITTING MACHINES
RT MILANESE KNITTING MACHINES
RASCHEL KNITTING MACHINES

SIMULTANEOUS DYEING AND FINISHING
BT DYEING (BY PROCESS FLOW)
SIMULTANEOUS FINISHING
PROCESSES
WET FINISHING
RT BATCH DYEING
CONTINUOUS DYEING
CONTINUOUS FINISHING
PAD DYEING
PAD FINISHING
REACTIVE DYEING
SEMI CONTINUOUS DYEING

SIMULTANEOUS FINISHING PROCESSES
NT SIMULTANEOUS DYEING AND
FINISHING
RT DESIZING
FINISHING PROCESS (GENERAL)
FULLING
SCOURING
WET FINISHING

SINGEING
BT DRY FINISHING
RT BURNING
FUZZ
GLAZE
HAIRINESS
MELTING
MELTING POINT
PATTERN DEFINITION
PILLS
SHEARING (FINISHING)
SINGEING MACHINES

SINGEING MACHINES
RT DRY FINISHING
OVENS
SHEARING MACHINES
SINGEING

SINGLE APRONS
BT APRONS
RT APRON CONTROL
CONTROL SURFACES
DOUBLE APRONS
DRAFTING (STAPLE FIBER)
FIBER CONTROL

SINGLE FILAMENT YARNS
USE MONOFILAMENT YARNS

SINGLE NEEDLE LOCKSTITCHES
BT LOCKSTITCHES
STITCHES (SEWING)

SINGLE THREAD BLINDSTITCHES
BT BLINDSTITCHES
STITCHES (SEWING)
RT BLIND STITCHING

SINGLE THREAD CHAIN STITCHES
BT CHAIN STITCHES
STITCHES (SEWING)

SINGLES
USE SINGLES YARNS

SINGLES TWIST
BT TWIST
RT HIGH TWIST
LOW TWIST
PLY TWIST
REAL TWIST

SINGLES YARNS
UF SINGLES
BT YARNS
RT BALANCED YARNS
COARSE YARNS
CONTINUOUS FILAMENT YARNS
CORDS
CORE SPUN YARNS
COUNT
DENIER
FILAMENT YARNS
FINE YARNS
MULTIPLIED YARNS
NOVELTY YARNS
PLIED YARNS
PLYING
S TWIST
S TWIST YARNS
SPUN YARNS
TWIST LIVELINESS
TWIST MULTIPLIER
TWO PLY YARNS
YARN GEOMETRY
YARN STRUCTURE
Z TWIST
Z TWIST YARNS

SINGLINGS
BT FABRIC DEFECTS
RT END BREAKAGES
FABRIC INSPECTION

SINKER BARS
RT GUIDE BARS
KNITTING MACHINES
MILANESE KNITTING MACHINES
NEEDLE BARS
PRESSER BARS
SINKER CAMS
SINKERS
SLURCOCKS

SINKER CAMS
BT CAMS
RT KNITTING
KNITTING MACHINES
SINKER BARS
SINKERS
SLURCOCKS
STITCH CAMS

SINKER LOOPS
RT DOUBLE KNIT FABRICS
KNITTED FABRICS
KNITTED STITCHES
KNITTING
LOOP DISTORTION
NEEDLE LOOPS
SINKERS
SINKING (KNITTING)
STITCH LENGTH

SINKERS
RT CIRCULAR KNITTING MACHINES
DIVIDERS (KNITTING)
FILLING KNITTING
(CONT.)

RT FILLING KNITTING MACHINES
FULLY FASHIONED KNITTING MACHINES
HOSIERY MACHINES
KNITTING
KNITTING MACHINES
MILANESE KNITTING MACHINES
MOCK SEAMS
NEEDLE BARS
SEAMLESS HOSIERY MACHINES
SINKER CAMS
SINKER LOOPS
SINKING (KNITTING)
WEBHOLDERS

SINKING (KNITTING)
RT FILLING KNITTING
FILLING KNITTING MACHINES
FULLY FASHIONED KNITTING MACHINES
KNITTING
LANDING LOOPS
LAYING IN (KNITTING)
SINKER LOOPS
SINKERS

SINTERED METAL RINGS
BT RINGS
RT ANTIWEDGE RINGS
BALLOON DYNAMICS
FRICTION
LUBRICATED RINGS
LUBRICATION
PLYING
POROUS MATERIALS
RING SPINNING
SPINNING
TRAVELERS
TWISTING
UNLUBRICATED RINGS
WEAR

SISAL
BT CELLULOSIC FIBERS
LEAF FIBERS
RT CORDAGE
MANILA
ROPES

SITES (DYEING)
NT ANIONIC SITES
CATIONIC SITES
RT ADSORPTION
DYEABILITY
DYEING
DYEING THEORY
ION EXCHANGE
OVERDYEING
SORPTION

SIVAL MACHINES
BT LACE MACHINES
RT BAR WARP MACHINES
BARMEN MACHINES
BOBBINET MACHINES
CURTAIN MACHINES
DOUBLE LOCKER MACHINES
GO THROUGH MACHINES
LEAVERS MACHINES
MECHLIN MACHINES
ROLLING LOCKER MACHINES
STRING WARP MACHINES
WARP LACE MACHINES

SIZE (MAGNITUDE)
USE DIMENSIONS

SIZE BOXES
RT DESIZING
SIZES (SLASHING)
SLASHERS
SLASHERS BEAMS
SLASHING
SPLIT RODS
SQUEEZE ROLLS
SQUEEZING
STRETCH ROLLS
WARP BEAMS
WARP PREPARATION
WEAVING

SIZES (SLASHING)
UF SIZING AGENTS
BT TEXTILE CHEMICALS
RT ABRASION
ABRASION RESISTANCE
ADHESIVES
BENDING RIGIDITY
CARBOXYMETHYLCELLULOSE
COMFORT
DEACETYLATED CHITIN
DELIQUESCENT AGENTS
DESIZING
DESIZING MACHINES
DEXTRINS
GELATINE
GLUES
HAIRINESS

RT KETTLES (HEATING)
LUBRICANTS
PRESERVATIVES
RESINS
ROSIN
SIZE BOXES
SIZING (PAPERMAKING)
SLASHERS
SLASHERS BEAMS
SLASHING
SQUEEZING
STARCH
STARCH GUMS
STIFFENERS (AGENTS)
STRETCH ROLLS
STYMER (TN)
WARP PREPARATION

SIZING
USE SLASHING

SIZING (PAPERMAKING)
RT PAPER MAKING
ROSIN
SIZES (SLASHING)
SLASHING
STARCH
STARCH GUMS

SIZING AGENTS
USE SIZES (SLASHING)

SKEIN DYEING
BT YARN DYEING
RT CHEESE DYEING
CONE DYEING
HANK DYEING
PACKAGE DYEING
SKEINS

SKEINS
RT BUNDLES
FIBER ASSEMBLIES
HANK NUMBER
HANKS
INDIRECT COUNT
PACKAGES
REELING
REELING MACHINES (WINDING)
ROVINGS
SILK
YORKSHIRE SKEIN WOOLEN COUNT

SKETCHING (APPAREL DESIGN)
RT APPAREL DESIGN
DESIGN
DRAFTING (PATTERN)

SKEW
BT FABRIC DEFECTS
RT BAGGINESS
BOW
BOW STRAIGHTENERS
DISTORTION
FABRIC PROPERTIES
FABRIC PROPERTIES (STRUCTURAL)
SHEAR (MODE OF DEFORMATION)
SKEW STRAIGHTENERS

SKEW STRAIGHTENERS
RT BOW STRAIGHTENERS
DRY FINISHING
EXPANDERS
FABRIC DEFECTS
FILLING STRAIGHTENING MACHINES
OPTICAL SCANNERS
OPTICAL SCANNING
SKEW

SKI PANTS
BT ATHLETIC CLOTHING
GARMENTS
SPORTSWEAR
RT LEOTARDS
SLACKS
STRETCH FABRICS

SKIN
RT ANTHROPOMETRIC KINEMATICS
CONTACT
FOLLICLE
HAND
PIGSKIN

SKIN (FIBER)
UF SHEATH (FIBER)
BT FINE STRUCTURE (FIBERS)
RT CORE (FIBER)
FIBER CROSS SECTIONS
FIBER SURFACE
RADIAL DISTRIBUTION
SURFACE PROPERTIES (PHYSICAL CHEMICAL)
VISCOSE RAYON

SKIN IRRITATION
UF IRRITATION (SKIN)
RT ABRASION

RT BIOCHEMICAL PROPERTIES
DERMATITIS
FRICTION
ROUGHNESS
TOXICITY

SKIN WOOL
USE PULLED WOOL

SKINEMATICS
USE ANTHROPOMETRIC KINEMATICS

SKIP DENT
BT FLAT WOVEN FABRICS
WOVEN FABRICS
RT LIGHT DENT
REED WIRES
REEDING
SWOLLEN DENT

SKIRTS
BT GARMENTS
RT BLOUSES
DRESSES

SKITTERINESS
BT DYEING DEFECTS
FABRIC DEFECTS
RT COLORFASTNESS
LEVELNESS (DYEING)
SPECKINESS
STREAKINESS
TIPPINESS

SKYLOFT (TN)
BT AIR JET TEXTURED YARNS
TEXTURED YARNS (FILAMENT)
RT TASLAN (TN)

SLACK COURSE
RT COURSES
KNITTED FABRICS
KNITTING
SLACK FEEDERS (KNITTING)
SLACK THREADS
WALES

SLACK FEEDERS (KNITTING)
RT FABRIC DEFECTS
KNITTING
SLACK COURSE
SLACK THREADS
STITCH LENGTH
TENSION CONTROL

SLACK MERCERIZING
BT MERCERIZING
RT STRETCH FABRICS
STRETCH YARNS (FILAMENT)
TENSION CONTROL

SLACK SELVEDGES
UF PUCKERED SELVEDGES
COCKLED SELVEDGES
BT FABRIC DEFECTS
RT CENTER SELVEDGES
CRACKED SELVEDGES
DOG LEGGED SELVEDGES
LENO SELVEDGES
LOOMS
LOOP SELVEDGES
PULLED IN SELVEDGES
SEALED SELVEDGES
SELVEDGE WARPS
SELVEDGES
TAPE SELVEDGES
TEMPLES
TIGHT SELVEDGES
WEAVING
WOVEN FABRICS

SLACK THREADS
BT FABRIC DEFECTS
RT FABRIC INSPECTION
SEAMS
SLACK COURSE
SLACK FEEDERS (KNITTING)
STITCHES (SEWING)
TENSION CONTROL

SLACKS
BT ATHLETIC CLOTHING
GARMENTS
SPORTSWEAR
WORK CLOTHING
RT BAGGINESS
LEOTARDS
SKI PANTS
STRETCH FABRICS
TROUSERS

SLANT SADDLE STITCHES
BT STITCHES (SEWING)

SLASHERS
RT CUT MARKS (SLASHING)
SIZE BOXES
SIZES (SLASHING)

RT SLASHERS BEAMS
SLASHING
SPLIT RODS
SQUEEZE ROLLS
SQUEEZING
STRETCH ROLLS
WARP BEAMS
WARP PREPARATION
YARN FINISH
YARN TENSION

SLASHERS BEAMS
RT WARP PREPARATION
SIZE BOXES
SIZES (SLASHING)
SLASHERS
SLASHING
SPLIT RODS
SQUEEZING
STRETCH ROLLS
WARP BEAMS
WARP ENDS

SLASHING
UF SIZING
NT CAKE SIZING
HANK SIZING
BT WARP PREPARATION
WET FINISHING
RT BATHS
BEAMS
COTTON SYSTEM PROCESSING
DESIZING
DESIZING MACHINES
GELATINE
KETTLES (HEATING)
LOOMS
ROSIN
SECTIONAL WARPING
SIZE BOXES
SIZES (SLASHING)
SIZING (PAPERMAKING)
SLASHERS
SLASHERS BEAMS
SPLIT RODS
SQUEEZING
STRETCH ROLLS
THROWING
WARP BEAMS
WARP ENDS
WARPING
YARN FINISH
YARN TENSION

SLAYS (LOOM)
USE SLEYS (LOOM)

SLEEVE BEARINGS
BT BEARINGS
RT AIR BEARINGS
BALL BEARINGS
BUSHINGS
CONICAL BEARINGS
FOOTSTEP BEARINGS
JOURNAL BEARINGS
NEEDLE BEARINGS
NYLON BEARINGS
TEFLON BEARINGS
THRUST BEARINGS

SLEEVES
BT GARMENT COMPONENTS
RT GARMENT MANUFACTURE
TAILORING

SLENDERNESS RATIO
UF LENGTH DIAMETER RATIO
RT BUCKLING
ENTANGLEMENTS
NEPS
PILLS
STRESS STRAIN PROPERTIES (COMPRESSIVE)
TORSIONAL BUCKLING
YARN BUCKLING

SLEYING
USE REEDING

SLEYS (LOOM)
UF BATTENS (LOOM)
FLY BEAMS (LOOM)
GOING PART (LOOM)
LATHES (LOOM)
LAYS (LOOM)
SLAYS (LOOM)
RT BEATING UP
FIXED REEDS
HEDDLES
LOOMS
LOOSE REEDS
PICKING (WEAVING)
QUILLS
RACE BOARDS
REEDS
SHUTTLE BINDERS
SHUTTLE BOXES
SHUTTLE CHECKING

RT SHUTTLE DEFLECTORS
SHUTTLE MARKS
SHUTTLE SMASH PROTECTORS
SHUTTLES
WEAVING

SLICKENERS
USE LUBRICANTS

SLICKNESS
BT AESTHETIC PROPERTIES
FINISH (PROPERTY)
SURFACE PROPERTIES (MECHANICAL)
RT FIBER FIBER FRICTION
FIBER FRICTION
HAND
LUBRICANTS
LUBRICATION
LUBRICITY
SLIDING
SLIDING FRICTION
SURFACE PROPERTIES (MECHANICAL)
YARN FINISH

SLIDE FASTENERS
USE ZIPPERS

SLIDE RULES
RT COMPUTATIONS
COMPUTERS
COUNTERS
NOMOGRAMS

SLIDING
UF SLIPPING
RT ABRASION
BOUNDARY LUBRICATION
DYNAMIC FRICTION
FRICTION
GALLING
HYDRODYNAMIC LUBRICATION
LUBRICATION
PLOUGHING
ROLLER SLIPPAGE
ROLLING
ROUGHNESS
RUBBING
SCRUBBING
SHEAR (MODE OF DEFORMATION)
SLICKNESS
SLIDING FRICTION
SLIPPAGE
SMOOTHNESS
SYTON (TN)
UNDERLAY (CARPETS)
WEAR

SLIDING FRICTION
BT FRICTION
RT BOUNDARY LUBRICATION
COEFFICIENT OF FRICTION
DYNAMIC FRICTION
GALLING
HYDRODYNAMIC LUBRICATION
LUBRICATION
PICKING (SNAGGING)
ROLLING FRICTION
RUBBING
SCORING
SCRATCHING
SCUFFING
SLICKNESS
SLIDING
SLIPPAGE
SNAGGING
VOLUME WEAR

SLIP
USE SLIPPAGE

SLIP ONS
BT GARMENTS
OUTERWEAR
SPORTSWEAR
RT CARDIGANS
CIRCULAR KNITTING
FULLY FASHIONED KNITTING
JERKINS
JUMPERS
KNITTED FABRICS
KNITTING
PULL OVERS
SWEATERS

SLIP PLANES

RT CRACKS (MECHANICAL)
CRYSTAL DEFECTS
CRYSTALLINE REGION
CRYSTALLINITY
DEFORMATION
DISLOCATIONS
SLIPPAGE

SLIP RESISTANCE
BT END USE PROPERTIES
FABRIC PROPERTIES
FABRIC PROPERTIES (MECHANICAL)
FINISH (PROPERTY)
RT ANTISLIP AGENTS
DIMENSIONAL STABILITY
FRICTION
LUBRICITY
ROUGHNESS
SLIPPAGE
YARN SLIPPAGE
YARN TO YARN FRICTION

SLIPE (WOOL)
USE PULLED WOOL

SLIPPAGE
UF SLIP
NT BELT SLIPPAGE
FIBER SLIPPAGE
ROLLER SLIPPAGE
SEAM SLIPPAGE
YARN SLIPPAGE
RT ABRASION
ALLOWANCE FACTOR (WARP PREPARATION)
DRAFTING (STAPLE FIBER)
DYNAMIC FRICTION
FIBER BREAKAGE
FRICTION
LUBRICATION
RUBBING
SHEAR (MODE OF DEFORMATION)
SLICING
SLICING FRICTION
SLIP PLANES
SLIP RESISTANCE
STICK SLIP FRICTION
VELOCITY (RELATIVE)

SLIPPERS
BT FOOTWEAR
RT BINDINGS (FOOTWEAR)
BOOTS
LACES (SHOES)
LEATHER
LININGS (FOOTWEAR)
PUMPS (FOOTWEAR)
SHOES
SNEAKERS
SOLES
STITCHING

SLIPPING
USE SLICING

SLIPS
BT GARMENTS
LINGERIE
UNDERWEAR
RT BRASSIERES
BRIEFS
FOUNDATION GARMENTS
GIRDLES
PANTIES
TRICOT KNITTED FABRICS

SLIT FILM YARNS
BT UNCONVENTIONAL YARNS
YARNS
RT BONDED YARNS
LAMINATED YARNS
PAPER YARNS

SLITTERS (CLOTH)
BT CUTTING MACHINES
GARMENT MANUFACTURING MACHINES
RT SCISSORS
SHEARING MACHINES

SLITTING
RT CRACK PROPAGATION
CUTTING (TAILORING)
SCRATCHING
SPLITTING
TEARING (TAILORING)

SLIVER CANS
UF COILER CANS
RT BALLING HEADS
CARD SLIVERS
CARD WEBS
CARDING
CARDS
COILERS
COMBED SLIVERS
CONDENSERS
COTTON CARDING
COTTON CARDS
DRAFTING (STAPLE FIBER)
DRAWFRAMES
FALSE TWIST
FINISHER PART
LEADING HOOKS
NOBLE COMBS
PACKAGES
SLIVER REVERSALS
SLIVERS
SLUBBINGS
(CONT.)

RT TRAILING HOOKS
TRUMPETS
WEBS
WORSTED CARDING

SLIVER KNITTING
RT ARTIFICIAL FURS
CARD SLIVERS
COMBED SLIVERS
KNITTING
PILE FABRICS (KNITTED)
SLIVERS

SLIVER LAP MACHINES
BT LAPPERS
RT SLIVERS

SLIVER NOIL RATIO
RT COMB TEAR
COMBER LAPS
COMBING
COMBING EFFICIENCY
DETACHING EFFICIENCY
NOBLE COMBING
NOIL PER CENT
NOILS
RECTILINEAR COMBS
TOP NOIL RATIO
YIELD (RETURN)

SLIVER REVERSALS
RT CARD SLIVERS
DRAFTING (STAPLE FIBER)
DRAFTING THEORY
DRAWFRAMES
FIBER HOOKS
FIBER ORIENTATION
HOOK REMOVAL
HOOKED FIBERS
LEADING HOOKS
SLIVER CANS
SLIVERS
TRAILING HOOKS

SLIVER TENDERNESS
RT CARD SLIVERS
COMBED SLIVERS
END BREAKAGES
SLIVERS
TENDERING
WEAK SPOTS

SLIVER TESTERS
BT TESTING EQUIPMENT
RT CARD SLIVERS
COMBED SLIVERS
SLIVERS

SLIVER TO YARN SPINNING
BT SPINNING
RT AUTOMATED SPINNING (STAPLE FIBER)
COTTON SYSTEM PROCESSING
SLIVERS

SLIVERS
NT CARD SLIVERS
COMBED SLIVERS
BT FIBER ASSEMBLIES
RT ACTUAL DRAFT
AUTOLEVELLERS
AUTOLEVELLING
BALLING HEADS
CAPACITANCE TYPE IRREGULARITY TESTERS
CARD WEBS
CARDING
CARDS
COILERS
COMBING
COTTON CARDING
COTTON SYSTEM PROCESSING
DETACHING ROLLS
DOFFERS (CARD)
DOFFING (OF A CARD)
DOUBLING (DRAFTING)
DRAFT
DRAFTING (STAPLE FIBER)
DRAFTING MACHINES
DRAFTING ROLLS
DRAFTING WAVES
DRAFTING ZONE
DRAWFRAMES
FIBER HOOKS
FIBER SLIPPAGE
FINISHER GILLING
FINISHER PART
GILLING
HALF LAP
HIGH DRAFT
HOOK REMOVAL
HOOKED FIBERS
IRREGULARITY
IRREGULARITY TESTERS
LEADING HOOKS
NEPS
NOBLE COMBS
NOILS
OVERLAP (FIBERS)
PNEUMATIC IRREGULARITY TESTERS
RECTILINEAR COMBS
ROVINGS
SLIVER CANS
SLIVER KNITTING
SLIVER LAP MACHINES
SLIVER REVERSALS
SLIVER TENDERNESS
SLIVER TESTERS
SLIVER TO YARN SPINNING
SLUBBINGS
STRANDS
THICK SPOTS
TOP COMBS
TOPS
TOW
TRAILING HOOKS
TRUMPETS
TUFTS (FIBER)
VELOCITY CHANGE POINT
WEBS
WOOLEN CARDING
WORSTED CARDING
YARNS

SLOPING
BT GARMENT MANUFACTURE
RT CUTTING (TAILORING)
PATTERN (APPAREL)
SEWING
SPREADING (TAILORING)
TAILORING

SLOTS
RT INSPECTION
KNOTS
SLUB CATCHERS
SLUB CATCHING
SLUBS
THICK SPOTS
WINDING
YARN DEFECTS

SLOUGHING
UF SHELL OFF
SLOUGHING OFF
RT DRUM WINDING
END BREAKAGES
ENTANGLEMENTS
FULL PACKAGES
LAYER LOCKING
PACKAGE DENSITY
PACKAGES
PIRN DENSITY
SNARLING
TAPER (BOBBIN)
UNWINDING
WINDING
YARN DYNAMICS

SLOUGHING OFF
USE SLOUGHING

SLUB CATCHERS
UF CLEARERS (YARN)
YARN CLEARERS
NT ELECTRONIC SLUB CATCHERS
MECHANICAL SLUB CATCHERS
RT CHEESE WINDERS
DETECTION
DRUM WINDERS
END BREAKAGES
INSPECTION
IRREGULARITY
KNOTS
KNOTTERS
KNOTTING
NEPS
OPTICAL SCANNERS
OPTICAL SCANNING
PRECISION WINDERS
PROCESS CONTROL
QUALITY CONTROL
SLOTS
SLUB CATCHING
SLUBS
THICK SPOTS
UNICONER (TN)
WINDERS
WINDING
YARN DEFECTS

SLUB CATCHING
UF CLEARING (YARN)
RT AUTOMATIC KNOTTING
CHEESE WINDING
CONING
DRUM WINDING
ELECTRONIC SLUB CATCHERS
END BREAKAGES
INSPECTION
QUALITY CONTROL
SLOTS
SLUB CATCHERS
SLUBS
TENSION CONTROL
WINDING

SLUB YARNS
BT NOVELTY YARNS
YARNS
RT BEAD YARNS
BOUCLES
FRILLS (NOVELTY YARNS)
KNICKERBOCKER YARNS
LOOPS (NOVELTY YARNS)
NEP YARNS
NEPS
NUB YARNS
SEEDS (NOVELTY YARNS)
SLUBS
SPIRALS (NOVELTY YARNS)
SPLASHES
TEXTURED YARNS (FILAMENT)
THICK AND THIN YARNS
YARN GEOMETRY
YARN STRUCTURE

SLUBBING DYEING
USE TOP DYEING

SLUBBING DYEING MACHINES
USE TOP DYEING MACHINES

SLUBBING FRAMES
USE ROVING FRAMES

SLUBBINGS
BT FIBER ASSEMBLIES
RT AUTOCOUNT (TN)
CARD SLIVERS
CARD WEBS
CARDING
CARDS
CARRIAGES (SPINNING MULE)
CONDENSED COUNT
CONDENSED YARNS
CONDENSERS
CONDENSING (YARN MANUFACTURING)
FINISHER PART
JACKSPOOLS
MULE DRAFT
MULE SPINNING
MULES
NEP COUNT
NEPS
ROVINGS
SLIVER CANS
SLIVERS
STRANDS
TOPS
TOW
WEB UNIFORMITY
WOOLEN CARDING
WOOLEN CARDS
WOOLEN SYSTEM PROCESSING
YARNS

SLUBINESS
RT AESTHETIC PROPERTIES
NEP COUNT
SLUBS
THICK SPOTS
THICK AND THIN YARNS

SLUBS
BT DEFECTS
FABRIC DEFECTS
YARN DEFECTS
RT ELECTRONIC SLUB CATCHERS
ENTANGLEMENTS
FABRIC INSPECTION
FLY (WASTE)
IMPURITIES
IRREGULARITY
KNOTTING
MECHANICAL SLUB CATCHERS
NEP YARNS
NEPS
NOVELTY YARNS
PILLS
SLOTS
SLUB CATCHERS
SLUB CATCHING
SLUB YARNS
SLUBINESS
THICK SPOTS
TUFTS (FIBER)
YARN DIAMETER

SLUDGE
USE WASTE TREATMENT

SLURCOCKS
RT FULLY FASHIONED KNITTING MACHINES
KNITTING MACHINES
SINKER BARS
SINKER CAMS
SINKERS

SLURGALLING
RT HOSIERY
KNITTING
STITCH LENGTH

SLURRY
    RT EMULSIONS
        PAPER MACHINES
        PAPER MAKING
        PULP
        WEB FORMING MACHINES
        WET PROCESS (WEB)

SMALL CIRCLE (COMB)
    RT COMB CIRCLES
        COMB FINS
        DABBING BRUSHES
        LARGE CIRCLE (COMB)
        NOBLE COMBS
        NOIL KNIVES
        SETOVER

SMASH
    USE SHUTTLE SMASH

SMOCKS
    BT GARMENTS
        WORK CLOTHING
    RT APRONS (CLOTHING)

SMOOTH DRYING FINISHES
    USE WASH WEAR FINISHES

SMOOTHNESS
    BT AESTHETIC PROPERTIES
        FINISH (PROPERTY)
        STRUCTURAL PROPERTIES
        SURFACE PROPERTIES
         (MECHANICAL)
    RT APPEARANCE
        BRIGHTNESS
        COMFORT
        DULLNESS
        FABRIC PROPERTIES
        FABRIC PROPERTIES (AESTHETIC)
        FABRIC PROPERTIES (MECHANICAL)
        GLOSS
        HAIRINESS
        HAND
        HARSHNESS
        LUSTER
        LUXURIOUSNESS
        PRICKLINESS
        ROUGHNESS
        SHEEN
        SLICING
        SOFTNESS
        SURFACE FRICTION
        TEXTURE
        WARMTH
        YARN PROPERTIES

SNAG RESISTANCE
    BT END USE PROPERTIES
        FABRIC PROPERTIES
        FABRIC PROPERTIES (MECHANICAL)
        MECHANICAL DETERIORATION
         PROPERTIES
        MECHANICAL PROPERTIES
    RT ABRASION
        ABRASION RESISTANCE
        ABRASION RESISTANCE FINISHES
        ASPERITIES
        CATCHING
        DURABILITY
        FATIGUE RESISTANCE
        FRICTION
        LADDERING
        MATRIX FREEDOM
        PICKING (SNAGGING)
        PICKING RESISTANCE
        PUNCTURE RESISTANCE
        PUNCTURING
        RUN RESISTANCE
        RUN RESISTANT HOSIERY
        RUNS (KNITTING)
        SNAG TESTERS
        SNAG TESTING
        SNAGGING
        SNAGS
        WEAR RESISTANCE
        YARN SLIPPAGE

SNAG RESISTANCE TESTING
    USE SNAG TESTING

SNAG TESTERS
    BT TESTING EQUIPMENT
    RT ABRASION TESTERS
        CATCHING
        PICKING (SNAGGING)
        PUNCTURE RESISTANCE
        PUNCTURING
        RUN RESISTANCE
        RUNS (KNITTING)
        SNAG RESISTANCE
        SNAG TESTING
        SNAGGING
        SNAGS

SNAG TESTING
    UF SNAG RESISTANCE TESTING
    BT TESTING
    RT ABRASION TESTING
        HOSIERY
        SNAG RESISTANCE
        SNAG TESTERS
        SNAGGING
        SNAGS

SNAGGING
    RT ABRASION
        ASPERITIES
        CATCHING
        DISTORTION
        ENTANGLEMENTS
        LADDERING
        PICKING (SNAGGING)
        PENETRATION
        PICKING RESISTANCE
        PILL WEAR OFF
        PILLING
        PUNCTURE RESISTANCE
        PUNCTURING
        RIPPING
        RUN RESISTANCE
        RUNS (KNITTING)
        SNAG RESISTANCE
        SNAG TESTERS
        SNAGS
        SNAP BACK
        STRESS CONCENTRATION

SNAGS
    RT CATCHING
        DISTORTION
        END BREAKAGES
        ENTANGLEMENTS
        KNITTED FABRICS
        NEPS
        PENETRATION
        PICKING (SNAGGING)
        PILL WEAR OFF
        PILLS
        PUNCTURE RESISTANCE
        PUNCTURING
        RUN RESISTANCE
        RUN RESISTANT HOSIERY
        RUN RESISTANT STITCHES
        RUNS (KNITTING)
        SNAG RESISTANCE
        SNAG TESTERS
        SNAG TESTING
        SNAGGING
        SNAP BACK
        STRESS CONCENTRATION

SNAP BACK
    RT BREAKING STRENGTH
        BUCKLING
        DYNAMIC BUCKLING
        ENTANGLEMENTS
        FIBER BREAKAGE
        FIBER STRENGTH
        LONGITUDINAL VIBRATIONS
        SLENDERNESS RATIO
        SNAGGING
        SNAGS
        SNARLING
        STRAIN WAVES
        STRESS PROPAGATION
        STRETCH BREAKING
        TORSIONAL BUCKLING
        YARN BUCKLING
        YARN DYNAMICS

SNAP FASTENERS
    BT FASTENERS
        GARMENT COMPONENTS
    RT HOOK CLOSURES
        JOINTS
        VELCRO (TN)
        ZIPPERS

SNARLING
    UF SNARLS
    RT DIMENSIONAL STABILITY
        ENTANGLEMENTS
        HIGH TWIST
        NEPS
        PILLING
        PILLS
        SLOUGHING
        SNAP BACK
        SNARLING TENDENCY
        STABILITY
        TORSIONAL BUCKLING
        TWIST LIVELINESS
        WILDNESS

SNARLING TENDENCY
    BT MECHANICAL PROPERTIES
    RT BALANCED YARNS
        DIMENSIONAL STABILITY
        LOOPING
        PILLING
        PILLS
        RESIDUAL TORQUE
        SNARLING
        TORSIONAL BUCKLING
        TORSIONAL ENERGY
        TWIST LIVELINESS
        WILDNESS

SNARLS
    USE SNARLING

SNEAKERS
    BT FOOTWEAR
    RT ATHLETIC CLOTHING
        BINDINGS (FOOTWEAR)
        BOOTS
        DUCK
        GALOSHES
        LACES (SHOES)
        LEATHER
        LININGS (FOOTWEAR)
        PUMPS (FOOTWEAR)
        RUBBER
        RUBBER FOOTWEAR
        SHOES
        SLIPPERS
        SOLES
        STITCHING
        SURGICAL STOCKINGS
        SYNTHETIC LEATHER
        VINYL COATED FABRICS

SOAKING
    RT BATHS
        LAUNDERING
        MOISTURE CONTENT
        RINSING
        SCOURING
        SOAPING
        SWELLING
        WETTING

SOAP SODA SCOURING
    BT SCOURING
    RT CARBONIZING
        CAUSTIC SCOURING
        DETERGENTS
        DRYING
        LANOLIN
        NEUTRALIZING
        OILING
        RINSING
        SCOURING BATHS
        SCOURING SHRINKAGE
        SCOURING TRAINS
        SCOURING WASTE
        SCOURING YIELD
        SOLVENT SCOURING
        SQUEEZING
        SUINT
        WOOL GREASE
        WOOL WAX

SOAPING
    UF AFTERWASH
    BT WET FINISHING
    RT AZOIC DYEING
        CROCKFASTNESS
        CRYSTALLIZATION
        SCOURING
        SOAKING
        SOAPS
        SURFACTANTS
        VAT DYEING

SOAPS
    NT ALKALI SOAPS
        AMINE SOAPS
        METALLIC SOAPS
        OLEATE SALTS
    BT CARBOXYLATE SURFACTANTS
    RT ALKANOLAMINES
        ANIONIC SOFTENERS
        ANTIFOAM AGENTS
        BLEACHING AGENTS
        DEGREASING
        DETERGENTS
        EMULSIFYING AGENTS
        FATTY ACIDS
        FOAMS (FROTH)
        LAUNDERING
        LEVELLING AGENTS
        LIME SOAP
        OLEIC ACID
        SARCOSIDE SURFACTANTS
        SCOURING
        SOAPING
        SOIL REMOVAL
        WETTING AGENTS

SOCKS
    BT ATHLETIC CLOTHING
        HOSIERY
    RT BOARDING
        CIRCULAR HOSIERY
        FASHIONING
        FOOTWEAR
        FULLY FASHIONED HOSIERY
        GALOSHES
        HALF HOSE
        HEAT SETTING (SYNTHETICS)
        KNITTED OUTERWEAR
        KNITTING
        PRESSING
        REINFORCED KNITTED FABRICS
        RUN RESISTANT HOSIERY
        SEAMLESS HOSIERY
        STOCKINGS
         (CONT.)

RT STRETCH HOSIERY
SUPPORT HOSIERY
SURGICAL STOCKINGS

SODA ASH
USE SODIUM CARBONATE

SODA CELLULOSE
USE ALKALI CELLULOSE

SODIUM ACETATE
BT ACETATE SALTS
RT HUMECTANTS
POTASSIUM ACETATE

SODIUM ACRYLATE
BT ACRYLIC COMPOUNDS
ACRYLIC SALTS

SODIUM BENZOATE
BT BENZOATE SALTS

SODIUM BICARBONATE
BT BICARBONATES
RT DYEING AUXILIARIES
PH CONTROL
SODIUM CARBONATE
SODIUM SESQUICARBONATE

SODIUM BISULFATE
BT BISULFATES
RT ACIDS
DYEING AUXILIARIES

SODIUM BISULFITE
BT BISULFITES
RT OXIDIZING AGENTS
REDUCING AGENTS

SODIUM BOROHYDRIDE
BT BORON COMPOUNDS
BORON HYDRIDES
RT REDUCING AGENTS

SODIUM BROMATE
RT DESIZING
DESIZING AGENTS
SODIUM BROMITE

SODIUM BROMIDE
BT BROMIDES
RT SODIUM BROMITE

SODIUM BROMITE
RT DESIZING
DESIZING AGENTS
SODIUM BROMATE
SODIUM BROMIDE

SODIUM CARBONATE
UF SODA ASH
BT CARBONATES
RT DYEING AUXILIARIES
PH CONTROL
SODIUM BICARBONATE
SODIUM SESQUICARBONATE

SODIUM CARBOXYMETHYLCELLULOSE
USE CARBOXYMETHYLCELLULOSE

SODIUM CHLORATE
BT CHLORATES
RT CHLORINE COMPOUNDS
OXIDIZING AGENTS
SODIUM CHLORITE

SODIUM CHLORIDE
UF SALT
BT CHLORIDES
RT DYEING AUXILIARIES
LEVELLING AGENTS
RETARDING AGENTS
SODIUM CHLORITE

SODIUM CHLORITE
RT SODIUM CHLORATE
SODIUM CHLORIDE

SODIUM COMPOUNDS (GENERAL)
NT SODIUM ACETATE
SODIUM ACRYLATE
SODIUM BENZOATE
SODIUM BICARBONATE
SODIUM BISULFITE
SODIUM BOROHYDRIDE
SODIUM BROMATE
SODIUM BROMIDE
SODIUM BROMITE
SODIUM CARBONATE
SODIUM CHLORATE
SODIUM CHLORIDE
SODIUM CUPRATE
SODIUM CYANIDE
SODIUM DICHROMATE
SODIUM DITHIONITE
SODIUM FLUORIDE
SODIUM HYDROXIDE
SODIUM HYPOCHLORITE
SODIUM IODIDE
NT SODIUM LAURYL SULFATE
SODIUM NITRATE
SODIUM NITRITE
SODIUM OLEATE
SODIUM PALMITATE
SODIUM PERBORATE
SODIUM PEROXIDE
SODIUM PERSULFATE
SODIUM POLYACRYLATE
SODIUM ROSINATE
SODIUM SESQUICARBONATE
SODIUM SILICATE
SODIUM SOAPS
SODIUM STEARATE
SODIUM SULFATE
SODIUM SULFIDE
SODIUM THIOSULFATE
TRISODIUM PHOSPHATE
TSPP
RT SALTS (GENERAL)

SODIUM CUPRATE
RT CUPRIC COMPOUNDS

SODIUM CYANIDE
BT CYANIDES

SODIUM DICHROMATE
BT CHROMIUM COMPOUNDS
DICHROMATES
RT MORDANT DYEING

SODIUM DITHIONITE
UF SODIUM HYDROSULFITE
BT DITHIONITES
RT REDUCING AGENTS
REDUCTION
VAT DYEING

SODIUM FLUORIDE
BT FLUORIDES

SODIUM HEXAMETAPHOSPHATE
USE CALGON (TN)

SODIUM HYDROSULFIDE
USE SODIUM SULFIDE

SODIUM HYDROSULFITE
USE SODIUM DITHIONITE

SODIUM HYDROXIDE
UF CAUSTIC (NOUN)
CAUSTIC SODA
BT HYDROXIDES
RT ALKALINE SOLUTIONS
DESIZING
DESIZING AGENTS
MERCERIZING
SWELLING AGENTS

SODIUM HYPOCHLORITE
BT HYPOCHLORITES
RT BLEACHING
CHLORINE BLEACHING AGENTS
HYDROGEN PEROXIDE
HYPOBROMITES
OXIDATION
OXIDIZING AGENTS
SODIUM PERBORATE
SODIUM PEROXIDE

SODIUM IODIDE
BT IODIDES

SODIUM LAURYL SULFATE
BT SULFATE SURFACTANTS
SULFATES

SODIUM NITRATE
BT NITRATES

SODIUM NITRITE
BT NITRITES

SODIUM OLEATE
BT ALKALI SOAPS
OLEATE SALTS
RT OLEIC ACID

SODIUM PALMITATE
BT ALKALI SOAPS

SODIUM PERBORATE
BT PERBORATES
RT PEROXIDE BLEACHING AGENTS
SODIUM HYPOCHLORITE
SODIUM PEROXIDE

SODIUM PEROXIDE
BT PEROXIDES
RT BLEACHING
HYDROGEN PEROXIDE
PEROXIDE BLEACHING AGENTS
SODIUM HYPOCHLORITE
SODIUM PERBORATE

SODIUM PERSULFATE
BT PERSULFATES

SODIUM POLYACRYLATE
RT ACRYLIC SALTS

SODIUM ROSINATE
BT ALKALI SOAPS
RT ROSIN

SODIUM SESQUICARBONATE
BT CARBONATES
RT DYEING AUXILIARIES
PH CONTROL
SODIUM BICARBONATE
SODIUM CARBONATE

SODIUM SILICATE
BT SILICATES

SODIUM SOAPS
BT ALKALI SOAPS

SODIUM STEARATE
BT ALKALI SOAPS

SODIUM SULFATE
NT GLAUBERS SALT
BT SULFATES
RT DYEING AUXILIARIES
LEVELLING AGENTS
RETARDING AGENTS
SULFURIC ACID

SODIUM SULFIDE
UF SODIUM HYDROSULFIDE
BT SULFIDES (INORGANIC)
RT DYEING AUXILIARIES
REDUCING AGENTS
SULFUR DYEING

SODIUM SULFOXYLATE FORMALDEHYDE
USE FORMALDEHYDE SULFOXYLATES

SODIUM THIOSULFATE
BT THIOSULFATES

SOFTENERS
UF SOFTENING AGENTS
NT ANIONIC SOFTENERS
CATIONIC SOFTENERS
NONIONIC SOFTENERS
BT FINISH (SUBSTANCE ADDED)
RT BENDING RIGIDITY
CREPING
DELIQUESCENT AGENTS
HAND
LUBRICANTS
LUBRICITY
NEEDLE HEATING
OCTADECYL ETHYLENE UREA
POLYETHYLENE
QUATERNARY AMMONIUM COMPOUNDS
SEWABILITY
SILICONES
SOFTENING
SOFTNESS

SOFTENING
BT WET FINISHING
RT LUBRICATION
SANDWICH ROLLING MILLS
SOFTENERS

SOFTENING AGENTS
USE SOFTENERS

SOFTENING POINT
UF SOFTENING TEMPERATURE
BT PHYSICAL PROPERTIES (EXCLUDING MECHANICAL)
THERMAL PROPERTIES
RT GLASS TRANSITION TEMPERATURE
MELTING POINT
PLASTIC FLOW
PLASTICITY
SOLIDIFICATION

SOFTENING TEMPERATURE
USE SOFTENING POINT

SOFTNESS
BT AESTHETIC PROPERTIES
END USE PROPERTIES
FINISH (PROPERTY)
MECHANICAL PROPERTIES
RT COMPRESSIBILITY
FABRIC PROPERTIES
FABRIC PROPERTIES (AESTHETIC)
FABRIC PROPERTIES (MECHANICAL)
HAND
HARDNESS
HARSHNESS
PACKAGE HARDNESS
PAPERINESS
ROUGHNESS
SMOOTHNESS
SOFTENERS

RT STIFFNESS
SURFACE CONTOUR
SURFACE FRICTION
TEXTURE
WADDINGS

SOIL
NT ARTIFICIAL SOIL
NATURAL SOIL
RT CLAYS
DIRT
IMPURITIES
SOIL REDEPOSITION
SOIL RELEASING FINISHES
SOIL REMOVAL
SOIL RESISTANCE
SOIL RESISTANCE FINISHES
SOILING

SOIL BURIAL TEST
BT TESTING
RT ANTIBACTERIAL FINISHES
ANTIMICROBIAL FINISHES
FUNGICIDES
MILDEW RESISTANCE FINISHES

SOIL DEPOSITION
USE SOILING

SOIL REDEPOSITION
RT DEPOSITION
DRY CLEANING
SOILING
STAIN REMOVAL
WHITENESS RETENTION

SOIL RELEASING FINISHES
BT FINISH (SUBSTANCE ADDED)
RT SOIL
SOIL REMOVAL

SOIL REMOVAL
UF DIRT REMOVAL
BT CLEANING
RT ANTIREDEPOSITION AGENTS
DEGREASING
DETERGENTS
DEWAXING (COTTON)
DRY CLEANING
LAUNDERING
REMOVAL
SCOURING
SOAPS
SOIL
SOIL RELEASING FINISHES
SOIL RESISTANCE
SOILING
STAIN REMOVAL
WASHING MACHINES

SOIL RESISTANCE
BT END USE PROPERTIES
FINISH (PROPERTY)
RT CROCKFASTNESS
CROCKING
DIRT
DRY CLEANING
FABRIC PROPERTIES
FABRIC PROPERTIES (PHYSICAL EXCLUDING MECHANICAL)
LAUNDERING
SERVICEABILITY
SOIL
SOIL REMOVAL
SOIL RESISTANCE FINISHES
SOILING
STAIN RESISTANCE
STAINING

SOIL RESISTANCE FINISHES
UF ANTISOILING FINISHES
BT FINISH (SUBSTANCE ADDED)
RT ANTIREDEPOSITION AGENTS
COLLOIDAL ALUMINA
OXIDES
SOIL
SOIL RELEASING FINISHES
SOIL RESISTANCE
STAIN RESISTANCE AGENTS

SOILING
UF SOIL DEPOSITION
RT DEFECTS
DETERGENCY
DIRT
FABRIC DEFECTS
FOGGY YARNS
FOGMARKING
IMPURITIES
OIL STAINS
SOIL
SOIL REDEPOSITION
SOIL REMOVAL
SOIL RESISTANCE
SPOTS
STAINING
STAINS
STATIC ELECTRICITY

SOLES
BT FOOTWEAR COMPONENTS
RT BINDINGS (FOOTWEAR)
BOOTS
FOOTWEAR
GALOSHES
LACES (SHOES)
LASTS
LEATHER
LEGGINGS
LININGS (FOOTWEAR)
PUMPS (FOOTWEAR)
RUBBER FOOTWEAR
SHOES
SLIPPERS
SNEAKERS
SPATS
STITCHING
SYNTHETIC LEATHER
TONGUES
UPPERS
VINYL COATED FABRICS
WELTS (FOOTWEAR)

SOLID BRAIDS
BT BRAIDS
RT CORE BRAIDS
FLAT BRAIDS
HOLLOW BRAIDS
NO CORE BRAIDS
REGULAR BRAIDS

SOLID COLORS
RT COLOR
MULTICOLORS
PIECE DYEING

SOLID SHADE
RT DYEING
TONE IN TONE DYEING

SOLID SHADE DYEING
USE UNION DYEING

SOLIDIFICATION
RT COAGULATING
CRYSTALLIZATION
DISSOLVING
FREEZING
FUSION (MELTING)
MELTING
SOFTENING POINT
SOLIDS
SOLUTES
SOLUTIONS (LIQUID)
TRANSITIONS (POLYMERS)

SOLIDS
RT LIQUIDS
SOLIDIFICATION
SOLUTES
SOLUTIONS (LIQUID)

SOLS
NT AEROSOLS
HYDROSOLS
BT COLLOIDS
RT GELS

SOLUBILITY
BT PHYSICAL CHEMICAL PROPERTIES
RT ACID SOLUBILITY
ALKALI SOLUBILITY
CONCENTRATION
CRYSTALLIZATION
DILUTION
DISSOLVING
EXTRACTION
PARTITION COEFFICIENT
PRECIPITATION
SATURATED SOLUTIONS
SATURATION (MATERIAL)
SOLUBILITY PARAMETER
SOLUTES
SOLUTIONS (LIQUID)
SOLVENTS
UREA BISULFITE SOLUBILITY

SOLUBILITY PARAMETER
RT DISPERSION FORCES
SOLUBILITY
SOLUTIONS (LIQUID)
SOLVENTS

SOLUBLE VAT DYES
USE LEUCO ESTER DYES

SOLUTES
RT DILUTION
DISSOLVING
SOLIDIFICATION
SOLIDS
SOLUBILITY
SOLUTIONS (LIQUID)
SOLVENTS

SOLUTION DYEING
USE DOPE DYEING

SOLUTION TYPE BINDERS
BT BINDERS (NONWOVENS)
RT BONDING
EMULSION BINDERS (NONWOVENS)
NONWOVEN FABRICS
THERMOPLASTIC POWDERS
THERMOSETTING POWDERS

SOLUTIONS (LIQUID)
NT ACID SOLUTIONS
ALKALINE SOLUTIONS
DYE SOLUTIONS
NONAQUEOUS SOLUTIONS
SATURATED SOLUTIONS
WATER SOLUTIONS
RT AQUEOUS SCOURING
COLLOIDS
DILUTION
DISSOLVING
EXTRACTION
GELS
LIQUIDS
MIXTURES
PRECIPITATION
SOLIDIFICATION
SOLIDS
SOLUBILITY
SOLUBILITY PARAMETER
SOLUTES
SOLVENTS
SUSPENSIONS

SOLUTIONS (MATHEMATICAL)
NT GENERAL SOLUTIONS
GRAPHICAL SOLUTIONS
NUMERICAL SOLUTIONS
PARTICULAR SOLUTIONS
TRIAL AND ERROR SOLUTIONS
RT COMPUTATIONS
DIFFERENTIATION
EQUATIONS
INTEGRATION
LINEAR DIFFERENTIAL EQUATIONS
MATHEMATICAL ANALYSIS
METHODS
RESULTS
ROOTS (MATHEMATICAL)

SOLVENT ASSISTED DYEING
BT DYEING (BY ENVIRONMENTAL CONDITIONS)
RT BENZYL ALCOHOL
CARRIER DYEING
CARRIERS (DYEING)
DYEING AUXILIARIES
EMULSIFYING AGENTS
LEVELLING AGENTS
LOW TEMPERATURE DYEING
SOLVENT DYEING
SOLVENT RECOVERY
SOLVENT TREATMENTS
SURFACTANTS
SWELLING
VACUUM DYEING
WETTING AGENTS

SOLVENT BONDING
BT BONDING
RT BONDED FIBER FABRICS
DIP BONDING
DRY POWDER BONDING
FLAME BONDING
FOAM BONDING
SOLVENT RECOVERY
SOLVENT TREATMENTS
SATURATION (MATERIAL)

SOLVENT DEGREASING
USE SOLVENT SCOURING

SOLVENT DYEING
NT VAPOCOL PROCESS (TN)
BT DYEING (BY ENVIRONMENTAL CONDITIONS)
RT CARRIER DYEING
HIGH TEMPERATURE DYEING
LOW TEMPERATURE DYEING
SOLVENT ASSISTED DYEING
SOLVENT RECOVERY
SOLVENT TREATMENTS
SWELLING
VACUUM DYEING
VAPOR PHASE DYEING

SOLVENT DYES
BT DYES (BY CHEMICAL CLASSES)
RT DYEING (BY DYE CLASSES)
DYES (BY FIBER CLASSES)
SOLVENT DYEING

SOLVENT EMULSIONS
BT EMULSIONS
RT EMULSIFYING AGENTS
EMULSION POLYMERS
LATICES
OIL IN WATER EMULSIONS
WATER IN OIL EMULSIONS

SOLVENT RECOVERY
BT RECOVERY (CHEMICAL)
RT SOLVENT ASSISTED DYEING
SOLVENT BONDING
SOLVENT DYEING
SOLVENT SCOURING
SOLVENT SPINNING
SOLVENT TREATMENTS
SOLVENTS

SOLVENT SCOURING
UF SOLVENT DEGREASING
BT SCOURING
SOLVENT TREATMENTS
RT CARBONIZING
GREASE RECOVERY
LANOLIN
SCOURING BATHS
SCOURING SHRINKAGE
SCOURING WASTE
SCOURING YIELD
SOAP SODA SCOURING
SOLVENT RECOVERY
SUINT
WOOL GREASE
WOOL WAX

SOLVENT SPINNING
UF DRY SPINNING
BT EXTRUSION
SPINNING (EXTRUSION)
RT COAGULATING BATHS
DRAWING (FILAMENT)
FILAMENT YARNS
GODET ROLLS
MAN MADE FIBERS
MELT SPINNING
MONOFILAMENT YARNS
MULTIFILAMENT YARNS
NEUTRALIZING (FIBER EXTRUSION)
POT SPINNING
SOLVENT RECOVERY
SOLVENT TREATMENTS
SPINNERETS
SYNTHETIC FIBERS
TOW
WET SPINNING

SOLVENT TREATMENTS
NT SOLVENT SCOURING
BT WET FINISHING
RT SOLVENT ASSISTED DYEING
SOLVENT BONDING
SOLVENT DYEING
SOLVENT RECOVERY
SOLVENT SPINNING
SOLVENTS
VAPOR PHASE TREATMENTS

SOLVENTS
NT CHLORINATED SOLVENTS
DRY CLEANING SOLVENTS
RT ACETONE
ALCOHOLS
ALKYLENE CARBONATES
BENZENE
CADOXEN
CARBON DISULFIDE
CHLORINATED HYDROCARBONS
CHLOROHYDRINS
CUPRAMMONIUM HYDROXIDE
CUPRIETHYLENEDIAMINE HYDROXIDE
DICHLOROBENZENE
DILUTION
DIMETHYL ACETAMIDE
DIMETHYL FORMAMIDE
DIMETHYL SULFOXIDE
DISSOLVING
DRY CLEANING
DRY CLEANING MACHINES
DYES
ESTERS
ETHERS
ETHYL ACETATE
EXTRACTION
FERRIC TARTRATE
KETONES
MARSCHALL SOLUTION
PHENOLS
SOLUBILITY
SOLUBILITY PARAMETER
SOLUTES
SOLUTIONS (LIQUID)
SOLVENT ASSISTED DYEING
SOLVENT BONDING
SOLVENT DYEING
SOLVENT RECOVERY
SOLVENT SCOURING
SOLVENT SPINNING
SOLVENT TREATMENTS
STAIN REMOVAL
THINNERS
TOLUENE
TRICHLOROBENZENE
VAPOCOL PROCESS (TN)

SONIC VELOCITY
RT DYNAMIC MODULUS
DYNAMIC MODULUS TESTERS
IMPULSES
LONGITUDINAL VIBRATIONS
PULSE PROPAGATION METER
SOUND
STRESS PROPAGATION
ULTRASONICS

SONICS
USE SOUND

SORBENTS
RT SORPTION
SORPTION OF DYES
SORPTION OF GASES
SORPTION OF LIQUIDS
SORPTION OF WATER

SORBITAN ESTER SURFACTANTS
UF POLYETHOXY ANHYDROSORBITOL ESTERS
NT SPANS (TN)
TWEENS (TN)
BT NONIONIC SURFACTANTS

SORBITOL
BT ALCOHOLS

SORPTION
NT ABSORPTION (MATERIAL)
ADSORPTION
CHEMISORPTION
DESORPTION
MOISTURE ABSORPTION
PREFERENTIAL ADSORPTION
SORPTION OF DYES
SORPTION OF GASES
SORPTION OF LIQUIDS
SORPTION OF WATER
RT ABSORBERS
ABSORPTION (RADIATION)
ABSORPTION RATE
ACQUISITION
ADSORPTIVITY
CAPILLARY WATER
CHEMICAL ENGINEERING
DIFFUSION
ENTHALPY OF SORPTION
EXTRACTION
FILTRATION
SITES (DYEING)
SORBENTS
SORPTION EQUILIBRIUM
SURFACE PROPERTIES (PHYSICAL CHEMICAL)
SWELLING

SORPTION OF DYES
UF DYE ABSORPTION
DYE ADSORPTION
BT SORPTION
RT ABSORPTION (MATERIAL)
ABSORPTION RATE
ADSORPTION
ADSORPTION ISOTHERM
ADSORPTION RATE
CHEMISORPTION
DESORPTION
DYE AFFINITY
DYE MIGRATION
DYE PENETRATION
DYEABILITY
EQUILIBRIUM (DYEING)
SORBENTS
SORPTION RATE

SORPTION EQUILIBRIUM
UF ABSORPTION EQUILIBRIUM
BT EQUILIBRIUM (CHEMICAL)
RT EQUILIBRIUM (DYEING)
SORPTION
SORPTION OF GASES
SORPTION OF LIQUIDS
SORPTION OF WATER
SORPTION RATE

SORPTION OF GASES
BT PHYSICAL CHEMICAL PROPERTIES
SORPTION
RT ABSORPTION (MATERIAL)
ADSORPTION
CHEMICAL PROPERTIES
CHEMISORPTION
DESORPTION
DIFFUSION
SORBENTS
SORPTION EQUILIBRIUM
SORPTION OF LIQUIDS
SORPTION OF WATER
SORPTION RATE
SORPTIVITY
SURFACE AREA
SWELLING

SORPTION OF LIQUIDS
BT PHYSICAL CHEMICAL PROPERTIES
SORPTION
RT ABSORPTION (MATERIAL)
ADSORPTION
CHEMICAL PROPERTIES
CHEMISORPTION
DESORPTION
DIFFUSION
DYEING PROPERTIES (GENERAL)
SORBENTS
SORPTION EQUILIBRIUM
SORPTION OF GASES
SORPTION OF WATER
SORPTION RATE
SORPTIVITY
SURFACE AREA
SWELLING
WIPING CLOTHS

SORPTION OF WATER
NT WATER ABSORPTION
BT PHYSICAL CHEMICAL PROPERTIES
SORPTION
RT ABSORPTION (MATERIAL)
ADSORPTION
CHEMICAL PROPERTIES
CHEMISORPTION
DESORPTION
DIFFUSION
DYEING PROPERTIES (GENERAL)
SORBENTS
SORPTION EQUILIBRIUM
SORPTION OF GASES
SORPTION OF LIQUIDS
SORPTION RATE
SORPTIVITY
SURFACE AREA
SWELLING

SORPTION RATE
NT ABSORPTION RATE
ADSORPTION RATE
BT RATE
RT SORPTION
SORPTION EQUILIBRIUM
SORPTION OF DYES
SORPTION OF GASES
SORPTION OF LIQUIDS
SORPTION OF WATER

SORPTIVITY
RT SORPTION
SORPTION OF GASES
SORPTION OF LIQUIDS
SORPTION OF WATER

SORTING
RT COTTON CLASSING
COTTON QUALITY
EFFECTIVE LENGTH (FIBER)
FIBER ARRAY
FIBER DIAGRAM
FIBER LENGTH DETERMINATION
FIBROGRAPH
FLEECES
GRADING
MATCHINGS
SHORT FIBERS INDEX
SPAN LENGTH
STAPLE FIBERS
STAPLING
SUTER WEBB TESTING

SOUND
UF SONICS
SOUND WAVES
RT ACOUSTIC BAFFLING
ACOUSTIC INSULATION
ACOUSTIC PROPERTIES
DYNAMIC MODULUS
ENERGY ABSORPTION
LONGITUDINAL VIBRATIONS
PULSE PROPAGATION METER
SONIC VELOCITY
TRANSVERSE VIBRATIONS
ULTRASONICS
VIBRATION
WAVE LENGTH

SOUND WAVES
USE SOUND

SOUTANES
BT CEREMONIAL CLOTHING

SOUTH AFRICAN MERINO WOOL
BT MERINO WOOL

SOUTH AMERICAN MERINO WOOL
BT MERINO WOOL

SOUTHDOWN WOOL
BT MEDIUM WOOL
WOOL

SPACE
RT ABSENCE
AREA
BULK
CAPACITY (GENERAL)
CORES (CENTER)
DENSITY
HOLLOWNESS

RT MATRIX FREEDOM
PACKING FACTOR
POROSITY
VOLUME

SPACE DRAWING
RT DRAWTWISTING
DRAWING (FILAMENT)
HEAT TRANSFER

SPACE DYEING
UF RANDOM DYEING
BT DYEING (FOR EFFECT)

SPAN LENGTH
RT COTTON CLASSING
COTTON QUALITY
EFFECTIVE LENGTH (FIBER)
FIBER ARRAY
FIBER DIAGRAM
FIBER LENGTH DETERMINATION
FIBER LENGTH DISTRIBUTION
FIBERS
FIBROGRAPH
GRADING
LENGTH UNIFORMITY RATIO (FIBER)
MEAN FIBER LENGTH
SHORT FIBERS INDEX
SORTING
STAPLING
SUTER WEBB TESTING
UPPER HALF MEAN LENGTH

SPANDELLE (TN)
BT SPANDEX FIBERS

SPANDEX
USE SPANDEX FIBERS

SPANDEX FIBER BLENDS
BT BLENDS (FIBERS)
RT SPANDEX FIBERS

SPANDEX FIBERS
UF SPANDEX
NT BLUE C SPANDEX (TN)
DURASPAN (TN)
GLOSPAN (TN)
INTERSPAN (TN)
LYCRA (TN)
NUMA (TN)
OROFIL (TN)
POLYTHANE (TN)
SPANDELLE (TN)
VYRENE (TN)
BT ELASTOMER FIBERS
FIBERS
MAN MADE FIBERS
SYNTHETIC FIBERS
RT CORE SPUN YARNS
COVERED YARNS
FOUNDATION GARMENTS
GIRDLES
ISOCYANATES
POLYURETHANES
POWER (FABRIC)
RUBBER
RUBBER FIBER (NATURAL)
SPANDEX FIBER BLENDS
STRETCH FABRICS
STRETCH YARNS (FILAMENT)
STRETCH YARNS (SPUN)
SURGICAL STOCKINGS
SYNTHETIC RUBBER FIBERS
URETHANE FOAMS
URETHANE RUBBERS

SPANISH (LACE)
BT HAND MADE LACES
LACES

SPANS (TN)
BT **SORBITAN ESTER SURFACTANTS**

SPARKLE
BT AESTHETIC PROPERTIES
OPTICAL PROPERTIES
RT APPEARANCE
BLOOM
BRIGHTNESS
COLOR
DRABNESS
DULLNESS
FABRIC PROPERTIES
FABRIC PROPERTIES (AESTHETIC)
FIBER CROSS SECTIONS
FINISH (PROPERTY)
GLOSS
LUSTER
RICHNESS
SHADE
SHEEN
TEXTURE

SPATS
BT FOOTWEAR
RT BOOTS
GALOSHES
LEATHER
LEGGINGS
SHOES
SOLES
SYNTHETIC LEATHER
TONGUES

SPECIALTY YARNS
USE NOVELTY YARNS

SPECIFIC GRAVITY
UF RELATIVE DENSITY
RT DENSITY
HYDROMETERS
PACKING FACTOR
SPECIFIC GRAVITY TESTING (FIBER IDENTIFICATION)
VOLUME
WEIGHT

SPECIFIC GRAVITY TESTING (FIBER IDENTIFICATION)
BT FIBER IDENTIFICATION
RT CHEMICAL ANALYSIS (FIBER IDENTIFICATION)
MICROSCOPIC ANALYSIS (FIBER IDENTIFICATION)
REAGENT TESTING (FIBER IDENTIFICATION)
SPECIFIC GRAVITY
STAINING TESTING (FIBER IDENTIFICATION)
SWELLING TESTING (FIBER IDENTIFICATION)
THERMAL TESTING (FIBER IDENTIFICATION)

SPECIFIC HEAT
BT PHYSICAL PROPERTIES (EXCLUDING MECHANICAL)
THERMAL PROPERTIES
RT BOILING POINT
CALORIMETRY
CONDUCTIVITY (THERMAL)
ENTHALPY
ENTROPY
HEAT CONTENT
LATENT HEAT
MELTING
MELTING POINT
PHYSICAL CHEMICAL PROPERTIES
TEMPERATURE
THERMAL TESTING
THERMODYNAMICS
WARMTH

SPECIFIC HUMIDITY
BT HUMIDITY
RT BOILING POINT
CRITICAL MOISTURE CONTENT
DEW POINT
DIFFUSION
DIFFUSIVITY
DRY BULB TEMPERATURE
DRYING
EQUILIBRIUM MOISTURE CONTENT
FREEZING POINT
MOISTURE CONTENT
PSEUDO WET BULB TEMPERATURE
PSYCHROMETRY
REGAIN
RELATIVE HUMIDITY
SATURATED VAPOR
VAPOR PRESSURE
WET BULB TEMPERATURE

SPECIFICATIONS
RT PRODUCTION
PRODUCTIVITY
QUALITY CONTROL
STANDARDS
TESTING

SPECIMEN LENGTH
USE GAUGE LENGTH

SPECIMEN PREPARATION
RT SAMPLES
SAMPLING
TESTING

SPECK DYEING
BT DYEING (BY ENVIRONMENTAL CONDITIONS)
RT FABRIC INSPECTION
VEGETABLE MATTER

SPECKELON (TN)
BT NYLON (POLYAMIDE FIBERS)
NYLON 66

SPECKINESS
BT DYEING DEFECTS
FABRIC DEFECTS
RT BRONZING
COLORFASTNESS
LEVELNESS (DYEING)
SKITTERINESS
SPECKLED FABRIC
STREAKINESS
TIPPINESS

SPECKING
BT DRY FINISHING
RT DEFECTS
FABRIC DEFECTS
FABRIC INSPECTION
LINT (WASTE)
SPECKS

SPECKLED FABRIC

RT DYEING (FOR EFFECT)
SPECKINESS

SPECKS
BT FABRIC DEFECTS
RT FABRIC INSPECTION
LINT (WASTE)
LINT BLADES
LINT CATCHERS
SPECK DYEING
SPECKING

SPECTRA
NT ABSORPTION SPECTRA
INFRARED SPECTRA
RT INFRARED SPECTROSCOPY
SPECTROPHOTOMETERS
SPECTROSCOPY
SPECTRUM ANALYSIS
ULTRAVIOLET SPECTROSCOPY
X RAY SPECTROSCOPY

SPECTROGRAPHY
RT MASS SPECTROMETRY
OPTICAL INSTRUMENTS

SPECTROPHOTOMETERS
NT INFRARED SPECTROPHOTOMETERS
BT MEASURING INSTRUMENTS
OPTICAL INSTRUMENTS
PHOTOMETERS
RT CHEMICAL ANALYSIS
COLORIMETERS
INFRARED SPECTROSCOPY
OPTICAL PROPERTIES
REFLECTOMETERS
SPECTRA
SPECTROPHOTOMETRY
SPECTROSCOPY
ULTRAVIOLET SPECTROSCOPY

SPECTROPHOTOMETRIC MEASUREMENTS
USE SPECTROPHOTOMETRY

SPECTROPHOTOMETRY
UF SPECTROPHOTOMETRIC MEASUREMENTS
RT ABSORPTION (RADIATION)
BRIGHTNESS
CHROMATICITY DIAGRAM
COLOR
COLOR MATCHING
COLORIMETRY
DULLNESS
DYEING
HUE
JUDD NBS COLOR SYSTEM
MUNSELL COLOR SYSTEM
OPTICAL INSTRUMENTS
OPTICS
SATURATION (COLOR)
SHADE
SPECTROPHOTOMETERS

SPECTROSCOPY
NT INFRARED SPECTROSCOPY
ULTRAVIOLET SPECTROSCOPY
X RAY SPECTROSCOPY
RT CHEMICAL ANALYSIS
ELECTRON SPIN RESONANCE
MASS SPECTROMETRY
MEASURING INSTRUMENTS
NUCLEAR MAGNETIC RESONANCE
OPTICAL INSTRUMENTS
OPTICS
SPECTRA
SPECTROPHOTOMETERS
SPECTRUM ANALYSIS

SPECTRUM ANALYSIS
RT ANALYZING
COLOR MATCHING
INFRARED SPECTRA
MASS SPECTROMETRY
SPECTRA
SPECTROPHOTOMETRY
SPECTROSCOPY
X RAY SPECTROSCOPY

SPEED
USE VELOCITY

SPEED CONTROL
RT AUTOMATIC CONTROL
CONSTANT SPEED
PROCESS CONTROL
STARTING TIME
VARIABLE SPEED
VELOCITY

SPEED FRAMES
USE ROVING FRAMES

SPHERICITY
BT SHAPE
RT CYLINDERS
DIAMETER
ELLIPTICITY
ROUNDNESS

SPHERULITES
BT FINE STRUCTURE (FIBERS)
RT CRYSTALLINE REGION
CRYSTALLINITY
CRYSTALLITES
CRYSTALLIZATION
FIBRIL REVERSALS
MICROFIBRILS
PARACRYSTALLINE REGION
SPHERULITIC CRYSTALLIZATION
SPHERULITIC FIBRILS

SPHERULITIC CRYSTALLIZATION
BT CRYSTALLIZATION
RT FIBRILS

SPHERULITIC FIBRILS
BT FIBRILS
FINE STRUCTURE (FIBERS)
RT CRYSTAL LATTICE
CRYSTALLINE REGION
CRYSTALLITES
FRINGED MICELLES
LONG CHAIN MOLECULES
SPHERULITES

SPIN DRYING
USE CENTRIFUGAL DRYING

SPIN DYEING
USE DOPE DYEING

SPIN FINISHES
UF SPINNING FINISH
NT SPINNING LUBRICANTS
BT FINISH (SUBSTANCE ADDED)
RT ANTISTATIC AGENTS
FILAMENT YARNS
SPINNING (EXTRUSION)
TOW

SPINDLE DRAFTING
USE MULE DRAFT

SPINDLE ECCENTRICITY
BT ECCENTRICITY
RT AXLES
BALANCING
BALLOON DYNAMICS
BEARINGS
HAND
ROLL ECCENTRICITY
SPINDLE SPEED
SPINDLE VIBRATION
SPINDLES
SPINNING
SPINNING FRAMES
TENSION CONTROL
TWISTERS
TWISTING
VIBRATION
YARN TENSION

SPINDLE HEAD
RT SPINDLE TAPES
SPINDLES
SPINNING FRAMES

SPINDLE RAILS
RT BUILDER MOTIONS
LIFT (WINDING)
RING FRAMES
RING RAILS
SPINNING FRAMES
TRAVERSE

SPINDLE SPEED
RT BALLOON TENSION
CONSTANT SPEED
ROTATION
SPINDLE ECCENTRICITY
SPINDLE TAPES
SPINDLE VIBRATION
SPINDLES
SPINNING
TENSION CONTROL
TRAVELER BURNOUT
VARIABLE SPEED
VELOCITY

SPINDLE TAPES
UF LISTINGS
RT BELT SLIPPAGE
BELT TENSION
BELTS
INDUSTRIAL FABRICS
SPINDLE HEAD
SPINDLE SPEED
SPINDLES
SPINNING FRAMES

SPINDLE VIBRATION
RT BALANCING
BEARINGS
BENDING RIGIDITY
BOLSTERS
DYNAMIC BALANCING
END BREAKAGES
RESONANT FREQUENCY
SPINDLE ECCENTRICITY
SPINDLE SPEED
SPINDLES
TENSION CONTROL

SPINDLES
NT DOUBLE TWIST SPINDLES
FALSE TWIST SPINDLES
RT AIR BEARINGS
AXLES
BALANCING
BALLOON COLLAPSE
BALLOON DYNAMICS
BALLOONS
BEARINGS
BOBBINS
BOLSTERS
CAP SPINNING FRAMES
CAPS (SPINNING)
CARRIAGES (SPINNING MULE)
CHEESES
CONES
DYNAMIC BALANCING
ECCENTRICITY
FLYER SPINNING FRAMES
FOOTSTEP BEARINGS
HEADSTOCKS
MULE CARRIAGES
NOVELTY TWISTERS
PACKAGES
QUILLS
RING FRAMES
RING SPINNING
RING TWISTERS
SPINDLE ECCENTRICITY
SPINDLE HEAD
SPINDLE SPEED
SPINDLE TAPES
SPINDLE VIBRATION
SPINNING
SPINNING FRAMES
TAPE DRIVES
TWISTERS
TWISTING
UPTWISTERS
VIBRATION
WINDING

SPINNABILITY (STAPLE FIBER)
USE SPINNING LIMIT

SPINNERETS
RT COAGULATING BATHS
DOPE (POLYMER)
EXTRUSION
FIBER CROSS SECTIONS
FIBERS
FILAMENT YARNS
FILAMENTS
HOLES
MAN MADE FIBERS
MELT SPINNING
METER PUMPS
MONOFILAMENT YARNS
MONOFILAMENTS
MULTIFILAMENT YARNS
NEUTRALIZING (FIBER EXTRUSION)
ORIFICES
POLYMERS
SOLVENT SPINNING
SPINNING (EXTRUSION)
SPINNING ASSEMBLIES
(EXTRUSION)
SPINNING FILTERS
SPINNING MACHINES (EXTRUSION)
SYNTHETIC FIBERS
TOW
WET SPINNING

SPINNING
NT AIR SPINNING
AUTOMATED SPINNING (STAPLE FIBER)
CAP SPINNING
COLLAPSED BALLOON SPINNING
COMB SPINNING
CORE SPINNING
CONTINUOUS SPINNING (STAPLE
FIBER)
COTTON SYSTEM SPINNING
NT FLUID SPINNING
FLYER SPINNING
HIGH DRAFT SPINNING
MULE SPINNING
POT SPINNING
RING SPINNING
SLIVER TO YARN SPINNING
WOOLEN SYSTEM SPINNING
WORSTED SYSTEM SPINNING
RT ACTUAL DRAFT
AMBLER SUPERDRAFT
AMERICAN SYSTEM PROCESSING
ANTIWEDGE RINGS
APRON CONTROL
APRONS
BALLOON COLLAPSE
BALLOON CONTROL
BALLOON CONTROL RINGS
BALLOON DYNAMICS
BALLOON SEPARATORS
BALLOON TENSION
BALLOONS
BRADFORD SYSTEM PROCESSING
BROKEN END COLLECTORS
BUILDER MOTIONS
CAP SPINNING FRAMES
CAPS (SPINNING)
CIRCULAR TRAVELERS
CONDENSED YARNS
CONTINENTAL SYSTEM PROCESSING
COTTON SYSTEM PROCESSING
DOFFING (PACKAGE)
DOUBLE ROVINGS
DOWNTWISTING
DRAFT
DRAFTING (STAPLE FIBER)
ELLIPTICAL TRAVELERS
EMPTY PACKAGES
FALSE TWIST SPINDLES
FALSE TWISTING (EXCEPT
TEXTURING)
FIBER HOOKS
FLYER SPINNING FRAMES
FLYERS
FULL PACKAGES
HIGH DRAFT
HOOKED FIBERS
INTERLACING (FILAMENT IN YARN)
IRREGULARITY
LAP UP
LARGE PACKAGES
LASHING END RADIUS
LIFT (WINDING)
LUBRICATED RINGS
MARTINDALE LIMIT
NEW BRADFORD SYSTEM PROCESSING
NYLON TRAVELERS
PACKAGE DENSITY
PACKAGING (OPERATION)
PATROLLING
PIECING
PLYING
PLYING TRAVELERS
POT SPINNING FRAMES
RADIAL DISTRIBUTION
RING FRAMES
RING RAILS
RINGS
ROLLER LAP
ROVING (OPERATION)
SINTERED METAL RINGS
SPINDLE ECCENTRICITY
SPINDLE SPEED
SPINDLES
SPINNING BALLOONS
SPINNING DYNAMICS
SPINNING EFFICIENCY
SPINNING FRAMES
SPINNING JENNYS
SPINNING LIMIT
SPINNING TRAVELERS
SPUN YARNS
STOP MOTIONS
TAPER (BOBBIN)
THREADLINE
TOW TO YARN CONVERSION
TRAVELER CHATTER
TRAVELERS
TRAVERSE
TWIST CONTROL
TWIST TRIANGLE
TWISTING
TWO FOR ONE TWISTING
UNLUBRICATED RINGS
UPTWISTING
WARP WIND
WINDING TENSION
WOOLEN DRAFTING
WOOLEN SYSTEM PROCESSING
WORSTED SPUN YARNS
WORSTED SYSTEM PROCESSING
YARN DYNAMICS
YARN TENSION
YARNS

SPINNING (EXTRUSION)
UF FIBER SPINNING

NT MELT SPINNING
SOLVENT SPINNING
WET SPINNING
BT EXTRUSION
RT BULGING (FILAMENT)
CAKES
CANDLE FILTERS
COAGULATING BATHS
CRYSTALLINITY
DOPE (POLYMER)
DRAWING (FILAMENT)
EXTRUSION PUMPS
FILAMENTS
FILTER PACKS
JET QUENCHING
MAN MADE FIBERS
NECKING (FILAMENT)
ORIENTATION
ORIFICES
POT SPINNING
QUENCHING (FILAMENT)
RELATIVE VISCOSITY
RING SPINNING
SPIN FINISHES
SPINNERETS
SPINNING ASSEMBLIES (EXTRUSION)
SPINNING FILTERS
SYNTHETIC FIBERS
VISCOSITY

SPINNING ASSEMBLIES (EXTRUSION)
RT EXTRUSION
FILTER PACKS
GODET ROLLS
METER PUMPS
SPINNERETS
SPINNING (EXTRUSION)
SPINNING FILTERS
SPINNING MACHINES (EXTRUSION)

SPINNING BALLOONS
BT BALLOONS
RT AIR RESISTANCE
BALLOON COLLAPSE
BALLOON CONTROL
BALLOON CONTROL RINGS
BALLOON DYNAMICS
BALLOON SEPARATORS
BALLOON TENSION
BALLOONS
DOWNTWISTING
RING SPINNING
SPINNING
TWIST RUN BACK
UNWINDING BALLOONS
YARN TENSION

SPINNING BATHS
USE COAGULATING BATHS

SPINNING BOBBINS
BT BOBBINS
PACKAGES
RT QUILLS
TWISTING BOBBINS

SPINNING DYNAMICS
NT BALLOON DYNAMICS
BT PROCESS DYNAMICS
YARN DYNAMICS
RT BALLOON COLLAPSE
BALLOON TENSION
CARDING DYNAMICS
CENTRIFUGAL FORCE
DRAFTING DYNAMICS
KNITTING DYNAMICS
MECHANISM (FUNDAMENTAL)
SPINNING
SPINNING TENSION (STAPLE YARN)
TRAVELERS
TWISTING DYNAMICS
TWISTING TENSION
WAVE LENGTH
WEAVING DYNAMICS
WINDING DYNAMICS
YARN TENSION

SPINNING EFFICIENCY
BT PROCESS EFFICIENCY
RT DOFFING (PACKAGE)
END BREAKAGES
KNOTTING
PATROLLING
PERFORMANCE
SPINNING
SPINNING LIMIT
STOPS
WEAVING EFFICIENCY
WINDING EFFICIENCY

SPINNING FILTERS
BT FILTER PACKS
RT METER PUMPS
SPINNERETS
SPINNING ASSEMBLIES (EXTRUSION)

SPINNING FINISH
USE SPIN FINISHES

SPINNING FRAMES
UF SPINNING MACHINERY (STAPLE FIBER)
NT CAP SPINNING FRAMES
FLYER SPINNING FRAMES
MULES
POT SPINNING FRAMES
RING FRAMES
WOOLEN SPINNING FRAMES
RT AMBLER SUPERDRAFT
BOBBINS
BUILDER MOTIONS
CAPS (SPINNING)
CASABLANCA SYSTEM
CLEARERS (LINT)
CONTROL SURFACES
COTTON SYSTEM PROCESSING
COTTON SYSTEM SPINNING
CREELING
DELIVERY ROLLS
DOFFING (PACKAGE)
DOUBLE TRUMPETS
DRAFTING (STAPLE FIBER)
DRAFTING ROLLS
DRAWFRAMES
DUO ROTH SYSTEM
FLYERS
FOOTSTEP BEARINGS
FRONT ROLLS
MAGNADRAFT (TN)
PACKAGES
REAL TWIST
RING RAILS
RINGS
ROTH SYSTEM
ROVING FRAMES
SHAW SYSTEM
SPINDLE RAILS
SPINDLES
SPINNING
SPINNING JENNYS
SPINNING MACHINES (EXTRUSION)
TAPE DRIVES
THREAD GUIDES
TRAVELERS
TRAVERSE
TWIST
TWIST GEARS
TWISTING
WOOLEN DRAFTING
WOOLEN SYSTEM PROCESSING
WOOLEN SYSTEM SPINNING
WORSTED SYSTEM PROCESSING
WORSTED SYSTEM SPINNING

SPINNING JENNYS
RT MULES
SPINNING
SPINNING FRAMES
SPINNING WHEELS

SPINNING LIMIT
UF SPINNABILITY (STAPLE FIBER)
RT ENTANGLEMENTS
FIBER DENIER
FIBER FIBER FRICTION
FIBER FINENESS
FIBER GEOMETRY
FIBER LENGTH
FIBER LENGTH DISTRIBUTION
FIBER MIGRATION
FIBER PROPERTIES
PROCESS EFFICIENCY
SPINNING
SPINNING EFFICIENCY
TWIST
TWIST MULTIPLIER

SPINNING LUBRICANTS
BT LUBRICANTS
SPIN FINISHES
RT CARDING OIL
DRAFTING (STAPLE FIBER)
FRICTION
HYDRODYNAMIC LUBRICATION
LUBRICATED RINGS
LUBRICATION
OILING
SLIDING
YARN FINISH

SPINNING MACHINERY (STAPLE FIBER)
USE SPINNING FRAMES

SPINNING MACHINES (EXTRUSION)
RT COAGULATING BATHS
EXTRUDERS
MELT SPINNING
SOLVENT SPINNING
SPINNERETS
SPINNING ASSEMBLIES (EXTRUSION)
SPINNING FRAMES
WET SPINNING

SPINNING MULES
USE MULES

SPINNING POTS
UF POTS (SPINNING)
RT POT SPINNING FRAMES

SPINNING RINGS
USE RINGS

SPINNING TENSION (EXTRUSION)
BT TENSION
RT DRAWING (FILAMENT)
SPINNING MACHINES (EXTRUSION)
WINDING TENSION
YARN TENSION

SPINNING TENSION (STAPLE YARN)
BT TENSION
RT BALLOON DYNAMICS
BALLOON TENSION
SPINNING DYNAMICS
TWISTING TENSION
WINDING TENSION
YARN TENSION

SPINNING TRAVELERS
NT CIRCULAR TRAVELERS
ELLIPTICAL TRAVELERS
NYLON TRAVELERS
STEEL TRAVELERS
BT TRAVELERS
RT PLYING TRAVELERS
RING SPINNING
RINGS
SPINNING
TRAVELER CHATTER

SPINNING WHEELS
RT SPINNING
SPINNING JENNYS

SPIRALITY (KNITTED FABRICS)
BT FABRIC DEFECTS
RT CIRCULAR KNITTING MACHINES
KNITTED FABRICS
PLAIN KNITTED FABRICS
TORSIONAL STABILITY

SPIRALS (NOVELTY YARNS)
BT NOVELTY YARNS
RT BEAD YARNS
BOUCLES
BUG YARNS
FLAKE YARNS
FLAMES (NOVELTY YARNS)
FRILLS (NOVELTY YARNS)
KNICKERBOCKER YARNS
LOOPS (NOVELTY YARNS)
NOVELTY TWISTERS
NUB YARNS
PLIED YARNS
RATINES
SEEDS (NOVELTY YARNS)
SLUB YARNS
SPLASHES

SPLASH VOILE
BT FLAT WOVEN FABRICS
WOVEN FABRICS
RT PLAIN WEAVES

SPLASHES
BT NOVELTY YARNS
RT BEAD YARNS
BOUCLES
BUG YARNS
FLAKE YARNS
FLAMES (NOVELTY YARNS)
FRILLS (NOVELTY YARNS)
KNICKERBOCKER YARNS
LOOPS (NOVELTY YARNS)
NUB YARNS
RATINES
SEEDS (NOVELTY YARNS)
SLUB YARNS
SPIRALS (NOVELTY YARNS)

SPLICING (KNITTING)
RT CORDAGE
JOINING
KNITTING
KNOTTING
PIECING
SEAMING
SUTURING

SPLIT DRUM WINDERS
BT DRUM WINDERS
RT WINDING

SPLIT RODS
RT LEASE RODS
SIZE BOXES
SLASHERS BEAMS
SLASHING

SPLIT SHED
UF CENTER SHED
BT SHED (WEAVING)
RT CLOSED SHED
HARNESSES
LOOMS
NEGATIVE SHEDDING
OPEN SHED
PICKING (WEAVING)
POSITIVE SHEDDING
SHEDDING
SHUTTLE SMASH
SHUTTLES
WARP ENDS
WEAVING

SPLITTING
RT BREAKING
CHIPPING
CRACK PROPAGATION
CUTTING
DELAMINATING
FLAKING
FRACTURING
GRINDING (MATERIAL REMOVAL)
SEPARATION
SLITTING
TEARING (TAILORING)

SPONGINESS
RT ABSORBENCY (MATERIAL)
COMPRESSIBILITY
HAND
HARSHNESS
LIVELINESS
RESILIENCE
SOFTNESS

SPONGING
BT DRY FINISHING
WET FINISHING
GARMENT MANUFACTURE
RT CRABBING
FABRIC RELAXATION
LONDON SHRINKING
STRESS RELAXATION
WET RELAXATION

SPOOLING
USE WINDING

SPOOLS
USE BOBBINS

SPORT SATIN
BT FLAT WOVEN FABRICS
SATIN
WOVEN FABRICS
RT SATIN WEAVES

SPORTSWEAR
NT BATHING SUITS
BEACH ROBES
BIKINIS
HALTERS
JACKETS
JERKINS
JUMPERS
KNICKERS
LEOTARDS
PARKAS
PLAYSUITS
PULL OVERS
SHORTS
SKI PANTS
SLACKS
SLIP ONS
SUN SUITS
SWEATSHIRTS
TRUNKS
BT GARMENTS
RT ATHLETIC CLOTHING
CARDIGANS
CHILDRENS WEAR
DUNGAREES
STRETCH WOVEN FABRICS
SWEATERS

SPOT WEAVES
BT WEAVE (TYPES)
RT FLAT WOVEN FABRICS

SPOTS
BT FABRIC DEFECTS
RT DIRT
FOGGY YARNS
FOGMARKING
OIL STAINS
SOIL RESISTANCE
SOILING
SPOTTINESS
SPOTTING
STAINING
STAINS

SPOTTINESS
RT BLOTCHINESS
BUNCHING COEFFICIENT
RT CARD WEBS
IRREGULARITY
MASS UNIFORMITY
NEPS
ORIENTATION UNIFORMITY
PATCHINESS
SPOTS
WEB ORIENTATION
WEB UNIFORMITY
WEBS
WIRA WEB LEVELNESS TESTER

SPOTTING
RT DISCOLORATION
DRY CLEANING
SPOTS
STAINING (DYEING DEFECTS)

SPRAY BONDING
BT BONDING
RT BONDED FIBER FABRICS
BONDED YARN FABRICS
DIP BONDING
FLAME BONDING

SPRAY TESTING
UF SPRAY TESTS
BT WATER REPELLENCY TESTING
RT WATER REPELLENCY
WATERPROOFING

SPRAY TESTS
USE SPRAY TESTING

SPRAYING
RT AEROSOLS
ATOMIZERS
ATOMIZING
JETS
PADDING
RINSING
SEALING
SOAKING

SPRAYING MACHINES
RT ATOMIZERS
COATING MACHINES
IMPREGNATING MACHINES
MANGLES
NOZZLES
PADDERS

SPREAD LOOPS
BT FABRIC DEFECTS
RT KNITTED FABRICS
STITCH LENGTH

SPREADERS (CLOTH)
BT GARMENT MANUFACTURING MACHINES
RT BOW STRAIGHTENERS
CALENDER ROLLS
SELVEDGE UNCURLERS
SCUTCHING (FINISHING)
SKEW STRAIGHTENERS
STRETCH ROLLS
TENTER FRAMES

SPREADING (TAILORING)
BT GARMENT MANUFACTURE
RT CUTTING (TAILORING)
MARKING
SCUTCHING (FINISHING)
SLOPING
SPREADING MACHINES (TAILORING)
SPREADING TABLES
TAILORING

SPREADING MACHINES (TAILORING)
RT GARMENT MANUFACTURE
SPREADING (TAILORING)
SCUTCHING (FINISHING)
SPREADING TABLES

SPREADING TABLES
RT GARMENT MANUFACTURE
SPREADING (TAILORING)
SPREADING MACHINES (TAILORING)
TAILORING

SPRING BEARD NEEDLES
USE SPRING NEEDLES

SPRING CONSTANT
RT COMPRESSIVE MODULUS
DEFLECTION
DEFORMATION
ELASTIC MODULUS (TENSILE)
MASS
MODULUS
RESONANT FREQUENCY
SPRINGS
STICK SLIP FRICTION
STIFFNESS
VIBRATION

SPRING LOADING
BT LOADING
RT DEAD WEIGHT LOADING
RT DRAWFRAMES
LET OFF MOTIONS
SPRINGS
TOP ROLLS

SPRING NEEDLES
UF BEARDED NEEDLES
SPRING BEARD NEEDLES
BT KNITTING NEEDLES
NEEDLES
RT COMPOUND NEEDLES
LATCH NEEDLES
MILANESE KNITTING MACHINES
NEEDLE BEARDS
NEEDLE EYE
PRESSER BARS
TRICOT KNITTING MACHINES
WARP KNITTING MACHINES

SPRINGINESS
USE RESILIENCE

SPRINGS
RT ACCESSORIES
DASHPOTS
DEVICES
ELASTICITY
ENERGY ABSORPTION
MAXWELL MODEL
MECHANISMS
SHOCK ABSORBERS
SPRING CONSTANT
SPRING LOADING
VIBRATION ISOLATORS
VISCOELASTICITY
VOIGT MODEL

SPUN SILK
BT SILK
RT SPUN YARNS
STAPLE FIBERS

SPUN YARNS
UF STAPLE FIBER YARNS
STAPLE YARNS
NT CARDED YARNS
COMBED YARNS
CONDENSED YARNS
COTTON SPUN YARNS
DIRECT SPUN YARNS
NOVELTY YARNS
TEXTURED SPUN YARNS
WOOLEN SPUN YARNS
WORSTED SPUN YARNS
BT YARNS
RT FIBER MIGRATION
FIBER SLIPPAGE
FILAMENT YARNS
MONOFILAMENT YARNS
REGENERATED FIBERS (EXCLUDING CELLULOSIC AND PROTEIN)
SINGLES YARNS
SPINNING
SPUN SILK
UNCONVENTIONAL YARNS
YARN STRUCTURE

SPUNBONDED (NONWOVENS)
BT BONDED FIBER FABRICS
NONWOVEN FABRICS
RT BACKING FABRICS
BONDING
BONDING STRENGTH
SELF BONDED FABRICS (NONWOVENS)
SELF BONDING FIBERS
TUFTED CARPETS

SPUNIZE (TN)
BT CRIMPED YARNS
TEXTURED YARNS (FILAMENT)
RT BANLON (TN)
STUFFER BOX CRIMPING
STUFFER BOXES

SQUEEGEES
RT DOCTOR BLADES
SCREEN PRINTING
SCREEN PRINTING MACHINES

SQUEEZE ROLLS
BT ROLLS
RT CALENDER ROLLS
CONTACT AREA
HYDROSTATIC LOADING
MANGLES
MOISTURE CONTENT
NIP
NIP PRESSURE
PADDERS
PRESS SECTION
PRESSURE ROLLS
ROLL COVERINGS
ROLLER NIP
SIZE BOXES
SLASHERS

RT SQUEEZERS (CLOTH)
SQUEEZING
SUCTION PRESSES
WET PICKUP

SQUEEZERS (CLOTH)
RT CALENDER ROLLS
MANGLES
PRESS SECTION
PRESSES
SQUEEZE ROLLS
SQUEEZING

SQUEEZING
RT CALENDERING
CARBONIZING
COMPRESSIBILITY
CONTACT AREA
DESIZING
DRYING
DYEING
LAUNDERING
MANGLING
MOISTURE CONTENT
NIP ROLLS
PADDING
PRESS SECTION
PRESSING
ROLLER NIP
SCOURING
SCOURING TRAINS
SIZE BOXES
SIZES (SLASHING)
SLASHERS
SLASHERS BEAMS
SLASHING
SOAP SODA SCOURING
SQUEEZE ROLLS
STARCH
STRETCH ROLLS
WARP PREPARATION
WET PICKUP

SRRL CARD
BT CARDS
RT COTTON CARDS
SRRL OPENER

SRRL OPENER
BT OPENERS
RT COTTON
DOFFING (OPENER)
DOFFER BRUSHES
SRRL CARD

STABILITY
RT AERODYNAMICS
BALLOON STABILITY
BUCKLING
CHEMICAL STABILITY
CREEP
CURL
DEGRADATION
DELAYED RECOVERY
DIMENSIONAL STABILITY
DRIFT
DURABILITY
DYNAMIC CHARACTERISTICS
DYNAMIC RESPONSE
HEAT STABILIZERS
PERMANENT DEFORMATION
PERMANENT SET
SNARLING
STABILIZATION
STABILIZERS (AGENTS)
THERMAL STABILITY
TORSIONAL STABILITY
WILDNESS

STABILITY TO CHEMICALS
USE CHEMICAL STABILITY

STABILIZATION
RT AGING (MATERIALS)
CONTROL SYSTEMS
HEAT STABILIZERS
INHIBITION
PARACHUTES
RECOVERY SYSTEMS (AERODYNAMIC)
STABILITY

STABILIZERS (AGENTS)
UF STABILIZING AGENTS
NT ANTIOXIDANTS
HEAT STABILIZERS
INFRARED ABSORBERS
ULTRAVIOLET ABSORBERS
ULTRAVIOLET STABILIZERS
RT CHEMICAL STABILITY
DEGRADATION
PHOTOCHEMICAL DEGRADATION
PRESERVATIVES
STABILITY
SUSPENDING AGENTS

STABILIZERS AGAINST THERMAL
DEGRADATION
USE HEAT STABILIZERS

STABILIZING AGENTS
USE STABILIZERS (AGENTS)

STAIN REMOVAL
BT CLEANING
RT DRY CLEANING
OIL STAINS
REMOVAL
SOIL REDEPOSITION
SOIL REMOVAL
SOLVENTS
STAIN RESISTANCE
STAIN RESISTANCE AGENTS
STAINING
STAINS
WATER SPOTS

STAIN REPELLENTS
USE STAIN RESISTANCE AGENTS

STAIN RESISTANCE
BT END USE PROPERTIES
FINISH (PROPERTY)
RT FABRIC PROPERTIES
FABRIC PROPERTIES (PHYSICAL
EXCLUDING MECHANICAL)
FLUOROCARBON POLYMERS
FLUOROCARBONS
LIPOPHILIC PROPERTY
OIL REPELLENCY
OIL REPELLENTS
OIL STAINS
PERSPIRATION
SOIL RESISTANCE
STAIN REMOVAL
STAIN RESISTANCE AGENTS
STAINING
STAINS
WATER REPELLENCY
WETTING

STAIN RESISTANCE AGENTS
UF STAIN REPELLENTS
BT FINISH (SUBSTANCE ADDED)
RT CLEANLINESS
EASY CARE FABRICS
SOIL RESISTANCE FINISHES
STAIN REMOVAL
STAIN RESISTANCE
STAINING

STAINING
UF MARK OFF
RT CROCKING
DIRT
DISCOLORATION
DYEING
OILS
SOIL RESISTANCE
SOILING
STAIN REMOVAL
STAIN RESISTANCE
STAIN RESISTANCE AGENTS
STAINING TESTING (FIBER
IDENTIFICATION)
STAINS

STAINING (DYEING DEFECTS)
BT DYEING DEFECTS
FABRIC DEFECTS
RT DISCOLORATION
SPOTTING
STAINING
WET CROCKING

STAINING TESTING (FIBER
IDENTIFICATION)
BT FIBER IDENTIFICATION
RT CHEMICAL ANALYSIS (FIBER
IDENTIFICATION)
MICROSCOPIC ANALYSIS (FIBER
IDENTIFICATION)
REAGENT TESTING (FIBER
IDENTIFICATION)
SPECIFIC GRAVITY TESTING
(FIBER IDENTIFICATION)
STAINING
THERMAL TESTING (FIBER
IDENTIFICATION)

STAINLESS STEEL
BT ALLOYS
METALLIC ALLOYS

STAINS
NT OIL STAINS
BT DEFECTS
FABRIC DEFECTS
RT DIRT
DYEING
FABRIC INSPECTION
FOGGY YARNS
FOGMARKING
OILS
PERSPIRATION
PERSPIRATION RESISTANCE
SOILING
SPOTS

RT STAIN REMOVAL
STAIN RESISTANCE
STAINING

STAINS DIRECT SPINNER
BT CONVERTERS (TOW)
TOW TO YARN MACHINES
RT DIRECT SPUN YARNS
DRAFT CUT
TOW
TOW TO YARN CONVERSION

STANDARD DEVIATION
BT STATISTICAL MEASURES
RT ACCURACY
COEFFICIENT OF VARIATION
CONFIDENCE COEFFICIENT
CONFIDENCE LIMITS
CORRELATION
CORRELATION COEFFICIENT
DEGREES OF FREEDOM
DISTRIBUTION
ERRORS
ESTIMATION
FLUCTUATIONS
KURTOSIS
LEVEL OF SIGNIFICANCE
MEAN
PROBABILITY
STANDARD ERROR
STATISTICS
VARIANCE

STANDARD ERROR
BT ERRORS
STATISTICAL MEASURES
RT COEFFICIENT OF VARIATION
CORRELATION
CORRELATION COEFFICIENT
DEGREES OF FREEDOM
DISTRIBUTION
ESTIMATION
MEAN
PROBABILITY
STANDARD DEVIATION
STATISTICS
VARIANCE

STANDARD HEDDLES
RT BACK CROSSING HEDDLES
DOUP
FRONT CROSSING HEDDLES
GROUND HEDDLES
HARNESSES
LENO WEAVES
LENO WEAVING

STANDARDS
RT COLOR STANDARDS
PRODUCTION
QUALITY CONTROL
SPECIFICATIONS
TESTING

STANDFAST (TN)
RT DYE FIXING
MOLTEN METAL BATHS

STANDS
UF PEDESTAL STANDS
RT ACCESSORIES
DEVICES

STANNOUS FLUORIDE
BT FLUORIDES
TIN COMPOUNDS

STAPLE ARRAY
USE FIBER ARRAY

STAPLE CUTTERS
BT CUTTERS
RT CUTTING
FIBER CUTTING
STAPLE FIBERS
THREAD CUTTERS

STAPLE DIAGRAM
USE FIBER DIAGRAM

STAPLE FIBER YARNS
USE SPUN YARNS

STAPLE FIBERS
BT FIBERS
RT ANIMAL FIBERS (GENERAL)
COEFFICIENT OF LENGTH
VARIATION
FIBER DIAGRAM
FIBER LENGTH
FIBER LENGTH DISTRIBUTION
FIBRO (TN)
FIBROGRAPH
MAN MADE FIBERS
MEAN FIBER LENGTH
NATURAL FIBERS
SEED HAIR FIBERS (GENERAL)
SORTING
SPUN SILK
(CONT.)

RT STAPLE CUTTERS
STAPLING
STOCK (FIBER)
STOCK DYEING
STOCK DYEING MACHINES
SYNTHETIC FIBERS
TOW CONVERSION
VEGETABLE FIBERS (GENERAL)

STAPLE LENGTH
USE FIBER LENGTH

STAPLE STRENGTH
USE FIBER STRENGTH

STAPLE YARNS
USE SPUN YARNS

STAPLES (WIRE)
RT CARD WIRES
CARDS
STAPLING (WIRE)

STAPLING
BT FIBER LENGTH DETERMINATION
RT COEFFICIENT OF LENGTH VARIATION
COTTON CLASSING
COTTON QUALITY
DIGITAL FIBROGRAPH
EFFECTIVE LENGTH (FIBER)
FIBER ARRAY
FIBER DIAGRAM
FIBER LENGTH
FIBER LENGTH DISTRIBUTION
FIBERS
FIBROGRAPH
FIBROGRAPH MEAN LENGTH
GRADING
LENGTH UNIFORMITY RATIO (FIBER)
MEAN FIBER LENGTH
SERVO FIBROGRAPH
SHORT FIBERS INDEX
SORTING
SPAN LENGTH
STAPLE FIBERS
SUTER WEBB TESTING
UPPER HALF MEAN LENGTH
UPPER QUARTILE LENGTH

STAPLING (WIRE)
BT GARMENT MANUFACTURE
RT JOINING
SEAMING
STAPLES (WIRE)

STAR WHEELS
RT NOBLE COMBS

STARCH
UF NATURAL STARCH
STARCHES
NT AMYLOPECTIN
AMYLOSE
CORN STARCH
DEXTRINS
DIALDEHYDE STARCH
MODIFIED STARCHES
OXIDIZED STARCHES
POTATO STARCH
STARCH ETHERS
STARCH GUMS
STARCH PHOSPHATE
BT CARBOHYDRATES
FINISH (SUBSTANCE ADDED)
NATURAL POLYMERS
POLYSACCHARIDES
RT CRINOLINE
DEXTRAN
GELATINE
HAND
LAUNDERING
LAWN FINISH
LEASE RODS
PAPERINESS
SIZE BOXES
SIZES (SLASHING)
SIZING (PAPERMAKING)
SLASHERS
SLASHERS BEAMS
SLASHING
SQUEEZING
STARCHING
STIFFENERS (AGENTS)
STIFFNESS
STRETCH ROLLS
WARP BEAMS
WARP PREPARATION

STARCH ETHERS
BT MODIFIED STARCHES
STARCH

STARCH GUMS
BT STARCH
RT DEXTRINS
PRINTING PASTES

RT SIZES (SLASHING)
SIZING (PAPERMAKING)
STARCHING
THICKENING AGENTS

STARCH PHOSPHATE
BT MODIFIED STARCHES
STARCH
RT CELLULOSE PHOSPHATE

STARCHES
USE STARCH

STARCHING
BT WET FINISHING
RT BENDING RIGIDITY
HAIR SETTING
LAUNDERING
SLASHING
STARCH
STIFFENERS (AGENTS)
STIFFENING TREATMENTS
STIFFNESS

START UP MARKS
USE STARTING MARKS

STARTING MARKS
UF START UP MARKS
STOPPING LINE
STOPPING MARKS
BT FABRIC DEFECTS
RT CLOTH FELL
FABRIC INSPECTION
KNITTING
STOP MOTIONS
STREAKINESS
WEAVING

STARTING TIME
RT CONSTANT SPEED
CONTROL SYSTEMS
DRIVES
MOTORS
PRODUCTION
SPEED CONTROL
TIME
VARIABLE SPEED
VELOCITY

STATIC
USE STATIC ELECTRICITY

STATIC ABSORPTION (WATER)
BT ABSORPTION (MATERIAL)
WATER ABSORPTION
RT ABSORBENCY (MATERIAL)
DYNAMIC ABSORPTION (WATER)
MOISTURE ABSORPTION
MOISTURE CONTENT
STATIC ABSORPTION TESTING
WATER ABSORPTION
WATER REPELLENCY TESTING
WET PICKUP
WETTING

STATIC ABSORPTION TESTING
UF STATIC ABSORPTION TESTS
BT WATER ABSORPTION TESTING
RT ABSORPTION (MATERIAL)
DYNAMIC ABSORPTION TESTING
IMMERSION TESTING
MOISTURE CONTENT
PERMEABILITY TESTING
STATIC ABSORPTION (WATER)
WATER RESISTANCE TESTING
WATERPROOFING
WET PICKUP

STATIC ABSORPTION TESTS
USE STATIC ABSORPTION TESTING

STATIC ELECTRICITY
UF ELECTROSTATICS
STATIC
RT ANTISTATIC AGENTS
ANTISTATIC BEHAVIOR
CAPACITANCE (ELECTRICAL)
CORONA
DISCHARGE (ELECTRIC)
DISSIPATION
ELECTROSTATIC FLOCKING
FOGGY YARNS
FOGMARKING
SOILING
SPOTS
STAINING
STATIC ELIMINATORS
STATIC RESISTANCE
TRIBOELECTRIC SERIES
TRIBOELECTRICITY

STATIC ELIMINATORS
UF RADIOACTIVE STATIC ELIMINATORS
RT ANTISTATIC AGENTS
ANTISTATIC BEHAVIOR
STATIC ELECTRICITY

STATIC FRICTION
UF LIMITING FRICTION
BT FRICTION
RT COEFFICIENT OF FRICTION
DYNAMIC FRICTION
FABRIC DEFECTS
FABRIC TO FABRIC FRICTION
FIBER FIBER FRICTION
FIBER MIGRATION
FRICTIONAL FORCE
FUZZ RESISTANCE
INTERNAL FRICTION
KNITTED FABRICS
LOOP EFFICIENCY
LOOP STRENGTH
PILL RESISTANCE
PILLING
PILLS
ROLLING FRICTION
RUBBING
SLIDING FRICTION
STICK SLIP FRICTION
YARN TO METAL FRICTION
YARN TO YARN FRICTION

STATIC MODULUS
USE ELASTIC MODULUS (TENSILE)

STATIC RESISTANCE
RT ANTISTATIC AGENTS
ANTISTATIC BEHAVIOR
FABRIC PROPERTIES
FABRIC PROPERTIES (PHYSICAL EXCLUDING MECHANICAL)
STATIC ELECTRICITY
STATIC ELIMINATORS
TRIBOELECTRIC SERIES
TRIBOELECTRICITY

STATICS
BT MECHANICS
RT DYNAMICS
KINEMATICS
KINETICS
STRUCTURAL MECHANICS

STATISTICAL ANALYSIS
RT ANALYSIS OF VARIANCE
ANALYZING
COEFFICIENT OF VARIATION
CONFIDENCE LIMITS
EMPIRICAL ANALYSIS
EXPERIMENTAL ANALYSIS
MARKOV CHAIN
MATHEMATICAL ANALYSIS
STATISTICAL INFERENCE
STATISTICAL MEASURES
STATISTICS
STOCHASTIC PROCESSES
THEORETICAL ANALYSIS
THEORIES
TIME SERIES
VARIANCE LENGTH CURVES

STATISTICAL DISTRIBUTION
USE DISTRIBUTION

STATISTICAL INFERENCE
RT COEFFICIENT OF VARIATION
DISTRIBUTION
MARKOV CHAIN
PROBABILITY
STATISTICAL ANALYSIS
STATISTICAL MEASURES
STOCHASTIC PROCESSES
TIME SERIES

STATISTICAL MEASURES
NT COEFFICIENT OF VARIATION
CONFIDENCE COEFFICIENT
CONFIDENCE LIMITS
CORRELATION COEFFICIENT
DEGREES OF FREEDOM
DISTRIBUTION
FREQUENCY DENSITY
KURTOSIS
LEVEL OF SIGNIFICANCE
MEAN
MEDIAN
MODE
RANGE
REGRESSION COEFFICIENT
STANDARD DEVIATION
STANDARD ERROR
VARIANCE
RT ANALYSIS OF VARIANCE
AUTOCORRELATION
COEFFICIENT OF FINENESS VARIATION
FREQUENCY
STATISTICS
STOCHASTIC PROCESSES
T DISTRIBUTION
TIME SERIES
UPPER CONTROL LIMIT
VARIANCE LENGTH CURVES

STATISTICAL MECHANICS
RT KINETICS (CHEMICAL)
MECHANICS
PROBABILITY
STATISTICS
THERMODYNAMICS

STATISTICAL METHODS
RT ANALYSIS OF VARIANCE
CORRELATION
DATA REDUCTION
DISTRIBUTION
EXPERIMENTAL DESIGN
STATISTICAL ANALYSIS
STATISTICAL MEASURES
STATISTICS
STOCHASTIC PROCESSES

STATISTICAL THEORY
USE STATISTICS

STATISTICS
UF STATISTICAL THEORY
RT ANALYSIS OF VARIANCE
COEFFICIENT OF VARIATION
DEGREES OF FREEDOM
DISTRIBUTION
ERRORS
ESTIMATION
EXPERIMENTAL DESIGN
FREQUENCY
MODE
POPULATION
PROBABILITY
STANDARD DEVIATION
STANDARD ERROR
STATISTICAL MEASURES
STATISTICAL MECHANICS
STOCHASTIC PROCESSES
TIME SERIES

STEAM
NT SUPERHEATED STEAM
RT AGERS
AGING (STEAMING)
HEATING EQUIPMENT
MOISTURE CONTENT
STEAM HEATING
STEAM JET TEXTURING
STEAMERS

STEAM HEATING
BT HEATING
RT STEAM
STEAMERS

STEAM JET TEXTURING
BT TEXTURING
RT AIR JET TEXTURING
AIR JET TEXTURING MACHINES
BELT CRIMPING
BULKING
CRIMPING
EDGE CRIMPING
GEAR CRIMPING
JET QUENCHING
JETS
STEAM
STUFFER BOX CRIMPING

TEAMERS
BT GARMENT MANUFACTURING MACHINES
RT AGERS
CONTINUOUS DYEING RANGE
DECATING MACHINES
DRYERS
DYEING MACHINES
DYEING MACHINES (BY PROCESS FLOW)
HEATING EQUIPMENT
OVENS
PIECE DYEING MACHINES
SEMI CONTINUOUS DYEING RANGE
STEAM
STEAM HEATING
TENTER FRAMES
WET FINISHING

TEAMING
NT HEAT SETTING (MOIST)
BT WET FINISHING
RT ACID SHOCK METHOD
AGERS
AGING (STEAMING)
CURING
HEAT SETTING (SYNTHETICS)
PAD STEAM FINISHING
STEAMERS
SUPERHEATED STEAM
TWIST SETTERS
WET FINISHING

EARATE ESTERS
BT ESTERS
RT STEARIC ACID

STEARIC ACID
BT FATTY ACIDS
RT STEARATE ESTERS

STEARYL ALCOHOL
BT FATTY ALCOHOLS

STEEL TRAVELERS
BT SPINNING TRAVELERS
TRAVELERS
RT CIRCULAR TRAVELERS
ELLIPTICAL TRAVELERS
NYLON TRAVELERS
PLYING TRAVELERS
RING SPINNING
RINGS
SPINNING

STEELON (TN)
BT NYLON (POLYAMIDE FIBERS)
NYLON 6

STEEP TWILLS
BT TWILLS
RT CHEVRON TWILLS
FANCY TWILLS
HERRINGBONE TWILLS
LOW TWILLS
REGULAR TWILLS

STELOMETER
BT TENSILE TESTERS
RT STELOMETER STRENGTH TESTING
STRAINOMETER

STELOMETER STRENGTH TESTING
UF STELOMETER STRENGTH TESTS
RT BREAKING STRENGTH
CLEMSON STRENGTH TESTING
FIBER STRENGTH
PRESSLEY TEST
STELOMETER
STRENGTH OF MATERIALS
TENACITY

STELOMETER STRENGTH TESTS
USE STELOMETER STRENGTH TESTING

STENTERING
USE TENTERING

STEREO SPECIFICITY
USE TACTIC POLYMERS

STEREOSCAN MICROSCOPY
UF SCANNING ELECTRON MICROSCOPY
BT ELECTRON MICROSCOPY
MICROSCOPY
RT INFRARED MICROSCOPY
INTERFERENCE MICROSCOPY
LIGHT MICROSCOPY (OPTICAL)
PHASE MICROSCOPY
POLARIZING MICROSCOPE
PROJECTION MICROSCOPY
STEREOSCOPIC MICROSCOPY
ULTRAVIOLET MICROSCOPY
X RAY MICROSCOPY

STEREOSCOPIC MICROSCOPY
BT MICROSCOPY
RT ELECTRON MICROSCOPY
INFRARED MICROSCOPY
INTERFERENCE MICROSCOPY
LIGHT MICROSCOPY (OPTICAL)
STEREOSCAN MICROSCOPY
ULTRAVIOLET MICROSCOPY
X RAY MICROSCOPY

STICK SHUTTLES
BT SHUTTLES
RT BLANKET SHUTTLES
BOAT SHUTTLES
FLY SHUTTLES
LOOMS
PICKER SHOES
PICKER STICKS
PICKING (WEAVING)
QUILLS
RACE BOARDS
RAPIERS (LOOMS)
REEDS
SHEDDING
SHUTTLE BINDERS
SHUTTLE BOXES
SHUTTLE DEFLECTORS
SHUTTLE MARKS
SHUTTLE SMASH
SHUTTLE SMASH PROTECTORS
SLEYS (LOOM)
WEAVING

STICK SLIP
USE STICK SLIP FRICTION

STICK SLIP FRICTION
UF STICK SLIP
STICTION
BT FRICTION
RT COEFFICIENT OF FRICTION
DYNAMIC FRICTION
FIBER FIBER FRICTION
FIBER SLIPPAGE
LUBRICATION
MASS
ROLLER SLIPPAGE
ROLLING FRICTION
SLICING FRICTION
SLIPPAGE
SPRING CONSTANT
STATIC FRICTION

STICKING (ADHESION)
USE ADHESION

STICTION
USE STICK SLIP FRICTION

STIFFENERS (AGENTS)
UF FABRIC STIFFENERS
BT FINISH (SUBSTANCE ADDED)
RT COLLAR STIFFENERS
HYDROXYALKYLCELLULOSE
PAPERINESS
SIZES (SLASHING)
STARCH
STARCHING
STIFFENING TREATMENTS
STIFFNESS
WRINKLE RECOVERY

STIFFENING TREATMENTS
BT WET FINISHING
RT STARCHING
STIFFENERS (AGENTS)

STIFFENINGS (GARMENT)
BT GARMENT COMPONENTS
RT FACING
FRONTS
GARMENT MANUFACTURE
INTERLININGS

STIFFNESS
UF RIGIDITY
BT END USE PROPERTIES
STRESS STRAIN PROPERTIES
STRESS STRAIN PROPERTIES (TENSILE)
RT BENDING
BENDING RIGIDITY
BOWING
COMPLIANCE
DRAPE
FABRIC PROPERTIES
FABRIC PROPERTIES (MECHANICAL)
HAND
HARDNESS
INITIAL MODULUS
MOMENT OF INERTIA
PAPERINESS
SOFTNESS
SPRING CONSTANT
STARCH
STARCHING
STIFFENERS (AGENTS)
WRINKLE RESISTANCE

STIRRERS
UF AGITATORS
RT HOMOGENIZERS
MIXING
STIRRING

STIRRING
RT BLENDING
CIRCULATION
COMPOUNDING
DISPERSING
EMULSIFYING
MIXING
STIRRERS

STITCH BONDED FABRICS
NT ARABEVA FABRIC (TN)
ARACHNE FABRIC (TN)
ARALOOP FABRIC (TN)
MALIMO FABRIC (TN)
MALIPOL FABRIC (TN)
MALIWATT FABRIC (TN)
VOLTEX FABRIC (TN)
BT FABRICS (ACCORDING TO STRUCTURE)
RT ARACHNE SYSTEMS (TN)
BONDED FIBER FABRICS
COMPOSITES
CROSS LAID YARN FABRICS
FLAT WOVEN FABRICS
KNITTED FABRICS
LAMINATED FABRICS
MALI PROCESSES (TN)
MULTIPLE LAYER FABRICS
NEEDLED FABRICS
NONWOVEN FABRICS
PILE FABRICS
STITCH BONDING
STITCH BONDING MACHINES
(CONT.)

RT STITCH REINFORCED FABRICS
STITCHED PILE FABRICS
STITCHING
TUFTED FABRICS
WOVEN FABRICS

STITCH BONDING
NT MALIMO PROCESS (TN)
MALIPOL PROCESS (TN)
MALIWATT PROCESS (TN)
BT BONDING
RT CHAIN STITCHES
NEEDLING
STITCH BONDED FABRICS
STITCH BONDING MACHINES
STITCH REINFORCED FABRICS
STITCHES (SEWING)
STITCHING
TAPING
WEAVING
WEBS

STITCH BONDING MACHINES
RT ARACHNE SYSTEMS (TN)
MALI PROCESSES (TN)
STITCH BONDED FABRICS
STITCH BONDING

STITCH CAMS
BT CAMS
RT AUXILIARY CAMS
CAM ANGLES
CIRCULAR KNITTING MACHINES
HOSIERY MACHINES
KNITTING
KNITTING MACHINES
KNITTING NEEDLES
MOCK SEAMS
NEEDLE CAMS
PATTERN WHEELS
RAISING CAMS
SEAMLESS HOSIERY MACHINES
SINKER CAMS

STITCH CLARITY
BT FABRIC PROPERTIES
FABRIC PROPERTIES (AESTHETIC)
FABRIC PROPERTIES (STRUCTURAL)
RT AESTHETIC PROPERTIES
APPEARANCE
BOTTOM COVER
COURSES
COVER
DESIGN
FINISH (PROPERTY)
KNITTED FABRICS
LOOP DISTORTION
NAP
OPTICAL PROPERTIES
PATTERN (FABRICS)
PATTERN DEFINITION
STITCH DENSITY
TEXTURE
TOP COVER
WEAVE

STITCH DENSITY
BT FABRIC PROPERTIES
FABRIC PROPERTIES (STRUCTURAL)
RT COURSES
COURSES PER INCH
COVER
COVER FACTOR
DENSITY
FABRIC GEOMETRY
FABRIC STRUCTURE
KNITTED FABRICS
KNITTING
LOOP DISTORTION
SET (WOVEN FABRIC)
STITCH BONDING
STITCH CLARITY
STITCH LENGTH
STITCHING
WALES PER INCH

STITCH DISTORTION
RT FABRIC DISTORTION (TAILORING)
KNITTED FABRICS
LOOP DISTORTION

STITCH ELASTICITY
UF SEAM ELASTICITY
SEAM ELONGATION
STITCH ELONGATION
STITCH RESILIENCE
RT ELASTICITY
SEAMING
SEAMS
SEWING
STITCHES (SEWING)
STITCHING

STITCH ELONGATION
USE STITCH ELASTICITY

STITCH LENGTH
UF LOOP LENGTH
RT COURSES
COURSES PER INCH
FABRIC GEOMETRY
KNITTED FABRICS
KNITTING
KNITTING TENSION
LOOP DISTORTION
MACHINE KNITTING
NEEDLE LOOPS
POSITIVE FEED
PROCESS VARIABLES
SINKER LOOPS
SPREAD LOOPS
STITCH DENSITY
STITCHES (SEWING)
WALES PER INCH

STITCH LENGTH CONTROL
RT AUTOMATIC CONTROL
KNITTING
POSITIVE FEED
PROCESS CONTROL
STITCH DENSITY
TENSION CONTROL

STITCH QUALITY
RT BASTING
GATHERING
GYMP
JOINING
KNITTING
LEATHER
RAWHIDE
SEAM EFFICIENCY
SEAMING
SERGING
SEWING
SHOES
STITCH ELASTICITY
STITCH SEQUENCE
STITCH SIZE (SEWING)
STITCH TENSION
STITCHES (SEWING)
STITCHING
TAILORING
THREADS
TRIMMING (OPERATION)
WEAVING

STITCH REINFORCED FABRICS
NT ARACHNE FABRIC (TN)
MALIWATT FABRIC (TN)
BT FABRICS (ACCORDING TO STRUCTURE)
RT ARACHNE SYSTEMS (TN)
BONDED YARN FABRICS
COMPOSITES
CROSS LAID YARN FABRICS
FLAT WOVEN FABRICS
MALI PROCESSES (TN)
NEEDLED FABRICS
NONWOVEN FABRICS
STITCH BONDED FABRICS
STITCHED PILE FABRICS
STITCHING
WOVEN FABRICS

STITCH RESILIENCE
USE STITCH ELASTICITY

STITCH SEQUENCE
RT SEAMING
SEAMS
SEWING
STITCH QUALITY
STITCHES (SEWING)
STITCHING

STITCH SIZE (SEWING)
RT SEAM EFFICIENCY
SEAM GRIN
SEAM QUALITY
SEAM SLIPPAGE
SEAM SPECIFICATIONS
SEAM STRENGTH
SEAMING
SEAMS
SEWING
STITCH QUALITY
STITCH TENSION
STITCHES (SEWING)
STITCHING
THREAD TENSION

STITCH TENSION
BT TENSION
RT PUCKERING
SEAM GRIN
SEAM PUCKER
SEAM SLIPPAGE
SEAMS
SEWING
STITCH QUALITY
STITCH SIZE (SEWING)
STITCHING
TENSION CONTROL
TENSION DEVICES

STITCH TRANSFER
RT FASHIONING
FASHIONING MARKS
FULLY FASHIONED HOSIERY
FULLY FASHIONED KNITTING
KNITTING
NARROWING (KNITTING)
NEEDLE TRANSFER
TRANSFERRING (MATERIAL)
WIDENING (KNITTING)

STITCHED PILE FABRICS
NT ARALOOP FABRIC (TN)
MALIPOL FABRIC (TN)
BT FABRICS (ACCORDING TO STRUCTURE)
PILE FABRICS
RT ARACHNE SYSTEMS (TN)
BONDED YARN FABRICS
COMPOSITES
CROSS LAID YARN FABRICS
FLOCKED FABRICS
MALI PROCESSES (TN)
NEEDLED FABRICS
NONWOVEN FABRICS
PILE FABRICS (TUFTED)
STITCH BONDED FABRICS
STITCH BONDING
STITCH REINFORCED FABRICS
STITCHING
TUFTED FABRICS

STITCHES (KNITTING)
USE KNITTED STITCHES

STITCHES (SEWING)
NT AIR TUCK STITCHES
BACKSTITCHES
BLINDSTITCHES
CHAIN STITCHES
COMBINATION STITCHES
COVER STITCHES
CROSS STITCHES
DIAGONAL STITCHES
DOT DASH STITCHES
DOUBLE DIAGONAL STITCHES
DOUBLE ROW STITCHES
FLATLOCK STITCHES
FOUR NEEDLE COVER STITCHES
GIANT SADDLE STITCHES
HAND PICK STITCHES
HAND SADDLE STITCHES
LADDER STITCHES
LOCKSTITCH BLINDSTITCHES
LOCKSTITCHES
MERROW STITCHES
OVERCAST STITCHES
OVEREDGE STITCHES
OVERLOCK STITCHES
OVERSEAM STITCHES
PICK STITCHES
PICCETTA FAGOTING (STITCH)
PIN POINT SADDLE STITCHES
REGULAR SADDLE STITCHES
SADDLE STITCHES
SERGING STITCHES
SHORT LONG STITCHES
SINGLE NEEDLE LOCKSTITCHES
SINGLE THREAD BLINDSTITCHES
SINGLE THREAD CHAIN STITCHES
SLANT SADDLE STITCHES
THREE NEEDLE COVER STITCHES
TWO THREAD CHAIN STITCHES
BT GARMENT COMPONENTS
RT BLIND STITCHING
BUTTED SEAMS
BUTTERFLY SEAMS
CHAIN SEWING
COVERED SEAMS
GATHERING
GATHERS (TAILORING)
HEMMING
LOOP CATCHING
NEEDLE EYE
PINCHING
PINKING (TAILORING)
PUCKERING
PUCKERS (TAILORING)
RIPPING
RUFFLING
SEAM STRENGTH
SEAMING
SEAMS
SERGING
SEWING
SEWING MACHINES
SEWING NEEDLES
SHIRRING
STITCH BONDING
STITCH ELASTICITY
STITCH LENGTH
STITCH QUALITY
STITCHING
TAILORED SEAMS
TAILORING
TRIMMING (OPERATION)

STITCHING
BT GARMENT MANUFACTURE
JOINING
RT ANGULAR SEAMING
ARACHNE PROCESS (TN)
BACK TACKING
BASTING
BINDINGS (FOOTWEAR)
BLIND STITCHING
BOBBIN THREADS (SEWING)
BUTTED SEAMS
BUTTERFLY SEAMS
CHAIN SEWING
COCKLING
COTTON THREADS
COVERED SEAMS
CURVED SEAMING
DRESSMAKING
EMBROIDERY
FABRIC DISTORTION (TAILORING)
FELLED SEAMS
FELLING
FRENCH SEAMS
FULL SEAM WIDTH
GATHERING
HAND SEWING
HEAT SEALING
HEM SEAMS
HEMMING
KNITTING
LAPPED SEAMS
LINEN THREADS
LOCKSTITCH BLINDSTITCHES
LOOP CATCHING
MACHINE SEWING
MALI PROCESSES (TN)
MALIMO PROCESS (TN)
MALIPOL PROCESS (TN)
MALIWATT PROCESS (TN)
NEEDLE EYE
NEEDLE GROOVE
NEEDLE HEATING
NEEDLE PENETRATION
NEEDLE POINT
NEEDLE SHAFT (SEWING)
NEEDLE SHANK (SEWING)
NEEDLING
NYLON THREADS
PINCHING
PUCKERING
QUILTING
RIPPLES
RUFFLING
SEAM DAMAGE
SEAM PUCKER
SEAM SLIPPAGE STRENGTH
SEAM STRENGTH
SEAMING
SEAMS
SEAMSTRESSES
SERGING
SEWABILITY
SEWING
SEWING DAMAGE
SEWING MACHINES
SEWING NEEDLES
SEWING THREADS
SHIRRING
STITCH BONDED FABRICS
STITCH ELASTICITY
STITCH QUALITY
STITCH REINFORCED FABRICS
STITCH SEQUENCE
STITCH SIZE (SEWING)
STITCH TENSION
STITCHED PILE FABRICS
STITCHES (SEWING)
TACKING
TAILORING
TAPING
THREADS
TRIMMING (OPERATION)
TUCKING (GARMENT MANUFACTURE)
YARN SEVERANCE

STOCHASTIC PROCESSES
RT DISTRIBUTION
EXPERIMENTAL DESIGN
STATISTICAL ANALYSIS
STATISTICAL INFERENCE
STATISTICAL MEASURES
STATISTICAL METHODS
STATISTICS

STOCK (FIBER)
BT FIBER ASSEMBLIES
RT BALE BREAKERS
BALES
HOPPER FEED
LAPS
OPENING
SCOURING
STAPLE FIBERS
STOCK DYEING
TINTING
WEBS

STOCK DYEING
UF RAW STOCK DYEING
LOOSE STOCK DYEING
BT DYEING (BY MATERIAL ASSEMBLY)
RT DOPE DYEING
PIECE DYED FABRICS
PIECE DYEING
STAPLE FIBERS
STOCK (FIBER)
STOCK DYEING MACHINES
TOP DYEING
TOW DYEING
YARN DYED FABRICS
YARN DYEING

STOCK DYEING MACHINES
UF LOOSE STOCK DYEING MACHINES
RAW STOCK DYEING MACHINES
BT DYEING MACHINES
DYEING MACHINES (BY MATERIAL ASSEMBLY)
RT BATCH DYEING MACHINES
DOPE DYEING EQUIPMENT
DYEING
DYEING (BY MATERIAL ASSEMBLY)
KETTLES (DYEING)
PACKAGE DYEING MACHINES
PIECE DYEING MACHINES
PRESSURE DYEING MACHINES
PROPELLERS
STAPLE FIBERS
STOCK DYEING
TOP DYEING MACHINES
TOW DYEING MACHINES
YARN DYEING MACHINES

STOCKINETTE
USE PLAIN KNITTED FABRICS

STOCKINETTE WARP KNIT FABRICS
USE TRICOT KNITTED FABRICS

STOCKINGS
NT SURGICAL STOCKINGS
BT HOSIERY
RT AFTERWELT
BOARDING
BOOTS
CIRCULAR HOSIERY
DENIER
FASHIONING
FOOTWEAR
FULLY FASHIONED HOSIERY
HALF HOSE
HEELS (STOCKINGS)
KNITTED OUTERWEAR
KNITTING
KNITTING MACHINES
LEGGINGS
MESH (KNITTING)
MOCK SEAMS
PRESSING
RUN RESISTANCE
RUN RESISTANT HOSIERY
SEAMLESS HOSIERY
SHEERNESS
SOCKS
STRETCH HOSIERY

STODDARD SOLVENT
RT DRY CLEANING
DRY CLEANING SOLVENTS
HYDROCARBONS

STOLL QM ABRASION TESTING
UF STOLL QM ABRASION TESTS
BT ABRASION TESTING
RT ABRASION RESISTANCE
SCHIEFER ABRASION TESTING
TABER ABRASION TESTING

STOLL QM ABRASION TESTS
USE STOLL QM ABRASION TESTING

STOP MOTIONS
NT AUTOMATIC STOP MOTIONS
DROP WIRES
RT AUTOMATIC CONTROL
BANGING OFF
CHEESE WINDERS
CREELING
DISC TENSION
END BREAKAGES
FEELER MECHANISMS
KNITTING
KNOCK OFF
LOOMS
LOOM STOPS
OPTICAL SCANNERS
PRECISION WINDERS
PROCESS CONTROL
QUILLING
SHUTTLE SMASH PROTECTORS
SPINNING
STARTING MARKS
STOPS
TENSION
TWISTING
WARP ENDS
WARPERS (MACHINE)
RT WARPING
WEAVING
WEFT FORKS
WINDERS
WINDING

STOPPAGES
USE STOPS

STOPPING LINE
USE STARTING MARKS

STOPPING MARKS
USE STARTING MARKS

STOPS
UF STOPPAGES
NT LOOM STOPS
RT DOWN TIME
END BREAKAGES
PATROLLING
PRODUCTION
SPINNING EFFICIENCY
STARTING MARKS
STOP MOTIONS
WEAVING EFFICIENCY

STORAGE AGING
USE AGING (STORAGE)

STORAGE VESSELS
BT VESSELS
RT AGING (STORAGE)
KETTLES (DYEING)
KETTLES (HEATING)
TRANSPORTING VESSELS

STRAIGHTENING
BT MODES OF DEFORMATION
RT BENDING
BENDING MOMENTS
BENDING RECOVERY
BENDING RIGIDITY
CURVATURE

STRAIN
UF ELONGATION
ELONGATION PERCENT
EXTENSION
PERCENTAGE STRAIN
NT BREAKING ELONGATION
ELASTIC STRAIN
PLASTIC STRAIN
TRUE STRAIN
BT MODES OF DEFORMATION
RT BEARING STRENGTH
BENDING
BREAKING ELONGATION
BUCKLING
BUNDLE STRENGTH
COMPLIANCE
COMPRESSIVE STRENGTH
CREASE RETENTION
CREEP
DEFLECTION
DEFORMATION
DELAYED RECOVERY
DIMENSIONAL STABILITY
DISPLACEMENT
DISTORTION
DRAPE
DRAWING (FILAMENT)
DRY RELAXATION
DUCTILITY
ELASTIC MODULUS (TENSILE)
ELASTICITY
EXTENSIBILITY
FABRIC RELAXATION
GAUGE LENGTH
GROWTH (FABRIC)
LOADING
LOOP EFFICIENCY
LOOP STRENGTH
PERMANENT SET
POSITION
RECOVERY (SELF RESTORATION TO ORIGINAL CONDITION)
RELAXATION SHRINKAGE
RELAXATION TIME (MECHANICAL)
RESIDUAL SHRINKAGE
RESIDUAL STRAIN
SAG
SHEAR (MODE OF DEFORMATION)
SHRINKAGE
STRAIN ENERGY
STRAIN GAUGES
STRAIN HARDENING
STRAIN RATE
STRAINOMETER
STRENGTH
STRENGTH ELONGATION TESTING
STRENGTH OF MATERIALS
STRESS
STRESS RELAXATION
STRESS STRAIN CURVES
STRESS STRAIN PROPERTIES (TENSILE)
STRETCH
(CONT.)

RT STRETCHING (YARNS)
TENSION
UNIAXIAL STRESS
WET RELAXATION
YIELD (MECHANICAL)
YIELD POINT

STRAIN ENERGY
UF ENERGY OF DEFORMATION
NT BENDING ENERGY
TORSIONAL ENERGY
BT ENERGY
RT BREAKING ENERGY
ENERGY ABSORPTION
INTERNAL ENERGY
KINETIC ENERGY
POTENTIAL ENERGY
STRAIN
TOUGHNESS

STRAIN GAUGES
BT MEASURING INSTRUMENTS
RT EXPERIMENTAL STRESS ANALYSIS
STRAIN
STRAINOMETER

STRAIN HARDENING
RT PLASTIC STRAIN
PLASTICITY
STRAIN
YIELD POINT

STRAIN RATE
BT RATE
RT DYNAMIC MODULUS
GAUGE LENGTH
IMPACT STRENGTH
RATE PROCESSES
SHOCK RESISTANCE
STRAIN
STRESS PROPAGATION
STRESS STRAIN CURVES
STRESS STRAIN PROPERTIES
STRESS STRAIN PROPERTIES (TENSILE)
TENSILE TESTING

STRAIN WAVES
RT DYNAMIC MODULUS
IMPULSES
LONGITUDINAL VIBRATIONS
PULSE PROPAGATION METER
SNAP BACK
SONIC VELOCITY
STRESS PROPAGATION

STRAINOMETER
BT MEASURING INSTRUMENTS
RT EXPERIMENTAL STRESS ANALYSIS

STELOMETER
STRAIN
STRAIN GAUGES

STRANDS
BT FIBER ASSEMBLIES
RT BUNDLES
CORDAGE
CORDS
FISHING LINES
HAWSERS
ROPES
ROVINGS
SLIVERS
SLUBBINGS
STRING
YARNS

STRAPPING
RT BALES
CORDAGE
NARROW FABRICS
PACKAGES
RIBBON
STRAPPING MACHINES
TAPE
TAPING
WEBS

STRAPPING MACHINES
RT BAGGING MACHINES
BALES
BALING MACHINES
PACKAGES
PACKAGING EQUIPMENT
STRAPPING

STREAKINESS
UF STRIPINESS
NT WEFT STREAKS
BT DEFECTS
DYEING DEFECTS
FABRIC DEFECTS
RT BARRE
BRONZING
COLORFASTNESS
CRACKS (FABRIC)
DYEING
FABRIC INSPECTION
FOGMARKING
LEVELNESS (DYEING)
MISPICKS
REED MARKS
SKITTERINESS
SPECKINESS
STARTING MARKS
TIPPINESS

STREAM SANITATION
USE WATER POLLUTION

STRENGTH
NT BEARING STRENGTH
BIAXIAL STRENGTH
BONDING STRENGTH
BREAKING STRENGTH
BUNDLE STRENGTH
BURSTING STRENGTH
COMPRESSIVE STRENGTH
COUNT STRENGTH PRODUCT
CREEP RUPTURE STRENGTH
CREEP STRENGTH
CUTTING RESISTANCE
DRY STRENGTH
FABRIC STRENGTH
FIBER STRENGTH
GRAB STRENGTH
IMPACT STRENGTH
KNOT STRENGTH
LEA STRENGTH PRODUCT
LOOP STRENGTH
PRESSLEY STRENGTH
RAVEL STRIP STRENGTH
SHEAR STRENGTH
SHOCK RESISTANCE
TEAR STRENGTH
TENACITY
UNIAXIAL STRENGTH
WET STRENGTH
YARN STRENGTH
BT STRESS STRAIN PROPERTIES
STRESS STRAIN PROPERTIES (TENSILE)
RT BREAKING
BREAKING LENGTH
ELASTICITY
END BREAKAGES
FORCE
FRACTURING
KNOT EFFICIENCY
LOOP EFFICIENCY
MOISTURE STRENGTH CURVES
PLASTICITY
PLASTICOVISCOSITY
PUNCTURING
RUPTURE
STRAIN
STRENGTH EFFICIENCY
STRENGTH ELONGATION TESTING
STRENGTH OF MATERIALS
STRENGTH TESTING
STRESS
STRESS HISTORY
STRESS STRAIN CURVES
TENDERING
TESTING EQUIPMENT
VISCOELASTICITY
VISCOSITY
YARN STRENGTH EFFICIENCY

STRENGTH AT RUPTURE
USE BREAKING STRENGTH

STRENGTH EFFICIENCY
NT YARN STRENGTH EFFICIENCY
BT EFFICIENCY (STRUCTURAL)
STRESS STRAIN PROPERTIES (TENSILE)
RT BINDER EFFICIENCY
BREAKING LENGTH
BREAKING STRENGTH
BUNDLE EFFICIENCY
BUNDLE STRENGTH
COUNT STRENGTH PRODUCT
FIBER SLIPPAGE
KNOT EFFICIENCY
LEA STRENGTH PRODUCT
LOOP EFFICIENCY
LOOP STRENGTH
SEAM EFFICIENCY
STRENGTH
STRENGTH ELONGATION TESTING
STRENGTH OF MATERIALS
TENACITY
WEAK SPOTS
YARN BUNDLE COHESION
YARN PROPERTIES
YARN STRENGTH

STRENGTH ELONGATION TESTING
UF STRESS STRAIN TESTING
BT TESTING
RT BEARING STRENGTH
BONDING STRENGTH
BREAKING STRENGTH
BUNDLE STRENGTH
COMPRESSIVE STRENGTH
COUNT STRENGTH PRODUCT
CREEP RUPTURE STRENGTH
DRY STRENGTH
FABRIC STRENGTH
FIBER STRENGTH
GAUGE LENGTH
GRAB STRENGTH
IMPACT STRENGTH
KNOT EFFICIENCY
KNOT STRENGTH
LEA STRENGTH PRODUCT
LOOP EFFICIENCY
LOOP STRENGTH
MECHANICAL PROPERTIES
MOISTURE STRENGTH CURVES
PRESSLEY STRENGTH
RAVEL STRIP STRENGTH
RUPTURE
SHEAR STRENGTH
STRAIN
STRENGTH
STRENGTH EFFICIENCY
STRENGTH OF MATERIALS
STRENGTH TESTING
STRESS
STRESS STRAIN CURVES
TEAR STRENGTH
TESTING EQUIPMENT
WET STRENGTH
YARN STRENGTH

STRENGTH OF MATERIALS
RT APPLIED ELASTICITY
BIAXIAL STRESS
BREAKING ELONGATION
ELASTICITY
IMPACT
KNOT EFFICIENCY
LOOP EFFICIENCY
MACHINE DESIGN
PLASTICITY
PRESSLEY TEST
RHEOLOGY
STRENGTH
STRENGTH EFFICIENCY
STRENGTH ELONGATION TESTING
STRESS ANALYSIS
STRESS HISTORY
STRESS OPTICAL COEFFICIENT
STRUCTURAL MECHANICS
VISCOELASTICITY

STRENGTH TESTING
UF STRENGTH TESTS
NT BURST TESTING
COMPRESSION TESTING
IMPACT TESTING
TEAR TESTING
TENSILE TESTING
BT TESTING
RT ABRASION TESTING
BEARING STRENGTH
BIAXIAL STRENGTH
BIAXIAL STRESS
BUNDLE STRENGTH
COMPRESSIVE STRENGTH
FABRIC STRENGTH
LOOP EFFICIENCY
LOOP STRENGTH
STRENGTH
STRENGTH ELONGATION TESTING
TESTING EQUIPMENT

STRENGTH TESTS
USE STRENGTH TESTING

STRESS
UF STRESSES
NT BIAXIAL STRESS
COMPRESSIVE STRESS
HOOP STRESS
HYDROSTATIC STRESS
SHEAR STRESS
TENSILE STRESS
UNIAXIAL STRESS
RT BEARING STRENGTH
BUNDLE STRENGTH
COLD DRAWING
COMPLIANCE
COMPRESSIVE STRENGTH
DELAYED RECOVERY
DRAWING (FILAMENT)
DRY RELAXATION
ELASTIC MODULUS (TENSILE)
ELASTIC STRAIN
EQUILIBRIUM (PHYSICAL)
EXPERIMENTAL STRESS ANALYSIS
FABRIC RELAXATION
FIBER STRENGTH
FORCE
HOT DRAWING
LOADING
LOOP EFFICIENCY
LOOP STRENGTH
PERMANENT SET
RELAXATION TIME (MECHANICAL)
STRAIN

RT STRENGTH
STRENGTH ELONGATION TESTING
STRESS ANALYSIS
STRESS OPTICAL COEFFICIENT
STRESS RELAXATION
STRESS STRAIN CURVES
TENSION
WET RELAXATION
YIELD POINT

STRESS (TRANSVERSE)
USE TRANSVERSE FORCE

STRESS ANALYSIS
NT EXPERIMENTAL STRESS ANALYSIS
RT ANALYZING
BIAXIAL STRESS
CHEMICAL ANALYSIS
COMPRESSIVE STRESS
MECHANICS
PHYSICAL ANALYSIS
SHEAR STRESS
STATISTICAL ANALYSIS
STRENGTH OF MATERIALS
STRESS
STRESS CONCENTRATION
STRESS DISTRIBUTION
STRESS HISTORY
STRESS OPTICAL COEFFICIENT
STRUCTURAL ANALYSIS
TENSILE STRESS
THERMAL ANALYSIS
X RAY ANALYSIS

STRESS CONCENTRATION
RT ABRASION
BIAXIAL STRESS
CRACK PROPAGATION
CUTTING RESISTANCE
EXPERIMENTAL STRESS ANALYSIS
FIBER BREAKAGE
IMPACT PENETRATION TESTING
KNOT EFFICIENCY
KNOT STRENGTH
LOOP STRENGTH
PENETRATION
PICKING (SNAGGING)
PILL WEAR OFF
PUNCTURING
RUPTURE
SHEAR (MODE OF DEFORMATION)
SNAGGING
SNAGS
STRESS DISTRIBUTION
STRETCH BREAKING
TEAR STRENGTH
UNIAXIAL STRESS

STRESS DISTRIBUTION
RT BIAXIAL STRESS
PRESSURE DISTRIBUTION
STRESS ANALYSIS
STRESS CONCENTRATION
TEMPERATURE DISTRIBUTION

STRESS HISTORY
RT EQUILIBRIUM (PHYSICAL)
HISTORY
MECHANICAL CONDITIONING
MECHANICAL HISTORY
STRENGTH
STRENGTH OF MATERIALS
STRESS ANALYSIS
STRESS OPTICAL COEFFICIENT
STRESS OPTICAL PROPERTIES
STRESS PROPAGATION
STRESS RELAXATION

STRESS OPTICAL COEFFICIENT
BT STRESS OPTICAL PROPERTIES
RT BIREFRINGENCE
GLASS RUBBER TRANSITION
GLASS TRANSITION TEMPERATURE
MELTING POINT
MOLECULAR CHAIN MOVEMENT
SECONDARY TRANSITION
TEMPERATURES
STRENGTH OF MATERIALS
STRESS ANALYSIS
STRESS HISTORY
SUBGROUP MOVEMENT (MOLECULAR)
TRANSITIONS (POLYMERS)
X RAY ANALYSIS

STRESS OPTICAL PROPERTIES
NT BIREFRINGENCE
STRESS OPTICAL COEFFICIENT
BT OPTICAL PROPERTIES
PHYSICAL PROPERTIES (EXCLUDING
MECHANICAL)
STRESS STRAIN PROPERTIES
RT ACOUSTIC PROPERTIES
AESTHETIC PROPERTIES
BIOCHEMICAL PROPERTIES
CHEMICAL PROPERTIES
ELECTRICAL PROPERTIES
EXPERIMENTAL STRESS ANALYSIS
FABRIC PROPERTIES

RT FABRIC PROPERTIES (MECHANICAL)
FIBER PROPERTIES
MECHANICAL DETERIORATION
PROPERTIES
OPTICS
ORIENTATION BIREFRINGENCE
PHOTOELASTICITY
PHYSICAL CHEMICAL PROPERTIES
POLARIZED LIGHT
REFRACTIVE INDEX
STRESS HISTORY
STRUCTURAL PROPERTIES
SURFACE PROPERTIES
(MECHANICAL)
THERMAL PROPERTIES
TRANSFER PROPERTIES

STRESS PROPAGATION
RT BUCKLING
CRACK PROPAGATION
DYNAMIC MODULUS
IMPULSES
LONGITUDINAL VIBRATIONS
SNAP BACK
SONIC VELOCITY
STRAIN RATE
STRAIN WAVES
STRESS ANALYSIS
STRESS CONCENTRATION
STRESS HISTORY
TRANSVERSE VIBRATIONS
WAVE LENGTH

STRESS RELAXATION
BT DRY FINISHING
RELAXATION
STRESS STRAIN PROPERTIES
STRESS STRAIN PROPERTIES
(TENSILE)
WET FINISHING
RT BENDING RECOVERY
BOARDING
CONDITIONING
CRABBING
CREEP RECOVERY
CRYSTALLIZATION
DELAYED RECOVERY
DRAWING (FILAMENT)
DRY RELAXATION
FABRIC RELAXATION
FABRIC SHRINKAGE
HEAT SETTING (SYNTHETICS)
LINEAR VISCOELASTICITY
LONDON SHRINKING
NONLINEAR VISCOELASTICITY
PERMANENT DEFORMATION
PERMANENT SET
RECOVERY (SELF RESTORATION TO
ORIGINAL CONDITION)
RELAXATION
RELAXATION SHRINKAGE
RELAXATION SPECTRUM
RELAXATION TIME (MECHANICAL)
RESIDUAL STRESS
SPONGING
STRAIN
STRESS
STRESS HISTORY
SUPERCONTRACTION
TWIST SETTERS
VISCOELASTICITY
WET RELAXATION

STRESS STRAIN BEHAVIOR
USE STRESS STRAIN CURVES

STRESS STRAIN CURVES
UF LOAD ELONGATION CURVES
LOAD EXTENSION CURVES
STRESS STRAIN BEHAVIOR
TENACITY ELONGATION CURVES
BT STRESS STRAIN PROPERTIES
STRESS STRAIN PROPERTIES
(TENSILE)
RT BEARING STRENGTH
BREAKING ELONGATION
BREAKING STRENGTH
BUNDLE STRENGTH
COLD DRAWING
COMPLIANCE
COMPRESSIVE STRENGTH
CRIMP INTERCHANGE
DELAYED RECOVERY
DRAWING (FILAMENT)
DYNAMIC MODULUS
ELASTIC MODULUS (TENSILE)
ELASTIC PERFORMANCE
ELASTIC RECOVERY
ELASTIC STRAIN
ELASTICITY
ENERGY OF RETRACTION
EXTENSIBILITY
FIBER BREAKAGE
FIBER STRENGTH
FILLINGWISE STRETCH
GAUGE LENGTH
HOOKES LAW
LOOP EFFICIENCY

RT LOOP STRENGTH
MOISTURE STRENGTH CURVES
NECKING (FILAMENT)
PERMANENT SET
PLASTIC FLOW
PLASTIC STRAIN
RESILIENCE
SECANT MODULUS
STRAIN
STRAIN RATE
STRENGTH
STRENGTH ELONGATION TESTING
STRESS
STRETCH
TANGENT MODULUS
TENACITY
TOUGHNESS
UNIAXIAL STRESS
VISCOELASTICITY
WARPWISE STRETCH
WORK RECOVERY
YARN STRENGTH
YIELD POINT

STRESS STRAIN PROPERTIES
UF RHEOLOGICAL PROPERTIES
NT ANISOTROPY (STRESS STRAIN)
BREAKING ELONGATION
BREAKING ENERGY
BRITTLENESS
BULKING
BULK MODULUS
COEFFICIENT OF CROSS VISCOSITY
COEFFICIENT OF VISCOSITY
COMPLIANCE
COMPRESSIBILITY
COMPRESSIVE MODULUS
COMPRESSIVE STRENGTH
CREEP
CREEP RECOVERY
CUTTING RESISTANCE
DILATANCY
DUCTILITY
DYNAMIC MODULUS
ELASTIC MODULUS (TENSILE)
ELASTIC RECOVERY
ELASTIC STRAIN
ELASTICITY
ENERGY ABSORPTION
ENERGY OF RETRACTION
FIRMOVISCOSITY
HARDNESS
HYSTERESIS (MECHANICAL)
IMAGINARY COMPONENT (DYNAMIC
MODULUS)
INITIAL MODULUS
KNOT EFFICIENCY
KNOT STRENGTH
LOOP EFFICIENCY
LOOP STRENGTH
MODULUS
MODULUS OF CROSS ELASTICITY
PERMANENT SET
PLASTIC FLOW
PLASTICITY
PLASTICOELASTICITY
REAL COMPONENT (DYNAMIC
MODULUS)
RECOVERY (SELF RESTORATION TO
ORIGINAL CONDITION)
RELAXATION TIME (MECHANICAL)
RESILIENCE
SECANT MODULUS
SEEPAGE (RHEOLOGY)
SHEAR MODULUS
SHEAR RESISTANCE
SHEAR STRENGTH
SHEAR STRESS
SHOCK RESISTANCE
STIFFNESS
STRENGTH
STRESS RELAXATION
STRESS STRAIN CURVES
STRESS STRAIN PROPERTIES
(COMPRESSIVE)
STRESS STRAIN PROPERTIES
(SHEAR)
STRESS STRAIN PROPERTIES
(TENSILE)
STRETCH
TANGENT MODULUS
TEAR STRENGTH
TENACITY
THIXOTROPY
TOUGHNESS
VISCOELASTICITY
VISCOSITY
VISCOUS FLOW
WORK RECOVERY
YIELD POINT
BT MECHANICAL PROPERTIES
RT AESTHETIC PROPERTIES
BIOCHEMICAL PROPERTIES
CHEMICAL PROPERTIES
COLD DRAWING
DRAWING (FILAMENT)
FABRIC PROPERTIES
FABRIC PROPERTIES (MECHANICAL)
(CONT.)

RT FIBER PROPERTIES
FREE VOLUME
LINEAR VISCOELASTICITY
MECHANICAL DETERIORATION PROPERTIES
NONLINEAR VISCOELASTICITY
PENETRATION
PHYSICAL CHEMICAL PROPERTIES
PHYSICAL PROPERTIES (EXCLUDING MECHANICAL)
RELAXATION SPECTRUM
STRAIN
STRAIN RATE
STRESS OPTICAL PROPERTIES
SURFACE PROPERTIES (MECHANICAL)
SURFACE PROPERTIES (PHYSICAL CHEMICAL)
TRANSFER PROPERTIES
YARN PROPERTIES

STRESS STRAIN PROPERTIES (COMPRESSIVE)
NT BUCKLING
COMPRESSIBILITY
COMPRESSIVE MODULUS
COMPRESSIVE STRENGTH
HARDNESS
BT STRESS STRAIN PROPERTIES
RT CRUSH RESISTANCE
STRESS STRAIN PROPERTIES (SHEAR)
STRESS STRAIN PROPERTIES (TENSILE)

STRESS STRAIN PROPERTIES (SHEAR)
NT CUTTING RESISTANCE
SHEAR (MODE OF DEFORMATION)
SHEAR MODULUS
SHEAR STRENGTH
SHEAR STRESS
BT STRESS STRAIN PROPERTIES
RT LOOP EFFICIENCY
STRESS STRAIN PROPERTIES (COMPRESSIVE)
STRESS STRAIN PROPERTIES (TENSILE)
TORSIONAL BEHAVIOR

STRESS STRAIN PROPERTIES (TENSILE)
UF TENSILE PROPERTIES
NT BREAKING ELONGATION
BREAKING ENERGY
BREAKING LENGTH
BREAKING STRENGTH
BRITTLENESS
BUNDLE EFFICIENCY
BUNDLE STRENGTH
BURSTING STRENGTH
COMPLIANCE
COUNT STRENGTH PRODUCT
CREEP
CREEP RECOVERY
CREEP RUPTURE STRENGTH
CREEP STRENGTH
DELAYED RECOVERY
DUCTILITY
DYNAMIC MODULUS
ELASTIC MODULUS (TENSILE)
ELASTIC RECOVERY
ELASTIC STRAIN
ELASTICITY
ENERGY ABSORPTION
ENERGY OF RETRACTION
INITIAL MODULUS
LEA STRENGTH PRODUCT
MODULUS
PERMANENT SET
PLASTIC FLOW
PLASTICITY
RECOVERY (SELF RESTORATION TO ORIGINAL CONDITION)
RESILIENCE
SECANT MODULUS
STIFFNESS
STRENGTH
STRENGTH EFFICIENCY
STRESS RELAXATION
STRESS STRAIN CURVES
TANGENT MODULUS
TENACITY
TOUGHNESS
VISCOUS FLOW
WET STRENGTH
WORK RECOVERY
YIELD POINT
BT STRESS STRAIN PROPERTIES
RT DRAWING TENSION (FILAMENT)
STRAIN RATE
STRESS STRAIN PROPERTIES (COMPRESSIVE)
STRESS STRAIN PROPERTIES (SHEAR)

STRESS STRAIN TESTING
USE STRENGTH ELONGATION TESTING

STRESSES
USE STRESS

STRETCH
NT FILLINGWISE STRETCH
TWO WAY STRETCH
WARPWISE STRETCH
BT MODES OF DEFORMATION
STRESS STRAIN PROPERTIES
RT BAGGINESS
BIAXIAL STRESS
BREAKING ELONGATION
BULK
COCKLE (YARN)
CREPING
CRIMP
DIMENSIONAL STABILITY
EDGE CRIMPING
ELASTIC STRAIN
ELASTICITY
EXTENSIBILITY
FABRIC PROPERTIES
FABRIC PROPERTIES (MECHANICAL)
FALSE TWIST YARNS
FIT
GROWTH (FABRIC)
KNITTED FABRICS
LOOP DISTORTION
POWER (FABRIC)
RECOVERY (SELF RESTORATION TO ORIGINAL CONDITION)
STRAIN
STRESS STRAIN CURVES
STRETCH FABRICS
STRETCH HOSIERY
STRETCH KNITTED FABRICS
STRETCH WOVEN FABRICS
STRETCH YARNS (FILAMENT)
STRETCHING (YARNS)
TEXTURE
TEXTURED YARNS (FILAMENT)
TEXTURING
UNIAXIAL STRESS

STRETCH BREAKING
RT BREAKING LENGTH
COLD DRAWING
CONVERTERS (TOW)
DRAFT CUT
FIBER BREAKAGE
FIBER CUTTING
FIBER STRENGTH
GREENFIELD CONVERTER
PERLOK SYSTEM
SACO LOWELL DIRECT SPINNER (TN)
SNAP BACK
STAINS DIRECT SPINNER
STRESS CONCENTRATION
TENSION
TOHO DIRECT SPINNER (TN)
TOW
TOW CONVERSION
TOW TO TOP CONVERSION
TOW TO YARN CONVERSION
TURBOSTAPLER

STRETCH FABRICS
UF ELASTIC FABRICS
NT STRETCH KNITTED FABRICS
STRETCH WOVEN FABRICS
BT FABRICS
RT ANTHROPOMETRIC KINEMATICS
ANTHROPOMETRIC MEASUREMENTS
BAGGINESS
BULKED YARNS
COMFORT
CORE SPUN YARNS
CREPING
CRIMPED YARNS
DIMENSIONAL STABILITY
FILLINGWISE STRETCH
FIT
FORM PERSUASIVE GARMENTS
FOUNDATION GARMENTS
GROWTH (FABRIC)
LYCRA (TN)
POWER (FABRIC)
SKI PANTS
SLACK MERCERIZING
SPANDEX FIBERS
STRETCH
STRETCH HOSIERY
STRETCH YARNS (FILAMENT)
STRETCH YARNS (SPUN)
TEXTURED YARNS (FILAMENT)
TEXTURING
TORSIONAL BUCKLING
TWIST LIVELINESS
TWO WAY STRETCH
WARPWISE STRETCH
WILDNESS

STRETCH HOSE
USE STRETCH HOSIERY

STRETCH HOSIERY
UF STRETCH HOSE
BT HOSIERY
RT BOARDING
FASHIONING
FULLY FASHIONED HOSIERY
HALF HOSE
HEAT SETTING (SYNTHETICS)
KNITTING
PRESSING
SOCKS
STOCKINGS
STRETCH
STRETCH FABRICS
STRETCH KNITTED FABRICS
STRETCH YARNS (FILAMENT)
SUPPORT HOSIERY

STRETCH KNITTED FABRICS
BT **KNITTED FABRICS**
STRETCH FABRICS
RT **BULKED YARNS**
COMFORT
CRIMPED YARNS
ELASTICITY
EXTENSIBILITY
FIT
FOUNDATION GARMENTS
POWER (FABRIC)
SPANDEX FIBERS
STRETCH
STRETCH HOSIERY
STRETCH WOVEN FABRICS
STRETCH YARNS (FILAMENT)
TEXTURED YARNS (FILAMENT)

STRETCH **PUCKERS**
RT **COCKLING**
GARMENT MANUFACTURE
PUCKERING
RIPPLES
SEAM PUCKER
STRETCHING (SEAMING)

STRETCH **ROLLS**
RT **DESIZING**
SIZE BOXES
SIZES (SLASHING)
SLASHERS
SLASHERS BEAMS
SLASHING
SPLIT RODS
SQUEEZING
WARP BEAMS
WARP PREPARATION

STRETCH WOVEN FABRICS
BT STRETCH FABRICS
WOVEN FABRICS
RT BAGGINESS
BULKED YARNS
COMFORT
CRIMPED YARNS
ELASTICITY
EXTENSIBILITY
FILLINGWISE STRETCH
FIT
GROWTH (FABRIC)
LEOTARDS
POWER (FABRIC)
SKI PANTS
SPANDEX FIBERS
SPORTSWEAR
STRETCH
STRETCH KNITTED FABRICS
STRETCH YARNS (FILAMENT)
TEXTURED YARNS (FILAMENT)
TWO WAY STRETCH
WARPWISE STRETCH
WEAVING

STRETCH YARNS (FILAMENT)
NT AGILON (TN)
BICOMPONENT FIBER YARNS (FILAMENT)
CRINKLE TYPE YARNS
DYNALOFT (TN)
EDGE CRIMPED YARNS
FALSE TWIST YARNS
FLUFLON (TN)
HELANCA (TN)
BT FILAMENT YARNS
TEXTURED YARNS (FILAMENT)
YARNS
RT AIR JET TEXTURING

BULKED YARNS
CORE SPUN YARNS
CREPING
CRIMP
CRIMPED YARNS
FALSE TWIST
FALSE TWIST SPINDLES
FILLINGWISE STRETCH
HEAT SETTING (SYNTHETICS)
HELICAL CRIMP
POWER (FABRIC)
SELF CRIMPING YARNS
SPANDEX FIBERS
STRETCH
STRETCH FABRICS
STRETCH HOSIERY
STRETCH KNITTED FABRICS

RT STRETCH WOVEN FABRICS
STRETCH YARNS (SPUN)
STUFFER BOXES
TEXTURING
TORSIONAL BUCKLING
TWIST
TWIST LIVELINESS
TWISTING
WARPWISE STRETCH
WILDNESS
YARN CRIMP
Z CRIMP

STRETCH YARNS (SPUN)
BT TEXTURED SPUN YARNS
RT BICOMPONENT FIBER YARNS (FILAMENT)
BICOMPONENT FIBER YARNS (STAPLE)
CORE SPUN YARNS
FALSE TWIST YARNS
FALSE TWISTING (TEXTURING)
FORM PERSUASIVE GARMENTS
HILOW BULKED YARNS
SLACK MERCERIZING
SPANDEX FIBERS
STRETCH FABRICS
STRETCH YARNS (FILAMENT)

STRETCHING (SEAMING)
BT FABRIC DISTORTION (TAILORING)
RT SEAMING
SEAMS
SEWING
STRETCH PUCKERS

STRETCHING (CORDS)
USE STRETCHING (YARNS)

STRETCHING (YARNS)
UF STRETCHING (CORDS)
RT BREAKING ELONGATION
HEAT SETTING (SYNTHETICS)
HEAT TREATMENT
STRAIN
STRETCH

STRICT GOOD MIDDLING (COTTON GRADE)
BT COTTON GRADE
GRADE (FIBERS)

STRICT GOOD ORDINARY (COTTON GRADE)
BT COTTON GRADE
GRADE (FIBERS)
RT GOOD ORDINARY (COTTON GRADE)

STRICT LOW MIDDLING (COTTON GRADE)
BT COTTON GRADE
GRADE (FIBERS)

STRICT MIDDLING (COTTON GRADE)
BT COTTON GRADE
GRADE (FIBERS)

STRING
BT CORDAGE
RT FISHING LINES
STRANDS
TWINE

STRING WARP MACHINES
BT LACE MACHINES
RT BAR WARP MACHINES
BARMEN MACHINES
BOBBINET MACHINES
CURTAIN MACHINES
DOUBLE LOCKER MACHINES
GO THROUGH MACHINES
LEAVERS MACHINES
MECHLIN MACHINES
ROLLING LOCKER MACHINES
SIVAL MACHINES
WARP LACE MACHINES

STRIP DYEING
BT DYEING (BY MATERIAL ASSEMBLY)
RT GARMENT DYEING
KNITTED FABRICS
SWEATERS

STRIPES
RT FABRIC DESIGN
RIB WEAVES
TWILLS
YARN DYED FABRICS

STRIPINESS
USE STREAKINESS

STRIPING UNIT (KNITTING)
USE YARN CHANGING UNIT (KNITTING)

STRIPPED SEAMS
BT SEAMS
RT FRENCH SEAMS

STRIPPERS
NT ANGLE STRIPPERS
BT CARD ROLLS
RT AERODYNAMIC CARDING
BOBBIN STRIPPING MACHINES
BREAKERS
CARD CLOTHING
CARD CYLINDERS
CARD GRINDING
CARD STRIPS
CARDING
CARDING ACTION
CARDING EFFICIENCY
CARDS
COTTON CARDS
DOFFERS (CARD)
DOFFING (OF A CARD)
FANCY (NOUN)
FETTLING
FINISHER PART
LICKER IN
STRIPPING ACTION
WOOLEN CARDING
WOOLEN CARDS
WORKERS
WORSTED CARDING
WORSTED CARDS

STRIPPING (CARDING)
USE FETTLING

STRIPPING (COLOR)
RT BATHS
BLEACHING
DECOLORIZING
DYEING
HYDROGEN PEROXIDE
PEROXIDES
REDUCING AGENTS
REDYEING
SCOURING
STRIPPING AGENTS

STRIPPING ACTION
RT AERODYNAMIC CARDING
ANGLE STRIPPERS
BRUSHING ACTION
CARD CLOTHING
CARD CYLINDERS
CARD FLATS
CARD WEBS
CARDING ACTION
CARDING EFFICIENCY
CARDS
COTTON CARDING
DOFFERS (CARD)
DOFFING (OF A CARD)
FANCY (NOUN)
FETTLING
FETTLING WASTE
LICKER IN
SCRIBBLING
STRIPPERS
WOOLEN CARDING
WORKERS
WORSTED CARDING

STRIPPING AGENTS
BT DYEING AUXILIARIES
RT POLY (VINYL PYRROLIDONE)
STRIPPING (COLOR)

STROBOSCOPES
BT ELECTRONIC INSTRUMENTS
RT BALLOON DYNAMICS
KINEMATICS
OPTICAL SCANNING
PHOTOGRAPHY
PROCESS CONTROL

STRUCTURAL ANALYSIS
RT ANALYZING
ANISOTROPIC MATERIALS
CHEMICAL ANALYSIS
DESIGN
FABRIC DESIGN
FABRIC GEOMETRY
FABRIC PROPERTIES (STRUCTURAL)
FABRIC STRUCTURE
FIBER GEOMETRY
FINE STRUCTURE (FIBERS)
ISOTROPIC MATERIALS
MATHEMATICAL ANALYSIS
MOLECULAR STRUCTURE
PHYSICAL ANALYSIS
QUALITATIVE ANALYSIS
QUANTITATIVE ANALYSIS
SPECTRUM ANALYSIS
STATISTICAL ANALYSIS
STRESS ANALYSIS
STRUCTURAL MECHANICS
THERMAL ANALYSIS
X RAY ANALYSIS
YARN GEOMETRY
YARN STRUCTURE

STRUCTURAL EFFICIENCY
USE EFFICIENCY (STRUCTURAL)

STRUCTURAL MECHANICS
BT MECHANICS
RT ANISOTROPIC MATERIALS
DIFFERENTIAL GEOMETRY
DYNAMICS
FABRIC GEOMETRY
FABRIC STRUCTURE
ISOTROPIC MATERIALS
KINEMATICS
STATICS
STRENGTH OF MATERIALS
STRUCTURAL ANALYSIS
YARN GEOMETRY
YARN STRUCTURE

STRUCTURAL PROPERTIES
UF GEOMETRICAL PROPERTIES
NT ANISOTROPY (STRESS STRAIN)
BULK
BULK DENSITY
CRIMP
DENSITY
DIMENSIONS
EFFICIENCY (STRUCTURAL)
MATRIX FREEDOM
ORIENTATION
PACKING FACTOR
PARALLELIZATION
POROSITY
ROUGHNESS
SMOOTHNESS
TEXTURE
BT PHYSICAL PROPERTIES (EXCLUDING MECHANICAL)
RT ACOUSTIC PROPERTIES
AESTHETIC PROPERTIES
ANISOTROPIC MATERIALS
BIOCHEMICAL PROPERTIES
CHEMICAL PROPERTIES
ELECTRICAL PROPERTIES
FABRIC GEOMETRY
FABRIC PROPERTIES
FABRIC PROPERTIES (PHYSICAL EXCLUDING MECHANICAL)
FABRIC PROPERTIES (STRUCTURAL)
FABRIC STRUCTURE
FIBER ASSEMBLIES
FIBER PROPERTIES
FINE STRUCTURE (FIBERS)
IRREGULARITY
ISOTROPIC MATERIALS
MECHANICAL PROPERTIES
MOLECULAR STRUCTURE
OPTICAL PROPERTIES
PHYSICAL CHEMICAL PROPERTIES
STRESS OPTICAL PROPERTIES
STRUCTURE OF MATERIALS
STRUCTURES
THERMAL PROPERTIES
TRANSFER PROPERTIES
YARN GEOMETRY
YARN STRUCTURE

STRUCTURE OF MATERIALS
RT CRYSTALLINITY
FIBRILLAR STRUCTURE
FINE STRUCTURE (FIBERS)
STRUCTURAL PROPERTIES

STRUCTURES
RT ASSEMBLIES
CONFIGURATION
EFFICIENCY (STRUCTURAL)
FABRIC GEOMETRY
FABRIC STRUCTURE
FIBER ASSEMBLIES
FINE STRUCTURE (FIBERS)
MOLECULAR STRUCTURE
NONWOVEN FABRIC STRUCTURE
STRUCTURAL ANALYSIS
STRUCTURAL PROPERTIES
YARN GEOMETRY
YARN STRUCTURE

STUDENTS DISTRIBUTION
USE T DISTRIBUTION

STUFFER BOX CRIMPED YARNS
BT CRIMPED YARNS
FILAMENT YARNS
TEXTURED YARNS (FILAMENT)
RT AIR JET TEXTURED YARNS
BICOMPONENT FIBER YARNS (FILAMENT)
BICOMPONENT FIBER YARNS (STAPLE)
EDGE CRIMPED YARNS
FALSE TWIST YARNS
GEAR CRIMPED YARNS
STUFFER BOXES
STUFFER BOX CRIMPING

STUFFER BOX CRIMPERS
USE STUFFER BOXES

STUFFER BOX CRIMPING
BT CRIMPING
TEXTURING
RT AIR JET TEXTURING
BANLON (TN)
BICOMPONENT FIBER YARNS (FILAMENT)
BICOMPONENT FIBER YARNS (STAPLE)
BICOMPONENT SPINNING
BULKED YARNS
BULKING
CELACRIMP (TN)
CRIMPED YARNS
EDGE CRIMPED YARNS
EDGE CRIMPING
FALSE TWISTING (TEXTURING)
GEAR CRIMPING
HEAT SETTING (SYNTHETICS)
NYLOFT (TN)
SPUNIZE (TN)
STUFFER BOX CRIMPED YARNS
STUFFER BOXES
TEXTRALIZED YARNS (TN)
TEXTURED YARNS (FILAMENT)

STUFFER BOXES
UF STUFFER BOX CRIMPERS
BT CRIMPING MACHINES
RT AIR JET TEXTURING MACHINES
BANLON (TN)
BULKED YARNS
CRIMPING
EDGE CRIMPING MACHINES
HEAT SETTING (SYNTHETICS)
SPUNIZE (TN)
STRETCH YARNS (FILAMENT)
STUFFER BOX CRIMPED YARNS
STUFFER BOX CRIMPING

STUFFING
RT BATTING
COMPRESSIBILITY
FILLER
FLOCK
INDUSTRIAL FABRICS
INSULATION (THERMAL)
PACKING
PADDING (MATERIAL)
WADDINGS
WARMTH

STYLES
USE STYLING (APPAREL)

STYLING (APPAREL)
UF STYLES
BT GARMENT MANUFACTURE
RT APPAREL DESIGN
FASHION (APPAREL)
MANNEQUINS
MARKETING
MODELS (FASHION)
PRODUCT DESIGN
PRODUCTION
RESEARCH

STYMER (TN)
RT SIZES (SLASHING)

STYRENE
BT VINYL COMPOUNDS
RT GRAFTING
POLYSTYRENE

STYROFOAM (TN)
BT FOAMS
RT FOAM RUBBER
RIGID FOAMS
URETHANE FOAMS
VINYL FOAMS

SUBASSEMBLIES
USE ASSEMBLIES

SUBGROUP MOVEMENT (MOLECULAR)
RT BIREFRINGENCE
GLASS RUBBER TRANSITION
GLASS TRANSITION TEMPERATURE
MELTING POINT
MOLECULAR CHAIN MOVEMENT
SECONDARY TRANSITION TEMPERATURES
STRESS OPTICAL COEFFICIENT
TRANSITIONS (POLYMERS)
X RAY ANALYSIS

SUBJECTIVE MECHANICAL PROPERTIES
USE FABRIC PROPERTIES (AESTHETIC)

SUBLIMATION
RT DISTILLATION
EVAPORATION
IRONING FASTNESS
VAPOR PRESSURE

SUBSTANTIVITY
RT DYE AFFINITY
EXHAUSTION (WET PROCESS)
PREFERENTIAL ADSORPTION

SUBSTANTIVITY (DYES)
USE DYEABILITY

SUBTRACTION
BT COMPUTATIONS
RT ADDITION
DIVISION
MULTIPLICATION

SUBSTRATES
RT COATING (PROCESS)
FABRICS
NONWOVEN FABRICS

SUCCINIC ACID
BT CARBOXYLIC ACIDS
DIBASIC ACIDS
ORGANIC ACIDS
RT ADIPIC ACID
OXALIC ACID
SULFOSUCCINAMIDES
SULFOSUCCINIC ESTERS

SUCROSE
BT CARBOHYDRATES
SUGARS
RT GLUCOSE

SUCROSE ESTER SURFACTANTS
BT NONIONIC SURFACTANTS

SUCTION PRESSES
BT ROLLS
RT CALENDER ROLLS
COMPACTING
DRAFTING ROLLS
FEED ROLLS
HYDROSTATIC LOADING
MANGLES
MANGLING
MOISTURE CONTENT
NIP
PAPER MACHINES
PAPER MAKING
PERALTA ROLLS
PERFORATED CYLINDERS
PRESS SECTION
PRESSING
PRESSURE DROP
PRESSURE ROLLS
ROLL COVERINGS
SQUEEZE ROLLS
VACUUM
WET PICKUP

SUEDE
RT CORFAM (TN)
KID LEATHER
LEATHER
SUEDING
SUEDING MACHINES
SYNTHETIC LEATHER

SUEDING
BT DRY FINISHING
RT BRUSHING
NAPPING
SHEARING (FINISHING)
SUEDE
SUEDING MACHINES

SUEDING MACHINES
RT CHINCHILLA MACHINES
SUEDE
SUEDING

SUFFOLK WOOL
BT MEDIUM WOOL
WOOL

SUGARS
UF MONOSACCHARIDES
NT GLUCOSE
SUCROSE
BT CARBOHYDRATES

SUINT
RT GREASE
GREASE RECOVERY
GREASY WOOL
IMPURITIES
LAMBSWOOL
LANOLIN
LUBRICATION
PULLED WOOL
SCOURED WOOL
SCOURING
SCOURING TRAINS
SCOURING WASTE
SCOURING YIELD
SOAP SODA SCOURING
SOLVENT SCOURING
WOOL
WOOL GREASE
WOOL WAX

SUITS
BT GARMENTS
OUTERWEAR
RT COATS
DRESSCLOTHING
DRESSES
FROCKS
JACKETS
SACKS (DRESSES)

SULFATE SURFACTANTS
NT ALKYL ETHOXY SULFATES
ALKYL GLYCEROSULFATES
ALKYL SULFATES
ALKYLAMIDE SULFATES
ALKYLARYL ETHOXY SULFATES
SODIUM LAURYL SULFATE
SULFATED OILS
BT ANIONIC SURFACTANTS
RT ANIONIC SOFTENERS
SULFONATE SURFACTANTS
SULFUR COMPOUNDS

SULFATED CASTOR OIL
USE SULFORICINOLEATE

SULFATED GLYCERIDES
USE ALKYL GLYCEROSULFATES

SULFATED OILS
UF SULFONATED OILS
NT SULFATED OLEIC ESTERS
SULFATED TALLOW
SULFORICINOLEATE
BT SULFATE SURFACTANTS

SULFATED OLEFINS
BT ALKYL SULFATES

SULFATED OLEIC ESTERS
BT SULFATED OILS

SULFATED TALLOW
UF SULFONATED TALLOW
BT SULFATED OILS

SULFATES
NT AMMONIUM SULFATE
CERIUM SULFATE
CUPRIC SULFATE
GLAUBERS SALT
MAGNESIUM SULFATE
SODIUM LAURYL SULFATE
SODIUM SULFATE
ZINC SULFATE
BT SULFUR COMPOUNDS
RT BISULFATES
PERSULFATES
SULFIDES (INORGANIC)
SULFITES
SULFURIC ACID
THIOSULFATES

SULFIDES (INORGANIC)
UF DISULFIDES (INORGANIC)
POLYSULFIDES (INORGANIC)
NT CARBON DISULFIDE
SODIUM SULFIDE
BT SULFUR COMPOUNDS
RT DISULFIDE BONDS
PERSULFATES
SULFATES
SULFITES
SULFUR DYES
THIOSULFATES

SULFIDES (ORGANIC)
USE THIOETHERS

SULFINATES
USE SULFINIC ACIDS

SULFINIC ACIDS
UF SULFINATES
BT ORGANIC ACIDS
SULFUR COMPOUNDS
RT REDUCING AGENTS
SULFONIC ACIDS
THIOSULFINIC ACIDS

SULFITES
UF MONOETHANOLAMINE SULFITE
BT SULFUR COMPOUNDS
RT BISULFITES
PERSULFATES
SULFATES
SULFIDES (INORGANIC)
SULFUR DIOXIDE
THIOSULFATES

SULFIX A (TN)
USE SULFONIUM REACTANTS

SULFOBETAINE SURFACTANTS
BT ZWITTERIONIC SURFACTANTS
RT BETAINE SURFACTANTS

SULFOFATTY ACID ESTERS
UF ALPHASULFOMETHYL PALMITATE
BT SULFONATE SURFACTANTS
RT SULFOFATTY ACIDS

SULFOFATTY ACIDS
UF ALPHASULFOPALMITIC ACID
BT SULFONATE SURFACTANTS
RT SULFOFATTY ACID ESTERS

SULFOGLYCERIDES
USE ALKYL GLYCEROSULFATES

SULFONAMIDE DYES
UF LEVAFIX (TN)
BT REACTIVE DYES

SULFONATE SURFACTANTS
NT ACYL ESTER SULFONATES
ALKYL SULFONATES
ALKYLAMIDE SULFONATES
ALKYLARYL OXYETHYL SULFONATES
ALKYLARYL SULFONATES
ALKYLETHER SULFONATES
LIGNIN SULFONATES
NAPHTHALENE FORMALDEHYDE SULFONATES
SULFOFATTY ACID ESTERS
SULFOFATTY ACIDS
SULFOPOLYCARBOXYLIC ESTERS
SULFOSUCCINAMIDES
TETRAHYDRONAPHTHALENE SULFONATES
BT ANIONIC SURFACTANTS
RT ANIONIC SOFTENERS
SULFATE SURFACTANTS
SULFUR COMPOUNDS

SULFONATED OILS
USE SULFATED OILS

SULFONATED TALLOW
USE SULFATED TALLOW

SULFONATES
USE SULFONIC ACIDS

SULFONES
NT BIS (HYDROXYETHYL) SULFONE
DIVINYL SULFONE
BT SULFUR COMPOUNDS
REACTANTS
RT POLYSULFONES
SULFONIUM REACTANTS

SULFONIC ACIDS
UF SULFONATES
BT ORGANIC ACIDS
SULFUR COMPOUNDS
RT SULFINIC ACIDS
SULFONIC END GROUPS
SULFURIC ACID
SULTONES
THIOSULFONATES

SULFONIC END GROUPS
BT ACID END GROUPS
END GROUPS
RT CARBOXYL END GROUPS
DEGREE OF POLYMERIZATION
POLYMERS
SULFONIC ACIDS

SULFONIUM COMPOUNDS
NT SULFONIUM REACTANTS
BT ONIUM COMPOUNDS
SULFUR COMPOUNDS
RT QUATERNARY AMMONIUM COMPOUNDS
SULFONIUM SURFACTANTS

SULFONIUM REACTANTS
UF SULFIX A (TN)
BT REACTANTS
SULFONIUM COMPOUNDS
RT BIS (HYDROXYETHYL) SULFONE
DIVINYL SULFONE
SULFONES
SULFONIUM COMPOUNDS

SULFONIUM SURFACTANTS
BT ONIUM SURFACTANTS
RT PHOSPHONIUM SURFACTANTS
QUATERNARY AMMONIUM SURFACTANTS
SULFUR COMPOUNDS

SULFOPOLYCARBOXYLIC ESTERS
NT SULFOSUCCINIC ESTERS
TRIHEXYL SULFOCARBALLYLATE
BT SULFONATE SURFACTANTS

SULFORICINOLEATE
UF SULFATED CASTOR OIL
BT SULFATED OILS

SULFOSUCCINAMIDES
NT AEROSOL 18 (TN)
AEROSOL 22 (TN)
BT SULFONATE SURFACTANTS
RT SUCCINIC ACID
SULFOSUCCINIC ESTERS

SULFOSUCCINIC ESTERS
NT DIOCTYL SULFOSUCCINATE
BT SULFOPOLYCARBOXYLIC ESTERS
RT SUCCINIC ACID
SULFOSUCCINAMIDES

SULFUR COMPOUNDS
NT BISULFATES
BISULFITES
DISULFIDES (ORGANIC)
DITHIOCARBAMATES
DITHIONITES
ISOTHIOCYANATES
MERCAPTANS
PERSULFATES
PYROSULFITES
RHODANINE
SULFATES
SULFIDES (INORGANIC)
SULFINIC ACIDS
SULFITES
SULFONES
SULFONIC ACIDS
SULFONIUM COMPOUNDS
SULFONIUM REACTANTS
SULFUR DIOXIDE
SULFURIC ACID
SULTONES
THIOCYANATES
THIOETHERS
THIOSULFATES
THIOSULFINIC ACIDS
THIOSULFONATES
THIOUREA DIOXIDE
RT SULFATE SURFACTANTS
SULFONAMIDE DYES
SULFONATE SURFACTANTS
THIOUREA

SULFUR DIOXIDE
BT OXIDES
SULFUR COMPOUNDS
RT SULFITES

SULFUR DYEING
BT DYEING (BY DYE CLASSES)
RT ACID DYEING
AZOIC DYEING
BASIC DYEING
DIRECT DYEING
DISPERSE DYEING
DYES
METALLIZED DYEING
MORDANT DYEING
NEUTRAL DYEING
PIGMENT DYEING
REACTIVE DYEING
SODIUM SULFIDE
SULFUR DYES
VAT DYEING

SULFUR DYES
BT DYES (BY CHEMICAL CLASSES)
RT AZOIC DYES
DIRECT DYES
DISULFIDE BONDS
DISULFIDES (ORGANIC)
DYEING (BY DYE CLASSES)
DYES (BY FIBER CLASSES)
SULFIDES (INORGANIC)
SULFUR DYEING
VAT DYES

SULFURIC ACID
BT DIBASIC ACIDS
INORGANIC ACIDS
SULFUR COMPOUNDS
RT BISULFATES
CARBONIZING
DYEING AUXILIARIES
SODIUM SULFATE
SULFATES
SULFONIC ACIDS

SULTONES
BT SULFUR COMPOUNDS
RT ESTERS
REACTANTS
SULFONIC ACIDS

SULZER LOOMS (TN)
BT LOOMS
SHUTTLELESS LOOMS
RT AIR JET LOOMS
DRAPER SHUTTLELESS LOOMS (TN)
JET LOOMS
OUTSIDE FILLING SUPPLY
RAPIER LOOMS
SHUTTLELESS WEAVING
WATER JET LOOMS
WEAVING

SUMMARIES
RT COMPILATIONS
HISTORY
LITERATURE SURVEYS
REVIEWS
SURVEYS

SUN SUITS
BT GARMENTS
SPORTSWEAR
RT BATHING SUITS
BIKINIS

SUNLIGHT
BT DAYLIGHT
LIGHT
RT ARTIFICIAL DAYLIGHT
COLORFASTNESS
FADING
FLUORESCENT LIGHT
ILLUMINATION
INCANDESCENT LIGHT
LIGHTFASTNESS (COLOR)
LIGHTFASTNESS (OF FINISH)
MONOCHROMATIC LIGHT
POLARIZED LIGHT
RADIATION

SUNLIGHT TESTS
USE LIGHTFASTNESS TESTING

SUPERCONTRACTION
RT CONTRACTION
DISULFIDE INTERCHANGE
KERATIN
SHRINKAGE
STRESS RELAXATION
WOOL

SUPERHEATED STEAM
BT STEAM
RT DRYING
HEATING
HIGH TEMPERATURE
STEAMING

SUPERLEVELLER (DRAFTING)
RT AUTODRAFTER
AUTOLEVELLERS
AUTOMATIC CONTROL
DRAFT CONTROL
DRAFTING (STAPLE FIBER)
IRREGULARITY CONTROL
PROCESS CONTROL

SUPERLOFT (TN)
BT FALSE TWIST YARNS
TEXTURED YARNS (FILAMENT)
RT BULKED YARNS
DYNALOFT (TN)
FALSE TWISTING (TEXTURING)
FLUFLON (TN)
HELANCA (TN)
SAABA (TN)

SUPERPOSED SEAMS
BT SEAMS
RT BUTTED SEAMS
BUTTERFLY SEAMS
COVERED SEAMS
SEAMING
SEWING

SUPERVISING
USE PATROLLING

SUPPORT BANDAGES
BT MEDICAL TEXTILES
RT POWER (FABRIC)
SPANDEX FIBERS
STRETCH FABRICS
STRETCH YARNS (FILAMENT)
STRETCH YARNS (SPUN)
SUPPORT HOSIERY
SURGICAL STOCKINGS

SUPPORT HOSIERY
BT HOSIERY
RT FOUNDATION GARMENTS
MEDICAL TEXTILES
SOCKS
SPANDEX FIBERS
STRETCH FABRICS
STRETCH HOSIERY
STRETCH YARNS (FILAMENT)
STRETCH YARNS (SPUN)
SUPPORT BANDAGES
SURGICAL STOCKINGS

SURFACE ACTIVE AGENTS
USE SURFACTANTS

SURFACE AREA
RT CHEMICAL PROPERTIES
CROSS SECTIONS
CROSS SECTIONAL AREA
DYEING PROPERTIES (GENERAL)
FINENESS
PERIMETER
SORPTION OF GASES
SORPTION OF LIQUIDS
SORPTION OF WATER
SURFACE CHEMISTRY
SWELLING

SURFACE CHEMISTRY
RT CHEMICAL ENGINEERING
CONTACT ANGLES
HYDROPHILIC PROPERTY
HYDROPHOBIC PROPERTY
INTERFACIAL TENSION
OIL REPELLENCY
SURFACE AREA
SURFACE FRICTION
SURFACE PROPERTIES (MECHANICAL)
SURFACE PROPERTIES (PHYSICAL CHEMICAL)
SURFACE TENSION
SURFACES
SURFACTANT APPLICATIONS
SURFACTANTS
WATER REPELLENCY
WETTING

SURFACE CONTOUR
RT AESTHETIC PROPERTIES
COMFORT
CONTACT AREA
DULLNESS
FABRIC PROPERTIES (PHYSICAL EXCLUDING MECHANICAL)
HAND
HARSHNESS
REFLECTANCE
ROUGHNESS
SOFTNESS
SURFACE FRICTION
TEXTURE
WARMTH

SURFACE DRUMS
RT CARRIAGES (SPINNING MULE)
DRAFTING ROLLS
HEADSTOCKS
JACKSPOOLS
MULES

SURFACE FRICTION
BT FRICTION
SURFACE PROPERTIES (MECHANICAL)
RT AESTHETIC PROPERTIES
COEFFICIENT OF FRICTION
COMFORT
DRAFTING (STAPLE FIBER)
FABRIC PROPERTIES
FABRIC PROPERTIES (MECHANICAL)
FABRIC PROPERTIES (PHYSICAL EXCLUDING MECHANICAL)
FIBER SURFACE
GROOVES
HAIRINESS
HAND
INTERNAL FRICTION
ROUGHNESS
SMOOTHNESS
SOFTNESS
SURFACE CHEMISTRY
SURFACE CONTOUR
SURFACE PROPERTIES (MECHANICAL)
TEXTURE

SURFACE PROPERTIES (MECHANICAL)
NT ABRASION RESISTANCE
COEFFICIENT OF FRICTION
HAIRINESS
HAND
PILL RESISTANCE
ROUGHNESS
ROUGHNESS COEFFICIENT
SLICKNESS
SMOOTHNESS
SURFACE FRICTION
TEXTURE
BT MECHANICAL PROPERTIES
RT AESTHETIC PROPERTIES
BIOCHEMICAL PROPERTIES
CHEMICAL PROPERTIES
COMFORT
DEGRADATION PROPERTIES
FABRIC PROPERTIES
FABRIC PROPERTIES (AESTHETIC)
FABRIC PROPERTIES (MECHANICAL)
FABRIC PROPERTIES (PHYSICAL EXCLUDING MECHANICAL)
FABRIC PROPERTIES (STRUCTURAL)
FIBER PROPERTIES
FIBER SURFACE
MECHANICAL DETERIORATION PROPERTIES
MECHANICAL PROPERTIES
PHYSICAL CHEMICAL PROPERTIES
PHYSICAL PROPERTIES (EXCLUDING MECHANICAL)
POTENTIAL (ELECTRICAL)
PROPERTIES
SHAPE
STRESS OPTICAL PROPERTIES
STRESS STRAIN PROPERTIES
SURFACE CHEMISTRY
SURFACE FRICTION
RT SURFACE PROPERTIES (PHYSICAL CHEMICAL)
SURFACES
TOXICITY
TRANSFER PROPERTIES

SURFACE PROPERTIES (PHYSICAL CHEMICAL)
NT HYDROPHILIC PROPERTY
HYDROPHOBIC PROPERTY
ZETA POTENTIAL
BT PHYSICAL CHEMICAL PROPERTIES
RT ADSORPTION
AESTHETIC PROPERTIES
BIOCHEMICAL PROPERTIES
CHEMICAL PROPERTIES
CONTACT ANGLES
CORROSION
FABRIC PROPERTIES
FABRIC PROPERTIES (MECHANICAL)
FABRIC PROPERTIES (PHYSICAL EXCLUDING MECHANICAL)
FIBER PROPERTIES
FIBER SURFACE
HYDROPHILIC FIBERS
HYDROPHOBIC FIBERS
INTERFACIAL TENSION
MECHANICAL PROPERTIES
PHYSICAL PROPERTIES (EXCLUDING MECHANICAL)
PROPERTIES
SKIN (FIBER)
SORPTION
SURFACE CHEMISTRY
SURFACE TENSION
SURFACES
TRANSFER PROPERTIES
WATERPROOFING
WETTING

SURFACE TENSION
RT CAPILLARITY
CAPILLARY PRESSURE
CAPILLARY TUBES
CAPILLARY WATER
CONTACT ANGLES
ENERGY
HYDROPHILIC FIBERS
HYDROPHILIC PROPERTY
HYDROPHOBIC FIBERS
HYDROPHOBIC PROPERTY
INTERFACIAL TENSION
METARHEOLOGY
ORIENTATION (SURFACE)
RHEOLOGY
SURFACE CHEMISTRY
SURFACE PROPERTIES (PHYSICAL CHEMICAL)
WATER REPELLENCY
WATERPROOFING
WETTING
WICKING

SURFACES
RT AREA
FIBER SURFACE
FINISH (PROPERTY)
GEOMETRY
INTERFACES
SURFACE CHEMISTRY
SURFACE PROPERTIES (MECHANICAL)
SURFACE PROPERTIES (PHYSICAL CHEMICAL)

SURFACTANT APPLICATIONS
NT DISPERSING AGENTS
EMULSIFYING AGENTS
FOAMING AGENTS
REWETTING AGENTS
SUSPENDING AGENTS
WETTING AGENTS
RT ANTIFOAM AGENTS
BUILDERS (DETERGENTS)
DETERGENTS
DRY CLEANING
DYEING AUXILIARIES
LAUNDERING
LEVELLING AGENTS
SCOURING
SURFACE CHEMISTRY
SURFACTANTS
TEXTILE CHEMICALS

SURFACTANTS
UF SURFACE ACTIVE AGENTS
NT AMPHOTERIC SURFACTANTS
ANIONIC SURFACTANTS
CATIONIC SURFACTANTS
NONIONIC SURFACTANTS
BT TEXTILE CHEMICALS
RT ANTIFOAM AGENTS
BIODEGRADABILITY
CARRIER DYEING
CARRIERS (DYEING)
CRITICAL MICELLE CONCENTRATION
DETERGENCY
DETERGENTS
DISPERSING AGENTS
RT DYEING AUXILIARIES
EMULSIFYING AGENTS
FOAMING AGENTS
FOAMS (FROTH)
LAUNDERING
LEVELLING AGENTS
REWETTING AGENTS
SCOURING
SOAPING
SOLVENT ASSISTED DYEING
STAIN RESISTANCE
SURFACE CHEMISTRY
SURFACTANT APPLICATIONS
SUSPENDING AGENTS
SYNTHETIC DETERGENTS
WATER REPELLENCY
WETTING
WETTING AGENTS

SURGICAL STOCKINGS
BT MEDICAL TEXTILES
STOCKINGS
RT FOOTWEAR
HALF HOSE
SNEAKERS
SOCKS
SPANDEX FIBERS
SUPPORT BANDAGES
SUPPORT HOSIERY

SURGING
RT AXIAL MOTION
DRAFTING (STAPLE FIBER)
DRAFTING WAVES
FLOW (MATERIALS)
FLUID FLOW
FREQUENCY
IMPULSES
IRREGULARITY
OSCILLATIONS
THICK SPOTS
VARIATION
VIBRATION
YARN DYNAMICS

SURVEYS
RT ANALYZING
BIBLIOGRAPHY
COMPILATIONS
HISTORY
LITERATURE SURVEYS
REVIEWS
THEORIES
YEARS COVERAGE (1)
YEARS COVERAGE (5)
YEARS COVERAGE (10)
YEARS COVERAGE (25)
YEARS COVERAGE (50)
YEARS COVERAGE (75)
YEARS COVERAGE (100)

SUSPENDERS
BT GARMENTS
RT BELTS (APPAREL)
BRACES
BRASSIERES
BRIEFS
GARTER BELTS
GARTERS
UNDERWEAR

SUSPENDING AGENTS
BT SURFACTANT APPLICATIONS
RT DISPERSING AGENTS
EMULSIFYING AGENTS
STABILIZERS (AGENTS)
SURFACTANTS
SUSPENSIONS

SUSPENSION POLYMERIZATION
BT POLYMERIZATION
RT BULK POLYMERIZATION
EMULSION POLYMERIZATION
SUSPENSION POLYMERS
SUSPENSIONS

SUSPENSION POLYMERS
RT EMULSION POLYMERS
SUSPENSION POLYMERIZATION
SUSPENSIONS

SUSPENSIONS
RT ANTIREDEPOSITION AGENTS
COLLOIDS
DISPERSING
DISPERSING AGENTS
FOAMS (FROTH)
SOLUTIONS (LIQUID)
SUSPENDING AGENTS
SUSPENSION POLYMERIZATION
SUSPENSION POLYMERS
TURBIDITY

SUTER WEBB TESTING
BT FIBER LENGTH DETERMINATION
RT COTTON CLASSING
COTTON QUALITY
EFFECTIVE LENGTH (FIBER)
FIBER ARRAY

RT FIBER DIAGRAM
FIBER LENGTH DISTRIBUTION
FIBERS
FIBROGRAPH
GRADING
MEAN FIBER LENGTH
SHORT FIBERS INDEX
SORTING
SPAN LENGTH
STAPLING
UPPER HALF MEAN LENGTH

SUTURING
RT COLLAGEN
COLLAGEN FIBERS
JOINING
KNITTING
MEDICAL TEXTILES
SEAMING
SPLICING (KNITTING)

SWEATERS
NT CARDIGANS
BT GARMENTS
KNITTED OUTERWEAR
OUTERWEAR
RT CIRCULAR KNITTING
FILLING KNITTED FABRICS
FULLY FASHIONED KNITTING
JERKINS
JUMPERS
KNITTED FABRICS
KNITTING
PULL OVERS
SLIP ONS
SPORTSWEAR
STRIP DYEING

SWEATSHIRTS
BT ATHLETIC CLOTHING
GARMENTS
INDUSTRIAL CLOTHING
SPORTSWEAR
WORK CLOTHING
RT SHIRTS

SWELLING
RT ABSORPTION (MATERIAL)
CHEMICAL PROPERTIES
DECRYSTALLIZATION
DECRYSTALLIZED COTTON
DIFFUSIVITY
DIMENSIONAL STABILITY
DISTORTION
DYEING PROPERTIES (GENERAL)
ENTHALPY OF SWELLING
HYDROPHILIC FIBERS
HYDROPHILIC PROPERTY
HYDROPHOBIC FIBERS
HYDROPHOBIC PROPERTY
MERCERIZED COTTON
OSMOSIS
PRIMARY WALLS
SHRINKAGE
SOAKING
SORPTION
SORPTION OF GASES
SORPTION OF LIQUIDS
SORPTION OF WATER
SURFACE AREA
SWELLING AGENTS

SWELLING AGENTS
RT CARRIERS (DYEING)
COLD SETTING (SYNTHETICS)
DIMETHYL FORMAMIDE
DIMETHYL SULFOXIDE
SODIUM HYDROXIDE
SWELLING
SWELLING METHOD (COTTON IMMATURITY)
UREA

SWELLING METHOD (COTTON IMMATURITY)
BT IMMATURITY TESTING (COTTON FIBER)
RT CROSS SECTIONAL SHAPE METHOD (COTTON IMMATURITY)
DIFFERENTIAL DYEING METHOD (COTTON IMMATURITY)
POLARIZED LIGHT METHOD (COTTON IMMATURITY)
SWELLING AGENTS

SWELLING TESTING (FIBER IDENTIFICATION)
BT FIBER IDENTIFICATION
RT CHEMICAL ANALYSIS (FIBER IDENTIFICATION)
MICROSCOPIC ANALYSIS (FIBER IDENTIFICATION)
REAGENT TESTING (FIBER IDENTIFICATION)
SPECIFIC GRAVITY TESTING (FIBER IDENTIFICATION)
STAINING TESTING (FIBER IDENTIFICATION)
THERMAL TESTING (FIBER IDENTIFICATION)

SWIFTS
USE CARD CYLINDERS

SWIM SUITS
USE BATHING SUITS

SWING DOOR MECHANISMS
RT IRREGULARITY CONTROL
PROCESS CONTROL

SWISS
BT FLAT WOVEN FABRICS
WOVEN FABRICS
RT MUSLIN

SWISS DOUBLE PIQUE
BT DOUBLE PIQUE
RT FRENCH DOUBLE PIQUE
KNITTED FABRICS
RIB KNITTED FABRICS

SWIVEL LOOMS
BT LOOMS
RT AUTOMATIC LOOMS
DOBBY LOOMS

SWIVEL WEAVES
BT WEAVE (TYPES)
RT EMBROIDERY
FLAT WOVEN FABRICS

SWOLLEN DENT
BT FABRIC DEFECTS
WRONG DRAW
RT DENTS
DRAWING IN
LIGHT DENT
LOOMS
REED MARKS
REED WIRES
REEDING
REEDS
SKIP DENT
SWOLLEN HEDDLE
WARP ENDS
WARP PREPARATION
WEAVING

SWOLLEN HEDDLE
BT FABRIC DEFECTS
WRONG DRAW
RT DRAWING IN
SWOLLEN DENT
WEAVING

SYMMETRY
UF ASYMMETRY
RT CONFIGURATION
CONGRUENCE
HOLLOWNESS
ROUNDNESS
SHAPE

SYNDIOTACTIC POLYMERS
BT TACTIC POLYMERS

SYNTHESIS
USE PREPARATION (CHEMICAL)

SYNTHETIC DETERGENTS
UF NON-SOAP DETERGENTS

BT TEXTILE CHEMICALS
RT ANTIFOAM AGENTS
BIODEGRADABILITY
DETERGENTS
FOAMS (FROTH)
LAUNDERING
SCOURING
SOAPS
SURFACTANTS
WETTING
WETTING AGENTS

SYNTHETIC DYEING
NT ACRYLIC DYEING
NYLON DYEING
OLEFIN DYEING
POLYESTER DYEING
BT DYEING (BY FIBER CLASSES)
RT ACETATE DYEING
BLEND DYEING
CELLULOSE DYEING
COTTON DYEING
DISPERSING
DOPE DYEING
MAN MADE FIBERS
RAYON DYEING
SILK DYEING
SYNTHETIC FIBERS
TOW DYEING
WOOL DYEING

SYNTHETIC FELTS
BT FELTS
RT FIBER WOVEN (TN)
MALIWATT FABRIC (TN)
MAN MADE FIBERS
NEEDLE PUNCHED FELTS
NEEDLING
NONWOVEN FABRICS
PAPERMAKERS FELTS
PRESSED FELTS
SHRINKAGE
SYNTHETIC FIBERS
WEBS
WOVEN FELTS

SYNTHETIC FIBER BLENDS
USE BLENDS (FIBERS)

SYNTHETIC FIBERS
UF SYNTHETICS
NT ACRYLIC FIBERS
BUTADIENE RUBBER FIBER
FLUOROCARBON FIBERS
MODACRYLIC FIBERS
NITRILE RUBBER FIBER
NYLON (POLYAMIDE FIBERS)
NYLON 11
NYLON 6
NYLON 610
NYLON 66
NYLON 7
NYTRIL FIBERS
OLEFIN FIBERS
POLYAMIDOESTER FIBERS
POLYESTER FIBERS
POLYETHYLENE FIBERS
POLYPROPYLENE FIBERS
POLYTETRAFLUOROETHYLENE FIBERS
POLYTRIFLUOROCHLOROETHYLENE FIBERS
POLYURETHANE FIBERS
SARAN POLY (VINYLIDENE CHLORIDE) FIBERS
SPANDEX FIBERS
SYNTHETIC RUBBER FIBERS
VINAL POLY (VINYL ALCOHOL) FIBERS
VINYL FIBERS (GENERAL)
VINYON POLY (VINYL CHLORIDE) FIBERS
BT FIBERS
MAN MADE FIBERS
RT ARTIFICIAL SILK (ARCHAIC)
AZLON (REGENERATED PROTEIN FIBERS)
BLENDS (FIBERS)
COAGULATING BATHS
COPOLYMERS
EXTRUSION
FIBRILS
FILAMENT YARNS
FILAMENTS
HIGH TEMPERATURE FIBERS
HYDROPHILIC FIBERS
HYDROPHOBIC FIBERS
INORGANIC FIBERS (MAN MADE)
MELT SPINNING
NATURAL FIBERS
NEUTRALIZING (FIBER EXTRUSION)
POLYAMIDES
POLYMERS
RAYON (REGENERATED CELLULOSE FIBERS)
REGENERATED FIBERS (EXCLUDING CELLULOSIC AND PROTEIN)
SOLVENT SPINNING
SPINNING (EXTRUSION)
SYNTHETIC DYEING
SYNTHETIC LEATHER
SYNTHETIC POLYMERS
SYNTHETIC RUBBER
VEGETABLE FIBERS (GENERAL)
WET SPINNING

SYNTHETIC LEATHER
NT CORFAM (TN)
BT LEATHER
RT ALLIGATOR
BOOTS
CALFSKIN
COATED FABRICS
COATINGS (SUBSTANCES)
COWHIDE
FIBERS
FLOCKED FABRICS
FOOTWEAR
HIDES
KID LEATHER
LAMINATES
LEGGINGS
LUGGAGE
MOISTURE VAPOR TRANSMISSION
NONWOVEN FABRICS
PIGSKIN
POLYMERS
RESINS
SHOES
SOLES
SUEDE
SYNTHETIC FIBERS
SYNTHETIC RUBBER
UPHOLSTERY FABRICS
UPHOLSTERY LEATHER
UPPERS
VINYL COATED FABRICS

SYNTHETIC POLYMERS
USE POLYMERS

SYNTHETIC RUBBER
NT SILICONE RUBBER
BT RUBBER
RT ELASTOMERS
NATURAL RUBBER
SYNTHETIC FIBERS
SYNTHETIC LEATHER
SYNTHETIC RUBBER FIBERS

SYNTHETIC RUBBER FIBER BLENDS
BT BLENDS (FIBERS)

SYNTHETIC RUBBER FIBERS
NT BUTADIENE RUBBER FIBER
NITRILE RUBBER FIBER
BT ELASTOMER FIBERS
FIBERS
MAN MADE FIBERS
SYNTHETIC FIBERS
RT POLYURETHANES
RAYON (REGENERATED CELLULOSE FIBERS)
RUBBER
RUBBER FIBER (NATURAL)
SPANDEX FIBERS
SYNTHETIC RUBBER

SYNTHETIC SOIL
USE ARTIFICIAL SOIL

SYNTHETICS
USE SYNTHETIC FIBERS

SYSTEM DESIGN
USE SYSTEMS ENGINEERING

SYSTEMS ENGINEERING
UF SYSTEM DESIGN
RT AUTOMATIC CONTROL
AUTOMATION
CONTROL SYSTEMS
DOUBLE SYSTEMS
INTEGRATED SYSTEMS
PROCESS CONTROL
PROCESS EFFICIENCY
PRODUCTION
SAFETY
TEXTILE MACHINERY (GENERAL)

SYTON (TN)
BT COLLOIDAL SILICA
SILICA
RT ANTISLIP AGENTS
FIBER FIBER FRICTION
FRICTION
LUDOX (TN)
SLICING

T DISTRIBUTION
UF STUDENTS DISTRIBUTION
RT BINOMIAL DISTRIBUTION
CHI SQUARED DISTRIBUTION
DISTRIBUTION
F DISTRIBUTION
NORMAL DISTRIBUTION
STATISTICAL MEASURES
STATISTICAL METHODS

TABBY WEAVES
BT WEAVE (TYPES)
RT FLAT WOVEN FABRICS

TABLE DAMASK
BT DAMASK
FLAT WOVEN FABRICS
RT TABLECLOTHS

TABLECLOTHS
RT HOUSEHOLD FABRICS
TABLE DAMASK

TABER ABRASION TESTING
UF TABER ABRASION TESTS
BT ABRASION TESTING
RT SCHIEFER ABRASION TESTING
STOLL QM ABRASION TESTING

TABER ABRASION TESTS
USE TABER ABRASION TESTING

TABLET WEAVING
BT WEAVING

TACK
BT MECHANICAL PROPERTIES
RT ADHESION
DUCTILITY
PEELING
PENETRATION
SEEPAGE (RHEOLOGY)
THIXOTROPY

TACKING
BT SEWING
RT ADHESION
BACK TACKING
BAR TACKS
BASTING
CHAIN SEWING
CURVED SEAMING
JOINING
PIN TACKS
SEAMING
STITCHING

TACTIC POLYMERS
UF STEREO SPECIFICITY
NT ISOTACTIC POLYMERS
SYNDIOTACTIC POLYMERS
RT ATACTIC POLYMERS
POLYMERS

TAFFETA
BT FLAT WOVEN FABRICS
WOVEN FABRICS
RT FAILLE
LININGS
PERCALINE
PLAIN WEAVES
POULT
TAFFETA WEAVES

TAFFETA WEAVES
BT WEAVE (TYPES)
RT FLAT WOVEN FABRICS
TAFFETA

TAILORED SEAMS
BT SEAMS
RT BUTTED SEAMS
BUTTERFLY SEAMS
COVERED SEAMS
FELLED SEAMS
FRENCH SEAMS
GARMENT MANUFACTURE
SEAMING
SEWING
STITCHES (SEWING)
STITCHING
TAILORING

TAILORING
NT BUNDLING (TAILORING)
BT GARMENT MANUFACTURE
RT ALTERATIONS
APPAREL DESIGN
BACK TACKING
BASTING
BLIND STITCHING
BUTTONHOLING
CHAIN SEWING
CLIPPINGS
COLLARS
COMFORT
COMPACTING
CURVED SEAMING
CUTTING (TAILORING)
CUTTING DRILLS
CUTTING MACHINES
DARTS
DECATING
DRESS DESIGN
DRESSMAKING
DURABLE PRESS
FIT
FITTING
FLAT PRESSING
GARMENT COMPONENTS
GARMENTS
GATHERING
HEMMING
LETTING OUT (GARMENTS)
LONDON SHRINKING
MARKERS
MARKING
MOLDING (TAILORING)
PATTERN (APPAREL)
PATTERN GRADING
PINCHING
PINKING (TAILORING)
PLEATING
POCKETS
PRESSING
PUCKERING
REVERES
RIPPING
RUFFLING
SEAM ALLOWANCE
SEAMING
SERGING
SEWING
SHADING (TAILORING)
SHIRRING
SHRINKING
SLEEVES
SLOPING
SPREADING (TAILORING)
STITCH QUALITY
STITCHES (SEWING)
STITCHING
TAILORED SEAMS
TAKING IN (GARMENTS)
TEARING (TAILORING)
TRIMMING (OPERATION)

TAILORING MACHINES
USE GARMENT MANUFACTURING MACHINES

TAKE DOWN
RT CASTING OFF
HOSIERY MACHINES
KNITTING
KNOCKOVER
MOCK SEAMS
TAKE DOWN TENSION
WEBHOLDERS

TAKE DOWN TENSION
BT TENSION
RT DISC TENSION
KNITTING
KNITTING TENSION
PROCESS VARIABLES
TAKE DOWN
TENSION CONTROL

TAKE UP
RT BEATING UP
FABRICS
FRICTION LET OFF
LET OFF
LET OFF BRACKETS
LET OFF MOTIONS
LOOMS
PICKING (WEAVING)
PICKS PER INCH
SHEDDING
TAKE UP ROLLS
WEAVING

TAKE UP (SEWING)
RT CHAIN SEWING
SEWING
SEWING CYCLES
SEWING MACHINES
TAKE UP MECHANISMS
TENSION CONTROL

TAKE UP MECHANISMS
RT CHECK SPRINGS
SEWING
SEWING MACHINES
TAKE UP (SEWING)

TAKE UP ROLLS
RT LET OFF MOTIONS
LOOMS
TAKE UP
WEAVING

TAKER IN
USE LICKER IN

TAKING IN (GARMENTS)
BT GARMENT MANUFACTURE
RT ALTERATIONS
LETTING OUT (GARMENTS)
TAILORING

TANGENT MODULUS
BT MODULUS
STRESS STRAIN PROPERTIES (TENSILE)
RT COMPLIANCE
ELASTIC MODULUS (TENSILE)
ELASTIC STRAIN
INITIAL MODULUS
SECANT MODULUS
STRESS STRAIN CURVES

TANGENTIAL FORCE
BT FORCE
RT FRICTIONAL FORCE
NORMAL FORCE
RESOLUTION (OF FORCES)
SHEAR STRESS
TRANSVERSE FORCE

TANKS
USE VESSELS

TANNIC ACID
BT CARBOXYLIC ACIDS
ORGANIC ACIDS
RT DYE FIXING AGENTS
MORDANT DYEING
PHENOLS
TANNING (LEATHER)
TANNING AGENTS

TANNING (LEATHER)
RT HIDES
LEATHER
TANNIC ACID
TANNING AGENTS

TANNING AGENTS
RT CHROMIUM COMPOUNDS
LEATHER
TANNIC ACID
TANNING (LEATHER)
TEXTILE CHEMICALS

TAPE
NT CONDENSER TAPES
BT FLAT WOVEN FABRICS
NARROW FABRICS
WOVEN FABRICS
RT BRAIDS
FLAT BRAIDS
INDUSTRIAL FABRICS
MEDICAL TEXTILES
RIBBON
TAPE DRIVES
TAPES (DATA MEDIA)

TAPE CONDENSERS
BT CONDENSERS
RT CARD WEBS
RING CONDENSERS
SLUBBINGS
WOOLEN CARDING
WOOLEN CARDS
WOOLEN SYSTEM PROCESSING

TAPE DRIVES
BT DRIVES
RT SPINDLES
SPINNING FRAMES
TAPE
TWISTERS

TAPE SELVEDGES
BT SELVEDGES
RT CENTER SELVEDGES
LENO SELVEDGES
LOOMS
LOOP SELVEDGES
SEALED SELVEDGES
SELVEDGE WARPS
TEMPLES
WEAVING
WOVEN FABRICS

TAPER (BOBBIN)
RT BOBBINS
BUILDER MOTIONS
CONES
LARGE PACKAGES
PACKAGE HARDNESS
PACKAGES
SHAPE
SLOUGHING
SPINNING
TWISTING
WIND
WINDING

TAPERED ROLLER BEARINGS
BT BEARINGS
ROLLER BEARINGS
(CONT.)

RT AIR BEARINGS
BALL BEARINGS
JOURNAL BEARINGS

TAPES (DATA MEDIA)
NT MAGNETIC TAPE
PUNCHED TAPE
RT COMPUTERS
CONTROL SYSTEMS
PUNCHED CARD EQUIPMENT
PUNCHED CARDS
TAPE

TAPESTRIES
NT ARRAS TAPESTRIES
AUBOSSON TAPESTRIES
BEAUVAIS TAPESTRIES
BRUSSELS TAPESTRIES
GOBELIN TAPESTRIES
GOTHIC TAPESTRIES
LILLE TAPESTRIES
SAVONNERIE TAPESTRIES
VERDURES TAPESTRIES
RT ART
AXMINSTER CARPETS
CARPETS
CHENILLE CARPETS
DECORATIVE FABRICS
DRAPES
FURNISHING FABRICS
FURNISHINGS
HAND WEAVING
TAPESTRY CARPETS
UPHOLSTERY
WILTON CARPETS
WOVEN CARPETS
WOVEN FABRICS

TAPESTRY CARPETS
BT WOVEN CARPETS
RT TAPESTRIES

TAPING
BT JOINING
RT ADHESION
ADHESIVES
BANDING
BONDING
BONDING STRENGTH
CEMENTING
FUSION (MELTING)
GLUING
HEAT SEALING
PEELING
SEALING
SEAMING
SEWING
STITCH BONDING
STITCHING
STRAPPING

TARPAULINS
BT INDUSTRIAL FABRICS
RT AGRICULTURAL FABRICS
DUCK
INFLATABLE FABRICS
MILITARY PRODUCTS
SHELTERS
TENTS

TASLAN (TN)
BT AIR JET TEXTURED YARNS
LOOP YARNS
TEXTURED YARNS (FILAMENT)
RT AIR JET TEXTURING
BULKED YARNS
COVER

TATTING LACES
BT HAND MADE LACES
LACES

THAR
USE TOP NOIL RATIO

TEAR RESISTANCE
USE TEAR STRENGTH

TEAR STRENGTH
UF TEAR RESISTANCE
NT ELMENDORF TEAR STRENGTH
TONGUE TEAR STRENGTH
TRAPEZOID TEAR STRENGTH
BT END USE PROPERTIES
FABRIC PROPERTIES
FABRIC PROPERTIES (MECHANICAL)
MECHANICAL PROPERTIES
STRENGTH
STRESS STRAIN PROPERTIES
RT BREAKING ELONGATION
BREAKING LENGTH
BREAKING STRENGTH
COMPRESSIVE STRENGTH
COUNT STRENGTH PRODUCT
COVER FACTOR
CUTTING RESISTANCE
EFFICIENCY (STRUCTURAL)
ELMENDORF TEAR TESTING
FABRIC STRENGTH
GRAB STRENGTH
IMPACT STRENGTH
INTERLACINGS (YARN IN FABRIC)
LOOP EFFICIENCY
LOOP STRENGTH
LUBRICATION
MATRIX FREEDOM
PICKING (SNAGGING)
PUNCTURE RESISTANCE
RAVEL STRIP STRENGTH
SHOCK RESISTANCE
SNAGGING
SOFTENERS
STRENGTH EFFICIENCY
STRENGTH ELONGATION TESTING
STRESS CONCENTRATION
TEAR TESTING
TEARING (TAILORING)
TENACITY
TIGHTNESS
TONGUE TEAR TESTING
TRAPEZOID TEAR TESTING
WEAVE (TYPES)
YARN SLIPPAGE

TEAR TESTERS
BT TESTING EQUIPMENT
RT ELMENDORF TEAR TESTING
TEAR STRENGTH
TEAR TESTING
TONGUE TEAR TESTING
TRAPEZOID TEAR TESTING

TEAR TESTING
UF TEAR TESTS
NT ELMENDORF TEAR TESTING
TONGUE TEAR TESTING
TRAPEZOID TEAR TESTING
BT STRENGTH TESTING
TESTING
RT ABRASION TESTING
BENDING TESTING
BURST TESTING
SNAG TESTING
STRESS CONCENTRATION
TEAR STRENGTH
TEAR TESTERS
TEARING (TAILORING)
TENSILE TESTING

TEAR TESTS
USE TEAR TESTING

TEARING (TAILORING)
BT GARMENT MANUFACTURE
RT CUTTING (TAILORING)
RIPPING
SLITTING
SPLITTING
TAILORING
TEAR STRENGTH
TEAR TESTING

TEAZLE GIG
RT DRAWN PILE FINISH
DRESS FACE FINISH
NAP
NAPPING
TEAZLES

TEAZLES
RT NAP
NAPPING
TEAZLE GIG

TECHNIQUES
RT ANALYZING
CONVENTIONAL PRACTICE
EXPERIMENTAL STRESS ANALYSIS
METHODS
PERFORMANCE

TEKJA (TN)
RT BONDED YARNS

TEFLON (TN)
BT FLUOROCARBON POLYMERS
RT POLYTETRAFLUOROETHYLENE
TEFLON BEARINGS
TEFLON FIBER (TN)

TEFLON BEARINGS
BT BEARINGS
RT ANTIFRICTION BEARINGS
BALL BEARINGS
BUSHINGS
FRICTION
JOURNAL BEARINGS
NYLON BEARINGS
PLAIN BEARINGS
SLEEVE BEARINGS
TEFLON (TN)

TEFLON FIBER (TN)
BT FLUOROCARBON FIBERS
HIGH TEMPERATURE FIBERS
POLYTETRAFLUOROETHYLENE FIBERS
RT FIBER FIBER FRICTION
FRICTION
FTORLON (TN)
NOMEX (TN)
POLYTETRAFLUOROETHYLENE
POLYTRIFLUOROCHLOROETHYLENE FIBERS
TEFLON (TN)

TEMPERATURE
NT AMBIENT TEMPERATURE
HIGH TEMPERATURE
LOW TEMPERATURE
ROOM TEMPERATURE
BT PHYSICAL PROPERTIES (EXCLUDING MECHANICAL)
THERMAL PROPERTIES
RT BOILING POINT
CONDUCTIVITY (THERMAL)
DENSITY
DEW POINT
DRY BULB TEMPERATURE
EMISSIVITY
ENTHALPY
ENTROPY
EQUILIBRIUM (PHYSICAL)
FLASH POINT
GLASS TRANSITION TEMPERATURE
HEAT
HEAT CONTENT
HEAT RESISTANCE
HEAT TRANSFER
HEAT TREATMENT
LATENT HEAT
PRESSURE
SECONDARY TRANSITION TEMPERATURES
SPECIFIC HEAT
TEMPERATURE CONTROL
TEMPERATURE DISTRIBUTION
TEMPERATURE GRADIENT
THERMODYNAMICS
THERMOMETERS
TRANSFER PROPERTIES
WET BULB TEMPERATURE

TEMPERATURE CONTROL
RT AIR CONDITIONING
AUTOMATIC CONTROL
DRYERS
HEAT SETTING (SYNTHETICS)
HEATING
HEATING EQUIPMENT
HUMIDITY CONTROL
MOISTURE CONTENT CONTROL
OVENS
PROCESS CONTROL
TEMPERATURE DISTRIBUTION
TEMPERATURE GRADIENT
TENTERING

TEMPERATURE DISTRIBUTION
RT DISTRIBUTION
HEAT TRANSFER
TEMPERATURE
TEMPERATURE CONTROL
TEMPERATURE GRADIENT

TEMPERATURE GRADIENT
RT CONDUCTIVITY (THERMAL)
HEAT TRANSFER
PRESSURE DROP
TEMPERATURE
TEMPERATURE CONTROL
TEMPERATURE DISTRIBUTION

TEMPLE MARKS
BT FABRIC DEFECTS
RT FABRIC INSPECTION
LOOMS
TEMPLES
WEAVING

TEMPLES
BT CLOTH GUIDES
RT CLOTH FELL
CRIMP INTERCHANGE
DOG LEGGED SELVEDGES
GUIDERS (CLOTH)
GUIDES
LENO SELVEDGES
LOOM TAKE UP
LOOMS
PULLED IN SELVEDGES
SELVEDGE WARPS
SELVEDGES
SLACK SELVEDGES
TAPE SELVEDGES
TEMPLE MARKS
TIGHT SELVEDGES
WEAVING
WOVEN FABRICS

TENACITY
UF GRAMS PER DENIER
BT STRENGTH
STRESS STRAIN PROPERTIES
STRESS STRAIN PROPERTIES (TENSILE)

RT BREAKING LENGTH
BREAKING STRENGTH
COUNT STRENGTH PRODUCT
DENIER
DENSITY
DRAWING (FILAMENT)
ELASTICITY
FIBER PROPERTIES
FIBER STRENGTH
FORCE
LEA STRENGTH PRODUCT
PILL RESISTANCE
PILLING
PRESSLEY TEST
STRENGTH EFFICIENCY
STRESS STRAIN CURVES
TEAR STRENGTH
YARN STRENGTH

TENACITY ELONGATION CURVES
USE STRESS STRAIN CURVES

TENASCO (TN)
BT CELLULOSIC FIBERS
RAYON (REGENERATED CELLULOSE FIBERS)
VISCOSE RAYON

TENAX (TN)
BT CELLULOSIC FIBERS
HIGH TENACITY VISCOSE RAYON
RAYON (REGENERATED CELLULOSE FIBERS)
VISCOSE RAYON

TENDERING
RT ACID DEGRADATION
ALKALI DEGRADATION
DAMAGE
DEGRADATION
SLIVER TENDERNESS
STRENGTH
THERMAL DEGRADATION

TENSILE BUCKLING
BT BUCKLING
RT COMPRESSIVE BUCKLING
DIMENSIONAL STABILITY
TORSIONAL BUCKLING

TENSILE PROPERTIES
USE STRESS STRAIN PROPERTIES (TENSILE)

TENSILE STRENGTH
USE BREAKING STRENGTH

TENSILE STRESS
NT BIAXIAL STRESS
UNIAXIAL STRESS
BT STRESS
RT BREAKING STRENGTH
COMPRESSIVE STRESS
SHEAR STRESS
TENACITY

TENSILE TESTERS
UF TENSILE TESTING MACHINES
NT CAMBRIDGE EXTENSOMETER (TN)
INSTRON TENSILE TESTER (TN)
SCOTT TENSILE TESTER (TN)
STELOMETER
BT TESTING EQUIPMENT
RT TENSILE TESTING

TENSILE TESTING
UF TENSILE TESTS
BT STRENGTH TESTING
TESTING
RT ABRASION TESTING
BENDING TESTING
BREAKING STRENGTH
BURST TESTING
CREEP TESTING
CYCLIC STRESS
FABRIC STRENGTH
FATIGUE TESTING
GAUGE LENGTH
IMPACT TESTING
IRREGULARITY
STRAIN RATE
STRENGTH
STRENGTH ELONGATION TESTING
TEAR TESTING
TENSILE TESTERS
TENSION
WEAK SPOTS
YARN STRENGTH

TENSILE TESTING MACHINES
USE TENSILE TESTERS

TENSILE TESTS
USE TENSILE TESTING

TENSIOMETERS
UF TENSION METERS
BT MEASURING INSTRUMENTS
RT TENSION
TENSION CONTROL
TENSION DEVICES
THREAD TENSION
YARN TENSION

TENSION
NT BALLOON TENSION
DRAWING TENSION (FILAMENT)
FILLING TENSION
KNITTING TENSION
SPINNING TENSION (EXTRUSION)
SPINNING TENSION (STAPLE YARN)
STITCH TENSION
TAKE DOWN TENSION
THREAD TENSION
TWISTING TENSION
WARP TENSION
WINDING TENSION
YARN TENSION
BT FORCE
RT ANGLE OF WRAP
BALL AND SOCKET TENSION
BENDING
BIAXIAL STRESS
BREAKING STRENGTH
COMPRESSION
DISC TENSION
DYNAMOMETERS
END BREAKAGES
FINGER TENSION
FORCE
FRICTION
GATE TENSION
NORMAL FORCE
SHEAR (MODE OF DEFORMATION)
STOP MOTIONS
STRAIN
STRESS
STRETCH BREAKING
TENSILE TESTING
TENSIOMETERS
TENSION CONTROL
TENSION DEVICES
TIGHTNESS
TORSION (GEOMETRIC)
UNIAXIAL STRESS
WEAK SPOTS
WINDING
WRAP
YARN DYNAMICS

TENSION CONTROL
UF TENSION REGULATION
RT AUTOMATIC CONTROL
BALL AND SOCKET TENSION
CHEESE WINDERS
CHEESE WINDING
CONING
DISC TENSION
DRUM WINDERS
DRUM WINDING
END BREAKAGES
FINGER TENSION
GATE TENSION
PAPER MACHINES
POSITIVE FEED
PRECISION WINDERS
PROCESS CONTROL
QUILLING
SLACK FEEDERS (KNITTING)
SLACK MERCERIZING
SLACK THREADS
SLUB CATCHING
SPINDLE ECCENTRICITY
SPINDLE SPEED
SPINDLE VIBRATION
STITCH LENGTH CONTROL
STITCH TENSION
TAKE DOWN TENSION
TAKE UP (SEWING)
TAKE UP MECHANISMS
TENSIOMETERS
TENSION
TENSION DEVICES
THREAD BREAKAGES
TWIST TRIANGLE
WARP ENDS
WARP PREPARATION
WARPERS (MACHINE)
WARPING
WINDERS
WINDING
WINDING TENSION
WRAP
YARN DYNAMICS
YARN TENSION

TENSION CONTROLLERS
USE TENSION DEVICES

TENSION DEVICES
UF TENSION CONTROLLERS
NT DIRECT TENSION DEVICES
HYDRAULIC TENSION DEVICES
INDIRECT TENSION DEVICES
MAGNETIC TENSION DEVICES
TENSION DISCS
TENSION SPRINGS
BT DEVICES
RT BALL AND SOCKET TENSION
CHECK SPRINGS
CHEESE WINDERS
DISC TENSION
ELECTROMECHANICAL DEVICES
MECHANICAL DEVICES
GATE TENSION
LET OFF MOTIONS
PRECISION WINDERS
SEWING MACHINES
STITCH TENSION
TAKE UP MECHANISMS
TENSION
TENSION CONTROL
WARP ENDS
WARP PREPARATION
WARPERS (MACHINE)
WASHERS
WINDERS
WINDING TENSION
WRAP
YARN TENSION

TENSION DISCS
BT TENSION DEVICES
RT GATE TENSION
HYDRAULIC TENSION DEVICES
MAGNETIC TENSION DEVICES
TENSION SPRINGS
WASHERS

TENSION METERS
USE TENSIOMETERS

TENSION MODULUS
USE ELASTIC MODULUS (TENSILE)

TENSION REGULATION
USE TENSION CONTROL

TENSION SPRINGS
BT TENSION DEVICES
RT GATE TENSION
HYDRAULIC TENSION DEVICES
LET OFF MOTIONS
MAGNETIC TENSION DEVICES
SEWING
TENSION DEVICES
TENSION DISCS
WASHERS

TENSOR ANALYSIS
USE TENSORS

TENSOR CALCULUS
USE TENSORS

TENSORS
UF TENSOR ANALYSIS
TENSOR CALCULUS
RT DIFFERENTIAL GEOMETRY
MATHEMATICAL ANALYSIS
MATRICES
SENSE (DIRECTION)
VECTORS

TENTER CLIPS
BT CLOTH GUIDES
RT GUIDERS (CLOTH)
GUIDES
PIN TENTERS
TENTER FRAMES
TENTER PINS
TENTERING

TENTER FRAMES
UF FRAMES (MACHINES)
NT CLIP TENTERS
PIN TENTERS
BT DRYERS
RT DRYING
EXPANDERS
GUIDERS (CLOTH)
GUIDES
HEAT SETTING MACHINES
HEATING EQUIPMENT
HOT AIR DRYERS
OVENS
SELVEDGE UNCURLERS
SPREADERS (CLOTH)
TENTER CLIPS
TENTER PINS
TENTERING

TENTER PINS
BT CLOTH GUIDES
RT PIN TENTERS
TENTER CLIPS
TENTER FRAMES
TENTERING

TENTERING
UF FRAMING
STENTERING
BT DRY FINISHING
RT CLIP TENTERS
(CONT.)

RT DRYING
DYEING
FILLING SHRINKAGE
HEAT SETTING (SYNTHETICS)
IRONING
OVENS
PIN TENTERS
TENTER FRAMES
TENTER PINS

TENTS
BT MILITARY PRODUCTS
RT DUCK
END USES
EQUIPAGE
INFLATABLE FABRICS
RADOMES
SHELTERS
TARPAULINS
VINYL COATED FABRICS

TEREPHTHALIC ACID
BT CARBOXYLIC ACIDS
DIBASIC ACIDS
ORGANIC ACIDS
RT ADIPIC ACID
AROMATIC COMPOUNDS
ISOPHTHALIC ACID
MONOMERS
PHTHALIC ACID
POLY (ETHYLENE GLYCOL TEREPHTHALATE)
POLY (HEXAMETHYLENE TEREPHTHALAMIDE)

TERGAL (TN)
BT POLYESTER FIBERS

TERMINAL VELOCITY
BT VELOCITY
RT AIR DRAG
HIGH SPEED
VARIABLE SPEED
VELOCITY (RELATIVE)
VELOCITY CHANGE POINT

TERMINOLOGY
USE NOMENCLATURE

TERON (TN)
BT POLYESTER FIBERS
RT POLY (ETHYLENE GLYCOL TEREPHTHALAMIDE)

TERPOLYMERS
RT COPOLYMERS
POLYMERS

TERRYCLOTH
BT PILE FABRICS (WOVEN)
WOVEN FABRICS
RT CRASH TOWELING
HOUSEHOLD FABRICS
TOWELS

TERTIARY ALKYLAMINE OXIDES
USE AMINE OXIDE SURFACTANTS

TERTIARY AMINE OXIDES
USE AMINE OXIDE SURFACTANTS

TERTIARY AMINE SURFACTANTS
USE ALKYLAMINE SURFACTANTS

TERTIARY AMINES
NT TRIETHANOLAMINE
BT AMINES

TERYLENE (TN)
BT POLYESTER FIBERS

TESTING
UF TESTS
NT ABRASION TESTING
AIR PERMEABILITY TESTING
BALLISTIC TESTING
BENDING TESTING
BURST TESTING
CHEMICAL TESTING
CLEMSON STRENGTH TESTING
COLOR TESTING
COLORFASTNESS TESTING
COMPRESSION TESTING
CORROSION TESTING
COUNT TESTING
CREEP TESTING
CROCK TESTING
CUTTING TESTING
DIAPHRAGM BURST TESTING
DRAVES TEST
FATIGUE TESTING
FIBER BUNDLE TESTING
FIBER IDENTIFICATION
FIBER LENGTH DETERMINATION
FINENESS TESTING
FIRE RESISTANCE TESTING
IMMATURITY TESTING (COTTON FIBER)
IMMERSION TESTING
NT IMPACT TESTING
IRREGULARITY TESTING
LAUNDERABILITY TESTING
LIGHTFASTNESS TESTING
MOISTURE CONTENT TESTING
PERMEABILITY TESTING
SCORCH TESTING
SHRINKAGE TESTING
SNAG TESTING
SOIL BURIAL TEST
STRENGTH ELONGATION TESTING
STRENGTH TESTING
TEAR TESTING
TENSILE TESTING
THERMAL TESTING
TWIST TESTING
WASHFASTNESS TESTING
WATER ABSORPTION TESTING
WATER RESISTANCE TESTING
WATER VAPOR TRANSMISSION TESTING
WEAR TESTING
WRINKLE RECOVERY TESTING

RT ANALYZING
CLASSING
CORING
EVALUATION
FIBER LENGTH DISTRIBUTION
GRADING
HYDROSTATIC PRESSURE TESTING
IRREGULARITY
SAMPLES
SAMPLING
SPECIFICATIONS
SPECIMEN PREPARATION
STANDARDS
STAPLING
STATISTICAL ANALYSIS
TESTING EQUIPMENT

TESTING APPARATUS
USE TESTING EQUIPMENT

TESTING EQUIPMENT
UF TESTING APPARATUS
TESTING INSTRUMENTS
NT ABRASION TESTERS
ABSORPTION TESTERS
ACCELEROMETERS
BURST TESTERS
COUNT TESTERS
CRIMP TESTERS
DRAFTING FORCE TESTERS
FINENESS TESTERS
FIRE RESISTANCE TESTERS
IMPACT TESTERS
IRREGULARITY TESTERS
LAUNDERABILITY TESTERS
LIGHTFASTNESS TESTERS
MOISTURE CONTENT TESTERS
NEP TESTERS
PERMEABILITY TESTERS
PULSE PROPAGATION METER
SCORCH TESTERS
SHRINKAGE TESTERS
SLIVER TESTERS
SNAG TESTERS
TEAR TESTERS
TENSILE TESTERS
TWIST TESTERS
WASHFASTNESS TESTERS
WATER RESISTANCE TESTERS
WRINKLE RECOVERY TESTERS
RT COMPONENTS
EQUIPMENT
EXPERIMENTATION
INSTRUMENTATION
MEASURING INSTRUMENTS
RECORDING INSTRUMENTS
SHIRLEY ANALYZER
STRENGTH
TESTING

TESTING INSTRUMENTS
USE TESTING EQUIPMENT

TESTS
USE TESTING

TETRAHYDRONAPHTHALENE
BT AROMATIC HYDROCARBONS
RT ALKYLARYL SULFONATES
TETRAHYDRONAPHTHALENE SULFONATES

TETRAHYDRONAPHTHALENE SULFONATES
NT ALKANOL S (TN)
BT SULFONATE SURFACTANTS
RT ALKYLNAPHTHALENE SULFONATES
NAPHTHALENE FORMALDEHYDE SULFONATES
TETRAHYDRONAPHTHALENE
WETTING AGENTS

TETRAKIS (HYDROXYMETHYL) PHOSPHONIUM CHLORIDE
USE THPC

TETRAPOTASSIUM PYROPHOSPHATE
BT PHOSPHATES
PYROPHOSPHATES
RT CHELATING AGENTS

TETRASODIUM PYROPHOSPHATE
USE TSPP

TETRONICS (TN)
USE POLYETHOXY POLYPROPOXY ETHYLENEDIAMINE

TEX
UF KILOTEX
BT COUNT
DIRECT COUNT
RT COUNT TESTING
DENIER
FIBER DENIER
GREX
INDIRECT COUNT
LINEAR DENSITY

TEXAS COTTON
BT COTTON

TEXTILE CHEMICALS
NT BLEACHING AGENTS
DELIQUESCENT AGENTS
DESIZING AGENTS
DETERGENTS
DYEING AUXILIARIES
DYES
FINISH (SUBSTANCE ADDED)
LUBRICANTS
SIZES (SLASHING)
SURFACTANTS
SYNTHETIC DETERGENTS
RT ADDITIVES (CHEMICAL)

DEODORANTS
SURFACTANT APPLICATIONS
TANNING AGENTS

TEXTILE COMPLIANCE
USE COMPLIANCE

TEXTILE MACHINERY (GENERAL)
NT AGERS
AIR JET TEXTURING MACHINES
BACK FILLING MACHINES
BACKWASHING MACHINES
BAGGING MACHINES
BALING MACHINES
BATT MAKING MACHINES
BLEACHING MACHINES
BLENDERS
BOBBIN STRIPPING MACHINES
BOIL OFF MACHINES
BOW CONTROL MACHINES
BOW STRAIGHTENERS
BRAIDERS
BRUSHING MACHINES
BULKING MACHINES
BURLING MACHINES
BURR CRUSHERS
BUTTON BREAKERS
CALENDERING MACHINES
CAP SPINNING FRAMES
CARBONIZERS
CARDS
CENTRIFUGAL DRYERS
CENTRIFUGES
CHEESE WINDERS
CHINCHILLA MACHINES
CIRCULAR KNITTING MACHINES
CLEANERS
COATING MACHINES
COMBER LAP MACHINES
COMBS
COMPENSATORS (CLOTH)
COMPRESSIVE SHRINKAGE MACHINES
CONVEYORS
CORDAGE MACHINES
COTTON GINS
CRABBING MACHINES
CRIMPING MACHINES
CROCHETING MACHINES
CURING OVENS
CUTTERS
CUTTING DRILLS
CUTTING MACHINES
DECATING MACHINES
DESIZING MACHINES
DRAWFRAMES
DRAWING IN MACHINES
DRY CLEANING MACHINES
DRYERS
DUSTING MACHINES
DYEING MACHINES
DYEING MACHINES (BY ENVIRONMENTAL CONDITIONS)
DYEING MACHINES (BY MATERIAL ASSEMBLY)
DYEING MACHINES (BY PROCESS FLOW)
EDGE CRIMPING MACHINES
EXPANDERS

NT FALSE TWISTERS
FELTING MACHINES
FILLING STRAIGHTENING MACHINES
FLYER SPINNING FRAMES
FOLDING MACHINES (FOR CLOTH)
FULLING MACHINES
FULLY FASHIONED KNITTING MACHINES
GARMENT MANUFACTURING MACHINES
GARNETTING MACHINES
GILL BOXES
HAIR PROCESSING MACHINES
HEAT SETTING MACHINES
HOMOGENIZERS
HOSIERY MACHINES
IMPREGNATING MACHINES
INSPECTION MACHINES
J BOXES
KETTLES (DYEING)
KIERS
KNOTTERS
LABELLING MACHINES
LACE MACHINES
LAPPERS
LOOMS
LOOPERS
LUSTERING MACHINES
MANGLES
MARKING MACHINES
MEASURING MACHINES
MECHANICAL MIXERS
MERCERIZING MACHINES
MILLS (COLLOID)
MILLS (FULLING ROTARY)
MILLS (KICKER)
MULES
NAPPING MACHINES
NONWOVEN FABRIC MACHINES
OPENERS
OVENS
PACIFIC CONVERTER
PACKAGING EQUIPMENT
PACCERS
PICKERS (OPENING)
PILERS
PINKING MACHINES
PLEATING MACHINES
POT SPINNING FRAMES
PRECISION WINDERS
PRESSES
PRINTING MACHINES
QUILLERS
RAG PULLING MACHINES
RING FRAMES
ROPE MACHINES
ROVING FRAMES
SANFORIZING MACHINES
SCOURING MACHINES
SELVEDGE UNCURLERS
SEWING MACHINES
SHEARING MACHINES
SINGEING MACHINES
SLASHERS
SLITTERS (CLOTH)
SPINNING FRAMES
SPRAYING MACHINES
SPREADERS (CLOTH)
SQUEEZERS (CLOTH)
STEAMERS
STITCH BONDING MACHINES
STRAPPING MACHINES
STUFFER BOXES
SUEDING MACHINES
TENTER FRAMES
THROWING MACHINES
TINTING MACHINES
TOW TO TOP MACHINES
TOW TO YARN MACHINES
TUFTING MACHINES
TURBOSTAPLER
TWIST SETTERS
TWISTERS
TYING IN MACHINES
UNWINDERS (CLOTH)
UPTWISTERS
WARP KNITTING MACHINES
WARP TYING IN MACHINES
WARPERS (MACHINE)
WASHING MACHINES
WASTE MACHINES
WEB FORMING MACHINES
WINDERS
WRAPPING MACHINES
RT AUTOMATIC CONTROL
COMPONENTS
COSTS
DRY FINISHING
EQUIPMENT
FINISHING MACHINERY (GENERAL)
FINISHING PROCESS (GENERAL)
INVENTORIES
INVESTMENTS
MACHINE DESIGN
MACHINERY
MOTORS
PLANT
PLANT LAYOUT
POWER CONSUMPTION
RT PROCESS DESIGN
PROCESSING MACHINERY
SCHREINERING
SYSTEMS ENGINEERING
TEXTILE PROCESSES (GENERAL)
TEXTILE MATERIALS
WET FINISHING

TEXTILE MATERIALS
UF FIBROUS MATERIALS
RT FABRICS
FIBERS
MATERIALS
TEXTILE MACHINERY (GENERAL)
TEXTILE PROCESSES (GENERAL)
YARNS

TEXTILE MODULUS
USE MODULUS

TEXTILE PROCESSES (GENERAL)
NT AMERICAN SYSTEM PROCESSING
ARACHNE SYSTEMS (TN)
BALING
BRADFORD SYSTEM PROCESSING
CARDING
COMBING
CONTINENTAL SYSTEM PROCESSING
COTTON SYSTEM PROCESSING
DRAFTING (STAPLE FIBER)
DRAWING (FILAMENT)
DRY FINISHING
DYEING
FLAX SYSTEM PROCESSING
GARMENT MANUFACTURE
GILLING
GINNING
INTERLACING (FILAMENT IN YARN)
KNITTING
MALI PROCESSES (TN)
NEEDLING
OPENING
PICKING (OPENING)
PRESSING
PRINTING
QUILLING
SILK SYSTEM PROCESSING
SLASHING
SPINNING
SPINNING (EXTRUSION)
TAILORING
TEXTURING
THROWING
TUFTING
TWISTING
WARPING
WEAVING
WET FINISHING
WINDING
WOOLEN SYSTEM PROCESSING
WORSTED SYSTEM PROCESSING
RT FINISHING PROCESS (GENERAL)
PLANT
POWER CONSUMPTION
PROCESS DESIGN
TEXTILE MACHINERY (GENERAL)
TEXTILE MATERIALS

TEXTRALIZED YARNS (TN)
BT BULKED YARNS
CRIMPED YARNS
TEXTURED YARNS (FILAMENT)
RT BANLON (TN)
BULKING
STUFFER BOX CRIMPING

TEXTURE
BT AESTHETIC PROPERTIES
FABRIC PROPERTIES
FABRIC PROPERTIES (AESTHETIC)
FABRIC PROPERTIES (STRUCTURAL)
STRUCTURAL PROPERTIES
SURFACE PROPERTIES (MECHANICAL)
RT AESTHETIC APPEAL
APPEARANCE
BULK
COMFORT
COURSES PER INCH
COVER
FABRICS
FINENESS
FRICTION
HAIRINESS
HAND
LUSTER
LUXURIOUSNESS
OPTICAL PROPERTIES
PACKING FACTOR
PATTERN DEFINITION
ROUGHNESS
SET (WOVEN FABRIC)
SMOOTHNESS
SOFTNESS
SPARKLE
STITCH CLARITY
STRETCH
SURFACE CONTOUR
RT SURFACE FRICTION
TEXTURED YARNS (FILAMENT)
TEXTURING
TRANSPARENCY
WALES PER INCH
WARMTH

TEXTURED SPUN YARNS
UF BULKED YARNS (STAPLE)
TEXTURED YARNS (SPUN)
NT BICOMPONENT FIBER YARNS (STAPLE)
HILOW BULKED YARNS
STRETCH YARNS (SPUN)
BT SPUN YARNS
YARNS
RT NOVELTY YARNS
TEXTURED YARNS (FILAMENT)

TEXTURED YARNS (FILAMENT)
UF MODIFIED CONTINUOUS FILAMENT YARNS
TEXTURIZED YARNS
NT AGILON (TN)
AIR JET TEXTURED YARNS
BANLON (TN)
BICOMPONENT FIBER YARNS (FILAMENT)
BULKED YARNS
BURMILIZED (TN)
CANTRECE (TN)
CRIMPED YARNS
CRINKLE TYPE YARNS
CUMULOFT (TN)
CUPREL (TN)
EDGE CRIMPED YARNS
FALSE TWIST YARNS
FLUFLON (TN)
GEAR CRIMPED YARNS
HELANCA (TN)
LOFTURA (TN)
LOOP YARNS
SAABA (TN)
SELF CRIMPING YARNS
SKYLOFT (TN)
SPUNIZE (TN)
STRETCH YARNS (FILAMENT)
STUFFER BOX CRIMPED YARNS
SUPERLOFT (TN)
TASLAN (TN)
TEXTRALIZED YARNS (TN)
TOPEL (TN)
TYCORA (TN)
BT FILAMENT YARNS
YARNS
RT AERATED YARNS
AIR JET TEXTURING
BICOMPONENT FIBERS
BULKING
CRIMP FREQUENCY
CRIMP INDEX
CRIMPING
EDGE CRIMPING
FALSE TWIST
FALSE TWIST SPINDLES
FALSE TWISTING (TEXTURING)
FILLINGWISE STRETCH
MAN MADE FIBERS
ORLON SAYELLE (TN)
PRODUCERS YARNS
SLUB YARNS
STRETCH
STRETCH FABRICS
STRETCH KNITTED FABRICS
STRETCH WOVEN FABRICS
STRETCH YARNS (SPUN)
STUFFER BOX CRIMPING
SYNTHETIC FIBERS
TEXTURE
TEXTURED SPUN YARNS
TEXTURING
TWISTING
WARPWISE STRETCH
YARN CRIMP
YARN GEOMETRY
YARN STRUCTURE
Z CRIMP

TEXTURED YARNS (SPUN)
USE TEXTURED SPUN YARNS

TEXTURING
NT AIR JET TEXTURING
BELT CRIMPING
BULKING
CRIMPING
EDGE CRIMPING
FALSE TWISTING (TEXTURING)
GEAR CRIMPING
STEAM JET TEXTURING
STUFFER BOX CRIMPING
RT BANLON (TN)
BICOMPONENT FIBER YARNS (FILAMENT)
BICOMPONENT FIBER YARNS (STAPLE)
BULK
BULKED YARNS
(CONT.)

RT BULKING MACHINES
CRIMP
CRIMPED YARNS
CRIMPING MACHINES
FALSE TWIST SPINDLES
FILAMENT YARNS
HEAT SETTING (SYNTHETICS)
HEAT TRANSFER
LATENT CRIMP
MAN MADE FIBERS
MOLECULAR ORIENTATION
RELAXATION TIME (MECHANICAL)
RESIDUAL STRAIN
RESIDUAL STRESS
RESIDUAL TORQUE
SELF CRIMPING YARNS
STRESS RELAXATION
STRETCH
STRETCH FABRICS
STRETCH YARNS (FILAMENT)
STUFFER BOX CRIMPING
SYNTHETIC FIBERS
TEXTURE
TEXTURED YARNS (FILAMENT)
THERMOPLASTICITY
THROWING
THROWSTERS
TORSIONAL BUCKLING
TWIST LIVELINESS
TWO FOR ONE TWISTING
UPTWISTERS
UPTWISTING
YARN CRIMP
Z CRIMP

TEXTURIZED YARNS
USE TEXTURED YARNS (FILAMENT)

THEORETICAL ANALYSIS
RT ANALYZING
EMPIRICAL ANALYSIS
EXPERIMENTAL ANALYSIS
MATHEMATICAL ANALYSIS
MECHANISM (FUNDAMENTAL)
QUALITATIVE ANALYSIS
QUANTITATIVE ANALYSIS
STATISTICAL ANALYSIS
THEORIES

THEORIES
RT ANALYZING
DRAFTING THEORY
DYEING THEORY
EXPERIMENTAL ANALYSIS
EXPERIMENTATION
FRICTION THEORY
HISTORY
MATHEMATICAL ANALYSIS
MECHANISM (FUNDAMENTAL)
REVIEWS
STATISTICAL ANALYSIS
SURVEYS
THEORETICAL ANALYSIS

THERMAL ANALYSIS
NT CALORIMETRY
DIFFERENTIAL THERMAL ANALYSIS
RT ANALYZING
CHEMICAL ANALYSIS
ENTHALPY OF REACTION
MEASURING INSTRUMENTS
PHYSICAL ANALYSIS
QUALITATIVE ANALYSIS
QUANTITATIVE ANALYSIS
SPECTRUM ANALYSIS
STATISTICAL ANALYSIS
STRESS ANALYSIS
STRUCTURAL ANALYSIS
X RAY ANALYSIS

THERMAL CONDUCTIVITY
USE CONDUCTIVITY (THERMAL)

THERMAL DEGRADATION
UF PYROLYSIS
BT DEGRADATION
RT CHEMICAL PROPERTIES
DRYING
HEAT RESISTANCE
HEAT SETTING (SYNTHETICS)
HEAT STABILIZERS
HEATING
LAUNDRY DEGRADATION
PHOTOCHEMICAL DEGRADATION
TENDERING
THERMAL STABILITY
TRAVELER BURNOUT

THERMAL PROPERTIES
NT ABSORPTIVITY (RADIATION)
BOILING POINT
CONDUCTIVITY (THERMAL)
EMISSIVITY
ENTHALPY
ENTROPY
FLASH POINT
HEAT CONTENT
HEAT OF FUSION
NT HEAT RESISTANCE
LATENT HEAT
MELTING POINT
REFLECTIVITY
SOFTENING POINT
SPECIFIC HEAT
TEMPERATURE
THERMAL STABILITY
TRANSMISSIVITY
WARMTH
BT PHYSICAL PROPERTIES (EXCLUDING MECHANICAL)
RT ABSORPTION (RADIATION)
ACOUSTIC PROPERTIES
AESTHETIC PROPERTIES
BIOCHEMICAL PROPERTIES
CHEMICAL PROPERTIES
COMFORT
DRYING
ELECTRICAL PROPERTIES
FABRIC PROPERTIES
FABRIC PROPERTIES (PHYSICAL EXCLUDING MECHANICAL)
FIBER PROPERTIES
FLAT SPOTTING (TIRES)
HEAT SETTING (SYNTHETICS)
HEAT TRANSFER
HEATING
MECHANICAL PROPERTIES
OPTICAL PROPERTIES
PHYSICAL CHEMICAL PROPERTIES
REFLECTANCE
STRESS OPTICAL PROPERTIES
STRUCTURAL PROPERTIES
THERMAL TESTING
THERMAL TRANSMITTANCE TESTING
THERMOPLASTICITY
TIRE FABRICS
TRANSFER PROPERTIES
TRANSLUCENCY
THERMODYNAMICS
TRANSMITTANCE
TRANSPARENCY

THERMAL RESISTANCE
USE CONDUCTIVITY (THERMAL)

THERMAL STABILITY
BT THERMAL PROPERTIES
RT ANTIOXIDANTS
CHEMICAL STABILITY
COLORFASTNESS
GELS
HEAT STABILIZERS
HEATING
OXIDATION
SHRINKAGE
STABILITY
THERMAL DEGRADATION
THERMODYNAMICS
WEATHER RESISTANCE

THERMAL TESTING
NT THERMAL TRANSMITTANCE TESTING
BT TESTING
RT CONDUCTIVITY (THERMAL)
DIFFERENTIAL THERMAL ANALYSIS
EMISSIVITY
FLAMMABILITY TESTING
REFLECTANCE
SPECIFIC HEAT
THERMAL PROPERTIES
THERMAL STABILITY
TRANSMITTANCE
WATER VAPOR TRANSMISSION TESTING

THERMAL TESTING (FIBER IDENTIFICATION)
BT FIBER IDENTIFICATION
RT CHEMICAL ANALYSIS (FIBER IDENTIFICATION)
MICROSCOPIC ANALYSIS (FIBER IDENTIFICATION)
REAGENT TESTING (FIBER IDENTIFICATION)
SPECIFIC GRAVITY TESTING (FIBER IDENTIFICATION)
STAINING TESTING (FIBER IDENTIFICATION)
SWELLING TESTING (FIBER IDENTIFICATION)

THERMAL TRANSMITTANCE TESTING
BT THERMAL TESTING
RT APPARENT CONTACT AREA
CONDUCTIVITY (THERMAL)
CONTACT AREA
THERMAL PROPERTIES
TRUE CONTACT AREA
WATER VAPOR TRANSMISSION TESTING

THERMODYNAMICS
RT AERODYNAMICS
BOILING POINT
DONNAN EQUILIBRIUM
ENTHALPY
ENTHALPY OF HYDRATION
RT ENTHALPY OF REACTION
ENTHALPY OF SWELLING
ENTHALPY OF WETTING
ENTROPY
EQUILIBRIUM (CHEMICAL)
FREE ENERGY
HEAT CONTENT
LATENT HEAT
SATURATED VAPOR
SPECIFIC HEAT
STABILITY
STATISTICAL MECHANICS
TEMPERATURE
THERMAL PROPERTIES
THERMAL STABILITY

THERMOFIXING (DYEING)
BT DYE FIXING
RT ACID SHOCK METHOD
DYEING
HEAT SETTING (SYNTHETICS)
HIGH TEMPERATURE DYEING
THERMOSOL PROCESS (TN)

THERMOMETERS
BT MEASURING INSTRUMENTS
RT HIGH TEMPERATURE
LOW TEMPERATURE
PYROMETERS
TEMPERATURE
THERMOPILES

THERMOPILES
BT ELECTRONIC INSTRUMENTS
MEASURING INSTRUMENTS
RT HIGH TEMPERATURE
THERMOMETERS

THERMOPLASTIC FIBER BONDING
BT BONDING
RT BONDED FIBER FABRICS
DIP BONDING
DRY POWDER BONDING
FOAM BONDING
NONWOVEN FABRICS
SATURATION BONDING
SOLVENT BONDING
SPRAY BONDING
THERMOPLASTIC FIBERS

THERMOPLASTIC FIBERS
RT NONWOVEN FABRICS
THERMOPLASTIC FIBER BONDING
THERMOPLASTIC POLYMERS
THERMOPLASTIC RESINS
THERMOPLASTICITY

THERMOPLASTIC POLYMERS
RT THERMOPLASTIC FIBERS
THERMOPLASTIC POWDERS
THERMOPLASTIC RESINS
THERMOPLASTICITY
THERMOSETTING POWDERS
THERMOSETTING RESINS

THERMOPLASTIC POWDERS
BT BINDERS (NONWOVENS)
RT BONDED FIBER FABRICS
SOLUTION TYPE BINDERS
THERMOPLASTIC POLYMERS
THERMOPLASTIC RESINS
THERMOPLASTICITY
THERMOSETTING POWDERS

THERMOPLASTIC RESINS
NT ACRYLIC RESINS
POLYESTER RESINS
BT RESINS
RT COATING RESINS
LAMINATING RESINS
POLYCARBONATES
THERMOPLASTIC FIBERS
THERMOPLASTIC POLYMERS
THERMOPLASTICITY
THERMOSETTING RESINS

THERMOPLASTICITY
RT ADHESION
BONDED FIBER FABRICS
BONDING
COHESION
CRYSTALLIZATION
CURING
FLAME BONDING
FUSION (MELTING)
HEAT OF FUSION
HEAT SEALING
HEAT SETTING (SYNTHETICS)
IRONING
JOINING
MELTING
MOLDING SHRINKAGE
PLASTICITY
SEALING
TEXTURING
THERMAL PROPERTIES
THERMOPLASTIC FIBERS
THERMOPLASTIC POLYMERS
THERMOPLASTIC POWDERS
THERMOPLASTIC RESINS

THERMOSETTING
USE CURING

THERMOSETTING POWDERS
BT BINDERS (NONWOVENS)
RT BONDED FIBER FABRICS
CURING
EMULSION BINDERS (NONWOVENS)
SOLUTION TYPE BINDERS
THERMOPLASTIC POWDERS

THERMOSETTING RESINS
NT EPOXY RESINS
MELAMINE RESINS
PHENOL-FORMALDEHYDE RESINS
PHENOLIC RESINS
BT RESINS
RT CURING
THERMOPLASTIC RESINS
BINDERS (PIGMENTS)
THERMOPLASTIC POLYMERS

THERMOSOL PROCESS (TN)
BT CONTINUOUS DYEING
DYEING (BY PROCESS FLOW)
HIGH TEMPERATURE DYEING
RT CURING
HEAT SETTING (SYNTHETICS)
PRESSURE DYEING
RADIATION
THERMOFIXING (DYEING)

THESAURI
RT DOCUMENTATION
INFORMATION RETRIEVAL
INFORMATION SYSTEMS
NOMENCLATURE

THICK AND THIN YARNS
BT NOVELTY YARNS
RT NEP YARNS
NUB YARNS
SLUB YARNS
THICK SPOTS

THICK SPOTS
BT DEFECTS
RT FILAMENTS
IRREGULARITY
IRREGULARITY (SHORT TERM)
NEPS
ROVINGS
SLIVERS
SLOTS
SLUB CATCHERS
SLUBINESS
SLUBS
SURGING
THICK AND THIN YARNS
THICKNESS
WEAK SPOTS
YARN DEFECTS
YARNS

THICKENING
RT COAGULATING
GELLING
HYDROXYETHYLCELLULOSE
THICKENING AGENTS

THICKENING AGENTS
RT ALGINIC ACID
CARBOXYMETHYLCELLULOSE
COAGULATING
COAGULATING BATHS
CORN STARCH
METHYLCELLULOSE
POTATO STARCH
PRINTING
PRINTING PASTES
STARCH GUMS
THICKENING

THICKNESS
BT DIMENSIONS
STRUCTURAL PROPERTIES
RT AMPLITUDE
AREA
BULK
BULK DENSITY
DIAMETER
FABRIC PROPERTIES
FABRIC PROPERTIES (STRUCTURAL)
HEIGHT
LENGTH
MATRIX FREEDOM
POROSITY
ROUGHNESS
SEAM THICKNESS
TEXTURE
THICK SPOTS
WARMTH
WEAK SPOTS
WEIGHT
WIDTH

THIN SPOTS
USE WEAK SPOTS

THINNERS
RT DILUTION
PIGMENT DYEING
SOLVENTS

THIOCYANATES
BT SULFUR COMPOUNDS
RT ISOTHIOCYANATES

THIOETHERS
UF SULFIDES (ORGANIC)
BT SULFUR COMPOUNDS
RT ALKYLATION
CYSTINE
DISULFIDE BONDS
DISULFIDES (ORGANIC)
ETHERS
KERATIN
MERCAPTANS

THIOGLYCOLIC ACID
BT CARBOXYLIC ACIDS
MERCAPTANS
ORGANIC ACIDS

THIOLS
USE MERCAPTANS

THIOSULFATES
NT SODIUM THIOSULFATE
BT SULFUR COMPOUNDS
RT SULFATES
SULFIDES (INORGANIC)
SULFITES
PERSULFATES

THIOSULFINIC ACIDS
BT ORGANIC ACIDS
SULFUR COMPOUNDS
RT SULFINIC ACIDS

THIOSULFONATES
UF BUNTE SALTS
BT SULFUR COMPOUNDS
RT ORGANIC ACIDS
SULFONIC ACIDS

THIOUREA
BT SULFUR COMPOUNDS
RT THIOUREA DIOXIDE
UREA

THIOUREA DIOXIDE
BT SULFUR COMPOUNDS
RT REDUCING AGENTS
THIOUREA

THIXOTROPY
BT STRESS STRAIN PROPERTIES
RT DUCTILITY
PENETRATION
RHEOLOGY
SEEPAGE (RHEOLOGY)
TACK

THPC
UF TETRAKIS (HYDROXYMETHYL) PHOSPHONIUM CHLORIDE
BT CHLORIDES
PHOSPHONIUM COMPOUNDS
PHOSPHORUS COMPOUNDS
REACTANTS
RT FIRE RETARDANCY AGENTS

THREAD BREAKAGES
RT END BREAKAGES
PROCESS EFFICIENCY
SEAM EFFICIENCY
SEAM STRENGTH
SEWING
SEWING EFFICIENCY
TENSION CONTROL
THREAD TENSION
THREADS

THREAD CUTTERS
BT CUTTERS
RT LOOMS
PICKING (WEAVING)
STAPLE CUTTERS
THREADS
WEAVING

THREAD GUIDES
UF EYELETS (GUIDES)
GUIDE EYES
PIGTAILS (GUIDES)
YARN GUIDES
BT GUIDES
RT CERAMIC GUIDES
CHEESE WINDERS
FLYER SPINNING FRAMES
PRECISION WINDERS
RING FRAMES
SEWING MACHINES
SPINNING FRAMES
THREADLINE
THREADS
TWISTERS
WARPERS (MACHINE)

THREAD PRESSERS
RT MILANESE KNITTING MACHINES

THREAD SLACK
RT CHECK SPRINGS
SEWING
THREAD TENSION
THREADS

THREAD SLIPPAGE
RT DIMENSIONAL STABILITY
PUCKER (DEFECT)
SEAM EFFICIENCY
SEAM GRIN
SEAM PUCKER
SEAM SLIPPAGE
SEAM SLIPPAGE STRENGTH
SEAM STRENGTH
THREADS
YARN SLIPPAGE

THREAD TENSION
BT TENSION
RT BALLOON TENSION
PUCKER (DEFECT)
RIPPLES
SEAM EFFICIENCY
SEAM GRIN
SEAM PUCKER
SEAM SLIPPAGE
SEAM STRENGTH
SEWING
STITCH SIZE (SEWING)
THREAD BREAKAGES
THREAD SLACK
THREADLINE
THREADS
YARN TENSION

THREADING
RT DRAWING IN
FILLING YARNS
NEEDLE EYE
NEEDLE THREADS
SELF THREADING DEVICES
SEWING NEEDLES
SHUTTLES
THREADS
TYING IN

THREADLINE
RT BALLOON DYNAMICS
SEWING
SPINNING
THREAD GUIDES
THREAD TENSION
THREADS
TWISTING
WINDING
YARN DYNAMICS
YARN TENSION
YARNS

THREADS
NT BOBBIN THREADS (SEWING)
COTTON THREADS
LINEN THREADS
NEEDLE THREADS
NYLON THREADS
SEWING THREADS
RT BOBBIN THREADS (SEWING)
GYMP
LACES (SHOES)
RAWHIDE
SEAMING
SELF THREADING DEVICES
SEWING
STITCH QUALITY
STITCHING
THREAD BREAKAGES
THREAD CUTTERS
THREAD GUIDES
THREAD SLACK
THREAD SLIPPAGE
THREAD TENSION
THREADING
THREADLINE

THREE EIGHTHS BLOOD (WOOL GRADE)
BT GRADE (FIBERS)
WOOL GRADE
RT WOOL

THREE NEEDLE COVER STITCHES
BT STITCHES (SEWING)

THREONINE
BT AMINO ACIDS

THROAT PLATES
RT FEED DOGS
NEEDLE PENETRATION
SEWING
SEWING MACHINES
YARN DAMAGE
YARN SEVERANCE

THROSTLE SPINNING FRAMES
USE FLYER SPINNING FRAMES

THROUGH PRINTING
BT PRINTING
RT BLOCK PRINTING
DISCHARGE PRINTING
DUPLEX PRINTING
MELANGE PRINTING
RESIST PRINTING
ROLLER PRINTING
SCREEN PRINTING

THROUGHPUT
RT CHARGING
CONVEYORS
DELIVERY RATE
FEED RATE
FEED ROLLS
FEEDERS
FLOW (MATERIALS)
FLOW CONTROL
HOPPER FEED
INPUT
LOADING (PROCESS)
OUTPUT
PROCESS VARIABLES
PRODUCTION
UNLOADING (PROCESS)

THROWING
RT BULKING
CRIMPING
CRINKLE TYPE YARNS
DOWNTWISTING
FALSE TWISTERS
FALSE TWISTING (TEXTURING)
FILAMENT YARNS
PACKAGING (OPERATION)
PLYING
SLASHING
TEXTURING
THROWING MACHINES
THROWSTERS
TOW
TOW CONVERSION
TWISTING
TWO FOR ONE TWISTING
UPTWISTING

THROWING MACHINES
NT BULKING MACHINES
CRIMPING MACHINES
FALSE TWISTERS
TWO FOR ONE TWISTERS
UPTWISTERS
RT THROWING
THROWSTERS

THROWOVER
USE SETOVER

THROWSTERS
RT DOWNTWISTERS
TEXTURING
THROWING
THROWING MACHINES
TWIST
TWISTERS
TWISTING
TWO FOR ONE TWISTERS
UPTWISTERS
UPTWISTING

THRUST BEARINGS
BT BEARINGS
RT AIR BEARINGS
BALL BEARINGS
CONICAL BEARINGS
FOOTSTEP BEARINGS
JOURNAL BEARINGS
NEEDLE BEARINGS
NYLON BEARINGS
ROLLER BEARINGS
SLEEVE BEARINGS

TICKING
BT FLAT WOVEN FABRICS
WOVEN FABRICS
RT CONVERTED TICKING
MATTRESSES
PLAIN WEAVES
SATIN
SATIN WEAVES

TIGHT SELVEDGES
BT FABRIC DEFECTS
RT CRACKED SELVEDGES
DOG LEGGED SELVEDGES
LOOMS
PULLED IN SELVEDGES
SELVEDGE WARPS
SELVEDGES
SLACK SELVEDGES
TEMPLES
WEAVING
WOVEN FABRICS

TIGHTNESS
RT COMFORT
COURSES PER INCH
COVER FACTOR
DENSITY
DOUBLE JAMMING
FABRIC GEOMETRY
FIT
JAMMING POINT
LOOSENESS
MATRIX FREEDOM
PACKING FACTOR
PICKS PER INCH
SET (WOVEN FABRIC)
SHEAR (MODE OF DEFORMATION)
TENSION
YARN SLIPPAGE

TIME
RT AGING (MATERIALS)
DRYING TIME
KINETICS
KINETICS (CHEMICAL)
RATE
RELAXATION TIME (MECHANICAL)
SETTING TIME
STARTING TIME
TIME AND MOTION STUDIES
TIME CONSTANT
TIME SERIES
YEARS COVERAGE (1)
YEARS COVERAGE (5)
YEARS COVERAGE (10)
YEARS COVERAGE (25)
YEARS COVERAGE (50)
YEARS COVERAGE (75)
YEARS COVERAGE (100)

TIME AND MOTION STUDIES
RT JOB ANALYSIS
MANAGEMENT
PERSONNEL
PROCESS EFFICIENCY
TIME

TIME CONSTANT
RT CAPACITANCE (ELECTRICAL)
DYNAMIC CHARACTERISTICS
DYNAMIC RESPONSE
RELAXATION TIME (MECHANICAL)
STRESS RELAXATION
TIME

TIME SERIES
RT DISTRIBUTION
STATISTICAL ANALYSIS
STATISTICAL INFERENCE
STATISTICAL MEASURES
STATISTICAL METHODS
STATISTICS
TIME

TIN
BT METALS
RT TIN COMPOUNDS
WHISKER FIBERS (METAL)

TIN COMPOUNDS
NT ORGANOTIN COMPOUNDS
STANNOUS FLUORIDE
RT ANTIMICROBIAL FINISHES
CATALYSTS
TIN

TINSEL
BT HIGH TEMPERATURE FIBERS
INORGANIC FIBERS (MAN MADE)
METALLIC FIBERS
METALLIC YARNS
RT BICOMPONENT FIBERS
FIBERS
FOIL
LAME (TN)
LUSTER
MINERAL FIBERS

TINTING
RT BLENDING
BLENDS (FIBERS)
DYEING
FIBER IDENTIFICATION
FUGITIVE DYES
QUALITY CONTROL
STOCK (FIBER)
TINTING MACHINES

TINTING MACHINES
RT DYEING MACHINES
TINTING

TIPPINESS
BT DYEING DEFECTS
FABRIC DEFECTS
RT COLORFASTNESS
LEVELNESS (DYEING)
SKITTERINESS
SPECKINESS
STREAKINESS

TIRE CORDS
BT CORDAGE
CORDS
RT CASING PLY
DYNAMIC MODULUS
ELASTIC MODULUS (TENSILE)
FATIGUE RESISTANCE
FLAT SPOTTING (TIRES)
PLIED YARNS
ROPES
THERMAL PROPERTIES
TIRE FABRICS
TIRES
V BELTS
YARN REINFORCED ELASTOMERS

TIRE FABRICS
UF BEAD FILLER FABRICS
BEAD WOVEN FABRICS
FLIPPER FABRICS
TYRE FABRICS
BT FLAT WOVEN FABRICS
RT CHAFER FABRICS
FLAT SPOTTING (TIRES)
INDUSTRIAL FABRICS
THERMAL PROPERTIES
TIRE CORDS
TIRES

TIRES
UF TYRES
BT YARN REINFORCED ELASTOMERS
RT BREAKER FABRICS
CASING PLY
FLAT SPOTTING (TIRES)
GROWTH (FABRIC)
INDUSTRIAL FABRICS
TIRE CORDS
TIRE FABRICS

TITANIUM COMPOUNDS
NT TITANIUM DIOXIDE
RT PIGMENTS

TITANIUM DIOXIDE
UF ANATASE
BT OXIDES
TITANIUM COMPOUNDS
RT DELUSTERING AGENTS
LUSTER
PIGMENTS
YARN LUSTER

TITRATION
USE VOLUMETRIC ANALYSIS

TMCEA
UF TRIS (METHYLOL CARBAMOYLETHYL) AMINE
BT METHYLOL AMIDES
REACTANTS

TOBACCO CLOTH
BT FLAT WOVEN FABRICS
WOVEN FABRICS
RT AGRICULTURAL FABRICS
CRINOLINE
PLAIN WEAVES
SHEETING (FABRIC)

TOES
RT BALLET TOES (KNITTING)
HEELS (STOCKINGS)
HOSIERY
KNITTING
REINFORCED KNITTED FABRICS

TOHO DIRECT SPINNER (TN)
BT CONVERTERS (TOW)
TOW TO YARN MACHINES
RT STRETCH BREAKING
TOW TO YARN CONVERSION

TOLERANCES
RT ADJUSTMENTS
ALIGNMENT
ALTERATIONS
CALIBRATION
CORRECTION
DRIFT
ECCENTRICITY
ERRORS

TOLUENE
BT AROMATIC HYDROCARBONS
RT CARRIERS (DYEING)
SOLVENTS

TOLUENE ISOCYANATES
BT ISOCYANATES
RT POLYURETHANES

TONE IN TONE DYEING
BT CROSS DYEING
DYEING (FOR EFFECT)
RT BLEND DYEING
CONTRAST EFFECT
RESERVE EFFECT

RT SOLID SHADE
TWO COLOR DYEING
TWO TONE DYEING
UNION DYEING

TONGUE TEAR STRENGTH
BT FABRIC PROPERTIES (MECHANICAL)
FABRIC PROPERTIES
TEAR STRENGTH
RT ELMENDORF TEAR STRENGTH
TRAPEZOID TEAR STRENGTH

TONGUE TEAR TESTING
UF TONGUE TEAR TESTS
BT TEAR TESTING
RT ELMENDORF TEAR TESTING
TEAR STRENGTH
TRAPEZOID TEAR TESTING

TONGUE TEAR TESTS
USE TONGUE TEAR TESTING

TONGUES
BT FOOTWEAR COMPONENTS
RT BINDINGS (FOOTWEAR)
BOOTS
FOOTWEAR
GALOSHES
LACES (SHOES)
LEATHER
LEGGINGS
LININGS (FOOTWEAR)
PUMPS (FOOTWEAR)
RUBBER FOOTWEAR
SHOES
SLIPPERS
SNEAKERS
SOLES
SPATS
SYNTHETIC LEATHER
UPPERS
VINYL COATED FABRICS

TOP COMBING (COMB CYCLE)
RT COMBING
CUTTER COMBING
DETACHING
NIPPING

TOP COMBS
RT COMB BRUSHES
COMB CYLINDERS
COMB PINS
COMB SEGMENTS
COMBED SLIVERS
COMBING
COMBS
DETACHING ROLLS
HALF LAPS
LAP PLATES
NIPPER JAWS
NIPPER KNIVES
NIPPER PLATES
NOILS
RECTILINEAR COMBS
SLIVERS

TOP COVER
BT COVER
RT BOTTOM COVER
FABRIC COVER
NAP
PATTERN DEFINITION
STITCH CLARITY
YARN COVER

TOP DYEING
UF SLUBBING DYEING
NT CONTINUOUS TOP DYEING
BT DYEING (BY MATERIAL ASSEMBLY)
RT DOPE DYEING
PIECE DYEING
RECOMBING
STOCK DYEING
TOP DYEING MACHINES
TOPS
TOW DYEING
WOOL DYEING
YARN DYEING

TOP DYEING MACHINES
UF SLUBBING DYEING MACHINES
BT DYEING MACHINES
DYEING MACHINES (BY MATERIAL ASSEMBLY)
RT BATCH DYEING MACHINES
DYEING
DYEING (BY MATERIAL ASSEMBLY)
PACKAGE DYEING MACHINES
PIECE DYEING MACHINES
STOCK DYEING MACHINES
TOP DYEING
TOPS
TOW DYEING MACHINES
YARN DYEING MACHINES

TOP MAKING
BT WORSTED SYSTEM PROCESSING
RT BACKWASHING
NOBLE COMBING
TOP NOIL RATIO
TOPS

TOP MATCHING
RT COLOR
COLOR MATCHING
COMBING
GRADING
SHADE
TOPS

TOP NOIL RATIO
UF TEAR
RT COMB TEAR
COMBING
NOBLE COMBING
NOIL PER CENT
NOILS
SLIVER NOIL RATIO
TOP MAKING
TOPS
YIELD (RETURN)

TOP ROLLS
BT DRAFTING ROLLS
ROLLS
RT BACK ROLLS
BOTTOM ROLLS
CLEARERS (LINT)
DEAD WEIGHT LOADING
DRIVEN ROLLS
DRIVING ROLLS
FLUTED ROLLS
FRONT ROLLS
INTERMEDIATE ROLLS
PLAIN ROLLS
SPRING LOADING
TUMBLERS (ROLLS)

TOP STITCHED SEAMS
BT SEAMS
RT FELLED SEAMS

[illegible]
BT TEXTURED YARNS (FILAMENT)

TOPOLOGY
RT CONFIGURATION
MATHEMATICAL ANALYSIS
SHAPE

TOPPING
RT BOTTOMING
DYEING
PRINTING

TOPS
UF COMBED TOPS
NT OIL COMBED TOPS
RECOMBED TOPS
BT FIBER ASSEMBLIES
RT AUTOLEVELLERS
BALLING HEADS
CARD SLIVERS
COMB TEAR
COMBED SLIVERS
COMBING
DRAFTING (STAPLE FIBER)
FIBER ORIENTATION
IRREGULARITY
LAPS
LEADING HOOKS
NOBLE COMBS
NOILS
ROVINGS
SLIVERS
SLUBBINGS
TOP DYEING
TOP DYEING MACHINES
TOP MAKING
TOP MATCHING
TOP NOIL RATIO
TOW
TRAILING HOOKS
WEBS
WORSTED SYSTEM PROCESSING
YARNS

TORCHON
BT HAND MADE LACES
LACES

TORCHWICK YARNS
RT BUMP YARNS
CANDLEWICK YARNS

TORQUE
USE TWISTING MOMENTS

TORSION (GEOMETRIC)
RT ANGULAR DEFLECTION
COUPLES
CURVATURE
DIFFERENTIAL GEOMETRY
HELIX ANGLE
POLAR MOMENT OF INERTIA
REAL TWIST
TENSION
TORSIONAL BUCKLING
TORSIONAL ENERGY
TORSIONAL RIGIDITY
TWIST
TWISTING
TWISTING MOMENTS

TORSION (MECHANICAL)
USE TWISTING MOMENTS

TORSIONAL BEHAVIOR
RT STRESS STRAIN PROPERTIES (SHEAR)
TORSIONAL BUCKLING
TORSIONAL ENERGY
TORSIONAL RIGIDITY
TORSIONAL STABILITY
TWIST
TWISTING MOMENTS

TORSIONAL BUCKLING
BT BUCKLING
RT COMPRESSIVE BUCKLING
CREPE
CREPE YARNS
CURL
DIMENSIONAL STABILITY
ENTANGLEMENTS
HIGH TWIST
LOOPS
RESIDUAL TORQUE
SNARLING
SNARLING TENDENCY
SLENDERNESS RATIO
SNAP BACK
STRETCH YARNS (FILAMENT)
TENSILE BUCKLING
TEXTURING
TORSION (GEOMETRIC)
TORSIONAL BEHAVIOR
TORSIONAL ENERGY
TORSIONAL RIGIDITY
TWIST
TWIST LIVELINESS
WILDNESS

TORSIONAL ENERGY
BT ENERGY
STRAIN ENERGY
RT BENDING ENERGY
HELIX ANGLE
INTERNAL ENERGY
KINETIC ENERGY
POLAR MOMENT OF INERTIA
POTENTIAL ENERGY
RESIDUAL TORQUE
SNARLING TENDENCY
TORSION (GEOMETRIC)
TORSIONAL BEHAVIOR
TORSIONAL BUCKLING
TWIST
TWIST LIVELINESS
TWISTING MOMENTS
WILDNESS

TORSIONAL MOMENT
USE TWISTING MOMENTS

TORSIONAL RIGIDITY
BT MECHANICAL PROPERTIES
RT ANGULAR DEFLECTION
COUPLES
HELIX ANGLE
MOMENT OF INERTIA
PILL RESISTANCE
PILLING
POLAR MOMENT OF INERTIA
SHEAR MODULUS
TORSION (GEOMETRIC)
TORSIONAL BEHAVIOR
TORSIONAL ENERGY
TORSIONAL RIGIDITY
TWIST
TWISTING
TWISTING MOMENTS

TORSIONAL STABILITY
RT CIRCULAR KNITTING
DIMENSIONAL STABILITY
SPIRALITY (KNITTED FABRICS)
STABILITY
TORSIONAL BEHAVIOR
TORSIONAL BUCKLING
TORSIONAL ENERGY
TORSIONAL RIGIDITY

TOUCH
USE HAND

TOUGHNESS
UF TOUGHNESS INDEX
BT END USE PROPERTIES
STRESS STRAIN PROPERTIES
STRESS STRAIN PROPERTIES (TENSILE)
RT BREAKING ELONGATION
(CONT.)

RT BREAKING ENERGY
ENERGY ABSORPTION
ENERGY OF RETRACTION
FATIGUE RESISTANCE
STRESS STRAIN CURVES

TOUGHNESS INDEX
USE TOUGHNESS

TOW
BT FIBER ASSEMBLIES
RT CARD SLIVERS
CIGARETTE FILTERS
CONVERTERS (TOW)
DRAFT CUT
EXTRUSION
FIBER CUTTING
FILAMENT YARNS
FILAMENTS
GREENFIELD CONVERTER
LAPS
MAN MADE FIBERS
PACIFIC CONVERTER
PERLOK SYSTEM
ROVINGS
SACO LOWELL DIRECT SPINNER (TN)
SLIVERS
SLUBBINGS
SOLVENT SPINNING
SPIN FINISHES
SPINNERETS
STAINS DIRECT SPINNER
STRETCH BREAKING
SYNTHETIC FIBERS
THROWING
TOPS
TOW CONVERSION
TOW CRIMPING
TOW DYEING
TOW DYEING MACHINES
TOW TO TOP CONVERSION
TOW TO YARN CONVERSION
TURBOSTAPLER
WEBS
YARNS

TOW CONVERSION
NT TOW TO TOP CONVERSION
TOW TO YARN CONVERSION
RT CONVERTERS (TOW)
DRAFT CUT
FIBER BREAKAGE
FIBER CUTTING
FIBER SHUFFLING
FIBERS
GREENFIELD CONVERTER
HALLE SEYDEL STRETCH BREAKER
PACIFIC CONVERTER
PERLOK SYSTEM
STAPLE FIBERS
STRETCH BREAKING
THROWING
TOW
TURBOSTAPLER

TOW CONVERTERS
USE CONVERTERS (TOW)

TOW CRIMPING
BT CRIMPING
RT BUCKLING
CRIMP AMPLITUDE
CRIMP FREQUENCY
GEAR CRIMPING
HEAT SETTING (SYNTHETICS)
STUFFER BOX CRIMPING
THERMOPLASTICITY
TOW
Z CRIMP

TOW DYEING
BT DYEING (BY MATERIAL ASSEMBLY)
RT ACETATE DYEING
DOPE DYEING
PIECE DYEING
RAYON DYEING
STOCK DYEING
SYNTHETIC DYEING
TOP DYEING
TOW
TOW DYEING MACHINES
YARN DYEING

TOW DYEING MACHINES
BT DYEING MACHINES
DYEING MACHINES (BY MATERIAL ASSEMBLY)
RT DOPE DYEING EQUIPMENT
DYEING
DYEING (BY MATERIAL ASSEMBLY)
PACKAGE DYEING MACHINES
PIECE DYEING MACHINES
STOCK DYEING MACHINES
TOP DYEING MACHINES
TOW
TOW DYEING
YARN DYEING MACHINES

TOW TO SLIVER CONVERSION
USE TOW TO TOP CONVERSION

TOW TO TOP CONVERSION
UF TOW TO SLIVER CONVERSION
BT TOW CONVERSION
RT CONVERTERS (TOW)
COURTAULDS CONVERTER
FIBER CUTTING
FIBER SHUFFLING
HALLE SEYDEL STRETCH BREAKER
PACIFIC CONVERTER
PERLOK SYSTEM
RIETER CONVERTER
STRETCH BREAKING
TOW
TURBOSTAPLER

TOW TO TOP MACHINES
NT COURTAULDS CONVERTER
HALLE SEYDEL STRETCH BREAKER
PACIFIC CONVERTER
PERLOK SYSTEM
RIETER CONVERTER
TURBOSTAPLER
BT CONVERTERS (TOW)

TOW TO YARN CONVERSION
UF DIRECT SPINNING
BT TOW CONVERSION
RT CONVERTERS (TOW)
DIRECT SPUN YARNS
DRAFT CUT
SACO LOWELL DIRECT SPINNER (TN)
STAINS DIRECT SPINNER
STRETCH BREAKING
TOHO DIRECT SPINNER (TN)
TOW
TOW TO YARN CONVERSION

TOW TO YARN MACHINES
UF DIRECT SPINNING MACHINES
NT SACO LOWELL DIRECT SPINNER (TN)
STAINS DIRECT SPINNER
TOHO DIRECT SPINNER (TN)
BT CONVERTERS (TOW)
RT DIRECT SPUN YARNS
TOW TO YARN CONVERSION

TOWELS
RT ABSORBENCY (MATERIAL)
CRASH TOWELING
HOUSEHOLD FABRICS
TERRYCLOTH

TOXICITY
UF POISONOUS
RT ACCIDENTS
BIOCHEMICAL PROPERTIES
DERMATITIS
OCCUPATIONAL HAZARDS
SAFETY
SKIN IRRITATION

TRACER FIBER TECHNIQUES
USE TRACER TECHNIQUES

TRACER FIBERS
UF DYED FIBER TRACERS
FIBER (TRACER)
BT TRACERS
RT BLENDING
BLENDS (FIBERS)
FIBER MIGRATION
HOOKED FIBERS
RADIAL DISTRIBUTION
RADIATION COUNTERS
RADIOACTIVE TRACERS
TRACER TECHNIQUES

TRACER TECHNIQUES
UF DYED FIBER TRACER TECHNIQUES
RADIOACTIVE TRACER TECHNIQUES
TRACER FIBER TECHNIQUES
RT RADIATION COUNTERS
RADIOACTIVE COMPOUNDS
RADIOACTIVE TRACERS
TRACER FIBERS
TRACERS

TRACERS
NT RADIOACTIVE TRACERS
TRACER FIBERS
RT TRACER TECHNIQUES

TRAILING HOOKS
BT FIBER HOOKS
HOOK DIRECTION
RT CARD SLIVERS
CARD WEBS
CARDING
DEGREE OF ORIENTATION
DRAFT
DRAFTING (STAPLE FIBER)
DRAFTING THEORY
FIBER ORIENTATION

RT HOOK FORMATION
HOOK REMOVAL
HOOKED FIBERS
LEADING HOOKS
LINDSLEY RATIOS
NEPS
PROCESSING DIRECTION
SLIVER CANS
SLIVER REVERSALS
SLIVERS
TOPS
WEB ORIENTATION

TRAINING
RT EVALUATION
JOB ANALYSIS
MAINTENANCE
MANAGEMENT
PATROLLING
PERSONNEL
PROCESS EFFICIENCY
PRODUCTION

TRAM
RT FILLING YARNS
ORGANZINE
SILK

TRANSDUCERS
NT ELECTRICAL INPUT TRANSDUCERS
ELECTRICAL OUTPUT TRANSDUCERS
RT AMPLIFIERS
CONTROL SYSTEMS
ELECTRONIC INSTRUMENTS
INSTRUMENTATION
OSCILLATORS

TRANSFER PROPERTIES
NT ABSORBENCY (MATERIAL)
ABSORPTIVITY (RADIATION)
CONDUCTIVITY (THERMAL)
DIFFUSIVITY
EMISSIVITY
REFLECTIVITY
TRANSMISSIVITY
BT MECHANICAL PROPERTIES
PHYSICAL CHEMICAL PROPERTIES
PHYSICAL PROPERTIES (EXCLUDING MECHANICAL)
RT ACOUSTIC PROPERTIES
AESTHETIC PROPERTIES
BIOCHEMICAL PROPERTIES
CHEMICAL PROPERTIES
ELECTRICAL PROPERTIES
FABRIC PROPERTIES
FABRIC PROPERTIES (MECHANICAL)
FIBER PROPERTIES
MECHANICAL DETERIORATION PROPERTIES
PROPERTIES
STRESS OPTICAL PROPERTIES
STRESS STRAIN PROPERTIES
STRUCTURAL PROPERTIES
SURFACE PROPERTIES (MECHANICAL)
SURFACE PROPERTIES (PHYSICAL CHEMICAL)
THERMAL PROPERTIES
TRANSPARENCY
TURBULENCE
WATER VAPOR TRANSMISSION

TRANSFER TAILS
RT KNITTING
KNITTING MACHINES
LOOMS
WEAVING
WINDERS
WINDING

TRANSFERRING (LOOPS)
USE LOOP TRANSFER

TRANSFERRING (MATERIAL)
RT CONVEYORS
DELIVERY
FLOW (MATERIALS)
MATERIAL PROCESSING
MATERIALS HANDLING
STITCH TRANSFER

TRANSIENT RESPONSE
BT DYNAMIC RESPONSE
RT DYNAMIC CHARACTERISTICS

TRANSITION TEMPERATURES
NT GLASS TRANSITION TEMPERATURE
MELTING POINT
SECOND ORDER TRANSITION TEMPERATURE
SECONDARY TRANSITION TEMPERATURE (AMORPHOUS PHASE)
SECONDARY TRANSITION TEMPERATURE (CRYSTALLINE PHASE)
SECONDARY TRANSITION TEMPERATURES
RT TRANSITIONS (POLYMERS)

TRANSITIONS (POLYMERS)
UF POLYMER TRANSITIONS
NT GLASS RUBBER TRANSITION
PRIMARY TRANSITIONS
SECONDARY TRANSITIONS
RT GLASS TRANSITION TEMPERATURE
MELTING POIN
SECONDARY TRANSITION TEMPERATURES
SOLIDIFICATION
TRANSITION TEMPERATURES

TRANSLATION
RT ACCELERATION (MECHANICAL)
AXIAL MOTION
FORCE
KINETIC ENERGY
MOMENTUM
ROTATION
VELOCITY

TRANSLUCENCY
BT AESTHETIC PROPERTIES
END USE PROPERTIES
OPTICAL PROPERTIES
RT APPEARANCE
COVER
FABRIC COVER
FABRIC PROPERTIES
FABRIC PROPERTIES (PHYSICAL EXCLUDING MECHANICAL)
FINISH (PROPERTY)
LUSTER
OPACITY
REFLECTANCE
TRANSMITTANCE
TRANSPARENCY

TRANSMISSIVITY
BT OPTICAL PROPERTIES
THERMAL PROPERTIES
TRANSFER PROPERTIES
RT ABSORPTIVITY (RADIATION)
CONDUCTIVITY (THERMAL)
EMISSIVITY
REFLECTIVITY
TRANSLUCENCY
TRANSMITTANCE
TRANSPARENCY

TRANSMITTANCE
UF LIGHT TRANSMISSION
RT ABSORBENCY (MATERIAL)
ABSORPTION (RADIATION)
OPTICAL PROPERTIES
THERMAL PROPERTIES
THERMAL TESTING
TRANSLUCENCY
TRANSMISSIVITY
TRANSPARENCY

TRANSPARENCY
BT AESTHETIC PROPERTIES
END USE PROPERTIES
OPTICAL PROPERTIES
RT ABSORPTION (RADIATION)
APPEARANCE
BRIGHTNESS
COVER
DULLNESS
FABRIC COVER
FABRIC PROPERTIES
FABRIC PROPERTIES (PHYSICAL EXCLUDING MECHANICAL)
FINISH (PROPERTY)
LUSTER
OPACITY
REFLECTANCE
SHEERNESS
TEXTURE
TRANSFER PROPERTIES
TRANSLUCENCY
TRANSMISSIVITY
TRANSMITTANCE

TRANSPORTING VESSELS
BT VESSELS
RT AGING (STORAGE)
KETTLES (DYEING)
KETTLES (HEATING)
STORAGE VESSELS

TRANSVERSE FORCE
UF STRESS (TRANSVERSE)
BT FORCE
RT CENTRIFUGAL FORCE
NORMAL FORCE
SHEAR FORCE
TANGENTIAL FORCE

TRANSVERSE VIBRATIONS
BT VIBRATION
RT AMPLITUDE
FREQUENCY
HIGH FREQUENCY
IMPACT PENETRATION TESTING
LONGITUDINAL VIBRATIONS
RESONANT FREQUENCY
SOUND
STRESS PROPAGATION
TRAVERSE
VIBRASCOPE
WAVE LENGTH
WINDING
YARN DYNAMICS

TRAP SORTING (WOOL)
USE WOOL SORTING

TRAPEZOID TEAR STRENGTH
BT FABRIC PROPERTIES
TEAR STRENGTH
RT ELMENDORF TEAR STRENGTH
TONGUE TEAR STRENGTH

TRAPEZOID TEAR TESTING
UF TRAPEZOID TEAR TESTS
BT TEAR TESTING
RT ELMENDORF TEAR TESTING
TEAR STRENGTH
TONGUE TEAR TESTING

TRAPEZOID TEAR TESTS
USE TRAPEZOID TEAR TESTING

TRASH
NT **BURRS**
MOTES
SEEDS (TRASH)
VEGETABLE MATTER
BT **IMPURITIES**
WASTE
RT **BOLLS**
CARD STRIPS
CLEANLINESS
CLEARERS (CARD)
CLIPPINGS
FLY (WASTE)
GINNING
LICKER IN SCREENS
LINT (WASTE)
MOTE KNIVES
PICKING (OPENING)
TRASH CONTENT
WASTE CONTROL KNIVES
WEB PURIFIERS

TRASH CONTENT
RT LINT CONTENT
NEP COUNT
TRASH
YIELD (RETURN)

TRAVELER BURNOUT
RT BALLOON DYNAMICS
END BREAKAGES
FRICTION
HEATING
LUBRICATION
RINGS
SPINDLE SPEED
SPINNING
THERMAL DEGRADATION
TRAVELER CHATTER
TRAVELERS
UNLUBRICATED RINGS
WEAR
WEDGING

TRAVELER CHATTER
RT BALLOON DYNAMICS
FRICTION
LUBRICATION
RING SPINNING
RINGS
SPINNING
SPINNING TRAVELERS
TRAVELER BURNOUT
TRAVELERS
TWIST
TWISTING
UNLUBRICATED RINGS
VIBRATION
WEAR
WEDGING
YARN TENSION

TRAVELERS
UF RING TRAVELERS
NT CIRCULAR TRAVELERS
ELLIPTICAL TRAVELERS
NYLON TRAVELERS
PLYING TRAVELERS
SPINNING TRAVELERS
STEEL TRAVELERS
RT ANTIWEDGE RINGS
BALLOON DYNAMICS
BALLOON TENSION
BALLOONS
CAPS (SPINNING)
CENTRIFUGAL FORCE
FLYERS
LUBRICATED RINGS
NOVELTY TWISTERS
RING FRAMES
RING RAILS
RT RING SPINNING
RING TWISTERS
RINGS
SINTERED METAL RINGS
SPINNING
SPINNING DYNAMICS
SPINNING FRAMES
TRAVELER BURNOUT
TRAVELER CHATTER
TWIST
TWISTERS
TWISTING
UNLUBRICATED RINGS
WEDGING
YARN TENSION

TRAVELING BLOWERS
RT BLOWERS
CLEANERS

TRAVERSE
UF CHASE
TRAVERSE MOTION
NT RECIPROCATING TRAVERSE
ROTARY TRAVERSE
RT BUILDER MOTIONS
CLOSE WINDING
COMBINATION WIND
DRUM WINDING
FILLING WIND
GAIN (WINDING)
LARGE PACKAGES
LIFT (WINDING)
NOSE BUNCHING
OPEN WINDING
PACKAGES
PACKAGING (OPERATION)
PRECISION WINDING
QUILLING
RING RAILS
SPINDLE RAILS
SPINNING
SPINNING FRAMES
TWISTERS
WARP WIND
WINDING

TRAVERSE MOTION
USE TRAVERSE

TREADLES
RT LOOMS
SHEDDING
WEAVING

TRELLIS MODEL
RT BIAXIAL STRESS
FABRIC GEOMETRY
FABRIC PROPERTIES
FABRIC PROPERTIES (STRUCTURAL)
GEOMETRY
MODELS (MATHEMATICAL)
POISSONS RATIO
UNIAXIAL STRESS

TRIACETATE
USE TRIACETATE FIBERS

TRIACETATE DYEING
BT DYEING (BY FIBER CLASSES)
RT ACETATE DYEING
DISPERSE DYEING

TRIACETATE FIBERS
UF TRIACETATE
NT ARNEL (TN)
TRICEL C (TN)
BT CELLULOSE ESTER FIBERS
MAN MADE FIBERS
RT ACETATE FIBERS
ARTIFICIAL SILK (ARCHAIC)
CELLULOSE ESTERS
CUPRAMMONIUM RAYON
VISCOSE RAYON

TRIAL AND ERROR SOLUTIONS
BT SOLUTIONS (MATHEMATICAL)
RT ANALYZING
APPROXIMATIONS
COMPUTATIONS
COMPUTERS
EXPERIMENTAL ANALYSIS
EXPERIMENTATION
GENERAL SOLUTIONS
GRAPHICAL SOLUTIONS
NUMERICAL SOLUTIONS
PARTICULAR SOLUTIONS

TRIAMINES
USE AMINES

TRIAZINES
NT CHLOROCYANURIC ACID
CYANURIC ACID
CYANURIC CHLORIDE
MELAMINE
BT HETEROCYCLIC COMPOUNDS
RT CHLOROTRIAZINE DYES

TRIAZONES
NT ETHYL TRIAZONE
METHYL TRIAZONE
METHYLOL TRIAZONES
BT HETEROCYCLIC COMPOUNDS

TRIBASIC ACIDS
NT CITRIC ACID
PHOSPHORIC ACID
BT ACIDS
RT INORGANIC ACIDS
ORGANIC ACIDS

TRIBOELECTRIC SERIES
RT ANTISTATIC AGENTS
ANTISTATIC BEHAVIOR
FRICTION
RUBBING
STATIC ELECTRICITY
TRIBOELECTRICITY

TRIBOELECTRICITY
RT ANTISTATIC AGENTS
ANTISTATIC BEHAVIOR
DISCHARGE (ELECTRIC)
FRICTION
RUBBING
STATIC ELECTRICITY
TRIBOELECTRIC SERIES

TRICEL C (TN)
BT TRIACETATE FIBERS

TRICHLOROACETIC ACID
USE CHLOROACETIC ACID

TRICHLOROBENZENE
BT CHLORINATED HYDROCARBONS
RT CARRIERS (DYEING)
CHLORINATED SOLVENTS
DICHLOROBENZENE
SOLVENTS

TRICHLOROCYANURIC ACID
USE CHLOROCYANURIC ACID

TRICHLOROETHYLENE
BT CHLORINATED HYDROCARBONS
RT CHLORINATED SOLVENTS
DRY CLEANING SOLVENTS
NAPHTHA
PERCHLOROETHYLENE
PERCLENE (TN)

TRICHLOROPYRIMIDINE DYES
USE CHLOROPYRIMIDINE DYES

TRICHROMATIC COEFFICIENTS
BT CIE SYSTEM
RT CHROMATICITY DIAGRAM
COLOR
COLOR MATCHING
COLORIMETRY
TRISTIMULUS VALUES

TRICOT
USE TRICOT KNITTED FABRICS

TRICOT FABRICS
USE TRICOT KNITTED FABRICS

TRICOT KNITTED FABRICS
UF JERSEY (WARP KNITTING)
STOCKINETTE WARP KNIT FABRICS
TRICOT
TRICOT FABRICS
BT KNITTED FABRICS
WARP KNITTED FABRICS
RT DOUBLE KNIT FABRICS
FILLING KNITTED FABRICS
FLAT KNITTING
LAPPING (WARP KNITTING)
MACHINE KNITTING
OVERLAP
RUNNER RATIO
SET (WARP KNITTING)
TRICOT KNITTING
TRICOT KNITTING MACHINES
TRICOT STITCHES
UNDERLAP
WARP KNITTING
WARP KNITTING MACHINES

TRICOT KNITTING
BT KNITTING
WARP KNITTING
RT COMPOUND NEEDLES
GUIDE BARS
HALF SET
LACE MAKING
OVERLAP
PRESSER BARS
RUNNER RATIO
SET (WARP KNITTING)
TRICOT KNITTED FABRICS
TRICOT KNITTING MACHINES
UNDERLAP
WARP KNITTING MACHINES

TRICOT KNITTING MACHINES
BT KNITTING MACHINES
WARP KNITTING MACHINES
RT CASTING OFF
COMPOUND NEEDLES
FULL SET
GUIDE BARS
HALF SET
LACE MACHINES
LAPPING (WARP KNITTING)
MILANESE KNITTING MACHINES
OVERLAP
PRESSER BARS
RASCHEL KNITTING MACHINES
RUNNER RATIO
RUNNERS
SET (WARP KNITTING)
SPRING NEEDLES
TRICOT KNITTED FABRICS
TRICOT KNITTING
TRICOT STITCHES
UNDERLAP
WEBHOLDERS

TRICOT STITCHES
RT CLOSED STITCHES
COMPOUND NEEDLES
FILLING KNITTED FABRICS
KNITTED FABRICS
LAPPING (WARP KNITTING)
OPEN STITCHES
OVERLAP
TRICOT KNITTED FABRICS
TRICOT KNITTING
TRICOT KNITTING MACHINES
UNDERLAP
WARP KNITTED FABRICS
WARP KNITTING
WARP KNITTING MACHINES

TRIETHANOLAMINE
BT ALKANOLAMINES
TERTIARY AMINES

TRIETHANOLAMINE SOAPS
BT ALKANOLAMINE SOAPS
RT AMINE SURFACTANTS

TRIGLYCERIDES
RT GLYCERYL ESTER SURFACTANTS

TRIHEXYL SULFOCARBALLYLATE
NT NEKAL NS (TN)
BT SULFOPOLYCARBOXYLIC ESTERS

TRILOBAL CROSS SECTION
BT FIBER CROSS SECTIONS

TRILOCK (TN)
BT MULTIPLE LAYER FABRICS
WOVEN FABRICS

TRIMETHYLOL MELAMINE
BT METHYLOL MELAMINES

TRIMMING (OPERATION)
BT GARMENT MANUFACTURE
RT BOARDING
CUTTING (TAILORING)
GATHERING
HEMMING
PINCHING
PUCKERING
RUFFLING
SEAMING
SERGING
SEWING
SHIRRING
STITCH QUALITY
STITCHES (SEWING)
STITCHING
TAILORING
TRIMMINGS

TRIMMINGS
BT GARMENT COMPONENTS
RT PIPING
TRIMMING (OPERATION)

TRIMNESS
RT BAGGINESS
COMFORT
DRAPE
FIT
STRETCH FABRICS

TRIS (AZIRIDINYL) PHOSPHINE OXIDE
USE APO

TRIS (METHYLAZIRIDINYL) PHOSPHINE OXIDE
USE MAPO

TRIS (METHYLOL CARBAMOYLETHYL) AMINE
USE TMCEA

TRISODIUM PHOSPHATE
BT PHOSPHATES
RT CALGON (TN)
TSPP

TRISTIMULUS VALUES
BT CIE SYSTEM
RT CHROMATICITY DIAGRAM
COLOR
COLOR MATCHING
COLORIMETRY
TRICHROMATIC COEFFICIENTS

TROPICAL FOOTWEAR
BT FOOTWEAR

TROPICAL WORSTED
BT FLAT WOVEN FABRICS
WORSTED FABRICS
WOVEN FABRICS
RT PLAIN WEAVES
WOOL
WORSTED SYSTEM PROCESSING

TROUSERS
BT GARMENTS
WORK CLOTHING
RT BAGGINESS
SLACKS

TRUE CONTACT AREA
BT CONTACT AREA
RT APPARENT CONTACT AREA
COEFFICIENT OF FRICTION
CONDUCTION
FRICTION
FRICTION THEORY
HEAT TRANSFER

PLASTIC FLOW
RESISTIVITY

TRUE STRAIN
BT STRAIN
RT ELASTIC STRAIN
PLASTIC STRAIN

TRUE TWIST
USE REAL TWIST

TRUMPETS
UF CONDENSING FUNNELS
RT CARD SLIVERS
CARD WEBS
CARDING
CARDS
COILERS
COTTON CARDING
COTTON CARDS
CROSROL WEB PURIFIER (TN)
DETACHING ROLLS
DOUBLE TRUMPETS
HALF LAP
RECTILINEAR COMBS
SLIVER CANS
SLIVERS
WORSTED CARDING

TRUNKS
BT ATHLETIC CLOTHING
OUTERWEAR
SPORTSWEAR
RT BRIEFS
DRAWERS
SHORTS

TRYPTOPHAN
BT AMINO ACIDS

TSATLEE SILK
BT SILK
RT TUSSAH SILK

TSPP
UF TETRASODIUM PYROPHOSPHATE
BT PHOSPHATES
PYROPHOSPHATES
RT CALGON (TN)
CHELATING AGENTS
DYEING AUXILIARIES
PH CONTROL
TRISODIUM PHOSPHATE

TUBES
RT COPS
PACKAGES
QUILLS

TUBULAR FABRICS
NT CIRCULAR KNITTED FABRICS
PILLOW TUBING
WOVEN TUBES
BT FABRICS
RT PAPERMAKERS FELTS

TUBULAR KNITTED FABRICS
USE CIRCULAR KNITTED FABRICS

TUCK LACES
BT HAND MADE LACES
LACES

TUCK STITCHES
BT KNITTED STITCHES
RT FALL PLATE FABRICS
FLOAT STITCHES (KNITTING)
KNITTING
KNOTTED STITCHES (KNITTING)
PURL STITCHES
TUCKING (KNITTING)

TUCKING (GARMENT MANUFACTURE)
NT AIR TUCKING
KNIFE TUCKING
BT GARMENT MANUFACTURE
RT BACK TACKING
BASTING
CHAIN SEWING
CURVED SEAMING
SEAMING
SEWING
STITCHING
TUCKS

TUCKING (KNITTING)
RT CASTING OFF
FASHIONING
FILLING KNITTING
FILLING KNITTING MACHINES
KNITTING
LANDING LOOPS
LAYING IN (KNITTING)
TUCK STITCHES
WELTING

TUCKS
RT DARTS
GARMENT COMPONENTS
GARMENT MANUFACTURE
PLEATS
TUCKING (GARMENT MANUFACTURE)

TUFT FORMERS
RT AIR BRIDGES
CARDS
CONDENSER SCREENS
CURLATORS (TN)
NONWOVEN FABRICS
RANDO FEEDER (TN)
RANDO WEBBER (TN)
WEB FORMING MACHINES
WEBS

TUFTED BEDSPREADS
BT BEDSPREADS
RT TUFTED FABRICS
TUFTING MACHINES

TUFTED CARPETS
BT CARPETS
FURNISHING FABRICS
PILE FABRICS (TUFTED)
TUFTED FABRICS
RT AXMINSTER CARPETS
BACKING FABRICS
CHENILLE CARPETS
CHENILLE FABRICS (TUFTED FABRIC)
KNITTED CARPETS
MALIPOL FABRIC (TN)
NONWOVEN CARPETS
SEWING
SPUNBONDED (NONWOVENS)
TUFTING
TUFTING MACHINES
TUFTS (FIBER)
WILTON CARPETS
WOVEN CARPETS

TUFTED FABRICS
UF FABRICS (TUFTED)
NT CANDLEWICK FABRICS (TUFTED FABRIC)
CHENILLE FABRICS (TUFTED FABRIC)
MALIPOL FABRIC (TN)
TUFTED CARPETS
BT FABRICS
FABRICS (ACCORDING TO STRUCTURE)
PILE FABRICS (TUFTED)
RT BACKING FABRICS
BRAIDS
CARPETS
FABRIC STRUCTURE
FLOCKED FABRICS
KNITTED FABRICS
LACES
LAMINATED FABRICS
MALIWATT FABRIC (TN)
NEEDLED FABRICS
NEEDLING
NETS
NONWOVEN FABRICS
SEWING
STITCHED PILE FABRICS
TUFTING
TUFTING MACHINES
WOVEN FABRICS

TUFTING
RT BACKING FABRICS
BEAMS
KNITTING
MALI PROCESSES (TN)
MALIPOL FABRIC (TN)
MALIPOL PROCESS (TN)
NEEDLE HEATING
NEEDLE PENETRATION
NEEDLE POINT
NEEDLE SHANK (SEWING)
NEEDLES
NEEDLING
PILE FABRICS (TUFTED)
PUNCHING
PUNCTURING
SEWING
SEWING DAMAGE
TUFTED CARPETS
TUFTED FABRICS
TUFTING MACHINES
TUFTING NEEDLES
TUFTS (FIBER)
WEAVING
YARN DAMAGE
YARN SEVERANCE
YARN SLIPPAGE

TUFTING NEEDLES
BT NEEDLES
RT KNITTING NEEDLES
NEEDLE HEATING
NEEDLE PENETRATION
NEEDLE POINT
NEEDLE SHANK (SEWING)
NEEDLING
TUFTING
TUFTING MACHINES

TUFTING MACHINES
NT CUT PILE TUFTING MACHINES
LOOP PILE TUFTING MACHINES
RT NEEDLE PENETRATION
NEEDLE POINT
NEEDLING
PILE FABRICS (TUFTED)
TUFTED BEDSPREADS
TUFTED CARPETS
TUFTED FABRICS
TUFTING
TUFTING NEEDLES

TUFTS (FIBER)
BT FIBER ASSEMBLIES
RT CARDING
FIBER ARRANGEMENT
FIBER CONFIGURATION
LAPS
NEPS
OPENING
SLIVERS
SLUBS
TUFTED CARPETS
TUFTING

TUMBLE DRYING
BT DRYING
RT BATCH DRYING
CENTRIFUGAL DRYING
DRYERS
DRYING CANS
FREEZE DRYING
LAUNDERING
LINE DRYING (EASY CARE GARMENTS)
TUMBLING
VACUUM DRYING

TUMBLERS (ROLLS)
BT DRAFTING ROLLS
ROLLS
RT BOTTOM ROLLS
DRAWFRAMES
INTERMEDIATE ROLLS
TOP ROLLS

TUMBLING
BT GARMENT MANUFACTURE
RT BULKING
KNITTED OUTERWEAR
TUMBLE DRYING

TUNGSTEN COMPOUNDS
RT GOLD COMPOUNDS
MOLYBDENUM COMPOUNDS
SILVER COMPOUNDS

TURBIDITY
BT OPTICAL PROPERTIES
RT OPACITY
SUSPENSIONS

TURBOSTAPLER
BT CONVERTERS (TOW)
TOW TO TOP MACHINES
RT DRAFT CUT
STRETCH BREAKING
TOW
TOW CONVERSION

TURBULENCE
RT AERODYNAMICS
AIR FLOW
AIR RESISTANCE
AIR STREAMS
BOUNDARY LAYER
FLUID FLOW
LAMINAR FLOW
MASS TRANSFER
MOMENTUM TRANSFER
TRANSFER PROPERTIES

TURBULENT FLOW
BT FLUID FLOW
RT AERODYNAMICS
AIR FLOW
AIR STREAMS
BOUNDARY LAYER
LAMINAR FLOW
MASS TRANSFER
VISCOUS FLOW

TURNS PER INCH
RT HIGH TWIST
LOW TWIST
REAL TWIST
TORSION (GEOMETRIC)
TWIST
TWIST LIVELINESS
TWIST MULTIPLIER
TWIST TESTING
WILDNESS

TUSSAH SILK
BT SILK
RT TSATLEE SILK

TWEENS (TN)
BT SORBITAN ESTER SURFACTANTS

TWILL ANGLES
RT CHEVRON TWILLS
FABRIC GEOMETRY
FABRIC PROPERTIES (STRUCTURAL)
PATTERN (FABRICS)
TWILL WEAVES
TWILLS
TWIST
TWIST SENSE

TWILL CHECKBOARD
BT FLAT WOVEN FABRICS
WOVEN FABRICS
RT TWILL WEAVES

TWILL SET
RT CARD CLOTHING
CARD CLOTHING FOUNDATION
CARD NAILING
CARD WIRES
CARDING
CARDS
COUNT CROWN AND GAUGE
FILLET CLOTHING
KNEE
METALLIC CARD CLOTHING
NOGG
NOGGS PER INCH
RIB SET
SHEET CLOTHING

TWILL WEAVES
BT FLAT WOVEN FABRICS
WEAVE (TYPES)
WOVEN FABRICS
RT ALBERT TWILL
ANGOLA (FABRIC)
BANNOCK BURN
BARATHEA
BLUETTE
BOLTON SHEETING
BROADCLOTH
CAVALRY TWILL
CHARMANETTE SATIN
CHEVRON TWILLS
COUTIL
COVERT
DENIMS
DOESKIN
ELASTIQUE
FLAT WOVEN FABRICS
FOULARD (FABRIC)
GABARDINE
HARVARD
ITALIAN CLOTH
JEANS
LASTING
SERGE
SHARKSKIN
TWILL ANGLES
TWILL CHECKBOARD
TWILLS
UMBRELLA CLOTH
WHIPCORD

TWILLS
NT CHEVRON TWILLS
FANCY TWILLS
(CONT.)

NT HERRINGBONE TWILLS
LOW TWILLS
REGULAR TWILLS
STEEP TWILLS
Z TWILLS
RT FABRIC GEOMETRY
FILLING EFFECT (TWILL)
S TWILLS
STRIPES
TWILL ANGLES
TWILLS
TWIST SENSE
WARP EFFECT (TWILL)

TWINE
NT FIBER ASSEMBLIES
BT CORDAGE
RT CORDS
FISHING LINES
STRANDS
STRING

TWIST
NT ALTERNATING TWIST
FALSE TWIST
FIBER TWIST
HIGH TWIST
LOW TWIST
PLY TWIST
PRODUCERS TWIST
REAL TWIST
S TWIST
SINGLES TWIST
Z TWIST
ZERO TWIST
BT MODES OF DEFORMATION
YARN PROPERTIES
RT ANGULAR DEFLECTION
BALANCED YARNS
CORDS
COUPLES
DIRECTION (TWILL)
DRAFTING TWIST
FALSE TWIST SPINDLES
FALSE TWIST TUBES
FALSE TWIST YARNS
FALSE TWISTING (EXCEPT TEXTURING)
FIBER MIGRATION
HELIX ANGLE
HELIX REVERSALS
OVEREND UNWINDING
PACKING FACTOR
PLYING
PROCESS VARIABLES
RING SPINNING
ROPES
ROVING (OPERATION)
ROVINGS
SENSE (DIRECTION)
SPINNING
SPINNING FRAMES
SPINNING LIMIT
STRETCH YARNS (FILAMENT)
TORSION (GEOMETRIC)
TORSIONAL BEHAVIOR
TORSIONAL BUCKLING
TORSIONAL ENERGY
TORSIONAL RIGIDITY
TRAVELERS
TURNS PER INCH
TWIST CONTRACTION
TWIST DISTRIBUTION
TWIST LIVELINESS
TWIST MULTIPLIER
TWIST SENSE
TWIST TESTING
TWIST TRIANGLE
TWISTERS
TWISTING
TWISTING MOMENTS
YARN CROSS SECTIONS
YARN GEOMETRY
YARN STRUCTURE
YARNS
Z TWIST YARNS

TWIST BREAK METHOD (TWIST TESTING)
BT TWIST TESTING
RT TWIST UNTWIST METHOD (TWIST TESTING)

TWIST CONSTANT
USE TWIST MULTIPLIER

TWIST CONTRACTION
RT CONTRACTION
HELIX ANGLE
REAL TWIST
TORSION (GEOMETRIC)
TORSIONAL BUCKLING
TWIST
TWISTING
YARN GEOMETRY

TWIST CONTROL
RT AMBLER SUPERDRAFT
APRON DRAFTING
CASABLANCA SYSTEM
CONTROL SURFACES
DRAFTING (STAPLE FIBER)
FALSE TWIST
FALSE TWIST TUBES
FIBER SLIPPAGE
HIGH DRAFT SPINNING
MULE SPINNING
OVEREND UNWINDING
PREFERENTIAL DRAFT
PROCESS CONTROL
SPINNING
TWIST DISTRIBUTION
TWISTING

TWIST DETERMINATION
USE TWIST TESTING

TWIST DISTRIBUTION
UF TWIST VARIATION
RT DRAFTING THEORY
OVEREND UNWINDING
REAL TWIST
THICK SPOTS
TWIST
TWIST CONTROL
TWIST RUN BACK
TWIST TRIANGLE
TWISTING
WEAK SPOTS

TWIST GEARS
BT GEARS
RT GEAR DRIVES
SPINNING FRAMES
TWISTERS

TWIST LIVELINESS
NT WILDNESS
BT LIVELINESS
YARN PROPERTIES
RT BALANCED YARNS
BUCKLING
CREPING
DIMENSIONAL STABILITY
MULTIPLIED YARNS
PLIED YARNS
PLY TWIST
PLYING
RESIDUAL TORQUE
S TWIST
SINGLES YARNS
SNARLING
SNARLING TENDENCY
STRETCH YARNS (FILAMENT)
TEXTURING
TORSIONAL BUCKLING
TORSIONAL ENERGY
TURNS PER INCH
TWIST
TWIST MULTIPLIER
TWISTING
TWO PLY YARNS
YARNS
Z TWIST

TWIST MULTIPLE
USE TWIST MULTIPLIER

TWIST MULTIPLIER
UF TWIST CONSTANT
TWIST MULTIPLE
RT [illegible]
HELIX ANGLE
HIGH TWIST
LOW TWIST
PACKING FACTOR
REAL TWIST
SPINNING LIMIT
TURNS PER INCH
TWIST
TWIST LIVELINESS
TWISTING
WILDNESS
YARN CROSS SECTIONS
YARN GEOMETRY
YARN STRUCTURE

TWIST RUN BACK
RT BALLOON DYNAMICS
BALLOONS
SPINNING
SPINNING BALLOONS
TWIST DISTRIBUTION
TWIST TRIANGLE
TWISTING

TWIST SENSE
BT YARN PROPERTIES
RT OVEREND UNWINDING
S TWIST
SENSE (DIRECTION)
TWILL ANGLES
TWILLS
TWIST
Z TWIST

TWIST SETTERS
RT ANNEALING
OVENS
PERMANENT DEFORMATION
STEAMING
STRESS RELAXATION

TWIST TESTERS
BT TESTING EQUIPMENT
RT TWIST
TWIST TESTING

TWIST TESTING
UF TWIST DETERMINATION
NT TWIST BREAK METHOD (TWIST TESTING)
TWIST UNTWIST METHOD (TWIST TESTING)
BT TESTING
RT HELIX ANGLE
TURNS PER INCH
TWIST
TWIST TESTERS

TWIST TRIANGLE
RT FIBER MIGRATION
RADIAL DISTRIBUTION
REAL TWIST
RING SPINNING
SPINNING
TENSION CONTROL
TWIST
TWIST RUN BACK
TWISTING

TWIST UNTWIST METHOD (TWIST TESTING)
BT TWIST TESTING
RT TWIST BREAK METHOD (TWIST TESTING)

TWIST VARIATION
USE TWIST DISTRIBUTION

TWISTERS
UF TWISTING MACHINES
NT DOWNTWISTERS
FALSE TWISTERS
NOVELTY TWISTERS
PLY TWISTERS
RING TWISTERS
TWO FOR ONE TWISTERS
UPTWISTERS
RT BOBBINS
BUILDER MOTIONS
CORDAGE MACHINES
DELIVERY ROLLS
DOFFERS (PACKAGE)
DOFFING (PACKAGE)
FALSE TWIST
FALSE TWIST SPINDLES
PACKAGES
PLIED YARNS
PLYING
REAL TWIST
RING RAILS
RINGS
SPINDLES
TAPE DRIVES
THREAD GUIDES
THROWSTERS
TRAVELERS
TRAVERSE
TWIST
TWIST GEARS
TWISTING

TWISTING
NT DOWNTWISTING
FALSE TWISTING (EXCEPT TEXTURING)
FLUID TWISTING
RING TWISTING
TWO FOR ONE TWISTING
UPTWISTING
RT AMERICAN SYSTEM PROCESSING
ANGULAR DEFLECTION
ANTIWEDGE RINGS
BALANCED YARNS
BALLOON COLLAPSE
BALLOON CONTROL
BALLOON CONTROL RINGS
BALLOON DYNAMICS
BALLOON SEPARATORS
BALLOON TENSION
BALLOONS
BRADFORD SYSTEM PROCESSING
CABLING
CIRCULAR TRAVELERS
CONTINENTAL SYSTEM PROCESSING
COTTON SYSTEM PROCESSING
DISTORTION
DOFFERS (PACKAGE)
DRAW TWISTING
ELLIPTICAL TRAVELERS
FALSE TWIST
FALSE TWIST SPINDLES
FALSE TWIST TUBES
INTERLACING (FILAMENT IN YARN)

RT FALSE TWIST YARNS
FIBER MIGRATION
FLYER SPINNING FRAMES
FLYERS
HELIX ANGLE
HIGH TWIST
LASHING END RADIUS
LOW TWIST
LUBRICATED RINGS
MULE SPINNING
MULTIPLIED YARNS
NEW BRADFORD SYSTEM PROCESSING
NYLON TRAVELERS
PACKAGING (OPERATION)
PLIED YARNS
PLY TWIST
PLYING
PLYING TRAVELERS
POT SPINNING FRAMES
RADIAL DISTRIBUTION
REAL TWIST
RING RAILS
RING SPINNING
RINGS
ROVING (OPERATION)
ROVINGS
S TWIST
SINTERED METAL RINGS
SPINDLES
SPINNING
SPINNING FRAMES
STOP MOTIONS
STRETCH YARNS (FILAMENT)
TAPER (BOBBIN)
TEXTURED YARNS (FILAMENT)
THREADLINE
THROWING
THROWSTERS
TORSION (GEOMETRIC)
TORSIONAL RIGIDITY
TRAVELERS
TURNS PER INCH
TWIST
TWIST CONTRACTION
TWIST CONTROL
TWIST DISTRIBUTION
TWIST LIVELINESS
TWIST MULTIPLIER
TWIST TRIANGLE
TWISTERS
TWISTING AT THE HEAD
TWISTING DYNAMICS
TWISTING MOMENTS
TWO FOR ONE TWISTERS
TWO PLY YARNS
UNLUBRICATED RINGS
WARP PREPARATION
WINDING
WINDING TENSION
WOOLEN SYSTEM PROCESSING
WORSTED SYSTEM PROCESSING
YARNS
Z TWIST
Z TWIST YARNS

TWISTING AT THE HEAD
RT COUNTERFALLERS
DRAFTING (STAPLE FIBER)
FALLERS (MULE)
MULE CRAFT
MULE SPINNING
MULES
PAYING OUT
TWISTING
WINDING
YARNS

TWISTING BOBBINS
BT BOBBINS
PACKAGES
RT SPINNING BOBBINS

TWISTING DYNAMICS
NT BALLOON DYNAMICS
BT PROCESS DYNAMICS
YARN DYNAMICS
RT BALLOON TENSION
CARDING DYNAMICS
DRAFTING DYNAMICS
KNITTING DYNAMICS
MECHANISM (FUNDAMENTAL)
SPINNING DYNAMICS
TWISTING
WEAVING DYNAMICS
WINDING DYNAMICS

TWISTING IN
USE TYING IN

TWISTING MACHINES
USE TWISTERS

TWISTING MOMENTS
UF TORQUE
TORSION (MECHANICAL)
TORSIONAL MOMENT
TWISTING TORQUE
BT MOMENTS
RT ANGULAR DEFLECTION
BENDING MOMENTS
BUCKLING
COUPLES
HELIX ANGLE
POLAR MOMENT OF INERTIA
REAL TWIST
SENSE (DIRECTION)
SHEAR FORCE
SHEAR STRESS
TORSION (GEOMETRIC)
TORSIONAL ENERGY
TORSIONAL RIGIDITY
TWIST
TWISTING

TWISTING TENSION
BT TENSION
RT BALLOON DYNAMICS
BALLOON TENSION
DISC TENSION
SPINNING DYNAMICS
SPINNING TENSION (STAPLE YARN)
UPTWISTING
WINDING TENSION
YARN TENSION

TWISTING TORQUE
USE TWISTING MOMENTS

TWISTLESS
USE ZERO TWIST

TWO BATH PROCESS
BT WET FINISHING
RT DYEING (BY PROCESS FLOW)
ONE BATH PROCESS

TWO COLOR DYEING
BT CROSS DYEING
DYEING (FOR EFFECT)
RT CONTRAST EFFECT
RESERVE EFFECT
TONE IN TONE DYEING
TWO TONE DYEING

TWO FOLD SINGLES
USE TWO PLY YARNS

TWO FOLD YARNS
USE TWO PLY YARNS

TWO FOR ONE TWISTERS
BT THROWING MACHINES
TWISTERS
RT DOUBLE SYSTEMS
DOUBLE TWIST SPINDLES
DOWNTWISTERS
NOVELTY TWISTERS
PACKAGING (OPERATION)
THROWSTERS
TWISTING
TWO FOR ONE TWISTING
UNWINDING
UPTWISTERS

TWO FOR ONE TWISTING
BT TWISTING
UPTWISTING
RT DOFFING (PACKAGE)
DOWNTWISTING
OVEREND UNWINDING
REAL TWIST
RING SPINNING
SPINNING
TEXTURING
THROWING
TWIST
TWO FOR ONE TWISTERS
UPTWISTERS

TWO PHASE PRINTING
USE TWO STAGE PRINTING

TWO PLY YARNS
UF TWO FOLD SINGLES
TWO FOLD YARNS
BT PLIED YARNS
RT BALANCED YARNS
MULTIPLIED YARNS
PLY TWIST
PLYING
S TWIST
SINGLES YARNS
TWIST LIVELINESS
TWISTING
Z TWIST

TWO POINT FIVE PERCENT SPAN LENGTH
RT FIBER ARRAY
FIBER LENGTH
FIBER LENGTH DETERMINATION
FIBER LENGTH DISTRIBUTION
FIBROGRAPH

TWO STAGE PRINTING
UF TWO PHASE PRINTING
BT PRINTING
RT FLASH AGING
REACTIVE PRINTING
ROLLER PRINTING
SCREEN PRINTING
VAT PRINTING

TWO THREAD CHAIN STITCHES
BT STITCHES (SEWING)

TWO TONE DYEING
BT DYEING (FOR EFFECT)
CROSS DYEING
RT CONTRAST EFFECT
RESERVE EFFECT
TONE IN TONE DYEING
TWO COLOR DYEING

TWO WAY STRETCH
BT STRETCH
RT ANTHROPOMETRIC KINEMATICS
BIAXIAL STRESS
FILLINGWISE STRETCH
FORM PERSUASIVE GARMENTS
STRETCH FABRICS
STRETCH WOVEN FABRICS
WARPWISE STRETCH

TYCORA (TN)
BT TEXTURED YARNS (FILAMENT)
RT BULKED YARNS

TYING
USE KNOTTING

TYING IN
UF TWISTING IN
RT DRAWING IN
KNOTTING
LEASING
LOOMS
THREADING
TYING IN MACHINES
WARP ENDS
WARP PREPARATION

TYING IN MACHINES
RT DRAWING IN MACHINES
KNOTTERS
SELF THREADING DEVICES
TYING IN

TYNEX (TN)
BT NYLON (POLYAMIDE FIBERS)
NYLON 66

TYPEWRITER RIBBONS
BT RIBBON
RT INDUSTRIAL FABRICS
NARROW FABRICS
TAPE

TYPP COUNT
BT COUNT
INDIRECT COUNT

TYRE FABRICS
USE TIRE FABRICS

TYRES
USE TIRES

TYROSINE
BT AMINO ACIDS

ULTIMATE ELONGATION
USE BREAKING ELONGATION

ULTIMATE STRENGTH
USE BREAKING STRENGTH

ULTIMATE TENACITY
USE BREAKING STRENGTH

ULTRACENTRIFUGES
RT CENTRIFUGAL FORCE
CENTRIFUGES
DEGREE OF POLYMERIZATION
INSTRUMENTATION
MOLECULAR WEIGHT
MOLECULAR WEIGHT DISTRIBUTION

ULTRAMICROSCOPY
BT MICROSCOPY
RT ELECTRON MICROSCOPY
INFRARED MICROSCOPY
INTERFERENCE MICROSCOPY
LIGHT MICROSCOPY (OPTICAL)
ULTRAVIOLET MICROSCOPY
X RAY MICROSCOPY

ULTRAMARINE BLUE
BT BLUEING AGENTS

ULTRASONICS
RT CLEANING
SONIC VELOCITY
SOUND

ULTRAVIOLET ABSORBERS
BT STABILIZERS (AGENTS)
RT INFRARED ABSORBERS
LIGHTFASTNESS (COLOR)
LIGHTFASTNESS (OF FINISH)
PHOTOCHEMICAL DEGRADATION
ULTRAVIOLET RADIATION
ULTRAVIOLET STABILIZERS

ULTRAVIOLET IRRADIATION
USE ULTRAVIOLET RADIATION

ULTRAVIOLET MICROSCOPY
BT MICROSCOPY
RT ELECTRON MICROSCOPY
FLUORESCENCE
INFRARED MICROSCOPY
INTERFERENCE MICROSCOPY
LIGHT MICROSCOPY (OPTICAL)
STEREOSCAN MICROSCOPY
ULTRAVIOLET RADIATION
X RAY MICROSCOPY

ULTRAVIOLET RADIATION
UF ULTRAVIOLET IRRADIATION
BT RADIATION
RT IRRADIATION
LIGHT
LIGHTFASTNESS (COLOR)
LIGHTFASTNESS (OF FINISH)
ULTRAVIOLET ABSORBERS
ULTRAVIOLET MICROSCOPY
ULTRAVIOLET SPECTROSCOPY
ULTRAVIOLET STABILIZERS
ULTRAVIOLET STERILIZERS

ULTRAVIOLET SPECTROSCOPY
BT SPECTROSCOPY
RT SPECTRA
SPECTROPHOTOMETERS
ULTRAVIOLET MICROSCOPY
ULTRAVIOLET RADIATION
X RAY SPECTROSCOPY

ULTRAVIOLET STABILIZERS
BT STABILIZERS (AGENTS)
RT HEAT STABILIZERS
LIGHTFASTNESS (COLOR)
LIGHTFASTNESS (OF FINISH)
PHOTOCHEMICAL DEGRADATION
RHODAMINE
ULTRAVIOLET ABSORBERS
ULTRAVIOLET RADIATION

ULTRAVIOLET STERILIZERS
RT ANTIMICROBIAL FINISHES
ANTIMICROBIAL TREATMENTS
ULTRAVIOLET RADIATION

UMBRELLA CLOTH
BT FLAT WOVEN FABRICS
WOVEN FABRICS
RT PLAIN WEAVES
RAINWEAR
SATIN WEAVES
TWILL WEAVES
UMBRELLAS

UMBRELLAS
RT RAINWEAR
UMBRELLA CLOTH

UNBALANCED SHED
BT SHED (WEAVING)
RT BALANCED SHED
CLOSED SHED
OPEN SHED
SHEDDING
SPLIT SHED

UNCONVENTIONAL YARNS
NT BONDED YARNS
LAMINATED YARNS
PAPER YARNS
SLIT FILM YARNS
BT YARNS
RT FILAMENT YARNS
SPUN YARNS

UNCUT PILE FABRICS
BT PILE FABRICS (WOVEN)
WOVEN FABRICS
RT CUT PILE FABRICS
LOOP PILE FABRICS
TERRYCLOTH

UNDERLAP
BT PROCESS VARIABLES
RT COMPOUND NEEDLES
GUIDE BARS
HALF SET
LAPPING (WARP KNITTING)
MILANESE KNITTING MACHINES
OVERLAP
RASCHEL KNITTING MACHINES
RUNNER RATIO
RUNNERS
SET (WARP KNITTING)
TRICOT KNITTED FABRICS
TRICOT KNITTING
TRICOT KNITTING MACHINES
TRICOT STITCHES
WARP KNITTING MACHINES

UNDERLAY (CARPETS)
RT BACKING (MATERIAL)
BACKING FABRICS
CARPETS
FOAM RUBBER
FOAMS
FRICTION
FURNISHING FABRICS
SLIDING

UNDERWEAR
NT BRASSIERES
BRIEFS
DRAWERS
FORM PERSUASIVE GARMENTS
FOUNDATION GARMENTS
GARTER BELTS
GARTERS
GIRDLES
KNITTED UNDERWEAR
LINGERIE
PANTIES
SHORTS
SLIPS
VESTS
BT **GARMENTS**
RT **KNITTED FABRICS**
SUSPENDERS

UNEVEN SELVEDGES
USE PULLED IN SELVEDGES

UNEVENNESS
USE IRREGULARITY

UNIAXIAL STRENGTH
BT FABRIC PROPERTIES
FABRIC PROPERTIES (MECHANICAL)
STRENGTH
RT BIAXIAL STRENGTH
MECHANICAL PROPERTIES
UNIAXIAL STRESS

UNIAXIAL STRESS
BT STRESS
TENSILE STRESS
RT BIAXIAL STRESS
BREAKING ELONGATION
BREAKING STRENGTH
COMPRESSION
COMPRESSIVE STRESS
CREEP
DYNAMIC MODULUS
ELASTIC MODULUS (TENSILE)
ELASTIC STRAIN
ELASTICITY
EXTENSIBILITY
FILLINGWISE STRETCH
GRAB STRENGTH
HOOP STRESS
NECKING (FILAMENT)
PLASTIC STRAIN
POISSONS RATIO
RAVEL STRIP STRENGTH
RECOVERY (SELF RESTORATION TO ORIGINAL CONDITION)
SHEAR STRESS
STRAIN
RT STRESS CONCENTRATION
STRESS STRAIN CURVES
STRETCH
TENSION
TRELLIS MODEL
UNIAXIAL STRENGTH
WARPWISE STRETCH

UNICONER (TN)
BT CONERS
DRUM WINDERS
WINDERS
RT AUTOMATIC KNOTTING
CONES
ELECTRONIC SLUB CATCHERS
SLUB CATCHERS

UNIDIRECTIONALLY ORIENTED WEBS
RT ANISOTROPIC MATERIALS
BONDED FIBER FABRICS
ISOTROPIC MATERIALS
WEB FORMING MACHINES
WEB ORIENTATION
WEBS

UNIFIL (TN)
BT QUILLERS
RT LOOMS
QUILLING
QUILLS
ROTARY MAGAZINES

UNIFORMITY
USE IRREGULARITY

UNIFORMITY RATIO
RT FIBER LENGTH DISTRIBUTION
FIBROGRAPH

UNIFORMS
BT CEREMONIAL CLOTHING
GARMENTS
INDUSTRIAL CLOTHING
MILITARY CLOTHING
WORK CLOTHING

UNION DYEING
UF SOLID SHADE DYEING
BT DYEING (FOR EFFECT)
RT BLEND DYEING
CROSS DYEING
LEVELLING AGENTS
RETARDING AGENTS
TONE IN TONE DYEING

UNIT CELLS (NONWOVEN FABRICS)
RT BUNCHING COEFFICIENT
FABRIC PROPERTIES (STRUCTURAL)
NONWOVEN FABRIC STRUCTURE
NONWOVEN FABRICS

UNIT OPERATIONS
RT CHEMICAL ENGINEERING
CONTINUOUS DYEING
CONTINUOUS FINISHING
FINISHING PROCESS (GENERAL)
PROCESSES

UNLEVEL DYEING
USE LEVELNESS (DYEING)

UNLOADING (PROCESS)
RT EJECTION
EMPTYING
FLOW (MATERIALS)
LOADING (PROCESS)
REMOVAL
THROUGHPUT
UNROLLING
UNWINDING

UNLUBRICATED RINGS
BT RINGS
RT ANTIWEDGE RINGS
LUBRICATED RINGS
LUBRICATION
PLYING
RING SPINNING
SINTERED METAL RINGS
SPINNING
TRAVELER BURNOUT
TRAVELER CHATTER
TRAVELERS
TWISTING
WEDGING

UNRAVELLING
RT CRINKLE TYPE YARNS
DIMENSIONAL STABILITY
KNITTED FABRICS
OVEREDGE STITCHES
SNAGGING
YARN SLIPPAGE

UNROLLING
RT EJECTION
ENTANGLEMENTS
INERTIA

RT OVEREND UNWINDING
REMOVAL
RESIDUAL STRESS
UNLOADING (PROCESS)
UNWINDERS (CLOTH)

UNSATURATED ALIPHATIC COMPOUNDS
USE UNSATURATED COMPOUNDS

UNSATURATED COMPOUNDS
UF UNSATURATED ALIPHATIC COMPOUNDS
NT ACETYLENIC COMPOUNDS
OLEFINIC COMPOUNDS
UNSATURATED HYDROCARBONS
BT ALIPHATIC COMPOUNDS
RT IODINE NUMBER
SATURATED COMPOUNDS
VINYL GROUPS

UNSATURATED HYDROCARBONS
BT HYDROCARBONS
UNSATURATED COMPOUNDS
RT ALICYCLIC HYDROCARBONS
ALIPHATIC HYDROCARBONS
AROMATIC HYDROCARBONS

UNWINDERS (CLOTH)
RT UNROLLING
UNWINDING
WINDERS

UNWINDING
NT OVEREND UNWINDING
RT BALLOON DYNAMICS
BALLOON TENSION
BALLOONS
CREELING
DRUM WINDING
EJECTION
ENTANGLEMENTS
NOSE BUNCHING
PRECISION WINDERS
REMOVAL
RESIDUAL STRESS
SLOUGHING
STRESS CONCENTRATION
TWO FOR ONE TWISTERS
UNLOADING (PROCESS)
UNWINDING ACCELERATORS
UNWINDING BALLOONS
WARP ENDS
WARP PREPARATION
WARP WIND
WINDERS
WINDING
WINDING DYNAMICS
YARN TENSION

UNWINDING ACCELERATORS
NT SCHLAFORST UNWINDING ACCELERATOR
RT BALLOON COLLAPSE
BALLOON CONTROL
BALLOON CONTROL RINGS
BALLOON DYNAMICS
BALLOON TENSION
BALLOONS
OVEREND UNWINDING
PRECISION WINDERS
UNWINDING
UNWINDING BALLOONS
WINDERS
WINDING
WINDING DYNAMICS

UNWINDING BALLOONS
BT BALLOONS
RT BALLOON COLLAPSE
BALLOON CONTROL
BALLOON DYNAMICS
BALLOON HEIGHT
BALLOON SEPARATORS
BALLOON TENSION
OVEREND UNWINDING
SPINNING BALLOONS
UNWINDING
UNWINDING ACCELERATORS
WINDING
WINDING DYNAMICS
YARN TENSION

UNWINDING DYNAMICS
USE WINDING DYNAMICS

UPHOLSTERY
RT AUTOMOTIVE FABRICS
DRAPES
END USES
FURNISHING FABRICS
FURNISHINGS
HOUSEHOLD FABRICS
RUCHE
TAPESTRIES
UPHOLSTERY FABRICS
UPHOLSTERY LEATHER
VINYL COATED FABRICS
WELTS (SEWN)

UPHOLSTERY FABRICS
BT FABRICS (BY END USES)
RT AUTOMOTIVE FABRICS
COATED FABRICS
FABRICS (ACCORDING TO STRUCTURE)
FURNISHING FABRICS
FURNISHINGS
HOUSEHOLD FABRICS
LAMINATED FABRICS
LEATHER
SYNTHETIC LEATHER
UPHOLSTERY
UPHOLSTERY LEATHER
VINYL COATED FABRICS
WELTS (SEWN)

UPHOLSTERY LEATHER
BT LEATHER
RT AUTOMOTIVE FABRICS
COATED FABRICS
SYNTHETIC LEATHER
UPHOLSTERY
UPHOLSTERY FABRICS
WELTS (SEWN)

UPKEEP
USE MAINTENANCE

UPLAND COTTON
USE AMERICAN UPLAND COTTON

UPPER CONTROL LIMIT
RT CONFIDENCE LIMITS
CONTROL CHARTS
LOWER CONTROL LIMIT
QUALITY CONTROL
STATISTICAL MEASURES

UPPER HALF MEAN LENGTH
RT COEFFICIENT OF LENGTH VARIATION
COTTON CLASSING
COTTON QUALITY
DIGITAL FIBROGRAPH
EFFECTIVE LENGTH (FIBER)
FIBER DIAGRAM
FIBER LENGTH
FIBER LENGTH DETERMINATION
FIBER LENGTH DISTRIBUTION
FIBROGRAPH
FIBROGRAPH MEAN LENGTH
MEAN FIBER LENGTH
SERVO FIBROGRAPH
SHORT FIBERS INDEX
SPAN LENGTH
STAPLING
SUTER WEBB TESTING
UPPER QUARTILE LENGTH

UPPER QUARTILE LENGTH
RT COEFFICIENT OF LENGTH VARIATION
COTTON CLASS
DIGITAL FIBROGRAPH
FIBER ARRAY
FIBER DIAGRAM
FIBER LENGTH
FIBER LENGTH DETERMINATION
FIBER LENGTH DISTRIBUTION
FIBROGRAPH
FIBROGRAPH MEAN LENGTH
MEAN FIBER LENGTH
SERVO FIBROGRAPH
STAPLING
UPPER HALF MEAN LENGTH

UPPERS
BT FOOTWEAR COMPONENTS
RT BINDINGS (FOOTWEAR)
BOOTS
FOOTWEAR
LACES (SHOES)
LASTS
LEATHER
LININGS (FOOTWEAR)
PUMPS (FOOTWEAR)
SHOES
SLIPPERS
SNEAKERS
SOLES
STITCHING
SYNTHETIC LEATHER
TONGUES
WELTS (FOOTWEAR)

UPTAKE
USE WET PICKUP

UPTWISTERS
UF UPTWISTING FRAMES
BT THROWING MACHINES
TWISTERS
RT BOBBINS
DOFFING (PACKAGE)
DOWNTWISTERS
FALSE TWISTERS
OVEREND UNWINDING
PACKAGES
SPINDLES
TEXTURING
THROWSTERS
TWIST
TWISTING DYNAMICS
TWO FOR ONE TWISTERS
TWO FOR ONE TWISTING
UPTWISTING

UPTWISTING
NT TWO FOR ONE TWISTING
BT TWISTING
RT DOFFING (PACKAGE)
DOWNTWISTING
OVEREND UNWINDING
PACKAGING (OPERATION)
REAL TWIST
SPINNING
TEXTURING
THROWING
THROWSTERS
TWIST
TWISTING DYNAMICS
TWISTING TENSION
UPTWISTERS
YARN TENSION

UPTWISTING FRAMES
USE UPTWISTERS

UREA
RT AMIDES
ISOCYANATES
METHYLOL ACETYLENE DIUREA
METHYLOL UREAS
SWELLING AGENTS
THIOUREA
UREA BISULFITE SOLUBILITY
UREA DERIVATIVES
URETHANE FOAMS
URETHANE RUBBERS
URETHANES

UREA BISULFITE SOLUBILITY
BT CHEMICAL PROPERTIES
DEGRADATION PROPERTIES
RT ACID SOLUBILITY
ALKALI SOLUBILITY
BISULFITES
DEGRADATION
KERATIN
SOLUBILITY
UREA
WOOL

UREA DERIVATIVES
UF ETHYLENE UREA
NT DMEU
METHYLOL UREAS
OCTADECYL ETHYLENE UREA
RT UREA

UREA-FORMALDEHYDE
USE METHYLOL UREAS

UREA-FORMALDEHYDE CONDENSATES
USE UREA-FORMALDEHYDE RESINS

UREA-FORMALDEHYDE PRECONDENSATES
USE METHYLOL UREAS

UREA-FORMALDEHYDE RESINS
UF UREA-FORMALDEHYDE CONDENSATES
BT AMINO RESINS
RT DMEU
MELAMINE FORMALDEHYDE RESINS
METHYLOL UREAS
PRECONDENSATES

URETHANE FOAMS
BT FOAMS
RT FOAM RUBBER
ISOCYANATES
ISOTHIOCYANATES
RIGID FOAMS
SPANDEX FIBERS
UREA
URETHANE RUBBERS
URETHANES
VINYL FOAMS

URETHANE RUBBERS
RT ISOCYANATES
SPANDEX FIBERS
UREA
URETHANE FOAMS
URETHANES

URETHANES
RT ISOCYANATES
ISOTHIOCYANATES
METHYLOL CARBAMATES
POLYURETHANES
UREA
URETHANE FOAMS
URETHANE RUBBERS

URON-FORMALDEHYDE CONDENSATES
USE METHYLOL URONS

USTER IRREGULARITY TESTER (TN)
BT CAPACITANCE TYPE IRREGULARITY TESTERS
IRREGULARITY TESTERS
RT CAPACITANCE (ELECTRICAL)
FIELDEN WALKER IRREGULARITY TESTER (TN)
IRREGULARITY
IRREGULARITY TESTING
PNEUMATIC IRREGULARITY TESTERS
QUALITY CONTROL
ROVINGS
SACO LOWELL IRREGULARITY TESTER (TN)
SLIVERS
TESTING
VARIANCE LENGTH CURVES
YARNS
ZELLWEGER TESTER (TN)

V BELTS
BT YARN REINFORCED ELASTOMERS
RT CREEP
REINFORCED HOSES
TIRE CORDS
TIRES

VACUUM
BT PRESSURE
RT ABSENCE
FREEZE DRYING
HOLLOWNESS
PRESENCE
SUCTION PRESSES
VACUUM COATING
VACUUM DRYING

VACUUM BAG MOLDING
BT MOLDING
RT COMPRESSION MOLDING
CONTACT MOLDING
FLEXIBLE PLUNGER MOLDING
INJECTION MOLDING
MATCHED DIE MOLDING
PRESSURE BAG MOLDING
REINFORCED COMPOSITES

VACUUM CLEANERS
RT BLOWERS
CLEANERS
VACUUM END COLLECTORS
VACUUM PUMPS

VACUUM COATING
BT COATING (PROCESS)
RT METALLIC FIBERS
VACUUM

VACUUM DRYERS
BT DRYERS
RT ATMOSPHERIC DRYERS
CENTRIFUGAL DRYERS
DRUM DRYERS
FLUIDIZED BED DRYERS
HOT AIR DRYERS
INFRARED DRYERS
LOOP DRYERS
VACUUM DRYING
VACUUM EXTRACTION
VACUUM PUMPS

VACUUM DRYING
BT DRYING
RT BATCH DRYING
CENTRIFUGAL DRYING
CONTINUOUS DRYING
DIELECTRIC DRYING
FREEZE DRYING
HOT AIR DRYING
INFRARED DRYING
LINE DRYING (EASY CARE GARMENTS)
TUMBLE DRYING
VACUUM
[illegible]

VACUUM DYEING
BT DYEING (BY ENVIRONMENTAL CONDITIONS)
RT CARRIER DYEING
HIGH TEMPERATURE DYEING
LOW TEMPERATURE DYEING
PRESSURE DYEING
SOLVENT ASSISTED DYEING
SOLVENT DYEING

VACUUM END COLLECTORS
UF PNEUMATIC BROKEN END COLLECTORS
BT BROKEN END COLLECTORS
RT PNEUMAFIL (TN)
VACUUM CLEANERS
VACUUM PUMPS

VACUUM EXTRACTION
BT EXTRACTION
RT DRYING
MOISTURE CONTENT
VACUUM DRYERS

VACUUM PUMPS
BT PUMPS
RT VACUUM DRYERS
VACUUM CLEANERS
VACUUM END COLLECTORS

VALENCIENNES
BT HAND MADE LACES
PILLOW LACES (BOBBIN LACES)

VALUE (MUNSELL)
BT MUNSELL COLOR SYSTEM
RT COLOR
HUE
SATURATION (COLOR)
WHITENESS

VALVES
RT ACCESSORIES
BEARINGS
CAMS
HYDRAULIC CYLINDERS
LINKAGES
MECHANISMS
PRESSURE CONTROL

VAMPS
BT FOOTWEAR COMPONENTS

VAN DER WAALS FORCES
BT FORCE
RT BOND ENERGY
CHEMICAL BONDS
DIPOLE MOMENT
DISPERSION FORCES
HYDROPHOBIC BONDS
MOLECULAR ATTRACTION

VAPOCOL PROCESS (TN)
BT SOLVENT DYEING
RT CONTINUOUS DYEING
SOLVENTS

VAPOR PHASE
RT DYEING
GAS PHASE
REACTIONS (CHEMICAL)
VAPOR PHASE DYEING
VAPOR PHASE TREATMENTS
VAPOR PRESSURE
WATER VAPOR

VAPOR PHASE DYEING
BT DYEING (BY ENVIRONMENTAL CONDITIONS)
RT SOLVENT DYEING
VAPOR PHASE
VAPOR PHASE TREATMENTS

VAPOR PHASE FINISHING
USE VAPOR PHASE TREATMENTS

VAPOR PHASE TREATMENTS
UF VAPOR PHASE FINISHING
BT WET FINISHING
RT SOLVENT TREATMENTS
VAPOR PHASE
VAPOR PHASE DYEING

VAPOR PRESSURE
RT BOILING POINT
CRITICAL MOISTURE CONTENT
DEW POINT
DIFFUSION
DIFFUSIVITY
DRY BULB TEMPERATURE
FLASH POINT
FREEZING POINT
EQUILIBRIUM MOISTURE CONTENT
MOISTURE CONTENT
PSEUDO WET BULB TEMPERATURE
[illegible]
REGAIN
RELATIVE HUMIDITY
SATURATED VAPOR
SPECIFIC HUMIDITY
SUBLIMATION
VAPOR PHASE
WATER VAPOR

VARIABILITY
USE IRREGULARITY

VARIABLE CAPACITORS
RT CAPACITANCE (ELECTRICAL)
FIXED CAPACITORS

VARIABLE SPEED
BT VELOCITY
RT ACCELERATION (MECHANICAL)
CONSTANT SPEED
DRIVES
HIGH SPEED
MOMENTUM
MOTION
MOTORS
ROTATION
SPEED CONTROL
SPINDLE SPEED
STARTING TIME
TERMINAL VELOCITY
VARIABLE SPEED DRIVES

VARIABLE SPEED DRIVES
BT DRIVES
RT BELT DRIVES
CHAIN DRIVES
GEAR DRIVES
HYDRAULIC DRIVES
MECHANICAL DRIVES
VARIABLE SPEED

VARIABLES
UF FACTORS (PHYSICAL)
RT EXPERIMENTAL DESIGN
FLUCTUATIONS
OSCILLATIONS
PROCESS VARIABLES

VARIANCE
BT STATISTICAL MEASURES
RT COEFFICIENT OF LENGTH VARIATION
COEFFICIENT OF VARIATION
CORRELATION COEFFICIENT
DEGREES OF FREEDOM
DISTRIBUTION
DOUBLING (DRAFTING)
DRAFTING THEORY
FIBER LENGTH
KURTOSIS
LEVEL OF SIGNIFICANCE
PROBABILITY
STANDARD DEVIATION
STANDARD ERROR
VARIANCE LENGTH CURVES

VARIANCE LENGTH CURVES
UF B (L) CURVES
RT ANALYSIS OF VARIANCE
CAPACITANCE TYPE IRREGULARITY TESTERS
IRREGULARITY
IRREGULARITY TESTERS
IRREGULARITY TESTING
PNEUMATIC IRREGULARITY TESTERS
STATISTICAL ANALYSIS
STATISTICAL MEASURES
USTER IRREGULARITY TESTER (TN)
VARIANCE

VARIATION
RT AUTOCORRELATION
FLUCTUATIONS
IRREGULARITY
PERIODICITY
REPEATS
SURGING
VIBRATION
WEAK SPOTS

VAT DYEING
BT DYEING (BY DYE CLASSES)
RT ACETALDEHYDE SULFOXYLATES
ACID DYEING
AZOIC DYEING
BASIC DYEING
DIRECT DYEING
DISPERSE DYEING
DITHIONITES
FORMALDEHYDE SULFOXYLATES
HYDROXYACETONE
LEUCO ESTER DYES
METALLIZED DYEING
MORDANT DYEING
NEUTRAL DYEING
OXIDIZING COMPARTMENT (DYEING)
PIGMENT DYEING
REACTIVE DYEING
SOAPING
SODIUM DITHIONITE
SULFUR DYEING
VAT DYES
VAT PRINTING
ZINC FORMALDEHYDE SULFOXYLATE

VAT DYES
UF INDIGOID DYES
NT LEUCO ESTER DYES
BT ANIONIC DYES
DYES (BY CHEMICAL CLASSES)
RT AZOIC DYES
DIRECT DYES
DYEING (BY DYE CLASSES)
DYES (BY FIBER CLASSES)
OXIDIZING COMPARTMENT (DYEING)
REDUCING AGENTS
SULFUR DYES
VAT DYEING

VAT ESTER DYES
USE LEUCO ESTER DYES

VAT PRINTING
BT PRINTING
RT ACETALDEHYDE SULFOXYLATES
AGERS
AGING (STEAMING)
EMULSION PRINTING
FORMALDEHYDE SULFOXYLATES
HYDROXYACETONE
PIGMENT PRINTING
REACTIVE PRINTING
TWO STAGE PRINTING
VAT DYEING
ZINC FORMALDEHYDE SULFOXYLATE

VECTORS
RT DIFFERENTIAL GEOMETRY
DYNAMICS
FORCE
KINEMATICS
MATHEMATICAL ANALYSIS
SENSE (DIRECTION)
TENSORS

VECTRA NYLON (TN)
    BT NYLON (POLYAMIDE FIBERS)
    RT NYLON 6
       NYLON 610

VECTRA POLYESTER (TN)
    BT POLYESTER FIBERS

VECTRA POLYPROPYLENE (TN)
    BT OLEFIN FIBERS
       POLYPROPYLENE FIBERS

VECTRA SARAN (TN)
    BT SARAN POLY (VINYLIDENE CHLORIDE) FIBERS

VEGETABLE FATS
    BT FATS
    RT ANIMAL FATS

VEGETABLE FIBERS (GENERAL)
    NT BAST FIBERS
       COIR
       COTTON
       FLAX
       FRUIT FIBERS
       HEMP
       HIBISCUS
       JUTE
       KAPOK
       LEAF FIBERS
       LINT (FIBER)
       MANILA
       RAFFIA
       RAMIE
       SEED HAIR FIBERS (GENERAL)
       SISAL
       WOOD FIBERS
    BT CELLULOSIC FIBERS
       FIBERS
       NATURAL FIBERS
    RT ANIMAL FIBERS (GENERAL)
       FIBRILS
       FILAMENTS
       FUR
       IMMATURE FIBERS
       MAN MADE FIBERS
       MINERAL FIBERS
       STAPLE FIBERS
       SYNTHETIC FIBERS

VEGETABLE MATTER
    NT BURRS
       SEEDS (TRASH)
    BT TRASH
       WASTE
    RT CARBONIZING
       IMPURITIES
       SCOURING
       SCOURING WASTE
       SCOURING YIELD
       SPECK DYEING

VEILING (LACE)
    BT LACES

VELCRO (TN)
    BT FASTENERS
       GARMENT COMPONENTS
    RT BUTTONS
       HOOK CLOSURES
       SNAP FASTENERS
       ZIPPERS

VELOCITY
    UF SPEED
    NT ANGULAR VELOCITY
       CONSTANT SPEED
       TERMINAL VELOCITY
       VARIABLE SPEED
       VELOCITY (RELATIVE)
    RT ACCELERATION (MECHANICAL)
       ACCELEROMETERS
       BRAKING
       CENTRIFUGAL FORCE
       CORIOLIS ACCELERATION
       DISPLACEMENT
       FLUID FLOW
       GRAVITATION
       INERTIA
       KINETIC ENERGY
       MOMENTUM
       MOTION
       MOTORS
       POSITION
       PROCESS VARIABLES
       RATE
       REVERSAL
       SPINDLE SPEED
       STARTING TIME
       TRANSLATION
       VELOCITY CHANGE POINT

VELOCITY (RELATIVE)
    NT HIGH SPEED
    BT VELOCITY
    RT ACCELERATION (MECHANICAL)
       ANGULAR VELOCITY
       FIBER FIBER FRICTION
       FIBER SLIPPAGE
       FRICTION
       MOMENTUM
       SLIPPAGE
       SPINDLE SPEED
       TERMINAL VELOCITY
       VELOCITY CHANGE POINT

VELOCITY CHANGE POINT
    RT ACCELERATION (MECHANICAL)
       APRON CONTROL
       BACK ROLLS
       DRAFT
       DRAFTING (STAPLE FIBER)
       DRAFTING THEORY
       DRAFTING WAVES
       FIBER CONTROL
       FLOATING FIBERS
       FRONT ROLLS
       IRREGULARITY
       RATCH
       SLIVERS
       TERMINAL VELOCITY
       VELOCITY
       VELOCITY (RELATIVE)

VELON (TN)
    BT SARAN POLY (VINYLIDENE CHLORIDE) FIBERS

VELOUR
    BT FLAT WOVEN FABRICS
       WOVEN FABRICS
    RT FACE GOODS
       NAPPING
       PILE FABRICS (WOVEN)
       PLAIN WEAVES
       VELVET

VELVET
    NT GENOA VELVET
    BT CUT PILE FABRICS
       PILE FABRICS (WOVEN)
       WOVEN FABRICS
    RT VELOUR

VELVET CARPETS
    BT CUT PILE FABRICS
       WOVEN CARPETS
    RT AXMINSTER CARPETS
       CHENILLE CARPETS
       PILE FABRICS
       WILTON CARPETS

VELVETEEN
    BT CUT PILE FABRICS
       PILE FABRICS (WOVEN)
       WOVEN FABRICS

VENICE (LACE)
    BT HAND MADE LACES
       NEEDLE POINT LACES

VENTILATING INSOLES
    BT FOOTWEAR COMPONENTS
       INSOLES (FOOTWEAR)

VERDURES TAPESTRIES
    BT TAPESTRIES

VEREL (TN)
    BT MODACRYLIC FIBERS

VERTAFORMERS
    BT PAPER MACHINES
    RT CYLINDER MACHINES
       FOURDRINIER MACHINES
       INVERFORM MACHINES
       KAMYR MACHINES
       ROTAFORMERS
       WET MACHINES (PAPERMAKING)

VESSELS
    UF TANKS
    NT AUTOCLAVES
       KETTLES (DYEING)
       KETTLES (HEATING)
       KIERS
       PRESSURE KETTLES
       STORAGE VESSELS
       TRANSPORTING VESSELS

VESTAN (TN)
    BT POLYESTER FIBERS
    RT KODEL (TN)

VESTMENTS
    BT CEREMONIAL CLOTHING
       GARMENTS

VESTS
    BT GARMENTS
       OUTERWEAR
       UNDERWEAR
    RT BRIEFS
       SHORTS

VIBRASCOPE
    BT COUNT TESTERS
    RT COUNT TESTING
       FIBER FINENESS
       FREQUENCY
       LINEAR DENSITY
       RESONANT FREQUENCY
       TRANSVERSE VIBRATIONS
       VIBRATION

VIBRATION
    NT TRANSVERSE VIBRATIONS
    RT ACCELEROMETERS
       AMPLITUDE
       AXIAL MOTION
       AXLES
       DAMPERS
       DAMPING
       DISPLACEMENT
       DYNAMIC BALANCING
       ECCENTRICITY
       ENERGY ABSORPTION
       FREQUENCY
       FREQUENCY RESPONSE
       HIGH FREQUENCY
       HIGH SPEED
       IMPACT
       IMPULSES
       LARGE PACKAGES
       MECHANICAL SHOCK
       MOTION
       OSCILLATIONS
       SAFETY
       SHAFTS
       SHOCK
       SPINDLES
       SPRING CONSTANT
       SOUND
       SURGING
       TRAVELER CHATTER
       VARIATION
       VIBRASCOPE
       VIBRATION DAMPERS
       VIBRATION ISOLATORS

VIBRATION DAMPERS
    UF VIBRATION DAMPING
    BT DAMPERS
    RT DAMPING
       DYNAMIC BALANCING
       ENERGY ABSORPTION
       FREQUENCY RESPONSE
       HYSTERESIS (MECHANICAL)
       IMPACT
       IMPULSES
       SHOCK
       SHOCK ABSORBERS
       VIBRATION
       VIBRATION ISOLATORS
       VISCOUS DAMPING

VIBRATION DAMPING
    USE VIBRATION DAMPERS

VIBRATION ISOLATION
    USE VIBRATION ISOLATORS

VIBRATION ISOLATORS
    UF VIBRATION ISOLATION
    RT DAMPERS
       DAMPING
       ENERGY ABSORPTION
       FREQUENCY RESPONSE
       IMPACT
       IMPULSES
       RESONANT FREQUENCY
       SHOCK
       SHOCK ABSORBERS
       SPRINGS
       VIBRATION
       VIBRATION DAMPERS

VICARA (TN)
    BT AZLON (REGENERATED PROTEIN FIBERS)
       PROTEIN FIBERS
    RT ARALAC (TN)
       ARDIL (TN)

VICTORIA LAWN
    BT FLAT WOVEN FABRICS
       WOVEN FABRICS
    RT LAWN

VICUNA
    BT HAIR
       KERATIN FIBERS
       PROTEIN FIBERS

VIGOUREUX PRINTING
    USE MELANGE PRINTING

VINAL POLY (VINYL ALCOHOL) FIBERS
    NT VINYLON (TN)
    BT MAN MADE FIBERS
       SYNTHETIC FIBERS
    RT POLY (VINYL ALCOHOL)
       SARAN POLY (VINYLIDENE CHLORIDE) FIBERS

RT VINYON POLY (VINYL CHLORIDE) FIBERS

VINYL ACETATE
BT ACETATE ESTERS
VINYL COMPOUNDS
RT POLY (VINYL ACETATE)

VINYL BROMIDE
BT BROMINE COMPOUNDS
VINYL COMPOUNDS

VINYL CHLORIDE
BT CHLORINE COMPOUNDS
VINYL COMPOUNDS
RT POLY (VINYL CHLORIDE)

VINYL COATED FABRICS
BT COATED FABRICS
COMPOSITES
RT ADHESION
AUTOMOTIVE FABRICS
BACKING FABRICS
BOOTS
BUNA N COATED FABRICS
BUTYL COATED FABRICS
COATINGS (SUBSTANCES)
FOOTWEAR
LACQUER COATED FABRICS
LAMINATING
LEATHER
LUGGAGE
NEOPRENE COATED FABRICS
POLYMERS
RAINWEAR
RESINS
RUBBER COATED FABRICS
RUBBER FOOTWEAR
SBR COATED FABRICS
SHOES
SNEAKERS
SYNTHETIC LEATHER
TENTS
UPHOLSTERY
UPHOLSTERY FABRICS
UPPERS

VINYL COMPOUNDS
NT STYRENE
VINYL ACETATE
VINYL BROMIDE
VINYL CHLORIDE
VINYL ETHERS
VINYL MORPHOLINE
VINYL PYRIDINE
VINYL PYRROLIDONE
BT MONOMERS
RT ACRYLIC COMPOUNDS
ADDITION POLYMERS
ALLYL COMPOUNDS
CHEMICAL MODIFICATION (FIBERS)
COPOLYMERIZATION
GRAFTING
POLYVINYLS
VINYL FIBERS (GENERAL)
VINYL GROUPS
VINYLATED COTTON
VINYLATION
VINYLIDENE COMPOUNDS

VINYL ETHERS
BT ETHERS
VINYL COMPOUNDS
RT POLYVINYLS
VINYLATED COTTON

VINYL FIBER BLENDS
NT ACRYLIC FIBER BLENDS
BT BLENDS (FIBERS)

VINYL FIBERS (GENERAL)
NT ACRYLIC FIBERS
MODACRYLIC FIBERS
NYTRIL FIBERS
SARAN POLY (VINYLIDENE CHLORIDE) FIBERS
VINAL POLY (VINYL ALCOHOL) FIBERS
VINYON POLY (VINYL CHLORIDE) FIBERS
BT FIBERS
MAN MADE FIBERS
SYNTHETIC FIBERS
RT ADDITION POLYMERS
OLEFIN FIBERS
POLYTETRAFLUOROETHYLENE FIBERS
POLYVINYLS
RAYON (REGENERATED CELLULOSE FIBERS)
VINYL COMPOUNDS

VINYL FOAMS
BT FOAMS
RT FOAM RUBBER
RIGID FOAMS
URETHANE FOAMS

VINYL GROUPS
RT ADDITION POLYMERIZATION
UNSATURATED COMPOUNDS
VINYL COMPOUNDS

VINYL MORPHOLINE
BT HETEROCYCLIC COMPOUNDS
VINYL COMPOUNDS

VINYL PYRIDINE
BT PYRIDINE DERIVATIVES
VINYL COMPOUNDS
RT POLY (VINYL PYRIDINE)

VINYL PYRROLIDONE
BT HETEROCYCLIC COMPOUNDS
VINYL COMPOUNDS
RT POLY (VINYL PYRROLIDONE)

VINYLATED COTTON
BT CHEMICALLY MODIFIED COTTON
RT ACETYLENE
CELLULOSE ETHERS
COTTON
ETHYLENE DIBROMIDE
VINYL COMPOUNDS
VINYL ETHERS

VINYLATED DYES
BT DYES (BY CHEMICAL CLASSES)
RT ADDITION POLYMERIZATION
POLYMERIC DYES
REACTIVE DYES
VINYLSULFONE DYES

VINYLATION
BT ETHERIFICATION
RT ACETYLENE
CHEMICALLY MODIFIED COTTON
VINYL COMPOUNDS

VINYLIDENE COMPOUNDS
BT MONOMERS
RT POLYMERS
POLY (VINYLIDENE CHLORIDE)
NYTRIL FIBERS

SARAN POLY (VINYLIDENE CHLORIDE) FIBERS
VINYL COMPOUNDS

VINYLON (TN)
BT VINAL POLY (VINYL ALCOHOL) FIBERS

VINYLSULFONE DYES
UF REMAZOL (TN)
BT REACTIVE DYES
RT AZIDOSULFONYL DYES
VINYLATED DYES

VINYON
USE VINYON POLY (VINYL CHLORIDE) FIBERS

VINYON POLY (VINYL CHLORIDE) FIBERS
UF VINYON
BT MAN MADE FIBERS
SYNTHETIC FIBERS
RT POLY (VINYL CHLORIDE)
SARAN POLY (VINYLIDENE CHLORIDE) FIBERS
VINAL POLY (VINYL ALCOHOL) FIBERS

VIRGIN WOOL
BT WOOL
RT LAMBSWOOL
MUNGO
PULLED WOOL
RECOVERED WOOL
SHEEP
SHODDY
WOOL CLASS

VISCOELASTICITY
NT LINEAR VISCOELASTICITY
NONLINEAR VISCOELASTICITY
BT STRESS STRAIN PROPERTIES
RT COLD DRAWING
COMPLIANCE
CREEP
CREEP RECOVERY
DAMPING
DEFORMATION
DELAYED RECOVERY
DRAWING (FILAMENT)
ELASTIC RECOVERY
ELASTICITY
FIRMOVISCOSITY
INTERNAL FRICTION
MAXWELL MODEL
PERMANENT SET
PHENOMENOLOGICAL RHEOLOGY
PLASTIC FLOW
PLASTICITY
PLASTICOVISCOELASTICITY
PLASTICOVISCOSITY
RT RELAXATION
RELAXATION TIME (MECHANICAL)
RHEOLOGY
RUBBER
SHOCK RESISTANCE
STRENGTH
STRENGTH OF MATERIALS
STRESS STRAIN CURVES
VISCOSITY
VOIGT MODEL
YIELD (MECHANICAL)

VISCOSE
USE VISCOSE RAYON

VISCOSE RAYON
UF VISCOSE
NT AVRIL (TN)
CORDURA (TN)
DURAFIL (TN)
FIBRO (TN)
HIGH TENACITY VISCOSE RAYON
HIGH WET MODULUS RAYON
POLYNOSIC FIBERS
TENASCO (TN)
TENAX (TN)
BT CELLULOSIC FIBERS
MAN MADE FIBERS
RAYON (REGENERATED CELLULOSE FIBERS)
RT ACETATE FIBERS
CARBON DISULFIDE
CELLOPHANE
CELLULOSE ESTER FIBERS
CUPRAMMONIUM RAYON
FORTISAN (TN)
JETSPUN (TN)
SKIN (FIBER)
TRIACETATE FIBERS

VISCOSE SOLUTION
USE DOPE (POLYMER)

VISCOSIMETERS
BT MEASURING INSTRUMENTS
RT INTRINSIC VISCOSITY
MOLECULAR WEIGHT
POLYMERS
RELATIVE VISCOSITY
VISCOSITY
VISCOUS FLOW

VISCOSITY
UF FLUIDITY
NT INTRINSIC VISCOSITY
MELT VISCOSITY
BT STRESS STRAIN PROPERTIES
RT AIR FLOW
CHEMICAL PROPERTIES
CUPRAMMONIUM FLUIDITY
ELASTICITY
FERRIC TARTRATE
FIRMOVISCOSITY
FLUID FLOW
FLUID FRICTION
HYDRODYNAMIC LUBRICATION
LINEAR VISCOELASTICITY
MARSCHALL SOLUTION
MOLECULAR WEIGHT CONTROL
MOTION
PLASTICITY
PLASTICOVISCOELASTICITY
PLASTICOVISCOSITY
RELATIVE VISCOSITY
RELAXATION TIME (MECHANICAL)
SPINNING (EXTRUSION)
STRENGTH
VISCOELASTICITY
VISCOSIMETERS
VISCOUS DAMPING
VISCOUS FLOW

VISCOUS DAMPING
BT DAMPING
RT DAMPERS
ENERGY ABSORPTION
HYSTERESIS (MECHANICAL)
VIBRATION DAMPERS
VISCOSITY
VISCOUS FLOW

VISCOUS FLOW
BT FLUID FLOW
STRESS STRAIN PROPERTIES
STRESS STRAIN PROPERTIES (TENSILE)
RT CREEP
ELASTIC STRAIN
INVISCID FLOW
LAMINAR FLOW
PEELING
PLASTIC FLOW
TURBULENT FLOW
VISCOSIMETERS
VISCOSITY
VISCOUS DAMPING

VOIGT MODEL
RT DELAYED RECOVERY
LINEAR VISCOELASTICITY
MAXWELL MODEL
NONLINEAR VISCOELASTICITY
PERMANENT SET
RELAXATION
RELAXATION SPECTRUM
RELAXATION TIME (MECHANICAL)
VISCOELASTICITY
VISCOUS FLOW
YIELD (MECHANICAL)

VITREOUS FIBERS
USE CERAMIC FIBERS

VOILE
BT FLAT WOVEN FABRICS
WOVEN FABRICS
RT PLAIN WEAVES

VULTEX FABRIC (TN)
BT STITCH BONDED FABRICS

VOLTAGE
RT CAPACITANCE (ELECTRICAL)
CURRENT (ELECTRICAL)
IMPEDANCE
REACTANCE (ELECTRICAL)
RESISTIVITY
VOLTAGE REGULATION

VOLTAGE REGULATION
RT CONTROL SYSTEMS
AUTOMATIC CONTROL
PROCESS CONTROL
VOLTAGE

VOLUME
RT BULK
CAPACITY (GENERAL)
DENSITY
DIMENSIONS
HOLLOWNESS
PACKING FACTOR
SPACE
SPECIFIC GRAVITY

VOLUME TEMPERATURE MEASUREMENTS
RT GLASS TRANSITION TEMPERATURE
SECONDARY TRANSITION TEMPERATURES

VOLUME WEAR
BT WEAR
RT ABRASION
FRICTION
SLIDING FRICTION
WEAR RESISTANCE

VOLUMETRIC ANALYSIS
UF TITRATION
BT CHEMICAL ANALYSIS
RT GRAVIMETRIC ANALYSIS

VYCRON (TN)
BT POLYESTER FIBERS

VYRENE (TN)
BT SPANDEX FIBERS

WADDINGS
RT BATTING
COMPOSITES
COMPRESSIBILITY
FILLER
PACKING
PADDING
SOFTNESS
STUFFING
VOLUME

WAGES
UF LABOR COST
RT COSTS
JOB ANALYSIS
MANAGEMENT
PERSONNEL

WALDRON SATURATOR
BT NONWOVEN FABRIC MACHINES
SATURATORS
RT BINDERS (NONWOVENS)
BONDED FIBER FABRICS
CURLATORS (TN)
PADDERS
RODNEY HUNT SATURATOR

WALE SPACING
USE WALES PER INCH

WALES
RT COURSES
COURSES PER INCH
DOUBLE KNIT FABRICS
KNITTED FABRICS
KNITTING
LOOP DISTORTION
MACHINE KNITTING
RUNS (KNITTING)
SLACK COURSE
WALES PER INCH

WALES PER INCH
UF WALE SPACING
BT FABRIC PROPERTIES
FABRIC PROPERTIES (STRUCTURAL)
RT COURSES
COURSES PER INCH
COVER FACTOR
DESIGN
ENDS PER INCH
FABRIC DESIGN
FABRIC GEOMETRY
FABRIC STRUCTURE
KNITTED FABRICS
KNITTING
LOOP DISTORTION
STITCH DENSITY
STITCH LENGTH
TEXTURE
WALES

WARMTH
BT END USE PROPERTIES
THERMAL PROPERTIES
RT BATTING
COMFORT
CONDUCTIVITY (THERMAL)
HAND
HEAT TRANSFER
HIGH TEMPERATURE
INSULATION (THERMAL)
LUXURIOUSNESS
PADDING
SMOOTHNESS
SPECIFIC HEAT
STUFFING
TEXTURE
THICKNESS

WARP BEAMING
USE WARPING

WARP BEAMS
RT DRAWING IN
FRICTION LET OFF
LEASE RODS
LET OFF
LET OFF BRACKETS
LET OFF CHAINS
LET OFF SPOOLS
LET OFF WEIGHTS
PACKAGES
SIZE BOXES
SLASHERS
SLASHERS BEAMS
SLASHING
STARCH
STRETCH ROLLS
WARP ENDS
WARP PREPARATION
WARPING

WARP DYEING
USE BEAM DYEING (YARN)

WARP DYEING MACHINES
BT PACKAGE DYEING MACHINES
YARN DYEING MACHINES
RT BEAM DYEING (YARN)
BOBBIN DYEING MACHINES
CAKE DYEING MACHINES
CHEESE DYEING MACHINES
CONE DYEING MACHINES
COP DYEING MACHINES
HANK DYEING MACHINES
SEMI CONTINUOUS DYEING RANGE

WARP EFFECT (TWILL)
RT FABRIC STRUCTURE
FILLING EFFECT (TWILL)
TWILLS
WARP FACE

WARP ENDS
UF ENDS (WEAVING)
WARP YARNS
WARPS
NT BACK WARP
RT BEER
CREELING
DRAWING IN
END BREAKAGES
ENDS PER INCH
FILLING YARNS
HARNESS DRAFT
HARNESSES
HEDDLES
JAMMING
LEASE RODS
LEASING
LET OFF
LET OFF BRACKETS
LET OFF CHAINS
LET OFF MOTIONS
LET OFF SPOOLS
LET OFF WEIGHTS
LIGHT DENT
LOOMS
OPEN SHED
PICKS
PIECING
REED DRAFT
REED HOOKS
REED WIDTH
REED WIRES
REEDING
REEDS
SECTIONAL WARPING
SELVEDGE WARPS
SHED (WEAVING)
SHEDDING
SLASHERS BEAMS
SLASHING
SPLIT SHED
STOP MOTIONS
SWOLLEN DENT
TENSION CONTROL
TENSION DEVICES
TYING IN
UNWINDING
WARP BEAMS
WARP PREPARATION
WARP TENSION
WARPERS (MACHINE)
WARPING
WARPING LEASE
WEAVING
WEAVING LEASE
WINDING
YARNS

WARP FACE
RT FABRIC STRUCTURE
FACE (FABRIC)
FILLING FACE
WARP EFFECT (TWILL)

WARP FILLING BLENDS
USE ORTHOBLENDS

WARP KNIT FABRICS
USE WARP KNITTED FABRICS

WARP KNITTED FABRICS
UF WARP KNIT FABRICS
NT FALL PLATE FABRICS
MILANESE FABRICS
RASCHEL FABRICS
SIMPLEX FABRICS
TRICOT KNITTED FABRICS
BT KNITTED FABRICS
RT DOUBLE KNIT FABRICS
LAPPING (WARP KNITTING)
MACHINE KNITTING
RASCHEL KNITTING MACHINES
TRICOT STITCHES
WARP KNITTING MACHINES

WARP KNITTING
NT TRICOT KNITTING
BT KNITTING
RT BEAMS
FILLING KNITTING
FLAT KNITTING
LACE MAKING
LAPPING (WARP KNITTING)
LAYING IN (KNITTING)
MALI PROCESSES (TN)
MILANESE KNITTING MACHINES
RASCHEL KNITTING MACHINES
SIMPLEX KNITTING MACHINES
TRICOT KNITTED FABRICS
TRICOT STITCHES
WARP KNITTING MACHINES
WARPING
WEAVING

WARP KNITTING MACHINES
NT MILANESE KNITTING MACHINES
RASCHEL KNITTING MACHINES
SIMPLEX KNITTING MACHINES
TRICOT KNITTING MACHINES
BT KNITTING MACHINES
RT CASTING OFF
CIRCULAR KNITTING MACHINES
COMPOUND NEEDLES
FILLING KNITTING MACHINES
FULL SET
GUIDE BARS
HALF SET
KNITTING
LACE MACHINES
LAPPING (WARP KNITTING)
LINKS LINKS MACHINES
NEEDLE BARS
OVERLAP
PRESSER BARS
RUNNER RATIO
RUNNERS
SET (WARP KNITTING)
SPRING NEEDLES
TRICOT KNITTED FABRICS
TRICOT KNITTING
TRICOT STITCHES
UNDERLAP
WARP KNITTED FABRICS
WARP KNITTING
WEBHOLDERS

WARP LACE MACHINES
BT LACE MACHINES
RT BAR WARP MACHINES
BARMEN MACHINES
BOBBINET MACHINES
CURTAIN MACHINES
GO THROUGH MACHINES
LEAVERS MACHINES
MECHLIN MACHINES
ROLLING LOCKER MACHINES
SIVAL MACHINES
STRING WARP MACHINES

WARP MILLS
USE BEAMS

WARP PREPARATION
UF YARN PREPARATION
NT CREELING
REEDING
SECTIONAL WARPING
SLASHING
WARPING
RT ALLOWANCE FACTOR (WARP PREPARATION)
BEAM DYEING (YARN)
BEAMS
BEER
DRAWING IN
ENDS PER INCH
HEDDLES
LEASE RODS
LEASING
LIGHT DENT
LOOMS
PACKAGES
PACKAGING (OPERATION)
QUALITY CONTROL
REED WIRES
REEDS
SHEDDING
SIZE BOXES
SIZES (SLASHING)
SLASHERS
SLASHERS BEAMS
SQUEEZING
STARCH
STRETCH ROLLS
SWOLLEN DENT
TENSION CONTROL
TENSION DEVICES
TWISTING
TYING IN
UNWINDING
WARP BEAMS
WARP ENDS
WARPERS (MACHINE)
WEAVING
WINDERS
WINDING
YARNS

WARP REPP
BT FLAT WOVEN FABRICS
WOVEN FABRICS
RT PLAIN WEAVES

WARP SHRINKAGE
BT SHRINKAGE
RT COMPACTING
COMPRESSIVE SHRINKING
CONTRACTION
DIMENSIONAL STABILITY
FELTING SHRINKAGE
FILLING SHRINKAGE
FUSION CONTRACTION SHRINKAGE
MOLDING SHRINKAGE
PRESHRINKING
RELAXATION SHRINKAGE
RESIDUAL SHRINKAGE
SHRINK GRADING
SHRINK RESISTANCE
SHRINKING

WARP STREAKS
BT FABRIC DEFECTS
RT BARRE
FILLING STREAKS
REED MARKS
STREAKINESS

WARP TENSION
BT TENSION
RT ABRASION
BEATING UP
BENDING RIGIDITY
BOUNCE (WEAVING)
CLOTH FELL
CRIMP INTERCHANGE
CYCLIC STRESS
DISC TENSION
DROP WIRES
ELASTIC MODULUS (TENSILE)
END BREAKAGES
FILLING TENSION
FRICTION LET OFF
JAMMING
LET OFF
LET OFF BRACKETS
LOOM TAKE UP
LOOMS
PROCESS VARIABLES
REEDS
SHEDDING
WARP ENDS
WEAVING
WEAVING DYNAMICS
YARN DAMAGE
YARN TENSION

WARP TYING IN MACHINES
RT LOOMS
TYING IN
WARP PREPARATION
WEAVING

WARP WIND
BT WIND
RT BUILDER MOTIONS
CHEESE WINDERS
CHEESES
CLOSE WIND
COMBINATION WIND
CONES
FILLING WIND
OPEN WIND
PACKAGES
QUILLS
ROVING WIND
SPINNING
TRAVERSE
UNWINDING
WINDERS
WINDING

WARP YARNS
USE WARP ENDS

WARPERS (MACHINE)
UF BEAMERS
RT BEAMS
CHEESE WINDERS
COMBS (LEASING)
CONERS
CONES
CREELING
CREELS
DRUM WINDERS
HACK STANDS
PACKAGES
PRECISION WINDERS
QUILLERS
QUILLS
REED WIDTH
REED WIRES
STOP MOTIONS
TENSION CONTROL
TENSION DEVICES
THREAD GUIDES
WARP ENDS
WARP PREPARATION
WINDERS

WARPING
UF BEAM WARPING
BEAMING
DRESSING (WARP)
WARP BEAMING
BT WARP PREPARATION
RT BEAM DYEING (YARN)
BEAMS
BEER
CONING
COTTON SYSTEM PROCESSING
CREELING
DENTS
DESIZING MACHINES
DRAWING IN
DRUM WINDING
JACKSPOOLS
LEASE RODS
LEASING
LOOMS
MALIMO PROCESS (TN)
PACKAGES
PACKAGING (OPERATION)
REED DRAFT
REEDING
REEDS
SECTION LEASE
SECTIONAL WARPING
SILK SYSTEM WARPING
SLASHING
STOP MOTIONS
TENSION CONTROL
WARP BEAMS
WARP ENDS
WARP KNITTING
WARPERS (MACHINE)
WEAVING
WINDING

WARPING LEASE
BT LEASE
RT COTTON SYSTEM WARPING
SILK SYSTEM WARPING
WARP ENDS

WARPS
USE WARP ENDS

WARPWISE STRETCH
BT STRETCH
RT ELASTIC STRAIN
ELASTICITY
EXTENSIBILITY
FILLINGWISE STRETCH
FORM PERSUASIVE GARMENTS
FOUNDATION GARMENTS
RECOVERY (SELF RESTORATION TO ORIGINAL CONDITION)
STRESS STRAIN CURVES
STRETCH FABRICS
STRETCH WOVEN FABRICS
STRETCH YARNS (FILAMENT)
TEXTURED YARNS (FILAMENT)
TWO WAY STRETCH
UNIAXIAL STRESS

WASH WEAR FABRICS
USE EASY CARE FABRICS

WASH WEAR FINISHES
UF ANTICREASE FINISHES
EASY CARE FINISHES
SMOOTH DRYING FINISHES
WRINKLE RECOVERY FINISHES
WRINKLE RESISTANCE FINISHES
BT FINISH (SUBSTANCE ADDED)
RT AZIRIDINE COMPOUNDS
CHLORINE RETENTION
DISULFIDE BONDS
DIVINYL SULFONE
DMEU
DMPU
EASY CARE FABRICS
FORMALDEHYDE
HALOHYDRINS
METHYLOL CARBAMATES
METHYLOL COMPOUNDS
ODORS
OXIRANE COMPOUNDS
REACTANTS
RESIN FINISHES
SHRINK RESISTANCE FINISHES
WASH WEAR PROPERTIES
WASH WEAR RATING
WASH WEAR TREATMENTS
WRINKLE RECOVERY
WRINKLE RESISTANCE

WASH WEAR PROPERTIES
RT DURABLE PRESS
EASY CARE FABRICS
LAUNDERABILITY TESTING
REACTANTS
WASH WEAR FINISHES
WASH WEAR RATING
WASH WEAR TREATMENTS
WRINKLE RESISTANCE

WASH WEAR RATING
RT BENDING RECOVERY
CREASE RETENTION
DURABLE PRESS
EASY CARE FABRICS
LAUNDERABILITY
LAUNDERABILITY TESTING
WASH WEAR FABRICS
WASH WEAR FINISHES
WASH WEAR PROPERTIES
WASH WEAR TREATMENTS
WRINKLE RECOVERY
WRINKLE RECOVERY ANGLE
WRINKLE RESISTANCE

WASH WEAR TREATMENTS
UF EASY CARE TREATMENTS
WRINKLE RESISTANCE TREATMENTS
NT BELFAST (TN)
BT WET FINISHING
RT CHEMICAL MODIFICATION (FIBERS)
CURING
DRY CURING
EASY CARE FABRICS
PAD BATCH CURE PROCESS
PAD DRY CURE PROCESS
WASH WEAR FINISHES
WASH WEAR PROPERTIES
WASH WEAR RATING
WET CURING
WRINKLE RECOVERY
WRINKLE RESISTANCE

WASHABILITY
USE LAUNDERABILITY

WASHERS
RT GASKETS
GATE TENSION
HYDRAULIC TENSION DEVICES
MAGNETIC TENSION DEVICES
SEALS
TENSION DEVICES
TENSION DISCS
TENSION SPRINGS

WASHFASTNESS (COLOR)
BT COLORFASTNESS
DEGRADATION PROPERTIES
END USE PROPERTIES
RT BLEEDING
CHLORINE FASTNESS (COLOR)
CROCKFASTNESS
CUPRIC SULFATE
DURABILITY
DURABLE FINISH (GENERAL)
DYEING
LAUNDERABILITY
LAUNDERING
LAUNDRY DEGRADATION
RESISTANCE TO BLEEDING
SEAWATER FASTNESS (COLOR)
WASHFASTNESS (OF FINISH)
WASHFASTNESS TESTERS
WASHFASTNESS TESTING

WASHFASTNESS (OF FINISH)
BT DEGRADATION PROPERTIES
END USE PROPERTIES
RT CHEMICALLY MODIFIED COTTON
CHLORINE FASTNESS (OF FINISH)
DMEU
DRY CLEANING FASTNESS (OF FINISH)
DURABLE FINISH (GENERAL)
DURABILITY
LAUNDERABILITY
LAUNDERING
REACTANTS
WASHFASTNESS (COLOR)
WASHFASTNESS TESTERS
WASHFASTNESS TESTING

WASHFASTNESS TESTERS
BT TESTING EQUIPMENT
RT LAUNDERABILITY
LAUNDERABILITY TESTING
LAUNDEROMETER
SHRINKAGE TESTERS
WASHFASTNESS (COLOR)
WASHFASTNESS (OF FINISH)
WASHING MACHINES

WASHFASTNESS TESTING
BT COLORFASTNESS TESTING
LAUNDERABILITY TESTING
TESTING
RT COLORFASTNESS
DYEING
LAUNDEROMETER
LIGHTFASTNESS TESTING
RESISTANCE TO BLEEDING
WASHFASTNESS (COLOR)
WASHFASTNESS (OF FINISH)

WASHING (LAUNDERING)
USE LAUNDERING

WASHING (PROCESSING)
USE SCOURING

WASHING MACHINES
UF WASHING MACHINES (LAUNDERING)
RT DRY CLEANING MACHINES
LAUNDERABILITY TESTERS
LAUNDERING
LAUNDEROMETER
LINT CATCHERS
SHRINKAGE TESTERS
SOIL REMOVAL
WASHFASTNESS TESTERS
WASHFASTNESS TESTING

WASHING MACHINES (LAUNDERING)
USE WASHING MACHINES

WASHING MACHINES (PROCESSING)
USE SCOURING MACHINES

WASTE
NT BURRS
LEAVES
MOTES
SEEDS (TRASH)
TRASH
VEGETABLE MATTER
RT BIODEGRADABILITY
BYPRODUCTS
CLEANLINESS
CLIPPINGS
CONSUMPTION (MATERIAL)
CONTAMINATION
COTTON GINS
CYLINDER SCREENS
EFFLUENTS
FLAT STRIPPING COMBS
FLY (WASTE)
GARNETTING
GINNING
IMPURITIES
LAP WASTE
NOIL REMOVAL
NOILS
PICKING (OPENING)
RECOVERY (WASTE)
SCOURING WASTE
WASTE DISPOSAL
WASTE MACHINES
WASTE TREATMENT
WATER POLLUTION
YIELD (RETURN)

WASTE CONTROL KNIVES
NT MOTE KNIVES
RT CARD WASTE
CARDS
COTTON CARDS
TRASH
WASTE

WASTE DISPOSAL
RT EFFLUENTS
POLLUTION
WASTE
WASTE TREATMENT
WATER POLLUTION

WASTE HEAT RECOVERY
RT HEAT
HEAT EXCHANGERS
HEAT TRANSFER
RECOVERY (CHEMICAL)

WASTE MACHINES
RT CLIPPINGS
GARNETTING MACHINES
RAG PULLING MACHINES
SHODDY
WASTE

WASTE TREATMENT
UF EFFLUENT TREATMENT
SLUDGE
RT EFFLUENTS
POLLUTION
PURIFICATION
RECOVERY (CHEMICAL)
RECOVERY (WASTE)
WASTE
WASTE DISPOSAL
WATER TREATMENT
WATER POLLUTION

WATER
NT DISTILLED WATER
HARD WATER
RT BOUND WATER
CALGON (TN)
CAPILLARY WATER
DEHYDRATION
FLUID FLOW
HUMIDITY
LIQUIDS
MOISTURE CONTENT
OXIDES
WATER ABSORPTION
WATER JET LOOMS
WATER POLLUTION
WATER REPELLENCY
RT WATER REPELLENTS
WATER RESISTANCE
WATER SOLUTIONS
WATER TREATMENT
WATER VAPOR

WATER ABSORPTION
NT DYNAMIC ABSORPTION (WATER)
STATIC ABSORPTION (WATER)
BT ABSORPTION (MATERIAL)
SORPTION OF WATER
RT MOISTURE ABSORPTION
WATER
WATER ABSORPTION TESTING
WATER IMBIBITION
WATER REPELLENCY
WATERPROOFING
WETTING
WIPING CLOTHS

WATER ABSORPTION TESTING
UF WATER ABSORPTION TESTS
NT DYNAMIC ABSORPTION TESTING
STATIC ABSORPTION TESTING
BT TESTING
RT DRAVES TEST
WATER REPELLENCY
WATER RESISTANCE TESTING

WATER ABSORPTION TESTS
USE WATER ABSORPTION TESTING

WATER BRAKES
BT BRAKES (FOR ARRESTING MOTION)
RT MAGNETIC BRAKES
MECHANICAL BRAKES

WATER IMBIBITION
RT CAPILLARY WATER
CROSSLINKING
WATER ABSORPTION
WATER RETENTION

WATER IN OIL EMULSIONS
BT EMULSIONS
RT BINDERS (PIGMENTS)
EMULSIFYING AGENTS
EMULSION POLYMERS
LATICES
OIL IN WATER EMULSIONS
SOLVENT EMULSIONS

WATER JET LOOMS
BT JET LOOMS
LOOMS
SHUTTLELESS LOOMS
RT AIR JET LOOMS
DRAPER SHUTTLELESS LOOMS (TN)
OUTSIDE FILLING SUPPLY
RAPIER LOOMS
SHUTTLELESS WEAVING
SULZER LOOMS (TN)
WATER
WEAVING

WATER PERMEABILITY
BT PERMEABILITY
RT AIR PERMEABILITY
MOISTURE VAPOR TRANSMISSION
PENETRATION
PERMEABILITY TESTING
WATER ABSORPTION
WATER REPELLENCY
WATER RESISTANCE
WATER RESISTANCE TESTING
WATER VAPOR TRANSMISSION

WATER PERMEABILITY TESTING
USE WATER RESISTANCE TESTING

WATER POLLUTION
UF STREAM SANITATION
BT POLLUTION
RT CONTAMINATION
WASTE
WASTE DISPOSAL
WASTE TREATMENT
WATER
WATER TREATMENT

WATER PROOFING
USE WATERPROOFING

WATER REPELLENCY
BT END USE PROPERTIES
FABRIC PROPERTIES
FABRIC PROPERTIES (PHYSICAL EXCLUDING MECHANICAL)
FINISH (PROPERTY)
RT ABSORBENCY (MATERIAL)
AIR PERMEABILITY
BUNDESMANN TESTING
CONTACT ANGLES
DYNAMIC ABSORPTION TESTING
FABRIC PROPERTIES
FABRIC PROPERTIES (PHYSICAL EXCLUDING MECHANICAL)
FLUOROCARBON POLYMERS
RT FLUOROCARBONS
HYDROPHILIC PROPERTY
HYDROPHOBIC PROPERTY
HYDROSTATIC PRESSURE TESTING
IMMERSION TESTING
IMPACT PENETRATION TESTING
PENETRATION
PERMEABILITY
PERMEABILITY TESTING
PORES
RAIN PENETRATION TESTING
RAINWEAR
SPRAY TESTING
STAIN RESISTANCE
SURFACE CHEMISTRY
SURFACE PROPERTIES (PHYSICAL CHEMICAL)
SURFACE TENSION
WATER
WATER ABSORPTION
WATER ABSORPTION TESTING
WATER PERMEABILITY
WATER REPELLENCY TESTING
WATER REPELLENCY TREATMENTS
WATER REPELLENTS
WATER RESISTANCE
WATER RESISTANCE TESTING
WATER VAPOR TRANSMISSION
WATERPROOFING
WETTING
WETTING TIME

WATER REPELLENCY TESTING
NT SPRAY TESTING
RT DROP PENETRATION TESTING
WATER REPELLENCY
WATER REPELLENCY TREATMENTS

WATER REPELLENCY TREATMENTS
BT WET FINISHING
RT HYDROPHILIC PROPERTY
HYDROPHOBIC PROPERTY
WATER REPELLENCY
WATER REPELLENCY TESTING
WATER REPELLENTS
WATERPROOFING

WATER REPELLENTS
BT FINISH (SUBSTANCE ADDED)
RT HYDROPHILIC PROPERTY
HYDROPHOBIC PROPERTY
METALLIC SOAPS
OCTADECYL ETHYLENE UREA
ORGANOTIN COMPOUNDS
PYRIDINE DERIVATIVES
SILICONES
WATER
WATER REPELLENCY
WATER REPELLENCY TESTING
WATER REPELLENCY TREATMENTS
WAX EMULSIONS
WAXES
ZIRCONIUM COMPOUNDS

WATER RESISTANCE
BT END USE PROPERTIES
FABRIC PROPERTIES
FABRIC PROPERTIES (PHYSICAL EXCLUDING MECHANICAL)
FINISH (PROPERTY)
RT PORES
WATER
WATER PERMEABILITY
WATER REPELLENCY
WATER RESISTANCE TESTING
WATER RESISTANT FOOTWEAR
WATER RESISTANT LEATHER
WATER VAPOR TRANSMISSION
WEATHERING

WATER RESISTANCE TESTERS
BT TESTING EQUIPMENT
RT WATER REPELLENCY TESTING
WATER RESISTANCE TESTING

WATER RESISTANCE TESTING
UF WATER PERMEABILITY TESTING
NT BUNDESMANN TESTING
DROP PENETRATION TESTING
HYDROSTATIC PRESSURE TESTING
IMPACT PENETRATION TESTING
RAIN PENETRATION TESTING
BT PERMEABILITY TESTING
TESTING
RT AIR PERMEABILITY TESTING
BURST TESTING
DYNAMIC ABSORPTION TESTING
SPRAY TESTING
STATIC ABSORPTION TESTING
WATER ABSORPTION TESTING
WATER PERMEABILITY
WATER REPELLENCY
WATER RESISTANCE

WATER RESISTANT FOOTWEAR
BT FOOTWEAR
RT WATER RESISTANCE

WATER RESISTANT LEATHER
BT LEATHER
RT WATER RESISTANCE

WATER RETENTION
RT CAPILLARY WATER
CROSSLINKING
MOISTURE CONTENT
WATER IMBIBITION

WATER SOLUTIONS
UF AQUEOUS SOLUTIONS
BT SOLUTIONS (LIQUID)
RT AQUEOUS SCOURING
DYE SOLUTIONS
NONAQUEOUS SOLUTIONS
WATER

WATER SPOTS
BT FABRIC DEFECTS
RT FOGMARKING
OIL STAINS
SOILING
STAIN REMOVAL
STAINING
STAINS

WATER TREATMENT
RT DEMINERALIZING
HARD WATER
ION EXCHANGE
PERMUTITE (TN)
PH CONTROL
PURIFICATION
WASTE TREATMENT
WATER
WATER POLLUTION

WATER VAPOR
RT VAPOR PHASE
VAPOR PRESSURE
WATER
WATER VAPOR TRANSMISSION

WATER VAPOR TRANSMISSION
UF MOISTURE VAPOR TRANSMISSION
RT AIR PERMEABILITY
AIR RESISTANCE
TRANSFER PROPERTIES
WATER ABSORPTION
WATER PERMEABILITY
WATER REPELLENCY
WATER RESISTANCE
WATER VAPOR
WATER VAPOR TRANSMISSION TESTING

WATER VAPOR TRANSMISSION TESTING
BT TESTING
RT THERMAL TRANSMITTANCE TESTING
WATER VAPOR
WATER VAPOR TRANSMISSION

WATERPROOFING
UF WATER PROOFING
BT WET FINISHING
RT AIR PERMEABILITY
BUNDESMANN TESTING
COATING (PROCESS)
CONTACT ANGLES
DYNAMIC ABSORPTION TESTING
HYDROPHILIC PROPERTY
HYDROPHOBIC PROPERTY
HYDROSTATIC PRESSURE TESTING
IMPACT PENETRATION TESTING
ORIENTATION (SURFACE)
PENETRATION
PERMEABILITY
RAIN PENETRATION TESTING
RAINWEAR
RESIN FINISHES
SEALING
SPRAY TESTING
STATIC ABSORPTION TESTING
SURFACE PROPERTIES (PHYSICAL CHEMICAL)
SURFACE TENSION
WATER ABSORPTION
WATER REPELLENCY
WATER REPELLENCY TREATMENTS
WATERPROOFING AGENTS
WAX EMULSIONS
WETTING

WATERPROOFING AGENTS
BT FINISH (SUBSTANCE ADDED)
RT COATINGS (SUBSTANCES)
WATER REPELLENTS
WATERPROOFING
WAX EMULSIONS

WAVE LENGTH
RT AMPLITUDE
BALLOON DYNAMICS
COLOR
CRIMP AMPLITUDE (FIBER)
CRIMP FREQUENCY
DRAFTING WAVES
FREQUENCY
LONGITUDINAL VIBRATIONS
PERIODICITY
SONIC VELOCITY
SOUND
SPECTRUM ANALYSIS
SPINNING DYNAMICS
STRESS PROPAGATION
TRANSVERSE VIBRATIONS
VIBRATION
WAVINESS

WAVINESS
RT BICOMPONENT FIBER YARNS (FILAMENT)
BICOMPONENT FIBER YARNS (STAPLE)
BUCKLING
CRIMP
CRIMP AMPLITUDE (FIBER)
CRIMP FREQUENCY
DRAPE
LATENT CRIMP
ROUGHNESS
SELF CRIMPING YARNS
SMOOTHNESS
WAVE LENGTH
YARN CRIMP
Z CRIMP

WAX EMULSIONS
RT SILICONES
WATER REPELLENTS
WATERPROOFING
WATERPROOFING AGENTS

WAXES
NT COTTON WAX
PARAFFIN WAXES
WOOL WAX
RT BATIK DYEING
COATINGS (SUBSTANCES)
FATS
FINISH (SUBSTANCE ADDED)
LUBRICANTS
OILS
WATER REPELLENTS

WEAK SPOTS
UF THIN SPOTS
RT BREAKING LENGTH
BREAKING STRENGTH
COUNT STRENGTH PRODUCT
DEFECTS
DOUBLING (DRAFTING)
DRAFTING WAVES
EFFICIENCY (STRUCTURAL)
END BREAKAGES
FIBER BREAKAGE
FIBER STRENGTH
FIBER TENDERNESS
IRREGULARITY
IRREGULARITY (SHORT TERM)
LEA STRENGTH PRODUCT
SLIVER TENDERNESS
STRENGTH EFFICIENCY
TENSION
THICK SPOTS
THICKNESS
VARIATION
WEAVING EFFICIENCY
YARN DEFECTS
YARN STRENGTH

WEAR
NT VOLUME WEAR
RT ABRASION
ABRASIVES
BOUNDARY LUBRICATION
BRAKES (FOR ARRESTING MOTION)
BREAKING
COEFFICIENT OF FRICTION
DAMAGE
DEGRADATION
DURABILITY
DYNAMIC FRICTION
EMERY
FRICTION
GALLING
HYDRODYNAMIC LUBRICATION
LUBRICATION
PLOUGHING
ROLLING
RUBBING
SCUFFING
SLICING
WEAR RESISTANCE
WEAR TESTING
WEDGING

WEAR RESISTANCE
BT END USE PROPERTIES
MECHANICAL DETERIORATION PROPERTIES
MECHANICAL PROPERTIES
RT ABRASION
ABRASION RESISTANCE
ABRASION RESISTANCE FINISHES
DURABILITY
FABRIC PROPERTIES
FABRIC PROPERTIES (MECHANICAL)
FATIGUE RESISTANCE
FRICTION
PERFORMANCE
SNAG RESISTANCE
VOLUME WEAR
WEAR
WEAR TESTING

WEAR TESTING
BT TESTING
RT ABRASION TESTING
DRY CLEANING
LAUNDERING
SAMPLING
WEAR
WEAR RESISTANCE
WEARING (OF APPAREL)

WEARING (OF APPAREL)
RT APPLICATIONS
END USES
WEAR TESTING

WEATHER RESISTANCE
BT DEGRADATION PROPERTIES
END USE PROPERTIES
FINISH (PROPERTY)
RT AIR POLLUTION
COLORFASTNESS
FABRIC PROPERTIES
PHOTOCHEMICAL DEGRADATION
THERMAL STABILITY
WEATHERING
WEATHERING TESTING

WEATHERING
UF EFFECT OF WEATHER
RT AGING (STORAGE)
AIR POLLUTION
CHEMICAL PROPERTIES
WATER RESISTANCE
WEATHER RESISTANCE
WEATHERING TESTING

WEATHERING TESTING
BT COLORFASTNESS TESTING
RT AIR POLLUTION
WEATHER RESISTANCE
WEATHERING

WEAVABILITY
USE WEAVING EFFICIENCY

WEAVE
BT FABRIC PROPERTIES
FABRIC PROPERTIES (STRUCTURAL)
RT AESTHETIC PROPERTIES
APPEARANCE
BALANCED WEAVES
DESIGN
DRAFTING (PATTERN)
DRAWING IN
FABRIC ANALYSIS
FABRIC DESIGN
FABRIC STRUCTURE
HARNESS DRAFT
PATTERN DEFINITION
PATTERN (FABRICS)
PICKING (WEAVING)
REPEATS
STITCH CLARITY
TEXTURE
WARPING
WEAVING
WOVEN FABRICS

WEAVE (TYPES)
NT BASKET WEAVES
BIRDS EYE WEAVES
CHECKER BOARD WEAVES
CLIPSPOT WEAVES
COMBINATION WEAVES
DIAMOND WEAVES
DOGSTOOTH WEAVES
FOLK WEAVES
GRANITE WEAVES
GRECIAN WEAVES
HONEYCOMB WEAVES
HUCKABACK WEAVES
IMPERIAL SATEEN
JACQUARD WEAVES
LAPPET WEAVES
LENO WEAVES
MOCK LENO WEAVES
OXFORD WEAVES
PIQUE WEAVES
PLAIN WEAVES
RIB WEAVES
SATIN WEAVES
SPOT WEAVES
SWIVEL WEAVES
TABBY WEAVES
TAFFETA WEAVES
TWILL WEAVES

WEAVERS KNOTS
BT KNOTS
RT JOINING
KNOTTING
LOOMS
PIECING
TYING IN
WARP ENDS
WEAVING

WEAVING
UF SHUTTLE WEAVING
NT DOBBY WEAVING
HAND WEAVING
JACQUARD WEAVING
SHUTTLELESS WEAVING
TABLET WEAVING
RT ALTERNATE PICKS
ARACHNE PROCESS (TN)
BACK CROSSING HEDDLES
BACK FILLING
BALANCED SHED
BALANCED WEAVES
BANGING OFF
BEAMS
BEATING UP
BEER
BLANKET SHUTTLES
BOAT SHUTTLES
BOUNCE (WEAVING)
BRAIDING
CAM LOOMS
CENTER SELVEDGES
CIRCULAR LOOMS
CLOSED SHED
CLOTH FELL
CLOTH ROLLS
CRACKED SELVEDGES
CRIMP INTERCHANGE
DOG LEGGED SELVEDGES
DOUBLE PICKS
DRAWING IN
DROP WIRES
END BREAKAGES
FELTING
FILLING KNITTING
FIXED REEDS
RT FLYSHOT LOOMS
FRICTION LET OFF
HARNESS DRAFT
HARNESSES
HEDDLES
INTERLACINGS (YARN IN FABRIC)
JAMMING
KNITTING
KNOTTING
LACE MAKING
LENO SELVEDGES
LET OFF
LET OFF BRACKETS
LET OFF CHAINS
LET OFF MOTIONS
LET OFF SPOOLS
LET OFF WEIGHTS
LOOM STOPS
LOOM TAKE UP
LOOMS
LOOSE REEDS
MALI PROCESSES (TN)
MALIMO PROCESS (TN)
MALIPOL PROCESS (TN)
MALIWATT PROCESS (TN)
MISPICKS
NEEDLING
NEGATIVE SHEDDING
OPEN SHED
PICK AND PICK
PICKERS (LOOM)
PICKING (WEAVING)
POSITIVE SHEDDING
PULLED IN SELVEDGES
QUILL CHANGING
QUILLS
RAPIER LOOMS
ROTARY MAGAZINES
SEALED SELVEDGES
SELVEDGE WARPS
SELVEDGES
SHED (WEAVING)
SHEDDING
SHUTTLE BINDERS
SHUTTLE CHECKING
SHUTTLE DEFLECTORS
SHUTTLE MARKS
SHUTTLE SMASH PROTECTORS
SHUTTLES
SLACK SELVEDGES
SLEYS (LOOM)
SPLIT SHED
STARTING MARKS
STICK SHUTTLES
STITCH BONDING
STOP MOTIONS
STRETCH WOVEN FABRICS
TAKE UP
TAPE SELVEDGES
TEMPLES
TIGHT SELVEDGES
TRANSFER TAILS
TREACLES
TUFTING
WARP ENDS
WARP KNITTING
WARP PREPARATION
WARP TENSION
WARPING
WATER JET LOOMS
WEAVE
WEAVING DYNAMICS
WEAVING EFFICIENCY
WEAVING LEASE
WEFT FORKS
WOVEN FABRICS
YARN DYNAMICS
YARN TENSION

WEAVING DYNAMICS
BT PROCESS DYNAMICS
RT ABRASION
BENDING
BIAXIAL STRESS
CARDING DYNAMICS
CRIMP INTERCHANGE
DRAFTING DYNAMICS
FATIGUE RESISTANCE
FILLING TENSION
FRICTION
IMPACT
IMPACT STRENGTH
JAMMING
KNITTING DYNAMICS
LET OFF MOTIONS
LOOM TAKE UP
MECHANISM (FUNDAMENTAL)
SPINNING DYNAMICS
TWISTING DYNAMICS
WARP TENSION
WEAVING
WEAVING EFFICIENCY
WINDING DYNAMICS
YARN DYNAMICS
YARN TENSION

WEAVING EFFICIENCY
UF WEAVABILITY
BT PROCESS EFFICIENCY
RT DOWN TIME
END BREAKAGES
IRREGULARITY
LOOM STOPS
PATROLLING
PERFORMANCE
PRODUCTION
SET UP TIME
SPINNING EFFICIENCY
STARTING MARKS
STOPS
WEAK SPOTS
WEAVING
WEAVING DYNAMICS
WINDING EFFICIENCY

WEAVING LEASE
BT LEASE
RT LOOMS
WARP ENDS
WEAVING

WEB FORMING MACHINES
RT BATT MAKING MACHINES
MALI PROCESSES (TN)
NONWOVEN FABRIC MACHINES
PAPER MACHINES
TUFT FORMERS

WEB ORIENTATION
RT ANISOTROPIC MATERIALS
BLOTCHINESS
BUNCHING COEFFICIENT
CARD WEBS
CARDING
CROSS LAID WEBS
DEGREE OF ORIENTATION
FIBER ARRANGEMENT
ISOTROPIC MATERIALS
ORIENTATION UNIFORMITY
ORTHOTROPIC MATERIALS
PATCHINESS
RANDO WEBS (TN)
SPOTTINESS
TRAILING HOOKS
UNIDIRECTIONALLY ORIENTED WEBS
WEB UNIFORMITY
WEBS
WIRA WEB LEVELNESS TESTER

WEB PURIFIERS
NT CROSROL WEB PURIFIER (TN)
PERALTA ROLLS
RT BURR CRUSHERS
CARD WEBS
CLEANLINESS
CROSS ROLLS
TRASH
WEBS

WEB UNIFORMITY
RT AUTOCOUNT (TN)
BLOTCHINESS
BUNCHING COEFFICIENT
CARD SLIVERS
CARD WEBS
CARDING
IRREGULARITY
MASS UNIFORMITY
NEPS
ORIENTATION UNIFORMITY
PATCHINESS
SLUBBINGS
SPOTTINESS
STREAKINESS
WEB ORIENTATION
WEBS
WIRA WEB LEVELNESS TESTER

WEBBING
NT LADDER WEBBING
BT FLAT WOVEN FABRICS
WOVEN FABRICS
NARROW FABRICS
RT CORDAGE
RIBBON
TAPE

WEBHOLDERS
RT CIRCULAR KNITTING MACHINES
FULLY FASHIONED KNITTING MACHINES
HOSIERY MACHINES
KNITTING
KNITTING MACHINES
KNOCKOVER BITS
SEAMLESS HOSIERY MACHINES
SINKERS
TAKE DOWN
TRICOT KNITTING MACHINES
WARP KNITTING MACHINES

WEBS
NT CARD WEBS
WET LAY RANDOM WEBS
BT FIBER ASSEMBLIES
RT AIR BRIDGES

RT AUTOCCUNT (TN)
BATTING
BONDEC FIBER FABRICS
BUNCHING COEFFICIENT
CARDING
CONDENSER SCREENS
CROSROL WEB PURIFIER (TN)
DOFFERS (CARD)
FABRICS
FELTS
FIBER ORIENTATION
FIBERS
FILMS
FOAMS
MALI PROCESSES (TN)
MALIMO PROCESS (TN)
MALIPOL PROCESS (TN)
MALIWATT PROCESS (TN)
NEEDLING
NEPPINESS INDEX
NEPS
NONWOVEN FABRICS
ORIENTATION
ORIENTATION UNIFORMITY
PATCHINESS
RANDO WEBBER (TN)
RANDO WEBS (TN)
ROVINGS
SCOTCH FEED
STITCH BONDING
SLIVER CANS
SLIVERS
SPOTTINESS
STOCK (FIBER)
STREAKINESS
TOPS
TOW
TUFT FORMERS
UNIDIRECTIONALLY ORIENTED WEBS
WEB ORIENTATION
WEB PURIFIERS
WEB UNIFORMITY
WET PROCESS (WEB)
WIRA WEB LEVELNESS TESTER
YARNS

WEDGE SCALE
RT FIBER DIAMETER
FIBER FINENESS
FINENESS TESTING
WOOL

WEDGING
RT ANTIWEDGE RINGS
FORCE
FRICTION
GALLING
PLOUGHING
RINGS
SCOURING
SCRATCHING
SCUFFING
TRAVELER BURNOUT
TRAVELER CHATTER
TRAVELERS
UNLUBRICATED RINGS
WEAR
WELDING
YARNS

WEFT
USE FILLING YARNS

WEFT FORKS
RT FEELER MECHANISMS
FILLING YARNS
LOOMS
STOP MOTIONS
WEAVING

WEFT INSERTION
USE PICKING (WEAVING)

WEFT KNITTED FABRICS
USE FILLING KNITTED FABRICS

WEFT KNITTING MACHINERY
USE FILLING KNITTING MACHINES

WEFT LOOPS
USE CENTER LOOPS

WEFT PIRNS
USE QUILLS

WEFT STRAIGHTENERS
USE FILLING STRAIGHTENING MACHINES

WEFT STREAKS
BT DYEING DEFECTS
STREAKINESS

WEFT YARNS
USE FILLING YARNS

WEIGHT
NT DRY WEIGHT
HEAVY WEIGHT
LIGHT WEIGHT
MEDIUM WEIGHT
RT DEAD WEIGHT LOADING
DENSITY
DIMENSIONS
FABRIC WEIGHT
GRAVITATION
LINEAR DENSITY
LOADING
MASS
ROLL WEIGHTING
SPECIFIC GRAVITY
THICKNESS
VOLUME
WEIGHT CONTROL

WEIGHT AVERAGE MOLECULAR WEIGHT
BT MOLECULAR WEIGHT
RT END GROUPS
MOLECULAR WEIGHT DISTRIBUTION
NUMBER AVERAGE MOLECULAR WEIGHT

WEIGHT CONTROL
RT AUTOLEVELLERS
AUTOMATIC CONTROL
CONTROL SYSTEMS
DRAFT CONTROL
DRAFTING (STAPLE FIBER)
HOPPER FEED
IRREGULARITY CONTROL
PROCESS CONTROL
RAPER AUTOLEVELLER (TN)
SCHUPP REGULATOR
WEIGHT

WEIGHT PER UNIT LENGTH
USE LINEAR DENSITY

WEIGHT UNIFORMITY
USE IRREGULARITY

WELDING
RT COEFFICIENT OF FRICTION
COHESION
FRICTION
FRICTION THEORY
FUSION (MELTING)
JOINING
PLOUGHING
WEDGING

WELSH MOUNTAIN WOOL
BT MOUNTAIN WOOL
WOOL

WELTING
RT FASHIONING
FILLING KNITTING
FILLING KNITTING MACHINES
FLOATING (KNITTING)
HOSIERY
KNITTING
KNITTING NEEDLES
LANDING LOOPS
LAYING IN (KNITTING)
RACKING
REVERSE WELTING
TUCKING (KNITTING)
WELTS (KNITTED)

WELTS (FOOTWEAR)
BT FOOTWEAR COMPONENTS
RT SOLES
UPPERS

WELTS (KNITTED)
NT INTURNED WELTS
RT HEELS (STOCKINGS)
KNITTING
WELTING
WELTS (SEWN)

WELTS (SEWN)
UF SEWN WELTS
RT OVEREDGE STITCHES
SEWING
UPHOLSTERY
UPHOLSTERY FABRICS
UPHOLSTERY LEATHER
WELTS (KNITTED)

WEST POINT COHESION TESTER
USE DRAFTING FORCE TESTERS

WET BULB DEPRESSION
USE WET BULB TEMPERATURE

WET BULB TEMPERATURE
UF WET BULB DEPRESSION
RT DEW POINT
DRY BULB TEMPERATURE
DRYING
HYGROMETERS
PSYCHROMETRY
RELATIVE HUMIDITY

WET CREASE RECOVERY
USE WET WRINKLE RECOVERY

WET CROCKING
BT CROCKING
RT BLEEDING
CROCKFASTNESS
FADING
RUBBING
SCRUBBING
STAINING
WET TESTING

WET CURING
BT CURING
WET FINISHING
RT DEFERRED CURE METHOD
DOUBLE CURE METHOD
DRY CURING
DURABLE PRESS TREATMENTS
HOT PRESS METHOD (DURABLE PRESS)
PAD BATCH CURE PROCESS
PRE-CURE METHOD
WASH WEAR TREATMENTS

WET DECATING
USE DECATING

WET FINISHING
UF WET PROCESSING
NT ANTIFELTING TREATMENTS
ANTIMICROBIAL TREATMENTS
BATCH FINISHING
BAULK FINISHING
BLEACHING
BLOWN FINISH
BLUEING
CARBONIZING
CARROTING
COLD SETTING (SYNTHETICS)
CONTINUOUS BLEACHING
CONTINUOUS FINISHING
CRABBING
CURING
DAMPENING
DELUSTERING
DEODORIZING
DESIZING
DURABLE PRESS TREATMENTS
DYE FIXING
FABRIC RELAXATION
FELTING
FIRE RETARDANCY TREATMENTS
FIRE PROOFING TREATMENTS
FULLING
GLAZING
GRASS BLEACHING
HEAT SETTING (MOIST)
INSECT RESISTANCE TREATMENTS
LONDON SHRINKING
MANGLING
MERCERIZING
MILDEW RESISTANCE TREATMENTS
OIL REPELLENT TREATMENTS
ONE BATH PROCESS
OPEN WIDTH BLEACHING
PAD BATCH CURE PROCESS
PAD DRY CURE PROCESS
PAD FINISHING
PAD JIG FINISHING
PAD ROLL FINISHING
PAD STEAM FINISHING
PADDING
PERMANENT PLEATING
PILL RESISTANCE TREATMENTS
PRESHRINKING
PRINTING
RELAXATION
RINSING
ROPE BLEACHING
SCOURING
SCUTCHING (FINISHING)
SETTING (EXCEPT SYNTHETICS)
SHRINK PROOFING
SHRINKING
SIMULTANEOUS DYEING AND FINISHING
SLASHING
SOAPING
SOFTENING
SPONGING
STARCHING
STIFFENING TREATMENTS
TWO BATH PROCESS
WASH WEAR TREATMENTS
WATER REPELLENCY TREATMENTS
WATERPROOFING
WET CURING
WET RELAXATION
BT FINISHING PROCESS (GENERAL)
RT AFTERTREATMENTS (GENERAL)
CRIMP FIXING
DRY CLEANING FASTNESS (OF FINISH)
DRY FINISHING
FINISHED FABRICS
HEAT TREATMENT
(CONT.)

RT OPEN WIDTH PROCESSING
PRETREATMENTS (GENERAL)
PROCESS DESIGN
ROPE PROCESSING
SETTING TREATMENTS (GENERAL)
SIMULTANEOUS FINISHING PROCESSES
SOLVENT TREATMENTS
STEAMERS
STEAMING
TEXTILE MACHINERY (GENERAL)
TEXTILE MATERIALS
TEXTILE PROCESSES (GENERAL)
VAPOR PHASE TREATMENTS

WET LAY RANDOM WEBS
BT WEBS
RT PAPER
PAPER MAKING
RANDO WEBS (TN)
SLURRY
WET MACHINES (PAPERMAKING)
WET PROCESS (WEB)

WET MACHINES (PAPERMAKING)
BT PAPER MACHINES
RT CYLINDER MACHINES
FOURDRINIER MACHINES
INVERFORM MACHINES
KAMYR MACHINES
ROTAFORMERS
VERTAFORMERS
WET LAY RANDOM WEBS
WET PROCESS (WEB)

WET MODULUS
BT MODULUS
RT AMORPHOUS REGION
ELASTIC MODULUS (TENSILE)
HYDROPHILIC PROPERTY
MOISTURE CONTENT
SWELLING
WET STRENGTH
WET TESTING

WET PICKUP
UF MANGLE EXPRESSION
UPTAKE
RT ABSORPTION (MATERIAL)
ADD ON
MANGLING
MOISTURE CONTENT
NIP
NIP PRESSURE
NIP ROLLS
PAD DYEING
PADDERS
PADDING
PRESS SECTION
SQUEEZE ROLLS
SQUEEZING
STATIC ABSORPTION (WATER)
STATIC ABSORPTION TESTING
SUCTION PRESSES

WET PROCESS (WEB)
RT AIR BRIDGES
NONWOVEN FABRICS
PAPER
PAPER MAKING
PULPING
RANDO WEBBER (TN)
RANDO WEBS (TN)
SLURRY
WEBS
WET LAY RANDOM WEBS

WET PROCESSING
USE WET FINISHING

WET PROCESSING RANGES
NT BLEACHING RANGES

WET RELAXATION
UF WET RELAXING
NT LONDON SHRINKING
BT FABRIC RELAXATION
RELAXATION
WET FINISHING
RT AMORPHOUS REGION
CONDITIONING
CRABBING
DRY RELAXATION
FABRIC SHRINKAGE
MOISTURE CONTENT
RELAXATION SHRINKAGE
RELAXATION TIME (MECHANICAL)
RESIDUAL STRESS
SPONGING
STRAIN
STRESS
STRESS RELAXATION
WET TESTING

WET RELAXING
USE WET RELAXATION

WET SPINNING
BT EXTRUSION
SPINNING (EXTRUSION)
RT CAKES
COAGULATING
COAGULATING BATHS
FILAMENT YARNS
GODET ROLLS
MAN MADE FIBERS
MELT SPINNING
MONOFILAMENT YARNS
MULTIFILAMENT YARNS
NEUTRALIZING (FIBER EXTRUSION)
POT SPINNING
SOLVENT SPINNING
SPINNERETS
SYNTHETIC FIBERS

WET STRENGTH
BT STRENGTH
STRESS STRAIN PROPERTIES (TENSILE)
RT AMORPHOUS REGION
BEARING STRENGTH
BREAKING STRENGTH
BUNDLE STRENGTH
COMPRESSIVE STRENGTH
DRY STRENGTH
EQUILIBRIUM MOISTURE CONTENT
FABRIC STRENGTH
HYDROGEN BONDS
HYDROPHILIC PROPERTY
HYDROPHOBIC PROPERTY
LOOP EFFICIENCY
MOISTURE CONTENT
MOISTURE STRENGTH CURVES
STRENGTH ELONGATION TESTING
SWELLING
WET MODULUS
WET TESTING

WET TESTING
RT WET CROCKING
WET MODULUS
WET RELAXATION
WET STRENGTH
WET WRINKLE RECOVERY
WETTING
WETTING TIME

WET WRINKLE RECOVERY
UF WET CREASE RECOVERY
BT WRINKLE RECOVERY
RT BENDING RECOVERY
DRY WRINKLE RECOVERY
EPICHLOROHYDRIN
MOISTURE CONTENT
WET TESTING
WRINKLE RECOVERY ANGLE
WRINKLE RECOVERY TESTING
WRINKLE RESISTANCE

WETTABILITY
USE WETTING

WETTING
UF WETTABILITY
RT AMPHOTERIC SURFACTANTS
ANIONIC SURFACTANTS
ANTIFOAM AGENTS
CAPILLARITY
CATIONIC SURFACTANTS
CONTACT ANGLES
DETERGENCY
DETERGENTS
DYEING AUXILIARIES
DYNAMIC ABSORPTION (WATER)
EMULSIFYING AGENTS
FOAMS (FROTH)
ENTHALPY OF WETTING
HYDROPHILIC FIBERS
HYDROPHILIC PROPERTY
HYDROPHOBIC FIBERS
HYDROPHOBIC PROPERTY
LAUNDERING
LEVELLING AGENTS
MOISTURE ABSORPTION
MOISTURE CONTENT
SCOURING
SOAPS
STATIC ABSORPTION (WATER)
SURFACE CHEMISTRY
SURFACE PROPERTIES (PHYSICAL CHEMICAL)
SURFACE TENSION
SURFACTANTS
STAIN RESISTANCE
SYNTHETIC DETERGENTS
WATER ABSORPTION
WATER REPELLENCY
WATERPROOFING
WET TESTING
WETTING AGENTS
WETTING TIME

WETTING AGENTS
UF PENETRANTS
BT SURFACTANT APPLICATIONS
RT ALKYLNAPHTHALENE SULFONATES
AMPHOTERIC SURFACTANTS
ANIONIC SURFACTANTS
ANTIFOAM AGENTS
CARRIER DYEING
CARRIERS (DYEING)
CATIONIC SURFACTANTS
DETERGENCY
DETERGENTS
DISPERSING AGENTS
DRAVES TEST
DYEING AUXILIARIES
EMULSIFYING AGENTS
FOAMS (FROTH)
HYDROPHILIC PROPERTY
HYDROPHOBIC PROPERTY
IMMERSION
LAUNDERING
LEVELLING AGENTS
MOISTURE CONTENT
NEKAL A (TN)
NEKAL B (TN)
NEKAL NS (TN)
NONIONIC SURFACTANTS
REWETTING AGENTS
SCOURING
SOAPS
SOLVENT ASSISTED DYEING
STAIN RESISTANCE
SURFACTANTS
SYNTHETIC DETERGENTS
TETRAHYDRONAPHTHALENE SULFONATES
WATER REPELLENCY
WETTING
WETTING TIME

WETTING TIME
RT HYDROPHILIC PROPERTY
IMMERSION
WATER REPELLENCY
WET TESTING
WETTING
WETTING AGENTS

WHEEL BRAKES
USE BRAKES (FOR ARRESTING MOTION

WHIP ROLLS
RT BACK REST (LOOM)
LOOMS

WHIPCORD
BT FLAT WOVEN FABRICS
WOVEN FABRICS
RT TWILL WEAVES

WHISKER FIBERS (METAL)
BT HIGH TEMPERATURE FIBERS
INORGANIC FIBERS (MAN MADE)
METAL FIBERS
METALLIC FIBERS
RT FOIL
METALLIC YARNS
MINERAL FIBERS
TIN

WHITE CHINA SILK
BT SILK
RT WHITE JAPAN SILK
YELLOW CHINA SILK

WHITE JAPAN SILK
BT SILK
RT WHITE CHINA SILK

WHITENESS
UF WHITENING
RT BLEACHING
BRIGHTNESS
FLUORESCENT BRIGHTENERS
VALUE (MUNSELL)
WHITENESS RETENTION

WHITENESS RETENTION
RT ANTIREDEPOSITION AGENTS
CHLORINE RETENTION
DISCOLORATION
LAUNDERING
SOIL REDEPOSITION
WHITENESS

WHITENING
USE WHITENESS

WICK LUBRICATED RINGS
USE LUBRICATED RINGS

WICKING
RT CAPILLARITY
CHROMATOGRAPHY
FIBER ASSEMBLIES
LUBRICATION
MASS TRANSFER
PORES
POROUS MATERIALS
STAINING
SURFACE TENSION
WIPING CLOTHS

WIDENING
RT BAGGINESS
DIMENSIONAL STABILITY
NARROWING
WIDTH

WIDENING (KNITTING)
BT FASHIONING
RT FASHIONING MARKS
FULLY FASHIONED HOSIERY
FULLY FASHIONED KNITTING
NARROWING (KNITTING)
WIDTH

WIDTH
UF BREADTH
BT DIMENSIONS
RT DIAMETER
HEIGHT
LENGTH
SHAPE
THICKNESS
WIDENING
WIDENING (KNITTING)

WIGAN
BT WOVEN FABRICS
RT LININGS
PRINT CLOTH
SHEETING (FABRIC)

WILD YARNS
USE WILDNESS

WILDNESS
UF WILD YARNS
BT TWIST LIVELINESS
YARN PROPERTIES
RT BALANCED YARNS
DIMENSIONAL STABILITY
ENTANGLEMENTS
MECHANICAL PROPERTIES
RESIDUAL TORQUE
SNARLING
SNARLING TENDENCY
STABILITY
STRETCH YARNS (FILAMENT)
TORSIONAL BUCKLING
TORSIONAL ENERGY
TURNS PER INCH
TWIST LIVELINESS
TWIST MULTIPLIER
YARN BUCKLING

WILLEYS
USE WILLOWS

WILLIAMS UNIT (TN)
RT CONTINUOUS DYEING RANGE
SEMI CONTINUOUS DYEING RANGE

WILLOWS
UF WILLEYS
BT DUSTING MACHINES
RT FEARNOUGHTS

WILLIAMSON DICKIE (TN)
BT DURABLE PRESS TREATMENTS
RT KORATRON (TN)

WILTON
USE WILTON CARPETS

WILTON CARPETS
UF WILTON
BT CARPETS
PILE FABRICS (WOVEN)
WOVEN CARPETS
WOVEN FABRICS
RT AXMINSTER CARPETS
CHENILLE CARPETS
FLOCKED CARPETS
NONWOVEN CARPETS
TAPESTRIES
TUFTED CARPETS
VELVET CARPETS
WILTON LOOMS

WILTON LOOMS
BT LOOMS
RT WILTON CARPETS

WINCH DYEING
USE BECK DYEING

WIND
NT CLOSE WIND
COMBINATION WIND
FILLING WIND
OPEN WIND
ROVING WIND
WARP WIND
RT CONING
DRUM WINDERS
DRUM WINDING
GAIN (WINDING)
PACKAGES
PRECISION WINDING
PROCESS VARIABLES
RIBBON BREAKING
RIBBONING
TAPER (BOBBIN)
WINDING
WINDING ANGLE

WIND TUNNELS
RT AERODYNAMICS
AIR DRAG
AIR FLOW
AIR PERMEABILITY
LIFT (FLUID)
PARACHUTES
PERMEABILITY TESTERS
RECOVERY SYSTEMS (AERODYNAMIC)
YARN DYNAMICS

WIND UP
USE WINDING

WINDERS
UF WINDING MACHINERY
NT CHEESE WINDERS
CONERS
DRUM WINDERS
PRECISION WINDERS
QUILLERS
UNICONER (TN)
RT BEAMS
COMBINATION WIND
CONES
CREELING
DOFFING (PACKAGE)
FILLING WIND
KNOTTERS
OPEN WIND
PACKAGES
PAPER MACHINES
QUILLS
ROTARY TRAVERSE
SLUB CATCHERS
STOP MOTIONS
TENSION CONTROL
TENSION DEVICES
TRANSFER TAILS
UNWINDING
UNWINDING ACCELERATORS
WARP PREPARATION
WARP WIND
WARPERS (MACHINE)
WINDING ANGLE
WINDING DYNAMICS
WINDING EFFICIENCY

WINDING
UF SPOOLING
WIND UP
WINDING ON
NT AUTOMATIC WINDING
CHEESE WINDING
CLOSE WINDING
CONING
CROSS WINDING
DRUM WINDING
OPEN WINDING
PARALLEL WINDING
PRECISION WINDING
QUILLING
RT AUTOMATIC KNOTTING
BALLOON DYNAMICS
BALLOON TENSION
BEAMS
BROKEN END COLLECTORS
BUILDER MOTIONS
CHEESES
CLOSE WIND
CONES
COTTON SYSTEM PROCESSING
CREELING
DOFFING (PACKAGE)
DRUM WINDERS
ELECTRONIC SLUB CATCHERS
FILLING WIND
FILLING YARNS
FINGER TENSION
FULL PACKAGES
GATE TENSION
HALCHING
IRREGULARITY
KNOTS
KNOTTING
LAP UP
LARGE PACKAGES
LASHING END RADIUS
LAYER LOCKING
LIFT (WINDING)
MECHANICAL SLUB CATCHERS
MULE SPINNING
NOSE BUNCHING
OPEN WINDING
OVEREND UNWINDING
PACKAGE DENSITY
PACKAGES
PACKAGING (OPERATION)
PIECING
QUILLS
RIBBON BREAKERS
RIBBON BREAKING
RING SPINNING
ROVING WIND
SECTIONAL WARPING
SLOTS
SLOUGHING
SLUB CATCHERS
SLUB CATCHING
SPINDLES
STOP MOTIONS
TAPER (BOBBIN)
TENSION
TENSION CONTROL
THREADLINE
TRANSFER TAILS
TRAVERSE
TWISTING
UNWINDING
UNWINDING ACCELERATORS
UNWINDING BALLOONS
WARP ENDS
WARP PREPARATION
WARP WIND
WARPING
WIND
WINDERS
WINDING ANGLE
WINDING DYNAMICS
WINDING EFFICIENCY
WINDING TENSION
YARN DYNAMICS
YARN TENSION

WINDING ANGLE
UF ANGLE OF WIND
RT ANGLE OF INCIDENCE
CLOSE WIND
CONING
DRUM WINDING
EMPTY PACKAGES
FULL PACKAGES
GAIN (WINDING)
OPEN WIND
PACKAGE DENSITY
PACKAGE HARDNESS
PACKAGES
PRECISION WINDERS
PRECISION WINDING
PROCESS VARIABLES
WIND
WINDERS
WINDING
WINDING DYNAMICS

WINDING DYNAMICS
UF UNWINDING DYNAMICS
NT BALLOON DYNAMICS
BT PROCESS DYNAMICS
YARN DYNAMICS
RT CARDING DYNAMICS
DRAFTING DYNAMICS
KNITTING DYNAMICS
MECHANISM (FUNDAMENTAL)
PACKAGE DENSITY
PACKAGE HARDNESS
SPINNING DYNAMICS
TWISTING DYNAMICS
UNWINDING
UNWINDING ACCELERATORS
UNWINDING BALLOONS
WEAVING DYNAMICS
WINDERS
WINDING
WINDING ANGLE
WINDING EFFICIENCY
WINDING TENSION

WINDING EFFICIENCY
BT PROCESS EFFICIENCY
RT DOWN TIME
END BREAKAGES
PACKAGE DENSITY
PACKAGE HARDNESS
PATROLLING
PERFORMANCE
PRODUCTION
SPINNING EFFICIENCY
WEAVING EFFICIENCY
WINDERS
WINDING
WINDING ANGLE
WINDING EFFICIENCY
WINDING TENSION

WINDING MACHINERY
USE WINDERS

WINDING ON
USE WINDING

WINDING TENSION
BT TENSION
RT BALLOON DYNAMICS
BALLOON TENSION
BALLOONS
DISC TENSION
EMPTY PACKAGES
END BREAKAGES
(CONT.)

RT FULL PACKAGES
PACKAGE DENSITY
PACKAGE HARDNESS
SPINNING
SPINNING TENSION (EXTRUSION)
SPINNING TENSION (STAPLE YARN)
TENSION CONTROL
TENSION DEVICES
TWISTING
TWISTING TENSION
WINDING
YARN TENSION
YARNS

WINDOW BLINDS
USE WINDOW HOLLAND

WINDOW HOLLAND
UF WINDOW BLINDS
WINDOW SHADES
BT FLAT WOVEN FABRICS
WOVEN FABRICS
RT HOUSEHOLD FABRICS
LAMPSHADES

WINDOW SHADES
USE WINDOW HOLLAND

WIPING CLOTHS
BT INDUSTRIAL FABRICS
RT OIL STAINS
OILS
SORPTION OF LIQUIDS
WATER ABSORPTION
WICKING

WIRA WEB LEVELNESS TESTER
BT IRREGULARITY TESTERS
RT IRREGULARITY
ORIENTATION UNIFORMITY
PATCHINESS
SPOTTINESS
TESTING
WEB ORIENTATION
WEB UNIFORMITY
WEBS
ZELLWEGER TESTER (TN)

WIRE DRAWING
UF DRAWING (WIRE)
RT DRAWTWISTING
DRAWING (FILAMENT)
METALLIC FIBERS
ROLLING
ROLLING MILLS
WIRES

WIRE FILLET CLOTHING
USE FILLET CLOTHING

WIRE ROPES
BT ROPES
RT CABLES
CORDAGE
LAY (ROPE)
METALLIC YARNS
PLIED YARNS
STRANDS
WIRES

WIRES
RT METALLIC FIBERS
WIRE DRAWING
WIRE ROPES

WOOD
RT CELLULOSE
LIGNIN
PULP
RAYON (REGENERATED CELLULOSE FIBERS)
WOOD FIBERS
WOOD FIBRILS

WOOD FIBERS
BT CELLULOSIC FIBERS
FIBERS
NATURAL FIBERS
RT CELLULOSE
HYDROGEN BONDS
PAPER
PRIMARY WALLS
PULP
RAYON (REGENERATED CELLULOSE FIBERS)
SECONDARY WALLS
WOOD
WOOD FIBRILS

WOOD FIBRILS
RT FIBRILS
PULP
WOOD
WOOD FIBERS

WOOD PULP
USE PULP

WOOL
UF WOOL FIBERS
NT CHEVIOT WOOL
COTSWOLD WOOL
CROSSBRED WOOL
DORSET WOOL
GREASY WOOL
HAMPSHIRE WOOL
LAMBSWOOL
LEICESTER WOOL
LINCOLN WOOL
LONG WOOL
MEDIUM WOOL
MERINO WOOL
MOUNTAIN WOOL
OXFORD WOOL
PULLED WOOL
RECOVERED WOOL
ROMNEY MARSH WOOL
RYELAND WOOL
SCOURED WOOL
SHORT WOOL
SHROPSHIRE WOOL
SOUTHDOWN WOOL
SUFFOLK WOOL
VIRGIN WOOL
BT FIBERS
KERATIN FIBERS
NATURAL FIBERS
PROTEIN FIBERS
RT ACID SOLUBILITY
ALKALI SOLUBILITY
AMINO ACIDS
ANIMAL HUSBANDRY
BICOMPONENT FIBERS
BICOMPONENT STRUCTURE
BRADFORD SYSTEM PROCESSING
BURRY WOOL
CHLORINATED WOOL
CHLORINATION (SHRINK PROOFING)
CONDENSED YARNS
CONDENSING (YARN MANUFACTURING)
CONTINENTAL SYSTEM PROCESSING
CRIMP
CYSTEINE
CYSTINE
DIFFERENTIAL FRICTION
DISULFIDE BONDS
DISULFIDE INTERCHANGE
DISULFIDES (ORGANIC)
ENDOCUTICLE
EPICUTICLE
EXOCUTICLE
FIBER LENGTH
FIBRILS
FOLLICLE
FUR
GRADE (FIBERS)
GRADING
KERATIN
LYSINE
MATCHINGS
MATRIX (FINE STRUCTURE)
MEDULLA
MICROFIBRILS
NEW BRADFORD SYSTEM PROCESSING
NOILS
NYLON (POLYAMIDE FIBERS)
ORTHOCORTEX
PARACORTEX
POLYPEPTIDES
PROTEINS
RESIDUAL FAT
SCALE MASKING
SCALES (WOOL FIBERS)
SETTING (EXCEPT SYNTHETICS)
SHEEP
SUINT
SUPERCONTRACTION
UREA BISULFITE SOLUBILITY
WOOL BALES
WOOL DYES
WOOL FABRICS
WOOL GRADE
WOOL GREASE
WOOL SORTING
WOOL WAX
WOOLEN FABRICS
WOOLEN SPUN YARNS
WOOLEN SYSTEM PROCESSING
WORSTED FABRICS
WORSTED SPUN YARNS
WORSTED SYSTEM PROCESSING
WORSTED SYSTEM SPINNING
WURLAN (TN)

WOOL BALES
BT BALES
RT COTTON BALES
WOOL

WOOL BLENDS
BT BLENDS (FIBERS)

WOOL CLASS
NT BABY DELAINE (WOOL CLASS)
CARDING WOOL
COMBING WOOL
DELAINE (WOOL CLASS)
FRENCH COMBING WOOL
PREPARING WOOL
BT CLASS (FIBERS)
RT CLASSING
GREASY WOOL
LAMBSWOOL
NOILS
PULLED WOOL
SCOURED WOOL
SHORN WOOL
VIRGIN WOOL
WOOL GRADE

WOOL CLASSING
RT WOOL CLASS
WOOL GRADE
WOOL SORTING

WOOL DYEING
BT DYEING (BY FIBER CLASSES)
RT ACETATE DYEING
ANIMALIZING
BLEND DYEING
CELLULOSE DYEING
CONTINUOUS TOP DYEING
COTTON DYEING
HAIR DYEING (HUMAN HAIR)
RAYON DYEING
SILK DYEING
SYNTHETIC DYEING
TOP DYEING

WOOL DYES
BT DYES (BY FIBER CLASSES)
RT ACETATE DYES
ACRYLIC DYES
CELLULOSE DYES
DYEING
DYES (BY CHEMICAL CLASSES)
NYLON DYES
POLYESTER DYES
SILK DYES
WOOL

WOOL FABRICS
UF WOOLENS AND WORSTEDS
NT WOOLEN FABRICS
WORSTED FABRICS
BT FABRICS (ACCORDING TO FIBER)
RT COTTON FABRICS
SETTING (EXCEPT SYNTHETICS)
WOOL
WOOLEN SPUN YARNS
WOOLEN SYSTEM PROCESSING
WORSTED SPUN YARNS
WORSTED SYSTEM PROCESSING
WURLAN (TN)

WOOL FIBERS
USE WOOL

WOOL FLEECES
USE FLEECES

WOOL GRADE
UF WOOL GRADING
NT BRAID (WOOL GRADE)
COMMON (WOOL GRADE)
FINE (WOOL GRADE)
FINE MEDIUM (WOOL GRADE)
LOW QUARTER BLOOD (WOOL GRADE)
MEDIUM (WOOL GRADE)
MEDIUM FINE (WOOL GRADE)
ONE HALF BLOOD (WOOL GRADE)
QUARTER BLOOD (WOOL GRADE)
THREE EIGHTHS BLOOD (WOOL GRADE)
BT GRADE (FIBERS)
GRADING
RT COTTON GRADE
CROSSBRED WOOL
FIBER FINENESS
LONG WOOL
SHORT WOOL
WOOL CLASS

WOOL GRADING
USE WOOL GRADE

WOOL GREASE
RT GREASE
GREASE RECOVERY
GREASY WOOL
LAMBSWOOL
LANOLIN
SCOURED WOOL
SCOURING
SCOURING TRAINS
SCOURING WASTE
SCOURING YIELD
SOAP SODA SCOURING
SOLVENT SCOURING
SUINT
WOOL
WOOL WAX

WOOL SCALES
    USE SCALES (WOOL FIBERS)

WOOL SHEARING
    UF SHEARING (WOOL)
    RT FLEECES

WOOL SORTING
    UF BENCH SORTING (WOOL)
       TRAP SORTING (WOOL)
    RT FIBER FINENESS
       FIBER LENGTH
       FLEECES
       GRADING
       LAMBSWOOL
       WOOL
       WOOL CLASSING
       WOOL GRADE

WOOL WAX
    BT WAXES
    RT GREASE
       GREASE RECOVERY
       GREASY WOOL
       LAMBSWOOL
       SCOURED WOOL
       SCOURING
       SCOURING WASTE
       SOAP SODA SCOURING
       SOLVENT SCOURING
       SUINT
       WOOL
       WOOL GREASE

WOOLEN CARDING
    BT WOOLEN SYSTEM PROCESSING
       CARDING
    RT AERODYNAMIC CARDING
       ANGLE STRIPPERS
       AUTOCOUNT (TN)
       BLAMIRE FEED
       BREAKERS
       BURR BEATERS
       BURR CRUSHERS
       CARD CLOTHING
       CARD CYLINDERS
       CARD SLIVERS
       CARD WASTE
       CARD WEBS
       CARDING ACTION
       CARDING EFFICIENCY
       CARDS
       CLEARERS (CARD)
       CONDENSERS
       COTTON CARDING
       DOFFERS (CARD)
       DOFFING (OF A CARD)
       FANCY (NOUN)
       FEED ROLLS
       FETTLING
       FINISHER PART
       HOPPER FEED
       INTERMEDIATE FEED
       INTERMEDIATE PART
       LICKER IN
       MOTE KNIVES
       NEP COUNT
       NEPS
       PERALTA ROLLS
       SLIVERS
       SLUBBINGS
       STRIPPERS
       STRIPPING ACTION
       WOOLEN CARDS
       WORKERS
       WORSTED CARDING

WOOLEN CARDS
    BT CARDS
    RT AERODYNAMIC CARDING
       ANGLE STRIPPERS
       BLAMIRE FEED
       BREAKERS
       BURR CRUSHERS
       CLEARERS (CARD)
       CONDENSERS
       CONDENSING (YARN
        MANUFACTURING)
       DUO CARDS
       FANCY (NOUN)
       FEED LATTICES
       FINISHER PART
       HOPPER FEED
       PERALTA ROLLS
       SCOTCH FEED
       SLUBBINGS
       STRIPPERS
       WOOLEN CARDING

WOOLEN DRAFTING
    BT DRAFTING (STAPLE FIBER)
       WOOLEN SYSTEM PROCESSING
    RT BACK ROLLS
       CONDENSED YARNS
       DRAFT
       DRAFTING ROLLS
       FALSE TWIST
       FALSE TWIST TUBES
       FALSE TWISTING (EXCEPT
        TEXTURING)
       FIBER LENGTH DISTRIBUTION
       FRONT ROLLS
       MULE DRAFT
       PREFERENTIAL DRAFT
       RATCH
       SPINNING
       SPINNING FRAMES
       WOOLEN SPINNING FRAMES

WOOLEN FABRICS
    UF WOOLENS
    BT WOOL FABRICS
    RT BAULK FINISHING
       WOOL
       WOOLEN SPUN YARNS
       WOOLEN SYSTEM PROCESSING
       WORSTED FABRICS

WOOLEN SPINNING FRAMES
    BT SPINNING FRAMES
    RT FALSE TWIST TUBES
       FALSE TWISTING (EXCEPT
        TEXTURING)
       WOOLEN DRAFTING
       WOOLEN SYSTEM SPINNING

WOOLEN SPUN YARNS
    UF WOOLEN YARNS
    NT ANGOLA (YARN)
    BT SPUN YARNS
       YARNS
    RT CARDED YARNS
       COMBED YARNS
       COTTON SPUN YARNS
       WOOL FABRICS
       WOOLEN FABRICS
       WOOLEN SYSTEM PROCESSING
       WOOLEN SYSTEM SPINNING
       WORSTED SPUN YARNS
       WORSTED SYSTEM SPINNING

WOOLEN SYSTEM PROCESSING
    NT CONDENSING (YARN
        MANUFACTURING)
       WOOLEN CARDING
       WOOLEN DRAFTING
       WOOLEN SYSTEM SPINNING
    RT AMERICAN SYSTEM PROCESSING
       BENDING
       BRADFORD SYSTEM PROCESSING
       BULK
       CARBONIZING
       CARDING
       CONDENSED YARNS
       CONTINENTAL SYSTEM PROCESSING
       COTTON SYSTEM PROCESSING
       FALSE TWIST TUBES
       FALSE TWISTING (EXCEPT
        TEXTURING)
       FLAX SYSTEM PROCESSING
       HOOKED FIBERS
       LAMBSWOOL
       NEW BRADFORD SYSTEM PROCESSING
       OILING
       PICKING (OPENING)
       RING FRAMES
       SCOURING
       SLUBBINGS
       SPINNING
       SPINNING FRAMES
       TWISTING
       WOOL
       WOOL FABRICS
       WOOLEN FABRICS
       WOOLEN SPUN YARNS
       WORSTED SYSTEM PROCESSING

WOOLEN SYSTEM SPINNING
    BT SPINNING
       WOOLEN SYSTEM PROCESSING
    RT COTTON SYSTEM SPINNING
       SPINNING FRAMES
       WOOLEN SPINNING FRAMES
       WOOLEN SPUN YARNS

WOOLEN YARNS
    USE WOOLEN SPUN YARNS

WOOLENS
    USE WOOLEN FABRICS

WOOLENS AND WORSTEDS
    USE WOOL FABRICS

WORK CLOTHING
    NT APRONS (CLOTHING)
       CAPS (HEADGEAR)
       COVERALLS
       DENIMS
       DUNGAREES
       INDUSTRIAL CLOTHING
       JEANS
       OVERALLS
       SAFETY CLOTHING
       SHIRTS
       SLACKS
       SMOCKS
       SWEATSHIRTS
       TROUSERS
       UNIFORMS
    BT GARMENTS
    RT COATS
       FATIGUES
       GLOVES
       HOSIERY
       JACKETS
       RAINWEAR

WORK RECOVERY
    BT RECOVERY (SELF RESTORATION TO
        ORIGINAL CONDITION)
       STRESS STRAIN PROPERTIES
       STRESS STRAIN PROPERTIES
        (TENSILE)
    RT CREEP RECOVERY
       CRIMP RECOVERY
       DELAYED RECOVERY
       ELASTIC PERFORMANCE
       ELASTIC RECOVERY
       ENERGY ABSORPTION
       ENERGY OF RETRACTION
       HYSTERESIS (MECHANICAL)
       LIVELINESS
       PERMANENT SET
       RESILIENCE
       STRESS STRAIN CURVES

WORK TO BREAK
    USE BREAKING ENERGY

WORKABLE FIBERS
    USE LINT (FIBER)

WORKERS
    BT CARD ROLLS
    RT BREAKERS
       CARD CLOTHING
       CARD CYLINDERS
       CARD GRINDING
       CARD STRIPS
       CARDING
       CARDING ACTION
       CARDING EFFICIENCY
       CARDS
       DOFFERS (CARD)
       FANCY (NOUN)
       FETTLING
       FINISHER PART
       INTERMEDIATE PART
       STRIPPERS
       STRIPPING ACTION
       WOOLEN CARDING
       WORSTED CARDING

WORKING ACTION
    USE CARDING ACTION

WORKLOADS
    RT MACHINE INTERFERENCE
       MANAGEMENT
       PATROLLING
       PERSONNEL
       PROCESS EFFICIENCY
       PRODUCTION
       PRODUCTIVITY
       QUALITY CONTROL

WORSTED CARDING
    BT CARDING
       WORSTED SYSTEM PROCESSING
    RT AERODYNAMIC CARDING
       ANGLE STRIPPERS
       BREAKERS
       BURR BEATERS
       BURR CRUSHERS
       CARD CLOTHING
       CARD CYLINDERS
       CARD SLIVERS
       CARD WASTE
       CARD WEBS
       CARDING ACTION
       CARDING EFFICIENCY
       CARDS
       COTTON CARDING
       DIVIDERS (CARD ROLLS)
       DOFFERS (CARD)
       DOFFING (OF A CARD)
       FANCY (NOUN)
       FEED ROLLS
       FETTLING
       FINISHER PART
       HOPPER FEED
       INTERMEDIATE PART
       LEADING HOOKS
       LICKER IN
       NEP COUNT
       NEPS
       PERALTA ROLLS
       RANDO WEBBER (TN)
       SLIVER CANS
       SLIVERS
       SLUBBINGS
       STRIPPERS
       STRIPPING ACTION
        (CONT.)

RT TRAILING HOOKS
TRUMPETS
WOOLEN CARDING
WORKERS
WORSTED CARDS
WORSTED SYSTEM SPINNING

WORSTED CARDS
BT CARDS
RT ANGLE STRIPPERS
COTTON CARDS
DIVIDERS (CARD ROLLS)
DOFFERS (CARD)
FEED LATTICES
HOPPER FEED
STRIPPERS
WOOLEN CARDS
WORKERS
WORSTED CARDING

WORSTED COMBING
BT COMBING
WORSTED SYSTEM PROCESSING
RT COTTON COMBING
NOBLE COMBING
RECTILINEAR COMBING
WORSTED SYSTEM SPINNING

WORSTED COUNT
BT COUNT
INDIRECT COUNT
RT DENIER
WORSTED SPUN YARNS

WORSTED FABRICS
UF WORSTEDS
NT TROPICAL WORSTED
BT WOOL FABRICS
RT NEW BRADFORD SYSTEM PROCESSING
WOOL
WOOLEN FABRICS
WORSTED SYSTEM PROCESSING
WORSTED SYSTEM SPINNING

WORSTED LASTING
BT LASTING
RT COTTON LASTING
SATIN WEAVES

WORSTED SPINNING
USE WORSTED SYSTEM SPINNING

WORSTED SPUN YARNS
UF WORSTED YARNS
BT SPUN YARNS
YARNS
RT CAP SPINNING
CARDED YARNS
CARDING WOOL
COMBED YARNS
COTTON SPUN YARNS
LONG STAPLE SPINNING
PREPARING WOOL
RING SPINNING
SPINNING
WOOL
WOOL FABRICS
WOOLEN SPUN YARNS
WORSTED COUNT
WORSTED SYSTEM PROCESSING
WORSTED SYSTEM SPINNING

WORSTED SYSTEM PROCESSING
NT AMERICAN SYSTEM PROCESSING
BACKWASHING
BRADFORD OPEN DRAWING
BRADFORD SYSTEM PROCESSING
CAP SPINNING
CONTINENTAL SYSTEM PROCESSING
GILLING
NEW BRADFORD SYSTEM PROCESSING
PIN DRAFTING
TOP MAKING
WORSTED CARDING
WORSTED COMBING
WORSTED SYSTEM SPINNING
RT AMBLER SUPERDRAFT
APRONS
BLENDING
CAP SPINNING FRAMES
CARBONIZING
CASABLANCA SYSTEM
COMBING
COTTON SYSTEM PROCESSING
DRAFTING (STAPLE FIBER)
FLAX SYSTEM PROCESSING
HALF LAP
HIGH DRAFT SPINNING
LONG STAPLE SPINNING
NOBLE COMBING
NOBLE COMBS
OILING
RAPER AUTOLEVELLER (TN)
RECTILINEAR COMBS
RING FRAMES
SCOURING
SPINNING
SPINNING FRAMES
RT TOPS .
TWISTING
WOOL
WOOL FABRICS
WOOLEN SYSTEM PROCESSING
WORSTED FABRICS
WORSTED SPUN YARNS

WORSTED SYSTEM SPINNING
UF WORSTED SPINNING
BT SPINNING
WORSTED SYSTEM PROCESSING
RT CAP SPINNING
COTTON SYSTEM SPINNING
GILL BOXES
GILLING
SPINNING FRAMES
WOOL
WOOLEN SPUN YARNS
WOOLEN SYSTEM SPINNING
WORSTED CARDING
WORSTED COMBING
WORSTED FABRICS
WORSTED SYSTEM SPINNING

WORSTED YARNS
USE WORSTED SPUN YARNS

WORSTEDS
USE WORSTED FABRICS

WOVEN CARPETS
NT AXMINSTER CARPETS
CHENILLE CARPETS
SCULPTURED CARPETS
TAPESTRY CARPETS
VELVET CARPETS
WILTON CARPETS
BT CARPETS
PILE FABRICS (WOVEN)
WOVEN FABRICS
RT BACKING FABRICS
CARPET STUFFER YARNS
HOUSEHOLD FABRICS
KNITTED CARPETS
NONWOVEN CARPETS
TAPESTRIES
TUFTED CARPETS
WOVEN FABRICS

WOVEN FABRIC STRUCTURE
BT FABRIC STRUCTURE
RT KNITTED FABRIC STRUCTURE (FILLING KNIT)
KNITTED FABRIC STRUCTURE (WARP KNIT)
NONWOVEN FABRIC STRUCTURE
YARN STRUCTURE

WOVEN FABRICS
UF FABRICS (WOVEN)
NT ALBERT TWILL
ALPACA FABRICS
ARTIFICIAL FURS
AWNING CLOTH
AXMINSTER CARPETS
BARATHEA
BATISTE
BEDFORD CORD
BOMBAZINE
BRASSIER CLOTH
BRILLIANT SPOT
BROADCLOTH
BROCADES
BROCATELLES
BUCKRAM
BUNTING
CALICO
CAMBRIC
CASEMENT
CAVALRY TWILL
CHAMBRAY
CHARMANETTE SATIN
CHECK SHIRTING
CHEESE CLOTH
CHENILLE CARPETS
CHEVIOT TWEED
CHIFFON
CHINTZ
CORDUROY
CORSET BATISTE
COUTIL
COVERT
CRASH TOWELING
CREPE
CREPON GEORGETTE
CRETONNE
CRINOLINE
CURL PILE FABRICS
CUT PILE FABRICS
DAMASK
DENIMS
DIMITY
DOESKIN
DOTTED MUSLIN
DRILLS (FABRICS)
DUCK
ELASTIQUE
NT EMBOSSED CREPE
FAILLE
FANCY LENO
FLANNEL
FLAT WOVEN FABRICS
FOULARD (FABRIC)
FURNISHING FABRICS
GABARDINE
GAUZE
GENOA VELVET
GEORGETTE
GINGHAM
HAND WOVEN FABRICS
HESSIAN
HIMALAYA CLOTH
HOPSACK
IMITATION GINGHAM
INDIA LINON
INDIANA CLOTH
ITALIAN CLOTH
JACONET
JEANS
KERSEY
LASTING
LAWN
LINGETTE
LONGCLOTH
LOOP PILE FABRICS
MAINSOOK CHECK
MARQUISETTE
MELTON
MESALINE
MOCK LENO
MULTIPLE LAYER FABRICS
MUSLIN
NETTING
OPALINE
ORGANDIE
OSNABURG
PAJAMA CHECK
PERCALE
PERCALINE
PILE FABRICS (WOVEN)
PILLOW TUBING
PIQUE
PLISSE
PLUSH
PONCEE
POPLIN
PRINT CLOTH
RAYON STRIPE
REPP
RIBBON
SATEEN
SATIN
SCREENING
SCRIM
SEERSUCKER
SERGE
SHADOW STRIPE
SHARKSKIN
SHEETING (FABRIC)
SHIRTING
SILESIA
SKIP DENT
SPLASH VOILE
SPORT SATIN
STRETCH WOVEN FABRICS
SWISS
TAFFETA
TAPE
TERRYCLOTH
TICKING
TOBACCO CLOTH
TRILOCK (TN)
TROPICAL WORSTED
TWILL CHECKBOARD
TWILL WEAVES
UMBRELLA CLOTH
UNCUT PILE FABRICS
VELOUR
VELVET
VELVETEEN
VICTORIA LAWN
VOILE
WARP REPP
WEBBING
WILTON CARPETS
WINDOW HOLLAND
WOVEN CARPETS
WOVEN TUBES
BT FABRICS
FABRICS (ACCORDING TO STRUCTURE)
RT BACK (FABRIC)
CENTER SELVEDGES
CLOTH FELL
COATED FABRICS
CROSS LAID YARN FABRICS
DESIGN
FABRIC STRUCTURE
KNITTED FABRICS
LACES
LENO SELVEDGES
LOOMS
MALI FABRICS (TN)
MALIPOL FABRIC (TN)
MISPICKS

RT NONWOVEN FABRICS
REED MARKS
REVERSIBLE FABRICS
SEALED SELVEDGES
SELVEDGE WARPS
SELVEDGES
STITCH BONDED FABRICS
STITCH REINFORCED FABRICS
SYNTHETIC LEATHER
TAPE SELVEDGES
TAPESTRIES
TEMPLES
TUFTED FABRICS
WEAVE
WEAVING
WOVEN CARPETS
WOVEN FELTS

WOVEN FELTS
BT FELTS
RT FELTING
NEEDLE PUNCHED FELTS
PAPERMAKERS FELTS
PRESSED FELTS
WOVEN FABRICS

WOVEN NETS
BT NETS

WOVEN TUBES
BT MULTIPLE LAYER FABRICS
TUBULAR FABRICS
WOVEN FABRICS

WRAP
RT ANGLE OF WRAP
BALL AND SOCKET TENSION
BELT TENSION
DISC TENSION
FINGER TENSION
FRICTION
GATE TENSION
TENSION
TENSION CONTROL
TENSION DEVICES
YARN TENSION

WRAP UP
USE LAP UP

WRAPPED YARNS
USE COVERED YARNS

WRAPPING MACHINES
RT CORE SPINNING
PACKAGING (OPERATION)
PACKAGING EQUIPMENT

WRINKLE RECOVERY
UF CREASE RECOVERY
NT DRY WRINKLE RECOVERY
WET WRINKLE RECOVERY
BT END USE PROPERTIES
FABRIC PROPERTIES
FABRIC PROPERTIES (MECHANICAL)
FINISH (PROPERTY)
RECOVERY (SELF RESTORATION TO ORIGINAL CONDITION)
RT BENDING RECOVERY
BENDING RIGIDITY
CREASE RESISTANCE
CREASE RETENTION
CREASES
CREEP
CREEP RECOVERY
DURABLE PRESS
EASY CARE FABRICS
ELASTIC RECOVERY
FIBER FIBER FRICTION
FRICTION
MATRIX FREEDOM
MECHANICAL PROPERTIES
RELAXATION
STIFFENERS (AGENTS)
WASH WEAR FINISHES
WASH WEAR TREATMENTS
WRINKLE RECOVERY ANGLE
WRINKLE RECOVERY TESTING
YARN TO YARN FRICTION
YIELD POINT

WRINKLE RECOVERY ANGLE
RT DRY WRINKLE RECOVERY
WASH WEAR RATING
WRINKLE RECOVERY
WRINKLE RESISTANCE

WRINKLE RECOVERY FINISHES
USE WASH WEAR FINISHES

WRINKLE RECOVERY TESTERS
UF MONSANTO CREASE RECOVERY
BT TESTING EQUIPMENT
RT WRINKLE RECOVERY
WRINKLE RECOVERY ANGLE
WRINKLE RESISTANCE

WRINKLE RECOVERY TESTING
RT DRY WRINKLE RECOVERY
WET WRINKLE RECOVERY
WRINKLE RECOVERY
WRINKLE RECOVERY ANGLE
WRINKLE RESISTANCE

WRINKLE RESISTANCE
UF CREASE RESISTANCE
MUSS RESISTANCE
BT END USE PROPERTIES
FABRIC PROPERTIES
FABRIC PROPERTIES (MECHANICAL)
FINISH (PROPERTY)
RT CREASE RETENTION
CREASING
DELAYED RECOVERY
DIMENSIONAL STABILITY
DRY WRINKLE RECOVERY
DURABLE PRESS
EASY CARE FABRICS
MECHANICAL PROPERTIES
PAPERINESS
PERFORMANCE
PRESSING
RECOVERY (SELF RESTORATION TO ORIGINAL CONDITION)
STIFFNESS
WASH WEAR FINISHES
WASH WEAR TREATMENTS
WET WRINKLE RECOVERY
WRINKLE RECOVERY
WRINKLE RECOVERY ANGLE
WRINKLE RECOVERY TESTERS
WRINKLE RECOVERY TESTING
WRINKLES
WRINKLING

WRINKLE RESISTANCE FINISHES
USE WASH WEAR FINISHES

WRINKLE RESISTANCE TREATMENTS
USE WASH WEAR TREATMENTS

WRINKLES
RT BENDING RECOVERY
CREASES
PLEATS
WRINKLE RECOVERY
WRINKLE RECOVERY ANGLE
WRINKLING

WRINKLING
RT BENDING
CREASE RETENTION
CREASES
CREASING
PERMANENT DEFORMATION
PERMANENT SET
PLEATING
PUCKER (DEFECT)
WRINKLE RESISTANCE
WRINKLES

WRONG DRAW
NT SWOLLEN DENT
SWOLLEN HEDDLE
BT FABRIC DEFECTS

WURLAN (TN)
BT ANTIFELTING AGENTS
ANTIFELTING TREATMENTS
SHRINK PROOFING
SHRINK RESISTANCE FINISHES
RT INTERFACIAL POLYMERIZATION
LAUNDERABILITY
WOOL
WOOL FABRICS

X RAY ANALYSIS
RT ALPHA BETA TRANSFORMATION
ANALYZING
BIREFRINGENCE
BRAGGS LAW
CHEMICAL ANALYSIS
CRYSTALLINITY
DIFFRACTION PATTERNS
FINE STRUCTURE (FIBERS)
GLASS RUBBER TRANSITION
GLASS TRANSITION TEMPERATURE
MELTING POINT
MOLECULAR CHAIN MOVEMENT
ORIENTATION
PHYSICAL ANALYSIS
QUALITATIVE ANALYSIS
QUANTITATIVE ANALYSIS
SECONDARY TRANSITION TEMPERATURES
SPECTRUM ANALYSIS
STATISTICAL ANALYSIS
STRESS ANALYSIS
STRESS OPTICAL COEFFICIENT
SUBGROUP MOVEMENT (MOLECULAR)
STRUCTURAL ANALYSIS
THERMAL ANALYSIS
TRANSITIONS (POLYMERS)
X RAY DIFFRACTION
X RAY MICROSCOPY
X RAY SPECTROSCOPY
X RAYS

X RAY DIFFRACTION
RT DEGREE OF ORIENTATION
DIFFRACTION
FINE STRUCTURE (FIBERS)
MOLECULAR STRUCTURE
X RAY ANALYSIS
X RAY MICROSCOPY
X RAY SPECTROSCOPY
X RAYS

X RAY MICROSCOPY
BT MICROSCOPY
RT ELECTRON MICROSCOPY
INFRARED MICROSCOPY
INTERFERENCE MICROSCOPY
LIGHT MICROSCOPY (OPTICAL)
ULTRAVIOLET MICROSCOPY
X RAY ANALYSIS
X RAY DIFFRACTION
X RAY SPECTROSCOPY
X RAYS

X RAY SPECTROSCOPY
BT SPECTROSCOPY
RT SPECTRA
SPECTRUM ANALYSIS
ULTRAVIOLET SPECTROSCOPY
X RAY MICROSCOPY
X RAYS

X RAYS
BT IONIZING RADIATION
RADIATION
RT GAMMA RAYS
IRRADIATION
X RAY ANALYSIS
X RAY DIFFRACTION
X RAY MICROSCOPY
X RAY SPECTROSCOPY

XENOTEST (TN)
BT LIGHTFASTNESS TESTERS
RT COLORFASTNESS TESTING
FADEOMETER (TN)
LIGHTFASTNESS (COLOR)
LIGHTFASTNESS (OF FINISH)
LIGHTFASTNESS TESTING

YAK
BT HAND MADE LACES
LACES

YARN BALLOONS
USE BALLOONS

YARN BUCKLING
BT BUCKLING
RT BENDING RIGIDITY
COMPRESSION
CREPING
FABRIC GEOMETRY
SLENDERNESS RATIO
SNAP BACK
TORSIONAL BUCKLING
WILDNESS
YARN CRIMP
YARN DYNAMICS
YARN STRUCTURE

YARN BREAKAGES
USE END BREAKAGES

YARN BUNDLE COHESION
RT EFFICIENCY (STRUCTURAL)
STRENGTH EFFICIENCY
YARN PROPERTIES
YARN STRUCTURE

YARN CARRIERS
RT BOBBINS
CARRIERS (DYEING)
CONES
FEEDERS
FLOW (MATERIALS)
FULLY FASHIONED KNITTING MACHINES
KNITTING
KNITTING MACHINES
KNITTING NEEDLES
YARNS

YARN CHANGING UNIT (KNITTING)
UF STRIPING UNIT (KNITTING)
RT CIRCULAR KNITTING MACHINES
KNITTING

YARN CLEARERS
USE SLUB CATCHERS

YARN COUNT
USE COUNT

YARN COVER
BT COVER
YARN PROPERTIES
RT BOTTOM COVER
COVER FACTOR
FABRIC COVER
HAIRINESS
TOP COVER
YARN DIAMETER

YARN CRIMP
BT CRIMP
RT BICOMPONENT FIBER YARNS (FILAMENT)
BICOMPONENT FIBER YARNS (STAPLE)
BULKED YARNS
BULKING
CRIMP BALANCE
CRIMP FREQUENCY
CRIMP INDEX
CRIMP INTERCHANGE
CRIMP PERCENT
CRIMP TENDENCY
CRIMPED YARNS
CRIMPING
FABRIC CRIMP
FIBER CRIMP
FOLLOW THE LEADER CRIMP
LATENT CRIMP
SELF CRIMPING YARNS
STRETCH YARNS (FILAMENT)
TEXTURED YARNS (FILAMENT)
TEXTURING
WAVINESS
YARN BUCKLING
Z CRIMP

YARN CROSS SECTIONS
RT COUNT
FABRIC CROSS SECTIONS
FIBER CROSS SECTIONS
DENIER
HELIX ANGLE
MATRIX FREEDOM
PACKING FACTOR
TWIST
TWIST MULTIPLIER
YARN DIAMETER
YARN GEOMETRY
YARN STRUCTURE

YARN DAMAGE
BT DAMAGE
RT NEEDLE HEATING
NEEDLE PENETRATION
NEEDLE POINT
SEAM EFFICIENCY
SEWABILITY
SEWING
SEWING DAMAGE
TUFTING
WARP TENSION
YARN DEFECTS
YARN SEVERANCE

YARN DEFECTS
NT FOGGY YARNS
KNOTS
NEPS
SLUBS
BT DEFECTS
RT DETECTION
ELECTRONIC SLUB CATCHERS
END BREAKAGES
ENTANGLEMENTS
IRREGULARITY
MARRIED YARNS
NEPS
NUBS
SLOTS
SLUB CATCHERS
THICK SPOTS
WEAK SPOTS
WILDNESS
YARN DAMAGE

YARN DENIER
USE DENIER

YARN DIAMETER
BT YARN PROPERTIES
RT COUNT
DIAMETER
HAIRINESS
MATRIX FREEDOM
PACKING FACTOR
ROUNDNESS
SLUBS
YARN COVER
YARN CROSS SECTIONS
YARN FLATTENING
YARN GEOMETRY
YARN STRUCTURE
YARNS

YARN DYED FABRICS
BT DYED FABRICS
RT DOPE DYEING
FABRICS
PIECE DYED FABRICS
PIECE DYEING
STOCK (FIBER)
STRIPES
YARN DYEING

YARN DYEING
NT BEAM DYEING (YARN)
BOBBIN DYEING
CAKE DYEING
CHEESE DYEING
CONE DYEING
COP DYEING
HANK DYEING
MUFF DYEING
PACKAGE DYEING
SKEIN DYEING
BT DYEING (BY MATERIAL ASSEMBLY)
RT DOPE DYEING
MULTICOLORS
PIECE DYED FABRICS
PIECE DYEING
STOCK DYEING
TOP DYEING
TOW DYEING
YARN DYEING MACHINES
YARNS

YARN DYEING MACHINES
NT BOBBIN DYEING MACHINES
BEAM DYEING MACHINES (YARN)
CAKE DYEING MACHINES
CHEESE DYEING MACHINES
CONE DYEING MACHINES
COP DYEING MACHINES
HANK DYEING MACHINES
PACKAGE DYEING MACHINES
WARP DYEING MACHINES
BT DYEING MACHINES
DYEING MACHINES (BY MATERIAL ASSEMBLY)
RT BATCH DYEING MACHINES
CIRCULATION PUMPS
DOPE DYEING EQUIPMENT
DYEING
DYEING (BY MATERIAL ASSEMBLY)
KETTLES (DYEING)
PERFORATED BEAMS
PERFORATED CAGES
PERFORATED CYLINDERS
PERFORATED PLATES
RT PERFORATED TUBES (DYEING)
PIECE DYEING MACHINES
PRESSURE DYEING MACHINES
STOCK DYEING MACHINES
TOP DYEING MACHINES
YARN DYEING
YARNS

YARN DYNAMICS
NT BALLOON DYNAMICS
SPINNING DYNAMICS
TWISTING DYNAMICS
WINDING DYNAMICS
BT PROCESS DYNAMICS
RT AIR DRAG
BALLOONS
BUCKLING
CARDING DYNAMICS
DRAFTING DYNAMICS
END BREAKAGES
KNITTING
KNITTING DYNAMICS
LONGITUDINAL VIBRATIONS
MECHANISM (FUNDAMENTAL)
SLOUGHING
SNAP BACK
SPINNING
SURGING
TENSION
TENSION CONTROL
THREADLINE
TRAVERSE
VIBRATION
WAVE LENGTH
WEAVING
WEAVING DYNAMICS
WIND TUNNELS
WINDING
YARN BUCKLING
YARN TENSION
YARNS

YARN FINISH
BT FINISH (SUBSTANCE ADDED)
RT CARDING OIL
COEFFICIENT OF FRICTION
FRICTION
HYDRODYNAMIC LUBRICATION
LUBRICATION
OILING
SOFTENERS
SPINNING LUBRICANTS
YARNS

YARN FLATTENING
RT FABRIC GEOMETRY
MATRIX FREEDOM
PACKING FACTOR
YARN DIAMETER
YARN GEOMETRY
YARN STRUCTURE
YARNS

YARN GEOMETRY
BT GEOMETRY
YARN PROPERTIES
RT CORDAGE
CORDS
CORE SPUN YARNS
COUNT
COVERED YARNS
CRIMP
CRIMP FREQUENCY
CRIMP INDEX
CRIMP PERCENT
CRIMP RADIUS
DENIER
DIFFERENTIAL GEOMETRY
FABRIC GEOMETRY
FABRIC STRUCTURE
FIBER MIGRATION
FILAMENT YARNS
HAIRINESS
HELIX ANGLE
MATRIX FREEDOM
NOVELTY YARNS
PACKING FACTOR
PLIED YARNS
SINGLES YARNS
SLUB YARNS
STRUCTURAL MECHANICS
STRUCTURAL PROPERTIES
STRUCTURES
TEXTURED YARNS (FILAMENT)
TWIST
TWIST CONTRACTION
TWIST MULTIPLIER
YARN CROSS SECTIONS
YARN DIAMETER
YARN FLATTENING
YARN STRUCTURE
YARNS

YARN GUIDES
USE THREAD GUIDES

YARN LUSTER
BT LUSTER
(CONT.)

RT APPEARANCE
BRIGHTNESS
OPACITY
TITANIUM DIOXIDE
YARN PROPERTIES
YARNS

YARN PREPARATION
USE WARP PREPARATION

YARN PROPERTIES
NT COUNT
COUNT STRENGTH PRODUCT
DENIER
FIBER CONTENT
HAIRINESS
TWIST
TWIST LIVELINESS
TWIST SENSE
WILDNESS
YARN COVER
YARN CRIMP
YARN DIAMETER
YARN GEOMETRY
YARN STRENGTH
RT AESTHETIC PROPERTIES
BENDING RIGIDITY
BIOCHEMICAL PROPERTIES
BULK
CHEMICAL PROPERTIES
DEGRADATION PROPERTIES
DYNAMIC MODULUS
ELASTICITY
ELECTRICAL PROPERTIES
FABRIC PROPERTIES
FIBER PROPERTIES
IRREGULARITY
MECHANICAL DETERIORATION PROPERTIES
MECHANICAL PROPERTIES
MODULUS
MONOFILAMENTS
OPTICAL PROPERTIES
PHYSICAL CHEMICAL PROPERTIES
PHYSICAL PROPERTIES (EXCLUDING MECHANICAL)
PROPERTIES
ROUNDNESS
SMOOTHNESS
STRENGTH EFFICIENCY
STRESS OPTICAL PROPERTIES
STRESS STRAIN PROPERTIES
STRESS STRAIN PROPERTIES (TENSILE)
STRUCTURAL PROPERTIES
SURFACE PROPERTIES (MECHANICAL)
SURFACE PROPERTIES (PHYSICAL CHEMICAL)
THERMAL PROPERTIES
TRANSFER PROPERTIES
YARN BUNDLE COHESION
YARN LUSTER
YARNS

YARN REINFORCED COMPOSITES
NT FILAMENT WOUND COMPOSITES
BT REINFORCED COMPOSITES
RT FIBER REINFORCED COMPOSITES
GLASS FIBERS
GLASS MAT
LAMINATED FABRICS

YARN REINFORCED ELASTOMERS
NT REINFORCED HOSES
TIRES
V BELTS
RT BELTS
FIBER REINFORCED COMPOSITES
FILAMENT WOUND COMPOSITES
INDUSTRIAL FABRICS
INFLATABLE FABRICS
TIRE CORDS

YARN SEVERANCE
RT BUTTED SEAMS
COVERED SEAMS
CUTTING
FUSION (MELTING)
GATHERING
PINCHING
PUCKERING
RUFFLING
SEAM EFFICIENCY
SEAMS
SEWING
STITCHING
YARN DAMAGE
YARN STRENGTH
YARN TENSION

YARN SLIPPAGE
BT SLIPPAGE
RT ABRASION
DIMENSIONAL STABILITY
FABRIC GEOMETRY
MATRIX FREEDOM
PERMANENT DEFORMATION
THREAD SLIPPAGE
TIGHTNESS

YARN STRENGTH
NT COUNT STRENGTH PRODUCT
LEA STRENGTH PRODUCT
BT STRENGTH
YARN PROPERTIES
RT BREAKING STRENGTH
COMPRESSIVE STRENGTH
DYNAMIC MODULUS
EFFICIENCY (STRUCTURAL)
END BREAKAGES
FABRIC STRENGTH
FIBER BREAKAGE
FIBER FIBER FRICTION
FIBER SLIPPAGE
FIBER STRENGTH
FRICTION
IRREGULARITY
LOOP EFFICIENCY
LOOP STRENGTH
STRENGTH EFFICIENCY
STRENGTH ELONGATION TESTING
STRESS STRAIN CURVES
TENACITY
TENSILE TESTING
WEAK SPOTS
YARN SEVERANCE
YARN STRENGTH EFFICIENCY
YARN STRUCTURE
YARN TENSION
YARNS

YARN STRENGTH EFFICIENCY
BT STRENGTH EFFICIENCY
RT BREAKING LENGTH
BREAKING STRENGTH
EFFICIENCY (STRUCTURAL)
FIBER STRENGTH
STRENGTH
YARN STRENGTH
YARN STRUCTURE
YARNS

YARN STRUCTURE
RT CORDAGE
CORDS
CORE SPUN YARNS
COUNT
COVERED YARNS
CRIMP
CRIMP FREQUENCY
CRIMP INDEX
CRIMP PERCENT
CRIMP RADIUS
DENIER
DIFFERENTIAL GEOMETRY
FABRIC GEOMETRY
FABRIC STRUCTURE
FIBER MIGRATION
FILAMENT YARNS
HAIRINESS
HELIX ANGLE
MIGRATION PERIOD
NOVELTY YARNS
PACKING FACTOR
PLIED YARNS
RADIAL DISTRIBUTION
SINGLES YARNS
SLUB YARNS
SPUN YARNS
STRUCTURAL MECHANICS
STRUCTURAL PROPERTIES
STRUCTURES
TEXTURED YARNS (FILAMENT)
TWIST
TWIST MULTIPLIER
YARN BUNDLE COHESION
YARN CROSS SECTIONS
YARN DIAMETER
YARN FLATTENING
YARN GEOMETRY
YARN STRENGTH
YARN STRENGTH EFFICIENCY
YARN TENSION
YARNS
Z TWIST

YARN TENSION
BT TENSION
RT BALLOON DYNAMICS
BALLOON HEIGHT
BALLOON TENSION
BALLOONS
CIRCULAR TRAVELERS
DISC TENSION
END BREAKAGES
FINGER TENSION
KNITTING
KNITTING TENSION
PROCESS VARIABLES
ROBBING BACK
SPINNING
SPINNING TENSION (EXTRUSION)
SPINNING TENSION (STAPLE YARN)
TENSIOMETERS
TENSION CONTROL
TENSION DEVICES
THREAD TENSION
THREADLINE
RT TRAVELERS
TWISTING TENSION
UNWINDING
UNWINDING BALLOONS
UPTWISTING
WARP TENSION
WEAVING
WEAVING DYNAMICS
WINDING
WINDING TENSION
WRAP
YARN DYNAMICS
YARN SEVERANCE
YARN STRENGTH
YARN STRUCTURE
YARNS

YARN TO METAL FRICTION
BT FRICTION
RT ABRASION
COEFFICIENT OF FRICTION
DYNAMIC FRICTION
FABRIC TO FABRIC FRICTION
FIBER FIBER FRICTION
FRICTIONAL FORCE
NORMAL FORCE
RUBBING
STATIC FRICTION
YARN TO YARN FRICTION

YARN TO YARN FRICTION
BT FRICTION
RT ABRASION
BENDING RIGIDITY
COEFFICIENT OF FRICTION
CONTACT AREA
DYNAMIC FRICTION
FABRIC TO FABRIC FRICTION
FIBER FIBER FRICTION
FRICTIONAL FORCE
NORMAL FORCE
RUBBING
SLIP RESISTANCE
STATIC FRICTION
WRINKLE RECOVERY
YARN TO METAL FRICTION

YARNS
NT BALANCED YARNS
BLENDED YARNS
BULKED YARNS
CARDED YARNS
CELLULAR YARNS
COARSE YARNS
COMBED YARNS
CONDENSED YARNS
CORE SPUN YARNS
COTTON SPUN YARNS
CRIMPED YARNS
CRINKLE TYPE YARNS
DIRECT SPUN YARNS
DOPE DYED YARNS
FALSE TWIST YARNS
FILAMENT YARNS
FILLING YARNS
FINE YARNS
FINGERING YARNS
KNITTING YARNS
LAMINATED YARNS
MONOFILAMENT YARNS
MULTIFILAMENT YARNS
NEP YARNS
NOVELTY YARNS
NUB YARNS
PAPER YARNS
PLIED YARNS
S TWIST YARNS
SINGLES YARNS
SLIT FILM YARNS
SLUB YARNS
SPUN YARNS
STRETCH YARNS (FILAMENT)
TEXTURED SPUN YARNS
TEXTURED YARNS (FILAMENT)
UNCONVENTIONAL YARNS
WOOLEN SPUN YARNS
WORSTED SPUN YARNS
Z TWIST YARNS
BT FIBER ASSEMBLIES
RT ALTERNATING TWIST
BLENDS (FIBERS)
CARD SLIVERS
CARRIER RODS
COUNT
DENIER
DRAFTING (STAPLE FIBER)
FABRICS
FIBERS
FILAMENTS
FILMS
FOAMS
HANKS
HIGH TWIST
INDIRECT COUNT
IRREGULARITY TESTERS
LOW TWIST
NEPS
PLY TWIST

RT REELING
REELING MACHINES (WINDING)
ROPES
ROVINGS
SAMPLES
SELF THREADING DEVICES
SLIVERS
SLUBBINGS
SPINNING
STRANDS
TEXTILE MATERIALS
THICK SPOTS
TOPS
TOW
TWIST
TWIST LIVELINESS
TWISTING
WARP ENDS
WARP PREPARATION
WEBS
WEDGING
YARN CARRIERS
YARN DIAMETER
YARN DYEING
YARN DYEING MACHINES
YARN DYNAMICS
YARN FINISH
YARN FLATTENING
YARN GEOMETRY
YARN LUSTER
YARN PROPERTIES
YARN STRENGTH
YARN STRENGTH EFFICIENCY
YARN STRUCTURE
YARN TENSION

YARNS (CONTINUOUS FILAMENT)
USE FILAMENT YARNS

YEARS COVERAGE (1)
RT COMPILATIONS
HISTORY
LITERATURE SURVEYS
REVIEWS
SURVEYS
TIME
YEARS COVERAGE (5)
YEARS COVERAGE (10)
YEARS COVERAGE (25)
YEARS COVERAGE (50)
YEARS COVERAGE (75)
YEARS COVERAGE (100)

YEARS COVERAGE (5)
RT COMPILATIONS
HISTORY
LITERATURE SURVEYS
REVIEWS
SURVEYS
TIME
YEARS COVERAGE (1)

YEARS COVERAGE (10)
RT COMPILATIONS
HISTORY
LITERATURE SURVEYS
REVIEWS
SURVEYS
TIME
YEARS COVERAGE (1)

YEARS COVERAGE (25)
RT COMPILATIONS
HISTORY
LITERATURE SURVEYS
REVIEWS
SURVEYS
TIME
YEARS COVERAGE (1)

YEARS COVERAGE (50)
RT COMPILATIONS
HISTORY
LITERATURE SURVEYS
REVIEWS
SURVEYS
TIME
YEARS COVERAGE (1)

YEARS COVERAGE (75)
RT COMPILATIONS
HISTORY
LITERATURE SURVEYS
REVIEWS
SURVEYS
TIME
YEARS COVERAGE (1)

YEARS COVERAGE (100)
RT COMPILATIONS
HISTORY
LITERATURE SURVEYS
REVIEWS
SURVEYS
TIME
YEARS COVERAGE (1)

YEAST
BT FUNGUS
RT ENZYMES

YELLOW CHINA SILK
BT SILK
RT WHITE CHINA SILK
YELLOW JAPAN SILK

YELLOW JAPAN SILK
BT SILK
RT WHITE JAPAN SILK
YELLOW CHINA SILK

YELLOWING
RT COLORANTS
DEGRADATION
DISCOLORATION
FADING
HYDROCELLULOSE
HYDROLYSIS
OXYCELLULOSE

YIELD (MECHANICAL)
UF YIELDING
RT CARD YIELD
COLD DRAWING
COLOR YIELD
DEFORMATION
DRAWTWISTING
DRAWING (FILAMENT)
ELASTIC STRAIN
HOT DRAWING
IMPACT STRENGTH
PLASTIC FLOW
PLASTIC STRAIN
RHEOLOGY
SCOURING YIELD
STRAIN
VISCOELASTICITY
VOIGT MODEL
YIELD POINT

YIELD (RETURN)
RT BYPRODUCTS
COMB TEAR
COMBING
FLY (WASTE)
MANAGEMENT
NOBLE COMBING
NOIL PER CENT
PRODUCTION
SLIVER NOIL RATIO
TOP NOIL RATIO
WASTE

YIELD POINT
UF ELASTIC LIMIT
YIELD STRESS
BT STRESS STRAIN PROPERTIES
STRESS STRAIN PROPERTIES (TENSILE)
RT BREAKING STRENGTH
COEFFICIENT OF CROSS VISCOSITY
COLD DRAWING
COMPLIANCE
DEFORMATION
ELASTIC MODULUS (TENSILE)
ELASTIC STRAIN
ELASTICITY
FIRMOVISCOSITY
HOT DRAWING
LOADING
MODULUS
NECKING (FILAMENT)
PERMANENT SET
PLASTIC STRAIN
RECOVERY (SELF RESTORATION TO ORIGINAL CONDITION)
RELAXATION SPECTRUM
RELAXATION TIME (MECHANICAL)
STRAIN
STRESS
STRESS STRAIN CURVES
STRESS STRAIN PROPERTIES (TENSILE)
WRINKLE RECOVERY

YIELD STRESS
USE YIELD POINT

YIELDING
USE YIELD (MECHANICAL)

YORKSHIRE SKEIN WOOLEN COUNT
BT COUNT
INDIRECT COUNT
RT SKEINS

YOUNGS MODULUS
USE ELASTIC MODULUS (TENSILE)

Z CRIMP
UF SAW TOOTH CRIMP
BT CRIMP
RT BICOMPONENT FIBER YARNS (FILAMENT)
BICOMPONENT FIBER YARNS (STAPLE)
BULKED YARNS
BULKING
CRIMP FREQUENCY
CRIMP INDEX
CRIMP INTERCHANGE
CRIMP TENDENCY
CRIMPED YARNS
CRIMPING
FIBER CRIMP
FOLLOW THE LEADER CRIMP
LATENT CRIMP
SELF CRIMPING YARNS
STRETCH YARNS (FILAMENT)
TEXTURED YARNS (FILAMENT)
TEXTURING
WAVINESS
YARN CRIMP

Z TWILLS
BT DIRECTION (TWILL)
TWILLS
RT CHEVRON TWILLS
FABRIC STRUCTURE
FANCY TWILLS
HERRINGBONE TWILLS
LOW TWILLS
REGULAR TWILLS
S TWILLS
SENSE (DIRECTION)

Z TWIST
UF LEFT HAND TWIST
BT TWIST
RT ALTERNATING TWIST
BALANCED YARNS
DIRECTION (TWILL)
HIGH TWIST
LOW TWIST
MULTIPLIED YARNS
PLIED YARNS
PLY TWIST
PLYING
REAL TWIST
S TWIST
SENSE (DIRECTION)
SINGLES YARNS
TWIST LIVELINESS
TWIST SENSE
TWISTING
TWO PLY YARNS
YARN STRUCTURE
ZERO TWIST

Z TWIST YARNS
BT YARNS
RT BALANCED YARNS
FILAMENT YARNS
GEAR CRIMPING
PLIED YARNS
S TWIST YARNS
RT SINGLES YARNS
SPINNING
TEXTURED YARNS (FILAMENT)
TWIST
TWISTING
Z TWIST

ZEFKROME (TN)
BT ACRYLIC FIBERS

ZEFRAN (TN)
BT ACRYLIC FIBERS

ZEIN
BT PROTEINS
RT AZLON (REGENERATED PROTEIN FIBERS)

ZELLWEGER TESTER (TN)
BT IRREGULARITY TESTERS
RT CAPACITANCE TYPE IRREGULARITY TESTERS
FIELDEN WALKER IRREGULARITY TESTER (TN)
PNEUMATIC IRREGULARITY TESTERS
SACO LOWELL IRREGULARITY TESTER (TN)
USTER IRREGULARITY TESTER (TN)
WIRA WEB LEVELNESS TESTER

ZERO TWIST
UF TWISTLESS
BT TWIST
RT ALTERNATING TWIST
CYCLOSET YARNS (TN)
FALSE TWIST
FLUID TWISTING
HIGH TWIST
INTERLACING (FILAMENT IN YARN)
LOW TWIST
NO TORQUE YARNS
REAL TWIST
ROTOSET YARNS (TN)
S TWIST
Z TWIST

ZETA POTENTIAL
UF ELECTROKINETIC POTENTIAL
BT ELECTRICAL PROPERTIES
SURFACE PROPERTIES (PHYSICAL CHEMICAL)
RT AGGREGATION
COAGULATING
COLLOIDS

ZIG ZAG MACHINES
BT SEWING MACHINES

ZINC CHLORIDE
BT CHLORIDES
ZINC COMPOUNDS
RT CATALYSTS
MAGNESIUM NITRATE
MARSCHALL SOLUTION
ZINC NITRATE
ZINC SULFATE

ZINC COMPOUNDS
NT ZINC CHLORIDE
ZINC FLUOBORATE
ZINC FORMALDEHYDE SULFOXYLATE
ZINC NITRATE
ZINC OXIDE
ZINC SOAPS
ZINC STEARATE
ZINC SULFATE

ZINC FLUOBORATE
BT BORON COMPOUNDS
ZINC COMPOUNDS
RT CATALYSTS

ZINC FORMALDEHYDE SULFOXYLATE
BT FORMALDEHYDE SULFOXYLATES
ZINC COMPOUNDS
RT ACETALDEHYDE SULFOXYLATES
REDUCING AGENTS
VAT DYEING
VAT PRINTING

ZINC NITRATE
BT NITRATES
ZINC COMPOUNDS
RT CATALYSTS
MAGNESIUM CHLORIDE
MAGNESIUM NITRATE
ZINC CHLORIDE

ZINC OXIDE
BT OXIDES
ZINC COMPOUNDS
RT PIGMENT PRINTING
PIGMENTS

ZINC SOAPS
BT ZINC COMPOUNDS
METALLIC SOAPS

ZINC STEARATE
BT METALLIC SOAPS
ZINC COMPOUNDS

ZINC SULFATE
BT SULFATES
ZINC COMPOUNDS
RT ZINC CHLORIDE

ZIPPERS
UF SLIDE FASTENERS
BT FASTENERS
GARMENT COMPONENTS
RT HOOK CLOSURES
JOINTS
POLYAMIDES
SNAP FASTENERS
VELCRO (TN)

ZIRCONIUM COMPOUNDS
RT WATER REPELLENTS

ZWITTERIONIC SURFACTANTS
NT BETAINE SURFACTANTS
SULFOBETAINE SURFACTANTS

ALPHABETICAL LIST OF KEYWORDS

ABRASION
ABRASION MARKS
ABRASION RESISTANCE
ABRASION RESISTANCE FINISHES
ABRASION TESTERS
ABRASION TESTING
ABRASIVES
ABSENCE
ABSORBATES
ABSORBENCY (MATERIAL)
ABSORBERS
ABSORPTION (MATERIAL)
ABSORPTION (RADIATION)
ABSORPTION EQUILIBRIUM
ABSORPTION ISOTHERM
ABSORPTION RATE
ABSORPTION SPECTRA
ABSORPTION TESTERS
ABSORPTIVITY (RADIATION)
ACCELERANTS (DYEING)
ACCELERATION (MECHANICAL)
ACCELEROMETERS
ACCESSIBILITY (INTERNAL)
ACCESSORIES
ACCIDENTS
ACCORDION FABRICS
ACCUMULATION
ACCURACY
ACETAL RESINS
ACETALDEHYDE SULFOXYLATES
ACETALS
ACETAMIDE
ACETATE
ACETATE DYEING
ACETATE DYES
ACETATE ESTERS
ACETATE FIBERS
ACETATE RAYON
ACETATE SALTS
ACETIC ACID
ACETIC ANHYDRIDE
ACETONE
ACETONITRILE
ACETYLATED COTTON
ACETYLATION
ACETYLENE
ACETYLENE DIUREA FORMALDEHYDE
ACETYLENIC COMPOUNDS
ACETYLENIC HYDROCARBONS
ACETYLTRIETHYL CITRATE
ACID AGING
ACID ANHYDRIDES
ACID BINDING
ACID CHLORIDES
ACID DAMAGE
ACID DEGRADATION
ACID DYEING
ACID DYES
ACID END GROUPS
ACID HYDROLYSIS
ACID MILLING DYES
ACID SHOCK METHOD
ACID SOLUBILITY
ACID SOLUTIONS
ACIDITY
ACIDS
ACOUSTIC BAFFLING
ACOUSTIC INSULATION
ACOUSTIC PROPERTIES
ACQUISITION
ACRILAN (TN)
ACROLEIN
ACRYLALDEHYDE
ACRYLAMIDE
ACRYLAMIDO DYES
ACRYLIC ACID
ACRYLIC COMPOUNDS
ACRYLIC DYEING
ACRYLIC DYES
ACRYLIC ESTERS
ACRYLIC FIBER BLENDS
ACRYLIC FIBERS
ACRYLIC RESINS
ACRYLIC SALTS
ACRYLICS
ACRYLONITRILE
ACTINIC DEGRADATION
ACTIVATED CARBON
ACTIVATION ENERGY
ACTIVATORS
ACTIVE SPORTSWEAR
ACTUAL DRAFT
ACYL ESTER SULFONATES
ACYLATION
ADAPTATION
ADD ON
ADDITION
ADDITION POLYMERIZATION
ADDITION POLYMERS
ADDITIVES (CHEMICAL)
ADHESION
ADHESIVE BACKING
ADHESIVE BONDED (NONWOVENS)
ADHESIVE BONDED FABRICS
ADHESIVES
ADHESIVES (FOOTWEAR)
ADIPALDEHYDE
ADIPIC ACID
ADJUSTMENTS
ADSORPTION
ADSORPTION ISOTHERM
ADSORPTION RATE
ADSORPTIVITY
ADVANTAGES
AERATED YARNS
AERESS (TN)
AERODYNAMIC CARDING
AERODYNAMIC CARDS

AERODYNAMIC CONFIGURATIONS
AERODYNAMIC HEATING
AERODYNAMICS
AEROSOL 18 (TN)
AEROSOL 22 (TN)
AEROSOLS
AESTHETIC APPEAL
AESTHETIC PROPERTIES
AFFINITY (DYES)
AFTERCHROME DYEING
AFTERCHROME DYES
AFTERTREATMENTS (GENERAL)
AFTERWASH
AFTERWELT
AGERS
AGGREGATION
AGILON (TN)
AGING (MATERIALS)
AGING (RAYON MANUFACTURE)
AGING (STEAMING)
AGING (STORAGE)
AGITATORS
AGRICULTURAL FABRICS
AIR
AIR BEARINGS
AIR BRAKES
AIR BRIDGES
AIR CONDITIONING
AIR DRAG
AIR FLOW
AIR FLOW TESTERS
AIR JET BULKED YARNS
AIR JET BULKERS
AIR JET BULKING
AIR JET LOOMS
AIR JET TEXTURED YARNS
AIR JET TEXTURING
AIR JET TEXTURING MACHINES
AIR LAY RANDOM WEB
AIR PERMEABILITY
AIR PERMEABILITY TESTING
AIR POLLUTION
AIR RESISTANCE
AIR SPINNING
AIR STREAMS
AIR TEXTURING
AIR TUCK STITCHES
AIR TUCKING
AIR TWISTING
ALBERT TWILL
ALBUMIN
ALCIAN BLUE (TN)
ALCOHOL SULFATES
ALCOHOLS
ALDEHYDE GROUPS
ALDEHYDES
ALENCON
ALEO
ALGINATE FIBERS
ALGINATES
ALGINIC ACID
ALHAMBRA QUILT
ALICYCLIC COMPOUNDS
ALICYCLIC HYDROCARBONS
ALIGNMENT
ALIGNMENT CHARTS
ALIPHATIC COMPOUNDS
ALIPHATIC HYDROCARBONS
ALKALI CELLULOSE
ALKALI DEGRADATION
ALKALI HYDROLYSIS
ALKALI SOAPS
ALKALI SOLUBILITY
ALKALI STEEPING
ALKALIES
ALKALINE SOLUTIONS
ALKALINITY
ALKANES
ALKANOL S (TN)
ALKANOLAMIDES
ALKANOLAMINE SOAPS
ALKANOLAMINES
ALKENES
ALKOXYL GROUPS
ALKYD RESINS
ALKYL CARBAMATE FORMALDEHYDE
ALKYL DIMETHYLAMINE SURFACTANTS
ALKYL ETHOXY PHOSPHATES
ALKYL ETHOXY SULFATES
ALKYL GLYCEROSULFATES
ALKYL PHENOXYETHYL SULFONATES
ALKYL PHOSPHATE SURFACTANTS
ALKYL SULFATES
ALKYL SULFONATES
ALKYLAMIDE SULFATES
ALKYLAMIDE SULFONATES
ALKYLAMINE SURFACTANTS
ALKYLAMINO CARBOXYLATE SURFACTANTS
ALKYLAMINO SULFATE SURFACTANTS
ALKYLAMINO SULFONATE SURFACTANTS
ALKYLARYL ETHOXY PHOSPHATES
ALKYLARYL ETHOXY SULFATES
ALKYLARYL OXYETHYL SULFONATES
ALKYLARYL POLYETHOXY ETHERS
ALKYLARYL SULFONATES
ALKYLATION
ALKYLBENZENE SULFONATES
ALKYLBIPHENYL SULFONATES
ALKYLDI(HYDROXYETHYL)AMINE OXIDE SURFACTANTS
ALKYLDIMETHYLAMINE OXIDE SURFACTANTS
ALKYLDIMETHYLAMINE SURFACTANTS
ALKYLENE CARBONATES
ALKYLETHER SULFONATES
ALKYLMORPHOLINE SURFACTANTS
ALKYLNAPHTHALENE SULFONATES

ALKYNES
ALL OVER (LACE)
ALLIGATOR
ALLOTMENT
ALLOWANCE FACTOR (WARP PREPARATION)
ALLOYS
ALLYL ALCOHOL
ALLYL BROMIDE
ALLYL CHLORIDE
ALLYL COMPOUNDS
ALLYL GLYCIDYL ETHER
ALLYLCELLULOSE
ALPACA
ALPACA FABRICS
ALPHA BETA TRANSFORMATION
ALPHA KERATIN
ALPHA PARTICLES
ALPHASULFOMETHYL PALMITATE
ALPHASULFOPALMITIC ACID
ALTERATIONS
ALTERING
ALTERNATE PICKS
ALTERNATING TWIST
ALUMINA
ALUMINUM
ALUMINUM ABIETATE
ALUMINUM COMPOUNDS
ALUMINUM FOIL
ALUMINUM OXIDE
ALUMINUM SILICATE FIBERS
ALUMINUM SOAPS
ALUMINUM STEARATE
AMBIENT TEMPERATURE
AMBLER SUPERDRAFT
AMERICAN CUT
AMERICAN EGYPTIAN COTTON
AMERICAN MERINO WOOL
AMERICAN RUN
AMERICAN SYSTEM PROCESSING
AMERICAN UPLAND COTTON
AMIDES
AMIDOAMINE SURFACTANTS
AMILAR (TN)
AMINE END GROUPS
AMINE GROUPS
AMINE OXIDE SURFACTANTS
AMINE OXIDES
AMINE SOAPS
AMINE SURFACTANTS
AMINES
AMINIZED COTTON
AMINO ACIDS
AMINO GROUPS
AMINO RESINS
AMINOPLASTS
AMMONIA
AMMONIUM COMPOUNDS
AMMONIUM SOAPS
AMMONIUM SULFATE
AMORPHOUS POLYMERS
AMORPHOUS REGION
AMPHOLYTIC SURFACTANTS
AMPHOTERIC SURFACTANTS
AMPLIFIERS
AMPLITUDE
AMYLNAPHTHALENE SULFONATES
AMYLOPECTIN
AMYLOSE
ANALOGUES
ANALYSIS
ANALYSIS (MATHEMATICAL)
ANALYSIS OF VARIANCE
ANALYZING
ANATASE
ANGLE OF INCIDENCE
ANGLE OF TWIST
ANGLE OF WIND
ANGLE OF WRAP
ANGLE STRIPPERS
ANGOLA (FABRIC)
ANGOLA (YARN)
ANGORA (GOAT)
ANGORA (RABBIT)
ANGULAR DEFLECTION
ANGULAR SEAMING
ANGULAR VELOCITY
ANHYDRIDES
ANILINE BLACK DYES
ANIMAL FATS
ANIMAL FIBERS (GENERAL)
ANIMAL HUSBANDRY
ANIMALIZING
ANIMALS
ANION EXCHANGE
ANION EXCHANGE RESINS
ANIONIC DYES
ANIONIC SITES
ANIONIC SOFTENERS
ANIONIC SURFACTANTS
ANIONS
ANISOTROPIC MATERIALS
ANISOTROPY (STRESS STRAIN)
ANNEALING
ANTHROPOMETRIC KINEMATICS
ANTHROPOMETRIC MEASUREMENTS
ANTIBACTERIAL AGENTS
ANTIBACTERIAL FINISHES
ANTIBACTERIAL TREATMENTS
ANTICLOCKWISE
ANTICREASE FINISHES
ANTIFELTING AGENTS
ANTIFELTING TREATMENTS
ANTIFOAM AGENTS
ANTIFRICTION BEARINGS
ANTILADDER BANDS
ANTILUBRICANTS
ANTIMICROBIAL FINISHES

ANTIMICROBIAL TREATMENTS
ANTIOXIDANTS
ANTIQUE (LACE)
ANTIREDEPOSITION AGENTS
ANTIRUN COURSES
ANTISEPTICS
ANTISHRINK COMPOUNDS
ANTISLIP AGENTS
ANTISOILING FINISHES
ANTISTATIC AGENTS
ANTISTATIC BEHAVIOR
ANTISTATIC FINISHES (SUBSTANCE ADDED)
ANTIWEDGE RINGS
ANTRON (TN)
APO
APPARATUS
APPAREL
APPAREL DESIGN
APPAREL FABRICS
APPARENT CONTACT AREA
APPARENT DENSITY
APPEARANCE
APPLICATIONS
APPLIED ELASTICITY
APPROXIMATIONS
APRON CONTROL
APRON DRAFTING
APRONS
APRONS (CLOTHING)
AQUEOUS SCOURING
AQUEOUS SOLUTIONS
ARABEVA FABRIC (TN)
ARABEVA PROCESS (TN)
ARACHNE FABRIC (TN)
ARACHNE PROCESS (TN)
ARACHNE SYSTEMS (TN)
ARAKNIT PROCESS (TN)
ARALAC (TN)
ARALOOP FABRIC (TN)
ARALOOP PROCESS (TN)
ARDIL (TN)
AREA
AREA MOMENT OF INERTIA
AREA OF CONTACT
ARGYLE GIMP
ARITHMETIC
ARMACS (TN)
ARMEENS (TN)
ARMENIAN (LACE)
ARMURE
ARNEL (TN)
AROMATIC COMPOUNDS
AROMATIC HYDROCARBONS
ARRAS TAPESTRIES
ARRESTERS
ARSONIUM COMPOUNDS
ART
ARTERIAL REPLACEMENTS
ARTIFICIAL DAYLIGHT
ARTIFICIAL FURS
ARTIFICIAL SILK (ARCHAIC)
ARTIFICIAL SOIL
ARYL COMPOUNDS
ASBESTOS
ASBESTOS FIBERS
ASPERITIES
ASSEMBLIES
ASTRAKHAN
ASYMMETRY
ATACTIC POLYMERS
ATHLETIC CLOTHING
ATMOSPHERIC DRYERS
ATOMIZERS
ATOMIZING
AUBOSSON TAPESTRIES
AUSTRALIAN MERINO WOOL
AUTOCLAVES
AUTOCORRELATION
AUTOCORRELATION COEFFICIENT
AUTOCOUNT (TN)
AUTODRAFTER
AUTOLEVELLERS
AUTOLEVELLING
AUTOMATED SPINNING (STAPLE FIBER)
AUTOMATIC CONTROL
AUTOMATIC DOFFING
AUTOMATIC KNOTTING
AUTOMATIC LOOMS
AUTOMATIC REGULATION
AUTOMATIC STOP MOTIONS
AUTOMATIC TRANSFER (KNITTING)
AUTOMATIC WINDING
AUTOMATION
AUTOMOTIVE FABRICS
AUXILIARY CAMS
AVRIL (TN)
AWL CUTTING DRILLS
AWNING CLOTH
AWNINGS
AXIAL MOTION
AXLES
AXMINSTER
AXMINSTER CARPETS
AXMINSTER LOOMS
AZIDES
AZIDOSULFONYL DYES
AZIRIDINE COMPOUNDS
AZLON (REGENERATED PROTEIN FIBERS)
AZLON BLENDS
AZOIC BASES
AZOIC DYEING
AZOIC DYES
B(L)CURVES
BABY DELAINE (WOOL CLASS)
BABY FLANNEL

BACK (FABRIC)
BACK AND FACE EFFECTS
BACK COATING
BACK CROSSING HEDDLES
BACK FILLING
BACK FILLING MACHINES
BACK GRAY
BACK REST (LOOM)
BACK ROLLS
BACK STANDARD
BACK TACKING
BACK TO BACK FABRICS (KNITTED)
BACK WARP
BACKED FABRICS
BACKING (MATERIAL)
BACKING CLOTHS
BACKING COMPOUNDS
BACKING FABRICS
BACKSTITCHES
BACKWASHING
BACKWASHING MACHINES
BACTERIA REPELLENCY
BACTERIA RESISTANCE
BACTERIAL DEGRADATION
BACTERIAL INHIBITION
BACTERICIDES
BACTERIOSTATIC AGENTS
BAGGINESS
BAGGING (PACKAGES)
BAGGING MACHINES
BAGGING TENDENCY
BAGS
BAIZE
BAKING
BALANCE
BALANCED SHED
BALANCED WEAVES
BALANCED YARNS
BALANCING
BALBRIGGAN
BALDNESS
BALE BREAKERS
BALE BREAKING
BALE PLUCKERS
BALES
BALING
BALING MACHINES
BALING PRESSERS
BALL AND SOCKET TENSION
BALL BEARINGS
BALL BURST TESTING
BALL BURST TESTS
BALL WARPS
BALLET TOES (KNITTING)
BALLING HEADS
BALLISTIC FABRICS
BALLISTIC LIMIT
BALLISTIC TESTING
BALLOON COLLAPSE
BALLOON CONTROL
BALLOON CONTROL RINGS
BALLOON CONTROLLERS
BALLOON DYNAMICS
BALLOON HEIGHT
BALLOON RINGS
BALLOON SEPARATORS
BALLOON SHAPE
BALLOON STABILITY
BALLOON TENSION
BALLOONS
BANCROFT PROCESS (TN)
BAND KNIFE CUTTING MACHINES
BANDAGES
BANDING
BANGING OFF
BANLON (TN)
BANNOCK BURN
BAR TACKS
BAR WARP MACHINES
BARATHEA
BARIUM ACTIVITY NUMBER
BARIUM COMPOUNDS
BARIUM HYDROXIDE
BARMEN MACHINES
BAROTOR MACHINES
BARRAS
BARRE
BASE BINDING
BASES
BASIC DYEING
BASIC DYES
BASKET WEAVES
BAST FIBERS
BASTARD COPS
BASTING
BATCH DRYING
BATCH DYEING
BATCH DYEING MACHINES
BATCH FINISHING
BATCH SCOURING
BATH RATIO
BATHING SUITS
BATHS
BATIK DYEING
BATIK DYEING MACHINES
BATINES
BATISTE
BATT MAKING MACHINES
BATTENBERG
BATTENS (LOOM)
BATTING
BATTLEDRESS
BAULK FINISHING
BAVE (SILK)
BEACH ROBES
BEAD FILLER FABRICS
BEAD WOVEN FABRICS
BEAD YARNS

BEADING (LACE)
BEAM DYEING (YARN)
BEAM DYEING MACHINES (FABRIC)
BEAM DYEING MACHINES (YARN)
BEAM WARPING
BEAMERS
BEAMING
BEAMS
BEARDED MOTES
BEARDED NEEDLES
BEARING ALLOYS
BEARING STRENGTH
BEARINGS
BEAT UP
BEATERS (OPENING)
BEATING UP
BEAUVAIS TAPESTRIES
BEAVER CLOTH
BECK DYEING
BECKS
BEDFORD CORD
BEDJACKETS
BEDSOCKS
BEDSPREADS
BEER
BEETLING
BELFAST (TN)
BELT CRIMPING
BELT DRIVES
BELT SLIPPAGE
BELT TENSION
BELTING
BELTS
BELTS (APPAREL)
BEMBERG FIBERS
BENCH SORTING (WOOL)
BENDING
BENDING ENERGY
BENDING LENGTH
BENDING MOMENT DIAGRAMS
BENDING MOMENTS
BENDING RECOVERY
BENDING RIGIDITY
BENDING STIFFNESS
BENDING STRENGTH
BENDING TESTING
BENZAMIDE
BENZENE
BENZOATE ESTERS
BENZOATE SALTS
BENZOIC ACID
BENZOQUINONE
BENZOYLATED COTTON
BENZYL ALCOHOL
BENZYL BENZOATE
BENZYL CHLORIDE
BENZYLATION
BENZYLCELLULOSE
BERYLLIUM CHLORIDE
BERYLLIUM COMPOUNDS
BERYLLIUM HYDROXIDE
BET ISOTHERM
BETA KERATIN
BETA PARTICLES
BETA RAYS
BETAINE SURFACTANTS
BHES
BIAXIAL STRENGTH
BIAXIAL STRESS
BIBLIOGRAPHY
BICARBONATES
BICOMPONENT FIBER YARNS (FILAMENT)
BICOMPONENT FIBER YARNS (STAPLE)
BICOMPONENT FIBERS
BICOMPONENT FILAMENTS
BICOMPONENT SPINNING
BICOMPONENT STRUCTURE
BICONSTITUENT FIBERS
BIKINIS
BILATERAL FIBER STRUCTURE
BILLIARD CLOTH
BINCHE
BINDER EFFICIENCY
BINDER-MIGRATION
BINDERS (NONWOVENS)
BINDERS (PIGMENTS)
BINDING (SEAMS)
BINDINGS (FOOTWEAR)
BINGHAM BODY
BINOMIAL DISTRIBUTION
BIOCHEMICAL OXYGEN DEMAND
BIOCHEMICAL PROPERTIES
BIOCIDES
BIODEGRADABILITY
BIOLOGICAL OXYGEN DEMAND
BIRDS EYE WEAVES
BIREFRINGENCE
BIS(HYDROXYETHYL) SULFONE
BISMUTH BROMIDE
BISMUTH CHLORIDE
BISMUTH COMPOUNDS
BISULFATES
BISULFITES
BLACKBODY
BLADES (ROLLER)
BLAMIRE FEED
BLANC DE BLANCS (TN)
BLANCOPHOR (TN)
BLANK NEEDLES
BLANKET (FINISHING)
BLANKET MARKS
BLANKET SHUTTLES
BLANKETS
BLEACH HOUSE
BLEACHERIES
BLEACHES
BLEACHING

BLEACHING AGENTS
BLEACHING MACHINES
BLEACHING RANGES
BLEEDING
BLEEDING TESTING
BLEMISHES
BLEND DRAFTING
BLEND DYEING
BLEND LEVEL
BLENDED FABRICS
BLENDED FIBER RATIO
BLENDED YARNS
BLENDED YARNS (FILAMENT)
BLENDED YARNS (STAPLE)
BLENDERS
BLENDING
BLENDS (FIBERS)
BLENDS (POLYMERS)
BLIND STITCHING
BLINDSTITCHES
BLOCK POLYMERS
BLOCK PRINTING
BLOCK PRINTING EQUIPMENT
BLONDE (LACE)
BLOOM
BLOTCHES
BLOTCHINESS
BLOUSES
BLOWERS
BLOWING EQUIPMENT
BLOWN FINISH
BLUE C NYLON (TN)
BLUE C POLYESTER (TN)
BLUE C SPANDEX (TN)
BLUEING
BLUEING AGENTS
BLUETTE
BOARDING
BOAT SHUTTLES
BOBBIN CASES
BOBBIN DYEING
BOBBIN DYEING MACHINES
BOBBIN HOOKS
BOBBIN LEAD
BOBBIN LOOPS
BOBBIN RINGS
BOBBIN STRIPPING MACHINES
BOBBIN THREADS (SEWING)
BOBBINET MACHINES
BOBBINETTE LACE
BOBBINS
BOD
BODICES
BODY KINEMATICS
BODY MOVEMENTS
BODY PLY
BOHEMIAN (LACE)
BOIL OFF
BOIL OFF MACHINES
BOILING
BOILING OFF
BOILING POINT
BOLLS
BOLSTERS
BOLT CAMS (KNITTING)
BOLTON SHEETING
BOMBAZINE
BOND DISSOCIATION ENERGY
BOND ENERGY
BONDED FABRICS (LAMINATED)
BONDED FABRICS (NONWOVENS)
BONDED FIBER FABRICS
BONDED YARN FABRICS
BONDED YARNS
BONDING
BONDING AGENTS (NONWOVENS)
BONDING STRENGTH
BOOKBINDING
BOOKS (SILK)
BOOSTER BOXES
BOOTS
BORON COMPOUNDS
BORON HYDRIDES
BORON NITRIDE
BORON NITRIDE FIBERS
BORON TRICHLORIDE
BOTANY WOOL
BOTTLE BOBBINS
BOTTOM COVER
BOTTOM FILLER (FOOTWEAR)
BOTTOM ROLLS
BOTTOMING
BOUCLES
BOUNCE (WEAVING)
BOUNCING
BOUND WATER
BOUNDARY LAYER
BOUNDARY LAYER CONTROL
BOUNDARY LUBRICATION
BOURDON
BOW
BOW CONTROL MACHINES
BOW STRAIGHTENERS
BOWED FABRICS
BOWED ROLLS
BOWING
BOWLS
BOX LININGS
BOX LOOMS
BOX TOES
BOXER SHORTS
BOXES
BRABANT
BRACE WEB
BRACES
BRADFORD OPEN DRAWING
BRADFORD SYSTEM PROCESSING
BRAGGS LAW

BRAID (WOOL GRADE)
BRAIDED FABRICS
BRAIDER BOBBINS
BRAIDERS
BRAIDING
BRAIDS
BRAKE DISCS
BRAKE DRUMS
BRAKE LININGS
BRAKES (FOR ARRESTING MOTION)
BRAKING
BRANCHED ALIPHATIC HYDROCARBONS
BRANCHED CHAIN ALIPHATIC
 HYDROCARBONS
BRANCHED CHAIN MOLECULES
BRASSIER CLOTH
BRASSIERES
BRATTICE CLOTH
BREADTH
BREAK DRAFT
BREAK SPINNING
BREAKAGES
BREAKER FABRICS
BREAKER PART
BREAKERS
BREAKING
BREAKING ELONGATION
BREAKING ENERGY
BREAKING EXTENSION
BREAKING LENGTH
BREAKING LOAD
BREAKING STRAIN
BREAKING STRENGTH
BREAKING STRESS
BREAKING TENACITY
BREAKING TENSION
BREAKS
BREAST BEAMS
BREAST SWIFTS
BRETON
BRI-NYLON (TN)
BRIEFS
BRIGHT PICKS
BRIGHT SPECKS
BRIGHTNESS
BRILLIANT SPOT
BRINS (SILK)
BRITTLE FRACTURE
BRITTLENESS
BROADCLOTH
BROCADES
BROCATELLES
BROKEN END COLLECTION
BROKEN END COLLECTORS
BROKEN FIBERS
BROKEN SELVEDGES
BROKEN THREADS
BROMIDES
BROMINE
BROMINE COMPOUNDS
BRONZING
BRUISED PLACES
BRUSHING
BRUSHING ACTION
BRUSHING MACHINES
BRUSSELS TAPESTRIES
BUCKET SPINNING
BUCKLES
BUCKLING
BUCKRAM
BUFFING WHEELS
BUG YARNS
BUILDER FABRICS
BUILDER MOTIONS
BUILDERS (DETERGENTS)
BULGING (FILAMENT)
BULK
BULK DENSITY
BULK MODULUS
BULK POLYMERIZATION
BULK YARNS (FILAMENT)
BULKED YARNS
BULKED YARNS (STAPLE)
BULKING
BULKING MACHINES
BULLET PROOF CLOTHING
BUMP GRAY
BUMP YARNS
BUNA N COATED FABRICS
BUNCHING COEFFICIENT
BUNDESMANN TESTING
BUNDLE EFFICIENCY
BUNDLE STRENGTH
BUNDLES
BUNDLING (TAILORING)
BUNTE SALTS
BUNTING
BURLING
BURLING IRONS
BURLING MACHINES
BURMILIZED (TN)
BURNING
BURR BEATERS
BURR CRUSHERS
BURR CRUSHING MACHINES
BURR PICKING
BURR REMOVAL
BURRS
BURRY WOOL
BURST TESTERS
BURST TESTING
BURSTING
BURSTING STRENGTH
BUSHINGS
BUSTS
BUTADIENE
BUTADIENE DIEPOXIDE
BUTADIENE RUBBER FIBER

BUTTED SEAMS
BUTTERFLY SEAMS
BUTTON BREAKERS
BUTTON BREAKING
BUTTONHOLES
BUTTONHOLING
BUTTONS
BUTTS (NEEDLES)
BUTYL ACRYLATE
BUTYL ALCOHOL
BUTYL COATED FABRICS
BUTYLNAPHTHALENE SULFONATES
BUTYROLACTONE
BYPRODUCTS
CABLE CORDS
CABLE STITCH
CABLED YARNS
CABLES
CABLING
CADMIUM CHLORIDE
CADMIUM COMPOUNDS
CADMIUM ETHYLENEDIAMINE
 HYDROXIDE
CADMIUM SELENIDE
CADON (TN)
CADOXEN
CAKE DYEING
CAKE DYEING MACHINES
CAKE SIZING
CAKES
CALCIUM CHLORIDE
CALCIUM COMPOUNDS
CALCIUM HYDROXIDE
CALCIUM SOAPS
CALCIUM STEARATE
CALCULATING
CALCULATIONS
CALENDER COATING
CALENDER PRINTING
CALENDER ROLLS
CALENDER SPREADING
CALENDERING
CALENDERING MACHINES
CALFSKIN
CALGON (TN)
CALIBRATION
CALICO
CALIFORNIA COTTON
CALORIMETRY
CAM ANGLES
CAM LOOMS
CAMBER
CAMBRIC
CAMBRIDGE EXTENSOMETER (TN)
CAMEL HAIR
CAMOUFLAGES
CAMS
CANDLE FILTERS
CANDLEWICK FABRICS (TUFTED
 FABRIC)
CANDLEWICK YARNS
CANOPIES
CANS (DRYING)
CANTON SILK
CANTRECE (TN)
CANVAS
CAP SPINNING
CAP SPINNING FRAMES
CAPACITANCE (ELECTRICAL)
CAPACITANCE BRIDGES
CAPACITANCE MOTORS
CAPACITANCE TYPE IRREGULARITY
 TESTERS
CAPACITY (ELECTRICAL)
CAPACITY (GENERAL)
CAPILLARITY
CAPILLARY EFFECTS
CAPILLARY PRESSURE
CAPILLARY TUBES
CAPILLARY WATER
CAPROLACTAM
CAPROLAN (TN)
CAPRYL DIETHANOLAMIDE
CAPS (HEADGEAR)
CAPS (SPINNING)
CARBAMOYLETHYLATED COTTON
CARBODIIMIDES
CARBOHYDRATES
CARBON
CARBON BLACK
CARBON DISULFIDE
CARBON TETRACHLORIDE
CARBONATES
CARBONIZED NOILS
CARBONIZERS
CARBONIZING
CARBONYL GROUPS
CARBOXYL CONTENT
CARBOXYL END GROUPS
CARBOXYL GROUPS
CARBOXYLATE SURFACTANTS
CARBOXYLIC ACIDS
CARBOXYMETHYLATED COTTON
CARBOXYMETHYLATION
CARBOXYMETHYLCELLULOSE
CARCASE
CARD CLEARERS
CARD CLOTHING
CARD CLOTHING FOUNDATION
CARD COUNT
CARD CROWN
CARD CYLINDERS
CARD CYLINDERS (JACQUARD
 KNITTING)
CARD DOFFERS
CARD FLATS
CARD GAUGE
CARD GRINDERS
CARD GRINDING

CARD NAILING
CARD ROLLERS
CARD ROLLS
CARD SLIVERS
CARD STRIPPING
CARD STRIPS
CARD WASTE
CARD WEBS
CARD WIRES
CARD YIELD
CARDED SLIVERS
CARDED YARNS
CARDER PART
CARDIGANS
CARDING
CARDING ACTION
CARDING DYNAMICS
CARDING EFFICIENCY
CARDING ENGINES
CARDING LEATHER
CARDING OIL
CARDING POWER
CARDING WOOL
CARDS
CARPET BACKING
CARPET DYEING
CARPET STUFFER YARNS
CARPETING
CARPETS
CARRIAGES (SPINNING MULE)
CARRICKMACROSS (LACE)
CARRIER (BRAIDS)
CARRIER DYEING
CARRIER RODS
CARRIERS (DYEING)
CARROTING
CARTONS
CARVED PILE FABRICS
CASABLANÇA SYSTEM
CASCADE WASHERS
CASEIN
CASEMENT
CASHMERE
CASING PLY
CASKET LININGS
CAST COATING
CASTING OFF
CATALYSIS
CATALYSTS
CATAPHORESIS
CATCH BARS
CATCHING
CATION EXCHANGE
CATION EXCHANGE RESINS
CATIONIC DYEING
CATIONIC DYES
CATIONIC SITES
CATIONIC SOFTENERS
CATIONIC SURFACTANTS
CATIONS
CAUCHY DISTRIBUTION
CAUSTIC (NOUN)
CAUSTIC POTASH
CAUSTIC SCOURING
CAUSTIC SODA
CAUSTICAIRE INDEX
CAUSTICAIRE SCALE
CAUSTICAIRE VALUE
CAVALRY TWILL
CAVITIES
CELACRIMP (TN)
CELAFIBRE (TN)
CELLOBIOSE
CELLOPHANE
CELLOTETRAOSE
CELLULAR FABRICS
CELLULAR PLASTICS
CELLULAR YARNS
CELLULASE
CELLULOSE
CELLULOSE ACETATE
CELLULOSE ACETATE BUTYRATE
CELLULOSE BLENDS
CELLULOSE DERIVATIVES
CELLULOSE DYEING
CELLULOSE DYES
CELLULOSE ESTER FIBER BLENDS
CELLULOSE ESTER FIBERS
CELLULOSE ESTERS
CELLULOSE ETHERS
CELLULOSE I
CELLULOSE II
CELLULOSE III
CELLULOSE IV
CELLULOSE NITRATE
CELLULOSE PHOSPHATE
CELLULOSE PULP
CELLULOSE REACTANTS
CELLULOSE XANTHATE
CELLULOSIC FIBERS
CELON (TN)
CEMENTING
CEMENTS
CENTER GUIDES
CENTER LOOPS
CENTER OF GRAVITY
CENTER OF MASS
CENTER SELVEDGES
CENTER SHED
CENTRALIZED CONTROL
CENTRIFUGAL DRYERS
CENTRIFUGAL DRYING
CENTRIFUGAL EXTRACTORS
CENTRIFUGAL FORCE
CENTRIFUGAL PUMPS
CENTRIFUGAL SPINNING
CENTRIFUGES
CENTRIPETAL FORCE

CERAMIC FIBERS
CERAMIC GUIDES
CERAMICS
CEREMONIAL CLOTHING
CERIUM COMPOUNDS
CERIUM SULFATE
CESIUM COMPOUNDS
CESIUM HYDROXIDE
CETYL ALCOHOL
CHAFE MARKS
CHAFER FABRICS
CHAIN DRAFT
CHAIN DRIVES
CHAIN FOLDING
CHAIN MERCERIZERS
CHAIN MOLECULES
CHAIN SEWING
CHAIN STITCHES
CHAINLESS MERCERIZERS
CHAINS
CHAMBRAY
CHANTILLY (LACE)
CHARACTER RECOGNITION
CHARACTERISTICS
CHARGE (ELECTRICAL)
CHARGING
CHARMANETTE SATIN
CHASE
CHECK SHIRTING
CHECK SPRINGS
CHECK STRAPS
CHECKER BOARD WEAVES
CHEESE CLOTH
CHEESE DYEING
CHEESE DYEING MACHINES
CHEESE WINDERS
CHEESE WINDING
CHEESES
CHELATES
CHELATING
CHELATING AGENTS
CHELATION
CHEMICAL ANALYSIS
CHEMICAL ANALYSIS (FIBER IDENTIFICATION)
CHEMICAL BONDS
CHEMICAL COMPOSITION
CHEMICAL ENGINEERING
CHEMICAL MODIFICATION (FIBERS)
CHEMICAL PROPERTIES
CHEMICAL REACTIONS
CHEMICAL RESISTANT LEATHER
CHEMICAL STABILITY
CHEMICAL TESTING
CHEMICALLY MODIFIED CELLULOSIC FIBERS
CHEMICALLY MODIFIED COTTON
CHEMICALLY MODIFIED PROTEIN FIBERS
CHEMICKING
CHEMILUMINESCENCE
CHEMISORPTION
CHENILLE (YARN)
CHENILLE AXMINSTER CARPETS
CHENILLE CARPETS
CHENILLE FABRICS (TUFTED FABRIC)
CHENILLE YARNS
CHEVIOT TWEED
CHEVIOT WOOL
CHEVRON TWILLS
CHI SQUARED DISTRIBUTION
CHIFFON
CHILDRENS WEAR
CHINCHILLA MACHINES
CHINTZ
CHIPPING
CHITIN
CHLORAL
CHLORATES
CHLORIDES
CHLORINATED HYDROCARBONS
CHLORINATED SOLVENTS
CHLORINATED WOOL
CHLORINATION (CHEMICAL REACTION)
CHLORINATION (SHRINK PROOFING)
CHLORINE
CHLORINE BLEACHING AGENTS
CHLORINE COMPOUNDS
CHLORINE DAMAGE
CHLORINE DIOXIDE
CHLORINE FASTNESS (COLOR)
CHLORINE FASTNESS (OF FINISH)
CHLORINE RETENTION
CHLORINE RETENTION DAMAGE
CHLORITES
CHLOROACETIC ACID
CHLOROACETYLAMINO DYES
CHLOROBENZOTHIAZOLE DYES
CHLOROCYANURIC ACID
CHLOROETHANOL
CHLOROFORM
CHLOROHYDRINS
CHLOROPROPANOL
CHLOROPYRIMIDINE DYES
CHLOROTRIAZINE DYES
CHLOROUS ACID
CHOPPING
CHROMA
CHROMATICITY DIAGRAM
CHROMATOGRAPHY
CHROME
CHROME DYEING
CHROME DYES
CHROME PRINTING COLORS
CHROMIUM COMPOUNDS
CHROMIUM FLUORIDE
CHROMOTROPISM
CHRYSOTILE

CIBACRON (TN)
CIBAPHASOL (TN)
CIE SYSTEM
CIGARETTE FILTERS
CINNAMIC ACID
CIRCULAR COMBING
CIRCULAR COMBS
CIRCULAR HOSIERY
CIRCULAR KNITTED FABRICS
CIRCULAR KNITTING
CIRCULAR KNITTING MACHINES
CIRCULAR LOOMS
CIRCULAR TRAVELERS
CIRCULARITY
CIRCULATION
CIRCULATION PUMPS
CIRCUMFERENCE
CITRIC ACID
CLASS (FIBERS)
CLASSING
CLAYS
CLEANERS
CLEANING
CLEANLINESS
CLEANNESS
CLEARANCES
CLEARERS (CARD)
CLEARERS (LINT)
CLEARERS (YARN)
CLEARING (YARN)
CLEARING LEASE
CLEARNESS
CLEMSON STRENGTH TESTING
CLEMSON STRENGTH TESTS
CLERICAL CLOTHES
CLICKERS
CLIP TENTERS
CLIPPINGS
CLIPSPOT WEAVES
CLOAKS
CLOCKWISE
CLOSE FACE FINISH
CLOSE WIND
CLOSE WINDING
CLOSED LOOP CONTROL
CLOSED SHED
CLOSED STITCHES
CLOSURES
CLOSURES (FOOTWEAR)
CLOTH
CLOTH ANALYSIS
CLOTH FELL
CLOTH GEOMETRY
CLOTH GUIDES
CLOTH INSPECTION
CLOTH ROLLS
CLOTH STRUCTURE
CLOTHING
CLUNY (LACE)
CLUTCHES
CMC
COACERVATE SYSTEMS
COAGULATING
COAGULATING BATHS
COAL TAR
COARSE YARNS
COARSENESS
COATED FABRICS
COATING (PROCESS)
COATING MACHINES
COATING RESINS
COATINGS (SUBSTANCES)
COATS
COBALT CHLORIDE METHOD
COBALT COMPOUNDS
COCKLE (DEFECT)
COCKLE (TAILORING)
COCKLE (YARN)
COCKLE YARNS
COCKLED BAR
COCKLED SELVEDGES
COCKLING
COCONUT DIETHANOLAMIDE
COCOONS
COEFFICIENT OF CROSS VISCOSITY
COEFFICIENT OF EXPANSION
COEFFICIENT OF FINENESS VARIATION
COEFFICIENT OF FRICTION
COEFFICIENT OF LENGTH VARIATION
COEFFICIENT OF PARTITION
COEFFICIENT OF VARIATION
COEFFICIENT OF VISCOSITY
COHESION
COILER CANS
COILERS
COIR
COLD CLIMATE FOOTWEAR
COLD DRAWING
COLD DYEING
COLD SETTING (SYNTHETICS)
COLLAGEN
COLLAGEN FIBERS
COLLAPSED BALLOON SPINNING
COLLAR STIFFENERS
COLLARS
COLLECTING
COLLECTION
COLLOID CHEMISTRY
COLLOIDAL ALUMINA
COLLOIDAL SILICA
COLLOIDS
COLOR
COLOR BLEEDING
COLOR BLINDNESS
COLOR CONTROL
COLOR DESIGNATION
COLOR FADING

COLOR GRADE
COLOR GRADING
COLOR MATCHING
COLOR MEASUREMENT
COLOR MEASUREMENT INSTRUMENTS
COLOR STANDARDS
COLOR TEMPERATURE
COLOR TESTING
COLOR TROUGH
COLOR VISION
COLOR YIELD
COLORANTS
COLORAY (TM)
COLORFASTNESS
COLORFASTNESS TESTING
COLORFASTNESS TO HEAT
COLORFASTNESS TO HOT PRESSING
COLORFASTNESS TO LIGHT TESTING
COLORIMETERS
COLORIMETRIC ANALYSIS
COLORIMETRY
COLORING
COMB BRUSHES
COMB CIRCLES
COMB CYLINDERS
COMB LEAD
COMB NOILS
COMB PINS
COMB SEGMENTS
COMB SPINNING
COMB TEAR
COMB TEARAGE
COMB WASTE
COMBAT SUITS
COMBED SLIVERS
COMBED TOPS
COMBED YARNS
COMBER LAP MACHINES
COMBER LAPS
COMBERS
COMBINATION STITCHES
COMBINATION WEAVES
COMBINATION WIND
COMBING
COMBING EFFICIENCY
COMBING SLIVERS
COMBING WOOL
COMBS
COMBS (LEASING)
COMFORT
COMMINUTION
COMMON (WOOL GRADE)
COMPACTING
COMPACTOR
COMPARISON
COMPATIBLE DYES
COMPENSATION
COMPENSATORS (CLOTH)
COMPILATIONS
COMPLETELY AUTOMATED SPINNING
COMPLEXING
COMPLIANCE
COMPONENTS
COMPOSITES
COMPOUND NEEDLES
COMPOUNDING
COMPRESSIBILITY
COMPRESSION
COMPRESSION MOLDING
COMPRESSION TESTING
COMPRESSIVE BUCKLING
COMPRESSIVE MODULUS
COMPRESSIVE SHRINKAGE
COMPRESSIVE SHRINKAGE MACHINES
COMPRESSIVE SHRINKING
COMPRESSIVE STRENGTH
COMPRESSIVE STRESS
COMPUTATIONS
COMPUTERS
COMPUTING
CONCENTRATION
CONCENTRATION GRADIENT
CONDENSATION (CHEMICAL)
CONDENSATION POLYMERIZATION
CONDENSATION POLYMERS
CONDENSED COUNT
CONDENSED NAPHTHALENE SULFONATES
CONDENSED YARNS
CONDENSER BOBBINS
CONDENSER SCREENS
CONDENSER SPUN YARNS
CONDENSER TAPES
CONDENSERS
CONDENSING (YARN MANUFACTURING)
CONDENSING FUNNELS
CONDITION (TEXTILES)
CONDITIONING
CONDUCTION
CONDUCTIVITY (ELECTRICAL)
CONDUCTIVITY (THERMAL)
CONDUCTOMETRY
CONDUCTOR BOXES
CONE DRAWING
CONE DYEING
CONE DYEING MACHINES
CONE WINDING
CONERS
CONES
CONFIDENCE COEFFICIENT
CONFIDENCE INTERVAL
CONFIDENCE LIMITS
CONFIGURATION
CONGRUENCE
CONICAL BEARINGS
CONING
CONJUGATE SPINNING
CONSTANT RATE PERIOD
CONSTANT SPEED

CONSUMPTION (MATERIAL)
CONTACT
CONTACT ANGLES
CONTACT AREA
CONTACT MOLDING
CONTAMINATION
CONTINENTAL COMBS
CONTINENTAL SYSTEM PROCESSING
CONTINUOUS BLEACHING
CONTINUOUS CONTROL
CONTINUOUS DRYING
CONTINUOUS DYEING
CONTINUOUS DYEING RANGE
CONTINUOUS FILAMENT YARNS
CONTINUOUS FINISHING
CONTINUOUS SCOURING
CONTINUOUS SPINNING (STAPLE FIBER)
CONTINUOUS TOP DYEING
CONTOURS
CONTRACTION
CONTRAST EFFECT
CONTROL
CONTROL APRONS
CONTROL CHARACTERISTICS
CONTROL CHARTS
CONTROL EQUIPMENT
CONTROL INSTRUMENTS
CONTROL PANELS
CONTROL POINTS
CONTROL SURFACES
CONTROL SYSTEMS
CONTROL THEORY
CONTROL VALVES
CONTROLLERS
CONTROLLING
CONVECTION
CONVENTIONAL PRACTICE
CONVERTED TICKING
CONVERTERS (TOW)
CONVEYOR BELTS
CONVEYOR BUCKETS
CONVEYORS
CONVOLUTIONS
COOLING
COOLING CAN
COP BUILD
COP DYEING
COP DYEING MACHINES
COP END EFFECT
COP WIND
COPES
COPOLYMERIZATION
COPOLYMERS
COPPER
COPPER 8-HYDROXYQUINOLATE
COPPER COMPOUNDS
COPPER FORMATE
COPPER NAPHTHENATE
COPPER NUMBER
COPPER PENTACHLOROPHENATE
COPS
CORDAGE
CORDAGE MACHINES
CORDS
CORDURA (TN)
CORDUROY
CORE (FIBER)
CORE BRAIDS
CORE SPINNING
CORE SPUN YARNS
CORED YARNS
CORES (CENTER)
CORFAM (TN)
CORING
CORING DEVICES
CORIOLIS ACCELERATION
CORIOLIS FORCE
CORK
CORN STARCH
CORONA
CORONIZING (TN)
CORRECTION
CORRELATION
CORRELATION COEFFICIENT
CORROSION
CORROSION INHIBITORS
CORROSION RESISTANCE
CORROSION TESTING
CORSET BATISTE
CORSETS
CORTEX
CORUNDUM
COST CONTROL
COSTS
COTS (ROLLERS)
COTSWOLD WOOL
COTTON
COTTON BALES
COTTON BLENDS
COTTON BOLLS
COTTON CARDING
COTTON CARDS
COTTON CLASS
COTTON CLASSING
COTTON COMBING
COTTON COMBS
COTTON COUNT
COTTON DYEING
COTTON DYES
COTTON FABRICS
COTTON FIBERS
COTTON GINNING
COTTON GINS
COTTON GRADE
COTTON GRADING
COTTON LASTING
COTTON LINT

COTTON LINTERS
COTTON PROCESSING
COTTON QUALITY
COTTON SEEDS
COTTON SPINNING
COTTON SPUN YARNS
COTTON SYSTEM PROCESSING
COTTON SYSTEM SPINNING
COTTON SYSTEM WARPING
COTTON THREADS
COTTON TYPE YARNS
COTTON WAX
COULIER MOTION (KNITTING)
COUNT
COUNT CROWN AND GAUGE
COUNT STRENGTH PRODUCT
COUNT TESTERS
COUNT TESTING
COUNTERCLOCKWISE
COUNTERFALLERS
COUNTERFLOW PROCESSES
COUNTERS
COUNTING
COUPLER CURVES
COUPLER POINTS
COUPLES
COUPLING (CHEMICAL)
COUPLINGS (MECHANICAL)
COURSE SPACING
COURSES
COURSES PER INCH
COURTAULDS CONVERTER
COUTIL
COVALENT BONDS
COVER
COVER FACTOR
COVER PLATES
COVER STITCHES
COVERALLS
COVERED ROLLS
COVERED SEAMS
COVERED YARNS
COVERT
COW HAIR
COWHIDE
COWS
CRABBING
CRABBING MACHINES
CRACK PROPAGATION
CRACKED SELVEDGES
CRACKS (FABRIC)
CRACKS (MECHANICAL)
CRADLES
CRANKS
CRASH TOWELING
CRATCH
CRAVATS
CREASE ACCEPTANCE
CREASE MARKS
CREASE RECOVERY
CREASE RESISTANCE
CREASE RESISTANT FABRICS
CREASE RETENTION
CREASES
CREASING
CREELING
CREELS
CREEP
CREEP RECOVERY
CREEP RESISTANCE
CREEP RUPTURE
CREEP RUPTURE STRENGTH
CREEP STRENGTH
CREEP TESTING
CREEP TESTS
CREPE
CREPE CORDS
CREPE EMBOSSING
CREPE FABRICS
CREPE SUZETTE
CREPE WEAVES
CREPE YARNS
CREPING
CREPON GEORGETTE
CREPSET (TN)
CRESLAN (TN)
CRETONNE
CRIMP
CRIMP AMPLITUDE (FIBER)
CRIMP BALANCE
CRIMP FIXING
CRIMP FIXING AGENTS
CRIMP FREQUENCY
CRIMP INDEX
CRIMP INTERCHANGE
CRIMP PERCENT
CRIMP RADIUS
CRIMP RATIO
CRIMP RECOVERY
CRIMP REMOVAL
CRIMP RETENTION
CRIMP TENDENCY
CRIMP TESTERS
CRIMPED YARNS
CRIMPERS
CRIMPING
CRIMPING MACHINES
CRIMPING POTENTIAL
CRIMPING TENDENCY
CRINKLE
CRINKLE TEXTURED YARNS
CRINKLE TYPE YARNS
CRINOLINE
CRITICAL MICELLE CONCENTRATION
CRITICAL MOISTURE CONTENT
CRITICAL VELOCITY
CROCHET NEEDLES
CROCHETED LACES

CROCHETING
CROCHETING MACHINES
CROCK TESTING
CROCK TESTS
CROCKFASTNESS
CROCKING
CROCKING RESISTANCE
CROPPING
CROSROL WEB CLEANERS
CROSROL WEB PURIFIER (TN)
CROSS DYEING
CROSS LAID WEBS
CROSS LAID YARN FABRICS
CROSS PLAITING
CROSS ROLLS
CROSS SECTIONAL AREA
CROSS SECTIONAL SHAPE METHOD
 (COTTON IMMATURITY)
CROSS SECTIONS
CROSS STITCHES
CROSS WINDING
CROSSBRED WOOL
CROSSLINK DENSITY
CROSSLINKING
CROSSLINKING AGENTS (FIBERS)
CROWNS
CROWNS (ROLL)
CRUSH RESISTANCE
CRUSH ROLLS
CRUSHING
CRYSTAL DEFECTS
CRYSTAL LATTICE
CRYSTALLINE ORIENTATION
CRYSTALLINE POLYMERS
CRYSTALLINE REGION
CRYSTALLINITY
CRYSTALLITES
CRYSTALLIZATION
CUEN
CUFFS
CUMULOFT (TN)
CUOXAM
CUPRAMMONIUM FLUIDITY
CUPRAMMONIUM HYDROXIDE
CUPRAMMONIUM RAYON
CUPREL (TN)
CUPRIC CHLORIDE
CUPRIC CHROMATE
CUPRIC COMPOUNDS
CUPRIC SULFATE
CUPRIC SULFITE
CUPRIETHYLENEDIAMINE HYDROXIDE
CUPROUS CHLORIDE
CUPROUS COMPOUNDS
CURING
CURING AGENTS
CURING OVENS
CURING RANGES
CURL
CURL PILE FABRICS
CURL TENDENCY
CURLATORS (TN)
CURLING
CURLING TENDENCY
CURRENT (ELECTRICAL)
CURRENTS
CURTAIN MACHINES
CURTAINS
CURVATURE
CURVED SEAMING
CURVILINEAR SCALE
CUSHIONS
CUT MARKS (SLASHING)
CUT PILE FABRICS
CUT PILE TUFTING MACHINES
CUT RUCHE
CUT SYSTEM
CUTICLE
CUTTERS
CUTTING
CUTTING (TAILORING)
CUTTING DRILLS
CUTTING MACHINES
CUTTING MARKERS
CUTTING RATIO
CUTTING RESISTANCE
CUTTING SHEARS
CUTTING TESTING
CYANATES
CYANIDES
CYANO COMPOUNDS
CYANOETHYLATED COTTON
CYANOETHYLATION
CYANOETHYLCELLULOSE
CYANURIC ACID
CYANURIC CHLORIDE
CYBERNETICS
CYCLIC COMPRESSION
CYCLIC STRESS
CYCLIC TENSION
CYCLOHEXANE, 1,4-
 BIS(METHYLAMINE)-
CYCLOSET YARNS (TN)
CYLINDER DRYERS
CYLINDER MACHINES
CYLINDER NEEDLES
CYLINDER PRINTING
CYLINDER SCREENS
CYLINDER STRIPS
CYLINDERS
CYLINDERS (CARD)
CYSTEINE
CYSTINE
CYSTINE BRIDGE
DABBING (COMBING)
DABBING BRUSHES
DACRON (TN)
DALEMBERTS FORCE

DAMAGE
DAMASK
DAMPENING
DAMPERS
DAMPING
DAMPING (MOISTURE)
DARLAN (TN)
DARNED LACES
DARNING
DARNING NEEDLES
DARTS
DARVAN (TN)
DASHPOTS
DATA REDUCTION
DAWSON WHEELS
DAYLIGHT
DEACETYLATED CHITIN
DEAD WEIGHT LOADING
DEAFNESS
DECATING
DECATING MACHINES
DECATIZING
DECELERATION
DECERESOL OT (TN)
DECOLORIZING
DECOLORIZING CARBON
DECOMPOSITION
DECORATIVE FABRICS
DECORTICATING
DECRYSTALLIZATION
DECRYSTALLIZED COTTON
DEFECTS
DEFERRED CURE METHOD
DEFLECTION
DEFOAMERS
DEFORMATION
DEFORMING
DEGRADATION
DEGRADATION PROPERTIES
DEGRADED CELLULOSE
DEGREASING
DEGREE OF CRYSTALLINITY
DEGREE OF ORIENTATION
DEGREE OF POLYMERIZATION
DEGREE OF SUBSTITUTION
 (CELLULOSE)
DEGREES OF FREEDOM
DEGUMMING
DEHAIRING
DEHYDRATION
DELAINE (WOOL CLASS)
DELAMINATING
DELAYED CURE
DELAYED ELASTIC RECOVERY
DELAYED RECOVERY
DELIQUESCENT AGENTS
DELIVERY
DELIVERY RATE
DELIVERY ROLLS
DELUSTERANTS
DELUSTERING
DELUSTERING AGENTS
DEMINERALIZING
DENIER
DENIMS
DENSITY
DENTS
DEODORANTS
DEODORIZING
DEODORIZING CARBON
DEPOLYMERIZATION
DEPOSITION
DERBY RIB
DERMATITIS
DESIGN
DESIGNING
DESIZING
DESIZING AGENTS
DESIZING MACHINES
DESORPTION
DETACHING
DETACHING EFFICIENCY
DETACHING ROLLS
DETECTION
DETERGENCY
DETERGENTS
DETERIORATION
DETERMINATION
DEUTERIUM
DEVELOPED DYES
DEVELOPERS (DYEING)
DEVELOPMENTS
DEVICES
DEW POINT
DEWAXING (COTTON)
DEWING
DEXTRAN
DEXTRINS
DEXTROSE
DHEU
DIACETALS
DIACETONE ACRYLAMIDE
DIACIDS
DIAGONAL STITCHES
DIAL NEEDLES
DIALDEHYDE CELLULOSE
DIALDEHYDE STARCH
DIALDEHYDES
DIALYSIS
DIAMETER
DIAMINES
DIAMOND BARRING
DIAMOND WEAVES
DIAPERS
DIAPHRAGM BURST TESTING
DIAPHRAGM BURST TESTS
DIAPHRAGMS
DIAZO COMPOUNDS

DIAZOBENZOYLATED COTTON
DIAZOMETHANE
DIAZOTIZATION
DIBASIC ACIDS
DIBORANE
DICHLOROBENZENE
DICHLOROCYANURIC ACID
DICHLOROHYDRIN
DICHLOROPROPANOL
DICHLOROPYRIMIDINE DYES
DICHLOROQUINOXALINE DYES
DICHLOROTRIAZINE DYES
DICHROIC RATIO
DICHROISM
DICHROMATES
DIE CUTTERS
DIELDRIN
DIELECTRIC CONSTANT
DIELECTRIC DRYERS
DIELECTRIC DRYING
DIELECTRIC HEAT SEALING
DIELECTRIC HEATING
DIELECTRIC PROPERTIES
DIELECTRIC STRENGTH
DIELECTRICS
DIENES
DIEPOXIDES
DIETHANOLAMIDE SURFACTANTS
DIETHANOLAMIDES
DIETHANOLAMINE
DIETHANOLAMINE SOAPS
DIETHYLAMINE
DIETHYLENEGLYCOL ESTER SURFACTANTS
DIFFERENCES
DIFFERENTIAL CONTROL
DIFFERENTIAL DISTRIBUTION (CROSSLINKING)
DIFFERENTIAL DYEING METHOD (COTTON IMMATURITY)
DIFFERENTIAL EQUATIONS
DIFFERENTIAL FRICTION
DIFFERENTIAL GEOMETRY
DIFFERENTIAL SHRINKAGE
DIFFERENTIAL THERMAL ANALYSIS
DIFFERENTIATION
DIFFRACTION
DIFFRACTION PATTERNS
DIFFUSION
DIFFUSION COEFFICIENT
DIFFUSION CONSTANT
DIFFUSIVITY
DIGITAL FIBROGRAPH
DIGLYCIDYL ETHERS
DIGLYCOL ESTER SURFACTANTS
DIHYDROXYETHYLENE UREA
DIISOCYANATES
DILATANCY
DILUTION
DIMENSIONAL ANALYSIS
DIMENSIONAL STABILITY
DIMENSIONS
DIMETHYL ACETAL
DIMETHYL ACETAMIDE
DIMETHYL FORMAMIDE
DIMETHYL SULFOXIDE
DIMETHYLOL ACETAMIDE
DIMETHYLOL DIHYDROXYETHYLENE UREA
DIMETHYLOL ETHYL CARBAMATE
DIMETHYLOL ETHYL TRIAZONE
DIMETHYLOL ETHYLENE UREA
DIMETHYLOL FORMAMIDE
DIMETHYLOL HYDROXYETHYLENE UREA
DIMETHYLOL HYDROXYPROPYLENE UREA
DIMETHYLOL ISOPROPYL CARBAMATE
DIMETHYLOL ITACONAMIDE
DIMETHYLOL METHYL CARBAMATE
DIMETHYLOL METHYL TRIAZONE
DIMETHYLOL PROPYLENE UREA
DIMETHYLOL UREA
DIMETHYLOL URON
DIMITY
DIOCTYL SULFOSUCCINATE
DIOLS
DIP BONDING
DIP COATING
DIPOLE MOMENT
DIPOLES
DIPPING
DIRECT COUNT
DIRECT DYEING
DIRECT DYES
DIRECT MOLDED SOLE (FOOTWEAR)
DIRECT PRINTING
DIRECT SPINNING
DIRECT SPINNING MACHINES
DIRECT SPUN YARNS
DIRECT TENSION DEVICES
DIRECTION (TWILL)
DIRT
DIRT REMOVAL
DISC TENSION
DISCHARGE (ELECTRIC)
DISCHARGE PRINTING
DISCOLORATION
DISCRETE DISTRIBUTION
DISINFECTANTS
DISINFECTION
DISLOCATIONS
DISORIENTATION
DISPERSANTS
DISPERSE DYEING
DISPERSE DYES
DISPERSE SYSTEMS
DISPERSING
DISPERSING AGENTS
DISPERSION (STATISTICAL)

DISPERSION FORCES
DISPERSIONS
DISPLACEMENT
DISSIPATION
DISSOLVING
DISTILLATION
DISTILLED WATER
DISTORTION
DISTRIBUTION
DISTRIBUTION FUNCTION
DISULFIDE BONDS
DISULFIDE INTERCHANGE
DISULFIDES (INORGANIC)
DISULFIDES (ORGANIC)
DITHIOCARBAMATES
DITHIONITES
DIVIDER ROLLS
DIVIDERS (CARD ROLLS)
DIVIDERS (KNITTING)
DIVIDING (KNITTING)
DIVINYL SULFONE
DIVISION
DMDHEU
DMEC
DMET
DMEU
DMHEU
DMHPU
DMPU
DMU
DMUR
DOBBIES
DOBBY HEADS
DOBBY LOOMS
DOBBY WEAVING
DOCTOR BLADES
DOCUMENTATION
DODECYL SULFATE
DOESKIN
DOFF
DOFFER BRUSHES
DOFFER COMBS
DOFFER STRIPS
DOFFERS (CARD)
DOFFERS (PACKAGE)
DOFFING (OF A CARD)
DOFFING (OPENER)
DOFFING (PACKAGE)
DOG LEGGED SELVEDGES
DOGBONE CROSS SECTION
DOGSTOOTH WEAVES
DOLLY
DOMESTIC SEWING MACHINES
DOMINANT WAVELENGTH
DONNAN EQUILIBRIUM
DONNAN THEORY
DOPE (POLYMER)
DOPE DYED YARNS
DOPE DYEING
DOPE DYEING EQUIPMENT
DORLON (TN)
DORSET WOOL
DOT DASH STITCHES
DOTTED MUSLIN
DOUBLE APRONS
DOUBLE BOW
DOUBLE BRAIDS
DOUBLE CLOTH
DOUBLE CURE METHOD
DOUBLE DIAGONAL STITCHES
DOUBLE FACED FABRICS (KNITTED)
DOUBLE HEADED NEEDLES
DOUBLE JAMMING
DOUBLE JERSEY
DOUBLE KNIT FABRICS
DOUBLE LOCKER MACHINES
DOUBLE PICKS
DOUBLE PIQUE
DOUBLE RING SYSTEM
DOUBLE ROVINGS
DOUBLE ROW STITCHES
DOUBLE SYSTEMS
DOUBLE TRUMPETS
DOUBLE TWIST SPINDLES
DOUBLED YARNS
DOUBLING (DRAFTING)
DOUBLING (TWISTING)
DOUBLING TRAVELERS
DOUP
DOUP ENDS
DOUP WEAVES
DOUP WEAVING
DOWN TIME
DOWNTWISTERS
DOWNTWISTING
DP
DRABNESS
DRAFT
DRAFT (PATTERN)
DRAFT CONSTANT
DRAFT CONTROL
DRAFT CONTROL SURFACES
DRAFT CUT
DRAFT RATIO
DRAFTING (PATTERN)
DRAFTING (STAPLE FIBER)
DRAFTING DYNAMICS
DRAFTING FORCE
DRAFTING FORCE TESTERS
DRAFTING MACHINES
DRAFTING RESISTANCE
DRAFTING ROLLS
DRAFTING SYSTEMS
DRAFTING THEORY
DRAFTING TWIST
DRAFTING WAVES
DRAFTING ZONE
DRAG

DRAPE
DRAPEOMETER
DRAPER SHUTTLELESS LOOMS (TN)
DRAPES
DRAVES TEST
DRAW
DRAW MECHANISM (KNITTING)
DRAW RATIO
DRAWERS
DRAWFRAMES
DRAWING (FILAMENT)
DRAWING (STAPLE FIBER)
DRAWING (WIRE)
DRAWING FRAMES
DRAWING IN
DRAWING IN MACHINES
DRAWING TENSION (FILAMENT)
DRAWN PIECE
DRAWN PILE FINISH
DRAWTWISTING
DRESDEN POINT
DRESS BANDINGS
DRESS DESIGN
DRESS FACE FINISH
DRESSCLOTHING
DRESSES
DRESSGOODS
DRESSING (WARP)
DRESSING SELVEDGES (LACE)
DRESSMAKING
DRESSWEAR
DRIFT
DRILLS (FABRICS)
DRIMARENE (TN)
DRIP DRY FABRICS
DRIP DRYING
DRIVEN ROLLS
DRIVES
DRIVING BANDS
DRIVING ROLLS
DROP BOX LOOMS
DROP PENETRATION TESTING
DROP PENETRATION TESTS
DROP WIRES
DROPPED STITCHES
DRUM (DYEING)
DRUM DRYERS
DRUM SCOURING MACHINES
DRUM WINDERS
DRUM WINDING
DRY BULB TEMPERATURE
DRY CLEANING
DRY CLEANING FASTNESS (COLOR)
DRY CLEANING FASTNESS (OF FINISH)
DRY CLEANING MACHINES
DRY CLEANING SOLVENTS
DRY CREASE RECOVERY
DRY CURING
DRY DECATING
DRY FINISHING
DRY IN BLEACHING
DRY POWDER BONDING
DRY RELAXATION
DRY RELAXING
DRY SPINNING
DRY STRENGTH
DRY WEIGHT
DRY WRINKLE RECOVERY
DRYERS
DRYING
DRYING CANS
DRYING MACHINES
DRYING TIME
DS (CELLULOSE)
DUCHESS (LACE)
DUCK
DUCTILITY
DUFFEL
DUFFLE
DULLNESS
DUMMY NEEDLES
DUMMY SLIDERS
DUNGAREES
DUNNAGE BAGS
DUO CARDS
DUO ROTH SYSTEM
DUOMEENS (TN)
DUPLEX PRINTING
DUPLEX PRINTING MACHINES
DUPLICATE PRINTING
DURABILITY
DURABLE FINISH (GENERAL)
DURABLE PRESS
DURABLE PRESS FINISHES
DURABLE PRESS TREATMENTS
DURAFIL (TN)
DURASPAN (TN)
DUSTING MACHINES
DYE ABSORPTION
DYE ADSORPTION
DYE AFFINITY
DYE BATHS
DYE FIXING
DYE FIXING AGENTS
DYE HOUSE
DYE KETTLES
DYE LIQUORS
DYE MIGRATION
DYE PENETRATION
DYE SOLUTIONS
DYE SPECKS
DYE UNIFORMITY
DYEABILITY
DYED FABRICS
DYED FIBER TRACER TECHNIQUES
DYED FIBER TRACERS
DYEFASTNESS

DYEING
DYEING (BY DYE CLASSES)
DYEING (BY ENVIRONMENTAL CONDITIONS)
DYEING (BY FIBER CLASSES)
DYEING (BY MATERIAL ASSEMBLY)
DYEING (BY PROCESS FLOW)
DYEING (FOR EFFECT)
DYEING ASSISTANTS
DYEING AUXILIARIES
DYEING DEFECTS
DYEING MACHINES
DYEING MACHINES (BY ENVIRONMENTAL CONDITIONS)
DYEING MACHINES (BY MATERIAL ASSEMBLY)
DYEING MACHINES (BY PROCESS FLOW)
DYEING PROPERTIES (GENERAL)
DYEING RATE
DYEING THEORY
DYES
DYES (BY CHEMICAL CLASSES)
DYES (BY FIBER CLASSES)
DYESTUFFS
DYNALOFT (TN)
DYNAMIC ABSORPTION (WATER)
DYNAMIC ABSORPTION TESTING
DYNAMIC ABSORPTION TESTS
DYNAMIC BALANCING
DYNAMIC BALANCING MACHINES
DYNAMIC BUCKLING
DYNAMIC CHARACTERISTICS
DYNAMIC DRAPE
DYNAMIC FRICTION
DYNAMIC MODULUS
DYNAMIC MODULUS TESTERS
DYNAMIC RECOVERY
DYNAMIC RESPONSE
DYNAMICS
DYNAMOMETERS
DYNEL (TN)
EAR MUFFS
EASY CARE FABRICS
EASY CARE FINISHES
EASY CARE TREATMENTS
ECCENTRICITY
ECONOMICS
EDDIES
EDGE CRIMPED YARNS
EDGE CRIMPING
EDGE CRIMPING MACHINES
EDGE GUIDES
EDGEING (LACE)
EDTA
EDUCATION
EFFECT DYEING
EFFECT OF WEATHER
EFFECTIVE LENGTH (FIBER)
EFFECTIVE POROSITY
EFFICIENCY (PROCESS)
EFFICIENCY (STRUCTURAL)
EFFLUENT TREATMENT
EFFLUENTS
EGYPTIAN COTTON
EIDERDOWNS
EIGHTLOCK (KNITTED FABRICS)
EJECTION
ELASTIC DEFORMATION
ELASTIC FABRICS
ELASTIC FOAMS
ELASTIC LIMIT
ELASTIC LIQUIDS
ELASTIC MODULUS (COMPRESSIVE)
ELASTIC MODULUS (TENSILE)
ELASTIC PERFORMANCE
ELASTIC RECOVERY
ELASTIC REGION
ELASTIC STRAIN
ELASTICITY
ELASTIQUE
ELASTOMER FIBERS
ELASTOMERS
ELECTRIC BLANKETS
ELECTRIC BRAKES
ELECTRIC CONTROLLERS
ELECTRIC CURRENT
ELECTRIC EQUIPMENT
ELECTRIC HEATING (RESISTANCE)
ELECTRICAL CONDUCTIVITY
ELECTRICAL INPUT TRANSDUCERS
ELECTRICAL OUTPUT TRANSDUCERS
ELECTRICAL PROPERTIES
ELECTRICAL RESISTIVITY
ELECTRO-OSMOSIS
ELECTRODEPOSITION
ELECTRODIALYSIS
ELECTROKINETIC POTENTIAL
ELECTROLYSIS
ELECTROLYTES
ELECTROMAGNETIC DEVICES
ELECTROMECHANICAL DEVICES
ELECTROMETERS
ELECTRON MICROSCOPE
ELECTRON MICROSCOPY
ELECTRON SPIN RESONANCE
ELECTRONIC INSTRUMENTS
ELECTRONIC MULES
ELECTRONIC SLUB CATCHERS
ELECTRONICS
ELECTROPHORESIS
ELECTROPLATING
ELECTROSTATIC BONDS
ELECTROSTATIC FLOCKING
ELECTROSTATICS
ELLIPTICAL TRAVELERS
ELLIPTICITY
ELMENDORF TEAR STRENGTH

ELMENDORF TEAR TESTING
ELMENDORF TEAR TESTS
ELONGATION
ELONGATION AT BREAK
ELONGATION PERCENT
EMBEDDING MEDIUM
EMBOSSED CREPE
EMBOSSED FABRICS
EMBOSSING
EMBOSSING CALENDERS
EMBROIDERY
EMBROIDERY FLOSS
EMBROIDERY MACHINES
EMERY
EMISSIVITY
EMPIRICAL ANALYSIS
EMPTY PACKAGES
EMPTYING
EMULSIFYING
EMULSIFYING AGENTS
EMULSION BINDERS (NONWOVENS)
EMULSION BINDERS (PIGMENTS)
EMULSION DYEING
EMULSION POLYMERIZATION
EMULSION POLYMERS
EMULSION PRINTING
EMULSION STABILIZERS
EMULSIONS
END BREAKAGES
END FENT
END GROUPS
END SPACING
END USE PROPERTIES
END USES
ENDOCUTICLE
ENDS (WEAVING)
ENDS DOWN
ENDS CUT
ENDS PER INCH
ENERGY
ENERGY ABSORPTION
ENERGY OF DEFORMATION
ENERGY OF RETRACTION
ENGRAVED CYLINDERS
ENGRAVING
ENKALOFT (TN)
ENKALON (TN)
ENKALURE (TN)
ENKATRON (TN)
ENTANGLEMENTS
ENTERING LEASE
ENTHALPY
ENTHALPY OF DYEING
ENTHALPY OF HYDRATION
ENTHALPY OF REACTION
ENTHALPY OF SOLUTION
ENTHALPY OF SORPTION
ENTHALPY OF SWELLING
ENTHALPY OF WETTING
ENTROPY
ENTROPY OF DYEING
ENTROPY OF REACTION
ENZYMATIC DEGRADATION
ENZYMES
EPICHLOROHYDRIN
EPICUTICLE
EPIHALOGENOHYDRIN
EPOXIDES
EPOXY COMPOUNDS
EPOXY RESINS
EQUALIZER (DRAFTING)
EQUALIZING (DRAFTING)
EQUATIONS
EQUILIBRIUM (CHEMICAL)
EQUILIBRIUM (DYEING)
EQUILIBRIUM (PHYSICAL)
EQUILIBRIUM MOISTURE CONTENT
EQUIPAGE
EQUIPMENT
EQUIVALENT COUNT
ERRORS
ESTERIFICATION
ESTERS
ESTIMATION
ESTRON (TN)
ETHANOL
ETHERALCOHOL SULFATES
ETHERIFICATION
ETHERS
ETHODUOMEENS (TN)
ETHOFATS (TN)
ETHOMEENS (TN)
ETHOMIDS (TN)
ETHOXY PROPOXY BLOCK POLYMERS
ETHOXYLATED ALKYL PHENOLS
ETHOXYLATION
ETHYL ACETATE
ETHYL ACRYLATE
ETHYL ALCOHOL
ETHYL MERCAPTAN
ETHYL TRIAZONE
ETHYLATION
ETHYLCELLULOSE
ETHYLENE CARBONATE
ETHYLENE CHLOROHYDRIN
ETHYLENE DIBROMIDE
ETHYLENE GLYCOL
ETHYLENE IMINE
ETHYLENE OXIDE
ETHYLENE UREA
ETHYLENE UREA FORMALDEHYDE
 CONDENSATES
ETHYLENEDIAMINE
ETHYLENEDIAMINE TETRAACETIC ACID
EULER BUCKLING
EVALUATION
EVAPORATION
EVENERS

EVENESS
EVEREVEN (DRAFTING)
EVERLASTING (LACE)
EWNN
EXAMINATION
EXHAUSTION (DYEING)
EXHAUSTION (MECHANICAL)
EXHAUSTION (WET PROCESS)
EXHAUSTION RATE (DYEING)
EXOCUTICLE
EXPANDED PLASTICS
EXPANDERS
EXPANSION TANKS
EXPERIMENTAL ANALYSIS
EXPERIMENTAL DESIGN
EXPERIMENTAL STRESS ANALYSIS
EXPERIMENTATION
EXPERIMENTS
EXPONENTIAL DISTRIBUTION
EXTENSIBILITY
EXTENSION
EXTENSION AT BREAK
EXTRACT PRINTING
EXTRACTION
EXTRAPOLATION
EXTRUDERS
EXTRUSION
EXTRUSION COATING
EXTRUSION PUMPS
EYELET EMBROIDERY
EYELETS (GUIDES)
F DISTRIBUTION
FABRIC ANALYSIS
FABRIC BOLTS
FABRIC COVER
FABRIC CRIMP
FABRIC CROSS SECTIONS
FABRIC DEFECTS
FABRIC DESIGN
FABRIC DISTORTION (TAILORING)
FABRIC GEOMETRY
FABRIC GUIDES
FABRIC INSPECTION
FABRIC INTERSTICES
FABRIC LAMINATES
FABRIC PROPERTIES
FABRIC PROPERTIES (AESTHETIC)
FABRIC PROPERTIES (DEGRADATION)
FABRIC PROPERTIES (MECHANICAL)
FABRIC PROPERTIES (PHYSICAL EXCLUDING MECHANICAL)
FABRIC PROPERTIES (STRUCTURAL)
FABRIC REINFORCED COMPOSITES
FABRIC RELAXATION
FABRIC SHRINKAGE
FABRIC STIFFENERS
FABRIC STRENGTH
FABRIC STRUCTURE
FABRIC TO FABRIC FRICTION
FABRIC TO FABRIC LAMINATES
FABRIC UPPERS (FOOTWEAR)
FABRIC WEIGHT
FABRICS
FABRICS (ACCORDING TO FIBER)
FABRICS (ACCORDING TO STRUCTURE)
FABRICS (BRAIDS)
FABRICS (BY END USES)
FABRICS (COATED)
FABRICS (COMPOSITES)
FABRICS (FLAT WOVEN)
FABRICS (FLOCKED)
FABRICS (KNITTED)
FABRICS (LACE)
FABRICS (MALI)
FABRICS (MULTIPLE LAYER)
FABRICS (NET)
FABRICS (NONWOVEN)
FABRICS (PILE)
FABRICS (TUFTED)
FABRICS (WOVEN)
FACE (FABRIC)
FACE FINISH
FACE FINISHED WOOLENS
FACE GOODS
FACE YARNS
FACILITIES
FACING
FACTORS (PHYSICAL)
FADEOMETER (TN)
FADING
FAILLE
FALL PLATE FABRICS
FALLER BARS
FALLER SCREWS
FALLER SPEED
FALLERS (GILL BOXES)
FALLERS (MULE)
FALLING RATE PERIOD
FALSE DRAFT
FALSE TWIST
FALSE TWIST SPINDLES
FALSE TWIST TUBES
FALSE TWIST YARNS
FALSE TWISTERS
FALSE TWISTING (EXCEPT TEXTURING)
FALSE TWISTING (TEXTURING)
FANCY (NOUN)
FANCY KNITTING
FANCY LENO
FANCY TWILLS
FANCY YARNS
FASHION (APPAREL)
FASHION MODELS
FASHIONING
FASHIONING MARKS
FASTENERS
FASTENING

FASTNESS (COLOR)
FATIGUE
FATIGUE RESISTANCE
FATIGUE RUPTURE
FATIGUE TESTING
FATIGUES
FATS
FATTY ACIDS
FATTY ALCOHOL SULFATES
FATTY ALCOHOLS
FATTY DIGLYCERIDES
FATTY KETONES
FATTY MONOGLYCERIDES
FAULTS
FEARNOUGHTS
FEED BOXES
FEED DOGS
FEED LATTICES
FEED PLATES
FEED RATE
FEED ROLLS
FEEDBACK CONTROL
FEEDER BOXES
FEEDERS
FEEL
FEELER MECHANISMS
FELL (CLOTH)
FELLED SEAMS
FELLING
FELT BOOTS
FELT PROOFING
FELTED FABRICS
FELTING
FELTING MACHINES
FELTING RESISTANCE
FELTING SHRINKAGE
FELTS
FERRIC CHLORIDE
FERRIC COMPOUNDS
FERRIC TARTRATE
FERROUS COMPOUNDS
FETTLING
FETTLING COMBS
FETTLING WASTE
FIBER (TRACER)
FIBER ANALYSIS
FIBER ARRANGEMENT
FIBER ARRAY
FIBER ASSEMBLIES
FIBER BLENDS
FIBER BREAKAGE
FIBER BUNDLE STRENGTH
FIBER BUNDLE TESTING
FIBER COARSENESS
FIBER CONFIGURATION
FIBER CONTENT
FIBER CONTROL
FIBER COUNT
FIBER CRIMP
FIBER CROSS SECTIONS
FIBER CUTTING
FIBER DENIER
FIBER DIAGRAM
FIBER DIAMETER
FIBER ENDS
FIBER ENTANGLEMENTS
FIBER EXTENT
FIBER FIBER FRICTION
FIBER FINENESS
FIBER FLOCK
FIBER FRICTION
FIBER G (TN)
FIBER GEOMETRY
FIBER HOOKS
FIBER IDENTIFICATION
FIBER IMMATURITY
FIBER LENGTH
FIBER LENGTH DETERMINATION
FIBER LENGTH DISTRIBUTION
FIBER LENGTH DISTRIBUTION TESTING
FIBER MATURITY
FIBER MIGRATION
FIBER OPTICS
FIBER ORIENTATION
FIBER PROPERTIES
FIBER PROPERTIES (GEOMETRIC)
FIBER QUALITY
FIBER REACTIVE COMPOUNDS
FIBER REACTIVITY
FIBER REINFORCED COMPOSITES
FIBER REINFORCED PLASTICS
FIBER SHUFFLING
FIBER SLIPPAGE
FIBER SOUNDNESS
FIBER SPINNING
FIBER STRENGTH
FIBER STRUCTURE
FIBER SURFACE
FIBER TENDERNESS
FIBER TO METAL FRICTION
FIBER TWIST
FIBER V (TN)
FIBER WEAKNESS
FIBER WITHDRAWAL FORCE
FIBER WOVEN (TN)
FIBERGLAS (TN)
FIBERS
FIBERS (BICOMPONENT)
FIBRIL REVERSALS
FIBRILLAR STRUCTURE
FIBRILLATION
FIBRILS
FIBRO (TN)
FIBROCETA (TN)
FIBROGRAPH
FIBROGRAPH MEAN LENGTH
FIBROIN (SILK)

FIBROLANE (TM)
FIBROUS MATERIALS
FICKIAN DIFFUSION
FIELD EMISSION MICROSCOPY
FIELDEN WALKER IRREGULARITY
 TESTER (TM)
FIGURED FABRICS
FIGURED REPP
FILAMENT COUNT
FILAMENT DENIER
FILAMENT WOUND COMPOSITES
FILAMENT YARNS
FILAMENTS
FILET (LACE)
FILLER
FILLET CLOTHING
FILLING
FILLING BANDS
FILLING BARS
FILLING EFFECT (TWILL)
FILLING FACE
FILLING FLOATS
FILLING KNITTED FABRICS
FILLING KNITTING
FILLING KNITTING MACHINES
FILLING SHRINKAGE
FILLING STRAIGHTENING MACHINES
FILLING STREAKS
FILLING TENSION
FILLING WARP BLENDS
FILLING WIND
FILLING YARNS
FILLINGWISE STRETCH
FILM READING
FILMS
FILTER CAKES
FILTER CLOTH
FILTER FABRICS
FILTER PACKS
FILTER PAPER
FILTERS (ELECTRICAL)
FILTERS (FLUID)
FILTRATION
FINDINGS
FINE (WOOL GRADE)
FINE MEDIUM (WOOL GRADE)
FINE STRUCTURE (FIBERS)
FINE WOOL
FINE YARNS
FINENESS
FINENESS TESTERS
FINENESS TESTING
FINGER TENSION
FINGERING YARNS
FINISH (AGENTS)
FINISH (PROCESS)
FINISH (PROPERTY)
FINISH (SUBSTANCE ADDED)
FINISHED FABRICS
FINISHER GILLING
FINISHER PART
FINISHING AGENTS
FINISHING MACHINERY (GENERAL)
FINISHING PROCESS (GENERAL)
FIRE PROOFING AGENTS
FIRE PROOFING TREATMENTS
FIRE RESISTANCE
FIRE RESISTANCE TESTERS
FIRE RESISTANCE TESTING
FIRE RETARDANCY
FIRE RETARDANCY AGENTS
FIRE RETARDANCY TREATMENTS
FIRING (REFRACTORY)
FIRMNESS
FIRMOVISCOSITY
FIRST ORDER TRANSITION
 TEMPERATURE
FISHING LINES
FIT
FITTERS (GARMENTS)
FITTING
FIXED CAPACITORS
FIXED REEDS
FIXING (DYES)
FLAGS
FLAKE YARNS
FLAKING
FLAME BACKING
FLAME BONDING
FLAME LAMINATING
FLAME PROOFING AGENTS
FLAME RESISTANCE
FLAME RETARDANTS
FLAMEPROOF (PROPERTY)
FLAMES (NOVELTY YARNS)
FLAMMABILITY
FLAMMABILITY TESTERS
FLAMMABILITY TESTING
FLANGES
FLANNEL
FLASH AGING
FLASH POINT
FLAT BRAIDS
FLAT CHAINS (CARD)
FLAT CLOTHING
FLAT CROSS SECTION
FLAT KNITTED FABRICS
FLAT KNITTING
FLAT KNITTING MACHINES
FLAT PRESSING
FLAT SCREEN PRINTING
FLAT SETTINGS (CARD)
FLAT SPOTTING (TIRES)
FLAT STRIPPING BRUSHES
FLAT STRIPPING COMBS
FLAT STRIPS
FLAT TOP CARDS
FLAT WOVEN FABRICS

FLATLOCK STITCHES
FLATS
FLAWS
FLAX
FLAX SYSTEM PROCESSING
FLEECES
FLEXIBILITY
FLEXIBLE FILM LAMINATING
FLEXIBLE PLUNGER MOLDING
FLEXING
FLEXURAL FATIGUE
FLEXURAL RIGIDITY
FLEXURAL STRENGTH
FLIPPER FABRICS
FLOAT (OPERATION)
FLOAT DESIGN
FLOAT LENGTH
FLOAT STITCHES (KNITTING)
FLOATING (KNITTING)
FLOATING FIBERS
FLOATS
FLOCK
FLOCK COATING
FLOCK COTS
FLOCK POWDERS
FLOCK PRINTING
FLOCKED CARPETS
FLOCKED FABRICS
FLOCKING
FLOUNCING (LACE)
FLOW (FLUIDS)
FLOW (MATERIALS)
FLOW (PLASTIC)
FLOW CONTROL
FLUCTUATIONS
FLUFLON (TN)
FLUID DRIVES
FLUID FLOW
FLUID FRICTION
FLUID MECHANICS
FLUID SPINNING
FLUID TWISTING
FLUIDICS
FLUIDITY
FLUIDIZED BED DRYERS
FLUIDIZED BEDS
FLUORESCENCE
FLUORESCENCE MICROSCOPY
FLUORESCENT BRIGHTENERS
FLUORESCENT DYES
FLUORESCENT LIGHT
FLUORIDES
FLUORINE COMPOUNDS
FLUOROCARBON COMPOUNDS
FLUOROCARBON CONTAINING POLYMERS
FLUOROCARBON FIBERS
FLUOROCARBON POLYMERS
FLUOROCARBONS
FLUTED ROLLS
FLUTES
FLY (GARMENT)
FLY (WASTE)
FLY BEAMS (LOOM)
FLY COMBS
FLY SHUTTLES
FLYER FRAMES
FLYER LEAD
FLYER LEG
FLYER SPINNING
FLYER SPINNING FRAMES
FLYERS
FLYSHOT LOOMS
FOAM BACKED FABRICS
FOAM BACKING
FOAM BONDING
FOAM LAMINATES
FOAM RUBBER
FOAM YARNS
FOAMING
FOAMING AGENTS
FOAMS
FOAMS (FROTH)
FOCUSING
FOGGY YARNS
FOGMARKING
FOIL
FOLDED YARNS
FOLDING (OF CLOTH)
FOLDING (TWISTING)
FOLDING MACHINES (FOR CLOTH)
FOLK WEAVES
FOLLICLE
FOLLOW THE LEADER CRIMP
FOOD DYES
FOOTSTEP BEARINGS
FOOTWEAR
FOOTWEAR COMPONENTS
FORCE
FORCED CONVECTION DRYERS
FORECASTING
FORM PERSUASIVE GARMENTS
FORMALDEHYDE
FORMALDEHYDE PRECONDENSATES
FORMALDEHYDE SULFOXYLATES
FORMALS
FORMATION
FORMIC ACID
FORMING
FORMULAS
FORMULATIONS
FORTE MOISTURE METER
FORTISAN (TN)
FORTREL (TN)
FOULARD (DYEING)
FOULARD (FABRIC)
FOUNDATION GARMENTS
FOUR NEEDLE COVER STITCHES
FOURDRINIER MACHINES

FRACTIONATION
FRACTURE
FRACTURING
FRAMES (MACHINES)
FRAMING
FRAZIER PERMEOMETER
FREE ENERGY
FREE ENERGY OF DYEING
FREE ENERGY OF REACTION
FREE ENERGY OF SOLUTION
FREE ENERGY OF SWELLING
FREE ENERGY OF WETTING
FREE RADICALS
FREE VOLUME
FREEZE DRYING
FREEZING
FREEZING POINT
FRENCH COMBING
FRENCH COMBING WOOL
FRENCH COMBS
FRENCH CREPE CORDS
FRENCH DOUBLE PIQUE
FRENCH SEAMS
FRENCH SYSTEM PROCESSING
FREQUENCY
FREQUENCY ANALYSERS
FREQUENCY DENSITY
FREQUENCY DISTRIBUTION
FREQUENCY DISTRIBUTION (FIBER LENGTH)
FREQUENCY RESPONSE
FREUNDLICH ISOTHERM
FRICTION
FRICTION CALENDERING
FRICTION COEFFICIENT
FRICTION LET OFF
FRICTION THEORY
FRICTIONAL CHARACTERISTICS
FRICTIONAL FORCE
FRILLS (NOVELTY YARNS)
FRINGE (NARROW FABRIC)
FRINGED FIBRILS
FRINGED MICELLES
FRL DRAPE TESTER
FROCKS
FRONT CROSSING HEDDLES
FRONT ROLLERS
FRONT ROLLS
FRONTS
FROSTING
FRUIT FIBERS
FTORLON (TN)
FUGITIVE DYES
FUGITIVE TINTS
FULL CARDIGAN FABRICS
FULL FACE FINISH
FULL PACKAGES
FULL SEAM WIDTH
FULL SET
FULLING
FULLING MACHINES
FULLY FASHIONED FABRICS
FULLY FASHIONED HOSIERY
FULLY FASHIONED HOSIERY MACHINE
FULLY FASHIONED KNITTING
FULLY FASHIONED KNITTING MACHINES
FULLY FASHIONED STOCKINGS
FUME FADING
FUNGICIDES
FUNGISTATS
FUNGUS
FUNGUS RESISTANCE
FUR
FURNISHING FABRICS
FURNISHINGS
FURNITURE
FUSING (MELTING)
FUSION (MELTING)
FUSION CONTRACTION SHRINKAGE
FUZZ
FUZZ RESISTANCE
FUZZINESS
FUZZING
FUZZY MOTES
GABARDINE
GADGETS
GAIN (WINDING)
GALLING
GALLOON
GALOSHES
GAMMA DISTRIBUTION
GAMMA RADIATION
GAMMA RAYS
GARDINOL (TN)
GARMENT COMPONENTS
GARMENT DESIGN
GARMENT DYEING
GARMENT DYEING MACHINES
GARMENT MANUFACTURE
GARMENT MANUFACTURING MACHINES
GARMENTS
GARNETT WIRES
GARNETTING
GARNETTING MACHINES
GARTER BANDS
GARTER BELTS
GARTERS
GAS (FUEL)
GAS FUME FASTNESS
GAS FUME FASTNESS TESTING
GAS HEATERS
GAS HEATING (GAS FUEL)
GAS PHASE
GASKETS
GATE TENSION
GATHERING
GATHERS (TAILORING)

GATTERWALKING (FULLING MACHINES)
GAUGE (WIRE)
GAUGE LENGTH
GAUSSIAN DISTRIBUTION
GAUZE
GAUZE WEAVES
GEAR CRIMPED YARNS
GEAR CRIMPING
GEAR DRIVES
GEARS
GEIGER COUNTERS
GELATINE
GELLATION
GELLING
GELS
GENERAL SOLUTIONS
GENOA VELVET
GEOMETRIC POROSITY
GEOMETRICAL PROPERTIES
GEOMETRY
GEORGETTE
GERMICIDES
GIANT SADDLE STITCHES
GIGGING
GILBERT-RIDEAL THEORY
GILL BOXES
GILL DRAFTING
GILL PREPARING
GILLING
GIN CUT
GIN FALL
GINGHAM
GINNING
GINS (COTTON)
GIRDLES
GLASS
GLASS CLOTH
GLASS FIBERS
GLASS FILAMENTS
GLASS MAT
GLASS RUBBER TRANSITION
GLASS TRANSITION TEMPERATURE
GLAUBERS SALT
GLAZE
GLAZING
GLOSPAN (TN)
GLOSS
GLOVE LININGS
GLOVES
GLUCOSE
GLUES
GLUING
GLUTARALDEHYDE
GLYCERIN
GLYCERIN DICHLOROHYDRIN
GLYCERIN MONOCHLOROHYDRIN
GLYCEROL
GLYCEROPHOSPHATES
GLYCERYL ESTER SURFACTANTS
GLYCIDYL (PREFIX)
GLYCIDYL COMPOUNDS
GLYCOL ESTER SURFACTANTS
GLYCOLIC ACID
GLYCOLS
GLYOXAL
GO THROUGH MACHINES
GOAT BREEDING
GOAT HAIR
GOATS
GOBELIN TAPESTRIES
GODET ROLLS
GOING PART (LOOM)
GOLD CHLORIDE
GOLD COMPOUNDS
GOLDEN CAPROLAN (TN)
GOOD MIDDLING (COTTON GRADE)
GOOD ORDINARY (COTTON GRADE)
GOTHIC TAPESTRIES
GOWNS
GRAB STRENGTH
GRADE (FIBERS)
GRADE STANDARDS
GRADING
GRAFT COPOLYMERS
GRAFT POLYMERIZATION
GRAFT POLYMERS
GRAFTING
GRAIN BOUNDARIES
GRAMS PER DENIER
GRANITE WEAVES
GRAPHICAL SOLUTIONS
GRAPHITE
GRAPHITE FIBERS
GRAPHS
GRASS BLEACHING
GRAVIMETRIC ANALYSIS
GRAVITATION
GREASE
GREASE RECOVERY
GREASY WOOL
GRECIAN ALHAMBRA
GRECIAN WEAVES
GREENFIELD CONVERTER
GREX
GREY GOODS
GREY SCALE
GRID BARS
GRILENE (TN)
GRINDING (COMMINUTION)
GRINDING (MATERIAL REMOVAL)
GRINDING ROLLS (CARDS)
GRIPPERS (YARN)
GRIT
GROOVES
GROUND HEDDLES
GROWTH (FABRIC)
GUARD HAIRS
GUARDING DEVICES

GUARDS (ATHLETIC CLOTHING)
GUIDE BARS
GUIDE EYES
GUIDERS (CLOTH)
GUIDES
GUIPURE
GUMS
GUNNY SACKS
GURLEY PERMEOMETER
GUSSETS (FOOTWEAR)
GUT (COLLAGEN)
GUT THREAD
GYMP
HACK STANDS
HACKLE COMBS
HAIR
HAIR DYEING (HUMAN HAIR)
HAIR PROCESSING MACHINES
HAIR SETTING
HAIRINESS
HALCHING
HALF CARDIGAN FABRICS
HALF HOSE
HALF HOSIERY
HALF LAP
HALF SET
HALIDES
HALLE SEYDEL STRETCH BREAKER
HALOGEN COMPOUNDS
HALOGENATION
HALOHYDRINS
HALTERS
HAMBURG (LACE)
HAMPSHIRE WOOL
HAND
HAND COMBING
HAND KNITTING
HAND KNITTING YARNS
HAND LOOMS
HAND MADE LACES
HAND PICK STITCHES
HAND SADDLE STITCHES
HAND SEWING
HAND WEAVING
HAND WOVEN FABRICS
HANDLE (TEXTILE)
HANK DYEING
HANK DYEING MACHINES
HANK NUMBER
HANK SIZING
HANKS
HARD WATER
HARDENERS (FELTS)
HARDENING (FELTS)
HARDNESS
HARES
HARNESS CHAINS
HARNESS CORDS
HARNESS DRAFT
HARNESS DROPS
HARNESS LEVERS
HARNESS SKIPS
HARNESS STRAPS
HARNESSES
HARROW FORKS (SCOURING)
HARSHNESS
HART MOISTURE METER
HARVARD
HATS
HAWSERS
HEADGEAR
HEADING (NARROW FABRIC)
HEADS (TOP)
HEADSTOCKS
HEALDS
HEART LOOP
HEART VALVES (FABRIC)
HEAT
HEAT BONDING
HEAT CAPACITY
HEAT CONDUCTIVITY
HEAT CONTENT
HEAT EXCHANGERS
HEAT OF DYEING
HEAT OF FUSION
HEAT OF HYDRATION
HEAT OF REACTION
HEAT OF SOLUTION
HEAT OF SWELLING
HEAT OF WETTING
HEAT RESISTANCE
HEAT RESISTANCE (HIGH TEMPERATURE)
HEAT SEALING
HEAT SETTING (DRY)
HEAT SETTING (MOIST)
HEAT SETTING (SYNTHETICS)
HEAT SETTING MACHINES
HEAT STABILIZERS
HEAT TRANSFER
HEAT TREATMENT
HEATING
HEATING EQUIPMENT
HEAVY METAL SOAPS
HEAVY WEIGHT
HEDDLE BARS
HEDDLE EYES
HEDDLE HOOKS
HEDDLE SMASH
HEDDLES
HEELS (FOOTWEAR)
HEELS (HOSIERY)
HEELS (STOCKINGS)
HEIGHT
HEILMANN COMBS
HELANCA (TN)
HELICAL CRIMP
HELIX ANGLE

HELIX REVERSALS
HELMETS
HEM SEAMS
HEMIACETALS
HEMICELLULOSE
HEMMING
HEMP
HEMS
HERCULON (TN)
HERRINGBONE TWILLS
HESSIAN
HETEROCYCLIC COMPOUNDS
HEXAMETHYLENE DIISOCYANATE
HEXAMETHYLENEDIAMINE
HEXAMETHYLOL MELAMINE
HIBISCUS
HIDES
HIGH DRAFT
HIGH DRAFT SPINNING
HIGH FREQUENCY
HIGH PRODUCTION CARDS
HIGH SPEED
HIGH TEMPERATURE
HIGH TEMPERATURE DYEING
HIGH TEMPERATURE FIBERS
HIGH TENACITY VISCOSE RAYON
HIGH TWIST
HIGH WET MODULUS RAYON
HILOW BULKED YARNS
HIMALAYA CLOTH
HISTORY
HOLDERS (PACKAGE)
HOLES
HOLES (FABRIC DEFECTS)
HOLLOW BRAIDS
HOLLOW FILAMENT YARNS
HOLLOWNESS
HOMOGENEOUS MATERIALS
HOMOGENIZERS
HOMOPOLYMERS
HONEYCOMB LAMINATES
HONEYCOMB WEAVES
HOODS
HOOK CLOSURES
HOOK DIRECTION
HOOK FORMATION
HOOK REMOVAL
HOOK SHAFTS
HOOKEAN REGION
HOOKEAN STRAIN
HOOKED FIBERS
HOOKES LAW
HOOKS (FIBER)
HOOP STRESS
HOPPER FEED
HOPPERS
HOPSACK
HORSE HAIR
HORSEHIDE
HOSE
HOSES (PIPE)
HOSIERY
HOSIERY DYEING
HOSIERY DYEING MACHINES
HOSIERY MACHINES
HOT AIR DRYERS
HOT AIR DRYING
HOT AIR HEATING
HOT DRAWING
HOT PRESS METHOD (DURABLE PRESS)
HOT SPOTS
HOUSEHOLD FABRICS
HUCKABACK WEAVES
HUE
HULLS
HUMAN HAIR
HUMAN HAIR SETTING
HUMECTANTS
HUMIDITY
HUMIDITY CONTROL
HUNTING
HUSKS
HYDRAULIC BRAKES
HYDRAULIC CONTROL
HYDRAULIC CYLINDERS
HYDRAULIC DRIVES
HYDRAULIC TENSION DEVICES
HYDRAULICS
HYDRAZIDES
HYDROCARBON SULFONATES
HYDROCARBONS
HYDROCELLULOSE
HYDROCHLORIC ACID
HYDRODYNAMIC LUBRICATION
HYDROGEN
HYDROGEN BONDS
HYDROGEN IODIDE
HYDROGEN ION CONCENTRATION
HYDROGEN PEROXIDE
HYDROLYSIS
HYDROMETERS
HYDROPHILIC FIBERS
HYDROPHILIC PROPERTY
HYDROPHOBIC BONDS
HYDROPHOBIC FIBERS
HYDROPHOBIC PROPERTY
HYDROSETTING
HYDROSOLS
HYDROSTATIC HEAD TESTING
HYDROSTATIC LOADING
HYDROSTATIC PRESSURE
HYDROSTATIC PRESSURE TESTING
HYDROSTATIC PRESSURE TESTS
HYDROSTATIC STRESS
HYDROSULFITES
HYDROXIDES
HYDROXYACETIC ACID
HYDROXYACETONE

HYDROXYALKYLCELLULOSE
HYDROXYETHYLCELLULOSE
HYDROXYLAMINE
HYDROXYMETHYL COMPOUNDS
HYGROMETERS
HYGROSCOPICITY
HYPERSONIC FLOW
HYPOBROMITES
HYPOCHLORITES
HYPODERMIC CUTTING DRILLS
HYSTERESIS (MECHANICAL)
ICI SYSTEM
IDENTIFICATION (FIBERS)
IGEPON A (TN)
IGEPON T (TN)
ILLUMINATION
IMAGINARY COMPONENT (DYNAMIC MODULUS)
IMBIBITION
IMIDAZOLIDINONES
IMIDAZOLINE SURFACTANTS
IMIDES
IMINES
IMITATION FRENCH PIPING
IMITATION GINGHAM
IMMATURE FIBERS
IMMATURITY (FIBER)
IMMATURITY TESTING (COTTON FIBER)
IMMEDIATE ELASTIC RECOVERY
IMMERSION
IMMERSION ROLLS
IMMERSION TESTING
IMPACT
IMPACT PENETRATION TESTING
IMPACT PENETRATION TESTS
IMPACT RESISTANCE
IMPACT STRENGTH
IMPACT TESTERS
IMPACT TESTING
IMPACT TESTS
IMPEDANCE
IMPEDANCE BRIDGES
IMPELLERS
IMPERFECTIONS
IMPERIAL SATEEN
IMPREGNATING MACHINES
IMPREGNATION
IMPROVEMENTS
IMPULSES
IMPURITIES
INCANDESCENT LIGHT
INCREASE
INDIA LINON
INDIAN COTTON
INDIANA CLOTH
INDICATORS (INSTRUMENTATION)
INDIGOID DYES
INDIGOSOLS
INDIRECT COUNT
INDIRECT TENSION DEVICES
INDUCTANCE
INDUSTRIAL CLOTHING
INDUSTRIAL FABRICS
INDUSTRIAL SEWING MACHINES
INERTIA
INERTIAL REFERENCE
INFANTS CLOTHING
INFLAMMABILITY
INFLATABLE FABRIC STRUCTURES
INFLATABLE FABRICS
INFORMATION RETRIEVAL
INFORMATION SYSTEMS
INFRARED ABSORBERS
INFRARED DRYERS
INFRARED DRYING
INFRARED HEATING
INFRARED MICROSCOPY
INFRARED RADIATION
INFRARED SPECTRA
INFRARED SPECTROPHOTOMETERS
INFRARED SPECTROSCOPY
INGRAIN CARPETS
INGRAIN DYES
INHIBITION
INITIAL MODULUS
INITIATION
INITIATORS
INJECTION MOLDING
INKS
INORGANIC ACIDS
INORGANIC COMPOUNDS (GENERAL)
INORGANIC FIBER BLENDS
INORGANIC FIBERS (MAN MADE)
INPUT
INSECT DEGRADATION
INSECT RESISTANCE
INSECT RESISTANCE FINISHES
INSECT RESISTANCE TREATMENTS
INSECTS
INSERTION
INSERTION (LACE)
INSOLES (FOOTWEAR)
INSOLUBLE AZO DYES
INSPECTION
INSPECTION MACHINES
INSTALLATION
INSTRON TENSILE TESTER (TN)
INSTRUMENTAL COLOR MATCHING
INSTRUMENTATION
INSTRUMENTS
INSULATED BOOTS
INSULATION (ACOUSTICAL)
INSULATION (ELECTRICAL)
INSULATION (THERMAL)
INTEGRAL CONTROL
INTEGRATED SYSTEMS
INTEGRATION

INTENSITY
INTERFACES
INTERFACIAL POLYMERIZATION
INTERFACIAL TENSION
INTERFERENCE FRINGES
INTERFERENCE MICROSCOPY
INTERFERENCE PATTERNS
INTERFEROMETRY
INTERLACING (FILAMENT IN YARN)
INTERLACINGS (YARN IN FABRIC)
INTERLININGS
INTERLOCK (KNITTED FABRICS)
INTERLOCK FABRICS
INTERLOCK STITCH
INTERMEDIATE FEED
INTERMEDIATE FRAMES
INTERMEDIATE PART
INTERMEDIATE ROLLS
INTERNAL ENERGY
INTERNAL FRICTION
INTERNAL REFLECTION
INTERPOLATION
INTERSECTING GILL BOXES
INTERSPAN (TN)
INTERSTICES
INTIMATE BLENDS
INTRINSIC VISCOSITY
INTURNED WELTS
INVENTIONS
INVENTORIES
INVENTORY CONTROL
INVERFORM MACHINES
INVERSION
INVESTMENTS
INVISCID FLOW
IODIDES
IODINE
IODINE ABSORPTION
IODINE COMPOUNDS
IODINE NUMBER
IODONIUM COMPOUNDS
ION EXCHANGE
ION EXCHANGE RESINS
IONIC BONDS
IONIZATION
IONIZING RADIATION
IONS
IRISH (LACE)
IRON
IRON COMPOUNDS
IRONING
IRONING FASTNESS
IRONS (BURLING MACHINE)
IRRADIATION
IRREGULAR CROSS SECTION
IRREGULARITY
IRREGULARITY (LONG TERM)
IRREGULARITY (PERIODIC)
IRREGULARITY (SHORT TERM)
IRREGULARITY CONTROL
IRREGULARITY TESTERS
IRREGULARITY TESTING
IRRITATION (SKIN)
ISOCYANATE POLYMERS
ISOCYANATES
ISOELECTRIC POINT
ISOIONIC POINT
ISOMERS (MOLECULAR)
ISOPHTHALIC ACID
ISOPROPYL ALCOHOL
ISOTACTIC POLYMERS
ISOTHERMAL PROCESSES
ISOTHIOCYANATES
ISOTOPES
ISOTROPIC MATERIALS
ITACONIC ACID
ITALIAN CLOTH
ITALIAN SILK
IWS FINISH 6 (TN)
J BOXES
JACKETS
JACKS (KNITTING)
JACKSPOOLS
JACONET
JACQUARD HARNESSES
JACQUARD HEADS
JACQUARD KNITTED FABRICS
JACQUARD KNITTING
JACQUARD KNITTING MACHINES
JACQUARD LOOMS
JACQUARD PUNCHED CARDS
JACQUARD WEAVES
JACQUARD WEAVING
JAMMING
JAMMING POINT
JEANS
JERKINS
JERSEY (FILLING KNITTING)
JERSEY (WARP KNITTING)
JET BULKING
JET CRIMPING
JET LOOMS
JET LOOPING
JET QUENCHING
JET SCOURING
JETS
JETSPUN (TN)
JIG DYEING
JIGGERS
JIGS (DYEING)
JOB ANALYSIS
JOINING
JOINTS
JOURNAL BEARINGS
JUDD NBS COLOR SYSTEM
JUMPERS
JUTE
KAMYR MACHINES

KAOLIN
KAPOK
KAPRON (TN)
KARL FISCHER TITRATION METHOD
KEEL (SLASHING)
KENAF
KERATIN
KERATIN FIBERS
KERSEY
KETONES
KETTLES (DYEING)
KETTLES (HEATING)
KID LEATHER
KIDDERMINSTER CARPETS
KIER BOILING
KIER SCOURING
KIERS
KILOTEX
KINEMATIC ANALYSIS
KINEMATIC SYNTHESIS
KINEMATICS
KINETIC ELASTICITY
KINETIC ENERGY
KINETIC FRICTION
KINETICS
KINETICS (CHEMICAL)
KINKING (KNITTING)
KINKY FILLING
KISS COATING
KISS ROLLS
KNEE
KNICKERBOCKER YARNS
KNICKERS
KNIFE BLADE GINNING
KNIFE COATING
KNIFE PLATES
KNIFE TUCKING
KNIFING
KNIT DEKNIT YARNS
KNITTED CARPETS
KNITTED FABRIC STRUCTURE (FILLING KNIT)
KNITTED FABRIC STRUCTURE (WARP KNIT)
KNITTED FABRICS
KNITTED LACE
KNITTED LOOPS
KNITTED NETS
KNITTED OUTERWEAR
KNITTED PILE FABRICS
KNITTED STITCHES
KNITTED UNDERWEAR
KNITTING
KNITTING DYNAMICS
KNITTING MACHINERY
KNITTING MACHINES
KNITTING NEEDLES
KNITTING STATIONS
KNITTING STITCHES
KNITTING TENSION
KNITTING YARNS
KNITWEAR
KNIVES
KNOCK OFF
KNOCKING OVER
KNOCKOVER
KNOCKOVER BITS
KNOP YARNS
KNOPPED YARNS
KNOPT YARNS
KNOT EFFICIENCY
KNOT STRENGTH
KNOTS
KNOTTED NETS
KNOTTED STITCHES (KNITTING)
KNOTTERS
KNOTTING
KNUCKLE YARNS
KODEL (TN)
KORATRON (TN)
KRITCHEVSKY PRODUCTS
KUBELKA-MUNK EQUATION
KURTOSIS
LABELLING
LABELLING MACHINES
LABOR
LABOR COST
LABORATORY APPARATUS
LACE FURNISHING MACHINES
LACE MACHINES
LACE MAKING
LACE NETS
LACES
LACES (SHOES)
LACING CORDS
LACQUER COATED FABRICS
LACQUERS
LACTAMS
LACTIC ACID
LACTONES
LADDER STITCHES
LADDER WEBBING
LADDERING
LADDERS (HOSIERY)
LAID IN YARNS
LAMBS
LAMBSWOOL
LAME (TN)
LAMEPONS
LAMINAR FLOW
LAMINARIN
LAMINATED FABRICS
LAMINATED JERSEY (FOAM)
LAMINATED YARNS
LAMINATES
LAMINATING
LAMINATING RESINS
LAMINATION

AMPSHADES
ANDING (LOOPS)
ANDING BARS
ANDING LOOPS
ANGMUIR ISOTHERM
ANITAL (TN)
ANOLIN
ANOLINE
ANOLISED NYLON
ANTHIONINE
AP ARBOURS
AP GUIDES
AP PLATES
AP RODS
AP ROLLS
AP STICKS
AP UP
AP WASTE
AP WEIGHT CONTROL
AP WINDING
APELS
APPED SEAMS
APPERS
APPET LOOMS
APPET WEAVES
APPING (OF ROLLS)
APPING (WARP KNITTING)
APS
ARGE CIRCLE (COMB)
ARGE PACKAGES
ARVAE
ARVICIDES
AS
ASHED IN WEFT
ASHING END RADIUS
ASHING IN
ASTING
ASTS
ASTS (FOOTWEAR)
ATCH GUARDS
ATCH NEEDLES
ATCHES
ATENT CRIMP
ATENT HEAT
ATEXES
ATHES (LOOM)
ATICES
AUNDERABILITY
AUNDERABILITY TESTERS
AUNDERABILITY TESTING
AUNDERING
AUNDEROMETER
AUNDRY DEGRADATION
AUNDRY NETS
AURIC ACID
AURYL ALCOHOL
AURYL DIETHANOLAMIDE
AWN
AWN CLOTH
LAWN FINISH
LAY (ROPE)
LAY (WIND)
LAYER LOCKING
LAYING IN (KNITTING)
LAYS (LOOM)
LAZIES
LEA COUNT
LEA STRENGTH PRODUCT
LEAD COMPOUNDS
LEADER CLOTH
LEADING HOOKS
LEAF FIBERS
LEAKAGE RESISTANCE (ELECTRICAL)
LEASE
LEASE RODS
LEASING
LEATHER
LEATHER BOOTS
LEATHER CLOTH
LEAVERS LACE
LEAVERS MACHINES
LEAVES
LECITHIN
LECITHIN SULFATE
LEFT HAND COORDINATE SYSTEMS
LEFT HAND RULE
LEFT HAND TWIST
LEGGINGS
LEICESTER WOOL
LENGTH
LENGTH DIAMETER RATIO
LENGTH MEASURING DEVICES
LENGTH PER UNIT WEIGHT
LENGTH UNIFORMITY
LENGTH UNIFORMITY RATIO (FIBER)
LENGTH VARIATION (FIBER)
LENO
LENO CELLULAR FABRICS
LENO LOOMS
LENO SELVEDGES
LENO WEAVES
LENO WEAVING
LEOTARDS
LET OFF
LET OFF BRACKETS
LET OFF CHAINS
LET OFF MOTIONS
LET OFF SPOOLS
LET OFF WEIGHT ARM
LET OFF WEIGHTS
LETTING OUT (GARMENTS)
LEUCO ESTER DYES
LEUCO VAT ESTER DYES
LEVAFIX (TN)
LEVAFIX E (TN)
LEVEL DYEING
LEVEL OF SIGNIFICANCE
LEVELLING

LEVELLING AGENTS
LEVELNESS (DYEING)
LEVELNESS (YARN)
LICKER IN
LICKER IN SCREENS
LICKING (YARNS)
LIFE RAFTS
LIFT (FLUID)
LIFT (WINDING)
LIFTER PLATES
LIGHT
LIGHT DENT
LIGHT METERS
LIGHT MICROSCOPY (OPTICAL)
LIGHT TRANSMISSION
LIGHT WEIGHT
LIGHTFASTNESS (COLOR)
LIGHTFASTNESS (OF FINISH)
LIGHTFASTNESS TESTERS
LIGHTFASTNESS TESTING
LIGHTING
LIGHTNESS
LIGNIN
LIGNIN SULFONATES
LILION (TN)
LILLE (LACE)
LILLE TAPESTRIES
LIME SOAP
LIMITING FRICTION
LINCOLN WOOL
LINDSLEY RATIOS
LINE DRYING (EASY CARE GARMENTS)
LINEAR ALKYLBENZENE SULFONATES
LINEAR CHAIN MOLECULES
LINEAR DENSITY
LINEAR DIFFERENTIAL EQUATIONS
LINEAR VISCOELASTICITY
LINEARITY
LINEN
LINEN THREADS
LINENS
LINGERIE
LINGETTE
LINGOE
LINING FABRICS
LINING FELT
LININGS
LININGS (FOOTWEAR)
LINKAGES
LINKING (KNITTING)
LINKS LINKS FABRICS
LINKS LINKS MACHINES
LINT (FIBER)
LINT (FLY)
LINT (FUZZ)
LINT (WASTE)
LINT BLADES
LINT CATCHERS
LINT COLOR
LINT CONTENT
LINTERS
LIPIDS
LIPOPHILIC PROPERTY
LIQUIDS
LIQUOR (DYEING)
LIQUOR RATIO
LISLE THREADS
LISTINGS
LITERATURE SURVEYS
LITHIUM BROMIDE
LITHIUM CHLORIDE
LITHIUM COMPOUNDS
LITHIUM HYDROXIDE
LITHIUM HYPOCHLORITE
LITHIUM STEARATE
LIVELINESS
LOAD BEARING CAPACITY
LOAD DEFLECTION CURVES
LOAD ELONGATION CURVES
LOAD EXTENSION CURVES
LOADING
LOADING (PROCESS)
LOCKING COURSES
LOCKKNIT
LOCKSTITCH BLINDSTITCHES
LOCKSTITCHES
LODEN COATS
LOFT
LOFTURA (TN)
LOGWOOD
LONDON FORCES
LONDON SHRINKING
LONDON-HEITLER FORCES
LONG CHAIN MOLECULES
LONG STAPLE COTTON
LONG STAPLE SPINNING
LONG WOOL
LONGCLOTH
LONGITUDINAL VIBRATIONS
LOOM LET OFF
LOOM STOPS
LOOM TAKE UP
LOOMS
LOOP CATCHERS
LOOP CATCHING
LOOP DISTORTION
LOOP DRYERS
LOOP EFFICIENCY
LOOP LENGTH
LOOP PILE FABRICS
LOOP PILE TUFTING MACHINES
LOOP SELVEDGES
LOOP STITCHES
LOOP STRENGTH
LOOP TRANSFER
LOOP TYPE YARNS
LOOP YARNS
LOOPED YARNS

LOOPER CLIPS
LOOPER POINTS
LOOPERS
LOOPING
LOOPS
LOOPS (NOVELTY YARNS)
LOOSE REEDS
LOOSE STOCK DYEING
LOOSE STOCK DYEING MACHINES
LOOSENESS
LOSS TANGENT
LOW MIDDLING (COTTON GRADE)
LOW QUARTER BLOOD (WOOL GRADE)
LOW TEMPERATURE
LOW TEMPERATURE DYEING
LOW TWILLS
LOW TWIST
LOWER CONTROL LIMIT
LUBRICANTS
LUBRICATED RINGS
LUBRICATING
LUBRICATION
LUBRICITY
LUDOX (TN)
LUG STRAPS
LUGGAGE
LUMEN
LUMINESCENCE
LUREX (TN)
LUSTER
LUSTERING MACHINES
LUXURIOUSNESS
LYCRA (TN)
LYSINE
MACARONI YARNS
MACHINE DESIGN
MACHINE INTERFERENCE
MACHINE KNITTING
MACHINE MADE LACES
MACHINE SEWING
MACHINERY
MACHINES
MACROFIBRILS
MACRORHEOLOGY
MAGNADRAFT (TN)
MAGNESIUM CHLORIDE
MAGNESIUM COMPOUNDS
MAGNESIUM HYDROXIDE
MAGNESIUM NITRATE
MAGNESIUM SULFATE
MAGNETIC BEARINGS
MAGNETIC BRAKES
MAGNETIC RINGS
MAGNETIC TAPE
MAGNETIC TENSION DEVICES
MAGNITUDE
MAIL (JACQUARD LOOM)
MAINSOOK
MAINSOOK CHECK
MAINTENANCE
MAKING UP
MALEIC ACID
MALEIC ANHYDRIDE
MALI FABRICS (TN)
MALI PROCESSES (TN)
MALIMO FABRIC (TN)
MALIMO PROCESS (TN)
MALINES
MALIPOL FABRIC (TN)
MALIPOL PROCESS (TN)
MALIWATT FABRIC (TN)
MALIWATT PROCESS (TN)
MALTESE (LACE)
MAN MACHINE CONTROL SYSTEMS
MAN MADE FIBER BLENDS
MAN MADE FIBERS
MAN POWER
MANAGEMENT
MANDRELS
MANGANESE COMPOUNDS
MANGLE EXPRESSION
MANGLES
MANGLING
MANILA
MANNEQUINS
MANUAL CONTROL
MANUFACTURE
MANUFACTURING
MAPO
MARIONETTE
MARK OFF
MARKER MAKING
MARKER STENCILS
MARKERS
MARKETING
MARKING
MARKING MACHINES
MARKING PAPER
MARKOV CHAIN
MAROCAIN
MARQUISETTE
MARRIED YARNS
MARSCHALL SOLUTION
MARTINDALE LIMIT
MASS
MASS DENSITY
MASS DYEING
MASS MOMENT OF INERTIA
MASS PER UNIT LENGTH
MASS SPECTROMETRY
MASS TRANSFER
MASS UNIFORMITY
MATCHED DIE MOLDING
MATCHINGS
MATERIAL CONSUMPTION
MATERIAL PROCESSING
MATERIALS
MATERIALS ENGINEERING

MATERIALS FLOW
MATERIALS HANDLING
MATERIALS SHAPING
MATHEMATICAL ANALYSIS
MATHEMATICAL MODEL
MATRICES
MATRIX (FINE STRUCTURE)
MATRIX FREEDOM
MATS
MATTING
MATTRESSES
MATURITY (FIBER)
MATURITY INDEX
MAXWELL MODEL
MAXWELLIAN DISTRIBUTION
MEAN
MEAN FIBER LENGTH
MEAN STAPLE LENGTH
MEANS
MEASUREMENTS
MEASURING
MEASURING INSTRUMENTS
MEASURING MACHINES
MECHANICAL BONDING
MECHANICAL BRAKES
MECHANICAL CONDITIONING
MECHANICAL DAMAGE
MECHANICAL DAMPING
MECHANICAL DETERIORATION
PROPERTIES
MECHANICAL DEVICES
MECHANICAL DRAFT
MECHANICAL DRIVES
MECHANICAL FINISHING
MECHANICAL FLOCKING
MECHANICAL HISTORY
MECHANICAL MIXERS
MECHANICAL PROPERTIES
MECHANICAL SHOCK
MECHANICAL SLUB CATCHERS
MECHANICS
MECHANISM (FUNDAMENTAL)
MECHANISMS
MECHLIN MACHINES
MEDIAN
MEDICAL TEXTILES
MEDIUM (WOOL GRADE)
MEDIUM FINE (WOOL GRADE)
MEDIUM WEIGHT
MEDIUM WOOL
MEDULLA
MELAMINE
MELAMINE FORMALDEHYDE
PRECONDENSATES
MELAMINE FORMALDEHYDE RESINS
MELAMINE RESINS
MELANGE PRINTING
MELANGE PRINTING EQUIPMENT
MELT CRYSTALLIZATION
MELT SPINNING
MELT VISCOSITY
MELTING
MELTING POINT
MELTON
MEMBRANES
MEMORY WHEELS
MENDING
MENSURATION
MERAKLON (TN)
MERCAPTANS
MERCERIZATION
MERCERIZED COTTON
MERCERIZED FINISH
MERCERIZERS
MERCERIZING
MERCERIZING MACHINES
MERCHANDISING
MERCURIC ACETATE
MERCURIC CHLORIDE
MERCURY COMPOUNDS
MERINO WOOL
MERROW STITCHES
MESALINE
MESH (KNITTING)
METACHROMASY
METACHROME DYEING
METACHROME DYES
METAL CHELATING DYES
METAL FIBERS
METAL FILAMENTS
METALLIC ALLOYS
METALLIC BONDS
METALLIC CARD CLOTHING
METALLIC FIBERS
METALLIC SOAPS
METALLIC YARNS
METALLIZED DYEING
METALLIZED DYES
METALLIZED FIBERS
METALLOGRAPHY
METALORGANIC COMPOUNDS
METALS
METAMERISM
METARHEOLOGY
METER PUMPS
METERS (INSTRUMENTS)
METHACROLEIN
METHACRYLAMIDE
METHACRYLIC ACID
METHACRYLIC ESTERS
METHACRYLONITRILE
METHALLYL CHLORIDE
METHANOL
METHAPHOSPHATES
METHODS
METHOXYMETHYL COMPOUNDS
METHOXYMETHYL MELAMINE
METHYL ACRYLATE

METHYL METHACRYLATE
METHYL TRIAZONE
METHYLATED METHYLOL MELAMINE
METHYLATED METHYLOL UREA
METHYLATION
METHYLCELLULOSE
METHYLENE BLUE
METHYLENE BLUE NUMBER
METHYLENE CHLORIDE
METHYLOL ACETAMIDE
METHYLOL ACETYLENE DIUREA
METHYLOL ACRYLAMIDE
METHYLOL ALKYL CARBAMATE
METHYLOL AMIDES
METHYLOL CARBAMATES
METHYLOL COMPOUNDS
METHYLOL DYES
METHYLOL ETHYL CARBAMATE
METHYLOL ETHYL TRIAZONE
METHYLOL FORMAMIDE
METHYLOL ISOPROPYL CARBAMATE
METHYLOL MELAMINES
METHYLOL METHACRYLAMIDE
METHYLOL METHYL CARBAMATE
METHYLOL METHYL TRIAZONE
METHYLOL TRIAZINES
METHYLOL TRIAZONES
METHYLOL UREAS
METHYLOL URONS
METHYLOLATION
METHYLTAURIDE SURFACTANTS
METLON (TN)
METRIC COUNT
MICELLES
MICROBIAL DEGRADATION
MICROBIOLOGICAL DEGRADATION
MICROFIBRILS
MICROMESH (KNITTING)
MICRONAIRE
MICRONAIRE FINENESS
MICRORADIOGRAPHY
MICRORHEOLOGY
MICROSCOPIC ANALYSIS (FIBER
IDENTIFICATION)
MICROSCOPY
MICROTOMES
MIDDLING (COTTON GRADE)
MIDDLING FAIR (COTTON GRADE)
MIGRATION (FIBER)
MIGRATION (SUBSTANCE)
MIGRATION PERIOD
MILAN (LACE)
MILANESE FABRICS
MILANESE KNITTING MACHINES
MILDEW
MILDEW RESISTANCE
MILDEW RESISTANCE FINISHES
MILDEW RESISTANCE TREATMENTS
MILDEW RESISTANT LEATHER
MILITARY CLOTHING
MILITARY PRODUCTS
MILLING
MILLING ACID DYES
MILLING DYES
MILLS (COLLOID)
MILLS (FULLING ROTARY)
MILLS (KICKER)
MINERAL FIBERS
MINERAL OIL
MINIMIZATION
MIRECOURT
MISPICKS
MISREED
MITRES
MIXING
MIXTURES
MOCK FASHIONING MARKS
MOCK LENO
MOCK LENO WEAVES
MOCK SEAMS
MODACRYLIC FIBERS
MODACRYLICS
MODE
MODELLING
MODELS (FASHION)
MODELS (MATHEMATICAL)
MODES
MODES OF DEFORMATION
MODIFICATION
MODIFIED CONTINUOUS FILAMENT
YARNS
MODIFIED STARCHES
MODULUS
MODULUS OF CROSS ELASTICITY
MODULUS OF ELASTICITY
MOHAIR
MOHRS CIRCLE
MOIRE PATTERNS
MOISTURE
MOISTURE ABSORPTION
MOISTURE CONTENT
MOISTURE CONTENT CONTROL
MOISTURE CONTENT TESTERS
MOISTURE CONTENT TESTING
MOISTURE REGAIN
MOISTURE STRENGTH CURVES
MOISTURE VAPOR TRANSMISSION
MOLDING
MOLDING (TAILORING)
MOLDING SHRINKAGE
MOLECULAR ATTRACTION
MOLECULAR CHAIN MOVEMENT
MOLECULAR ORIENTATION
MOLECULAR STRUCTURE
MOLECULAR WEIGHT
MOLECULAR WEIGHT CONTROL
MOLECULAR WEIGHT DISTRIBUTION
MOLECULES

MOLTEN METAL BATHS
MOLYBDENUM COMPOUNDS
MOMENT CURVATURE CURVES
MOMENT OF INERTIA
MOMENTS
MOMENTUM
MOMENTUM TRANSFER
MONFORTS REACTOR
MONOALKYL CARBAMATE FORMALDEHYDE
MONOCHLOROHYDRIN
MONOCHLOROTRIAZINE DYES
MONOCHROMATIC LIGHT
MONOETHANOLAMIDES
MONOETHANOLAMINE
MONOETHANOLAMINE BISULFITE
MONOETHANOLAMINE CARBONATE
MONOETHANOLAMINE SULFITE
MONOFILAMENT YARNS
MONOFILAMENTS
MONOGLYCERIDE SULFATES
MONOMERS
MONOMETHYLOL ETHYL CARBAMATE
MONOMETHYLOL METHYL CARBAMATE
MONOMETHYLOL UREA
MONOSACCHARIDES
MONOSHEER (TN)
MONOUNSATURATED COMPOUNDS
MONSANTO CREASE RECOVERY
MONTMORILLONITE
MORDANT DYEING
MORDANT DYES
MORPHOLINE SOAPS
MORPHOLINIUM COMPOUNDS
MORPHOLOGY
MOSQUITO RESISTANCE
MOTE KNIVES
MOTES
MOTH PROOFING
MOTH RESISTANCE
MOTH RESISTANCE FINISHES
MOTHS
MOTIFS (LACE)
MOTION
MOTORS
MOUNTAIN WOOL
MOVEMENT
MUFF DYEING
MUFTI
MULE CARRIAGES
MULE DRAFT
MULE DRAFTING
MULE DRAW
MULE FRAMES
MULE SPINNING
MULE SPINNING FRAMES
MULES
MULLEN BURST TESTING
MULTICOLORS
MULTIFILAMENT YARNS
MULTIFOLD YARNS
MULTIPLE LAYER FABRICS
MULTIPLES
MULTIPLICATION
MULTIPLIED YARNS
MUNGO
MUNSELL COLOR SYSTEM
MUSLIN
MUSS RESISTANCE
MYLAR (TN)
MYRISTYL ALCOHOL
N FADING
N-METHYLOL COMPOUNDS
NACCONOL NR (TN)
NAP
NAPHTHA
NAPHTHALENE FORMALDEHYDE
SULFONATES
NAPHTHOL DYES
NAPHTHOLS
NAPKINS
NAPPIES
NAPPING
NAPPING MACHINES
NARROW FABRIC LOOMS
NARROW FABRICS
NARROWING
NARROWING (KNITTING)
NARROWING COMBS
NASMITH COMBS
NATURAL CONVECTION DRYERS
NATURAL CRIMP
NATURAL DRAW RATIO
NATURAL DYES
NATURAL FIBER BLENDS
NATURAL FIBERS
NATURAL FREQUENCY
NATURAL POLYMERS
NATURAL RUBBER
NATURAL SOIL
NATURAL STARCH
NECK FORMATION
NECKING (FILAMENT)
NECKTIES
NEEDLE BARS
NEEDLE BEARDS
NEEDLE BEARINGS
NEEDLE BEDS
NEEDLE BUTTS
NEEDLE CAMS
NEEDLE EYE
NEEDLE GROOVE
NEEDLE HEATING
NEEDLE HOOK
NEEDLE JACKS
NEEDLE LATCHES
NEEDLE LEADS
NEEDLE LOOMS (NARROW FABRICS)
NEEDLE LOOMS (NEEDLING)

NEEDLE LOOPS
NEEDLE PENETRATION
NEEDLE POINT
NEEDLE POINT LACES
NEEDLE PRESSER BARS
NEEDLE PUNCHED FELTS
NEEDLE PUNCHING
NEEDLE PUNCHING MACHINES
NEEDLE SCARF
NEEDLE SHAFT (SEWING)
NEEDLE SHANK (SEWING)
NEEDLE THREADS
NEEDLE TRANSFER
NEEDLE TRICKS
NEEDLED FABRICS
NEEDLES
NEEDLEWORK
NEEDLING
NEGATIVE SHEDDING
NEGLIGEES
NEKAL A (TN)
NEKAL B (TN)
NEKAL NS (TN)
NEOPATE PROCESS (TN)
NEOPRENE COATED FABRICS
NEP COUNT
NEP COUNTERS
NEP FORMATION
NEP POTENTIAL
NEP SIZE
NEP TESTERS
NEP YARNS
NEPPINESS
NEPPINESS INDEX
NEPS
NEPTEX (TN)
NEPTOMETER (TN)
NEPTOMETER GRADE
NETS
NETT SILK
NETTING
NEUTRAL AXIS
NEUTRAL DYEING
NEUTRAL DYES
NEUTRALIZING
NEUTRALIZING (FIBER EXTRUSION)
NEW BRADFORD SYSTEM PROCESSING
NEWTONIAN LIQUIDS
NICKEL
NICKEL COMPOUNDS
NIGHTCAPS (TEXTILE)
NIGHTSHIRTS
NIGHTWEAR
NIP
NIP COATING
NIP PRESSURE
NIP ROLLS
NIPPER JAWS
NIPPER KNIVES
NIPPER PLATES
NIPPING
NITRATES
NITRIC ACID
NITRILE RUBBER FIBER
NITRILES
NITRITES
NITRO COMPOUNDS
NITROCELLULOSE
NITROGEN OXIDES
NITRON (TN)
NITROUS ACID
NMR
NO CORE BRAIDS
NO TORQUE YARNS
NOBLE COMBING
NOBLE COMBS
NOGG
NOGGS PER INCH
NOIL CHUTES
NOIL KNIVES
NOIL PER CENT
NOIL REMOVAL
NOILS
NOISE
NOMENCLATURE
NOMEX (TN)
NOMOGRAMS
NOMOGRAPHS
NON LINT CONTENT
NON-FROSTING
NON-SOAP DETERGENTS
NONAQUEOUS SOLUTIONS
NONIONIC SOFTENERS
NONIONIC SURFACTANTS
NONLINEAR DIFFERENTIAL EQUATIONS
NONLINEAR VISCOELASTICITY
NONUNIFORMITY
NONWOVEN CARPETS
NONWOVEN FABRIC MACHINES
NONWOVEN FABRIC STRUCTURE
NONWOVEN FABRICS
NONWOVEN SCRIMS
NONWOVENS
NONYLBENZENE SULFONATES
NONYLNAPHTHALENE SULFONATES
NORMAL ALIPHATIC HYDROCARBONS
NORMAL DISTRIBUTION
NORMAL FORCE
NORMANDIE
NOSE BUNCHING
NOTTINGHAM LACE
NOVELTY TWISTERS
NOVELTY YARNS
NOZZLES
NUB YARNS
NUBS
NUCLEAR MAGNETIC RESONANCE
NULL METHODS

NUMA (TN)
NUMBER AVERAGE MOLECULAR WEIGHT
NUMBERED DUCK
NUMBERS
NUMERICAL ANALYSIS
NUMERICAL INTEGRATION
NUMERICAL SOLUTIONS
NYLEX (TN)
NYLOFT (TN)
NYLON
NYLON (POLYAMIDE FIBERS)
NYLON 6
NYLON 7
NYLON 11
NYLON 66
NYLON 610
NYLON BEARINGS
NYLON BLENDS
NYLON DYEING
NYLON DYES
NYLON THREADS
NYLON TRAVELERS
NYLTEST (TN)
NYPEL (TN)
NYTRIL FIBERS
O FACING
OCCUPATIONAL HAZARDS
OCTADECYL ETHYLENE UREA
OCTYL SULFATE
OCTYLBENZENE SULFONATES
ODOR CONTROL
ODORS
OFF SHADE
OIL BATHS
OIL COMBED TOPS
OIL COMBING
OIL HEATING (OIL FUEL)
OIL IN WATER EMULSIONS
OIL PROOFING
OIL REPELLENCY
OIL REPELLENT FINISHES
OIL REPELLENT TREATMENTS
OIL REPELLENTS
OIL STAINS
OILING
OILS
OLEATE SALTS
OLEFIN COPOLYMERS
OLEFIN DYEING
OLEFIN FIBER BLENDS
OLEFIN FIBERS
OLEFINIC COMPOUNDS
OLEFINIC HYDROCARBONS
OLEFINS
OLEIC ACID
OLEYL ALCOHOL
OLIGOMERS
ON OFF CONTROL
ONE BATH PROCESS
ONE HALF BLOOD (WOOL GRADE)
ONIUM COMPOUNDS
ONIUM SURFACTANTS
OPACITY
OPALINE
OPEN DRAWING
OPEN END SPINNING
OPEN LOOP CONTROL
OPEN SHED
OPEN STITCHES
OPEN WIDTH BLEACHING
OPEN WIDTH DYEING
OPEN WIDTH FINISHERS (FULLING)
OPEN WIDTH PROCESSING
OPEN WIDTH SCOURING
OPEN WIND
OPEN WINDING
OPENERS
OPENING
OPENING MACHINES
OPERATIONS RESEARCH
OPTICAL BLEACHES
OPTICAL BLEACHING
OPTICAL BRIGHTENING
OPTICAL DENSITY
OPTICAL INSTRUMENTS
OPTICAL PROPERTIES
OPTICAL SCANNERS
OPTICAL SCANNING
OPTICAL WHITENERS
OPTICS
OPTIMIZATION
ORDINARY DIFFERENTIAL EQUATIONS
ORGANDIE
ORGANIC ACIDS
ORGANIC COMPOUNDS (GENERAL)
ORGANIC HALOGEN COMPOUNDS
ORGANOBISMUTH COMPOUNDS
ORGANOMERCURY COMPOUNDS
ORGANOMETALLIC COMPOUNDS (EXCLUDING SILICONES)
ORGANOSILICON COMPOUNDS
ORGANOSOLS
ORGANOTIN COMPOUNDS
ORGANZINE
ORIENTAL LACE
ORIENTATION
ORIENTATION (SURFACE)
ORIENTATION BIREFRINGENCE
ORIENTATION UNIFORMITY
ORIENTED WEBS
ORIFICES
ORLON (TN)
ORLON SAYELLE (TN)
OROFIL (TN)
ORTHOBLENDS
ORTHOCORTEX
ORTHOPEDIC FOOTWEAR
ORTHOTROPIC MATERIALS

OSCILLATIONS
OSCILLATORS
OSCILLOSCOPES
OSMOSIS
OSNABURG
OUTERWEAR
OUTLINE
OUTPUT
OUTSIDE FILLING SUPPLY
OVEN METHOD (MOISTURE TESTING)
OVENS
OVERALLS
OVERCAST STITCHES
OVERCOATS
OVERDYEING
OVEREDGE STITCHES
OVEREND UNWINDING
OVERFED PIN TENTERS
OVERHAULING
OVERLAP
OVERLAP (FIBERS)
OVERLOCK STITCHES
OVERSEAM STITCHES
OXALIC ACID
OXAZOLINE SURFACTANTS
OXFORD WEAVES
OXFORD WOOL
OXIDANTS
OXIDATION
OXIDATION DYES
OXIDES
OXIDIZED STARCHES
OXIDIZING AGENTS
OXIDIZING COMPARTMENT (DYEING)
OXIRANE COMPOUNDS
OXIRANE DYES
OXYCELLULOSE
OXYGEN
OZONE
OZONE FADING
PACIFIC CONVERTER
PACKAGE DENSITY
PACKAGE DYEING
PACKAGE DYEING MACHINES
PACKAGE HARDNESS
PACKAGE HOLDERS
PACKAGE SECURING
PACKAGES
PACKAGING (MATERIAL)
PACKAGING (OPERATION)
PACKAGING EQUIPMENT
PACKING
PACKING DENSITY
PACKING FACTOR
PAD BATCH CURE PROCESS
PAD DRY CURE PROCESS
PAD DYEING
PAD FINISHING
PAD FIX MACHINES
PAD JIG DYEING
PAD JIG FINISHING
PAD ROLL DYEING
PAD ROLL FINISHING
PAD ROLL MACHINES
PAD STEAM DYEING
PAD STEAM FINISHING
PAD STEAM MACHINES
PADDERS
PADDING
PADDING (MATERIAL)
PADDING (SLEEVES)
PADDING MANGLES
PADS
PAINTS
PAJAMA CHECK
PAJAMAS
PANTIES
PAPER
PAPER MACHINES
PAPER MAKING
PAPER YARNS
PAPERBOARD
PAPERINESS
PAPERMAKERS FELTS
PARACHUTE CORDS
PARACHUTES
PARACORTEX
PARACRYSTALLINE REGION
PARAFFIN WAXES
PARAFORMALDEHYDE
PARAGUAY (LACE)
PARALLEL MOTION
PARALLEL WINDING
PARALLELIZATION
PARKAS
PARTIAL DIFFERENTIAL EQUATIONS
PARTIAL PRESSURE
PARTICLE SIZE
PARTICULAR SOLUTIONS
PARTITION COEFFICIENT
PASTES
PATCHINESS
PATCHING
PATROLLING
PATTERN (APPAREL)
PATTERN (FABRICS)
PATTERN AREA
PATTERN CHAINS
PATTERN DEFINITION
PATTERN GRADING
PATTERN MAKING
PATTERN REEDS
PATTERN REPEATS
PATTERN WHEEL JACKS
PATTERN WHEEL KNITTING
PATTERN WHEELS
PAYING OUT
PDC PROCESS

PECTINS
PEDESTAL STANDS
PEELING
PEG BOARD KNITTING
PELERINE STITCHES
PENETRANTS
PENETRATING
PENETRATION
PENTACHLOROPHENYL LAURATE
PENTAERYTHRITOL
PEPTIDES
PERACETIC ACID
PERALTA ROLLS
PERBORATES
PERCALE
PERCALINE
PERCENTAGE CRIMP
PERCENTAGE PLATES
PERCENTAGE STRAIN
PERCHING
PERCHLORATES
PERCHLORIC ACID
PERCHLOROETHYLENE
PERCLENE (TN)
PERFLUOROALKYL COMPOUNDS
PERFLUOROCARBONS
PERFORATED BEAMS
PERFORATED BELTS
PERFORATED CAGES
PERFORATED CYLINDERS
PERFORATED PLATES
PERFORATED TUBES (DYEING)
PERFORMANCE
PERFORMANCE PROPERTIES
PERFORMIC ACID
PERIMETER
PERIODATES
PERIODICITY
PERLOK SYSTEM
PERLON (TN)
PERLON U (TN)
PERMANENT CREASING
PERMANENT DEFORMATION
PERMANENT FINISHES (GENERAL)
PERMANENT PLEATING
PERMANENT PRESS
PERMANENT SET
PERMANENT SETTING
PERMANGANATES
PERMEABILITY
PERMEABILITY TESTERS
PERMEABILITY TESTING
PERMEABILITY TESTS
PERMEOMETERS
PERMUTITE (TN)
PEROXIDE BLEACHES
PEROXIDE BLEACHING AGENTS
PEROXIDES
PEROXYGEN COMPOUNDS
PERSONNEL
PERSPIRATION
PERSPIRATION RESISTANCE
PERSULFATES
PERSULFURIC ACID
PETROLEUM SULFONATES
PH
PH CONTROL
PH METERS
PHASE BOUNDARY CROSSLINKING
PHASE MICROSCOPY
PHENOL-FORMALDEHYDE RESINS
PHENOLIC PLASTICS
PHENOLIC RESINS
PHENOLS
PHENOMENOLOGICAL RHEOLOGY
PHENYLMERCURIC ACETATE
PHENYLMERCURIC
DIOCTYLSULFOSUCCINATE
PHENYLMERCURIC SUCCINATE
PHOSPHATE SURFACTANTS
PHOSPHATED GLYCERIDES
PHOSPHATES
PHOSPHITES
PHOSPHONIC ACIDS
PHOSPHONIUM COMPOUNDS
PHOSPHONIUM SURFACTANTS
PHOSPHORIC ACID
PHOSPHORUS COMPOUNDS
PHOTOCHEMICAL DEGRADATION
PHOTOCHEMICAL REACTIONS
PHOTOCHROMY
PHOTOELASTICITY
PHOTOELECTRIC CELLS
PHOTOELECTRIC COUNTERS (NEPS)
PHOTOELECTRIC SCANNING
PHOTOGRAPHY
PHOTOMETERS
PHOTOMICROGRAPHS
PHOTONS
PHOTOTROPY
PHTHALATE ESTERS
PHTHALIC ACID
PHTHALIC ANHYDRIDE
PHTHALOCYANINE COMPOUNDS
PHYSICAL ANALYSIS
PHYSICAL CHEMICAL PROPERTIES
PHYSICAL PROPERTIES (EXCLUDING
MECHANICAL)
PHYSICAL PROPERTIES (MECHANICAL)
PIANO MOTIONS
PICK AND PICK
PICK COUNTERS
PICK GLASS
PICK INSERTION
PICK SPACING
PICK STITCHES
PICKER EVENERS
PICKER LAPS

PICKER POINTS
PICKER SHOES
PICKER STICKS
PICKER STRAPS
PICKERS (LOOM)
PICKERS (OPENING)
PICKING (HARVESTING)
PICKING (OPENING)
PICKING (SNAGGING)
PICKING (WEAVING)
PICKING MOTIONS
PICKING RESISTANCE
PICKING TONGS
PICKS
PICKS PER INCH
PICOETTA FAGOTING (STITCH)
PICOT
PIECE DYED FABRICS
PIECE DYEING
PIECE DYEING MACHINES
PIECED UP SECTIONS
PIECENINGS
PIECING
PIGMENT DISPERSIONS
PIGMENT DYEING
PIGMENT PRINTING
PIGMENTS
PIGSKIN
PIGTAILS (GUIDES)
PILE
PILE DENSITY
PILE FABRIC LOOMS
PILE FABRICS
PILE FABRICS (FLOCKED)
PILE FABRICS (KNITTED)
PILE FABRICS (TUFTED)
PILE FABRICS (WOVEN)
PILE HEIGHT
PILE LENGTH
PILERS
PILGRIMSTEP (FULLING MACHINES)
PILL RESISTANCE
PILL RESISTANCE FINISHES
PILL RESISTANCE TREATMENTS
PILL WEAR OFF
PILLING
PILLING TESTING
PILLOW LACES (BOBBIN LACES)
PILLOW TUBING
PILLOWCASES
PILLOWS
PILLS
PIMA COTTON
PIN DRAFTERS
PIN DRAFTING
PIN POINT SADDLE STITCHES
PIN TACKS
PIN TENTERS
PINCHING
PINKED SEAMS
PINKING (TAILORING)
PINKING MACHINES
PINKING SHEARS
PIPE COVERINGS
PIPED SEAMS
PIPES
PIPING
PIQUE
PIQUE WEAVES
PIRN BARRE
PIRN DENSITY
PIRN WINDERS
PIRN WINDING
PIRNS
PISTONS
PLAIN BEARINGS
PLAIN KNITTED FABRICS
PLAIN KNITTING
PLAIN NET MACHINES
PLAIN ROLLS
PLAIN STITCHES (KNITTING)
PLAIN WEAVES
PLAITING (BRAIDING)
PLAITING (FOLDING)
PLAITING (KNITTING)
PLAITING MACHINES (BRAIDING)
PLAITING MACHINES (FOLDING)
PLANE MOTION
PLANKING (FULLING MACHINES)
PLANNING
PLANT
PLANT LAYOUT
PLASTIC DEFORMATION
PLASTIC FLOW
PLASTIC REGION
PLASTIC STRAIN
PLASTICITY
PLASTICIZERS
PLASTICOELASTICITY
PLASTICOVISCOELASTICITY
PLASTICOVISCOSITY
PLASTICS
PLATEN HARDENERS
PLATINUM
PLAYSUITS
PLEATING
PLEATING MACHINES
PLEATS
PLIED YARNS
PLISSE
PLOUGH KNIVES
PLOUGHING
PLUNGER CAMS
PLURONICS (TN)
PLUSH
PLY TWIST
PLY TWISTERS
PLYING

PLYING TRAVELERS
PNEUMAFIL (TN)
PNEUMATIC BROKEN END COLLECTORS
PNEUMATIC CONTROL
PNEUMATIC INSTRUMENTS
PNEUMATIC IRREGULARITY TESTERS
PNEUMATIC LAP WEIGHT CONTROL
POCKET LININGS
POCKETS
POINT DE PARIS
POINT DESPRIT
POISONOUS
POISSON DISTRIBUTION
POISSONS RATIO
POLAR COMPOUNDS
POLAR MOMENT OF INERTIA
POLARITY
POLARIZED LIGHT
POLARIZED LIGHT METHOD (COTTON IMMATURITY)
POLARIZING MICROSCOPE
POLAROGRAPHY
POLIAFIL (TN)
POLISHING
POLLUTION
POLY(ETHYL ACRYLATE)
POLY(ETHYLENE GLYCOL TEREPHTHALATE)
POLY(ETHYLENE GLYCOL)
POLY(ETHYLENE OXIDE)
POLY(HEXAMETHYLENE ADIPAMIDE)
POLY(HEXAMETHYLENE SEBACAMIDE)
POLY(HEXAMETHYLENE TEREPHTHALAMIDE)
POLY(M-XYLYLENE ADIPAMIDE)
POLY(METHYL ACRYLATE)
POLY(METHYL METHACRYLATE)
POLY(METHYL VINYL KETONE)
POLY(PROPYLENE GLYCOL)
POLY(PROPYLENE OXIDE)
POLY(VINYL ACETAL)
POLY(VINYL ACETATE)
POLY(VINYL ALCOHOL)
POLY(VINYL BUTYRAL)
POLY(VINYL CHLORIDE)
POLY(VINYL OXAZOLIDINONE)
POLY(VINYL PYRIDINE)
POLY(VINYL PYRROLIDINE)
POLY(VINYL PYRROLIDONE)
POLY(VINYLIDENE CHLORIDE)
POLY(VINYLIDENE CYANIDE) FIBERS
POLYACROLEIN
POLYACRYLAMIDE
POLYACRYLATES
POLYACRYLIC ACID
POLYACRYLONITRILE
POLYALKYLAMINE SURFACTANTS
POLYALKYLENEOXIDE BLOCK COPOLYMERS
POLYAMIDE DYES
POLYAMIDES
POLYAMIDOESTER FIBERS
POLYAMINES
POLYBENZIMIDAZOLES
POLYBUTADIENE
POLYCAPROAMIDE
POLYCAPROLACTAM
POLYCARBONATES
POLYELECTROLYTES
POLYENES
POLYESTER DYEING
POLYESTER DYES
POLYESTER FIBER BLENDS
POLYESTER FIBERS
POLYESTER RESINS
POLYESTERS
POLYETHERS
POLYETHOXY ALCOHOLS
POLYETHOXY ALKYL PHENOLS
POLYETHOXY ALKYL POLYAMINES
POLYETHOXY ALKYLAMINE SURFACTANTS
POLYETHOXY ALKYLARYL ETHERS
POLYETHOXY AMIDES
POLYETHOXY AMINE SURFACTANTS
POLYETHOXY ANHYDROSORBITOL ESTERS
POLYETHOXY ESTERS
POLYETHOXY ETHERS
POLYETHOXY FATTY ACIDS
POLYETHOXY GLYCERIDES
POLYETHOXY POLYOL ESTERS
POLYETHOXY POLYPROPOXY ETHYLENEDIAMINE
POLYETHOXY POLYPROPOXY SURFACTANTS
POLYETHOXY THIOETHERS
POLYETHOXYLATED CASTOR OIL
POLYETHYLENE
POLYETHYLENE (HIGH DENSITY)
POLYETHYLENE FIBERS
POLYETHYLENE GLYCOL ESTERS
POLYETHYLENE GLYCOL ETHERS
POLYETHYLENEIMINE
POLYGLYCERYL ESTERS
POLYGLYCOL ESTERS
POLYKETONES
POLYMER ALLOYS
POLYMER DEPOSITION
POLYMER TRANSITIONS
POLYMER TYPES (GENERAL)
POLYMERIC DYES
POLYMERIZATION
POLYMERS
POLYMERS (NATURAL)
POLYMETHACRYLATES
POLYMETHACRYLIC ACID
POLYNOSIC FIBERS

POLYOLEFINS
POLYOLS
POLYOXYALKYL COMPOUNDS
POLYOXYETHYLENE
POLYOXYETHYLENE ALCOHOLS
POLYOXYETHYLENE ALKYLAMIDES
POLYOXYETHYLENE ALKYLAMINES
POLYOXYETHYLENE ALKYLPHENOLS
POLYOXYETHYLENE COMPOUNDS
POLYOXYETHYLENE ESTERS
POLYOXYETHYLENE THIOETHERS
POLYOXYMETHYLENE
POLYPEPTIDES
POLYPROPYLENE
POLYPROPYLENE FIBERS
POLYSACCHARIDES
POLYSILOXANES
POLYSTYRENE
POLYSULFIDES (INORGANIC)
POLYSULFONES
POLYTETRAFLUOROETHYLENE
POLYTETRAFLUOROETHYLENE FIBER BLENDS
POLYTETRAFLUOROETHYLENE FIBERS
POLYTHANE (TN)
POLYTHENE
POLYTRIFLUOROCHLOROETHYLENE FIBERS
POLYUNSATURATED COMPOUNDS
POLYUREAS
POLYURETHANE FIBERS
POLYURETHANES
POLYVINYL DERIVATIVES
POLYVINYLS
PONCHOS
PONGEE
POPLIN
POPULATION
PORCUPINE DRAWING
PORCUPINE ROLLS
PORES
POROSITY
POROUS MATERIALS
POROUS MEDIA
PORTER
POSITION
POSITIONING
POSITIONING MARKING
POSITIVE DISPLACEMENT PUMPS
POSITIVE FEED
POSITIVE FEED MECHANISMS
POSITIVE SHEDDING
POST CURE
POST FORMED LAMINATES
POST FORMING (LAMINATES)
POT SPINNING
POT SPINNING FRAMES
POTASH
POTASSIUM ACETATE
POTASSIUM BICARBONATE
POTASSIUM BROMIDE
POTASSIUM CARBONATE
POTASSIUM CHLORIDE
POTASSIUM COMPOUNDS (GENERAL)
POTASSIUM CYANATE
POTASSIUM CYANIDE
POTASSIUM FLUORIDE
POTASSIUM HYDROXIDE
POTASSIUM IODIDE
POTASSIUM NITRITE
POTASSIUM OLEATE
POTASSIUM PERSULFATE
POTASSIUM SOAPS
POTASSIUM STEARATE
POTATO STARCH
POTENTIAL (ELECTRICAL)
POTENTIAL ENERGY
POTS (SPINNING)
POULT
POWDERING
POWER (FABRIC)
POWER BRAKES
POWER CONSUMPTION
POWER LOOMS
PRE-CURE METHOD
PREBOARDING
PRECIPITATION
PRECISION WIND
PRECISION WINDERS
PRECISION WINDING
PRECONDENSATES
PREFERENTIAL ADSORPTION
PREFERENTIAL DRAFT
PREMETALLIZED DYES
PREORIENTATION
PREPARATION (CHEMICAL)
PREPARING (WORSTED)
PREPARING WOOL
PRESENCE
PRESERVATIVES
PRESHRINKING
PRESS FINISHING
PRESS SECTION
PRESSED FELTS
PRESSER BARS
PRESSER FOOT
PRESSER KNIVES
PRESSES
PRESSING
PRESSING (BEARDED NEEDLES)
PRESSING MACHINES
PRESSING OFF
PRESSLEY INDEX
PRESSLEY STRENGTH
PRESSLEY TEST
PRESSURE
PRESSURE BAG MOLDING
PRESSURE BECKS

PRESSURE CONTAINERS
PRESSURE CONTROL
PRESSURE DISTRIBUTION
PRESSURE DROP
PRESSURE DYEING
PRESSURE DYEING MACHINES
PRESSURE FINGERS
PRESSURE JET DYEING MACHINES
PRESSURE JIGS (DYEING)
PRESSURE KETTLES
PRESSURE ROLLS
PRESSURE SUITS
PRESSURE VESSELS
PRETREATMENTS (GENERAL)
PREVENTIVE MAINTENANCE
PRICKLINESS
PRIMARY AMINE SURFACTANTS
PRIMARY AMINES
PRIMARY CREEP
PRIMARY TRANSITION TEMPERATURES
PRIMARY TRANSITIONS
PRIMARY WALLS
PRIMAZIN (TN)
PRINCESS LACE
PRINT CLOTH
PRINT ROLLS
PRINTED FABRICS
PRINTERS BLANKETS
PRINTING
PRINTING DEFECTS
PRINTING MACHINES
PRINTING PASTES
PRINTING ROLLS
PRINTING SCREENS
PRINTING SHELLS
PRINTS
PROBABILITY
PROBABILITY DENSITY FUNCTION
PROBAN PROCESS (TN)
PROCESS CONTROL
PROCESS DESIGN
PROCESS DYNAMICS
PROCESS EFFICIENCY
PROCESS VARIABLES
PROCESSES
PROCESSING
PROCESSING DIRECTION
PROCESSING MACHINERY
PROCILAN (TN)
PROCION (TN)
PROCION H (TN)
PRODUCERS TWIST
PRODUCERS YARNS
PRODUCT DESIGN
PRODUCTION
PRODUCTION CONTROL
PRODUCTION RATE
PRODUCTIVITY
PRODUCTS
PROJECTION MICROSCOPY
PROPANEDIOL ESTER SURFACTANTS
PROPELLANTS (AEROSOLS)
PROPELLERS
PROPERTIES
PROPIOLACTONE
PROPIONIC ANHYDRIDE
PROPORTIONATING FEEDERS
PROPYLENE CARBONATE
PROPYLENE IMINE
PROTECTIVE COLLOIDS
PROTEIN FATTY ACID CONDENSATES
PROTEIN FIBERS
PROTEINS
PROTEOLYTIC ENZYMES
PROXIMITY TESTERS
PSEUDO WET BULB TEMPERATURE
PSYCHROMETRIC CHARTS
PSYCHROMETRY
PUCKER (DEFECT)
PUCKERED SELVEDGES
PUCKERING
PUCKERS (FOR EFFECT)
PUCKERS (TAILORING)
PULL OVERS
PULLED IN SELVEDGES
PULLED WOOL
PULP
PULPING
PULSE PROPAGATION METER
PULVERIZING
PUMICE
PUMPS
PUMPS (FOOTWEAR)
PUNCH BALLS (COMBING)
PUNCHED CARD EQUIPMENT
PUNCHED CARDS
PUNCHED TAPE
PUNCHING
PUNCTURE RESISTANCE
PUNCTURING
PURIFICATION
PURL FABRICS
PURL KNIT FABRICS
PURL KNITTED FABRICS
PURL KNITTING
PURL STITCHES
PVC
PYRIDINE
PYRIDINE DERIVATIVES
PYRIDINIUM COMPOUNDS
PYROLYSIS
PYROMETERS
PYROPHOSPHATES
PYROSULFITES
PYRUVIC ACID
QUADRANTS (MULE)
QUALITATIVE ANALYSIS
QUALITY

QUALITY CONTROL
QUALITY IMPROVEMENT
QUANTITATIVE ANALYSIS
QUARTER BLOOD (WOOL GRADE)
QUASI HOMOGENEOUS MATERIALS
QUATERNARY AMMONIUM COMPOUNDS
QUATERNARY AMMONIUM SURFACTANTS
QUENCHING (FILAMENT)
QUILL CHANGING
QUILL STRIPPERS
QUILLERS
QUILLING
QUILLING MACHINES
QUILLS
QUILTING
QUILTS
QUINONE
R C CIRCUITS
RABBIT HAIR
RABBITS
RACE (LOOM)
RACE BOARDS
RACE PLATES
RACEWAYS
RACKED STITCHES
RACKING
RADIAL BEARINGS
RADIAL DISTRIBUTION
RADIANT HEATING
RADIATION
RADIATION COUNTERS
RADIATION GRAFTING
RADIATION TREATMENT
RADICALS (CHEMICAL)
RADIOACTIVE COMPOUNDS
RADIOACTIVE STATIC ELIMINATORS
RADIOACTIVE TRACER TECHNIQUES
RADIOACTIVE TRACERS
RADIUS
RADIUS OF CURVATURE
RADOMES
RAFFIA
RAG PULLING MACHINES
RAIN PENETRATION TESTING
RAINCOATS
RAINWEAR
RAISING
RAISING CAMS
RAMBOUILLET WOOL
RAMIE
RAMS
RANDO FEEDER (TN)
RANDO WEBBER (TN)
RANDO WEBS (TN)
RANDOM DISTRIBUTION
RANDOM DYEING
RANDOM LAID WEBS
RANGE
RAPER AUTOLEVELLER (TN)
RAPIER LOOMS
RAPIERS (LOOMS)
RASCHEL FABRICS
RASCHEL KNITTING MACHINES
RASCHEL LOOMS
RATCH
RATE
RATE PROCESSES
RATINES
RAVEL STRIP STRENGTH
RAW STOCK DYEING
RAW STOCK DYEING MACHINES
RAWHIDE
RAYON
RAYON (CUPRAMMONIUM)
RAYON (REGENERATED CELLULOSE FIBERS)
RAYON BLENDS
RAYON DYEING
RAYON DYES
RAYON STRIPE
RE-CURE
REACTANCE (ELECTRICAL)
REACTANTS
REACTION KINETICS
REACTION RATE
REACTIONS (CHEMICAL)
REACTIVE DYEING
REACTIVE DYES
REACTIVE PRINTING
REACTIVITY
REACTONE (TN)
REAGENT TESTING (FIBER IDENTIFICATION)
REAGENTS
REAL COMPONENT (DYNAMIC MODULUS)
REAL TWIST
RECIPROCATING CUTTING MACHINES
RECIPROCATING TRAVERSE
RECOMBED TOPS
RECOMBING
RECORDING INSTRUMENTS
RECOVERABILITY
RECOVERED WOOL
RECOVERING
RECOVERY (CHEMICAL)
RECOVERY (SELF RESTORATION TO ORIGINAL CONDITION)
RECOVERY (WASTE)
RECOVERY SYSTEMS (AERODYNAMIC)
RECRYSTALLIZATION
RECTILINEAR COMBING
RECTILINEAR COMBS
REDOX CATALYSIS
REDOX POTENTIAL
REDOX REACTIONS
REDUCING (WORSTED PROCESS)
REDUCING AGENTS
REDUCTION

REDUCTION PADDERS
REDYEING
REED DRAFT
REED HOOKS
REED MARKS
REED OMBRE
REED PLAN
REED RAKE
REED SPACING
REED WIDTH
REED WIRES
REEDING
REEDS
REEDY FABRICS
REEL MACHINE (DYEING)
REELING
REELING MACHINES (WINDING)
REELS
REFLECTANCE
REFLECTIVITY
REFLECTOMETERS
REFRACTANCE
REFRACTIVE INDEX
REFRACTOMETRY
REFRACTORY FIBERS
REFRIGERATION
REFULLING MACHINES
REGAIN
REGENERATED CELLULOSE
REGENERATED CELLULOSE FIBERS
REGENERATED FIBERS (EXCLUDING CELLULOSIC AND PROTEIN)
REGENERATED PROTEIN FIBERS
REGISTERED PRINTING
REGRESSION
REGRESSION ANALYSIS
REGRESSION COEFFICIENT
REGULAR BRAIDS
REGULAR SADDLE STITCHES
REGULAR TWILLS
REGULARITY
REGULATION
REGULATORS
REINFORCED COMPOSITES
REINFORCED HOSES
REINFORCED KNITTED FABRICS
REINFORCED PLASTICS
RELATIVE DENSITY
RELATIVE HUMIDITY
RELATIVE VISCOSITY
RELAXATION
RELAXATION SHRINKAGE
RELAXATION SPECTRUM
RELAXATION TIME (MECHANICAL)
RELAXING
REMAZOL (TN)
REMODELLING
REMOTE CONTROL
REMOVAL
RENAISSANCE (LACE)
REPAIRS
REPAIRS (MACHINERY)
REPEATS
REPP
RESEARCH
RESERVE EFFECT
RESERVE PRINTING
RESIDUAL FAT
RESIDUAL SHRINKAGE
RESIDUAL STRAIN
RESIDUAL STRESS
RESIDUAL TORQUE
RESILIENCE
RESIN CURING MACHINES
RESIN FINISHES
RESIN PRECONDENSATES
RESINOGRAPHY
RESINS
RESIST PRINTING
RESISTANCE (ELECTRICAL)
RESISTANCE (HEAT)
RESISTANCE TO BLEEDING
RESISTANCE TO DELAMINATION
RESISTIVITY
RESOLUTION (OF FORCES)
RESONANCE
RESONANT FREQUENCY
RESPONSE TIME
RESTRAINING AGENTS
RESTYLING
RESULTS
RETARDATION
RETARDERS
RETARDING AGENTS
RETARDING AGENTS
RETTING
REVERES
REVERSAL
REVERSE PLAITING
REVERSE WELTING
REVERSIBLE FABRICS
REVERSIBLE GARMENTS
REVIEWING
REVIEWS
REVISIONS
REVOLVING FLAT CARDS
REWETTING AGENTS
RHEOLOGICAL PROPERTIES
RHEOLOGY
RHODANINE
RIB KNITTED FABRICS
RIB KNITTING
RIB SET
RIB TRANSFER
RIB WEAVES
RIBBON
RIBBON BREAKERS
RIBBON BREAKING

RIBBON CROSS SECTION
RIBBON LAP MACHINES
RIBBON LAPS
RIBBONING
RICHNESS
RIETER CONVERTER
RIGHT HAND COORDINATE SYSTEMS
RIGHT HAND RULE
RIGHT HAND TWILLS
RIGHT HAND TWIST
RIGID FOAMS
RIGIDITY
RIGMEL PROCESS
RILSAN (TN)
RING CONDENSERS
RING FRAME SPINNING
RING FRAMES
RING RAILS
RING SPINNING
RING SPINNING FRAMES
RING TRAVELERS
RING TWISTERS
RING TWISTING
RINGS
RINGS (BOBBINS)
RINSING
RINSING BATHS (SCOURING)
RIPPING
RIPPLES
RIVETS
ROBBING BACK
ROBES
RODNEY HUNT SATURATOR
ROLL BOSSES
ROLL CLEARERS
ROLL COVERINGS
ROLL CROWNS
ROLL ECCENTRICITY
ROLL WEIGHTING
ROLLER BEARINGS
ROLLER COATING
ROLLER DRAFT
ROLLER DRAFTING
ROLLER GINNING
ROLLER HARDENERS
ROLLER LAP
ROLLER LEATHERS
ROLLER NIP
ROLLER PRINTING
ROLLER PRINTING MACHINES
ROLLER SHAFTS
ROLLER SLIP
ROLLER SLIPPAGE
ROLLER TOP CARDS
ROLLERS
ROLLING
ROLLING FRICTION
ROLLING LOCKER MACHINES
ROLLING MILLS
ROLLS
ROMNEY MARSH WOOL
ROMPERS
RONGALITE (TN)
ROOM TEMPERATURE
ROOTS (MATHEMATICAL)
ROPE BLEACHING
ROPE DYEING
ROPE FORM PROCESSING
ROPE MACHINES
ROPE MARKS
ROPE PROCESSING
ROPE SCOURING
ROPE SCOURING MACHINES
ROPES
ROSE POINT
ROSIN
ROSIN DERIVATIVES
ROT RESISTANCE
ROT RESISTANCE FINISHES
ROT RESISTANCE TREATMENTS
ROTAFORMERS
ROTAMETERS
ROTARY BATTERIES
ROTARY CUTTING MACHINES
ROTARY DYEING MACHINES
ROTARY MAGAZINES
ROTARY SCREEN PRINTING
ROTARY TRAVERSE
ROTATION
ROTATIONAL VELOCITY
ROTH SYSTEM
ROTOSET YARNS (TN)
ROTTING
ROUGHNESS
ROUGHNESS COEFFICIENT
ROUND CROSS SECTION
ROUNDNESS
ROVANA (TN)
ROVING (OPERATION)
ROVING FRAMES
ROVING MACHINES
ROVING WIND
ROVINGS
ROVOMATIC (TN)
RUBBER
RUBBER COATED FABRICS
RUBBER ELASTICITY
RUBBER FIBER (NATURAL)
RUBBER FOOTWEAR
RUBBER SHRINKING BLANKETS
RUBBING
RUBBING LEATHERS
RUBFASTNESS
RUBIDIUM COMPOUNDS
RUBIDIUM HYDROXIDE
RUCHE
RUFFLES
RUFFLING

RUGS
RUN OUT
RUN RESISTANCE
RUN RESISTANT HOSIERY
RUN RESISTANT STITCHES
RUN SYSTEM
RUNNER RATIO
RUNNERS
RUNS (KNITTING)
RUPTURE
RUPTURE STRENGTH
RUPTURE TENACITY
RUPTURING
RUSSIAN POINT
RUST RESISTANCE
RUSTING
RYELAND WOOL
S TWILLS
S TWIST
S TWIST YARNS
SAABA (TN)
SACKCLOTH
SACKS (BAGS)
SACKS (DRESSES)
SACO LOWELL DIRECT SPINNER (TN)
SACO LOWELL IRREGULARITY TESTER (TN)
SADDLE STITCHES
SAFETY
SAFETY BELTS
SAFETY CLOTHING
SAFETY FOOTWEAR
SAG
SAGGING
SAICATEX PROCESS (TN)
SAILCLOTH
SALES
SALICYLANILIDE
SALICYLATE SALTS
SALICYLIC ACID
SALT
SALTS (GENERAL)
SAMPLES
SAMPLING
SANDWICH BLENDING
SANDWICH LAMINATES
SANDWICH ROLLING MILLS
SANDWICH SEAMS
SANFORIZING (TN)
SANFORIZING MACHINES
SANITARY NAPKINS
SANITARY TOWELS
SANITIZING AGENTS
SAPONIFICATION
SARAN
SARAN POLY(VINYLIDENE CHLORIDE) FIBERS
SARCOSIDE SURFACTANTS
SATEEN
SATIN
SATIN WEAVES
SATURATED ALIPHATIC COMPOUNDS
SATURATED ALIPHATIC HYDROCARBON
SATURATED COMPOUNDS
SATURATED HYDROCARBONS
SATURATED SOLUTIONS
SATURATED VAPOR
SATURATION (COLOR)
SATURATION (MATERIAL)
SATURATION BONDING
SATURATORS
SAVONNERIE TAPESTRIES
SAW BLADE GINNING
SAW TOOTH CLOTHING
SAW TOOTH CRIMP
SAW TOOTHED ROLLS
SAXONY WOOL
SBR COATED FABRICS
SCALE (CORROSION)
SCALE MASKING
SCALES (WOOL FIBERS)
SCANNING
SCANNING ELECTRON MICROSCOPY
SCARVES
SCHAPPE SILK
SCHAPPE SYSTEM PROCESSING
SCHIEFER ABRASION TESTING
SCHIFFLE EMBROIDERY
SCHIFFLI LACE
SCHLAFORST UNWINDING ACCELERATOI
SCHREINERING
SCHREINERIZING
SCHUPP REGULATOR
SCISSORS
SCORCH TESTERS
SCORCH TESTING
SCORCHING
SCORCHING MARKS
SCORING
SCOTCH BLACKFACE WOOL
SCOTCH CARPETS
SCOTCH FEED
SCOTT TENSILE TESTER (TN)
SCOURED WOOL
SCOURING
SCOURING BATHS
SCOURING BOWLS
SCOURING MACHINES
SCOURING SETS
SCOURING SHRINKAGE
SCOURING TRAINS
SCOURING WASTE
SCOURING YIELD
SCRATCHING
SCREEN PRINTING
SCREEN PRINTING MACHINES
SCREENING
SCREENS (PRINTING)

SCRIBBLER PART
SCRIBBLING
SCRIM
SCROOP
SCRUBBING
SCUFFING
SCULPTURED CARPETS
SCULPTURED PILE FABRICS
SCUTCHER LAPS
SCUTCHING (FINISHING)
SCUTCHING (FLAX)
SCUTCHING (PICKING)
SEA ISLAND COTTON
SEALED SELVEDGES
SEALERS
SEALING
SEALS
SEAM ABRASIVE STRENGTH
SEAM ALLOWANCE
SEAM BASTING
SEAM BINDING
SEAM DAMAGE
SEAM DETECTORS
SEAM EFFICIENCY
SEAM ELASTICITY
SEAM ELONGATION
SEAM GRIN
SEAM GRINNING
SEAM HEADING
SEAM LET OUT
SEAM PIPING
SEAM PUCKER
SEAM QUALITY
SEAM SIZE
SEAM SLIPPAGE
SEAM SLIPPAGE STRENGTH
SEAM SPECIFICATIONS
SEAM STRENGTH
SEAM TAPING
SEAM TENSILE STRENGTH
SEAM TENSILE STRENGTH EFFICIENCY
SEAM THICKNESS
SEAM WIDTH
SEAMING
SEAMLESS HOSIERY
SEAMLESS HOSIERY MACHINES
SEAMLESS STOCKINGS
SEAMS
SEAMSTRESSES
SEAWATER FASTNESS (COLOR)
SEBUM
SECANT MODULUS
SECOND ORDER TRANSITION TEMPERATURE
SECONDARY AMINE SURFACTANTS
SECONDARY AMINES
SECONDARY CREEP
SECONDARY TRANSITION TEMPERATURE (AMORPHOUS PHASE)
SECONDARY TRANSITION TEMPERATURE (CRYSTALLINE PHASE)
SECONDARY TRANSITION TEMPERATURES
SECONDARY TRANSITIONS
SECONDARY WALLS
SECONDS (FABRICS)
SECTION LEASE
SECTIONAL BEAMING
SECTIONAL DRESSING
SECTIONAL MOMENT OF INERTIA
SECTIONAL WARPING
SEED COAT NEPS
SEED HAIR FIBERS (GENERAL)
SEEDS (NOVELTY YARNS)
SEEDS (TRASH)
SEEPAGE (RHEOLOGY)
SEERSUCKER
SELF ACTOR MULES
SELF ADJUSTING ACTION
SELF BONDED FABRICS (NONWOVENS)
SELF BONDING FIBERS
SELF CRIMPING YARNS
SELF SMOOTHING FABRICS
SELF SUPPORTING PACKAGES
SELF THREADING DEVICES
SELVAGES
SELVEDGE UNCURLERS
SELVEDGE WARPS
SELVEDGES
SEMI CONTINUOUS DYEING
SEMI CONTINUOUS DYEING RANGE
SEMIPERMEABLE MEMBRANES
SENSE (DIRECTION)
SEPARATING
SEPARATION
SEPARATION (SOLUTION)
SEPARATOR SPACING
SEPARATORS
SEQUESTERING
SEQUESTERING AGENTS
SERGE
SERGING
SERGING STITCHES
SERICIN
SERRATED CROSS SECTION
SERVICEABILITY
SERVO DRAFT
SERVO FIBROGRAPH
SERVOMECHANISMS
SET (WARP KNITTING)
SET (WOVEN FABRIC)
SET UP TIME
SETOVER
SETTING (ADJUSTMENTS)
SETTING (EXCEPT SYNTHETICS)
SETTING AGENTS
SETTING TIME
SETTING TREATMENTS (GENERAL)

SETTLING TANKS
SEWABILITY
SEWING
SEWING BOBBINS
SEWING CYCLES
SEWING DAMAGE
SEWING EFFICIENCY
SEWING MACHINES
SEWING NEEDLES
SEWING THREADS
SEWING YARNS
SEWN WELTS
SEYDEL MACHINE
SHADE
SHADE BARS
SHADING (TAILORING)
SHADOW LACE
SHADOW STRIPE
SHADOW WELT
SHAFTS
SHANKS
SHANTUNG
SHAPE
SHAPE FACTOR
SHAPING
SHARKSKIN
SHARPENING
SHAW SYSTEM
SHEAR (MODE OF DEFORMATION)
SHEAR FORCE
SHEAR FORCE DIAGRAMS
SHEAR MODULUS
SHEAR RESISTANCE
SHEAR STRAIN
SHEAR STRENGTH
SHEAR STRESS
SHEARING (FINISHING)
SHEARING (MODE OF DEFORMATION)
SHEARING (TAILORING)
SHEARING (WOOL)
SHEARING MACHINES
SHEATH (FIBER)
SHED (WEAVING)
SHED OPENING
SHEDDING
SHEDDING (FIBER)
SHEEN
SHEEP
SHEEP BREEDING
SHEEP BREEDS
SHEERNESS
SHEET CLOTHING
SHEETING (FABRIC)
SHELL OFF
SHELTERS
SHINE
SHINERS
SHINGUARDS
SHIRLEY ANALYZER
SHIRLEY MOISTURE METER
SHIRRER BLADES
SHIRRING
SHIRTING
SHIRTS
SHOCK
SHOCK ABSORBERS
SHOCK RESISTANCE
SHODDY
SHOE LININGS
SHOE SIZES
SHOEPAC
SHOES
SHOGGING
SHORN WOOL
SHORT FIBERS INDEX
SHORT LONG STITCHES
SHORT STAPLE COTTON
SHORT STAPLE SPINNING
SHORT WOOL
SHORTS
SHRINK GRADING
SHRINK PROOFING
SHRINK RESISTANCE
SHRINK RESISTANCE FINISHES
SHRINKAGE
SHRINKAGE RESISTANCE
SHRINKAGE TESTERS
SHRINKAGE TESTING
SHRINKING
SHROPSHIRE WOOL
SHUTTLE ACCESSORIES
SHUTTLE BINDERS
SHUTTLE BOXES
SHUTTLE BREAKING
SHUTTLE CHANGING
SHUTTLE CHECKING
SHUTTLE DEFLECTORS
SHUTTLE FEELERS
SHUTTLE GUARDS
SHUTTLE MARKS
SHUTTLE SMASH
SHUTTLE SMASH PROTECTORS
SHUTTLE WEAVING
SHUTTLELESS LOOMS
SHUTTLELESS WEAVING
SHUTTLES
SIDE CHAINS (MOLECULAR)
SIDE TO SIDE SHADING
SILANES
SILESIA
SILESIAN WOOL
SILICA
SILICATES
SILICOFLUORIDES
SILICON CARBIDE FIBERS
SILICON COMPOUNDS
SILICONE FLUIDS
SILICONE RESINS

SILICONE RUBBER
SILICONES
SILK
SILK BLENDS
SILK DYEING
SILK DYES
SILK LAP (WARP KNITTING)
SILK NOILS
SILK SYSTEM PROCESSING
SILK SYSTEM WARPING
SILK THROWING
SILKWORMS
SILOXANES
SILVER COMPOUNDS
SIMPLEX FABRICS
SIMPLEX KNITTING MACHINES
SIMULTANEOUS DYEING AND FINISHING
SIMULTANEOUS FINISHING PROCESSES
SINGEING
SINGEING MACHINES
SINGLE APRONS
SINGLE FILAMENT YARNS
SINGLE NEEDLE LOCKSTITCHES
SINGLE THREAD BLINDSTITCHES
SINGLE THREAD CHAIN STITCHES
SINGLES
SINGLES TWIST
SINGLES YARNS
SINGLINGS
SINKER BARS
SINKER CAMS
SINKER LOOPS
SINKERS
SINKING (KNITTING)
SINTERED METAL RINGS
SISAL
SITES (DYEING)
SIVAL MACHINES
SIZE (MAGNITUDE)
SIZE BOXES
SIZES (SLASHING)
SIZING
SIZING (PAPERMAKING)
SIZING AGENTS
SKEIN DYEING
SKEINS
SKETCHING (APPAREL DESIGN)
SKEW
SKEW STRAIGHTENERS
SKI PANTS
SKIN
SKIN (FIBER)
SKIN IRRITATION
SKIN WOOL
SKINEMATICS
SKIP DENT
SKIRTS
SKITTERINESS
SKYLOFT (TN)
SLACK COURSE
SLACK FEEDERS (KNITTING)
SLACK MERCERIZING
SLACK SELVEDGES
SLACK THREADS
SLACKS
SLANT SADDLE STITCHES
SLASHERS
SLASHERS BEAMS
SLASHING
SLAYS (LOOM)
SLEEVE BEARINGS
SLEEVES
SLENDERNESS RATIO
SLEYING
SLEYS (LOOM)
SLICKENERS
SLICKNESS
SLIDE FASTENERS
SLIDE RULES
SLIDING
SLIDING FRICTION
SLIP
SLIP ONS
SLIP PLANES
SLIP RESISTANCE
SLIPE (WOOL)
SLIPPAGE
SLIPPERS
SLIPPING
SLIPS
SLIT FILM YARNS
SLITTERS (CLOTH)
SLITTING
SLIVER CANS
SLIVER KNITTING
SLIVER LAP MACHINES
SLIVER NOIL RATIO
SLIVER REVERSALS
SLIVER TENDERNESS
SLIVER TESTERS
SLIVER TO YARN SPINNING
SLIVERS
SLOPING
SLOTS
SLOUGHING
SLOUGHING OFF
SLUB CATCHERS
SLUB CATCHING
SLUB YARNS
SLUBBING DYEING
SLUBBING DYEING MACHINES
SLUBBING FRAMES
SLUBBINGS
SLUBINESS
SLUBS
SLUDGE
SLURCOCKS

SLUBGALLING
SLURRY
SMALL CIRCLE (COMB)
SMASH
SMOCKS
SMOOTH DRYING FINISHES
SMOOTHNESS
SNAG RESISTANCE
SNAG RESISTANCE TESTING
SNAG TESTERS
SNAG TESTING
SNAGGING
SNAGS
SNAP BACK
SNAP FASTENERS
SNARLING
SNARLING TENDENCY
SNARLS
SNEAKERS
SOAKING
SOAP SODA SCOURING
SOAPING
SOAPS
SOCKS
SODA ASH
SODA CELLULOSE
SODIUM ACETATE
SODIUM ACRYLATE
SODIUM BENZOATE
SODIUM BICARBONATE
SODIUM BISULFATE
SODIUM BISULFITE
SODIUM BOROHYDRIDE
SODIUM BROMATE
SODIUM BROMIDE
SODIUM BROMITE
SODIUM CARBONATE
SODIUM CARBOXYMETHYLCELLULOSE
SODIUM CHLORATE
SODIUM CHLORIDE
SODIUM CHLORITE
SODIUM COMPOUNDS (GENERAL)
SODIUM CUPRATE
SODIUM CYANIDE
SODIUM DICHROMATE
SODIUM DITHIONITE
SODIUM FLUORIDE
SODIUM HEXAMETAPHOSPHATE
SODIUM HYDROSULFIDE
SODIUM HYDROSULFITE
SODIUM HYDROXIDE
SODIUM HYPOCHLORITE
SODIUM IODIDE
SODIUM LAURYL SULFATE
SODIUM NITRATE
SODIUM NITRITE
SODIUM OLEATE
SODIUM PALMITATE
SODIUM PERBORATE
SODIUM PEROXIDE
SODIUM PERSULFATE
SODIUM POLYACRYLATE
SODIUM ROSINATE
SODIUM SESQUICARBONATE
SODIUM SILICATE
SODIUM SOAPS
SODIUM STEARATE
SODIUM SULFATE
SODIUM SULFIDE
SODIUM SULFOXYLATE FORMALDEHYDE
SODIUM THIOSULFATE
SOFTENERS
SOFTENING
SOFTENING AGENTS
SOFTENING POINT
SOFTENING TEMPERATURE
SOFTNESS
SOIL
SOIL BURIAL TEST
SOIL DEPOSITION
SOIL REDEPOSITION
SOIL RELEASING FINISHES
SOIL REMOVAL
SOIL RESISTANCE
SOIL RESISTANCE FINISHES
SOILING
SOLES
SOLID BRAIDS
SOLID COLORS
SOLID SHADE
SOLID SHADE DYEING
SOLIDIFICATION
SOLIDS
SOLS
SOLUBILITY
SOLUBILITY PARAMETER
SOLUBLE VAT DYES
SOLUTES
SOLUTION DYEING
SOLUTION TYPE BINDERS
SOLUTIONS (LIQUID)
SOLUTIONS (MATHEMATICAL)
SOLVENT ASSISTED DYEING
SOLVENT BONDING
SOLVENT DEGREASING
SOLVENT DYEING
SOLVENT DYES
SOLVENT EMULSIONS
SOLVENT RECOVERY
SOLVENT SCOURING
SOLVENT SPINNING
SOLVENT TREATMENTS
SOLVENTS
SONIC VELOCITY
SONICS
SORBENTS
SORBITAN ESTER SURFACTANTS
SORBITOL

SORPTION
SORPTION EQUILIBRIUM
SORPTION OF DYES
SORPTION OF GASES
SORPTION OF LIQUIDS
SORPTION OF WATER
SORPTION RATE
SORPTIVITY
SORTING
SOUND
SOUND WAVES
SOUTANES
SOUTH AFRICAN MERINO WOOL
SOUTH AMERICAN MERINO WOOL
SOUTHDOWN WOOL
SPACE
SPACE DRAWING
SPACE DYEING
SPAN LENGTH
SPANDELLE (TN)
SPANDEX
SPANDEX FIBER BLENDS
SPANDEX FIBERS
SPANISH (LACE)
SPANS (TN)
SPARKLE
SPATS
SPECIALTY YARNS
SPECIFIC GRAVITY
SPECIFIC GRAVITY TESTING (FIBER IDENTIFICATION)
SPECIFIC HEAT
SPECIFIC HUMIDITY
SPECIFICATIONS
SPECIMEN LENGTH
SPECIMEN PREPARATION
SPECK DYEING
SPECKELON (TN)
SPECKINESS
SPECKING
SPECKLED FABRIC
SPECKS
SPECTRA
SPECTROGRAPHY
SPECTROPHOTOMETERS
SPECTROPHOTOMETRIC MEASUREMENTS
SPECTROPHOTOMETRY
SPECTROSCOPY
SPECTRUM ANALYSIS
SPEED
SPEED CONTROL
SPEED FRAMES
SPHERICITY
SPHERULITES
SPHERULITIC CRYSTALLIZATION
SPHERULITIC FIBRILS
SPIN DRYING
SPIN DYEING
SPIN FINISHES
SPINDLE DRAFTING
SPINDLE ECCENTRICITY
SPINDLE HEAD
SPINDLE RAILS
SPINDLE SPEED
SPINDLE TAPES
SPINDLE VIBRATION
SPINDLES
SPINNABILITY (STAPLE FIBER)
SPINNERETS
SPINNING
SPINNING (EXTRUSION)
SPINNING ASSEMBLIES (EXTRUSION)
SPINNING BALLOONS
SPINNING BATHS
SPINNING BOBBINS
SPINNING DYNAMICS
SPINNING EFFICIENCY
SPINNING FILTERS
SPINNING FINISH
SPINNING FRAMES
SPINNING JENNYS
SPINNING LIMIT
SPINNING LUBRICANTS
SPINNING MACHINERY (STAPLE FIBER)
SPINNING MACHINES (EXTRUSION)
SPINNING MULES
SPINNING POTS
SPINNING RINGS
SPINNING TENSION (EXTRUSION)
SPINNING TENSION (STAPLE YARN)
SPINNING TRAVELERS
SPINNING WHEELS
SPIRALITY (KNITTED FABRICS)
SPIRALS (NOVELTY YARNS)
SPLASH VOILE
SPLASHES
SPLICING (KNITTING)
SPLIT DRUM WINDERS
SPLIT RODS
SPLIT SHED
SPLITTING
SPONGINESS
SPONGING
SPOOLING
SPOOLS
SPORT SATIN
SPORTSWEAR
SPOT WEAVES
SPOTS
SPOTTINESS
SPOTTING
SPRAY BONDING
SPRAY TESTING
SPRAY TESTS
SPRAYING
SPRAYING MACHINES
SPREAD LOOPS

SPREADERS (CLOTH)
SPREADING (TAILORING)
SPREADING MACHINES (TAILORING)
SPREADING TABLES
SPRING BEARD NEEDLES
SPRING CONSTANT
SPRING LOADING
SPRING NEEDLES
SPRINGINESS
SPRINGS
SPUN SILK
SPUN YARNS
SPUNBONDED (NONWOVENS)
SPUNIZE (TN)
SQUEEGEES
SQUEEZE ROLLS
SQUEEZERS (CLOTH)
SQUEEZING
SRRL CARD
SRRL OPENER
STABILITY
STABILITY TO CHEMICALS
STABILIZATION
STABILIZERS (AGENTS)
STABILIZERS AGAINST THERMAL DEGRADATION
STABILIZING AGENTS
STAIN REMOVAL
STAIN REPELLENTS
STAIN RESISTANCE
STAIN RESISTANCE AGENTS
STAINING
STAINING (DYEING DEFECTS)
STAINING TESTING (FIBER IDENTIFICATION)
STAINLESS STEEL
STAINS
STAINS DIRECT SPINNER
STANDARD DEVIATION
STANDARD ERROR
STANDARD HEDDLES
STANDARDS
STANDFAST (TN)
STANDS
STANNOUS FLUORIDE
STAPLE ARRAY
STAPLE CUTTERS
STAPLE DIAGRAM
STAPLE FIBER YARNS
STAPLE FIBERS
STAPLE LENGTH
STAPLE STRENGTH
STAPLE YARNS
STAPLES (WIRE)
STAPLING
STAPLING (WIRE)
STAR WHEELS
STARCH
STARCH ETHERS
STARCH GUMS
STARCH PHOSPHATE
STARCHES
STARCHING
START UP MARKS
STARTING MARKS
STARTING TIME
STATIC
STATIC ABSORPTION (WATER)
STATIC ABSORPTION TESTING
STATIC ABSORPTION TESTS
STATIC ELECTRICITY
STATIC ELIMINATORS
STATIC FRICTION
STATIC MODULUS
STATIC RESISTANCE
STATICS
STATISTICAL ANALYSIS
STATISTICAL DISTRIBUTION
STATISTICAL INFERENCE
STATISTICAL MEASURES
STATISTICAL MECHANICS
STATISTICAL METHODS
STATISTICAL THEORY
STATISTICS
STEAM
STEAM HEATING
STEAM JET TEXTURING
STEAMERS
STEAMING
STEARATE ESTERS
STEARIC ACID
STEARYL ALCOHOL
STEEL TRAVELERS
STEELON (TN)
STEEP TWILLS
STELOMETER
STELOMETER STRENGTH TESTING
STELOMETER STRENGTH TESTS
STENTERING
STEREO SPECIFICITY
STEREOSCAN MICROSCOPY
STEREOSCOPIC MICROSCOPY
STICK SHUTTLES
STICK SLIP
STICK SLIP FRICTION
STICKING (ADHESION)
STICTION
STIFFENERS (AGENTS)
STIFFENING TREATMENTS
STIFFENINGS (GARMENT)
STIFFNESS
STIRRERS
STIRRING
STITCH BONDED FABRICS
STITCH BONDING
STITCH BONDING MACHINES
STITCH CAMS
STITCH CLARITY

STITCH DENSITY
STITCH DISTORTION
STITCH ELASTICITY
STITCH ELONGATION
STITCH LENGTH
STITCH LENGTH CONTROL
STITCH QUALITY
STITCH REINFORCED FABRICS
STITCH RESILIENCE
STITCH SEQUENCE
STITCH SIZE (SEWING)
STITCH TENSION
STITCH TRANSFER
STITCHED PILE FABRICS
STITCHES (KNITTING)
STITCHES (SEWING)
STITCHING
STOCHASTIC PROCESSES
STOCK (FIBER)
STOCK DYEING
STOCK DYEING MACHINES
STOCKINETTE
STOCKINETTE WARP KNIT FABRICS
STOCKINGS
STODDARD SOLVENT
STOLL QM ABRASION TESTING
STOLL QM ABRASION TESTS
STOP MOTIONS
STOPPAGES
STOPPING LINE
STOPPING MARKS
STOPS
STORAGE AGING
STORAGE VESSELS
STRAIGHTENING
STRAIN
STRAIN ENERGY
STRAIN GAUGES
STRAIN HARDENING
STRAIN RATE
STRAIN WAVES
STRAINOMETER
STRANDS
STRAPPING
STRAPPING MACHINES
STREAKINESS
STREAM SANITATION
STRENGTH
STRENGTH AT RUPTURE
STRENGTH EFFICIENCY
STRENGTH ELONGATION TESTING
STRENGTH OF MATERIALS
STRENGTH TESTING
STRENGTH TESTS
STRESS
STRESS (TRANSVERSE)
STRESS ANALYSIS
STRESS CONCENTRATION
STRESS DISTRIBUTION
STRESS HISTORY
STRESS OPTICAL COEFFICIENT
STRESS OPTICAL PROPERTIES
STRESS PROPAGATION
STRESS RELAXATION
STRESS STRAIN BEHAVIOR
STRESS STRAIN CURVES
STRESS STRAIN PROPERTIES
STRESS STRAIN PROPERTIES (COMPRESSIVE)
STRESS STRAIN PROPERTIES (SHEAR)
STRESS STRAIN PROPERTIES (TENSILE)
STRESS STRAIN TESTING
STRESSES
STRETCH
STRETCH BREAKING
STRETCH FABRICS
STRETCH HOSE
STRETCH HOSIERY
STRETCH KNITTED FABRICS
STRETCH PUCKERS
STRETCH ROLLS
STRETCH WOVEN FABRICS
STRETCH YARNS (FILAMENT)
STRETCH YARNS (SPUN)
STRETCHING (CORDS)
STRETCHING (SEAMING)
STRETCHING (YARNS)
STRICT GOOD MIDDLING (COTTON GRADE)
STRICT GOOD ORDINARY (COTTON GRADE)
STRICT LOW MIDDLING (COTTON GRADE)
STRICT MIDDLING (COTTON GRADE)
STRING
STRING WARP MACHINES
STRIP DYEING
STRIPES
STRIPINESS
STRIPING UNIT (KNITTING)
STRIPPED SEAMS
STRIPPERS
STRIPPING (CARDING)
STRIPPING (COLOR)
STRIPPING ACTION
STRIPPING AGENTS
STROBOSCOPES
STRUCTURAL ANALYSIS
STRUCTURAL EFFICIENCY
STRUCTURAL MECHANICS
STRUCTURAL PROPERTIES
STRUCTURE OF MATERIALS
STRUCTURES
STUDENTS DISTRIBUTION
STUFFER BOX CRIMPED YARNS
STUFFER BOX CRIMPERS
STUFFER BOX CRIMPING

STUFFER BOXES
STUFFING
STYLES
STYLING (APPAREL)
STYMER (TN)
STYRENE
STYROFOAM (TN)
SUBASSEMBLIES
SUBGROUP MOVEMENT (MOLECULAR)
SUBJECTIVE MECHANICAL PROPERTIES
SUBLIMATION
SUBSTANTIVITY
SUBSTANTIVITY (DYES)
SUBSTRATES
SUBTRACTION
SUCCINIC ACID
SUCROSE
SUCROSE ESTER SURFACTANTS
SUCTION PRESSES
SUEDE
SUEDING
SUEDING MACHINES
SUFFOLK WOOL
SUGARS
SUINT
SUITS
SULFATE SURFACTANTS
SULFATED CASTOR OIL
SULFATED GLYCERIDES
SULFATED OILS
SULFATED OLEFINS
SULFATED OLEIC ESTERS
SULFATED TALLOW
SULFATES
SULFIDES (INORGANIC)
SULFIDES (ORGANIC)
SULFINATES
SULFINIC ACIDS
SULFITES
SULFIX A (TN)
SULFOBETAINE SURFACTANTS
SULFOFATTY ACID ESTERS
SULFOFATTY ACIDS
SULFOGLYCERIDES
SULFONAMIDE DYES
SULFONATE SURFACTANTS
SULFONATED OILS
SULFONATED TALLOW
SULFONATES
SULFONES
SULFONIC ACIDS
SULFONIC END GROUPS
SULFONIUM COMPOUNDS
SULFONIUM REACTANTS
SULFONIUM SURFACTANTS
SULFOPOLYCARBOXYLIC ESTERS
SULFORICINOLEATE
SULFOSUCCINAMIDES
SULFOSUCCINIC ESTERS
SULFUR COMPOUNDS
SULFUR DIOXIDE
SULFUR DYEING
SULFUR DYES
SULFURIC ACID
SULTONES
SULZER LOOMS (TN)
SUMMARIES
SUN SUITS
SUNLIGHT
SUNLIGHT TESTS
SUPERCONTRACTION
SUPERHEATED STEAM
SUPERLEVELLER (DRAFTING)
SUPERLOFT (TN)
SUPERPOSED SEAMS
SUPERVISING
SUPPORT BANDAGES
SUPPORT HOSIERY
SURFACE ACTIVE AGENTS
SURFACE AREA
SURFACE CHEMISTRY
SURFACE CONTOUR
SURFACE DRUMS
SURFACE FRICTION
SURFACE PROPERTIES (MECHANICAL)
SURFACE PROPERTIES (PHYSICAL CHEMICAL)
SURFACE TENSION
SURFACES
SURFACTANT APPLICATIONS
SURFACTANTS
SURGICAL STOCKINGS
SURGING
SURVEYS
SUSPENDERS
SUSPENDING AGENTS
SUSPENSION POLYMERIZATION
SUSPENSION POLYMERS
SUSPENSIONS
SUTER WEBB TESTING
SUTURING
SWEATERS
SWEATSHIRTS
SWELLING
SWELLING AGENTS
SWELLING METHOD (COTTON IMMATURITY)
SWELLING TESTING (FIBER IDENTIFICATION)
SWIFTS
SWIM SUITS
SWING DOOR MECHANISMS
SWISS
SWISS DOUBLE PIQUE
SWIVEL LOOMS
SWIVEL WEAVES
SWOLLEN DENT
SWOLLEN HEDDLE

SYMMETRY
SYNDIOTACTIC POLYMERS
SYNTHESIS
SYNTHETIC DETERGENTS
SYNTHETIC DYEING
SYNTHETIC FELTS
SYNTHETIC FIBER BLENDS
SYNTHETIC FIBERS
SYNTHETIC LEATHER
SYNTHETIC POLYMERS
SYNTHETIC RUBBER
SYNTHETIC RUBBER FIBER BLENDS
SYNTHETIC RUBBER FIBERS
SYNTHETIC SOIL
SYNTHETICS
SYSTEM DESIGN
SYSTEMS ENGINEERING
SYTON (TN)
T DISTRIBUTION
TABBY WEAVES
TABER ABRASION TESTING
TABER ABRASION TESTS
TABLE DAMASK
TABLECLOTHS
TABLET WEAVING
TACK
TACKING
TACTIC POLYMERS
TAFFETA
TAFFETA WEAVES
TAILORED SEAMS
TAILORING
TAILORING MACHINES
TAKE DOWN
TAKE DOWN TENSION
TAKE UP
TAKE UP (SEWING)
TAKE UP MECHANISMS
TAKE UP ROLLS
TAKER IN
TAKING IN (GARMENTS)
TANGENT MODULUS
TANGENTIAL FORCE
TANKS
TANNIC ACID
TANNING (LEATHER)
TANNING AGENTS
TAPE
TAPE CONDENSERS
TAPE DRIVES
TAPE SELVEDGES
TAPER (BOBBIN)
TAPERED ROLLER BEARINGS
TAPES (DATA MEDIA)
TAPESTRIES
TAPESTRY CARPETS
TAPING
TARPAULINS
TASLAN (TN)
TATTING LACES
TEAR
TEAR RESISTANCE
TEAR STRENGTH
TEAR TESTERS
TEAR TESTING
TEAR TESTS
TEARING (TAILORING)
TEAZLE GIG
TEAZLES
TECHNIQUES
TEFLON (TN)
TEFLON BEARINGS
TEFLON FIBER (TN)
TEKJA (TN)
TEMPERATURE
TEMPERATURE CONTROL
TEMPERATURE DISTRIBUTION
TEMPERATURE GRADIENT
TEMPLE MARKS
TEMPLES
TENACITY
TENACITY ELONGATION CURVES
TENASCO (TN)
TENAX (TN)
TENDERING
TENSILE BUCKLING
TENSILE PROPERTIES
TENSILE STRENGTH
TENSILE STRESS
TENSILE TESTERS
TENSILE TESTING
TENSILE TESTING MACHINES
TENSILE TESTS
TENSIOMETERS
TENSION
TENSION CONTROL
TENSION CONTROLLERS
TENSION DEVICES
TENSION DISCS
TENSION METERS
TENSION MODULUS
TENSION REGULATION
TENSION SPRINGS
TENSOR ANALYSIS
TENSOR CALCULUS
TENSORS
TENTER CLIPS
TENTER FRAMES
TENTER PINS
TENTERING
TENTS
TEREPHTHALIC ACID
TERGAL (TN)
TERMINAL VELOCITY
TERMINOLOGY
TERON (TN)
TERPOLYMERS
TERRYCLOTH

TERTIARY ALKYLAMINE OXIDES
TERTIARY AMINE OXIDES
TERTIARY AMINE SURFACTANTS
TERTIARY AMINES
TERYLENE (TN)
TESTING
TESTING APPARATUS
TESTING EQUIPMENT
TESTING INSTRUMENTS
TESTS
TETRAHYDRONAPHTHALENE
TETRAHYDRONAPHTHALENE SULFONATES
TETRAKIS(HYDROXYMETHYL) PHOSPHONIUM CHLORIDE
TETRAPOTASSIUM PYROPHOSPHATE
TETRASODIUM PYROPHOSPHATE
TETI NICS (TN)
TEX
TEXAS COTTON
TEXTILE CHEMICALS
TEXTILE COMPLIANCE
TEXTILE MACHINERY (GENERAL)
TEXTILE MATERIALS
TEXTILE MODULUS
TEXTILE PROCESSES (GENERAL)
TEXTRALIZED YARNS (TN)
TEXTURE
TEXTURED SPUN YARNS
TEXTURED YARNS (FILAMENT)
TEXTURED YARNS (SPUN)
TEXTURING
TEXTURIZED YARNS
THEORETICAL ANALYSIS
THEORIES
THERMAL ANALYSIS
THERMAL CONDUCTIVITY
THERMAL DEGRADATION
THERMAL PROPERTIES
THERMAL RESISTANCE
THERMAL STABILITY
THERMAL TESTING
THERMAL TESTING (FIBER IDENTIFICATION)
THERMAL TRANSMITTANCE TESTING
THERMODYNAMICS
THERMOFIXING (DYEING)
THERMOMETERS
THERMOPILES
THERMOPLASTIC FIBER BONDING
THERMOPLASTIC FIBERS
THERMOPLASTIC POLYMERS
THERMOPLASTIC POWDERS
THERMOPLASTIC RESINS
THERMOPLASTICITY
THERMOSETTING
THERMOSETTING POWDERS
THERMOSETTING RESINS
THERMOSOL PROCESS (TN)
THESAURI
THICK AND THIN YARNS
THICK SPOTS
THICKENING
THICKENING AGENTS
THICKNESS
THIN SPOTS
THINNERS
THIOCYANATES
THIOETHERS
THIOGLYCOLIC ACID
THIOLS
THIOSULFATES
THIOSULFINIC ACIDS
THIOSULFONATES
THIOUREA
THIOUREA DIOXIDE
THIXOTROPY
THPC
THREAD BREAKAGES
THREAD CUTTERS
THREAD GUIDES
THREAD PRESSERS
THREAD SLACK
THREAD SLIPPAGE
THREAD TENSION
THREADING
THREADLINE
THREADS
THREE EIGHTHS BLOOD (WOOL GRADE)
THREE NEEDLE COVER STITCHES
THREONINE
THROAT PLATES
THROSTLE SPINNING FRAMES
THROUGH PRINTING
THROUGHPUT
THROWING
THROWING MACHINES
THROWOVER
THROWSTERS
THRUST BEARINGS
TICKING
TIGHT SELVEDGES
TIGHTNESS
TIME
TIME AND MOTION STUDIES
TIME CONSTANT
TIME SERIES
TIN
TIN COMPOUNDS
TINSEL
TINTING
TINTING MACHINES
TIPPINESS
TIRE CORDS
TIRE FABRICS
TIRES
TITANIUM COMPOUNDS
TITANIUM DIOXIDE
TITRATION

TMCEA
TOBACCO CLOTH
TOES
TOHO DIRECT SPINNER (TN)
TOLERANCES
TOLUENE
TOLUENE ISOCYANATES
TONE IN TONE DYEING
TONGUE TEAR STRENGTH
TONGUE TEAR TESTING
TONGUE TEAR TESTS
TONGUES
TOP COMBING (COMB CYCLE)
TOP COMBS
TOP COVER
TOP DYEING
TOP DYEING MACHINES
TOP MAKING
TOP MATCHING
TOP NOIL RATIO
TOP ROLLS
TOP STITCHED SEAMS
TOPEL (TN)
TOPOLOGY
TOPPING
TOPS
TORCHON
TORCHWICK YARNS
TORQUE
TORSION (GEOMETRIC)
TORSION (MECHANICAL)
TORSIONAL BEHAVIOR
TORSIONAL BUCKLING
TORSIONAL ENERGY
TORSIONAL MOMENT
TORSIONAL RIGIDITY
TORSIONAL STABILITY
TOUCH
TOUGHNESS
TOUGHNESS INDEX
TOW
TOW CONVERSION
TOW CONVERTERS
TOW CRIMPING
TOW DYEING
TOW DYEING MACHINES
TOW TO SLIVER CONVERSION
TOW TO TOP CONVERSION
TOW TO TOP MACHINES
TOW TO YARN CONVERSION
TOW TO YARN MACHINES
TOWELS
TOXICITY
TRACER FIBER TECHNIQUES
TRACER FIBERS
TRACER TECHNIQUES
TRACERS
TRAILING HOOKS
TRAINING
TRAM
TRANSDUCERS
TRANSFER PROPERTIES
TRANSFER TAILS
TRANSFERRING (LOOPS)
TRANSFERRING (MATERIAL)
TRANSIENT RESPONSE
TRANSITION TEMPERATURES
TRANSITIONS (POLYMERS)
TRANSLATION
TRANSLUCENCY
TRANSMISSIVITY
TRANSMITTANCE
TRANSPARENCY
TRANSPORTING VESSELS
TRANSVERSE FORCE
TRANSVERSE VIBRATIONS
TRAP SORTING (WOOL)
TRAPEZOID TEAR STRENGTH
TRAPEZOID TEAR TESTING
TRAPEZOID TEAR TESTS
TRASH
TRASH CONTENT
TRAVELER BURNOUT
TRAVELER CHATTER
TRAVELERS
TRAVELING BLOWERS
TRAVERSE
TRAVERSE MOTION
TREADLES
TRELLIS MODEL
TRIACETATE
TRIACETATE DYEING
TRIACETATE FIBERS
TRIAL AND ERROR SOLUTIONS
TRIAMINES
TRIAZINES
TRIAZONES
TRIBASIC ACIDS
TRIBOELECTRIC SERIES
TRIBOELECTRICITY
TRICEL C (TN)
TRICHLOROACETIC ACID
TRICHLOROBENZENE
TRICHLOROCYANURIC ACID
TRICHLOROETHYLENE
TRICHLOROPYRIMIDINE DYES
TRICHROMATIC COEFFICIENTS
TRICOT
TRICOT FABRICS
TRICOT KNITTED FABRICS
TRICOT KNITTING
TRICOT KNITTING MACHINES
TRICOT STITCHES
TRIETHANOLAMINE
TRIETHANOLAMINE SOAPS
TRIGLYCERIDES
TRIHEXYL SULFOCARBALLYLATE
TRILOBAL CROSS SECTION

TRILOCK (TN)
TRIMETHYLOL MELAMINE
TRIMMING (OPERATION)
TRIMMINGS
TRIMNESS
TRIS(AZIRIDINYL) PHOSPHINE OXIDE
TRIS(METHYLAZIRIDINYL) PHOSPHINE OXIDE
TRIS(METHYLOL CARBAMOYLETHYL) AMINE
TRISODIUM PHOSPHATE
TRISTIMULUS VALUES
TROPICAL FOOTWEAR
TROPICAL WORSTED
TROUSERS
TRUE CONTACT AREA
TRUE STRAIN
TRUE TWIST
TRUMPETS
TRUNKS
TRYPTOPHAN
TSATLEE SILK
TSPP
TUBES
TUBULAR FABRICS
TUBULAR KNITTED FABRICS
TUCK LACES
TUCK STITCHES
TUCKING (GARMENT MANUFACTURE)
TUCKING (KNITTING)
TUCKS
TUFT FORMERS
TUFTED BEDSPREADS
TUFTED CARPETS
TUFTED FABRICS
TUFTING
TUFTING MACHINES
TUFTING NEEDLES
TUFTS (FIBER)
TUMBLE DRYING
TUMBLERS (ROLLS)
TUMBLING
TUNGSTEN COMPOUNDS
TURBIDITY
TURBOSTAPLER
TURBULENCE
TURBULENT FLOW
TURNS PER INCH
TUSSAH SILK
TWEENS (TN)
TWILL ANGLES
TWILL CHECKBOARD
TWILL SET
TWILL WEAVES
TWILLS
TWINE
TWIST
TWIST BREAK METHOD (TWIST TESTING)
TWIST CONSTANT
TWIST CONTRACTION
TWIST CONTROL
TWIST DETERMINATION
TWIST DISTRIBUTION
TWIST GEARS
TWIST LIVELINESS
TWIST MULTIPLE
TWIST MULTIPLIER
TWIST RUN BACK
TWIST SENSE
TWIST SETTERS
TWIST TESTERS
TWIST TESTING
TWIST TRIANGLE
TWIST UNTWIST METHOD (TWIST TESTING)
TWIST VARIATION
TWISTERS
TWISTING
TWISTING AT THE HEAD
TWISTING BOBBINS
TWISTING DYNAMICS
TWISTING IN
TWISTING MACHINES
TWISTING MOMENTS
TWISTING TENSION
TWISTING TORQUE
TWISTLESS
TWO BATH PROCESS
TWO COLOR DYEING
TWO FOLD SINGLES
TWO FOLD YARNS
TWO FOR ONE TWISTERS
TWO FOR ONE TWISTING
TWO PHASE PRINTING
TWO PLY YARNS
TWO POINT FIVE PERCENT SPAN LENGTH
TWO STAGE PRINTING
TWO THREAD CHAIN STITCHES
TWO TONE DYEING
TWO WAY STRETCH
TYCORA (TN)
TYING
TYING IN
TYING IN MACHINES
TYNEX (TN)
TYPEWRITER RIBBONS
TYPP COUNT
TYRE FABRICS
TYRES
TYROSINE
ULTIMATE ELONGATION
ULTIMATE STRENGTH
ULTIMATE TENACITY
ULTRACENTRIFUGES
ULTRAMARINE BLUE
ULTRAMICROSCOPY

ULTRASONICS
ULTRAVIOLET ABSORBERS
ULTRAVIOLET IRRADIATION
ULTRAVIOLET MICROSCOPY
ULTRAVIOLET RADIATION
ULTRAVIOLET SPECTROSCOPY
ULTRAVIOLET STABILIZERS
ULTRAVIOLET STERILIZERS
UMBRELLA CLOTH
UMBRELLAS
UNBALANCED SHED
UNCONVENTIONAL YARNS
UNCUT PILE FABRICS
UNDERLAP
UNDERLAY (CARPETS)
UNDERWEAR
UNEVEN SELVEDGES
UNEVENNESS
UNIAXIAL STRENGTH
UNIAXIAL STRESS
UNICONER (TN)
UNIDIRECTIONALLY ORIENTED WEBS
UNIFIL (TN)
UNIFORMITY
UNIFORMITY RATIO
UNIFORMS
UNION DYEING
UNIT CELLS (NONWOVEN FABRICS)
UNIT OPERATIONS
UNLEVEL DYEING
UNLOADING (PROCESS)
UNLUBRICATED RINGS
UNRAVELLING
UNROLLING
UNSATURATED ALIPHATIC COMPOUNDS
UNSATURATED COMPOUNDS
UNSATURATED HYDROCARBONS
UNWINDERS (CLOTH)
UNWINDING
UNWINDING ACCELERATORS
UNWINDING BALLOONS
UNWINDING DYNAMICS
UPHOLSTERY
UPHOLSTERY FABRICS
UPHOLSTERY LEATHER
UPKEEP
UPLAND COTTON
UPPER CONTROL LIMIT
UPPER HALF MEAN LENGTH
UPPER QUARTILE LENGTH
UPPERS
UPTAKE
UPTWISTERS
UPTWISTING
UPTWISTING FRAMES
UREA
UREA BISULFITE SOLUBILITY
UREA DERIVATIVES
UREA-FORMALDEHYDE
UREA-FORMALDEHYDE CONDENSATES
UREA-FORMALDEHYDE PRECONDENSATES
UREA-FORMALDEHYDE RESINS
URETHANE FOAMS
URETHANE RUBBERS
URETHANES
URON-FORMALDEHYDE CONDENSATES
USTER IRREGULARITY TESTER (TN)
V BELTS
VACUUM
VACUUM BAG MOLDING
VACUUM CLEANERS
VACUUM COATING
VACUUM DRYERS
VACUUM DRYING
VACUUM DYEING
VACUUM END COLLECTORS
VACUUM EXTRACTION
VACUUM PUMPS
VALENCIENNES
VALUE (MUNSELL)
VALVES
VAMPS
VAN DER WAALS FORCES
VAPOCOL PROCESS (TN)
VAPOR PHASE
VAPOR PHASE DYEING
VAPOR PHASE FINISHING
VAPOR PHASE TREATMENTS
VAPOR PRESSURE
VARIABILITY
VARIABLE CAPACITORS
VARIABLE SPEED
VARIABLE SPEED DRIVES
VARIABLES
VARIANCE
VARIANCE LENGTH CURVES
VARIATION
VAT DYEING
VAT DYES
VAT ESTER DYES
VAT PRINTING
VECTORS
VECTRA NYLON (TN)
VECTRA POLYESTER (TN)
VECTRA POLYPROPYLENE (TN)
VECTRA SARAN (TN)
VEGETABLE FATS
VEGETABLE FIBERS (GENERAL)
VEGETABLE MATTER
VEILING (LACE)
VELCRO (TN)
VELOCITY
VELOCITY (RELATIVE)
VELOCITY CHANGE POINT
VELON (TN)
VELOUR
VELVET
VELVET CARPETS

VELVETEEN
VENICE (LACE)
VENTILATING INSOLES
VERDURES TAPESTRIES
VEREL (TN)
VERTAFORMERS
VESSELS
VESTAN (TN)
VESTMENTS
VESTS
VIBRASCOPE
VIBRATION
VIBRATION DAMPERS
VIBRATION DAMPING
VIBRATION ISOLATION
VIBRATION ISOLATORS
VICARA (TN)
VICTORIA LAWN
VICUNA
VIGOUREUX PRINTING
VINAL POLY(VINYL ALCOHOL) FIBERS
VINYL ACETATE
VINYL BROMIDE
VINYL CHLORIDE
VINYL COATED FABRICS
VINYL COMPOUNDS
VINYL ETHERS
VINYL FIBER BLENDS
VINYL FIBERS (GENERAL)
VINYL FOAMS
VINYL GROUPS
VINYL MORPHOLINE
VINYL PYRIDINE
VINYL PYRROLIDONE
VINYLATED COTTON
VINYLATED DYES
VINYLATION
VINYLIDENE COMPOUNDS
VINYLON (TN)
VINYLSULFONE DYES
VINYON
VINYON POLY(VINYL CHLORIDE) FIBERS
VIRGIN WOOL
VISCOELASTICITY
VISCOSE
VISCOSE RAYON
VISCOSE SOLUTION
VISCOSIMETERS
VISCOSITY
VISCOUS DAMPING
VISCOUS FLOW
VITREOUS FIBERS
VOIGT MODEL
VOILE
VOLTAGE
VOLTAGE REGULATION
VOLTEX FABRIC (TN)
VOLUME
VOLUME TEMPERATURE MEASUREMENTS
VOLUME WEAR
VOLUMETRIC ANALYSIS
VYCRON (TN)
VYRENE (TN)
WADDINGS
WAGES
WALDRON SATURATOR
WALE SPACING
WALES
WALES PER INCH
WARMTH
WARP BEAMING
WARP BEAMS
WARP DYEING
WARP DYEING MACHINES
WARP EFFECT (TWILL)
WARP ENDS
WARP FACE
WARP FILLING BLENDS
WARP KNIT FABRICS
WARP KNITTED FABRICS
WARP KNITTING
WARP KNITTING MACHINES
WARP LACE MACHINES
WARP MILLS
WARP PREPARATION
WARP REPP
WARP SHRINKAGE
WARP STREAKS
WARP TENSION
WARP TYING IN MACHINES
WARP WIND
WARP YARNS
WARPERS (MACHINE)
WARPING
WARPING LEASE
WARPS
WARPWISE STRETCH
WASH WEAR FABRICS
WASH WEAR FINISHES
WASH WEAR PROPERTIES
WASH WEAR RATING
WASH WEAR TREATMENTS
WASHABILITY
WASHERS
WASHFASTNESS (COLOR)
WASHFASTNESS (OF FINISH)
WASHFASTNESS TESTERS
WASHFASTNESS TESTING
WASHING (LAUNDERING)
WASHING (PROCESSING)
WASHING MACHINES
WASHING MACHINES (LAUNDERING)
WASHING MACHINES (PROCESSING)
WASTE
WASTE CONTROL KNIVES
WASTE DISPOSAL
WASTE HEAT RECOVERY

WASTE MACHINES
WASTE TREATMENT
WATER
WATER ABSORPTION
WATER ABSORPTION TESTING
WATER ABSORPTION TESTS
WATER BRAKES
WATER IMBIBITION
WATER IN OIL EMULSIONS
WATER JET LOOMS
WATER PERMEABILITY
WATER PERMEABILITY TESTING
WATER POLLUTION
WATER PROOFING
WATER REPELLENCY
WATER REPELLENCY TESTING
WATER REPELLENCY TREATMENTS
WATER REPELLENTS
WATER RESISTANCE
WATER RESISTANCE TESTERS
WATER RESISTANCE TESTING
WATER RESISTANT FOOTWEAR
WATER RESISTANT LEATHER
WATER RETENTION
WATER SOLUTIONS
WATER SPOTS
WATER TREATMENT
WATER VAPOR
WATER VAPOR TRANSMISSION
WATER VAPOR TRANSMISSION TESTING
WATERPROOFING
WATERPROOFING AGENTS
WAVE LENGTH
WAVINESS
WAX EMULSIONS
WAXES
WEAK SPOTS
WEAR
WEAR RESISTANCE
WEAR TESTING
WEARING (OF APPAREL)
WEATHER RESISTANCE
WEATHERING
WEATHERING TESTING
WEAVABILITY
WEAVE
WEAVE (TYPES)
WEAVERS KNOTS
WEAVING
WEAVING DYNAMICS
WEAVING EFFICIENCY
WEAVING LEASE
WEB FORMING MACHINES
WEB ORIENTATION
WEB PURIFIERS
WEB UNIFORMITY
WEBBING
WEBHOLDERS
WEBS
WEDGE SCALE
WEDGING
WEFT
WEFT FORKS
WEFT INSERTION
WEFT KNITTED FABRICS
WEFT KNITTING MACHINERY
WEFT LOOPS
WEFT PIRNS
WEFT STRAIGHTENERS
WEFT STREAKS
WEFT YARNS
WEIGHT
WEIGHT AVERAGE MOLECULAR WEIGHT
WEIGHT CONTROL
WEIGHT PER UNIT LENGTH
WEIGHT UNIFORMITY
WELDING
WELSH MOUNTAIN WOOL
WELTING
WELTS (FOOTWEAR)
WELTS (KNITTED)
WELTS (SEWN)
WEST POINT COHESION TESTER
WET BULB DEPRESSION
WET BULB TEMPERATURE
WET CREASE RECOVERY
WET CROCKING
WET CURING
WET DECATING
WET FINISHING
WET LAY RANDOM WEBS
WET MACHINES (PAPERMAKING)
WET MODULUS
WET PICKUP
WET PROCESS (WEB)
WET PROCESSING
WET PROCESSING RANGES
WET RELAXATION
WET RELAXING
WET SPINNING
WET STRENGTH
WET TESTING
WET WRINKLE RECOVERY
WETTABILITY
WETTING
WETTING AGENTS
WETTING TIME
WHEEL BRAKES
WHIP ROLLS
WHIPCORD
WHISKER FIBERS (METAL)
WHITE CHINA SILK
WHITE JAPAN SILK
WHITENESS
WHITENESS RETENTION
WHITENING
WICK LUBRICATED RINGS
WICKING

WIDENING
WIDENING (KNITTING)
WIDTH
WIGAN
WILD YARNS
WILDNESS
WILLEYS
WILLIAMS UNIT (TN)
WILLIAMSON DICKIE (TN)
WILLOWS
WILTON
WILTON CARPETS
WILTON LOOMS
WINCH DYEING
WIND
WIND TUNNELS
WIND UP
WINDERS
WINDING
WINDING ANGLE
WINDING DYNAMICS
WINDING EFFICIENCY
WINDING MACHINERY
WINDING ON
WINDING TENSION
WINDOW BLINDS
WINDOW HOLLAND
WINDOW SHADES
WIPING CLOTHS
WIRA WEB LEVELNESS TESTER
WIRE DRAWING
WIRE FILLET CLOTHING
WIRE ROPES
WIRES
WOOD
WOOD FIBERS
WOOD FIBRILS
WOOD PULP
WOOL
WOOL BALES
WOOL BLENDS
WOOL CLASS
WOOL CLASSING
WOOL DYEING
WOOL DYES
WOOL FABRICS
WOOL FIBERS
WOOL FLEECES
WOOL GRADE
WOOL GRADING
WOOL GREASE
WOOL SCALES
WOOL SHEARING
WOOL SORTING
WOOL WAX
WOOLEN CARDING
WOOLEN CARDS
WOOLEN DRAFTING
WOOLEN FABRICS
WOOLEN SPINNING FRAMES
WOOLEN SPUN YARNS
WOOLEN SYSTEM PROCESSING
WOOLEN SYSTEM SPINNING
WOOLEN YARNS
WOOLENS
WOOLENS AND WORSTEDS
WORK CLOTHING
WORK RECOVERY
WORK TO BREAK
WORKABLE FIBERS
WORKERS
WORKING ACTION
WORKLOADS
WORSTED CARDING
WORSTED CARDS
WORSTED COMBING
WORSTED COUNT
WORSTED FABRICS
WORSTED LASTING
WORSTED SPINNING
WORSTED SPUN YARNS
WORSTED SYSTEM PROCESSING
WORSTED SYSTEM SPINNING
WORSTED YARNS
WORSTEDS
WOVEN CARPETS
WOVEN FABRIC STRUCTURE
WOVEN FABRICS
WOVEN FELTS
WOVEN NETS
WOVEN TUBES
WRAP
WRAP UP
WRAPPED YARNS
WRAPPING MACHINES
WRINKLE RECOVERY
WRINKLE RECOVERY ANGLE
WRINKLE RECOVERY FINISHES
WRINKLE RECOVERY TESTERS
WRINKLE RECOVERY TESTING
WRINKLE RESISTANCE
WRINKLE RESISTANCE FINISHES
WRINKLE RESISTANCE TREATMENTS
WRINKLES
WRINKLING
WRONG DRAW
WURLAN (TN)
X RAY ANALYSIS
X RAY DIFFRACTION
X RAY MICROSCOPY
X RAY SPECTROSCOPY
X RAYS
XENOTEST (TN)
YAK
YARN BALLOONS
YARN BREAKAGES
YARN BUCKLING
YARN BUNDLE COHESION

YARN CARRIERS
YARN CHANGING UNIT (KNITTING)
YARN CLEARERS
YARN COUNT
YARN COVER
YARN CRIMP
YARN CROSS SECTIONS
YARN DAMAGE
YARN DEFECTS
YARN DENIER
YARN DIAMETER
YARN DYED FABRICS
YARN DYEING
YARN DYEING MACHINES
YARN DYNAMICS
YARN FINISH
YARN FLATTENING
YARN GEOMETRY
YARN GUIDES
YARN LUSTER
YARN PREPARATION
YARN PROPERTIES
YARN REINFORCED COMPOSITES
YARN REINFORCED ELASTOMERS
YARN SEVERANCE
YARN SLIPPAGE
YARN STRENGTH
YARN STRENGTH EFFICIENCY
YARN STRUCTURE
YARN TENSION
YARN TO METAL FRICTION
YARN TO YARN FRICTION
YARNS
YARNS (CONTINUOUS FILAMENT)
YEARS COVERAGE (100)
YEARS COVERAGE (10)
YEARS COVERAGE (1)
YEARS COVERAGE (25)
YEARS COVERAGE (50)
YEARS COVERAGE (5)
YEARS COVERAGE (75)
YEAST
YELLOW CHINA SILK
YELLOW JAPAN SILK
YELLOWING
YIELD (MECHANICAL)
YIELD (RETURN)
YIELD POINT
YIELD STRESS
YIELDING
YORKSHIRE SKEIN WOOLEN COUNT
YOUNGS MODULUS
Z CRIMP
Z TWILLS
Z TWIST
Z TWIST YARNS
ZEFKROME (TN)
ZEFRAN (TN)
ZEIN
ZELLWEGER TESTER (TN)
ZERO TWIST
ZETA POTENTIAL
ZIG ZAG MACHINES
ZINC CHLORIDE
ZINC COMPOUNDS
ZINC FLUOBORATE
ZINC FORMALDEHYDE SULFOXYLATE
ZINC NITRATE
ZINC OXIDE
ZINC SOAPS
ZINC STEARATE
ZINC SULFATE
ZIPPERS
ZIRCONIUM COMPOUNDS
ZWITTERIONIC SURFACTANTS

REFERENCES USED IN PREPARATION OF THE THESAURUS

1. Science, Government and Information, A Report of the President's Science Advisory Committee, The White House, Washington, D. C., January 10, 1963.
2. Current Research and Development in Scientific Documentation, No. 13, National Science Foundation, November 1964. (See also Nos. 1 through 12 of the same publication.)
3. Specialized Science Information Services in the United States, National Science Foundation, NSF-61-68, November 1961.
4. Scientific Information Services for the Textile Industry, Proceedings of a Special Seminar held at the School of Textiles of North Carolina State College of the University of North Carolina at Raleigh, October 1963.
5. Textracts, Published by J. B. Goldberg, 11 West 42nd Street, New York, New York.
6. Thesaurus of Engineering Terms, 1st Edition, Engineers Joint Council, New York, May 1964.
7. Costello, J.C. Jr., "Storage and Retrieval of Chemical Research and Patent Information by Links and Roles in du Pont", American Documentation, Vol. 12, pp. 111-120, April 1961.
8. Opler, A., "Dow Refines Structural Searching", Chemical and Engineering News, Vol. 35, No. 33, pp. 92-96, August 1957.

9. Ranganathan, S.R., "Self-perpetuating Scheme of Classification", Journal of Documentation, Vol. 4, pp. 223-250, March 1949.

10. Luhn, H. P., "Keyword-in-context Index for Technical Literature (KWIC Index)", IBM Advanced Systems Development Division, Yorktown Heights, New York, 1959.

11. Kessler, M.M., "Concerning some Problems of Intra-Science Communication", Lincoln Laboratory Group Report 45-35, Lexington, Massachusetts, December 1958.

12. Holm, Bart E., "Information Retrieval--The Problem.... Coordinate Indexing--A Solution", Engineers Joint Council, New York, August 1963.

13. Frank, O. (Editor), Modern Documentation and Information Practices, International Federation for Documentation, The Hague, FID Publ. 334, 1961.

14. Costello, J. C. Jr., Training Manual and Workbook for use in Abstracting and Coordinate Indexing, Information Systems Research Division, Battelle Memorial Institute, July 1, 1964.

15. Classification, Class T, Technology, Fourth Edition, Subject Cataloging Division, Library of Congress, Washington, D.C., 1948, Reprinted 1964, also

Subject Headings, Sixth Edition, Subject Cataloging Division, Library of Congress, 1957 + Annual Supplements through 1964.

16. Annual Subject Index, The Journal of the Textile Institute Abstracts, Vol. 47, Page 228, 1956.

17. Annual Subject Index, The Journal of the Textile Institute Abstracts, Vol. 53, page 114, 1962.

18. Information Publications of the Engineers Joint Council, 345 East 47th Street, New York, New York 10017.

(a) Study of Engineering Terminology and Relationships among Engineering Terms, Final Report, August 1963.

(b) Proposed EJC Course of Action in the Field of Information Systems, September 21, 1962.

(c) Information Systems--Essential Tools in Engineering Application of Science for the Needs of Society. Proceedings of a Panel Program jointly sponsored by EJC and Section M (Engineering) - AAAS, December 27, 1962.

(d) Information Retrieval Thesauri by Eugene Wall, 1962.

(e) Small Scale Information Retrieval Systems by Eugene Wall, August, 1962.

(f) The Mechanization of Information Dissemination by Eugene Wall, October 1962.

(g) Proceedings of the Engineering Information Symposium, 1962.

(h) Information Handling Systems and Technical Information Flow, 1962.

(i) Information Retrieval--The Problem.....Coordinate Indexing--A Solution, by Bart E. Holm, August 1963.

(j) Thesaurus of Engineering Terms, 1st Edition, May 1964.

19. Taube, M., "Specificity in Subject Headings and Coordinate Indexing", Library Trends, Vol. 1, No. 2, pp. 219-223, October 1952.

20. Jaster, J.J., Murray, B.R. and Taube, M., The State of the Art of Coordinate Indexing, Preliminary Edition. Prepared for the Office of Science Information Service, National Science Foundation, February 1962.

21. Annual Editorial Index, Hydrocarbon Processing and Petroleum Engineer, Vol. 42, No. 12, December 1963.

22. Morse, R., "Information Retrieval", Chemical Engineering Progress, Vol. 57, No. 5, pp. 55-58, May 1961.

23. Costello, J.C. Jr., "A Basic Theory of Roles as Syntactical Control Devices in Coordinate Indexes", Journal of Chemical Documentation, Vol. 4, pp. 116-124, 1964.

24. Holm, Bart E., "Information Retrieval--A Solution", Chemical Engineering Progress, Vol. 57, No. 6, pp. 73-78, June 1961.

25. Cleverdon, C. W., "Report on the Testing and Analysis of an Investigation into the Comparative Efficiency of Indexing Systems", ASLIB Cranfield Research Project, Cranfield, England, October 1962.

26. Overhage, C.F.J., and Harman, R.J., "INTREX: Report of a Planning Conference on Information Transfer Experiments", M.I.T. Press, Cambridge, Massachusetts, 1965.

27. Hamby, D.S., "The American Cotton Handbook", Interscience Publishers, New York, 1966.

28. Burlington Industries, Inc., "Textile Fibers and Their Properties", Greensboro, North Carolina, 1965.

29. Linton, G.E., "The Modern Textile Dictionary", Little Brown and Company, Boston, 1954.

30. Marks, S.S., "Fairchild's Dictionary of Textiles",

Fairchild Publications Inc., New York, 1959.

31. Kaswell, E.R., "Wellington Sears Handbook of Industrial Textiles", Wellington Sears Company, New York, 1963.

32. "Textile Terms and Definitions", The Textile Institute, Manchester, England, 1963 (5th Edition).

33. Solinger, J., "Apparel Manufacturing Analysis", Interscience Publishers, Inc., New York, 1961.

34. Grover, E.B., and Hamby, D.S., "Handbook of Textile Testing and Quality Control", Interscience Publishers Inc., New York, 1960.

35. Goldberg, J.B., "Fabric Defects", McGraw-Hill Book Company, New York, 1950.

36. Moncrieff, R.W., "Man-Made Fibres", John Wiley & Sons Inc., New York, 1957.

37. Wray, G.R., "Modern Yarn Production", Columbine Press, Manchester, 1960.

38. Marsh, J.T., "An Introduction to Textile Finishing", Chapman & Hall Ltd., London, 1951.

39. "Saco-Lowell Handbook of Engineering and Technical Data", Saco-Lowell Shops, Boston, Massachusetts, 1958.

40. Hoye, S., "Staple Cotton Fabrics", McGraw-Hill Company, New York, 1942.

41. Nielson, L.E., "Mechanical Properties of Polymers", Reinhold Publishing Corp., New York, 1962.

42. Roff, W.J., "Fibres Plastics and Rubbers", Butterworths Scientific Publications, London, 1956.

43. Flory, P.J., "Principles of Polymer Chemistry", Cornell University Press, Ithaca, New York 1953.

44. Coulson, J.M., and Richardson, J.F., "Chemical Engineering", Pergamon Press, New York, 1956.

45. Harris, M., "Handbook of Textile Fibers", Harris Research Laboratories, Washington, D. C., 1954.

46. Stephenson, J.N., "Manufacture and Testing of Paper and Board", McGraw-Hill Book Company, New York, 1953.

47. "The Science of Color", Committee on Colorimetry, Optical Society of America, Thomas Crowell Company, New York, 1953.

48. "International Dictionary of Applied Mathematics", D. Van Nostrand Company, Princeton, New Jersey, 1960.

49. Lennox-Kerr, P., "Index to Man-Made Fibres of the World", Harlequin Press, Manchester, 1964.

50. Thesaurus of Pulp and Paper Terms, Pulp and Paper Research Institute of Canada, Pointe Claire, Quebec, 1965.

51. Classification Definitions, U. S. Patent Office, Washington, D. C. , 1961, 1962 (Textile and Apparel Terms).

52. Susskind, C., "The Encyclopedia of Electronics", Reinhold Publishing Company, New York, 1962.

53. Clark, G.L., "The Encyclopedia of Microscopy", Reinhold Publishing Company, New York, 1961.

54. Meredith, R. and Hearle, J.W.S., "Physical Methods of Investigating Textiles", Interscience Publishers Inc., New York, 1959.

55. Hearle, J.W.S., and Peters, R.H., "Fibre Structure", Butterworths, Manchester, 1963.

56. Von Bergen, W., Wool Handbook, Interscience Publishers Inc., New York, 1963.

57. Courtaulds Vocabulary of Textile Terms, Courtaulds Limited, Compiled by R.J.E. Savill, London, 1964.

58. Burgers, J.M. and Scott Blair, G.W., Report on the Principles of Rheological Nomenclature, North Holland Publishing Co., Amsterdam, 1949.

59. Glossary of Plastics Terms, A Consensus, Plastics Technical Evaluation Center, Picatinny Arsenal, Dover, New Jersey, December 1966.

60. Reichman, Charles, Knitting Dictionary, National Knitted Outerwear Association, New York, 1966.

61. Thesaurus of Engineering and Scientific Terms, Engineers Joint Council, New York, 1967.

62. Sharp, J.R., Some Fundamentals of Information Retrieval, Andre Deutsch Ltd., London, 1965.

APPENDIX A

Problems of Textile Information Retrieval

INTRODUCTION 340

NEEDS OF THE INFORMATION USER 342

INPUT TO THE INFORMATION STORE 345

COORDINATION WITH OTHER SYSTEMS 353

COORDINATE INDEXING 354

THE THESAURUS AND ITS OBJECTIVES 358

LINKS AND ROLES IN COORDINATE INDEXING 363

DEPTH OF INDEXING 366

SEARCHING BY REFERENCES 372

BROWSING 375

ACCESS TO DOCUMENTS 377

INFORMATION FLOW VELOCITY 378

SELECTION OF A COMPUTER FILE STRUCTURE 381

COURSE OF FUTURE ACTIVITY 389

REFERENCES 391

APPENDIX A

Problems of Textile Information Retrieval

INTRODUCTION

Is the current concern with the technical information explosion a legitimate worry? Does the research man have too much or too little information?

The size of both industrial and government budgets for research designed to develop new information suggests that the research man has too little information; whereas the magnitude of national spending on information retrieval suggests that he may have too much. The fact is that there is too little usable pertinent information located near and accessible to the technical working areas. Research managers are concerned about the effectiveness and efficiency of individual research efforts in given problem areas, and not knowing of nor being sure of prior work done by others in the same or related fields and publicly reported.

Textile technology, in common with other fields of applied and pure science, requires adequate communication. There is a vast store of human knowledge in the theory and techniques of information transfer quite outside the field of textiles that should be freely utilized in approaching our own textile problems. Similar informational efforts can be found around the world, in proportion to human effort devoted to science and technology.

The textile field can benefit from the experience of other disciplines. This may be derived through the borrowing of techniques relating to specific informational problems, those relating to "hardware" such as filing systems, microfilm devices, and computers to "software"-filing procedures, film organization and computer programs and to the theory of information-retrieval, for example, probability theories. In studying the experience and designs in other areas, we must be aware of the differences which exist between other subjects and the textile field, relating to the scope of information encountered, types of documents available, different users of the information, varying uses to which it will be put, and accumulated experience and habits of information transfer in textiles. An effective textile information system, therefore, must tie in with local (national) information networks in the other disciplines as well as with international textile networks.

We are concerned here with recorded information, with emphasis on publicly available information. This means the collection of knowledge, documents, and facts, that are conveyed by books, journals, patents, translations, catalogs, drawings, specifications, lectures, seminars, speeches, films, and magnetic tapes. Clearly, the physical format for information transfer often dictates the style and arrangement of the information.

The audience, too, is a determinant in establishing the type of information formulated in each of the transfer media mentioned, just as it is a key factor in designing an information center with a view towards optimum consumer satisfaction.

NEEDS OF THE INFORMATION USER

Satisfaction for the information user is, of course, a relative matter. Once he has asked a question of an information system, the user wants assurance of a high probability that the documents containing the answers to his question will be retrieved and delivered to him. He wants retrieval and delivery of documents in a reasonable time and with a minimum of cost. And he wants a minimum of nonpertinent documents included in the retrieval.

Obviously, the importance ranking of these factors in assessing satisfaction will vary with the needs of the consumer. The plant engineer who needs a few ideas on how to overcome the causes of today's process stoppage wants a few pertinent documents--but quick. He does not want to wade through a mass of irrelevant papers. The patent lawyer, on the other hand, puts a premium on complete retrieval of pertinent documents, he does not mind a reasonable wait, and he is willing to accept a larger share of nonpertinent material if the assurance of receiving useful material is maximized. The satisfaction specifications of the research scientist or engineer can be said to parallel those of the patent lawyer, though the type of documents retrieved for the researcher will often differ considerably. The needs of the technical service specialist are likely to be intermediate to those of the plant engineer and of the research scientist. The design of a textile information retrieval system must take into account these varying needs of the information consumers, and therefore we have a real need for a count and for a classification of active or potentially active users of an industrywide textile information system. Professional textile societies can perform an important service in studying the information needs of the industry and in determining the size of each user class in the industry. But whoever undertakes such a survey must keep the following in mind:

(1) Most technical personnel in industry and in the government agencies fill but a small part of their information needs directly from the open literature. If one probes further, one observes the tendency for people to consult other staff members of their own organization rather than to turn directly to the professional literature. But in each organization there exist a limited number of technical experts who closely follow the professional literature and, in passing information to others, serve in effect as technological gatekeepers[1] between the outside literature and their company associates. Clearly, the information needs of these specialists must be given particular attention in assessing information needs of the industry, and more weight should be given to their requirements than to the projected needs of the infrequent user of the literature.

(2) Technical personnel in industry will hesitate to use an industrywide information center if there is a chance of security leaks revealing the areas of their interest.

(3) Industry will not make much use of an information

center if the boundaries of its data base are too limited or are poorly defined,

(4) The potential market for an efficient, sophisticated, computer-based information service cannot be estimated from questionnaires submitted to people who are unaware of the power and versatility of a well-designed retrieval system. Experience in estimating the need for computers in large organizations shows that people simply cannot project their requirements until they have had some experience in using actual hardware and in developing software to solve current technical problems. And once this experience permeates the organization, the expressed need for computers tends to increase exponentially.

The results of such a survey will also provide a rational basis for specifying the type of information to be included in a textile information system and the subject matter to be covered. In short, once consumer specifications are at hand, one can proceed to a detailed determination of suitable input to the information store. We can, for the time, consider the potential users of the system to include all fiber producers, textile manufacturers, textile finishers, manufacturers of textile machinery, textile auxiliaries, and dyestuffs, clothing manufacturers and distributors, industrial-textile-user industries (such as tire manufacturers), as well as the individual consumer. The textile manufacturing and finishing industry stands at the center of interest in the system.

On the basis of this general user profile one can attempt an initial approach to the question of input to the information store.

INPUT TO THE INFORMATION STORE

Provision should, of course, be made for inclusion of primary literature sources that touch on textile matters in their numerous facets. Identification of the central textile theme of current interest offers no difficulty in considering input to the system. A simple indicator of such "central" material is the presence of the word "textile" in the title of the publication, or the name of textile or clothing product or machine or company. There are over 125 such journals published monthly in the world and so the question of pure textile input is dealt with almost automatically.

On the other hand, many difficult problems arise in identifying items or channels of information that bear on peripheral themes of _current_ textile interest and still more difficult, in tagging peripheral information packages of _future_ potential value on the textile scene. Some peripheral themes can be tagged as being of current textile interest either because of their similarity in composition to textiles, or because of their concern with fibrous materials. Thus one can expand the list of journals to be included in a hypothetical information warehouse by incorporating titles that contain such words as polymers, plastics, paper, wood, or even dyestuffs. The question remains as to how much of this peripheral material should be incorporated into the textile system.

An even more difficult question relates to the material published in general fields of science, engineering, technology, and management where research, technological advance, and management techniques are reported without any special reference to textile materials and processes. Yet this material can be shown to be of general interest to many segments of the textile industry and, in some cases, it may include applications of systems or techniques to actual textile problems.

The problem is to determine how much of this expansive field is to be introduced into the hypothetical store without inundating the system and its users as well. There are said to be well over 25,000 primary technical journals published in the world and 1,700 secondary journals (such as abstract publications). It remains to be determined how many of these primary and secondary journals belong in the proposed textile system. Should the input encompass nontextile journals which nevertheless frequently contain articles about metal and ceramic fibers, composites and plastics, strings, rods and beams, membranes and composite structures, and fiber reinforced structures? One can quickly compile a list of some 1000 journals with apparent possible utility for the textile industry.[2] How many of these journals should be subscribed to?

The specialized library staff located in different parts of the information system must assume major responsibility for deciding on suitable material input to the store.

But there is a need to share such responsibility with the professional societies in studying material in peripheral textile areas and in recommending its inclusion (or exclusion) in the industry's information system. Such a study will require a survey of world-wide titles in specified areas and an evaluation of their contents over a period of several years to ascertain their potential textile contribution. And once a journal or publication channel has been adopted, its contribution to the textile store must be reviewed from time to time to insure pertinence of title inclusion and continuity of input.

Still another method of judging the desirability of including a given journal in the store is to measure the nature of its interaction with the recognized textile literature. In setting up an information store at M.I.T. for the literature of physics, Kessler[3] developed a journal reference matrix in order to measure the flow of information between journals through the medium of references or citations. In the reference matrix, $J_m (m = 0,1,2,3....)$ represents a list of journals and J_{mn} represents the percentage or number of references in journal m to journal n. As Kessler points out, when J_{mn} is sizable and roughly equals J_{nm}, it means that journals m and n maintain a strong, two-way communication. If J_{mn} is large and J_{nm} small, then it can be said that journal m uses journal n as a source but does not contribute to journal n as a significant reference. If both J_{mn} and J_{nm} are small, it follows that journals m and n have little interaction with one another, as judged through the medium of reference flow.

The brief study reported here is based on the **Kessler** matrix limiting the reference counts to the following **five** principal journals: Textile Research Journal (TRJ) = **0;** Journal of the Textile Institute--Transactions (JTI) = **1;** Journal of the Textile Machinery Society of Japan **(TMSJ) =** 2; Journal of the Society of Dyers and Colourists (SDC) = **3;** Transactions of the Society of Rheology (TSRh) = 4. Preliminary results for 1963-1965 for these journals are shown in Table I in actual rather than relative counts. The data indicate

TABLE I. Textile Journal Reference Matrix

From → / To ↓	TRJ J_0	JTI J_1	TMSJ J_2	SDC J_3	TSRh J_4
J_0 TRJ	1542	269	26	53	0
J_1 (JTI)	562	435	36	57	0
J_2 (TMSJ)	4	3	66	0	0
J_3 (SDC)	272	22	2	253	0
J_4 (TSRh)	0	0	0	0	87

strong reference **interactions** between TRJ, JTI, and SDC. The reference flow **between TMSJ** and TRJ or JTI is somewhat one-sided at present, **and the** TMSJ interaction with SDC is negligible. Difference **in** subject coverage of these journals often accounts for apparent lack of information exchange, but sometimes it is simply a matter of poor communications. TSRh has essentially no reaction with the prime textile journals, a rather surprising finding considering the important contributions made to the field of rheology by textile researchers.

Reference counts from the primary textile research journals include data on numerous other journals. In Table II is listed a large number of those journals cited more than a few times by the key journals--TRJ, JTI, SDC, TMSJ, and TSRh, Careful study of Table II will provide considerable insight on the flow of information to our primary textile research journals from different sources, including the following:

(a) Journals of the classical science (e.g., Organic Chemistry);

(b) Journals of peripheral sciences (e.g., Polymer Science);

(c) Journals of peripheral technology (e.g., TAPPI);

(d) Textile journals printed in foreign languages (e.g., Institut Textile de France); and

(e) Journals with textile mill orientation (e.g., Textile World).

In this reference count evidence is noted of but a limited flow from textile trade journals to the selected research journals. This is to be expected, for the normal flow tends to be from the primary (or research) journal level to the secondary (trade, review, or evaluative) journals.

TABLE II. Selected References in Key Textile Journals

	U. S. A. T.R.J.	British J.T.I.	Japanese T.M.S.J.	U. S. A. Rheol.	British D. & C.
American Dyestuff Reporter	247	15	1	0	75
Am. Soc. Testing Materials	80	10	1	2	0
Analytical Chemistry	32	4	0	2	8
Annales Scien. Textile Belges	8	6	1	0	0
Australian J. Agricul. Res.	3	4	0	0	0
Australian J. Biological Scien.	38	2	0	0	0
Australian J. of Chemistry	21	2	0	0	0
Biochemica et Biophysica Acta	48	6	0	0	0
Biochemical Journal	60	15	0	0	4
British J. of Appl. Physics	11	9	5	3	0
Bull.de l'Institut Text.de France	33	6	1	0	0
Canadian J. of Chemistry	3	1	0	0	10
Canadian Textile Journal	5	0	0	0	3
Chemical Abstracts	40	3	0	0	13
Chem. & Engineering News	3	1	0	0	0
Chemistry and Industry	15	4	0	0	5
Chem. of High Polymers, Japan	6	1	2	3	0
Dyer	5	0	0	0	15
Faraday Soc. (Trans. & Disc.)	71	12	0	9	33
Faserforschung & Textil-technik, Berlin	14	1	0	0	0
Helvetica Chemica Acta	9	1	0	0	12
Indian Central Cotton (tech. bull.)	5	0	0	0	0
Indian Cotton Grow. Rev.	4	1	0	0	0
Indian Textile Journal	3	4	0	0	0
Industrial & Eng'g. Chemistry	65	3	0	18	7
Ind. & Eng'g. Chem. Anal. Ed.	14	2	0	4	0
Investigacieninformacion Text., Barcelona	2	3	0	0	0
J. Acoustical Soc., Japan	0	0	4	0	0
J. Am. Chemical Society	128	5	0	15	28
J. Am. Oil Chemist Soc.	5	4	0	0	0
J. Applied Chemistry, London	6	0	0	2	0
J. Applied Mechanics	6	0	0	9	0
J. Applied Physics	37	8	5	74	4
J. Applied Poly. Sci., N. Y.	63	5	0	20	11
J. of Bacteriology	40	0	0	0	0
J. of Biological Chemistry	31	15	0	0	7
J. Chemical Physics	29	3	0	36	12
J. Chemical Society	47	4	0	0	17
J. Colloid Science	16	4	0	29	0
J. Optical Soc. of America	12	2	0	0	41
J. Polymer Science A.B.C.	162	15	0	14	16
J. Res. Nat'l. Bu. of Stands.	66	8	2	3	8
J. Soc. of Dyers & Colourists	272	22	2	0	253
J. Soc. Text. & Cellulose Inds. Japan	22	5	18	0	0
J. Textile Institute	562	435	36	0	57
J. Text. Machy. Soc., Japan	5	3	66	0	0
Kolloid-Z & Zeitschrift für Polymere	35	5	1	14	0
Méilliand Textilberichte	67	15	3	0	42
Modern Textiles Magazine	3	1	1	0	0
Nature—London	96	123	0	16	20
Organic Chemistry	13	0	0	0	0
Paper and Timber	3	0	0	0	0
Physic Chem.	37	4	0	0	7
Polymer—London	7	2	0	2	0
Proceedings Phys. Society	11	8	8	3	0
Proceeding Royal Soc. A.B.	52	16	7	17	9
Shirley Institute Memo	8	47	0	0	0
Svensk Papperstidning	25	2	0	0	0
Tappi	30	1	0	0	4
Teintex	4	1	0	0	4
Textile Bulletin, Charlotte, N. C.	19	0	0	0	0
Textile Industries	20	0	0	0	0
Text. Inst. & Industry, Manchester, England	0	4	0	0	0
Textile J. of Australia	4	4	0	0	0
Textile Manufacturer, Manchester, England	7	4	0	0	0
Textil—Praxis	15	9	1	0	9
Textile Recorder	3	9	0	0	0
Textile Research Journal	1542	269	26	0	53
Textil—Rundschau	18	1	0	0	15
Textile Weekly	3	0	0	0	0
Textile World	5	2	2	0	0
Wool Science Review	5	2	0	0	0
Z-Ges-Textilind . . .	11	3	0	0	0

However, the existence of this normal flow is difficult to verify because of the failure on the part of many secondary journals to provide references to primary sources.

A breakdown of the Table II count into 5-year reference periods for selected journals provides additional information of value. For now it can be seen how far back the textile scientist goes to collect information for his current work, e.g., that published during 1963-1965. Table III demonstrates that a considerable portion of his references for recent papers come from sources 5 to 10 years old. An even more surprising trend is evident in the large number of references that come from literature more than 10 years old. Such data can be used as a basis for fixing the chronological boundaries of literature coverage in the hypothetical information system. These would be minimum boundaries, of course, since the data of Table III reflect the usage of references obtained by relatively inefficient methods of searching. It is likely that, given better retrieval methods, the textile scientist will make better use of literature pertinent to his field of inquiry, whether it appeared 1 year or 25 years ago.

For the present, one concludes that a comprehensive textile information system should include literature dating back more than 10 years (perhaps 15) Similar results from Kessler[4] for three major journals in the physics literature suggest that the average physicist relies for about 80% of his references on material that is less than 8 years old. No doubt, the useful age of references will vary from field to field in science, engineering, and technology.

TABLE III. Selected Textile References by Year

From → / To	TRJ 63-65 Issues			JTI 63-65 Issues		
	1960-65	1955-59	-1954	1960-65	1955-59	-1954
American Dyestuff Reporter	126	59	62	1	5	9
Am. Soc. Testing Materials	60	15	5	8	2	0
Biochemical Journal	10	15	35	2	6	7
Chemical Abstracts	6	12	22	0	0	3
Faraday Soc. (Trans. & Disc.)	21	11	39	3	0	9
Industrial & Engineering Chem.	4	12	49	0	0	3
J. of the Am. Chem. Soc.	20	25	83	0	0	5
J. Applied Polymer Sci.	61	2	0	5	0	0
J. Polymer Sci., A.B.C.	90	56	16	5	9	1
J. Soc. Dyers & Colourists	53	55	164	5	6	11
J. Textile Institute	194	104	265	170	104	161
Mélliand Textilberichte	19	18	30	2	5	8
Nature—London	28	30	38	19	9	95
Proceedings Royal Society	3	5	44	1	0	15
Tappi	13	12	5	0	1	0
Textile Research Journal	922	359	261	90	74	105

COORDINATION WITH OTHER SYSTEMS

Obviously, decisions for input selection cannot be made in an information vacuum. While decisions on incorporating core subject matter will flow easily, those choices relating to peripheral subjects must take into account the channels of communications in peripheral areas and the existence of information centers in outside fields possessing either major or minor textile implications. There is little point in building into the textile system extensive coverage on a subject such as wood pulp because of the textile industry's interest in nonwovens and paper, or on metal because of its interest in metallic fibers--not when the paper industry and the metals societies have organized information centers in their specialties. Clearly, the information activities of the Engineers' Joint Council must be studied in deciding on the input to the textile system, and particularly the chemical information system of the American Chemical Society, including the Polymer Science and Technology (POST) service. Likewise, textile planning should include consideration of the collections and fields of interest of the large government information centers, such as the National Library of Medicine, the Agricultural Library, the Smithsonian Institute, National Aeronautics and Space Administration, the Department of Defense Documentation Center, the Atomic Energy Commission, and the Clearing House for Federal Scientific and Technical Information. Many of these nontextile organizations collect documents (of both journal and report types) in given fields, issue abstract and announcement literature, and provide active search services which are available to industry and academic institutions. There is little need to undertake expensive duplication of these facilities. But it is imperative that their boundaries of interest and information coverage be clearly defined vis-à-vis textile subject interests. Similar provision must be made for coordination of input planning with the interests and experience of textile information centers overseas.

Implicit in this planned input to the hypothetical information center is the assumption that continuous effective communication will be maintained with the many information centers external to the textile industry. And the proposed system must take an active role in answering the peripheral subject needs of the textile industry by the aggressive pursuit of information normally flowing through external channels and centers.

COORDINATE INDEXING

Once we have decided what literature we wish to receive and store and what we wish to keep under surveillance, we must then set up a storage and retrieval system under compression ratios that stagger the imagination. By compression ratio is meant the ratio between the size of the information store and the size of the index used to point to its existence and to its contents. For example, in the case of a 5- or 600-page book we may find only four or five subject headings used to identify the entire informational content of this book, whereas in the case of journal articles ranging from two to twenty pages, we may have two to three subject headings per article. Clearly, it is impossible under such compressional restraints to reveal the detailed contents of our traditional documents using our traditional indices as identifiers of the information and pointers to its location.

It is possible to reduce the excessive compressive ratios which exist today, but this would mean expansion of the information contained in the indexing system. If this expansion were to be carried too far, the index would become unwieldy and the user might be tempted to avoid the gross index and go directly to the actual store of documents, or to a less detailed subindex. Further, it takes so long for the publication of current indexes in their present form that an expansion of these would, in many cases, be accompanied by an unacceptable delay in availability of informational pointers for storage and retrieval of current information. However, the fact remains that the information compression ratio must be reduced and more information is required in our pointer systems than has been available in the past.

As Stevens[5] points out, the operations known as indexing involve first choosing clues that will serve to identify, for purposes of later retrieval, a particular recorded item, and second, either marking on the item itself or recording as a separate item the tags, labels, or codes representing these clues. The second of these two steps can be purely clerical in nature but the first has been, to date, primarily the result of human intellectual efforts in subject content analysis.

In choosing clues to identify documents for the purpose of later retrieval, one can use two general types of indexing. The first is based on words, or catch words, the second on subject indexing, concept indexing, or control indexing. In terms of operational definition[5] the contrast is best expressed as the distinction between index entries that are derived from the text of an item itself, and those that are assigned to it from a list or schedule of subject categories, descriptors, and the like, which exist independently of the text.

The coordinate indexing method suggested by Taube[6] has recently received considerable attention, and several information systems based on this approach are currently in operation in the U.S.A.[7,8,9] In this approach documents are indexed by terms representing the concepts expressed in the document. A small or large number of indexing terms may be used to describe the informational content depending upon the nature of the document itself and on the depth of indexing desired. In the inverted arrangement of the file organization a record such as an index, or catalog card, is created for each term derived from the document indexing, and the cards or files have document identification numbers marked on them to identify all documents that have been indexed under the terms in question. Thus, for example, a paper on nep formation in carding might typically be indexed under such terms, or key words, as neps, lickerin, velocity, carding, cotton, fineness, micronaire fineness, etc. An identification or accession number is assigned to each document, this being the one that is posted to the appropriate cards in the file as illustrated in Figure 1. The numbered document is then transferred to its appropriate shelf, its information content having been noted in the key to the information system.

In a typical case, we would require that this document be retrieved in response to a request for information on the effect of lickerin velocity on neps. We must first determine the concept coordinates of the inquiry, and here the coordinates velocity, lickerin, neps, might be considered appropriate. The files corresponding to these key words are withdrawn from the index system and the accession numbers on these files are compared. Any accession numbers common

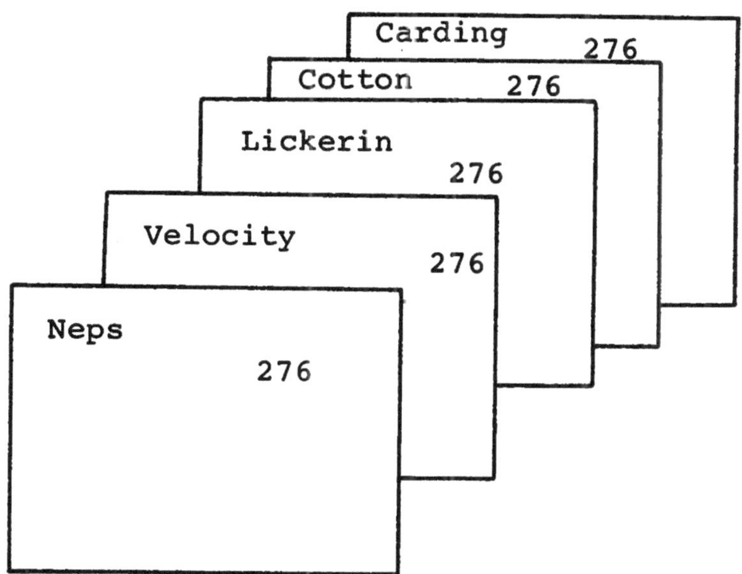

Fig.1: Keyword file, each card bearing the same identification or accession number

to all three identify documents with information on the subject of interest.

It is evident that this approach to index organization does not recognize a preconceived hierarchy of terms. Each key word enjoys equal status in the card file and proper names can be indexed as readily as more abstract concepts. The system allows indexing to be done in terms of the natural language of the literature, and data, as well as information, can be indexed within the general framework of the single file. There are, however, some difficulties inherent in this system which relate to language control and which need to be resolved before it can be used effectively for information retrieval.

THE THESAURUS AND ITS OBJECTIVES

There is considerable lack of uniformity in the natural language of textile technology as it is used by individuals, and agreement on the definition of terms needs to be established if effective retrieval of information is required. Control of the system vocabulary is one of the more critical requirements in coordinate indexing. The thesaurus is a logical language control for information storage and retrieval, providing a list of words or key terms to be used in the indexing of documents, storing the index data, and questioning or searching the store at a later date.

The purpose of a technical thesaurus is to reduce ambiguity of meaning to a minimum, to eliminate the multiplicity of equal terms wherever possible, and to suggest concepts related to the given subject. It presents synonyms and antonyms as does Roget's well-known thesaurus, but it usually directs the reader to use but one word or term in a set of terms, if this is possible. The technical thesaurus also lists terms related hierarchically for each recommended key word to indicate the existence of concepts which either overlap or border on the subject of it and the nature of these hierarchical relations is spelled out in a definite way.

As an illustration, we can take the subject of variability. The set of terms having to do with variability in a textile product could conceivably include: variability, uniformity, nonuniformity, homogeneity, nonhomogeneity, regularity and irregularity. These are not, of course, completely synonymous terms. In fact, antonyms are obviously included, but they do relate to the general concept of variability in a given property or form of textile material.

In this example, the thesaurus might arbitrarily suggest the use of the word irregularity to represent the indicated concept and further suggest suppression of the remaining words in the set. This suppression should not be taken as an example of thought control arbitrarily imposed by the compiler of the thesaurus. The choice is made simply to avoid multiple labels being put on the same file of information to the puzzlement of indexer and searcher. It allows the author to use the term inhomogeneous in his title if he wishes, but suggests that the information be stored and searched for under the concept of irregularity.

Another case in which the thesaurus may attempt to suppress usage of a word is when the word is so narrowly defined that it is unlikely to be used with any frequency in the literature. Since the compiler of the thesaurus, with one eye on the information store, seeks to eliminate categories or labels that will receive negligible use, he may combine this narrow word with a term expressing a broader function, advising the user not to use the narrower term. This concept is designated diagrammatically in Figure 2a where a is the narrower concept term unlikely to be used in the given literature, and A is the broader concept more likely to be used.

Thus, the thesaurus has been employed to reduce the use of multiple terms for similar concepts, and it has suppressed usage of terms judged too narrow to include in the language of the system with provision of a pointer to the preferred broad term.

Fig. 2: (a) Shows the broader and narrower term relationship; (b) shows the relationship between given term, and broader and narrower terms.

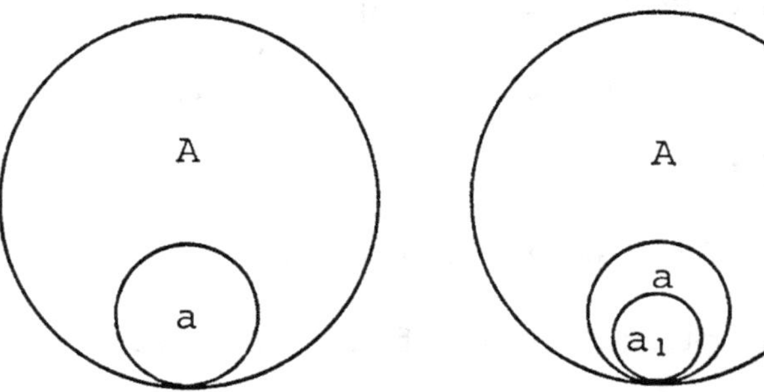

(a) A = Generally used term
a = Rarely used term

(b) A = Broader term
a = Given term
a_1 = Narrower term

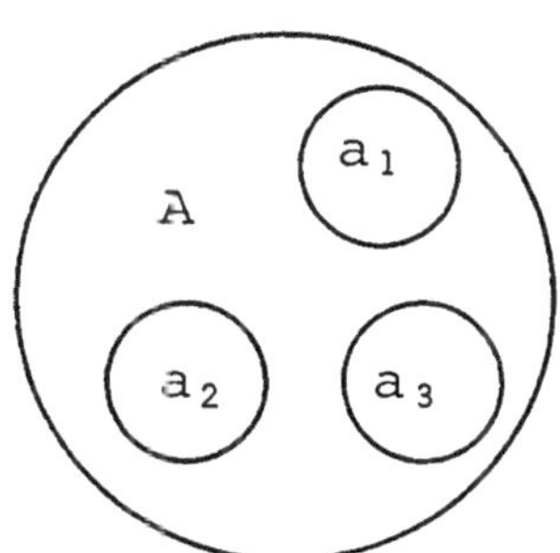

(a) A = Broad Term

a_1, a_2, a_3 = Related Terms

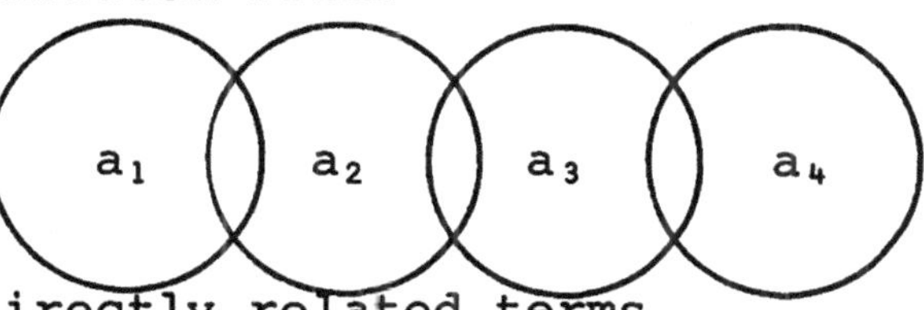

(b) a_1, a_4 = Indirectly related terms

a_2, a_3 = Directly related terms

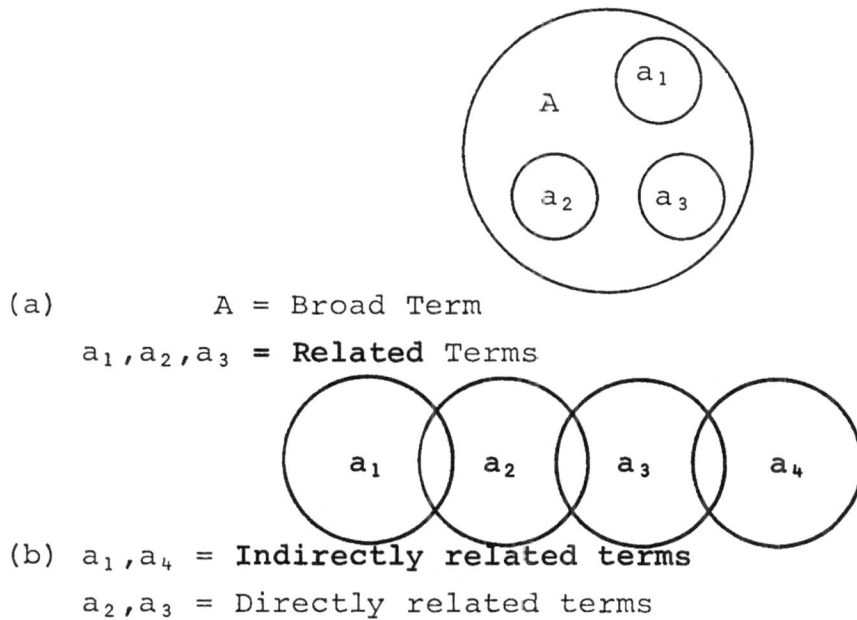

Fig. 3 General relationship, both direct and indirect, between terms.

The broader and narrower terms can be designated diagrammatically, as in Fig. 2b, for the given concept as follows. Here a is the given concept, A is the broader term, and a_1 is the narrower concept. This facet of term relationship assumes a "father-son-grandfather" relationship. The only difficulty occurs in cases where there may be different ways of relating the terms. For example, two terms may be related by composition, or by function, or by use in a technological sense, and each of these different types of relationships may promote a broader-narrower term connection between concepts.

The generally related terms are frequently more numerous than the broader-narrower relationships for a given concept.

In terms of Figure 3a the concept term a_1 is related to concept term a_2 and concept term a_3, etc. as brothers in the family A. But the relationship can be extended as well to step-brothers and distant cousins. Figure 3b indicates that concepts a_2 and a_3 are clearly related to one another, but a relationship can also be argued for a_1 and a_4, via the $a_2 \cdot a_4$ coupling.

In addition to the semantic or logical relationship developed in accordance with the diagram above, a connection between words can be developed in terms of positions in a process or in terms of mechanistic associations, and they can also be developed in terms of connections evidenced in the literature itself. Word or concept relationships thus derived provide another means for tracing information in the literature. Establishing such a relationship is an extremely difficult task at the start, because it depends upon the technological background and experience of the compiler of the thesaurus. But with use of the thesaurus for indexing over a period of time, and with feedback of information on the interrelationship between concepts in many documents, it becomes possible to update and expand the relationships between terms in it, and this updating provides for continual improvement of operation of the system.

The thesaurus is expected to undertake three major functions: first, to eliminate, or at least reduce, ambiguity in meaning of words and key terms; second, to determine or suggest the level of refinement or detail to be used in concepts and key terms; and third, to point out routes and alternate routes of search and/or of indexing. These functions relate to all stages of information storage and retrieval. They relate to the processes of indexing, storing or posting, searching, and retrieval.

The indexer is well advised to refer constantly to the thesaurus in the process of indexing his abstracts or original documents. Reliance on a key word list alone during the indexing process is unwise in that the indexer will be swayed towards the language and terminology of the author, or will be limited in the use of terms which may describe the concepts of the document. In posting the indexed document to the system, it is of utmost importance that the exact words of the thesaurus be used, otherwise there will be no guidance for the searcher at a later date to retrieve information, for the vocabulary of the input will vary from that of subsequent searches. Likewise, the searcher must resort to the thesaurus to govern the terminology that he uses. Furthermore, if the searcher is dissatisfied with the documentation returned in his early searches, the thesaurus undertakes the important function of suggesting alternate terms that he may use in subsequent questions put to the system.

LINKS AND ROLES IN COORDINATE INDEXING

Coordinate indexing allows for the indexing of documents by terms representing the concepts expressed in that document, and a small or large number of indexing terms may be used. In the primary form of coordinate indexing, the terms are in an unmodified form and no attempt is made to separate or group the terms which appear under a given document.

However, as the number of terms used to index a given document grows quite large, the danger arises that any or all of the words used will be considered in subsequent searches to have equal value, or equal relationships with one another.

To avoid ambiguity and confusion arising from this situation, it has been found advisable to group those terms that relate to individual concepts within the given document. Thus, a document that deals both with the marketing of tops and with the manufacture of worsted fabrics would have the terms related to marketing and tops linked together, while the terms relating to manufacture and worsted fabrics would be linked together. The link is provided as a simple subscript letter to the document accession number, and thus it becomes, in effect, the designation of a subdocument. That portion of the document dealing with marketing of tops is given an arbitrary subscript "a", and that dealing with manufacture of worsted fabrics, an arbitrary subscript "b" in the document numbering system. Thus, the searcher for information on the marketing of worsted fabrics would not want or expect this particular document to be returned to him, and it would not be returned if he were careful to link marketing with worsted fabrics. In short, the proper use of these document links, or subdocument identifications, makes it possible to deal separately with individual concepts within a given document.

A second refinement in the system of coordinate indexing relates to roles which are used to show the syntactical relationship of terms to one another. A paper on the packaging of Mylar would be entered in the index file under the key words packaging and Mylar (among others), but if roles were not assigned to these key words, this document might also be retrieved in response to a question for information on Mylar to be used for packaging, clearly a false retrieval in terms of the intent of the searcher's question. This problem is resolved through use of role indicators. The Engineers Joint Council in the U.S.A. has developed a system of numerically designated roles for this purpose, and these are described in the publication from the Council and also in Costello[10] and Holm.[11] These roles, which in effect designate the grammatical or functional association between related key words, are indicated by assigning single numbers to the key words.

The coordinate system may be operated either on a manual or on a mechanized or computerized basis. The indexing input to the manual system requires posting of document numbers on various key word cards, and this represents the storage function. The searching function is conducted by comparing the cards, listing the key words desired in the search, and determining what document numbers appear in common on two or more desired key term cards. In a computer system the same processes are involved, including indexing, or selection of appropriate key words with links and roles, followed by posting of the indexed information that may be retained in the computer on magnetic tapes, discs, or drums, or printed out in "hard copy" in a so-called inverted file that lists individual key words and the document numbers posted under each such key word.

Clearly, a computer search of the posted files is much faster and more accurate than that of files kept manually and if the information is correct when inserted, there is less opportunity for error in the search of the computer tape than there is in the search of the manual file.

The supporters of the coordinate indexing method are quite confident as to the potential and efficiency of their system in handling very large stores of information, and it is being adopted by the Engineers Joint Council (U.S.A.) and by such societies as the American Society for Metals in the operation of their information systems. But it should be emphasized that the coordinate indexing system for information storage and retrieval is a subject-based system. As such, it is open to all the weaknesses and bottlenecks involved in the intellectual activity of translating the contents of the information store item by item into the select language adopted for a given field of technology, engineering, or science.

DEPTH OF INDEXING

A few words should be said about language refinement (or detail) to be used in indexing and its influence on retrieval effectiveness and efficiency. If one views the hierarchical structure of a typical page in the M.I.T. Thesaurus, one can pick out, for example, the subject of "yarn balloons" such as occur in ring spinning. The sole logical subject heading for this topic might have been simply "balloons". But a study of the textile literature showed considerable differentiation in terminology in this area, and so the subject was developed to permit a rapid identification of different kinds of balloons, such as spinning and unwinding balloons, and different balloon parameters such as balloon control, balloon control rings, balloon dynamics, balloon separators, balloon shape, balloon stability, and balloon tension.

Obviously, a chemist would be satisfied with the more general term "balloons" or even "ring spinning" or "spinning." But the engineer trying to optimize package size in a given spinning or twisting system will want his information delivered with a more detailed label than simply "balloons." There may be 15 to 20 articles per year relating to balloons, and so over a 15-20-year search period, the system output under balloons might approach an unmanageable 400 items. Breakdown into the subconcepts or "precoordinated" detail seems necessary.

With this detailed breakdown, one increases the danger that the indexer will place the item in the wrong box, so to speak, or the searcher will look for it under a different box or key term. The article may discuss balloon stability as a function of tension in the yarn. One man may index the item under "balloon stability" and "tension" and "yarns." A second man may index the same item under "balloon tension" and "balloon collapse." The third man may search for it under "yarn tension" and "balloon shape" and "stability"--with obvious failure to retrieve the article indexed by either of the first two men. Stated simply, the searcher uses perfectly accurate and approved terminology to describe the concept on which he was searching, yet he fails to retrieve this pertinent paper. The textile system must avoid such an occurrence.

The situation is not unlike that which faces the small boy who favors hard candies and who seeks to pick some out from a bag of mixed sweets. If he consumes the whole bag, he will get all the hard candies at hand, but with some danger of a stomachache. If he reaches deep into the bag and tries to detect the harder items by judicious pinching, his retrieval will consist mostly of hard candies, but he will probably miss many of those he favors.

Figure 4 shows the outline of a hypothetical hierarchical structure of terms or concepts ranging from subject A to subject R7. The solid lines represent broader-narrower term relationships between terms and the dotted lines indicate the more general term relationships (based on semantic, mechanistic, or process associations). This structure is used to establish the file labels used for both indexing and filing and searching the information collection.

The information store resembles the bag of candies, for the searcher reaches into it without prior knowledge of its contents. Let us assume now that the searcher's primary interest lies in the region of the term (or concept) A_1d. The expression "in the region of" is used to indicate that the searcher's interest does not exactly match the concept A_1d. If he asks for the broadest subject in the group A_1, he hopes to retrieve all items of interest including A_1d; but the answer will contain also an infinitely large surplus of information.

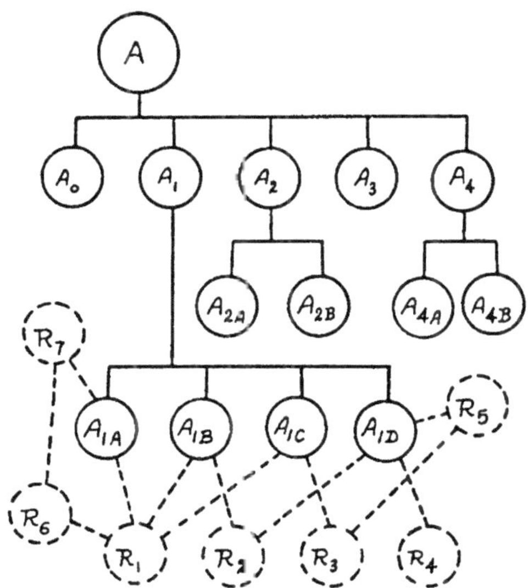

Fig. 4 A hierarchical structure of concept relationships

If he seeks information in the more detailed category labelled A_1d, he may miss valuable pertinent information contained in A_1c or in R_3 or R_4 or R_5. The presence of this valuable datum of interest in files other than A_1d may be due to the slight mismatch between the searcher's actual subject of interest and the system's file title or to the indexer's slight variation in assignment of index titles. It seems desirable, then, to ask for several items at the desired depth of inquiry, with the request being guided by the thesaurus. It, likewise, seems desirable to provide for liberal indexing at that depth to insure that the detailed documents sought are caught in the net. At the same time, it is useful to index up the generic tree one or more steps to insure that the search directed at concept A_1 does indeed scoop up all documents containing concepts A_1a **to A_1d,** inclusive. In other words, if a document treats **the** subject A_1d, it should be indexed under A_1d, but also under A_1 and even under A.

Most of the time upindexing (or upposting of **a document** along the concept hierarchy) is desirable. **For example,** in Figure 5 concepts A and B are represented as **independent** key words, while A_1 and A_1a and B_1 and B_1a are **shown (in a** somewhat different form than in Figure 4) to represent the first and second generations of their narrower terms. A hypothetical document is pictured as encompassing completely the two concepts A_1a and B_1a. The document may include other coverage, but as pictured, it is concerned to a major degree with A_1a and B_1a. Here it would be logical to post or index the document to A_1 and B_1, as well as to A_1a and B_1a since it deals in large measure with a significant part of subjects A_1 and B_1. One might carry the posting further, to A and B, but the value of this extension is open to question.

The situation in Figure 6 differs considerably. For here the document is devoted, for the most part, to subject matter outside of concepts A_1a and B_1a. Posting of the document to these narrow concepts is nonetheless desirable, but there is some question as to whether upposting to the broader subjects A_1 and B_1 is worthwhile. For example, if the document makes a brief comment about A_1a in passing, it will hardly be of interest to the man searching at the

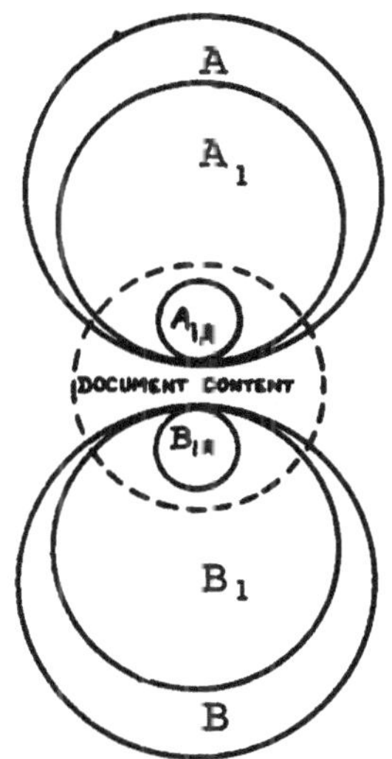

Fig. 5 Matching document content to a concept hierarchy.

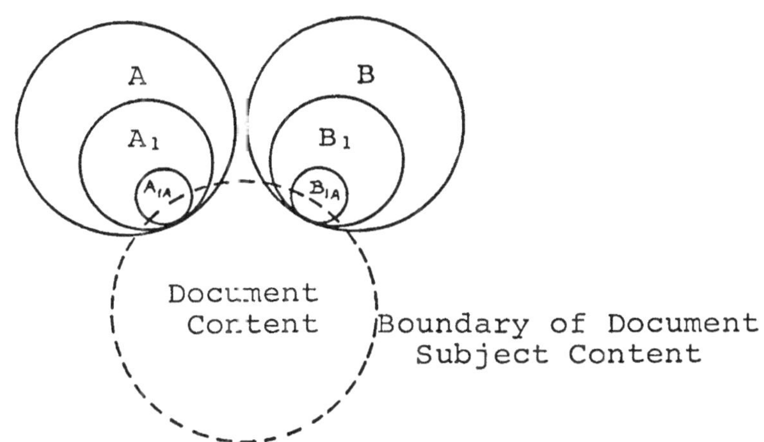

Fig. 6 Matching document content to a concept hierarchy.

level A_1 and less so in answer to questions on A. A decision not to post to A_1 and A is intended to avoid clogging the answers on A with material having very little to do with the general subject of A. Yet the item is not lost in the system, for the searcher interested in the refinements of A_1a will detail his questions accordingly.

The chances of document recovery with single or multiple key word searches can be dealt with in terms of probability theory. But the proof of the system still remains in its testing--which lies immediately ahead in this program. The approach thus far is to rely heavily on the thesaurus for development of related key terms to describe document content and to post automatically the document to the next broader term. Thus, the document is given an expanded coverage in breadth and height. By stressing the same philosophy for subject searching, there is hope of developing retrieval efficiency at a suitable level.

SEARCHING BY REFERENCES

Tables I and II show how reference tracing can be used to evaluate the interaction between journals as a measure of information flow between countries, between scientific disciplines, between science and engineering and technology, between research and production, between the past literature and the present. The availability in the computer of references used in the principal textile journals in the U.S.A. and in the key textile journals of other technically advanced countries will greatly facilitate the formulation of quantitative conclusions concerning information flow through formal journal channels. A similar study could be made of patent references with the promise of interesting results.

Just as important a use for entry of bibliographic data in the system is its potential as a searching channel. Here, too, Kessler[12] points the way in his studies of the physics literature, showing that the results of title word searches and of bibliographic couplings are correlated to a considerable degree. Bibliographic coupling refers to the association between individual articles and/or authors sharing one or more common references. **For example,** if Smith in paper A and Brown in paper B, **both referred** to paper C by Green, **they would be associated with a coupling** magnitude of 1. **Clearly, if Brown and Smith over the years** used many of the **same references,** their record would show a strong bibliographic coupling and one would expect that they are working in the same field or subject. If paper A and paper B shared many references, one would expect that they dealt with the same or very similar subjects. The reference evaluation has to do with how a given author, or authors, use the past literature.

On the other hand, when one lists the later citations by other authors of paper A and of author Smith in the file of A by Smith, one can, in effect, look into the future as viewed from the date of publication of A by Smith. For example, if Smith wrote A in 1945 and the system appropriately filed bibliographic data for 1946-1966 inclusive, one could look back in the Smith-A file and determine how many people in later years used this work as a citation. And one could compare A by Smith with B by Brown to see if they were used (or cited) by the same authors or in the same papers at a later date. This would be another form of bibliographic coupling.

Now, a few words as to how a citation search can apply to a textile information system. The search may start with a given author or document (without a definition of subject content) and a request that the system furnish a collection of papers which have couplings at or above a specified value. On the other hand, it may be desirable to start with a given subject, then identify authors who have written in this area, then evaluate the relative number of citations each man has experienced in the years following each of his publications. This could lead to a rough (admittedly very rough) estimate of the level of expertise in the given area. Information of this nature could conceivably be of considerable value to the manager of a plant more interested in hiring an expert to solve his problem than in reading a number of papers which might be over his head.

Finally, one may use the coupling of bibliographic data as a means of tracing the chronological development of ideas and of technical effort. Starting with a given document (given author and subject matter), one can determine the chronological distribution of papers that share one, two, or more references with the given paper. Plots of this nature have been reported by Kessler[13] and show the rise and fall of human interest in a given subject. A similar indication of technical interests can be characterized by determining the chronological frequency distribution of papers according to a given set of key words or concepts. Such data can serve to separate the subject of significance from the fad. Another such study could be directed towards the rise and fall of interest in a given research topic in the literature of different countries. It would be interesting to have such chronological frequency distributions on subjects such as stitch-bonded structures, or high speed carding, or self-crimping fibers, or permanent press. While all this discussion has been directed at printed journal, trade paper, and book categories, one can easily picture the fascinating data which could come from conducting some of these last-named studies on the programs of the technical meetings of the textile professional societies, both here and abroad, and in the area of patents.

The point to be made is that the primary value of such studies will be in the design of an overall textile information network compatible and interlocked with the textile networks of other countries and with the general information networks both here and abroad. Without a grasp of the nature of information flow, planning of a textile information network becomes somewhat academic.

BROWSING

Some object to the complexity of information storage and retrieval systems which have been devised in today's technical society. Objectors may point at the inefficiencies of such systems, for they cannot be perfect. But few persons will advocate their elimination. A reasonable view on this point is that each segment of technical interest, or each separate industry must work towards setting up its system designed for maximum efficiency, but at the same time, it should advertise clearly the system limitations so the consumer is forewarned.

Others object to the development of a too efficient

system for fear that it will eliminate the pleasurable experience of browsing. Browsing sometimes serves as a substitute for boredom. In a way, it represents educational prospecting. It provides for joy of learning. However, in terms of an information system, it more often represents blunt edge searching. When a directed inquiry into an efficient system gives an exact answer, there is little inclination to browse. But when an inquiry evokes a lot of irrelevant material, reviewing this output may be considered a form of browsing. While browsing offers promise of unplanned discovery, one must remember that it is generally limited by the arbitrary ordering of the information store. If one browses on the front shelves of a library, one will have no chance of finding the material shelved in the back room. With the computer store, however, one can browse in a variety of ways, slicing the collection by authors, by journals, by subjects, by categories of information, by origin of work, by common references, or completely at random.

The difference between this type of computerized browsing and that involving musty shelves is that the conventional form provides immediate access of full text. The computerized systems envisioned at present do not, but one can provide for limited forms of access, say the title or the list of key words used in indexing the document, or the abstract. And if photoimage storage is provided, the table of contents or first page of the article can easily be furnished and, ultimately, the entire text.

ACCESS TO DOCUMENTS

While immediate access to all or part of the document is desirable for the browsing function, the access requirements of an information network may vary considerably. It is sometimes enough to identify the location of the original article or of its abstract and let the user do the rest. There is much to be said for providing immediate title and key word information to the user as he queries the store. And identification of the item in one of the classical abstract systems (JTI or TTD) will permit him to consult quickly his local library collection to view the abstracts of interest. Hard copies of the abstracts can easily be furnished by regional centers. Many will find the availability of the principal textile abstracts on microfilm to be particularly useful and efficient for searching, and their examination on commercially available microfilm equipment is found to be most convenient.

It is also important that the original documents be made readily available to interested users at reasonable cost. To facilitate location of original journals of interest to users in the U. S. textile industry, a list[2] has been compiled of about 1000 journals indicating the abstract services which incorporate the various journals in their coverage, reporting the university and textile school libraries that subscribe, the journal languages, and the publishers' addresses. This list was prepared with the aid of the computer-edited program and it can be updated with minimum clerical effort. It is hoped that such listings will insure better usage of existing textile library .ities and will also point up the occasional lack of :age in textile abstract systems and in local library

A word of caution is needed in considering the abstract service coverage of the journals on such a list. The fact that an abstract system includes a given journal in its list does not mean that all items from that journal will be abstracted in the system. The abstract input from many primary journals is a matter of editorial decision. Needless to say, the editor's decision may differ from that of the user as to what is pertinent on the textile scene.

INFORMATION FLOW VELOCITY

Few will argue against speeding up the flow of information in technologies such as that of the textile industry. In planning an information network for the industry one does well to give serious study to the matter of flow rates. At the same time, review of the paths of flow may reveal unnecessary circuits which slow up the overall flow process, even though local flow rates in parts of the circuit are very high.

The time delay which the present system of handling textile information engenders can be calculated roughly as follows: two to three (or more) years from the filing of patents to their issuance in the U.S.A., two to three years (or more) between the time of completion of industrial research to the submission of a manuscript for publication.

This delay will, of course, vary with the nature of the research, with the market situation, and with company policy. A time lag of up to one year for refereeing, editing, and publishing research manuscripts is not uncommon as is a time lag of a month to a year for publication of the abstract of the original research article. There is often a time lag of up to a year or more for publication of a substantial abstract subject index or of a comprehensive review of textile progress covering the given abstract or the original article. The total lag from the time of completion of the work to the time of its insertion in the "currently best" form for efficient information retrieval may run to six or more years.

How to decrease the time lag and release the results of completed research is a subject for more extensive discussion within the industry. The merits of early release of information should be argued on the basis of cooperative give and take.

The problem of decreasing the time lag in the formal primary publication is a worthwhile subject for study in information councils of the industry. As an individual, one is often a contributor to such delay in one's failure to meet refereeing schedules, but the journals can help in the overall process by such techniques as releasing abstracts and even text (in advance of publication) to the abstract services. The journals can require their authors, referees, and the technical editors to contribute key words for subsequent indexing. The J.T.I. journal is doing this at present.

One way to decrease the time lag in publication of secondary journals is best illustrated by the current program[8] of the Chemical Abstracts System which shows that it is possible to reduce the time span from primary journal publication date to the posting date of the abstract journal to an overall 14 weeks for the Chemical Abstracts with traditional publication methods, and to an overall 4 weeks for Chemical-Biological Activities and Chemical Titles with computer-based publication methods.

Clearly, a textile-based system cannot hope for the speed and coverage of a system like the Chemical Abstracts Service, but it can do a great deal to reduce the time span between release of research results for publication and appearance of said results in the primary and secondary information channels. It can develop better coordination between publications at all levels, reduce duplications and introduce valid short cuts, and it can try to short-circuit the practice of passing information from one country to another, to another, and to another through abstracts of abstracts of abstracts.

Finally, the industry should differentiate between different forms of primary information in developing its flow of secondary information. Present systems have a tendency to process abstract coverage of daily newspapers, monthly trade journals, monthly research journals, patents, yearly seminars, and even books in single monthly abstract publications. It would appear more fruitful to publish abstracts of items, with current-awareness emphasis on a weekly or biweekly basis. Such a publication could include announcement of journal titles (sometimes in advance of publication) and also recent patents. The monthly abstract services could then concentrate on the information of more lasting value that is likely to be processed in the long-term information store. Certain companies are doing just this on an internal basis and their formula of success might well be applied across the industry. In any case, this is a topic worthy of serious discussion in the information councils of the industry.

SELECTION OF A COMPUTER FILE STRUCTURE

So much has been said above concerning the use of computers in storing and retrieving information that one is almost led to believe that selection and setting up of the computer elements of an information retrieval system is a simple matter of making selections from the manufacturer's catalogue and plugging in the wall socket. This is hardly the case, for selection of computer components must be based on the overall design of the information system. And despite the availability of some packaged programs from the manufacturer of the hardware, it is more than likely that establishment of a computerized retrieval system will require extensive programming, or softwear development.

This is not the place to treat the details of various hardware systems. But it is appropriate here to describe the general types of file systems which have been devised for information storage and retrieval with a view towards meeting different criteria of use and different types of data collections. We will touch on seven different systems and attempt to illustrate their advantage and/or drawbacks.

(1) Serial Document File. The organization of a computer retrieval system often parallels that of a manual retrieval system. In the simplest manual system, one may take each document as it comes into the store, label it with appropriate descriptive keywords, or subject headings, and then store it in order of its receipt. When information is retrieved under a given subject or combination of subjects, one goes through the entire store of documents, picking out those whose labels include the sought-after word combinations. It should be pointed out that the development of a suitable key word list is a prerequisite to the establishment of such a system, simple as it is.

This kind of linear search may be facilitated by substituting ordinary cards for the actual documents, or by use of punched cards amenable to mechanized or optical sorting. Or the data corresponding to each document may be chronologically recorded on magnetic tape in digital form and then searched linearly to select the document numbers which meet the searcher's subject criteria. The data can likewise be recorded in graphic form on microfilm with digital subject identification on each frame. One advantage of microfilm storage resides in the fact that the linear search of the film may provide a copy of the full document as its reward rather than simply the document number. But whatever hardware is used to implement this filing system, the result is the same, for the entire document store must be reviewed in each search. And if a document is inserted with an incorrect subject label, its chance of correct retrieval is negligible.

(2) Inverted Keyword File. The inverted file in which the document identification is listed under one or more subject headings, can be utilized both in manual and in computer systems. It permits the user to move through the file cards or through the file tapes until he comes to a desired key word below which are listed the document numbers of interest. The search can be repeated for second, third, etc., key words. Linear search of the inverted file requires only half the effort of the serially filed documents, since in the serial case the entire file must be searched, while in the inverted file, an average of only half the file need be searched. But it should be noticed that the data must be handled more carefully during input into an inverted file. The data cannot be simply tacked on to the end of the file.

(3) Indexed Inverted Keyword File on Magnetic Tape. If the inverted files are organized on more than one tape, a simple index can be provided on a small tape designating the first and last key words on each main tape. Use of this index permits further reduction in search time, since there is no need to scan any tapes other than those containing the key words of interest. But, again, more effort is required during input.

(4) Use of a Computer with Random Access. If one stores the inverted file in its organized or structured form on a

disk, drum, or data cell, one can enter the store of data at any point, making a quick reading to identify the key word where entry occurred. This is the case of ordered input, but with random access during the search. Now comparing this key word alphabetically with the desired key word, one knows in which section of the store the desired word resides. The next entry is made accordingly in that section, that is, another sample is taken, etc., until the desired word is bracketed and identified, and its associated document numbers read out. The sampling and reading of key words has now been reduced from N (the total number of key words in the store) to $\log_2 N$. This is called the binary search technique.

(5) Employing an Index with Random Input and Random Access. If the input is stored in inverted form at random, an index should be used to identify its location. And the random access feature may be used to reach in anywhere in storage to retrieve it. Now instead of using the binary search technique, one can reduce the time of search simply to the reading of the index and to the readout of the specific word and its associated document numbers. It should be pointed out that while this involves another step (i.e. reading in the index, which itself must be searched for the appropriate key word/pointer), it results in a substantial saving in computer charges due to the relatively high cost per access on high capacity, random access storage devices.

(6) The Interactive System. When search time is reduced sufficiently by use of proper file systems, indices,

computer programs, and appropriate computer hardware, one can use the rapidly delivered answer to a given question as the basis for quick formulation of a second, then a third question until, in effect, a dialogue is developed between man and machine, between searcher and the information system. A system which responds this quickly is called an interactive system. One such system will be discussed briefly below. It has been described earlier in some detail[14]

(7) An Interactive Time-Shared System. Such a system provides such rapid answers to a given question by user A that the computer can turn and answer the questions of 5, 10, or 30 other users before user A can even think to ask another question. User A may be located in another city or continent, asking his questions without knowing that 25 or so other people are using the same computer, in effect, simultaneously.

The basic systems for storage and retrieval of subject matter can generally be categorized under one or more of these types:

(1) Serial document file,

(2) Inverted key word file in linear form,

(3) Indexed inverted file, segmented and in linear form,

(4) Inverted key word file and random access storage,

(5) Indexed random input and random access storage,

(6) An interactive system operating on condition (5),

(7) An interactive time-shared system operating on condition (5).

Notice that as we move from system type (1) to system type (7), several factors are changing simultaneously. First, the hardware goes from simple to complex and its processing speed goes from slow to fast. Second, the amount of time and effort spent for input processing of the data increases, for in type (1) one simply adds each new item at the end of the file, while in type (7) one fits it in place with appropriate location tags much like a small piece of stone being set into a large mosaic as it is constructed "by the numbers." Third, in going from type (2) to type (7), the opportunity for effectively storing and retrieving bibliographic data, together with subject identified material, increases. For example, one can treat authors' names as subject matter. One may also include in the file system all the references listed in a given document. Inclusion of such data provides for interconnecting links between documents based on common subjects, shared authors, common references or common citations. Thus, the movement from type (2) to type (7) allows for greater flexibility in searching and retrieving desired information. (Of course, system type (1) can be constructed as a file of actual, i.e., original or microfilmed documents, and such a file offers all the information available as a result of the first retrieval step.) Fourth, in going from type (1) to type (7) the amount of scanning (that is, the percent of the store which must be scanned to locate a given document) is decreased. Fifth, the cost of both the hardware and the programming software increases in going from system type (1) to type (7). But if the data store is large and the questions tend to increase in frequency, the cost per inquiry will soon decrease.

At this point a question should be raised as to which system operates at the least cost per document handled. This is the sort of question that interests industry management. The answer is not a simple one and it must be based on system specifications dictated by the intended use conditions. Some of the system parameters which must be specified include:

(1) the number of documents to be stored,

(2) the amount of information to be stored for each document: how many subject descriptions, how much bibliographic data, etc.,

(3) the storage requirement (product of 1 and 2),

(4) how many users are to be expected per hour,

(5) how fast a response does each user expect?

(6) how many questions will the various users pose to find the desired information?

(7) how sophisticated will the user be?

(8) is remote input potential required?

(9) is remote search potential required?

This list can be expanded to cover about 30 parameters which enter into consideration of optimum system design. But if we concentrate on a simple group of such parameters (as the number of documents, the information per document, the number of users per unit time, and the acceptable delay in providing answers), we can do some rough estimating of system requirements. For example, if you have but 50 documents for your personal use, you can label them manually, then drop them in your desk drawer. Here the input effort can be kept to a minimum since you are the only user and it takes but a few minutes to scan the labels of the whole store.

If we move to a collection of 5,000 internal company reports with infrequent usage, the serial storage on magnetic tape or microfilm or even punch cards with optical mechanical search capacity would probably provide a satisfactory system at a minimum cost. But when the document store rises to 200,000 or more, which would correspond to an industrywide collection, the selection of the more sophisticated systems becomes mandatory. And at this stage, the search frequency and the nature of the searches loom as critical factors dictating system design.

We keep mentioning the nature of the search, implying thereby that searching need not follow a fixed pattern. There is an important distinction between asking a question of a system when you know its answer is there, and asking a question when you do not know whether an answer exists at all, and you are not even sure the question is correctly worded, or if the answer you receive is pertinent to your inquiry. Information retrieval from a vast store of documents is not a cut and dried matter of picking beans out of a bottle. It is more like fishing in a muddy stream. The successful fisherman fishes at all levels, frequently moves his boat, and changes his bait.

COURSE OF FUTURE ACTIVITY

Throughout this report we have referred to the needs of users of textile information and to the complex problems facing information specialists in establishing systems to meet user demands. Clearly, the best approach in developing optimum information systems for the textile industry is one based on cooperative planning and joint study by users and information specialists. A joint venture of consumers and specialists could, in time, move to:

(1) Establish industry councils on information, responsible for gathering support and promoting cooperation in the development of an information network.

(2) Develop liaison between the information activities of the U.S.A. textile industry and those of other countries.

(3) Develop a textile information program to mesh with the information centers of other industries and disciplines and the networks being formed in various government agencies.

(4) Survey and recommend input fields and channels for the textile network.

(5) Recommend specifications for control of input coverage within accepted fields and channels.

(6) Recommend specifications for indexing and searching techniques on the basis of research results.

(7) Survey the information needs in the industry and study the range of consumer requirements as a basis for formulating modifications in the network.

(8) Study the patterns of information flow into, within, and out of the industry.

(9) Establish liaison channels between the publishers of primary and secondary journals and the information centers of the industry with a view towards reducing time lag in dissemination of information.

(10) Plan different treatments for current awareness and retrospective search material.

(11) Point out the strengths and weaknesses of the industry's network.

(12) Encourage research on information transfer and the use of advanced computer systems in the textile industry.

(13) Encourage preparation of reviews and state of the art reports as a third level of publication.

(14) Encourage the discussion of information transfer problems and research within the professional societies of the industry.

(15) Study the feasibility of developing a computerized data bank for the industry.

REFERENCES

1. Allen, T.J., Communications in the Research and Development Laboratory, Technology Review, 31-37 (October/November, 1967).

2. Jacobs, P., and Backer, S., Journals of Interest to the Textile Industry, M.I.T., Cambridge, Mass. (1967).

3. Kessler, M.M., The M.I.T. Technical Information Project, Physics Today, 28-36 (Mar. 1965).

4. Kessler, M.M., Technical Information Flow Patterns, Lincoln Laboratory Report, M.I.T.

5. Stevens, M.E., Automatic Indexing, A State of The Art Report, NBS Monograph 91, U.S. Dept. of Commerce, Washington, D.C., 1965.

6. Taube, M., Specificity in Subject Headings and Coordinate Indexing, Library Trends, 1, 2, pp. 219-223, October 1952.

7. Jaster, J.J., Murray, B.R., and Taube, M., The State of the Art of Coordinate Indexing, Preliminary Edition, Prepared for the Office of Science Information Service, National Science Foundation, February 1962.

8. Annual Editorial Index, Hydrocarbon Processing and Petroleum Engineer, 42, 12 Dec. 1963.

9. Morse, R., Information Retrieval, Chemical Engineering Progress, 57, 5, pp. 55-58, May 1961.

10. Costello, J.C., Jr., A Basic Theory of Roles as Syntactical Control Devices in Coordinate Indexes, Journal of Chemical Documentation, 4, pp. 116-124, 1964.

11. Holm, B.E., Information Retrieval--A Solution, Chemical Engineering Progress, 57, 6, pp. 73-78, June 1961.

12. Kessler, M.M., Comparison of the Results of Bibliographic Coupling and Analytic Subject Searching, American Documentation 16, No. 3, 223 (July 1965).

13. Kessler, M.M., Bibliographic Coupling Extended in Time--Ten Case Histories, "Information Storage and Retrieval", 169-187 (1963).

14. Sheldon, R.C., Roach, R.A., and Backer, S., Design of an On-Line Computer-Based Textile Information Retrieval System, Textile Research Journal 38, 81-100 (1968).

APPENDIX B

GROUND RULES FOR USE OF THE TEXTILE THESAURUS IN INDEXING AND SEARCHING

INTRODUCTION

During the process of compiling this thesaurus numerous questions of procedure arose requiring ad hoc decisions by the editors. There were questions concerning the meaning of terms, concerning the hierarchical relationships between key words, and concerning the choice of a preferred term among two or more synonyms. Decisions had to be made as to the detail of terminology to be included and the degree of pre-coordination to be employed in the thesaurus. (Precoordination is the process of combining two or more independent terms into a single key word concept, reflecting common usage in the technical literature.)

When these questions touched on problems which, it was anticipated, would frequently arise during the process of compilation, the ad hoc decisions made were recorded as tentative ground rules of procedure. Such ground rules were reviewed several times with the recurrence of each specific difficulty, and were frequently revised as compiling experience broadened. Considerable effort was directed towards minimization of the requisite number of ground rules and maintenance of consistency with the COSATI guidelines* for

*Guidelines for the Development of Information Retrieval Thesauri, Committee on Scientific and Technical Information, Federal Council for Science and Technology, 1 September 1967.

the formulation of information retrieval thesauri.

As suggested in the COSATI guidelines, subject index terms were selected on the basis of their relative significance in the textile literature and their effectiveness in connoting useful retrieval concepts. In evaluating candidate terms, consideration was given to (1) their anticipated frequency in indexing and searching, (2) their relationships to other terms in the vocabulary, and (3) their acceptability as authentic terminology. Acceptability of terms has been determined by reference to textile dictionaries and encyclopedias as well as to periodicals and technical monographs in the field.

CLASSES OF ASSOCIATION BETWEEN TERMS

The classes of association between key words were established on the basis of the COSATI guidelines as follows:

Narrower Term (NT) One which is related to the given key word in a hierarchical organization of concepts, but at a lower generic level.

Example: Key word Texturing
Narrower Term False Twisting

Broader Term (BT) One which is related to the given key word in a hierarchical organization of concepts, but at a higher generic level.

Example: Key word Quilling
Broader Term Winding

Synonyms Terms whose meaning is identical in usage with the key word under consideration.

Example: Irregularity, Uneveness, Non-Uniformity, Variability

We have attempted to avoid usage of synonyms in indexing by designating a single key word to which the others are referred. The key word to be used in indexing and retrieval is indicated by the notation USE, while the words to be omitted in indexing will be marked UF (USED FOR).

Example:	Key word	Uneveness
		USE Irregularity
	Key word	Irregularity
		UF Uneveness
		UF Non-Uniformity
		UF Variability
		etc.

Where the meaning of one term completely overlaps that of another term, but the reverse does not hold, the associative designation may be either USE-UF, or BT-NT. The USE-UF variant is appropriate in cases where it is desirable to suppress the (indexing usage of the) narrower meaning term because of its limited utility and to employ in its place the broader-meaning term

Example:	Key word	Cellular Fabrics
		UF Leno Cellular Fabrics
	Key word	Leno Cellular Fabrics
		USE Cellular Fabrics

The BT-NT variant is appropriate in cases where the indexing of the narrower meaning term is considered desirable.

Example:	Key word	Packages
		NT Cones
	Key word	Cones
		BT Packages

Related Terms (RT). Where there is a partial overlap in meaning between a given key word and an associated second term, the second term will be designated as a related term. When the second term is taken as a separate key word, the first key word will now be related to it by the designation RT, if it is considered appropriate.

Example: Key word Seaming

RT Stitching

Key word Stitching

RT Seaming

In this particular case, the reverse relationship was appropriate. In the case of the terms slub catching and quality control, the reverse relationship (key word QUALITY CONTROL, RT SLUB CATCHING) might not be appropriate.

The RT designation was also used for antonyms to assist the user of the thesaurus to detect all terms of direct interest. It was employed to show mechanistic connection between terms, for example, cause and effect relationships. And, finally, the RT designation was used to identify "upstream" and "downstream" processing components, materials, etc., associated with a given key word. This use of related terms is seen by the authors as a means of guiding the user in the selection of entry points into the indexing system.

Scope Notes. Finally, the limiting or specifying of term meaning was accomplished by adding a scope note in parentheses.

Example: Key word Picking (Harvesting)

Key word Picking (Opening)

Example: Key word <u>Picking (Snagging)</u>

Key word <u>Picking (Weaving)</u>

GROUND RULES

1. <u>Purpose of Thesaurus</u>. This document is intended to provide language control in retrieval of information from the textile research, and technological literature. It contains terms relating to all segments of the textile industrial complex from fiber producer to machinery maker to consumer goods testing laboratory. However, the focal point of the compilation is the language and literature of the textile mill and/or finishing plant, i.e., the "core" industry. The detail of terminology contained in the thesaurus is for the most part determined from the viewpoint of this "core" industry.

2. <u>Depth of Indexing</u>. In selecting key words to designate the concepts contained in a given document, the indexer should first seek those terms that represent each given concept in the most specific manner possible. The meaning and scope of the indexing term should closely match that of the document concept. For example, in the paper by J. W. S. Hearle entitled "Structural Mechanics of Torque-Stretch Yarns: The Mechanism of Snarl Formulation," the terms which best match the principal concepts of the document are "False Twist Yarns," "Structural Mechanics," "Snarling," and "Theoretical Analysis" rather than the broader terms "Yarns," "Mechanics," Stretch Yarns (Filament)," and "Theories." (In addition, there are several other terms which can be used to represent more detailed concepts of the paper, including "Torsional Buckling," "Strain Energy," "Contraction," and "Tension.")

Having identified the most specific key words to represent the most pertinent document concepts, the indexer must now decide whether to include appropriate broader terms for these key words. The reason for including terms further up the hierarchical tree for each of the key words chosen thus far is to improve the probability of recall. For it is recognized that the searcher who is unaware of the existence of this paper, but who seeks the information it contains, will likely use different key words in his query of the information store. He may wish to know about the "Mechanics" of "Textured Yarns (Filament)" or about the "Mechanical Properties" of "Stretch Yarns (Filament)."

The indexer's decision on the employment of key words which are broader in scope and meaning than the specific concepts of the document must be guided by the philosophy and design of the information system for which the indexing is being conducted. There are several alternatives.

(a) If the searcher of the information store understands that no provision is made for indexing terms broader than those of the specific document concepts, then the indexer need not expand his key word list beyond the first group of very specific terms. And it will be up to the searcher to modify his successive queries as he moves up or down the hierarchical tree of textile terminology, going from narrow concepts to broad concepts, or vica versa. But it should be recognized that the principal burden in this case is on the searcher.

(b) If the information system itself automatically indexes the broader term or terms for each designated key word, it is no longer necessary for the indexer to do so explicitly. Or, if the information system automatically reforms each inquiry concerning specific key words into a multiple search involving all broader and all narrower terms thereto, again the indexer may omit any reference to concepts of wider scope than those he encounters in each document. In short, if the information system design is intended to remove the reasons for multiple hierarchical indexing of document concepts, there is no point in having the indexer consciously expanding the concept coverage in his task of keywording each item of literature.

(c) If a document is indexed at the source, for example, in the case of an original paper in the Journal of the Textile Institute, the ultimate use of its key word list is rather uncertain. It may be used in a computerized information system possessing the automatic features mentioned above. Or, it may find its way into a mechanical optical system whose users are accustomed to formulating multiple searches based on manipulation of the hierarchical tree. It may be used by companies or groups whose interests are peripheral to the subject matter of the document, in which case use of the more general terminology is desirable. It may be used by personnel who are unfamiliar with the detailed terminology of the textile industry but who rely on the more general terms of the Engineers Joint Council Thesaurus of Engineering and Scientific Terms* for

*First Edition, December 1967

classifying documents containing textile information. Where the ultimate use of the indexed terminology is not known, it would appear logical to provide the additional key words relating to the broader concepts, and these can be discarded later, if the system does not require them.

3. Precoordination of Terms. In many cases it is desirable to use a precoordinated arrangement of terms to represent a unit key word in the textile thesaurus.

Example: SPINNING FRAMES, STRESS STRAIN CURVES, JACQUARD LOOMS, TEXTURED YARNS, FIBER LENGTH. Precoordination is avoided except in those cases where the combination of terms as entered in the thesaurus is peculiar to the textile field. Where such precoordination is desirable, the individual elements of the multiterm key word will be considered for entry as separate key terms in the thesaurus and may be so entered.

Examples:

(a) VELOCITY CHANGE POINT will be precoordinated, but CHANGE and POINT will not appear as individual thesaurus key words.

(b) SPINDLE VIBRATION will be precoordinated. SPINDLE and VIBRATION will appear as individual thesaurus key words.

4. Broader-Narrower Term Relationships. The generic trees of individual terms have been developed from standard reference sources and should be used as a guide to selection of associated generic terms. See, for examples, the tables of Appendix C. In certain cases, a decision will be necessary as to a specific entry in the thesaurus. For example, the term

TEXTILES is a logical broader term of FABRICS and YARNS, but will not be so indicated in the thesaurus, since the total thesaurus relates to textiles.

In certain other cases, for example "Textile Processes (General)," the list of narrower terms is so long and the field of coverage is so extensive that the term loses its usefulness for indexing and information retrieval purposes except for those documents which consider or survey textile processes in general. Yet, it is not only retained in the thesaurus, but its narrower terms are listed, explicitly, for the sole purpose of instructing the user of the thesaurus concerning the many different types of processes which are available. But the general term, in this case Textile Processes (General), is not shown under each individual process as a broader term, i.e. the relationship is not shown in reversed form. This departure from the conventional procedure for thesaurus formulation is intended to suppress the manual or automatic broader term "upposting" from a specific textile process to the concept of textile processes in general. This special rule should cause no difficulty in retrieval if the searcher notes the presence and implication of the scope note "(General)."

In certain other cases, for example "Fibers," a similar problem occurs. But here the scope note (General) is not added for the decision is made that the term Fibers, broad as it is, has use in gathering under a single indexing term all those documents dealing with the many different generic classes of fibers. In addition, the term Fibers refers to a wide range of materials whose geometric specifications are understood to include a high slenderness ratio, a high surface-to-volume ratio, etc., and many papers are devoted to properties and behavior patterns which are dominated by geometric rather than chemical-material considerations.

Finally, in some cases where large numbers of narrower terms and related concepts occur under a single word, it is desirable to provide scope-noted subdivisions of the word in question to enhance proper grouping of closely related terms. An example of this is the term Dyeing with subclasses:

Dyeing (By Dye Classes)

Dyeing (By Environmental conditions)

Dyeing (By Fiber Classes)

Dyeing (By Material Assembly)

Dyeing (By Process Flow)

Dyeing (For Effect).

While these scope-noted subdivisions do not represent terminology common to the technological literature, they do provide logical groupings for trade designations of dyeing methods. Thus, they fulfill a teaching function for the benefit of both indexer and retriever.

5. Terminology for Chemical Compounds. The periodic and patent literature of the synthesis and manufacture of chemical compounds used in the textile industry, including that of dyes, finishes, and their intermediates, is very extensive. Thousands of different compounds are used commercially as dyes alone. Because of the enormous task involved in the detailed indexing of this literature and since the literature of chemical

compounds is adequately covered in Chemical Abstracts and its associated retrieval system, the present thesaurus is not designed to serve this purpose. However, information on the application of chemical compounds in processing, dyeing, and finishing is covered in the thesaurus with effort made to limit the number of chemical terms used. To provide such a limitation, the following rules were observed.

(a) In general, compounds are listed whenever the literature deals with their use in the textile industry.

(b) Common names, rather than those of the Geneva or ACS nomenclature, are used, preference being given to the names used in the textile literature itself.

(c) General names are used in preference to a series of individual compound names, unless the literature deals with the differences in the behavior of the individual compounds in textile applications.

(d) In general, no effort is exerted to give a complete set of hierarchical and related terms of chemical compounds.

(e) Whenever possible, dyes are listed in broad groups, according to chemical classes or fiber classes to which they are applied.

(f) Surface active agents are listed in an extensive hierarchical system tabulated in Appendix C.

(g) Cross-linking agents are identified as "reactants" rather than "resins", although the latter term is still frequently used in a technically obsolete meaning. Because of the difficulty of unequivocal decision, in general, no effort has been made to distinguish between monofunctional and multifunctional reactants, although strictly speaking, only the latter are cross-linking agents.

(h) Documents dealing with the absorption of dyes by fibers are indexed under the appropriate terms of "dyeing" rather than "dyes," e.g., "acid dyeing," "wool dyeing" rather than "acid dyes," "wool dyes." Literature dealing with the properties and behavior of dyes, apart from their absorption by fibers, e.g.,conductivity of dyes, is indexed under the appropriate terms of "dyes," e.g., "acid dyes," "wool dyes."

6. Testing and Test Instruments. For most literature dealing with the testing of fibers, yarns, fabrics, and finishes, the preferred term treats the specific kind of testing such as "Abrasion Testing." The use of terms denoting a testing instrument should be restricted to literature dealing with the details of the apparatus, or the design of the instrument, or with a comparison between results of two instruments where the emphasis is on the devices and their mechanisms rather than on the materials subject to test. Where both materials and instruments are emphasized, the test method, as well as the testing device, should be indexed.

7. Fiber, Yarn and Fabric Properties. To avoid excessive precoordination of terms denoting properties of textiles in various states of assembly (such as fiber modulus, yarn modulus, fabric modulus, web modulus, etc.) the following procedure was followed. The narrow terms under "Fiber Properties" are generally limited to those properties which are unique to material in a fiber form, likewise for "Yarn Properties" and "Fabric Properties." For example, "Fiber Length" is a common

measurement for fibers, but is meaningless as a yarn or fabric measurement, hence it is listed under "Fiber Properties." "Twist Liveliness" is a common yarn property, but has no meaning when related to fibers, fabrics, webs, nets, laces, etc. And "Grab Strength" is meaningful only in a fabric context. (There are a few exceptions as in the case of fiber yarn and fabric strength which are included under fiber, yarn, and fabric properties because of their frequency of appearance.)

Generally, each property which relates to any material, regardless of its form, will be listed in a hierarchical tree of pure properties, such as:

Aesthetic Properties

Biochemical Properties

Chemical Properties

Mechanical Properties

Physical Chemical Properties

Physical Properties (excluding Mechanical).

The indexer encountering any of the 'pure' properties in connection with a particular fiber assembly, simply coordinates the two appropriate terms, such as "Fabrics" and "Compressive Modulus," or "Yarns" and "Breaking Elongation."

8. Process, Machine and Product. In cases where a process, its related machine, and related product all provide distinct nouns, it is not always necessary to have a key word for each. For example, wrapping or covering, wrapping or covering machines, and wrapped or covered yarns. An arbitrary decision was made here to use "Covered Yarns" and "Wrapping Machines" as key words. The indexer should therefore use the former to denote the product and the latter to denote both process and

machines. If it turns out that the process and the machine are written about extensively, separate words will be added at a later date. For the time being, decisions of this sort are made with a view towards limiting the size of the thesaurus.

9. Histories, Surveys, and Reviews. Surveys and reviews generally cover a given period of time. This coverage is denoted by use of the terms Years Coverage (1), Years Coverage (5), (10), (25), (50), (75), (100).

10. Polymer Terms. Polymer terms imply polymers or resins, but not fibers of those polymers. Separate fiber terms are provided. For example, "Nylon" is the polymer, "Nylon (Polyamide Fibers)" is the fiber.

11. Keyword A without Keyword B. Many new processes are described in terms contrasting with conventional processes. And the contrast is often the absence of some part of device used in the conventional system. For example, twisting without rings, travelers, flyers, or spindles. It is often difficult to index the new system (example: comb spinning) in a positive way. Use of the term "absence" plus missing parts will help to describe the novel system.

12. Double Systems. The adjective double occurs so often--double trumpets, double ring system, duo-cards, etc.--that we propose the term "double system" be used with the term which is doubled. These are kept together in the simplest link form.

13. Felting vs. Fulling. Although fiber migration takes place in both cases, the purpose and sometimes the equipment for felting differs from that of fulling. Felting should be

used in describing the development of a felt material. Fulling may be an intermediate process in felting. Or it may be a terminal finishing process for woven goods or knitted to improve bulk and body. (Fulling is used for milling.)

14. Dye Uniformity refers to "piece to piece" or bobbin to bobbin, etc., comparison of color or shade uniformity. Levelness (dyeing) is reserved for color uniformity in a single piece, a single beam, etc.

15. Mangles vs. Padders. Padders include a mangle, but a mangle may be operated as a separate unit to remove water from textiles. Padders generally introduce a solution into (i.e., impregnate) the fabric, then remove the excess.

16. Blends are handled as follows. Say, Dacron-cotton blends. Use generic blend terms such as:

POLYESTER FIBER BLENDS

COTTON BLENDS

then add in same link the trade name (if given) for the individual component, in this case

DACRON (TN).

It is not necessary to add the term cotton here.

We also have terms to indicate where blend takes place, i.e,

BLENDS (FIBERS)

BLENDED YARNS

BLENDED FABRICS

BLENDS (POLYMERS).

If an article deals with cotton-Dacron fabrics, say dyeing thereof, index all in one link.

BLENDED FABRICS (This is a case where one does not know where blending took place)

COTTON BLENDS

POLYESTER FIBER BLENDS

DACRON (TN).

If the article is on strength of blended yarns (case 1) or if the article is on blended fabrics (case 2), and one knows the yarns here are blended, then index as follows:

<u>Case 1</u>	<u>Case 2</u>
BLENDED YARNS	BLENDED YARNS
COTTON BLENDS	BLENDED FABRICS
Etc.	COTTON BLENDS
	Etc.

If emphasis is on method or process, etc., location of blend may need terms as:

INTIMATE BLENDS (Stock blends or draw frame blends)

ORTHOBLENDS (If different yarns in fabric are of different fibers).

For stock blending use:

INTIMATE BLENDS

STOCK

BLENDS (FIBERS)

and, if necessary,

BLENDING.

For draw frame blends, use:

INTIMATE BLENDS

SLIVERS

DRAFTING, or

PIN DRAFTING, or

GILLING, as the case may be.

Polymer Blends--use for blending that takes place prior to spinnaret. If two polymers join at or after spinnaret, this is Bicomponent Fiber and not a true polymer blend. Fiber made from two-phase polymer blend is termed Biconstituent Fiber.

17. Glass Laminates. For glass or Fiberglas reinforced laminates, use:

FIBER REINFORCED COMPOSITES

or

YARN REINFORCED COMPOSITES

or

FABRIC REINFORCED COMPOSITES

together with

GLASS FIBERS

or

GLASS CLOTH.

18. General Science Terms are not structured in detail, such as:

BIOLOGY

CHEMISTRY

CHEMICAL ENGINEERING

MECHANICAL ENGINEERING

ELECTRICAL ENGINEERING

PHYSICS

PHYSICAL CHEMISTRY

MATHEMATICS

PHYSIOLOGY

STATISTICAL MECHANICS.

19. Spun Yarns of Different Fibers.

For: Cotton Type Yarns, use: Cotton Spun Yarns

For: Woolen Yarns, use: Woolen Spun Yarns

For: Worsted Yarns, use: Worsted Spun Yarns.

If the yarn is definitely made of cotton, use in addition to Cotton Spun Yarns the identifying fiber term, i.e., Cotton. Similarly, for a Woolen Spun Yarn, add the term Wool, if that is the fiber used. And for Worsted Spun Yarns, add the term Wool, if it is the fiber. A Dacron yarn spun on the worsted system would be Worsted Spun Yarns and Dacron (TN).

20. Fiber Classification. The development of a hierarchical structure of generic fiber terms is generally a simple task, but considerable complexity is introduced with superposition of the commonly used categories of "Cellulosic Fibers," "Protein Fibers" and "High Temperature Fibers." Reference to the fiber classification table of Appendix C will provide the necessary guidelines to consistent indexing of the subject matter.

APPENDIX C

TABLES OF HIERARCHICAL RELATIONSHIPS

FIBERS

	CELLULOSE FIBERS	PROTEIN FIBERS	HIGH TEMP. FIBERS
NATURAL FIBERS			
VEGETABLE FIBERS (GENERAL)	X		
Bast Fibers	X		
Flax	X		
Hemp, Hibiscus	X		
Jute, Ramie	X		
Fruit Fibers	X		
Coir	X		
Leaf Fibers	X		
Manila, Raffia, Sisal	X		
Seed Hair Fibers (General)	X		
Cotton	X		
Kapok	X		
Lint (fiber)	X		
Wood Fibers	X		
MINERAL FIBERS			X
Asbestos			X
ANIMAL FIBERS (GENERAL)		X	
Silk		X	
Fur		X	
Keratin Fibers		X	
Wool		X	
Lambswool, Short Wool, Medium Wool,		X	
Long Wool, Merino Wool,		X	
Mountain Wool, Recovered Wool,		X	
Crossbred Wool		X	
Hair		X	
Guard Hairs, Mohair, Alpaca,		X	
Camel Hair, Cashmere, Cow Hair,		X	
Goat Hair, Horse Hair, Human Hair,		X	
Rabbit Hair		X	

	C	P	H.T.
MAN MADE FIBERS			
— **SYNTHETIC FIBERS**			
—— Nylon (Polyamide Fibers)			
——— Nylon 6, Nylon 66, Nylon 7, Nylon 11,			
——— Nylon 610			
——— Nomex (TN)			X
—— Polyester Fibers			
——— Blue C Polyester (TN), Dacron (TN),			
——— Fortrel (TN), Grilene (TN), Kodel (TN),			
——— Tergal (TN), Terylene (TN),			
——— Vestan (TN), Vycron (TN),			
——— Vectra Polyester (TN)			
—— Vinyl Fibers (General)			
——— Acrylic Fibers			
———— Acrilan (TN), Creslan (TN),			
———— Nitron (TN), Orlon (TN),			
———— Orlon Sayelle (TN),			
———— Zefkrome (TN), Zefran (TN)			
——— Modacrylic Fibers			
———— Acress (TN), Dynel (TN),			
———— Verel (TN)			
——— Nytril Fibers			
———— Darvan (TN)			
——— Vinal Poly(vinyl alcohol) Fibers			
———— Vinylon (TN)			
——— Vinyon (PVC) Fibers			
——— Saran Poly(vinylidene chloride) Fibers			
———— Rovana (TN), Vectra Saran (TN),			
———— Velon (TN)			
—— Olefin Fibers			
——— Polyethylene Fibers			
——— Polypropylene Fibers			
———— Herculon (TN), Meraklon (TN),			
———— Vectra Polypropylene (TN)			

	C	P	H.T.
Spandex Fibers			
Blue C Spandex (TN), Duraspan (TN),			
Glospan (TN), Interspan (TN),			
Lycra (TN), Numa (TN), Orofil (TN),			
Polythane (TN), Spandelle (TN),			
Vyrene (TN)			
Polyurethane Fibers			
Perlon U (TN), Dorlon (TN)			
Fluorocarbon Fibers			
Polytetrafluoroethylene Fibers			X
Teflon Fiber (TN)			X
Polytrifluorochloroethylene Fibers			X
Ftorlon (TN)			X
Synthetic Rubber Fibers			
Butadiene Rubber Fiber			
Nitrile Rubber Fiber			
REGENERATED FIBERS			
Rayon (Regenerated Cellulosic Fibers)	X		
Viscose Rayon	X		
High Tenacity Viscose Rayon	X		
High Wet Modulus Rayon	X		
Cuprammonium Rayon	X		
Azlon (Regenerated Protein Fibers)		X	
Aralac (TN), Ardil (TN),		X	
Collagen Fibers, Lanital (TN),		X	
Vicara (TN), Fibrolane (TN)		X	
Regenerated Fibers (excluding Cellulosic and Protein)			
Alginate Fibers, Rubber Fiber (Natural)			
Cellulose Ester Fibers			
Acetate Fibers			
Celacrimp (TN), Celafibre (TN),			
Estron (TN), Fibroceta (TN),			
Loftura (TN)			
Triacetate Fibers			
Arnel (TN), Tricel C (TN)			

FIBERS

	C	P	H.T.
ARTIFICIAL SILK (ARCHAIC)			
INORGANIC FIBERS (MAN MADE)			X
Ceramic Fibers			X
Aluminum Silicate Fibers			X
Boron Nitride Fibers			X
Silicon Carbide Fibers			X
Graphite Fibers			X
Metallic Fibers			X
Foil, Lamé (TN), Lurex (TN),			X
Metlon (TN), Metal Fibers, Tinsel,			X
Whisker Fibers, Aluminum Foil			X
Glass Fibers			X
Fiberglas (TN)			X
CHEMICALLY MODIFIED CELLULOSIC FIBERS	X		
CHEMICALLY MODIFIED COTTON	X		
CHEMICALLY MODIFIED PROTEIN FIBERS		X	

- YARNS
    - BLENDED YARNS
        - Blended Yarns (Filament)
        - Blended Yarns (Spun)
    - DOPE DYED YARNS
    - FILAMENT YARNS
        - Monofilament Yarns
        - Multifilament Yarns
        - Textured Yarns (Filament)
            - Bulked Yarns
                - Air Jet Textured Yarns
                - Bicomponent Fiber Yarns (Filament)
                - Crinkle Type Yarns
                - Edge Crimped Yarns
                - Loop Yarns
                - Textralized Yarns (TN)
            - Crimped Yarns
                - Air Jet Textured Yarns
                - Bicomponent Fiber Yarns (Filament)
                - Celacrimp (TN)
                - Edge Crimped Yarns
                    - Agilon (TN)
                - False Twist Yarns
                - Gear Crimped Yarns
                - Nyloft (TN)
                - Self Crimping Yarns
                - Spunize (TN)
                - Stuffer Box Crimped Yarns
                    - Banlon (TN)
            - Stretch Yarns (Filament)
                - Bicomponent Fiber Yarns (Filament)
                - Crinkle Type Yarns
                - Edge Crimped Yarns
                    - Agilon (TN)
                - False Twist Yarns

Stretch Yarns (Filament) - continued

- False Twist Yarns
  - Fluflon (TN)
  - Helanca (TN)
  - Dynaloft (TN)

SPUN YARNS

- [By Spinning System]*
  - Cotton Spun Yarns
  - Woolen Spun Yarns
  - Worsted Spun Yarns
- [By Process Designation]*
  - Carded Yarns
  - Combed Yarns
  - Condensed Yarns
  - Direct Spun Yarns
- Textured Spun Yarns
  - Bicomponent Fiber Yarns (Staple)
  - Hilow Bulked Yarns
  - Stretch Yarns (Spun)

UNCONVENTIONAL YARNS

- Bonded Yarns
- Cellular Yarns
- Laminated Yarns
- Paper Yarns
- Slit Film Yarns

[YARNS BY STRUCTURE]*

- Balanced Yarns
- Coarse Yarns
- Core Spun Yarns
- Crepe Yarns
- Fine Yarns
- Interlaced Yarns
- Novelty Yarns
  - Batines
  - Bead Yarns
  - Boucles
  - Bug Yarns

[YARNS BY STRUCTURE]* - continued

- Novelty Yarns
    - Cockle (Yarn)
    - Flake Yarns
    - Flames (Novelty Yarns)
    - Frills (Novelty Yarns)
    - Intermittently Bulked Yarns
    - Knickerbocker Yarns
    - Loops (Novelty Yarns)
    - Nep Yarns
    - Nub Yarns
    - Ratines
    - Seeds (Novelty Yarns)
    - Slub Yarns
    - Spirals (Novelty Yarns)
    - Splashes
    - Thick and Thin Yarns
- Plied Yarns
- Singles Yarns
- S Twist Yarns
- Z Twist Yarns

[YARNS BY USE]*

- Filling Yarns
- Fingering Yarns
- Knitting Yarns
- Warp Ends

*These terms are included in this chart to aid in its structuring. They are not present in the Thesaurus.

FABRICS

FABRICS*

- APPAREL FABRICS
- AUTOMOTIVE FABRICS
- BACKED FABRICS
- BRAIDS
- CARPETS
- DOUBLE KNIT FABRICS
- FELTS
- FILLING KNIT FABRICS
- FLAT BRAIDS
- FLAT WOVEN FABRICS
- FLOCKED FABRICS
- HOUSEHOLD FABRICS
- INDUSTRIAL FABRICS
- KNITTED FABRICS
- KNITTED NETS
- KNOTTED NETS
- LACES
- LAMINATED FABRICS
- MALI FABRICS (TN)
- NEEDLED FABRICS
- NETTING
- NONWOVEN FABRICS
- PILE FABRICS
- PRINT CLOTH
- RIBBON
- SCREENING
- SCRIM
- STRETCH FABRICS
- TAPE
- TIRE FABRICS
- TRICOT KNITTED FABRICS

*The terms selected for inclusion in this chart under "Fabrics" represent the more general type fabrics. The Thesaurus actually contains many additional narrow terms under "Fabrics"

FABRICS - continued

- UPHOLSTERY FABRICS
- WARP KNITTED FABRICS
- WEBBING
- WOVEN FABRICS

FABRICS (ACCORDING TO FIBER)

- COTTON FABRICS
- WOOL FABRICS

FABRICS (ACCORDING TO STRUCTURE)

- BONDED YARN FABRICS
- BRAIDS
  - Core Braids
  - Double Braids
  - Flat Braids
  - Hollow Braids
  - No Core Braids
  - Parachute Cords
  - Regular Braids
  - Solid Braids
- COATED FABRICS
  - Buna N Coated Fabrics
  - Butyl Coated Fabrics
  - Lacquer Coated Fabrics
  - Neoprene Coated Fabrics
  - Rubber Coated Fabrics
  - SBR Coated Fabrics
  - Vinyl Coated Fabrics
- CROSS LAID YARN FABRICS
  - Malimo Fabric (TN)
- FABRIC REINFORCED COMPOSITES
- FELTS
  - Needle Punched Felts
  - Paper Makers Felts
  - Pressed Felts
  - Synthetic Felts
  - Woven Felts

FABRICS (ACCORDING TO STRUCTURE) - continued

FLAT WOVEN FABRICS

- Albert Twill
- Alpaca Fabrics
- Awning Cloth
- Barathea
- Batiste
- .
- .
- .
- .
- .

For detailed listing of additional terms, see Thesaurus

- Victoria Lawn
- Voile
- Warp Repp
- Webbing
- Whipcord
- Window Holland

FLOCKED FABRICS

- Flocked Carpets

KNITTED FABRICS

- Double Faced Fabrics (Knitted)
- Double Pique
- Filling Knitted Fabrics
    - Accordion Fabrics
    - Astrakhan
    - Circular Knitted Fabrics
    - Double Knit Fabrics
    - Eightlock (Knitted Fabrics)
    - Flat Knitted Fabrics
    - Full Cardigan Fabrics
    - Fully Fashioned Fabrics
    - Half Cardigan Fabrics
    - Interlock (Knitted Fabrics)
    - Jacquard Knitted Fabrics
    - Knitted Pile Fabrics

FABRICS (ACCORDING TO STRUCTURE) - continued

- KNITTED FABRICS
  - Filling Knitted Fabrics
    - Plain Knitted Fabrics
    - Purl Knitted Fabrics
    - Rib Knitted Fabrics
  - Knitted Lace
  - Reinforced Knitted Fabrics
  - Stretch Knitted Fabrics
  - Warp Knitted Fabrics
    - Milanese Fabrics
    - Raschel Fabrics
    - Simplex Fabrics
    - Tricot Knitted Fabrics
      - Lockknit
- LACES
  - All Over (Lace)
  - Armenian (Lace)
  - Beading (Lace)
  - .
  - .
  - .
  - .
  - . } For detailed listing of additional terms, see Thesaurus.
  - Tuck **Laces**
  - Veiling (Lace)
  - Yak
- LAMINATED FABRICS
  - Fabric to Fabric Laminates
  - Foam Backed Fabrics
  - Laminated Jersey (Foam)
- MULTILAYER FABRICS
  - Trilock (TN)
  - Woven Tubes
- NEEDLED FABRICS
  - Fiber Woven (TN)
  - Needle Punched Felts

FABRICS (ACCORDING TO STRUCTURE) - continued

- NETS
  - Knitted Nets
  - Knotted Nets
  - Lace Nets
  - Laundry Nets
  - Woven Nets
- NONWOVEN FABRICS
  - Arabeva Fabric (TN)
  - Arachne Fabric (TN)
  - Araloop Fabric (TN)
  - Bonded Fiber Fabrics
    - Adhesive Bonded (Nonwovens)
    - Paper
    - Self Bonded Fabrics (Nonwovens)
    - Spunbonded (Nonwovens)
  - Fiber Woven (TN)
  - Malipol Fabric (TN)
  - Maliwatt Fabric (TN)
  - Nonwoven Scrims
  - Pressed Felts
- PILE FABRICS
  - Curl Pile Fabrics
  - Knitted Pile Fabrics
    - Plush
  - Loop Pile Fabrics
  - Pile Fabrics (Flocked)
    - Flocked Carpets
    - Flocked Fabrics
  - Pile Fabrics (Tufted)
    - Tufted Carpets
    - Tufted Fabrics
      - Candlewick Fabrics (Tufted Fabric)
      - Chenille Fabrics (Tufted Fabric)
      - Malipole Fabric (TN)

FABRICS (ACCORDING TO STRUCTURE) - continued

- PILE FABRICS
  - Pile Fabrics (Woven)
    - Artificial Furs
    - Corduroy
    - Curl Pile Fabrics
    - Cut Pile Fabrics
      - Velvet
      - Velvet Carpets
      - Velveteen
    - Loop Pile Fabrics
    - Plush
    - Terrycloth
    - Uncut Pile Fabrics
    - Woven Carpets
      - Axminster Carpets
      - Chenille Carpets
      - Ingrain Carpets
      - Sculptured Carpets
      - Tapestry Carpets
      - Velvet Carpets
      - Wilton Carpets
    - Sculptured Pile Fabrics
    - Stitched Pile Fabrics
      - Araloop Fabric (TN)
      - Malipol Fabric (TN)
- STITCH BONDED FABRICS
  - Arabeva Fabric (TN)
  - Arachne Fabric (TN)
  - Araloop Fabric (TN)
  - Mali Fabrics (TN)
  - Malimo Fabric (TN)
  - Malipol Fabric (TN)
  - Maliwatt Fabric (TN)
  - Voltex Fabric (TN)
- STITCH REINFORCED FABRICS
  - Arachne Fabric (TN)
  - Maliwatt Fabric (TN)

FABRICS (ACCORDING TO STRUCTURE) - continued

- STITCHED PILE FABRIC
  - Araloop Fabric (TN)
  - Malipol Fabric (TN)
- WOVEN FABRICS
  - Albert Twill
  - Alpaca Fabrics
  - Artificial Furs
  - Awning Cloth
  - .
  - .
  - . For detailed listing of additional terms, see Thesaurus.
  - .
  - .
  - Window Holland
  - Woven Carpets
  - Woven Tubes

FABRICS (BY END USES)

- AGRICULTURAL FABRICS
- APPAREL FABRICS
- AUTOMOTIVE FABRICS
- DECORATIVE FABRICS
- FURNISHING FABRICS
- HOUSEHOLD FABRICS
- INDUSTRIAL FABRICS
- UPHOLSTERY FABRICS

COMPOSITES

- COATED FABRICS
  - Buna N Coated Fabrics
  - Butyl Coated Fabrics
  - Lacquer Coated Fabrics
  - Neoprene Coated Fabrics
  - Rubber Coated Fabrics
  - SBR Coated Fabrics
  - Vinyl Coated Fabrics

LAMINATES

- FABRIC TO FABRIC LAMINATES
- FOAM BACKED FABRICS
- HONEYCOMB LAMINATES
- LAMINATED FABRICS
- LAMINATED JERSEY (FOAM)
- POST FORMED LAMINATES
- SANDWICH LAMINATES

REINFORCED COMPOSITES

- FABRIC REINFORCED COMPOSITES
- FIBER REINFORCED COMPOSITES
- FILAMENT WOUND COMPOSITES
- HONEYCOMB LAMINATES
- POST FORMED LAMINATES
- YARN REINFORCED COMPOSITES

PROPERTIES

AESTHETIC PROPERTIES

Aesthetic Appeal
Appearance
Bloom
Brightness
Comfort
Compressibility
Drabness
Drape
Dullness
Hairiness
Hand
Liveliness
Luster
Luxuriousness
Pattern Definition
Richness
Roughness
Slickness
Smoothness
Softness
Sparkle
Texture
Translucency
Transparency

BIOCHEMICAL PROPERTIES

Biodegradability

CHEMICAL PROPERTIES

Acid Binding
Acid Solubility
Alkali Solubility
Biodegradability
Chemical Stability
Copper Number
Cuprammonium Fluidity

*Terms listed are structured as related terms in the Thesaurus.

CHEMICAL PROPERTIES - continued

- Dyeing Properties (General)
- Finish (Property)
- Isoelectric Point
- Isoionic Point
- Solubility
- Urea Bisulfite Solubility

DEGRADATION PROPERTIES

- Acid Solubility
- Alkali Solubility
- Bacterial Inhibition
- Biodegradability
- Colorfastness
- Chlorine Fastness (Color)
- Chlorine Fastness (of Finish)
- Copper Number
- Cuprammonium Fluidity
- Dry Cleaning Fastness (Color)
- Dry Cleaning Fastness (of Finish)
- Gas Fume Fastness
- Insect Resistance
- Lightfastness (Color)
- Lightfastness (of Finish)
- Mildew Resistance
- Perspiration Resistance
- Urea Bisulfite Solubility
- Washfastness (Color)
- Washfastness (of Finish)
- Weather Resistance

END USE PROPERTIES

- Abrasion Resistance
- Absorbency (Material)
- Antistatic Behavior
- Appearance
- Bacterial Inhibition
- Breaking Strength
- Chlorine Fastness (Color)

END USE PROPERTIES - continued

- Chlorine Fastness (of Finish)
- Colorfastness
- Comfort
- Crease Acceptance
- Crease Retention
- Crockfastness
- Crush Resistance
- Dimensional Stability
- Dry Cleaning Fastness (Color)
- Dry Cleaning Fastness (of Finish)
- Durable Press
- Durability
- Drape
- Fatigue Resistance
- Felting Resistance
- Finish (Property)
- Fire Resistance
- Fire Retardancy
- Gas Fume Fastness
- Insect Resistance
- Ironing Fastness
- Lightfastness (Color)
- Lightfastness (of Finish)
- Liveliness
- Mildew Resistance
- Oil Repellency
- Perspiration Resistance
- Resilience
- Roughness
- Run Resistance
- Sewability
- Sheen
- Shrink Resistance
- Slickness
- Slip Resistance
- Smoothness

END USE PROPERTIES - continued

- Snag Resistance
- Softness
- Soil Resistance
- Stain Resistance
- Stiffness
- Tear Strength
- Toughness
- Translucency
- Transparency
- Warmth
- Washfastness (Color)
- Washfastness (of Finish)
- Water Repellency
- Water Resistance
- Wear Resistance
- Weather Resistance
- Wrinkle Recovery
- Wrinkle Resistance

FABRIC PROPERTIES

- Aesthetic Appeal
- Appearance
- Bagginess
- .
- . for detailed listing of additional terms,
- . see Thesaurus.
- **Wrinkle** Recovery
- Wrinkle Resistance

FABRIC PROPERTIES (AESTHETIC)

- Aesthetic Appeal
- Appearance
- Bagginess
- Bloom
- Brightness
- Comfort
- Drabness
- Drape

FABRIC PROPERTIES (AESTHETIC)

- Dullness
- Finish (Property)
- Hand
- Harshness
- Luxuriousness
- Pattern (Fabrics)
- Pattern Definition
- Prickliness
- Richness
- Scroop
- Sheen
- Stitch Clarity
- Texture

FABRIC PROPERTIES (MECHANICAL)

- Bagginess
- Biaxial Strength
- Bursting Strength
- Crease Acceptance
- Crease Retention
- Crush Resistance
- Curling Tendency
- Drape
- Elmendorf Tear Strength
- Fabric Relaxation
- Fabric Shrinkage
- Fabric Strength
- Fabric to Fabric Friction
- Finish (Property)
- Grab Strength
- Liveliness
- Pill Resistance
- Puncture Resistance
- Ravel Strip Strength
- Scroop
- Seam Efficiency
- Seam Strength
- Sewability

FABRIC PROPERTIES (MECHANICAL) - continued

- Shrink Resistance
- Slip Resistance
- Snag Resistance
- Tear Strength
- Tongue Tear Strength
- Trapezoid Tear Strength
- Wrinkle Recovery
- Wrinkle Resistance

FABRIC PROPERTIES (PHYSICAL EXCLUDING MECHANICAL)

- Brightness
- Comfort
- Fabric Weight
- Finish (Property)
- Water Repellency
- Water Resistance

FABRIC PROPERTIES (STRUCTURAL)

- Courses per Inch
- Cover
- Cover Factor
- Ends per Inch
- Fabric Crimp
- Fabric Geometry
- Float Length
- Pattern (Fabrics)
- Pattern Definition
- Picks per Inch
- Set (Woven Fabric)
- Stitch Clarity
- Stitch Density
- Texture
- Wales per Inch
- Weave

FIBER PROPERTIES

- Coefficient of Fineness Variation
- Coefficient of Length Variation
- Crimp Amplitude (Fiber)

FIBER PROPERTIES - continued

- Crimp Frequency
- Crimp Index
- Fiber Array
- Fiber Bundle Strength
- Fiber Crimp
- Fiber Cross Sections
- Fiber Denier
- Fiber Diagram
- Fiber Diameter
- Fiber Extent
- Fiber Fineness
- Fiber Friction
- Fiber Geometry
- Fiber Length
- Fiber Length Distribution
- Fiber Strength
- Fiber Tenderness
- Fiber Twist
- Fine Structure (Fibers)
- Grade (Fibers)
- Immaturity (Fiber)
- Mean Fiber Length
- Micronaire Fineness
- Natural Draw Ratio
- Pressley Strength

MECHANICAL PROPERTIES

- Abrasion Resistance
- Adhesion
- Bearing Strength
- Bending Recovery
- Bending Rigidity
- Coefficient of Friction
- Cohesion
- Crimp
- Cutting Resistance
- Dimensional Stability

MECHANICAL PROPERTIES - continued

- Durability
- Efficiency (Structural)
- Fatigue Resistance
- Flexural Strength
- Frictional Characteristics
- Hardness
- Knot Efficiency
- Liveliness
- Loop Efficiency
- Loop Strength
- Mechanical Deterioration Properties
- Modulus
- Paperiness
- Permeability
- Pill Resistance
- Puncture Resistance
- Resistance to Delamination
- Shear Resistance
- Shock Resistance
- Shrink Resistance
- Snag Resistance
- Snarling Tendency
- Softness
- Stress Strain Properties
  - Stress Strain Properties (Compressive)
    - Buckling
    - Compressibility
    - Compressive Modulus
    - Compressive Strength
    - Hardness
  - Stress Strain Properties (Shear)
    - Cutting Resistance
    - Shear (Mode of Deformation)
    - Shear Modulus
    - Shear Strength
    - Shear Stress

MECHANICAL PROPERTIES - continued

Stress Strain Properties

Stress Strain Properties (Tensile)

Breaking Elongation

Breaking Energy

Breaking Length

.
.
. For detailed listing of additional terms,
. see Thesaurus
.

Wet Strength

Work Recovery

Yield Point

Surface Properties (Mechanical)

Tack

Tear Strength

Texture

Torsional Rigidity

Transfer Properties

Wear Resistance

PHYSICAL CHEMICAL PROPERTIES

Absorbency (Material)

Activation Energy

Boiling Point

Diffusion Coefficient

Fire Retardancy

Flash Point

Freezing Point

Glass Transition Temperature

Hydrophilic Property

Hydrophobic Property

Hygroscopicity

Isoelectric Point

Isoionic Point

Latent Heat

Melting Point

Molecular Attraction

PHYSICAL CHEMICAL PROPERTIES - continued

- Secondary Transition Temperatures
- Soil Resistance
- Solubility
- Specific Heat
- Surface Properties (Physical Chemical)
- Transfer Properties

PHYSICAL PROPERTIES (EXCLUDING MECHANICAL)

- Acoustic Properties
- Bulk
- Bulk Density
- Crimp
- Electrical Properties
- Electron Spin Resonance
- Enthalpy
- Entropy
- Flash Point
- Freezing Point
- Heat Resistance
- Latent Heat
- Melting Point
- Nuclear Magnetic Resonance
- Optical Properties
- Packing Factor
- Softening Point
- Stress Optical Properties
- Structural Properties
- Temperature
- Thermal Properties
- Warmth

TRANSFER PROPERTIES

- Absorbency (Material)
- Absorptivity (Radiation)
- Conductivity (Thermal)
- Diffusivity
- Emissivity
- Reflectivity
- Transmissivity

PROPERTIES - continued

YARN PROPERTIES

Count

Count Strength Product

Denier

Fiber Content

Hairiness

Twist

Twist Liveliness

Twist Sense

Wildness

Yarn Cover

Yarn Crimp

Yarn Diameter

Yarn Geometry

Yarn Strength

DYEING

- DYEING (BY DYE CLASSES)
  - Acid Dyeing
  - Azoic Dyeing
  - Basic Dyeing
  - Direct Dyeing
  - Disperse Dyeing
  - Metallized Dyeing
  - Mordant Dyeing
  - Neutral Dyeing
  - Pigment Dyeing
  - Reactive Dyeing
  - Sulfur Dyeing
  - Vat Dyeing
- DYEING (BY ENVIRONMENTAL CONDITIONS)
  - Carrier Dyeing
  - High Temperature Dyeing
  - Low Temperature Dyeing
  - Pressure Dyeing
  - Solvent Assisted Dyeing
  - Solvent Dyeing
  - Speck Dyeing
  - Vacuum Dyeing
  - Vapor Phase Dyeing
- DYEING (BY FIBER CLASSES)
  - Acetate Dyeing
  - Acrylic Dyeing
  - Blend Dyeing
  - Cellulose Dyeing
  - Cotton Dyeing
  - Hair Dyeing (Human Hair)
  - Nylon Dyeing
  - Olefin Dyeing
  - Polyester Dyeing
  - Rayon Dyeing

DYEING (BY FIBER CLASSES) - continued

- Silk Dyeing
- Synthetic Dyeing
- Triacetate Dyeing
- Wool Dyeing

DYEING (BY MATERIAL ASSEMBLY)

- Dope Dyeing
- Garment Dyeing
- Piece Dyeing
- Stock Dyeing
- Strip Dyeing
- Top Dyeing
- Tow Dyeing
- Yarn Dyeing

DYEING (BY PROCESS FLOW)

- Batch Dyeing
- Continuous Dyeing
- Pad Dyeing
- Semi Continuous Dyeing
- Simultaneous Dyeing and Finishing

DYEING (FOR EFFECT)

- Batik Dyeing
- Cross Dyeing
- Space Dyeing
- Union Dyeing

DYES

- DYES (BY CHEMICAL CLASSES)
  - Acid Dyes
  - Aniline Black Dyes
  - Anionic Dyes
  - Azoic Dyes
  - Basic Dyes
  - Direct Dyes
  - Disperse Dyes
  - Fluorescent Dyes
  - Ingrain Dyes
  - Metallized Dyes
  - Mordant Dyes
  - Natural Dyes
  - Neutral Dyes
  - Oxidation Dyes
  - Polymeric Dyes
  - Reactive Dyes
  - Solvent Dyes
  - Sulfur Dyes
  - Vat Dyes
  - VINYLATED DYES
- DYES (BY FIBER CLASSES)
  - Acetate Dyes
  - Acrylic Dyes
  - Cellulose Dyes
  - Nylon Dyes
  - Polyester Dyes
  - Silk Dyes
  - Wool Dyes
- FOOD DYES

SOME PROPERTIES OFTEN IMPROVED THROUGH CHEMICAL FINISHING PROCESSES

FINISH (PROPERTY)	FINISH (SUBSTANCE ADDED)	FINISHING PROCESS (GENERAL)
ABRASION RESISTANCE	ABRASION RESISTANCE FINISHES	-----
ABSORBENCY (MATERIAL)	ABSORBENT FINISHES*	-----
ANTISTATIC BEHAVIOR	ANTISTATIC AGENTS	-----
BACTERIAL INHIBITION	ANTIBACTERIAL FINISHES	ANTIBACTERIAL TREATMENTS
	ANTIMICROBIAL FINISHES	ANTIMICROBIAL TREATMENTS
CREASE RETENTION	DURABLE PRESS FINISHES	DURABLE PRESS TREATMENTS
DURABLE PRESS	DURABLE PRESS FINISHES	DURABLE PRESS TREATMENTS
FELTING RESISTANCE	ANTIFELTING AGENTS	ANTIFELTING TREATMENTS
FIRE RESISTANCE	FIRE PROOFING AGENTS	FIRE PROOFING TREATMENTS
FIRE RETARDANCY	FIRE RETARDANCY AGENTS	FIRE RETARDANCY TREATMENTS
INSECT RESISTANCE	INSECT RESISTANCE FINISHES	INSECT RESISTANCE TREATMENTS
LUSTER	DELUSTERING AGENTS	DELUSTERING
MERCERIZED FINISH	-----	MERCERIZING
MILDEW RESISTANCE	MILDEW RESISTANCE FINISHES	MILDEW RESISTANCE TREATMENTS
	FUNGICIDES	
OIL REPELLENCY	OIL REPELLENTS	OIL REPELLENT TREATMENTS
PILL RESISTANCE	PILL RESISTANCE FINISHES	PILL RESISTANCE TREATMENTS
SEWABILITY	SOFTENERS	SOFTENING
	LUBRICANTS	
SHRINK RESISTANCE	SHRINK RESISTANCE FINISHES	SHRINK PROOFING

FINISH (PROPERTY)	FINISH (SUBSTANCE ADDED)	FINISHING PROCESS (GENERAL)
SLICKNESS	LUBRICANTS	-----
SLIP RESISTANCE	ANTISLIP AGENTS	-----
SOFTNESS	SOFTENERS	SOFTENING
SOIL RESISTANCE	SOIL RESISTANCE FINISHES	SOIL RESISTANCE* TREATMENTS
SOIL RELEASE*	SOIL RELEASE FINISHES	SOIL RELEASE* TREATMENTS
STAIN RESISTANCE	STAIN RESISTANCE AGENTS	STAIN RESISTANCE* TREATMENTS
STIFFNESS	STIFFENERS (AGENTS)	STIFFENING TREATMENTS
WATER REPELLENCY	WATER REPELLENTS	WATER REPELLENCY TREATMENTS
WATER RESISTANCE	WATERPROOFING AGENTS	WATERPROOFING
WEATHER RESISTANCE	{ WEATHER RESISTANCE FINISHES* PRESERVATIVES STABILIZERS (AGENTS)	WEATHER RESISTANCE TREATMENTS*
WASH WEAR PROPERTIES	WASH WEAR FINISHES	WASH WEAR TREATMENTS
WRINKLE RECOVERY	WASH WEAR FINISHES	WASH WEAR TREATMENTS

*This term not included as of completion of 2nd edition of Thesaurus.

- SURFACTANTS
  - ANIONIC SURFACTANTS
    - Carboxylate Surfactants
      - Soaps
        - Alkali Soaps
          - Ammonium Soaps, Potassium Oleate, Potassium Soaps, Potassium Stearate, Sodium Oleate, Sodium Palmitate, Sodium Soaps, Sodium Rosinate, Sodium Stearate
        - Amine Soaps
          - Alkanolamine Soaps
            - Ethanolamine Soaps
            - Triethanolamine Soaps
          - Morpholine Soaps
        - Metallic Soaps
          - Aluminum Soaps, Aluminum Stearate, Calcium Soaps, Calcium Stearate, Heavy Metal Soaps, Lithium Stearate, Zinc Soaps, Zinc Stearate
      - Sarcosides
      - Lamepons (UF Protein--Fatty Acid Condensates)
    - Sulfonate Surfactants
      - Alkyl Sufonates (UF Petroleum Sulfonates, Hydrocarbon Sulfonates)
      - Acyl Ester Sulfonates
        - Igepon A (TN)
      - Alkylamide Sulfonates
        - Methyltauride Surfactants
          - Igepon T (TN)
      - Alkylaryl Sulfonates
        - Alkylbiphenyl Sulfonates
        - Alkylnaphthalene Sulfonates
          - Amylnaphthalene Sulfonates, Butylnaphthalene Sulfonates, Nekal A (TN), Nekal B (TN), Nonylnaphthalene Sulfonates
        - Alkylbenzene Sulfonates
          - Linear Alkylbenzene Sulfonates, Nacconol NR (TN), Nonylbenzene Sulfonates, Octylbenzene Sulfonates

- Sulfonate Surfactants (Continued)
  - Alkylether Sulfonates
  - Sulfofatty Acids
  - Sulfofatty Acid Esters
  - Sulfopolycarboxylic Esters
    - Trihexyl Sulfocarballylate
      - Nekal NS (TN)
    - Sulfosuccinic Esters
      - Dioctyl Sulfosuccinate
        - Deceresol OT (TN)
  - Naphthalene Formaldehyde Sulfonates
  - (UF Condensed Naphthalene Sulfonates)
  - Tetrahydronaphthalene Sulfonates
    - Alkanol S (TN)
  - Alkylaryl Oxyethyl Sulfonates
    - Alkyl Phenoxyethyl Sulfonates
  - Lignin Sulfonates
  - Sulfosuccinamides
    - Aerosol 18 (TN)
    - Aerosol 22 (TN)
- Sulfate Surfactants
  - Alkyl Sulfates
  - (UF Alcohol Sulfates, Fatty Alcohol Sulfates)
    - Gardinol (TN), Octyl Sulfates, Dodecyl Sulfates
    - Sodium Lauryl Sulfate
    - Sulfated Olefins
  - Sulfated Oils (UF Sulfonated Oils)
    - Sulfated Oleic Esters
    - Sulfated Tallow (UF Sulfonated Tallow)
    - Sulforicinoleates (UF Sulfated Castor Oil)
  - Alkylamide Sulfates
  - Alkyl Ethoxy Sulfates (UF Ether Alcohol Sulfates)
  - Alkylaryl Ethoxy Sulfates
  - Alkyl Glycerosulfates (UF Monoglyceride Sulfates)
  - Sulfated Glycerides, Sulfoglycerides)
    - Lecithin Sulfate

Phosphate Surfactants
Alkyl Phosphate Surfactants, Alkyl Ethoxy Phosphates,
Alkylaryl Ethoxy Phosphates, Glycerophosphates
(UF Lecithin, UF Phosphated Glycerides)
CATIONIC SURFACTANTS
Amine Surfactants
Alkylamine Surfactants
(UF Alkyldimethylamine Surfactants)
Armeens (TN)
Armacs (TN)
Alkylmorpholine Surfactants
Imidazoline Surfactants
Polyalkylamine Surfactants
Duomeens (TN)
Oxazoline Surfactants
Amine Oxide Surfactants
(UF Tertiary Alkylamine Oxides, UF Tertiary Amine Oxides)
Alkyldimethylamine Oxide Surfactants
Alkyldi(hydroxyethyl)amine Oxide Surfactants
Polyethoxy Polypropoxy Ethylenediamine
(UF Tetronics [TN])
Polyethoxy Alkylamine Surfactants
(UF Ethomeens [TN], UF Polyethoxy Alkyl Polyamines,
UF Ethoduomeens [TN], UF Polyethoxy Amine Surfactants)
Onium Surfactants
Phosphonium Surfactants
Sulfonium Surfactants
Quaternary Ammonium Surfactants
Polyalkoxy Quaternary Ammonium Surfactants
NONIONIC SURFACTANTS
Diethanolamide Surfactants (UF Kritchevsky Products)
Lauryl Diethanolamide, Capryl Diethanolamide,
Coconut Diethanolamide
Glucoside Surfactants
Glyceryl Ester Surfactants
Fatty Monoglycerides, Fatty Diglycerides,
Polyglyceryl Esters

SURFACTANTS

NONIONIC SURFACTANTS (CONTINUED)

- Glycol Ester Surfactants
  - Diglycol Ester Surfactants
  - Propanediol Ester Surfactants
- Polyethoxy Alkylaryl Ethers
  (UF Polyethoxy Alkyl Phenols)
  (UF Ethoxylated Alkyl Phenols)
- Polyethoxy Amides (UF Polyoxyethylene Alkylamides)
  - Ethomids (TN)
- Polyethoxy Esters (UF Polyethoxy Fatty Acids, Polyethoxy Polyol Esters, Polyethylene Glycol Esters, Polyglycol Esters, Polyoxyethylene Esters)
  - Ethofats (TN)
- Polyethoxy Ethers (UF Polyethoxy Alcohols, Polyethyleneglycol Ethers, Polyoxyethylene Alcohols)
- Polyethoxy Glycerides
  - Polyethoxylated Castor Oil
- Polyethoxy Polypropoxy Surfactants
  (UF Ethoxy Propoxy Block Polymers)
  - Pluronics (TN)
- Polyethoxy Thioethers (UF Polyoxyethylene Thioethers)
- Polyglycidyl Alkylaryl Ethers
- Sorbitan Ester Surfactants
  (UF Polyethoxy Anhydrosorbitol Ester)
  - Spans (TN), Tweens (TN)
- Sucrose Ester Surfactants

AMPHOTERIC SURFACTANTS (UF AMPHOLYTIC SURFACTANTS)

- Alkylamino Carboxylate Surfactants
- Alkylamino Sulfate Surfactants
- Alkylamino Sulfonate Surfactants
- Zwitterionic Surfactants
  - Betaine Surfactants
  - Sulfobetaine Surfactants

*UF = Used For

RECENTLY RECOMMENDED KEYTERMS*

ACCELEROTOR ABRASION TESTING
  BT ABRASION TESTING
ACETHYDRAZIDE DISULFIDE
  BT DISULFIDES (ORGANIC)
  BT HYDRAZIDES
  BT REACTANTS
ACRIBEL (TN)
  BT ACRYLIC FIBERS
ACYLATED COTTON
  BT CHEMICALLY MODIFIED COTTON
ALGINIC ACID
  BT CARBOXYLIC ACIDS
  BT ORGANIC ACIDS
  BT POLYSACCHARIDES
ALANINE
  BT AMINO ACIDS
ALKYLARYL COMPOUNDS
ALKYLARYL HYDROCARBONS
  BT HYDROCARBONS
  BT ALKYLARYL COMPOUNDS
ALLYL SULFONIC ACID
  BT ALLYL COMPOUNDS
  BT SULFONIC ACIDS
ALLYLAMINE
  BT ALLYL COMPOUNDS
  BT AMINES
ALUMINUM SULFATE
  BT ALUMINUM COMPOUNDS
  BT SULFATES
AMILAN (TN)
  BT NYLON 6
  BT NYLON (POLYAMIDE FIBERS)
AMMONIUM PHOSPHATE
  BT AMMONIUM COMPOUNDS
  BT PHOSPHATES
ANTHRANILATED COTTON
  BT CHEMICALLY MODIFIED COTTON
ANTIBIOTICS
ANTIMONY COMPOUNDS
ANTIMONY OXIDE
  BT ANTIMONY COMPOUNDS
  BT OXIDES
ARSENIC COMPOUNDS
ARSENIC TRIOXIDE
  BT OXIDES
  BT ARSENIC COMPOUNDS
ASCORBIC ACID
  BT ORGANIC ACIDS
ASH CONTENT
  BT CHEMICAL COMPOSITION
ASPIRATORS
BETA (TN)
  BT GLASS FIBERS
BORAX
  BT BORON COMPOUNDS
BROMINATION
  BT HALOGENATION
  BT REACTIONS (CHEMICAL)
BROMATES
BROMITES
BUFFERS
BUTYL RUBBER
  BT SYNTHETIC RUBBER
CALCIUM CARBONATE
  BT CALCIUM COMPOUNDS
  BT CARBONATES
CALCIUM PEROXIDE
  BT PEROXIDES
  BT CALCIUM COMPOUNDS
CARBONIZATION
CASHMILLON (TN)
  BT ACRYLIC FIBERS
CELLULOSE ACETATE SULFATE
  BT CELLULOSE ESTERS
CELLULOSE ACETATE VALERATE
  BT CELLULOSE ESTERS
CELLULOSE BUTYRATE
  BT CELLULOSE ESTERS
CELLULOSE BUTYRATE VALERATE
  BT CELLULOSE ESTERS
CELLULOSE PROPIONATE
  BT CELLULOSE ESTERS
CELLULOSE PROPIONATE VALERATE
  BT CELLULOSE ESTERS
CELLULOSE SULFATE
  BT CELLULOSE ESTERS
CELLULOSE VALERATE
  BT CELLULOSE ESTERS
CHLOROACRYLIC ACID
  BT ACRYLIC COMPOUNDS
  BT CARBOXYLIC ACIDS
  BT ORGANIC ACIDS
CHLOROMETHYLSTYRENE
  BT VINYL COMPOUNDS
COERCIVE COUPLES
  BT COUPLES
COORDINATION POLYMERS
CORRIEDALE WOOL
  BT WOOL
CORVAL (TN)
  BT CELLULOSIC FIBERS
  BT RAYON (REGENERATED CELLULOSE FIBERS)
  BT VISCOSE RAYON
COURTELLE (TN)
  BT ACRYLIC FIBERS
CRIMPLENE (TN)
  BT FALSE TWIST YARNS
  BT TEXTURED YARNS (FILAMENT)
  BT POLYESTER FIBERS
CROCK FASTNESS (OF FINISH)
  BT END USE PROPERTIES
CROTONALDEHYDE
  BT ALDEHYDES
  BT VINYL COMPOUNDS
CROTONIC ACID
  BT CARBOXYLIC ACIDS
  BT ORGANIC ACIDS
  BT VINYL COMPOUNDS
CRYLOR (TN)
  BT ACRYLIC FIBERS
CRYPOANIONIC SURFACTANTS
  BT ANIONIC SURFACTANTS
CUPRESA (TN)
  BT CUPRAMMONIUM RAYON
  BT RAYON (REGENERATED CELLULOSE FIBERS)
  BT CELLULOSIC FIBERS
DIALDEHYDE COTTON
  BT CHEMICALLY MODIFIED COTTON
DICEL (TN)
  BT ACETATE FIBERS
DDT
DIESTERS
  BT ESTERS
DIET
DIETHANOLAMINE CARBONATE
  BT CARBONATES
DIHYDROXYETHYLENE THIOUREA
  BT THIOUREA DERIVATIVES
DIKETENE
  BT VINYL COMPOUNDS
DIMETHYLOL ALLOXAN
  BT METHYLOL COMPOUNDS
DIMETHYLOL ETHYLENE THIOUREA
  BT METHYLOL THIOUREAS
DIMETHYLOL THIOUREA
  BT METHYLOL THIOUREAS
DIOLEN (TN)
  BT POLYESTER FIBERS
DOLAN (TN)
  BT ACRYLIC FIBERS
DORLASTAN (TN)
  BT SPANDEX FIBERS
DRALON (TN)
  BT ACRYLIC FIBERS
ELASTOMER FIBER BLENDS
  BT BLENDS (FIBERS)
ELECTRIFYING (PILE FABRICS)
  BT DRY FINISHING
ELECTROSTATIC PRINTING
  BT PRINTING
EULAN (TN)
  BT RAYON (REGENERATED CELLULOSE FIBERS)
  BT VISCOSE RAYON
  BT CRIMPED YARNS
EXLAN (TN)
  BT ACRYLIC FIBERS
EXTRACTS
FERROUS SULFATE
  BT FERROUS COMPOUNDS
  BT SULFATES
  BT IRON COMPOUNDS
FIBER END DENSITY
  BT STRUCTURAL PROPERTIES
FIBER SHRINKAGE
  BT SHRINKAGE
  BT FIBER PROPERTIES
FIBERS (BY CHARACTERISTICS)
FLUOROFATTY ACIDS
  BT FLUOROCARBONS
  BT ORGANIC ACIDS
  BT CARBOXYLIC ACIDS
FORMYLATED COTTON
  BT CHEMICALLY MODIFIED COTTON
FURAN DERIVATIVES
  BT HETEROCYCLIC COMPOUNDS
FURANDIMETHANOL
  BT ALCOHOLS
  BT FURAN DERIVATIVES
FURFURYL ALCOHOL
  BT ALCOHOLS
  BT FURAN DERIVATIVES
GLYCINE
  BT AMINO ACIDS
GLUCOSIDE SURFACTANTS
  BT NONIONIC SURFACTANTS
GLUTAMIC ACID
  BT AMINO ACIDS
GLUTEN
  BT PROTEINS
HETERO FIBERS
HIGH SHRINKAGE FIBERS
  BT FIBERS (BY CHARACTERISTICS)
HISTIDINE
  BT AMINO ACIDS
HOMOFIBERS
HOUSEHOLD DYEING
  BT DYEING
HYDROGEN SULFIDE
  BT INORGANIC ACIDS
HYDROXYETHYLATED COTTON
  BT CHEMICALLY MODIFIED COTTON
HYDROXYMETHYL SULFONIC ACID
  BT SULFONIC ACIDS
HYDROXYPROPYLMETHYL CELLULOSE
  BT CELLULOSE ETHERS
HYDROXYQUINOLINE
  BT PHENOLS
INSTANTANEOUS LINEAR DENSITY
  BT LINEAR DENSITY
IRON OXIDE
  BT IRON COMPOUNDS
  BT OXIDES
ISATOIC ANHYDRIDE
  BT ACID ANHYDRIDES
  BT REACTANTS
LACTONITRILE
  BT NITRILES
LASTRALENE (TN)
  BT SPANDEX FIBERS
LAVSAN (TN)
  BT POLYESTER FIBERS
LEACRYL N (TN)
  BT ACRYLIC FIBERS
LEADING ENDS
  BT FIBER ENDS
LLAMA FIBER
  BT HAIR
  BT KERATIN FIBERS
  BT PROTEIN FIBERS
LOOP RATIO (RIB TO PLAIN)
  BT KNITTED FABRIC STRUCTURE (FILLING KNIT)
LOW SHRINKAGE FIBERS
  BT FIBERS (BY CHARACTERISTICS)
METHIONINE
  BT AMINO ACIDS
  BT THIOETHERS
METHYLENEDISULFATE
  BT SULFATES
  BT SULFUR COMPOUNDS
METHYLOL THIOUREAS
  BT METHYLOL COMPOUNDS
  BT THIOUREA DERIVATIVES
MITIN (TN)
MOIRE FINISH
  BT FINISH (PROPERTY)
MORPHOLINE
  BT HETEROCYCLIC COMPOUNDS
MORPHOLINE DERIVATIVES
  BT HETEROCYCLIC COMPOUNDS
NEOMYCIN
  BT ANTIBIOTICS
NITROGEN
NOMINAL LINEAR DENSITY
  BT LINEAR DENSITY
NOMINAL STRAIN
  BT STRAIN
NOMINAL STRESS
  BT STRESS
NTA
  BT CARBOXYLIC ACIDS
  BT ORGANIC ACIDS

NYLON 4
BT NYLON (POLYAMIDE FIBERS)
BT MAN MADE FIBERS
BT SYNTHETIC FIBERS
NYLSUISSE (TN)
BT NYLON (POLYAMIDE FIBERS)
BT NYLON 66
NYMCRYLON (TN)
BT ACRYLIC FIBERS
OIL SOLUBLE DYES
BT DYES (BY CHEMICAL CLASSES)
ORGANOANTIMONY COMPOUNDS
BT ANTIMONY COMPOUNDS
BT ORGANOMETALLIC COMPOUNDS (EXCLUDING SILICONES)
ORGANOARSENIC COMPOUNDS
BT ARSENIC COMPOUNDS
BT ORGANOMETALLIC COMPOUNDS (EXCLUDING SILICONES)
ORGANOTITANIUM COMPOUNDS
BT ORGANOMETALLIC COMPOUNDS (EXCLUDING SILICONES)
BT TITANIUM COMPOUNDS
PENICILLIN
BT ANTIBIOTICS
PHASE RETARDATION
PHOSPHATED COTTON
BT CHEMICALLY MODIFIED COTTON
PHOSPHINIC ACIDS
BT ORGANIC ACIDS
BT PHOSPHORUS COMPOUNDS
PHOSPHONIC ACIDS
BT PHOSPHORUS COMPOUNDS
BT ORGANIC ACIDS
PHOSPHOROUS ACID
BT PHOSPHORUS COMPOUNDS
BT INORGANIC ACIDS
POLY (VINYL ETHERS)
BT ETHERS
BT POLYVINYLS
POLY (VINYL IMIDAZOLE)
BT POLYVINYLS
POLY (VINYL STEARATE)
BT ESTERS
BT POLYVINYLS
POLYBENZOXAZOLES
BT POLYMERS
POLYCARBONATE FIBERS
BT FIBERS
BT MAN MADE FIBERS
BT SYNTHETIC FIBERS
POLYGLYCIDYL ALKYLARYL ETHERS
BT NONIONIC SURFACTANTS
POLYMETHACRYLONITRILE
BT NITRILES
BT POLYVINYLS
PLUGS (FIBER)
BT FIBER ASSEMBLIES
POTENTIOMETRY
BT CHEMICAL ANALYSIS
PPX (TN)
BT POLYPROPYLENE FIBERS
BT OLEFIN FIBERS
PRESTWICK (TN)
BT WASH WEAR TREATMENTS
PRINTING SHARPNESS
PROTECTING AGENTS
BT DYEING AUXILIARIES
PULLING (WOOL)
PUNCHING RESISTANCE
BT MECHANICAL PROPERTIES
PYROSULFATES
BT SULFUR COMPOUNDS
PYRROLIDINE
BT HETEROCYCLIC COMPOUNDS
PNC
BT PHOSPHORUS COMPOUNDS
RADIATION RESISTANCE
BT END USE PROPERTIES
BT FINISH (PROPERTY)
BT DEGRADATION PROPERTIES
RADIATION RESISTANCE FINISHES
BT FINISH (SUBSTANCE ADDED)
RADIATION RESISTANCE TREATMENTS
BT WET FINISHING
REDON (TN)
BT ACRYLIC FIBERS
ROOTS (FIBERS)
SACCHARIDE ETHER SURFACTANTS
BT NONIONIC SURFACTANTS
SERINE
BT AMINO ACIDS
SHAMPOOING (CARPETS)
BT CLEANING
SODIUM SULFITE
BT SULFITES
SOIL RELEASE TREATMENTS
BT WET FINISHING
SOIL RESISTANCE TREATMENTS
BT WET FINISHING
SPANZELLE (TN)
BT SPANDEX FIBERS
SPECIFIC SURFACE
BT SURFACE PROPERTIES (MECHANICAL)
SPRAY PRINTING
BT PRINTING
STABILIZERS (FOR PEROXYGEN COMPOUNDS)
STAIN RESISTANCE TREATMENTS
BT WET FINISHING
STANNIC CHLORIDE
BT CHLORIDES
BT TIN COMPOUNDS
STATIC CHARACTERISTICS
STENCIL PRINTING
BT PRINTING
STRETCHEVER (TN)
BT SPANDEX FIBERS
STRONTIUM COMPOUNDS
STYRENE SULFONIC ACID
BT SULFONIC ACIDS
BT VINYL COMPOUNDS
SUBLIMATION FASTNESS
BT COLORFASTNESS
SULFAMIC ACIDS
BT ORGANIC ACIDS
BT SULFUR COMPOUNDS
SULFONAMIDES
BT AMIDES
BT SULFUR COMPOUNDS
SULFUR CONTENT
BT CHEMICAL COMPOSITION
SULFUR TRIOXIDE
BT OXIDES
BT SULFUR COMPOUNDS
SUPERMOLECULAR STRUCTURE
BT FINE STRUCTURE (FIBERS)
TACRYL (TN)
BT ACRYLIC FIBERS
TEKLAN (TN)
BT MODACRYLIC FIBERS
TETRON (TN)
BT POLYESTER FIBERS
THERMAL SHRINKAGE
BT SHRINKAGE
BT THERMAL PROPERTIES
THERMOCOUPLES
BT MEASURING INSTRUMENTS
THIOUREA DERIVATIVES
THIOUREA-FORMALDEHYDE RESINS
BT AMINO RESINS
TIPS (FIBERS)
TORAYLON (TN)
BT ACRYLIC FIBERS
TRAILING ENDS (FIBERS)
BT FIBER ENDS
TRANSFER PRINTING
BT PRINTING
TREVIRA (TN)
BT POLYESTER FIBERS
TRIALLYL PHOSPHATE
BT ALLYL COMPOUNDS
BT PHOSPHORUS COMPOUNDS
TRIMESIC ACID
BT ORGANIC ACIDS
BT CARBOXYLIC ACIDS
BT TRIBASIC ACIDS
TRUE STRESS
BT STRESS
ULTRAVIOLET SPECTRA
BT OPTICAL PROPERTIES
BT SPECTRA
UNCRIMPING RESISTANCE
VANNEL (TN)
BT ACRYLIC FIBERS
VARYING CONDITIONS
VENTURI TUBES
VINCEL (TN)
BT POLYNOSIC FIBERS
BT HIGH WET MODULUS RAYON
BT VISCOSE RAYON
BT RAYON (REGENERATED CELLULOSE FIBERS)
BT CELLULOSIC FIBERS
VINITRON (TN)
BT SARAN POLY (VINYLCHLORIDE) FIBERS
VINYL CHLOROACETATE
BT VINYL COMPOUNDS
VINYL ESTERS
BT ESTERS
BT VINYL COMPOUNDS
VINYL IMIDAZOLE
BT VINYL COMPOUNDS
BT HETEROCYCLIC COMPOUNDS
VINYL STEARATE
BT VINYL COMPOUNDS
BT STEARATE ESTERS
BT VINYL ESTERS
VINYL SULFONIC ACID
BT SULFONIC ACIDS
BT VINYL COMPOUNDS
VINYLIDENE CHLORIDE
BT VINYLIDENE COMPOUNDS
VORTICES
WASHING EFFICIENCY
BT PROCESS EFFICIENCY
WEATHER RESISTANCE TREATMENTS
BT WET FINISHING
WEATHER RESISTANCE FINISHES
BT FINISH (SUBSTANCE ADDED)
WET FIXATION PROCESS (DURABLE PRESS)
BT DURABLE PRESS TREATMENTS
YARN SHRINKAGE
BT SHRINKAGE
BT YARN PROPERTIES
YIELD DETERMINATION
BT TESTING

* The keyterms listed in the last two pages have been recommended for inclusion in the thesaurus by various groups on the basis of recent indexing experience. These terms have been fully structured although space limitations permit listing of broad terms only at this time. It is expected that the thesaurus will continue to expand as keywording practices spread through the textile literature.

www.ingramcontent.com/pod-product-compliance
Lightning Source LLC
Chambersburg PA
CBHW080131220326
41598CB00032B/5025
* 9 7 8 0 2 6 2 5 2 3 5 5 4 *